Classical Complex Analysis

A Geometric Approach —Vol.1

Classical Complex Analysis

A Geometric Approach —Vol.1

I-Hsiung Lin

National Taiwan Normal University, Taiwan

World Scientific

NEW JERSEY · LONDON · SINGAPORE · BEIJING · SHANGHAI · HONG KONG · TAIPEI · CHENNAI

Published by

World Scientific Publishing Co. Pte. Ltd.

5 Toh Tuck Link, Singapore 596224

USA office: 27 Warren Street, Suite 401-402, Hackensack, NJ 07601

UK office: 57 Shelton Street, Covent Garden, London WC2H 9HE

British Library Cataloguing-in-Publication Data
A catalogue record for this book is available from the British Library.

CLASSICAL COMPLEX ANALYSIS — Volume 1
A Geometric Approach

Copyright © 2011 by World Scientific Publishing Co. Pte. Ltd.

ISBN-13 978-981-4261-22-7
ISBN-13 978-981-4261-23-4 (pbk)

Typeset by Stallion Press
Email: enquiries@stallionpress.com

To my wife Hsiou-O and my children
Zing, Ting, Ying and Fei

Preface

Complex analysis, or roughly equivalently the theory of analytic functions of one complex variable, budded in the early ages of Gauss, d'Alembert and Euler as a main branch of mathematical analysis. In the 19th century, Cauchy, Riemann, and Weierstrass laid a rigorous mathematical foundation for it (see Ref. 47). Nourished by joint effort of generations of brilliant mathematicians, it grows up into one of the remarkable branches of exact science, and serves as a prototype or model of other theories concerned with generalizations of analytic functions such as Riemann surfaces, analytic functions of several complex variables, quasiconformal and quasiregular mappings and complex dynamics, etc. Its methods and theory are widely used in many branches of mathematics, ranging from analytic number theory to fluid mechanics, elasticity theory, electrodynamics and string theory, etc.

Elementary complex analysis stands as a discipline to the whole mathematical training. This book is designed for beginners in this direction, especially for upper level undergraduate and graduate mathematics majors, and to those physics (or engineering) students who are interested in more theoretically oriented introduction to the subject rather than only in computational skills. The content is thus selective and its level of difficulty should be then adequately arranged.

Besides its strong intuitive flavor, it is the geometric (mapping) properties, derived from or characterized the analytic properties, that makes the theory of analytic functions differ so vehemently from that of real analysis and so special yet restrictive in applications. This is the reason why I favor a geometric approach to the basics. The degree of difficulty, as a whole, is not higher than that of Ahlfors's classic *Complex Analysis* (see Ref. 1). But I try my best to give as detailed and clear explanations to the theory as possible. I hope that the presentation will be less arduous in order to be more available to not-so-well-prepared or not-so-gifted students and be easier for self-study. Neither the greatest possible generality nor the most up-to-date terminologies is our purpose. Please refer to Ref. 9 for those purposes. I would consider my purpose fulfilled if the readers are able to acquire elementary yet solid fundamental classical results and techniques concerned.

Knowledge of elementary analysis, such as a standard calculus course including some linear algebra, is assumed. In many situations, mathematical maturity seems more urgent than purely mathematical prerequisites. Apart from these, the work is self-contained except for some difficult theorems to which references have been indicated. Yet for clearer and thorough understanding where one stands for the present in the whole mathematical realm and for the ability to compare with real analysis, I suggest readers get familiar with the theory of functions of two real variables.

Sketch of the Contents

If one takes a quick look at the Contents or read over Sketch of the Content at the beginning of each chapter, then s/he will have an overall idea about the book.

A complex number is not only a plane vector, but also carries by itself the composite motion of a one-way stretch and rotation, and hence, is a two-dimensional "number". They constitute a field, but cannot be ordered. Mathematics based on them is the one about similarity in global geometric sense and is the one about conformality in local infinitesimal sense. *Chapter 1* lays the algebraic, geometric, and point-set foundations barely needed in later chapters.

Just as one experienced in calculus, we need to know some standard *elementary* complex-valued *functions* of a complex variable before complex differentiation and integration are formally introduced. It is the *isolated-zero principle* (see (3) and (4) in (3.4.2.9)) that makes many of their *algebraic properties* or algebraic identities similar to their real counterparts. Owing to the complex plane $\mathbf{C}$ having the same topological structure as the Euclidean plane $\mathbf{R}^2$, their *point-set properties* (such as continuity and convergence) are the same as the real ones, too. It is the *geometric mapping properties* owned by these elementary functions that distinguishes them from the real ones and make one feel that complex analysis is not just a copy of the latter. In particular, the local and global single-valued continuous branches of arg z are deliberately studied, and then, prototypes of Riemann surfaces are introduced. *Chapter 2* tries to figure out, though loosely and vaguely organized, the common analytic and geometric properties owned by these individual elementary functions and then, to foresee what properties a general analytic function might have.

A complex-valued function $f(z)$, defined in a domain (or an open set) Ω, is called *analytic* if any one of the following equivalent conditions is satisfied:

(1) $f(z)$ is differentiable everywhere in Ω (Chap. 3).
(2) $f(z)$ is continuous in Ω and $\int_{\partial \Delta} f(z)dz = 0$ for any triangle Δ contained in Ω (Chaps. 3 and 4).
(3) For each fixed point $z_0 \in \Omega$, $f(z)$ can be expressed as a convergent power series $\sum_{n=0}^{\infty} a_n(z - z_0)^n$ in a neighborhood of z_0 (Chap. 5).

An analytic function $f(z)$ infinitesimally, via the Cauchy–Riemann equations, appears as a conformal mapping in case $f'(z) \neq 0$ and $\overline{f(z)}$ can be interpreted as the velocity field of a solenoidal, irrotational flow (see (3.2.4.3)). *Chapter 3* develops the most fundamental and important analytic and geometric properties, both locally and globally, which an analytic function might possess. The most subtle one, among all, is that a function $f(z)$ analytic at z_0 can always be written as $f(z) = f(z_0) + (z - z_0)\varphi(z)$ where $\varphi(z)$ is another function analytic at z_0. From this, many properties, such as the isolated-zero principle, maximum–minimum modulus principle and the open mapping property, inverse and implicit function theorems, can be either directly or indirectly deduced. This chapter also studies some global theorems such as Schwarz's lemma, symmetry principle, argument principle and Rouché's theorem and their illustrative examples.

After proving homotopic and homologous forms of Cauchy integral theorem, most part of *Chap. 4* is devoted to the residue theorem and its various applications in evaluating integrals, summation of series, and the Fourier and Laplace transforms.

Chapter 5 starts with various local power series representations of an analytic function and analytic continuation of power series which eventually lead to the monodromy theorem. Besides power series representation, an entire function can also be factorized as an infinite product of its zeros such as polynomials do, and meromorphic function can be expanded into partial fractions via its poles as rational functions. The most remarkable example is Euler's gamma function $\Gamma(z)$ and its colorful properties. Riemann zeta function is only sketched. The essential limit process in the whole chapter is the method of *local uniform convergence.* Weierstrass's theorem and Montel's normality criterion are two of the most fundamental results in this direction. Both are used to prove Picard's theorems via the elliptic modular function. These classical theorems can also be obtained by Schwarz–Ahlfors's lemma, a geometric theorem.

Riemann mapping theorem initiated the study of global geometric mapping properties of univalent analytic functions, simply called univalent mappings. Schwarz–Christoffel formulas provide fruitful examples for the theorem. As a consequence, *Chap. 6* solves the Dirichlet's problem for an open disk, a Jordan domain and hence, a class of general domains via Perron's method. This, in turn, is adopted to determine the canonical mappings and canonical domains for finitely connected domains.

Based on our intuitive and descriptive knowledge about Riemann surfaces of particularly chosen multiple-valued functions, scattered from Chaps. 2 to 6, *Chap. 7* tries to give a formal, rigorous yet concise introduction to *abstract* Riemann surfaces. We will cover the fundamental group, covering surfaces and covering transformations, and finally highlight the proof of the uniformization theorem of Riemann surfaces via available classical methods, even though most recently it admits a purely differential geometric one-page proof (see Ref. 17).

Almost all sections end up with an *Exercise A* for getting familiar with the basic techniques and applications; most of them also have an *Exercise B* for practice of combining techniques and deeper thinking or applications; few of them have an *Exercise C* for extra readings of a paper or *Appendices A, B, and C.*

As far as starred sections are concerned, see "How to Use the Book" below.

Features of the Book

(1) Style of writing. As a textbook for beginners, I try to introduce the concepts clearly and the whole theory gradually, by giving definite explanations and accompanying their geometric interpretations whenever possible. Geometric points of view are emphasized. There are about 546 figures and many of them are particularly valid or meaningful only for complex variable, but not for reals. Most definitions come out naturally in the middle of discussions, while main results obtained after a discussion are summarized and are numbered along with important formulas.

(2) Balance between theory and examples. As an introductory text or reference book to beginners, how to grasp and consolidate the basic theory and techniques seems more important and practical than to go immediately to deeper theories concerned. Therefore, there are sufficient amount of examples to practice main ideas or results. Exercises are usually divided into Part A and Part B; the former is designed to familiarize the readers with the established results, while the latter contains challenging exercises for mature and minded readers. Both examples and exercises are classic and are benefited very much from Refs. $31, 52, 58, 60$, and 80. What should be mentioned is that many exercise problems in Ref. 1 have been adopted as illustrative examples in this text.

(3) Careful treatment of multiple-valued functions. Owing to historical and pedagogical reasons, complex analysis is conventionally carried out in the (one-layer) classical complex plane. Later development shows that the most natural place to do so is multiple-layer planes, the so-called Riemann surfaces or one-dimensional complex manifolds in its modern terminology. Multiple-valuedness is a difficult subject to most beginners and most introductory books just avoid or sketch it by focusing on $\sqrt[n]{z}$ and $\log z$ only. To provide intuitive feeling toward abstract Riemann surfaces in Chap. 7, Chaps. 2–6 take no hesitation to treat multiple-valued analytic functions whenever possible in the theory and in the illustrative examples. Once the troublemaker $\arg z$, the origin of multiple-valuedness, is tamed (see Sec. 2.7.1), what is left is much easier to handle with. Also we construct many (purely descriptive and nonrigorous) Riemann surfaces or their line complexes of specified multiple-valued functions, merely for purposes of clearer illustration, wherever we feel worthy to do so.

(4) Emphasis on the difference between real and complex analyses. The origin of all these differences comes from the very character of what a complex number is (see the second paragraph inside the title Sketch of the Contents). This fact reflects, upon differentiating process, in the aspects of algebra, analysis as well as in geometry (see (3.2.2) for short; Secs. 3.2.1, 3.2.2, and 3.2.3 for details).

(5) Paving the way to advanced study. The contents chosen are so arranged that they will provide solid background knowledge to further study in fields mentioned in the first paragraph of this Preface. Besides, the book contains more materials than what is required in a Ph.D. qualifying examination for complex analysis.

How to Use the Book (A Suggestion to the Readers)?

The book is rich in contents, examples, and exercises when comparing to other books on complex analysis of the same level. It is designed for a variety of usages and motivations for advanced studies concerned.

The whole content is divided into two volumes: Vol. 1 contains Chaps. 1–4 plus Appendix A, while Vol. 2 contains Chaps. 5–7 plus Appendices B and C.

I may have the following suggestions for different proposes:

Chapters 1 and 2 are preparatory. Except those basic concepts such as limits and functions needed, topics in these two chapters could be selective, up to one's taste.

(1) As an introductory text for undergraduates.

Sections 2.5.2, 2.5.4, 2.6, 2.7.2, and 2.7.3 (sketch only), Sec. 2.9 (sketch only); Secs. 3.2.2, 3.3.1 (only basic examples and $\sqrt[n]{z}$, $\log z$), Secs. 3.3.2, 3.4.1–3.4.4, 3.5.1–3.5.5, 3.5.7 (sketch only); Secs. 4.8, 4.9, and 4.10 (sketch only), Sec. 4.11 (sketch only), Secs. 4.12.1–4.12.3; Secs. 5.3.1, 5.4.1, and 5.5.2 (optional and sketch).

As a whole, Examples and Exercise A should be selective. Minded readers should try more, both Examples and Exercises, and pay attention to more elementary multiple-valued functions and their Riemann surfaces, if possible.

In a class, the role played by a lecture to select topics is crucial.

(2) As a beginning graduate text.

With a solid understanding of materials in (1), the following topics are added: Secs. 2.8; 3.4.5, 3.4.6; 4.1–4.7, 4.12.3A–4.15 (selective); Secs. 5.1.3, 5.2, 5.3.2, 5.5–5.6 (selective), Secs. 5.8.1–5.8.3; Chap. 6 except Sec. 6.6.4.

Examples and Exercise A (even B) should be emphasized. Of course, the adding or deleting of some topics are still possible.

(3) To readers who are interested in Riemann surfaces.

Pay more attention to multiple-valued functions and their descriptive Riemann surfaces such as Secs. 2.7, 3.3.3, 3.4.7, 3.5.6; 5.1, 5.2, and end up with the whole Chap. 7.

(4) Several complex variables.

Sections 3.4.7, 3.5.6; Chap. 7.

(5) Quasiconformal mappings and complex dynamical systems.
    Section 3.2.3, Example 2 in Secs. 3.5.5; 5.3.4, 5.8; 6.6.4; Chap. 7 and
    Appendix C.
(6) As a general reference book supplement to other books on complex
    analysis.

Acknowledgments

The following students in Mathematics Department helped type my hand-written manuscript:

Jing-ya Shui; Ya-ling Zhan; Yu-hua Weng; Ming-yang Kao; Wei-ming Su; Wen-jie Li; Shuen-hua Liang; Shi-wei Lin; D. C. Peter Hong; Hsuan-ya Yu; Yi-hsuan Lin; Ming-you Chin; Che-wei Wu; Cheng-han Yang; Kuo-han Tseng; Yi-ting Tsai; Yi-chai Li; Po-tsu Lin; Hsin-han Huang.

Yan-yu Chen graphed all the Figures appearing in this book. Yan-yu Chen, Aileen Lin, Wen-jie Li, and Ming-yang Kao helped edit the final manuscript for printing. Here may I pay my sincerest thanks to all of them. Without their unselfish dedication, this book definitely cannot be published so soon.

Also, teaching assistant Jia-ming Ying helped improve and correct partially my English writing. My colleagues Prof. Tian-yu Tsai proof-read the entire manuscript, and Prof. Yu-lin Chang proof-read Chaps. 5–7 and adopted parts of the content in his graduate course on complex analysis. Both of them pointed out many mistakes and gave me valuable suggestions. It would be my pleasure to express my gratitude toward them for their kindest help.

As usual, teaching assistant Ching-yu Yang did all the computer work for the several editions of the manuscript. And Ms. Tan Rok Ting, an editor in World Scientific, copyedited the whole book with carefulness and expertise. Thank them so much.

I-hsiung Lin
21 January 2009
Taipei, Taiwan

Contents

CHAPTER 1

Complex Numbers

This opening chapter is to introduce the basic knowledge about the complex numbers in three aspects and the interactions among them.

Sketch of the Content

Algebraic: The imaginary solution $i = \sqrt{-1}$ of the equation $t^2 + 1 = 0$ creates the complex numbers $z = x + iy$ which obey the same basic laws of arithmetics as the reals do (Sec. 1.3). The main distinction between them lies on the fact that the complexes cannot be ordered, but compensated by the conjugate operation $z \to \bar{z}$ (Sec. 1.4.1). This enables us to interrelate both by the operation $|z|^2 = z\bar{z}$ and to introduce the inequalities among various $|z|$ (Secs. 1.4.1 and 1.4.2). An instant consequence is that every polynomial with complex coefficients is always solvable; in particular, $z^n - 1 = 0$ has exactly n-distinct roots (Sec. 1.5).

Geometric: $z = x + iy$ can be understood not only as a point or a vector in the Euclidean plane $\mathbf{R}^2$, but also as a planar motion composed of one-way stretch and rotation (Secs. 1.1, 1.2, and 1.4). It comes naturally that $z = x + iy$ can be represented as the point (x, y) in $\mathbf{R}^2$ which is thus renamed as the *complex plane* $\mathbf{C}$, or in the polar form $re^{i\theta}$ ($r = |z|, \theta = \arg z$) with the origin 0 as pole and the positive x-axis as the polar axis. And then, they can be used effectively to describe planar geometric objects or to solve geometric problems (Secs. 1.4.3 and 1.4.4). Therefore, almost every algebraic operation about complex numbers has its illuminative geometric meaning.

Topological (point-set properties): Owing to $|x|, |y| \leq |z| = \sqrt{x^2 + y^2} \leq |x| + |y|$ ($z = x + iy$), the limit concepts of real sequences and series and their properties can be carried verbatim over to the complex ones except the one appeared in (3) of (1.7.3), where $\arg z_n \to \arg z_0$ needs to be treated carefully. So do the concepts of open, closed, compact, and connected sets for both $\mathbf{R}^2$ and $\mathbf{C}$ (Sec. 1.8). The addition of the point at infinity ∞ to $\mathbf{C}$

to obtain the *extended complex plane* $\mathbf{C}^*$, realized as the *Riemann sphere* S in $\mathbf{R}^3$, seems more naturally for the need of the limit concept than the algebraic operations (Secs. 1.6 and 1.9). The most importance of all is that $\mathbf{C}$ is a complete metric space in which every Cauchy sequence is convergent (Sec. 1.9).

1.1.   How to Visualize Geometrically the Existence of the So-Called Complex Numbers in Our Daily Life

As already well known, man creates

(1) the natural number system $\mathbf{N}$ for counting;
(2) the integer number system $\mathbf{Z}$ for the negative of a quantity;
(3) the rational number system $\mathbf{Q}$ for fractions of a whole quantity or for measurement such as length, and
(4) the real number system $\mathbf{R}$ for measurement (see Appendix A).

These numbers are vivid in our daily life. Though we cannot see what irrational numbers such as e and π would exactly look like, nowadays we are able to approximate them as accurately as we want by rational numbers.

Numbers of the form $a + b\sqrt{-1}$, where a, $b \in \mathbf{R}$, the so-called complex numbers, appeared as early as 16th century when mathematicians tried to solve quadratic equations.

Man *imagines* that there exists a "number" i satisfying $x^2 + 1 = 0$. This i denotes $\sqrt{-1}$. As a consequence, a quadratic equation $ax^2 + bx + c = 0$ with a, b, $c \in \mathbf{R}$ and $b^2 - 4ac < 0$ would have two roots $\frac{-b \pm \sqrt{4ac - b^2}i}{2a}$. They are not real but only imaginary. They seem so freaky because man does not even know how to approximate them by known numbers from the well-established number systems. It was C. Gauss and I. Argand who interpreted the numbers $a + bi$ (also denoted as $a + ib$) geometrically as the point (a, b) on the plane and hence, laid a firm basis for the development of the complex function theory. For historical account, see Ref. [47].

In what follows, we assume the readers are familiar with basic knowledge about the Euclidean plane $\mathbf{R}^2$.

Fix the rectangular xy-coordinate system on the plane $\mathbf{R}^2$. The point (x, y) can be considered as *the position vector* from $(0, 0)$ pointed to (x, y) itself. See Fig. 1.1. Under this circumstance, *vector operations* are applied to (x, y) as follows to form *the two-dimensional real space* $\mathbf{R}^2$:

(1) Addition

$$(x_1, y_1) + (x_2, y_2) = (x_1 + x_2, y_1 + y_2)$$

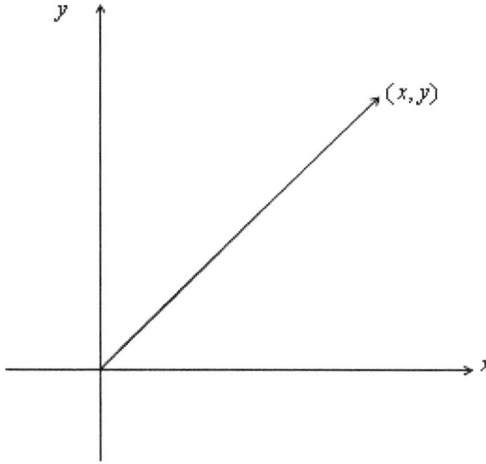

Fig. 1.1

satisfying the following properties:

(i) (commutative) $(x_1, y_1) + (x_2, y_2) = (x_2, y_2) + (x_1, y_1)$;

(ii) (associative) $((x_1, y_1) + (x_2, y_2)) + (x_3, y_3) = (x_1, y_1) + ((x_2, y_2) + (x_3, y_3))$;

(iii) (zero vector) $(x, y) + (0, 0) = (x, y)$;

(iv) (inverse vector) For each (x, y),

$$(x, y) + (-x, -y) = (0, 0).$$

(2) Scalar multiplication

$$\alpha(x, y) = (\alpha x, \alpha y), \quad \alpha \in \mathbf{R}$$

satisfying:

(i) $1(x, y) = (x, y)$.

(ii) $(\alpha\beta)(x, y) = \alpha(\beta(x, y))$, $\alpha, \beta \in \mathbf{R}$.

(3) Distributive law

$$(\alpha + \beta)(x, y) = \alpha(x, y) + \beta(x, y), \quad \alpha, \beta \in \mathbf{R};$$

$$\alpha((x_1, y_1) + (x_2, y_2)) = \alpha(x_1, y_1) + \alpha(x_2, y_2).$$

(1.1.1)

Owing to lack of product operation among vectors, conceptually it is not enough to identify fully the imaginary number $x + iy$ (also denoted alternatively as $x + yi$) with the point (x, y) or the vector it induces. The *point* is that we have to define what the *product*

$$(x_1, y_1)(x_2, y_2)$$

means properly so that it still represents a number $a + bi$ and satisfies nice operational properties such as commutative and associative laws, etc.

Hence, we designate the notation

$z = x + yi$, where $x, y \in \mathbf{R}$ and are not both equal to zero.

$\Leftrightarrow$ (1) z is the point (x, y) or the vector (x, y) in $\mathbf{R}^2$, and
   (2) z represents the *one-way stretch* along the x-axis with scale factor $r = \sqrt{x^2 + y^2}$ and then followed by a *rotation* with center at $(0, 0)$ through an angle in the counterclockwise direction so that the x-axis will coincide with the line generated by the

$$\text{vector } (x, y). \tag{1.1.2}$$

Note that in condition 2, we can perform rotation first and one-way stretch second. Both are commutative. See Fig. 1.2.

In particular,

$$i = (0, 1) = 0 + 1i, \tag{1.1.3}$$

represents the counterclockwise rotation of $90°$ *of the point* $(1, 0)$ *or any nonzero vector.* Let $i^2 = i \cdot i$ denote two such consecutive motions, etc.

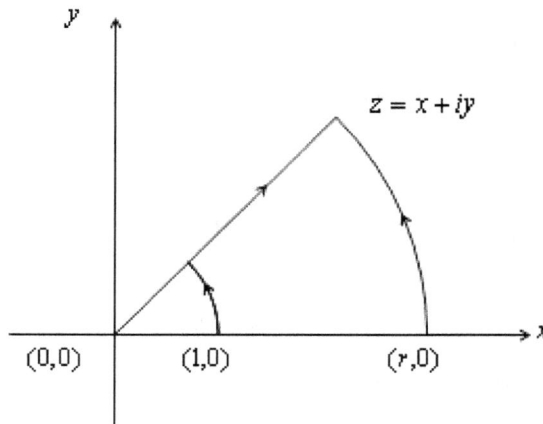

Fig. 1.2

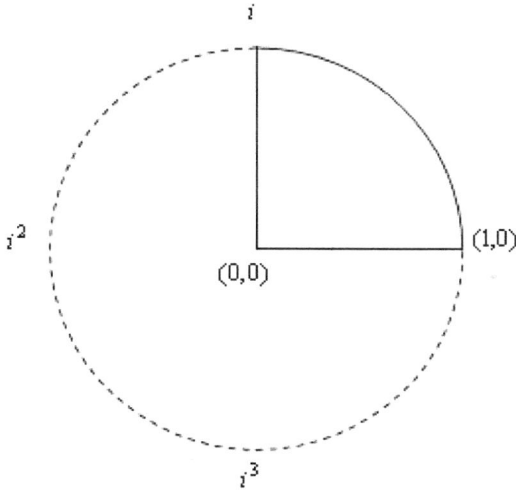

Fig. 1.3

Then (see Fig. 1.3)

$$i^2 = (-1,0) = -(1,0) = -1,$$
$$i^3 = (0,-1) = -(0,1) = -i, \qquad (1.1.4)$$
$$i^4 = (1,0) = 1 + i0 = 1.$$

Finally, we designate

$$0 = 0 + i0, \qquad (1.1.5)$$

where the two zeros 0 on the right are the zero number in **R**. Hence, it is reasonable to define

$$z = x + iy = 0 = 0 + i0 \Leftrightarrow x = 0 \quad \text{and} \quad y = 0. \qquad (1.1.6)$$

Also, we define, for $z \neq 0$,

the *modulus* or *absolute value* $|z| = \sqrt{x^2 + y^2} = r,$

and

the *principal argument* $\operatorname{Arg} z = \theta$ shown in Fig. 1.2. $\qquad (1.1.7)$

Let $z_1 = x_1 + iy_1$ and $z_2 = x_2 + iy_2$. We try to define the *product* $z_1 z_2$ properly. Knowing what z_1 means as in (1.1.2), we define, assuming $z_1 \neq 0$

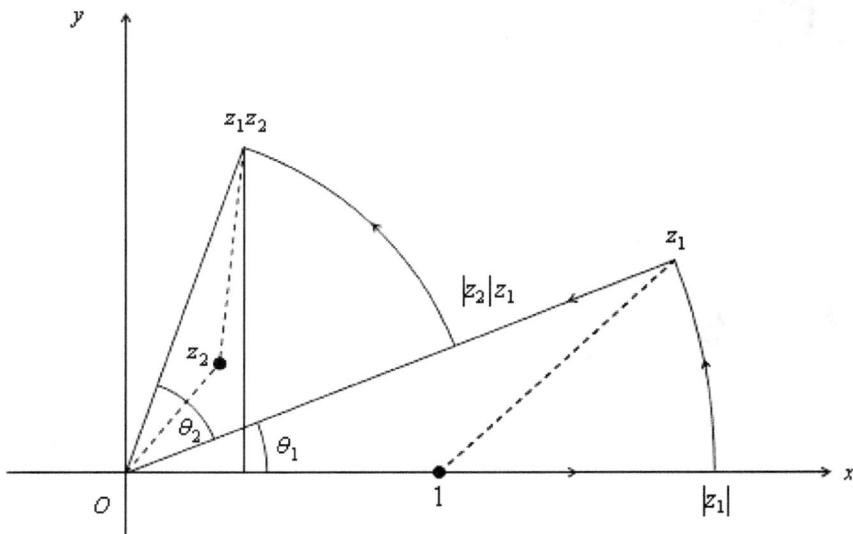

Fig. 1.4

and $z_2 \neq 0$,

> $z_1 z_2 =$ the point obtained by one-way stretch along the vector z_1
> with scale factor $|z_2|$ and then followed by a counterclockwise
> rotation through the principal argument $\text{Arg}\, z_2$ of z_2.   (1.1.8)

See Fig. 1.4.

To pinpoint the coordinate of $z_1 z_2$, let $\theta_1 = \text{Arg}\, z_1$ and $\theta_2 = \text{Arg}\, z_2$ for simplicity. Then $z_1 z_2$ has coordinate

$$|z_1|\,|z_2|(\cos(\theta_1 + \theta_2), \sin(\theta_1 + \theta_2)),$$

where

$$\cos(\theta_1 + \theta_2) = \cos\theta_1 \cos\theta_2 - \sin\theta_1 \sin\theta_2 = \frac{x_1}{|z_1|} \cdot \frac{x_2}{|z_2|} - \frac{y_1}{|z_1|} \cdot \frac{y_2}{|z_2|},$$

and

$$\sin(\theta_1 + \theta_2) = \sin\theta_1 \cos\theta_2 + \cos\theta_1 \sin\theta_2 = \frac{y_1}{|z_1|} \cdot \frac{x_2}{|z_2|} + \frac{x_1}{|z_1|} \cdot \frac{y_2}{|z_2|}.$$

Hence

$$z_1 z_2 = (x_1 x_2 - y_1 y_2, x_1 y_2 + x_2 y_1) = x_1 x_2 - y_1 y_2 + i(x_1 y_2 + x_2 y_1).$$

$$(1.1.9)$$

Note that, in Fig. 1.4, the triangle with vertices at points 0, 1, and z_1 is similar to the triangle with vertices at 0, z_2, and z_1z_2. *Suppose the symbol defined by* (1.1.2) *enjoys the operational properties such as commutative, associative, and distributive laws* which we are used to in the real number system **R**, then (1.1.9) is nothing but the usual product as

$$z_1z_2 = (x_1 + iy_1)(x_2 + iy_2)$$
$$= x_1x_2 + x_1(iy_2) + (iy_1)x_2 + (iy_1)(iy_2)$$
$$= x_1x_2 + ix_1y_2 + iy_1x_2 + i^2y_1y_2$$
$$= x_1x_2 - y_1y_2 + i(x_1y_2 + x_2y_1),$$

by using (1.1.4).

In case either $z_1 = 0$ or $z_2 = 0$ or $z_1 = z_2 = 0$, we just define

$$z_1z_2 = 0. \tag{1.1.10}$$

We call the symbol defined in (1.1.2) a *complex number*. *It is not only a plane vector, but also carries with itself the composite motion of one-way stretch and rotation, and hence, is a two-dimensional "number" which cannot be approximated by any number we knew before.* Mathematics based on the complex numbers is, geometrically, the one about similarity in global sense and is the one about conformality in local sense via limit processes.

Exercises A

(1) Let $z = x + iy \neq 0$. Try to use (1.1.2) and Fig. 1.2 to show that the unique w satisfying $zw = 1$ is

$$\frac{x - iy}{x^2 + y^2}$$

which is denoted as z^{-1} or $\frac{1}{z}$.

(2) Try to locate the following complex numbers:

$$1 - 2i; \quad \frac{1}{1+i}; \quad (3 + 2i)(1 - i); \quad \frac{1 - i}{3 + 2i}.$$

(3) Try to use (1.1.2) to show that the product operation satisfies

   (i) the commutative law $z_1z_2 = z_2z_1$;
   (ii) the associative law $(z_1z_2)z_3 = z_1(z_2z_3)$; and
   (iii) the distributive law $z_1(z_2 + z_3) = z_1z_2 + z_1z_3$.

(4) Let z_1, z_2, and z_3 be three complex numbers, noncollinear when considered as points. Fix any z_0 and let $w_1 = z_0z_1$, $w_2 = z_0z_2$, and $w_3 = z_0z_3$.

(a) Show that w_1, w_2, and w_3 are noncollinear.

(b) Graph the triangles $\triangle z_1 z_2 z_3$ and $\triangle w_1 w_2 w_3$. Compute their corresponding angles and areas.

Exercises B

(1) Try to design a kind of chess game based on the idea shown in (1.1.2).

1.2.   Complex Number and Its Geometric Representations

Section (1) The imaginary unit i

Suppose x is a real number. Then $x^2 \geq 0$ and hence, $x^2 + 1 \geq 1 > 0$ holds. Hence the equation $t^2 + 1 = 0$ does not have any solution in **R**.

By imagination, suppose there exists a "number", denoted by

$$i = \sqrt{-1} \tag{1.2.1}$$

satisfying that the "product" of i with itself is -1, namely,

$$i^2 = -1.$$

Then $i^2 + 1 = 0$ holds and i is a solution of $x^2 + 1 = 0$. Another solution is $-i$. We call i the *imaginary unit*.

Section (2) The complex number

We do formal "addition" and "multiplication" of two real numbers x, y and i into the symbol, denoted as

$$z = x + iy \quad \text{or} \quad x + yi, \tag{1.2.2}$$

and is called a *complex number* formed with

$$\text{the } real \ part \ \mathrm{Re}\, z = x,$$

and

$$\text{the } imaginary \ part \ \mathrm{Im}\, z = y. \tag{1.2.3}$$

For example, $1 + 0i$, $0 + 2i$, $\sqrt{2} + \frac{1}{2}i$, etc. are complex numbers.

A complex number $z = x + iy$ with its $\mathrm{Im}\, z = y = 0$ is specifically denoted as

$$x + i0 = x, \tag{1.2.4}$$

and is considered as real number in many occasions. For example, $-1 + 0i = -1$.

A complex number $z = x + iy$ with its $\operatorname{Re} z = x = 0$ is denoted as

$$0 + iy = iy, \tag{1.2.5}$$

and is called a *pure imaginary*. For example, $0 + 2i = 2i$. Hence, a complex number $x + iy$ is said to be *imaginary* if $y \neq 0$.

In particular, only the *zero* complex number

$$0 = 0i = 0 + 0i, \tag{1.2.6}$$

is both real and pure imaginary.

Section (3) As a point in the Euclidean plane $\mathbf{R}^2$

Fix a rectangular xy-coordinate system in the plane. We designate the *point* (x, y) as the complex number $z = x + iy$ and vice versa. This sets up a one-to-one and onto correspondence between complex numbers and points in the plane which is thus called a *complex plane*. In particular, real numbers x are in one-to-one and onto correspondence with points in the x-axis, hence called the *real axis*; pure imaginaries iy and the points y in the y-axis are in one-to-one and onto correspondence and hence the y-axis is called the *imaginary axis*. See Fig. 1.5.

Henceforth, no distinction will be made between the set of all complex numbers

$$\mathbf{C} = \{x + iy \mid x, y \in \mathbf{R}\} \tag{1.2.7}$$

and the complex plane, also denoted as $\mathbf{C}$.

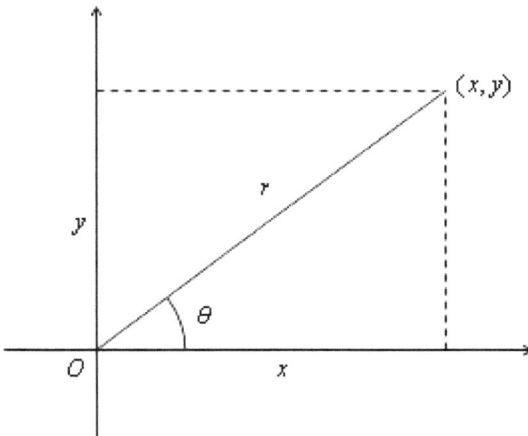

Fig. 1.5

Section (4) As a plane vector pointed from $(0,0)$ to (x,y)

$z = x + iy$ can also be considered as a *plane vector* pointed from $(0,0)$ toward the point (x,y). In this case, x and y are orthogonal projections of the vector on the real and imaginary axes, respectively. See Fig. 1.5.

The length r of the vector, denoted as

$$|z| = r = \sqrt{x^2 + y^2}, \qquad (1.2.8)$$

is called the *modulus* or the *absolute value* of the complex number z. Note that $|z| \geq 0$, and $|z| = 0 \Leftrightarrow z = 0$.

In the case $z = x + iy \neq 0$, we call the angle, denoted as

$$\arg z, \qquad (1.2.9)$$

between the vector z and the positive direction of the real axis *an argument* of z. Usually, we define

$$\arg z > 0 \quad \text{if } \arg z \text{ is obtained by counterclockwise rotation}$$

and

$$\arg z < 0 \quad \text{if obtained by clockwise rotation.} \qquad (1.2.10)$$

Figure 1.6 shows that $\arg z$ is multiple-valued. In the case $z = 0$, $\arg z$ is not defined.

We summarize the above as

The multiple-valuedness of $\arg z$ *($z \neq 0$). The value of* $\arg z$ *that lies on* $-\pi < \theta \leq \pi$ *(or $0 \leq \theta < 2\pi$) is called the principal values of* $\arg z$ *or the principal argument of z and is denoted as*

$$\text{Arg } z.$$

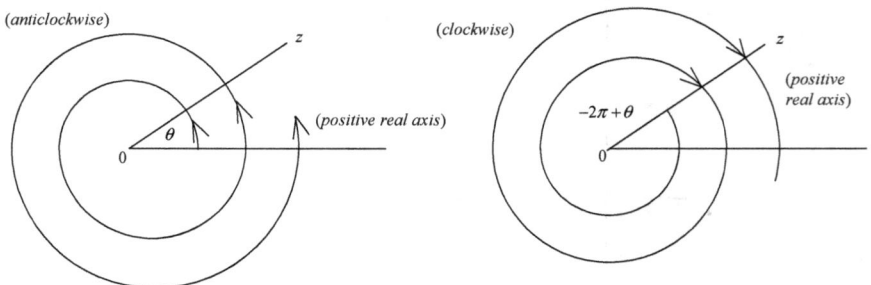

Fig. 1.6

Then

$$\arg z = \operatorname{Arg} z + 2n\pi, \quad n = 0, \pm 1, \pm 2, \ldots . \tag{1.2.11}$$

The argument $\arg z$ is the main trouble-maker in generating multiple-valued functions in complex analysis. For a detailed treatment, see Sec. 2.7.1.

We illustrate an

Example 1. For each of the following z, compute $|z|$, $\operatorname{Arg} z$ and $\arg z$.

(1) $x \in \mathbf{R}$ and $x \neq 0$;
(2) iy, $y \in \mathbf{R}$ and $y \neq 0$;
(3) $-1 - i$;
(4) $-\frac{1}{2} + \frac{\sqrt{3}}{2} i$.

Solution. (1) If $x > 0$, $|x| = x$, $\operatorname{Arg} x = 0$ and $\arg x = 2n\pi$, $n \in \mathbf{Z}$; if $x < 0$, $|x| = -x$, $\operatorname{Arg} x = \pi$ and $\arg x = \pi + 2n\pi$, $n \in \mathbf{Z}$;

(2) If $y > 0$, $|iy| = y$, $\operatorname{Arg}(iy) = \frac{\pi}{2}$, and $\arg(iy) = \frac{\pi}{2} + 2n\pi$, $n \in \mathbf{Z}$; if $y < 0$, $\operatorname{Arg}(iy) = -\frac{\pi}{2}$, and $\arg(iy) = -\frac{\pi}{2} + 2n\pi$, $n \in \mathbf{Z}$;

(3) $|z| = \sqrt{2}$, $\operatorname{Arg} z = -\frac{3\pi}{4}$, and $\arg z = -\frac{3\pi}{4} + 2n\pi$, $n \in \mathbf{Z}$;

(4) $|z| = 1$, $\operatorname{Arg} z = \frac{2\pi}{3}$, and $\arg z = \frac{2\pi}{3} + 2n\pi$, $n \in \mathbf{Z}$.

Section (5) Trigonometric or polar representation

Consider the origin $(0,0)$ as the pole and the positive x-axis as the polar axis in a polar coordinate system. Then, the modulus $|z|$ and the argument $\arg z$ of a complex number z are, respectively, the *polar radius* r and *polar angle* θ of the vector z in the polar coordinate system (see Fig. 1.5):

$$x = \operatorname{Re} z = r \cos \theta,$$
$$y = \operatorname{Im} z = r \sin \theta,$$

and then,

$$z = x + iy = r(\cos\theta + i\sin\theta), \quad \text{where } r = \sqrt{x^2 + y^2} \text{ and } \theta = \tan^{-1}\frac{y}{x}, \tag{1.2.12}$$

which is called *the trigonometric* or *polar form* of z. For simplicity, we introduce the *Euler's formula*

$$e^{i\theta} = \cos\theta + i\sin\theta, \quad \theta \in \mathbf{R}, \tag{1.2.13}$$

which is to be justified in Sec. 2.6.1. Then (1.2.12) can be rewritten as the concise form $re^{i\theta}$.

Remark (The relation between $\operatorname{Arg} z$ and $\tan^{-1} \frac{y}{x}$). Suppose $-\pi < \operatorname{Arg} z \le \pi$ and $-\frac{\pi}{2} < \operatorname{Arc\,tan} \frac{y}{x} < \frac{\pi}{2}$, the principle value of $\tan^{-1} \frac{y}{x}$.

Then

$$\operatorname{Arg} z = \begin{cases} \operatorname{Arc\,tan} \dfrac{y}{x}, & z \text{ in the first or the fourth quadrant,} \\[2mm] \operatorname{Arc\,tan} \dfrac{y}{x} + \pi, & z \text{ in the second quadrant,} \\[2mm] \operatorname{Arc\,tan} \dfrac{y}{x} - \pi, & z \text{ in the third quadrant.} \end{cases}$$

If $0 \le \operatorname{Arg} z < 2\pi$, then the above relations should be replaced, respectively, by

$$\operatorname{Arg} z = \begin{cases} \operatorname{Arc\,tan} \dfrac{y}{x}, & z \text{ in the first quadrant,} \\[2mm] \operatorname{Arc\,tan} \dfrac{y}{x} + \pi, & z \text{ in the second or the third quadrant,} \\[2mm] \operatorname{Arc\,tan} \dfrac{y}{x} + 2\pi, & z \text{ in the fourth quadrant.} \end{cases}$$

Recall that $\tan^{-1} \frac{y}{x} = \operatorname{Arc\,tan} \frac{y}{x} + n\pi$, $n \in \mathbf{Z}$.   □

Section (6) Some applications

Let $z_1, z_2 \in \mathbf{C}$. When we view both z_1 and z_2 as vectors, the *sum* $z_1 + z_2$ is defined as the addition of the vectors z_1 and z_2; on the other hand, $z_1 - z_2 = z_1 + (-z_2)$ as the difference of the vector z_1 from the vector z_2. See Fig. 1.7.

In case $z_1 \ne 0$ and $z_2 \ne 0$, the polar forms

$$z_k = |z_k|(\cos\theta_k + i\sin\theta_k), \quad k = 1, 2 \tag{1.2.14}$$

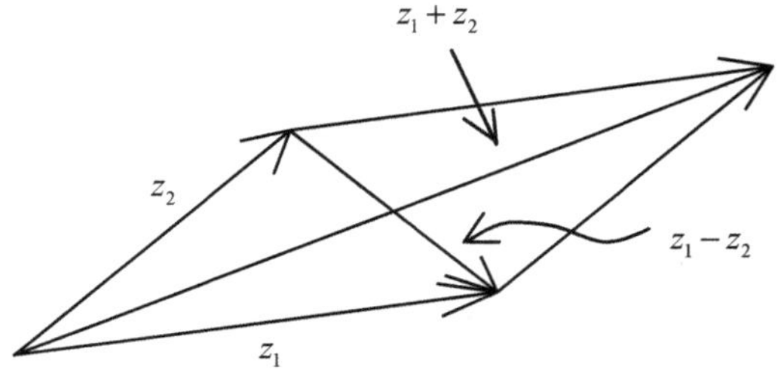

Fig. 1.7

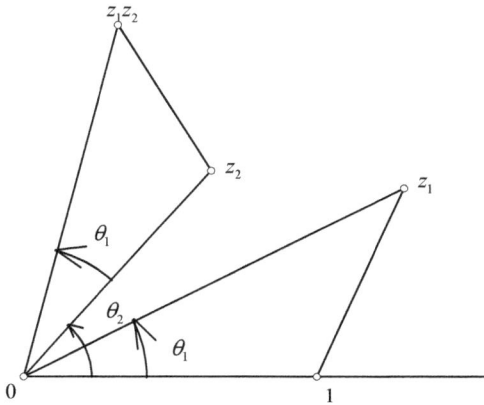

Fig. 1.8

enable us to *define* the *product* of z_1 and z_2 as

$$z_1 z_2 = |z_1| |z_2| (\cos(\theta_1 + \theta_2) + i \sin(\theta_1 + \theta_2)). \qquad (1.2.15)$$

This is what we did in (1.1.9). Figure 1.8 shows the geometric meaning behind this definition. The triangle with vertices at 0, 1, and z_1 is similar to the triangle with vertices at 0, z_2, and $z_1 z_2$. Rotate z_2 through the angle θ_1 and then truncate a vector of length $|z_1| |z_2|$. This resulting vector will be $z_1 z_2$.

Similarly, the division of z_1 by z_2 is defined as

$$\frac{z_2}{z_1} = \frac{|z_2|}{|z_1|} (\cos(\theta_2 - \theta_1) + i \sin(\theta_2 - \theta_1)). \qquad (1.2.15)'$$

Figure 1.9 indicates its geometric interpretation.

We summarize the above as

The moduli and arguments of product and division of two complex numbers.

(1)
$$|z_1 z_2| = |z_1| |z_2|,$$

$$\left| \frac{z_2}{z_1} \right| = \frac{|z_2|}{|z_1|} \quad (z_1 \neq 0).$$

(2) *Suppose* $z_1 \neq 0$ *and* $z_2 \neq 0$. *Then*

$$\arg z_1 z_2 = \arg z_1 + \arg z_2,$$

$$\arg \frac{z_2}{z_1} = \arg z_2 - \arg z_1. \qquad (1.2.16)$$

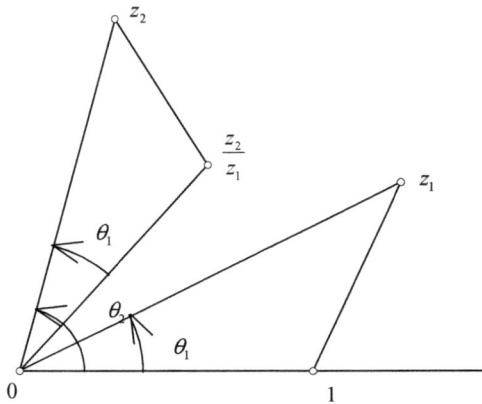

Fig. 1.9

It is understood that the last two relations should be interpreted by
any one of the following statements:

(i) consider both sides as sets of arguments and then treat both sides
as identical of sets;

(ii) preassign any one value to each of $\arg z_1$ and $\arg z_2$, then there
exists a value of $\arg z_1 z_2$ (or $\arg \frac{z_2}{z_1}$) so that both sides are equal;

(iii) $\arg z_1 z_2$ (or $\arg \frac{z_2}{z_1}$) is considered as the set of the sums (or differences) of all possible values of $\arg z_1$ and $\arg z_2$.

Two numerical examples are provided to illustrate the points made in
the last paragraph.

Example 2. Let $z_1 = 1 + \sqrt{3}i$ and $z_2 = 1 - i$. Compute $\text{Arg}\, z_1 z_2$, $\arg z_1 z_2$,
$\text{Arg}\, \frac{z_1}{z_2}$, and $\arg \frac{z_1}{z_2}$.

Solution. $z_1 z_2 = (1 + \sqrt{3}) + (-1 + \sqrt{3})i$ and $\frac{z_1}{z_2} = \frac{1}{2}[(1 - \sqrt{3}) + (1 + \sqrt{3})i]$
(see (1.1.9)).

Then, compute $\arg z_1 z_2$, etc., directly.

On the other hand, $\text{Arg}\, z_1 = \frac{\pi}{3}$, $\arg z_1 = \frac{\pi}{3} + 2n\pi$, $n \in \mathbf{Z}$, and $\text{Arg}\, z_2 = -\frac{\pi}{4}$, $\arg z_2 = -\frac{\pi}{4} + 2m\pi$, $m \in \mathbf{Z}$. Then,

$$\text{Arg}\, z_1 z_2 = \text{Arg}\, z_1 + \text{Arg}\, z_2 = \frac{\pi}{3} - \frac{\pi}{4} = \frac{\pi}{12},$$

$$\arg z_1 z_2 = \arg z_1 + \arg z_2 = \frac{\pi}{12} + 2(n+m)\pi = \frac{\pi}{12} + 2p\pi, \quad p \in \mathbf{Z};$$

$$\text{Arg}\,\frac{z_1}{z_2} = \text{Arg}\,z_1 - \text{Arg}\,z_2 = \frac{7\pi}{12},$$

$$\arg\frac{z_1}{z_2} = \arg z_1 - \arg z_2 = \frac{7\pi}{12} + 2(n-m)\pi = \frac{7\pi}{12} + 2q\pi, \quad q \in \mathbf{Z}.$$

Example 3. Compute $(5-i)^4(1+i)$ and then, prove the *Machin formula*

$$\frac{\pi}{4} = 4\,\text{Arc}\tan\frac{1}{5} - \text{Arc}\tan\frac{1}{239}.$$

Solution. $(5-i)^4(1+i) = (476 - 480i)(1+i) = 4(239 - i)$. Since

$$\text{Arg}(5-i)^4 = -4\,\text{Arc}\tan\frac{1}{5},$$

$$\text{Arg}(1+i) = \frac{\pi}{4},$$

$$\text{Arg}(239 - i) = -\text{Arc}\tan\frac{1}{239},$$

$$\Rightarrow \text{ (by using (1.2.15))} \ -4\,\text{Arc}\tan\frac{1}{5} + \frac{\pi}{4} = -\text{Arc}\tan\frac{1}{239}.$$

The result follows. This trigonometric identity can be used to compute approximate values of π.

Exercises A

(1) Locate the following complex numbers as points in the complex plane.
  (a) $-3 + \sqrt{2}i$.   (b) $\pm 1 \pm i$.   (c) $\sqrt{3} + 6i$.   (d) $(-2 + 3i)^4$.
  (e) $(-1 + 2i)(3 - i)^2$.   (f) $(\frac{1}{2} + i)^3 - (\frac{1}{2} - i)^3$.

(2) Write each of the following z in its polar form and compute $|z|$, $\text{Arg}\,z$, $\arg z$:
  $1 + i$; $1 - i$; $-1 - i$; $-1 + i$; $1 + \sqrt{3}i$; $-1 + \sqrt{3}i$; $\sqrt{3} - i$; $2 + \sqrt{3} + i$.

(3) Choose any two complex numbers z_1 and z_2 out of these in Exercise (1). Compute: $|z_1 z_2|$, $|\frac{z_1}{z_2}|$: $\text{Arg}\,z_1 z_2$, $\arg z_1 z_2$; $\text{Arg}\,\frac{z_1}{z_2}$, $\arg\frac{z_1}{z_2}$.

(4) (a) Compute $(2+i)(3+i)$ and prove that $\frac{\pi}{4} = \text{Arc}\tan\frac{1}{2} + \text{Arc}\tan\frac{1}{3}$.
  (b) Show that $\frac{\pi}{4} = 3\,\text{Arc}\tan\frac{1}{4} + \text{Arc}\tan\frac{1}{20} + \text{Arc}\tan\frac{1}{1985}$.

1.3. Complex Number System (Field) C

Repeat (1.2.7) and denote

$$\mathbf{C} = \{x + iy \,|\, x, y \in \mathbf{R}\},$$

the set of all complex numbers. We try to endow $\mathbf{C}$ with suitable operations of addition and multiplication (see (1.2.15)) so that $\mathbf{C}$ becomes *formally* a number system and hence, a field.

Let $z_1 = x_1 + iy_1$ and $z_2 = x_2 + iy_2$. Define

equality: $z_1 = z_2 \Leftrightarrow x_1 = x_2$ and $y_1 = y_2$;

addition: $z_1 + z_2 = (x_1 + x_2) + i(y_1 + y_2)$, called the *sum* of z_1 and z_2;

multiplication: $z_1 z_2 = (x_1 x_2 - y_1 y_2) + i(x_1 y_2 + x_2 y_1)$,

$$\text{called the } \textit{product} \text{ of } z_1 \text{ and } z_2. \tag{1.3.1}$$

The complex field **C**

(1) Addition has the following properties:

   (i) (commutative) $z_1 + z_2 = z_2 + z_1$;

   (ii) (associative) $(z_1 + z_2) + z_3 = z_1 + (z_2 + z_3)$;

   (iii) (zero element) there exists a unique element $0 = 0 + i0 \in$ **C** so that $z + 0 = z$, $z \in$ **C**;

   (iv) (inverse element) for each $z = x + iy$, there corresponds a unique element

$$-z = -(x + iy) = -x + i(-y)$$

   so that

$$z + (-z) = 0.$$

(2) Multiplication has the following properties:

   (i) (commutative) $z_1 z_2 = z_2 z_1$;

   (ii) (associative) $(z_1 z_2) z_3 = z_1 (z_2 z_3)$;

   (iii) (unit element) there exists a unique element $1 = 1 + i0$ so that $1z = z$, $z \in$ **C**;

   (iv) (inverse element) for each nonzero element $z = x + iy \in$ **C**, there exists a unique element denoted as

$$z^{-1} = \frac{1}{z} = \frac{1}{x + iy} = \frac{x - iy}{x^2 + y^2}$$

   so that

$$zz^{-1} = 1.$$

(3) Addition and multiplication satisfy

$$\text{(distributive law) } z_1(z_2 + z_3) = z_1 z_2 + z_1 z_3. \tag{1.3.2}$$

As a whole, **C** with two *such* operations is called *the complex field*; while properties (1) 1 and 2, (2) 1 and 2, and (3) together say that **C** indeed is a *number system*.

Remark 1 (The real field R as a subfield of C). The subset

$$\widetilde{\mathbf{R}} = \{x + i0 \,|\, x \in \mathbf{R}\},$$

has all the properties listed in (1.3.2) for its elements and hence, is called a *subfield* of **C**. Since the correspondence

$$x \in \mathbf{R} \rightarrow x + i0 \in \widetilde{\mathbf{R}},$$

is one-to-one and onto, and preserves all the operations concerned, we identify **R** with $\widetilde{\mathbf{R}}$ (see Section (3) in Sec. 1.2). Under this circumstance, the real number system (field)

$$\mathbf{R} \subseteq \mathbf{C}, \tag{1.3.3}$$

holds as a *subfield* of **C**.   □

Remark 2 (C is not an ordered field). Recall that **R** is an ordered field with respect to the operation "< (less than)" (refer to Appendix A).

Now, suppose **C** was an ordered field. Since $i \neq 0$, so either $i > 0$ or $i < 0$ holds. Both imply that

$$i^2 = -1 > 0,$$
$$\Rightarrow (\text{Multiply both sides by } -1 > 0)1 > 0; \text{ while}$$

(Add both sides by $1 > 0$) $0 > 1$, which is a contradiction.

When computing with complex numbers, one often needs the circular operational properties of i: $i^1 = i$, $i^2 = -1$, $i^3 = -i$, and $i^4 = 1$. In general,

$$i^{4n+1} = i, \quad i^{4n+2} = -1, \quad i^{4n+3} = -i, \quad i^{4n+4} = 1, \quad n \in \mathbf{Z}.$$

We give three examples.

Example 1. Write each of the following complex numbers z in the form $x + iy$.
(1) $\left(\frac{2+i}{3-2i}\right)^2$. (2) $\frac{(1-i)^5-1}{(1+i)^5+1}$. (3) $(1+i)^n + (1-i)^n$, $n \in \mathbf{Z}$. (4) $(1+i)^n - (1-i)^n$, $n \in \mathbf{Z}$.

Solution. (1)

$$\frac{2+i}{3-2i} = (2+i)\left(\frac{3}{13} + \frac{2}{13}i\right) = \frac{4}{13} + \frac{7}{13}i$$

$$\Rightarrow \left(\frac{2+i}{3-2i}\right)^2 = \frac{1}{169}(-33+56i).$$

(2)

$$(1-i)^5 = 1^5 + 5 \cdot 1^4 \cdot (-i) + 10 \cdot 1^3 \cdot (-i)^2 + 10 \cdot 1^2 \cdot (-i)^3$$

$$+ 5 \cdot 1 \cdot (-i)^4 + (-i)^5$$

$$= 1 - 5i - 10 + 10i + 5 - i = -4(1-i)$$

and

$$(1+i)^5 = -4(1+i).$$

Or,

$$(1-i)^2 = -2i \Rightarrow (1-i)^4 = -4 \Rightarrow (1-i)^5 = -4(1-i).$$

Hence,

$$\frac{(1-i)^5 - 1}{(1+i)^5 + 1} = \frac{-4(1-i) - 1}{-4(1+i) + 1}$$

$$= \frac{-5+4i}{-3-4i} = \frac{(5-4i)(3-4i)}{3^2+4^2} = \frac{1}{25}(-1-32i).$$

(3) and (4), by (2), for $m = 0, 1, 2, \ldots$

$$(1-i)^{4m+1} = ((1-i)^4)^m(1-i) = (-4)^m(1-i);$$

$$(1-i)^{4m+2} = -2(-4)^m i;$$

$$(1-i)^{4m+3} = -2(-4)^m(1+i);$$

$$(1-i)^{4m+4} = (-4)^{m+1},$$

and

$$(1+i)^{4m+1} = (-4)^m(1+i);$$

$$(1+i)^{4m+2} = 2(-4)^m i;$$

$$(1+i)^{4m+3} = 2(-4)^m(-1+i);$$

$$(1+i)^{4m+4} = (-4)^{m+1}.$$

The final results follow easily.

Example 2. Let

$$\omega = \frac{1}{2}\left(-1 + \sqrt{3}i\right).$$

(1) Show that $\omega^2 + \omega + 1 = 0$ and $\omega^3 = 1$;
(2) Evaluate $(a + b\omega + c\omega^2)(a + b\omega^2 + c\omega)$ and $(a + b)(a + b\omega)(a + b\omega^2)$.

Solution. (1) By direct computation,

$$\omega^2 = \frac{1}{2}\left(-1 - \sqrt{3}i\right) = -\omega - 1$$

$$\Rightarrow \omega^2 + \omega + 1 = 0.$$

Also,

$$\omega^3 = \omega^2 \cdot \omega = \frac{1}{2}\left(-1 - \sqrt{3}i\right) \cdot \frac{1}{2}\left(-1 + \sqrt{3}i\right) = \frac{4}{4} = 1.$$

Or,

$$\omega^3 = \omega^2 \cdot \omega = (-\omega - 1)\omega = -\omega^2 - \omega = 1.$$

(2) By using (1), direct computation shows that

$$(a + b\omega + c\omega^2)(a + b\omega^2 + c\omega)$$
$$= a^2 + b^2 + c^2 + ab(\omega^2 + \omega) + bc(\omega^4 + \omega^2) + ca(\omega^2 + \omega)$$
$$= a^2 + b^2 + c^2 - ab - bc - ca,$$

and

$$(a+b)(a+b\omega)(a+b\omega^2) = a^3 + b^3 + a^2 b(\omega^2 + \omega + 1) + ab^2(\omega^2 + \omega + 1) = a^3 + b^3.$$

By the way, how can one factorize $a^2 - ab + b^2$ over the complex field **C**? Just treat $a^2 - ab + b^2 = 0$ as a quadratic equation in a, and then its solutions are

$$a = \frac{b \pm \sqrt{b^2 - 4b^2}}{2} = \frac{1 \pm \sqrt{3}i}{2} b = -b\omega \quad \text{or} \quad -b\omega^2.$$

Hence $a^2 - ab + b^2 = (a + b\omega)(a + b\omega^2)$. See Exercise B(2).

Example 3. Solve the following equations, namely, try to find $z = x + iy$ so that

(1) (square root) $z^2 = -6 + \sqrt{5}i$;
(2) (cube root) $z^3 = 1$.

Solution. (1) $z^2 = x^2 - y^2 + 2xyi$. Hence

$$z^2 = -6 + \sqrt{5}i,$$
$$\Leftrightarrow x^2 - y^2 = -6 \quad \text{and} \quad 2xy = \sqrt{5},$$
$$\Rightarrow (x^2 + y^2)^2 = (x^2 - y^2)^2 + (2xy)^2 = 41,$$
$$\Rightarrow \text{(take the positive square root)} \ x^2 + y^2 = \sqrt{41}.$$

Solving this last equation with $x^2 - y^2 = -6$, we get

$$x^2 = \frac{1}{2}\left(-6 + \sqrt{41}\right) \quad \text{and} \quad y^2 = \frac{1}{2}\left(6 + \sqrt{41}\right)$$
$$\Rightarrow x = \pm\sqrt{\frac{1}{2}\left(-6 + \sqrt{41}\right)} \quad \text{and} \quad y = \pm\sqrt{\frac{1}{2}\left(6 + \sqrt{41}\right)}.$$

In appearance, we would get four solutions. Owing to the restrained condition that $xy > 0$, we have only two solutions left, namely,

$$z = \pm\left(\sqrt{\frac{1}{2}\left(-6 + \sqrt{41}\right)} + i\sqrt{\frac{1}{2}\left(6 + \sqrt{41}\right)}\right).$$

(2) $z^3 = x^3 - 3xy^2 + i(3x^2y - y^3)$. Then

$$z^3 = 1$$
$$\Rightarrow x^3 - 3xy^2 = 1 \quad \text{and} \quad 3x^2y - y^3 = y(3x^2 - y^2) = 0.$$

In case $y = 0$, then $x^3 = 1$ and we choose the only real root $x = 1$. If $3x^2 - y^2 = 0$, then $x^3 - 3xy^2 = x^3 - 9x^3 = -8x^3 = 1$ and we choose $x = -\frac{1}{2}$ and hence, $y^2 = 3x^2 = \frac{3}{4}$ which, in turn, results in $y = \pm\frac{\sqrt{3}}{2}$. Therefore, $z^3 = 1$ has solutions $z = 1$, w and w^2 where $w = \frac{1}{2}(-1 + \sqrt{3}i)$.

Exercises A

(1) Express each of the following complex numbers z in the form $z = x + iy$, where $x, y \in \mathbf{R}$, and then compute $|z|$ and $\arg z$.

(a) $(4 + 3i)(4 + 2i)(3 - i)(1 - i)$. (b) $\left(\sqrt{3} - i\right)^6$.

(c) $\dfrac{i}{(i-1)(i-2)(i-3)}$. (d) $\dfrac{(1+2i)^3 - (1-i)^3}{(3+2i)^3 - (2+i)^3}$. (e) $\dfrac{(i+1)^9}{(1-i)^7}$.

(f) $\dfrac{a + bi}{a - bi}$, $a, b \in \mathbf{R}$. (g) $\dfrac{(a+bi)^2}{(a-bi)^2} - \dfrac{(a-bi)^2}{(a+bi)^2}$, $a, b \in \mathbf{R}$.

(2) Find the real and the imaginary parts of the following complex numbers $z = x + iy$.

    (a) $\dfrac{z-1}{z+1}$.   (b) $\dfrac{1}{z^2}$.   (c) z^5.   (d) z^n, $n \in \mathbf{N}$.

(3) Let $\omega = \frac{1}{2}(-1 + \sqrt{3}i)$. Compute:

    (a) $(1-\omega)(1-\omega^2)(1-\omega^4)(1-\omega^8)$.   (b) $(1-\omega+\omega^2)(1+\omega-\omega^2)$.

    (c) $(a\omega^2 + b\omega)(a\omega + b\omega^2)$.   (d) $(a + b\omega + c\omega^2)^3 + (a + b\omega^2 + c\omega)^3$.

(4) Let $z = x + iy$, $x, y \in \mathbf{R}$. Solve the following equations.

    (a) $z^2 = -i$.   (b) $z^2 = \frac{1}{2}\left(1 - \sqrt{3}i\right)$.   (c) $z^3 = \dfrac{1+i}{1-i}$.   (d) $z^4 = -1$.

    (e) $z^2 + (6+7i) + \sqrt{2} + 5i = 0$.   (f) $z^2 - (3+2i)z + (1+3i) = 0$.

Exercises B

(1) Suppose $a, b \in \mathbf{R}$. Show that, in the complex field $\mathbf{C}$,

$$\sqrt{a + bi} = \begin{cases} \pm\left(\sqrt{\dfrac{a + \sqrt{a^2 + b^2}}{2}}\right. \\ \qquad \left. + (\mathrm{sgn}\, b)\sqrt{\dfrac{-a + \sqrt{a^2 + b^2}}{2}}\, i\right), & \text{if } a \neq 0, b \neq 0 \\ \pm\sqrt{a}, & \text{if } a \geq 0, b = 0, \\ \pm\sqrt{-a}\, i, & \text{if } a < 0, b = 0, \end{cases}$$

where $\mathrm{sgn}\, b = \frac{b}{|b|} = 1$ if $b > 0$, $= -1$ if $b < 0$ and the square root of a positive real number is chosen to be positive.

(2) Suppose $a, b, c \in \mathbf{C}$, and $a \neq 0$. Show that

$$az^2 + bz + c = a\left(z + \frac{b}{2a}\right)^2 + \frac{4ac - b^2}{4a}$$

$$= a\left(z + \frac{b + \sqrt{b^2 - 4ac}}{2a}\right)\left(z + \frac{b - \sqrt{b^2 - 4ac}}{2a}\right)$$

where $\sqrt{b^2 - 4ac}$ is as in Exercise (1). Note that, as long as $b^2 - 4ac \neq 0$, $\sqrt{b^2 - 4ac}$ always have two values with positive and negative sign, respectively, and $\sqrt{b^2 - 4ac}$ could be any one of them in the above expression. So the *quadratic equation* $az^2 + bz + c = 0$ has exactly two solutions

$$z = \frac{-b \pm \sqrt{b^2 - 4ac}}{2a},$$

as in the real case.

(3) Let $\omega = \frac{1}{2}(-1 + \sqrt{3}i)$. Suppose $a, b, c \in \mathbf{C}$. Show that

    (a) $a^3 - b^3 = (a - b)(a^2 + ab + b^2) = (a - b)(a - b\omega)(a - b\omega^2)$.

    (b) $a^3 + b^3 + c^3 - 3abc = (a + b + c)(a^2 + b^2 + c^2 - ab - bc - ca)$

$$= (a + b + c)(a + b\omega + c\omega^2)(a + b\omega^2 + c\omega).$$

(4) For simplicity, let $e^{i\alpha} = \cos\alpha + i\sin\alpha$ for real α (see (1.2.13)). Suppose $a \neq 0$. Show that the cubic equation $z^3 = a$ always has three distinct solutions

$$|a|^{\frac{1}{3}} e^{i\frac{\theta}{3}}, \quad |a|^{\frac{1}{3}} e^{i\frac{\theta}{3}}\omega, \quad |a|^{\frac{1}{3}} e^{i\frac{\theta}{3}}\omega^2,$$

where $\theta = \operatorname{Arg} a$ and $\omega = \frac{1}{2}(-1 + \sqrt{3}i)$. Try to find solutions of $z^n = a$.

(5) Suppose $a, b \in C$. Show that $z^3 - 3abz + (a^3 + b^3) = 0$ has solutions

$$-(a + b), \quad -(a\omega + b\omega^2), \quad -(a\omega^2 + b\omega),$$

where $\omega = \frac{1}{2}(-1 + \sqrt{3}i)$.

(6) (*Cardano formula* for cubic equations) Given a cubic equation

$$z^3 + a_2 z^2 + a_1 z + a_0 = 0,$$

with complex coefficients. Substitute $z = w - \frac{1}{3}a_2$ into the equation to obtain

$$w^3 + pw + q = 0.$$

    (a) By comparing to the cubic equation in Exercise (5), let $p = -3ab$ and $q = a^3 + b^3$. Show that a^3 and b^3 are roots of the quadratic equation

$$t^2 - qt - \frac{1}{27}p^3 = 0.$$

By Exercise (2), we may suppose that

$$a^3 = \frac{q}{2} + \sqrt{\frac{q^2}{4} + \frac{p^3}{27}} \quad \text{and} \quad b^3 = \frac{q}{2} - \sqrt{\frac{q^2}{4} + \frac{p^3}{27}}.$$

(b) Choose a and b as suitable cubic roots of

$$\sqrt[3]{\frac{q}{2} + \sqrt{\frac{q^2}{4} + \frac{p^3}{27}}} \quad \text{and} \quad \sqrt[3]{\frac{q}{2} - \sqrt{\frac{q^2}{4} + \frac{p^3}{27}}},$$

respectively, *subject to the constrained condition* $ab = -\frac{1}{3}p$. Then, according to Exercise (5), $w^3 + pw + q = 0$ has roots $-(a+b)$, $-(a\omega + b\omega^2)$, $-(a\omega^2 + b\omega)$. Therefore, the original equation has roots $-\frac{a_2}{3} - (a + b)$, $-\frac{a_2}{3} - (a\omega^2 + b\omega)$, and $-\frac{a_2}{3} - (a\omega + b\omega^2)$.

(c) Solve $z^3 + 3z^2 - 3z - 14 = 0$.

(7) Suppose $z = x + iy$ is not a negative real number and $z \neq 0$. Show that there exists a unique w, $\operatorname{Re} w > 0$, so that $w^2 = z$.

Exercises C

(1) In (1.1.1), replace the real scalars α, β by complex numbers α, β and then, we can view $\mathbf{R}^2$ as *one-dimensional complex vector space* over the field $\mathbf{C}$. We denote this vector space by $\mathbf{C}$ itself.

(a) Show that any linear transformation $T : \mathbf{C} \to \mathbf{C}$ is of the form

$$T(z) = \alpha z, \quad z \in \mathbf{C}$$

where α is a complex scalar. Let $\alpha = a + bi$ and $z = x + iy$. Then $\mathbf{T}(z)$ can be rewritten as the vector form $T_\alpha(x, y) = (ax - by, bx + ay)$ or as the matrix form, with respect to the natural basis for $\mathbf{R}^2$,

$$T_\alpha(x, y) = (x \ y) \begin{bmatrix} a & b \\ -b & a \end{bmatrix}$$

which is a special kind of linear transformations on $\mathbf{R}^2$.

(b) Conversely, given a linear transformation on $\mathbf{R}^2$ as

$$T(x, y) = (x \ y) \begin{bmatrix} a & b \\ c & d \end{bmatrix}, \quad a, b, c, d \in \mathbf{R}.$$

Let $z = (x \ y) = x + iy$. Suppose this T turns out to be a linear transformation $T_\alpha(z) = \alpha z$ on C. Then

$$T(z) = (x \ y) \begin{bmatrix} a & b \\ c & d \end{bmatrix} = \alpha z \quad \text{for all } z \in \mathbf{C}$$

$$\Rightarrow (\text{Let } z = 1.) \ (1 \ 0) \begin{bmatrix} a & b \\ c & d \end{bmatrix} = (a \ b) = \alpha \cdot 1 = \alpha \text{ or } \alpha = a + bi;$$

(Let $z = i$.) $(0\ 1)\begin{bmatrix} a & b \\ c & d \end{bmatrix} = (c\ d) = \alpha i$

or $\alpha = -i(c + di) = d - ic$

$\Rightarrow \alpha = a + bi = d - ci$ or $a = d$ and $b = -c$.

Hence, T should be of the form

$$T(x, y) = (x\ y)\begin{bmatrix} a & b \\ -b & a \end{bmatrix} = T_\alpha(z), \quad \text{where } \alpha = a + bi.$$

(a) and (b) suggest that the following peculiar set of matrices

$$\widetilde{SO}(2, \mathbf{R}) = \left\{ \begin{bmatrix} a & b \\ -b & a \end{bmatrix} \middle| a, b \in \mathbf{R} \right\},$$

is worthy being emphasized.

(c) Under the operations of matrix addition and multiplication, show that $\widetilde{SO}(2, \mathbf{R})$ is a field (namely, having properties listed in (1.3.2)).

(d) Show that the mapping $\Phi : \widetilde{SO}(2, \mathbf{R}) \to \mathbf{C}$ defined by

$$\Phi\left(\begin{bmatrix} a & b \\ -b & a \end{bmatrix} \right) = a + bi,$$

is a *field isomorphism* (namely, Φ is one-to-one, onto and preserves operations of addition and multiplication).

Therefore, one can treat

$$\begin{bmatrix} a & b \\ -b & a \end{bmatrix} = r\begin{bmatrix} \cos\theta & \sin\theta \\ -\sin\theta & \cos\theta \end{bmatrix}, \quad \text{where } r = \sqrt{a^2 + b^2} \text{ and } \tan\theta = \frac{b}{a}$$

as a complex number, especially in its polar form $re^{i\theta}$ (see (1.2.14)). Since the expression on the right side represents, geometrically, a stretch of vectors followed by a rotation through the angle θ (for details, refer to Ref. [56], Vol. 2), *a complex number is a two-dimensional number, taking care of both stretch and rotation at one time by its very nature.* See the end of Sec. 1.1.

(2) Fix a point (α, β), with $\beta \neq 0$, in $\mathbf{R}^2$. Then $(x, y) = (x - \frac{\alpha y}{\beta})(1, 0) + \frac{y}{\beta}(\alpha, \beta)$ always holds. Define: On $\mathbf{R}^2$,

equality: $(x_1, y_1) = (x_2, y_2) \Leftrightarrow x_1 = x_2$ and $y_1 = y_2$;
real number: $(x_1, 0)$;
addition: $(x_1, y_1) + (x_2, y_2) = (x_1 + x_2, y_1 + y_2)$, and

multiplication:

$$(x_1, y_1) \odot (x_2, y_2) = \left[\left(x_1 - \frac{\alpha y_1}{\beta} \right) \left(x_2 - \frac{\alpha y_2}{\beta} \right) - \frac{y_1 y_2}{\beta^2} \right] (1, 0)$$
$$+ \left[\left(x_1 - \frac{\alpha y_1}{\beta} \right) \frac{y_2}{\beta} + \left(x_2 - \frac{\alpha y_2}{\beta} \right) \frac{y_1}{\beta} \right] (\alpha, \beta).$$

Let $\widetilde{\mathbf{C}}(i)$ denote $\mathbf{R}^2$ with these two operations.

(a) Show that $\widetilde{\mathbf{C}}(i)$ is a field.

(b) Show that $\widetilde{\mathbf{C}}(i)$ is field isomorphic to $\mathbf{C}$.

Hence, conceptually we can view $\widetilde{\mathbf{C}}(i)$ as a complex field with $i = (\alpha, \beta)$ acting as the imaginary unit.

(3) Let

$$A = \begin{bmatrix} 1 & 2 \\ 2 & -1 \end{bmatrix}.$$

Show that $A^2 + I_2 = O$ where $I_2 = \begin{bmatrix} 1 & 0 \\ 0 & 1 \end{bmatrix}$ and O is the zero matrix.

Can we construct a field $\mathbf{F}$ with A as the imaginary unit so that $\mathbf{F}$ is field isomorphic to $\mathbf{C}$? If affirmative, try it (see Sec. 2.7.8 in Ref. [56], Vol. 1).

1.4. Algebraic Operations and Their Geometric Interpretations (Applications)

The section is divided into four subsections.

Section 1.4.1 introduces conjugate complex numbers, a unique operation particularly owned by the complex number system $\mathbf{C}$.

Section 1.4.2 discusses the relations among real and complex inequalities.

Section 1.4.3 uses examples to show how complex numbers can be adopted to solve planar geometrical problems.

Finally, the important concept of symmetric points with respect to a circle or a line will be discussed in Sec. 1.4.4.

1.4.1. *Conjugate complex numbers*

The symmetric point of a complex number $z = x + yi$ with respect to the real axis, denoted as

$$\bar{z} = x - iy, \tag{1.4.1.1}$$

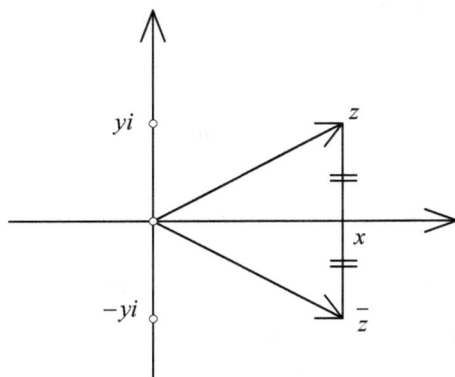

Fig. 1.10

is called the (complex) *conjugate* of z. See Fig. 1.10. Note that

$$z = \bar{z} \Leftrightarrow z \text{ is real, i.e., Im } z = 0. \tag{1.4.1.2}$$

This indicates that the *conjugate operation* $z \rightarrow \bar{z}$ is a two-dimensional operation, particularly owned by **C** but not by **R**. Also,

$$z\bar{z} = |z|^2 \geq 0, \tag{1.4.1.3}$$

shows that how a complex number z and its conjugate $\bar{z}$ can be so related, via multiplication, to produce a nonnegative real number $|z|^2$, whose positive square root $|z|$ is just the absolute value of itself and can be considered as *the length of the vector z or the distance of the point z to* 0. This compensates somewhat the fact that **C** is no more an ordered field.

It is easy to prove the following

Operational properties.

(1) $\operatorname{Re} z = \frac{1}{2}(z + \bar{z}), \quad \operatorname{Im} z = \frac{1}{2i}(z - \bar{z}).$
(2) $\overline{(\bar{z})} = z$ (in short, $\bar{\bar{z}} = z$).
(3) $\overline{z_1 \pm z_2} = \bar{z}_1 \pm \bar{z}_2, \quad \overline{z_1 z_2} = \bar{z}_1 \bar{z}_2, \quad \overline{\left(\frac{z_1}{z_2}\right)} = \frac{\bar{z}_1}{\bar{z}_2}(z_2 \neq 0).$
(4) $|\bar{z}| = |z|, \operatorname{Arg} \bar{z} = -\operatorname{Arg} z(-\pi < \operatorname{Arg} z < \pi).$
(5) $|z|^2 = z\bar{z} \geq 0, |z| = \sqrt{z\bar{z}}$ (positive square root as a real number).
(6) $z_1 = z_2 \Leftrightarrow |z_1| = |z_2|, \quad \operatorname{Arg} z_1 = \operatorname{Arg} z_2 \Leftrightarrow \bar{z}_1 = \bar{z}_2. \tag{1.4.1.4}$

Readers are urged to give these relations their geometric interpretations.

In what follows, we discuss two important applications of the relation

$$z\bar{z} = |z|^2.$$

It provides an easy way to compute the multiplicative inverse of a nonzero complex number. Suppose $z = x + iy \neq 0$. Then

$$\frac{1}{z} = \frac{\bar{z}}{|z|^2} = \frac{x - iy}{x^2 + y^2}. \tag{1.4.1.5}$$

Figure 1.11 also shows that

$$z \underset{(i)}{\longrightarrow} \frac{1}{\bar{z}} \underset{(ii)}{\longrightarrow} \overline{\left(\frac{1}{\bar{z}}\right)} = \frac{1}{z}$$

where (i) is called the *symmetric motion* or *reflection* with respect to the unit circle $|z| = 1$, while (ii) the one with respect to the real axis. Therefore,

$$w = \frac{1}{z}, \tag{1.4.1.6}$$

as a mapping from z to $\frac{1}{z}$, is the composite of two such reflections.

In computation involving absolute values of complex numbers, the relation $|z|^2 = z\bar{z}$ plays an essential role. For example,

$$|z_1 \pm z_2|^2 = (z_1 \pm z_2)(\bar{z}_1 \pm \bar{z}_2) = z_1\bar{z}_1 \pm (z_1\bar{z}_2 + \bar{z}_1 z_2) + z_2\bar{z}_2,$$

$$\Rightarrow |z_1 \pm z_2|^2 = |z_1|^2 \pm 2\operatorname{Re}(z_1\bar{z}_2) + |z_2|^2. \tag{1.4.1.7}$$

Hence, it follows immediately that

$$|z_1 + z_2|^2 + |z_1 - z_2|^2 = 2(|z_1|^2 + |z_2|^2), \tag{1.4.1.8}$$

which reflects the fact that the sum of the square of two diagonals of a parallelogram is equal to the sum of the square of its four sides (see Fig. 1.7).

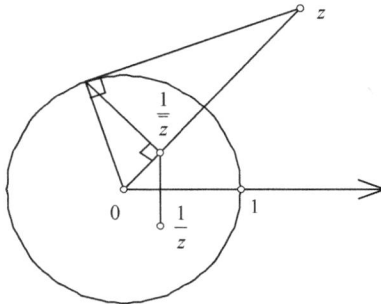

Fig. 1.11

In (1.4.1.7), let $z_1 = a_1 b_1$ and $z_2 = a_2 b_2$. Then

$$|a_1 b_1 + a_2 b_2|^2 = |a_1 b_1|^2 + |a_2 b_2|^2 + 2 \operatorname{Re} a_1 b_1 \overline{a_2} \overline{b_2}$$
$$= (|a_1|^2 + |a_2|^2)(|b_1|^2 + |b_2|^2)$$
$$- (|a_1|^2 |\overline{b_2}|^2 - 2 \operatorname{Re} a_1 \overline{b_2} a_2 \overline{b_1} + |a_2|^2 |\overline{b_1}|^2)$$
$$\Rightarrow |a_1 b_1 + a_2 b_2|^2 = (|a_1|^2 + |a_2|^2)(|b_1|^2 + |b_2|^2) - |a_1 \overline{b_2} - a_2 \overline{b_1}|^2.$$
$$\text{(1.4.1.9)}$$

This is a special case of the *Lagrange identity* (see Exercise A(11)).

Exercises A

(1) Prove (1.4.1.4) in detail and interpret them geometrically.

(2) For each of the following complex numbers z, compute $\operatorname{Re} z$, $\operatorname{Im} z$, $|z|$, $\operatorname{Arg} z$ and $\bar{z}$.

    (a) $\frac{1+2i}{3-4i} + \frac{2-i}{5i}$.   (b) $\frac{i}{(i-1)(i-2)(i-3)}$.   (c) $(\sqrt{3}+i)^{-3}$.   (d) $\frac{(\sqrt{3}+i)^5}{(1-\sqrt{3}i)^{10}}$.

(3) Let $w = \frac{az+b}{cz+d}$. Compute $\operatorname{Re} w$, $\operatorname{Im} w$, $|w|$, and $\bar{w}$.

(4) Let $z = \cos\theta + i\sin\theta$ or suppose $|z| = 1$ and $z + \frac{1}{z} = 2\cos\theta$. Show that

$$z^n + \frac{1}{z^n} = 2\cos n\theta \quad \text{and} \quad z^n - \frac{1}{z^n} = 2\sin n\theta, \quad n \in \mathbf{Z}.$$

(5) Let $(1 - \sqrt{3}i)^n = x_n + iy_n$, where $x_n, y_n \in \mathbf{R}$ for $n = 1, 2, 3, \ldots$.
    (a) Show that $x_n y_{n-1} - x_{n-1} y_n = 4^{n-1} \cdot \sqrt{3}$.
    (b) Compute $x_n x_{n-1} + y_n y_{n-1} = ?$

(6) Suppose $|z_1| = \lambda |z_2|, \lambda > 0$, show that $|z_1 - \lambda^2 z_2| = \lambda |z_1 - z_2|$. Conversely, if $|z_1 - \lambda^2 z_2| = \lambda |z_1 - z_2|$ for $\lambda > 0$ and $\lambda \neq 1$, then $|z_1| = \lambda |z_2|$ holds.

(7) (a) Show that, for $z \neq 0$,

$$|z| = 1 \Leftrightarrow z = \frac{1}{\bar{z}} \Leftrightarrow z = \frac{\zeta}{\bar{\zeta}} \quad \text{for some } \zeta \neq 0.$$

    (b) Suppose $|z| = 1$ but $z \neq -1$. Show that there exists a unique real number t so that $z = \frac{1+ti}{1-ti}$. Try to express t in terms of z.

    (c) Hence, the set $\{z \in \mathbf{C} \,||\, z| = 1 \text{ but} z \neq -1\}$ can be put in one-to-one correspondence onto the real axis. Try to deduce the polar form of a complex number.

(8) Prove the following identities.
  (a) $|z_1(1+|z_2|^2) - z_2(1+|z_1|^2)| = |z_1 - z_2|^2|1 - z_1\bar{z}_2|^2 - (z_1\bar{z}_2 - \bar{z}_1 z_2)^2$.
  (b) $|1 + z_1\bar{z}_2|^2 + |z_1 - z_2|^2 = (1 + |z_1|^2)(1 + |z_2|^2)$.
  (c) $|1 - z_1\bar{z}_2|^2 - |z_1 - z_2|^2 = (1 - |z_1|^2)(1 - |z_2|^2)$.
(9) Suppose $a, b, z \in \mathbf{C}$.
  (a) Show that

$$\left|\frac{z-a}{1-\bar{a}z}\right| = 1 \Leftrightarrow |a| = 1 \quad \text{or} \quad |z| = 1.$$

  Discuss what happens if $|a| = |z| = 1$.
  (b) Suppose $|z| = 1$. Show that $\left|\frac{az+b}{bz+\bar{a}}\right| = 1$ where $|a|^2 - |b|^2 \neq 0$.
(10) Use (1.4.1.8) to show that

$$\left|a + \sqrt{a^2 - b^2}\right| + \left|a - \sqrt{a^2 - b^2}\right| = |a + b| + |a - b|,$$

and hence, deduce that

$$|z_1| + |z_2| = \left|\frac{1}{2}(z_1 + z_2) + \sqrt{z_1 z_2}\right| + \left|\frac{1}{2}(z_1 + z_2) - \sqrt{z_1 z_2}\right|.$$

(11) For $a_j, b_j \in \mathbf{C}$ for $1 \leq j \leq n$, prove the *Lagrange identity*

$$\left|\sum_{j=1}^{n} a_j b_j\right|^2 = \left(\sum_{j=1}^{n} |a_j|^2\right)\left(\sum_{j=1}^{n} |b_j|^2\right) - \sum_{1 \leq j < k \leq n} |a_j \bar{b}_k - a_k \bar{b}_j|^2.$$

Exercises B

(1) Suppose z_0 is a *zero* of a polynomial $p(z) = a_n z^n + a_{n-1}z^{n-1} + \cdots + a_1 z + a_0$ with *real* coefficients, i.e., $p(z_0) = 0$. Show that $\bar{z}_0$ is also a zero of $p(z)$, i.e., $p(\bar{z}_0) = 0$. In this case, z_0 and $\bar{z}_0$ are called the *conjugate roots* of $p(z) = 0$.
(2) Given a quadratic equation $z^2 + az + b = 0$ with complex coefficients a and b. Determine the necessary and sufficient conditions so that the equation has

  (i) coincident roots;
  (ii) at least one real root; and
  (iii) two conjugate complex roots,

  respectively.

(3) Find a necessary condition so that the equation $z^3 + (a + ib)z^2 + (c + id)z + 1 = 0$, where a, b, c, and d are real numbers, has at least one real root.

(4) Find necessary and sufficient conditions so that the polynomial $z^3 + pz + q = 0$, where $p, q \in \mathbf{R}$, has

   (i) three distinct real roots,
   (ii) one real root and two conjugate complex roots, and
   (iii) three real roots with two of them coincident,

   respectively.

1.4.2. *Inequalities*

To each complex number z, there correspond three numbers $\operatorname{Re} z$, $\operatorname{Im} z$, and $|z|$ whose absolute values constitute the three side lengths of a right triangle (see Fig. 1.5). Hence, it follows immediately the inequalities

$$\left.\begin{array}{c} |\operatorname{Re} z|, |\operatorname{Im} z| \\ \dfrac{|\operatorname{Re} z| + |\operatorname{Im} z|}{\sqrt{2}} \end{array}\right\} \leq |z| \leq |\operatorname{Re} z| + |\operatorname{Im} z|. \qquad (1.4.2.1)$$

These are the most important elementary inequalities involving real and complex numbers. And above all, they connect the real and complex limit processes together (see Sec. 1.7).

By using (1.4.1.7) and (1.4.2.1),

$$\begin{aligned} |z_1 + z_2|^2 &= |z_1|^2 + |z_2|^2 + 2\operatorname{Re}(z_1 \bar{z}_2) \\ &\leq |z_1|^2 + |z_2|^2 + 2|z_1 \bar{z}_2| = (|z_1| + |z_2|)^2 \end{aligned}$$

$$\Rightarrow |z_1 + z_2| \leq |z_1| + |z_2|, \text{ which is called a } \textit{triangle inequality}$$

$$(\text{see Fig. 1.7}). \qquad (1.4.2.2)$$

Equality in (1.4.2.2) holds if and only if $|z_1 \bar{z}_2| = \operatorname{Re}(z_1 \bar{z}_2)$, and hence

$$|z_1 + z_2| = |z_1| + |z_2| \Leftrightarrow z_1 \bar{z}_2 \geq 0 \Leftrightarrow \operatorname{Arg} z_1 = \operatorname{Arg} z_2 \Leftrightarrow \frac{z_1}{z_2} \geq 0 \quad \text{if } z_2 \neq 0.$$

$$(1.4.2.2)'$$

Similarly, we have another *triangle inequality*

$$\big|\,|z_1| - |z_2|\,\big| \leq |z_1 + z_2|$$

and

the equality "=" holds $\Leftrightarrow z_1 \bar{z}_2 \leq 0 \Leftrightarrow \dfrac{z_1}{z_2} \leq 0 \quad$ if $z_2 \neq 0$. $\qquad (1.4.2.3)$

This can also be proved by observing, via (1.4.2.2), that $|z_1| = |(z_1 + z_2) - z_2| \leq |z_1 + z_2| + |-z_2| = |z_1 + z_2| + |z_2|$, with equality holds if and only if $(z_1 + z_2)(-\bar{z}_2) = -z_1\bar{z}_2 - |z_2|^2 \geq 0$ or $z_1\bar{z}_2 \leq 0$; and $|z_2| \leq |z_1 + z_2| + |z_1|$ with equality if and only if $\bar{z}_1 z_2 \leq 0$.

We illustrate three examples.

Example 1 (Cauchy–Schwarz inequality). Let $a_j, b_j \in \mathbf{C}$ for $1 \leq j \leq n$. Then

$$\left| \sum_{j=1}^{n} a_j b_j \right| \leq \left(\sum_{j=1}^{n} |a_j|^2 \right)^{\frac{1}{2}} \left(\sum_{j=1}^{n} |b_j|^2 \right)^{\frac{1}{2}} \tag{1.4.2.4}$$

with equality if and only if, as long as $a_j \neq 0, b_j \neq 0, \frac{a_j}{b_j}, 1 \leq j \leq n$, are all equal.

Proof. This follows immediately by using Lagrange identity (see Exercise A(11) of Sec. 1.4.1). A direct proof is as follows. We may suppose that $\sum_{j=1}^{n} |b_j|^2 \neq 0$.

For any complex number λ, (1.4.1.7) shows that

$$\sum_{j=1}^{n} |a_j - \lambda \bar{b}_j|^2 = \sum_{j=1}^{n} |a_j|^2 + |\lambda|^2 \sum_{j=1}^{n} |b_j|^2 - 2\operatorname{Re}\bar{\lambda} \sum_{j=1}^{n} a_j b_j$$

$$= \sum_{j=1}^{n} |a_j|^2 + \left(\sum_{j=1}^{n} |b_j|^2 \right) \left| \lambda - \frac{\sum_{j=1}^{n} a_j b_j}{\sum_{j=1}^{n} |b_j|^2} \right|^2$$

$$- \frac{\left| \sum_{j=1}^{n} a_j b_j \right|^2}{\sum_{j=1}^{n} |b_j|^2} \geq 0.$$

When choosing $\lambda = \frac{\sum_{j=1}^{n} a_j b_j}{\sum_{j=1}^{n} |b_j|^2}$, the left side will get the minimum value

$$\sum_{j=1}^{n} |a_j|^2 - \frac{\left| \sum_{j=1}^{n} a_j b_j \right|^2}{\sum_{j=1}^{n} |b_j|^2} \geq 0,$$

with equality if and only if $\sum_{j=1}^{n} |a_j - \lambda \bar{b}_j|^2 = 0 \Leftrightarrow \frac{a_j}{b_j} = \lambda$ for $1 \leq j \leq n$. $\square$

In general, there does not exist a constant $M > 0$ so that $|z_1 - z_2| \leq M||z_1| - |z_2||$ holds for any $z_1, z_2 \in \mathbf{C}$. But, we do have

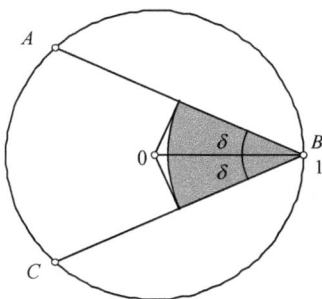

Fig. 1.12

Example 2. Fix δ, $0 \le \delta < \frac{\pi}{2}$. Denoted by D_δ the shaded part in Fig. 1.12. Show that

$$\frac{|1-z|}{1-|z|} \le \frac{2}{\cos\delta}, \quad z \in D_\delta.$$

The part of the disc inside the angle $\angle ABC$ is called a *Stolz domain*. Conversely, if $\frac{|1-z|}{1-|z|}$ is bounded when $|z| < 1$, then z should lie in some D_δ for $0 \le \delta < \frac{\pi}{2}$.

Proof. Take any $z \in D_\delta$ and let $1 - z = r(\cos\theta + i\sin\theta)$, where $|\theta| \le \delta$. Then $\cos\theta \ge \cos\delta$ and $r = |1-z| \le \cos\delta$. Therefore,

$$\frac{|1-z|}{1-|z|} = \frac{r(1+|z|)}{1-|z|^2} = \frac{r(1+|z|)}{1-(1-2r\cos\theta+r^2)}$$

$$= \frac{1+|z|}{2\cos\theta - r} \le \frac{2}{2\cos\delta - \cos\delta} = \frac{2}{\cos\delta}.$$

Conversely,

$$\frac{|1-z|}{1-|z|} = \frac{1+|z|}{2\cos\theta - r} = \frac{\cos\theta}{2\cos\theta - r} \cdot (1+|z|) \cdot \frac{1}{\cos\theta}.$$

If $|z| < 1$ and $|z|$ is close to 1, both $\frac{\cos\theta}{2\cos\theta - r}$ and $1 + |z|$ are bounded. Hence $\frac{|1-z|}{1-|z|}$ is bounded $\Leftrightarrow \frac{1}{\cos\theta}$ is bounded,

$$\Rightarrow |\theta| \le \delta < \frac{\pi}{2} \quad \text{for some } \delta \ge 0. \qquad \square$$

Example 3. Show that

$$|z - 1| \le \big||z| - 1\big| + |z||\mathrm{Arg}\,z|,$$

and interpret its geometric meaning.

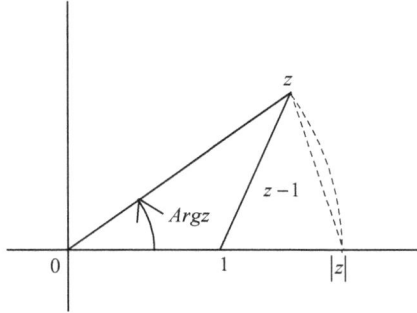

Fig. 1.13

Proof. In Fig. 1.13, the triangle with vertices 1, $|z|$ and z shows that $|z - 1| \le ||z| - 1| + |z - |z||$. But the secant $|z - |z||$ is not larger than the circular arc connecting $|z|$ to z whose length is equal to $|z| |\operatorname{Arg} z|$.

For an analytic proof, let $z = |z|(\cos\theta + i\sin\theta), \theta = \operatorname{Arg} z$. Then

$$|z - 1| = |(z - |z|) + (|z| - 1)| \le |z - |z|| + ||z| - 1| \le |z| |(\cos\theta + i\sin\theta)$$

$$-1| + ||z| - 1| = |z| \left| 2\sin\frac{\theta}{2} \right| + ||z| - 1|$$

$$\le ||z| - 1| + |z| |\theta| \left(\text{using} \left| \sin\frac{\theta}{2} \right| \le \frac{|\theta|}{2} \text{ here} \right). \qquad \square$$

Exercises A

(1) Show that

$$|z_1 + \cdots + z_n| \le |z_1| + |z_2| + \cdots + |z_n|$$

with equality if and only if, as long as $z_j z_k \ne 0$, then $\frac{z_j}{z_k} \ge 0$ holds.

(2) Suppose $|z_k| < 1, \lambda_k \ge 0$ for $1 \le k \le n$. If $\lambda_1 + \cdots + \lambda_n = 1$, show that

$$|\lambda_1 z_1 + \cdots + \lambda_n z_n| < 1.$$

Try to give a geometric interpretation if $n = 2$ or 3. How about for general n?

(3) (a) Show that

$$\left| \frac{z - a}{1 - \bar{a} z} \right| < 1 \Leftrightarrow |a| < 1, |z| < 1 \quad \text{or} \quad |a| > 1, |z| > 1.$$

(b) Find the necessary and sufficient condition so that $|\frac{z-a}{1-\bar{a}z}| = 1$ or >1.

(4) Suppose $|a| < 1, |b| < 1$. Show that

$$\frac{|a| - |b|}{1 - |a|\,|b|} \leq \left| \frac{a-b}{1-\bar{a}b} \right| \leq \frac{|a| + |b|}{1 + |a|\,|b|} \leq 1.$$

Try to figure out when equalities hold.

(5) Suppose $z_1 + z_2 = 1$. Show that $1 \leq |z_1| + |z_2|$ and, in case $z_1 z_2 \neq 0$, equality holds if and only if both z_1 and z_2 are positive real numbers.

(6) Suppose $\mathrm{Re}\, a > 0$ and $\mathrm{Re}(\sqrt{a^2 - 1}) \geq 0$ (see Exercise B(1) of Sec. 1.3).

   (a) Show that $\mathrm{Re}\{\bar{a}\sqrt{a^2 - 1}\} \geq 0$.
   (b) Show that $|a + \sqrt{a^2 - 1}| \geq 1$ with equality if and only if a is real and $0 < a \leq 1$.

(7) Consider the equation $z^2 - 2az + 1 = 0$.

   (a) In case a is real and $-1 \leq a \leq 1$, show that the roots z of the equation satisfy $|z| = 1$.
   (b) Otherwise, the equation has one root z_1 satisfying $|z_1| < 1$ and another root z_2 satisfying $|z_2| > 1$.

(8) Prove the following special case of *Minkowski inequality*

$$\left(\sum_{j=1}^{n} |a_j + b_j|^2 \right)^{\frac{1}{2}} \leq \left(\sum_{j=1}^{n} |a_j|^2 \right)^{\frac{1}{2}} + \left(\sum_{j=1}^{n} |b_j|^2 \right)^{\frac{1}{2}}$$

with equality if and only if $\frac{a_j}{b_j}, 1 \leq j \leq n$, are equal.

Exercises B

(1) Show that the roots z of the equation $az^2 + bz + c = 0$, where $ac \neq 0$, satisfy

$$\frac{|c|}{|b| + \sqrt{|a|\,|c|}} \leq |z| \leq \frac{|b| + \sqrt{|a|\,|c|}}{|a|}.$$

(2) Suppose $0 < a_n \leq a_{n-1} \leq \cdots \leq a_1 \leq a_0$. If $|z| < 1$, show that

$$a_n z^n + a_{n-1} z^{n-1} + \cdots + a_1 z + a_0 \neq 0.$$

(3) (a) (*Young inequality*) Suppose $p \geq 0, q \geq 0$, and $p + q = 1$. Show that

$$|a|^p |b|^q \leq p|a| + q|b|, \quad a, b \in \mathbf{C}$$

with equality if and only if $|a| = |b|$.

(b) (*Hölder inequality*) Suppose $p \geq 1, q \geq 1$, and $\frac{1}{p} + \frac{1}{q} = 1$. Then

$$\sum_{j=1}^{n} |a_j b_j| \leq \left(\sum_{j=1}^{n} |a_j|^p \right)^{\frac{1}{p}} \left(\sum_{j=1}^{n} |b_j|^q \right)^{\frac{1}{q}}.$$

When does equality hold?

(c) (*Minkowski inequality*) Suppose $p \geq 1$, then

$$\left(\sum_{j=1}^{n} |a_j + b_j|^p \right)^{\frac{1}{p}} \leq \left(\sum_{j=1}^{n} |a_j|^p \right)^{\frac{1}{p}} + \left(\sum_{j=1}^{n} |b_j|^p \right)^{\frac{1}{p}}.$$

When does equality hold?

1.4.3. *Applications in (planar) Euclidean geometry*

In this section, we try to use algebraic operational properties of complex numbers and their geometric meanings introduced so far to realize how complex numbers can be used to handle geometric problems.

Section (1) Lines, angles, triangles, circles, and domains determined by them, etc.

Let $z = x + iy$, where $x, y \in \mathbf{R}$.

A line in the plane $\mathbf{R}^2$ has the equation
$$\alpha x + \beta y + \gamma = 0, \quad \text{where } \alpha, \beta, \gamma \in \mathbf{R}$$
$$\Rightarrow \text{(by (1) in (1.4.1.4))} \quad \frac{\alpha}{2}(z + \bar{z}) - \frac{\beta i}{2}(z - \bar{z}) + \gamma = 0$$

or

$$\frac{\alpha - \beta i}{2} z + \frac{\alpha + \beta i}{2} \bar{z} + \gamma = 0.$$

We list this result as part of the following

Complex equations of a line.

(1) $\bar{a} z + a \bar{z} + b = 0$, where $a \in \mathbf{C}$ and $a \neq 0$, $b \in \mathbf{R}$.

(2) The line passing two *distinct* points z_1 and z_2 has the following expressions:

    (i) $z = z_1 + t(z_2 - z_1), t \in \mathbf{R}$ (passing the *point* z_1 with *direction* $z_2 - z_1$ and real *parameter* t). Note that the line *segment* $\overline{z_1 z_2}$ is $z = z_1 + t(z_2 - z_1), 0 \leq t \leq 1$.

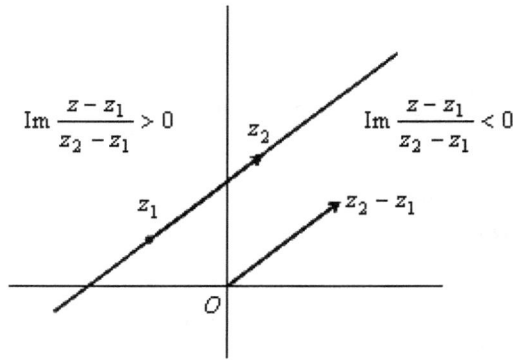

Fig. 1.14

(ii) $\text{Im} \frac{z-z_1}{z_2-z_1} = 0.$

(iii) $\begin{vmatrix} z & \bar{z} & 1 \\ z_1 & \overline{z_1} & 1 \\ z_2 & \overline{z_2} & 1 \end{vmatrix} = 0.$

See Fig. 1.14.

If walking along the line in the direction $z_2 - z_1$, then:

$$\text{left open half plane}: \left\{ z \in \mathbf{C} \,\middle|\, \text{Im} \frac{z-z_1}{z_2-z_1} > 0 \right\},$$

$$\text{simply denoted as } \text{Im} \frac{z-z_1}{z_2-z_1} > 0;$$

$$\text{right open half plane}: \text{Im} \frac{z-z_1}{z_2-z_1} < 0;$$

$$\text{left closed half plane}: \text{Im} \frac{z-z_1}{z_2-z_1} \geq 0$$

$$\left(\text{including the boundary line } \text{Im} \frac{z-z_1}{z_2-z_1} = 0 \right);$$

$$\text{right closed half plane}: \text{Im} \frac{z-z_1}{z_2-z_1} \leq 0. \qquad (1.4.3.1)$$

In particular, the real axis $\text{Im}\, z = 0$, pointed to right, separates $\mathbf{C}$ into

$$\text{(open) upper half plane}: \text{Im}\, z > 0 \text{ and}$$

$$\text{(open) lower half plane}: \text{Im}\, z < 0; \qquad (1.4.3.2)$$

while the imaginary axis $\operatorname{Re} z = 0$, pointed upward, separates $\mathbf{C}$ into

(open) right half plane : $\operatorname{Re} z > 0$ and

(open) left half plane : $\operatorname{Re} z < 0.$ (1.4.3.3)

Also we have

The relative positions of two lines (segments). Let $z = z_1 + t(z_2 - z_1)$ and $z = z_1' + t(z_2' - z_1')$ be two lines. Then, they are

(i) coincident $\Leftrightarrow z_1' - z_1$ and $z_2' - z_1'$ are real multiples of $z_2 - z_1$, i.e., $\frac{z_1' - z_1}{z_2 - z_1}$ and $\frac{z_2' - z_1'}{z_2 - z_1}$ are real;

(ii) parallel $\Leftrightarrow z_2' - z_1'$ is a real multiple of $z_2 - z_1$, i.e., $\frac{z_2' - z_1'}{z_2 - z_1}$ is real;

(iii) parallel and having the *same* direction $\Leftrightarrow z_2' - z_1'$ is a positive real multiple of $z_2 - z_1$, i.e., $\frac{z_2' - z_1'}{z_2 - z_1} > 0$;

(iv) intersecting at a point $\Leftrightarrow \operatorname{Im} \frac{z_2' - z_1'}{z_2 - z_1} \neq 0.$

The *angle* from the line $z = z_1' + t(z_2' - z_1')$ to the line $z = z_1 + t(z_2 - z_1)$ is

$$\operatorname{Arg} \frac{z_2 - z_1}{z_2' - z_1'}.$$

See Fig. 1.15. Hence, they are

(v)

$$\text{perpendicular} \Leftrightarrow \operatorname{Re} \frac{z_2 - z_1}{z_2' - z_1'} = 0. \qquad (1.4.3.4)$$

Proofs are left to the readers as Exercise A(3).

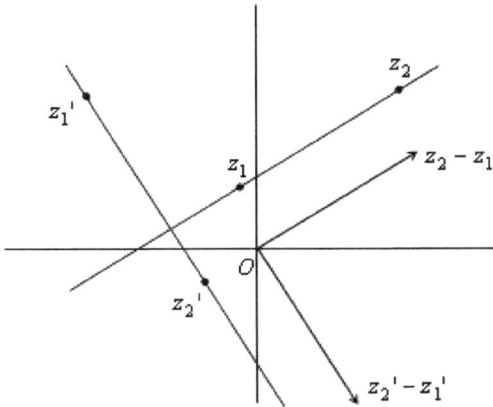

Fig. 1.15

Suppose z_1, z_2, and z_3 are not collinear. Then, the consecutive segments $\overline{z_1 z_2}$, $\overline{z_2 z_3}$, and $\overline{z_3 z_1}$ form the sides of a *triangle* $\Delta z_1 z_2 z_3$ with z_1, z_2, and z_3 as *vertices*. The ordering $z_1 \to z_2 \to z_3$ is said to determine an *orientation* of the triangle, usually called *positive* if counterclockwise and *negative* if clockwise.

Some facts about a triangle

(1) $\Delta z_1 z_2 z_3$ and $\Delta z_1' z_2' z_3'$, having the same orientation, are similar if and only if (see Fig. 1.16)

$$\frac{z_3 - z_1}{z_2 - z_1} = \frac{z_3' - z_1'}{z_2' - z_1'} \quad \text{or} \quad \begin{vmatrix} z_1 & z_1' & 1 \\ z_2 & z_2' & 1 \\ z_3 & z_3' & 1 \end{vmatrix} = 0.$$

If having opposite orientation, then they are similar if and only if (see Fig. 1.17)

$$\frac{z_3 - z_1}{z_2 - z_1} = \frac{\overline{z_3' - z_1'}}{\overline{z_2' - z_1'}} \quad \text{or} \quad \begin{vmatrix} z_1 & \overline{z_1'} & 1 \\ z_2 & \overline{z_2'} & 1 \\ z_3 & \overline{z_3'} & 1 \end{vmatrix} = 0.$$

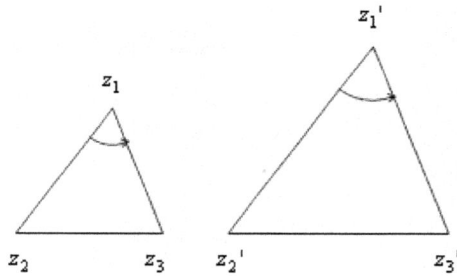

Fig. 1.16

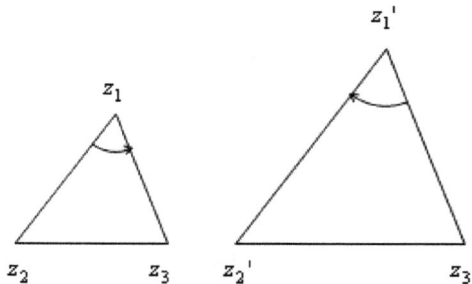

Fig. 1.17

(2) $\Delta z_1 z_2 z_3$ has *area*

$$\frac{1}{2}\mathrm{Im}(z_1 - z_3)(\overline{z_3} - \overline{z_2}) = \frac{1}{2}|z_3 - z_2|^2\mathrm{Im}\frac{z_1 - z_3}{z_3 - z_2} = \frac{1}{2}\mathrm{Im}(\overline{z_1}z_2 + \overline{z_2}z_3 + \overline{z_3}z_1)$$

if the triangle is *positively oriented*; or,

$$\pm\frac{i}{4}\begin{vmatrix} z_1 & \overline{z_1} & 1 \\ z_2 & \overline{z_2} & 1 \\ z_3 & \overline{z_3} & 1 \end{vmatrix},$$

where $\pm$ is to be chosen so that the area is nonnegative. $\qquad$ (1.4.3.5)

The details are left as Exercise A(4).

In two real variables x and y, a circle has the equation

$$\alpha(x^2 + y^2) + \beta x + \gamma y + \delta = 0$$

$$\Rightarrow \left(\text{Let } x = \frac{1}{2}(z + \bar{z}) \text{ and } y = -\frac{i}{2}(z - \bar{z}). \text{ Note that } z\bar{z} = |z|^2 = x^2 + y^2.\right)$$

$$\alpha z\bar{z} + \frac{\beta}{2}(z + \bar{z}) - \frac{i\gamma}{2}(z - \bar{z}) + \delta = 0, \quad \text{or}$$

$$\alpha z\bar{z} + \frac{1}{2}(\beta - i\gamma)z + \frac{1}{2}(\beta + i\gamma)\bar{z} + \delta = 0.$$

We summarize the above as

The complex equations of a circle. A circle has the equation

$$a|z|^2 + \bar{b}z + b\bar{z} + c = 0, \quad a, c \in \mathbf{R}, \quad \text{and} \quad b \in \mathbf{C}.$$

(1) If $a = 0$ and $b \neq 0$, it degenerates into a line $\bar{b}z + b\bar{z} + c = 0$.

(2) In case $a \neq 0$;

    (i) $|b|^2 < ac$: an imaginary circle;

    (ii) $|b|^2 = ac$: a point circle $-\frac{b}{a}$;

    (iii) $|b|^2 > ac$: a real circle

$$|z - z_0| = r$$

with *center* $z_0 = -\frac{b}{a}$ and *radius* $r = \frac{\sqrt{|b|^2 - ac}}{|a|}$ or, in polar form,

$$z = z_0 + r(\cos\theta + i\sin\theta) = z_0 + re^{i\theta}, \quad 0 \leq \theta \leq 2\pi.$$

See Fig. 1.18.

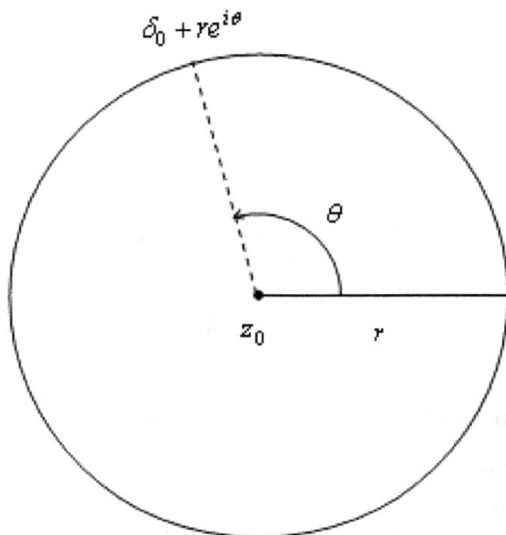

$$\delta_0 + re^{i\theta}$$

Fig. 1.18

Walking along the circle, we usually call the circle *positive-oriented* if counterclockwise and *negative-oriented* if clockwise. The circle separates the plane $\mathbf{C}$ into

open disk : $|z - z_0| < r$,

closed disk : $|z - z_0| \leq r$ (including the circle $|z - z_0| = r$ itself),

outside of the closed disk : $|z - z_0| > r$ and

outside of the open disk : $|z - z_0| \geq r$ (including $|z - z_0| = r$).   (1.4.3.6)

Section (2) Some illustrative examples

Example 1. A triangle $\triangle z_1 z_2 z_3$ is an equilateral triangle if and only if

$$z_1^2 + z_2^2 + z_3^2 - z_1 z_2 - z_2 z_3 - z_3 z_1 = 0.$$

Proof. *The necessity*: See Fig. 1.19. Then

$$\frac{z_3 - z_1}{z_2 - z_1} = \frac{z_1 - z_2}{z_3 - z_2} = \frac{z_2 - z_3}{z_1 - z_3}.$$ $(*_1)$

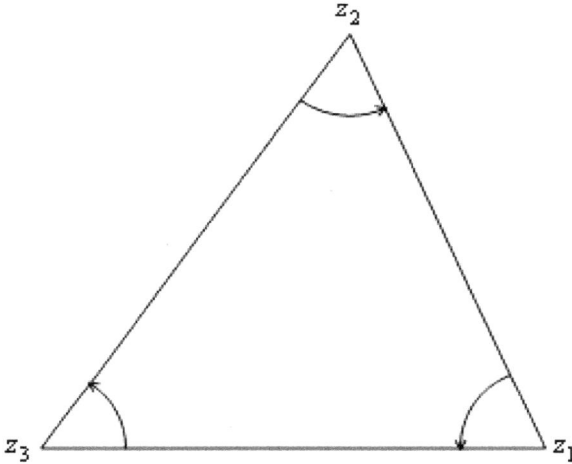

Fig. 1.19

After cross multiplication of the first equality, we have

$$(z_3 - z_1)(z_3 - z_2) = (z_2 - z_1)(z_1 - z_2) = -(z_1 - z_2)^2$$
$$\Rightarrow z_1^2 + z_2^2 + z_3^2 - z_1 z_2 - z_2 z_3 - z_3 z_1 = 0.$$

The sufficiency: Reversing the above process, we will recapture $(*_1)$ and hence, $\Delta z_1 z_2 z_3$ is equilateral. Or we can do this as follows. The given identity can be rewritten as

$$(z_1 - z_2)^2 + (z_2 - z_3)^2 + (z_3 - z_1)^2 = 0.$$

Now

$$(z_1 - z_2) + (z_2 - z_3) + (z_3 - z_1) = 0$$
$$\Rightarrow (z_1 - z_2)^2 + (z_2 - z_3)^2 + 2(z_1 - z_2)(z_2 - z_3)$$
$$= [-(z_3 - z_1)]^2 = (z_3 - z_1)^2$$
$$\Rightarrow (\text{Substitute } (z_1 - z_2)^2 + (z_2 - z_3)^2$$
$$= -(z_3 - z_1)^2.)(z_1 - z_2)(z_2 - z_3) = (z_3 - z_1)^2$$
$$\Rightarrow \frac{z_3 - z_1}{z_2 - z_1} = \frac{z_2 - z_3}{z_1 - z_3}.$$

And so on. □

Example 2. Given four distinct points z_1, z_2, z_3, and z_4 in the complex plane $\mathbf{C}$. Show that the identity

$$(z_1 - z_2)(z_3 - z_4) + (z_1 - z_3)(z_4 - z_2) + (z_1 - z_4)(z_2 - z_3) = 0,$$

and hence, deduce the inequality

$$|z_1 - z_2||z_3 - z_4| + |z_2 - z_3||z_1 - z_4| \geq |(z_3 - z_1)(z_4 - z_2)|,$$

with equality if and only if z_1, z_2, z_3, and z_4, in this ordering, lie on a circle or on a line. This is the classical *Ptolemy theorem*. See Fig. 1.20(a) and (c).

Proof. The identity can be rewritten as

$$-\frac{(z_2 - z_1)(z_4 - z_3)}{(z_4 - z_1)(z_2 - z_3)} + 1 = \frac{(z_4 - z_2)(z_3 - z_1)}{(z_3 - z_2)(z_4 - z_1)}, \qquad (*_2)$$

which we simply denoted as $-A + 1 = B$, where

$$A = \frac{(z_2 - z_1)(z_4 - z_3)}{(z_4 - z_1)(z_2 - z_3)}$$

and

$$B = \frac{(z_4 - z_2)(z_3 - z_1)}{(z_3 - z_2)(z_4 - z_1)}.$$

The necessity: Suppose these four points lie on a circle as shown in Fig. 1.20(a) or on a line as shown in Fig. 1.20(c). Set

$$\theta_1 = \operatorname{Arg} \frac{z_2 - z_1}{z_4 - z_1},$$

$$\theta_2 = \operatorname{Arg} \frac{z_4 - z_3}{z_2 - z_3},$$

$$\theta_3 = \operatorname{Arg} \frac{z_4 - z_2}{z_3 - z_2},$$

and

$$\theta_4 = \operatorname{Arg} \frac{z_3 - z_1}{z_4 - z_1}.$$

Then,

$$\operatorname{Arg} A = \theta_1 + \theta_2 = -\pi \quad \text{or} \quad \pi,$$

$$\operatorname{Arg} B = \theta_3 + \theta_4 = 0$$

$$\Rightarrow \operatorname{Arg}(-A) = 0.$$

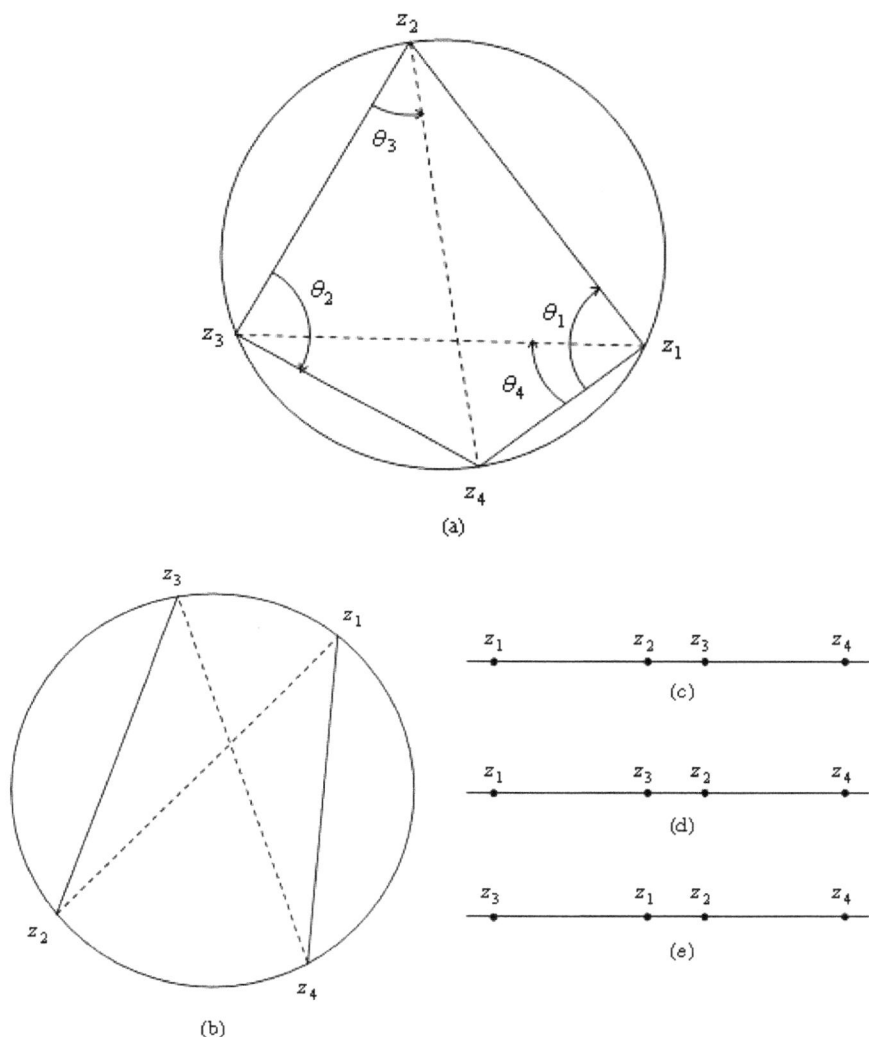

Fig. 1.20

Hence $-A > 0$ and $B > 0$. $(*_2)$ means $-A + 1 = B$ which is equivalent to $|-A| + 1 = |B|$, i.e., $|z_1 - z_2| \, |z_3 - z_4| + |z_2 - z_3| \, |z_1 - z_4| = |(z_3 - z_1)| \, |(z_4 - z_2)|$ holds.

The sufficiency: Reverse the above process. The identity is equivalent to $|-A| + 1 = |B|$. When comparing to $(*_2)$, namely, $-A + 1 = B$, we have

$-A > 0$ and hence $B > 0$. These mean that $\operatorname{Arg} A = \theta_1 + \theta_2 = -\pi$ and $\operatorname{Arg} B = \theta_3 + \theta_4 = 0$. Therefore, z_1, z_2, z_3, and z_4 lie on a circle. □

Remark. Suppose we disregard the ordering of appearance of z_1, z_2, z_3, and z_4 on a circle or on a line. Then, in addition to Figs. 1.20(a) and 1.20(c), we also have to consider the cases shown in Figs. 1.20(b)–1.20(e). In all cases, *the ratio of ratios*

$$\frac{z_1 - z_3}{z_2 - z_3} : \frac{z_1 - z_4}{z_2 - z_4}$$

will be a real number; as a matter of fact, it is positive if z_3 and z_4 does not separate z_1 and z_2, and it is negative if z_3 and z_4 separate z_1 and z_2.

Example 3. Let a and b be two distinct points in **C**.

(1) The set of points z satisfying

$$\left|\frac{z - a}{z - b}\right| = \lambda, \quad 0 \le \lambda \le \infty$$

represents

   (i) the point circle a, if $\lambda = 0$;
   (ii) the point circle b, if $\lambda = \infty$;
  (iii) the perpendicular bisector of the segment $\overline{ab}$, if $\lambda = 1$, and
  (iv) in case $0 < \lambda < \infty$ and $\lambda \ne 1$, the circle

$$\left|z - \frac{a - \lambda^2 b}{1 - \lambda^2}\right| = \frac{\lambda|a - b|}{|1 - \lambda^2|}.$$

See Fig. 1.21: If λ varies from 0 to ∞, the family of circles varies from the point a, via circles with centers lying on the line ab, called *Apollonius circles*, to the pointe b.

(2) The set of the points z satisfying

$$\operatorname{Arg}\frac{z - a}{z - b} = \theta, \quad -\pi < \theta \le \pi$$

represents

   (i) the segment $\overline{ab}$, if $\theta = \pi$;
  (ii) the two outward rays eminating from a and b along the line connecting a and b, if $\theta = 0$, and

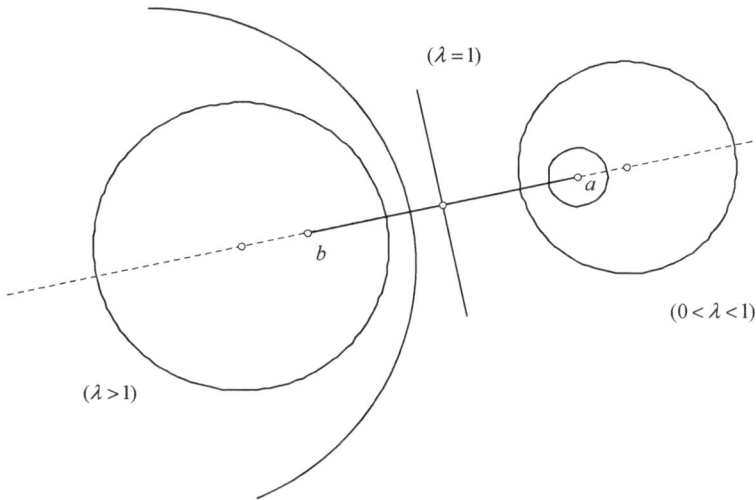

Fig. 1.21

(iii) in case $-\pi < \theta < \pi$ and $\theta \neq 0$, the circle $\left|z - \left(\frac{a+b}{2} - \frac{a-b}{2\tan\theta}i\right)\right| = \frac{|a-b|}{2}|\csc\theta|$.

See Fig. 1.22: If θ varies from $-\pi$ to π, the family of circles are *coaxial circles* with the segment $\overline{ab}$ as the coaxis.

(3) Any one of the circles in (1) is orthogonal to any one of the circles in (2). They together form the so-called *Steiner's circles*.

Proofs are left as Exercise A(5). For further discussions concerned, see Sec. 1.4.4. □

Example 4. Describe the plane curve

$$|z^2 - a^2| = \lambda \quad \text{where } a > 0 \text{ is a constant and } 0 \leq \lambda < \infty.$$

Try to find out the point set of z satisfying $|z^2 - a^2| < \lambda$ $(0 < \lambda < \infty)$.

Solution. $|z^2 - a^2| = \lambda$ has the following rectangular equation and polar equation

$$[(x-a)^2 + y^2][(x+a)^2 + y^2] = \lambda^2, \quad \text{where } z = x + iy;$$

$$(r^2 + a^2)^2 = \lambda^2 + 4a^2r^2\cos^2\theta, \quad \text{where } z = re^{i\theta},$$

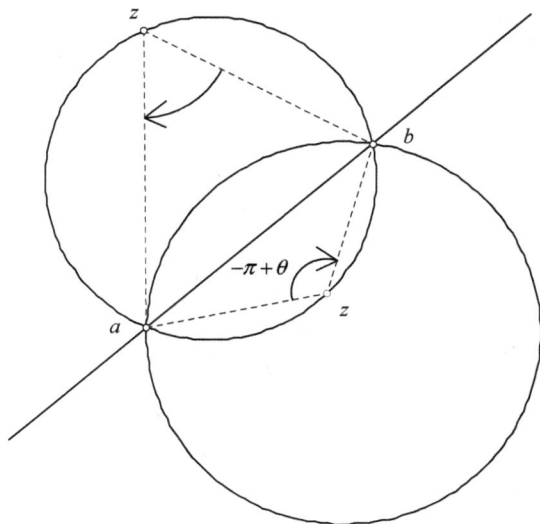

Fig. 1.22

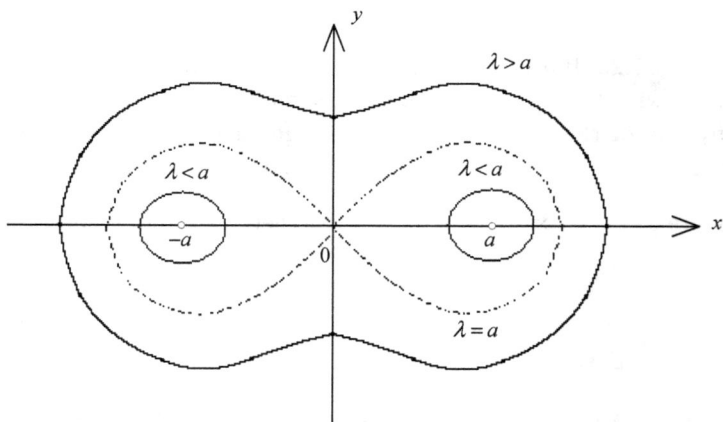

Fig. 1.23

respectively. It is a *Cassini oval* if $\lambda > a$, a *Bernoulli lemniscate* if $\lambda = a$, a *Cassini oval of two branches* if $0 < \lambda < a$, and the point set $\{a, -a\}$ if $\lambda = 0$. See Fig. 1.23. What is the point set defined by $|z^2 - a^2| < \lambda$? □

Example 5. Let $w = z^3 - 3z$. Try to describe the following point sets:

(1) $\operatorname{Re} w \gtreqless 0$, i.e., either $\operatorname{Re} w > 0, = 0$ or < 0.
(2) $\operatorname{Im} w \gtreqless 0$.

Solution. Let $z = x + iy$ and $w = u + iv$. Then

$$u + iv = (x + iy)^3 - 3(x + iy) = (x^3 - 3xy^2 - 3x) + i(3x^2 y - y^3 - 3y)$$
$$\Rightarrow u = x^3 - 3xy^2 - 3x = x(x^2 - 3y^2 - 3),$$
$$v = -y^3 + 3x^2 y - 3y = y(3x^2 - y^2 - 3).$$

Then

$$u = \operatorname{Re} w = 0 \Leftrightarrow x = 0 \text{ or } x^2 - 3y^2 = 3 \text{ (a hyperbola)};$$
$$u = \operatorname{Re} w > 0 \Leftrightarrow x > 0 \text{ and } x^2 - 3y^2 > 3 \text{ or } x < 0 \text{ and } x^2 - 3y^2 < 3;$$
$$u = \operatorname{Re} w < 0 \Leftrightarrow x > 0 \text{ and } x^2 - 3y^2 < 3 \text{ or } x < 0 \text{ and } x^2 - 3y^2 > 3.$$

Similarly,

$$v = \operatorname{Im} w \begin{cases} > 0 \Leftrightarrow y > 0 \text{ and } 3x^2 - y^2 > 3 \text{ or } y < 0 \text{ and } 3x^2 - y^2 < 3, \\ = 0 \Leftrightarrow y = 0 \text{ or } 3x^2 - y^2 = 3 \text{ (a hyperbola)}, \\ < 0 \Leftrightarrow y > 0 \text{ and } 3x^2 - y^2 < 3 \text{ or } y < 0 \text{ and } 3x^2 - y^2 > 3. \end{cases}$$

See Fig. 1.24.

Exercises A

(1) Do the following problems.

    (a) Prove that the direction vector of a line $\bar{a}z + a\bar{z} + b = 0$ is orthogonal to a.

    (b) Two points z_1 and z_2, considered as vectors, are orthogonal if and only if $z_1 \bar{z_2} + \bar{z_1} z_2 = 0$.

    (c) Three points $z_1, z_2,$ and z_3 are collinear if and only if, there exist real scalars $t_1, t_2,$ and t_3, not all equal to zero, so that

$$t_1 z_1 + t_2 z_2 + t_3 z_3 = 0 \quad \text{and} \quad t_1 + t_2 + t_3 = 0.$$

    (d) Suppose $z_1 + z_2 + z_3 = z_1 z_2 z_3$ holds. Show that $z_1, z_2,$ and z_3 cannot lie on the same side of the real axis.

(2) Prove (1.4.3.1) in detail.
(3) Prove (1.4.3.4) in detail.

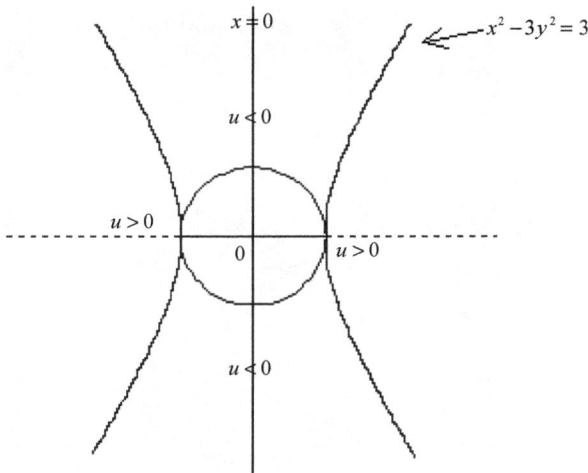

Fig. 1.24

(4) Prove (1.4.3.5) in detail.

(5) Prove Example 3 in detail.

(6) (a) Find the other two vertices of a square with two known vertices a_1 and a_2.

    (b) Find the fourth vertex of a parallelogram with three known vertices a_1, a_2, and a_3.

(7) Fix any three noncollinear points a_1, a_2, and a_3 in the plane $\mathbf{C}$. Show that any point z in $\mathbf{C}$ can be expressed uniquely as

$$z = t_1 a_1 + t_2 a_2 + t_3 a_3, \quad t_1 + t_2 + t_3 = 1.$$

Usually, we call (t_1, t_2, t_3) the *barycentric coordinate* of z with respect to the *affine basis* $\{a_1, a_2, a_3\}$ for the plane. Show that a point z lies in the interior of the triangle $\triangle a_1 a_2 a_3$ if and only if it has barycentric coordinate (t_1, t_2, t_3) where $t_1 > 0, t_2 > 0, t_3 > 0$, and $t_1 + t_2 + t_3 = 1$. In particular, show that

$$\frac{1}{3}(a_1 + a_2 + a_3),$$

is the *barycenter* of $\triangle a_1 a_2 a_3$.

(8) Any three vertices of the four vertices of a quadrilateral form a triangle. Suppose the locations of the barycenters of such four triangles are known, try to find the quadrilateral.

(9) Construct $\Delta a_1 a_2 a_3'$, $\Delta a_2 a_3 a_1'$ and $\Delta a_3 a_1 a_2'$ outward the triangle $\Delta a_1 a_2 a_3$ so that these three triangles are similar and have the same orientation. Show that both $\Delta a_1' a_2' a_3'$ and $\Delta a_1 a_2 a_3$ have the same barycenter.

(10) Show that $\Delta a_1 a_2 a_3$ is an isosceles triangle with vertex a_3 if and only if there exists a positive number k so that

$$\frac{a_3 - a_1}{a_2 - a_1} : \frac{a_3 - a_2}{a_1 - a_2} = k.$$

(11) Suppose $|z_k| = 1$, $1 \le k \le 4$. Show that z_1, z_2, z_3, and z_4 form a rectangle if and only if $z_1 + z_2 + z_3 + z_4 = 0$.

(12) Suppose $|a_1| = |a_2| = |a_3| = 1$. Show that $\Delta a_1 a_2 a_3$ is an equilateral triangle, inscribed to the unit circle, if and only if $a_1 + a_2 + a_3 = 0$.

(13) (a) Show that three distinct points z_1, z_2, and z_3 are collinear if and only if

$$\text{Im} \, \frac{z_1 - z_3}{z_2 - z_3} = 0.$$

(b) Show that four distinct points z_1, z_2, z_3, and z_4 are collinear or lie on a circle if and only if

$$\text{Im} \left(\frac{z_1 - z_3}{z_2 - z_3} \cdot \frac{z_2 - z_4}{z_1 - z_4} \right) = 0.$$

(14) Four distinct lines $\overline{a_1 b_1}$, $\overline{a_2 b_2}$, $\overline{a_3 b_3}$, and $\overline{a_4 b_4}$ meet at a point z_0, where a_1, a_2, a_3, a_4 and b_1, b_2, b_3, b_4 are collinear, respectively. See Fig. 1.25. Show that

$$\frac{a_1 - a_3}{a_1 - a_4} : \frac{a_2 - a_3}{a_2 - a_4} = \frac{b_1 - b_3}{b_1 - b_4} : \frac{b_2 - b_3}{b_2 - b_4}.$$

(15) Show that the equation of the circle passing three noncollinear points z_1, z_2, and z_3 is

$$\begin{vmatrix} |z|^2 & z & \bar{z} & 1 \\ |z_1|^2 & z_1 & \bar{z}_1 & 1 \\ |z_2|^2 & z_2 & \bar{z}_2 & 1 \\ |z_3|^2 & z_3 & \bar{z}_3 & 1 \end{vmatrix} = 0.$$

Determine the center and the radius. What happens if z_1, z_2, and z_3 are collinear?

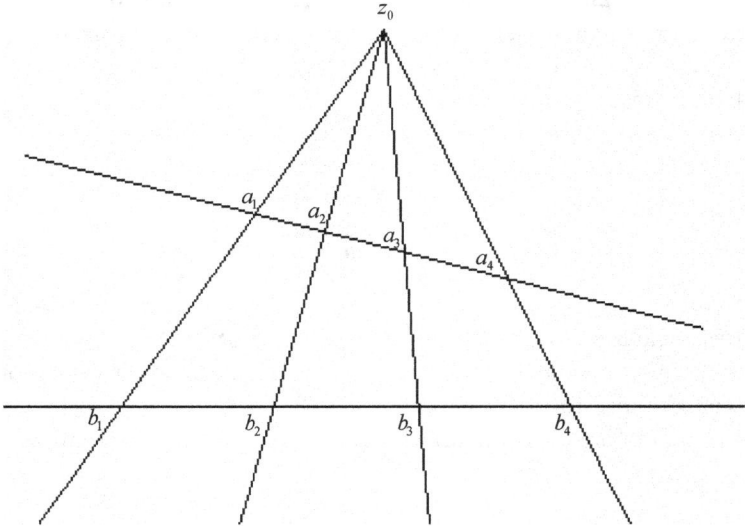

Fig. 1.25

(16) Suppose $z_1 \neq 0, 1$ and $0, z_1, z_2$ are not collinear. Show that the center and the radius of the circle passing z_1, z_2 and $\overline{z_1}^{-1}$ are

$$\frac{z_1(1 + |z_2|^2) - z_2(1 + |z_1|^2)}{z_1\overline{z_2} - \overline{z_1}z_2} \quad \text{and} \quad \frac{|z_1 - z_2|\,|1 - \overline{z_1}z_2|}{|z_1\overline{z_2} - \overline{z_1}z_2|},$$

respectively. Prove that $\overline{z_2}^{-1}$ also lies on the circle.

(17) Fix $a > 0$.

(a) Denote by Γ the circle passing $\pm a$ and with the center bi, where $b \in \mathbf{R}$. Show that a point z lies inside, on, or outside Γ if and only if

$$|z|^2 - 2b \operatorname{Im} z < a^2, \quad = a^2, \quad \text{or} \ > a^2,$$

respectively.

(b) For any point z with $\operatorname{Im} z \neq 0$. Show that $-a, a, z$, and $\frac{a^2}{z}$ lie on a circle.

(c) If a circle passes through $-a, z(\operatorname{Im} z \neq 0)$ and $\frac{a^2}{z}$, then it should pass a.

(d) Let Γ be as in (a). Show that z lies inside $\Gamma \Leftrightarrow \frac{a^2}{z}$ lies outside Γ.

(18) Describe the following point sets or curves and try to graph them.

    (a) $\alpha < \operatorname{Re} z < \beta$, where $\alpha, \beta \in \mathbf{R}$.

    (b) $\alpha < \operatorname{Im} z < \beta$, where $\alpha, \beta \in \mathbf{R}$.

    (c) $\alpha < \operatorname{Arg} z < \beta$, where $\alpha, \beta \in \mathbf{R}$ and $\alpha < \beta < \alpha + 2\pi$.

    (d) $\alpha < |z - z_0| < \beta$, where z_0 is a fixed point and $0 \le \alpha < \beta$.

    (e) $\frac{\pi}{4} < \operatorname{Arg} \frac{z+i}{z-i} \le \frac{\pi}{2}$.

    (f) $|z| \gtrless |\operatorname{Re} z| + 1$.

    (g) $(a\bar{z} + \bar{a}z)^2 \gtrless 2(b\bar{z} + \bar{b}z) + c$, where c is real.

    (h) $|z + a| + |z - a| \gtrless 2\lambda$, where $\lambda > 0$ and $|a| < \lambda$.

    (i) $||z + a| - |z - a|| \gtrless 2\lambda$ where $\lambda > 0$ and $|a| < \lambda$.

    (j) $\operatorname{Re} \frac{z(z+i)}{z-i} \gtrless 0$.

    (k) $|z| < 2$ and $0 < \operatorname{Arg} z < \frac{\pi}{4}$.

    (l) $0 < \operatorname{Arg}(z - 1) < \frac{\pi}{4}$ and $2 < \operatorname{Re} z \le 3$.

    (m) Let $\overline{ab}$ and $\overline{a'b'}$ be two line segments. Try to locate these points z so that Δzab and $\Delta za'b'$ both have the same orientation and are similar.

Exercises B

(1) Suppose $z_1, \ldots, z_n$ are distinct nonzero points and they all lie on the same side of a line passing 0.

    (a) Show that $\frac{1}{z_k}$, $1 \le k \le n$, all lie on the same side of a certain line passing 0.

    (b) $z_1 + \cdots + z_n \ne 0$ and $\frac{1}{z_1} + \cdots + \frac{1}{z_n} \ne 0$ hold.

This indicates that, as long as $z_1 + \cdots + z_n = 0$ and $z_1, \ldots, z_n$ do not lie on the same line, then any line passing 0 will *separate* the points $z_1, \ldots, z_n$, i.e., some lie on one side of the line while others on the other side.

(2) Let $z_1, \ldots, z_n$ be distinct points in $\mathbf{C}$. Denote the *convex closure* spanned by $z_1, \ldots, z_n$ as

$$\operatorname{Con}(z_1, \ldots, z_n)$$

$$= \left\{ \sum_{j=1}^{n} \lambda_j z_j \,\middle|\, \lambda_j \ge 0 \text{ for } 1 \le j \le n \text{ and } \lambda_1 + \cdots + \lambda_n = 1 \right\},$$

with $z_1, \ldots, z_n$ as *vertices*. It is a *convex set*, namely, the line segment connecting any two of its points lies entirely in the set. In case $\lambda_j > 0$ for $1 \le j \le n$ and $\lambda_1 + \cdots + \lambda_n = 1$, then $\sum_{j=1}^{n} \lambda_j z_j$ is called an *interior point* of $\mathrm{Con}(z_1, \ldots, z_n)$. If $z_1, \ldots, z_n$ are collinear, then $\mathrm{Con}(z_1, \ldots, z_n)$ is the smallest line segment containing $z_1, \ldots, z_n$.

(a) Suppose $z_1, \ldots, z_n$ are not collinear. Then any line passing an interior point of $\mathrm{Con}(z_1, \ldots, z_n)$ will separate $z_1, \ldots, z_n$.

(b) If $\sum_{j=1}^{n} \frac{1}{z - z_j} = 0$ holds, then z should lie on the set $\mathrm{Con}(z_1, \ldots, z_n)$.

1.4.4. *Steiner circles and symmetric points with respect to a circle (or line)*

We try to give another proof of the fact that an Apollonius circle C_1 : $\left| \frac{z-a}{z-b} \right| = \lambda \, (0 \leq \lambda \leq \infty)$ and a coaxial circle C_2 : $\mathrm{Arg} \frac{z-a}{z-b} = \theta \, (-\pi < \theta \leq \pi)$ will intersect orthogonally (refer to Example 3 in Sec. 1.4.3). And hence, we introduce how two points are said to be symmetric with respect to a circle (or line).

In Fig. 1.26, fix any point z_0 on a C_2 circle and draw the tangent to the circle at z_0 so that it intersects the extended line $\overline{ab}$ at a point p. Now,

$$\Delta p z_0 a \sim \Delta p b z_0 \text{ (similar)} \Rightarrow \frac{|z_0 - a|}{|z_0 - b|} = \frac{|p - a|}{|p - z_0|} = \frac{|p - z_0|}{|p - b|}.$$

Note that z_0 lies on a C_1 circle, namely,

$$\left| \frac{z - a}{z - b} \right| = \lambda = \left| \frac{z_0 - a}{z_0 - b} \right|.$$

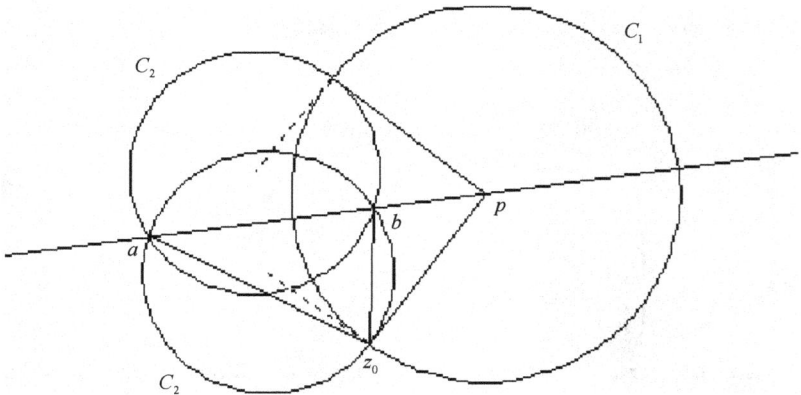

Fig. 1.26

Hence,

$$\frac{|p-a|}{|p-z_0|} = \frac{|p-z_0|}{|p-b|} = \lambda \Rightarrow \frac{|p-a|}{|p-b|} = \lambda^2, \quad \text{a constant.}$$

This means that, as long as a point z_0 lying on a C_1 circle: $\left|\frac{z-a}{z-b}\right| = \lambda$, then the tangent to the C_2 circle: $\text{Arg}\,\frac{z-a}{z-b} = \theta = \text{Arg}\,\frac{z_0-a}{z_0-b}$ will always intersect the line ab at a *fixed* point p. Moreover,

$$|p-z_0|^2 = |p-a|\,|p-b|,$$

shows that $|p - z_0|$ is also a *constant*. And we prove that such a C_1 circle has its center at p and radius $|p - z_0|$, and is orthogonal to any possible C_2 circle.

We formally summarize the above as

Symmetric or reflection points with respect to a circle (or a line). Fix a circle C with center $\mathbf{z_0}$ (including the degenerated case, a line L) and two points $z, z^* \in \mathbf{C}$. Then, the following are equivalent:

(1) z and z^* are situated on the same half line from z_0, namely $\frac{z^*-z_0}{z-z_0} > 0$, and $|z - z_0|\,|z^* - z_0| = r^2$.
(2) Any circle (or line) passing through z and z^* will intersect the fixed circle C (or line L) orthogonally.
(3) Any circle (or line) passing z and intersecting C orthogonally will also pass the point z^*.

In such a case, we called z and z^* *symmetric* with respect to C (or L), or *reflection* points of C (or L). In fact:

(a) If C has equation $|z - z_0| = r$, then z and z^* are symmetric w.r.t. C if and only if

$$z^* = z_0 + \frac{r^2}{\bar{z} - \bar{z}_0}.$$

In this case, the circle C can be expressed as $\left|\frac{\zeta-z}{\zeta-z^*}\right| = \frac{r}{r_1} = \frac{r_2}{r}$ where $z = z_0 + r_1 e^{i\varphi}$ and $z^* = z_0 + r_2 e^{i\varphi}$. If C has equation $az\bar{z} + \bar{b}z + b\bar{z} + c = 0$, then

$$az^*\bar{z} + \bar{b}z^* + b\bar{z} + c = 0.$$

(b) If L has equation $\bar{a}z + a\bar{z} + b = 0$, then z and z^* are symmetric w.r.t. L if and only if

$$a\bar{z} + \bar{a}z^* + b = 0;$$

if L has equation $z = a + bt(|b| = 1, t \in \mathbf{R})$, then

$$z^* = a + b^2(\bar{z} - \bar{a});$$

if L is the line passing two distinct points a_1 and a_2, then

$$z^* = \frac{1}{\overline{a_2} - \overline{a_1}} \left[(a_2 - a_1)\bar{z} + a_1\overline{a_2} - \overline{a_1}a_2 \right]. \tag{1.4.4.1}$$

See Fig. 1.27

Note that z lies inside $C \Leftrightarrow z^*$ lies outside C; z lies on $C \Leftrightarrow z^*$ lies on C and $z = z^*$ holds. In case of a line L, z and z^* lie on different sides of L.

By the way, we obtain partial results of the following

Steiner's circles. Fix two distinct points a and b in the plane $\mathbf{C}$. Then,

$$\text{Apollonius circles } C_1 : \left| \frac{z-a}{z-b} \right| = \lambda, \quad 0 \le \lambda \le \infty \quad \text{and}$$

$$\text{Coaxial circles } C_2 : \operatorname{Arg} \frac{z-a}{z-b} = \theta, \quad -\pi < \theta \le \pi,$$

together form the so-called *Steiner's circles*: a and b are called the *limit point* of C_1 circles, while the line segment $\overline{ab}$ the *coaxis* of C_2 circles. See Fig. 1.28.

They own the following basic properties:

(1) Except the limit points a and b, any point in $\mathbf{C}$ lies on exactly one C_1 circle and only one C_2 circle.

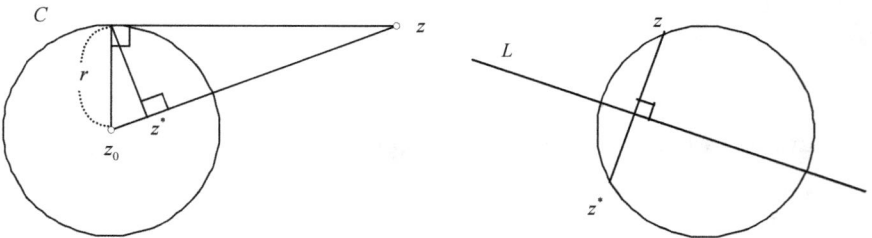

Fig. 1.27

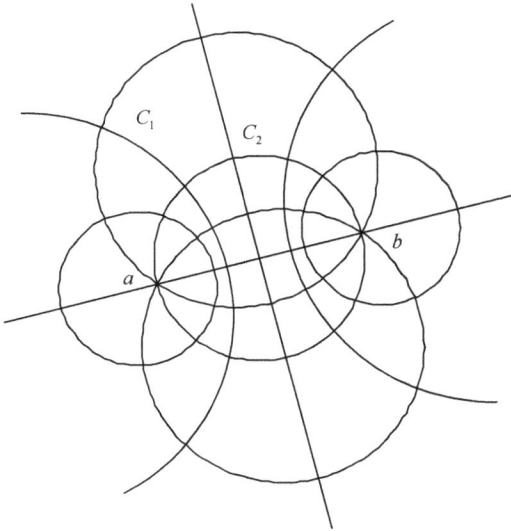

Fig. 1.28

(2) Every C_1 circle is orthogonal to every C_2 circle.
(3) A *reflection* or *symmetric motion* $(z \to z^*)$ with respect to a C_1 circle maps every C_2 circle onto itself, and maps a C_1 circle onto another C_1 circle. A *reflection* or *symmetric motion* with respect to a C_2 circle maps every C_1 circle onto itself, and maps a C_2 circle onto another C_2 circle. See Exercise A(7).
(4) The limit points a and b are symmetric with respect to every C_1 circle, but not to any other circles.   (1.4.4.2)

As a matter of fact, (1.4.4.1) and (1.4.4.2) are easy consequences of the conformality of the bilinear transformations $w = \frac{z - z_0}{z - z_0^*}$ and $w = \frac{z - a}{z - b}$, respectively. (See Sec. 2.5.4; in particular, Examples 1 and 2 there.)

Exercises A

(1) Prove (a) and (b) in (1.4.4.1) in detail.
(2) Prove (1.4.4.2). See also Exercise (7) below.
(3) Suppose a circle C_1 intersects another circle C_2 orthogonally and a line passing the center of C_1 will intersect C_2 at two points a and b. Show that a and b are symmetric with respect to C_1.

(4) (a) Show that a point z symmetric with respect to both $|z - a_1| = r_1$ and $|z - a_2| = r_2$ satisfies

$$r_1^2(z - a_1)^{-1} - r_2^2(z - a_2)^{-1} = \bar{a}_2 - \bar{a}_1.$$

(b) Show that a point z symmetric with respect to both a circle $|z - a| = r$ and a line passing a_1 and a_2 satisfies

$$\bar{a} + r^2(z - a)^{-1} = (a_2 - a_1)^{-1}[(\bar{a}_2 - \bar{a}_1)z + \bar{a}_1 a_2 - a_1 \bar{a}_2].$$

(5) Try to use Exercise (4) to do the following problems.

(a) Find out all the circles orthogonal to both $|z| = 1$ and $\left|z - \frac{1}{4}\right| = \frac{1}{4}$.
(b) Find out all the circles orthogonal to both $|z| = 1$ and the line $x = 2$.

(6) Discuss the following families of curves and graph them.

(a) $\operatorname{Re}\frac{1}{z} = \lambda_1$ and $\operatorname{Im}\frac{1}{z} = \lambda_2$, are they orthogonal to each other?
(b) $\operatorname{Re} z^2 = \lambda_1$ and $\operatorname{Im} z^2 = \lambda_2$, are they orthogonal to each other?

(7) (Refer to (3) in (1.4.4.2))

(a) Suppose z and z^* are symmetric with respect to a C_1 circle: $\left|\frac{z-a}{z-b}\right| = \lambda_0$ $(0 \le \lambda_0 \le \infty)$. Then

$$\left|\frac{z-a}{z-b}\right| = \lambda \Leftrightarrow \left|\frac{z^* - a}{z^* - b}\right| = \frac{\lambda_0^2}{\lambda} \quad (0 \le \lambda \le \infty);$$

$$\operatorname{Arg}\frac{z-a}{z-b} = \operatorname{Arg}\frac{z^* - a}{z^* - b} = \theta \quad (-\pi < \theta \le \pi).$$

(b) Suppose z and z^* are symmetric with respect to a C_2 circle. Then,

$$\left|\frac{z-a}{z-b}\right| = \left|\frac{z^* - a}{z^* - b}\right| = \lambda \quad (0 \le \lambda \le \infty);$$

$$\operatorname{Arg}\frac{z^* - a}{z^* - b} = \theta \Leftrightarrow \operatorname{Arg}\frac{z-a}{z-b} = -\theta \quad (-\pi < \theta \le \pi).$$

(8) Consider the graph of the circle $|z| = 1$ under the translation $z \to 1+z$. Show that, if $|z| = 1$ and $z \ne -1$, then $2\operatorname{Arg}(1 + z) = \operatorname{Arg} z$.

1.5. De Moivre Formula and nth Roots of Complex Numbers

In (1.2.15), let $z_1 = z_2 = z$, then

$$z^2 = |z|^2(\cos 2\theta + i\sin 2\theta), \quad \theta = \arg z;$$

in (1.2.15)$'$, let $z_2 = 1$ and $z_1 = z$, then

$$z^{-1} = |z|^{-1}(\cos(-\theta) + i\sin(-\theta)), \quad \theta = \arg z.$$

Based on these two identities, we get inductively the following

$$z^n = r^n(\cos n\theta + i\sin n\theta), \quad r = |z|, \quad \theta = \arg z$$

$$= r^n e^{in\theta}, \quad n = 0, \pm 1, \pm 2, \dots . \tag{1.5.1}$$

In case $|z| = r = 1$, we have the

De Moivre formula.

$$(\cos \theta + i\sin \theta)^n = \cos n\theta + i\sin n\theta, \quad n = \pm 0, \pm 1, \pm 2, \dots,$$

or,

$$(e^{i\theta})^n = e^{in\theta}. \tag{1.5.2}$$

One of the main advantages of this formula is that it provides an easy way to compute the nth roots of a complex number, while the other one is that, via binomial expansion, we can express both $\cos n\theta$ and $\sin n\theta$ as polynomials in $\cos \theta$ and $\sin \theta$ (see Exercise B(2)–(4)).

Let n be a fixed positive integer and $z \in \mathbf{C}$. Then, *the nth roots* of z, denoted as

$$z^{\frac{1}{n}} \quad \text{or} \quad \sqrt[n]{z}, \tag{1.5.3}$$

are defined as any complex numbers w such that $w^n = z$.

If $z = 0$, designate $z^{\frac{1}{n}} = 0$.

Now, suppose $z \neq 0$. In $w^n = z$, let

$$z = re^{i\theta}, \quad r = |z|, \quad \text{and} \quad \theta = \operatorname{Arg} z;$$

$$w = \rho e^{i\varphi}, \quad \rho = |w|, \quad \text{and} \quad \varphi = \arg w$$

$$\Rightarrow (\text{By } (1.5.1)) \; \rho^n e^{in\varphi} = re^{i\theta}$$

$$\Rightarrow \rho^n = r \quad \text{and} \quad n\varphi = \theta + 2k\pi, \quad k = 0, \pm 1, \pm 2, \dots$$

$$\Rightarrow \rho = r^{\frac{1}{n}} = |z|^{\frac{1}{n}} \quad (\text{as a positive real number}),$$

$$\varphi = \frac{\theta + 2k\pi}{n} = \frac{1}{n}(\operatorname{Arg} z + 2k\pi), \quad k = 0, \pm 1, \pm 2, \dots .$$

Hence, the roots of $w^n = z$ are

$$w_k = |z|^{\frac{1}{n}} e^{\frac{i}{n}(\text{Arg } z + 2k\pi)}, \quad k = 0, \pm 1, \pm 2, \ldots.$$

In case $k < 0$ or $k \geq n$, $w_k = w_m \Leftrightarrow k = pn + m$ for some integer p and $0 \leq m \leq n-1$. Hence, $w^n = z$ has only n *distinct* roots w_k for $0 \leq k \leq n-1$.

We summarize the above as

nth roots ($n \geq 2$) of a nonzero complex number.

(1) *The nth roots of the unit* 1. Let

$$\omega = \cos \frac{2\pi}{n} + i \sin \frac{2\pi}{n} = e^{i\frac{2\pi}{n}}.$$

Then $\omega^n = 1$ and nth roots of 1 are

$$1, \omega, \omega^2, \ldots, \omega^{n-1}$$

which form vertices of a regular n-gon inscribed in the unit circle. See Fig. 1.29 (for $n = 6$).

(2) *The nth roots of $z \neq 0$.* Let $\theta = \text{Arg } z$. Then, the nth roots of z are

$$\zeta_k = |z|^{\frac{1}{n}} e^{\frac{i}{n}(\theta + 2k\pi)}$$

$$= \zeta_0 \omega^k, \quad k = 0, 1, 2, \ldots, n-1,$$

ζ_0 is usually called the *principal value* of $\sqrt[n]{z}$. Note that $\zeta_0, \zeta_0\omega, \ldots, \zeta_0\omega^{n-1}$ form vertices of a regular n-gon inscribed in the circle with center at 0 and radius $|z|^{\frac{1}{n}}$.   (1.5.4)

A complex number ζ, such as ω, satisfying $\zeta^n = 1$ but $\zeta^m \neq 1$ for any $1 \leq m \leq n-1$, is called a *primitive nth root* of 1. If ζ is a primitive nth

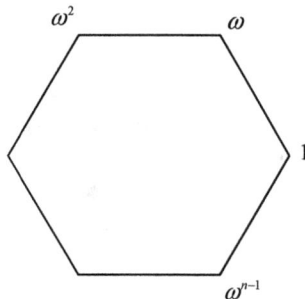

Fig. 1.29

root of 1, then $1, \zeta, \zeta^2, \ldots, \zeta^{n-1}$ will be all the nth roots of 1. $\zeta = \omega^k$ is a primitive nth root if and only if $1 \leq k \leq n-1$ and k is relatively prime to n.

Equation (1.5.3) can be extended as follows. Let n and m be integers so that $n > 0$ is relatively prime with $|m|$. Then, we define

$$z^{\frac{m}{n}} = \begin{cases} (z^{\frac{1}{n}})^m, & m > 0, \\ \dfrac{1}{(z^{\frac{1}{n}})^{-m}}, & m < 0 \quad \text{and} \quad z \neq 0. \end{cases} \tag{1.5.5}$$

By using (1.5.4), we have

$$z^{\frac{m}{n}} = |z|^{\frac{m}{n}} e^{i\frac{m(\theta + 2k\pi)}{n}}, \quad \theta = \mathrm{Arg}\, z, \quad k = 0, 1, 2, \ldots, n-1. \tag{1.5.6}$$

Note that, within this formula, $|z|^{\frac{m}{n}} = \sqrt[n]{|z|^m} = (\sqrt[n]{|z|})^m$ is chosen to be a positive number. For further discussion, see Exercise A(10).

We give four examples.

Example 1. Solve $z^6 - 2z^3 + 2 = 0$.

Solution. $z^6 - 2z^3 + 1 = -1 \Rightarrow (z^3 - 1)^2 = -1 = e^{\pi i}$. Hence

$$z^3 - 1 = e^{\frac{i}{2}(\pi + 2k\pi)}, \quad k = 0, 1.$$

In case $k = 0 : z^3 - 1 = e^{\frac{1}{2}\pi i} = i \Rightarrow z^3 = 1 + i = \sqrt{2} e^{\frac{1}{4}\pi i}$. Hence

$$z_l = 2^{\frac{1}{6}} e^{\frac{i}{3}\left(\frac{1}{4}\pi + 2l\pi\right)}, \quad l = 0, 1, 2$$

$$\Rightarrow z_0 = 2^{\frac{1}{6}} e^{\frac{1}{12}\pi i} = 2^{\frac{1}{6}} \left(\frac{\sqrt{6} + \sqrt{2}}{4} + i\frac{\sqrt{6} - \sqrt{2}}{4} \right);$$

$$z_1 = 2^{\frac{1}{6}} e^{\frac{3}{4}\pi i} = 2^{\frac{1}{6}} \left(-\frac{1}{\sqrt{2}} + i\frac{1}{\sqrt{2}} \right);$$

$$z_2 = 2^{\frac{1}{6}} e^{\frac{17}{12}\pi i} = 2^{\frac{1}{6}} \left(\frac{-\sqrt{6} + \sqrt{2}}{4} - i\frac{\sqrt{6} + \sqrt{2}}{4} \right).$$

In case $k = 1 : z^3 - 1 = -i \Rightarrow z^3 = 1 - i = \sqrt{2} e^{-\frac{1}{4}\pi i}$. Hence

$$z_{l+3} = 2^{\frac{1}{6}} e^{\frac{i}{3}\left(-\frac{1}{4}\pi + 2l\pi\right)}, \quad l = 0, 1, 2$$

$$\Rightarrow z_3 = \overline{z_0}, \quad z_4 = \overline{z_2} \quad \text{and} \quad z_5 = \overline{z_1}.$$

See Fig. 1.30.

Example 2. Compute (1) $[(1 + i)^{\frac{1}{2}}]^3$, $[(1 + i)^3]^{\frac{1}{2}}$;

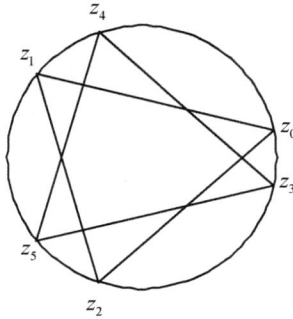

Fig. 1.30

(2) $[(1+i)^{\frac{1}{2}}]^2$, $[(1+i)^2]^{\frac{1}{2}}$, and compare them.

Solution. (1) $1+i = \sqrt{2}e^{\frac{1}{4}\pi i}$. So

$$(1+i)^{\frac{1}{2}} = 2^{\frac{1}{4}}e^{\frac{i}{2}\left(\frac{1}{4}\pi+2k\pi\right)}, \quad k = 0, 1,$$

$$\Rightarrow [(1+i)^{\frac{1}{2}}]^3 = 2^{\frac{3}{4}}e^{\frac{i}{2}\left(\frac{3}{4}\pi+6k\pi\right)}, \quad k = 0, 1,$$

namely, $z_0 = 2^{\frac{3}{4}}e^{\frac{3}{8}\pi i}$, $z_1 = -2^{\frac{3}{4}}e^{\frac{3}{8}\pi i}$.

On the other hand, $(1+i)^3 = 2(-1+i) = 2\sqrt{2}e^{\frac{3}{4}\pi i}$ and $[(1+i)^3]^{\frac{1}{2}}$ has two values z_0 and z_1.

(2) By (1),

$$[(1+i)^{\frac{1}{2}}]^2 = 2^{\frac{1}{2}}e^{i\left(\frac{1}{4}\pi+2k\pi\right)}, \quad k = 0, 1,$$

and we get only one value $\sqrt{2}e^{\frac{1}{4}\pi i} = 1+i$. While, $(1+i)^2 = 2i = 2e^{\frac{1}{2}\pi i}$ and hence

$$[(1+i)^2]^{\frac{1}{2}} = 2^{\frac{1}{2}}e^{\frac{i}{2}\left(\frac{1}{2}\pi+2k\pi\right)}, \quad k = 0, 1,$$

namely, $\pm(1+i)$, which has one more value than $[(1+i)^{\frac{1}{2}}]^2$.

Remark. Suppose $f_1(z)$ and $f_2(z)$ are two multi-valued functions. If, for each z in their common domain of definitions, the set of values of $f_1(z)$ is equal to the set of values of $f_2(z)$, then we *designate* $f_1(z) = f_2(z)$, otherwise $f_1(z) \neq f_2(z)$. According to this convention, $[(1+i)^{\frac{1}{2}}]^3 = [(1+i)^3]^{\frac{1}{2}}$ while $[(1+i)^{\frac{1}{2}}]^2 \neq [(1+i)^2]^{\frac{1}{2}}$. For general setting, see Exercises A(9) and (10). Also, refer to (1.2.16).

Example 3. Let $n \geq 2$ be an integer and $w_n = e^{\frac{2\pi i}{n}}$. Show that

(1) $w_n^{nm+k} = w_n^k$, $m \in \mathbf{Z}$ and $k = 0, 1, \ldots, n-1$;
(2) $w_n^{n-k} = \bar{w}_n^k$, $k = 0, 1, \ldots, n-1$; and
(3) $1 + w_n + w_n^2 + \cdots + w_n^{n-1} = 0$ (for geometric meaning, see Fig. 1.29)

Proof. (1) and (2) are obvious. (3) Note that $w_n \neq 1$. Now

$$w_n^n - 1 = (w_n - 1)(1 + w_n + \cdots + w_n^{n-1}) = 0$$

$$\Rightarrow 1 + w_n + \cdots + w_n^{n-1} = 0.$$

Or, let $p = 1 + w_n + \cdots + w_n^{n-1}$. Then $pw_n = w_n + w_n^2 \cdots + w_n^{n-1} + w_n^n = 1 + w_n + \cdots + w_n^{n-1} = p$. Hence $p(w_n - 1) = 0$ and it follows that $p = 0$. □

Example 4. Factorize the polynomial

$$x^{12} + x^9 + x^6 + x^3 + 1$$

over the integer, the real and the complex number systems, respectively.

Analysis: The original polynomial can be rewritten as $(x^3)^4 + (x^3)^3 + (x^3)^2 + x^3 + 1$. Recall the formula $a^5 - 1 = (a-1)(a^4 + a^3 + a^2 + a^1 + 1)$. Hence

$$\frac{x^{15} - 1}{x^3 - 1} = x^{12} + x^9 + x^6 + x^3 + 1$$

and try to factor the left side. Or, observe that $12 = 5 \times 2 + 2$, $9 = 5 \times 1 + 4$, $6 = 5 \times 1 + 1$. If we choose w as a primitive 5th root of 1, then

$$w^{12} + w^9 + w^6 + w^3 + 1 = w^{10}w^2 + w^5w^4 + w^5w + w^3 + 1$$

$$= w^2 + w^4 + w + w^3 + 1 = 0,$$

which means the original polynomial has a factor $x^4 + x^3 + x^2 + x + 1$.

Solution. In **C**, $z^{15} - 1 = 0$ has roots $e^{\frac{2k\pi}{15}i}$, $0 \leq k \leq 14$; while the roots corresponding to $k = 0, 5, 10$ are roots of $z^3 - 1 = 0$. Hence,

$$x^{12} + x^9 + x^6 + x^3 + 1 = \prod_{\substack{k=0 \\ k \neq 0,5,10}}^{14} (x - e^{\frac{2k\pi}{15}i}).$$

By use of Example 3(2), $e^{\frac{2\pi i}{15}(15-k)} = e^{-\frac{2k\pi i}{15}}$ for $k = 0, 1, 2, \ldots, 7$. Hence, for such k,

$$(x - e^{\frac{2k\pi}{15}i})(x - e^{-\frac{2k\pi}{15}i}) = x^2 - \left(2\cos\frac{2k\pi}{15}\right)x + 1.$$

Therefore, in $\mathbf{R}$,

$$x^{12} + x^9 + x^6 + x^3 + 1 = \prod_{\substack{k=1 \\ k \neq 5}}^{7} \left[x^2 - \left(2\cos\frac{2k\pi}{5}\right)x + 1\right].$$

In $\mathbf{Z}$

$$x^{12} + x^9 + x^6 + x^3 + 1 = \frac{x^{15} - 1}{x^3 - 1} = \frac{(x^5)^3 - 1}{x^3 - 1} = \frac{x^5 - 1}{x - 1} \cdot \frac{x^{10} + x^5 + 1}{x^2 + x + 1}$$

$$= (x^4 + x^3 + x^2 + x + 1)$$

$$\times (x^8 - x^7 + x^5 - x^4 + x^3 - x + 1),$$

where the two factor polynomials in the right cannot be factored any more over $\mathbf{Z}$ (why?). Or, according to the *Analysis* above, divide $x^{12} + x^9 + x^6 + x^3 + 1$ by $x^4 + x^3 + x^2 + x + 1$ via long division.

Exercises A

(1) Try to set $z = \cos\theta + i\sin\theta$ in identities such as $1 + z + \cdots + z^n = \frac{1 - z^{n+1}}{1 - z}$, $z \neq 1$, to prove the following identities.

(a) $1 + \sum_{k=1}^{n} \cos k\theta = \frac{1}{2} + \frac{\sin(n + \frac{1}{2})\theta}{2\sin\frac{\theta}{2}}$, $0 < \theta < 2\pi$.

(b) $\sum_{k=1}^{n} \sin k\theta = \frac{\sin\frac{n\theta}{2}\sin\frac{(n+1)}{2}\theta}{\sin\frac{\theta}{2}}$, $0 < \theta < 2\pi$.

(c) $\sum_{k=1}^{n} \cos(2k-1)\theta = \frac{\sin 2n\theta}{2\sin\theta}$, $0 < \theta < \pi$.

(d) $\sum_{k=1}^{n} \sin(2k-1)\theta = \frac{\sin^2 n\theta}{\sin\theta}$, $0 < \theta < \pi$.

(2) Solve the following equations.
   (a) $z^8 + z = 0$. (b) $z^8 - 4z^4 + 3 = 0$. (c) $z^5 + z + 1 = 0$.
   (d) $z^8 + z^4 + 1 = 0$.

(3) Solve $z^{n-1} = \bar{z}$.

(4) Compute $\left(\frac{\sqrt{3}}{2} + \frac{1}{2}i\right)^{\frac{3}{4}}$ and $i^{\frac{3}{2}}$.

(5) Show that all the roots of $(z+1)^5 + z^5 = 0$ lie on the line $\operatorname{Re} z = -\frac{1}{2}$.

(6) Show that $64z^6 = (z-1)^6$ has four complex roots, where two of them lie on the imaginary axis, and the other two lie on the first and the fourth quadrant. Also, show that all the roots lie on the circle $\left(x + \frac{1}{3}\right)^2 + y^2 = \left(\frac{2}{3}\right)^2$.

(7) (a) Show that the real part of $\sqrt[n]{z} + \sqrt[n]{\bar{z}}$, where $z = re^{i\theta}$, is

$$2\sqrt[n]{r}\cos\frac{\theta + 2k\pi}{n}, \qquad 0 \le k \le n-1.$$

(b) Write out the real part of $\sqrt[4]{-1+i} + \sqrt[4]{-1-i}$.

(8) Let z_1, z_2, and z_3 be the three roots of $z^3 - 3pz^2 + 3qz - r = 0$. Show that the barycenter of $\triangle z_1 z_2 z_3$ is p and, if $\triangle z_1 z_2 z_3$ is an equilateral triangle, then $p^2 = q$.

(9) Compute the following pairs of complex numbers and compare them.

(a) $[(1+i)^{\frac{1}{3}}]^5$; $[(i+i)^5]^{\frac{1}{3}}$.

(b) $[(1+i)^{\frac{1}{3}}]^6$; $[(1+i)^6]^{\frac{1}{3}}$.

(10) Let n and m be integers, where n is positive.

(a) Show that $z^{\frac{m}{n}} = (z^{\frac{1}{n}})^m$ has $\frac{n}{(n,|m|)}$ distinct values, where $z \ne 0$ and $(n, |m|)$ denotes the greatest common divisor of n and $|m|$.

(b) Show that $(z^{\frac{1}{n}})^m = (z^m)^{\frac{1}{n}} \Leftrightarrow (n, |m|) = 1$.

(11) If $n \ge 2$, show that

$$\sum_{k=0}^{n-1} \cos\frac{2k\pi}{n} = 0 \quad \text{and} \quad \sum_{k=0}^{n-1} \sin\frac{2k\pi}{n} = 0.$$

(12) Show that

$$\prod_{k=1}^{n-1} \sin\frac{k\pi}{n} = \frac{n}{2^{n-1}}.$$

(13) In Fig. 1.29, fix any vertex of the regular n-gon. Show that the product of the distances from that vertex to the other $(n-1)$ vertices is the constant n.

(14) Let k be an integer and $0 \le k \le n-1$. Set $\omega = e^{\frac{2k\pi i}{n}}$.

(a) The smallest positive integer p satisfying $\omega^p = 1$ is called the *order* of ω. Find the order of ω.

(b) For each positive integer l, show that

$$1 + \omega^l + \omega^{2l} + \cdots + \omega^{l(n-1)} = \begin{cases} n, & \text{if } n \text{ divides } lk \\ 0, & \text{if otherwise} \end{cases}.$$

(15) Let ω be a primitive nth root of 1. Compute the following values.

(a) $1 - \omega^l + \omega^{2l} - \cdots + (-1)^{n-1}\omega^{(n-1)l}$, where l is an integer but not a multiple of n.

(b) $1 + 2\omega + 3\omega^2 + \cdots + n\omega^{n-1}$.

(c) $\omega + 2\omega^2 + 3\omega^3 + \cdots + (n-1)\omega^{n-1}$.

(d) $1 + 4\omega + 9\omega^2 + \cdots + n^2\omega^{n-1}$.

(16) Let ω be a primitive nth root of 1 and $f(z) = a_0 + a_1 z + \cdots + a_k z^k$. Show that

$$\frac{1}{n}\{f(z) + f(\omega z) + f(\omega^2 z) + \cdots + f(\omega^{n-1} z)\}$$

$$= a_0 + a_n z^n + a_{2n} z^{2n} + \cdots + a_{\lambda n} z^{\lambda n},$$

where λn is the largest multiple of n, not larger than k.

(17) (a) Suppose ω is a primitive 5th root of 1. Show that

$$(x + y + z)(x + \omega y + \omega^4 z)(x + \omega^2 y + \omega^3 z)(x + \omega^3 y + \omega z)$$
$$\times (x + \omega^4 y + \omega z) = x^5 + y^5 + z^5 - 5x^3 yz + 5xy^2 z^2.$$

(b) Use (a) to solve $x^5 - 5ax^3 + 5a^2 x + (a^5 + 1) = 0$.

(18) Prove the following factorizations, where $a \geq 0$.

(a) $x^{2m} - a^{2m} = (x^2 - a^2) \prod_{k=1}^{m-1} \left(x^2 - 2ax \cos \frac{k\pi}{m} + a^2\right)$.

(b) $x^{2m+1} - a^{2m+1} = (x - a) \prod_{k=1}^{m} \left(x^2 - 2ax \cos \frac{2k\pi}{2m+1} + a^2\right)$.

(c) $x^{2m} + a^{2m} = \prod_{k=0}^{m-1} \left(x^2 - 2ax \cos \frac{(2k+1)\pi}{2m} + a^2\right)$.

(d) $x^{2m+1} + a^{2m+1} = (x + a) \prod_{k=0}^{m-1} \left(x^2 - 2ax \cos \frac{(2k+1)\pi}{2m+1} + a^2\right)$.

(e) $x^{2m} - 2a^m x^m \cos \theta + a^{2m} = \prod_{k=0}^{m-1} \left(x^2 - 2ax \cos \frac{\theta + 2k\pi}{m} + a^2\right)$.

Exercises B

(1) Factorize the following polynomials over **Z**, **R**, and **C**, respectively:

(a) $x^5 + x^4 + 1$.

(b) $x^7 + x^6 + x^4 + 2x^2 + 1$.

(c) $x^8 + x^6 + x^4 + x^2 + 1$.

(d) $a + (a + b)x + (a + 2b)x^2 + (a + 3b)x^3 + 3bx^4 + 2bx^5 + bx^6$.

(2) By expanding $(\cos x + i \sin x)^m$ binomially and comparing the real and the imaginary parts of both sides, show that

$$\cos mx = \sum_{k=0}^{\left[\frac{m}{2}\right]} (-1)^k C_{2k}^m \cos^{(m-2k)} x \sin^{2k} x,$$

$$\sin mx = \sum_{k=0}^{\left[\frac{m-1}{2}\right]} (-1)^k C_{2k+1}^m \cos^{(m-2k-1)} x \sin^{(2k+1)} x,$$

where $\left[\frac{m}{2}\right]$ denotes the largest integer not larger than $\frac{m}{2}$, etc.

(3) (a) If m is an even integer and $x \in \mathbf{R}$, show that

$$\cos mx = \prod_{k=1}^{\frac{m}{2}} \left(1 - \frac{\sin^2 x}{\sin^2 \frac{(2k+1)\pi}{2m}}\right),$$

$$\frac{\sin mx}{\cos x} = m \sin x \prod_{k=1}^{\frac{m}{2}-1} \left(1 - \frac{\sin^2 x}{\sin^2 \frac{k\pi}{m}}\right).$$

(b) If m is an odd integer and $x \in \mathbf{R}$, show that

$$\frac{\cos mx}{\cos x} = \prod_{k=1}^{\frac{m-1}{2}} \left(1 - \frac{\sin^2 x}{\sin^2 \frac{(2k-1)\pi}{m}}\right),$$

$$\sin mx = m \sin x \prod_{k=1}^{\frac{m-1}{2}} \left(1 - \frac{\sin^2 x}{\sin^2 \frac{k\pi}{m}}\right).$$

Note. These identities can be used to derive the infinite product expressions for $\cos x$ and $\sin x$:

$$\cos x = \prod_{n=0}^{\infty} \left(1 - \frac{x^2}{(n+\frac{1}{2})^2 \pi^2}\right); \quad \sin x = x \prod_{k=1}^{\infty} \left(1 - \frac{x^2}{n^2 \pi^2}\right).$$

(4) Let $z = re^{i\theta}$. Then $z^n + z^{-n} = 2\cos n\theta$ and $z^n - z^{-n} = 2i\sin n\theta$. Show that:

(a) If n is an odd integer,

$$\cos^n \theta = \frac{1}{2^{n-1}} \sum_{k=0}^{\frac{n-1}{2}} C_k^n \cos(n - 2k)\theta,$$

$$\sin^n \theta = \frac{1}{2^{n-1}} (-1)^{\frac{n-1}{2}} \sum_{k=0}^{\frac{n-1}{2}} (-1)^k C_k^n \sin(n - 2k)\theta.$$

(b) If n is an even integer,

$$\cos^n \theta = \frac{1}{2^{n-1}} \sum_{k=0}^{\frac{n-2}{2}} C_k^n \cos(n-2k)\theta + \frac{1}{2^n} C_{\frac{n}{2}}^n,$$

$$\sin^n \theta = \frac{1}{2^{n-1}} (-1)^{\frac{n}{2}} \sum_{k=0}^{\frac{n-2}{2}} (-1)^k C_k^n \cos(n-2k)\theta + \frac{1}{2^n} C_{\frac{n}{2}}^n.$$

(5) (a) If $0 < \theta < \frac{\pi}{2}$, show that

$$\sin(2m+1)\theta = \sin^{2m+1}\theta P_m(\cot^2 \theta)$$

where $P_m(x) = \sum_{k=0}^m (-1)^k C_{2k+1}^{2m+1} x^{m-k}$.

(b) Show that

$$\sum_{k=1}^m \cot^2 \frac{k\pi}{2m+1} = \frac{m(2m-1)}{3}.$$

1.6.   Spherical Representations of Complex Numbers: Riemann Sphere and Extended Complex Plane

In the Euclidean space $\mathbf{R}^3$, consider the *unit sphere*

$$S : x_1^2 + x_2^2 + x_3^2 = 1$$

with $N = (0,0,1)$ as the *north pole*. Designate the $x_1 x_2$-plane as the complex plane which intersects with S along the equator.

Pick any point z in $\mathbf{C}$. The line connecting z to N intersects S at a unique point Q, where $Q \neq N$, and vice versa. Note that $|z| < 1 \Leftrightarrow Q$ is in the lower hemisphere, $|z| > 1 \Leftrightarrow Q$ is in the upper hemisphere, and $|z| = 1 \Leftrightarrow Q$ lies on the equator. See Fig. 1.31.

This sets up a one-to-one and onto correspondence

$$\Phi : Q \in S \backslash \{N\} \leftrightarrow \Phi(Q) = z \in \mathbf{C}, \qquad (1.6.1)$$

between S (except N) and $\mathbf{C}$, called the *stereographic projection* from S onto $\mathbf{C}$ with N as *center* in which z is called the *projective point* of Q in $\mathbf{C}$

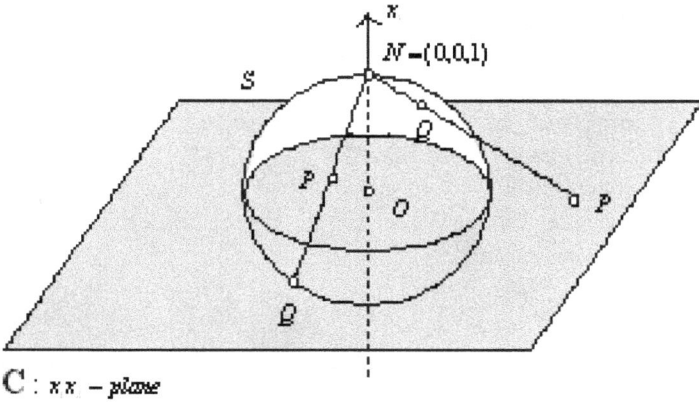

Fig. 1.31

and Q the spherical *image* or *representation* of z on S. In this sense, S is particularly called the *Riemann sphere*.

Section (1) The infinite point ∞

No point in $\mathbf{C}$ corresponds to the north pole N of the Riemann sphere S.

Observe that

$$|z| \text{ (the distance of } z \text{ from } 0) \to +\infty$$

$$\Leftrightarrow Q \in S \to N, \text{ i.e., the distance } |Q - N| \to 0.$$

Hence, it is natural to imagine that there exists a unique point, specifically denoted as

$$\infty, \tag{1.6.2}$$

beyond any disk $|z| < \mathrm{R}$ (no matter how large R is), corresponding to the north pole. We call ∞ the *infinite point* or *the point at infinity* of the *finite complex plane* $\mathbf{C}$ and

$$\mathbf{C}^* = \mathbf{C} \cup \{\infty\}, \tag{1.6.3}$$

the *extended complex plane*. Hence, points in $\mathbf{C}^*$ are in one-to-one and onto correspondence with points in S. Sometimes, we denote

$$\mathbf{C} \text{ by } |z| < +\infty, \quad \text{and} \quad \mathbf{C}^* \text{ by } |z| \le +\infty, \tag{1.6.4}$$

where $+\infty$ is the positive infinity, in the extended real number system.

Remark (The geometric properties and algebraic operational properties of ∞). Rotate the sphere S with x_1-axis as axis through $180°$. Then the upper and the lower hemispheres are interchanged so that (x_1, x_2, x_3) into $(x_1, -x_2, -x_3)$, and $N = (0,0,1)$ into the south pole $(0,0,-1)$ and conversely. Note that $\Phi(0,0,-1) = 0$. This motion is equivalent to the reflection $z \to \frac{1}{z}$ of the plane $\mathbf{C}$ (see Fig. 1.11 and (1.6.7))), which interchanges $0 < |z| < 1$ and $1 < |z| < \infty$, and 0 and ∞ at the same time. Hence, we designate

$$\frac{1}{0} = \infty \quad \text{and} \quad \frac{1}{\infty} = 0,$$

$$\frac{a}{0} = \infty \quad (a \in \mathbf{C} \text{ and } a \neq 0) \quad \text{and} \quad \frac{a}{\infty} = 0 \quad (a \in \mathbf{C}). \tag{1.6.5}$$

Note that the inverse mapping $\Phi^{-1} : \mathbf{C} \to S\backslash\{N\}$ does not preserve operations of addition and multiplication. On the other hand, ∞ is the *only* point in $\mathbf{C}^*$ which has the *infinite distance* $+\infty$ to the origin 0. Hence, it seems reasonable to define:

$$a + \infty = \infty + a = \infty, \quad a \in \mathbf{C};$$

$$a - \infty = \infty - a = \infty, \quad a \in \mathbf{C}; \tag{1.6.6}$$

$$a \cdot \infty = \infty \cdot a = \infty, \quad a \in \mathbf{C} \quad \text{and} \quad a \neq 0.$$

Yet $\arg \infty$, $\infty \pm \infty$, $0 \cdot \infty$, $\frac{\infty}{\infty}$, $\frac{0}{0}$, ∞^0 and 1^∞ cannot be defined definitely, especially from the viewpoint of the limit processes (see Exercise A(3) of Sec. 1.7).

In short, the infinite point ∞ is sometimes like a certain kind of mathematical concept rather than an ordinary complex number. Its appearance seems to be more natural in the eyes of the limit concept such as completeness (see Sec. 1.9). The study of ∞ and its neighborhood $(|z| > R)$ is usually transformed to the study of the origin 0 and its neighborhood $\left(|z| < \frac{1}{R}\right)$ via the reflection $z \to \frac{1}{z}$. It is from the geometric and the point-set aspects (Sec. 1.9), but not from the algebraic one, that we view $\mathbf{C}^*$ as the Riemann sphere S.   □

Section (2) The analytic expression of the stereographic projection Φ

In Fig. 1.31, let $Q = (x_1, x_2, x_3) \in S$ and $z = (x, y, 0) = x + iy$. Recall that the north pole $N = (0,0,1)$. Then

$$z, \ Q, \text{ and } N \text{ are collinear.}$$

$$\Leftrightarrow \frac{x_1 - 0}{x - 0} = \frac{x_2 - 0}{y - 0} = \frac{x_3 - 1}{0 - 1}$$

$$\Leftrightarrow x_1 = \lambda x, \ x_2 = \lambda y \quad \text{and} \quad x_3 = 1 - \lambda \text{ for some scalar } \lambda$$

$$\Rightarrow x_1^2 + x_2^2 + x_3^2 = \lambda^2(x^2 + y^2) + (1 - \lambda)^2 = 1$$

$$\Rightarrow \lambda = \frac{2}{1 + |z|^2}.$$

Summarizing, we have

$$\Phi(x_1, x_2, x_3) = \begin{cases} \dfrac{x_1 + ix_2}{1 - x_3}, & (x_1, x_2, x_3) \in S\backslash\{N\}, \text{ namely, } \ x_3 \neq 1 \\[2mm] \infty, & (x_1, x_2, x_3) \in S \text{ and } x_3 = 1, \text{ namely,} \\ & (x_1, x_2, x_3) = (0, 0, 1) \end{cases},$$

and

$$\Phi^{-1}(z) = \begin{cases} \left(\dfrac{z + \bar{z}}{|z|^2 + 1}, \ \dfrac{z - \bar{z}}{i(|z|^2 + 1)}, \ \dfrac{|z|^2 - 1}{|z|^2 + 1} \right), & z \in \mathbf{C} \\[2mm] (0, 0, 1), & z = \infty. \end{cases} \qquad (1.6.7)$$

Note that $\Phi^{-1} : \mathbf{C}^* \to S$ is the inverse of $\Phi : S \to \mathbf{C}^*$.

Section (3) The spherical distance on $\mathbf{C}^*$

"$\|$" cannot be adopted as the distance on $\mathbf{C}^*$ since each finite complex number $z \in \mathbf{C}$ has the *same* $+\infty$ distance to ∞, namely, $|z - \infty| = +\infty$.

Now, for any two points z_1 and z_2 on $\mathbf{C}^*$, we define the *distance* between them as

$$d(z_1, z_2) = |\Phi^{-1}(z_1) - \Phi^{-1}(z_2)|$$
$$= \{(x_1 - y_1)^2 + (x_2 - y_2)^2 + (x_3 - y_3)^2\}^{\frac{1}{2}},$$

where $\Phi^{-1}(z_1) = (x_1, x_2, x_3)$ and $\Phi^{-1}(z_2) = (y_1, y_2, y_3)$, and it is called the *spherical (chord) distance* of z_1 and z_2. In fact, it is the length of the chord connecting the points $\Phi^{-1}(z_1)$ and $\Phi^{-1}(z_2)$.

Since $x_1^2 + x_2^2 + x_3^2 = y_1^2 + y_2^2 + y_3^2 = 1$, via (1.6.7), it is easy to see that

$$d(z_1, z_2) = \begin{cases} \dfrac{2|z_1 - z_2|}{\sqrt{(1 + |z_1|^2)(1 + |z_2|^2)}}, & z_1, z_2 \in \mathbf{C}, \\[4mm] \dfrac{2}{\sqrt{1 + |z_1|^2}}, & z_1 \in \mathbf{C} \text{ and } z_2 = \infty \end{cases} \qquad (1.6.8)$$

satisfying the following properties: for $z_1, z_2, z_3 \in \mathbf{C}^*$

(1) $d(z_1, z_2) \geq 0$ and $=0 \Leftrightarrow z_1$ and z_2.
(2) $d(z_1, z_2) = d(z_2, z_1)$.
(3) $d(z_1, z_3) \leq d(z_1, z_2) + d(z_2, z_3)$.

Hence, d$(,)$ defines a *metric* on $\mathbf{C}^*$ and, endowed with d$(,)$, $\mathbf{C}^*$ becomes a *metric space* (refer to Exercise B(1) of Sec. 1.8). Note that $d(0, \infty) = 2$.

Section (4) Circle-preserving under Φ

The plane $a_1 x_1 + a_2 x_2 + a_3 x_3 = a_0$, where $a_0, a_1, a_2, a_3 \in \mathbf{R}$ and $a_1^2 + a_2^2 + a_3^2 = 1$, in space $\mathbf{R}^3$ has a nonempty intersection with the interior $x_1^2 + x_2^2 + x_3^2 < 1$ of S, if and only if the distance from (0, 0, 0) to the plane is not greater than 1, namely, $|a_0| \leq 1$. Hence, *the circle on the sphere S has the equation*

$$\begin{cases} a_1 x_1 + a_2 x_2 + a_3 x_3 = a_0, & \text{where } a_1^2 + a_2^2 + a_3^2 = 1 \text{ and } |a_0| \leq 1, \\ x_1^2 + x_2^2 + x_3^2 = 1. \end{cases}$$

In particular, it is

(1) a great circle (a circle whose center is at $(0,0,0)$) $\Leftrightarrow a_0 = 0$;
(2) a circle passing the north pole $N = (0,0,1) \Leftrightarrow a_3 = a_0$; and
(3) a circle passing the south pole $(0,0,-1) \Leftrightarrow a_3 = -a_0$.

A great circle passing both the north and the south poles is called a *meridian*, and called the *principal meridian* if it lies on the $x_3 x_1$-plane.

Via (1.6.7), Φ transforms a circle on S into a circle on $\mathbf{C}^*$:

$$(a_0 - a_3)|z|^2 + (a_1 - ia_2)z + (a_1 + ia_2)\bar{z} + a_0 + a_3 = 0. \qquad (1.6.9)$$

Since $|a_1 + ia_2|^2 = a_1^2 + a_2^2 \geq (a_0 - a_3)(a_0 + a_3) = a_0^2 - a_3^2$, (2) in (1.4.3.6) guarantees that it is either a point circle or a real circle.

A reverse process says that, a circle or line on $\mathbf{C}^*$ is mapped, via Φ^{-1}, onto a circle on the sphere S.

We summarize the above as

Circle-preserving of the stereographic projection.

(1) *The stereographic projection Φ maps circles on the Riemann sphere S onto circles or lines on the extended complex plane $\mathbf{C}^*$, and conversely.*
(2) *In particular, a circle on S passes through the north pole $N \Leftrightarrow$ its stereographic image on $\mathbf{C}^*$ is a line.*   (1.6.10)

Hence, *any line in the plane $\mathbf{C}$ should pass through the infinite point ∞,* and *every open half-plane (see (1.4.3.1)) does not contain the point ∞, which is merely a boundary point (see (1.8.6)). Φ also preserves angles (see Exercises B(1)).*

Exercises A

(1) Prove (1.6.7) in detail.
(2) Prove (1.6.8) in detail.
(3) Let Q_1 and Q_2 be two points on the sphere S, and $P_1 = z_1$ and $P_2 = z_2$ are their stereographic images under Φ, respectively. See Fig. 1.32.

   (a) Show that, as lengths of segments in $\mathbf{R}^3$,

   $$NP_j = (1 + |z_j|^2)^{\frac{1}{2}} \quad \text{and} \quad NQ_j = 2(1 + |z_j|^2)^{-\frac{1}{2}}$$

   for $j = 1, 2$.
   (b) Show that $\triangle NQ_1Q_2$ is similar to $\triangle NP_1P_2$. Then, give a geometric proof of (1.6.8).

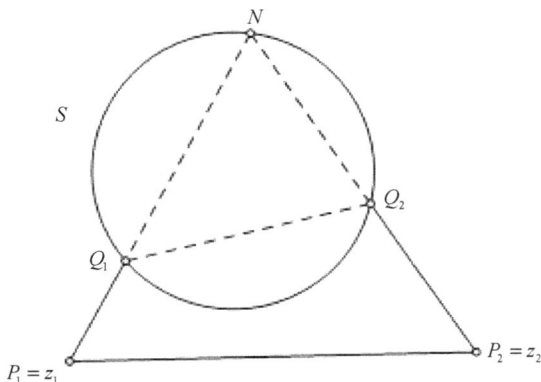

Fig. 1.32

(4) Construct the stereographic projection of the sphere $x_1^2 + x_2^2 + \left(x_3 - \frac{1}{2}\right)^2 = \frac{1}{4}$, with its center at the north pole $N = (0,0,1)$ and the coordinate plane $x_3 = 0$ as the complex plane. Recapture (1.6.7)–(1.6.10) in this case.

(5) Introduce the spherical coordinate

$$x_1 = \cos\varphi\cos\psi, \quad x_2 = \cos\varphi\sin\psi,$$

$$x_3 = \sin\varphi, \quad -\frac{\pi}{2} < \varphi \le \frac{\pi}{2}, \quad -\pi < \psi \le \pi$$

in the sphere S: $x_1^2 + x_2^2 + x_3^2 = 1$, where φ represents the latitude and ψ the longitude of a point (x_1, x_2, x_3) on S.

   (a) Show that a point $P = (\varphi, \psi)$ on S has its image, under the stereographic projection, on $\mathbf{C}$ the point

$$z = (\cos\psi + i\sin\psi)\tan\left(\frac{\pi}{4} + \frac{\varphi}{2}\right).$$

   (b) Use (a) to prove (1.6.7).

(6) Show that the spherical images, under Φ^{-1}, of z and $\frac{1}{z}$, on the sphere S are symmetric with respect to the $x_1 x_2$-plane.

(7) Pinpoint the spherical images, under Φ^{-1}, of the following points:

$$z, \ -z, \ -z^{-1}, \ z^{-1}, \ \bar{z}, \ -\bar{z}, \ -\bar{z}^{-1}, \ \bar{z}^{-1}$$

and compute their coordinates.

(8) Find the coordinates of the vertices of the following figures under Φ:

   (a) A cube inscribed to the sphere S, with sides parallel to the coordinate axis.

   (b) A regular tetrahedron inscribed to S.

(9) (a) Determine the center and the radius of the circle on S whose projection, under Φ, has the equation $|z - z_0| = r$.

   (b) Show that the circle on S in (a) is a great circle $\Leftrightarrow r^2 = 1 + |z_0|^2$.

(10) Find the spherical image of the ray $\operatorname{Arg} z = \theta$ (constant) under Φ^{-1}.

Exercises B

(1) *Angle-preserving of the stereographic projection* (calculus is needed). A continuous mapping $\gamma : (-1,1) \to S$ is said to *define a curve* on S. The point set $\gamma((-1,1))$ is usually called a *curve* on S and is still denoted

by γ. Let

$$\gamma(t) = (x_1(t), x_2(t), x_3(t)), \quad x_1(t)^2 + x_2(t)^2 + x_3(t)^2 = 1, \quad -1 < t < 1.$$

In case $x_j'(t) = \frac{d}{dt}x_j(t)$ for $j = 1, 2, 3$ do not equal to zero simultaneously, then γ is said to have *tangent* at t (or $\gamma(t)$) with *direction*

$$\gamma'(t) = (x_1'(t), \ x_2'(t), \ x_3'(t)).$$

Under Φ, the projective image of γ,

$$\widetilde{\gamma} : (-1, \ 1) \to \mathbf{C}^* \text{ defined by } \Phi(\gamma(t)) = \frac{x_1(t) + ix_2(t)}{1 - x_3(t)} = x(t) + iy(t),$$

thus defines a curve on $\mathbf{C}$. In short, we denote $\widetilde{\gamma} = \widetilde{\gamma}((-1,1))$. See Fig. 1.33.

(a) Show that $|\gamma'(t)| \neq 0 \Leftrightarrow |\widetilde{\gamma}'(t)| \neq 0$. In particular, show that

$$|\widetilde{\gamma}'(t)| = \frac{1}{1 - x_3(t)}|\gamma'(t)|$$

or, in differential form,

$$dr = \frac{1}{1 - x_3}ds \quad \text{or} \quad ds = \frac{2dr}{1 + |z|^2}$$

where $dr = \sqrt{dx^2 + dy^2}$ denotes the arc length element on $\mathbf{C}$, while $ds = \sqrt{dx_1^2 + dx_2^2 + dx_3^2}$ the one on S. Try to explain the factor $\frac{1}{1-x_3}$ geometrically.

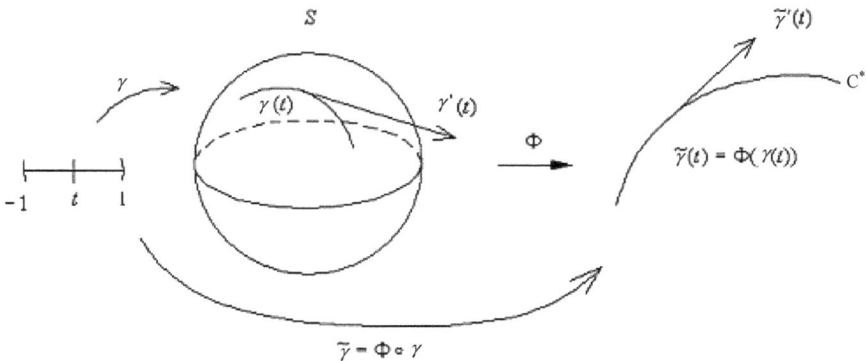

Fig. 1.33

If two curves γ_1 and γ_2 on S pass the same point Q and both
have the tangents at Q, then the *angle θ* between γ_1 and γ_2 at Q
is defined to be the angle formed by these tangents at Q. Under
Φ, suppose Q is mapped onto the point P, and γ_1 and γ_2 are
mapped onto the curves $\widetilde{\gamma_1} = \Phi \circ \gamma_1$ and $\widetilde{\gamma_2} = \Phi \circ \gamma_2$ passing P. By
(a), $\widetilde{\gamma_1}$ and $\widetilde{r}_2$ have tangents at P and the *angle θ'* between them
is defined to be the one between their respective tangents at P.
See Fig. 1.34.

In case $Q = N$, the north pole, then $P = \infty$, the infinite point.
Via the reflection $z \to \frac{1}{z}$, $\widetilde{\gamma_1}$, and $\widetilde{\gamma_2}$ are mapped onto two curves
$\widetilde{\widetilde{\gamma_1}}$ and $\widetilde{\widetilde{\gamma_2}}$ passing 0. Then, the *angle θ'* between $\widetilde{\widetilde{\gamma_1}}$ and $\widetilde{\widetilde{\gamma_2}}$ at 0 is
defined to be the angle θ' between $\widetilde{\gamma_1}$ and $\widetilde{\gamma_2}$ at ∞. See Fig. 1.35.

(b) Show that $\theta = \theta'$.

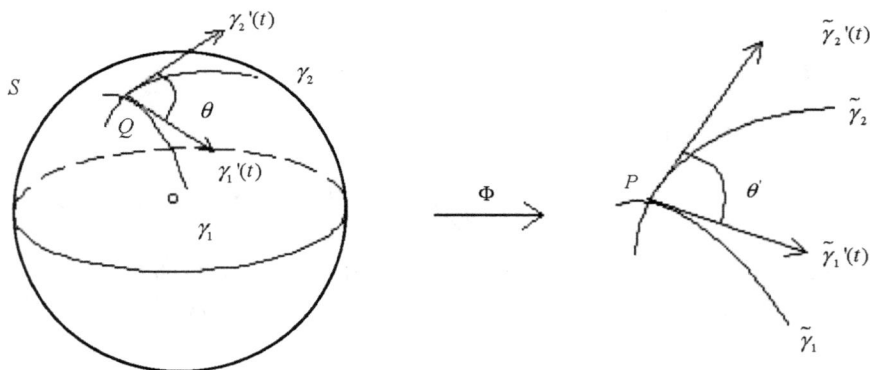

Fig. 1.34

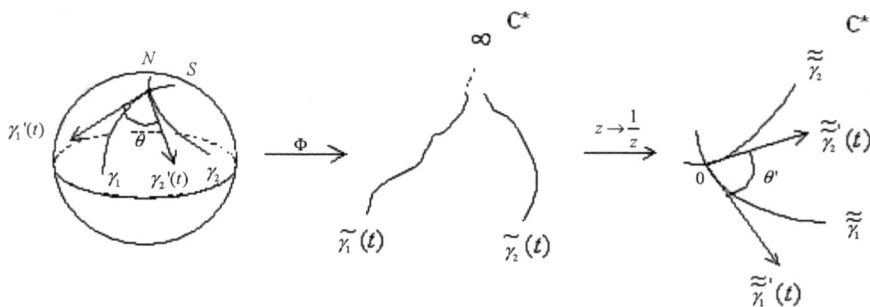

Fig. 1.35

As a conclusion, this means that *the stereographic projection Φ preserves angles between curves lying on the Riemann sphere S. It is a kind of conformal mapping.*

(2) (a) Let z_1 and z_2 be the projective points of two points Q_1 and Q_2 on S, respectively. Show that Q_1 and Q_2 are antipodal points $\Leftrightarrow z_1\overline{z_2} = -1$.

(b) Rotate the sphere S with the diameter Q_1Q_2 as the axis and through the angle θ. Suppose a point Q on S is then mapped into a point Q' on S and its projective point z into w. See Fig. 1.36. Suppose $\Phi(Q_1) = z_0$ and $\Phi(Q_2) = -\overline{z_0}^{-1}$. Explain why both

$$d(z, z_0) = d(w, z_0) \quad \text{and} \quad d(z, -\overline{z_0}^{-1}) = d(w, -\overline{z_0}^{-1})$$

hold. Then, try to show that $\Phi(Q) = z$ and $\Phi(Q') = w$ are related as

$$\frac{w - z_0}{1 + \overline{z_0}w} = e^{i\theta}\frac{z - z_0}{1 + \overline{z_0}z}$$

or,

$$w = \frac{az - b}{\overline{b}z + \overline{a}}.$$

Try to express a and b in terms of z_0 and θ. Is $|a|^2 + |b|^2 \neq 0$ true?

(3) (continued from Exercise (2)) Let z_1 and z_2 be two points in $\mathbf{C}$. Denote $Q_j = \Phi^{-1}(z_j) \in S$ for $j = 1, 2$. Suppose Q_1 and Q_2 are *not* antipodal points, namely, $z_1\overline{z_2} \neq -1$. Construct a great circle C passing Q_1 and

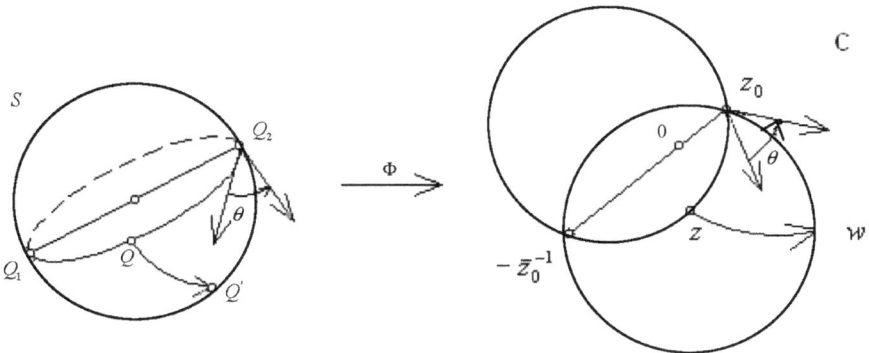

Fig. 1.36

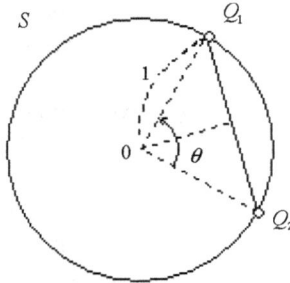

Fig. 1.37

Q_2 so that Q_1 and Q_2 divide C into two arcs. Let

$$\rho(z_1, z_2) = \text{the length of the smaller arc, } \widehat{Q_1 Q_2}$$

called the *spherical surface distance* of z_1 and z_2. See Fig. 1.37.

(a) Use $d(z_1, z_2) = 2 \sin \frac{1}{2}\rho(z_1, z_2)$ to prove that

$$\rho(z_1, z_2) = \begin{cases} 2\operatorname{Arc}\tan \dfrac{|z_1 - z_2|}{|1 + \overline{z_2} z_1|}, & z_1, z_2 \in \mathbf{C}, \\[3mm] 2\operatorname{Arc}\tan \dfrac{1}{|z_1|}, & z_1 \in C \text{ and } z_2 = \infty. \end{cases}$$

Note that $0 \le \rho(z_1, z_2) \le \pi$, and $\rho(z_1, z_2) = \pi \Leftrightarrow z_1 \overline{z_2} = -1$.

(b) Show that, for $z_1, z_2, z_3 \in \mathbf{C}^*$,

   (i) $\rho(z_1, z_2) \ge 0$ with equality $\Leftrightarrow z_1 = z_2$;

  (ii) $\rho(z_1, z_2) = \rho(z_2, z_1)$;

 (iii) $\rho(z_1, z_2) \le \rho(z_1, z_3) + \rho(z_3, z_2)$, with equality $\Leftrightarrow$ the spherical images $\Phi^{-1}(z_1)$, $\Phi^{-1}(z_2)$ and $\Phi^{-1}(z_3)$, in this ordering, lie on a great circle.

1.7. Complex Sequences

We will give a concise introduction to complex sequences without regard to details similar to real sequences.

   A *sequence* $\{z_n\}_{n=1}^{\infty}$ of complex numbers, simply denoted as z_n, $n \ge 1$ or just z_n, is said to *converge to a limit (point)* $z_0 \in \mathbf{C}$ as $n \to \infty$ and is denoted as

$$\lim_{n\to\infty} z_n = z_0 \quad \text{or} \quad \lim z_n = z_0 \quad \text{or} \quad z_n \to z_0, \tag{1.7.1}$$

if for any $\varepsilon > 0$, there exists a positive integer $N = N(\varepsilon)$ so that $|z_n - z_0| < \varepsilon$ as long as $n \geq N$. A complex sequence $z_n \in \mathbf{C}$ is said to *diverge to ∞* or *converge to ∞* in $\mathbf{C}^*$ as $n \to \infty$ and is denoted as

$$\lim_{n\to\infty} z_n = \infty \quad \text{or} \quad \lim z_n = \infty \quad \text{or} \quad z_n \to \infty, \qquad (1.7.2)$$

if for any $R > 0$, there exists a positive integer $N = N(R)$ so that $|z_n| > R$, $n \geq N$.

If a sequence $z_n \in \mathbf{C}$ does not converge to any point $z_0 \in \mathbf{C}$ or diverge to ∞, then z_n is called *divergent*.

A convergent sequence z_n has a unique limit and is *bounded*, i.e., there exists $M \geq 0$ so that $|z_n| \leq M$ for $n \geq 1$.

Also, a convergent sequence z_n is *Cauchy*, i.e., for any $\varepsilon > 0$, there exists a positive integer $N = N(\varepsilon)$ so that for all $m, n \geq N$, $|z_m - z_n| < \varepsilon$ always holds. By use of (1.4.2.1), namely,

$$|\mathrm{Re}\, z_m - \mathrm{Re}\, z_n|, \quad |\mathrm{Im}\, z_m - \mathrm{Im}\, z_n| \leq |z_m - z_n| \leq |\mathrm{Re}\, z_m - \mathrm{Re}\, z_n|$$
$$+ |\mathrm{Im}\, z_m - \mathrm{Im}\, z_n|$$

and the completeness of the real $\mathbf{R}$ (see Appendix A), it follows easily that *every Cauchy sequence does converge (to a point) in $\mathbf{C}$*. Hence, *the complex field $\mathbf{C}$ is complete as a metric space endowed with the metric $\|$* (refer to (1) in (1.9.3)).

Now, here comes

The necessary and sufficient conditions for convergent sequence. Let $z_n \in \mathbf{C}$ and $z_0 \in \mathbf{C}$. Then

(1) $\lim_{n\to\infty} z_n = z_0$;
$\Leftrightarrow$ (2) $\lim_{n\to\infty} \mathrm{Re}\, z_n = \mathrm{Re}\, z_0$ *and* $\lim_{n\to\infty} \mathrm{Im}\, z_n = \mathrm{Im}\, z_0$;
$\Leftrightarrow$ (3) *In case* $z_0 \neq 0$, $\lim_{n\to\infty} |z_n| = |z_0|$ *and* $\lim_{n\to\infty} \arg z_n = \arg z_0$.

Note: The last means that, for any preassigned value φ_0 of $\arg z_0$, a value φ_n of $\arg z_n$, $n \geq 1$, can be chosen so that

$$\lim_{n\to\infty} \varphi_n = \varphi_0 \qquad (1.7.3)$$

holds.

Proof. (1) $\Leftrightarrow$ (2) is an easy consequence of (1.4.2.1).

By using polar forms of z_n, (3) $\Rightarrow$ (1) follows immediately. For (1) $\Rightarrow$ (3), $\lim |z_n| = |z_0|$ follows from the inequality $||z_n| - |z_0|| \leq |z_n - z_0|$. What remains is to prove that $\arg z_n \to \arg z_0$.

Let us start from a concrete example. Set $z_n = -1 + i\frac{(-1)^n}{n}$. Then $z_n \to z_0 = -1$. Since $\text{Arg}\, z_{2n} = \pi - \text{Arc}\tan\frac{1}{2n}$ and $\text{Arg}\, z_{2n+1} = -\pi + \text{Arc}\tan\frac{1}{2n+1}$. $\text{Arg}\, z_n$ does not converge and neither does $\arg z_n$. Choose $\varphi_0 = \pi + 2m_0\pi$ (m_0 is a fixed integer), a value of $\arg(-1)$. Then, we choose

$$\varphi_n = \begin{cases} \text{Arg}\, z_{2m} + 2m_0\pi, & \text{if } n = 2m, \\ \text{Arg}\, z_{2m+1} + 2\pi + 2m_0\pi, & \text{if } n = 2m+1. \end{cases}$$

See Fig. 1.38. Under this circumstance $\varphi_n \to \varphi_0$.

In the general case, since $z_0 \neq 0, \infty$, then for all sufficiently small $\varepsilon > 0$, the open disk $|z - z_2| < |z_0| \sin \varepsilon$ does not contain 0. See Fig. 1.39.

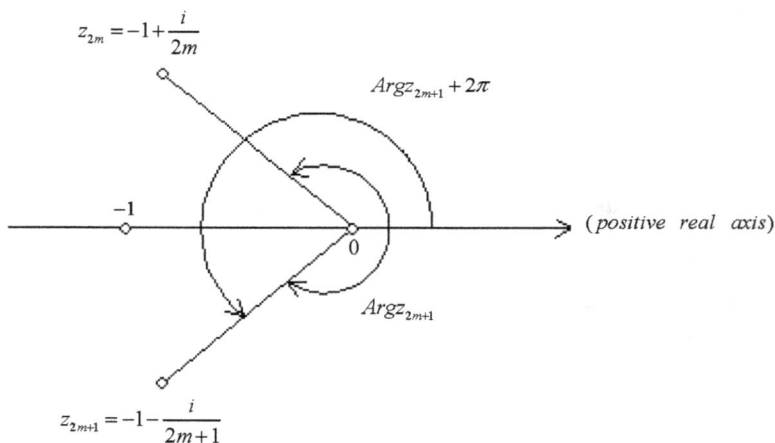

Fig. 1.38

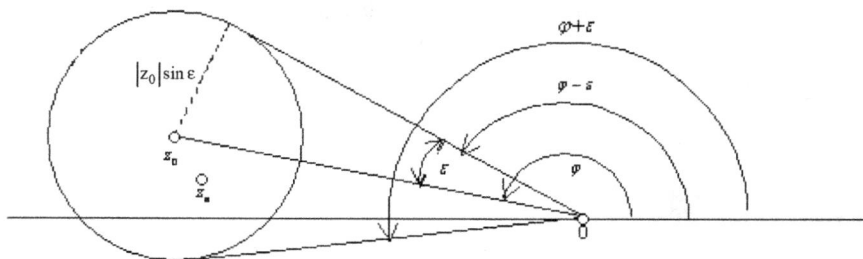

Fig. 1.39

By assumption $z_n \to z_0$, there exists n_0 so that $|z_n - z_0| < |z_0| \sin \varepsilon$, $n \geq n_0$. In case $0 \leq \mathrm{Arg}\, z_0 \leq \pi$ and $\varphi_0 = \mathrm{Arg}\, z_0 + 2m_0\pi$ (m_0 fixed): Choose

$$\varphi_n = \begin{cases} \mathrm{Arg}\, z_n + 2m_0\pi, & \text{if } 0 \leq \mathrm{Arg}\, z_n \leq \pi, \\ \mathrm{Arg}\, z_n + 2\pi + 2m_0\pi, & \text{if } -\pi < \mathrm{Arg}\, z_n < 0. \end{cases}$$

Then $|\varphi_n - \varphi_0| < \varepsilon$ if $n \geq n_0$. In case $-\pi < \mathrm{Arg}\, z_0 < 0$ and $\varphi_0 = \mathrm{Arg}\, z_0 + 2m_0\pi$: Choose

$$\varphi_n = \begin{cases} \mathrm{Arg}\, z_n - \pi + 2m_0\pi, & \text{if } 0 \leq \mathrm{Arg}\, z_n \leq \pi, \\ \mathrm{Arg}\, z_n + 2m_0\pi, & \text{if } -\pi < \mathrm{Arg}\, z_n < 0. \end{cases}$$

Then $|\varphi_n - \varphi_0| < \varepsilon$ if $n \geq n_0$. Hence, $\varphi_n \to \varphi_0$. $\qquad\square$

We list the following

operational properties of convergent sequences.

(1) Let $z_n \in \mathbf{C} \to z_0 \in \mathbf{C}$ and $z_n' \in \mathbf{C} \to z_0' \in \mathbf{C}$

    (i) $z_n \pm z_n' \to z_0 \pm z_0'$;

    (ii) $z_n z_n' \to z_0 z_0'$;

    (iii) If $z_n' \neq 0$ and $z_0' \neq 0$, $\dfrac{z_n}{z_n'} \to \dfrac{z_0}{z_0'}$.

(2) *Some exceptional cases* (see (1.6.5) and (1.6.6)):

    (i) $z_n \to z_0 \in \mathbf{C}$, $z_n' \to \infty \Rightarrow z_n + z_n' \to \infty$;

    (ii) $z_n \to z_0 \in \mathbf{C}$ and $z_0 \neq 0$, $z_n' \to \infty \Rightarrow z_n z_n' \to \infty$;

    (iii) $z_n \to z_0 \in \mathbf{C}^*$ and $z_0 \neq 0$, $z_n' \to 0 \Rightarrow \dfrac{z_n}{z_n'} \to \infty$;

$$z_n \to z_0 \in \mathbf{C}, \quad z_n' \to \infty \Rightarrow \frac{z_n}{z_n'} \to 0. \qquad (1.7.4)$$

Proofs are left as Exercise A(1).

We illustrate two examples.

Example 1. Prove that, if $z = x + iy$,

$$\lim_{n \to \infty} \left(1 + \frac{z}{n}\right)^n = e^x(\cos y + i \sin y) \underset{\text{(def.)}}{=} e^z. \qquad (1.7.5)$$

Proof. Since $\lim_{n \to \infty}\left(1 + \frac{z}{n}\right) = 1$, for any $1 > \varepsilon > 0$, there exists n_0 so that $n \geq n_0$ implies that $\left|\mathrm{Arg}\left(1 + \frac{z}{n}\right)\right| < \varepsilon$ always holds.

For simplicity, let $z_n = \left(1 + \frac{z}{n}\right)^n$. Then

$$|z_n| = \left[\left(1 + \frac{x}{n}\right)^2 + \frac{y^2}{n^2}\right]^{\frac{n}{2}}$$

$\Rightarrow$ (by using L'Hospital's rule on n)

$$\lim_{n \to \infty} \log|z_n| = \lim_{n \to \infty} \frac{1}{\frac{-2}{n^2}} \cdot \frac{-\frac{x}{n^2}\left(1 + \frac{x}{n}\right) - \frac{2y^2}{n^3}}{\left(1 + \frac{x}{n}\right)^2 + \frac{y^2}{n^2}} = x$$

$$\Rightarrow \lim_{n \to \infty} |z_n| = e^x.$$

On the other hand, if $n \geq n_0$,

$$\text{Arg } z_n = n\text{Arctan} \frac{\frac{y}{n}}{1 + \frac{x}{n}} \to y \text{ as } n \to \infty \quad \left(\text{recall that } \lim_{\theta \to 0} \frac{\text{Arctan } \theta}{\theta} = 1\right).$$

Hence,

$$\lim_{n \to \infty} \left(1 + \frac{z}{n}\right)^n = \lim_{n \to \infty} |z_n| e^{i\text{Arg } z_n} = e^x e^{iy} = e^x(\cos y + i \sin y). \qquad \square$$

Example 2. Prove that, if $z = x + iy \neq 0$,

$$\lim_{n \to \infty} n\left(\sqrt[n]{z} - 1\right) = \log|z| + i(\text{Arg } z + 2k\pi) \underset{(\text{def.})}{=} \log z, k = 0, \pm 1, \pm 2, \ldots.$$

$$(1.7.6)$$

Proof. $\sqrt[n]{z} = |z|^{\frac{1}{n}} e^{\frac{i(\theta + 2k\pi)}{n}}, \theta = \text{Arg } z$ and $k = 0, \pm 1, \pm 2, \ldots$. For any fixed k,

$$\sqrt[n]{z} - 1 = |z|^{\frac{1}{n}} \cos \frac{\theta + 2k\pi}{n} - 1 + i|z|^{\frac{1}{n}} \sin \frac{\theta + 2k\pi}{n}$$

$$\Rightarrow \text{Re } n(\sqrt[n]{z} - 1) = n\left(|z|^{\frac{1}{n}} \cos \frac{\theta + 2k\pi}{n} - 1\right) \text{ and}$$

$$\text{Im } n(\sqrt[n]{z} - 1) = n|z|^{\frac{1}{n}} \sin \frac{\theta + 2k\pi}{n}.$$

From the known result $\lim_{\alpha \to 0} \frac{a^\alpha - 1}{\alpha} = \log a$ ($\alpha \in \mathbf{R}$ and $a > 0$) or by using L'Hospital's rule,

$$\lim_{n \to \infty} \text{Re } n(\sqrt[n]{z} - 1) = \lim_{n \to \infty} \left\{\frac{1}{\frac{1}{n}}(|z|^{\frac{1}{n}} - 1) + \frac{1}{\frac{1}{n}}|z|^{\frac{1}{n}}\left(\cos \frac{\theta + 2k\pi}{n} - 1\right)\right\}$$

$$= \log|z| + 0 = \log|z| \text{ (real logarithm of } |z|\text{),}$$

while

$$\lim_{n\to\infty} \operatorname{Im} n(\sqrt[n]{z} - 1) = \lim_{n\to\infty} |z|^{\frac{1}{n}} \cdot (\theta + 2k\pi) \cdot \frac{\sin\frac{\theta+2k\pi}{n}}{\frac{\theta+2k\pi}{n}}$$

$$= \theta + 2k\pi = \operatorname{Arg} z + 2k\pi.$$

The result follows from (1.7.3).   □

Let $n_k, k \geq 1$, be a sequence of positive integers satisfying $1 \leq n_1 < n_2 < \cdots < n_k < \cdots$ and $\lim_{k\to\infty} n_k = \infty$. Then the sequence $z_{n_k}, k \geq 1$, is called a *subsequence* of the sequence $z_n, n \geq 1$.

A bounded sequence, such as $z_n = i^n$, is not necessarily convergent. But we do have

two basic facts about sequences.

(1) *A sequence z_n converges to a point z_0* $\Leftrightarrow$ *all the subsequence $z_{n_k}, k \geq 1$, of z_n converge to the same point z_0.*

(2) *(Completeness of $\mathbf{C}$) Any bounded sequence has a convergent subsequence (converging to some point in $\mathbf{C}$).*   (1.7.7)

For (2), refer to Sec. 1.9. Proofs are left as Exercise A(2).

A point $z \in \mathbf{C}^*$ is called a *limit point* of a complex sequence z_n, if there exists a subsequence z_{n_k} of z_n which converges to z.

Example 3. Show that every point on the unit circle $|z| = 1$ is a limit point of the sequence

$$z_n = e^{in}, \quad n = 0, \pm 1, \pm 2, \ldots.$$

Proof. Let α be any irrational number. It is well known from elementary real analysis that the set $\{n + m\alpha \mid n, m = 0, \pm 1, \ldots\}$ is *dense* in $\mathbf{R}$ (refer to Section (5) in Sec. 1.8), namely, for each $x \in \mathbf{R}$, there exists a sequence x_k ($x_k \neq x_l$, if $k \neq l$) from the set converging to x.

In our case, choose $\alpha = 2\pi$. By Example 1, $e^z = e^x(\cos y + i \sin y)$ is a continuous function of x and y. Let $x_k = n_k + 2m_k\pi, k \geq 1$ and $x_k \to x$. Then $e^{ix_k} = e^{in_k} \to e^{ix}$ as $k \to \infty$.   □

Exercises A

(1) Prove (1.7.4) in detail.

(2) Prove (1.7.7) in detail.

(3) Try to use complex sequences to explain why we cannot define each of the following expressions as a definite value:

$$\infty + \infty, \infty - \infty, 0 \cdot \infty, \frac{0}{0}, \frac{\infty}{\infty}, \infty^0, 1^\infty.$$

(4) Try to use $z_n = i^n n$ to explain that $z_n \to \infty$ does not necessarily imply that

$$|\operatorname{Re} z_n| \to +\infty \quad \text{and} \quad |\operatorname{Im} z_n| \to +\infty.$$

(5) Show that it is only if $z_0 = 0$ or ∞ that $\lim_{n \to \infty} |z_n| = |z_0|$ would imply $\lim_{n \to \infty} z_n = z_0$.

(6) Suppose $z_n \to z_0 \neq 0$. Let φ_n be any value of $\arg z_n$ satisfying the condition that $|\varphi_m - \varphi_n| < \pi$ for $m, n \geq N$ (a positive integer). Show that φ_n converges.

(7) Suppose $z_n \to z_0 \neq 0$ and z_0 is not a negative real number. Show that

$$\operatorname{Arg} z_n \to \operatorname{Arg} z_0.$$

(8) For each of the following sequences, find its limit if it converges; otherwise, find the set of all its limit points.

(a) $z_n = \sqrt[n]{n} + inr^n (|r| < 1)$. (b) $z_n = \dfrac{1}{n} + (-1)^n i^n$.

(c) $z_n = \dfrac{n}{n+1} \cos \dfrac{n\pi}{2} + \dfrac{i^n}{n}$. (d) $z_n = \left(1 + \dfrac{z}{n}\right)^{-n} + \dfrac{ni}{n+1}$.

(e) $z_n = n^{(-1)^n} + \sqrt[n]{3}i$. (f) $z_n = \dfrac{\log n!}{n} + \dfrac{n!}{2^n} i$.

(g) $z_n = e^{in\theta}$ (θ is an irrational number which is not of the form $k\pi, k = 0, \pm 1, \pm 2, \ldots$), $n = 0, \pm 1, \pm 2, \ldots$.

(9) Suppose $z_n \to z_0 \in \mathbf{C}$ and $w_n \to w_0 \in \mathbf{C}$.

(a) Prove that

$$\lim_{n \to \infty} \frac{z_1 + z_2 + \cdots + z_n}{n} = z_0.$$

What happens if $z_0 = \infty$?

(b) Prove that

$$\lim_{n \to \infty} \frac{z_1 w_n + z_2 w_{n-1} + \cdots + z_n w_1}{n} = z_0 w_0.$$

What happens if $z_0 \in \mathbf{C}$ and $w_0 = \infty$?

(c) Suppose $\lambda_1 > 0, \ldots, \lambda_n > 0$ and $\lim_{n \to \infty} (\lambda_1 + \cdots + \lambda_n) = \infty$. Show that

$$\lim_{n \to \infty} \frac{\lambda_1 z_1 + \lambda_2 z_2 + \cdots + \lambda_n z_n}{\lambda_1 + \lambda_2 + \cdots + \lambda_n} = z_0.$$

(10) Suppose a complex sequence z_n satisfying $\lim_{n\to\infty}(z_n - z_{n-2}) = 0$.

    (a) Show that

$$\lim_{n\to\infty} \frac{z_n - z_{n-1}}{n} = 0.$$

    Is the converse true?

    (b) Show that

$$\lim_{n\to\infty} \frac{z_n}{n} = 0.$$

    Is the converse true?

(11) In elementary *real* analysis, we learned the following limit processes:

    (a) $\lim_{n\to\infty} na^n = 0$ ($a \in \mathbf{R}$ and $|a| < 1$).

    (b) $\lim_{n\to\infty} \sqrt[n]{a} = 1$ ($a \in \mathbf{R}$ and $a > 0$).

    (c) $\lim_{n\to\infty} \frac{a^n}{n!} = 0$ ($a \in \mathbf{R}$ and $a > 0$).

    (d) $\lim_{n\to\infty} \frac{n^k}{a^n} = 0$ ($a \in \mathbf{R}$ and $a > 1, k$ a fixed integer).

    Try to prove these statements by the $\varepsilon - N$ process. In case a is replaced by a fixed suitable complex number z, where in (d), z is supposed that $|z| > 1$, do the following questions.

    (i) Use the $\varepsilon - N$ process to prove that these statements are still valid. Be careful that, in (b), $\sqrt[n]{z}$ is a multiple-valued function.

    (ii) Use (1.7.3) to prove them again. Then, try to compare it to the method in 1.

(12) If $|z_{n+1} - z_n| \le \lambda|z_n - z_{n-1}|, n \ge 1$, where $0 < \lambda < 1$ is a constant, then z_n converges.

Exercises B

Let $z_n \in \mathbf{C}, n \ge 1$ and $S_n = z_1 + z_2 + \cdots + z_n, n \ge 1$. If $\lim_{n\to\infty} S_n = S \in \mathbf{C}$, then the *complex series*

$$\sum_{n=1}^{\infty} z_n = z_1 + z_2 + \cdots + z_n + \cdots \quad \text{or} \quad \sum z_n$$

is said to *converge* to the *sum S* and is denoted as

$$\sum_{n=1}^{\infty} z_n = S \quad \text{or} \quad \sum z_n = S. \tag{1.7.8}$$

Otherwise, we call $\sum z_n$ a *divergent* series. If $\sum |z_n|$ converges (in this case, $\sum z_n$ is necessarily convergent), we call $\sum z_n$ *absolutely convergent*; if $\sum |z_n|$ diverges but $\sum z_n$ converges, we call $\sum z_n$ *conditionally convergent*.

Do the following problems.

(1) *Basic criteria for convergence*

$$\sum z_n \text{ converges.}$$

$\Leftrightarrow \sum \operatorname{Re} z_n$ and $\sum \operatorname{Im} z_n$ converge.

$\Leftrightarrow$ (*Cauchy condition*) For any $\varepsilon > 0$, there exists an integer $N = N(\varepsilon) > 0$ so that

$$|z_{n+1} + \cdots + z_{n+p}| < \varepsilon \text{ for all } n \geq N \text{ and all } p \geq 1.$$

$\Rightarrow \lim_{n \to \infty} z_n = 0.$

(2) *Basic properties*

    (a) For any integer $N \geq 1$, $\sum_{n=1}^{\infty} z_n$ and $\sum_{n=N}^{\infty} z_n$ converge or diverge together.

    (b) If $\sum z_n = S \Rightarrow \sum(\alpha z_n) = \alpha S, \alpha \in \mathbf{C}$.

    (c) If $\sum z_n = S$, then for any sequence of integers $0 < k_1 < k_2 < \cdots < k_n < k_{n+1} < \cdots$,

$$(z_1 + \cdots + z_{k_1}) + (z_{k_1+1} + \cdots + z_{k_2}) + \cdots + (z_{k_n+1} + \cdots + z_{k_{n+1}}) + \cdots = S.$$

    (d) If $\sum z_n = S$ and $\sum w_n = S' \Rightarrow \sum(z_n \pm w_n) = S \pm S'$.

    (e) If both $\sum z_n = S$ and $\sum w_n = S'$ *absolutely*, then $\sum_{n=1}^{\infty}(z_1 w_n + z_2 w_{n-1} + \cdots + z_n w_1) = SS'$ *absolutely*.

(3) *Some criteria for convergence*

    (a) (*comparison test*) Suppose $z_n \neq 0, w_n \neq 0, n \geq 1$. If

$$0 < \underline{\lim} \frac{|z_n|}{|w_n|} \leq \overline{\lim} \frac{|z_n|}{|w_n|} < \infty,$$

then both $\sum |z_n|$ and $\sum |w_n|$ converge or diverge together.

    (b) (*ratio test*) Suppose $z_n \neq 0, n \geq 1$.

$$\overline{\lim} \left| \frac{z_{n+1}}{z_n} \right| < 1 \Rightarrow \sum |z_n| \text{ converges;}$$

$$\underline{\lim} \left| \frac{z_{n+1}}{z_n} \right| > 1 \Rightarrow \sum |z_n| \text{ diverges.}$$

    (c) (*root test*)

$$\overline{\lim} \sqrt[n]{|z_n|} < 1 \Rightarrow \sum |z_n| \text{ converges;}$$

$$\underline{\lim} \sqrt[n]{|z_n|} > 1 \Rightarrow \sum |z_n| \text{ diverges.}$$

(d) (*Abel test*) Suppose $\sum |z_n - z_{n+1}|$ and $\sum w_n$ converges, then $\sum z_n w_n$ converges.

(e) (*Dirichlet test*) Let $\lambda_n \geq \lambda_{n+1} \geq 0$ for $n \geq 1$ and $\lim_{n \to \infty} \lambda_n = 0$. If the partial sum $S_n = z_1 + \cdots + z_n, n \geq 1$, of $\sum z_n$ is a bounded sequence, namely, $\sup_{n \geq 1} |z_1 + \cdots + z_n| < \infty$, then $\sum \lambda_n z_n$ converges.

(f) (*Leibniz test*) Let $\lambda_n \geq \lambda_{n+1} \geq 0$ for $n \geq 1$ and $\lim_{n \to \infty} \lambda_n = 0$, then $\sum_{n=1}^{\infty} (-1)^{n-1} \lambda_n$ converges.

(4) (*Absolute convergence*) $\sum_{n=1}^{\infty} z_n = S$ absolutely $\Leftrightarrow$ For any permutation $\sigma : \mathbf{N} \to \mathbf{N}$ (a one-to-one and onto mapping), the series $\sum_{n=1}^{\infty} z_{\sigma(n)} = S$ absolutely.

(5) (*Conditional convergence*) Suppose $\sum_{n=1}^{\infty} z_n$ converges conditionally and $S \in \mathbf{C}^*$ is any fixed point, then there exists a permutation $\sigma : \mathbf{N} \to \mathbf{N}$ so that $\sum_{n=1}^{\infty} z_{\sigma(n)} = S$.

(6) Test if the following series are convergent or divergent. In case of convergence, does it converge absolutely or conditionally?

(a) $\displaystyle\sum_{n=0}^{\infty} \frac{1}{(1+i)^n}$.

(b) $\displaystyle\sum_{n=0}^{\infty} \left(\cos \frac{n\pi}{4} + i \sin \frac{n\pi}{4} \right)$.

(c) $\displaystyle\sum_{n=0}^{\infty} \frac{1}{(1+i)^n} \left(\cos \frac{n\pi}{4} + i \sin \frac{n\pi}{4} \right)$.

(d) $\displaystyle\sum_{n=0}^{\infty} \frac{(1+i)^n}{i(1+i)^n + 2}$.

(e) $\displaystyle\sum_{n=1}^{\infty} \left(\frac{3+4i}{5} \right)^{n^2} i^n$.

(f) $\displaystyle\sum_{n=1}^{\infty} \left(\frac{3+4i}{5} \right)^{n} i^n$.

(7) Suppose $\sum_{n=1}^{\infty} |z_n|$ converges. Show that $\sum_{n=1}^{\infty} |z_n|^2$ converges. Is the converse true?

(8) Suppose $|\operatorname{Arg} z_n| \leq \alpha < \frac{\pi}{2}, n \geq 1$. If $\sum z_n$ converges, show that $\sum |z_n|$ converges, too.

(9) Recall that $\mathbf{C}^*$ with the spherical chord distance $d(,)$ is a metric space (see (1.6.8) and Exercise B(1) of Sec. 1.8). Define:

a sequence $z_n \in \mathbf{C}^*$ *converges* to a point $z_0 \in \mathbf{C}^* \Leftrightarrow d(z_n, z_0) \to 0$ as $n \to \infty$.

Show that this definition includes (1.7.1) and (1.7.2) as special cases, and conversely, if we use the spherical chord distance in (1.7.1) and (1.7.2), then they can be combined into the definition above.

1.8.   Elementary Point Sets

We will give a concise introduction to these point sets in $\mathbf{C}$ and $\mathbf{C}^*$, barely needed in our discussion of elementary complex analysis, without going into details.

Section (1) ε-neighborhood

Fix a point $a \in \mathbf{C}$. For $\varepsilon > 0$, define
   ε-neighborhood of a : the open disk $|z-a| < \varepsilon$, simply denoted as $D_\varepsilon(a)$;

$$\textit{deleted } \varepsilon\text{-}\textit{neighborhood of } a : 0 < |z - a| < \varepsilon. \tag{1.8.1}$$

For $R > 0$, define
   R-neighborhood of ∞ : $R < |z| \le +\infty$, simply denoted as $D_R(\infty)$;

$$\textit{deleted } R\text{-}\textit{neighborhood of } \infty : R < |z| < \infty. \tag{1.8.2}$$

Refer to (1.4.3.6) and Fig. 1.18.

Section (2) Open set

A nonempty subset O of $\mathbf{C}$ (or $\mathbf{C}^*$) is called *open* if O contains an ε-neighborhood $D_\varepsilon(a)$ of each point a belonging to O itself.

In particular, ε-neighborhood and deleted ε-neighborhood of a point are open sets. $\mathbf{C}$ (or $\mathbf{C}^*$) itself is open. We designate empty set as an open set. Note that $R < |z| < +\infty$ is open both in $\mathbf{C}$ and $\mathbf{C}^*$, yet $R < |z| \le +\infty$ is not a subset of $\mathbf{C}$ but is an open set in $\mathbf{C}^*$.

The intersection of *finitely* many open sets is open, while the union of *arbitrarily* many open sets is open.

We usually call an open set *an open neighborhood* of each of its points and a set a *neighborhood* of a point if the set contains an open neighborhood of that point.

Section (3) Limit (or accumulation or cluster) point of a set

We have the following

Characteristic properties of a limit point. Let $A \subseteq \mathbf{C}$ be a nonempty set and z_0 be a point in $\mathbf{C}^*$. Then:

   (1) *Any neighborhood of z_0 contains a point of A, other than z_0 itself.*
$\Leftrightarrow$ (2) *Any neighborhood of z_0 contains infinitely many points of A.*

$\Leftrightarrow$ (3) *There exists a sequence* $z_n \in A, z_n \neq z_0, n \geq 1$, *so that* $z_n \to z_0$.

$\Leftrightarrow$ (4) *There exists a sequence* $z_n \in A, z_n \neq z_m$ *if* $n \neq m$, *so that* $z_n \to z_0$.

Note: If $z_0 \in \mathbf{C}$, *one might impose that* $|z_1 - z_0| > |z_2 - z_0| > \cdots > |z_n - z_0| > \cdots \to 0$; *if* $z_0 = \infty$, *then* $|z_1| < |z_2| < \cdots < |z_n| < \cdots \to +\infty$.

Such a point z_0 *is called a limit (or accumulation or cluster) point of A.*

$$(1.8.3)$$

Proofs are left as Exercise A(1). Note that a limit point of a set A may not be in the set.

Empty set or finite set does not have limit point.

If a point set $\{z_n | n \geq 1\}$ has a limit point z_0, then the *sequence* $z_n, n \geq 1$, has a subsequence $z_{n_k} \to z_0$, i.e., z_0 is a limit point of the sequence (Sec. 1.7). But, the converse is not true, in general. For example, the sequence $z_n = i^n, n \geq 1$, has four limit points ± 1 and $\pm i$, yet as a point set, $\{z_n | n \geq 1\} = \{1, -1, i, -i\}$ does not have any limit point.

An infinite set in $\mathbf{C}$, for example $\{n \cdot i^n | n \geq 1\}$, does not necessarily have a limit point in $\mathbf{C}$ itself; but definitely has at least one limit point in $\mathbf{C}^*$ (why?). Anyway, a *bounded infinite subset* of $\mathbf{C}$, i.e., a set containing in a circle, *has at least one limit point in* $\mathbf{C}$ (compare to (2) in (1.7.7) and see (3) in (1.9.3)).

The set of all limit points of a set A is called the *derived set* of A and denoted as A'.

The set $\overline{A} = A \cup A'$ is called the *closure of A*. A point $z_0 \in \overline{A}$ if and only if for $D_\varepsilon(z_0) \cap A \neq \phi$ for $\varepsilon > 0$, or equivalently, if and only if there exists a sequence $z_n \in A \to z_0$. The closure of the open disk $|z - z_0| < r$ is the closed disk $|z - z_0| \leq r$.

Section (4) Closed set, compact set, etc.

Suppose $A \subseteq C$. Define A as

$$\begin{aligned}
&\text{a } \textit{closed set} \Leftrightarrow A' \subseteq A; \\
&\text{an } \textit{isolated set} \Leftrightarrow A \neq \phi \quad \text{and} \quad A' \cap A = \phi; \\
&\text{a } \textit{dense set by itself} \Leftrightarrow A \subseteq A', \text{ and} \\
&\text{a } \textit{perfect set} \Leftrightarrow A = A'.
\end{aligned} \qquad (1.8.3)$$

$\overline{A}$ is obviously a closed set. Point in $A - A'$ is called an *isolated point* of A. The set of all isolated points of a set is a countable set. Recall that a set is called *countable* if there exists a one-to-one and onto correspondence between the set and $\{1, 2, \ldots, n\}$ for some positive integer n or the set N of natural numbers; otherwise, it is called *uncountable*. Countable sets which are not finite and uncountable sets are called *infinite sets*.

We designate the empty set as a closed set (recall that it is also an open set). $\mathbf{C}$ itself is a closed set. A closed disk and a closed half-plane (see (1.4.3.1)) are basic closed sets.

The union of finitely many closed sets and the intersection of arbitrarily many closed sets are closed sets.

Indeed, we have

Characteristic properties of a closed set. Let $A \subseteq C$. Then:

(1) *A is a closed set, i.e., $A' \subseteq A$.*
$\Leftrightarrow$ (2) *If a sequence $z_n \in A \to z_0$, then it is necessary that $z_0 \in A$.*
$\Leftrightarrow$ (3) *If the distance from z_0 to A*

$$\text{dist}(z_0, A) = \inf_{z \in A} |z_0 - z|$$

is equal to zero, then $z_0 \in A$.
$\Leftrightarrow$ (4) *A is the intersection of all these closed sets containing A as a subset.*
$\Leftrightarrow$ (5) *$A = \overline{A}$.*
$\Leftrightarrow$ (6) *The complementary set $A^\sim = \mathbf{C} - A$ is an open set.*   (1.8.5)

Proofs are left as Exercise A(2).

A set is called *bounded* if it is contained in a disk $|z| < R$; otherwise, it is *unbounded*. A bounded closed set is called *compact* (for details, see (1.9.4)).

Section (5) Dense set

Let A and B be two sets in $\mathbf{C}$. If $A \subseteq \overline{B}$ holds, we say that B is *dense in* A; in case $A = \mathbf{C}$, namely, $\overline{B} = \mathbf{C}$, we call B a *dense set* in $\mathbf{C}$.

Even B is dense in A, it is possible that $A \cap B = \phi$, *the empty set*. For instance, let $A = \{z \in \mathbf{C}|$ at least one of $\text{Re}\,z$ and $\text{Im}\,z$ is an irrational number$\}$ and $B = \{z \in \mathbf{C}|$ both $\text{Re}\,z$ and $\text{Im}\,z$ are rational numbers$\}$, then both A and B are dense in $\mathbf{C}$; also B is dense in A, i.e., $A \subseteq \overline{B} = \mathbf{C}$ yet $A \cap B = \phi$.

On the contrary, B is called a *nowhere dense set* if its complement $B^\sim = \mathbf{C} - B$ is dense in $\mathbf{C}$.

Section (6) Interior, boundary, and exterior of a set

Let $A \subseteq \mathbf{C}$ and $z_0 \in \mathbf{C}$. Define z_0 as

an *interior point* of $A \Leftrightarrow$ there exists an $\varepsilon > 0$ so that $D_\varepsilon(z_0) \subseteq A$;
a *boundary point* of $A \Leftrightarrow$ for *any* $\varepsilon > 0$, $D_\varepsilon(z_0) \cap A \neq \phi$ and $D_\varepsilon(z_0) \cap$

$$A^\sim \neq \phi \text{ hold, and}$$

an *exterior point* of $A \Leftrightarrow$ there exists an $\varepsilon > 0$ so that $D_\varepsilon(z_0) \cap A = \phi$.

$$\Leftrightarrow z_0 \in \overline{A}^\sim . \tag{1.8.6}$$

The whole plane $\mathbf{C}$ is divided into the union of the following pairwise disjoint sets:

the *interior* of $A : A^0$ or Int $A = \{$interior points of $A\}$;
the *boundary* of $A : \partial A$ or Bdry $A = \{$boundary points of $A\}$;
the *exterior* of $A : \overline{A}^\sim$ or Ext $A = \{$exterior points of $A\}$. $\tag{1.8.7}$

They enjoy the following properties:

$$\bar{A}^\sim = (A^\sim)^0;$$
$$\overline{(A^\sim)} = (A^0)^\sim; \tag{1.8.8}$$
$$\partial A = \overline{A} \cap \overline{(A^\sim)} = \overline{A} - A^0 = \partial A^\sim.$$

Section (7) Connected set, domain

Let $A \subseteq \mathbf{C}$. If there exists two *nonempty* subsets A_1 and A_2 of A so that

$$A = A_1 \cup A_2, \quad \overline{A_1} \cap A_2 = A_1 \cap \overline{A_2} = \phi, \tag{1.8.9}$$

then we call A_1 and A_2 *separate* the set A and A is called a *disconnected set*; otherwise, A is called a *connected set*.

Remark (A closer look at (1.8.9)). Let $B_j = A - \bar{A}_j, j = 1, 2$.
Both $B_1 \neq \phi$ and $B_2 \neq \phi$: $\bar{A}_1 \cap A_2 = \phi \Rightarrow A_2 \subseteq \bar{A}_1^\sim \Rightarrow A \cap A_2 = A_2 \subseteq A \cap \bar{A}_1^\sim = A - \overline{A_1} = B_1 \Rightarrow B_1 \neq \phi$. Similarly, $B_2 \neq \phi$.
$B_1 \cap B_2 = \phi$: In case $z_0 \in B_1 \cap B_2 \Rightarrow z_0 \in A$ but $z_0 \notin \overline{A_j}$ for $j = 1, 2 \Rightarrow z_0 \in A_1 \cap A_2$, contradicting to $A_1 \cap A_2 = \phi$.
Both B_1 and B_2 are *open sets in A* (or *relatively open in A*): This means that there exist open sets O_1 and O_2 in $\mathbf{C}$ so that

$$B_1 = A \cap O_1, \quad B_2 = A \cap O_2.$$

Just take $O_j = \mathbf{C} - \overline{\mathbf{A}}_j$ for $j = 1, 2$.

Note that $A = B_1 \cup B_2$.

Both B_1 and B_2 are *closed sets in A* (or *relatively closed* in A): This means that there exist closed sets E_1 and E_2 in $\mathbf{C}$ so that

$$B_1 = A \cap E_1, \quad B_2 = A \cap E_2.$$

Since $B_1 = A - B_2 = A - (A \cap O_2) = A - O_2 = A \cap O_2^{\sim}$, choose $E_1 = O_2^{\sim} = \mathbf{C} - O_2$; similarly, take $E_2 = O_1^{\sim}$, where O_1 and O_2 are as above.   □

Hence, we can rephrase

The definition of a connected set.

(1) *Let $A \subseteq \mathbf{C}$. Then*

> (a) *A is a connected set (see (1.8.9)).*
> ⇔ (b) *A cannot be separated by two nonempty relatively open (or closed) subsets of A itself. Namely, there do not exist open sets B_1 and B_2 in A so that*

$$A = B_1 \cup B_2, B_1 \neq \phi, B_2 \neq \phi \quad \text{and} \quad B_1 \cap B_2 = \phi.$$

> ⇔ (c) *If a set B is both open and closed in A, then either $B = \phi$ or $B = A$.*

(2) *Therefore, an open (or closed) set A in $\mathbf{C}$ is connected if and only if A cannot be separated by two open (or closed) sets in $\mathbf{C}$.*   (1.8.10)

By a *curve* in the complex plane $\mathbf{C}$, we mean a continuous mapping $z = z(t) = x(t) + iy(t) : [0,1] \to \mathbf{C}$ where $x(t)$ and $y(t)$ are continuous real-valued function on the closed interval $[0,1]$ of the real line. If a curve is composed of finitely many line segments, connecting end-to-end, it is called a *polygonal curve* (for details, see Sec. 2.4).

It is well known that the only connected sets in $\mathbf{R}$ are intervals. While, in $\mathbf{C}$, we have the basic concept of a

Domain. Let $\Omega \subseteq \mathbf{C}$ be a nonempty open set. Then:

(1) *Ω is connected.*
⇔ (2) *Any two points of Ω can be joined by a curve lying entirely in Ω. This curve can be chosen to be a polygonal curve with its composed segments all parallel to the axes.*

We call a nonempty open connected subset Ω of $\mathbf{C}$ a *domain*. The closure $\overline{\Omega}$ of a domain Ω in $\mathbf{C}$ is called a *closed domain* in $\mathbf{C}$.   (1.8.11)

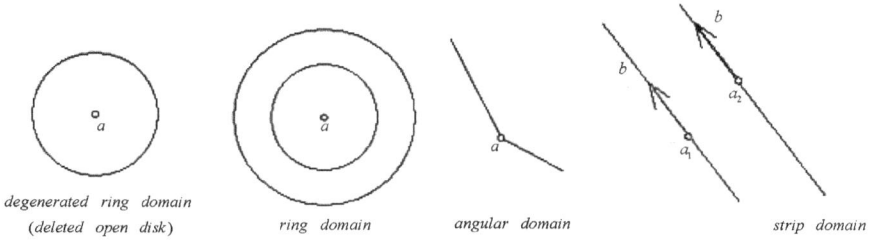

degenerated ring domain
(deleted open disk)   ring domain   angular domain   strip domain

Fig. 1.40

Commonly used domains in complex analysis are as follows:

open disk: $|z - a| < r$; $R < |z| \leq \infty$.
deleted open disk: $0 < |z - a| < r$; $R < |z| < \infty$.
open half-plane: $\operatorname{Im} \frac{z-a}{b} > 0$ or < 0.
ring domain: $0 < r_1 < |z - a| < r_2$ or (degenerated) $0 < |z - a| < r$.
angular domain: $\alpha < \operatorname{Arg}(z - a) < \beta$.
strip domain: $\{\operatorname{Im} \frac{z-a_1}{b} < 0\} \cap \{\operatorname{Im} \frac{z-a_2}{b} > 0\}$, where $\operatorname{Arg} a_1 - \pi < \operatorname{Arg} a_2 < \operatorname{Arg} a_1$.

Of course, $\mathbf{C}$ itself is a domain. See Fig. 1.40.

Proof of (1.8.11). (1) $\Rightarrow$ (2): Fix a point $a \in \Omega$. Let

$$B = \{z \in \Omega \mid \text{There exists a curve } in \ \Omega \text{ connecting a to } z.\}$$

Try to show that B is both open and closed in Ω, and $B \neq \phi$, and hence $B = \Omega$. This will prove the validity of (2), by (2) in (1.8.10).

Pick any point $z_0 \in B$. Since Ω is open, there exists a $\delta > 0$ so that the open disk $D_\delta(z_0) \subseteq \Omega$. Now, any point in $D_\delta(z_0)$ can be joined to z_0 by a radial segment or by a polygonal curve composed of two line segments, each parallel to the real or the imaginary axis. See Fig. 1.41. Hence $D_\delta(z_0) \subseteq B$ and B is open.

As a consequence, $a \in B$ and $B \neq \phi$.

Take any point $z_0 \in \Omega - B$. There exists a $\delta > 0$ so that $D_\delta(z_0) \subseteq \Omega$. Then, any point in $D_\delta(z_0)$ cannot be joined by a curve in A to a, otherwise z_0 can, too, a contradiction. Hence, $D_\delta(z_0) \subseteq \Omega \backslash B$ and hence, $\Omega \backslash B$ is open in Ω which, in turn, implies that B is closed in Ω.

(2) $\Rightarrow$ (1) (*We do not have to presume that Ω is open.*) In case Ω is not connected, then there exist nonempty open sets O_1 and O_2 in Ω so that $O_1 \cap O_2 = \phi$ and $\Omega = O_1 \cup O_2$ hold. By assumption, there exists a polygonal

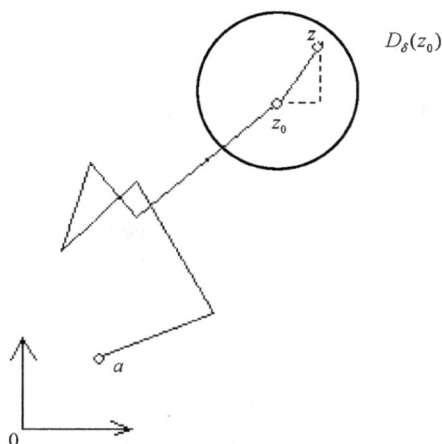

Fig. 1.41

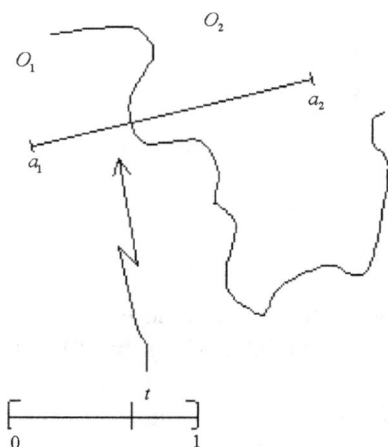

Fig. 1.42

curve in Ω connecting a fixed point $a_1 \in O_1$ to a fixed point $a_2 \in O_2$. One end point of a composed line segment will be in O_1, while the other end point will be in O_2. Hence, we may suppose that a_1 and a_2 are end points of a single line segment. See Fig. 1.42.

Now, the segment $\overline{a_1 a_2}$ can be expressed as $(1-t)a_1 + ta_2$, $0 \le t \le 1$. Let

$$B_1 = \{t \in [0,1] | (1-t)a_1 + ta_2 \in O_1\}, \text{ and}$$

$$B_2 = \{t \in [0,1] | (1-t)a_1 + ta_2 \in O_2\}.$$

Then $B_1 \neq \phi$, $B_2 \neq \phi$ and $B_1 \cap B_2 = \phi$. B_1 and B_2 are both open in $[0, 1]$ and $[0, 1] = B_1 \cup B_2$, contradicting to the connectedness of $[0, 1]$. Hence (1) holds. □

Section (8) Component of a set

Let $A \subseteq \mathbf{C}$ be a nonempty set.

The largest connected subset of A, i.e., a connected subset of A which is not contained in any other connected subset of A, is called a *component* of A.

For $a \in A$, let

$C(a) = $ the union of all connected subsets of A which contain a.

Since $\{a\}$ is connected, then $C(a) \neq \phi$.

$C(a)$ is connected: If not, there exist open sets O_1 and O_2 in $C(a)$ so that $C(a) = O_1 \cup O_2$ where $O_1 \neq \phi$, $O_2 \neq \phi$ and $O_1 \cap O_2 = \phi$. Suppose $a \in O_1$. Pick $b \in O_2$. Since $b \in C(a)$, there exists a connected subset B of A that contains b. Now, $B \subseteq C(a)$ implies that $B = (O_1 \cap B) \cup (O_2 \cap B)$, contradicting to the connectedness of B.

$C(a)$ is a closed set in A: It is known that $\overline{C(a)}$ is connected (see Exercise A(4)). Then $C(a) = A \cap \overline{C(a)}$ shows that $C(a)$ is relatively closed in A.

$C(a) = C(b)$ or $C(a) \cap C(b) = \phi$: If $z_0 \in C(a) \cap C(b)$, by definition, $C(a) \subseteq C(z_0)$ holds and, in particular, $a \in C(z_0)$ which, in turns, implies that $C(z_0) \subseteq C(a)$. Hence $C(a) = C(z_0)$. Similarly, $C(b) = C(z_0)$. Thus, $C(a) = C(b)$ holds.

Such a $C(a)$ is called the *component containing a*.

We summarize the above as

The components of a planar set.

(1) *Every nonempty set in* $\mathbf{C}$ *can be expressed as the disjoint union of components; each component is closed in the set.*

(2) *Every nonempty open set in* $\mathbf{C}$ *can be expressed as the disjoint union of countably many components; each component is an open set in* $\mathbf{C}$. (1.8.12)

Note that (1.8.10), the arguments about components of a set and (1.8.12) are still valid in $\mathbf{C}^*$.

A domain Ω in $\mathbf{C}$ is called *simply connected* or *n-connected* $(n \geq 2)$ if its complement $\mathbf{C}^* - \Omega$ in $\mathbf{C}^*$ has only one component or n components as subsets of $\mathbf{C}^*$, respectively.

We illustrate two examples.

Example 1. Let

$$B = \{z \in \mathbf{C} \mid \text{both } \operatorname{Re} z \text{ and } \operatorname{Im} z \text{ are rational numbers}\}.$$

Explanation. Let $A = \mathbf{C} - B$. Then $A \cap B = \phi$, but $\overline{A} = \overline{B} = \mathbf{C}$.

B is a countable set, while A is not. A and B are not bounded sets, nor are they closed and hence are not compact.

The interior $A^0 = B^0 = \phi$, yet $\partial A = \partial B = \mathbf{C}$.

B is not a connected set, while A is (why?). □

Example 2. Let

$$A = \{z \mid |z| < 1\} \cup \{z \mid 1 < |z| < r \text{ and } \operatorname{Arg} z \text{ is rational.}\}$$

Explanation. A is neither open nor closed. A is not connected.

$A' = \overline{A}$, both are $|z| \le r$.

A^0 is $|z| < 1$; ∂A is $1 \le |z| \le r$; $\overline{A}^\sim$ is $|z| > r$. A has countably infinitely many components: $|z| < 1$ and the line segments $\operatorname{Arg} z = $ rational, with $1 < |z| < r$. Note that $1 < |z| < r$ is not a component of A. □

Section (9) Point sets in $\mathbf{C}^*$

In the extended complex plane $\mathbf{C}^*$, we adopt the spherical chord distance $d(,)$ (see (1.6.8)) and use the *ball* with z_0 as *center* and $\varepsilon > 0$ as *radius*:

$$B_\varepsilon(z_0) = \{z \in \mathbf{C}^* \mid d(z, z_0) < \varepsilon\}, \tag{1.8.13}$$

to replace $D_\varepsilon(z_0)$ and $D_R(\infty)$ as indicated in (1.8.1) and (1.8.2). We can define various point sets in $\mathbf{C}^*$ exactly like the ways we did so far in $\mathbf{C}$ and obtain the same properties. *Hereafter, we will feel free to use them if needed.*

Caution: The following three facts should be noted.

(1) If z_1 and z_2 are any two points in $|z| \le R$, then

$$\frac{2}{1+R^2}|z_1 - z_2| \le d(z_1, z_2) \le 2|z_1 - z_2|.$$

This indicates that, *in a compact subset of* $\mathbf{C}$, *the limit process using* $(\mathbf{C}, \|\|)$ *will be coincident with the limit process using* $(\mathbf{C}^*, d(,))$.

(2) Special attention should be paid to the difference when an unbounded set is viewed as a subset of $\mathbf{C}$ or of $\mathbf{C}^*$. For example, open (closed) sets in $\mathbf{C}$ are still open (closed) in $\mathbf{C}^*$, and vice versa, if one disregards the point at infinity ∞.

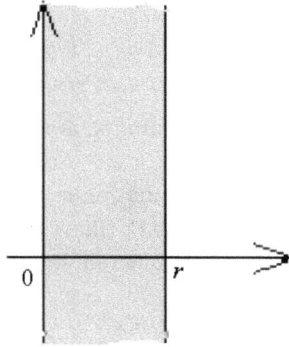

Fig. 1.43

The vertical strip $0 < \operatorname{Re} z < r$ has two boundary components $\operatorname{Re} z = 0$ and $\operatorname{Re} z = r$ when viewed as a subset of $\mathbf{C}$, yet it has only one boundary component $\{\operatorname{Re} z = 0\} \cup \{\infty\} \cup \{\operatorname{Re} z = r\}$ when viewed as a subset of $\mathbf{C}^*$. Hence, it is a simply connected domain. See Fig. 1.43. What is the boundary of $|z| > R$, when considered as a subset of $\mathbf{C}$ or $\mathbf{C}^*$?

(3) $(\mathbf{C}, \|)$ is a complete metric space but is not compact (see (1.9.4)), while $(\mathbf{C}^*, d(,))$ is both complete and compact. See Exercise B(1) and Sec. 1.9.

Exercises A

(1) Prove (1.8.3) in detail.
(2) Prove (1.8.5) in detail.
(3) Prove (1.8.8) in detail, plus the following:

    (a) $\bar{A} - \partial A = A^0$.
    (b) If $\bar{A} \cap \bar{B} = \phi$, then $\partial(A \cup B) = \partial A \cup \partial B$.
    (c) $(A')' \subseteq A'$, i.e., the derived set A' of a set A is closed.

(4) Suppose A is a connected set and $A \subseteq E \subseteq \bar{A}$. Show that E is also connected.

(5) Find the limit points of each of the following sets.

    (a) $\left\{ 1 + \frac{n}{n+1} i^n \,\middle|\, n \geq 1 \right\}$.
    (b) $\left\{ \frac{1}{m} + \frac{1}{n} i \,\middle|\, m, n = \pm 1, \pm 2, \ldots \right\}$.
    (c) $\left\{ \frac{p}{m} + \frac{q}{n} i \,\middle|\, m, n = \pm 1, \pm 2, \ldots; p, q = 0, \pm 1, \pm 2, \ldots \right\}$.

(d) $\left\{ \left(\frac{1}{m} + \frac{1}{n} \right)^{mn} + i^n \,\middle|\, m, n = 1, 2, 3, \dots \right\}$.

(e) $\left\{ \left(1 + \frac{1}{n} \right) \left(\cos \frac{n\pi}{\alpha} + i \sin \frac{n\pi}{\alpha} \right) \,\middle|\, n = 1, 2, 3, \dots \right\}$. Consider the cases that α is rational and irrational, respectively.

(6) Show that the union of two domains is a domain if and only if their intersection is nonempty.

(7) (a) Suppose A and B are disjoint connected subsets of a set E and C is a connected subset of $A \cup B$. Show that either $C \subseteq A$ or $C \subseteq B$.

(b) Suppose A and B are connected sets so that $A \cap B \neq \phi$, then $A \cup B$ is connected.

(8) Find a domain Ω so that $\partial\Omega \neq \partial\overline{\Omega}$.

(9) Let

$$A = \{z \in \mathbf{C} | \operatorname{Re} z = 0 \text{ and } |\operatorname{Im} z| \leq 1\}, \quad \text{and}$$

$$B = \left\{ z \in \mathbf{C} \,\middle|\, \operatorname{Re} z > 0 \text{ and } \operatorname{Im} z = \frac{1}{\sin(\operatorname{Re} z)} \right\}.$$

Is $A \cup B$ connected?

(10) Let

$$A = \left\{ z \in \mathbf{C} \,\middle|\, 0 \leq \operatorname{Re} z \leq 1 \text{ and } \operatorname{Im} z = 0 \text{ or } \operatorname{Im} z = \frac{1}{n}, n \geq 1 \right\}.$$

Find the components of A. Are they closed? Are they open in A? Show that A is *not locally connected*.

Note: A point set A in $\mathbf{C}$ is called *locally connected* if every neighborhood of each point a in A contains another connected neighborhood of a.

(11) Let

$$A = \{z \,|\, \operatorname{Im} z > 0\} - \left\{ \frac{1}{n} + it \,\middle|\, n = \pm 1, \pm 2, \dots; 0 \leq t \leq 1 \right\}.$$

Find ∂A. Is A a domain? Simply connected?

Exercises B

(1) Let X be a nonempty set. Suppose, for each pair x_1, x_2 of elements in X, it is associated with a nonnegative number $d(x_1, x_2)$ (could be $+\infty$), satisfying:

1. $d(x_1, x_2) \geq 0$ and $= 0 \Leftrightarrow x_1 = x_2$;
2. $d(x_1, x_2) = d(x_2, x_1)$; and
3. $d(x_1, x_2) \leq d(x_1, x_3) + d(x_3, x_2), x_1, x_2, x_3 \in X$;

then, we call X, endowed with the *metric $d(,)$*, a *metric space* and X is denoted as $(X, d(,))$ or simply as X. Elements in X are called *points*. For example,

$$(\mathbf{C}, \|), (\mathbf{C}^*, d(,))(\text{see } (1.6.8)) \text{ and } (\mathbf{C}^*, \rho(,))$$

(see Exercise B(3) of Sec. 1.6)

are such examples. The set, for $\delta > 0$ and $x \in X$,

$$B_\delta(x) = \{y \in X | d(y, x) < \delta\}$$

is called an *open sphere* with *center x* and *radius δ*. Try to use this $B_\delta(x)$ to define all point sets introduced in the text for X and derive their possible properties.

(2) Let T be a nonempty set. Let τ be a family of subsets of T, satisfying:

(i) $T \in \tau$ and $\phi \in \tau$;
(ii) any finite intersection of sets in τ is still in τ; and
(iii) any arbitrary union of sets in τ is still in τ;

then, we say τ defines a *topology* on T and call (T, τ) or simply T a *topological space*. Elements in τ are called *open* sets. If, for any two distinct points x and y in T, there exist two open sets $O(x)$ and $O(y)$ satisfying

$$x \in O(x), y \in O(y) \quad \text{and} \quad O(x) \cap O(y) = \phi,$$

then, call T a *Hausdorff space*. Try to use open sets to define all possible point sets for an abstract topological space or a Hausdorff space. Try to derive their possible properties and compare to Exercise (1).

Note. For planar point sets, refer to Ref. [65]; for general topology, refer to Ref. [27].

1.9. Completeness of the Complex Field C

Here, we try to extend the completeness of the real field $\mathbf{R}$ (see Appendix A) to the complex field $\mathbf{C}$.

Let $A \subseteq \mathbf{C}$. Suppose $\mathcal{F}$ is a family of *open* sets in $\mathbf{C}$ satisfying the property that the union of sets in $\mathcal{F}$ covers A, i.e.,

$$A \subseteq \bigcup_{O \in \mathcal{F}} O, \tag{1.9.1}$$

then we call $\mathcal{F}$ an *open covering* of A. A subfamily of $\mathcal{F}$ is called a *subcovering* if the union of open sets from the subfamily still covers A, and it is called *a finite* (or *countable*) *subcovering* if the number of open sets in the subfamily is finite (or countable).

The *diameter* of a set A in $\mathbf{C}$ is defined as

$$\text{diam } A = \sup_{z_1, z_2 \in A} |z_1 - z_2|. \tag{1.9.2}$$

Obviously, A is bounded if and only if $\text{diam } A < +\infty$.

Here comes the main result (refer to Appendix A).

The completeness of the complex field $\mathbf{C}$. *In* $\mathbf{C}$, *any one of the following statement holds and they are equivalent.*

(1) *Every Cauchy sequence in* $\mathbf{C}$ *converges (see Sec. 1.7).*
(2) *(Cantor's nested sets theorem) Let* A_n *be a sequence of closed sets in* $\mathbf{C}$ *satisfying*

   (i) $A_1 \supseteq A_2 \supseteq \cdots \supseteq A_n \supseteq A_{n+1} \supseteq \cdots$, *and*
   (ii) $\text{diam } A_n \to 0$,

   then there exists a unique point $z_0 \in \mathbf{C}$ *so that* $\cap_{n=1}^{\infty} A_n = \{z_0\}$.
(3) *(Bolzano–Weierstrass) Any bounded infinite set in* $\mathbf{C}$ *has at least one limit point (see* (1.8.3)*).*
(4) *(Bolzano–Weierstrass) Any bounded complex sequence has a convergent subsequence (see* (2) *in* (1.7.7)*).*
(5) *(Heine–Borel) Suppose* $A \subseteq \mathbf{C}$ *is a bounded closed set, i.e., a compact set. Then, any open covering* $\mathcal{F}$ *of* A *has a finite subconvering, i.e., there exist finitely many open sets* $O_1, \ldots, O_n$ *in* $\mathcal{F}$ *so that*

$$A \subseteq \bigcup_{j=1}^{n} O_j. \tag{1.9.3}$$

Proof. We show (1) holds and then, prove that (1)–(5) are equivalent.

The validity of (1): Let $z_n, n \geq 1$, be a Cauchy sequence. By inequality (1.4.2.1), both $\text{Re } z_n$ and $\text{Im } z_n, n \geq 1$, are real Cauchy sequences. Completeness of the real field $\mathbf{R}$ (see Appendix A) implies that $\text{Re } z_n \to x_0 \in \mathbf{R}$ and $\text{Im } z_n \to y_0 \in \mathbf{R}$. Let $z_0 = x_0 + iy_0$. Then same inequality implies that $z_n \to z_0 \in \mathbf{C}$.

(1) $\Rightarrow$ (2): Take any fixed point $z_n \in A_n, n \geq 1$. In case $m > n$, by $A_n \supseteq A_m$ it follows that $z_n, z_m \in A_n$. Hence,

$$|z_n - z_m| \leq \text{diam } A_n \to 0 \quad \text{as } n \to \infty.$$

Therefore, $z_n, n \geq 1$, is Cauchy. By (1), $z_n \to z_0 \in \mathbf{C}$.

Fix any $n \geq 1$. Then $z_m \in A_n$ if $m \geq n$. Recall that A_n is a closed set. Hence, $z_m \to z_0 \in A_n$ for $n \geq 1$ and $z_0 \in \cap_{n=1}^{\infty} A_n$.

In case there exists a point $z_0' \in \cap_{n=1}^{\infty} A_n$. Then, $z_0, z_0' \in A_n$ for $n \geq 1$. Hence

$$|z_0 - z_0'| \leq \operatorname{diam} A_n \to 0 \quad \text{as } n \to \infty$$

implies that $z_0' = z_0$. This proves that $\cap_{n=1}^{\infty} A_n = \{z_0\}$.

$(2) \Rightarrow (3)$: Suppose $A \subseteq \mathbf{C}$ is a bounded infinite set.

We may suppose that A is contained in a rectangle R_1. See Fig. 1.44. Divide R_1 into four equal parts. Then, at least one of these smaller rectangles, denoted as R_2, contains infinitely many points of A. Then,

(1) $R_1 \supseteq R_2$;
(2) $\operatorname{diam} R_2 = \frac{1}{2} \operatorname{diam} R_1$; and
(3) $R_2 \cap A$ is an infinite set.

Again, divide R_2 into four equal parts and pick up such a subrectangle R_3 so that

(1) $R_2 \supseteq R_3$;
(2) $\operatorname{diam} R_3 = \frac{1}{2} \operatorname{diam} R_2$; and
(3) $R_3 \cap A$ is an infinite set.

Continue this process to R_3. Then, by induction, there exists a sequence $R_1, R_2, \ldots, R_n, \ldots$ of *closed* rectangles satisfying

(1) $R_1 \supseteq R_2 \supseteq \cdots \supseteq R_n \supseteq R_{n+1} \supseteq \cdots$;
(2) $\operatorname{diam} R_n = \frac{1}{2} \operatorname{diam} R_{n-1} = \cdots = \frac{1}{2^{n-1}} \operatorname{diam} R_1 \to 0$ as $n \to \infty$; and
(3) $R_n \cap A$ is an infinite set, $n \geq 1$.

By 1 and 2, there exists a point $z_0 \in \mathbf{C}$ so that $\cap_{n=1}^{\infty} R_n = \{z_0\}$.

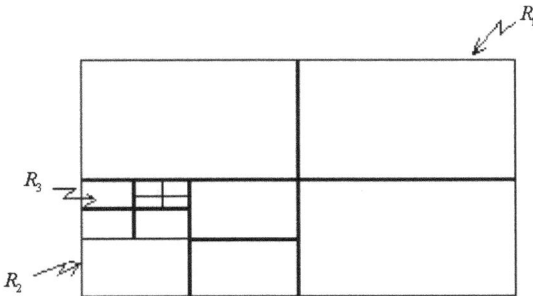

Fig. 1.44

For any $\varepsilon > 0$, since diam $R_n \to 0$, then $R_n \subseteq D_\varepsilon(z_0)$ for all sufficiently large n. Now, $R_n \cap A \subseteq D_\varepsilon(z_0) \cap A$ holds. So, by 3, $D_\varepsilon(z_0)$ contains infinitely many points of A. Therefore, z_0 is a limit point of A.

(3) $\Rightarrow$ (4): Given a bounded sequence $z_n, n \geq 1$. Consider the bounded set $A = \{z_n \mid n \geq 1\}$.

In case A is a finite set: Then, there exists a sequence of positive integers $n_1 < n_2 < \cdots < n_k < \cdots$ so that $\lim_{k \to \infty} n_k = +\infty$ and $z_{n_1} = z_{n_2} = \cdots = z_{n_k} = \cdots$, denoted by z_0 this common number. Then the subsequence $z_{n_k} \to z_0$.

In case A is an infinite set: Then, the result follows from (3) and (1.8.3).

(4) $\Rightarrow$ (1): Suppose $z_n, n \geq 1$, is Cauchy.

$z_n, n \geq 1$, is then a bounded sequence: There exists a positive integer n_0 so that $|z_m - z_n| < 1$ if $m > n \geq n_0$. Hence,

$$|z_n| \leq \max\{|z_1|, \ldots, |z_{n_0-1}|, |z_{n_0}| + 1\}, \quad n \geq 1.$$

By assumption, there exists a subsequence $z_{n_k} \to z_0 \in \mathbf{C}$.

To prove the original sequence $z_n \to z_0$: For any $\varepsilon > 0$, there exists a positive integer k_0 such that $|z_{n_k} - z_0| < \varepsilon$ if $k \geq k_0$. On the other hand, there exists a positive integer N such that $|z_m - z_n| < \varepsilon$ if $m, n \geq N$. Since $\lim_{k \to \infty} n_k = +\infty$, there exists a k_1 so that $n_k \geq N$ if $k \geq k_1$. Finally, for $n \geq N$, we may choose $k \geq \max\{k_0, k_1\}$ and then $n_k \geq N$ holds; in this case,

$$|z_n - z_0| \leq |z_n - z_{n_k}| + |z_{n_k} - z_0| < \varepsilon + \varepsilon = 2\varepsilon.$$

Hence $z_n \to z_0$.

(2) $\Rightarrow$ (5): Let $\mathcal{F}$ be an open covering of the compact set A.

Suppose A is contained in a rectangle R_1. If (5) is false, then the same process as indicated in the proof of (2) $\Rightarrow$ (3) shows that there exists a sequence $R_1, R_2, \ldots, R_n, \ldots$ of rectangles satisfying

(1) $R_1 \supseteq R_2 \supseteq \cdots \supseteq R_n \supseteq \cdots$;
(2) diam $R_n = \frac{1}{2^{n-1}}$ diam $R_1 \to 0$ as $n \to \infty$, and
(3) $R_n \cap A$ cannot be covered by finitely many open sets in $\mathcal{F}, n \geq 1$.

By assumption, $\bigcap_{n=1}^{\infty} R_n = \{z_0\}$ holds.

Since A is closed, $z_0 \in A$. Hence, there exists an open set $O \in \mathcal{F}$ so that $z_0 \in O$; and, in turn, there exists an $\varepsilon > 0$ so that $D_\varepsilon(z_0) \subseteq O$. Now, $\lim_{n \to \infty}$ diam $R_n = 0$ implies that

$$R_n \subseteq D_\varepsilon(z_0) \subseteq O \text{ for all sufficiently large } n \Rightarrow R_n \cap A \subseteq O,$$

contradicting to 3. Thus, (5) follows.

$(5) \Rightarrow (1)$: Let $z_n, n \geq 1$, be Cauchy. Note that the set $A = \{z_n \mid n \geq 1\}$ is bounded (see $(4) \Rightarrow (1)$).

If A has a limit point z_0, then A has a sequence $z_{n_k} \to z_0$ (see $(1.8.3)$). This $z_{n_k}, k \geq 1$, is a subsequence of $z_n, n \geq 1$ and $z_n \to z_0$ (see $(4) \Rightarrow (1)$).

If A is a finite set, then z_n converges (see $(3) \Rightarrow (4)$ and $(4) \Rightarrow (1)$).

In case A is an infinite set without a limit point: Then $A' = \phi \subseteq A$ and A is then closed. Hence, A is a closed bounded set without a limit point. For any $z_n \in A$, there exists an open neighborhood O_n of z_n so that $A \cap O_n$ is a *finite* set. Now, $\mathcal{F} = \{O_n \mid n \geq 1\}$ is an open covering of A. By assumption, only finitely many sets in $\mathcal{F}$, say $O_{n_1}, \ldots, O_{n_k}$, are enough to cover A, namely,

$$A \subseteq O_{n_1} \cup \cdots \cup O_{n_k}$$

$$\Rightarrow A = \bigcup_{j=1}^{k} (A \cap O_{n_j})$$

$$\Rightarrow A \text{ is a finite set.}$$

This is a contradiction. Hence, this case cannot happen.   □

Owing to its importance, we list

The characteristic properties of a compact set. Let $A \subseteq \mathbf{C}$. Then

(1) *A is closed and bounded.*

⇔ (2) *Any infinite subset (or sequence) of A has at least one limit point in A (or a subsequence converging to a point in A).*

⇔ (3) *Any open covering of A has a finite subcovering.*

⇔ (4) *A is a totally bounded closed set.*

Note: A set A is totally bounded if for any $\varepsilon > 0$, there exist finitely many points $z_1, \ldots, z_n$ in A so that $A \subseteq \cup_{j=1}^{n} D_\varepsilon(z_j)$.

Such a point set A is called compact. $(1.9.4)$

In $\mathbf{C}$, a set is bounded if and only if it is totally bounded. Yet in a general metric space, a totally bounded set is bounded but not conversely (see Exercise B(1)).

Sketch of proof. From $(1.9.3)$, it is easy to see that $(1) \Rightarrow (2) \Rightarrow (3)$ and $(4) \Rightarrow (1)$.

$(3) \Rightarrow (4)$: Fix $\varepsilon > 0$, then the open disks $D_\varepsilon(z), z \in A$, form an open covering of A. Hence A is totally bounded.

$(\bar{A}) - (A)$: Suppose A is not closed. Then, pick a point $z_0 \in \bar{A} - A$. Now

$$\bigcap_{n=1}^{\infty} \bar{D}_{\frac{1}{n}}(z_0)$$

$$= \{z_0\}, \quad \text{where } \bar{D}_{\frac{1}{n}}(z_0) \text{ is the closed disk } |z - z_0| \leq \frac{1}{n}, \quad n \geq 1.$$

$$\Rightarrow A \subseteq \bigcup_{n=1}^{\infty} \bar{D}_{\frac{1}{n}}(z_0)^{\sim}, \quad \text{where } \bar{D}_{\frac{1}{n}}(z_0)^{\sim} \text{ is } |z - z_0| > \frac{1}{n}, \quad n \geq 1.$$

Hence, there exists a positive integer n_0 so that

$$A \subseteq \bigcup_{n=1}^{n_0} \bar{D}_{\frac{1}{n}}(z_0)^{\sim} = \bar{D}_{\frac{1}{n_0}}(z_0)^{\sim}$$

$$\Rightarrow A \cap D_{\frac{1}{n_0}}(z_0) = \phi$$

contradicting to the fact that z_0 is a limit point of A.   □

Remark. (1.9.3) and (1.9.4) show that $(\mathbf{C}, \|)$ is a *noncompact, complete metric space*.

Since any infinite set in $\mathbf{C}^*$ has at least one limit point in itself, $(\mathbf{C}^*, d(,))$ is a *compact, complete metric space*, where $d(,)$ is the spherical chord distance defined in (1.6.8). By Exercise B(3) of Sec. 1.6, $(\mathbf{C}^*, \rho(,))$ is also a *compact, complete metric space*.

The adjoining of a point ∞ to $\mathbf{C}$ makes $(\mathbf{C}^*, d(,))$ a compact space. We usually call $\mathbf{C}^*$ a *one-point compactification* of $\mathbf{C}$ (refer to Section (9) in Sec. 1.8).   □

Exercises A

(1) Let $A_n, n \geq 1$, be a sequence of compact sets in $\mathbf{C}$ satisfying $A_1 \supseteq A_2 \supseteq \cdots \supseteq A_n \supseteq \cdots$. Show that $\cap_{n=1}^{\infty} A_n \neq \phi$. What happens if each A_n is merely a closed set? In case each A_n is a compact connected set, then so is $\cap_{n=1}^{\infty} A_n$. Prove this.

(2) (a) Suppose $A \subseteq \mathbf{C}$ is compact. Show that there exist two points z_1, z_2 in A so that

$$\text{diam} \, A = |z_1 - z_2|.$$

(b) The *distance* between two sets A and B in $\mathbf{C}$ is defined as

$$\text{dist}(A, B) = \inf_{z_1 \in A, z_2 \in B} |z_1 - z_2|.$$

If $B = \{z_0\}$, denote $d(A, B) = \mathrm{dist}(z_0, A)$ (see (3) in (1.8.5)). If both A and B are compact and $A \cap B = \phi$, show that there exist a point $z_1 \in A$ and a point $z_2 \in B$ so that

$$\mathrm{dist}(A, B) = |z_1 - z_2| > 0.$$

Show that this fact is still valid if A is compact and B is closed and $A \cap B = \phi$. What happens if both A and B are just closed sets?

Exercises B

Here, it is supposed that readers are familiar with basic knowledge about metric space (X, d). See Exercise B(1) of Sec. 1.8, for instance. A sequence $x_n, n \geq 1$, in X is called *Cauchy* if $d(x_n, x_m) \to 0$ as $m, n \to \infty$. If every Cauchy sequence in X converges to a point in X, then X is called a *complete metric space*. A subset A of X is called a *complete subspace* if (A, d) is complete as a metric space; in this case, A is a closed subset of X. Any closed subset of a complete metric space is a complete subspace.

Do the following problems.

(1) Let A be a subset of a metric space (X, d). Show that

    (a) (2) in (1.9.4).
⇔ (b) (3) in (1.9.4).
⇔ (c) (i) A is a complete subspace and
       (ii) A is totally bounded (see (4) in (1.9.4)).
⇒ (d) A is closed and bounded.

Such a set A satisfying (a) or (b) or (c) is called a *compact set*.

(2) Extend Exercise A(2) to general metric space (X, d).

(3) In a metric space (X, d), show that

    (a) X has a countable dense subset.
⇔ (b) X has a countable family of open sets $\{O_n\}_{n=1}^{\infty}$, so that for every nonempty open set O in X and each point $x \in O$, there exists an n so that $x \in O_n \subseteq O$.
⇔ (c) Any open covering of X has a countable subcovering.

We call such a space X a *separable space*. Show that both $\mathbf{C}$ and $\mathbf{C}^*$ are separable.

(4) (*Baire category theorem*) Suppose $A_n, n \geq 1$, is a sequence of closed sets in a complete metric space (X, d) such that $X = \cup_{n=1}^{\infty} A_n$ holds. Show that, there exists at least one A_n with nonempty interior, i.e., $A_n^0 \neq \phi$.

CHAPTER 2

Complex-Valued Functions of a Complex Variable

Introduction

Suppose A is a nonempty set in $\mathbf{C}$. If for each number z in A, there corresponds a unique complex number w, denoted as $f(z)$, then we call

$$f : A \to \mathbf{C}; \quad w = f(z) : A \to \mathbf{C}, \quad \text{or simply} \quad f(z) \text{ or } f \qquad (2.1)$$

a (single) *complex-valued function* of *one complex variable* z, with A as its *domain of definition* and $f(A)$ the *image* of A. In case, for some $z \in A$, there corresponds two or more complex numbers w, then we call f a *multiple-valued function* of z in A (for details, refer to Sec. 2.7). *Unless otherwise stated, all functions mentioned would be supposed to be single-valued.* When geometric flavor, such as correspondence between points or configurations, is emphasized, a function is also called a *mapping*.

Functions mentioned above are divided into four types:

Type one: If both the domain A of the definition and the image $f(A)$ are subsets of the real number (or real axis), then f is just a real-valued function of a real variable, which we studied in calculus. Properties of this type of functions will be supposed to be familiar to the readers and will be used to compare to or to be imitated by the functions of type four described below.

Type two: The cases that A is *not* a subset of $\mathbf{R}$ and $f(A) \subseteq \mathbf{R}$ are real-valued functions of one complex variable $z + x + iy$ or two real variables x and y, such as

$$\operatorname{Re} z, \operatorname{Im} z, \text{ and } \operatorname{Arg} z$$

and their composite functions.

Type three: The cases that $A \subseteq \mathbf{R}$ and $f(A) \subseteq \mathbf{C}$ are complex-valued functions of a real variable, usually denoted as

$$z = z(t) = x(t) + iy(t), \quad t \in A(\subseteq \mathbf{R}).$$

They are used to define curves in the plane. See Sec. 2.4.

Type four: Both A and $f(A)$ are *not* subsets of the real number (real axis). These are truly *complex-valued functions of one complex variable* to be investigated in the course of complex analysis. Usually, we set $z = x + iy$ and $w = u + iv$. Then $w = f(z)$ has the

$$\text{real part: Re } f(z) = u(z) = u(x,y);$$
$$\text{imaginary part: Im } f(z) = v(z) = v(x,y),$$
(2.2)

and the function itself can be expressed as

$$w = f(z) = u(z) + iv(z) = u(x,y) + iv(x,y).$$
(2.3)

Henceforth, *all functions concerned will be this type unless specified otherwise*.

By use of operational properties (1.3.2) of complex numbers, we have *operational properties of functions*. Suppose f and g have a common domain A of definition. Then the following are functions on A:

(1) (addition) $f + g$: $(f+g)(z) = f(z) + g(z)$, $z \in A$;
(2) (subtraction) $f - g$: $(f-g)(z) = f(z) - g(z)$, $z \in A$;
(3) (multiplication) fg: $(fg)(z) = f(z)g(z)$, $z \in A$;
(4) (division) $\frac{f}{g}$: $\left(\frac{f}{g}\right)(z) = \frac{f(z)}{g(z)}$, $z \in A$ and $g(z) \neq 0$.
(2.4)

The most simple functions are:

constant function: $c (c \in \mathbf{C})$, and
identity function: z.
(2.5)

By successive applications of multiplication, one obtains *power functions* $z^2, z^3, \ldots, z^n$ $(n \in \mathbf{N})$ and, then, *polynomials*

$$p(z) = a_n z^n + a_{n-1} z^{n-1} + \cdots + a_1 z + a_0,$$

$$\text{where } a_j \in \mathbf{C} \quad \text{for } 0 \leq j \leq n \text{ and } a_n \neq 0.$$
(2.6)

Their domains of definition are $\mathbf{C}$, so are their images (to be proved later on).

The division of two polynomials $p(z)$ and $q(z)$,

$$\frac{p(z)}{q(z)} = \frac{a_n z^n + a_{n-1} z^{n-1} + \cdots + a_1 z + a_0}{b_m z^m + b_{m-1} z^{m-1} + \cdots + b_1 z + b_0}, \quad a_n \neq 0,\ b_m \neq 0$$
(2.7)

is a *rational function* of degree $\max(m,n)$. Its domain of definition contains these points at which the denominator is not equal to zero and its range the extended complex plane $\mathbf{C}^*$ if zero points of the denominator are permitted.

Under limit process, some further functions may be introduced, such as exponential function e^z (Sec. 2.6.1), trigonometric functions $\sin z, \cos z,$ and $\tan z$ (Sec. 2.6.2) and hyperbolic functions $\sinh z, \cosh z,$ and $\tanh z$ (Sec. 2.6.2).

The most elementary *multiple-valued* functions are

$\arg z$ (see (1.2.11) and Sec. 2.7.1);

$z^{\frac{1}{n}}$ or $\sqrt[n]{z}$, $n \geq 2$ (see (1.5.4) and Sec. 2.7.2);

$z^{\frac{m}{n}}$, where m and n are relatively prime (see (1.5.5) and   (2.8)
    Sec. 2.7.2), and

$\log z$ (see Sec. 2.7.3).

Then comes the inverse trigonometric functions $\cos^{-1} z$, etc. and the inverse hyperbolic functions $\cosh^{-1} z$, etc. (see Sec. 2.7.4).

In general, if $p_n(z), \ldots, p_0(z)$ are polynomials which are not identically equal to zero, then the algebraic equation

$$p_n(z)w^n + p_{n-1}(z)w^{n-1} + \cdots + p_1(z)w + p_0(z) = 0 \qquad (2.9)$$

defines $w = f(z)$ as a (n-valued) multiple-valued function of z, called an *algebraic function*. A function which is not algebraic is called *transcendental* (see Sec. 2.6).

The study of algebraic, analytic, and geometric properties of these functions mentioned above constitutes one of the main topics in elementary complex analysis. Also, they motivate the development of abstract and general theory in this field.

Sketch of the Content

The concepts and general properties of the limit at a point and the continuity on a set of a function are sketched in Secs. 2.1 and 2.2 for reference. Section 2.3 devotes to a concise study of the criteria for uniform convergence of a sequence or series of functions, plus some illustrative examples. Common terminologies about curves or paths are in Sec. 2.4.

Section 2.5 will focus on basic algebraic and geometric mapping properties of rational functions as a prelude to analytic and meromorphic functions to be fully developed in later chapters. Detailed mapping properties of the power function $w = z^n$ (Sec. 2.5.2), the bilinear fractional transformation $w = \frac{az+b}{cz+d}$ (Sec. 2.5.4) and the Joukowski function (Sec. 2.5.5) are studied with the emphasis on their conformal characters.

Elementary transcendental functions such as $e^z, \sin z, \cos z, \tan z,$ $\sinh z,$ etc. are in Sec. 2.6. It is the limit process (less than the process of differentiation and integration) and the geometric mapping properties, rather then the algebraic ones, that distinguish them completely from the corresponding real functions of a real variable.

The most difficult thing that a beginner in complex analysis might probably encounter is the multiple-valuedness originated from that of $\arg z$ (Sec. 2.7.1). What are the main differences in algebraic, geometric, and analytic aspects between the well-known real function $\sqrt{x}$, $(x \geq 0)$ and the complex function $\sqrt{z}$ $(z \in \mathbf{C})$? Section 2.7 tries to describe the local and the global branches of the multiple-valued functions, in general, and to construct the Riemann surfaces (even though descriptive, informal and non-rigorous) of some elementary functions such as $\sqrt[n]{z}$ (Sec. 2.7.2), $\log z$ (Sec. 2.7.3), $\cos^{-1} z, \sin^{-1} z$ (Sec. 2.7.4), in particular.

What the differentiation of a plane-valued function of two real variables and the line integral along a planar curve might look like in the complex notation? Sections 2.8 and 2.9 answer these questions. By the way, the ∂_z and $\partial_{\bar{z}}$ operators, the differential forms, the Green, Stokes, and the generalized Cauchy integral formulas are introduced, too.

2.1. Limits of Functions

Just like complex sequences (Sec. 1.7), we will give a concise introduction to the concept of limit of a function at a point, disregarding of details similar to the real cases.

Let $f : A \to \mathbf{C}$ be a function and z_0 be a *limit* point of A. Suppose $w_0 \in \mathbf{C}$. If for each $\varepsilon > 0$, there exists a $\delta = \delta(z_0, \varepsilon) > 0$ so that for all points $z \in A$ satisfying $0 < |z - z_0| < \delta$, $|f(z) - w_0| < \varepsilon$ always holds, then we say that the *limit* of f as $z \in A$ approaches z_0 is equal to w_0 and is denoted as

$$\lim_{z \in A \to z_0} f(z) = w_0 \quad \text{or} \quad f(z) \to w_0 \quad \text{as} \quad z \in A \to z_0. \tag{2.1.1}$$

Such a limit is *unique* if it exists. In this case, f is *bounded* in a deleted neighborhood of z_0.

It follows easily the following

Characteristic properties of the existence of the limit of a function at a point.

Let $f \colon A \to \mathbf{C}$, z_0 a limit point of A and $w_0 \in \mathbf{C}$.

(1) $\lim_{z \in A \to z_0} f(z) = w_0$.

$\Leftrightarrow$ (2) $\lim_{z \in A \to z_0} \operatorname{Re} f(z) = \operatorname{Re} w_0$ and $\lim_{z \in A \to z_0} \operatorname{Im} f(z) = \operatorname{Im} w_0$.

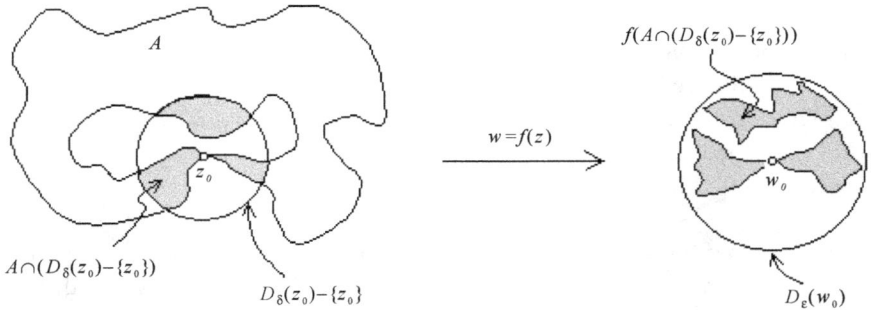

Fig. 2.1

$\Leftrightarrow$ (3) $\lim_{z \in A \to z_0} |f(z) - w_0| = 0$.

$\Leftrightarrow$ (4) For each sequence $z_n \in A \to z_0, f(z_n) \to w_0$ always holds.

$\Leftrightarrow$ (5) For all $\varepsilon > 0$, there exists $\delta = \delta(z_0, \varepsilon) > 0$ such that

$$f\left(A \cap (D_\delta(z_0) - \{z_0\})\right) \subseteq D_\varepsilon(w_0),$$

or

$$A \cap (D_\delta(z_0) - \{z_0\}) \subseteq f^{-1}(D_\varepsilon(w_0)). \qquad (2.1.2)$$

See Fig. 2.1, where the shaded part on the left is the part of the *deleted* neighborhood $D_\delta(z_0) - \{z_0\}$ that lies inside A, the one on the right side is its image $f(A \cap (D_\delta(z_0) - \{z_0\}))$ under f which lies inside the disk $D_{\varepsilon(w_0)}$.

Remark (Other forms of limits of functions). Similarly, we can define:

$$\lim_{z \in A \to z_0} f(z) = \infty;$$

$$\lim_{z \in A \to \infty} f(z) = w_0 \quad \text{where } w_0 \in \mathbf{C} \text{ and } \infty \text{ is a limit point of } A; \text{ and}$$

$$(2.1.3)$$

$$\lim_{z \in A \to \infty} f(z) = \infty.$$

Note that, in these definitions, f may not be defined at z_0 beforehand. "$z \in A \to z_0$" means that, in the process of approaching z_0, z always remains in A and approaches z_0 in all possible manners. In case z_0 is an interior point of A, $z \in A \to z_0$ is always rewritten as $z \to z_0$ in short.

Using spherical chord distance $d(\ ,\ )$ (see (1.6.8)), definitions in (2.1.1) and (2.1.3) can be combined into a single one as follows. Let $z_0 \in \mathbf{C}^*$ be a limit point of a set $A \subseteq \mathbf{C}$ and $f : A \to \mathbf{C}^*$, then

$$\lim_{z \in A \to z_0} f(z) = w_0, \quad \text{where} \quad w_0 \in \mathbf{C}^*$$

$$\Leftrightarrow \lim_{z \in A \to z_0} d(f(z), w_0) = 0 \tag{2.1.4}$$

Concerned with (2.1.1), we have

The operational properties of limits. Suppose $\lim_{z \to z_0} f(z) = w_0$ and $\lim_{z \to z_0} g(z) = w_0'$, where $z \to z_0$ means $z \in A \to z_0$.

(1) $\lim_{z \to z_0} (f(z) \pm g(z)) = w_0 \pm w_0'$;
   $\lim_{z \to z_0} f(z)g(z) = w_0 w_0'$;
   $\lim_{z \to z_0} \frac{f(z)}{g(z)} = \frac{w_0}{w_0'}$ if $w_0' \neq 0$ (in this case, $g(z) \neq 0$ on $0 < |z - z_0| < \delta$
   for some $\delta > 0$);
(2) $\lim_{z \to z_0} \overline{f(z)} = \overline{w_0}$ and $\lim_{z \to z_0} |f(z)| = |w_0|$. $\tag{2.1.5}$

Similar properties hold for (2.1.3). We remind the readers that $\infty \pm \infty$, $0 \cdot \infty$, $\frac{\infty}{\infty}$ and $\frac{0}{0}$ are not defined (see Exercise A (3) of Sec. 1.7).
   We illustrate an example.

Example.

(1) For the polynomial $p(z)$ in (2.6), where $n \geq 1$ and $a_n \neq 0$, show that
   $\lim_{z \to \infty} p(z) = \infty$.
(2) For the rational function $\frac{p(z)}{q(z)}$ in (2.7), show that

$$\lim_{z \to \infty} \frac{p(z)}{q(z)} = \begin{cases} 0, & n < m \\ \dfrac{a_n}{b_m}, & n = m \ (a_n \neq 0, b_m \neq 0). \\ \infty, & n > m \end{cases}$$

In case $p(z)$ and $q(z)$ do not have common zeros, then for each zero $z_0 \in q(z)$, $\lim_{z \to z_0} \frac{p(z)}{q(z)} = \infty$.

Proof. (1) Note that

$$|p(z)| \geq |\,|a_n z^n| - (|a_{n-1} z^{n-1}| + \cdots + |a_1 z| + |a_0|)|$$

$$= |z|^n \left| |a_n| - \left(\frac{|a_{n-1}|}{|z|} + \cdots + \frac{|a_1|}{|z|^{n-1}} + \frac{|a_0|}{|z|^n} \right) \right| \quad \text{if } |z| > 0.$$

Using the fact that $\lim_{z \to \infty} \frac{a}{z^m} = 0$ where $a \in \mathbf{C}$ and $m > 1$, the result follows easily.

(2) Rewrite

$$\frac{p(z)}{q(z)} = \frac{a_n z^n}{b_m z^m} \cdot \frac{1 + \frac{a_{n-1}}{a_n z} + \cdots + \frac{a_1}{a_n z^{n-1}} + \frac{a_0}{a_n z^n}}{1 + \frac{b_{m-1}}{b_m z} + \cdots \frac{b_1}{b_m z^{m-1}} + \frac{b_0}{b_m z^m}}$$

and let $z \to \infty$ according to $n < m$, $n = m$, and $n > m$, respectively.
If $q(z_0) = 0$ and $p(z_0) \neq 0$, then

$$\lim_{z \to z_0} \frac{p(z)}{q(z)} = \frac{\lim_{z \to z_0} p(z)}{\lim_{z \to z_0} q(z)} = \frac{p(z_0)}{0} = \infty. \qquad \square$$

Exercises A

(1) Use $\varepsilon - \delta$ process to prove the statements in the example.
(2) Prove (2.1.2) in detail.
(3) Define (2.1.3) and prove that, together with (2.1.1), they are equivalent to (2.1.4).
(4) Prove (2.1.5) in detail.
(5) Test if the following limits exist, where $z = x + iy$. If it does, find the limit.

(a) $\displaystyle\lim_{z \to 0} \frac{\operatorname{Re} z}{z}$.

(b) $\displaystyle\lim_{z \to 0} \frac{\operatorname{Re} z \cdot \operatorname{Im} z}{|z|^2}$.

(c) $\displaystyle\lim_{z \to 0} \frac{z}{|z|}$.

(d) $\displaystyle\lim_{z \to 0} \frac{\operatorname{Re} z^2}{|z|^2}$.

(e) $\displaystyle\lim_{z \to 0} \frac{z \operatorname{Re} z}{|z|}$.

(f) $\displaystyle\lim_{z \to i}(z^3 - 3\bar{z} + 2)$.

(g) $\displaystyle\lim_{z \to 2i} \frac{z^3 + 8i}{z^2 + 4}$.

(h) $\displaystyle\lim_{z \to i}(2z^2 - iz^3 + z \operatorname{Arg} \bar{z})$.

(i) $\displaystyle\lim_{z \to i} \frac{z^4 + 1}{z^2 + i}$.

(j) $\displaystyle\lim_{z \to 2i} \frac{z^2 - iz + 2}{z^2 + 4}$.

(k) $\displaystyle\lim_{z \to 0} \frac{x(\cos y - 1) + iy \sin x}{|z|}$.

(l) $\displaystyle\lim_{z \to 0} \frac{e^x \sin y - e^y \cos x}{y}$.

(6) Suppose $|z| < 1$ and $\lim_{z \to 1} \frac{|1-z|}{1-|z|} = 1$. Show that $\lim_{z \to 1} \operatorname{Arg}(1-z) = 0$.
(7) Suppose z_0 is a limit point of A. If a function $f : A \to \mathbf{C}$ satisfies $\lim_{\zeta \to z_0}\{\lim_{\eta \to z_0}[f(\zeta) - f(\eta)]\} = 0$, show that $\lim_{z \to z_0} f(z)$ exists.

2.2.   Continuous Functions

Suppose $z_0 \in A$ is a limit point of a set A. Call a function $f : A \to \mathbf{C}$
continuous at z_0 if

$$\lim_{z \in A \to z_0} f(z) = f(z_0), \qquad (2.2.1)$$

namely, for each $\varepsilon > 0$, there exists a $\delta = \delta(z_0, \varepsilon) > 0$ so that for all $z \in A$ with $|z - z_0| < \delta$, $|f(z) - f(z_0)| < \varepsilon$ always holds. Note that, for f to be continuous at z_0, it is necessary to assume that f is already defined at z_0 as a complex number $f(z_0)$. In this case, w_0 appeared from (1) to (5) in (2.1.2) should be replaced by $f(z_0)$; while in (5), the deleted neighborhood $D_\delta(z_0) - \{z_0\}$ should be replaced by the original neighborhood $D_\delta(z_0)$ and hence, $f(A \cap D_\delta(z_0)) \subseteq D_\varepsilon(f(z_0))$ holds.

Designate f to be continuous at isolated points (i.e., points in $A - A'$) of A.

If f is continuous at every point of a set A, then we call f a *continuous function on A*.

A function $f : A \to f(A) \subseteq \mathbf{C}$ is said to be a *homeomorphism* if f is a *one-to-one continuous* function from A onto $f(A)$ and the *inverse function* $f^{-1} : f(A) \to A$ is also *continuous*.

Remark 1. (Continuity in the generalized sense). Still suppose $A \subseteq \mathbf{C}$ and z_0 is a limit point of A. In case $z_0 \in \mathbf{C}$ and $\lim_{z \in A \to z_0} f(z) = \infty$, we may designate $f(z_0) = \infty$; if $z_0 = \infty$ and $\lim_{z \in A \to \infty} f(z) = w_0 \in \mathbf{C}^*$, we designate $f(\infty) = w_0$. Under these circumstances, we say f is *continuous* at z_0 in the *generalized sense*. Adopt spherical chord distance $d(\ ,\ )$ in $\mathbf{C}^*$.

Then f is continuous at z_0 or continuous at z_0 in the generalized sense.

$$\Leftrightarrow \lim_{d(z,z_0) \to 0} d(f(z), f(z_0)) = 0. \tag{2.2.2}$$

Unless stated otherwise, we would *not* use this definition. $\square$

In what follows, four sections are divided.

Section (1) Section Characteristic and operational properties

Based on (2.1.2) and the explanation after (2.2.1), we have

The characteristic properties of a continuous function. Let $f : A \to \mathbf{C}$ be a function. Then,

(1) f is continuous on A.
$\Leftrightarrow$ (2) For each open set O in $\mathbf{C}$, $f^{-1}(O)$ is open in A (see Section (7) in Sec. 1.8).
$\Leftrightarrow$ (3) For each closed set K in $\mathbf{C}$, $f^{-1}(K)$ is closed in A (see Section (7) in Sec. 1.8). (2.2.3)

Proofs are left as Exercise A (2). Based on (2.4) and (2.1.5), we have

The operational properties of continuous functions

(1) *Algebraic operations*: Suppose f and g are continuous on the same set A. Then, $f \pm g$, fg, and $\frac{f}{g}$ (at points $z \in A$ where $g(z) \neq 0$) are continuous.

(2) *Composite operation*: Suppose $f : A \to \mathbf{C}$ is continuous at z_0 and $g : B \to \mathbf{C}$ is continuous at $f(z_0) \in f(A) \subseteq B$. Then $g \circ f : A \to \mathbf{C}$ is continuous at z_0.

(3) *Others*: Suppose $f : A \to \mathbf{C}$ is continuous on A. Then, so are the following functions:

$$\operatorname{Re} f; \operatorname{Im} f;$$

$$modules \ function \ |f| \ (\text{defined as } |f|(z) = |f(z)|, \ z \in A) \text{ and}$$

$$conjugate \ function \ \bar{f} \ (\text{defined as } \bar{f}(z) = \overline{f(z)}, \ z \in A). \qquad (2.2.4)$$

Proofs are left as Exercise A (3). Constant and identity functions (see (2.5)) are continuous everywhere on $\mathbf{C}$ and hence, so are polynomials (see (2.6)). While a rational function (see (2.7)) is continuous at points where denominator is not equal to zero. Beyond these elementary functions, limit processes are usually needed in constructing new types of continuous functions (see Secs. 2.6 and 2.7).

Suppose F is a multiple-valued function on a set A. Let $f : A \to \mathbf{C}$ be a function satisfying:

(1) f is single-valued on A;
(2) f is continuous on A; and
(3) for each $z \in A$, $f(z)$ is one of the many values of F at z,   (2.2.5)

then f is called a *single-valued branch* of F on A, or simply a *branch* of F on A. A point z_0 is called a *singular point* of a branch f of a multiple-valued function F if f fails to be continuous at z_0 but is continuous at some point in every neighborhood of z_0.

We illustrate the following example. For more complicated examples in this direction, see Sec. 2.7.

Example 1. The principal argument function $\operatorname{Arg} z : \mathbf{C} - \{0\} \to (-\pi, \pi]$ fails to be continuous along the negative real axis $(-\infty, 0]$ and is continuous elsewhere in $\mathbf{C}$. Hence, the multiple-valued function $\arg z$ has infinitely many branches

$$f_n(z) = \operatorname{Arg} z + 2n\pi, \quad n = 0, \pm 1, \pm 2, \ldots$$

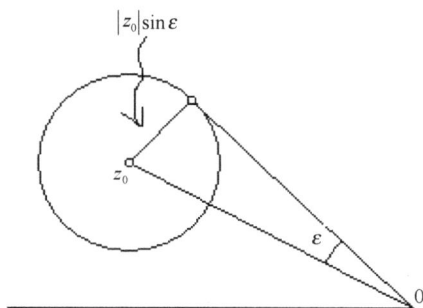

Fig. 2.2

on $\mathbf{C} - (\infty, 0] = \{z \in \mathbf{C} \mid z \neq 0$ and $-\pi < \operatorname{Arg} z < \pi\}$. Each of them has every point on the negative real axis as its singular point (see Fig. 2.2).

Proof. Take any point $z_0 \in \mathbf{C} - (\infty, 0]$. In Fig. 2.2, choose $\varepsilon > 0$ so small so that the open disk $|z - z_0| < \delta = |z_0| \sin \varepsilon$ has empty intersection with $(-\infty, 0]$. Then,

$$|z - z_0| < \delta \Rightarrow |\operatorname{Arg} z - \operatorname{Arg} z_0| < \varepsilon.$$

This proves the continuity of $\operatorname{Arg} z$ at z_0.

In case $z_0 = x_0 < 0$ is a point on $(-\infty, 0]$, take sequence $z_n = x_0 + \frac{i}{n}$ and $z'_n = x_0 - \frac{i}{n}$ for $n \geq 1$. Then $z_n \to x_0$ and $z'_n \to x_0$, yet (refer to the proof of (3) in (1.7.3))

$$\operatorname{Arg} z_n = \operatorname{Arc\,tan} \frac{1}{nx_0} + \pi \to \pi, \ \ n \to \infty \quad \text{and}$$

$$\operatorname{Arg} z'_n = \operatorname{Arc\,tan} \left(-\frac{1}{nx_0} \right) - \pi \to -\pi, \ \ \text{as } n \to \infty.$$

Hence, $\operatorname{Arg} z$ is not continuous at x_0.   □

Section (2) Continuity and connectedness

The main result is

The continuity of a function and the connectedness of a set. A continuous function maps a connected set onto a connected set.

$$(2.2.6)$$

Proof. Suppose $f : A \to \mathbf{C}$ is continuous and A is a connected set. If $f(A)$ fails to be connected, then there exist nonempty yet disjoint open sets O_1

and O_2 in $\mathbf{C}$ so that $f(A) \subseteq O_1 \cup O_2$, $f(A) \cap O_1 \neq \phi$, and $f(A) \cap O_2 \neq \phi$ and then $A \subseteq f^{-1}(O_1) \cup f^{-1}(O_2)$, $A \cap f^{-1}(O_1) \neq \phi$, and $A \cap f^{-1}(O_2) \neq \phi$. This means that A is not connected, a contradiction.

A slight modification of the argument: Suppose U_1 and U_2 are disjoint open sets in $f(A)$ so that $f(A) = U_1 \cup U_2$. Now, $A = f^{-1}(U_1) \cup f^{-1}(U_2)$, where $f^{-1}(U_1)$ and $f^{-1}(U_2)$ are disjoint open sets in A. Since A is connected, one of $f^{-1}(U_1)$ and $f^{-1}(U_2)$ must be empty. Hence, one of U_1 and U_2 is empty and $f(A)$ is then connected.

Remark 2 (How to use (2.2.6) in complex analysis?). The ways of arguments above are standard in complex analysis.

If a real-valued function defined on a connected set takes both positive and negative values, then it should take the value zero; if it takes only integer values, then it should be a constant function.   □

Two examples are provided as follows.

Example 2. Suppose Ω is a domain and $f : \Omega \to \mathbf{C}$ is a continuous function satisfying $|f(z)^2 - 1| < 1$, $z \in \Omega$. Show that either

$$|f(z) - 1| < 1, \;\; z \in \Omega \quad \text{or} \quad |f(z) + 1| < 1, \;\; z \in \Omega.$$

Proof. $f(\Omega)$ is connected and is contained in $|w^2 - 1| < 1$. According to Example 4 in Sec. 1.4.3 and Fig. 1.23 with $a = 1$, $|w^2 - 1| < 1$ has two components (see (1.8.12)): $|w - 1| < 1$ and $|w + 1| < 1$. Hence, $f(\Omega)$ should lie exactly in one of the components.   □

Example 3. Suppose a continuous function $f : \mathbf{C} \to \mathbf{C}$ satisfying the conditions:

(1) $|f(z)| \to \infty$ as $|z| \to \infty$; and
(2) $f(\mathbf{C})$ is open in $\mathbf{C}$.

Show that f is onto $\mathbf{C}$, namely, $f(\mathbf{C}) = \mathbf{C}$.

Proof. If not, then $D = f(\mathbf{C})$ is a *proper* open connected subset of $\mathbf{C}$.

As a proper subdomain D of $\mathbf{C}$, $\mathbf{C} - D \neq \phi$ and hence,

$$\phi \neq \partial D = \bar{D} \cap \overline{\mathbf{C} - D} \subseteq \bar{D} \subseteq \bar{\mathbf{C}} = \mathbf{C}.$$

(In general, if D and G are domains such that $D \subseteq G$, then either $D = G$ or $\partial D \cap G \neq \phi$). Fix a point $w_0 \in \partial D$ and choose a sequence $w_n \in D \to w_0$.

Pick a point $z_n \in f^{-1}(w_n)$ for $n \geq 1$. If z_n is a bounded sequence, then it has a subsequence z_{n_k} converging to a point $z_0 \in \mathbf{C}$. Hence, $f(z_{n_k}) = w_{n_k} \to f(z_0) = w_0 \in D$, contradicting to the fact that $w_0 \in \partial D$. Hence, z_n is an *unbounded* sequence and has a subsequence $z_{n_k} \to \infty$. Therefore, $|f(z_{n_k})| = |w_{n_k}| \to |w_0|$ and $+\infty$ simultaneously. Thus $|w_0| = +\infty$ and then, $w_0 = \infty$, contradicting to $w_0 \in \mathbf{C}$.

This proves that $f(\mathbf{C}) = \mathbf{C}$. $\qquad\qquad\qquad\qquad\qquad\qquad\qquad\qquad\square$

Section (3) Continuity and compactness

The main result is

The continuity of a function and the compactness of a set. Suppose A is a compact set in $\mathbf{C}$ and $f : A \to \mathbf{C}$ is a *continuous* function. Then:

(1) f is *bounded* on A, i.e., there exists a constant $M > 0$ so that $|f(z)| \leq M$ for all $z \in A$.
(2) $f(A)$ is compact in $\mathbf{C}$, i.e., a continuous function maps a compact set onto a compact set.
(3) The modular function $|f|$ attains its maximum and minimum values on A, i.e., there exist points z_1 and z_2 in A so that
$$|f(z_1)| \leq |f(z)| \leq |f(z_2)|, \quad z \in A.$$
(4) If f is one-to-one, then $f : A \to f(A)$ is a homeomorphism.
(5) f is *uniformly* continuous on A (see Section (4) below).   (2.2.7)

Sketch of Proof

(1) is contained in (2). The readers are asked to give a direct poof of (1).
(2) Take a sequence $w_n \in f(A)$ and pick a point $z_n \in f^{-1}(w_n)$ for $n \geq 1$. Then z_n has a subsequence $z_{n_k} \to z_0 \in A$. By continuity of f, $f(z_{n_k}) = w_{n_k} \to f(z_0) \in f(A)$. Hence $f(A)$ is compact.

Or, let $\{O_\lambda\}$ be an open covering of $f(A)$. Then $\{f^{-1}(O_\lambda)\}$ is an open covering of A. A finite subcovering of it, say $f^{-1}(O_1), \ldots, f^{-1}(O_n)$, still covers A. Then
$$A \subseteq \bigcup_{j=1}^{n} f^{-1}(O_j),$$
$$\Rightarrow f(A) \subseteq \bigcup_{j=1}^{n} O_j.$$

This shows that $f(A)$ is compact.

(3) Just like (2), $|f|(A)$ is a compact set in $\mathbf{R}$, i.e., a closed and bounded set. So $|f|(A)$, as a set in $\mathbf{R}$, has the smallest and largest elements.

(4) $f^{-1} : f(A) \to A$ exists as a single-valued function. All one needs to prove is that f^{-1} is continuous on $f(A)$, or equivalently, f maps each closed set K in A onto a closed set $f(K)$ in $f(A)$. Since A is closed, so is K in $\mathbf{C}$ and hence, K is compact in $\mathbf{C}$. By (2), $f(K)$ is compact and hence, is closed in $f(A)$ since $f(A)$ is known to be closed.

(5) The proof is postponed to Section (4). □

Section (4) Uniform continuity

To prove (5) in (2.2.7): Take any $\varepsilon > 0$. For each $z \in A$, by continuity of f at z, there exists a $\delta(z) > 0$ so that for each $\zeta \in A \cap D_{\delta(z)}(z), |f(\zeta) - f(z)| < \frac{\varepsilon}{2}$ holds. Then

$$\{D_{\frac{\delta(z)}{2}}(z)\}$$

forms an open covering of A. Hence, there exist finitely many points $z_1, \ldots, z_n$ in A so that

$$A \subseteq \bigcup_{j=1}^{n} D_{\frac{\delta(z_j)}{2}}(z_j).$$

Let $\delta = \min\{\frac{1}{2}\delta(z_j) | 1 \le j \le n\}$. Now, for any two points z and $\zeta \in A$ satisfying $|z - \zeta| < \delta$, there is some $j, 1 \le j \le n$, so that $z \in D_{\frac{\delta(z_j)}{2}}(z_j)$. Thus,

$$|\zeta - z_j| \le |\zeta - z| + |z - z_j| < \delta + \frac{1}{2}\delta(z_j) \le \frac{1}{2}\delta(z_j) + \frac{1}{2}\delta(z_j) = \delta(z_j)$$

$$\Rightarrow \zeta \in D_{\delta(z_j)}(z_j)$$

$$\Rightarrow |f(z) - f(\zeta)| \le |f(z) - f(z_j)| + |f(z_j) - f(\zeta)| < \frac{1}{2}\varepsilon + \frac{1}{2}\varepsilon = \varepsilon.$$

In short for any $\varepsilon > 0$, there is a $\delta = \delta(\varepsilon) > 0$, depending only on ε and independent of any point in A, so that for any two points z and ζ in A satisfying $|\zeta - z| < \delta$, $|f(\zeta) - f(z)| < \varepsilon$ always holds. Any function f with this property is said to be *uniformly continuous* on the *set A*.

We formally summarize as

The uniform continuity of a function on a set A. Let $f : A \to \mathbf{C}$ be a function.

(1) The following are equivalent:

(a) For each $\varepsilon > 0$, there exists a $\delta = \delta(\varepsilon) > 0$ so that for all $z, \zeta \in A$ with $|\zeta - z| < \delta$, then $|f(\zeta) - f(z)| < \varepsilon$ holds.

(b) (Assume that f is continuous on A) Fix any $\varepsilon > 0$. For each $z \in A$, let

$$\delta(z, \varepsilon) = \inf\{\delta' > 0| \text{ for } \zeta \in A \text{ and } |\zeta - z| < \delta', |f(\zeta) - f(z)| < \varepsilon\}.$$

Then

$$\inf_{z \in A} \delta(z, \varepsilon) = \delta(\varepsilon) > 0.$$

In this case, we call f *uniformly continuous* on A. Note that a uniformly continuous function on a set is automatically continuous on the same set.

(2) The following are equivalent:

   (a) There exists an $\varepsilon_0 > 0$ so that, for all $\delta > 0$, there exists a pair of points ζ_δ, z_δ in A satisfying $|\zeta_\delta - z_\delta| < \delta$, but $|f(\zeta_\delta) - f(z_\delta)| \geq \varepsilon_0$.

   (b) (continued from (b) in (1)) $\inf_{z \in A} \delta(z, \varepsilon) = 0$.

   (c) There exists an $\varepsilon_0 > 0$ and two sequences $z_n, \zeta_n \in A$ satisfying $|z_n - \zeta_n| \to 0$ but $|f(z_n) - f(\zeta_n)| \geq \varepsilon_0$ or $|f(\zeta_n) - f(z_n)| \nrightarrow 0$.

   In each such case, f is said to be *not* uniformly continuous on A.

$$(2.2.8)$$

Proofs are left as Exercise A (4). For further properties of uniform continuity, see Exercises A(13)–(16).

Three examples concerned are as follows.

Example 4. The distance function $d(z, A)$ from a point z to a fixed set A is uniformly continuous on $\mathbf{C}$. In fact

$$|d(z, A) - d(\zeta, A)| \leq |z - \zeta|, \quad z, \zeta \in \mathbf{C}.$$

Sketch of Proof

For fixed $z, \zeta \in \mathbf{C}, |z - \eta| \leq |z - \zeta| + |\zeta - \eta|$ implies that $\operatorname{dist}(z, A) \leq |\zeta - z| + \operatorname{dist}(\zeta, A)$. Similarly, $\operatorname{dist}(\zeta, A) \leq |z - \zeta| + \operatorname{dist}(z, A)$.

Example 5. Discuss the uniform continuity of the function

$$f(z) = \frac{1}{1 + z^2}.$$

Solution. $f(z)$ is continuous everywhere in C except at $\pm i$. $f(z)$ is *not* uniformly continuous on $\mathbf{C} - \{\pm i\}$. Let

$$z_n = i + \frac{1}{n} \quad \text{and} \quad \zeta_n = i + \frac{1}{n^2}, \quad n > 1 \quad \text{with} \quad z_n - \zeta_n \to 0.$$

$$\Rightarrow \lim_{n \to \infty} (f(z_n) - f(\zeta_n)) = \lim_{n \to \infty} n^2 \left\{ \frac{1}{1 + 2ni} - \frac{n^2}{1 + 2n^2 i} \right\} = \infty.$$

So the claim follows. As a matter of fact, as long as D is any open set in $\mathbf{C}$ containing $\pm i$, then $f(z)$ is not uniformly continuous on $D - \{\pm i\}$.

$f(z)$ is not uniformly continuous on $|z| < 1$. Choose $z_n = \left(1 - \frac{1}{n}\right)i$ and $\zeta_n = \left(1 - \frac{1}{n^2}\right)i$ for $n \geq 2$. Then $|z_n| < 1$, $|\zeta_n| < 1$, and $z_n - \zeta_n \to 0$, but $f(z_n) - f(\zeta_n) \to \infty$.

Take $0 < r < 1$. Then $f(z)$ is uniformly continuous on $\mathbf{C} - D_r(i) - D_r(-i)$, the closed set obtained by deleting two open disks $D_r(i)$ and $D_r(-i)$ from the plane $\mathbf{C}$. It is because

$$\lim_{z \to \infty} f(z) = 0.$$

$\Rightarrow$ For each $\varepsilon > 0$, there exists $R > 0$ so that $|z| \geq R$ implies that

$$|f(z)| < \frac{\varepsilon}{3}.$$

$\Rightarrow$ if $|\zeta| \geq R$, $|z| \geq R$ (no matter how large $|\zeta - z|$ is)

$$|f(\zeta) - f(z)| \leq |f(\zeta)| + |f(z)| < \frac{\varepsilon}{3} + \frac{\varepsilon}{3} = \frac{2\varepsilon}{3} < \varepsilon.$$

Let $R \geq 2$. Then $f(z)$ is uniformly continuous on the compact set $D = \{|z| \leq R\} - D_r(i) - D_r(-i)$, i.e., for $z, \zeta \in D$ and $|z - \zeta| < \delta$, then $|f(\zeta) - f(z)| < \frac{\varepsilon}{3} < \varepsilon$. In case $|z| > R$ and $\zeta \in D$ are such that $|z - \zeta| < \delta$ we may suppose ζ is an interior point of D. Let η be the point of intersection of the line segment $\overline{z\zeta}$ with the circle $|z| = R$. Then

$$|f(z) - f(\zeta)| \leq |f(z) - f(\eta)| + |f(\eta) - f(\zeta)| < \frac{\varepsilon}{3} + \frac{\varepsilon}{3} < \varepsilon. \qquad \square$$

According to Example 1 above, the two-valued function $\sqrt{z}$ has two (single-valued continuous) branches

$$f_1(z) = |z|^{\frac{1}{2}}\left(\cos\frac{\operatorname{Arg}z}{2} + i\sin\frac{\operatorname{Arg}z}{2}\right) \underset{(\text{def.})}{=} +\sqrt{z}(\sqrt{1} = +1) \quad \text{and}$$

$$f_2(z) = |z|^{\frac{1}{2}}\left(-\cos\frac{\operatorname{Arg}z}{2} - i\sin\frac{\operatorname{Arg}z}{2}\right) \underset{(\text{def.})}{=} -\sqrt{z}(\sqrt{1} = -1).$$

on $\mathbf{C} - (\infty, 0]$. Since $|f_1(z)| = f_2(z)| = |z|^{\frac{1}{2}} \to 0$ as $z \in \mathbf{C} - (\infty, 0) \to 0$, we may define $f_1(0) = f_2(0) = 0$ so that both of them are continuous on $\mathbf{C} - (\infty, 0)$ with points on $(\infty, 0)$ as their singular points (see (2.2.5)).

Example 6. Let $f(z) = \sqrt{z}$ be any one of the two branches of the two-valued function $\sqrt{z}$ on $\mathbf{C} - (\infty, 0]$. Let the angular (closed) domain $\Omega = \{z \in \mathbf{C} | z = 0 \text{ or } |\operatorname{Arg}z| \leq \alpha\}$, where $0 < \alpha < \pi$. See Fig. 2.3. Then

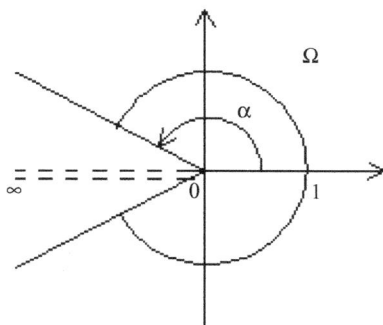

Fig. 2.3

(1) $f(z)$ is uniformly continuous on Ω.
(2) $f(z)$ is not uniformly continuous on $\mathbf{C} - (\infty, 0]$.

Proof. (1) For z and ζ in the set $D = \Omega \cap \{|z| \geq 1\}$,

$$|f(z) - f(\zeta)| = \frac{|z - \zeta|}{|\sqrt{z} + \sqrt{\zeta}|}.$$

As far as the denominator is concerned, observe that $|\sqrt{z} + \sqrt{\zeta}|^2 \geq 2(1 + \mathrm{Re}\sqrt{z}\overline{\sqrt{\zeta}})$. Let $\sqrt{z} = |z|^{\frac{1}{2}} e^{\frac{i\mathrm{Arg}\, z}{2}}$ and $\sqrt{\zeta} = |\zeta|^{\frac{1}{2}} e^{\frac{i\mathrm{Arg}\, \zeta}{2}}$. Since, $|\mathrm{Arg}\, z| \leq \alpha$, $|\mathrm{Arg}\, \zeta| \leq \alpha$, and $|\mathrm{Arg}\, z - \mathrm{Arg}\, \zeta| \leq 2\alpha$, we have

$$1 + \mathrm{Re}\ \sqrt{z}\overline{\sqrt{\zeta}}$$

$$= 1 + |z|^{\frac{1}{2}} |\zeta|^{\frac{1}{2}} \mathrm{Re}\ e^{i(\mathrm{Arg}\, z - \mathrm{Arg}\, \zeta)/2}$$

$$= 1 + |z|^{\frac{1}{2}} |\zeta|^{\frac{1}{2}} \cos \frac{1}{2}(\mathrm{Arg}\, z - \mathrm{Arg}\, \zeta)$$

$$\geq 1 + \cos \frac{2\alpha}{2} = 1 + \cos \alpha = 2 \cos^2 \frac{\alpha}{2}$$

$$\Rightarrow |\sqrt{z} + \sqrt{\zeta}|^2 \geq 2 \times 2 \cos^2 \frac{\alpha}{2} = 4 \cos^2 \frac{\alpha}{2}$$

$$\Rightarrow |\sqrt{z} + \sqrt{\zeta}| \geq 2 \cos \frac{\alpha}{2}$$

$$\Rightarrow |f(z) - f(\zeta)| \leq M|z - \zeta|, \quad \text{where } M = \frac{1}{2 \cos \frac{\alpha}{2}} \text{ and } z,\ \zeta \in D.$$

Hence $f(z)$ is uniformly continuous on D.

On the other hand, $\Omega \cap \{|z| \leq 1\}$ is compact on which $f(z)$ is uniformly continuous.

Since $\Omega \cap \{|z| \geq 1\}$ and $\Omega \cap \{|z| \leq 1\}$ have a nonempty common boundary along $|z| = 1$, $f(z)$ would be uniformly continuous on Ω (see the argument in Example 5).

(2) Suppose $f(z)$ is uniformly continuous on $\mathbf{C} - (\infty, 0]$. Fix a boundary point $x_0 < 0$. Then, for any sequence $z_n \in \mathbf{C} - (\infty, 0] \to x_0$, it follows that $f(z_n)$ always converges to the same limit (see Exercise A (15)(b)). Now, choose $z_n = x_0 + \frac{i}{n}$ and $z'_n = x_0 - \frac{i}{n}$ for $n \geq 1$. Then $f(z_n) \to |x_0|^{\frac{1}{2}} i$ (or $-|x_0|^{\frac{1}{2}} i$), while $f(z'_n) \to -|x_0|^{\frac{1}{2}} i$ (or $|x_0|^{\frac{1}{2}} i$), a contradiction.   □

Exercises A

(1) Prove (2.2.2) in detail.

(2) Prove (2.2.3) in detail.

(3) Prove (2.2.4) in detail.

(4) Prove (2.2.8) in detail.

(5) Show that the function $f : \mathbf{C} \to \mathbf{C}$ defined by

$$f(z) = \begin{cases} 0, & |z| \text{ is a rational number} \\ 1, & \text{otherwise} \end{cases}$$

is discontinuous everywhere in $\mathbf{C}$.

(6) Define f by

$$f(z) = \begin{cases} 0, & z = 0 \\ \dfrac{(\operatorname{Re} z^2)^2}{|z|^2}, & z \neq 0 \end{cases}.$$

Show that f is continuous at 0 and hence, is continuous everywhere in $\mathbf{C}$.

(7) For each of the following multiple-valued functions, do the following questions.

   (i) Find all possible branches, including domains of definition and their singular points;

   (ii) Discuss the uniform continuity of each branch.

     (a) $f(z) = \arg(1 - 2z)$;   (b) $f(z) = \arg iz^2$;   (c) $f(z) = \arg z^3$;

     (d) $f(z) = \sqrt{1 - 2z}$;   (e) $f(z) = \sqrt[3]{z}$;   (f) $f(z) = \arg \dfrac{z+i}{z-i}$;

     (g) $f(z) = \sqrt{\dfrac{z+i}{z-i}}$;   (h) $f(z) = \sqrt{z^2 - 1}$.

(8) Construct a homeomorphism mapping $|z| < 1$ onto $\mathbf{C}$.

(9) Let $A \subseteq \mathbf{C}$ be a compact set. Suppose a continuous function $f : A \to \mathbf{C}$ satisfies the condition that $f(z) \neq 0$ for all $z \in A$. Show that the image $f(A)$ lies outside of some closed disk $|w| \leq \delta$.

(10) If a continuous function $f : \mathbf{C} \to \mathbf{C}$ has the property that $\lim_{|z| \to \infty} f(z) = 0$, show that there exists a point $z_0 \in \mathbf{C}$ so that $|f(z)| \leq |f(z_0)|$ for all $z \in \mathbf{C}$.

(11) Let Ω be a domain and $f : \Omega \to \mathbf{C}$ be a continuous function satisfying $f(z)^2 = z$ for all $z \in \Omega$. Suppose $f(z)$ is never a pure imaginary number for each $z \in \Omega$. Show that either $f(z) = +\sqrt{z}$ on Ω or $f(z) = -\sqrt{z}$ on Ω.

(12) Discuss the continuity and uniform continuity of each of the following functions.

(a) $\dfrac{z}{1+z^2}$. (b) $\dfrac{z^3}{1+z^2}$. (c) $\dfrac{z^5}{z^3-1}$. (d) $\dfrac{z^4-z}{z^3+1}$.

(13) Suppose $f : \mathbf{C} \to \mathbf{C}$ is uniformly continuous. Show that, there exist constants $M > 0$ and $N > 0$ so that

$$|f(z)| \leq M|z| + N, \quad z \in \mathbf{C}.$$

Give an example to show that this is not a sufficient condition for uniform continuity.

(14) Suppose $f : \mathbf{C} \to \mathbf{C}$ is a continuous function satisfying $\lim_{|z| \to \infty} f(z) = 0$. Show that f is uniformly continuous on $\mathbf{C}$.

(15) Suppose $f : A \to \mathbf{C}$ is uniformly continuous.

(a) Suppose $z_n \in A, n \geq 1$, is a Cauchy sequence (so that its limit is in $\overline{A}$ but may not be in A itself). Show that $f(z_n), n \geq 1$, is Cauchy in $\mathbf{C}$.

(b) Let $z_0 \in A'$. Show that

$$\lim_{z \in A \to z_0} f(z)$$

exists. This limit may be defined as $f(z_0)$.

(c) There exists a unique continuous function $g : \overline{A} \to \mathbf{C}$ so that $g(z) = f(z)$ for all $z \in A$.

(d) Show that the extended function g in (c) is uniformly continuous on $\overline{A}$.

(16) Operational properties of uniform continuity.

(a) Suppose f and g are uniformly continuous on A. Then:

(i) $f \pm g$ are uniformly continuous on A;

(ii) Raise an example to show that fg is not necessarily uniformly continuous. But it is if both f and g are bounded on A;

(iii) In case f is bounded on A and $\inf_{z \in A} |g(z)| = \delta > 0$, then $\frac{f}{g}$ is uniformly continuous on A.

(b) Suppose $f : A \to \mathbf{C}$ and $g : B \subseteq f(A) \to \mathbf{C}$ are uniformly continuous. Then, so is $g \circ f$ on A.

(c) $f : A \to \mathbf{C}$ is uniformly continuous $\Leftrightarrow$ both $\operatorname{Re} f$ and $\operatorname{Im} f$ are so on A. $\Leftrightarrow \overline{f}$ is so on A. $\Rightarrow |f|$ is also uniformly continuous on A but the converse is not true.

(17) Let Ω be a domain. Suppose $f : \Omega \to \mathbf{C}$ is a continuous function satisfying $e^{f(z)} = 1$ for all $z \in \Omega$. Show that f is a constant function whose unique value could be any one in the set $\{2n\pi i \,|\, n = 0, \pm 1, \pm 2, \ldots\}$.

Exercises B

(1) Suppose a continuous function $f : \mathbf{C} \to \mathbf{C}$ satisfies the following conditions:

(i) $\lim_{z \to \infty} f(z) = \infty$; and

(ii) f is a *local homeomorphism*, i.e., for each $z_0 \in \mathbf{C}$, there exists a neighborhood of z_0 on which f is a homeomorphism.

Show that $f(\mathbf{C}) = \mathbf{C}$ and $f : \mathbf{C} \to \mathbf{C}$ is a (global) homeomorphism.

(2) Let $p(z)$ be a polynomial of degree $n(n \geq 1)$. Suppose $z_0 \in \mathbf{C}$ such that $p(z_0) \neq 0$. Then for every open neighborhood O of z_0, there exist two points z_1 and z_2 in O so that

$$|p(z_1)| < |p(z_0)| < |p(z_2)|.$$

This means that $|p(z)|$ does not obtain its local maximum and minimum at z_0. Try to do the following steps.

(a) Suppose $p(z) = a_0 + a_n(z - z_0)^n, a_0 \neq 0, a_n \neq 0$. Setting $z - z_0 = re^{i\theta}$, $0 \leq \theta \leq 2\pi$, we have

$$p(z) = a_0 \left\{ 1 + \left| \frac{a_n}{a_0} \right| r^n e^{i(n\theta + \varphi)} \right\}, \qquad \varphi = \operatorname{Arg} \frac{a_n}{a_0}.$$

Try to choose suitable θ_1 and θ_2, and the corresponding z_1 and z_2.

(b) In general, rewrite $p(z)$ as $a_0 + a_k(z - z_0)^k(1 + q(z))$, where $a_k \neq 0$ and $q(z)$ is a polynomial so that $q(z_0) = 0$. Try to use (a).

Then, prove the *fundamental theorem of algebra*: Any polynomial of positive degree has at least one zero.

2.3. Uniform Convergence of a Sequence or Series of Functions

Let $f_n : A \to \mathbf{C}$, $n \geq 1$, be a sequence of functions defined on the same set A. If to each fixed point $z \in A$, $f_n(z)$ converges, as a complex sequence, to a limit, denoted by $f(z)$. Then, we say that the sequence, f_n, $n \geq 1$ *converges pointwise* on A to the *limit function* $f : A \to \mathbf{C}$, denoted as

$$\lim_{n \to \infty} f_n(z) = f(z) \text{ on } A \quad \text{or} \quad f_n \to f \text{ on } A. \tag{2.3.1}$$

This means that, for each $z \in A$ and $\varepsilon > 0$, there exists a positive integer $n_0 = n_0(z, \varepsilon)$, depending on both z and ε, so that if $n \geq n_0$, then $|f_n(z) - f(z)| < \varepsilon$. We illustrate it by the following example.

Example 1. Show that $f_n(z) = nz^n$, $n > 1$, converges pointwise on $|z| < 1$ to the limit function $f(z) = 0$. What happens if on $|z| \leq r$ where $0 < r < 1$?

Solution. Fix z, $|z| < 1$. In order to have $|f_n(z) - f(z)| = n|z|^n < \varepsilon$, we may choose $n > 1 + \frac{2|z|^2}{\varepsilon(1-|z|)^3}$ (since $n|z|^n < \frac{2|z|^2}{(n-1)(1-|z|)^3}$ if $|z| < 1$). Then, take n_0 to be the smallest positive integer greater than $1 + \frac{2|z|^2}{\varepsilon(1-|z|)^3}$. This n_0 depends on z and ε. Observe that $n_0 \to \infty$ if $|z| \to 1$.

If we restrict the domains of the definition of f_n, $n \geq 1$, to $|z| \leq r$, then $n|z|^n \leq nr^n < \varepsilon$ holds, if $n_0 \geq 1 + \frac{2r^2}{\varepsilon(1-r)^3}$. In this case, this n_0 depends only on ε and $n|z|^n < \varepsilon$ is valid *for all z* in $|z| \leq r$, if $n \geq n_0$. We say that $f_n(z) \to f(z) = 0$ *uniformly* on $|z| \leq r$. □

This section is divided into two subsections.

Section (1) Uniform convergences of a sequence of functions

For each $\varepsilon > 0$, if there exists a positive integer $n_0 = n_0(\varepsilon)$, *depending only on* ε, so that $|f_n(z) - f(z)| < \varepsilon$ for all $z \in A$ if $n \geq n_0$ then f_n is said to *converge uniformly on* A to the *limit function* f.

We have

Basic criteria for uniform convergence.

(1) f_n converges uniformly on A to f;
$\Leftrightarrow$ (2) (*Supnorm method*) Let

$$\|f_n - f\|_\infty = \sup_{z \in A} |f_n(z) - f(z)|;$$

Then

$$\lim_{n \to \infty} \|f_n - f\|_\infty = 0.$$

$\Leftrightarrow$ (3) (*Cauchy condition*) For each $\varepsilon > 0$, there exists a positive integer $n_0 = n_0(\varepsilon)$, so that

$$|f_m(z) - f_n(z)| < \varepsilon \quad \text{for all } z \in A \text{ if } m, n \geq n_0. \tag{2.3.2}$$

Readers should try to figure out what "f_n does not converge to f uniformly on A" means. Proofs are left as Exercise A (1). Of course, uniform convergence implies pointwise converges but not conversely. In Example 1, $\|f_n - f\|_\infty = n \to \infty$ so that f_n does not converge uniformly on $|z| < 1$ to $f = 0$, but only on every smaller disk $|z| \leq r < 1$ and hence, on every compact subset of $|z| < 1$ (why?).

One of the main features of uniform convergence is to construct continuous functions from the known ones. This is *the uniform convergence and continuity.* Suppose each function $f_n : A \to \mathbf{C}$ is continuous on A and the sequence f_n, $n \geq 1$, *converges uniformly* on *every compact subset* of A to a function f. Then $f : A \to \mathbf{C}$ is continuous. $\tag{2.3.3}$

Sketch of Proof.

Fix any point $z_0 \in A$ and $\rho > 0$ so that $f_n \to f$ uniformly on $|z - z_0| \leq \rho$. Note that

$$|f(z) - f(z_0)| \leq |f(z) - f_n(z)| + |f_n(z) - f_n(z_0)| + |f_n(z_0) - f(z_0)|.$$

If z is such that $|z - z_0| \leq \rho$ and n is large enough, then $|f(z) - f_n(z)| < \frac{\varepsilon}{3}$ and $|f_n(z_0) - f(z_0)| < \frac{\varepsilon}{3}$ for a given $\varepsilon > 0$. Take $\delta > 0$ small enough, say $0 < \delta < \rho$ and n is as before, then $|f_n(z) - f_n(z_0)| < \frac{\varepsilon}{3}$ if $|z - z_0| < \delta$. $\quad \square$

Section (2) Uniform convergence of a series of functions

Let $f_n : A \to \mathbf{C}$, $n \geq 1$, be a sequence of functions. If the partial sums $S_n(z) = \sum_{k=1}^n f_k(z)$, $n \geq 1$, converge uniformly on A to a function $f(z)$, then we say the series $\sum_{n=1}^\infty f_n(z)$ *converges uniformly* on A to the function $f(z)$, denoted as

$$\sum_{n=1}^\infty f_n(z) = f(z) \quad \text{or} \quad \sum f_n(z) \quad \text{uniformly on } A. \tag{2.3.4}$$

In addition to those stated in Section (2) with f_n replaced by S_n, we also have

Criteria for uniform convergence of series of functions.

(1) **(Weierstrass)** If $\sum \|f_n\|_\infty$ converges, then $\sum f_n(z)$ converges uniformly (and absolutely) on A.

(2) **(Abel)** Suppose

  (i) $\sum f_n(z)$ converges uniformly on A; and
  (ii) $g_n : A \to \mathbf{R}$, $n > 1$, is a monotone sequence of real-valued functions (i.e., $g_n \geq g_{n+1}$ or $g_n \leq g_{n+1}$ for $n \geq 1$) satisfying $\sup_{n\geq1} \|g_n\|_\infty < \infty$.

  Then $\sum f_n(z)g_n(z)$ converges uniformly on A.

(3) **(Dirichlet)** Suppose

  (i) the sequence $f_n : A \to \mathbf{C}$, $n \geq 1$ satisfying $\sup_{n\geq1}\|f_1+\cdots+f_n\|_\infty < \infty$ and
  (ii) $g_n : A \to [0,\infty)$, $n \geq 1$, is a nonincreasing sequence of positive functions satisfying $\lim_{n\to\infty}\|g_n\|_\infty = 0$.

Then $\sum f_n(z)g_n(z)$ converges uniformly on A. $\hspace{2cm}$ (2.3.5)

Note that both Abel's and Dirichlet's methods are usually used to test those series that converges uniformly but not absolutely. Refer to Exercises B of Sec. 1.7 for complex series.

Sketch of Proof

(1) This follows by using Cauchy criterion.
  For (2) and (3), we need the following

  Abel's summation law: For $a_j, b_j \in \mathbf{C}$, $1 \leq j \leq m$, let $B_j = b_1 + b_2 + \cdots + b_j, 1 \leq j \leq m$. Then:
  (a) $\sum_{j=1}^m a_j b_j = \sum_{j=1}^{m-1}(a_j - a_{j+1})B_j + a_m B_m$.
  (b) In case a_j, $1 \leq j \leq m$, is monotone and $|B_j| \leq M$ for $1 \leq j \leq m$, it follows from (a) that

$$\left| \sum_{j=1}^m a_j b_j \right| \leq M(|a_1| + 2|a_m|). \hspace{1cm} (2.3.6)$$

Proofs are left as Exercises A(2).

(2) By Cauchy criterion, for $\varepsilon > 0$, there exists n_0 so that $\|f_n + \cdots + f_{n+p}\|_\infty < \varepsilon$ for $n \geq n_0$ and all $p \geq 0$. By (b) in (2.3.6), if

$n \geq n_0$ and $p \geq 0$

$$\|f_n g_n + \cdots + f_{n+p} g_{n+p}\|_\infty \leq \varepsilon(\|g_n\|_\infty + 2\|g_{n+p}\|_\infty).$$

(3) Let $M = \sup_{n \geq 1} \|f_1 + \cdots + f_n\|_\infty$. Then

$$\|f_n g_n + \cdots + f_{n+p} g_{n+p}\|_\infty \leq 2M(\|g_n\|_\infty + 2\|g_{n+p}\|_\infty) \to 0 \quad \text{as } n \to \infty. \square$$

Better illustrations are the examples.

Example 2. Show that

$$\sum_{n=1}^{\infty} \frac{(-1)^{n-1}}{z+n}$$

converges uniformly on every compact subset of $\Omega = \mathbf{C} - \{n \,|\, n = 0, -1, -2, \ldots\}$ but converges absolutely nowhere.

Solution. Let $z = x + iy$. Then

$$\frac{1}{z+n} = \frac{x+n}{(x+n)^2 + y^2} - i\frac{y}{(x+n)^2 + y^2}, \quad z \in \Omega.$$

It is sufficient to consider that uniform convergence of $\sum \mathrm{Re}\left(\frac{(-1)^{n-1}}{z+n}\right)$ and $\sum \mathrm{Im}\left(\frac{(-1)^{n-1}}{z+n}\right)$ on every closed disk $K : |z - z_0| \leq r$, where $K \subseteq \Omega$. Note that every compact subset of Ω can be covered by finitely many such disks.

The real part series: Observe that the function $\frac{t}{t^2+y^2}$ is monotone decreasing in t if $t \geq |y|$. Hence, if n is large enough so that $t = x+n \geq |y|$, by Leibniz test (see Exercises B (3)(f) of Sec. 1.7),

$$\sum_{n=n_0}^{\infty} \frac{(-1)^{n-1}(x+n)}{(x+n)^2 + y^2} \quad \text{converges on } K \text{ to a function } u(z),$$

where n_0 is a certain positive integer. On the other hand, if $n \geq n_0$, then

$$\left| u(z) - \sum_{k=n_0}^{n} \frac{(-1)^{k-1}(x+k)}{(x+k)^2 + y^2} \right| \leq \frac{|x+n+1|}{(x+n+1)^2 + y^2}$$

$$\leq \frac{1}{|x+n+1|} \leq \frac{1}{n+1-|Re z_0| - r}$$

$$\to 0 \quad \text{as } n \to \infty \text{ (independent of points in } K).$$

Therefore, the real part series converges uniformly on K.

The imaginary part series: If $z \in K$, then

$$\left| \frac{-(-1)^{n-1}y}{(x+n)^2 + y^2} \right| \leq M \cdot \frac{1}{n^2}, \quad n \geq 1 \quad \text{and} \quad M > 0 \text{ a constant.}$$

By Weierstrass test, the imaginary part series converges uniformly on K.
Since

$$\frac{\left| \frac{(-1)^{n-1}}{z+n} \right|}{\frac{1}{n}} = \frac{n}{\sqrt{(x+n)^2 + y^2}} \to 1 \quad \text{as } n \to \infty,$$

by comparison test (see Exercise B (3)(a) of Sec. 1.7), the series $\sum \frac{(-1)^{n-1}}{z+n}$ does not converge absolutely anywhere in Ω.

Example 3. Show that

$$\sum_{n=0}^{\infty} \frac{z}{(1+z^2)^n}$$

converges absolutely on the set $|\operatorname{Arg} z| \leq \frac{\pi}{4}$ but not uniformly.

Solution. Let $z = re^{i\theta}$, $|\theta| \leq \frac{\pi}{4}$ and $r > 0$. Then, $1 + z^2 = (1 + r^2 \cos 2\theta) + ir^2 \sin 2\theta$ and hence, $|1 + z^2|^2 = 1 + 2r^2 \cos 2\theta + r^4 \geq 1 + r^4 \geq r^4$. Therefore,

$$\sum_{n=0}^{\infty} \left| \frac{z}{(1+z^2)^n} \right| \leq \sum_{n=0}^{\infty} \frac{r}{(1+r^4)^{\frac{n}{2}}} \quad \text{which converges if } r > 0.$$

So the series converges absolutely at every point of $|\operatorname{Arg} z| \leq \frac{\pi}{4}$. Note that this series also converges absolutely at $z = 0$ and points z where $|1 + z^2| > 1$.

Since the positive real axis is contained in the set $|\operatorname{Arg} z| \leq \frac{\pi}{4}$, all we need to do is to show that if $z = x > 0$, the series $\sum \frac{x}{(1+x^2)^n}$ does not converge uniformly. It is known that

$$\sum_{n=0}^{\infty} \frac{x}{(1+x^2)^n} = \frac{1+x^2}{x}, \quad x > 0$$

$$\Rightarrow \sum_{n=0}^{k} \frac{x}{(1+x^2)^n} - \frac{1+x^2}{x} = -\frac{1}{x(1+x^2)^k} = f_k(x), \quad x > 0.$$

Now, $\|f_k\|_{\infty} = \sup_{x>0} |f_k(x)| = \infty$ for $k \geq 1$, so $\sum \frac{x}{(1+x^2)^n}$ does not converge uniformly on $x > 0$.

Exercises A

(1) Prove (2.3.2) in detail.

(2) Prove (2.3.6) in detail.

(3) Find the set of uniform convergence for each of the following sequence of functions or series of functions.

(a) $f_n(z) = \dfrac{1}{1 + z^n}.$ (b) $f_n(z) = \dfrac{z^n}{1 + z^{2n}}.$

(c) $f_n(z) = \dfrac{z^n}{1 + n^2 z}.$ (d) $\displaystyle\sum_{n=1}^{\infty} \dfrac{(-1)^{n-1}}{n} \cdot \dfrac{z^n}{1 + z^n}.$

(e) $\displaystyle\sum_{n=1}^{\infty} \dfrac{1}{n^2}\left(z^2 + \dfrac{1}{z^2}\right).$ (f) $\displaystyle\sum_{n=0}^{\infty} e^{-nz}.$

(g) $\displaystyle\sum_{n=1}^{\infty} e^{-z\log n}.$ (h) $\displaystyle\sum_{n=1}^{\infty} \dfrac{z^n}{1 + z^{2n}}.$

(4) Show that the series $\sum_{n=1}^{\infty} \frac{z^n}{n}$ converges uniformly on $|z| < \delta (0 \le \delta < 1)$, also uniformly on the open interval $(-1, 0)$ but $\sum_{n=1}^{\infty} \frac{|z|^n}{n}$ does converge on $(-1, 0)$, yet not uniformly. This shows that *Weierstrass test is a sufficient but not a necessary criterion for uniform convergence.*

(5) Show that

$$\sum_{n=1}^{\infty} \frac{(-1)^n z}{(1 + z^2)^n}$$

converges both absolutely and uniformly on the set $|\operatorname{Arg} z| \le \frac{\pi}{4}$, yet $\sum_{n=1}^{\infty} \frac{|z|}{|1+z^2|^n}$ does not converge uniformly on the same set. What happens to $\sum_{n=1}^{\infty} \frac{z}{(1+z^2)^n}$ on the same set?

(6) Show that

$$\sum_{n=1}^{\infty} \frac{z^n}{(1 - z^n)(1 - z^{n+1})}$$

converges uniformly on every compact subset of $|z| < 1$ to the function $\frac{z}{(1-z)^2}$, and uniformly on every compact subset of $|z| > 1$ to $\frac{1}{(1-z)^2}$.

(7) Show that $f_n(z) = ze^{-\frac{1}{2}n^2 z^2}$ converges uniformly on the real axis but does not converge uniformly on any disk $|z| \le r\ (r > 0)$.

2.4.   Curves

In short, a continuous complex-valued function of a single real variable is called a (planar) **curve** or **path**. It is a supplementary tool in complex analysis. We need it in complex integration, winding number, geometric interpretation of conformal mapping and approximating value along a curve, etc.

This section is divided into four subsections.

Section (1) Curve and its related terminologies

Let $[a, b]$, $-\infty < a < b < \infty$, be a finite closed interval on the real line $\mathbf{R}$. A pair of continuous functions $x = x(t), y = y(t) : [a, b] \to \mathbf{R}$ is said to define a (continuous) **curve** γ (Fig. 2.4) with parametric representation

$$z(t) = x(t) + iy(t) : [a, b] \to \mathbf{C} \tag{2.4.1}$$

on the plane $\mathbf{C}$. It is also denoted as

$$\gamma(t) = x(t) + iy(t).$$

Some related terminologies are as follows.

parameter: t (may be considered as time).
parameter interval: $[a, b]$.
point on the curve: t or $z(t), t \in [a, b]$.
locus of the curve: the point set $\{z(t) \,|\, a \le t \le b\}$, both compact and connected, also called a curve for simplicity.
initial point: $z(a)$.
terminal point: $z(b)$.

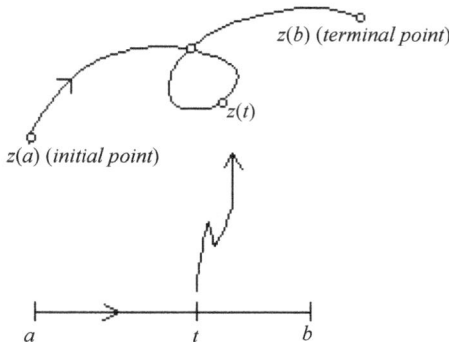

Fig. 2.4

orientation or *direction*: the direction induced by the increasing of t from a to b. For details, see Exercise B (1).

inverse curve $-\gamma$ of γ: $z(t) = x(a+b-t) + iy(a+b-t)$, $t \in [a,b]$, with direction induced by decreasing t from b to a.

multiple point: a point $z(t_1) = z(t_2)$ where $a \leq t_1 < t_2 \leq b$.

closed curve: $z(a) = z(b)$.

simple or *Jordan curve* (or *arc*): $z(t_1) \neq z(t_2)$ whenever $a < t_1 < t_2 < b$.

simple or *Jordan closed curve*: $z(t_1) = z(t_2)$ only if t_1 is one of a and b and t_2 is the other one of a and b.   (2.4.2)

It can be shown that any simple closed curve is homeomorphic to the unit circle $|z| = 1$ (see Ref. [65]). Note that $z_1(t) = a + re^{it}$, $t \in [0, 2\pi]$, and $z_2(t) = a + re^{it}$, $t \in [0, 3\pi]$ are different curves even though they have the same locus, the circle $|z-a| = r$.

For examples:

directed line segment $\overrightarrow{ab}$: $g(t) = a + (b-a)t, t \in [0,1]$ with the inverse curve $\overrightarrow{ba} : z(t) = b + (a-b)t, t \in [0,1]$. Refer to (1.4.3.1). *directed polygonal curve*: Let $a_1, \ldots, a_n$ be n distinct points in **C**. Then

$$z(t) = a_k + (a_{k+1} - a_k)(t - k + 1), \ t \in [k-1, k], \ k = 1, 2, \ldots, n-1$$

is a directed polygonal curve, denoted as $\overrightarrow{a_1 a_2} + \overrightarrow{a_2 a_3} + \cdots + \overrightarrow{a_{n-1} a_n} = \sum_{k=1}^{n-1} \overrightarrow{a_k a_{k+1}}$. See Fig. 2.5:

directed circle: $z(t) = a + r(\cos t + i \sin t) = a + re^{it}$, $t \in [0, 2\pi]$, in anti-clockwise direction, while the inverse curve is $z(t) = a + re^{-it}$, $t \in [-2\pi, 0]$, in clockwise direction. See Fig. 2.6.

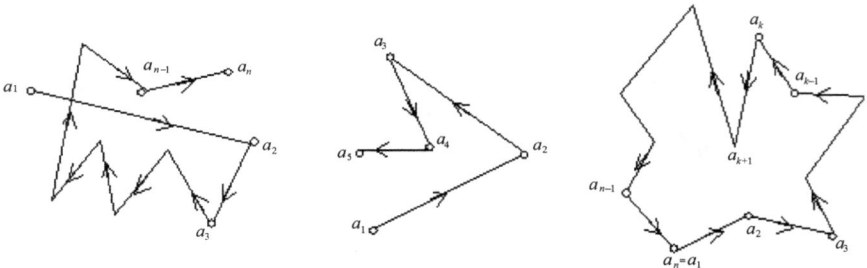

Fig. 2.5

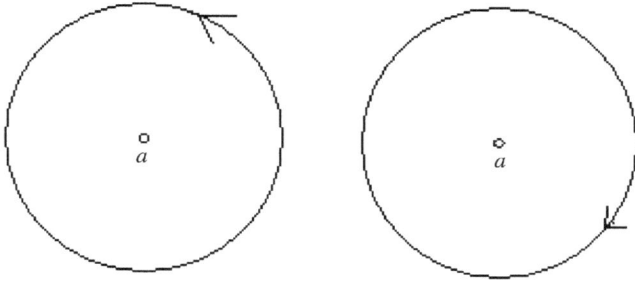

Fig. 2.6

Remark 1.   (Curve in $\mathbf{C}^*$). Fix a and b in the extended real number system (see Appendix A) so that $-\infty \le a < b \le \infty$.

A continuous function in the generalized sense (see (2.2.2))

$$z(t) = x(t) + iy(t) : [a, b] \to \mathbf{C}^* \tag{2.4.3}$$

is said to define a *curve* γ^* in $\mathbf{C}^*$. Conventionally, the locus of points $z(t)$, $t \in [a, b]$, in $\mathbf{C}^*$ is also called a *curve* and is denoted as γ^* for simplicity. Terminologies presented in (2.4.2) are still valid in this case.

If there exists a $t_0 \in [a, b]$ so that $\lim_{t \to t_0} z(t) = z(t_0) = \infty$, then γ^* is said to pass the point at infinity ∞.

Under stereographic projection (see Sec. 1.6), the preimage of a curve γ^* passing ∞ is a curve passing the north pole which, in turn, is mapped into a curve passing the south pole under the rotation of the sphere with x-axis as axis and through the angle $180°$. In the complex plane $\mathbf{C}$ (or $\mathbf{C}^*$), this is equivalent to say that, under the mapping $z \to \frac{1}{z}$, γ^* is pulled back to a curve passing the origin o. Therefore, in this text, the tangent at ∞ of a curve passing ∞ or the angle at ∞ between two such curves is defined as the one at o of the image curve or curves under $z \to \frac{1}{z}$, respectively.

If γ^* passes through ∞, then each component of the complement $\mathbf{C}^* - \gamma^*$ of γ^* in $\mathbf{C}^*$ does not contain ∞ anymore.

Suppose γ^* passes through ∞ and $\gamma^* - \{\infty\}$ is contained in the domain A of definition of a function $f : A \to \mathbf{C}$. If the limit

$$\lim_{z \in \gamma^* \to \infty} f(z) = w_0, \quad w_0 \in \mathbf{C}^* \tag{2.4.4}$$

exists, then call w_0 *the approximate value of f at ∞ along γ^**. $\square$

Notice that one may define a curve on the interval (a, b) or $[a, b)$ or $(a, b]$ if needed. For instance, on (a, b), a curve is open ending, namely, no end points. We will feel free to use these kinds of curves later if necessary.

Section (2) Rectifiable curve

Let a plane curve γ be defined as in (2.4.1). Consider a partition $P : a = t_0 < t_1 < \cdots < t_{n-1} < t_n = b$ of $[a, b]$ and construct the inscribed polygonal curve $\overline{z_0 z_1} + \cdots + \overline{z_{n-1} z_n}$ where $z_j = z(t_j), 0 \le j \le n$. See Fig. 2.7. This polygonal curve has total length $l(P) = \sum_{j=1}^{n} |z_j - z_{j-1}|$.

If for all partitions P of $[a, b]$, the corresponding lengths $l(P)$ have an upper bound, then γ is called *rectifiable* and its *length* is defined as

$$l(\gamma) = \sup_P l(P). \tag{2.4.5}$$

Otherwise, γ is said to be *nonrectifiable*.

Suppose γ is rectifiable and $z(t)$ is *not* a constant function on any non-degenerated subinterval of $[a, b]$. Let $s(t)$ be the length of γ restricted to the subinterval $[a, t]$, $a \le t \le b$. Then

$$s = s(t) : [a, b] \rightarrow [0, l(\gamma)]$$

is a strictly increasing continuous function called the *length function* of γ. Its inverse function $t = t(s) : [0, l(\gamma)] \rightarrow [a, b]$ is also a strictly increasing continuous function. The composite function

$$\tilde{z}(s) = z(t(s)) : [0, l(\gamma)] \rightarrow \mathbf{C} \tag{2.4.6}$$

is called the *length parameter equation (representation)* of γ. It is uniquely defined.

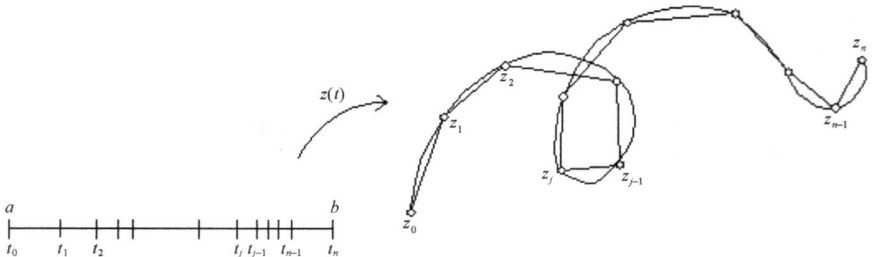

Fig. 2.7

Section (3) Differentiable curve

In case both $x'(t_0)$ and $y'(t_0)$ exist as finite real numbers for some $t_0 \in [a, b]$, then designate

$$z'(t_0) = x'(t_0) + iy'(t_0) \qquad (2.4.7)$$

as the *derivative* of $z(t)$ at t_0 and $z(t)$ is called *differentiable* at t_0. Note that if $t_0 = a$ or b, both $x'(t_0)$ and $y'(t_0)$ are right or left derivatives, respectively; if γ is closed, the right and left derivatives should coincide at a and b. $z'(t)$ can be defined directly as

$$z'(t) = \lim_{h \to 0} \frac{z(t+h) - z(t)}{h} = \lim_{h \to 0} \frac{\{x(t+h) - x(t)\} + i\{y(t+h) - y(t)\}}{h}.$$
$$(2.4.8)$$

If $z'(t) \neq 0$, we say the curve γ has a *tangent* at t (or at $z(t)$):

tangent vector: $z'(t)$.
direction of the tangent vector: Arg $z'(t)$.

Figure 2.8 says everything about these concepts.

If two curves γ_1 and γ_2, with respective equations $z_1(t)$ and $z_2(t)$, both have tangents at their point of intersection t_0, then the *angle* from γ_1 to γ_2 at t_0 is defined as Arg $z_2'(t_0) -$ Arg $z_1'(t_0)$ (see (1.4.3.4)).

If $z'(t)$ both exists and is continuous everywhere on $[a, b]$, then $z(t)$ is said to define a (*continuously*) *differentiable curve* γ on **C**. In addition, if $z'(t) \neq 0$ for all $t \in [a, b]$, then γ is called a *smooth* or *regular curve*. For instance, a circle is such a curve. A smooth curve γ has continuous tangent vectors $z'(t)$.

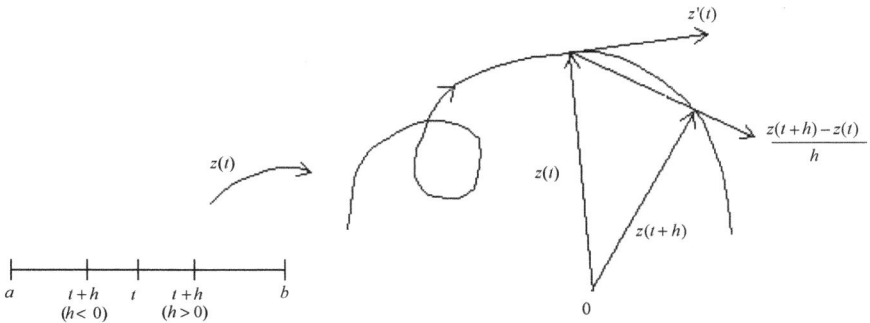

Fig. 2.8

Suppose:

(1) Except for finitely many points $t_1, \ldots, t_n$ on $[a, b]$, say $a \leq t_1 < \cdots < t_n \leq b$, the restriction of γ to each of the subintervals (a, t_1), $(t_1, t_2), \ldots, (t_n, b)$ is a differentiable (or smooth) curve;

(2) $z(t)$ is still continuous at these exceptional points $t_1, \ldots, t_n$. Also, for each $j, 1 \leq j \leq n$, left and/or right derivatives of $z(t)$ exist at t_j and are equal to the respective left and/or right limits of $z'(t)$ at t_j (which are not equal to zero in the case of a smooth curve), namely

$$\lim_{h \to 0^-} \frac{z(t_j + h) - z(t_j)}{h} = \lim_{h \to 0^-} z'(t_j + h) \underset{(\text{def.})}{=} z'(t_j^-) \ (\neq 0),$$

(2.4.9)

$$\lim_{h \to 0^+} \frac{z(t_j + h) - z(t_j)}{h} = \lim_{h \to 0^+} z'(t_j + h) \underset{(\text{def.})}{=} z'(t_j^+) \ (\neq 0).$$

Then $z(t)$ is said to define a *piecewise differentiable* (or *piecewise smooth (regular)*) *curve* γ. See Fig. 2.9.

A piecewise differentiable curve γ is rectifiable and has its length function: $s(t) = \int_a^t |z'(t)| dt = \int_a^t \sqrt{x'(t)^2 + y'(t)^2} dt$, $a \leq t \leq b$;

arc length element: $ds = \sqrt{dx^2 + dy^2} = |dz|$. (2.4.10)

Section (4) Simply and multiply connected domains (revisited)

Here, we try to redefine these two kinds of domains (see Sec. 1.8) via the concept of simple closed curve.

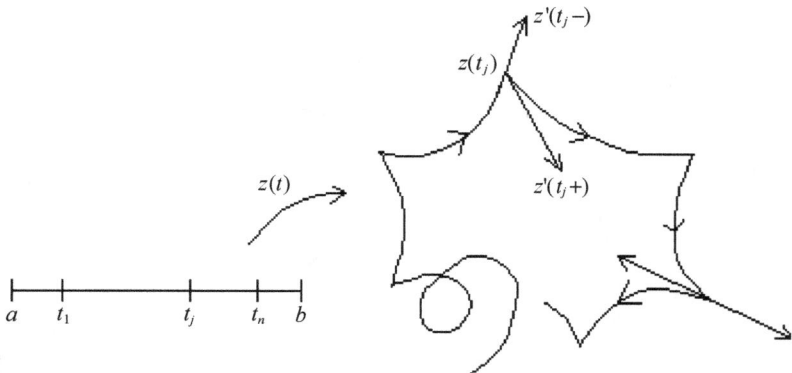

Fig. 2.9

Firstly, we have the famous and important

Jordan (1892) *Curve Theorem.* Suppose γ is a Jordan closed curve in $\mathbf{C}$. Then there exists two disjoint domains Ω_1 and Ω_2 satisfying:

(1) $\mathbf{C} - \gamma = \Omega_1 \cup \Omega_2$;
(2) exact one of Ω_1 and Ω_2 is a bounded set; and
(3) $\partial \Omega_1 = \partial \Omega_2 = \gamma$.

The bounded one is called the *interior* of γ and is denoted as $\text{Int}\,\gamma$, while the unbounded one the *exterior* of γ, denoted as $\text{Ext}\,\gamma$.   (2.4.11)

   For proof, see Refs. [23] and [65].

   Suppose Ω is a domain in $\mathbf{C}$. If, for each Jordan closed curve γ lying entirely inside Ω, Ω always contains the interior $\text{Int}\,\gamma$ of γ, then Ω is called a *simply connected domain*, otherwise *multiply connected*. $\text{Int}\,\gamma$ is always simply connected for any Jordan closed curve γ. Intuitively, a simply connected domain is a domain without a hole or some point or set within removed.

Remark 2.   (Relations between domains in C and C*) Domains in $\mathbf{C}$ might be either bounded or unbounded.

   For an unbounded domain Ω in $\mathbf{C}$, as a subset of $\mathbf{C}^*$, there are two cases as follows.

(1) In case the set $\Omega^* = \Omega \cup \{\infty\}$ is still a domain (namely, a nonempty open connected set) in $\mathbf{C}^*$: Then, the boundary $\partial \Omega$ is a bounded set in $\mathbf{C}$ and Ω itself contains an open neighborhood of ∞, say a set $\{z \in C \,|\, |z| > r\}$.
(2) Otherwise, the boundary $\partial^* \Omega$ of Ω in $\mathbf{C}^*$ should pass the infinite point ∞ and the boundary $\partial \Omega$ of Ω in $\mathbf{C}$ is an unbounded set. In this case, as a subset of $\mathbf{C}^*$, Ω is still a domain.

Conversely, for a domain Ω^* in $\mathbf{C}^*$, as a subset $\Omega = \Omega^* - \{\infty\}$ of $\mathbf{C}$, there are three cases:

(1) Ω is a bounded domain: then $\infty \notin \Omega^*$ and $\Omega^* = \Omega$.
(2) Ω is an unbounded domain but its boundary $\partial \Omega$ in $\mathbf{C}$ is a bounded set: then $\infty \in \Omega^*$.
(3) Ω is an unbounded domain but its boundary $\partial \Omega$ in $\mathbf{C}$ is an unbounded set: then $\infty \in \partial^* \Omega$ (the boundary in $\mathbf{C}^*$) and $\infty \notin \Omega^*$ and hence, $\Omega^* = \Omega$. $\square$

Secondly, we have

The topological characteristic properties of a simply connected domain. Let Ω be a domain in $\mathbf{C}$. Then

    (1) Ω is simply connected;

$\Leftrightarrow$ (2) If Ω contains a Jordan closed polygonal curve γ, then it should contain the interior $\operatorname{Int}\gamma$;

$\Leftrightarrow$ (3) The boundary $\partial^*\Omega$ of Ω in $\mathbf{C}^*$ is connected;

$\Leftrightarrow$ (4) The complement $\mathbf{C}^* - \Omega$ of Ω in $\mathbf{C}^*$ is connected;

$\Leftrightarrow^*$(5) The fundamental group of Ω is zero.   (2.4.12)

In case $\Omega \subseteq \mathbf{C}$ is a *bounded* domain, in properties 3 and 4, it is just enough to take boundary and complement in $\mathbf{C}$; if *unbounded*, then it should be done in $\mathbf{C}^*$. It is property 5 that characterizes the simply connectedness of a domain being independent of the space wherein the domain situates. For proofs, see Refs. [65] and [57] or refer to (7.4.6). Later in Chap. 4, we will use analytic function to characterize if a domain is simply connected or not.

    Therefore, if the boundary $\partial^*\Omega$ of a domain Ω in $\mathbf{C}^*$ has n components (see (1.8.12)), where $n \geq 1$, it is called *n-connected*; otherwise, *infinitely connected*. Note that a 1-connected domain is a simply connected domain.

    For examples:

Simply connected domain: The interior $\operatorname{Int}\gamma$ of a Jordan closed curve γ is specifically called a *Jordan domain*. Figures 2.10(a)–2.10(f) are bounded simply connected domains but they are not Jordan ones.

    If Ω is an unbounded simply connected domain in $\mathbf{C}$, then its boundary $\partial^*\Omega$ in C^* should pass ∞. For instance, fix a point $z_0 \in \mathbf{C}$ and construct a Jordan curve γ connecting z_0 to ∞. Then, $\mathbf{C} - \gamma$ is such a domain. Figures 2.11(a)–2.11(d) are unbounded simply connected domains.

Doubly connected (*2-connected*) *domain*, also called a *ring domain*: Fix $0 < r < R < \infty$, then $r < |z - z_0| < R$ is such a domain. If γ is a Jordan closed curve, then both $\operatorname{Ext}\gamma - \{\infty\}$ and $\operatorname{Int}\gamma - \{z_0\}$, where $z_0 \in \operatorname{Int}\gamma$, are ring domains.

    Readers are urged to construct domains with their boundaries in $\mathbf{C}^*$ having countably and uncountably infinitely many components, respectively.

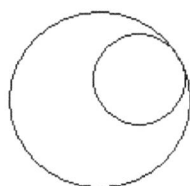

(a)the part between two
tangent circles

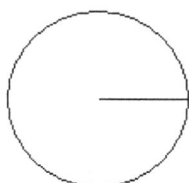

(b)a circle with a
radius deleted

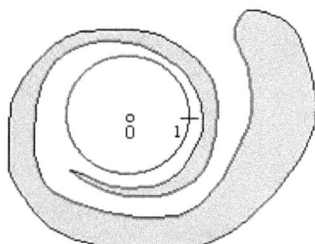

(c)the shaded part with
the spiral boundary
approaching $|z|=1$

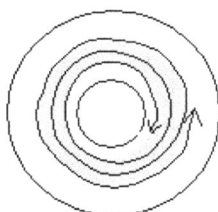

(d)the shaded part
whose boundary
consists of $|z|=1$,
$|z|=2$ and the spiral
$r=\dfrac{3}{2}+\dfrac{\tan^{-1}\theta}{\pi}$.

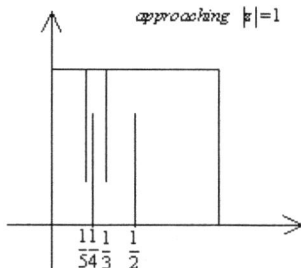

(e)$\Omega=(0,1)\times(0,1)-\{(x,y)\,|\,x=\dfrac{1}{2n},0<y\le\dfrac{3}{4},n\ge1\}$

$-\{(x,y)\,|\,x=\dfrac{1}{2n+1},\dfrac{1}{4}\le y<1,n\ge1\}$

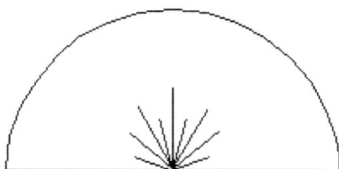

(f)$\Omega=\{z\,|\,0<Arg z<\pi,0<|z|<1\}-\{z\,|\,Arg z=\dfrac{q}{p},0<|z|<\dfrac{1}{p},p,q$ relatively prime positive integers$\}$

Fig. 2.10

Finally, we state

Brouwer's theorem on invariance of domain (1911). Suppose $\Omega\subseteq\mathbf{C}$ is a domain and $f:\Omega\to\mathbf{C}$ is a one-to-one and continuous mapping. Then

(1) $f(\Omega)$ is a domain in $\mathbf{C}$; and

(2) $f:\Omega\to f(\Omega)$ is a homeomorphism.   (2.4.13)

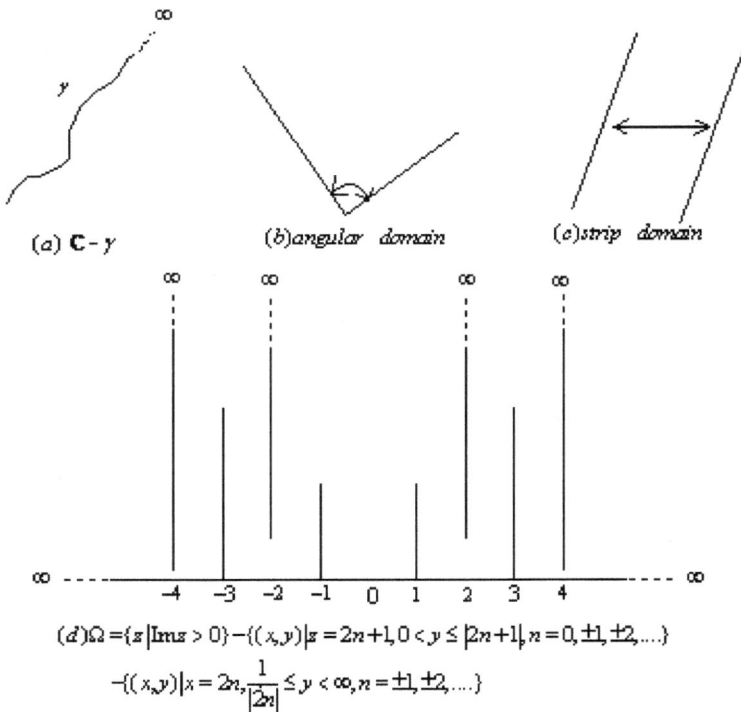

(a) **C** - γ (b)*angular domain*   (c)*strip domain*

(d)$\Omega = \{z \,|\, \mathrm{Im}\, z > 0\} - \{(x,y)\,|\, z = 2n+1, 0 < y \leq 2n+1, n = 0, \pm 1, \pm 2, \ldots\}$

$- \{(x,y)\,|\, x = 2n, \frac{1}{|2n|} \leq y < \infty, n = \pm 1, \pm 2, \ldots\}$

Fig. 2.11

For proofs, see Ref. [58], Vol. 1, pp. 94–99 or Ref. [65], pp. 120–123, pp. 135–137. If, in addition, $f : \bar{\Omega} \to \mathbf{C}^*$ is continuous in the generalized sense, then it follows that $f(\partial\Omega) \cap f(\Omega) = \phi$ and $f(\partial\Omega) = \partial f(\Omega)$ hold. As a consequence, *a homeomorphism preserves simple or multiple connectedness of a domain.*

Remark 3. (More about domains (for details see Ref. [65])). It is well-known that $f(z) = \frac{z}{1-|z|}$ maps $|z| < 1$ onto $\mathbf{C}$ homeomorphically. And it can be shown that *any simply connected domain in* $\mathbf{C}$ *is homeomorphic to the unit disk* $|z| < 1$. Hence, in $\mathbf{C}^*$, there are exactly two kinds of simply connected domains:

(1) $\mathbf{C}^*$;
(2) $\mathbf{C}$ (or $|z| < 1$).

Moreover, any two n-connected domains ($n \geq 1$) are homeomorphic.

The coming question is: Can a homeomorphism between two simply connected domains Ω_1 and Ω_2 always be extended to a homeomorphism between their closure $\bar{\Omega}_1$ and $\bar{\Omega}_2$? It is not true, in general. It could happen that $\partial\Omega_1$ is not homeomorphic to $\partial\Omega_2$. For instance, $|z| < 1$ and Fig. 2.10(b) are such cases. Even if $\partial\Omega_1$ and $\partial\Omega_2$ are homeomorphic, the result might still be invalid. For example, adopting polar coordinate, the homeomorphism

$$f : |z| < 1 \to |z| < 1 \text{ defined by } f(r, \theta) = \left(r, \frac{\theta}{1-r}\right), \quad 0 \le r < 1, \ 0 \le \theta \le 2\pi,$$

is such a case. All we can say is that *the closure domain $\bar{\Omega}$ of a Jordan domain Ω is homeomorphic to the closed disk $|z| \le 1$* (Ref. [65], pp. 169–173).

In complex analysis, via one-to-one analytic (namely, univalent) function, simply connected domains in $\mathbf{C}^*$ are divided into three standard types:

$$\mathbf{C}^*, \mathbf{C} \text{ and } |z| < 1$$

which are not conformally equivalent among them. This is the beginning of the geometric theory in complex analysis. See Chaps. 6 and 7.   □

Exercises A

(1) Sketch the following curves: What are their directions? Do they have any multiple point? Are they closed? Do they pass ∞?

    (a) $z(t) = (1 + it)^{-1}$, $-\infty < t < \infty$.

    (b) $z(t) = at + be^{i\theta t}$, $a > 0$, $b > 0$, $\theta > 0$, $-\infty < t < \infty$.

    (c) (parabola) $z(t) = at^2 + bt + c$, $a \in \mathbf{R}$ and $a \ne 0, b, c \in \mathbf{C}$ and Im $b \ne 0$, $-\infty < t < \infty$.

    (d) (hyperbola) $z(t) = a\frac{1+t^2}{1-t^2} + ib\frac{2t}{1-t^2}$, $a, b \in \mathbf{R}$, $-1 \le t \le 1$.

(2) For each curve in Exercise A (1), find its length function and the length on a preassigned closed interval.

(3) Suppose a smooth curve $z(t) : [a, b] \to \mathbf{C}$ is twice continuously differentiable (i.e., $z''(t)$ exists and is continuous). Adopt notations in (2.4.6).

    (a) Show that the curve has the *unit tangent vector* at s

$$\frac{d\tilde{z}}{ds}(s) = \frac{z'(t)}{|z'(t)|} = \frac{dz}{dt}(t) \left(\frac{ds}{dt}(t)\right)^{-1},$$

    where $\tilde{z}(s)$ is defined in (2.4.6).

    (b) Show that

$$\text{Re} \, \frac{\overline{d\tilde{z}}}{ds} \frac{d^2\tilde{z}}{ds^2} = 0$$

(refer to 5 in (1.4.3.4)). We call $\frac{d^2\tilde{z}}{ds^2}$ the *principal normal vector* or *inward normal* of the curve at s, and *curvature* at s: $\kappa(s) = \left|\frac{d^2\tilde{z}}{ds^2}(s)\right|$,

radius of curvature at s: $\rho(s) = \dfrac{1}{|\kappa(s)|}$,

center of curvature at s: $\tilde{z}(s) + \rho(s)\dfrac{d^2\tilde{z}}{ds^2}(s)$,

circle of curvature at s: $\left|z - (\tilde{z}(s) + \rho(s)\dfrac{d^2\tilde{z}}{ds^2}(s))\right| = \rho(s)$.

See Fig. 2.12.

(c) Show that

$$\frac{d^2\tilde{z}}{ds^2}(s) = \frac{d}{dt}\left(\frac{d\tilde{z}}{ds}\right)\left(\frac{ds}{dt}\right)^{-1} = \frac{1}{z'(t)}\frac{d}{dt}\left\{\frac{z'(t)}{|z'(t)|}\right\}.$$

(4) Let $z(t)$ be as in Exercise A (3) and the parameter t be interpreted as the time variable. Then $\frac{dz}{dt}(t)$ and $\frac{d^2z}{dt^2}(t)$ are, respectively, the *velocity* and *acceleration* (vectors) of the particle $z(t)$ at the time t. We adopt notations in (2.4.6).

(a) Show that, via $z(t) = z(s(t))$,

$$\frac{dz}{dt} = \frac{dz}{ds}\frac{ds}{dt} = e^{i\alpha}\frac{ds}{dt},$$

where $\frac{dz}{ds} = e^{i\alpha}$ is the unit tangent vector $\frac{d\tilde{z}}{ds}$ in Exercise A(3)(a), where $\tilde{z}(s) = z(t(s))$.

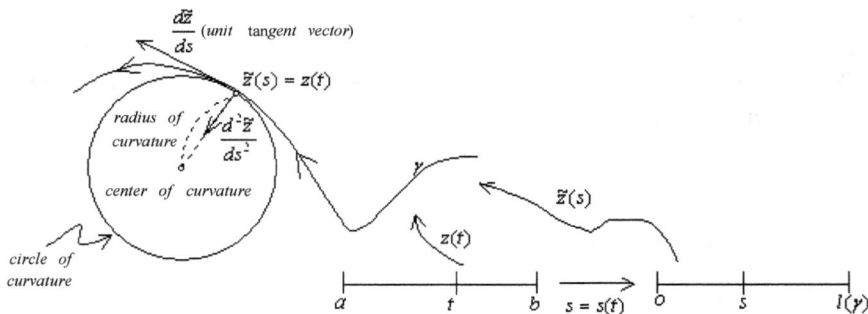

Fig. 2.12

(b) Show that

$$\frac{d^2z}{dt^2} = e^{i\alpha}\frac{d^2s}{dt^2} + ie^{i\alpha}\left(\frac{ds}{dt}\right)^2\frac{d\alpha}{ds},$$

where $ie^{i\alpha}$ is the unit inward normal and $\frac{d\alpha}{ds} = \kappa(s)$ is the curvature.

(5) Now we adopt polar coordinate (r,θ). Suppose the particle $z(t) = r(t)e^{i\theta(t)}$ is a function of the time variable t.

(a) Show that

$$\frac{dz}{dt} = e^{i\theta(t)}\frac{dr}{dt} + ir(t)e^{i\theta(t)}\frac{d\theta}{dt},$$

where $e^{i\theta(t)}$ is the unit *radial vector*.

(b) Show that

$$\frac{d^2z}{dt^2} = e^{i\theta(t)}\left[\frac{d^2r}{dt^2} - r(t)\left(\frac{d\theta}{dt}\right)^2\right] + ie^{i\theta(t)}\frac{1}{r(t)}\frac{d}{dr}\left[r(t)^2\frac{d\theta}{dt}\right].$$

Exercises B

(1) A curve might have many parametric representations. Suppose a curve γ_1 is defined by $z_1(t) : [a,b] \to \mathbf{C}$ and another curve γ_2 by $z_2(t) : [c,d] \to \mathbf{C}$. We say that γ_1 and γ_2 are *equivalent of the first* (or *second*) *kind* if there exists an increasing (or decreasing) function $\varphi : [a,b] \to [c,d]$ so that $\varphi(a) = c$, $\varphi(b) = d$ (or $\varphi(a) = d$, $\varphi(b) = c$) and $z_1 = z_2 \circ \varphi$. Equivalent curves have the same locus in the plane.

(a) Show that any two parametric representations of a Jordan arc are equivalent.

(b) Two parametric representations $z_1(t)$ and $z_2(t)$ of a Jordan closed curve are not equivalent except $z_1(a) = z_2(c)$. For example, the unit circle $|z| = 1$ has parametric representations $z_1(t) = e^{it}, 0 \leq t \leq 2\pi$, and $z_2(t) = e^{it}, \frac{\pi}{2} \leq t \leq \frac{5\pi}{2}$, which are not equivalent. In case, $z_1(a) \neq z_2(c)$, there exists a unique point $t_0 \in [c,d]$ so that $z_1(a) = z_2(t_0)$. Define $\tilde{z}_2(t) : [t_0, d - c + t_0] \to \mathbf{C}$ by

$$\tilde{z}_2(t) = \begin{cases} z_2(t), \, t \in [t_0, d] \\ z_2(t - d + c), \, t \in [d, d - c + t_0] \end{cases}.$$

Then $z_1(t)$ and $\tilde{z}_2(t)$ are equivalent. This means that *any points in a Jordan closed curve can be treated as the initial and terminal points.* Under this circumstance, we still say that $z_1(t)$ and $z_2(t)$ are equivalent. Hence, *two parametric representations define the same Jordan closed curve if and only if they are equivalent.*

For a fixed curve γ, we say each equivalent class of its parametric representations determine as *orientation* for it. There are two such orientations, opposite to each other. γ with a designated orientation is called an *oriented curve.*

(2) Construct a domain Ω in $\mathbf{C}$ so that its boundary in $\mathbf{C}^*$ has countably and uncountably infinitely many components, respectively.

Exercise C

Try to read a book on topology and see how to prove (2.4.11), (2.4.12), and (2.4.13).

2.5.   Elementary Rational Functions

Here, we will treat polynomials (see (2.6)) as a special case of rational functions (see (2.7)). In this section, we try to figure out some basic properties, basically without recourse to differentiation and integration (Chap. 3), of rational functions such as

(1) *Algebraic properties*: Zeros and factorizations of polynomials (Sec. 2.5.1); zeros, poles, and partial fraction expansions of rational functions (Sec. 2.5.3).
(2) *Topological properties*: Local maximum and minimum, openness, fundamental theorem of algebra. These topics are arranged in Exercises B.
(3) *Geometric mapping properties*: Conformality, domain of univalence, mapping properties of power function z^n (Sec. 2.5.2), linear transformation $\frac{az+b}{cz+d}$ (Sec. 2.5.4) and Joukowski function $\frac{1}{2}\left(z + \frac{1}{z}\right)$ (Sec. 2.5.5). For more concrete examples, refer to Sec. 3.5.7.

This section is divided into five subsections.

2.5.1.   *Polynomials*

Based on Exercise B (2) of Sec. 2.2, we can prove

The fundamental theorem of algebra: Any polynomials of positive degree has at least one zero.

$$(2.5.1.1)$$

Proof. Let $p(z) = a_n z^n + \cdots + a_1 z + a_0$ be a polynomial, where $n \geq 1$ and $a_n \neq 0$. By Example (1) in Sec. 2.1, $\lim_{z \to \infty} p(z) = \infty$. Hence, there exists an $R > 0$ so that $|p(z)| \geq 1$ on $|z| \geq R$; in particular, $p(z) \neq 0$ on $|z| \geq R$. Suppose

$$m = \min_{|z| < \infty} |p(z)|.$$

Then $0 \leq m < \infty$. There exists $r \geq R$ so that $|p(z)| \geq m + 1$ on $|z| \geq r$. Then, there exists z_0, $|z_0| \leq r$ so that $m = |p(z_0)|$. In case $m \neq 0$, it is necessary that $|z_0| < r$. For $0 < \delta < r - |z_0|$, by Exercise B(2) of Sec. 2.2, there exists a point z_1, $|z_1 - z_0| < \delta$ so that $|p(z_1)| < |p(z_0)| = m$, a contradiction. So $m = 0$ should hold and $p(z_0) = 0$.

Refer to Exercises B (1), B (2), and B (3) for further information.   □

Designate $p(\infty) = \infty$. *Then $p(z)$ is a continuous function on* $\mathbf{C}$ *and* also, continuous in the generalized sense on $\mathbf{C}^*$.

By the fundamental theorem of algebra, for each $w \in \mathbf{C}$, $p(z) = w$ has at least one solution and at most n distinct solutions. Hence $p(\mathbf{C}^*) = \mathbf{C}^*$ and indeed, $p(\mathbf{C}^*)$ covers $\mathbf{C}^*$ n-times (counting multiplicities) if $p(z)$ is a polynomial of degree n. If we use the fundamental theorem repeatedly, then a polynomial $p(z)$ of degree n has exactly n zeros $z_1, \ldots, z_n$ and $p(z)$ can be factorized as

$$p(z) = a_n(z - z_1)(z - z_2) \cdots (z - z_n).$$

If some z_j appears r_j times in the above factorization, then z_j is called a *zero of multiplicity* r_j of $p(z)$ or *a root of multiplicity* r_j of the equation $p(z) = 0$. Suppose $p(z)$ has k *distinct* zeros $z_1, \ldots, z_k$ with respective multiplicities $r_1, \ldots, r_k$. Then

$$p(z) = a_n(z - z_1)^{r_1} \cdots (z - z_k)^{r_k}, \quad r_j \geq 1$$

$$\text{for } 1 \leq j \leq k \text{ and } r_1 + \cdots + r_k = n. \qquad (2.5.1.2)$$

It is easy to see the following

Characterization of a zero z_0 of multiplicity (or order) r of a polynomial $p(z)$.

(1) z_0 is a zero of multiplicity r of $p(z)$, where $r \geq 1$;
⇔ (2) $p(z) = (z - z_0)^r q(z)$, where $q(z)$ is a polynomial and $q(z_0) \neq 0$;
⇔ (3) $p(z_0) = p'(z_0) = \cdots = p^{(r-1)}(z_0) = 0$ but $p^{(r)}(z_0) \neq 0$ (see Remark 1 below).

In case $r = 1$, z_0 is called a *simple zero* (*root*) of $p(z)$ $(p(z) = 0)$.

$$(2.5.1.3)$$

Remark 1. (Differentiation of polynomials). Let $p(z) = a_n z^n + \cdots + a_1 z + a_0$, where $n \geq 1$ and $a_n \neq 0$. The *derivative* of $p(z)$ at z_0 is defined as, if the limit exists as a finite complex number,

$$\lim_{z \to z_0} \frac{p(z) - p(z_0)}{z - z_0} = p'(z_0) = a_n \cdot n z_0^{n-1} + \cdots + a_2 \cdot 2z_0 + a_1, \quad z_0 \in \mathbf{C}$$

$$(2.5.1.4)$$

and thus, $p(z)$ is said to be *differentiable* at z_0. Exactly like in calculus, they enjoy the same properties such as:

$$(p \pm q)'(z) = p'(z) \pm q'(z); \quad (\alpha p)'(z) = \alpha p'(z), \quad \alpha \in \mathbf{C},$$

$$(pq)'(z) = p'(z)q(z) + p(z)q'(z), \quad (p \circ q)'(z) = p'(q(z))q'(z), \quad (2.5.1.5)$$

$$\left(\frac{p}{q}\right)'(z) = \frac{p'(z)q(z) - p(z)q'(z)}{(q(z))^2}, \quad q(z) \neq 0.$$

For general setting, refer to (3.1.6) in Chap. 3. In particular, if $p(z) = (z - z_0)^r q(z)$, then

$$p'(z) = r(z - z_0)^{r-1} q(z) + (z - z_0)^r q'(z).$$

This is what we want in proving (2) $\Leftrightarrow$ (3) in (2.5.1.3).   □

Remark 2 (Conformality of a polynomial $p(z)$ at a point z where $p'(z) \neq 0$). Let $z(t)$ represent a differentiable curve (see (2.4.7)). Then, for any polynomial $p(z), p(z(t))$ is a differentiable curve and

$$p(z(t))' = p'(z(t))z'(t).$$

Now, let $z_1(t)$ and $z_2(t)$ represent two differentiable curves so that $z_1(t_0) = z_2(t_0) = z_0$ and $z_1'(t_0) \neq 0, z_2'(t_0) \neq 0$. Then, if $p'(z_0) \neq 0$,

$$p(z_j(t_0))' = p'(z_j(t_0))z_j'(t_0) = p'(z_0)z_j'(t_0) \neq 0, \quad j = 1, 2.$$

$$\Rightarrow \operatorname{Arg} p(z_j(t_0))' = \operatorname{Arg} p'(z_0) + \operatorname{Arg} z_j'(t_0), \quad j = 1, 2.$$

$$\Rightarrow \operatorname{Arg} p(z_2(t_0))' - \operatorname{Arg} p(z_1(t_0))' = \operatorname{Arg} z_2'(t_0) - \operatorname{Arg} z_1'(t_0).$$

$$(2.5.1.6)$$

This means that a polynomial $w = p(z)$, as a *mapping* from $\mathbf{C}$ onto $\mathbf{C}$, *preserves angle and its direction at point z where $p'(z) \neq 0$*. Then, we say a polynomial is *conformal* at points where its first derivative is nonzero. For details, see Section (1) of Sec. 3.5.1.

We will extend properties (2.5.1.2)–(2.5.1.6) to analytic functions (see Chap. 3).   □

The proof of the fundamental theorem (2.5.1.1) cannot locate where zeros of a polynomial are. It provides only a possible range that zeros might be situated. For example, if $p(z) = a_n z^n + \cdots + a_1 z + a_0$, $n \geq 1$, and $a_n \neq 0$, then

$$|p(z)| \geq |a_n z^n| \left\{ 1 - \left| \frac{a_{n-1}}{a_n z} + \cdots + \frac{a_0}{a_n z^n} \right| \right\}$$

$$\geq |a_n||z|^n \left\{ 1 - \frac{1}{|z|} \left(\left| \frac{a_{n-1}}{a_n} \right| + \cdots + \left| \frac{a_0}{a_n} \right| \right) \right\}, \quad |z| \geq 1.$$

Choose $r \geq 1$ so that

$$\frac{1}{r} \left(\left| \frac{a_{n-1}}{a_n} \right| + \cdots + \left| \frac{a_0}{a_n} \right| \right) \leq \frac{1}{2}$$

$$\Rightarrow |p(z)| \geq \frac{1}{2} |a_n||z|^n \quad \text{if } |z| \geq r.$$

This means that all the zeros of $p(z)$ lie inside $|z| \leq r$, namely, all its zeros z_j satisfy

$$|z_j| \leq \max \left\{ 1, 2 \left(\left| \frac{a_{n-1}}{a_n} \right| + \cdots + \left| \frac{a_0}{a_n} \right| \right) \right\}, \quad 1 \leq j \leq n. \tag{2.5.1.7}$$

Therefore we have (see also Example 3 in Sec. 3.5.3) the Eneström–Kakeya theorem.

Eneström–Kakeya theorem. *Suppose $a_0 > a_1 > a_2 > \cdots > a_{n-1} > a_n > 0$. Then the polynomial $p(z) = a_0 + a_1 z + \cdots + a_{n-1} z^{n-1} + a_n z^n$ does not have zeros on $|z| \leq 1$. Hence, all its zeros lie on the ring domain*

$$1 < |z| < \frac{2n a_0}{a_n}. \tag{2.5.1.8}$$

Proof. Observe that $|(1 - z)p(z)| = |a_0 - (a_0 - a_1)z - \cdots - (a_{n-1} - a_n)z_n - a_n z^{n+1}| > a_0 - \{(a_0 - a_1)|z| + \cdots + (a_{n-1} - a_n)|z|^n + |a_n||z|^{n+1}\} \geq a_0 - (a_0 - a_1) - \cdots - (a_{n-1} - a_n) - a_n = 0$ if $|z| \leq 1$ and $z \neq 1$. Hence, $p(z) \neq 0$ if $|z| \leq 1$ and $z \neq 1$. In case $z = 1$, then $p(1) = a_0 + a_1 + \cdots + a_n > 0$. $\square$

If $f(x)$ is a differentiable real-valued function of a real variable x, Rolle's theorem says that between two distinct zeros of $f(x)$ lies a zero of $f'(x)$. This is no more true for complex-valued functions. For instance, $f(z) = e^z - 1$ (see Sec. 2.6.1) has its zeros $2n\pi i$ ($n = 0, \pm 1, \pm 2, \ldots$), yet $f'(z) = e^z$ (why?) $\neq 0$ everywhere. However, we have the Gauss–Lucas theorem.

Gauss–Lucas theorem. *The zeros of $p'(z)$ of a polynomial $p(z)$ lie on the convex closure generated by the zeros of $p(z)$. Moreover, these zeros of $p'(z)$ that are not zeros of $p(z)$ are interior points of the convex closure.*

(2.5.1.9)

For terminologies concerned, see Exercise B (2) of Sec. 1.4.3.

Proof. Let $p(z) = a_n(z - z_1) \cdots (z - z_n)$. Then, a direct computation shows that

$$\frac{p'(z)}{p(z)} = \frac{1}{z - z_1} + \cdots + \frac{1}{z - z_n}.$$

Suppose z_0 is a point so that $p(z_0) \neq 0$ and $p'(z_0) = 0$. Since $z_0 - z_k \neq 0$ for $1 \leq k \leq n$, then

$$\sum_{k=1}^{n} \frac{1}{z_0 - z_k} = \sum_{k=1}^{n} \frac{\bar{z}_0 - \bar{z}_k}{|z_0 - z_k|^2} = 0$$

$$\Rightarrow \sum_{k=1}^{n} \frac{z_0 - z_k}{|z_0 - z_k|^2} = 0 \quad \text{or}$$

$$z_0 = \left(\sum_{k=1}^{n} \frac{1}{|z_0 - z_k|^2} \right)^{-1} \left(\sum_{k=1}^{n} \frac{1}{|z_0 - z_k|^2} z_k \right).$$

Thus, z_0 is an interior point of the convex closure $\mathrm{Con}(z_1, \ldots, z_k)$. On the other hand, if $z \notin \mathrm{Con}(z_1, \ldots, z_n)$, then $p(z) \neq 0$. We try to show that $p'(z) \neq 0$. Since $\mathrm{Con}(z_1, \ldots, z_n)$ is a convex set, there exists a real θ so that

$$\theta - \pi < \mathrm{Arg}(z - z_k) < \theta, \quad 1 \leq k \leq n$$

$$\Rightarrow \pi - \theta > \mathrm{Arg}\, \frac{1}{z - z_k} > -\theta, \quad 1 \leq k \leq n$$

$$\Rightarrow 0 < \mathrm{Arg}\, \frac{e^{i\theta}}{z_k - z} < \pi \text{ and hence Im}\, \frac{e^{i\theta}}{z_k - z} > 0, \quad 1 \leq k \leq n$$

$$\Rightarrow \frac{p'(z)}{p(z)} = \sum_{k=1}^{n} \frac{1}{z - z_k} = -e^{-i\theta} \sum_{k=1}^{n} \frac{e^{i\theta}}{z_k - z} \neq 0.$$

Hence $p'(z) \neq 0$. $\square$

Exercises A

(1) Suppose $a_k \geq 0$, $0 \leq k \leq n - 1$, and $\sum_{k=0}^{n-1} a_k > 0$. Show that $x^a - \sum_{k=0}^{n-1} a_k x^k = 0$ has only one positive real root which is not larger

than $\max\left\{1, \sum_{k=0}^{n-1} a_k\right\}$. Try to use this result to define the positive nth **root** $\sqrt[n]{a}$ of $a > 0$.

(2) Suppose $\sum_{k=0}^{n-1} |a_k| > 0$. Show that the absolute values of zeros of $z^n + a_{n-1}z^{n-1} + \cdots + a_1 z + a_0 = 0$ are not larger than the only positive real root of $x^n - |a_{n-1}|x^{n-1} - \cdots - |a_1|x - |a_0| = 0$.

(3) In the polynomial $p(z) = a_0 + a_1 z + \cdots + a_n z^n$, suppose $a_0 \neq 0$ and $a_n \neq 0$.

  (a) Show that $|a_n|x^n + |a_{n-1}|x^{n-1} + \cdots + |a_1|x - |a_0| = 0$ has only one positive real root ρ.

  (b) Show that $p(z) \neq 0$ on $|z| < \rho$.

(4) Let $p(z) = a_n z^n + \cdots + a_1 z + a_0$ have positive coefficients $a_k > 0$, $0 \leq k \leq n$. Show that all the zeros of $p(z)$ lie on the closed ring domain

$$\min_{1 \leq k \leq n} \frac{a_{k-1}}{a_k} \leq |z| \leq \max_{1 \leq k \leq n} \frac{a_{k-1}}{a_k}.$$

(5) Suppose $p(z)$ is a polynomial with real coefficients.

  (a) If $p(z)$ has only real zeros, than $p'(z)$ does, too.

  (b) If $p(z)$ has an imaginary zero z_0, then $\bar{z}_0$ is also a zero of $p(z)$. A circle with the segment $\overline{z_0 \bar{z}_0}$ as its diameter is called a *Jensen circle*. If $p'(z)$ has an imaginary root, show that this root should lie inside or on a Jensen circle.

(6) Suppose $0 \leq a_k \leq M$ for $0 \leq k \leq n$ and $a_n \geq 1$. Show that the zeros z of $p(z) = a_n z^n + \cdots + a_1 z + a_0$ satisfy

  (i) $\operatorname{Re} z \leq 0$, or

  (ii) $\operatorname{Re} z > 0$ and $|z| \leq \frac{1}{2}(1 + \sqrt{1 + 4M})$.

(7) Suppose $p(z) = \sum_{k=0}^{n} a_k z^k$ with complex coefficients $a_k = \alpha_k + i\beta_k, \alpha_k, \beta_k \in \mathbf{R}$ for $0 \leq k \leq n$. Let

$$p_1(z) = \sum_{k=0}^{n} \alpha_k z^k \quad \text{and} \quad p_2(z) = \sum_{k=0}^{n} \beta_k z^k.$$

In case all the zeros of $p(z)$ are on the half-plane $\operatorname{Im} z < 0$, show that all the zeros of both $p_1(z)$ and $p_2(z)$ are real.

(8) Let $p_n(z) = 1 + 2z + 3z^2 + \cdots + nz^{n-1}$ and $0 < r < 1$. Show that, there exists a positive integer n_0 so that, if $n \geq n_0$, then $p_n(z) \neq 0$ on $|z| \leq r$.

(9) Prove (2.5.1.5) in detail.

(10) (*Lagrange interpolation formula*) For n *distinct* points $z_1, \ldots, z_n \in \mathbf{C}$ and n complex numbers $c_1, \ldots, c_n$ (which might be coincident among them), show that there exists a unique polynomial $p(z)$ of degree $n-1$ so that $p(z_k) = c_k$ for $1 \le k \le n$, and

$$p(z) = \sum_{k=1}^{n} c_k \frac{(z - z_1) \cdots (z - z_{k-1})(z - z_{k+1}) \cdots (z - z_n)}{(z_k - z_1) \cdots (z_k - z_{k-1})(z_k - z_{k+1}) \cdots (z_k - z_n)}, \quad z \in \mathbf{C}.$$

Exercises B

(1) (continued from Exercise B (2) of Sec. 2.2) Let $p(z)$ be a polynomial of degree n, $n \ge 1$.

(a) Fix $r > 0$. Show that $|p(z)|$ attains its maximum on $|z - z_0| \le r$ at some point on the circle $|z - z_0| = r$, namely,

$$\max_{|z-z_0| \le r} |p(z)| = \max_{|z-z_0| = r} |p(z)| \underset{\text{(def.)}}{=} M(r).$$

Also, show that $M(r)$ is a strictly increasing continuous function of r.

(b) Suppose $p(z) \ne 0$ on $|z - z_0| \le \rho$. For any $0 < r < \rho$, show that

$$\min_{|z-z_0| \le r} |p(z)| = \min_{|z-z_0| = r} |p(z)| \underset{\text{(def.)}}{=} m(r).$$

Also, show that $m(r)$ is a strictly decreasing continuous function of $r(0 < r < \rho)$.

(c) Suppose $p(z)$ and $q(z)$ are two polynomials so that $p(z) \equiv q(z)$ on some circle $|z - z_0| = r > 0$. Show that $p(z) = q(z), z \in \mathbf{C}$.

(2) Let $p(z)$ be polynomial of degree $n, n \ge 1$. Show that $p(z)$ is *open*, i.e., $w = p(z)$ maps open sets in $\mathbf{C}$ onto open sets. In particular, $p(\mathbf{C})$ is a domain in $\mathbf{C}$.

(3) Try to use the following two methods to prove the *fundamental theorem* (2.5.1.1).

(a) Show that $p(\mathbf{C})$ is both open and closed in $\mathbf{C}$.

(b) Show that $p(\mathbf{C})$ is open in $\mathbf{C}$ and $\lim_{z \to \infty} p(z) = \infty$. Then, try to use Example 3 in Sec. 2.2.

2.5.2.   The power function $w = z^n$ $(n \ge 2)$

According to (2.5.1.3), $z = 0$ is a zero of order n of z^n.

For each $w \ne 0$, De Moivre formula (1.5.2) shows that $z^n = w$ has n distinct solutions $|w|^{\frac{1}{n}} e^{i\theta_k}$, where $\theta_k = \frac{1}{n}(\text{Arg } w + 2k\pi), k = 0, 1, 2, \ldots, n-1$.

Hence, $w = z^n$ *covers the plane* $\mathbf{C}$ n *times* (counting multiplicity at $z = 0$ and even, at $z = \infty$).

Which part of $\mathbf{C}$ will cover $\mathbf{C}$ only once under $w = z^n$? Namely, we try to figure out some subsets of $\mathbf{C}$ which will be mapped one-to-one and onto $\mathbf{C}$ under $w = z^n$.

For this purpose, let $z = re^{i\theta}$, where $r = |z|$ and $\theta = \operatorname{Arg} z \in [0, 2\pi)$. Then if $w = z^n$, we have

$$w = (re^{i\theta})^n = r^n e^{in\theta}$$
$$\Rightarrow |w| = r^n \quad \text{and} \quad \operatorname{Arg} w = n\theta \ (\text{mod } 2\pi).$$

This indicates that $w = z^n$ maps the ray $\operatorname{Arg} z = \theta$ onto the ray $\operatorname{Arg} w = n\theta$ and the circle $|z| = r$ onto the circle $|w| = r^n$. For any α, $-\pi < \alpha \leq \pi$,

$$\alpha \leq n\theta < \alpha + 2\pi \Leftrightarrow \frac{\alpha}{n} \leq \theta < \frac{\alpha}{n} + \frac{2\pi}{n}.$$

Hence, $w = z^n$ maps the *sector* $\frac{\alpha}{n} \leq \operatorname{Arg} z < \frac{\alpha}{n} + \frac{2\pi}{n}$ one-to-one and onto the whole plane $\mathbf{C}$ (with $z = 0$ corresponding to 0). And at the same time, $w = z^n$ maps the arc of the circle $|z| = r$ within this sector one-to-one and onto the circle $|w| = r^n$. As a whole, $w = z^n$ maps $|z| = r$ onto $|w| = r^n$, covered n times. See Fig. 2.13: Ω_k is the open sector $\frac{2k\pi}{n} < \operatorname{Arg} z < \frac{2(k+1)\pi}{n}$, $0 \leq k \leq n-1$. Then

(1) $w = z^n$ maps each open sector Ω_k one-to-one and onto the domain $\mathbf{C} - [0, \infty)$, the plane $\mathbf{C}$ with the nonnegative real axis $[0, \infty)$ cut off. Domains such as Ω_k, $0 \leq k \leq n-1$, are called *domains of univalence* or *fundamental domains* (see (2.7.4)) for $w = z^n$;

(2) $w = z^n$ maps the upper edge $\operatorname{Arg} z = \frac{2k\pi}{n}$ one-to-one and onto the upper side of the cut $[0, \infty)$, i.e., $\operatorname{Arg} w = 0$; the lower edge $\operatorname{Arg} z =$

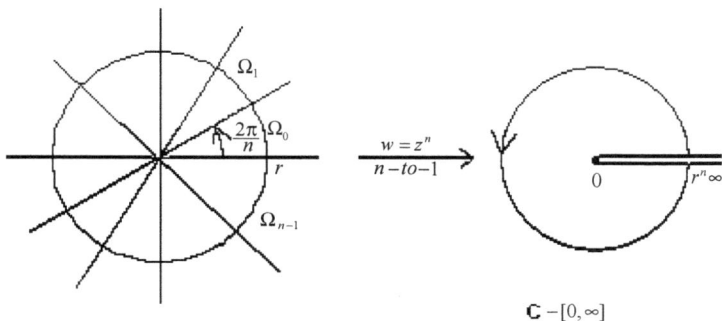

Fig. 2.13

$\frac{2(k+1)\pi}{n}$ one-to-one and onto the lower side of the cut $[0, \infty)$, i.e.,
$\operatorname{Arg} w = 2\pi$;

(3) $w = z^n$ maps either $\frac{2k\pi}{n} \le \theta < \frac{2(k+1)\pi}{n}$ or $\frac{2k\pi}{n} < \theta \le \frac{2(k+1)\pi}{n}$ one-to-one
and onto the whole plane $\mathbf{C}$ ($z = 0$ to $w = 0$);

(4) $|z| = r$ covers $|w| = r^n$ n-times via $w = z^n$.   (2.5.2.1)

These results still hold for the sector $\varphi < \operatorname{Arg} z < \varphi + \frac{2\pi}{n}$ with its image
under $w = z^n$ the domain $\mathbf{C} - \{w | \arg w = n\varphi\}$ (including 0). Remember
that $w = z^n$ is conformal everywhere except at $z = 0$ (see (2.5.1.6)).

Can you figure out what the inverse function $z = \sqrt[n]{w}$ is? For details,
see Sec. 2.7.2. We give an example below.

Example Let $w = z^2$ where $z = x + iy$ and $w = u + iv$.

(a) Find the image of the family of orthogonal lines $x = c_1$ (constant) and
$y = c_2$.

(b) Determine the preimage of the family of orthogonal lines $u = c_1$ and
$v = c_2$.

Solution. $u + iv = (x + iy)^2 = x^2 - y^2 + 2xyi$ shows that $u = x^2 - y^2$ and
$v = 2xy$.

(a) The images of $x = c_1$ have parametric equations $u = c_1^2 - y^2$ and
$v = 2c_1 y$ with $y \in \mathbf{R}$ as parameter. If $c_1 = 0$, then $x = 0$ is mapped
onto $u = -y^2, y \in \mathbf{R}$ and $v = 0$, namely, the nonpositive real axis
$(-\infty, 0]$ in a two-to-one manner except at $z = 0$. Suppose $c_1 \ne 0$. By
eliminating y from $u = c_1^2 - y^2$ and $v = 2c_1 y$, we have the parabola
$v^2 = 4c_1^2(c_1^2 - u)$.

While, the images of $y = c_2$ have equations $u = x^2 - c_2^2$ and $v = 2c_2 x$.
If $c_2 = 0$, then $y = 0$ is mapped onto $u = x^2, x \in \mathbf{R}$ and $v = 0$, which
is $[0, \infty)$, the nonnegative real axis, in a two-to-one manner except at
$z = 0$. If $c_2 \ne 0$, the image is $u = x^2 - c_2^2$ and $v = 2c_2 x$ which is the
parabola $v^2 = 4c_2^2(c_2^2 + u)$.

Note that these two families of parabolas $v^2 = 4c_1^2(c_1^2 - u)$ and $v^2 = 4c_2^2(c_2^2 + u)$ are orthogonal to each other wherever they meet except at
$w = 0$ because $w = z^2$ is conformal everywhere except at $z = 0$. See
Fig. 2.14.

(b) The preimages of $u = c_1$ are hyperbolas $x^2 - y^2 = c_1$, while the preimages of $v = c_2$ are hyperbolas $2xy = c_2$. These two families of hyperbolas are orthogonal everywhere except at $z = 0$. See Fig. 2.15.

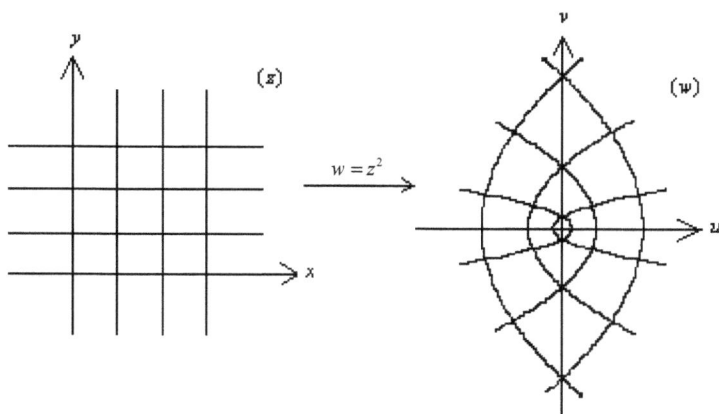

Fig. 2.14

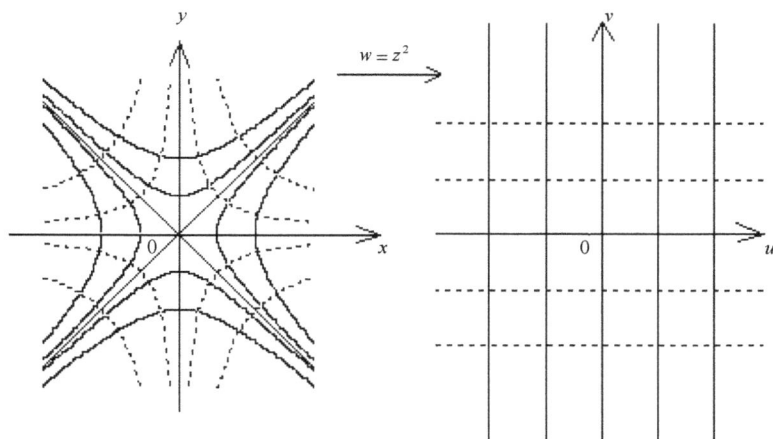

Fig. 2.15

Exercises A

(1) Do the following problems, $z = x + iy$ and $w = u + iv$.

    (a) $w = z^2$ maps the domain $\operatorname{Re} z > x_0 > 0$ *univalently* (namely, one-to-one) onto the domain $v^2 > 4x_0^2(x_0^2 - u)$. What is the image of $\operatorname{Re} z < x_0 < 0$, where $w = z^2$?

    (b) Find the image domains of $\operatorname{Im} z > y_0 > 0$ and $\operatorname{Im} z < y_0 < 0$, under $w = z^2$.

(c) Find the image domains of the strips $0 < x_1 < \operatorname{Re} z < x_2$ and $y_2 < \operatorname{Im} z < y_1 < 0$, under $w = z^2$.

(d) Find the image domains of $|z^2 - a^2| = \lambda (\lambda > 0, a > 0)$, under $w = z^2$.

(2) (a) Map the interior of the right branch of the hyperbola $x^2 - y^2 = a^2$ $(a > 0)$ *univalently* (and hence, *conformally*) onto the upper half plane $\operatorname{Im} w > 0$.

*(b) Map the exterior of the parabola $y^2 = 4px$ $(p > 0)$ univalently onto the right half-plane $\operatorname{Re} w > 0$.

(3) Show that $w = z^2$ maps the circle $|z \pm z_0| = |z_0|$ $(z_0 \neq 0)$ univalently onto the cardioid $w = 2z_0^2(1+\cos\theta)e^{i\theta}$, $-\pi < \theta \le \pi$; the circle $|z \pm z_0| = r$ $(r > 0$, and $z_0 \neq 0)$ univalently onto the curve $w = w_1 + 2r(z_0 + r\cos\theta)e^{i\theta}$, $-\pi < \theta \le \pi$. What is w_1? Graph both curves.

(4) Model after Figs. 2.13 and 2.14 for the mappings $w = z^3$ and $w = z^4$.

(5) Find domains of univalence for the mappings $w = (z - a)^n$, $w = z^n + b$, and $w = (z - a)^n + b$, respectively. Try to graph these domains as in Fig. 2.13.

2.5.3.   *Rational functions*

Let $R(z) = \frac{p(z)}{q(z)}$ be a rational function as in (2.7), where $p(z) = a_n z^n + \cdots + a_1 z + a_0$ $(n \ge 0, a_n \neq 0)$ and $q(z) = b_m z^m + \cdots + b_1 z + b_0$ $(m \ge 0, b_m \neq 0)$ *do not* have common factors.

According to (2.5.1.5), $R(z)$ is differentiable at points z where $q(z) \neq 0$. Hence, $R(z)$ is *analytic* (see Chap. 3) everywhere on $\mathbf{C}$ except points z where $q(z) = 0$ and is conformal at points z where $R'(z) \neq 0$.

$R(z) = 0 \Leftrightarrow p(z) = 0$ and both have the same order (multiplicity) at such z.

Now, suppose z_0 is a zero of order k of $q(z)$. Then, $q(z) = (z - z_0)^k r(z)$, where $r(z)$ is a polynomial and $r(z_0) \neq 0$ (see (2.5.1.3)). Hence

$$R(z) = \frac{\varphi(z)}{(z - z_0)^k}, \quad \text{where } \varphi(z) = \frac{p(z)}{r(z)} \text{ is rational with}$$

$$\varphi(z_0) \neq 0, \infty.$$

$$\Rightarrow \lim_{z \to z_0} R(z) = \infty. \tag{2.5.3.1}$$

Designate $R(z_0) = \infty$ and call z_0 a *pole* of *order* k of $R(z)$, and a *simple pole* if $k = 1$. As a consequence, $w = R(z) : \mathbf{C} \to \mathbf{C}^*$ is continuous in the generalized sense. Redefine (see Example (2) in Sec. 2.1)

$$R(\infty) = \begin{cases} 0, & n < m \\ \infty, & n > m \\ \dfrac{a_n}{b_n}, & n = m. \end{cases}$$

and call ∞ a *zero* of order $m - n$ if $n < m$, a *pole* of order $n - m$ if $n > m$. Therefore, $w = R(z) : \mathbf{C}^* \to \mathbf{C}^*$ is continuous, of course, in the generalized sense.

What is the range of $w = R(z)$? Observe that

the number of zeros of $R(z)$ (counting multiplicities)

$$= \text{the number of poles of } R(z)$$

(counting multiplicities)

$$= \max(m, n). \tag{2.5.3.2}$$

Now, for each $a \in \mathbf{C}^*$, then $R(z) = a \Leftrightarrow z$ is a *zero* of $R(z) - a$ if $a \in \mathbf{C}$ or z is a *pole* of $R(z)$ if $a = \infty$; moreover, $R(z) - a$ and $R(z)$ have the same set of poles. Hence, $w = R(z)$ takes each point $a \in \mathbf{C}^*$ the same number of times, namely, $\max(m, n)$. This means $w = R(z)$ covers $\mathbf{C}^* \max(m, n)$ times (counting multiplicities) and hence, the rational function $w = R(z)$ is said to be of *degree* $\max(m, n)$.

Another expression for $w = R(z)$ is the so-called *partial fraction expansion* according to the poles of it. This can be achieved by *coefficient-comparison* method in high school algebra course. In what follows, we suggest an alternative one, the so-called *method of change of variables* which is of interest by itself.

Suppose $R(z) = \frac{p(z)}{q(z)}$ has a pole at ∞. By long division,

$$R(z) = g(z) + r(z), \quad r(z) = \frac{h(z)}{q(z)} \tag{2.5.3.3}$$

where $g(z)$ is polynomial *without* the constant term whose degree is the order of pole ∞, while $h(z)$ is the remainder polynomial whose degree is *at most* equal to that of $q(z)$. Call $g(z)$ the *essential or principal part* of $R(z)$ at the pole ∞. Note that $r(\infty)$ is a finite complex number.

Next, suppose $q(z)$ has k distinct zeros $z_1, \ldots, z_k$ which are finite poles of $R(z)$. For each fixed j, $1 \le j \le k$, the change of variables

$$z \to \zeta = \frac{1}{z - z_j}$$

maps z_j into ∞ in the ζ-plane. Substitute $z = z_j + \zeta^{-1}$ into $R(z)$. As a rational function of ζ, the rational function $R(z_j + \zeta^{-1})$ has a pole at $\zeta = \infty$.

By (2.5.3.3),

$$R(z_j + \zeta^{-1}) = g_j(\zeta) + r_j(\zeta)$$

$$\Rightarrow R(z) = g_j\left(\frac{1}{z - z_j}\right) + r_j\left(\frac{1}{z - z_j}\right), \quad 1 \le j \le k$$

$$(2.5.3.4)$$

where $g_j((z - z_j)^{-1})$ is a polynomial in $(z - z_j)^{-1}$, without the constant term, and is called the *essential* or *principal part* of $R(z)$ at the pole z_j, while $r_j((z - z_j)^{-1})$ is a rational function in $(z - z_j)^{-1}$ and has a finite value at $z = z_j$.

Now, consider the rational function

$$f(z) = R(z) - g(z) - \sum_{j=1}^{k} g_j\left(\frac{1}{z - z_j}\right)$$

which is defined on $\mathbf{C}^* - \{\infty, z_1, \ldots, z_k\}$. Observe that

(1) at ∞, $R(z) - g(z) = r(z)$ is a finite complex number and each $g_j((z - z_j)^{-1})$ is equal to zero, and

(2) at each fixed z_j $(1 \le j \le k)$, $R(z) - g((z - z_j)^{-1}) = r_j((z - z_j)^{-1})$ is a finite complex function, while $g(z)$ and $g((z - z_l)^{-1})$, $l \ne j$, $1 \le l \le k$, are finite complex numbers, too.

Hence, $f(z)$ is a rational function without poles and hence, $f(z)$ is a constant c. Therefore, $R(z)$ has the *partial fraction expansion*

$$R(z) = g(z) + c + \sum_{j=1}^{k} g_j\left(\frac{1}{z - z_j}\right). \qquad (2.5.3.5)$$

As comparing to zeros, poles of a rational function characterize the function fully. Extension of rational functions are *meromorphic functions*. See Chap. 5, in particular Sec. 5.4, for details.

We give an example below

Example. A rational function $w = \frac{az+b}{cz+d}$ $(ad - bc \ne 0)$ of degree 1 satisfying the condition $|w| = 1 \Leftrightarrow |z| = 1$ should be of the form

(1) $e^{i\theta}\frac{z-z_0}{1-\bar{z}_0 z}$, $z_0 \in \mathbf{C}$ and $|z_0| \ne 1$ and $\theta \in \mathbf{R}$, or

(2) $\frac{e^{i\theta}}{z}$ in case $z_0 = \infty$.

Solution. For sufficiency, see Exercise A (9) of Sec. 1.4.1 and Exercise A (3) of Sec. 1.4.2. For necessity, firstly suppose $a = 0$. Let $z = 1, -1$ and i, then

$$|c + d| = |c - d| = |ic + d| = |b| \qquad (*)$$

hold. In case $d = 0$, then $bc \neq 0$. Now $|c| = |b| \Rightarrow b = e^{i\theta}c$ for some $\theta \in \mathbf{R}$ and then $w = \frac{e^{i\theta}}{z}$. If $d \neq 0$, then $bc \neq 0$ still holds. Now $|c + d| = |c - d| \Rightarrow c$ and d (as vectors) are perpendicular to each other and hence, d and ic are collinear. Then $|ic + d| > |b|$ or $< |b|$, contradicting to $(*)$.

Similarly, if $c = 0$, then $w = e^{i\theta}z$ should hold.

Now suppose $ac \neq 0$. Dividing numerator and denominator by a, we may rewrite w as

$$w = \frac{z + \alpha}{\beta z + \gamma}, \alpha, \beta, \gamma \in \mathbf{C} \quad \text{and} \quad \gamma - \alpha\beta \neq 0.$$

$|\alpha| \neq 1$, otherwise $|\alpha| = 1$ implies that $w(-\alpha) = 0$, contradicting to $|w(-\alpha)| = 1$. Let $z = 1, -1$ and i (these three distinct points uniquely determine the unit circle $|z| = 1$), then

$$|1 + \alpha| = |\beta + \gamma| \Rightarrow 1 + 2\text{Re}\,\alpha + |\alpha|^2 = |\beta|^2 + 2\text{Re}\,\beta\bar{\gamma} + |\gamma|^2;$$

$$|-1 + \alpha| = |-\beta + \gamma| \Rightarrow 1 - 2\text{Re}\,\alpha + |\alpha|^2 = |\beta|^2 - 2\text{Re}\,\beta\bar{\gamma} + |\gamma|^2;$$

$$|i + \alpha| = |i\beta + \gamma| \Rightarrow 1 + 2\text{Im}\,\alpha + |\alpha|^2 = |\beta|^2 - 2\text{Im}\,\beta\bar{\gamma} + |\gamma|^2,$$

$$\Rightarrow \text{Re}\,\alpha = \text{Re}\,\beta\bar{\gamma} \text{ and } \text{Im}\,\alpha = -\text{Im}\,\beta\bar{\gamma} \text{ and hence } \alpha$$

$$= \overline{(\beta\bar{\gamma})} = \bar{\beta}\gamma; \text{ and } 1 + |\alpha|^2 = |\beta|^2 + |\gamma|^2$$

$$\text{and hence } (1 - |\beta|^2)(1 - |\gamma|^2) = 0.$$

Therefore $|\beta| = 1$ or $|\gamma| = 1$. Note that $|\beta| = |\gamma| = 1$ cannot be true, otherwise $|\alpha| = 1$. In case $|\beta| = 1$ and $|\gamma| \neq 1$, let $\beta = e^{i\theta} (\theta \in \mathbf{R})$. Then $\gamma = e^{i\theta}\alpha$ contradicting to $\gamma - \alpha\beta \neq 0$. It follows that $|\gamma| = 1$ and $|\beta| \neq 1$. Since $\gamma^{-1} = \bar{\gamma}$, $\beta = \bar{\alpha}\gamma$ holds and hence

$$w = \frac{z + \alpha}{\gamma + \bar{\alpha}\gamma z} = \frac{1}{\gamma}\frac{z + \alpha}{1 + \bar{\alpha}z}, \quad |\gamma| = 1 \text{ and } |\alpha| \neq 1.$$

And the result follows. $\square$

As an application of both the Example and the method of change of variables, see Exercise B (1). For detailed account for $w = \frac{az+b}{cz+d}$, see Sec. 2.5.4.

As for a rational function of degree 2

$$w = f(z) = \frac{az^2 + bz + c}{a'z^2 + b'z + c'}$$

where a and a' are not equal to zero simultaneously, it can be shown that $w = f(z)$ is decomposed as either

(1) $z \to \eta = \dfrac{az + b}{cz + d}(ad - bc \neq 0) \to \zeta = \eta^2 \to w = \dfrac{A\zeta + B}{C\zeta + D}(AD - BC \neq 0)$; or

(2) $z \to \eta = \dfrac{az + b}{cz + d}(ad - bc \neq 0) \to \zeta = \dfrac{1}{2}\left(\eta + \dfrac{1}{\eta}\right) \to w = A\zeta + B.$

$$(2.5.3.6)$$

So, $w = z^2$ (see Sec. 2.5.2) and $w = \frac{1}{2}\left(z + \frac{1}{z}\right)$ (see Sec. 2.5.5) are two standard rational functions of degree 2.

Exercises A

(1) Expand the following rational functions into partial fractions.

(a) $\dfrac{z^4}{z^3 - 1}$, (b) $\dfrac{1}{z(z + 1)^2(z - i)^3}$.

(2) Suppose $R(z) = \dfrac{p(z)}{q(z)}$, where $\deg p = n$ and $\deg q = m$ and $n < m$. If $q(z)$ has m distinct zeros $z_1, \ldots, z_m$, show that

$$\frac{p(z)}{q(z)} = \sum_{j=1}^{m} \frac{p(z_j)}{q'(z_j)(z - z_j)}.$$

(3) Decompose the following rational functions of order 2 according to (2.5.3.6).

(a) $\dfrac{-4z}{(z + 1)^2}$. (b) $\dfrac{z - i}{(z + i)^2}$. (c) $\dfrac{z^2 - z + 1}{z^2 + z + 1}$. (d) $\dfrac{z^2 - 2cz + c^2}{z^2 - 2cz - c^2}$, $c > 0$.

(4) Show that

$$w = \frac{z^3 - z + 1}{z^3 + z + 1}$$

covers $\mathbf{C}^*$ three times and determines its algebraic branch points in the w-plane.

Note: Let $R(z) = \frac{p(z)}{q(z)}$ be a rational function. A point $w_0 = f(z_0) \in \mathbf{C}^*$ is called an *algebraic branch point of order* $k - 1$ ($k \geq 2$) of $R(z)$ or its inverse function $z = R^{-1}(w)$ if either z_0 is a pole of order k (in this case, $w_0 = \infty$) or z_0 is a zero of order k of $R(z) - w_0$ if $w_0 \in \mathbf{C}$.

(5) Show that

$$w = \frac{z^4 - z^2 + 1}{z^4 + z^2 + 1}$$

covers $\mathbf{C}^*$ four times and determines its algebraic branch points in the w-plane.

Exercises B

(1) (See Ref. [1], p. 33) A rational function $w = R(z)$ of positive degree satisfying $|R(z)| = 1$ if $|z| = 1$ should be of the form

$$e^{i\theta} z^{\alpha} \prod_{j=1}^{k} \left(\frac{z - a_j}{1 - \bar{a}_j z} \right)^{\alpha_j}, \quad z \in \mathbf{C}$$

where $\theta \in \mathbf{R}, \alpha$ is an integer, $\alpha_j \geq 1$ are integer for $1 \leq j \leq k$, and $a_j \in \mathbf{C}, |a_j| \neq 0, 1$ are distinct for $1 \leq j \leq k$. What happens if $|R(z)| = \rho$ on $|z| = r$?

(2) Determine these rational functions $w = R(z)$ satisfying that $|R(z)| = 1$ if z is real.

(3) Determine these rational functions $w = R(z)$ satisfying that $R(z)$ is real if z is real.

2.5.4.   *Linear fractional (or bilinear or Möbius) transformations*

This section will focus on the basic geometric mapping properties of a rational function of degree 1

$$w = \frac{az + b}{cz + d}, \quad ad - bc \neq 0 \qquad (2.5.4.1)$$

specifically called a *linear fractional* or *bilinear* or *Möbius transformation*, simply called a *linear transformation*.

Section (1) Conformality

According to (2.5.1.4) and (2.5.1.5),

$$w' = \frac{ad - bc}{(cz + d)^2}, \quad z \in \mathbf{C}^* - \left\{ \infty, -\frac{d}{c} \right\}. \qquad (2.5.4.2)$$

For such point z, $w'(z) \neq 0$, and hence, via (2.5.1.6), $w = w(z)$ is conformal at z.

Suppose $c \neq 0$. Define $w(\infty) = \frac{a}{c}$. Then (see Remark in Sec. 1.6)

$$w \left(\frac{1}{z} \right) = \frac{a + bz}{c + dz} \underset{\text{(def.)}}{=} g(z)$$

$$\Rightarrow g'(z) = \frac{bc - ad}{(c + dz)^2} \quad \text{and hence,} \quad g'(0) = \frac{bc - ad}{c^2} \neq 0.$$

Thus, $w = w(z)$ is conformal at ∞. On the other hand, $w\left(-\frac{d}{c}\right) = \infty$. Then

$$\frac{1}{w(z)} = \frac{cz + d}{az + b} \underset{\text{(def.)}}{=} h(z),$$

$$\Rightarrow h'(z) = \frac{bc - ad}{(az + b)^2} \quad \text{and hence,} \quad h'\left(-\frac{d}{c}\right) = \frac{c^2}{bc - ad} \neq 0.$$

Also, this means that $w = w(z)$ is conformal at $z = -\frac{d}{c}$.

In case $c = 0$, then $w = \frac{a}{d}z + \frac{b}{d}$ $(d \neq 0)$. Now, $w(\infty) = \infty$ and

$$\frac{1}{w\left(\frac{1}{z}\right)} = \frac{dz}{a + bz} \underset{\text{(def.)}}{=} k(z),$$

$$\Rightarrow k'(z) = \frac{ad}{(a + bz)^2} \quad \text{and hence,} \quad k'(0) = \frac{d}{a} \neq 0.$$

Therefore, the original $w = w(z) = \frac{a}{d}z + \frac{b}{d}$ is conformal at ∞. As a whole, $w = w(z)$ is conformal everywhere on $\mathbf{C}^*$.

For any two points $z_1, z_2 \in \mathbf{C}$, set

$$w_j = \frac{az_j + b}{cz_j + d}, \quad j = 1, 2$$

$$\Rightarrow w_1 - w_2 = \frac{(z_1 - z_2)(ad - bc)}{(cz_1 + d)(cz_2 + d)}. \tag{2.5.4.3}$$

This indicates that $w = w(z)$ is one-to-one on $\mathbf{C}^*$. Also, $w = w(z)$ is onto $\mathbf{C}^*$. This is because

$$\frac{az + b}{cz + d} = w$$

$$\Rightarrow z = \frac{-dw + b}{cw - a}, \quad (-d)(-a) - bc = ad - bc \neq 0, \tag{2.5.4.4}$$

where, $w = \infty$ corresponds to $z = -\frac{d}{c}$ and $w = \frac{a}{c}$ to $z = \infty$.

As a conclusion, $w = w(z)$ maps $\mathbf{C}^*$ one-to-one and conformally onto $\mathbf{C}^*$, i.e., $w = w(z)$ is an univalent function on $\mathbf{C}^*$. It can be shown that $w = w(z)$ is the only such univalent function (see Example 6 and Exercise A (6) of Sec. 4.10.2).

Section (2) The image of a circle under w = w(z)

The image of a circle under $w = w(z)$ is still a circle (or a line if degenerated). There are various ways to justify this claim.

One way is to observe that

$$w = w(z) = \begin{cases} \dfrac{a}{d}z + \dfrac{b}{d}, & \text{if } c = 0 \\[2ex] \dfrac{a}{c} + \dfrac{bc - ad}{c^2} \cdot \dfrac{1}{z + \frac{d}{c}}, & \text{if } c \neq 0. \end{cases} \tag{2.5.4.5}$$

Hence, a general mapping $w = w(z)$ is the composite of the following elementary ones:

(1) Translations: $w = z + a$, where a is fixed.
(2) Rotation: $w = e^{i\theta}z$, $\theta \in \mathbf{R}$.
(3) Stretch: $w = rz$, $r > 0$.
(4) Reflection: $w = \frac{1}{z} = (\overline{\frac{1}{\bar{z}}})$ (see Fig. 1.11).

It is easy to check that each of these four mappings preserves circles (or lines), so $w = w(z)$ does, too.

Another way is to recall that a circle has the equation $\alpha z\bar{z} + \bar{\beta}z + \beta\bar{z} + \gamma = 0$, where $\alpha, \gamma \in \mathbf{R}$ and $\beta \in \mathbf{C}$, and α and β cannot be both equal to zero (see (1.4.3.6)). Note that, if $\alpha = 0$ and $\beta \neq 0$, then it reduces to a line. Now, substitute $z = z(w)$ in (2.5.4.4) into this equation and we get

$$\alpha\frac{-dw+b}{cw-a}\frac{-\bar{d}\bar{w}+\bar{b}}{\overline{cw}-\bar{a}} + \bar{\beta}\frac{-dw+b}{cw-a} + \beta\frac{-\bar{d}\bar{w}+\bar{b}}{\overline{cw}-\bar{a}} + \gamma = 0,$$

$$\Rightarrow (\alpha|d|^2 - \bar{\beta}\bar{c}d - \beta c\bar{d} + \gamma|c|^2)|w|^2 + (-\alpha\bar{b}d + \bar{\beta}ad + \beta\bar{b}c - \gamma\bar{a}c)w$$

$$+ (-\alpha b\bar{d} + \beta a\bar{d} + \bar{\beta}b\bar{c} - \gamma a\bar{c})\bar{w}$$

$$+ \alpha|b|^2 - \bar{\beta}\bar{a}b - \beta a\bar{b} + \gamma|a|^2 = 0. \tag{$*_1$}$$

Note that the coefficient of $|w|^2$ and the constant term are real, and the coefficients of w and $\bar{w}$ are conjugate complex numbers. Also, the coefficients of $|w|^2$ and w cannot be both equal to zero, otherwise this will lead to $ad - bc = 0$ (try it!), a contradiction. Hence, this equation represents a circle in general. In particular, $(*_1)$ represents a line

$$\Leftrightarrow \alpha|d|^2 - \bar{\beta}\bar{c}d - \beta c\bar{d} + \gamma|c|^2 = 0$$

$$\Leftrightarrow (\text{Suppose } c \neq 0.)\ \alpha\left(-\frac{d}{c}\right)\left(-\frac{\bar{d}}{\bar{c}}\right) + \bar{\beta}\left(-\frac{d}{c}\right) + \beta\left(-\frac{\bar{d}}{\bar{c}}\right) + \gamma = 0.$$

This indicates that *only when the point* $-\frac{d}{c}$, where $w = w(z)$ assumes ∞, lies on the original circle, then $w = w(z)$ will map the circle (or line) into a line.

A third way is to observe that a circle (line) can always be expressed as

$$\left|\frac{z-z_1}{z-z_2}\right| = \lambda \quad (\lambda > 0), \tag{$*_2$}$$

where z_1 and z_2 are two points symmetric w.r.t. the circle (see Example 3 in Sec. 1.4.3 and (1.4.4.1)). Recall that it represents a line if $\lambda = 1$. Substitute

$z = z(w)$ in (2.5.4.4) into $(*_2)$ and we have

$$\left| \frac{(cz_1 + d)w - (az_1 + b)}{(cz_2 + d)w - (az_2 + b)} \right| = \lambda.$$

$\Rightarrow$ (1) $\left| \frac{w - w_1}{w - w_2} \right| = \lambda'$, if $cz_1 + d \neq 0$, $cz_2 + d \neq 0$, where $w_1 = w(z_1)$ and $w_2 = w(z_2)$ are symmetric w.r.t. the image circle.

(2) $|w - w_2| = \frac{|az_1 + b|}{\lambda |cz_2 + d|}$, if $cz_1 + d = 0$, where $w_1 = w(z_1) = \infty$ and $w_2 = w(z_2)$.

(3) $|w - w_1| = \lambda \frac{|az_2 + b|}{|cz_1 + d|}$, if $cz_2 + d = 0$, where $w_1 = w(z_1)$ and $w_2 = w(z_2) = \infty$.

This proves the claim. By the way, we obtain the fact that $w = w(z)$ *preserves symmetric points w.r.t. circles (lines).*

Conversely, we pose a *question: Does there exist a transformation* $w = w(z)$ *mapping a given circle (line) onto a preassigned circle (line)?* Certainly, it does exist, and there are many of them.

To answer this question, give two sets of distinct points z_2, z_3, z_4 and w_2, w_3, w_4 (point z_1 will appear later on with specific meaning). Repeating (2.5.4.3) to $w - w_3, w - w_4$ and $w_2 - w_3, w_2 - w_4$ will lead to

$$\frac{(w - w_3)(w_2 - w_4)}{(w - w_4)(w_2 - w_3)} = \frac{(z - z_3)(z_2 - z_4)}{(z - z_4)(z_2 - z_3)}. \tag{2.5.4.6}$$

This is a required transformation *mapping the circle (line) passing z_2, z_3, and z_4 onto the circle (line) passing w_2, w_3, and w_4.* It is understood that, in (2.5.4.6),

(1) if $z_2 = \infty$, the right side is $\dfrac{z - z_3}{z - z_4}$;

(2) if $z_3 = \infty$, the right side is $\dfrac{z_2 - z_4}{z - z_4}$;

(3) if $z_4 = \infty$, the right side is $\dfrac{z - z_3}{z_2 - z_3}$. $\tag{2.5.4.7}$

Similar explanation applies to w_2, w_3 and w_4.

If there exists another linear transformation $w = T(z)$ satisfying $w_j = T(z_j)$ for $2 \leq j \leq 4$. Then, the composite transformation $z = T^{-1}(w) = T^{-1}(w(z))$ is still a linear transformation (why?) and preserves z_j, $2 \leq j \leq 4$, fixed, i.e., $z_j = T^{-1}(w(z_j))$, $j = 2, 3, 4$. Stop here and we have to face a *question: How many fixed points could a linear transformation* $w = w(z)$

have? Solve

$$w(z) = z = \frac{az+b}{cz+d},$$

$$\Rightarrow cz^2 + (d-a)z - b = 0,$$

$$\Rightarrow z = \frac{(a-d) \pm \sqrt{(a-d)^2 + 4bc}}{2c} \quad (c \neq 0). \qquad (2.5.4.8)$$

This shows that $w = w(z)$ has *at most two distinct fixed points, except it is the identical mapping* $w = z$. If $c = 0$ and $a \neq d, \infty$ and $\frac{b}{d-a}$ are fixed points; if $c = 0$ and $a = d$, ∞ is the only fixed point.

Returning to the original problem, since $z = T^{-1}(w(z))$ has three distinct fixed points z_2, z_3, and z_4, it is the identical mapping and hence $w = w(z) = T(z)$, $z \in \mathbf{C}^*$. Therefore, *there exists a unique linear transformation, namely,* (2.5.4.6), *mapping three distinct points into three distinct points.*

Section (3) Cross ratio of four points (three of them are distinct.)

Equations (2.5.4.6) and (2.5.4.7) have further implications. Let z_2, z_3, and z_4 be *distinct* and denote the linear transformation

$$S(z) = \frac{(z - z_3)(z_2 - z_4)}{(z - z_4)(z_2 - z_3)}. \qquad (2.5.4.9)$$

If one of z_2, z_3, and z_4 is ∞, $S(z)$ should be interpreted as (2.5.4.7). This $S(z)$ is nothing but $w = w(z)$ in (2.5.4.6) mapping z_2, z_3, and z_4 onto $1, 0$, and ∞, respectively. Take any point $z_1 \in \mathbf{C}^*$ (z_1 could be one of z_2, z_3, and z_4). Define the *cross ratio* of z_1, z_2, z_3, and z_4, *in this ordering*, as

$$(z_1, z_2; z_3, z_4) = \text{the image } S(z_1) \text{ of } z_1 \text{ under } S(z) \text{ in } (2.5.4.9)$$

$$= \frac{(z_1 - z_3)(z_2 - z_4)}{(z_1 - z_4)(z_2 - z_3)}, \qquad (2.5.4.10)$$

subject to the *convention* stated in (2.5.4.7). See Fig. 2.16. In general, cross ratio $(z_1, z_2; z_3, z_4)$ is a complex number and it is a *real number if and only if* z_1, z_2, z_3, and z_4 lie on a circle (*line*).

Let us return to (2.5.4.6). Suppose z_1 is mapped to $w_1 = w(z_1)$ by (2.5.4.6). Then, it follows that

$$(z_1, z_2; z_3, z_4) = (w_1, w_2; w_3, w_4). \qquad (2.5.4.11)$$

This means that *linear transformation preserves cross ratio of four points.*

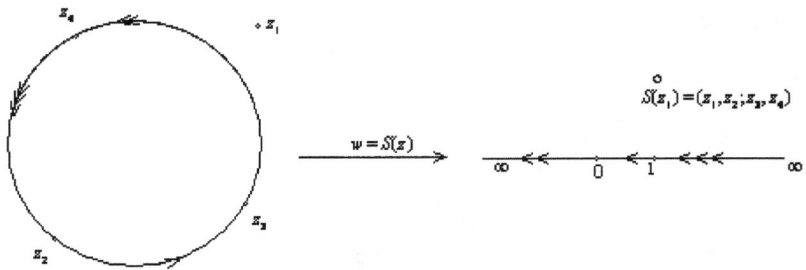

Fig. 2.16

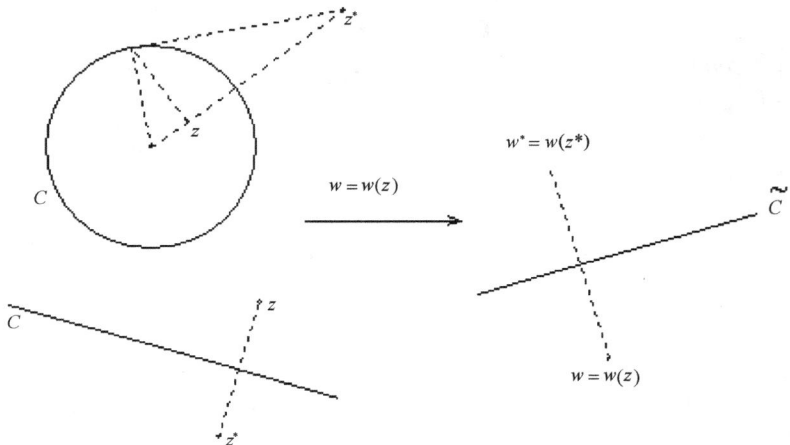

Fig. 2.17

Section (4) Invariance of symmetric points with respect to a circle under $w = w(z)$

Near the end of Section (2), we have shown that $w = w(z)$ preserves symmetric points with respect to a circle (line). See Fig. 2.17.

Now, we give a *geometric proof*. According to (1.4.4.1),
z and z^* are symmetric with respect to a circle (line) C.

$\Leftrightarrow$ Any circle passing z and z^* is orthogonal to C.

$\Leftrightarrow$ (Since $w = w(z)$ is conformal on $\mathbf{C}^*$). Any circle passing $w = w(z)$ and $w^* = w(z^*)$ is orthogonal to the image circle (line) $\tilde{C} = w(C)$.

$\Leftrightarrow$ w and w^* are symmetric *with* respect to $\tilde{C}$.

Yet, another *analytic proof*: According to (1.4.3.6), a circle $\alpha z\bar{z} + \bar{\beta}z + \beta\bar{z} + \gamma = 0$ can be written as $|z + \frac{\beta}{\alpha}| = \frac{\sqrt{|\beta|^2 - \alpha\gamma}}{|\alpha|}$ (only $\alpha \neq 0$ is considered here). Then

z and z^* are symmetric with respect to C.

$\Leftrightarrow$ (See (a) in (1.4.4.1).) $\left(z + \dfrac{\beta}{\alpha}\right)\left(\bar{z}^* + \dfrac{\bar{\beta}}{\alpha}\right) = \dfrac{|\beta|^2 - \alpha\gamma}{\alpha^2}$,

$\Leftrightarrow \alpha z\bar{z}^* + \bar{\beta}z + \beta\bar{z}^* + \gamma = 0$,

$\Leftrightarrow z^* = \dfrac{-\bar{\beta}\bar{z} - \gamma}{\alpha\bar{z} + \bar{\beta}}$.   (2.5.4.12)

Even if $\alpha = 0$, (2.5.4.12) is still true. Try to substitute $z = z(w)$ in (2.5.4.4) into (2.5.4.12) and get $(*_1)$ with $\bar{w}$ there replacing by w^*. Once again, (2.5.4.12) will show that w and w^* are symmetric with respect to the image circle.

Section (5) Summaries

We summarize Sections (1)–(4) as

The basic properties of linear transformations. For linear transformation $w = \frac{az+b}{cz+d}$, $ad - bc \neq 0$, the complex matrix of order 2

$$\begin{bmatrix} a & b \\ c & d \end{bmatrix}$$

is called the *coefficient matrix* of $w = w(z)$ and its determinant is $ad - bc$.

(1) *Operational properties.* The composite of $w = w(z)$ followed by $\tilde{w} = \frac{\tilde{a}w + \tilde{b}}{\tilde{c}w + \tilde{d}}$, $\tilde{a}\tilde{d} - \tilde{b}\tilde{c} \neq 0$, is still a linear transformation

$$\tilde{w} = \frac{\alpha z + \beta}{\gamma z + \delta}, \qquad \begin{bmatrix} \alpha & \beta \\ \gamma & \delta \end{bmatrix} = \begin{bmatrix} \tilde{a} & \tilde{b} \\ \tilde{c} & \tilde{d} \end{bmatrix}\begin{bmatrix} a & b \\ c & d \end{bmatrix}$$

and the inverse (2.5.4.4) $z = z(w)$ of $w = w(z)$ is also a linear transformation. Hence, the set of all linear transformations forms a *group* under composite operation.

(2) *Analytic (or topological) properties.* $w = w(z)$ maps $\mathbf{C}^*$ one-to-one and *conformally* onto $\mathbf{C}^*$. They are the only univalent functions from $\mathbf{C}^*$ onto $\mathbf{C}^*$.

(3) *Geometric mapping properties.* $w = w(z)$ *preserves*

    (i) *circle*: see Section (2), in part., (2.5.4.6) and (2.5.4.7);

    (ii) *cross ratio*: see Section (3), in part., (2.5.4.11);

    (iii) *symmetric points with respect to a circle (line)*: see Section (4), in part.,

        z and z^* are symmetric with respect to a circle (line) C.

  $\Leftrightarrow$ For any three distinct points z_2, z_3, and z_4 on C,

$$(z^*, z_2; z_3, z_4) = \overline{(z, z_2; z_3, z_4)}.$$

And hence, $w = w(z)$ maps points symmetric with respect to C onto points symmetric with respect to the image circle $\tilde{C} = w(C)$. Also, $(z_1, z_2; z_3, z_4)$ is real

$\Leftrightarrow z_1$, z_2, z_3, and z_4 are on a circle or a line.

In addition, $w = w(z)$ has at most two distinct fixed points except the identical mapping

$$w = z. \tag{2.5.4.13}$$

Note that $w = w(z)$ is the nonhomogeneous representation of a projective transformation on the two-dimensional projective complex plane $\mathbf{P}^2(\mathbf{C})$: $w_1' = \sigma(aw_1 + bw_2)$, $w_2' = \sigma(cw_1 + dw_2)$ where $\sigma \neq 0$ is a scalar. For extension of Möbius transformations to several variables, see Ref. [5].

As an application, we raise the *question*: How can we map a circle (line) onto a circle (line) so that the interior or exterior of the circle are mapped respectively, onto those of the image circle, or the half-planes determined by the line onto those of the image line, etc.? There are two aspects.

One: Three distinct points z_2, z_3, and z_4 on a circle (line) are known. And they are required to map into w_2, w_3, and w_4, respectively. Then

$$(w, w_2; w_3, w_4) = (z, z_2; z_3, z_4), \tag{2.5.4.14}$$

namely (2.5.4.6), is the required one. In case z_2, z_3, and z_4, in this ordering, is counterclockwise, while w_2, w_3, and w_4, in this ordering, is clockwise on the image circle $\tilde{C}$, then $w = w(z)$ maps the interior and exterior of C onto the exterior and interior of $\tilde{C}$, respectively. This is because of the conformality $w = w(z)$. See Fig. 2.18.

The other: One point on a circle and a pair of symmetric points with respect to the circle are known. Let point z_2 be on the circle C, and

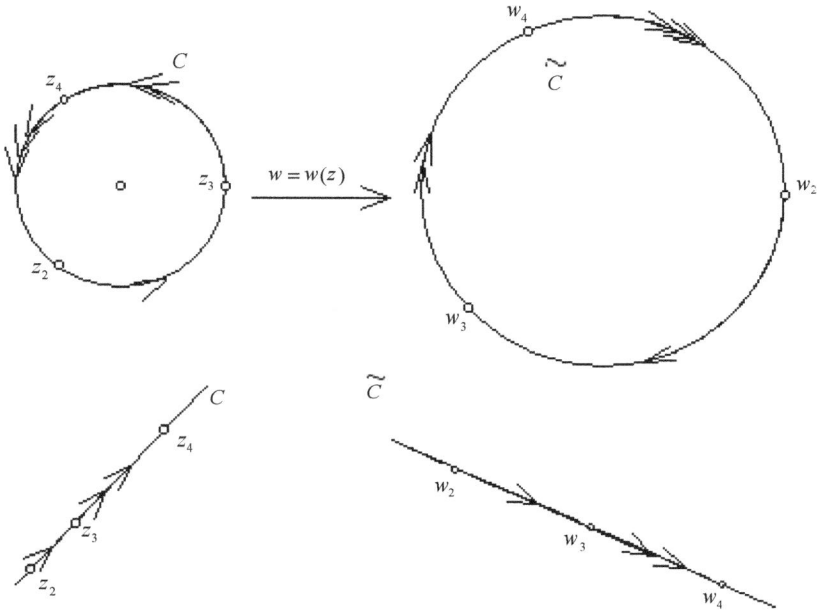

Fig. 2.18

z_3 and $z_4 = z_3^*$ be symmetric points with respect to C. We may suppose z_3 does not lie on C. From Example 3 in Sec. 1.4.3 and (1.4.4.1), each circle in the family

$$\left| \frac{z - z_3}{z - z_3^*} \right| = \lambda \ (0 < \lambda < \infty) \quad \text{or} \quad \left| z - \frac{z_3 - \lambda^2 z_3^*}{1 - \lambda^2} \right| = \frac{\lambda |z_3 - z_3^*|}{|1 - \lambda^2|}$$

has z_3 and z_3^* as its symmetric points. Hence, a pair of symmetric points cannot locate a circle *uniquely*. If a point z_2 on a circle is pinpointed, then the center and radius of that circle can be figured out precisely via the above equation and the circle is then well-determined. In this situation, the required $w = w(z)$ is given by, owing to (3)2 in (2.5.4.13),

$$(w(z), w(z_2); w(z_3), w(z_3^*)) = (z, z_2; z_3, z_3^*). \tag{2.5.4.15}$$

In case z_3 lies inside C and the image point $w(z_3)$ also lies inside the image circle $\tilde{C}$, then $w = w(z)$ maps the interior and exterior of C onto those of $\tilde{C}$, respectively. See Fig. 2.19.

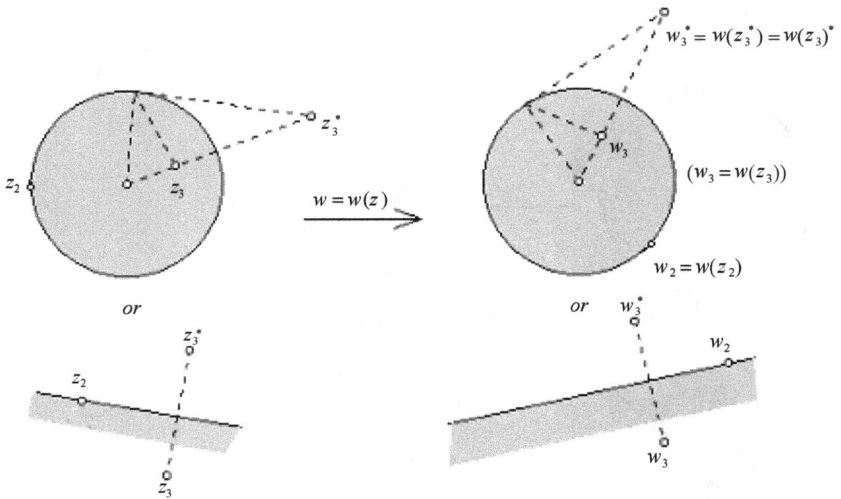

Fig. 2.19

Section (6) Examples

We give five examples in what follows.

Example 1. The most general linear transformation mapping Im $z > 0$ onto $|w| < 1$ is of the form

$$w = e^{i\theta}\frac{z - z_0}{z - z_0}, \quad \theta \in \mathbf{R} \text{ (a constant) and } z_0 \text{ is fixed with Im } z_0 > 0.$$

Solution. A point in the upper half-plane Im $z > 0$, say z_0, should be mapped into the zero point 0. Then, its symmetric points $\bar{z}_0$ should go into the symmetric point ∞ of 0 with respect to $|w| = 1$. So $w = w(z)$ is of the form

$$w = k\frac{z - z_0}{z - \bar{z}_0}.$$

Since $z = 0$ is mapped into the point $w(0)$ on the unit circle, namely, $|w(0)| = 1$, then

$$|w(0)| = |k| = 1.$$

Hence, $k = e^{i\theta}$ for some real θ and the result follows.

Owing to conformality, $w = w(z)$ maps

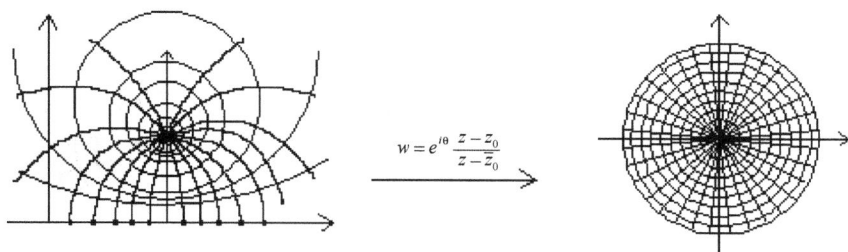

$$w = e^{i\theta} \frac{z - z_0}{z - \bar{z}_0}$$

Fig. 2.20

Apollonius circles $\left|\frac{z-z_0}{z-\bar{z}_0}\right| = \lambda \ (0 < \lambda \le 1) \to$ concentric circles $|w| = \lambda$;
Conjugate circles $\operatorname{Arg} \frac{z-z_0}{z-\bar{z}_0} = \theta \ (-\pi < \theta \le \pi)$ that lie on $\operatorname{Im} z > 0 \to$ the
diameters in $|w| = 1$.

See Fig. 2.20.

Note that

$$w'(z) = e^{i\theta} \frac{z_0 - \bar{z}_0}{(z - \bar{z}_0)^2} = 2ie^{i\theta} \frac{\operatorname{Im} z_0}{(z - \bar{z}_0)^2}.$$

If $z = \bar{z}_0 + i\alpha$ lies on the line $\operatorname{Re} z = \operatorname{Re} \bar{z}_0$, then $|z - \bar{z}_0| = |i\alpha| = |\alpha|$. This suggests that $|w'(z)| = \frac{2\operatorname{Im} z_0}{\alpha^2}$ is larger if $\alpha \to 0^+$ and is smaller if $\alpha > 1 \to \infty$. These facts reflect in Fig. 2.20 (refer to (3.2.3.1) and Fig. 3.1(a)).

Example 2. The most general linear transformation mapping $|z| < 1$ onto $|w| < 1$ is of the form (compare to the Example in Sec. 2.5.3)

$$w = e^{i\theta} \frac{z - z_0}{1 - \bar{z}_0 z}, z_0 \text{ is fixed and } |z_0| < 1, \theta \in \mathbf{R} \text{ (a constant)}.$$

Solution. A point $z_0, |z_0| < 1$, is mapped into 0, while its symmetric point $\frac{1}{\bar{z}_0}$ with respect to $|z| < 1$ is mapped into ∞. So

$$w = k \frac{z - z_0}{1 - \bar{z}_0 z},$$

where k is some constant. Since $|z| = 1 \Rightarrow |w| = 1$, it follows that $|k| = 1$ (see Exercise A (9) of Sec. 1.4.1). Set $k = e^{i\theta}$ for $\theta \in \mathbf{R}$ and the result follows.

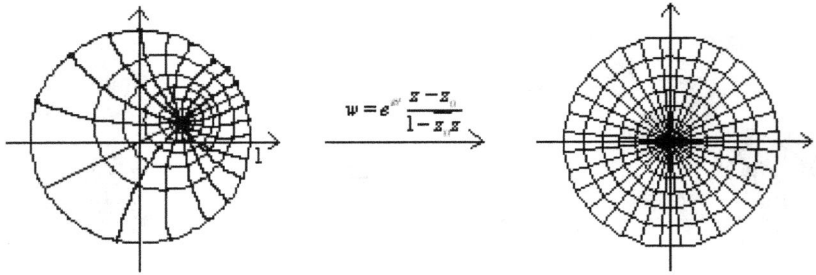

$$w = e^{i\theta} \frac{z - z_0}{1 - \bar{z}_0 z}$$

Fig. 2.21

On the other hand, $w = w(z)$ maps

Apollonius circles $\left| \frac{z - z_0}{z - \bar{z}_0^{-1}} \right| = \lambda \ (0 < \lambda \le |z_0|) \to$ concentric circles $|w| = \frac{\lambda}{|z_0|}$; Conjugate circles $\text{Arg} \frac{z - z_0}{z - \bar{z}_0^{-1}} = \theta \ (0 < \pi < \theta \le \pi)$ that lie inside $|z| < 1 \to$ the diameters in $|w| = 1$.

See Fig. 2.21.

Note that

$$w'(z) = e^{i\theta} \frac{1 - |z_0|^2}{(1 - \bar{z}_0 z)^2}, \quad |z| < 1, \quad |z_0| < 1,$$

$$\Rightarrow w'(z_0) = e^{i\theta} \frac{1}{1 - |z_0|^2} \quad \text{with } \text{Arg} \, w'(z_0) = \theta.$$

Now $1 - |z_0| < 1 - |z_0||z| \le |1 - \bar{z}_0 z| \le 1 + |z||z_0| < 1 + |z_0|$, the equality in the left "$\le$" holds $\Leftrightarrow 1 \ (\overline{-\bar{z}_0 z}) = -z_0 \bar{z} \le 0$ (see (1.4.2.3)), namely, $z = \alpha z_0$ and $0 < \alpha < \frac{1}{|z_0|}$; the equality in the right "$\le$" holds $\Leftrightarrow 1 \cdot \overline{(-\bar{z}_0 z)} = -z_0 \bar{z} \ge 0$ (see (1.4.2.2)), namely, $z = \alpha z_0$ and $-\frac{1}{|z_0|} < \alpha \le 0$. Hence

$$|w'(z)| = \frac{1 - |z_0|^2}{|1 - \bar{z}_0 z|^2} \begin{cases} \le \dfrac{1 - |z_0|^2}{(1 - |z_0||z|)^2} \le \dfrac{1 + |z_0|}{1 - |z_0|}, & |z| < 1 \\[3mm] \ge \dfrac{1 - |z_0|^2}{(1 + |z_0||z|)^2} \ge \dfrac{1 - |z_0|}{1 + |z_0|}, & |z| < 1 \end{cases}.$$

This indicates that, $|w'(z)|$ will attain larger values along the radius having the same direction as z_0; will attain smaller values along the radius having the opposite direction against that of z_0. These facts also reflect in Fig. 2.21 (again, refer to (3.2.3.1) and Fig. 3.1(a)).

Example 3. The most general linear transformation mapping $\operatorname{Im} z > 0$ onto $\operatorname{Im} w > 0$ is of the form

$$w = \frac{az+b}{cz+d}, \quad a,b,c,d \in \mathbf{R} \quad \text{and} \quad ad - bc > 0.$$

Solution. Take real numbers z_2, z_3, z_4 so that $z_2 < z_3 < z_4$. The linear transformation $w = w(z)$ mapping z_2, z_3, z_4 onto $0, 1, \infty$, respectively, is

$$(w, 1; 0, \infty) = (z, z_3; z_2, z_4)$$

$$\Rightarrow w = \frac{z_3 - z_4}{z_3 - z_2} \cdot \frac{z - z_2}{z - z_4} = \frac{az+b}{cz+d},$$

where $a = z_3 - z_4$, $b = -z_2(z_3 - z_4)$, $c = z_3 - z_2$, and $d = -z_4(z_3 - z_2)$ are reals, and $ad - bc = (z_2 - z_4)(z_3 - z_4)(z_3 - z_2) > 0$.

Or, pick up two sets of distinct real numbers z_2, z_3, z_4 and w_2, w_3, w_4. Then

$$(w, w_2; w_3, w_4) = (z, z_2; z_3, z_4)$$

$$\Rightarrow w = \frac{az+b}{cz+d}, \quad a,b,c,d \in \mathbf{R} \text{ (depending only on } z_2, \ldots, w_4 \text{)},$$

$$\Rightarrow \bar{w} = \frac{a\bar{z}+b}{c\bar{z}+d},$$

$$\Rightarrow w - \bar{w} = \frac{(ad-bc)(z-\bar{z})}{|cz+d|^2} \quad \text{or} \quad \operatorname{Im} w = \frac{ad-bc}{|cz+d|^2} \operatorname{Im} z.$$

The result follows easily.

Example 4. There exists a linear transformation mapping the concentric ring domain $\Omega_1; 0 < r_1 < |z - z_0| < r_2$ onto another one $\Omega_2 : 0 < R_1 < |w - w_0| < R_2$ if and only if

$$\frac{r_2}{r_1} = \frac{R_2}{R_1}.$$

This ratio is conformally invariant and is called the *modulus* of Ω_1.

Solution. See Fig. 2.22.

For sufficiency: $w = w_0 + \frac{R_1}{r_1}(z - z_0)$ is a required one.

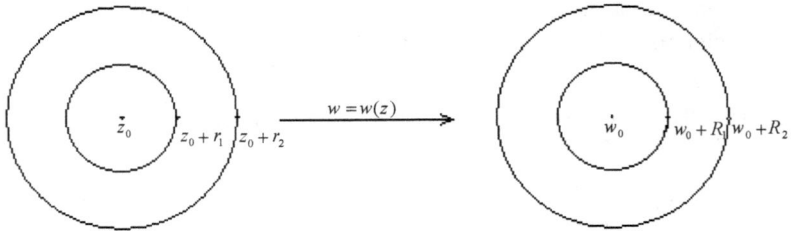

Fig. 2.22

For necessity: Construct a linear transformation mapping z_0 to w_0 and hence, ∞ to ∞, and $z_0 + r_1$ to $w_0 + R_1$. Then

$$(w, w_0 + R_1; w_0, \infty) = (z, z_0 + r_1; z_0, \infty),$$

$$\Rightarrow w = w_0 + \frac{R_1}{r_1}(z - z_0).$$

For this $w = w(z)$ to map $|z - z_0| = r_2$ onto $|w - w_0| = R_2$, it is necessary that

$$R_2 = \frac{R_1}{r_1}r_2 \quad \text{or} \quad \frac{r_2}{r_1} = \frac{R_2}{R_1}.$$

Similarly, $w = w_0 + \frac{R_2}{r_1}(z - z_0)$ maps $|z - z_0| = r_1$ onto $|w - w_0| = R_2$. In order to map $|z - z_0| = r_2$ onto $|w - w_0| = R_1$, it must be $R_1 = \frac{R_2}{r_1}r_2$ or $\frac{r_1}{r_2} = \frac{R_2}{R_1}$ (and $R_1 > R_2$).

Suppose there exists a mapping $w = \frac{az+b}{cz+d}$ that maps Ω_1 onto Ω_2. Since z_0 and ∞ are symmetric to both $|z - z_0| = r_1$ and $|z - z_0| = r_2$, so are their images $w(z_0)$ and $w(\infty)$ with respect to $|w - w_0| = R_1$ and $|w - w_0| = R_2$. Hence

$$w(\infty) = \infty \Rightarrow c = 0 \text{ and hence, } w(z) = \alpha z + \beta;$$

$$w(z_0) = w_0 \Rightarrow w_0 = \alpha z_0 + \beta \text{ or } \beta = w_0 - \alpha z_0,$$

$$\Rightarrow w(z) = w_0 + \alpha(z - z_0).$$

If the image of $|z - z_0| = r_1$ under this $w = w(z)$ is $|w - w_0| = R_1$, then $\alpha = \frac{R_1}{r_1}$; if the image is $|w - w_0| = R_2$, then $\alpha = \frac{R_2}{r_1}$. These are the cases considered in the last two paragraphs. Of course, it is possible that $w(\infty) = w_0$ and $w(z_0) = \infty$, and we leave this case to the readers.

Example 5.   Find a linear transformation mapping $|z| = 1$ and $\operatorname{Re} z = 2$ onto concentric circles and determine the ratio of their radii.

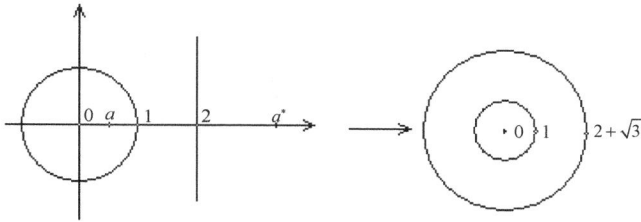

Fig. 2.23

Solution. See Fig. 2.23.

We try to find a real point a whose symmetric points with respect to both $|z| = 1$ and $\operatorname{Re} z = 2$ coincide, say a^*. This point a is going to be mapped into the center, say o, of the concentric circles. Hence, a^* is to ∞. Now

$$\frac{1}{a}(\text{symmetric point of a with respect to } |z| = 1)$$

$$= 4 - \bar{a} \ (\text{with respect to } \operatorname{Re} z = 2),$$

$$\Rightarrow a^2 - 4a + 1 = 0,$$

$$\Rightarrow a = 2 \pm \sqrt{3}.$$

Note that if $a = 2 + \sqrt{3}$, then $a^* = 2 - \sqrt{3}$, and vice versa.

If the point $z_0 = 1$ is to a point $w_0 \neq 0$, then

$$(w, w_0; 0, \infty) = (z, 1; 2 - \sqrt{3}, 2 + \sqrt{3}),$$

$$\Rightarrow w = -w_0(2 + \sqrt{3})\frac{z - (2 - \sqrt{3})}{z - (2 + \sqrt{3})}$$

maps $|z| = 1$ onto $|w| = |w_0|$, $\operatorname{Re} z = 2$ onto $|w| = |w_0|(2 + \sqrt{3})$ with the ratio of their radii equal to $2 + \sqrt{3}$.

Exercises A

In general, a linear transformation is denoted, in short, as $w = w(z)$.

(1) Find the image of the domain Ω under the given $w = w(z)$.

(a) $w = \dfrac{1}{z}$, $\quad \Omega = \{|z + i| < \sqrt{2}\} \cap \{|z + \sqrt{3}i| > 2\}$.

(b) $w = \dfrac{z - 1}{z - i}$, $\quad \Omega = \{|z| < 1\} \cap \{|z - (1 + i)| < 1\}$.

(c) $w = \dfrac{z - i}{z + i}$, $\Omega = \{\operatorname{Re} z > 0\} \cap \{\operatorname{Im} z > 0\}$.

(d) $w = \dfrac{z + 1}{z - 2}$, $\Omega = \{0 < \operatorname{Re} z < 1\}$.

(2) Determine $w = w(z)$ satisfying $w(z_j) = w_j$, $j = 2, 3, 4$, and find the image under $w = w(z)$ of the interior of the circle (or one side of the line) passing z_2, z_3, z_4.

    (a) $z_2 = 1, z_3 = i, z_4 = -i$ and $w_2 = 1, w_3 = 0, w_4 = -1$.

    (b) $z_2 = 1, z_3 = i, z_4 = -1$ and $w_2 = \infty, w_3 = -1, w_4 = 0$.

    (c) $z_2 = \infty, z_3 = 0, z_4 = 1$ and $w_2 = 0, w_3 = 1, w_4 = \infty$.

(3) Find $w = w(z)$, mapping $\operatorname{Im} z > 0$ onto $|w - w_0| < R$ and satisfying $w(i) = w_0$, $w'(i) > 0$.

(4) Find $w = w(z)$, mapping $\operatorname{Im} z > 0$ onto $|w| < 1$ and satisfying $w(i) = 0$ and $w'(i) > 0$, $\operatorname{Arg} w'(i) = \frac{\pi}{2}$, $w'(i) = 1$, respectively.

(5) Find $w = w(z)$, mapping $|z - 4i| < 2$ onto the half-plane $\operatorname{Im} w > \operatorname{Re} w$ and satisfying $w(4i) = -4$, $w(2i) = 0$.

(6) Determine $R > 0$ so that $w = w(z)$ maps $\operatorname{Im} z > 0$ onto $|w| < R$ and satisfies $w(i) = 0$, $w'(i) = 1$.

(7) Find $w = w(z)$, mapping $|z| < 2$ onto $\operatorname{Re} w > 0$ so that $w(0) = 1$ and $\operatorname{Arg} w'(0) = \frac{\pi}{2}$.

(8) Find $w = w(z)$, mapping $\operatorname{Im} z > 0$ onto $\operatorname{Im} w < 0$ so that $w(z_0) = \bar{z}_0$ $(\operatorname{Im} z_0 > 0)$ and $\operatorname{Arg} w'(z_0) = -\frac{\pi}{2}$.

(9) Given $w = w(z)$.

    (a) If it has two distinct fixed points z_1 and z_2, then $w = w(z)$ is of the form

$$\frac{w - z_1}{w - z_2} = k \frac{z - z_1}{z - z_2}, \qquad \text{where } k \text{ is a nonzero constant.}$$

    (b) If it has only one finite fixed point z_0, then

$$\frac{1}{w - z_0} = \frac{1}{z - z_0} + c, \qquad \text{where } c \text{ is a nonzero constant.}$$

Determine $w = w(z)$ so that $w(2) = 2$, $w(\frac{1}{2}) = \frac{1}{2}$ and $w(\frac{1}{4}(5 + 3i)) = 0$ and another $w = w(z)$ so that $w(1) = \infty$ and i is the only fixed point.

(10) Find $w = w(z)$, mapping the domain bounded by $|z - 3| = 9$ and
$|z - 8| = 16$ onto a concentric domain, with center at o and outer
radius equal to 1.

(11) (a) Find $w = w(z)$, mapping $\{|z - 3i| > 2\} \cap \{|z - 4| > 2\}$ onto
$r < |w| < 1$. Determine r.

(b) Find $w = w(z)$, mapping $\{\operatorname{Re} z > 0\} - \{|z - h| \leq 1\}$, where $h > 1$,
onto $1 < |w| < 2$. What is h?

(12) Any $w = w(z)$, mapping a triangle onto another triangle, should be
of the form $w = az + b$, where $a \neq 0$.

(13) Find $w = w(z)$, mapping $|z| < 1$ onto itself so that a segment
$0 \leq \operatorname{Re} z \leq a$ $(a < 1)$ lying on the real axis is mapped onto a line
segment lying on the real axis and symmetric with respect to the ori-
gin. Determine the length of this line segment. See Fig. 2.24.

(14) Find $w = w(z) = \frac{az+b}{cz+d}$ so that $w(w(z)) = z, z \in \mathbf{C}^*$.

(15) If $w = \frac{az+b}{cz+d}$ maps $|z| = 1$ onto a line; find conditions on a, b, c, and d.

(16) If $w = \frac{az+b}{cz+d}$ maps a circular triangle formed by three circular arcs onto
a triangle formed by line segments, find the necessary and sufficient
condition.

(17) Find all circles orthogonal to both $|z| = 1$ and $|z-1| = 4$.

(18) Let γ be any circle passing -1 and 1. Suppose z_1 and z_2 are any two
points not on γ but satisfy $z_1 z_2 = 1$. Show that one of z_1 and z_2 lies
inside γ and the other outside γ.

(19) Suppose $w = w(z)$ has two distinct fixed points z_1 and z_2. Show that
$w'(z_1)w'(z_2) = 1$ holds.

(20) Under the *symmetric motion* $(z \rightarrow z^*)$ with respect to a circle D
with radius R, a circle C of radius r is mapped onto a circle $\tilde{C}$ with

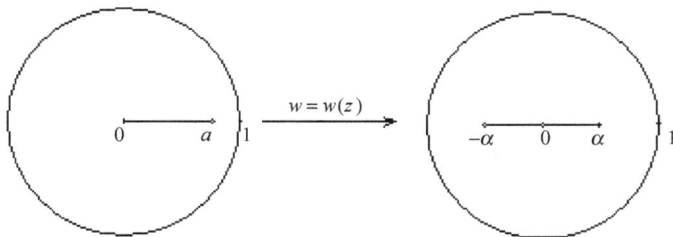

Fig. 2.24

radius ρ. Show that

$$\rho = \frac{R^2 r}{|\sigma^2 - r^2|},$$

where σ is the distance between the center of C and the center of D.

(21) Transformation $w = w(z)$ satisfying $(w, z_2; z_3, z_4) = \overline{(z, z_2; z_3, z_4)}$ is called a *symmetric motion*. Show that the composite of two such motions is a linear transformation.

(22) Suppose $w = w(z)$ has two distinct fixed points z_1 and z_2. Show that $w = w(z)$ maps Apollonius circles: $\left|\frac{z-z_1}{z-z_2}\right| = \lambda$ $(0 \le \lambda \le \infty)$; conjugate circles: Arg $\frac{z-z_1}{z-z_2} = \theta(-\pi < \theta \le \pi)$, onto themselves, respectively.

(23) Given four distinct points z_1, z_2, z_3 and z_4. Try to use $\lambda = (z_1, z_2; z_3, z_4)$ to express the 24 cross ratios caused by permutations of $z_1, z_2, z_3,$ and z_4.

(24) Let z_2 and z_4 be symmetric with respect to a circle C, and z_1 and z_3 be on the circle C. Show that $|(z_1, z_2; z_3, z_4)| = 1$.

(25) Try to use cross ratio to prove Ptolemy theorem stated in Example 2 of Sec. 1.4.3.

(26) Suppose z_1, z_2, z_3, and z_4, in this ordering, lie on a circle. Then z_1, z_3, z_4 and z_2, z_3, z_4 determine the same orientation of the circle $\Leftrightarrow (z_1, z_2; z_3, z_4) > 0$.

(27) Suppose a simply closed curve γ has its image $\Gamma = w(\gamma)$ under $w = w(z)$.

(a) Show that Γ is also a simply closed curve.
(b) Int γ is mapped, via $w = w(z)$, one-to-one and onto Int Γ or Ext Γ.

Exercises B

(1) (continued from Exercise A (9)). If $w = w(z)$ has two distinct fixed points and $k > 0$, $w = w(z)$ is called a linear transformation of *hyperbolic type*; if $|k| = 1$, an *elliptic type*. If $w = w(z)$ has only one fixed point, then it is called a *parabolic type*. Otherwise, $w = w(z)$ is called a *loxodromic type*.

(a) Classify the following according to their types:

$$w = \frac{z}{z-1}; \quad w = \frac{2z}{3z-1}; \quad w = \frac{3z-4}{z-1}; \quad w = \frac{z}{2-z}.$$

(b) Suppose $ad - bc = 1$ and *redefine* k by $k + \frac{1}{k} = (a + d)^2 - 2$. Show that $w = w(z)$ is

   (i) a hyperbolic type $\Leftrightarrow a + d$ is real and $|a + d| > 2$;

$$\Leftrightarrow k > 0, \quad k \neq 1;$$

   (ii) an elliptic type $\Leftrightarrow a + d$ is real and $|a + d| < 2$;

$$\Leftrightarrow |k| = 1 \quad \text{and} \quad k \neq 1.$$

   (iii) a parabolic type $\Leftrightarrow a + d = \pm 2$;

$$\Leftrightarrow k = 1$$

   (iv) a loxodromic type $\Leftrightarrow a + d$ is a nonreal complex number;

$$\Leftrightarrow k = |k|e^{i\theta}, \quad \theta \neq 2n\pi, \quad |k| > 0 \quad \text{and} \quad |k| \neq 1.$$

(2) Show that there exists a linear transformation mapping any presumed four distinct points z_1, z_2, z_3, z_4 onto 1, -1, k, $-k$, respectively, where k depends on these four points. How many are there? Any relations among them?

(3) Suppose $w = w(z)$ is a linear transformation mapping $|z| < 1$ onto itself.

   (a) Let $|z_1| < 1, |z_2| < 1$ and $w_1 = w(z_1), w_2 = w(z_2)$. Show that

$$\left| \frac{z_1 - z_2}{1 - \bar{z}_1 z_2} \right| = \left| \frac{w_1 - w_2}{1 - \bar{w}_1 w_2} \right|.$$

   (b) Show that

$$\frac{|dz|}{1 - |z|^2} = \frac{|dw|}{1 - |w|^2}, \quad |z| < 1.$$

Exercise C

Read Appendix B in Vol. 2 and do exercises within.

2.5.5. *Joukowski function* $w = \frac{1}{2}\left(z + \frac{1}{z}\right)$

The function

$$w = f(z) = \frac{1}{2}\left(z + \frac{1}{z}\right), \quad z \in \mathbf{C}^* \tag{2.5.5.1}$$

is specifically called a *Joukowski function* where $w(0) = w(\infty) = \infty$. For other functions of this type, see exercises in this section.

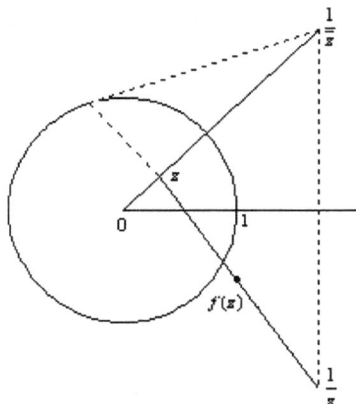

Fig. 2.25

Figure 2.25 shows that $f(z)$ is the midpoint of the segment connecting z to z^{-1}. In particular,

(1) In case $|z| = 1$, then $\frac{1}{z} = \bar{z}$ and $f(z)$ is real. Thus, $w = f(z)$ maps both the upper half-circle and the lower half-circle one-to-one and onto $[-1, 1]$, namely, $w = f(z)$ covers $[-1, 1]$ twice where -1 and 1 covered once with multiplicity 2.

(2) Since $f(z) = f(z^{-1})$, $w = f(z)$ maps $|z| < 1$ and $|z| > 1$ to the same domain.

We will see the domain in 2 is $\mathbf{C}^* - [-1, 1]$. Therefore, $w = f(z)$ covers the w-plane twice.

$w = f(z)$ has simple poles (see Sec. 2.5.3) at 0 and ∞.

Since $w' = \frac{1}{2}\left(1 - \frac{1}{z^2}\right) = 0 \Leftrightarrow z = \pm 1$, $w = f(z)$ ceases to be conformal at $z = \pm 1$ which correspond to the point $w = \pm 1$.

To find the *domain of univalence* (or *fundamental domain*) for $w = f(z)$, suppose $z_1, z_2 \in \mathbf{C}$ so that

$$\frac{1}{2}\left(z_1 + \frac{1}{z_1}\right) = \frac{1}{2}\left(z_2 + \frac{1}{z_2}\right),$$

$$\Rightarrow (z_1 - z_2)\left(1 - \frac{1}{z_1 z_2}\right) = 0.$$

Hence, a domain Ω is a domain of univalence if and only if Ω does not contain two points z_1 and z_2 satisfying $z_1 z_2 = 1$. Therefore, both $|z| < 1$ and $|z| > 1$ in $\mathbf{C}^*$ are domains of univalence for $w = f(z)$, on which

$f(z)$ are *univalent*, namely, one-to-one. Note that $f(z) = f(z^{-1})$. In case $|z_1| = |z_2| = 1$ and $z_1 \neq z_2$, then $1 - \frac{1}{z_1 z_2} = 0 \Leftrightarrow z_2 = \bar{z}_1$. This shows that $f(z)$ is *not* one-to-one on $|z| = 1$, but two-to-one with multiplicity 2 at $z = \pm 1$.

What are the images of $|z| = 1, |z| < 1$, and $|z| > 1$ under $w = f(z)$?

On $|z| = 1$, $f(z) = \frac{1}{2}\left(z + \frac{1}{z}\right) = \frac{1}{2}(z + \bar{z}) = \operatorname{Re} z$ holds. So $w = f(z)$ maps $|z| = 1$ onto $[-1, 1]$ in a two-to-one manner.

Fix any $w \in \mathbf{C}^* - [-1, 1]$. Since $w(0) = w(\infty) = \infty$, we may suppose $w \neq \infty$. Solve

$$\frac{1}{2}\left(z + \frac{1}{z}\right) = w,$$

$$\Rightarrow z = w \pm \sqrt{w^2 - 1} \quad \text{(refer to Exercises B (1) and (2) of Sec. 1.3).}$$

$\sqrt{w^2 - 1}$ is double-valued (for detailed account, see Example 2 in Sec. 2.7.2). We designate $z_1 = w - \sqrt{w^2 - 1}$ to be 0 at $w = \infty$ while $z_2 = w + \sqrt{w^2 - 1}$ to be ∞ at $w = \infty$. Then $z_1 z_2 = 1$ and $|z_1| < 1, |z_2| > 1$. Hence $w = f(z)$ maps $|z| < 1$ and $|z| > 1$ one-to-one and onto $\mathbf{C}^* - [-1, 1]$, respectively.

For geometric mapping properties, let $z = re^{i\theta}$, $w = f(z) = u + iv$. Then

$$u + iv = \frac{1}{2}\left(re^{i\theta} + \frac{1}{r}e^{-i\theta}\right)$$

$$\Rightarrow u = \frac{1}{2}\left(r + \frac{1}{r}\right)\cos\theta, \quad v = \frac{1}{2}\left(r - \frac{1}{r}\right)\sin\theta, \quad r \geq 0, \ 0 \leq \theta \leq 2\pi.$$

By eliminating the parameters θ and r, respectively, we get

$$\frac{u^2}{\left[\frac{1}{2}(r + r^{-1})\right]^2} + \frac{v^2}{\left[\frac{1}{2}(r - r^{-1})\right]^2} = 1, \qquad (2.5.5.2)$$

$$\frac{u^2}{\cos^2\theta} - \frac{v^2}{\sin^2\theta} = 1. \qquad (2.5.5.3)$$

We separate it into two cases.

Case 1. Suppose $r = \text{const.}$

$r = 0$: the point circle $|z| = r = 0$ corresponds to $w = \infty$ under $w = f(z)$.

$0 < r < 1$: The circle $|z| = r$ is mapped, via $w = f(z)$, one-to-one and onto the ellipse (2.5.5.2) in the w-plane. They ($0 < r < 1$) together form a *family of coaxial ellipses*:

$$\text{semiaxes:} \quad \frac{1}{2}(r + r^{-1}), \quad \frac{1}{2}|r - r^{-1}|$$

$$\text{foci:} \quad \sqrt{\left[\frac{1}{2}(r + r^{-1})\right]^2 - \left[\frac{1}{2}(r - r^{-1})\right]^2} = \pm 1.$$

Note that the interior $|z| < r$ is mapped onto the outside of the corresponding ellipse.

If r increases from 0 to 1, the corresponding ellipses decrease from ∞ to the degenerated segment $[-1, 1]$.

What we obtained so far are still valid for $1 \le r \le \infty$, except that as r ranges from 1 to ∞, then the ellipse increases from $[-1, 1]$ to ∞. See Fig. 2.26.

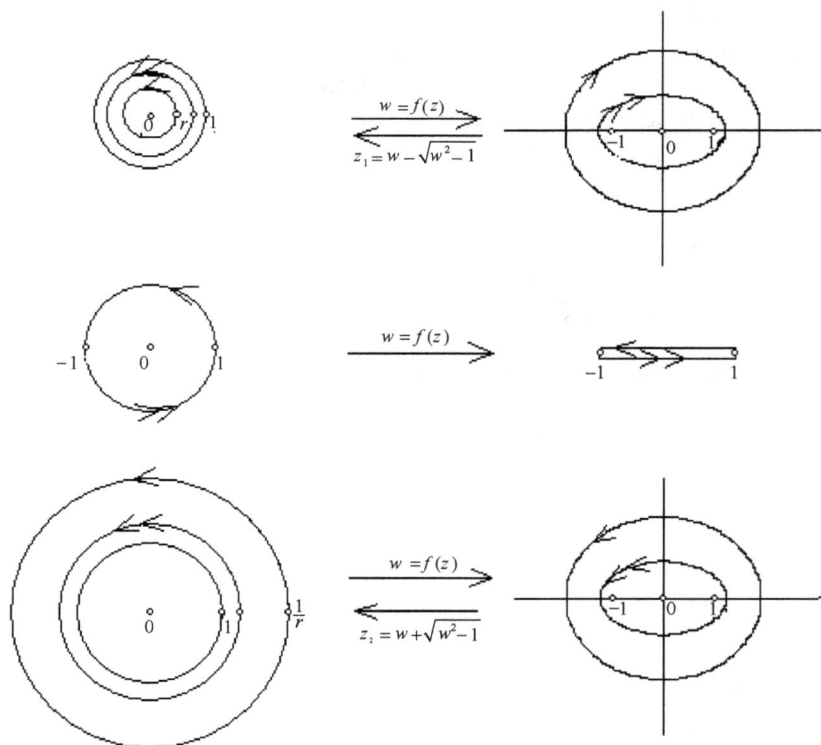

Fig. 2.26

Case 2. Suppose $\theta = \text{const.}$ $0 < \theta < \frac{\pi}{2}$:

① $0 < r \le 1$: Via $u = \frac{1}{2}(r + r^{-1})\cos\theta$ and $v = \frac{1}{2}(r - r^{-1})\sin\theta$, $w = f(z)$
maps the ray $\text{Arg}\, z = \theta$, $0 < r \le 1$, onto the part of the hyperbola
(2.5.5.3) in the fourth quadrant; if r increases from 0 to 1, the image
point $w = f(z)$ moves from ∞ along that part to the vertex $u = \cos\theta$
on $[-1, 1]$.

② $1 \le r < \infty$: The ray $\text{Arg}\, z = \theta$, $1 \le r < \infty$, is mapped onto the part
of the hyperbola in the first quadrant; as r increases from 1 to ∞, the
point $w = f(z)$ moves from $u = \cos\theta$ along this part to ∞.

③ $1 \le r < \infty$: The ray $\text{Arg}\, z = \theta + \pi$, $1 \le r < \infty$, is mapped onto the part
of the hyperbola in the third quadrant.

④ $0 < r \le 1$: The ray $\text{Arg}\, z = \theta + \pi$, $0 < r \le 1$, is mapped into the second
quadrant.

As a whole, the ray $\text{Arg}\, z = \theta$ is mapped to the right branch of a hyperbola, while $\text{Arg}\, z = \theta + \pi$ to the left branch of the same hyperbola. See
Fig. 2.28(b).

$\theta = 0$: $u = \frac{1}{2}(r + r^{-1})$ and $v = 0$. If r increases from 0 to 1 and then
to ∞, the image point $w = f(z)$ decreases from ∞, along the positive real
axis, to 1 and back to ∞. See Fig. 2.28(a).

$\theta = \frac{\pi}{2}$: $u = 0$, $v = \frac{1}{2}(r - r^{-1})$. If the point $z = iy$ moves from o $(r = 0)$,
along the imaginary axis, to i $(r = 1)$ and then to $i\infty = \infty$ $(r = \infty)$, the
image point $w = f(z)$ moves from $-i\infty = \infty$, along the imaginary axis, to
0 and then to $i\infty = \infty$. See Fig. 2.28(c).

$\theta = \pi$ or $\frac{3}{2}\pi$: Both are left to the readers.

Since $f(z) = f(\frac{1}{z})$, $|z| < 1$ and $\text{Im}\, z > 0$ is mapped by $w = f(z)$ one-
to-one and onto $\text{Im}\, w < 0$, so is $|z| > 1$ and $\text{Im}\, z < 0$. Hence, $|z| > 1$ and
$\text{Im}\, z > 0$ is mapped by $w = f(z)$ one-to-one and onto $\text{Im}\, w > 0$, so is $|z| < 1$
and $\text{Im}\, z < 0$. See Fig. 2.27.

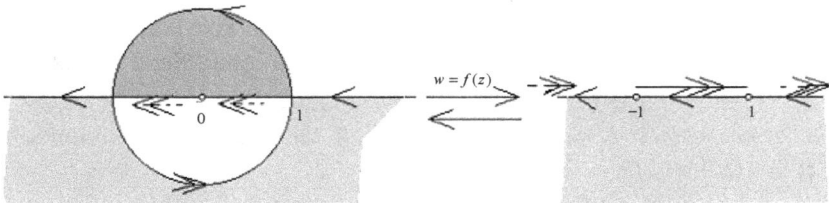

Fig. 2.27

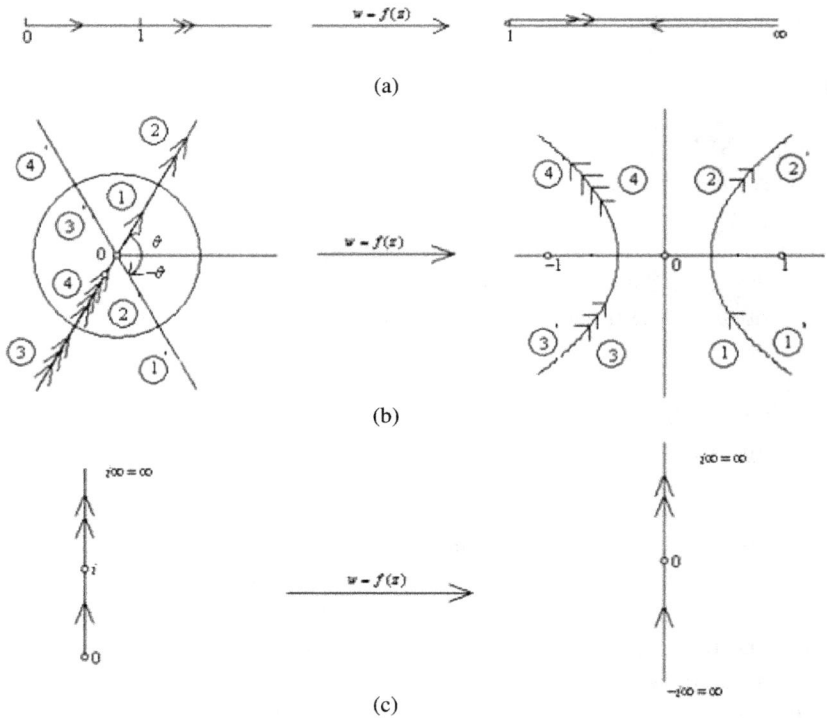

(a)

(b)

(c)

Fig. 2.28

$-\frac{\pi}{2} < \theta < 0$: Via the last paragraph,

①′ $\operatorname{Arg} z = \theta$, $1 \leq r < \infty$: Its image is the same as ① above.
②′ $\operatorname{Arg} z = \theta$, $0 < r \leq 1$: as ②.
③′ $\operatorname{Arg} z = \theta + \pi$, $0 < r \leq 1$: as ③.
④′ $\operatorname{Arg} z = \theta + \pi$, $1 \leq r < \infty$: as ④.

See Fig. 2.28(b).

Therefore, $w = f(z)$ maps lines passing the origin o into a family of hyperbolas (2.5.5.3) with foci at ± 1, the same as the family of ellipses (2.5.5.2). The two families are orthogonal to each other. See Fig. 2.29.

We summarize the above as

The basic geometric mapping properties of the Joukowski function $w = f(z) = \frac{1}{2}(z + z^{-1})$.

(1) $w = f(z)$ is conformal everywhere on $\mathbf{C}^*$ except at ± 1 with simple poles at $z = 0$ and ∞.

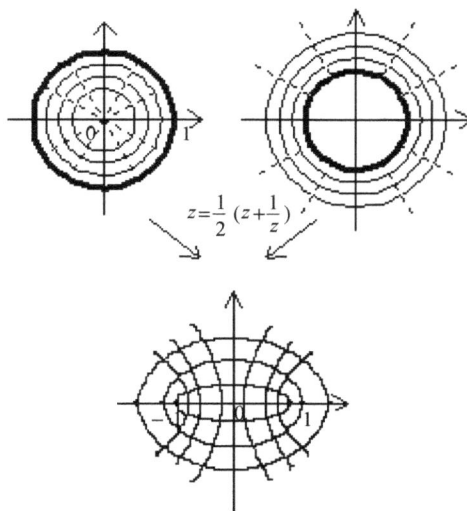

Fig. 2.29

(2) $z = \pm 1$ produces two algebraic branch points $f(\pm 1) = \pm 1$ in the w-plane, each of order 1 (see Exercise A (4) of Sec. 2.5.3).

(3) $w = f(z) : \{|z| = 1\} \to [-1, 1]$ two-to-one and onto. See Fig. 2.26.

　　: $\{\operatorname{Re} z \geq 0$ and $\operatorname{Im} z = 0\} \to [1, \infty]$ two-to-one, onto. See Fig. 2.28(a).

　　: $\{\operatorname{Re} z \leq 0$ and $\operatorname{Im} z = 0\} \to [\infty, -1]$ two-to-one, onto.

　　: $\{\operatorname{Re} z = 0$ and $\operatorname{Im} z \geq 0\} \to [-i\infty, i\infty]$, imaginary axis, one-to-one. See Fig. 2.28(c).

　　: $\{\operatorname{Re} z = 0$ and $\operatorname{Im} z \leq 0\} \to [-i\infty, i\infty]$, one-to-one.

(4) *Domain of univalence.* In general, $f(z) = f\left(\frac{1}{z}\right), f(\bar{z}) = \overline{f(z)}$, and $f(-\bar{z}) = -\overline{f(z)}$.

　(a) $w = f(z) : \{|z| < 1\} \to \mathbf{C}^* - [-1, 1]$ univalently. See Fig. 2.26.

　　(i) $[0, 1) \to (1, \infty]; (-1, 0] \to [\infty, -1);$

　　(ii) Upper half-disk $\{|z| < 1$ and $\operatorname{Im} z > 0\} \to \{\operatorname{Im} w < 0\};$

　　(iii) Lower half-disk $\{|z| < 1$ and $\operatorname{Im} z < 0\} \to \{\operatorname{Im} w > 0\}.$

　(b) $w = f(z) : \{|z| > 1\} \to \mathbf{C}^* - [-1, 1]$ univalently. See Fig. 2.26.

　　(i) $(1, \infty] \to (1, \infty]; [\infty, -1) \to [\infty, -1);$

　　(ii) $\{|z| > 1$ and $\operatorname{Im} z > 0\} \to \{\operatorname{Im} w > 0\};$

　　(iii) $\{|z| > 1$ and $\operatorname{Im} z < 0\} \to \{\operatorname{Im} w < 0\}.$

  (c) $w = f(z) : \{\operatorname{Im} z > 0\} \to \mathbf{C}^* - [\infty, -1] - [1, \infty]$ univalently.
  (d) $w = f(z) : \{\operatorname{Im} z < 0\} \to \mathbf{C}^* - [\infty, -1] - [1, \infty]$ univalently.

(5) (a) $w = f(z)$ maps $|z| = r$ $(0 < r < 1)$ onto a contracting family
        (2.5.5.2) of coaxial, confocal ellipses with foci at ± 1; it maps $|z| = r$ $(1 \leq r < \infty)$ onto an expanding family (2.5.5.2) of coaxial, confocal ellipses with foci at ± 1. See Figs. 2.29 and 2.28(b).
   (b) $w = f(z)$ maps $\operatorname{Arg} z = \theta \, (-\pi < \theta \leq \pi)$ onto a family (2.5.5.3) of coaxial, confocal hyperbolas with foci at ± 1. See Figs. 2.29 and 2.28(b).
   The two families are orthogonal to each other. See Fig. 2.29.
*(6) The inverse function $z = f^{-1}(w)$ on $\mathbf{C}^* - [-1, 1]$ of $w = f(z)$ has two single-valued analytic branches (refer to Example 2 of Sec. 2.7.2, Fig. 2.61, and Example 3(1) of Sec. 3.3.1).

   (a) $z_1 = w - \sqrt{w^2 - 1}(z_1(\infty) = 0) : \mathbf{C}^* - [-1, 1] \to \{|z| < 1\}$, mapping the outside of the ellipse (2.5.5.2) univalently onto $|z| < r$.
   (b) $z_2 = w + \sqrt{w^2 - 1}(z_2(\infty) = \infty) : \mathbf{C}^* - [-1, 1] \to \{|z| > 1\}$, mapping the outside of the ellipse (2.5.5.2) univalently onto $|z| > r$.

Also, the inverse function $z = f^{-1}(w)$ on $\mathbf{C}^* - [\infty, -1] - [1, \infty]$ has two single-valued analytic branches, determined by $\sqrt{-1} = i$ and $\sqrt{-1} = -i$, respectively. Such an inverse function maps the domain between two branches of a hyperbola (2.5.5.3) onto a sector domain in the z-plane. See Example 2 in Sec. 2.7.2.   (2.5.5.4)

Exercises A

(1) Let

$$w = z + \frac{R^2}{z}, \quad \text{where } R > 0 \text{ is a constant.}$$

Observe that $z = \pm R$ produce two algebraic branch points $w = \pm 2R$ in the w-plane. For two distinct points z_1 and z_2, $w(z_1) = w(z_2) \Leftrightarrow z_1 z_2 = R^2$. Also, $w(\bar{z}) = \overline{w(z)}$ and $w(-\bar{z}) = -\overline{w(z)}$, namely, $w = w(z)$ preserves points symmetric to the real and imaginary axes, respectively.

   (a) Prove results corresponding to those stated in (2.5.5.4).
   (b) $w = w(z)$ maps the circle $|z - iy_0| = \sqrt{y_0^2 + R^2}$ $(y_0 \geq 0)$ two-to-one onto a circular arc γ connecting $\pm 2R$ in the upper half-plane $\operatorname{Im} w > 0$. See Fig. 2.30. Also, the circular arcs $R \to i(y_0 + $

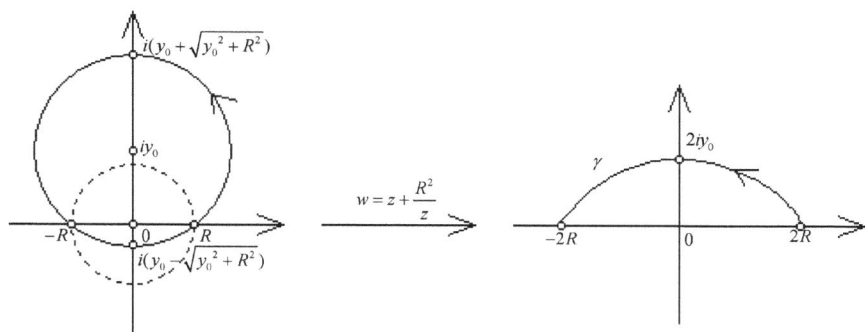

Fig. 2.30

$\sqrt{y_0^2 + R^2}) \rightarrow (-R)$ and $R \rightarrow i(y_0 - \sqrt{y_0^2 + R^2}) \rightarrow (-R)$ are mapped, respectively, one-to-one and onto γ. $w = w(z)$ maps the domain $|z - iy_0| > \sqrt{y_0^2 + R^2}$ univalently onto $\mathbf{C}^* - \gamma$.

(c) Let $z_0 \in \mathbf{C}$ be fixed so that $|\mathbf{R} - z_0| < |\mathbf{R} + z_0|$. $w = w(z)$ maps the circle $|z - z_0| = |z_0 + R|$ univalently (namely, one-to-one) onto a Jordan closed curve Γ, passing $-2R$ and having a cusp there (namely, the curve tangent to itself at $-2R$), and lying entirely in $\mathbf{C}^* - \gamma$ except $-2R$ (γ is as in (b)). Call Γ a *Joukowski airfoil*. Also, $w = w(z)$ maps $|z - z_0| > |z_0 + R|$ univalently onto Ext Γ. See Fig. 2.31.

(d) Fix x_0, $-\infty < x_0 < R$, and $x_0 \neq 0$, $\frac{R}{2}$. $w = w(z)$ maps

  (i) the circle $|z + x_0| = R - x_0$ one-to-one and onto a Jordan closed curve Γ, symmetric with respect to the real axis, passing a cusp

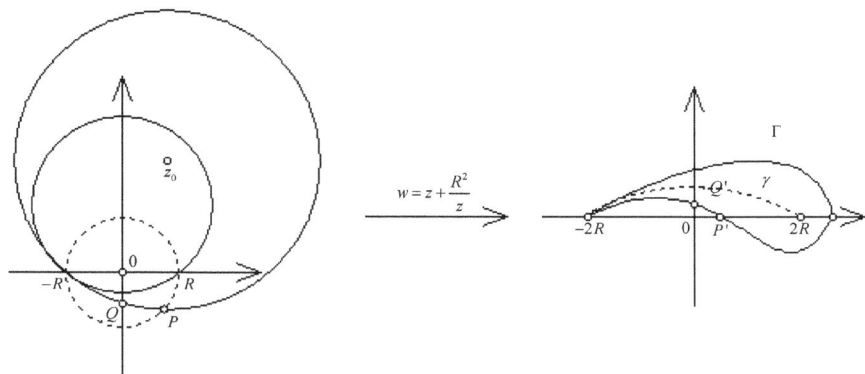

Fig. 2.31

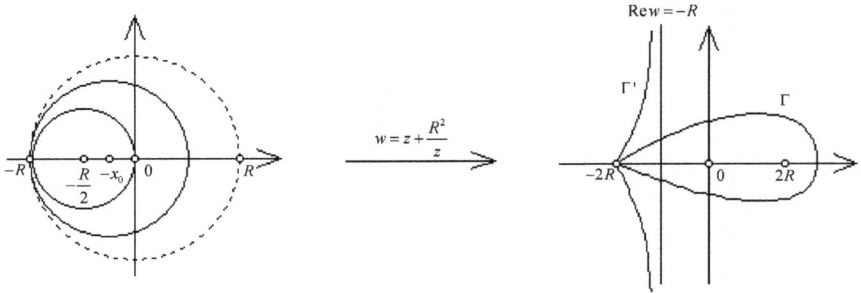

$$w = z + \frac{R^2}{z}$$

Fig. 2.32 $\left(0 < x_0 < \frac{R}{2}\right)$

at $-2R$. We call Γ a *symmetric Joukowski airfoil*. See Fig. 2.32 $\left(0 < x_0 < \frac{R}{2}\right)$;

(ii) the disk $|z + x_0| < R - x_0$ univalently onto Ext Γ;

(iii) the circle $|z + \frac{R}{2}| = \frac{R}{2}$ one-to-one and onto a curve Γ' in Fig. 2.32; and

(iv) the line $\text{Re}\, z = -R$ one-to-one and onto Γ' with the line $\text{Re}\, w = -R$ as its asymptote.

(2) Decompose $w = z + \frac{R^2}{z}$ $(R > 0)$ as the composite of the following mappings:

$$z \to \zeta = \frac{z}{R} \to \eta = \frac{1}{2}\left(\zeta + \frac{1}{\zeta}\right) \to w = 2R\eta.$$

Try to use (2.5.5.4) to redo Exercise (1)(a). Note that $w = z + \frac{R^2}{z}$ can be obtained by the process: $z \to \frac{R^2}{\bar{z}}$ (symmetric point of z with respect to $|z| = R$) $\to \frac{R^2}{z}$ (conjugate point of $\frac{R^2}{\bar{z}}$) $\to w$ (twice the middle point of the line segment joining z and $\frac{R^2}{z}$). See Fig. 2.33 (compare to Fig. 2.25). Can this process be helpful in doing this problem?

(3) Let $w = z + \frac{R^2}{z}$ $(R > 0)$.

(a) Suppose a curve γ has a tangent at $z = R$ or $z = -R$. Show that its image curve under $w = w(z)$ has a cusp at $w = 2R$ or $w = -2R$.

(b) Show that a half-circle of $|w| = 1$ is the image of $|z - i| = \sqrt{2}$ under $w = w(z)$, and another half is the image of $|z + i| = \sqrt{2}$.

(4) Decompose $w = z - \frac{R^2}{z}$ $(R > 0)$ as the composite of the following mappings:

$$z \to \zeta = iz \to \eta = \zeta + \frac{R^2}{\zeta} \to w = -i\eta.$$

Try to obtain results similar to (2.5.5.4).

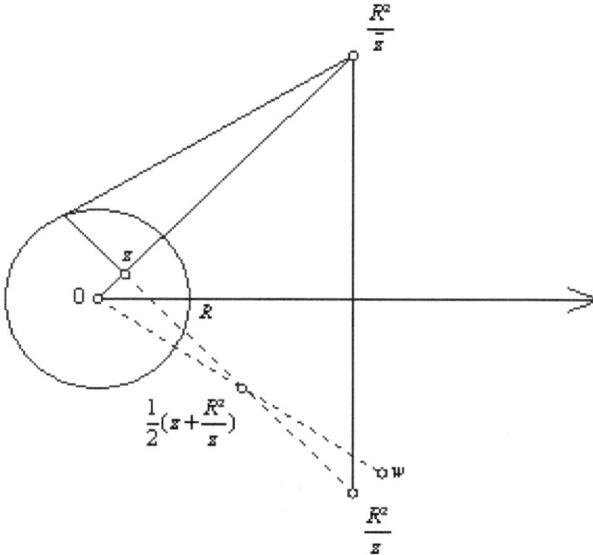

Fig. 2.33

Exercise B

(1) Suppose α and β are nonzero complex numbers. Then

$$w = \alpha z + \frac{\beta}{z}$$

can be rewritten as

$$\frac{w + 2R}{w - 2R} = \left(\frac{z + \frac{R}{\alpha}}{z - \frac{R}{\alpha}} \right)^2, \quad R = \sqrt{\alpha\beta} \neq 0, \quad \text{and} \quad -\frac{\pi}{2} < \operatorname{Arg} R \leq \frac{\pi}{2}.$$

Also, $w = w(z)$ can be decomposed as

$$z \to \zeta = \alpha z \to w = \zeta + \frac{R^2}{\zeta}.$$

Try to redo Exercise A (1) for this $w = w(z)$.

2.6.   Elementary Transcendental Functions

Using the limit process (see Example 1 in Sec. 1.7), Sec. 2.6.1 introduces the exponential function e^z, $z \in \mathbf{C}$. Based on Euler's formula for $\cos\theta$ and $\sin\theta$, Sec. 2.6.2 discusses trigonometric functions $\sin z$, $\cos z$ and $\tan z$, and then, hyperbolic functions $\sinh z$, $\cosh z$ and $\tanh z$.

We emphasize the following aspects of these elementary functions:

(1) *Algebraic properties*: Algebraic identities, periods, etc. They are similar to the real functions $e^x, \sin x$, etc. in most cases.
(2) *Limit properties*: Unboundedness, asymptotic values at ∞. These *analytic* properties concerned with differentiation and integration will be fully developed in Chaps. 3–5.
(3) *Geometric mapping properties*: Domain for univalence, fundamental periodic region, etc. For more concrete examples, refer to Sec. 3.5.7.

It is the limit and geometric properties that distinguish vehemently complex exponential and trigonometric functions from their corresponding real ones.

2.6.1. The exponential function e^z

Based on Example 1 of Sec. 1.7, we define *the exponential function e^z* ($e = \lim_{n \to \infty} \left(1 + \frac{1}{n}\right)^n$): For $z = x + iy \in \mathbf{C}$,

$$e^z \underset{\text{(def.)}}{=} \lim_{n \to \infty} \left(1 + \frac{z}{n}\right)^n = e^x(\cos y + i \sin y),$$

where the polynomial sequence $\left(1 + \frac{z}{n}\right)^n, n \geq 1$, converges to e^z uniformly on every compact subset of $\mathbf{C}$. Hence,

(1) $|e^z| = e^x$, $\operatorname{Arg} e^z = y$, if $z = x + iy$.
(2) *Euler identity*: $e^{i\theta} = \cos\theta + i\sin\theta, \theta \in \mathbf{R}$. In particular, $e^{\pi i} = -1$ connects important numbers e, π, i, and 1 in mathematics.
(3) *Euler formulas*: For $\theta \in \mathbf{R}$,

$$\cos\theta = \frac{1}{2}(e^{i\theta} + e^{-i\theta}), \quad \sin\theta = \frac{1}{2i}(e^{i\theta} - e^{-i\theta}). \qquad (2.6.1.1)$$

Remark. Conformality of e^z (see Remark 2 of Sec. 2.5.1.). Let $z(t) = x(t) + iy(t)$ represent any differentiable curve. Then

$$e^{z(t)} = e^{x(t)}(\cos y(t) + i \sin y(t))$$
$$\Rightarrow (\text{see } (2.4.8)) \frac{d}{dt} e^{z(t)} = x'(t)e^{x(t)}(\cos y(t) + i \sin y(t))$$
$$+ e^{x(t)}(-\sin y(t) + i \cos y(t))y'(t),$$
$$= (x'(t) + iy'(t))e^{x(t)}(\cos y(t) + i \sin y(t))$$
$$= e^{z(t)}z'(t).$$

Denote this fact simply by

$$(e^{z(t)})' = e^{z(t)} z'(t) \quad \text{or} \quad \frac{d}{dz} e^z = e^z \text{ (see Sec. 3.1).}$$

Just like (2.5.1.6), we also have

$$\text{Arg}(e^{z_2(t_0)})' - \text{Arg}(e^{z_1(t_0)})' = \text{Arg} z_2'(t_0) - \text{Arg} z_1'(t_0).$$

This means $w = e^z$, as a mapping from $\mathbf{C}$ into $\mathbf{C}$, *preserves angle and its direction* at $z_1(t_0) = z_2(t_0)$ of any two intersecting curves $z_1(t)$ and $z_2(t)$.   □

Via basic properties of real-valued functions e^x, $\cos y$, and $\sin y$, it follows easily that $e^{z_1 + z_2} = e^{z_1} e^{z_2}$ for $z_1, z_2 \in \mathbf{C}$. In case $z_2 = -z_1$, $e^z e^{-z} = 1$ implies both $e^z \neq 0$ and $(e^z)^{-1} = e^{-z}$ for any $z \in \mathbf{C}$.

The main difference between e^x, $x \in \mathbf{R}$, and e^z, $z \in \mathbf{C}$ comes from the fact that the latter enjoys the property $\text{Arg } e^z = \text{Im } z$. Recall that e^x is a strictly increasing function of $x \in \mathbf{R}$ and hence, it is one-to-one on $\mathbf{R}$.

$e^{z+2n\pi i} = e^z$, $n = 0, \pm 1, \pm 2, \ldots$, shows that e^z is not one-to-one on $\mathbf{C}$. Suppose $z_k = x_k + iy_k$, $k = 1, 2$ so that

$$e^{z_1} = e^{z_2} \text{ or } e^{z_1 - z_2} = 1,$$

$$\Rightarrow |e^{z_1 - z_2}| = e^{x_1 - x_2} = 1 \quad \text{and} \quad \arg e^{z_1 - z_2} = \arg 1 = 2n\pi = y_1 - y_2$$

$$\Rightarrow x_1 - x_2 = 0 \quad \text{or} \quad x_1 = x_2 \quad \text{and} \quad y_1 = y_2 + 2n\pi, \quad n = 0, \pm 1, \pm 2, \ldots.$$

Hence, $z_1 - z_2 = 2n\pi i$ or $z_1 = z_2 + 2n\pi i$, $n = 0, \pm 1, \pm 2, \ldots$.

For each $\alpha \in \mathbf{R}$, last paragraph shows that $w = e^z$ is one-to-one on the parallel strip $\alpha < \text{Im } z \leq \alpha + 2\pi$. Fix any $w \in \mathbf{C} - \{0\}$ and let $w = re^{i\varphi}$ with $r = |w|$ and $\varphi = \text{Arg } w$. Suppose

$$e^z = w,$$

$$\Rightarrow e^x = r \quad \text{and} \quad e^{iy} = e^{i\varphi},$$

$$\Rightarrow x = \log r \text{ (real logarithm)} \quad \text{and} \quad y = \varphi + 2n\pi, \quad n = 0, \pm 1, \pm 2 \ldots.$$

$$(2.6.1.2)$$

This shows that infinitely many points $z_n = \log |w| + i(\text{Arg } w + 2n\pi)$, $n = 0, \pm 1, \ldots$, are mapped, under $w = e^z$, onto a single point w. There exists a unique integer n_0 so that $\alpha < \text{Arg } w + 2n_0\pi \leq \alpha + 2\pi$. Then, the corresponding z_{n_0} lies on the strip $\alpha < \text{Im } z \leq \alpha + 2\pi$ and $e^{z_{n_0}} = w$.

We summarize the above as

Basic properties of e^z: e^z is a continuous function on $\mathbf{C}$.

(1) Properties similar to the real exponential:

$$e^{z_1+z_2} = e^{z_1}e^{z_2}, z_1, z_2 \in \mathbf{C}.$$

In particular,

  (i) $e^z \neq 0, z \in \mathbf{C}$;
  (ii) $(e^z)^{-1} = e^{-z}, z \in \mathbf{C}$;
  (iii) $(e^z)^n = e^{nz}, z \in \mathbf{C}$, n integers.

(2) Properties different from the real exponential:

  (a) e^z is *not* defined at $z = \infty$. For instance, suppose the curve $z(t) = x(t) + iy(t)\colon [0, \infty) \to \mathbf{C}$ satisfying $\lim_{t\to\infty} |x(t)| = \infty$, then

  $$\lim_{t\to\infty} e^{z(t)} = \begin{cases} \infty, & \text{if } x(t) \to \infty \\ 0, & \text{if } x(t) \to -\infty \end{cases}.$$

  (b) $w = e^z$ is an infinite-to-one function (originated from the fact that $\operatorname{Arg} e^z = \operatorname{Im} z$).

   (i) $e^{z_1} = e^{z_2} \Leftrightarrow z_2 = z_1 + 2n\pi i, n = 0, \pm 1, \pm 2, \ldots$;
   (ii) $w = e^z$ is a *simple-periodic* function with $2\pi i$ as its *primitive period*. All other periods of e^z are integer multiples of $2\pi i, 2n\pi i$.

  Hence, $w = e^z$ maps $z + 2n\pi i$, $n = 0, \pm 1, \ldots$, onto the same point e^z. In part;

   (iii) $w = e^z$ takes real values only on the horizontals $z = x + n\pi i, n = 0, \pm 1, \ldots, x \in \mathbf{R}$;
   (iv) $w = e^z$ takes pure imaginaries only on the horizontals $z = x + \left(n + \frac{1}{2}\right)\pi i, n = 0, \pm 1, \pm 2, \ldots, x \in \mathbf{R}$.

  (c) For any fixed $\alpha \in \mathbf{R}$, $w = e^z$ maps the parallel strip $\alpha < \operatorname{Im} z \leq \alpha + 2\pi$ one-to-one and onto $\mathbf{C} - \{0\}$. See Fig. 2.34. It maps the open strip $\alpha < \operatorname{Im} z < \alpha + 2\pi$, a *fundamental domain*, onto the slit domain $\mathbf{C} - \{w \mid \operatorname{Arg} w = \alpha\} \cup \{0\}$. In particular, we call $(\alpha = -\pi)$ the horizontal strip

  $$-\pi < \operatorname{Im} z \leq \pi$$

  the *fundamental periodic strip* of e^z.   (2.6.1.3)

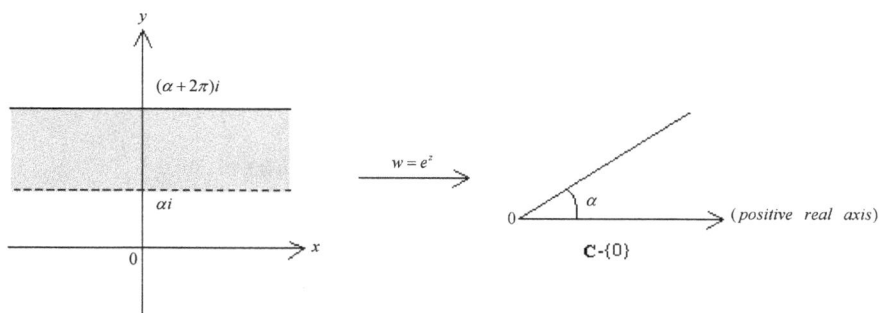

Fig. 2.34

Exercises A

(1) Show that $\left(1 + \frac{z}{n}\right)^n$ converges uniformly to e^z on every compact subset of **C**.

(2) Find $|z|$ and $\arg z$ of each of the following complex numbers z.

$$ e^{e^i}; e^{3-2i}; -ae^{-i\theta} \ (a \geq 0, -\pi < \theta < \pi); \frac{e^{i\theta} - 1}{e^{-i\theta} - 1} (\theta \in \mathbf{R}); $$

$$ e^{e^\alpha} \text{ where } \alpha \in \mathbf{C}. $$

(3) Let $z = x + iy \neq 0$. Compute $\mathrm{Re}\,(e^{\frac{1}{z}})$.

(4) Solve the equations: $e^z = 1 + \sqrt{3}i$; $e^z + 1 = 0$.

(5) (a) Letting $z \to \infty$ along the ray $\mathrm{Arg}\, z = \theta(-\pi < \theta \leq \pi)$, try to find the limits of e^z if they exist.

(b) Suppose $z_n \to \infty$. Try to find all possible limit points of e^{z_n}.

(6) Let $f(z) = e^{-\frac{1}{z}}, z \in \mathbf{C} - \{0\}$.

(a) Show that $f(z)$ is bounded on the set $0 < |z| \leq 1$, $|\mathrm{Arg}\, z| \leq \frac{\pi}{2}$. Also $f(z)$ is continuous but not uniformly there.

(b) Show that $f(z)$ is uniformly continuous on $0 < |z| \leq 1$, $|\mathrm{Arg}\, z| \leq \alpha < \frac{\pi}{2}$.

(7) (a) Suppose $a_j \in \mathbf{C}$ and $\alpha_j \in \mathbf{R}$ for $1 \leq j \leq n$ and $\alpha_1 < \alpha_2 < \cdots < \alpha_n$. If $a_1 e^{\alpha_1 z} + \cdots + a_n e^{\alpha_n z} = 0$ for all z, show that $a_1 = \cdots = a_n = 0$.

(b) Suppose $p_j(z), 1 \leq j \leq n$, are polynomials. If there exist n distinct complex numbers $\alpha_1, \ldots, \alpha_n$ so that $p_1(z)e^{\alpha_1 z} + \cdots + p_n(z)e^{\alpha_n z} = 0$ for all z, show that $p_j(z) \equiv 0, 1 \leq j \leq n$.

(c) Prove that e^z is a *transcendental function*, i.e., not an algebraic function. That is to say, if there exist polynomials $p_n(z), \ldots,$ $p_1(z), p_0(z)$ so that $p_n(z)e^{nz} + \cdots + p_1(z)e^z + p_0(z) = 0$ for all $z \in \mathbf{C}$, then $p_j(z) \equiv 0$ for $0 \le j \le n$.

(8) Treat $w = e^z$ as a mapping from the z-plane to w-plane.

   (a) Graph the image of the line $\operatorname{Re} z = x_0$ (a constant) under $w = e^z$.
   (b) Graph the image of the line $\operatorname{Im} z = y_0$ (a constant) under $w = e^z$.

Exercises B

(1) Try to show that

$$e^z = \sum_{n=0}^{\infty} \frac{z^n}{n!}, \quad z \in \mathbf{C},$$

where the series on the right converges to e^z uniformly on every compact subset of $\mathbf{C}$. Also, show that $|e^z - 1| \le e^{|z|} - 1 \le |z|e^{|z|}$, $z \in \mathbf{C}$.

(2) Let $\Omega \subseteq \mathbf{C}^*$ be a domain. A function $f : \Omega \to \mathbf{C}^*$ is called **open** if f maps each open set O in Ω onto an open set $f(\mathrm{O})$ in $\mathbf{C}^*$, and is called a **local homeomorphism** on Ω if each point z_0 in Ω has a neighborhood $N(z_0)$ on which f is a homeomorphism. Show that $w = e^z : \mathbf{C} \to \mathbf{C}$ is open and is locally homeomorphic.

(3) Let Ω be a domain in $\mathbf{C}^*$ and $F : \Omega \to \mathbf{R}$ be a real-valued function. If $z_0 \in \Omega$ has a neighborhood $\mathrm{O} \subseteq \Omega$ so that $F(z_0) \ge F(z)$ (or $F(z_0) \le F(z)$) for all $z \in \mathrm{O}$, then we say F attains a *local maximum* $F(z_0)$ (or a *local minimum* $F(z_0)$) at z_0. Show that $F(z) = |e^z|$ does *not* attain local maximum or minimum at any point of $\mathbf{C}$.

(4) Let $\Omega \subseteq \mathbf{C}$ be a domain, *symmetric with respect to the real axis*, i.e., $z \in \Omega \Leftrightarrow \bar{z} \in \Omega$ holds. A function $f : \Omega \to \mathbf{C}^*$ satisfying

$$f(\bar{z}) = \overline{f(z)}, \quad z \in \Omega$$

is said to *preserve symmetry* with respect to the real axis. Such a function has the following properties:

   (i) f preserves reals, i.e., $f(x) = \overline{f(x)}$, if $x \in \Omega \cap \mathbf{R}$, and
   (ii) zeros of f are conjugate in pairs, i.e., $f(z) = 0 \Leftrightarrow f(\bar{z}) = 0$.

For instance, a polynomial with real coefficients is such a function. Show that $w = e^z$ is another such function.

2.6.2. *Trigonometric functions* cos z, sin z, *and* tan z

Based on Euler's formula (see (2.6.1.1)), we define complex *cosine* and *sine* functions, respectively, as

$$\cos z = \frac{1}{2}(e^{iz} + e^{-iz}),$$

$$\sin z = \frac{1}{2i}(e^{iz} - e^{-iz}), \quad z \in \mathbf{C}. \tag{2.6.2.1}$$

Both are rational functions of e^{iz}.

Remark. Conformality of cosz and sinz (see Remark 2 of Sec. 2.5.1 and Remark of Sec. 2.6.1). It can be shown that, for any differentiable curve $z(t)$,

$$\frac{d}{dt}\cos z(t) = (-\sin z(t))z'(t), \text{ or simply as } \frac{d}{dz}\cos z = -\sin z;$$

$$\frac{d}{dt}\sin z(t) = (\cos z(t))z'(t), \text{ or simply as } \frac{d}{dz}\sin z = \cos z \text{ (see Sec. 3.1).}$$

Just like (2.5.1.6), $w = \cos z$ *preserves angles and their directions at points z where* $\frac{d}{dz}\cos z = -\sin z \neq 0$. So do $w = \sin z$ and $w = \tan z$ in a similar manner. □

By these very definitions, *Euler identity* (see (2.6.1.1)) can be generalized as

$$e^{iz} = \cos z + i \sin z, \quad z \in \mathbf{C}. \tag{2.6.2.2}$$

So are many identities concerned with $\cos z$ and $\sin z$, which are similar to the real cosine and sine functions. See (2.6.2.6) below.

According to $e^{z+2n\pi i} = e^z, z \in \mathbf{C}$, it follows that

$$\cos(z + 2n\pi) = \cos z, \quad \sin(z + 2n\pi) = \sin z, \quad z \in \mathbf{C}, \quad n = 0, \pm 1, \dots.$$

Hence, both $\cos z$ and $\sin z$ have periods $2n\pi$, $n = \pm 1, \pm 2, \dots$.

Do $\cos z$ and $\sin z$ have any other periods? As for $\cos z$, suppose $\omega \neq 0$ is a period of $\cos z$. Then

$$\cos(z + \omega) = \cos z, \quad z \in \mathbf{C},$$

$$\Rightarrow \left(\text{Let } z = 0 \text{ and } z = -\frac{\pi}{2}\right) \cos \omega = 1 \quad \text{and} \quad \sin \omega = 0.$$

$$\Rightarrow e^{i\omega} = \cos \omega + i \sin \omega = 1,$$

$$\Rightarrow i\omega = 2n\pi i \quad \text{or} \quad \omega = 2n\pi, \ n = \pm 1, \pm 2, \dots.$$

Hence, the only periods $\cos z$ has are $2n\pi, n = \pm 1, \pm 2, \dots$. So is $\sin z$.

Since $\cos(-z) = \cos z$, $w = \cos z$ is two-to-one on the vertical strip

$$-\pi \leq \operatorname{Re} z < \pi,$$

called *the fundamental periodic strip* of $\cos z$. Since $\cos\left(z - \frac{\pi}{2}\right) = \sin z$, so $w = \sin z$ is also two-to-one on its *fundamental periodic strip* $-\frac{\pi}{2} \leq \operatorname{Re} z < \frac{3\pi}{2}$.

What is the image of $-\pi \leq \operatorname{Re} z < \pi$ under $w = \cos z$? Suppose $w_0 \in \mathbf{C}$ so that

$$\cos z = \frac{1}{2}(e^{iz} + e^{-iz}) = w_0,$$

$$\Rightarrow (e^{iz})^2 - 2w_0 e^{iz} + 1 = 0,$$

$$\Rightarrow e^{iz} = w_0 + \sqrt{w_0^2 - 1} \text{ (remind that } \sqrt{w_0^2 - 1} \text{ is double-valued)}.$$

$$(2.6.2.3)$$

According to (2)(c) in (2.6.1.3), to each value of $\sqrt{w_0^2 - 1}$, there exists a unique z_0, $-\pi \leq \operatorname{Im}(iz_0) < \pi$ so that $e^{iz_0} = w_0 + \sqrt{w_0^2 - 1}$ except $w_0 = \pm 1$; equivalently, $-\pi \leq \operatorname{Re}(z_0) < \pi$ and $e^{iz_0} = w_0 + \sqrt{w_0^2 - 1}$ holds. Therefore, $w = \cos z$ takes each point in $\mathbf{C} - \{\pm 1\}$ twice on the strip $-\pi \leq \operatorname{Re} z < \pi$ and -1 at $z = -\pi$; 1 at $z = 0$ only once. A similar result holds for $w = \sin z$ on $-\frac{\pi}{2} \leq \operatorname{Re} z < \frac{3\pi}{2}$.

Where does $w = \cos z$ take real values?

Suppose that w_0 in (2.6.2.3) is real. If $|w_0| \geq 1$, then

$$e^{iz} = w_0 + \sqrt{w_0^2 - 1} \text{ (both values are real numbers)},$$

$$\Rightarrow (\text{let } z = x + iy) \ \sin x = 0,$$

$$\Rightarrow x = n\pi, n = 0, \pm 1, \pm 2, \ldots .$$

Hence, on the vertical lines $\operatorname{Re} z = n\pi, n = 0, \pm 1, \ldots$, $w = \cos z$ takes reals of absolute values not less than 1. If $|w_0| < 1$, then

$$e^{iz} = w_0 \pm i\sqrt{1 - w_0^2} \ (\text{here, } \sqrt{1} = 1),$$

$$\Rightarrow (\text{let } z = x + iy) \ e^{-y} = 1,$$

$$\Rightarrow y = 0.$$

This means that, only on the real axis $\operatorname{Im} z = 0$, $w = \cos z$ takes reals of absolute values less than 1 except at $x = n\pi (n = 0, \pm 1, \ldots)$ where it assumes the values ± 1. In particular,

$$\cos z = 0,$$
$$\Leftrightarrow e^{iz} = \pm i,$$
$$\Leftrightarrow y = 0 \quad \text{and} \quad x = \left(n + \frac{1}{2}\right)\pi, \quad n = 0, \pm 1, \pm 2, \ldots .$$

Therefore, only when $z = \left(n + \frac{1}{2}\right)\pi, n = 0, \pm 1, \pm 2, \ldots$, $\cos z = 0$ happens. This fact is identical with the real cosine function (refer to Exercise B (3) of Sec. 1.5).

It is well-known that $|\cos x| \leq 1$ and $|\sin x| \leq 1$ for all real x. As a very striking contrast, both $\cos z$ and $\sin z$ are *not* bounded for complex z since $\cos z$ takes all complex numbers and so does $\sin z$ via $\cos(z - \frac{\pi}{2}) = \sin z$.

In (2.6.2.1), let $z = iy$ $(y \in \mathbf{R})$, then

$$\cos iy = \frac{1}{2}(e^y + e^{-y}) = \cosh y,$$

$$\sin iy = \frac{1}{2i}(e^{-y} - e^y) = \frac{i}{2}(e^y - e^{-y}) = i \sinh y.$$

Hence, for $z = x + iy$,

$$\cos z = \cos(x + iy) = \cos x \cos iy - \sin x \sin iy,$$
$$= \cos x \cosh y - i \sin x \sinh y, \tag{2.6.2.4}$$
$$\Rightarrow |\cos z|^2 = \cos^2 x \cosh^2 y + \sin^2 x \sinh^2 y,$$
$$= \sinh^2 y + \cos^2 x = \cosh^2 y - \sin^2 x = \frac{1}{2}(\cosh 2y + \cos 2x), \tag{2.6.2.5}$$
$$\Rightarrow |\cos z| \geq |\sinh y|, \quad z = x + iy.$$

Since $\lim_{y \to \infty} \sinh y = \infty$, $\cos z$ is unbounded. So is $\sin z$.

We summarize the above as

Basic properties of $\cos z$ *and* $\sin z$. Both are continuous functions on $\mathbf{C}$.

(1) Properties similar to real cosine and sine: $z \in \mathbf{C}$,

$$\cos^2 z + \sin^2 z = 1;$$
$$\cos(-z) = \cos z, \quad \sin(-z) = -\sin z;$$
$$\cos\left(z - \frac{\pi}{2}\right) = \sin z;$$

$$\cos(z_1 \pm z_2) = \cos z_1 \cos z_2 \mp \sin z_1 \sin z_2;$$

$$\sin(z_1 \pm z_2) = \sin z_1 \cos z_2 \pm \cos z_1 \sin z_2.$$

Moreover, 2π is the *primitive period* of both $\cos z$ and $\sin z$, and $2n\pi, n = \pm 1, \pm 2, \ldots$, are all their periods. Both are *simple* periodic functions.

(2) Properties different from real cosine and sine:

(a) $w = \cos z$ maps its *fundamental periodic strip*

$$-\pi \le \operatorname{Re} z < \pi$$

two-to-one onto $\mathbf{C}$, except ± 1 which are covered only once. See Fig. 2.35. In particular,

(i) $\cos z$ are reals $\Leftrightarrow \operatorname{Im} z = 0$ (then, $|\cos z| \le 1$);

$$\operatorname{Re} z = n\pi, \quad n = 0, \pm 1, \ldots \quad \text{(then, } |\cos z| \ge 1\text{)};$$

(ii) $\cos z = 0 \Leftrightarrow z = \left(n + \frac{1}{2}\right)\pi, n = 0, \pm 1, \ldots$;

(iii) $\cos z$ are pure imaginaries $\Leftrightarrow \operatorname{Re} z = (n + \frac{1}{2})\pi, n = 0, \pm 1, \ldots$;

(iv) $w = \cos z$ maps the *domains of univalence (or fundamental domains)*: half-strip $0 < \operatorname{Re} z < \pi$ one-to-one onto $\mathbf{C} - (-\infty, -1] \cup [1, +\infty)$; half-strip $-\pi < \operatorname{Re} z < 0$ one-to-one onto $\mathbf{C} - (-\infty, -1] \cup [1, +\infty)$.

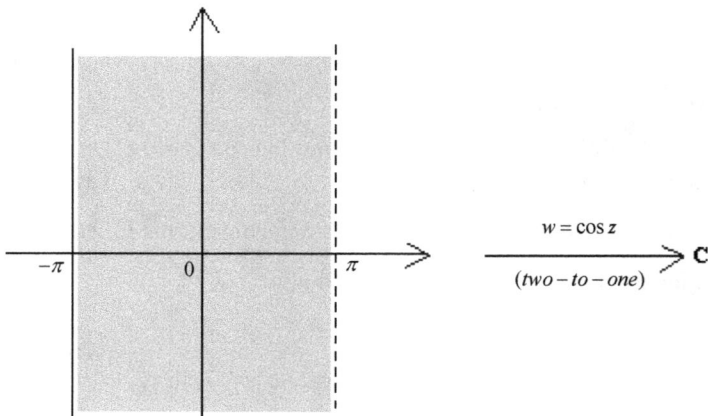

Fig. 2.35

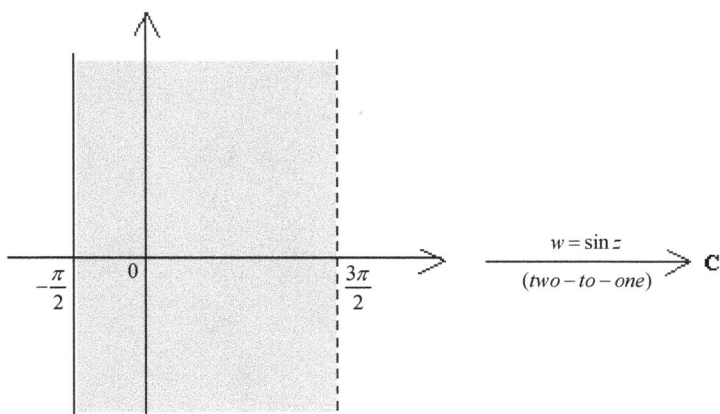

Fig. 2.36

(b) $w = \sin z$, via $\cos\left(z - \frac{\pi}{2}\right) = \sin z$, maps its *fundamental periodic strip*

$$-\frac{1}{2}\pi \le \operatorname{Re} z < \frac{3}{2}\pi$$

two-to-one onto **C**, except ± 1 which are covered only once. See Fig. 2.36. In particular,

(i) $\sin z$ are reals $\Leftrightarrow \operatorname{Im} z = 0$ (then, $|\sin z| \le 1$);

$$\operatorname{Re} z = \left(n + \frac{1}{2}\right)\pi, \quad n = 0, \pm 1, \ldots \quad (\text{then, } |\sin z| \ge 1);$$

(ii) $\sin z = 0 \Leftrightarrow z = n\pi, n = 0, \pm 1, \ldots$;

(iii) $\sin z$ are pure imaginaries $\Leftrightarrow \operatorname{Re} z = n\pi, n = 0, \pm 1, \ldots$;

(iv) $w = \sin z$ maps the *domains of univalence:* $\frac{\pi}{2} < \operatorname{Re} z < \frac{3\pi}{2}$ and $-\frac{\pi}{2} < \operatorname{Re} z < \frac{\pi}{2}$, respectively, one-to-one onto the same set $\mathbf{C} - (-\infty, -1] \cup [1, +\infty)$.

(3) $\cos z$, $\sin z$ and $\cos x$, $\sin x$, $\cosh x$, $\sinh x$ are related as follows.

(i) $\cos ix = \cosh x, \sin ix = i \sinh x, x \in \mathbf{R}$;

(ii) $\cos(x + iy) = \cos x \cosh y - i \sin x \sinh y$;
$\sin(x + iy) = \sin x \cosh y + i \cos x \sinh y, x, y \in \mathbf{R}$;

(iii) $|\cos(x + iy)|^2 = \sinh^2 y + \cos^2 x = \cosh^2 y - \sin^2 x = \frac{1}{2}(\cosh 2y + \cos 2x)$;
$|\sin(x + iy)|^2 = \sinh^2 y + \sin^2 x = \cosh^2 y - \cos^2 x = \frac{1}{2}(\cosh 2y - \cos 2x), x, y \in \mathbf{R}$.

(iv) Hence,

$$|\sinh y| \leq \begin{cases} |\cos(x+iy)| \\ |\sin(x+iy)| \end{cases} \leq \cosh y.$$

(4) $\cos z$ and $\sin z$ are *unbounded* functions on $\mathbf{C}$.   (2.6.2.6)

Imitating real trigonometric functions, define

$$\tan z = \frac{\sin z}{\cos z} = -i\frac{e^{2iz}-1}{e^{2iz}+1};$$

$$\cot z = \frac{\cos z}{\sin z} = i\frac{e^{2iz}+1}{e^{2iz}-1} = \frac{1}{\tan z};$$

$$\sec z = \frac{1}{\cos z} = \frac{2e^{iz}}{e^{2iz}+1};$$

$$\csc z = \frac{1}{\sin z} = \frac{2ie^{iz}}{e^{2iz}-1}.$$ (2.6.2.7)

They are rational functions of e^{iz}.

Model after (2.6.2.6), we have

Basic properties of $\tan z$ *and* $\cot z$.

(1) Domains of definition:

(i) $\tan z$ is continuous on $\mathbf{C} - \{(n+\frac{1}{2})\pi \mid n = 0, \pm 1, \ldots\}$; it takes ∞ at points $(n+\frac{1}{2})\pi, n = 0, \pm 1, \ldots$;

(ii) $\cot z$ is continuous on $\mathbf{C} - \{n\pi \mid n = 0, \pm 1, \ldots\}$; it takes ∞ at points $n\pi, n = 0, \pm 1, \ldots$.

(2) Properties similar to real tan and cot:

$$\tan\left(z - \frac{\pi}{2}\right) = -\cot z; \tan(-z) = -\tan z; \cot(-z) = -\cot z;$$

$$\tan(z_1 \pm z_2) = \frac{\tan z_1 \pm \tan z_2}{1 \mp \tan z_1 \tan z_2}; \text{etc.}$$

Moreover, both $\tan z$ and $\cot z$ are *simple* periodic functions with *primitive period* π, and all other periods are of the form $n\pi, n = \pm 1, \pm 2, \ldots$.

(3) Properties different from real tan and cot:

(a) $w = \tan z$ maps its *fundamental periodic strip*

$$0 \leq \operatorname{Re} z < \pi$$

one-to-one and onto $\mathbf{C}^* - \{\pm i\}$. See Fig. 2.37. In particular, for $n = 0, \pm 1, \pm 2, \ldots$, wherein

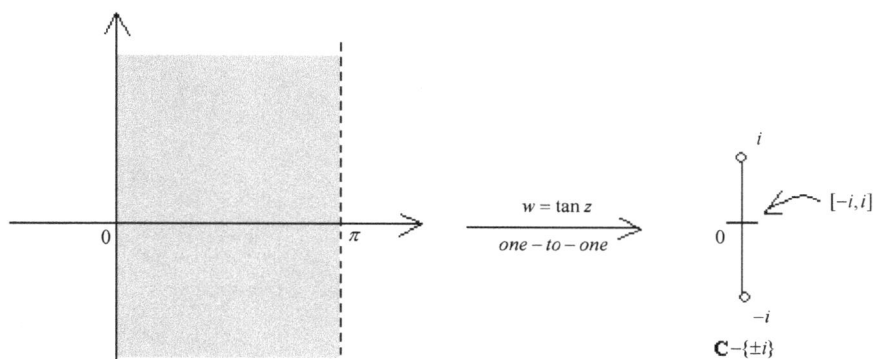

$w = \tan z$

$one - to - one$

$[-i, i]$

$\mathbf{C} - \{\pm i\}$

Fig. 2.37

(i) $\tan z$ are reals $\Leftrightarrow \operatorname{Im} z = 0$ and $n\pi \leq \operatorname{Re} z \leq \left(n + \frac{1}{2}\right)\pi$ (then, $0 \leq \tan z < \infty$); $\operatorname{Im} z = 0$ and $\left(n + \frac{1}{2}\right)\pi \leq \operatorname{Re} z \leq (n+1)\pi$ (then, $-\infty \leq \tan z \leq 0$).

(ii) $\tan z$ are pure imaginaries $\Leftrightarrow \operatorname{Re} z = n\pi$ (then, $\tan z \in [-i, i]$: $+i\infty \to i, n\pi \to 0, -i\infty \to -i$); $\operatorname{Re} z = (n+\frac{1}{2})\pi$ (then, $\tan z \in (-i\infty, -i) \cup (i, +i\infty)$: $+i\infty \to i, (n+\frac{1}{2})\pi \to i\infty, -i\infty \to -i$).

(iii) $w = \tan z$ maps the *domains of univalence*: the upper half $\operatorname{Im} z > 0$ of the strip $n\pi < \operatorname{Re} z < (n+1)\pi$ one-to-one and onto $\{\operatorname{Im} w > 0\} - [0, i]$; the lower half $\operatorname{Im} z < 0$ of the strip one-to-one and onto $\{\operatorname{Im} w < 0\} - [-i, 0]$.

(b) $w = \cot z$ maps its *fundamental periodic strip*

$$-\frac{\pi}{2} \leq \operatorname{Re} z < \frac{\pi}{2}$$

one-to-one and onto $\mathbf{C}^* - \{\pm i\}$. See Fig. 2.38. In particular, via $\cot z = -\tan\left(z - \frac{\pi}{2}\right)$ or $\tan z = -\cot\left(z + \frac{\pi}{2}\right)$, and for $n = 0, \pm 1, \pm 2, \ldots$ wherein,

(i) $\cot z$ are reals $\Leftrightarrow \operatorname{Im} z = 0$ and $(n - \frac{1}{2})\pi \leq \operatorname{Re} z \leq n\pi$ (then, $-\infty \leq \cot z \leq 0$); $\operatorname{Im} z = 0$ and $n\pi \leq \operatorname{Re} z \leq (n+\frac{1}{2})\pi$ (then, $0 \leq \cot z \leq +\infty$).

(ii) $\cot z$ are pure imaginaries $\Leftrightarrow \operatorname{Re} z = (n - \frac{1}{2})\pi$ (then, $\cot z \in [-i, i]$: $+i\infty \to -i, \left(n - \frac{1}{2}\right)\pi \to 0, -i\infty \to i$); $\operatorname{Re} z = n\pi$ (then, $\cot z \in (-i\infty, -i) \cup (i, +i\infty)$: $+i\infty \to -i, n\pi \to i\infty, -i\infty \to i$).

(iii) $w = \cot z$ maps the *domains of univalence*: the upper half $\operatorname{Im} z > 0$ of the strip $\left(n - \frac{1}{2}\right)\pi < \operatorname{Re} z < \left(n + \frac{1}{2}\right)\pi$ one-to-one

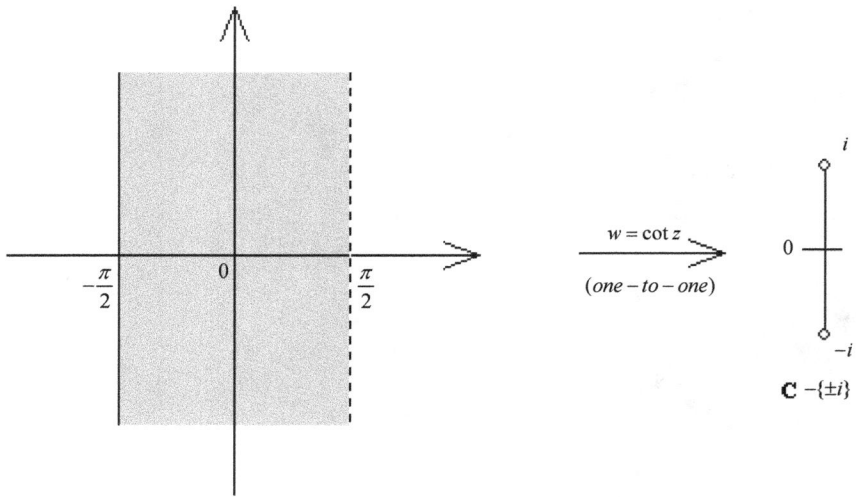

Fig. 2.38

and onto $\{\operatorname{Im} w < 0\} - [-i, 0]$; the lower half $\operatorname{Im} z < 0$ of the strip one-to-one and onto $\{\operatorname{Im} w > 0\} - [0, i]$.

(c) Let γ^* be a curve $z(t) = x(t) + iy(t) : [0, \infty) \to \mathbf{C}^*$ such that $\lim_{t \to \infty} y(t) = \infty$ or $-\infty$. Then

$$\lim_{t \to \infty} \tan z(t) = \begin{cases} -i, & \text{if } y(t) \to -\infty \\ i, & \text{if } y(t) \to \infty \end{cases};$$

$$\lim_{t \to \infty} \cot z(t) = \begin{cases} i, & \text{if } y(t) \to -\infty \\ -i, & \text{if } y(t) \to \infty \end{cases}.$$

Namely, $\tan z$ and $\cot z$ have approximate values $\pm i$ at ∞ along γ^*.

$$(2.6.2.8)$$

Similar to real hyperbolic functions, we define *complex hyperbolic functions* as follows:

$$\cosh z = \frac{1}{2}(e^z + e^{-z}), \quad z \in \mathbf{C};$$

$$\sinh z = \frac{1}{2}(e^z - e^{-z}), \quad z \in \mathbf{C};$$

$$\tanh z = \frac{\sinh z}{\cosh z} = \frac{e^z - e^{-z}}{e^z + e^{-z}}, \quad z \in \mathbf{C} - \left\{\left(n + \frac{1}{2}\right)\pi i \mid n = 0, \pm 1, \dots\right\};$$

$$\coth z = \frac{\cosh z}{\sinh z} = \frac{e^z + e^{-z}}{e^z - e^{-z}}, \quad z \in \mathbf{C} - \{n\pi i \mid n = 0, \pm 1, \dots\};$$

$$\operatorname{sech} z = \frac{1}{\cosh z} = \frac{2}{e^z + e^{-z}}, \quad z \in \mathbf{C} - \left\{\left(n + \frac{1}{2}\right)\pi i \mid n = 0, \pm 1, \dots\right\};$$

$$\operatorname{csch} z = \frac{1}{\sinh z} = \frac{2}{e^z - e^{-z}}, \quad z \in \mathbf{C} - \{n\pi i \mid n = 0, \pm 1, \dots\}.$$

$$(2.6.2.9)$$

These are rational functions of e^z and are continuous on their respective domains of definition. They enjoy properties similar to those stated in (2.6.2.6) and (2.6.2.8). Refer to Exercise A (3).

Exercises A

(1) Determine $|z|, \operatorname{Arg} z, \operatorname{Re} z$ and $\operatorname{Im} z$ for each of the following complex numbers z:
$\cos\left(\frac{\pi}{4} + \frac{i}{2}\right)$, $\cos e^i$, $\sin 2i$, $\sin\left(\frac{\pi}{6} + i\right)$, $\tan(2 - i)$, $\tan e^{-i}$, $\cot\left(\frac{\pi}{4} - i\right.$
$\log 2\big)$, $\cot(-1 + i)$, $\tanh\left(\log 3 + \frac{\pi}{4}i\right)$, $\sinh(-2i), \coth(2 + i)$.

(2) Solve the following equations.
   (a) $\sin z = 2$.   (b) $\cosh z = 0$.   (c) $\cos z + \sin z = 0$.   (d) $\tan z = 1 + 2i$.
   (e) $\cot z = 3$.

(3) Prove the following identities (when domains of definition concerned are permitted).

   (a) $\sin 2z = 2\sin z \cos z$; $\cos 2z = \cos^2 z - \sin^2 z$.

   (b) $\cos z_2 - \cos z_1 = -2\sin\dfrac{z_2 + z_1}{2}\sin\dfrac{z_2 - z_1}{2}$; $\sin z_2 - \sin z_1 =$
   $2\cos\dfrac{z_2 + z_1}{2}\sin\dfrac{z_2 - z_1}{2}$.

   (c) $\cosh^2 z - \sinh^2 z = 1$.

   (d) $\sinh\left(\dfrac{\pi}{2}i - z\right) = i\cosh z$.

   (e) What are $\sinh(z_1 \pm z_2), \cosh(z_1 \pm z_2), \sinh 2z$ and $\cosh 2z$?

   (f) $(1 + i)\cot z + (1 - i)\cot \bar{z} = 2\dfrac{\sin(2x) + \sinh(2y)}{\cosh(2y) - \cos(2x)}$, $z = x + iy$.

   (g) $\operatorname{sech}^2 z + \tanh^2 z = 1$.

(4) Prove (2.6.2.6) in detail, especially about $\sin z$.

(5) Show that $|\operatorname{Im} z| \le |\sin z| \le e^{|\operatorname{Im} z|}$.

(6) Prove (2.6.2.8) in detail.

(7) Model after (2.6.2.6) and (2.6.2.8) to obtain similar results for $\sec z$, $\csc z$ and hyperbolic functions in (2.6.2.9).

(8) (a) On each square with vertices $n\pi(\pm 1 \pm i), n = 1, 2, \ldots$, show that $|\cos z| \geq 1$ always holds.

 (b) On each square with vertices $\left(n + \frac{1}{2}\right)\pi(\pm 1 \pm i), n = 1, 2, \ldots$, show that $|\sin z| \geq 1$.

 (c) On $|\operatorname{Im} z| \geq \delta > 0$, show that

$$\begin{cases} |\tan z| \\ |\cot z| \end{cases} \leq (1 + \sinh \delta)^{-2})^{\frac{1}{2}}.$$

Exercises B

(1) Show that complex trigonometric and hyperbolic functions are symmetry-preserving functions (refer to Exercise B (4) of Sec. 2.6.1). Locate their zeros.

(2) Show that complex trigonometric and hyperbolic functions are open mapping and are local homeomorphisms in their respective domains of definition (refer to Exercise B (2) of Sec. 2.6.1).

(3) Except those points where the functions assume 0 or ∞, show that the following functions $|\cos z|, |\sin z|, \ldots, |\operatorname{csch} z|$ do not obtain local maximum or minimum in their respective domains of definition (refer to Exercise B (3) of Sec. 2.6.1).

2.7.   Elementary Multiple-Valued Functions

Functions mentioned so far are single-valued except $\arg z$ $(z \neq 0)$, $\sqrt[n]{z}$ $(n \geq 2$, see Sec. 1.5) and $\log z$ $(z \neq 0$, see Example 2 in Sec. 1.7), which were informally introduced.

If a (single-valued) function $z = f(z)$ is *not* univalent (namely, one-to-one) on its domain of definition, then its inverse function $w = f^{-1}(z)$ is definitely multiple-valued. The following are elementary multiple-valued functions.

(1) The inverse of the nth power function $z = w^n (n \geq 2)$: the n-th *root function*

$$w = \sqrt[n]{z} = \sqrt[n]{|z|}e^{\frac{i(\operatorname{Arg} z + 2k\pi)}{n}}, \quad k = 0, 1, 2, \ldots, n-1.$$

(2) The inverse of the Joukowski function $z = \frac{1}{2}\left(w + \frac{1}{w}\right)$:

$$w = z + \sqrt{z^2 - 1}.$$

(3) The inverse of the exponential function $z = e^w$: *the logarithmic function*

$$w = \log z = \log|z| + i(\operatorname{Arg} z + 2n\pi), \quad n = 0, \pm1, \pm2, \ldots, z \neq 0.$$

(4) The inverse of $\cos w$: *The arccosine function*

$$w = \cos^{-1} z = -i\log(z + \sqrt{z^2 - 1}).$$

(5) The inverse of $z = \sin w$: *The arcsine function*

$$w = \sin^{-1} z = -i\log i(z + \sqrt{z^2 - 1}).$$

(6) The inverse of $z = \tan w$: *The arctangent function*

$$w = \tan^{-1} z = \frac{i}{2}\log\frac{i+z}{i-z} = \frac{1}{2i}\log\frac{1+iz}{1-iz}.$$

(7) The inverse of $z = \cot w$: *The arccotangent function*

$$w = \cot^{-1} z = \frac{i}{2}\log\frac{z-i}{z+i}.$$

(8) The inverse of $z = \cosh w$: *The arc hyperbolic cosine function*

$$w = \cosh^{-1} z = \log(z + \sqrt{z^2 - 1}) \text{ (see Example 3 of Sec. 2.7.3)}$$

$$= i\cos^{-1} z.$$

(9) The inverse of $z = \sinh w$: *The arc hyperbolic sine function*

$$w = \sinh^{-1} z = \log(z + \sqrt{z^2 + 1}) = i\sin^{-1} iz.$$

(10) The inverse of $z = \tanh w$: *The arc hyperbolic tangent function*

$$w = \tanh^{-1} z = \frac{1}{2}\log\frac{1+z}{1-z} = -i\tan^{-1} iz.$$

(11) The inverse of $z = \coth w$: *The arc hyperbolic cotangent function*

$$w = \coth^{-1} z = \frac{1}{2}\log\frac{z+1}{z-1} \quad \text{(see Example 4 of Sec. 2.7.3).} \quad (2.7.1)$$

So are $\sec^{-1} z$, $\csc^{-1} z$, $\operatorname{sech}^{-1}z$ and $\operatorname{csch}^{-1}z$. They all originate from the multi-valuedness of $\arg z$, and $\sqrt[n]{z}$ (including $z + \sqrt{z^2 - 1}$) and $\log z$ are the most fundamental among all.

Our main purpose in this text is to study analytic properties of functions via the limit processes of differentiation (Chap. 3) and integration (Chaps. 3 and 4). Before we are able to perform these processes to multiple-valued functions, we need to make sure on which parts of the plane such functions have *local* single-valued branches (see (2.2.5), mainly for differentiation) and *global* ones (mainly, for integration).

Hence, in this section, we will focus on the following aspects for *elementary* multiple-valued functions.

(1) *Local single-valued branch*

   (a) To find out these points in the plane **C** (or **C***), on a certain open *disk* neighborhood of each such point where a single-valued branch can be defined.

   Note: For clearer purpose, by a *local* single-valued *branch* we always mean a single-valued branch defined on *open disk neighborhood* unless otherwise stated.

   (b) Could there be any explicit expressions for single-valued branches?

   (c) How many of them are there?

   (d) How are they related to each other?   (2.7.2)

Suppose $w = F(z) : G$ (domain in **C***) $\to$ **C*** is a multiple-valued function. Let H be a subdomain of G and z_0 in H is a point so that $F(z)$ has single-valued branches on H after deleting some ray or rays starting at z_0. Suppose z_0 has the following properties.

   (i) $w = F(z)$ may not be defined at z_0 (such as $\arg z$ and $\log z$ at $z_0 = 0$). In case $F(z)$ is defined at z_0, at least two of the so many values of $F(z_0)$ coincide (such as $\sqrt{z}$ at $z_0 = 0$).

   (ii) If z moves once along a closed Jordan curve γ in H, containing z_0 in Int γ, and back to its starting point, F takes a value at the starting point other than any preassigned value to it, namely, a single-valued branch is changed continually into another single-valued branch. (2.7.3)

Then, we call z_0 a *branch point* of $w = F(z)$. A Jordan curve connecting two or more branch points is called a *branch cut* for $w = F(z)$. Usually, points on a branch cut are *singular points* of branches of $w = F(z)$ (see (2.2.5) and Example 1 that follows there). Condition 2 above can be replaced by the following

(2′) Let γ be as in 2. If z moves along γ once and back to its starting point, the image curve $F(\gamma)$ is *not* a closed curve, i.e., $w = F(z)$ does not trace a closed curve. See Fig. 2.39.

Under this circumstance, if z moves along γ in the same direction a finite number of times, say n $(n \geq 2)$, and back to its starting point, $w = F(z)$ will take back its original value, i.e., $w = F(z)$ will trace a Jordan closed curve, then we say z_0 is a *branch point of order* $n - 1$. For instance, if $w = F(z)$ takes $w_0 = F(z_0)$ at z_0 n times, that is to say, $F(z) - F(z_0)$ has

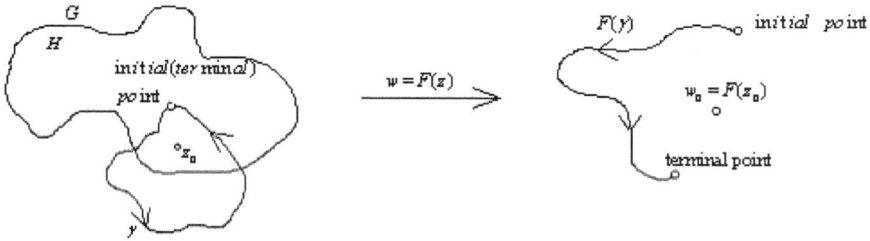

Fig. 2.39   Phenomenon caused by w branch point.

a zero of order n at z_0, then z_0 would be a branch point of order $n - 1$. Suppose z_0 is a branch point of order $n - 1$, and as $z \to z_0$, $F(z)$ has a limit in $\mathbf{C}^*$, then z_0 is called an *algebraic branch point of order* $n - 1$; otherwise, a *transcendental branch point*. A branch point which is not of finite-order is called a *logarithmic branch point*; in this case, no matter how many times z goes around z_0 along γ, the corresponding $F(z)$ will never take its original value back. For example, 0 and ∞ are algebraic branch points of $n - 1$ of $w = \sqrt[n]{z}$ ($n \geq 2$), while 0 and ∞ are logarithmic branch points of $w = \log z$. On the other hand, if z_0 satisfies only Condition 1 but not 2 in (2.7.3), then z_0 is called a *node* (point) of $w = F(z)$. For example, ∞ is a node of $\sqrt{(z - a)(z - b)}$ ($a \neq b$; see Example 2 in Sec. 2.7.2) but not for $\sqrt{(z - a)(z - b)(z - c)}$ ($a \neq b \neq c$) of which it is an algebraic branch point of order 1; ∞ is a node of $\sqrt[3]{(1 - z)(1 + z)^2}$ but not for $\sqrt[3]{(1 - z)(1 + z)}$ of which it is an algebraic branch point of order 2 (refer to Exercises, B (3) and (4) in Sec. 2.7.2).

(2) *Global single-valued branch*

Suppose $w = F(z) : G$ (domain in $\mathbf{C}^*$) $\to \mathbf{C}^*$ is multiple-valued and there exists a local single-valued branch $f(z)$ in an open disk neighborhood O_{z_0} of a point z_0 in G. How can O_{z_0} be expanded to a largest possible subdomain G_f in G so that the resulted $f(z)$ is still single-valued on G_f? Usually, G_f is obtained from G by deleting some branch cuts in it. And we can still ask questions such as b, c, and d in (2.7.2).

This phenomenon does happen in the following situation. Suppose $z = g(w) : \Omega$ (domain in the w − plane $\mathbf{C}^*$) $\to \mathbf{C}^*$ is a continuous function in the generalized sense so that its inverse function $w = F(z) : G \to \mathbf{C}^*$ is multiple-valued. If there exists a subdomain $D \subseteq \Omega$ so that:

(1) $g : D \to \mathbf{C}^*$ is univalent (namely, one-to-one); and
(2) $g(D)$ is a domain obtained from $\mathbf{C}^*$ by deleting one or more disjoint branch cuts in it.   (2.7.4)

then D is called a *fundamental domain* or *domain of univalence* for $z = g(w)$. Therefore, the restriction $f(z)$ of $w = F(z)$ to $g(D) = G_f$, namely, $w = f(z) : G_f \to D$ is such a *global* single-valued branch described in the last paragraph. For instance, $\theta < \operatorname{Arg} w < \theta + \frac{2\pi}{n}$ (θ fixed and $-\pi < \theta < \pi$) is a domain of univalence for $z = w^n$ (see (2.5.2.1)); both $|w| < 1$ and $|w| > 1$ are the ones for $z = \frac{1}{2}\left(w + \frac{1}{w}\right)$ (see (2.5.5.4)); $\alpha < \operatorname{Im} w < \alpha + \pi$ (α fixed) is one for $z = e^w$ (see (2.6.1.3)); $0 < \operatorname{Re} w < \pi$ for $z = \cos w$ (see (2.6.2.6)), etc.

(3) *Riemann surface*

Even though we can define single-valued branches on G, $w = F(z)$ is still multiple-valued on G because G itself is "single-layer" only. Is it possible to construct an ideal surface R_F on which $w = F(z)$ turns out to be single-valued? A rough idea is like this: To each global single-valued branch $f(z)$ of $w = F(z)$, pick up its largest possible domain G_f of definition. Then, paste these G_f together along the upper and lower edges of their common branch cut or cuts to form a "*multiple-layer*" G, namely, R_F, the so-called *Riemann surface* of the multiple-valued function $w = F(z)$. Note that the process of pasting G_f together should depend on the behavior of $w = F(z)$ at its branch points.

The simplest example of a Riemann surface is provided by $w = \sqrt{z}$. See Sec. 2.7.2.

This section is divided into four subsections.

Section 2.7.1 characterizes the existence of local and global single-valued branches of $\arg z$, and rough ideas about the winding number of a closed curve about a point not lying on it. These will lay the foundation for later subsections.

$w = \sqrt[n]{z}$ ($n \geq 2$), $w = \sqrt[n]{z^m}$, and $w = z + \sqrt{z^2 - 1}$, etc. will be in Sec. 2.7.2.

The logarithm function $w = \log z$ and other exponential functions are in Sec. 2.7.3, while Sec. 2.7.4 is for $w = \cos^{-1} z, \tan^{-1} z$, etc.

For more concrete examples concerning their geometric mapping properties, refer to Sec. 3.5.7.

2.7.1.   *The origin of multiple-valuedness: arg z*

Recall that $\arg z = \operatorname{Arg} z + 2n\pi, n = 0, \pm 1, \pm 2, \ldots$ and $-\pi < \operatorname{Arg} z \leq \pi$ for $z \neq 0, \infty$. Figure 1.6 and arguments in Example 1 of Sec. 2.2 show that it is not possible to define (single-valued) branches of $\arg z$ in any neighborhood of either 0 or ∞. Note that a Jordan closed curve oriented in counterclockwise direction and surrounding 0, when viewed from ∞, is the one oriented in clockwise direction and vice versa. This is so because $z \to \frac{1}{z}$ transforms 0 to ∞ and $\operatorname{Arg} z$ into $\operatorname{Arg} \frac{1}{z} = -\operatorname{Arg} z$.

Section (1) Local branches

Fix $z_0 \in \mathbf{C}^*$ and $z_0 \neq 0, \infty$.

Case 1. The disk $|z - z_0| < |z_0|$ does not intersect (but might be tangent to) the negative real axis, see Fig. 2.40(a). Then $|z - z_0| < |z_0|$ lies entirely within $\mathbf{C} - (-\infty, 0]$ to which Example 1 in Sec. 2.2 applies. Namely, $f(z) = \operatorname{Arg} z$ is the principal branch of $\arg z$ and any other branches are of the forms $f(z) + 2n\pi, n = 0, \pm 1, \ldots$.

Case 2. The disk $|z - z_0| < |z_0|$ intersects the negative real axis, see Fig. 2.40(b). Define

$$f(z) = \begin{cases} \operatorname{Arg} z, & |z - z_0| < |z_0| \text{ and } \operatorname{Im} z \geq 0 \\ \operatorname{Arg} z + 2\pi, & |z - z_0| < |z_0| \text{ and } \operatorname{Im} z < 0 \end{cases}. \qquad (2.7.1.1)$$

Then $f(z)$ is a branch of $\arg z$ on $|z - z_0| < |z_0|$, so are $f(z) + 2n\pi$, $n = 0, \pm 1, \ldots$.

Suppose $g(z)$ is a possible branch of $\arg z$ on $|z - z_0| < |z_0|$. Then

$$h(z) = \frac{1}{2\pi}(g(z) - f(z)), \quad |z - z_0| < |z_0| \qquad (2.7.1.2)$$

is a continuous function taking only integer values. Hence (see Remark 2 in Sec. 2.2), there exists a unique integer n so that $h(z) \equiv n$ on $|z - z_0| < |z_0|$ and thus, $g(z) = f(z) + 2n\pi$.

Let $g(z)$ be any *fixed* branch of $\arg z$ on $|z - z_0| < |z_0|$ and ζ be any *fixed* point in $|z - z_0| < |z_0|$. *Then $g(z)$ is completely determined on the entire disk $|z - z_0| < |z_0|$ by the value $g(\zeta)$.* Indeed, since $\operatorname{Arg} z$ is uniquely defined on $|z - z_0| < |z_0|$, $g(\zeta)$ would decide what integer n should be so that $g(z) = \operatorname{Arg} z + 2n\pi$ or $\operatorname{Arg} z + (2n + 2)\pi$.

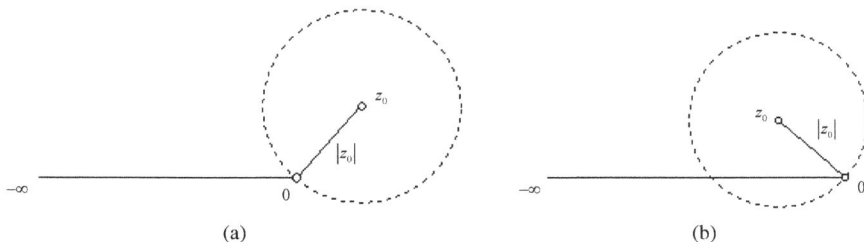

(a)   (b)

Fig. 2.40

Remark. Fix θ_0, $0 \le \theta_0 < 2\pi$, and designate the principal value $\operatorname{Arg} z$ of $\arg z$ in the range $\theta_0 \le \operatorname{Arg} z < \theta_0 + 2\pi$. Then $\operatorname{Arg} z$ is continuous on $\mathbf{C} - (\{0\} \cup \{z \mid \operatorname{Arg} z = \theta_0\})$ and results similar to what we obtained so far are still valid. We will feel free to use them if needed.

Section (2) Global branches

Firstly, we try to continue $\arg z$ along a curve not passing through 0.

Let γ be a *Jordan curve* with the representation $z(t) = x(t) + iy(t)$, $t \in [a, b]$ and $z(t) \ne 0$ for all t. For each fixed $t \in [a, b]$, take $0 < \varepsilon \le |z(t)|$. Then

(1) It is possible to define branches of $\arg z$ on $|z - z(t)| < \varepsilon$.
(2) There exists a $\delta_t = \delta(t, \varepsilon) > 0$ so that, if $t' \in [a, b]$ and $|t' - t| < \delta_t$, then $|z(t') - z(t)| < \varepsilon$ always holds. In particular, the relatively open interval $I_t = (t - \delta_t, t + \delta_t) \cap [a, b]$ has its image

$$z(I_t) \subseteq \{|z - z(t)| < \varepsilon\}.$$

For an assigned branch of $\arg z$ on $|z - z(t)| < \varepsilon$, designate

$$\arg z(t)$$

the restriction of the branch along $z(t), t \in I_t$. Then $\arg z(t)$ *is continuous on I_t and is called a* (*continuous*) *branch of* $\arg z$ *along I_t.* Note that, $\arg z(t)$ is also completely determined by its value at a point of I_t; and the difference of any two such branches of $\arg z$ along I_t is an integral multiple of 2π. See Fig. 2.41.

Now, I_t, $t \in [a, b]$, form an open covering of $[a, b]$. By Heine–Borel theorem (see (1.9.3)), there exist finitely many points $t_1, \ldots, t_n \in [a, b]$

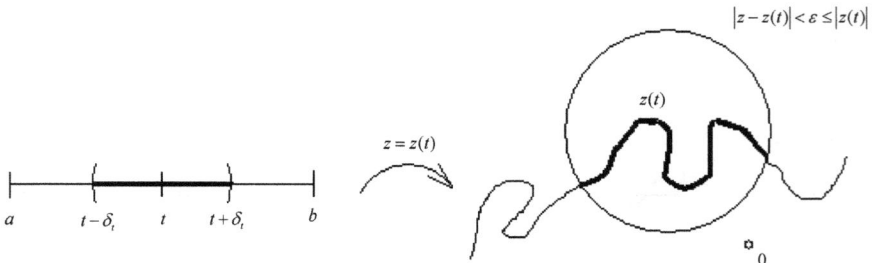

Fig. 2.41

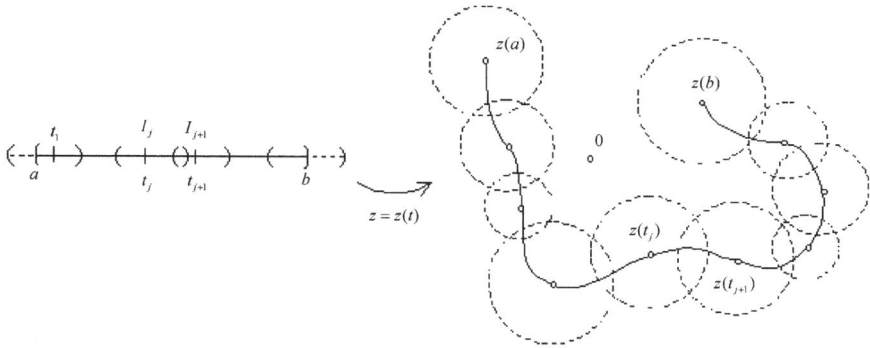

Fig. 2.42

so that

(1) $I_j \cap I_{j+1} \neq \phi, 1 \leq j \leq n-1$, where $I_j = I_{t_j}$ for short;
(2) $[a, b] \subseteq \bigcup_{j=1}^{n} I_j$.

See Fig. 2.42 (assume $a \leq t_1 < t_2 < \cdots < t_{n-1} < t_n \leq b$ and $I_j = (t_j - \delta_{t_j}, t_j + \delta_{t_j})$ rather than its intersection with $[a, b]$). Presume a fixed value of $\arg z(a)$. This value determines a unique branch $f_1(t)$ of $\arg z$ along I_1. For any fixed point $t \in I_1 \cap I_2, f_1(t)$ determines a unique branch $f_2(t)$ of $\arg z$ along I_2 so that

$$f_2(t) = f_1(t), \quad t \in I_1 \cap I_2.$$

Continuing this process, after $n-2$ similar steps and for any fixed point $t \in I_{n-1} \cap I_n, f_{n-1}(t)$ determines a unique branch $f_n(t)$ of $\arg z$ along I_n so that

$$f_n(t) = f_{n-1}(t), \quad t \in I_{n-1} \cap I_n.$$

Then, define a function $f : [a, b] \to \mathbf{R}$ by

$$f(t) = f_j(t), \quad t \in I_j, \quad 1 \leq j \leq n. \tag{2.7.1.3}$$

This f is well-defined on $[a, b]$ and hence, single-valued and continuous there.

   Note that, in case γ is *not* a Jordan curve, the process leading to (2.7.1.3) still works. Under this circumstance, f is well-defined on $[a, b]$ but is not more single-valued, in general, and is changed continuously on $[a, b]$. And we call such f an *analytic continuation of an assigned value of* $\arg z(a)$ *along* γ. For details, see Sec. 5.2.1.

We summarize the above as

The analytic continuation of $\arg z$ *along a curve.* Let γ, $z(t) = x(t) + iy(t)$: $[a, b] \to \mathbf{C}$, be a curve in $\mathbf{C}$ not passing 0.

(1) Each value of $\arg z(a)$ uniquely determines an analytic continuation f of $\arg z$ along γ; in case γ is a *Jordan curve but is not closed*, f is a single-valued continuous *branch* of $\arg z$ along γ.

    (a) If $f(t)$ is a branch of $\arg z$ along a Jordan curve γ, then any branch of $\arg z$ along γ is of the form $f(t) + 2n\pi, n = 0, \pm 1, \ldots$ (see (2.7.1.2)).

    (b) If $z = z(t)$ continuously moves along γ from $z(a)$ to $z(b)$, then $\arg z$ (of any branch of $\arg z$ along γ) changes continuously from $\arg z(a)$ to $\arg z(b)$. The *variation* $\Delta_\gamma \arg z$ of $\arg z$ along γ from $\arg z(a)$ to $\arg z(b)$ is defined by

$$\arg z(b) = \arg z(a) + \Delta_\gamma \arg z.$$

Note that $\Delta_\gamma \arg z$ is dependent only on γ and independent of the value of $\arg z(a)$ and the representations of γ.

(2) Let γ be a *closed* curve in $\mathbf{C}$, not passing 0, then $\Delta_\gamma \arg z$ is an integer multiple of 2π. In this case, the integer

$$n(\gamma; 0) = \frac{1}{2\pi} \Delta_\gamma \arg z$$

is called the *winding number* of γ around 0.   (2.7.1.4)

For an example, see Exercise A (2); for more explanation about winding number, see Exercises B and Sec. 3.5.2.

Suppose γ_1 and γ_2 are two curves, not passing 0 and having the same initial and terminal points, with the representations $z_j = z_j(t) : [a, b] \to \mathbf{C}$, $j = 1, 2$. Then $z_1(a) = z_2(a)$ and $z_1(b) = z_2(b)$. See Fig. 2.43. Also (see (1)b in (2.7.1.4))

$$\arg z_j(b) = \arg z_j(a) + \Delta_{\gamma_j} \arg z, \quad j = 1, 2.$$

If $\arg z_1(a) = \arg z_2(a)$ is assumed, then

$$\arg z_2(b) - \arg z_1(b) = \Delta_{\gamma_2} \arg z - \Delta_{\gamma_1} \arg z \underset{\text{(def.)}}{=} \Delta_{\gamma_2 - \gamma_1} \arg z$$

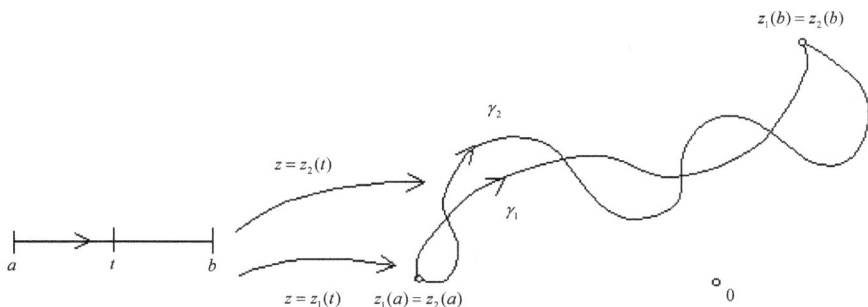

Fig. 2.43

where $\gamma_2 - \gamma_1$ denotes the closed curve starting from b, tracing back to a along γ_1 and then again to b along γ_2. Hence

$$\arg z_2(b) = \arg z_1(b),$$

$$\Leftrightarrow \Delta_{\gamma_2 - \gamma_1} \arg(z) = 0,$$

$$\Leftrightarrow \text{The winding number } n(\gamma_2 - \gamma_1; 0)$$

$$\text{of } \gamma_2 - \gamma_1 \text{ around } 0 \text{ is equal to zero.} \qquad (2.7.1.5)$$

That is to say, if $\arg z$ assumes the same value at $z_1(a) = z_2(a)$ and moves continuously along different curves γ_1 and γ_2 to the same point $z_2(a) = z_2(b)$, it might assume distinct values at the terminal point; and *it will assume the same value if and only if the winding number of the closed curve $\gamma_2 - \gamma_1$ around 0 is equal to zero.*

Finally, let $\Omega \subseteq \mathbf{C}$ be a domain, not containing 0.

Fix a point $z_0 \in \Omega$ and presume any value of $\arg z_0$.

Take any other point $z^* \in \Omega$ so that $z^* \neq z_0$. Let γ be a curve in Ω, connecting z_0 to z^*, with the representation $z = z(t)$, $t \in [a, b]$. Then, for an analytic continuation of $\arg z$ along γ,

$$\arg z^* = \arg z(b) = \arg z(a) + \Delta_\gamma \arg z = \arg z_0 + \Delta_\gamma \arg z.$$

In case $\arg z^* = \arg z(b)$ is independent of the curve γ (in Ω) connecting z_0 to z^* but dependent only on the assigned value $\arg z_0 = \arg z(a)$, then a unique *single-valued continuous branch* of $\arg z$ can be defined on the whole Ω.

We summarize the above as

The existence of global branches of $\arg z$. Let Ω be a domain in $\mathbf{C}$ so that $0 \notin \Omega$.

(1) Ω is a domain on which a (single-valued, continuous) *branch* of $\arg z$ can be defined. Such a Ω is called a *single-valued domain of* $\arg z$, for simplicity.

$\Leftrightarrow$ For any closed curve γ in Ω, the winding number $n(\gamma; 0) = 0$.

(2) Let Ω be a single-valued domain for $\arg z$.

    (a) Fix any $z_0 \in \Omega$. Then, any presumed value of $\arg z_0$ determines uniquely a branch of $\arg z$ on Ω.

    (b) If $f(z)$ is any branch of $\arg z$ on Ω, then all the branches of $\arg z$ on Ω are of the forms $f(z) + 2n\pi, n = 0, \pm 1, \pm 2, \ldots$.

(3) Single-valued domains for $\arg z$.

    (a) Any simply connected domain in $\mathbf{C}$, not containing 0 (or, any simply connected domain in $\mathbf{C}^*$, not containing 0 and ∞) is such a domain.

    (b) A domain Ω in $\mathbf{C}$ (or $\mathbf{C}^*$) is such a domain if and only if 0 and ∞ belongs to the same component of $\mathbf{C}^* - \Omega$. In this case, Ω is contained in a simply connected domain in $\mathbf{C}$ (or $\mathbf{C}^*$), not containing 0 (and ∞).

For example, if γ^* is a Jordan curve connecting 0 to ∞, then $\mathbf{C}^* - \gamma^*$ is a single-valued domain for $\arg z$. 0 and ∞ are *branch points* and γ^* *branch cut* of $\arg z$. Each point on γ^* is a *singular point* of $\arg z$ (refer to Example 1 of Sec. 2.2). See Fig. 2.44.   (2.7.1.6)

Explanation about (3): If $\Omega \subseteq C$ is a single-valued domain for $\arg z$, then $o \notin \Omega$ and hence, 0 and ∞ are in $\mathbf{C}^* - \Omega$. In case 0 and ∞ are in distinct components of $\mathbf{C}^* - \Omega$. Then there exists a Jordan closed curve γ so that $0 \in \operatorname{Int} \gamma$ and $\infty \in \operatorname{Ext} \gamma$ (see (2.4.11)). Since, then, $n(\gamma; 0) = 1$ or -1 (see Exercise A (2) or Exercise B (1)), this contradicts to our assumption that Ω is a single-valued domain. Hence, 0 and ∞ should be in the same component of $\mathbf{C}^* - \Omega$. If $\Omega \subseteq \mathbf{C}$ is a simply connected domain so that $0 \notin \Omega$, then 0 and ∞ are in the only component $\mathbf{C}^* - \Omega$ (see (2.4.12)). Therefore, for any closed curve γ in Ω, 0 lies in the unbounded component (that one containing ∞ of $\mathbf{C}^* - \gamma$), and thus (see Exercise B (2)(c)) $n(\gamma; 0) = 0$. Hence, Ω is a single-valued domain for $\arg z$. Similar arguments are still

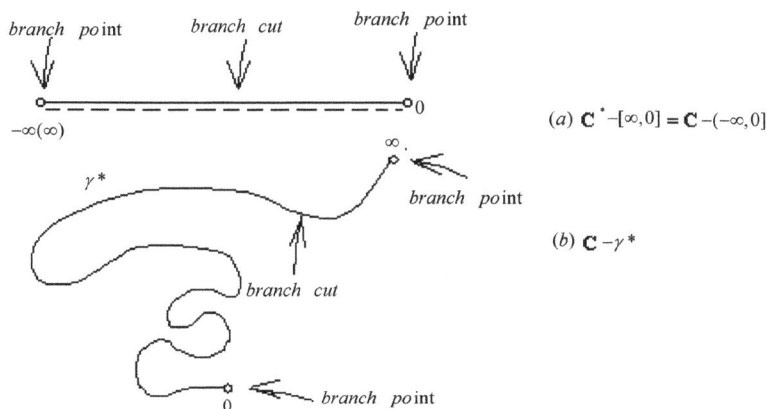

Fig. 2.44

valid for domain Ω in $\mathbf{C}$, where 0 and ∞ are in the same component of $\mathbf{C}^* - \Omega$.

Note that what we have obtained so far for $\arg z$ are still valid for $\arg(z - z_0)$, where $z_0 \in \mathbf{C}$, subject to minor yet obvious modifications. We will feel no hesitation to use them if needed. Refer to Exercise B for a somewhat different treatment.

Exercises A

(1) Prove (2.7.1.6) for $\arg(z - z_0)$, where $z_0 \in \mathbf{C}$ is a fixed point.

(2) For the unit circle $\gamma : z(t) = e^{it}$, $0 \le t \le 2\pi$, $z(0) = z(2\pi) = 1$. Designate $\arg z(0) = 2n_0\pi$ (n_0 is a fixed integer). Then $\arg z(2\pi) = \arg z(0) + \Delta_\gamma \arg z = 2n_0\pi + 2\pi = (2n_0 + 2)\pi$ and $n(\gamma; 0) = \frac{1}{2\pi} 2\pi = 1$.

  (a) Let $\gamma : z(t) = e^{2it}, 0 \le t \le 2\pi$. Designate $\arg z(0) = 0$. Compute $\arg z(2\pi), \Delta_\gamma \arg z$ and $n(\gamma; o)$.

  (b) For curves γ in Fig. 2.45, compute $n(\gamma; 0)$.

Exercise B

The winding number $n(\gamma; z_0)$.

Let $\gamma : z(t) = x(t) + iy(t), 0 \le t \le 1$, be a curve which does not pass a fixed point $z_0 \in \mathbf{C}$. Choose d so that

$$\inf_{t \in [0,1]} |z(t) - z_0| = \text{dist}(\gamma, z_0) > d > 0.$$

Since $z(t)$ is uniformly continuous on $[0, 1]$, there is a partition $0 = t_0 < t_1 < \cdots < t_{n-1} < t_n = 1$ of $[0, 1]$ so that, for each $t \in [t_j, t_{j+1}]$, $0 \le j \le$

Fig. 2.45

$n - 1$, $|z(t) - z(t_j)| < d$ always holds. Hence,

$$\left| \frac{z(t) - z_0}{z(t_j) - z_0} - 1 \right|$$

$$< \frac{d}{|z(t_j) - z_0|} \leq \frac{d}{d} = 1, \quad t \in [t_j, t_{j+1}], \quad 0 \leq j \leq n - 1$$

$$\Rightarrow \operatorname{Re} \frac{z(t) - z_0}{z(t_j) - z_0} > 1 - 1 = 0, \quad t \in [t_j, t_{j+1}], \quad 0 \leq j \leq n - 1.$$

$$\Rightarrow -\frac{\pi}{2} < \operatorname{Arg} \frac{z(t) - z_0}{z(t_j) - z_0} < \frac{\pi}{2}, \quad t \in [t_j, t_{j+1}], \quad 0 \leq j \leq n - 1.$$

$$(2.7.1.7)$$

By Example 1 of Sec. 2.2, $\operatorname{Arg} \frac{z(t) - z_0}{z(t_j) - z_0}$ is continuous for $t \in [t_j, t_{j+1}]$. Thus, *fix* a value of $\arg(z(0) - z_0)$ and *define*

$$\arg(z(t) - z_0) = \arg(z(t_j) - z_0) + \operatorname{Arg} \frac{z(t) - z_0}{z(t_j) - z_0},$$

$$t \in [t_j, t_{j+1}], \quad 0 \leq j \leq n - 1. \qquad (2.7.1.8)$$

Then, $\arg(z(t) - z_0)$ is well-defined and is continuously changed along $[0, 1]$. We call *such* $\arg(z(t) - z_0)$ an *analytic continuation* of $\arg(z - z_0)$ *along* γ, uniquely determined by any presumed value of $\arg(z(0) - z_0)$.

In case γ is a *closed* curve around $z_0(z_0 \notin \gamma)$ and $\arg(z(t) - z_0)$ is a continuation of $\arg(z - z_0)$ along γ, the *variation* of $\arg(z(t) - z_0)$ along γ from $z(o)$ to $z(1) = z(0)$ is defined as

$$\arg(z(1) - z_0) - \arg(z(0) - z_0) = \Delta_\gamma \arg(z - z_0), \qquad (2.7.1.9)$$

which is an integral multiple of 2π. This integer depends only on the relative position of z_0 and γ, and is independent of the initial values chosen for $\arg(z(0) - z_0)$ and the parametric representation of γ. Indeed, suppose

$\varphi(t), t \in [0,1]$ is another representation of γ and $\arg(\varphi(t) - z_0)$ is a continuation of $\arg(z - z_0)$ along γ subject to an initial value of $\arg(\varphi(0) - z_0)$. Then, the integral-valued continuous function

$$\frac{1}{2\pi}[\arg(\varphi(t) - z_0) - \arg(z(t) - z_0)], \quad 0 \le t \le 1$$

on $[0, 1]$ should be a fixed integer k (see Remark 2 in Sec. 2.2). As a consequence,

$$\frac{1}{2\pi}[\arg(z(1) - z_0) - \arg(z(0) - z_0)]$$

$$= \frac{1}{2\pi}[\arg(\varphi(1) - z_0) - \arg(\varphi(0) - z_0)] = \text{an integer.}$$

This proves the claim. We denote this common integer as

$$n(\gamma; z_0) = \frac{1}{2\pi}\Delta_\gamma \arg(z - z_0) \qquad (2.7.1.10)$$

and call it the *winding number* or *index* of γ *around* z_0. Intuitively, $|n(\gamma; z_0)|$ is the number of times the curve γ winds around a point z_0 not lying on it. An equivalent formulation of $n(\gamma; z_0)$ is

$$n(\gamma; z_0) = \frac{1}{2\pi}\sum_{j=1}^{n-1} \text{Arg} \frac{z(t_{j+1}) - z_0}{z(t_j) - z_0}, \qquad (2.7.1.11)$$

which is independent of the choices of partitions $0 = t_0 < t_1 < \cdots < t_{n-1} < t_n = 1$ as long as (2.7.1.7) holds. This follows immediately from (2.7.1.8). Equation (2.7.1.11) provides a way of computing $n(\gamma; z_0)$.

In case γ is a piecewise differentiable closed curve (see (2.4.9)), it can be shown that

$$n(\gamma; z_0) = \frac{1}{2\pi i} \int_\gamma \frac{dz}{z - z_0}. \qquad (2.7.1.12)$$

For details, see Sec. 3.5.2.

Try to do the following problems.

(1) Use (2.7.1.11) to show that, if γ is a circle,

$$n(\gamma; z_0) = \begin{cases} 1, & \text{if } z \in \text{Int } \gamma \text{ and } \gamma \text{ is oriented in counterclockwise} \\ & \quad \text{direction} \\ -1, & \text{if } z \in \text{Int } \gamma \text{ and } \gamma \text{ is oriented in clockwise direction} \\ 0, & \text{if } z \in \text{Ext } \gamma \end{cases}.$$

Results are still valid for a Jordan closed curve (see Ref. [66]).

(2) *Basic properties* of winding numbers.

    (a) If γ is an oriented closed curve, then $n(-\gamma; z_0) = -n(\gamma; z_0)$.

    (b) Considering $n(\gamma; z_0)$ as a function of z_0, then $n(\gamma; z_0)$ is continuous on each component of $\mathbf{C}^* - \gamma$ and hence, is a constant.

    (c) $n(\gamma; z_0) = 0$ if z_0 lies on the unbounded component (the one containing ∞) of $\mathbf{C}^* - \gamma$.

    (d) If a closed curve γ is deformed continuously into another closed curve γ' so that z_0 is kept untouched during the deformation, then

$$n(\gamma; z_0) = n(\gamma'; z_0).$$

In short, winding number is an *invariant under homotopy*. See Sec. 4.2.3.

Note: Suppose γ and γ' are represented by $z = z(t)$ and $z = \varphi(t), t \in [0, 1]$, respectively. If there exists a continuous function $f : [0, 1] \times [0, 1] \to \mathbf{C}$ satisfying

$$f(t, 0) = z(t) \quad \text{and} \quad f(t, 1) = \varphi(t), \quad t \in [0, 1],$$

then f is called a *homotopy*, deforming γ continuously into γ', and γ and γ' are said to be *homotopic*.

2.7.2. $w = \sqrt[n]{z}$ $(n \geq 2)$ *and its Riemann surface (etc.)*

The function $z = w^n : \mathbf{C}^* \to \mathbf{C}^*$ is n-to-one except at $w = 0$ and ∞, which are the zero and the pole of order n of $z = w^n$, respectively (see Sec. 2.5.2). Therefore, its inverse function $w = \sqrt[n]{z} : \mathbf{C}^* \to \mathbf{C}^*$ is multiple-valued; in fact, it is one-to-n except at $z = 0$ and ∞.

Usually, there are two ways to determine local and global branches of $w = \sqrt[n]{z}$ and then, to construct an ideal surface on which $w = \sqrt[n]{z}$ is single-valued.

Section (1) (Based on Sec. 2.5.2 and the method explained in (2.7.4))

We start from $z = w^n$ and try to find domains of univalence for it.

Fix φ, $-\pi < \varphi \leq \pi$ and define the *principal argument* $\varphi \leq \operatorname{Arg} w < \varphi + 2\pi$. Then $z = w^n$ maps

(1) the sector $\varphi \leq \operatorname{Arg} w < \varphi + \frac{2\pi}{n}$ with the upper edge $\operatorname{Arg} w = \varphi$ (including 0) one-to-one and onto entire plane $\mathbf{C}$, and

(2) the sector domain $\Omega_0 : \varphi < \operatorname{Arg} w < \varphi + \frac{2\pi}{n}$ (0 is excluded) one-to-one and onto the domain $C_\varphi = \mathbf{C} - (\{0\} \cup \{\arg z = n\varphi\})$, obtained by deleting the ray $\arg z = n\varphi$ (including 0) from $\mathbf{C}$.   (2.7.2.1)

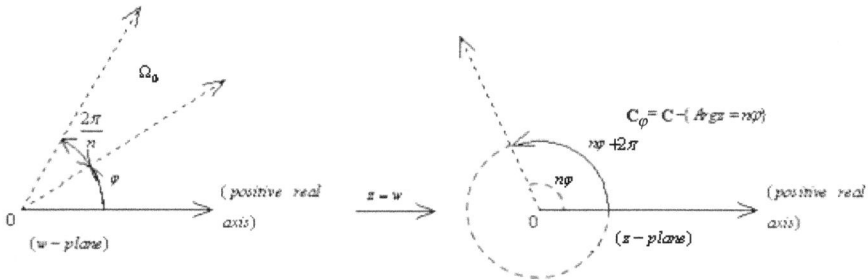

Fig. 2.46

The ray $\arg z = n\varphi$ or $n\varphi + 2\pi$ is, for convenience, supposed to have two *sides* with $\arg z = n\varphi$, the *upper side* and $\arg z = n\varphi + 2\pi$, the *lower side*. Therefore, the *upper edge* $\operatorname{Arg} w = \varphi$ is mapped onto the upper side and the *lower edge* $\operatorname{Arg} w = \varphi + \frac{2\pi}{n}$ onto the lower side (compare to (2.5.2.1)). See Fig. 2.46.

Let the sector domains

$$\Omega_k : \varphi + \frac{2k\pi}{n} < \operatorname{Arg} w < \varphi + \frac{2(k+1)\pi}{n}, \quad k = 0, 1, 2, \ldots, n-1.$$

$$(2.7.2.2)$$

They are *domains of univalence* for $z = w^n$. The restricted map $z = w^n$: $\Omega_k \to C_\varphi$ has a one-to-one and onto inverse function

$$w_k = |z|^{\frac{1}{n}} e^{i\frac{\operatorname{Arg} z + 2k\pi}{n}} : C_\varphi \to \Omega_k, \quad k = 0, 1, 2, \ldots, n-1 \qquad (2.7.2.3)$$

which are *branches* of $w = \sqrt[n]{z}$ on C_φ. Note that $w_k = w_0 e^{\frac{2k\pi i}{n}}$, $k = 0, 1, 2, \ldots, n-1$ (see Exercise B (1) for more information). 0 and ∞ are

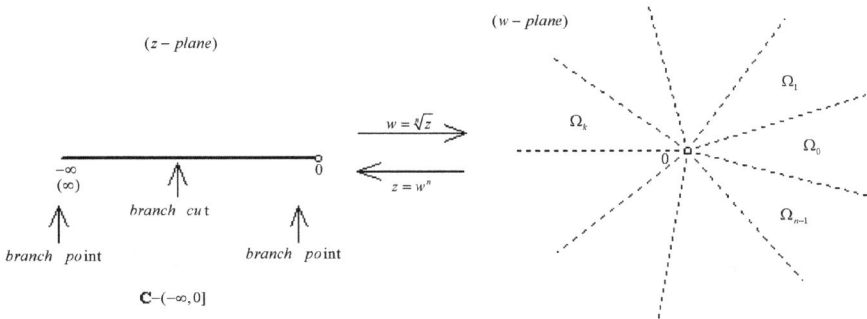

Fig. 2.47

algebraic branch points of *order* $n-1$ and the ray $\arg z = n\varphi$, connecting 0 to ∞, is a *branch cut*. See Fig. 2.47 (here $\varphi = -\frac{\pi}{n}$; compare to Fig. 2.13 where $\varphi = 0$).

Riemann surface for $w = \sqrt[n]{z}$.

For simplicity, set $\varphi = -\frac{\pi}{n}$. So $C_{-\frac{\pi}{2}} = \mathbf{C} - (\{0\} \cup \{\arg z = -\pi\}) = \mathbf{C} - (-\infty, 0]$ is the domain obtained from $\mathbf{C}$ by cutting off the nonpositive real axis $(-\infty, 0]$. Designate

the *upper side* of $\mathbf{C} - (\infty, 0]$: the negative real axis $(-\infty, 0]$ and $\operatorname{Arg} z = -\pi$;

the *lower side* of $\mathbf{C} - (\infty, 0]$: the negative real axis $(-\infty, 0]$ and $\operatorname{Arg} z = \pi$;

and the *upper* (or initial) *edge* of Ω_k : $\operatorname{Arg} w = \frac{(2k-1)\pi}{n}$; the *lower* (or terminal) *edge* of Ω_k : $\operatorname{Arg} w = \frac{(2k+1)\pi}{n}, k = 0, 1, 2, \ldots, n-1$.

Observe that $w = w_k(z)$ in (2.7.2.3) maps, respectively, the lower and upper sides of $\mathbf{C} - (\infty, 0]$ onto the lower and upper edges of Ω_k. See Fig. 2.48.

Now, take n replicas of $\mathbf{C} - (\infty, 0]$ so that the one corresponding to Ω_k under $w = w(z)$ is named the kth *sheet* for $k = 0, 1, \ldots, n-1$. Deposit these n-sheets, parallel to each other, in the three-dimensional Euclidean space so that the points over the same point in $\mathbf{C}$ always lie on the same vertical line perpendicular to these n-sheets on those points. On the w-plane, the lower edge of Ω_k is coincident with the upper edge of Ω_{k+1} for $0 \le k \le n-2$, while the lower edge of Ω_{n-1} is coincident with the upper edge of Ω_0. See the right figure in Fig. 2.47. Then, in imagination, paste these n replicas of $\mathbf{C} - (\infty, 0]$ together along the upper and lower sides according to the manner on how $\Omega_k, 0 \le k \le n-1$, are pasted together in the w-plane. Namely, the lower side of the kth sheet is pasted together with the upper side of the $(k+1)$th sheet for $0 \le k \le n-2$; the lower side of the $(n-1)$th-sheet with the upper side the 0th sheet. The resulted surface by pasting together these n replicas is denoted by $R_{\sqrt[n]{z}}$, called the *Riemann surface* of $w = \sqrt[n]{z}$. See Fig. 2.49.

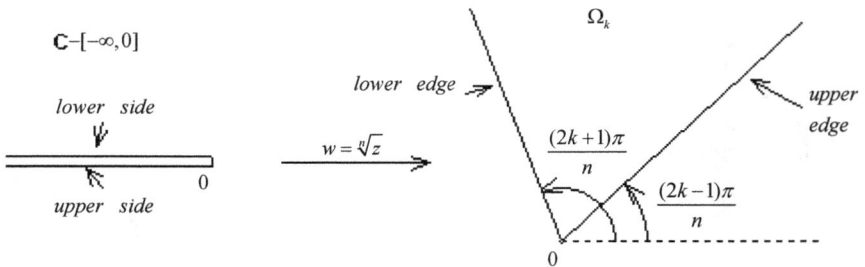

Fig. 2.48

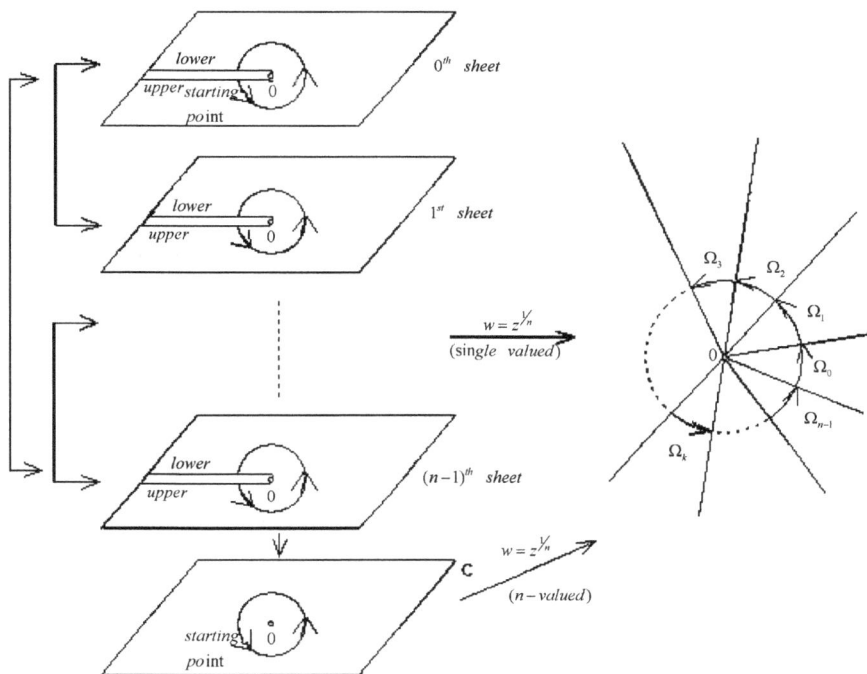

Fig. 2.49

Explanation

(1) Only points o and ∞ are common points to all the n replicas $\mathbf{C} - (\infty, 0] = \mathbf{C}^* - [\infty, 0]$ in $R_{\sqrt[n]{z}}$. They correspond to $z = 0$ and $z = \infty$, respectively, and are the algebraic branch points of order $n - 1$ of the multiple-valued function $w = \sqrt[n]{z}$.

(2) Now, $w = \sqrt[n]{z}$ is $single$-$valued$ on $R_{\sqrt[n]{z}}$. Its kth branch $w = w_k(z)$ has its domain the kth sheet $\mathbf{C} - (\infty, 0]$ and the range Ω_k for $0 \leq k \leq n-1$. As a whole, $w = \sqrt[n]{z}$ is a "homeomorphism" from $R_{\sqrt[n]{z}}$ (including 0 and ∞) onto $\mathbf{C}^*$.

(3) Each point z in the z-plane other than 0 and ∞ is covered n-times by n distinct "points" (one on each replica $\mathbf{C} - (\infty, 0]$) lying over it in $R_{\sqrt[n]{z}}$.

(4) If one starts from a fixed point (other than 0 and ∞) in the 0th sheet and goes around 0 in the counterclockwise direction to reach a point in the lower side, then one crosses into the 1th sheet from the same point on the upper side of this sheet. Similar process is continued until one

reaches a point in the lower side of the $(n-1)$th sheet and then comes out from the same point in the upper side of the 0th sheet to return to its starting point. See the left figure in Fig. 2.49. This amounts to one circling around "o" in the w-plane and n circlings around "o" in the original z-plane.

The process in (4) can be visualized by the following configuration. Use "x" and "o" to represent the lower and upper half-plane (or just a point in each half-plane), respectively. The action of crossing a branch cut (the positive and negative real axes) is denoted by a line segment connecting "x" to "o" or "o" to "x". The graph formed by "x" and "o" with line segments connecting them is called the *line complex* of $R_{\sqrt[n]{z}}$. See Fig. 2.50 ($n = 2$ and $\varphi = -\frac{\pi}{2}$).

Observe the following correspondence:

Riemann surface $R_{\sqrt[n]{z}}$	The w-plane	Line complex
Upper half-plane	Half of fundamental domain Δ	o
Lower half-plane	Half of fundamental domain Δ'	×
Branch cuts (positive and negative real axes)	sides of half-fundamental domain	Line segment—
Branch point o		Interior of line complex
Branch point ∞		Exterior of line complex

See Fig. 2.51. Finally, the "homeomorphism" between $R_{\sqrt[n]{z}}$ and $\mathbf{C}^*$ (the w-plane) in (2) suggests that the Riemann sphere (see Sec. 1.6) can be treated, via stereographic projection, as a *topological model* for the Riemann surface $R_{\sqrt[n]{z}}$ of $w = \sqrt[n]{z}$.

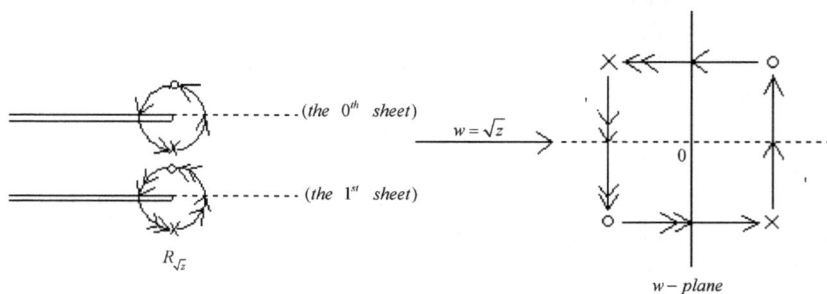

Fig. 2.50

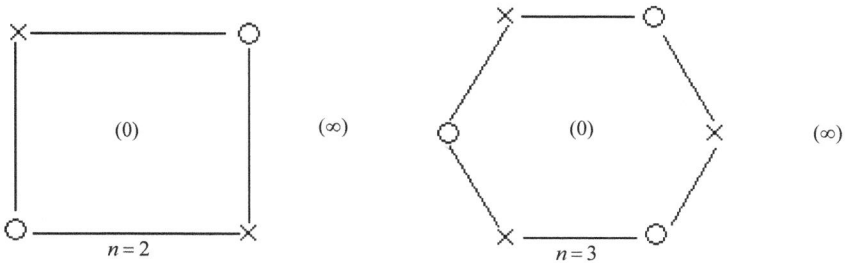

Fig. 2.51

Note: The cuts on the surface $R_{\sqrt[n]{z}}$ are never distinguished lines, but the nomination of specific cuts is necessary for descriptive purposes. Hence, the cut $[0, \infty]$ can be replaced by a cut along any Jordan curve from 0 to ∞, and the Riemann surface obtained using this new cut should be considered identical with the original $R_{\sqrt[n]{z}}$.

Section (2) (Based on Sec. 2.7.1 and the method explained in (2.7.2) and (2.7.3))

Since $\sqrt[n]{z} = |z|^{\frac{1}{n}} e^{i \frac{\text{Arg} z + 2k\pi}{n}}$, $k = 0, 1, 2, \ldots, n-1$, where $-\pi < \text{Arg} z \leq \pi$ and $z \neq 0, \infty$, the multi-valuedness of $w = \sqrt[n]{z}$ origins from that of $\arg z = \text{Arg} z + 2k\pi$. As a consequence, $w = \sqrt[n]{z}$ has exactly the same single-valued domains for its local and global branches as $\arg z$ (see (2.7.1.6)), with 0 and ∞ as its algebraic branch points of order $n-1$.

In order to verify intuitively the concept of branch points and branch cut described in (2.7.3) and to set up a model for later imitation, we try to study the behavior of $w = \sqrt[n]{z}$ around o in what follows.

Take a Jordan closed curve γ so that o is in the interior Int γ. Pick an arbitrarily fixed point z_0 on γ and a presumed value θ_0 of $\arg z_0$, which is equivalent to presuming a value $|z_0|^{\frac{1}{n}} e^{i \frac{\theta_0}{n}}$ of $\sqrt[n]{z_0}$. Let z start from z_0 and circle along γ once in counterclockwise direction. When z returns to z_0, the values taken by $w = \sqrt[n]{z}$ change continually as

$$|z|^{\frac{1}{n}} : |z_0|^{\frac{1}{n}} \to |z_0|^{\frac{1}{n}};$$

$$\arg z : \theta_0 \to \theta_0 + 2\pi;$$

$$\sqrt[n]{z} : w_0 = |z_0|^{\frac{1}{n}} e^{i \frac{\theta_0}{n}} \to |z_0|^{\frac{1}{n}} e^{i \frac{\theta_0 + 2\pi}{n}} = w_0 e^{i \frac{2\pi}{n}}. \qquad (2.7.2.4)$$

This shows that $w = \sqrt[n]{z}$ does not take back its original value w_0 but a value belonging to another branch. See Fig. 2.52. If z circles along γ twice and (back to z_0), then $w = \sqrt[n]{z}$ takes the values $w_0 e^{i \frac{4\pi}{n}}$. It takes z

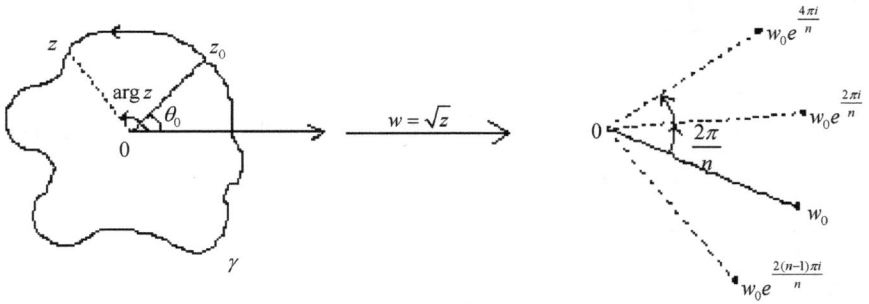

Fig. 2.52

to consecutively circle along γ exactly n times for $w = \sqrt[n]{z}$ to take back its original value $w_0 e^{i\frac{2n\pi}{n}} = w_0$. It is in this sense that we call o (and hence ∞), a *branch point* of order $n - 1$. To avoid such a phenomenon happening, construct a *Jordan arc* γ^* *connecting the branch point* o *to another branch point* ∞ so that, in $\mathbf{C}^* - \gamma^*$, no more such γ around o (∞) can exist.

Next, suppose o lies in the exterior of a Jordan closed curve γ. If z starts from z_0 and circles once along γ and back to z_0, the values taken by $w = \sqrt[n]{z}$ change continually as

$$|z|^{\frac{1}{n}} : |z_0|^{\frac{1}{n}} \rightarrow |z_0|^{\frac{1}{n}};$$

$$\arg : \theta_0 \rightarrow \theta_0;$$

$$\sqrt[n]{z} : w_0 = |z_0|e^{i\frac{\theta_0}{n}} \rightarrow |z_0|e^{i\frac{\theta_0}{n}} = w_0. \qquad (2.7.2.5)$$

Hence, no matter how many times z circles along γ, $w = \sqrt[n]{z}$ always takes back its original value. See Fig. 2.53. Therefore, any point, other than o and ∞, is not a branch point.

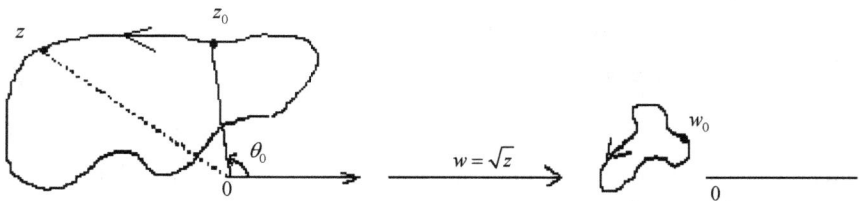

Fig. 2.53

As a summary, we have

Single-valued domains for $w = \sqrt[n]{z}$. A domain $\Omega \subseteq \mathbf{C}$ is a single-valued domain for $w = \sqrt[n]{z}$ if and only if o and ∞ belong to the same component of $\mathbf{C}^* - \Omega$. In particular,

(1) A simply connected domain in $\mathbf{C}$, not containing o, is such a domain.
(2) Let γ^* be a Jordan curve connecting o to ∞, then $\mathbf{C}^* - \gamma^*$ is such a
    domain.   (2.7.2.6)

The manner described in Fig. 2.52 suggests how to construct the Riemann surface of $\sqrt[n]{z}$, exactly like the one we knew in Section (1).

Section (3) Some further examples

In what follows, we illustrate some related examples and leave the details for the readers.

Example 1. Suppose m and n are positive integers, relatively prime. Then

$$w = \sqrt[n]{z^m} = (z^m)^{\frac{1}{n}} = (z^{\frac{1}{n}})^m$$

$$= |z|^{\frac{m}{n}} e^{i\frac{m(\operatorname{Arg} z + 2k\pi)}{n}}, \quad k = 0, 1, 2, \ldots, n-1,$$

which has the same single-valued domains as $\arg z = \operatorname{Arg} z + 2k\pi$, $k = 0, \pm 1, \ldots$, where $-\pi < \operatorname{Arg} z \leq \pi$ (see (2.7.1.6)).

Explanation. Suppose $m = 3$ and $n = 2$ for simplicity.
Treat $z^{\frac{3}{2}}$ as $(z^{\frac{1}{2}})^3$. So $w = z^{\frac{3}{2}}$ may be decomposed as $z \to \zeta = z^{\frac{1}{2}} \to w = \zeta^3$. It has two single-valued branches as indicated in Fig. 2.54. Note that both $\zeta_{-\frac{\pi}{2}} = |z|^{\frac{1}{2}} e^{i\frac{\operatorname{Arg} z}{2}}$ and $\zeta_{\frac{\pi}{2}} = |z|^{\frac{1}{2}} e^{i\frac{\operatorname{Arg} z + 2\pi}{2}}$ are one-to-one and onto, while $w = (\zeta_{-\frac{\pi}{2}})^3$ and $w = (\zeta_{\frac{\pi}{2}})^3$ are just onto but not one-to-one.
Treat $z^{\frac{3}{2}}$ as $(z^3)^{\frac{1}{2}}$. So $w = z^{\frac{3}{2}}$ may be decomposed as $z \to \zeta = z^3 \to w = \zeta^{\frac{1}{2}}$. Two single-valued branches are shown in Fig. 2.55, where $D_k = \{\frac{2k\pi}{3} < \operatorname{Arg} z < \frac{2(k+1)\pi}{3}\}$, $k = 0, 1, 2$ and $C_1 = C_2 = C_3 = \mathbf{C} - [0, \infty)$.
In general, $w = \sqrt[n]{z^m}$ has n branches on $\mathbf{C} - (\infty, 0]$, namely, for $k = 0, 1, \ldots, n-1$,

$$w_k = |z|^{\frac{m}{n}} e^{i\frac{m(\operatorname{Arg} z + 2k\pi)}{n}} : \mathbf{C} - (\infty, 0]$$

$$\to \left\{ \frac{m(2k-1)\pi}{n} < \arg w < \frac{m(2k+1)\pi}{n} \right\}.$$

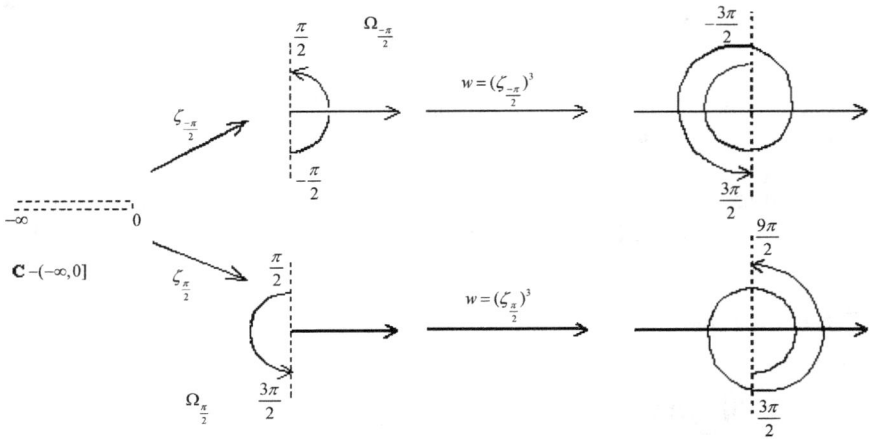

Fig. 2.54

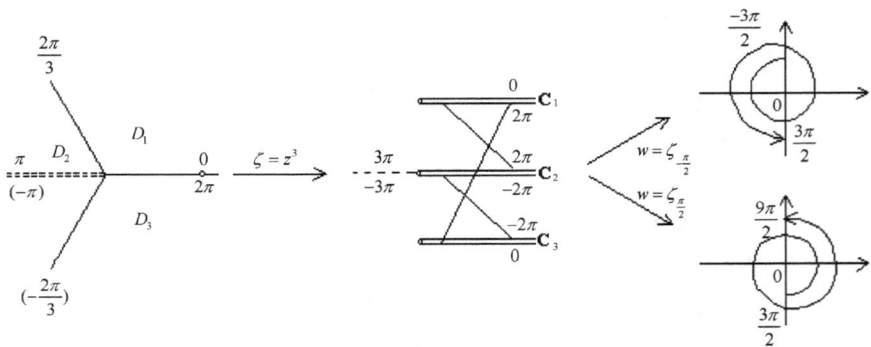

Fig. 2.55

They all are onto, and also one-to-one if $m < n$ but not if $m > n$. 0 and ∞ are branch points of order $n - 1$, and $[0, \infty)$ is a branch cut. The Riemann surface for $w = \sqrt[n]{z^m}$ is the same as $w = \sqrt[n]{z}$ (see Fig. 2.49).

Example 2. Let $w = \sqrt{z^2 - 1}$.

(1) Find branch points and branch cut for $w = w(z)$.
(2) Find local and global single-valued domains, branches, and their analytic expressions.

(3) What are the values on the upper and lower sides of $(-1,\ 1)$ of the branch determined by the condition $w(-2) = -\sqrt{3}$?

(4) Find the values at $z = -2$ and $z = -i$ of the branch determined by $w(i) = \sqrt{2}i$.

(5) Let $f(z)$ be the branch determined by $w(0) = -i$ in $|z| < 1$. As z changes continually along γ_1 and γ_2, respectively, and arrives at 2 (see Fig. 2.58), find possible values at 2 for $f(z)$ if $f(z)$ varies continually as a function of z.

(6) Construct Riemann surface for $w = \sqrt{z^2 - 1}$.

Solution. $w = w(z)$ is one-to-one twice except at $z = \pm 1$ and $z = \infty$, where $w(\pm 1) = 0$ and $w(\infty) = \infty$.

(1) Let z vary continuously in counterclockwise direction along a Jordan closed curve γ_1 not passing ± 1, according to the following cases:

(a) $\pm 1 \in \operatorname{Ext}\gamma$.   (b) $1 \in \operatorname{Int}\gamma$ and $-1 \in \operatorname{Ext}\gamma$.

(c) $-1 \in \operatorname{Int}\gamma$ and $1 \in \operatorname{Ext}\gamma$.   (d) $\pm 1 \in \operatorname{Int}\gamma$.

As z completes one revolution and returns to its starting point, observe if $w = w(z)$ will take back its original presumed value or not, just as in (2.7.2.4) and (2.7.2.5). Then we will see that both $z = 1$ and $z = -1$ are branch points of order 1, while ∞ is not but is a node point. $[-1, 1]$ or $[\infty, -1] \cup [1, \infty]$ can be treated as branch cut. See (2) below.

(2) Let

$$z - 1 = r_1 e^{i\theta_1}, \quad r_1 = |z - 1| \quad \text{and} \quad \theta_1 = \operatorname{Arg}(z - 1), \quad \text{and}$$
$$z + 1 = r_2 e^{i\theta_2}, \quad r_2 = |z + 1| \quad \text{and} \quad \theta_2 = \operatorname{Arg}(z + 1).$$

See Fig. 2.56. Then

$$w^2 = z^2 - 1 = r_1 r_2 e^{i(\theta_1 + \theta_2)}$$
$$\Rightarrow w = \sqrt{z^2 - 1} = \sqrt{r_1 r_2}\, e^{i\frac{\theta_1 + \theta_2 + 2k\pi}{2}}, \quad k = 0, 1.$$

By imposing the restrictions (on principal arguments θ_1 and θ_2)

$$-\pi < \theta_1 \leq \pi \quad \text{and} \quad 0 \leq \theta_2 < 2\pi$$
$$\Rightarrow w_1 = \sqrt{r_1 r_2}\, e^{i\frac{\theta_1 + \theta_2}{2}} \quad \text{and} \quad w_2 = -w_1.$$

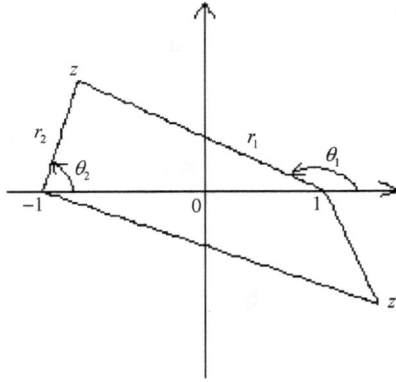

Fig. 2.56

As z approaches any point on $(1, \infty)$, the real axis to the right of the point $z = 1$, then

$$\lim \theta_1(z) = 0;$$

$$\lim \theta_2(z) = \begin{cases} 0, & \text{if } z \text{ approaches from above} \\ 2\pi, & \text{if } z \text{ approaches from below} \end{cases};$$

$$\Rightarrow \lim \frac{\theta_1 + \theta_2}{2} = \begin{cases} 0, & \text{if from above} \\ \pi, & \text{if from below} \end{cases};$$

$$\Rightarrow \text{Both } w_1 \text{ and } w_2 \text{ are discontinuous on } (1, \infty).$$

Similar arguments show that w_1 and w_2 are discontinuous on $(\infty, -1)$ but continuous on $(-1, 1)$. So two branches, namely w_1 and w_2, can be defined on $\mathbf{C}^* - [\infty, -1] \cup [1, \infty]$. If we impose the restrictions $-\pi < \theta_1, \theta_2 \leq \pi$, then it turns out that two branches, w_1 and w_2, can be defined on $\mathbf{C}^* - [-1, 1]$. Note that $w_1(\infty) = w_2(\infty) = \infty$, yet ∞ is not a branch point of $w = w(z)$.

For $z_0 \neq -1, 1, \infty, w = w(z)$ has two local branches on $|z - z_0| < \min\{|z_0 - 1|, |z_0 + 1|\}$: one is the negative of the other. A domain $\Omega \subseteq \mathbf{C}$ is a single-valued domain for $w = w(z)$ if and only if -1 and 1 belong to the same component of $\mathbf{C}^* - \Omega$. For instance, $\mathbf{C}^* - [-1, 1]$ is such a domain.

(3) By (2), the condition $w(-2) = -\sqrt{3}$ defines uniquely the branch $w_1 = \sqrt{r_1 r_2} e^{i\frac{\theta_1 + \theta_2}{2}}$, $-\pi < \theta_1, \theta_2 \leq \pi$ (see Fig. 2.57). Fix a point $x \in (-1, 1)$.

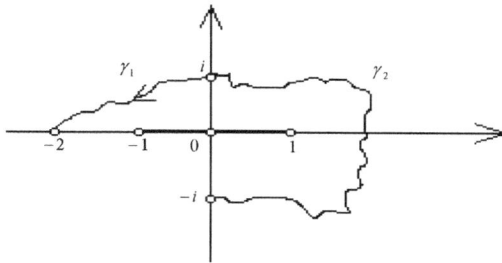

Fig. 2.57

Let z approach x. Then

$$\lim_{z \to x} \theta_1(z) = \begin{cases} \pi, & \text{from above} \\ -\pi, & \text{from below} \end{cases}; \quad \lim_{z \to x} \theta_2(z) = 0.$$

$$\Rightarrow \lim_{z \to x} w_1(z) = \begin{cases} |x^2 - 1|^{\frac{1}{2}} e^{i\frac{\pi+o}{2}} = |x^2 - 1|^{\frac{1}{2}} i, & \text{from above} \\ |x^2 - 1|^{\frac{1}{2}} e^{i\frac{-\pi+o}{2}} = -|x^2 - 1|^{\frac{1}{2}} i, & \text{from below} \end{cases}.$$

(4) Both -2 and $-i$ are in $\mathbf{C}^* - [-1, 1]$. So we need to consider the single-valued domain $\mathbf{C}^* - [-1, 1]$. Since $w_1(i) = \sqrt{2}e^{i\frac{1}{2}(\frac{3\pi}{4} + \frac{\pi}{4})} = \sqrt{2}i$, we choose the branch $w_1 = \sqrt{r_1 r_2}\, e^{i\frac{\theta_1 + \theta_2}{2}}$ with $-\pi < \theta_1, \theta_2 \leq \pi$. By computation, $w_1(-2) = -\sqrt{3}$ and $w_1(-i) = -\sqrt{2}i$.

In case we do not know the branch $w = w_1(z)$ beforehand, we can use the continuous variation $\Delta_\gamma \operatorname{Arg} f(z)$ of a branch $f(z)$ varying continuously along a curve γ to achieve our purpose. Let γ_1 be a Jordan curve connecting i to -2 and lying entirely in $\mathbf{C}^* - [-1, 1]$. See Fig. 2.57. Then (see (1) in (2.7.1.4)),

$$\Delta_{\gamma_1} \operatorname{Arg}(z - 1) = \operatorname{Arg}(-2 - 1) - \operatorname{Arg}(i - 1) = \pi - \frac{3\pi}{4} = \frac{\pi}{4};$$

$$\Delta_{\gamma_1} \operatorname{Arg}(z + 1) = \operatorname{Arg}(-2 + 1) - \operatorname{Arg}(i + 1) = \pi - \frac{\pi}{4} = \frac{3\pi}{4};$$

$$\Rightarrow \Delta_{\gamma_1} \operatorname{Arg} f(z) = \frac{1}{2}\{\Delta_{\gamma_1} \operatorname{Arg}(z - 1) + \Delta_{\gamma_2} \operatorname{Arg}(z + 1)\} = \frac{\pi}{2};$$

$$\Rightarrow \left(\text{Since } f(i) = \sqrt{2}i \text{ and } \operatorname{Arg} f(i) = \frac{\pi}{2}\right) \operatorname{Arg} f(-2)$$

$$= \Delta_{\gamma_1} \operatorname{Arg} f(z) + \operatorname{Arg} f(i) = \pi;$$

$$\Rightarrow f(-2) = |f(-2)| e^{i \operatorname{Arg} f(-2)} = \sqrt{3}e^{\pi i} = -\sqrt{3}.$$

On the other hand, let γ_2 be a Jordan curve in $\mathbf{C}^* - [-1, 1]$, connecting i to $-i$ as shown in Fig. 2.57. Then

$$\Delta_{\gamma_2}\mathrm{Arg}(z - 1) = \mathrm{Arg}(-i - 1) - \mathrm{Arg}(i - 1) = -\frac{3\pi}{4} - \frac{3\pi}{4} = \frac{-3\pi}{2};$$

$$\Delta_{\gamma_2}\mathrm{Arg}(z + 1) = -\frac{\pi}{4} - \frac{\pi}{4} = \frac{-\pi}{2};$$

$$\Rightarrow \Delta_{\gamma_2}\mathrm{Arg}\, f(z) = \frac{1}{2}\left\{-\frac{3\pi}{2} - \frac{\pi}{2}\right\} = -\pi;$$

$$\Rightarrow \mathrm{Arg}\, f(-i) = \Delta_{\gamma_2}\mathrm{Arg}\, f(z) + \mathrm{Arg}\, f(i) = -\pi + \frac{\pi}{2} = -\frac{\pi}{2};$$

$$\Rightarrow f(-i) = |f(-i)|e^{i\,\mathrm{Arg}\, f(-i)} = \sqrt{2}e^{-i\frac{\pi}{2}} = -\sqrt{2}i.$$

(5) By (2), since $w(0) = -i$, the branch so determined in $|z| < 1$ is $f(z) = w_2(z) = -\sqrt{r_1 r_2}e^{i\frac{\theta_1 + \theta_2}{2}}$, $-\pi < \theta_1 \le \pi$ and $0 \le \theta_2 < 2\pi$. Let γ_1 and γ_2 be as in Fig. 2.58. Along γ_1,

$$\Delta_{\gamma_1}\mathrm{Arg}(z + 1) = \mathrm{Arg}(2 + 1) - \mathrm{Arg}(1) = 2\pi - 0 = 2\pi;$$

$$\Delta_{\gamma_1}\arg(z - 1) = \arg(2 - 1) - \arg(-1) = 0 - \pi = -\pi$$

(why use arg instead of Arg?);

$$\Rightarrow \Delta_{\gamma_1}\mathrm{Arg}\, f(z) = \frac{1}{2}(\Delta_{\gamma_1}\mathrm{Arg}(z + 1) + \Delta_{\gamma_1}\arg(z - 1))$$

$$= \frac{1}{2}(2\pi - \pi) = \frac{1}{2}\pi;$$

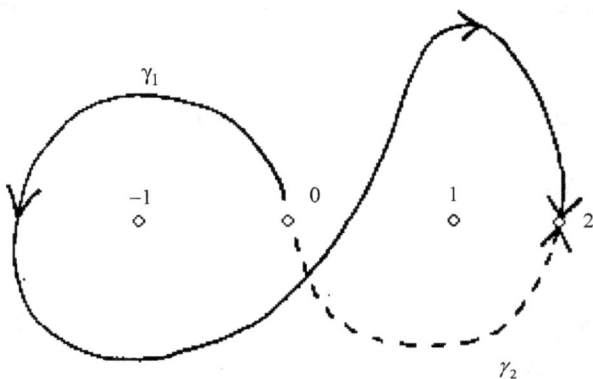

Fig. 2.58

$$\Rightarrow \left(\text{Since } f(0) = w_2(0) = -i \text{ and } \operatorname{Arg} f(0) = \frac{1}{2}(0 + \pi + 2\pi) = \frac{3}{2}\pi\right);$$

$$f(2) = |2^2 - 1|^{\frac{1}{2}} e^{i(\operatorname{Arg} f(0) + \Delta_{\gamma_1} \operatorname{Arg} f(z))} = \sqrt{3} e^{2\pi i} = \sqrt{3}.$$

Along γ_2,

$$\Delta_{\gamma_2} \operatorname{Arg}(z + 1) = \operatorname{Arg} 3 - \operatorname{Arg} 1 = 2\pi - 2\pi = 0;$$

$$\Delta_{\gamma_2} \operatorname{Arg}(z - 1) = \operatorname{Arg} 1 - \operatorname{Arg}(-1) = 0 - \pi = -\pi;$$

$$\Rightarrow \Delta_{\gamma_2} \operatorname{Arg} f(z) = \frac{1}{2}(0 - \pi) = -\frac{1}{2}\pi;$$

$$\Rightarrow f(2) = |2^2 - 1|^{\frac{1}{2}} e^{i(\operatorname{Arg} f(0) + \Delta_{\gamma_2} \operatorname{Arg} f(z))} = \sqrt{3} e^{\pi i} = -\sqrt{3}.$$

(6) Let $w_1 = \sqrt{r_1 r_2}\, e^{i \frac{\theta_1 + \theta_2}{2}}$ and $w_2 = -w_1$, where $-\pi < \theta_1 = \operatorname{Arg}(z-1) \leq \pi$ and $-\pi < \theta_2 = \operatorname{Arg}(z+1) \leq \pi$. By actual computation or by observing the composition $z \to \zeta = z^2 \to \eta = \zeta - 1 \to w = \sqrt{\eta}$, it can be shown that

$$w_1 = w_1(z) : \mathbf{C} - [-1, 1] \to \{\operatorname{Im} w \geq 0\} - [0, i], \text{ two-to-one and onto;}$$

$$w_2 = w_2(z) : \mathbf{C} - [-1, 1] \to \{\operatorname{Im} w \leq 0\} - [-i, 0], \text{ two-to-one and onto.}$$

See Fig. 2.59. Note that, the orientation of the boundary of $\{\operatorname{Im} w > 0]\} - [0, i]$, viewed as the image under $w = w_1(z)$ of $\operatorname{Im} z > 0$, is opposite to that of the boundary of $\{\operatorname{Im} w > 0]\} - [0, i]$, when viewed as the image under $w = w_1(z)$ of $\operatorname{Im} z < 0$. A similar explanation holds for $w = w_2(z)$. Now, paste two copies of $\mathbf{C} - [-1, 1]$ crosswise along the upper and lower sides of the cuts $[-1, 1]$, according to how $\operatorname{Im} w > 0$ and $\operatorname{Im} w < 0$ can be pasted together along the same direction indicated in their common boundary, the real axis. The resulted surface is the Riemann surface $R_{\sqrt{z^2-1}}$ of $w = \sqrt{z^2 - 1}$. See Fig. 2.60: $z = 1$ and $z = -1$ are algebraic branch points of order 1, while $z = \infty$ is a node point.

Remark. The Riemann surface for $z = w + \sqrt{w^2 - 1}$, the inverse of the Joukowski function $w = \frac{1}{2}\left(z + \frac{1}{z}\right)$.
    Recall (2.5.5.4), in particular, (6) within. Fig. 2.27 indicates how to construct this Riemann surface. See Fig. 2.61 see Ref. [1], p. 99). Its line complex is shown in Fig. 2.62. Note that both $z = 0$ and $z = \infty$ are

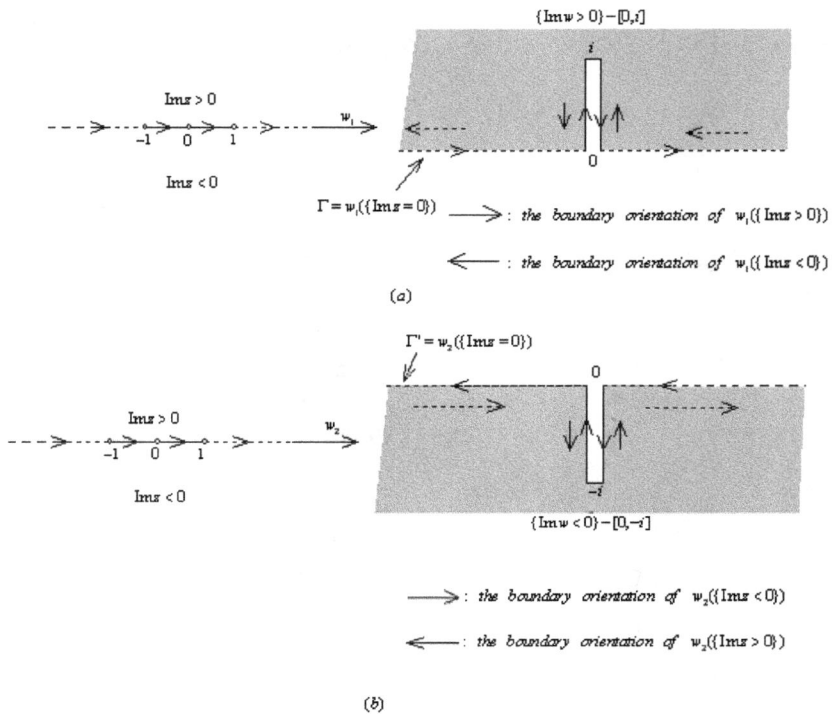

$\{\operatorname{Im} w > 0\} - [0, i]$

$\operatorname{Im} z > 0$

$-1 \quad 0 \quad 1$

w_1

$\operatorname{Im} z < 0$

$\Gamma = w_1(\{\operatorname{Im} z = 0\})$ ────⟶ : the boundary orientation of $w_1(\{\operatorname{Im} z > 0\})$

⟵──── : the boundary orientation of $w_1(\{\operatorname{Im} z < 0\})$

(a)

$\Gamma' = w_2(\{\operatorname{Im} z = 0\})$

$\operatorname{Im} z > 0$

$-1 \quad 0 \quad 1$

w_2

$\operatorname{Im} z < 0$

$\{\operatorname{Im} w < 0\} - [0, -i]$

────⟶ : the boundary orientation of $w_2(\{\operatorname{Im} z < 0\})$

⟵──── : the boundary orientation of $w_2(\{\operatorname{Im} z > 0\})$

(b)

Fig. 2.59

regular points (namely, points other than branch points) and are denoted symbolically by two-lines (a polygon with two sides).   □

Example 3. $w = \sqrt[3]{z^2 - 1}$.

Explanation. $w = w(z)$ is one-to-three except at $z = \pm 1, \infty$.

(1) Let $z - 1 = r_1 e^{i\theta_1}$ and $z + 1 = r_2 e^{i\theta_2}$, where $\theta_1 = \operatorname{Arg}(z-1)$, $0 \le \theta_1 < 2\pi$ and $\theta_2 = \operatorname{Arg}(z+1)$, $0 \le \theta_2 < 2\pi$. Then, for $z \ne \pm 1$, $w = \sqrt[3]{z^2 - 1}$ has three distinct values

$$w_k = \sqrt[3]{r_1 r_2}\, e^{i\frac{1}{3}(\theta_1 + \theta_2 + 2k\pi)}, \quad k = 0, 1, 2.$$

Let γ be a Jordan closed curve so that $1 \in \operatorname{Int}\gamma$ and $-1 \in \operatorname{Ext}\gamma$. As z encircles γ once, then, $r_1 \to r_1$, $r_2 \to r_2$ and $\theta_1 \to \theta_1 +$

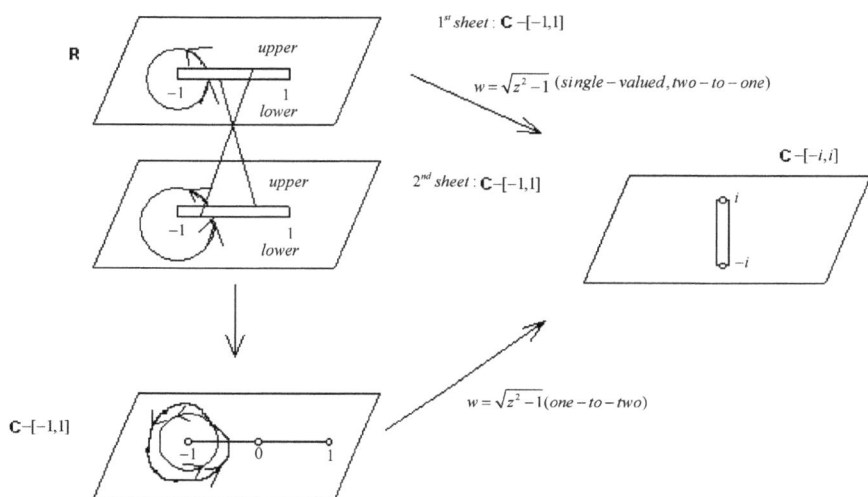

Fig. 2.60

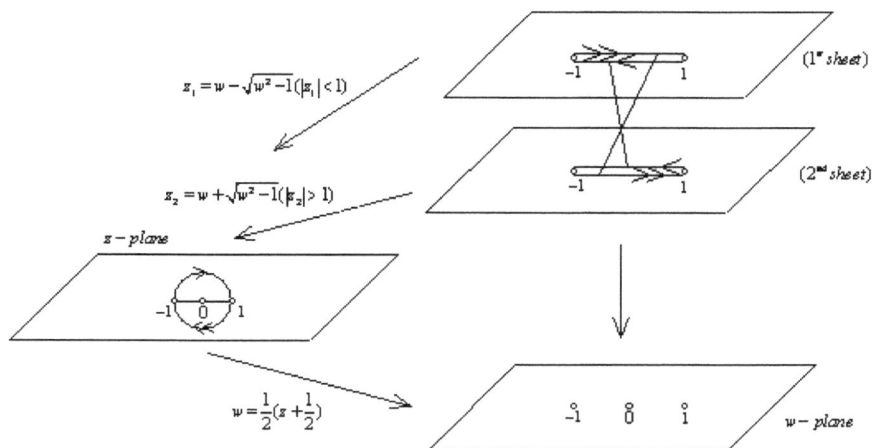

Fig. 2.61

$2\pi, \theta_2 \rightarrow \theta_2$ and hence, $w_1 \rightarrow w_2, w_2 \rightarrow w_3, w_3 \rightarrow w_1$. If z encircles γ two more times then w_1 returns to w_1 via w_2 and w_3. So do w_2 and w_3. Hence, 1 is an algebraic branch point of order 2. Similar arguments show that -1 and ∞ are algebraic branch points of order 2.

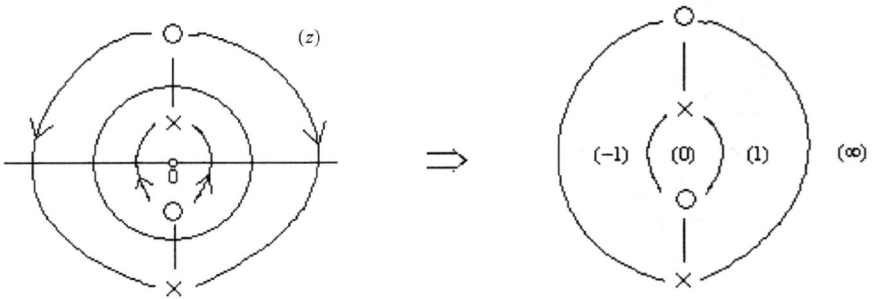

Fig. 2.62

(2) A domain $\Omega \subseteq \mathbf{C}$ is a single-valued domain for $w = w(z)$ if $1, -1$ and ∞ all lie in the same component of $\mathbf{C}^* - \Omega$. With $[\infty, -1]$ and $[1, \infty]$ as branch cuts, we can define three branches on $\mathbf{C}^* - [\infty, -1] - [1, \infty]$.

(3) What is the value at $-i$ of the branch f defined uniquely by $f(i) = \sqrt[3]{2}e^{\frac{\pi i}{3}}$?

Since $r_1 = |i - 1| = \sqrt{2}, \theta_1 = \operatorname{Arg}(i - 1) = \frac{3\pi}{4}$ and $r_2 = |i + 1| = \sqrt{2}, \theta_2 = \operatorname{Arg}(i + 1) = \frac{\pi}{4}$,

$$\sqrt[3]{r_1 r_2}\, e^{i\frac{1}{3}(\theta_1 + \theta_2 + 2k\pi)} = \sqrt[3]{2}\, e^{i\frac{1}{3}\left(\frac{3\pi}{4} + \frac{\pi}{4} + 2k\pi\right)} = \sqrt[3]{2}\, e^{\frac{\pi i}{3}},$$

$$\Rightarrow e^{\frac{2k\pi}{3}i} = 1 \quad \text{and hence } k = 0.$$

Hence, $f(z) = w_0(z)$. At $z = -i, r_1 = r_2 = \sqrt{2}$, and $\theta_1 = \frac{5\pi}{4}, \theta_2 = \frac{7\pi}{4}$, so $f(-i) = -\sqrt[3]{2}$.

An alternative way is as follows. Let γ be Jordan curve in $\mathbf{C}^* - [\infty, -1] - [1, \infty]$, connecting i to $-i$. Then

$$\Delta_\gamma \arg(z - 1) = \arg(-i - 1) - \arg(i - 1) = \frac{5\pi}{4} - \frac{3\pi}{4} = \frac{\pi}{2};$$

$$\Delta_\gamma \arg(z + 1) = \operatorname{Arg}(-i + 1) - \operatorname{Arg}(i + 1) = \frac{7\pi}{4} - \frac{\pi}{4} = \frac{3\pi}{2};$$

$$\Rightarrow \Delta_\gamma \operatorname{Arg} f(z) = \frac{1}{3}(\Delta_\gamma \arg(z - 1) + \Delta_\gamma \arg(z + 1)) = \frac{2\pi}{3};$$

$$\Rightarrow (\text{since } f(i) = \sqrt[3]{2}e^{\frac{\pi i}{3}})f(-i)$$

$$= |f(-i)|e^{i(\operatorname{Arg} f(i) + \Delta_\gamma \operatorname{Arg} f(z))} = -\sqrt[3]{2}.$$

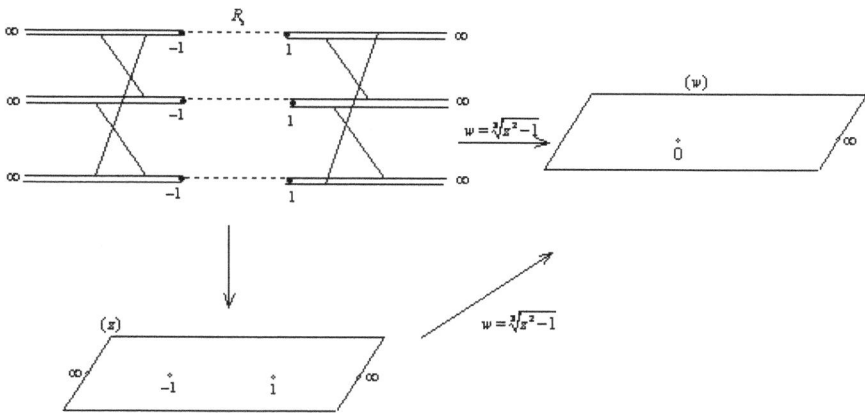

Fig. 2.63

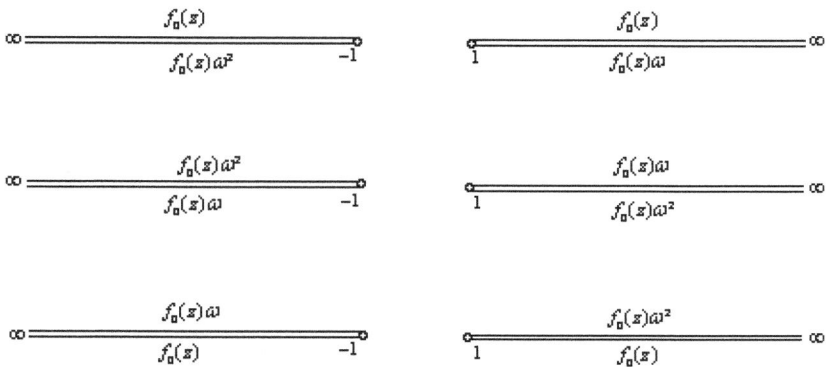

Fig. 2.64

(4) Let $f_0(z) = w_0(z)$ and $\omega = e^{\frac{2\pi i}{3}}$. Then $w_1 = f_0\omega$ and $w_2 = f_0\omega^2$. The Riemann surface for $w = w(z)$ is as Fig. 2.63 and, for simplicity, as in Fig. 2.64.

Example 4. (See Ref. [1], p. 99) $w = z^3 - 3z$.

Explanation. $w = w(z)$ ceases to be conformal (see (2.5.1.6)) at points z where $w'(z) = 3(z^2 - 1) = 0$, namely, $z = \pm 1$. If $z = 1$, then $w = 1 - 3 = -2$ and $z^3 - 3z + 2 = (z-1)^2(z+2)$ shows that $w(z) - w(1)$ has a zero of order 2 at $z = 1$. Similarly, $z^3 - 3z - 2 = (z+1)^2(z-2)$ shows that $w(z) - w(-1)$ has a zero of order 2 at $z = -1$ (see (2.5.1.3)). Hence, $w = 2$ and $w = -2$

are algebraic branch points of order 1 for the inverse function of $w = w(z)$ (refer to Exercise A (4) of Sec. 2.5.3). Also, $w = \infty$ is an algebraic branch point of order 2.

To figure out the fundamental domains for $w = w(z)$:

Let $z = x + iy$ and $w = u + iv$. From Example 5 in Sec. 1.4.3, we knew that

$$v = \operatorname{Im} w \begin{cases} > 0 \Leftrightarrow y > 0 \text{ and } 3x^2 - y^2 > 3 \text{ or } y < 0 \text{ and } 3x^2 - y^2 < 3 \\ = 0 \Leftrightarrow y = 0 \text{ or } 3x^2 - y^2 = 3 \\ < 0 \Leftrightarrow y > 0 \text{ and } 3x^2 - y^2 < 3 \text{ or } y < 0 \text{ and } 3x^2 - y^2 > 3 \end{cases}.$$

$$(*)$$

while $u = x(x^2 - 3y^2 - 3)$.

Where are the exact images of $y = 0$ and $3x^2 - y^2 = 3$ under $w = w(z)$? In case $y = 0$, $u = x(x^2 - 3) = (x+\sqrt{3})x(x-\sqrt{3})$ and $u'(x) = 3(x-1)(x+1)$, $u''(x) = 6x$. By calculus, we obtain the following facts that $w = w(z)$ maps

$$(-\infty, -1] \text{ onto } (-\infty, \ 2], \text{strictly increasing};$$
$$[-1, \ 1] \text{ onto } [-2, \ 2], \text{ strictly decreasing, and}$$
$$[1, \ +\infty) \text{ onto } [-2, \ +\infty), \text{ strictly increasing}.$$

In case $3x^2 - y^2 = 3$, $u = -8x^3 + 6x < 0$ for $x \geq 1$ and > 0 for $x \leq -1$. Then $w = w(z)$ maps

the upper half of the right branch: $3x^2 - y^2 = 3, \ x \geq 1$,
$$y > 0 \to (-\infty, -2];$$
the lower half of the right branch: $3x^2 - y^2 = 3, \ x \geq 1$,
$$y < 0 \to (-\infty, -2];$$
the upper half of the left branch: $3x^2 - y^2 = 3, \ x \leq -1$,
$$y > 0 \to [2, \infty), \text{ and}$$
the lower half of the left branch: $3x^2 - y^2 = 3, \ x \leq -1$,
$$y < 0 \to [2, +\infty).$$

See Fig. 2.65.

$w = w(z)$ maps the domains $3x^2 - y^2 > 3$, $x > 1$, $y > 0$ and $3x^2 - y^2 > 3$, $x > 1$, $y < 0$ univalently onto the domains $\operatorname{Im} w > 0$ and $\operatorname{Im} w < 0$, respectively(see $(*)$). Therefore, $w = w(z)$ maps the domain $3x^2 - y^2 > 3$, $x > 1$ univalently onto the domain $\mathbf{C}^* - [-\infty, -2]$. Similarly, $w = w(z)$ maps $3x^2 - y^2 < 3$, $-1 < x < 1$, univalently onto $\mathbf{C}^* - [-\infty, -2] - [2, \infty]$ and $3x^2 - y^2 > 3$, $x < -1$ univalently onto $\mathbf{C}^* - [2, \infty]$. These three domains

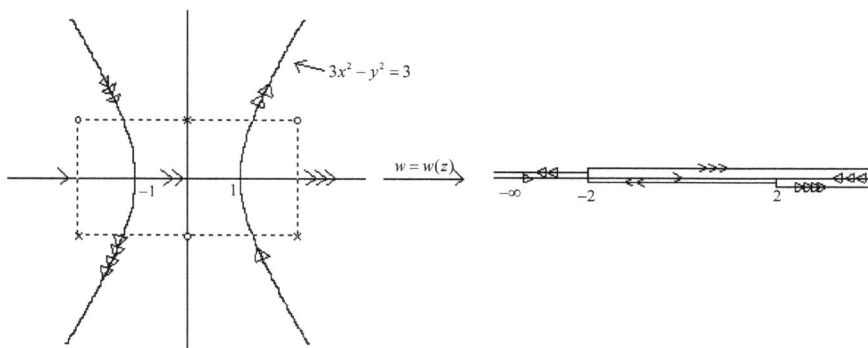

Fig. 2.65

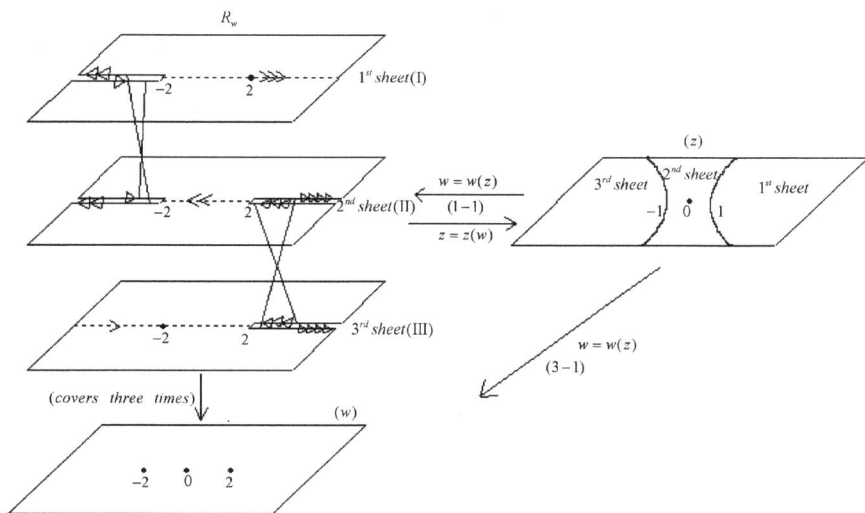

Fig. 2.66

in the z-plane are fundamental domains for $w = w(z)$. For univalence, refer to Sec. 3.3.5.

Take three copies of w-plane and paste them crosswise along the branch cuts $[\infty, -2]$ and $[2, \infty]$ in the manner in which $y = 0$ and $3x^2 - y^2 = 3$ are related to each other. The resulted surface R_w is the Riemann surface for the inverse function $z = z(w)$. See Fig. 2.66. The line complex for R_w is shown in the left figure in Fig. 2.65.

Exercises A

(1) Let $a \in \mathbf{C}$ be a fixed point. Do the content in Section. 2.6.2.2 for $w = \sqrt[n]{z - a}$.

(2) Do Example 1 for $w = z^{\frac{2}{3}}$. Try to construct a Riemann surface R_z over the z-plane and a Riemann surface R_w over the w-plane so that $w = z^{\frac{2}{3}} : R_z \to R_w$ is one-to-one and onto.

(3) Designate $-\pi < \operatorname{Arg} z \leq \pi$. Let $f : \mathbf{C} - [0, \infty) \to \mathbf{C}$ be defined as

$$f(z) = \begin{cases} |z|^{\frac{1}{3}} e^{i\frac{1}{3}\operatorname{Arg} z}, & \operatorname{Im} z \geq 0 \\ |z|^{\frac{1}{3}} e^{i\frac{1}{3}\operatorname{Arg} z} \omega, & \operatorname{Im} z < 0 \end{cases},$$

where $\omega = \frac{1}{2}(-1 + \sqrt{3}i)$. Show that f is a branch of $w = z^{\frac{1}{3}}$ on $\mathbf{C} - [0, \infty)$. Find the other two branches of $w = z^{\frac{1}{3}}$ on $\mathbf{C} - [0, \infty)$.

(4) Designate $-\pi < \operatorname{Arg} z \leq \pi$. Find the branch of $w = z^{\frac{1}{4}}$ on $\mathbf{C} - [0,\infty)$, determined uniquely by $f(-1) = \frac{1}{2}(\sqrt{2} + i\sqrt{2})$. Find the other three branches there.

(5) Let γ be an eight-figure closed curve starting at 0 as shown in Fig. 2.67. Suppose z winds along γ and back to 0. Find the change of values at 0 for each of the following functions $w = w(z)$.

   (a) $w = \sqrt{z^2 - 1}$.   (b) $w = \sqrt[3]{z - 1}$.

   (c) $w = \sqrt{\dfrac{z - 1}{z + 1}}$.   (d) $w = \sqrt[3]{(z + 1)(z - 1)^2}$.

   (e) $w = \sqrt{(1 - z^2)(1 - k^2 z^2)}$ $(0 < k < 1)$ but γ does not pass $\pm\frac{1}{k}$.

(6) Suppose $\arg w(1) = 0$. Find $\arg w(1)$ if z winds along $|z| = 1$ in the anticlockwise direction.

   (a) $w = \sqrt{z - a}$ $(0 < a < 1; a > 1)$.
   (b) $w = \sqrt[n]{z - a}$ $(0 < a < 1; a > 1)$.

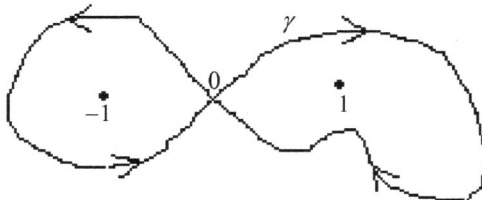

Fig. 2.67

(c) $w = \sqrt[3]{\dfrac{z-a}{z-b}}$ $\quad (0 < a, b < 1; 0 < a < 1, b > 1; a > 1, b > 1)$.

(d) $w = \sqrt{z^2 + 2z - 1}$.

(e) $w = \sqrt{(z - a_1) \cdots (z - a_n)}$ $\quad (0 < a_j < 1, 1 \leq j \leq k; a_j > 1,$
$k < j \leq n)$.

(7) Let $w = \sqrt{z - 1} + \sqrt{z + 1}$. Note that $w^4 - 4w^2 z + 4 = 0$.

(a) Show that $z = \pm 1, \infty$ are algebraic branch points of order 1.

(b) Show that a domain $\Omega \subseteq \mathbf{C}$ is a single-valued domain for $w = w(z)$ if ± 1 and ∞ are in the same component of $\mathbf{C}^* - \Omega$.

(c) On $\Omega_0 = \mathbf{C}^* - [\infty, -1] - [1, \infty]$, designate $-\pi < \text{Arg}(z + 1) \leq \pi$ and $0 \leq \text{Arg}(z - 1) < 2\pi$. Let

$$w_1^{(k)} = |z - 1|^{\frac{1}{2}} e^{i\frac{1}{2}[\text{Arg}(z-1) + 2k\pi]}, \quad k = 0, 1;$$

$$w_{-1}^{(l)} = |z + 1|^{\frac{1}{2}} e^{i\frac{1}{2}[\text{Arg}(z+1) + 2l\pi]}, \quad l = 0, 1.$$

Show that $w_1^{(k)} + w_{-1}^{(l)}, k, l = 0, 1$, are branches of $w = w(z)$ on Ω_0.

(d) Find the values along the upper and lower sides of $(\infty, -1)$ and $(1, \infty)$ of the branch f determined by $w(0) = 1 + i$.

(e) Let g be the branch determined by $w(0) = 1 - i$ on $|z| < 1$. Find the terminal values $g(0)$ if z winds once along γ_1, γ_2, and γ_3 from 0, respectively, as shown in Fig. 2.68.

(f) Try to construct the Riemann surface for $w = w(z)$. Note that

	$w_1^{(0)} + w_{-1}^{(0)}$ (1st sheet)	$w_1^{(1)} + w_{-1}^{(1)}$ (2nd)	$w_1^{(0)} + w_{-1}^{(1)}$ (3rd)	$w_1^{(1)} + w_{-1}^{(0)}$ (4th)
$z = 1$	$\sqrt{2}$	$-\sqrt{2}$	$\sqrt{2}$	$-\sqrt{2}$
$z = -1$	$\sqrt{2i}$	$-\sqrt{2i}$	$-\sqrt{2i}$	$\sqrt{2i}$
$z = \infty$	∞	∞	0	0

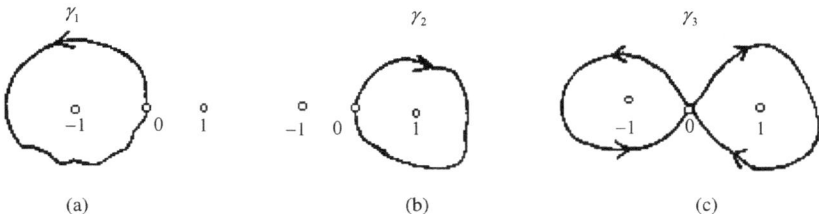

Fig. 2.68

This Table suggests that, on the Riemann surface R_w, the 1st sheet and the 3rd sheet paste crosswise along $[1, \infty)$, while the 2nd and the 4th along $[1, \infty)$; 1st and 4th along $[\infty, -1)$, while 2nd and 3rd along $[\infty, -1)$. There are two points on R_w which lie over $z = 1$, where two sheets meet at each of them, thus $z = 1$ is called a *double* branch point of order 1. So are $z = -1$ and $z = \infty$.

(8) Let $w = \sqrt[6]{(z-i)^2(z+i)^3}$.

    (a) Show that i is a double algebraic branch point of order 2, $-i$ is a triple one of order 1 and ∞ is a branch point of order 5.

    (b) Find single-valued domains for $w = w(z)$ and branches.

    (c) Find the value at 1 of the branch $f(z)$ uniquely determined by $w(0) = e^{\frac{13\pi}{12}i}$.

    (d) Construct the Riemann surface for $w = w(z)$.

(9) Let $w = \sqrt[3]{\frac{z-1}{z^2}} + \sqrt{z-i}$.

    (a) Show that 0 is a double branch point of order 2, so is 1, i is a triple one of order 1 and ∞ is a branch point of order 5.

    (b) Find single-valued domains for $w = w(z)$ and branches.

    (c) Find the value at -1 of the branch $f(z)$ uniquely determined by $w(1+i) = \frac{1}{\sqrt[3]{2}} e^{\frac{2\pi i}{3}} - 1$.

    (d) Construct the Riemann surface for $w = w(z)$.

(10) Show that $w = \sqrt[4]{(1-z)^3(1+z)}$ has four branches on $\mathbf{C} - [-1, 1]$. Find the values at $\pm i$ of the branch which takes positive values along the upper side of the cut $[-1, 1]$. Construct the Riemann surface for $w = w(z)$.

(11) Let $w = \sqrt{\frac{(1-z)^3}{z}} \cdot \frac{1}{1+z}$.

    (a) Show that $w = w(z)$ has two branches on $\mathbf{C} - [0, 1]$. Write them out.

    (b) Let $z = r_1 e^{i\theta_1}$, $0 \le \theta_1 < 2\pi$ and $1 - z = r_2 e^{i\theta_2}$, $-\pi < \theta_2 \le \pi$. Let $f(z)$ be the branch which has positive values along the upper side of $(0, 1)$. Show that

$$f(x) = \begin{cases} \sqrt{\dfrac{(1-x)^3}{x}} \cdot \dfrac{1}{1+x}, & x \text{ in the upper side of } (0, 1) \\[4mm] -\sqrt{\dfrac{(1-x)^3}{x}} \cdot \dfrac{1}{1+x}, & x \text{ in the lower side of } (0, 1) \end{cases}.$$

    (c) Let $f(z)$ be as in (b). Find $f(i)$ and $f(-i)$.

(12) (See Ref. [1], p. 99) Construct the Riemann surface for the inverse function of $w = (z^2 - 1)^2$.

Exercises B

(1) Suppose f is a branch of $w = z^{\frac{1}{n}}$ on a domain Ω. Show that $w = z^{\frac{1}{n}}$ has n branches on Ω and any such branch is of the form ωf, where ω is an nth root of 1.

(2) Let $f : \Omega(\text{domain}) \to \mathbf{C}$ be continuous so that $f(z_0) \neq 0$ for some $z_0 \in \Omega$. For $n \geq 2$, there exists an open neighborhood O of z_0 so that $\sqrt[n]{f(z)}$ has branches on O. Here we suppose that $f(O)$ is also an open set. This will be the case if f is *analytic* at z_0 (see (3.3.1.5)).

(3) Let $w = \sqrt{(z - a_1)(z - a_2) \cdots (z - a_n)}$, where $a_1, a_2, \ldots, a_{n-1}$ and a_n are distinct points in $\mathbf{C}$.

   (a) If $n = 2m$ is an even integer, show that

   (i) $a_1, a_2, \ldots, a_n$ are algebraic branch points of order 1,

   (ii) pairwise disjoint Jordan curves $\widehat{a_1 a_2}, \widehat{a_3 a_4}, \ldots, \widehat{a_{n-1} a_n}$ are branch cuts; and

   (iii) ∞ is a node point.

   There are two branches of $w = w(z)$ on $\mathbf{C}^* - \widehat{a_1 a_2} - \cdots - \widehat{a_{n-1} a_n}$.

   (b) If $n = 2m + 1$ is an odd integer, show that

   (i) branch points: $a_1, a_2, \ldots, a_n,\ \infty$, algebraic branch points of order 1; and

   (ii) branch cuts: pairwise disjoint Jordan curves $\widehat{a_1 a_2}, \widehat{a_3 a_4}, \ldots,$ $\widehat{a_{n-2} a_{n-1}}, \widehat{a_n \infty}$.

   There are two branches of $w = w(z)$ on $\mathbf{C}^* - \widehat{a_1 a_2} - \cdots - \widehat{a_{n-2} a_{n-1}} - \widehat{a_n \infty}$.

   (c) Try to construct the Riemann surface for $w = w(z)$. Note that a sphere with handles can be viewed as a topological model in case $n \geq 3$.

(4) Let $w = \sqrt[n]{(z - a_1)^{r_1}(z - a_2)^{r_2} \cdots (z - a_k)^{r_k}}, a_1 \neq a_2 \neq \cdots \neq a_k$ and $r_j \geq 1$ for $1 \leq j \leq k$. Show that

   (a) Branch points:

   a_j is a branch point $\Leftrightarrow n$ cannot divide $r_j, 1 \leq j \leq k$;

   ∞ is a branch point $\Leftrightarrow n$ cannot divide $r_1 + r_2 + \cdots + r_k$.

   If r_j and n are relatively prime, then a_j is an algebraic branch point of order $n - 1$; if $n = d_j r_j$, then a_j is an r_j multiple branch point of order $d_j - 1$.

(b) Branch cuts: If n divides $r_{j_1} + \cdots + r_{j_k}$, $1 \le j_1 < \cdots < j_k < k$, then the corresponding $a_{j_1}, \ldots, a_{j_k}$ can be connected to form a branch cut γ, namely, the value $w = w(z)$ keeps unchanged if z winds along a Jordan closed curve, having $a_{j_1}, \ldots, a_{j_k}$ in its interior, and back to its starting point.

There will be n branches on the domain obtained by deleting all possible branch cuts from $\mathbf{C}^*$.

(5) Construct the Riemann surface for each of the following multiple-valued functions.

(a) $w = \sqrt{(1 - z^2)(1 - k^2 z^2)}$, $0 < k < 1$.

(b) $w = \sqrt{z} + \sqrt[3]{z - 1}$.

(c) $\sqrt[3]{z^4 - 1}$.

(d) $w = \sqrt{\sqrt[3]{z} - 1}$.

(e) $w = \sqrt{\dfrac{z - a}{z - b}} + \sqrt[3]{z - c}$, $a \neq b \neq c$.

(f) $w = \sqrt[4]{(z - a)(z - b)}$, $a \neq b$.

(g) $w = \sqrt[4]{(z - a_1)(z - a_2)(z - a_3)(z - a_4)}$ according to $a_1 = a_2 = a_3 \neq a_4$, $a_1 = a_2 \neq a_3 = a_4$, $a_1 = a_2 \neq a_3 \neq a_4$, and $a_1 \neq a_2 \neq a_3 \neq a_4$.

(h) $w = \sqrt[5]{(z + 1)(z - 1)^2} + \sqrt[3]{z}$.

(i) $w = \sqrt{z + \sqrt{1 - z^2}}$.

(j) $w = \sqrt{z + \sqrt{z^2 - 1}}$. See Fig. 2.69 for its line complex.

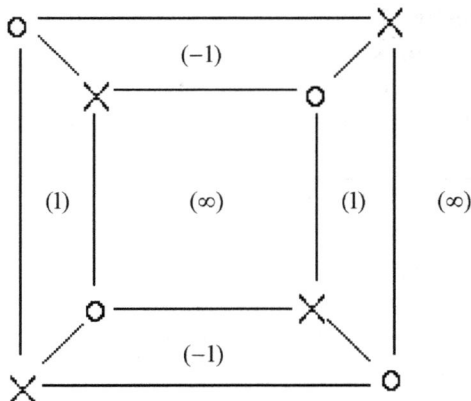

Fig. 2.69

(k) $w = \sqrt{\dfrac{z^2 - z + 1}{z^2 + z + 1}}$.

(l) $w = \sqrt{\dfrac{z^4 - z^2 + 1}{z^4 + z^2 + 1}}$.

(6) Construct the Riemann surface for the inverse function of each of the following functions.

(a) $w = \dfrac{z^2 - z + 1}{z^2 + z + 1}$. See Fig. 2.70.

(b) $w = \dfrac{z^4 - z^2 + 1}{z^4 + z^2 + 1}$. See Fig. 2.71.

(c) $w = \dfrac{z^{2n} - z^n + 1}{z^{2n} + z^n + 1}$ $(n \geq 1,$ integers$)$.

What we have talked about Riemann surfaces for multiple-valued functions is descriptive, informal and, of course, nonrigorous. For a little more

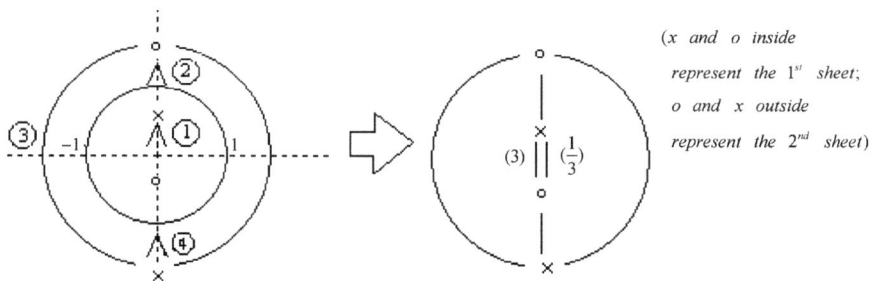

Fig. 2.70

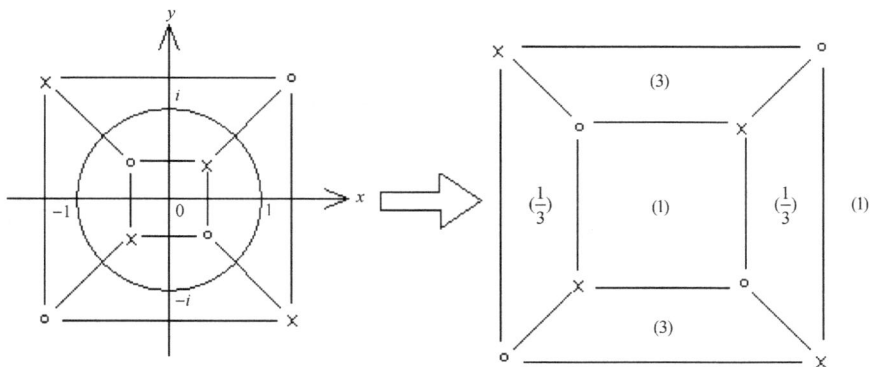

Fig. 2.71

unified and clearer concept about Riemann surfaces, we give the following definitions.

Definition 1. A connected Hausdorff space M is called a two-demensional (topological) *manifold* if each point of M has an open neighborhood which is homeomorphic to an open set in the Euclidean plane $\mathbf{R}^2$ or the complex plane $\mathbf{C}$.

$$(2.7.2.7)$$

Each manifold is arcwise connected, and has a countable base if and only if it has a countable covering by parametric disks (refer to Sec. 1.8 and Exercise B (2) there, and Sec. 1.9).

Definition 2. A manifold R is called an (abstract) *Riemann surface* if

(1) there is a collection $\{(O_\lambda, \varphi_\lambda)\}_{\lambda \in \Lambda}$, where $\{O_\lambda\}_{\lambda \in \Lambda}$ forms an open covering of M and each φ_λ is a homeomorphism of O_λ onto an open set in the complex plane $\mathbf{C}$, and
(2) in case $O_{\lambda_1} \cap O_{\lambda_2} \neq \phi$, the composite mapping

$$\varphi_{\lambda_2} \circ \varphi_{\lambda_1}^{-1} : \varphi_{\lambda_1}\left(O_1 \bigcap O_2\right) \to \varphi_{\lambda_2}\left(O_1 \bigcap O_2\right)$$

is a one-to-one and onto analytic function (see Chap. 3). $(2.7.2.8)$

Usually, $(O_\lambda, \varphi_\lambda)$ is called a *coordinate system* on R, where O_λ is a *coordinate neighborhood* and the associated φ_λ defines a *local coordinate* or *local uniformizing parameter*. It can be shown that the Riemann surfaces mentioned in our text are Riemann surfaces in this abstract sense. See Sec. 3.3.3 and Ref. [75], p. 73–74. For a formal and rigorous approach to this topic, see Refs. [81, 6, 32, 33 and 75]. We will come back to this topic in Chap. 7.

Definition 3. A *covering surface* N of a two-dimensional manifold M is another a two-dimensional manifold N together with a continuous mapping F from N into M so that, for any $P_0 \in N$ and $Q_0 = F(P_0) \in M$, then

(1) there exists a coordinate system (U, ψ) on N: $w = \psi(P)$, $P \in U$ and $\psi(P_0) = 0$; and
(2) there exists a coordinate system (O, φ) on M: $z = \varphi(Q)$, $Q = F(P) \in O$ if $P \in U$ and $\varphi(Q_0) = 0$ such that
(3) in terms of (U, ψ) and (O, φ), $\varphi \circ F \circ \psi^{-1}(w) = w^n = z$ for some integer $n \geq 1$. $(2.7.2.9)$

In this case, Q_0 *is the projection* of P_0 on M and P_0 is said to *lie over* Q_0, and F is the *projection* of N into M. If $n > 1$, P_0 is called a *branch point of order* $n-1$ of N with respect to M; if $n=1$, P_0 is called a *regular point*. For instance, in Fig. 2.49, $z=w^n$, so $w=0$ is a branch point of $n-1$, so is ∞ since $z=w^{-n}$ in a neighborhood of ∞. And the w-plane or $R_{\sqrt[n]{z}}$ is a covering surface of the z-plane.

2.7.3. $w = \log z$ (*the natural logarithm function with base e*) *and its Riemann surface*

Example 2 in Sec. 1.7 showed that, if $z = z + iy \neq 0$ then

$$\lim_{k\to\infty} k(\sqrt[k]{z}-1) = \log|z| + i(\mathrm{Arg}\,z + 2n\pi) \underset{\text{(def.)}}{=} \log z, \quad n = 0, \pm 1, \dots .$$

We can adopt this as the definition for the complex logarithm $\log z$ with base $e = \lim_{n\to\infty}\left(1 + \frac{1}{n}\right)^n$. But it would be much better in many aspects when we treat $w = \log z$ as the multiple-valued inverse of the exponential function $z = e^w$.

Three sections are divided in the sequel.

Section (1) w = log z and its branches

Let $z = re^{i\theta} \neq 0$, where $r = |z|$ and $\theta = \mathrm{Arg}\,z$. Let $w = u + iv$. Then

$$e^w = e^u e^{iv} = z = re^{i\theta},$$

$$\Rightarrow e^u = r \quad \text{and} \quad e^{iv} = e^{i\theta},$$

$$\Rightarrow u = \log r = \log|z| \quad \text{(this always means real logarithm of } |z|.);$$

$$v = \theta + 2n\pi, \quad n = 0, \pm 1, \dots .$$

This means that e^w takes the value z at points $w = \log r + i(\theta + 2n\pi)$, $n = 0, \pm 1, \dots$. Hence, we define, for $z \neq 0$,

$$\log z = \log|z| + i(\mathrm{Arg}\,z + 2n\pi) = \log|z| + i\arg z, \quad n = 0, \pm 1, \pm 2, \dots .$$
$$(2.7.3.1)$$

and call it the *logarithmic function* of z, inverse to $z = e^w$.

The multi-valuedness of $w = \log z$ originates from that of $\arg z$. Therefore, (2.7.1.6) is fully applicable to $\log z$. We summarize

The existence of local and global branches of $w = \log z$.

(1) Local branches. No branch can be defined in open neighborhoods of either 0 or ∞. For any $z_0 \in \mathbf{C}$ and $z_0 \neq 0$, countably infinitely many branches of $\log z$ can be defined on $|z - z_0| < |z_0|$, and each of them is

uniquely determined by its value at z_0. If $f_0(z)$ is such a branch, then any other branches are of the form

$$f_0(z) + 2n\pi i, \quad n = \pm 1, \pm 2, \dots .$$

(2) Global branches. A domain $\Omega \subseteq \mathbf{C}$ is a single-valued domain for $\log z$ if and only if 0 and ∞ lie on the same component of $\mathbf{C}^* - \Omega$. For instance:

(i) If γ^* is a Jordan curve connecting 0 and ∞, then $\mathbf{C}^* - \gamma^*$ is such a domain.

(ii) Any simply connected domain in $\mathbf{C}$, not containing 0, is such a domain.

(3) Let $\gamma^* = [\infty, 0]$ be the nonpositive real axis. We call

$$\operatorname{Log} z = \log |z| + i \operatorname{Arg} z : \mathbf{C} - \{0\} \to \{-\pi < \operatorname{Im} w \le \pi\}$$

the *principal branch* of $\log z$. $w = \operatorname{Log} z$ maps $\mathbf{C}^* - [\infty, 0]$ one-to-one and onto $-\pi < \operatorname{Im} w < \pi$ which is a fundamental domain of $z = e^w$ (see (2.6.1.3)). In general, each of the branches

$$w_n = \operatorname{Log} z + 2n\pi i : \mathbf{C}^* - [\infty, 0] \to \{(2n-1)\pi < \operatorname{Im} w < 2n\pi\} = \Omega_n$$

is one-to-one and onto, for $n = 0, \pm 1, \pm 2, \dots .$ (2.7.3.2)

The principal branch of $\log z$ is subject to change, if needed, and is dependent on what principal branch of $\arg z$ is designated. For instance, if we define $0 \le \operatorname{Arg} z < 2\pi$, then $\operatorname{Log} z = \log |z| + i \operatorname{Arg} z : \mathbf{C} - \{0\} \to \{0 < \operatorname{Im} w < w\pi\}$ is a favorable choice for the *principle branch*, too. In this case, $w = \operatorname{Log} z$ maps $\mathbf{C}^* - [0, \infty]$ univalently onto $0 < \operatorname{Im} w < 2\pi$.

We provide a computational example.

Example 1. Compute (1) $\log(-1)$, (2) $\log(i)$, (3) $\log \sqrt{-1 + 2i}$, and (4) $\log \sqrt[3]{-1}$.

Solution.

(1) $\operatorname{Arg}(-1) = \pi$. So $\operatorname{Log}(-1) = \log|-1| + \pi i$ and $\log(-1) = \pi i + 2n\pi i = (2n+1)\pi i$, $n = 0, \pm 1, \pm 2, \dots .$

(2) $\operatorname{Arg} i = \frac{\pi i}{2}$. $\operatorname{Log} i = \frac{\pi i}{2}$ and $\log i = \left(2n + \frac{1}{2}\right)\pi i$, $n = 0, \pm 1, \pm 2, \dots .$

(3) $\sqrt{-1 + 2i} = \sqrt[4]{5} e^{i\frac{1}{2}(\theta + 2k\pi)} = w_k$, $k = 0, 1$, where $\theta = \operatorname{Arg}(-1 + 2i) = \tan^{-1}(-2)$. Then $\log w_0 = \log \sqrt[4]{5} e^{i\frac{\theta}{2}} = \log \sqrt[4]{5} + i\left(\frac{\theta}{2} + 2n\pi\right)$ and $\log w_1 = \log \sqrt[4]{5} e^{-i\frac{\theta}{2}} = \log \sqrt[4]{5} + i\left(-\frac{\theta}{2} + 2n\pi\right)$, $n = 0, \pm 1, \pm 2, \dots .$

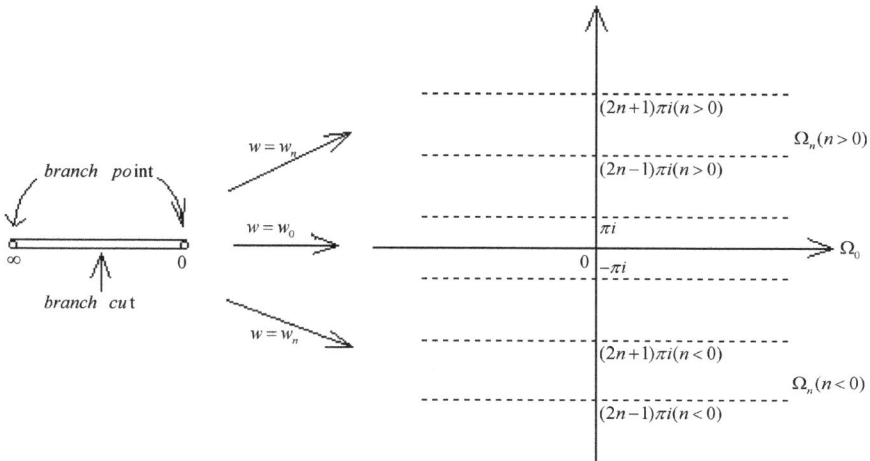

Fig. 2.72

(4) $\sqrt[3]{-1} = e^{i\frac{1}{3}(\pi + 2k\pi)} = w_k$, $k = 0, 1, 2$. Then $\log w_0 = i\left(2n + \frac{1}{3}\right)\pi$, $\log w_1 = i(2n + 1)\pi$, and $\log w_2 = i\left(2n - \frac{1}{3}\right)\pi$, $n = 0, \pm 1 \pm 2, \ldots$.

Note: Even though $\log(-z)^2 = \log z^2$, $\log \sqrt{-1 + 2i} \neq \frac{1}{2}\log(-1 + 2i)$.

The Riemann surface for log z

Figure 2.72 indicates how $z = e^w$ maps each fundamental domain $\Omega_n = \{(2n - 1)\pi < \operatorname{Im} w < (2n + 1)\pi\}$ onto the same domain $\mathbf{C}^* - [\infty, 0]$, called the nth sheet. Equivalently, the nth branch $w_n = \operatorname{Log} z + 2n\pi i$ of $w = \log z$ maps $\mathbf{C}^* - [\infty, 0]$ one-to-one and onto the parallel strip Ω_n, for $n = 0, \pm 1, \pm 2, \ldots$. Observe that the upper edge $\operatorname{Im} w = (2n - 1)\pi$ of Ω_{n-1} is the lower edge of Ω_n. Now, $z = e^w : \Omega_{n-1} \to \mathbf{C}^* - [\infty, 0]$ maps $\operatorname{Im} w = (2n - 1)\pi$ onto the upper side of the cut $[\infty, 0]$, while $z = e^w : \Omega_n \to \mathbf{C}^* - [\infty, 0]$ maps $\operatorname{Im} w = (2n-1)\pi$ onto the lower side of the cut $[\infty, 0]$. Hence, paste the upper side of the cut $[\infty, 0]$ of the $(n-1)$th sheet $\mathbf{C}^* - [\infty, 0]$ and the lower side of the cut $[\infty, 0]$ of the nth sheet $\mathbf{C}^* - [\infty, 0]$ pointwise together, just as how Ω_{n-1} and Ω_n have the common boundary $\operatorname{Im} w = (2n - 1)\pi$, for $n = 0, \pm 1, \pm 2, \ldots$. The resulted surface $R_{\log z}$ is the Riemann surface for $w = \log z$. See Fig. 2.73 for $R_{\log z}$ and Fig. 2.74 for the corresponding line complex. Note that if z winds around $|z| = 1$ continually in the anticlockwise direction, the corresponding $w = \log z$ will change values $(\operatorname{Arg} z + 2n\pi)i$ to $(\operatorname{Arg} z + (2n + 1)\pi)i$ for $n = 0, \pm 1, \pm 2, \ldots$ and will

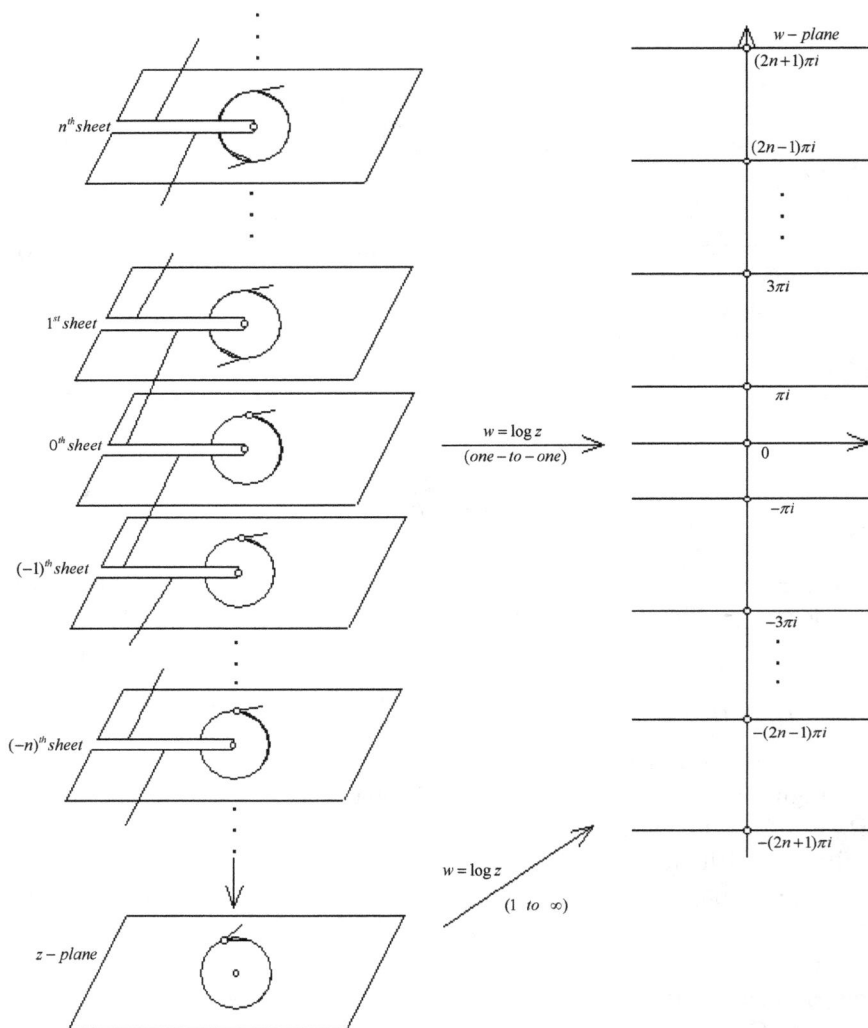

Fig. 2.73

never come back to its initial value. This amounts to say that, on $R_{\log z}$, the circle $|z| = 1$ forms an endless screw. In this sense, o and ∞ are called the *logarithmic branch points* of $w = \log z$. The origin 0 and the infinite point ∞ will *not* be points on $R_{\log z}$, since $z = e^w \neq 0$ and ∞ for $w \in \mathbf{C}^*$. If we replace the cut $[\infty, 0]$ by any Jordan curve γ^* from 0 to ∞, the resulted

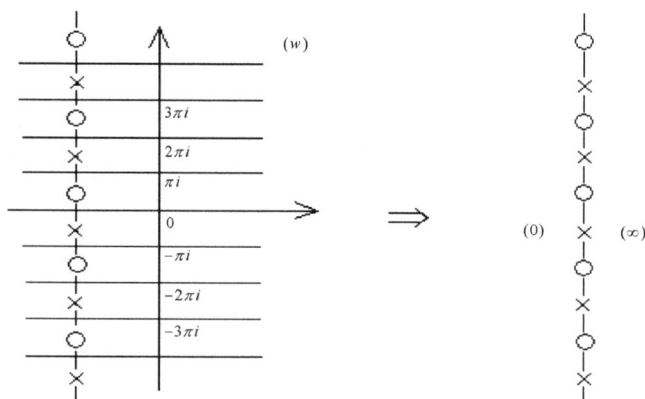

Fig. 2.74

surface based on $\mathbf{C}^* - \gamma^*$ is considered as identical with $R_{\log z}$ constructed above. And the Riemann sphere with a point punctured, or equivalently the finite complex plane, is the topological model for $R_{\log z}$.

Section (2) z^α and a^z

Fix a number $\alpha \in \mathbf{C}$. The general *power function* z^α is defined by

$$z^\alpha = e^{\alpha \log z}, \quad z \neq 0, \infty. \tag{2.7.3.3}$$

Let $\alpha = a + bi$. Then

$$z^\alpha = e^{(a+bi)[\log |z|+i(\operatorname{Arg} z+2k\pi)]}$$

$$= \rho_k e^{i\varphi_k},$$

where

$$\rho_k = e^{a \log |z|-b(\operatorname{Arg} z+2k\pi)},$$

$$\varphi_k = b \log |z| + a(\operatorname{Arg} z + 2k\pi), \quad k = 0, \pm 1, \pm 2, \ldots.$$

Note that $w = z^\alpha$ is multiple-valued if $b \neq 0$ or $b = 0$ and a is an irrational number. Also,

$$|z^\alpha| = |z|^\alpha \Leftrightarrow b = 0, \quad \text{namely, } \alpha = a \text{ is a real number.}$$

In case $\alpha = n$ is an integer, then $z^n = (|z|e^{i \operatorname{Arg} z})^n$ is single-valued; if $\alpha = \frac{m}{n}(0 < |m| < n$, and m and n are relatively prime), $z^\alpha = |z|^{\frac{m}{n}} e^{i\frac{m}{n}(\operatorname{Arg} z+2k\pi)}, k = 0, 1, 2, \ldots, n-1$, is n-valued.

We summarize the above as

The branches of z^α (α is irrational or complex but $\operatorname{Im}\alpha \neq 0$).

(1) The single-valued domains for $w = z^\alpha$ are exactly the same as these for $w = \log z$ (see (1) and (2) in (2.7.3.2)). 0 and ∞ are *logarithmic branch points* for $w = z^\alpha$. A Jordan curve from 0 to ∞ is a *branch cut*.

(2) Suppose f is a branch of $w = z^\alpha$ on a single-valued domain Ω. Then any branch of z^α on Ω is of the form

$$f(z)e^{2\alpha k\pi i}, \quad k = 0, \pm 1, \pm 2, \ldots .$$

In case $\Omega = \mathbf{C}^* - [\infty, 0]$, call $e^{\alpha \operatorname{Log} z}$ the *principal branch* of z^α on $\mathbf{C}^* - [\infty, 0]$.

(3) Suppose $\alpha > 0$. Let Ω be the sector domain $0 < \operatorname{Arg} z < \theta_0$ where $0 < \theta_0 \leq \frac{2\pi}{\alpha}$. Note that $e^{\alpha \operatorname{Log} z}$ is the composite of the following mappings:

$$z \to \zeta = \alpha \operatorname{Log} z \to w = e^\zeta.$$

So, $w = e^{\alpha \operatorname{Log} z}$ maps Ω univalently onto the sector domain $0 < \operatorname{Arg} w < \alpha\theta_0$. See Fig. 2.75.   (2.7.3.4)

Let $a \neq 0$ be a fixed number. Define the general *exponential function* a^z with *base a* as

$$a^z = e^{z \log a}, \quad z \neq \infty.   (2.7.3.5)$$

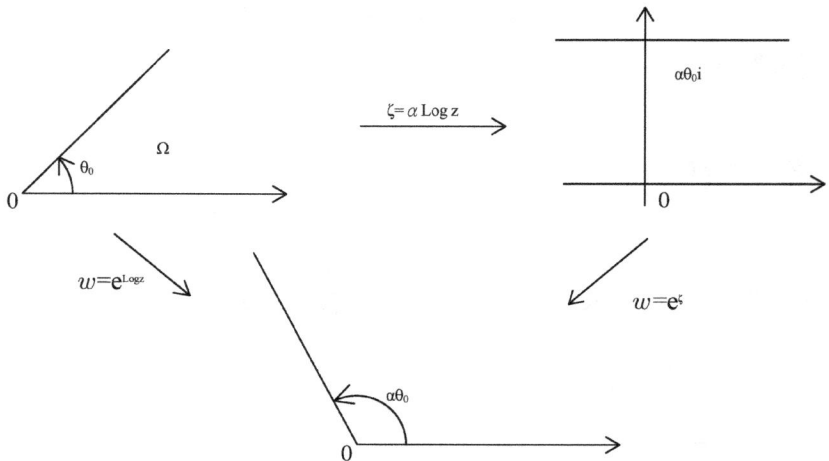

Fig. 2.75

If $z = n$ is an integer, then $a^z = a^n$; if $z = \frac{m}{n}$ $(0 < |m| < n$, and m and n are relatively prime), then $a^z = \sqrt[n]{a^m}$ as in Sec. 2.7.2. The remaining cases are summarized as

The branches of a^z (a is irrational or complex but $\operatorname{Im} a \neq 0$).

(1) The multi-valuedness of a^z originates from that of $\log a$. Choosing a definite value $\log |a| + i(\operatorname{Arg} a + 2n\pi)$, $n = 0, \pm 1, \ldots$, of $\log a$, then a branch

$$e^{z[\log |a| + i(\operatorname{Arg} a + 2n\pi)]}$$

is then defined on the whole plane **C**.

(2) Any two values of a^z on **C** differs by a factor of the form $e^{2n\pi i}$, $n = 0, \pm 1, \pm 2, \ldots$.

(3) There is no branch point for a^z on **C**.   (2.7.3.6)

Remark. Reconfirmation of e^z (defined in Sec. 2.6.1) and log z (defined in (2.7.3.1)). According to (2.7.3.5), e^z is multiple-valued with branches $e^{z \log e} = e^{z(1 + 2n\pi i)}$, $n = 0, \pm 1, \ldots$, while e^z defined in Sec. 2.6.1 is just one of its branches, namely, the one with $n = 0$. Also, e^i has value $\cos 1 + i \sin 1$ when e^z is defined as in Sec. 2.6.1, while $e^i = e^{i \log e} = e^{-2n\pi}(\cos 1 + i \sin 1)$, $n = 0, \pm 1, \ldots$, according to (2.7.3.5). In order to avoid this confusion, e^z *will always be adopted as in Sec. 2.6.1 in this text.*

Let $a \in \mathbf{C}$ and $a \neq 0, 1$. If $z \neq 0, \infty$, what values of w will satisfy $z = a^w$? Fix any value b of $\log a$. Then, $z = a^w = e^{w \log a} = e^{wb} \Rightarrow wb = \log z \Rightarrow w = \frac{\log z}{w}$. Hence, define

$$\log_a z = \frac{\log z}{\log a}, \quad z \neq 0, \infty. \tag{2.7.3.7}$$

But, for $a = e$, $\log z$ *will always be defined as in* (2.7.3.1).   □

We give an example:

Example 2. Compute (1) i^i, (2) $(-2)^{\sqrt{2}}$, and (3) $(-i)^{\sqrt{-i}}$.

Solution.

(1) $i^i = e^{i \log i} = e^{i[\log |i| + i(\frac{\pi}{2} + 2k\pi)]} = e^{-(\frac{\pi}{2} + 2k\pi)}$, $k = 0, \pm 1, \pm 2, \ldots$.

(2) $(-2)^{\sqrt{2}} = e^{\sqrt{2} \log(-2)} = e^{\sqrt{2}[\log |-2| + i(\pi + 2k\pi)]} = 2^{\sqrt{2}} e^{i\sqrt{2}(\pi + 2k\pi)}$, $k = 0, \pm 1, \pm 2, \ldots$.

(3) Since

$$\sqrt{-i} = e^{i\frac{1}{2}\left(-\frac{\pi}{2}+2k\pi\right)} = \begin{cases} \dfrac{1}{\sqrt{2}}(1-i), & k=0 \\[2mm] \dfrac{1}{\sqrt{2}}(-1+i), & k=1 \end{cases}.$$

$(-i)^{\sqrt{-i}}$ has two set of values: for $k = 0, \pm 1, \pm 2, \dots,$

$$(-i)^{\frac{1}{\sqrt{2}}(1-i)} = e^{\frac{1}{\sqrt{2}}(1-i)\left(-\frac{\pi}{2}+2k\pi\right)} \quad \text{and} \quad (-i)^{\frac{1}{\sqrt{2}}(-1+i)}$$

$$= e^{-\frac{1}{\sqrt{2}}(1-i)\left(-\frac{\pi}{2}+2k\pi\right)}.$$

Section (3) Some further examples

The details in what follows will be left to the readers.

Example 3. Let $w = \log(z + \sqrt{z^2 - 1})$, the inverse of $z = \cosh w = \frac{1}{2}(e^w + e^{-w})$.

(1) ± 1 are algebraic branch points of order 1; ∞ is a logarithmic branch point. $[\infty, -1]$ and $[1, \infty]$ (along the real axis) are branch cuts.

(2) Show that both $\mathbf{C}^* - [\infty, -1] - [1, \infty]$ and $\mathbf{C}^* - [\infty, -1] - [-1, 1] = \mathbf{C}^* - [\infty, 1]$ are single-valued domains for $w = w(z)$. On each of them, infinitely many branches may be defined.

(3) Find the value at i of the branch $f(z)$ uniquely determined by $w(0) = \frac{\pi}{2}i$.

(4) Construct the Riemann surface for $w = w(z)$.

Solution.

(1) By considering a Jordan closed curve γ and the cases: $\pm 1 \in \text{Ext}\,\gamma$, $1 \in \text{Int}\,\gamma$ and $-1 \in \text{Ext}\,\gamma$, $-1 \in \text{Int}\,\gamma$ and $1 \in \text{Ext}\,\gamma$, $\pm 1 \in \text{Int}\,\gamma$, the results follow just like the arguments shown in Figs. 2.52 and 2.53.

(2) According to (6) in (2.5.5.4), $\zeta = z + \sqrt{z^2 - 1}$ has two branches on $\Omega = \mathbf{C}^* - [\infty, -1] - [1, \infty]$:

   (i) the branch ζ_1 uniquely determined by $\zeta(0) = i$ maps Ω univalently onto $\text{Im}\,\zeta > 0$; and

   (ii) the branch ζ_2 uniquely determined by $\zeta(0) = -i$ maps Ω univalently onto $\text{Im}\,\zeta < 0$.

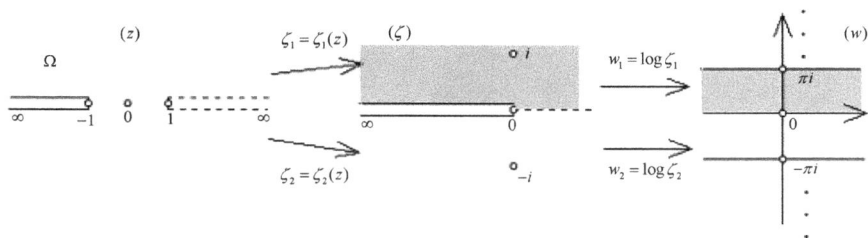

Fig. 2.76

See Fig. 2.76.

Since $\zeta_1(\Omega)$ and $\zeta_2(\Omega)$ are contained in the domain $\mathbf{C}^* - [\infty, 0]$ in the ζ-plane, the composite maps, for $n = 0, \pm 1, \pm 2, \ldots$,

$$w_1 = \log \zeta_1 = \operatorname{Log} \zeta_1 + 2n\pi i : \Omega \to \{2n\pi < \operatorname{Im} w < (2n+1)\pi\};$$
$$w_2 = \log \zeta_2 = \operatorname{Log} \zeta_2 + 2n\pi i : \Omega \to \{(2n-1)\pi < \operatorname{Im} w < 2n\pi\},$$

are well-defined branches of $w = w(z)$ on Ω. They differ by a sign or by an integral multiple of $2\pi i$.

Similarly, $\eta = z + \sqrt{z^2 - 1}$ defines two branches on $\mathbf{C}^* - [-1, 1]$:

(i) the branch η_1 determined by $\eta(\infty) = 0$ maps $\mathbf{C}^* - [-1, 1]$ univalently onto $|\eta| < 1$; and
(ii) the branch η_2 determined by $\eta(\infty) = \infty$ maps $\mathbf{C}^* - [-1, 1]$ univalently onto $|\eta| > 1$.

But both $|\eta| < 1$ and $|\eta| > 1$ are not contained in the domain $\mathbf{C}^* - [\infty, 0]$ in the η-plane. Based on (4) in (2.5.5.4), η_1 and η_2 maps $[\infty, -1]$ in the z-plane univalently onto $[-1, 0]$ and $[\infty, -1]$ in the η-plane, respectively. Hence, let $D = \mathbf{C}^* - [\infty, 1]$, a domain in the z-plane. Then $\eta_1(D) = \{|\eta| < 1\} - (-1, 0]$ and $\eta_2(D) = \{|\eta| > 1\} - (\infty, -1]$. See Fig. 2.77. Now, on D, there are infinitely many branches of $w = w(z)$: for $n = 0, \pm 1, \pm 2, \ldots$,

$$w_1 = \log \eta_1 : D \to \{(2n-1)\pi < \operatorname{Im} w < (2n+1)\pi \text{ and } \operatorname{Re} w < 0\}, \text{ and}$$

$$w_2 = \log \eta_2 : D \to \{(2n-1)\pi < \operatorname{Im} w < (2n+1)\pi \text{ and } \operatorname{Re} w > 0\}.$$

They differ by a sign or by an integral multiple of $2\pi i$.

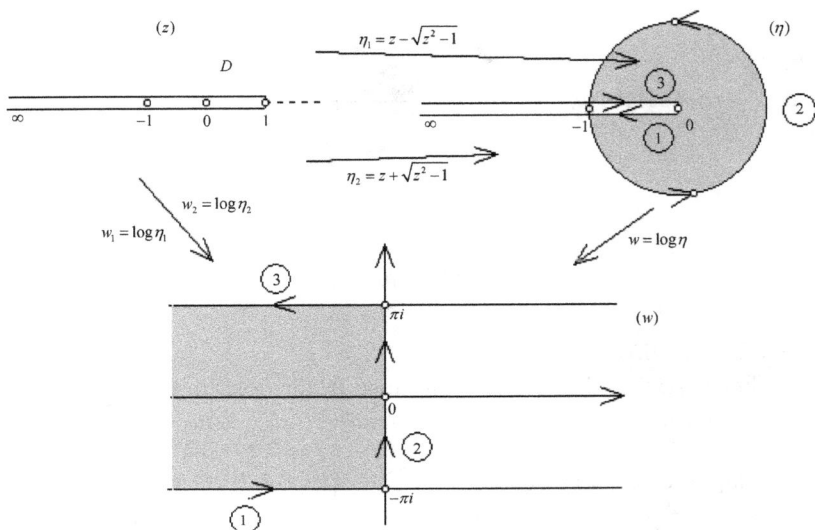

Fig. 2.77

(3) The branch $f(z) = \operatorname{Log}\zeta_1 : \Omega \to \{0 < \operatorname{Im} w < \pi\}$ is uniquely defined by $w(0) = \frac{\pi}{2}i$. Then, $f(i) = \operatorname{Log}(i + \sqrt{2}i) = \operatorname{Log}(1 + \sqrt{2})i = \log(1 + \sqrt{2}) + \frac{\pi}{2}i$.

(4) $\zeta = z + \sqrt{z^2 - 1}$ maps Ω onto $\mathbf{C}^* - [\infty, 0]$ in the one-to-two manner. So, based on Fig. 2.76, we construct a surface $\tilde{\Omega}$, composed of two copies of $\Omega = \mathbf{C}^* - [-\infty, -1] - [1, \infty]$ and crosswise pasted along $[1, \infty]$, so that $\zeta = \zeta(z)$ maps $\tilde{\Omega}$ univalently onto $\mathbf{C}^* - [\infty, 0]$. See Fig. 2.78 (compare to

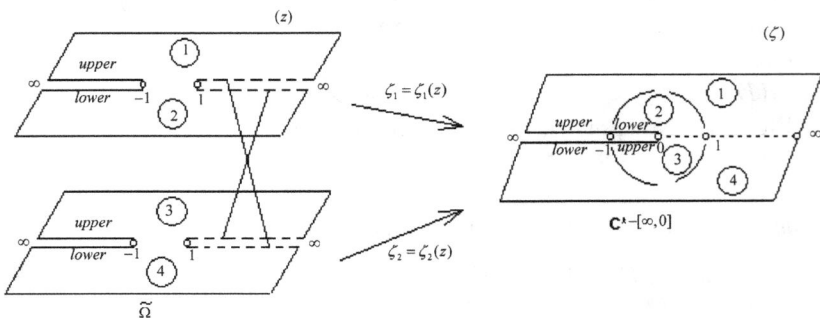

Fig. 2.78

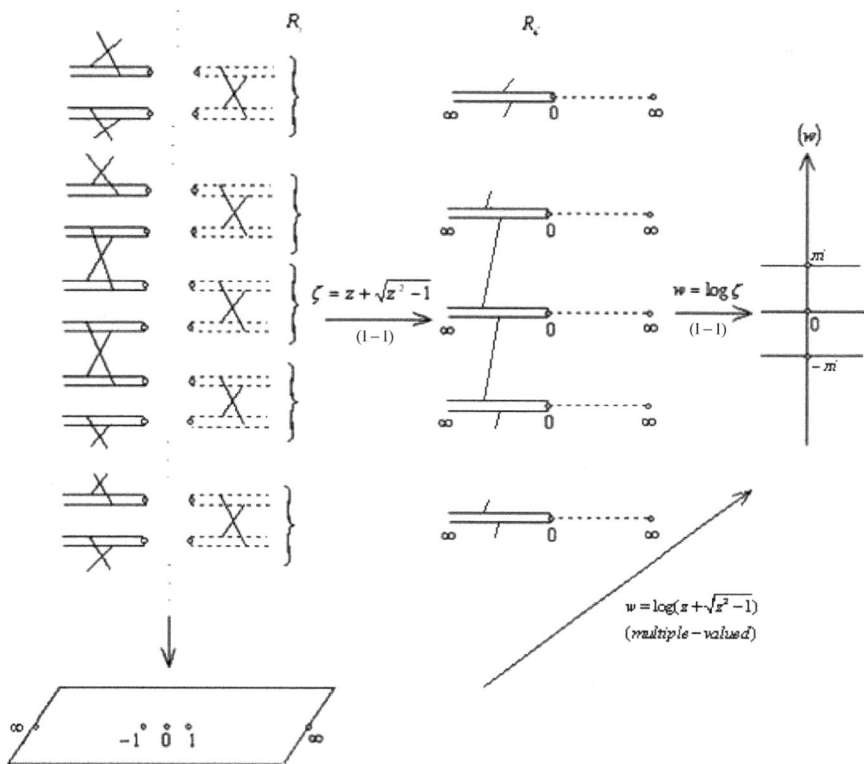

Fig. 2.79

Fig. 2.61). When combined with the Riemann surface of $w = \log \zeta$ (see Fig. 2.73), we obtain the Riemann surface R_z for $w = \log(z + \sqrt{z^2 - 1})$. See Fig. 2.79.

There are infinitely many branch points of order 1 over $z = \pm 1$ and two logarithmic branch points over ∞.

Example 4. Suppose a and b are distinct complex numbers. Let $w = \log \frac{z-a}{z-b}$.

(1) a and b are logarithmic branch points. A Jordan arc $\widehat{ab}$ in $\mathbf{C}^*$ from a to b is a branch cut. Note that ∞ is not a branch point.
(2) Infinitely many branches may be defined on the single-valued domain $\mathbf{C}^* - \widehat{ab}$.

(3) In case $a = 1$ and $b = -1$, find the value at $-1 + 2i$ of the branch $f(z)$ determined by $w(i) = \left(\frac{1}{2} + 2n_0\right)\pi i$, where n_0 is a fixed integer.

(4) Construct the Riemann surface of $w = w(z)$.

Solution.

(1) Routine checks will show the claim.

(2) $\zeta = \frac{z-a}{z-b}$ maps a, $\frac{1}{2}(a+b)$, and b into 0, -1 and ∞, respectively. Hence, $\zeta = \zeta(z)$ maps $\mathbf{C}^* - [a,b]$ univalently onto $\mathbf{C}^* - [\infty, 0]$ in the ζ-plane, on which $w = \log \zeta$ may be defined infinitely many branches. Hence, $w = w(z)$ has the branches: for $n = 0, \pm 1, \pm 2, \ldots$,

$$w_n = \operatorname{Log}\frac{z-a}{z-b} + 2n\pi i : \mathbf{C}^* - [a,b] \to \{(2n-1)\pi < \operatorname{Im} w < (2n+1)\pi\}.$$

(3) $w(i) = \log i = i\left(\frac{\pi}{2} + 2n\pi\right), n = 0, \pm 1, \ldots$. Now, choose $w(i) = w_{n_0}(i)$. Hence

$$w(-1+2i) = w_{n_0}(-1+2i) = \log\left|\frac{-2+2i}{2i}\right| + i\left(\operatorname{Arg}\frac{-2+2i}{2i} + 2n_0\pi\right)$$

$$= \log\sqrt{2} + i\left(\frac{\pi}{4} + 2n_0\pi\right).$$

An alternative method is as follows. Choose a Jordan curve γ from i to $-1 + 2i$ which lies in $\mathbf{C}^* - [-1, 1]$. Now

$$\Delta_\gamma \operatorname{Arg}(z-1) = \operatorname{Arg}(-2+2i) - \operatorname{Arg}(i-1) = 0,$$

$$\Delta_\gamma \operatorname{Arg}(z+1) = \operatorname{Arg}(2i) - \operatorname{Arg}(i+1) = \frac{\pi}{2} - \frac{\pi}{4} = \frac{\pi}{4},$$

$$\Rightarrow \Delta_\gamma \operatorname{Arg}\frac{z-1}{z+1} = 0 - \frac{\pi}{4} = -\frac{\pi}{4},$$

$$\Rightarrow \text{(note that } \log f(z_2)$$

$$= \log|f(z_2)| + i\{\Delta_\gamma \operatorname{Arg} f(z) + \operatorname{Im} \log f(z_1)\})$$

$$w(-1+2i) = \log\left|\frac{-2+2i}{2i}\right| + i\left\{-\frac{\pi}{4} + \frac{\pi}{2} + 2n_0\pi\right\}$$

$$= \log\sqrt{2} + i\left(\frac{\pi}{4} + 2n_0\pi\right).$$

(4) Via the linear transformation $\frac{\zeta-1}{\zeta+1} = \frac{z-a}{z-b}$, we may suppose $a = 1$ and $b = -1$. The decomposition $z \to \zeta = \frac{z-1}{z+1} \to w = \log\zeta$ of $w = \log\frac{z-1}{z+1}$ shows how to construct the Riemann surface R_z of $w = w(z)$. See Fig. 2.80.

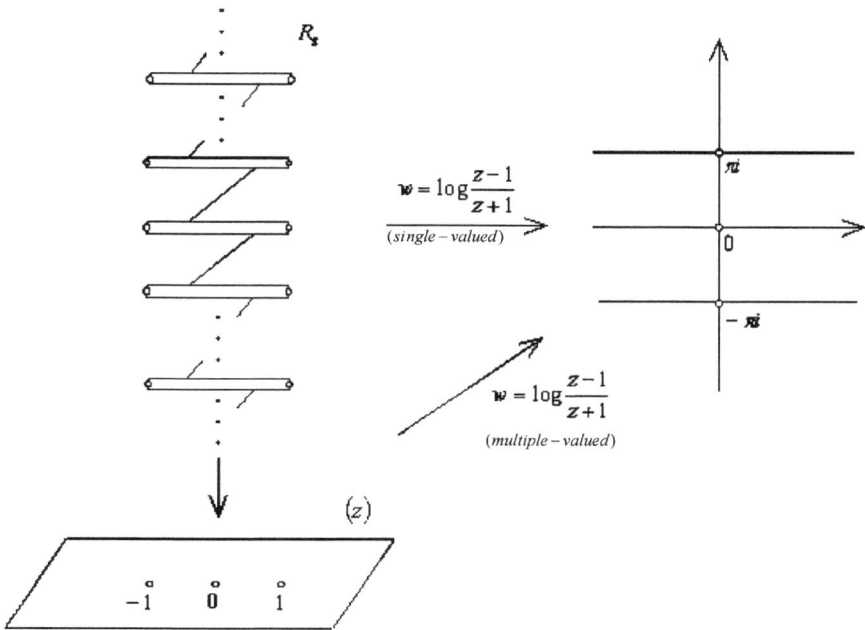

Fig. 2.80

Example 5. Let a and b be distinct complex numbers, and α and β be complex numbers but not integers. Let $w = (z-a)^\alpha (z-b)^\beta$.

(1) a and b are branch points. If α is an irrational number or $\operatorname{Im}\alpha \neq 0$, then a is a logarithmic branch point, etc.

(2) If $\alpha + \beta$ is an integer, ∞ is a node point; if $\alpha + \beta$ is not an integer, ∞ is a branch point.

(3) If $\alpha + \beta$ is an integer, then $\mathbf{C}^* - \gamma$ is a single-valued domain where γ is a Jordan curve from a to b; if $\alpha + \beta$ is not an integer, then $\mathbf{C}^* - \gamma^*$ is a single-valued domain where γ^* is a Jordan curve connecting a, b and ∞.

Solution. By definition (2.7.3.3),

$$w = e^{\alpha \log(z-a)} e^{\beta \log(z-a)}$$

$$= |z-a|^\alpha |z-b|^\beta e^{[\alpha \operatorname{Arg}(z-a)+\beta \operatorname{Arg}(z-b)]i} e^{(2\alpha k\pi + 2\beta l\pi)i},$$

$$k, l = 0, \pm 1, \pm 2, \ldots .$$

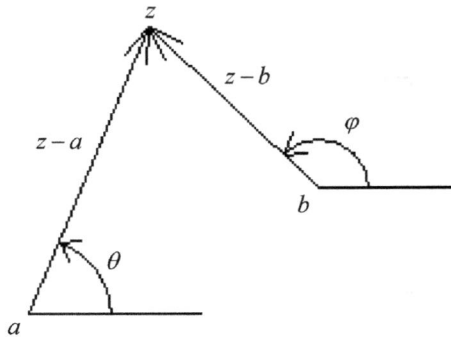

Fig. 2.81

See Fig. 2.81.

Let γ be a Jordan closed curve, passing neither a nor b. Use $\Delta_\gamma g(z)$ to denote the *variation* of $g(z) = \alpha \log(z - a) + \beta \log(z - b)$ along γ if z winds along γ from a given point z_0 on it and back to z_0 in the anticlockwise direction. Namely, $\Delta_\gamma g(z)$ is equal to the terminal value at z_0 minus the initial value at z_0 of $g(z)$. Four cases are considered as follows.

(1) If $a, b \in \text{Ext}\,\gamma$: then $\Delta_\gamma g(z) = 0$ and $w = w(z)$ assumes its original value.

(2) If $a \in \text{Int}\,\gamma$ and $b \in \text{Ext}\,\gamma$: then $\Delta_\gamma g(z) = 2\alpha\pi i$. Thus, the terminal value of $w = w(z)$ is $e^{2\alpha\pi i}$ times its initial value (note that α is not an integer) and a branch of $w = w(z)$ will be changed continually into another branch. Hence, a is a branch point.

(3) If $a \in \text{Ext}\,\gamma$ and $b \in \text{Int}\,\gamma$: b is branch point.

(4) If a and b are in $\text{Int}\,\gamma$: Then

$$\Delta_\gamma g(z) = \Delta_\gamma\{\alpha \log(z - a)\} + \Delta_\gamma\{\beta \log(z - b)\}$$

$$= 2\alpha\pi i + 2\beta\pi i = 2(\alpha + \beta)\pi i.$$

Therefore, the terminal value of $w = w(z)$ is equal to $e^{2(\alpha+\beta)\pi i}$ times its initial value, and equal to its initial value if and only if $\alpha + \beta$ is an integer.

Exercises A

(1) Let $a \in \mathbf{C}$ be a fixed point. Prove results similar to those stated in (2.7.3.2) for $\log(z - a)$.

(2) Determine if a branch of $w = \log z$ can be defined on each of the following domains Ω. If yes, try to define all possible branches on Ω.

  (a) $\Omega = \{z | \operatorname{Re} z + \operatorname{Im} z < 0\}$.
  (b) $\Omega = \{z | 1 < |z| < e\} - \{ti | 1 \le t \le e\}$.
  (c) $\Omega = \{z | 0 < \operatorname{Re} z - \operatorname{Im} z < 1\}$.
  (d) $\Omega = \{z | 0 < |\operatorname{Re} z| + |\operatorname{Im} z| < 1\}$.

(3) The curve $r = \theta$ is the Archimedean spiral γ in the polar coordinate (r, θ). Then $\Omega = \mathbf{C} - \gamma$ is a simply connected domain, not containing 0. Try to find the branch $f(z)$ of $w = \log z$ on Ω, uniquely defined by $w(1) = 0$. Find all other branches of $w = \log z$ on Ω.

(4) Sketch the curve $\gamma : z(t) = e^{t+it}, t \in \mathbf{R}$. Determine the branch $f(z)$ of $w = \log z$ on $\mathbf{C} - \gamma$, defined by $w(e) = 1$. Find the range of $f(z)$ and compute $f(e^6)$ and $f(ie^{n\pi})$, where n is any integer.

(5) Sketch the domain

$$\Omega = C - \{|z - 1| \le 1\} - \bigcup_{k=1}^{\infty} \{|z - 1| = 2k + 1, \ \operatorname{Im} z \ge 0\}$$

$$- \bigcup_{k=1}^{\infty} \{|z| = 2k, \ \operatorname{Im} z \le 0\}.$$

Determine the branch $f(z)$ of $w = \log z$ on Ω, defined by $w(1) = 0$. Compute $f(e^2)$ and $f(2n + 1)$, where n is any integer.

(6) Compute the following: 1^i; $(-i)^i$; $(-i)^{(-i)}$; $i^{\sqrt{-i}}$; $(3 - 4i)^{1+i}$; $\left(\frac{1-i}{\sqrt{2}}\right)^{1+i}$; $(1 + i)^i$.

(7) (a) Show that $|a^b| = |a|^b$ if $a \in \mathbf{C}$ and $a \ne 0$, and b is real.
  (b) When does $|a^b| = |a|^{|b|}$ hold if $a, b \in \mathbf{C}$ and $a \ne 0$?

(8) When will $\log a^b = b \log a$ be true?

(9) (a) Try to use branches of $\log z$ to show that $a^b a^c = a^{b+c}$, where $a, b, c \in \mathbf{C}$ and $a \ne 0$.
  (b) If $\log ab = \log a + \log b$ for a suitable branch of $\log z$, show that $(ab)^c = a^c b^c$ holds.

(10) When will $(a^b)^c = a^{bc}$ be true?

(11) Show that $z = \tan\{\frac{1}{i} \log \left(\frac{1+iz}{1-iz}\right)^{\frac{1}{2}}\}$.

(12) Let $w = (1 - z)^{-\alpha} z^{\alpha-1}$ $(0 < \alpha < 1)$.

  (a) Find the variation $\Delta_\gamma \arg w$ of $\arg w$ and the variation $\Delta_\gamma w(z)$ of $w(z)$ if z traces the curve γ respectively: $z_1(t) = \frac{1}{2} e^{it}, \ 0 \le t \le 2\pi$; $z_2(t) = 1 + \frac{1}{2} e^{it}, \ 0 \le t \le 2\pi$; $z_3(t) = 2e^{it}, 0 \le t \le 2\pi$.
  (b) Find single-valued domains for $w = w(z)$.

(13) Compute $\Delta_\gamma w$ for each of the following functions $w = w(z)$, where γ is a Jordan closed curve surrounding 0.
   (a) $\log \frac{z}{1+z}$.   (b) $\log z^2$.   (c) $(\log z^2)^2$.   (d) $\log \sqrt{z}$.

(14) Compute $\Delta_\gamma w$, where $w = \log(z^2 - 1)$ and $w = \log\sqrt{z^2 - 1}$, respectively, and γ is as in Fig. 2.67.

(15) Let $w = \log \frac{z-i}{z+i}$.
   (a) Find the value at 1 of the branch determined by $w(0) = \pi i$ along a Jordan curve γ from 0 to 1, not passing $\pm i$.
   (b) If γ is the curve γ_1 and γ_2 shown in Fig. 2.82, respectively, find $w(1)$ in (a).
   (c) $w(1) = i(2n - \frac{1}{2})\pi$, $n = 0, \pm 1, \pm 2, \ldots$. What value should $w(0)$ take so that, if z varies along r_1 and r_2 in Fig. 2.82, respectively, then $w(1)$ will have the value $i(2n - \frac{1}{2})\pi$.
   (d) Find the value at $z = 1$ of the branch determined by $w(0) = \pi i$ in $\mathbf{C}^* - [i, +i\infty] - [-i\infty, -i]$.
   (e) Find the value on the left and the right sides of the cut $[-i, i]$ of the branch determined by $w(1) = -\frac{1}{2}\pi i$ in $\mathbf{C}^* - [-i, i]$.

(16) Consider the branch of $\log z$, defined by $\log 1 = 0$ on $\mathbf{C}^* - [\infty, 0]$. Compute, where $x \in \mathbf{R}$,
$$\lim_{x \to 0^-} \log \frac{x-i}{x+i} \quad \text{and} \quad \lim_{x \to 0^+} \log \frac{x-i}{x+i}.$$

(17) Let $w = \log(z - a)(z - b)$, $a \neq b$. Construct the Riemann surface of R_z. Note that there is a logarithmic branch point over each of a and b, and there are two logarithmic branch points over $z = \infty$. Also, note that $\mathbf{C}^* - \widehat{a\infty} - \widehat{b\infty}$ is a single-valued domain.

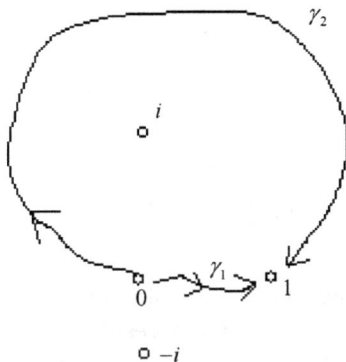

Fig. 2.82

(18) Decompose $w = \log(1 + z^3)$ as $z \to \eta = 1 + z^3 \to w = \log \eta$.

    (a) Show that $\Omega = \mathbf{C} - \{re^{\frac{\pi i}{3}} \mid r \geq 1\} - \{re^{\pi i} \mid r \geq 1\} - \{re^{\frac{-\pi i}{3}} \mid r \geq 1\}$ is a single-valued domain for $w = w(z)$.

    (b) Construct the Riemann surface for $w = w(z)$. Note that there is a logarithmic branch point over each of the points $e^{\frac{\pi i}{3}}$, $e^{\pi i}$, and $e^{-\frac{\pi i}{3}}$, and there are three logarithmic branch points over ∞.

(19) Construct the Riemann surface for $w = \log(z - \frac{1}{z})$. Note that there is a logarithmic branch point over each of the points $z = \pm 1, 0, \infty$. $\mathbf{C}^* - [\infty, -1] - [0, 1]$ is a single-valued domain.

(20) Find single-valued domains for each of the following functions $w = w(z)$ and determine all branches on such domain. (a) $(z^2 + 1)^z$. (b) $z^{\sqrt{z}}$. (c) $(e^z + 1)^{z^2}$. (d) $(\log z)^z$. (e) $(z + \sqrt{z^2 + 1})^{\log z}$.

Exercises B

(1) Suppose $f : S \subseteq \mathbf{C} \to \mathbf{C} - \{0\}$ is a continuous function. If there exist continuous functions $g : S \to \mathbf{C}$ and $h : S \to \mathbf{R}$ so that $f = e^g$ and $\frac{f}{|f|} = e^{ih}$, respectively, then g is called a *branch* of $\log f$ on S and h a *branch* of $\arg f$ on S.

    (a) Show that g is a branch of $\log f \Leftrightarrow \mathrm{Re}\, g = |f|$, and $\mathrm{Im}\, g$ is a branch of $\arg f$.

    (b) Suppose S is a connected set. If g_1 and g_2 are two branches of $\log f$ on S, then there exists an integer n so that $g_1(z) - g_2(z) = 2n\pi i$ on S. Hence, if g is a branch of $\log f$ on S, then $\{g + 2n\pi i \mid n = 0, \pm 1, \ldots\}$ is the set of all branches of $\log f$ on S.

    *(c) Let S be an open set and g be a branch of $\log f$ on S. If f is analytic on S (see (3.3.1.8)), then g is also analytic on S, called an *analytic single-valued branch* of $\log f$.

(2) For each of the following multiple-valued functions $w = w(z)$, do the following problems.

    (i) Find its single-valued domains and branches wherein;

    (ii) Determine the ranges of branches, if possible;

    (iii) Construct the Riemann surface for $w = w(z)$, including branch points and cuts.

    (a) $w = \log(z - a)(z - b)(b - c), \quad a \neq b \neq c.$

    (b) $w = \log \dfrac{(z - a)(z - b)}{z - c}, \quad a \neq b \neq c.$

    (c) $w = \log(z + \sqrt{z^2 + 1}).$

(d) $w = \log(z^4 - 1)$.

(e) $w = \log(1 + \sqrt{z})$.

(f) $w = \log(e^z - 1)$.

(g) $w = \sqrt[n]{e^z}$.

(h) $w = e^{\sqrt[n]{z}}$.

(i) $z = e^{\frac{1}{2}\left(w + \frac{1}{w}\right)}$.

(j) $z = \frac{1}{2}(e^w + e^{-w})$.

(k) $w = \sqrt[n]{\log z}$. See Fig. 2.83(a) for $n = 2$.

(l) $w = \log(1 + z^n)$.

(m) $w = \dfrac{1}{2}\left(\log z + \dfrac{1}{\log z}\right)$.

(n) $w = e^{\sqrt[n]{\log z}}$.

(o) (See Ref. [1], p. 72) $w = \log\log z$. See Fig. 2.83(b)).

(p) (See Ref. [64], p. 293) $z = we^{-w}$. See Fig. 3.58.

(q) $w = \log\dfrac{1}{2}\left(\sqrt{z} + \dfrac{1}{\sqrt{z}}\right)$.

2.7.4. $w = \cos^{-1} z$ and $w = \tan^{-1} z$ and their Riemann surfaces

We divide the content into four sections.

Section (1) $w = \cos^{-1}z$ and its Riemann surface

Since $z = \cos w = \frac{1}{2}(e^{iw} + e^{-iw})$, $w = \cos^{-1} z = -i\log(z + \sqrt{z^2 - 1})$. There are three ways to handle this topic:

(1) Direct usage of Example 3 in Sec. 2.7.3: Rotate the right figure (in the w-plane) of Fig 2.79 by $90°$ in the clockwise direction. The resulted configuration is a description of the Riemann surface for $w = \cos^{-1} z$.

(2) Decomposition as the composite of elementary functions: $w = \cos^{-1} z$ is the composite of the inverses of the following mappings: $w \to \zeta = iw \to \eta = e^{\zeta} \to z = \frac{1}{2}(\eta + \frac{1}{\eta})$.

(3) Description of fundamental domains of $z = \cos w$, based on (2) in (2.6.2.6).

Here, we try to adopt method 3 in order to practise more about elementary geometric mapping properties of $z = \cos w$. Interchange of z and w are needed when applying (2.6.2.6) to our case now.

Recall that $(n - 1)\pi \le \operatorname{Re} w < (n + 1)\pi, n = 0, \pm 1, \ldots$, are periodic strips for $z = \cos w$. We try to figure out what is the image, under $z =$

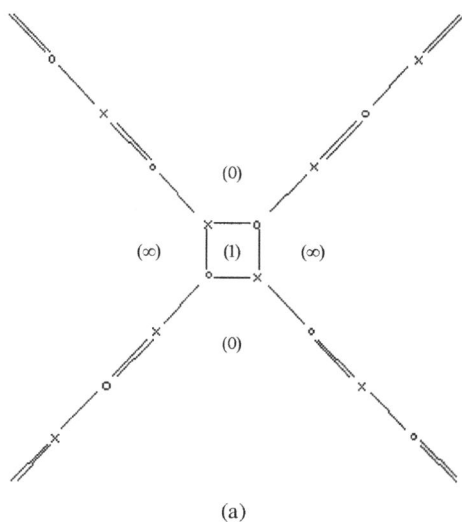

(a)

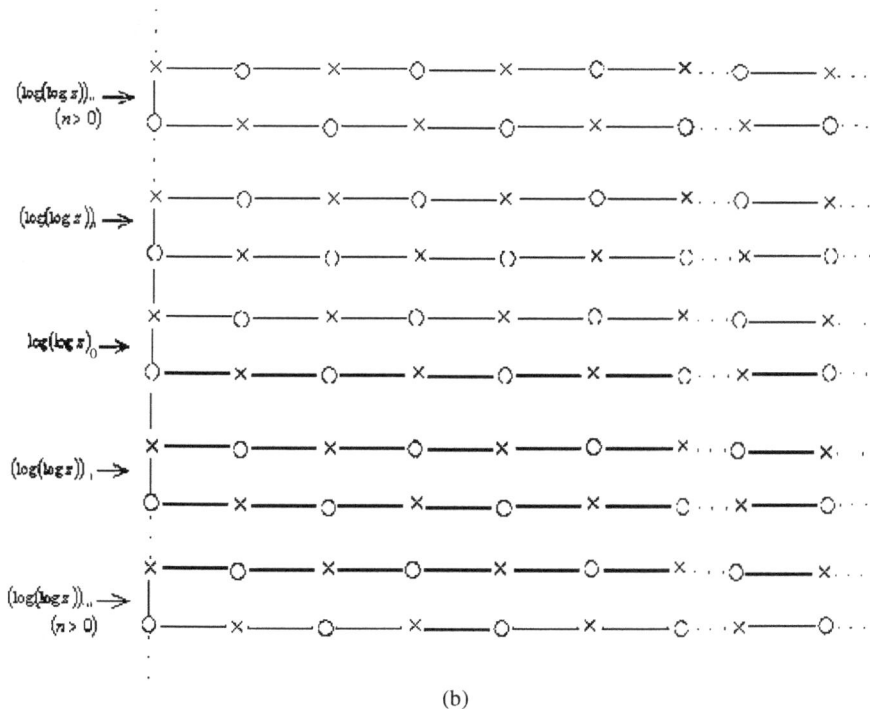

(b)

Fig. 2.83

$\cos w$, of the open half-strip $k\pi < \mathrm{Re}\,w < (k+1)\pi$, denoted as A_k, for $k = 0, \pm 1, \pm 2, \dots$.

Let $z = x + iy$ and $w = u + iv$. Then

$$z = \cos w$$

$$\Rightarrow x = \frac{1}{2}(e^{-v} + e^v)\cos u \quad \text{and} \quad y = \frac{1}{2}(e^{-v} - e^v)\sin u.$$

Four cases are considered in what follows: $k = 0, \pm 1, \pm 2, \dots$.

Case 1. If $w = 2k\pi + iv$ and $-\infty < v < \infty$, then $x = \frac{1}{2}(e^{-v} + e^v)$ and $y = 0$. This indicates that, if v decreases from $+\infty$ to 0 and then to $-\infty$, the image of $\mathrm{Re}\,w = 2k\pi$ is the segment $[1, \infty)$ along the x-axis, decreasing from ∞ to 1 and then increasing back to $+\infty$.

Case 2. If $w = (2k-1)\pi + iv$ and $-\infty < v < \infty$, then $x = -\frac{1}{2}(e^{-v} + e^v)$ and $y = 0$: Thus, the image of $\mathrm{Re}\,w = (2k-1)\pi$ is the segment $(-\infty, -1]$ along the x-axis such that $x : -\infty \to -1 \to -\infty$ as $v : +\infty \to 0 \to -\infty$.

Case 3. If $w = u, (2k-1)\pi < u < 2k\pi$, then $x = \cos u$ and $y = 0$. If $u : (2k-1)\pi \to 2k\pi$, the image of $[(2k-1)\pi, 2k\pi]$ is the segment $[-1, 1]$ along the x-axis, increasing from -1 to 1.

Case 4. Which part of $k\pi < \mathrm{Re}\,w < (k+1)\pi$ is mapped onto the upper or the lower half-space? Observe that

$$y = \frac{1}{2}(e^{-v} - e^v)\sin u > 0,$$

$$\Leftrightarrow e^{-v} - e^v > 0 \text{ and } \sin u > 0, \quad \text{or} \quad e^{-v} - e^v < 0 \text{ and } \sin u < 0.$$

$$\Leftrightarrow k\pi < u < (k+1)\pi \quad \text{and} \quad v < 0, \quad k = 0, \pm 2, \pm 4, \dots,$$

or

$$k\pi < u < (k+1)\pi \text{ and } v > 0, \quad k = \pm 1, \pm 3, \dots.$$

Combining these results together (plus some arguments in Sec. 2.6.2), it follows that *each open strip* $A_k : k\pi < \mathrm{Re}\,w < (k+1)\pi, k = 0, \pm 1, \dots$, *is a fundamental domain for* $z = \cos w$, *whose image is the domain* $\mathbf{C}^* - [\infty, -1] - [1, \infty]$. See Fig. 2.84.

$z = \cos w$ *maps its open fundamental periodic strip* $-\pi < \mathrm{Re}\,w < \pi$ *onto* $\mathbf{C}^* - [\infty, -1]$ *in two-to-one manner except at* $w = 0$ *where* $z = \cos w$ *assumes the value 1 twice*. Similarly, so does for each open periodic strip $(n-1)\pi < \mathrm{Re}\,w < (n+1)\pi$. See Fig. 2.85.

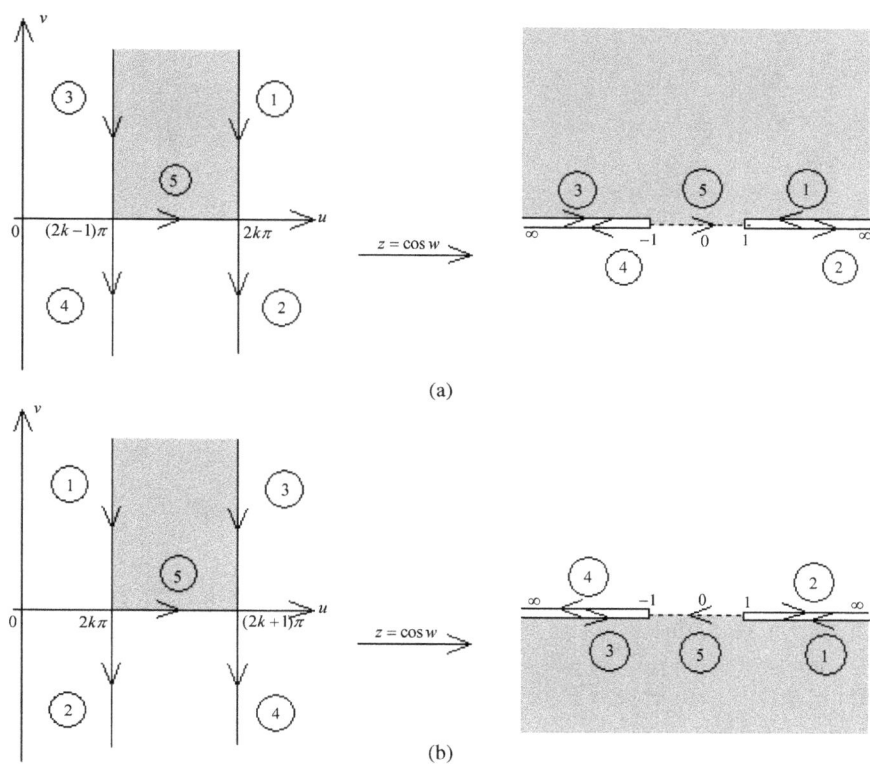

(a)

(b)

Fig. 2.84

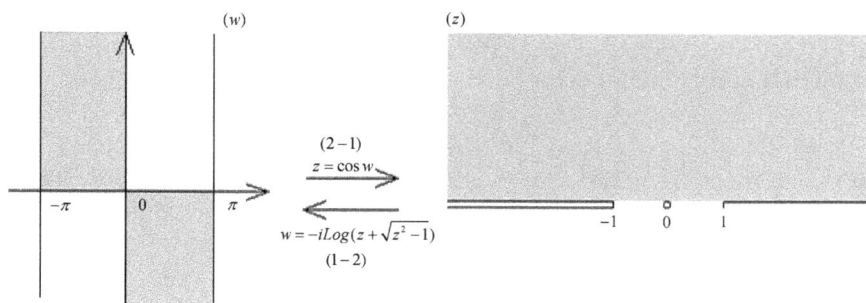

Fig. 2.85

Therefore, for each $n = 0, \pm 1, \pm 2, \ldots$, *the inverse function* $w = \cos^{-1} z$ *has a* "*branch*"

$$(\cos^{-1} z)_n = w_n = 2n\pi - i \operatorname{Log}(z + \sqrt{z^2 - 1})$$

$$\mathbf{:C^*} - [\infty, -1] - [1, \infty] \;\begin{array}{l} \nearrow \{(2n-1)\pi < \operatorname{Re} w < 2n\pi\}, \text{ if } \sqrt{-1} = -i \\[2ex] \searrow \{2n\pi < \operatorname{Re} w < (2n+1)\pi\}, \text{ if } \sqrt{-1} = i \end{array}$$

$$(2.7.4.1)$$

composed of two single-valued branches. In particular, we call

$$(\cos^{-1} z)_0 = w_0 = -i \operatorname{Log}(z + \sqrt{z^2 - 1})$$

$$\mathbf{:C^*} - [\infty, -1] - [1, \infty] \;\begin{array}{l} \nearrow \{-\pi < \operatorname{Re} w < 0\}, \text{ if } \sqrt{-1} = -i \\[2ex] \searrow \{0 < \operatorname{Re} w < \pi\}, \quad \text{ if } \sqrt{-1} = i \end{array}$$

$$(2.7.4.2)$$

the principal "*branch*" *of* $w = \cos^{-1} z$. It is one-to-two.

More precisely, $z = \cos w : D = \{w \,|\, 0 < \operatorname{Re} w < \pi; \operatorname{Re} w = 0$ and $\operatorname{Im} w \geq 0; \operatorname{Re} w = \pi$ and $\operatorname{Im} w \leq 0\} \to \mathbf{C}$ is one-to-one and onto. Then, we call the inverse function

$$\operatorname{Arc} \cos z = -i \operatorname{Log}(z + \sqrt{z^2 - 1}) : \mathbf{C} \to \mathrm{D} \qquad (2.7.4.3)$$

with $\sqrt{-1} = i$ the *positive branch* of the principal "*branch*" w_0 described above. This $\operatorname{Arc} \cos z$ maps $\mathbf{C^*} - [\infty, -1] - [1, \infty]$ *univalently* onto $0 < \operatorname{Re} w < \pi$. See (2.7.4.2). What is the *negative branch* of w_0?

The Riemann surface for $w = \cos^{-1} z$: From Fig. 2.84, it follows easily the Riemann surface for $w = \cos^{-1} z$ as shown in Fig. 2.86, where $\Omega_n = \mathbf{C^*} - [\infty, -1] - [1, \infty]$ is the image of A_n under $z = \cos w$ for $n = 0, \pm 1, \pm 2, \ldots$. The corresponding line complex is in Fig. 2.87. *There are infinitely many branch points of order 1 over $z = \pm 1$ and two logarithmic branch points over ∞.*

Section (2) $\sin^{-1} z$ and its Riemann surface

Let $z = \sin w = \cos\left(w - \frac{\pi}{2}\right)$. Also, set

$$\zeta = w - \frac{\pi}{2} \quad \text{or} \quad w = \zeta + \frac{\pi}{2}.$$

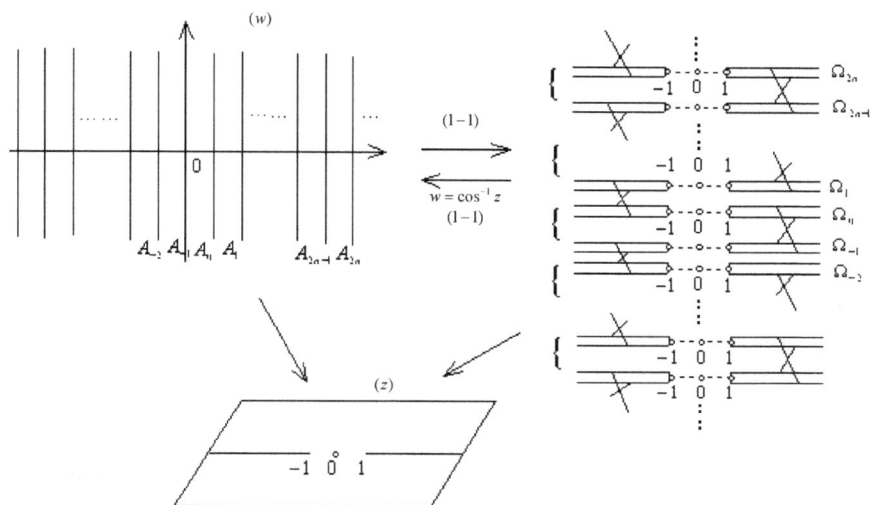

Fig. 2.86

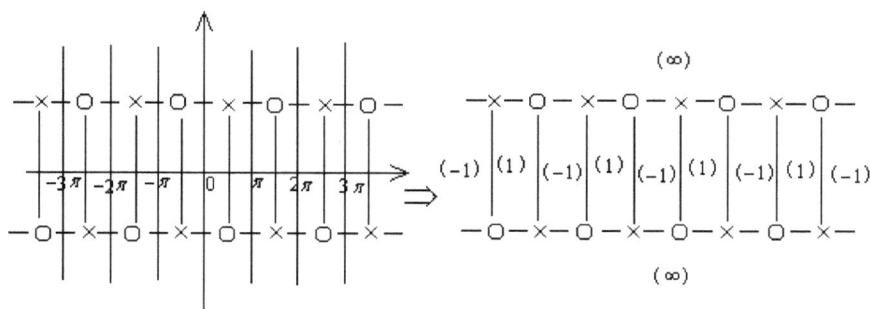

Fig. 2.87

Then $z = \sin w = \cos \zeta$. All the figures in the ζ-plane, after translating $\frac{\pi}{2}$ unit to the right, will become the corresponding figures in the w-plane. This amounts to say that $\zeta = \cos^{-1} z$ and $w = \sin^{-1} z$ have the relation

$$w = \sin^{-1} z = \frac{\pi}{2} + \cos^{-1} z = \frac{\pi}{2} - i \log(z + \sqrt{z^2 - 1}). \qquad (2.7.4.4)$$

And what we obtained for $\zeta = \cos^{-1} z$ in Section (1) can now be easily transformed to $w = \sin^{-1} z$. We list these results as

The basic mapping properties of $w = \sin^{-1} z$.

(1) $z = \sin w$ maps, for $n = 0, \pm 1, \pm 2, \ldots$,

  (i) $\operatorname{Re} w = \left(2n + \frac{1}{2}\right)\pi \to [1, \infty)$ and $\operatorname{Re} w = \left(2n - \frac{1}{2}\right)\pi \to (\infty, -1]$,
  two-to-one and onto, and

  (ii) $\left[\left(2n - \frac{1}{2}\right)\pi, \left(2n + \frac{1}{2}\right)\pi\right] \to [-1, 1]$, one-to-one.

(2) $z = \sin w$ maps, for $n = 0, \pm 1, \pm 2, \ldots$,

  (i) $\left(2n - \frac{1}{2}\right)\pi < \operatorname{Re} w < \left(2n + \frac{1}{2}\right)\pi \to \Omega = \mathbf{C}^* - [\infty, -1] - [1, \infty]$
  univalently, with $\operatorname{Im} w > 0$ onto $\operatorname{Im} z > 0$ and $\operatorname{Im} w < 0$ onto
  $\operatorname{Im} z < 0$, and

  (ii) $\left(2n + \frac{1}{2}\right)\pi < \operatorname{Re} w < \left(2n + \frac{3}{2}\right)\pi \to \Omega$ univalently, with $\operatorname{Im} w > 0$
  and $\operatorname{Im} w < 0$ onto $\operatorname{Im} z < 0$ and $\operatorname{Im} z > 0$, respectively.

  See Fig. 2.88.

(3) $z = \sin w$ maps $D = \left\{w \mid \frac{\pi}{2} < \operatorname{Re} w < \frac{3\pi}{2}; \operatorname{Re} w = \frac{\pi}{2} \text{ and } \operatorname{Im} w \geq 0; \right.$
$\left. \operatorname{Re} w = \frac{3\pi}{2} \text{ and } \operatorname{Im} w \leq 0 \right\}$ one-to-one and onto $\mathbf{C}$. The inverse function

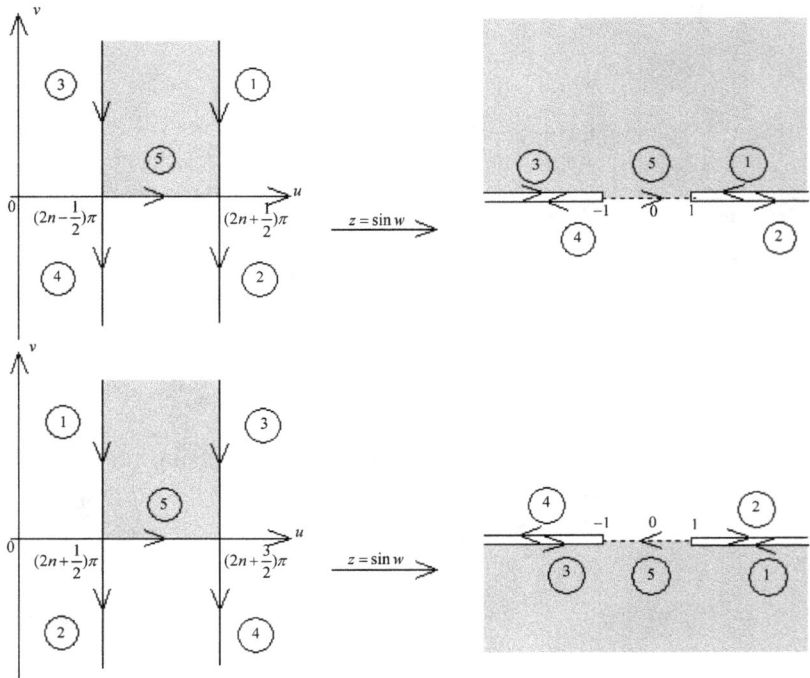

Fig. 2.88

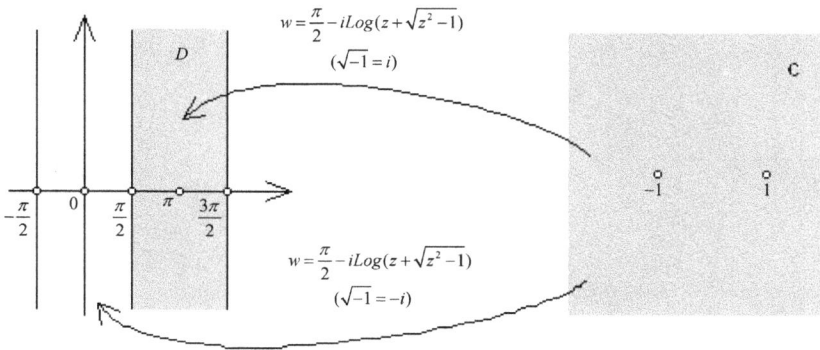

$$w = \frac{\pi}{2} - i Log(z + \sqrt{z^2 - 1})$$
$$(\sqrt{-1} = i)$$

$$w = \frac{\pi}{2} - i Log(z + \sqrt{z^2 - 1})$$
$$(\sqrt{-1} = -i)$$

Fig. 2.89

$$z = \operatorname{Arc} \sin w = \frac{\pi}{2} - i \operatorname{Log}(z + \sqrt{z^2 - 1}) : \mathbf{C} \to \mathbf{D}$$

with $\sqrt{-1} = i$, is called the *positive branch* of the principal "branch" of $w = \sin^{-1} z$; that defined by $\sqrt{-1} = -i$ is called the *negative branch*. See Fig. 2.89. In general, we call $(\sin^{-1} z)_0 = \frac{\pi}{2} - i \operatorname{Log}(z + \sqrt{z^2 - 1})$: $\Omega \to \left\{ -\frac{\pi}{2} < \operatorname{Re} w < \frac{3\pi}{2} \right\}$ the *principal "branch"* of $w = \sin^{-1} z$. It is one-to-two.

(4) $w = \sin^{-1} z$ has infinitely many "branches", for $n = 0, \pm 1, \pm 2, \ldots$,

$$(\sin^{-1} z)_n = 2n\pi - \left\{ \frac{\pi}{2} - i \operatorname{Log}(z + \sqrt{z^2 - 1}) \right\}$$

$$:\Omega \begin{cases} \left\{ \left(2n - \tfrac{1}{2}\right)\pi < \operatorname{Re} w < \left(2n + \tfrac{1}{2}\right)\pi \right\}, & \text{if } \sqrt{-1} = -i \\[2mm] \left\{ \left(2n + \tfrac{1}{2}\right)\pi < \operatorname{Re} w < \left(2n + \tfrac{3}{2}\right)\pi \right\}, & \text{if } \sqrt{-1} = i \end{cases}$$

$$(2.7.4.5)$$

The Riemann surface for $w = \sin^{-1} z$ is exactly the same as that for $w = \cos^{-1} z$. See Figs. 2.86 and 2.87.

Section (3) w = tan⁻¹ z and its Riemann surface

Decompose $z = \tan w = i \frac{1 - e^{2iw}}{1 + e^{2iw}}$ as the composite of the following elementary functions:

$$w \xrightarrow{f_1} \zeta = 2iw \xrightarrow{f_2} \eta = e^\zeta \xrightarrow{f_3} \lambda = \frac{1 - \eta}{1 + \eta} \xrightarrow{f_4} z = i\lambda.$$

See Fig. 2.90.

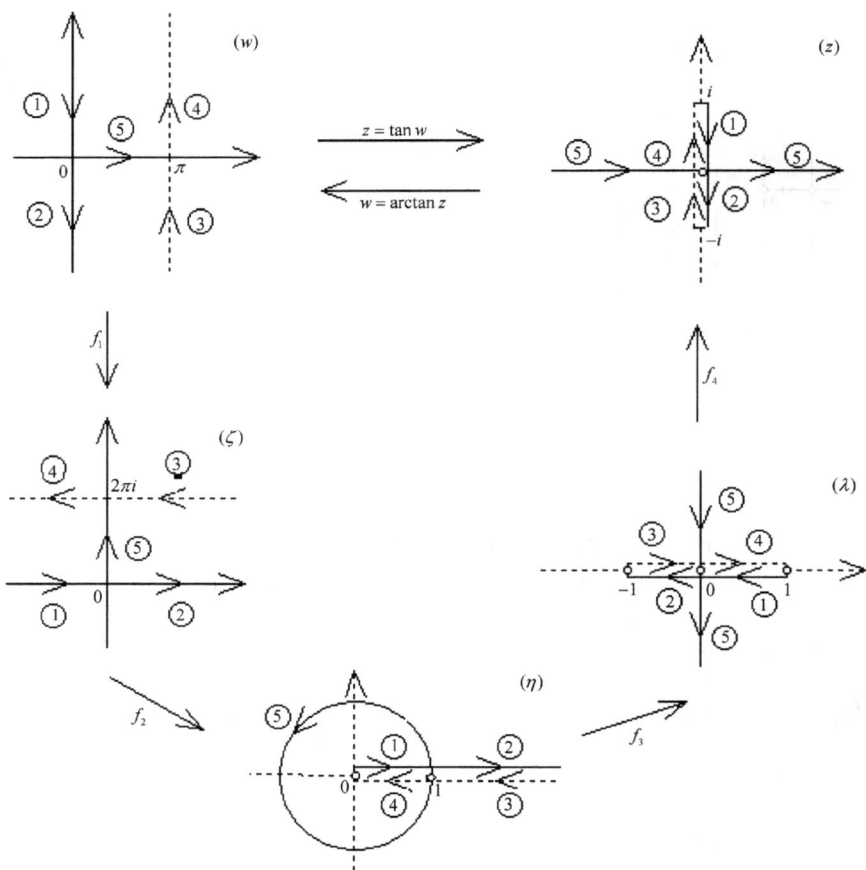

Fig. 2.90

Since $z = \tan w$ has periods $n\pi, n = \pm 1, \pm 2, \ldots,$ it maps $0 \leq \operatorname{Re} w < \pi$ one-to-one and onto $\mathbf{C}^* - \{\pm i\}$. We call the inverse function

$$\text{Arc } \tan z = \frac{-i}{2} \operatorname{Log} \frac{1 + iz}{1 - iz} : \mathbf{C}^* - \{\pm i\} \to \{0 \leq \operatorname{Re} w < \pi\}, \quad \text{or}$$

$$(2.7.4.6)$$

$$: \mathbf{C}^* - [-i, i] \to \{0 < \operatorname{Re} w < \pi\}$$

the *principal branch* of the multiple-valued function

$$w = \tan^{-1} z = \frac{-i}{2} \log \frac{1 + iz}{1 - iz}.$$

Combining the above result with (2.6.2.8) and Example 4 in Sec. 2.7.3, we have,

The basic mapping properties of $w = \tan^{-1} z$.

(1) $z = \tan w$ takes $\pm i$ at $w = \pm i\infty$. $\pm i$ are logarithmic branch points for $w = \tan^{-1} z$. There is no algebraic branch point.

(2) $z = \tan w$ maps, for $n = 0, \pm 1, \pm 2, \ldots$,

   (i) $\operatorname{Re} z = n\pi \to [-i, i]$, one-to-one and onto: $+i\infty \to i$, $n\pi \to 0$, $-i\infty \to -i$;

  (ii) the segment $[n\pi, (n+1)\pi] \to$ the real axis, one-to-one and onto (except the end points) : $\left[n\pi, \left(n + \frac{1}{2}\right)\pi\right] \to [0, +\infty]$; $\left[\left(n + \frac{1}{2}\right)\pi, (n+1)\pi\right] \to [-\infty, 0]$, and

 (iii) $\operatorname{Re} z = \left(n + \frac{1}{2}\right)\pi \to$ the imaginary axis minus $[-i, i]$, one-to-one and onto: $+i\infty \to i$, $\left(n + \frac{1}{2}\right)\pi \to i\infty$, $-i\infty \to -i$.

(3) $z = \tan w$ maps, for $n = 0, \pm 1, \pm 2, \ldots$,

$$n\pi < \operatorname{Re} w < (n+1)\pi \to \Omega = \mathbf{C}^* - [-i, i] \text{ univalently:}$$
$$\operatorname{Im} w > 0 \to \{\operatorname{Im} z > 0\} - [0, i],$$
$$\operatorname{Im} w < 0 \to \{\operatorname{Im} w < 0\} - [-i, 0], \quad \text{and}$$
$$n\pi < \operatorname{Re} w < \left(n + \frac{1}{2}\right)\pi \to \operatorname{Re} z > 0,$$

$$\left(n + \frac{1}{2}\right)\pi < \operatorname{Re} w < (n+1)\pi \to \operatorname{Re} z < 0.$$

Observe that $z(x) = \tan x$ is real if x is real. See Fig. 2.91.

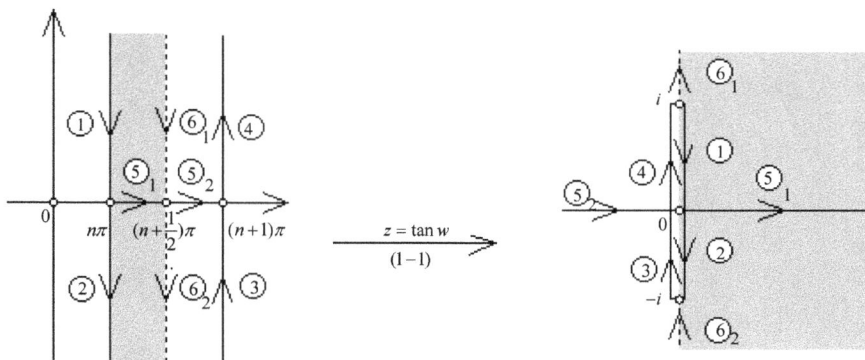

Fig. 2.91

(4) $w = \tan^{-1} z$ has infinitely many branches, for $n = 0, \pm 1, \pm 2, \ldots,$

$$(\tan^{-1} z)_n = n\pi + \operatorname{Arc} \tan w = n\pi - \frac{i}{2} \operatorname{Log} \frac{1 + iz}{1 - iz} : \Omega$$

$$= \mathbf{C}^* - [-i, i] \to \{n\pi < \operatorname{Re} w < (n+1)\pi\}$$

where $(\tan^{-1} z)_0 = \operatorname{Arc} \tan w$ is the *principal branch* $((2.7.4.6))$.

Note that, instead, one can use $[+i, +i\infty] \cup [-i, -i\infty]$ as the branch cut.   (2.7.4.7)

The Riemann surface for $w = \tan^{-1} z$: To each domain $D_n : n\pi < \operatorname{Re} w < (n+1)\pi$, $n = 0, \pm 1, \pm 2, \ldots$, take a copy of $\mathbf{C}^* - [-i, i]$ and denote it as Ω_n. Then, it follows from Fig. 2.91 the desired Riemann surface R_z for $w = \tan^{-1} z$. See Fig. 2.92.

Use "0" and "x" to denote the right and the left half-plane, respectively. This surface has the line complex as shown in Fig. 2.93.

Note that $\cot w = -\tan(w - \frac{\pi}{2}) = \tan(\frac{\pi}{2} - w)$.

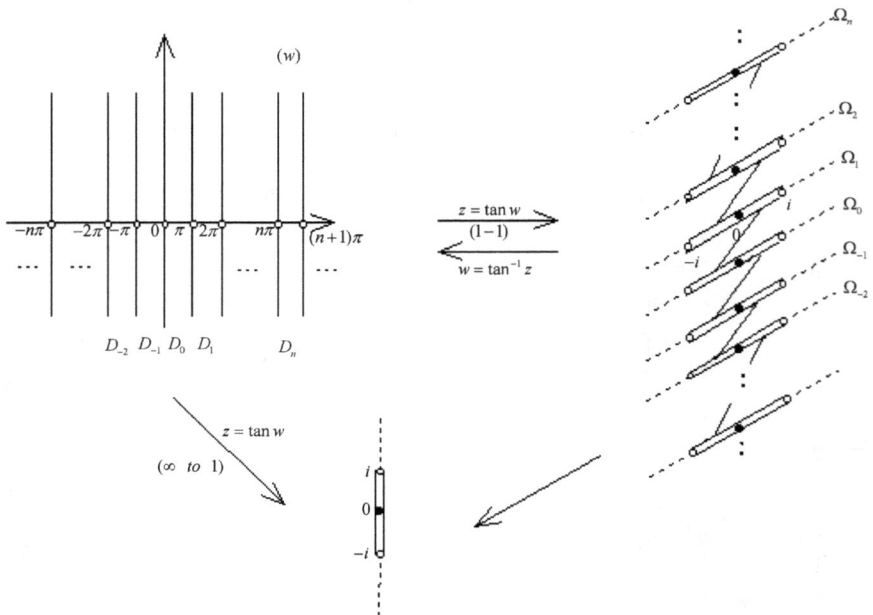

Fig. 2.92

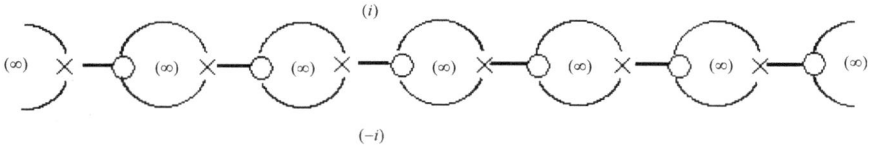

Fig. 2.93

Hence,

$$\cot^{-1} z = \frac{\pi}{2} - \tan^{-1} z. \tag{2.7.4.8}$$

Just like $\sin^{-1} z$ to $\cos^{-1} z$, it is easy to transform results in (2.7.4.7) to corresponding results for $w = \cot^{-1} z$. The details are left to the readers.

Section (4) Some further examples

Example 1. Show that $w = \cos \sqrt{z}$ is a single-valued continuous function on **C**, and $w = \sin \sqrt{z}$ is a double-valued function on $\Omega = \mathbf{C} - \{n^2\pi^2 \mid n = 0, 1, 2, \ldots\}$. Find the single-valued domains for $w = \sin \sqrt{z}$.

Solution. $\zeta = \sqrt{z}$ is double-valued on $\mathbf{C} - \{0\}$ and has two branches on $\mathbf{C}^* - [0, \infty] : \zeta_1 = \sqrt{z}$ with $\sqrt{-1} = i$ and $\zeta_2 = \sqrt{z}$ with $\sqrt{-1} = -i$.

Now, $\cos \sqrt{0} = \cos 0 = 1$ and $\cos \zeta_1 = \cos \zeta_2 = \cos \sqrt{z}$ for $z \in \mathbf{C} - \{0\}$. Thus, $w = \cos \sqrt{z}$ is single-valued and continuous on **C**. In Chap. 3, we can show that $w = \cos \sqrt{z}$ is analytic everywhere on **C**.

As for $w = \sin \sqrt{z}$, $\sin \zeta_1 \neq \sin \zeta_2$ holds for $z \in \Omega$. Hence, $w = \sin \sqrt{z}$ is double-valued on Ω. On $\mathbf{C}^* - [0, \infty]$, it has two branches $w_1 = \sin \zeta_1$ and $w_2 = \sin \zeta_2$.

Example 2. Find single-valued domains for $w = \frac{\sin \sqrt{iz+1}}{z^2+1}$.

Solution. $\sqrt{iz+1}$ has branches if $iz + 1$ does not lie on the set $[\infty, 0]$, namely, z does not lie on the set $[i, +i\infty]$. Hence, $\sqrt{iz+1}$ has two branches on $\mathbf{C}^* - [i, +i\infty]$.

As a consequence, $w = w(z)$ has two branches on $\mathbf{C}^* - [i, +i\infty]$.

Example 3. Find the single-valued domains for $w = \sqrt{\tan z}$ and construct its Riemann surface.

Solution. Let $\zeta = \tan z$. Then $w = \sqrt{\zeta}$.

It is known that $\zeta = \tan z$ is continuous on $\mathbf{C} - \{(n + \frac{1}{2})\pi \,|\, n = 0, \pm 1, \pm 2, \dots\}$. Since $w = \sqrt{\zeta}$ has two branches on ζ-plane excluding $[\infty, 0]$, we try to find points on the z-plane whose images, under $\zeta = \tan z$, will lie on the negative real axis $[\infty, 0]$. According to Fig. 2.91, $\zeta = \tan z$ maps each interval $\left[(n + \frac{1}{2})\pi, (n + 1)\pi\right]$ one-to-one and onto $[\infty, 0]$. Hence, let

$$\Omega = \mathbf{C} - \bigcup_{n=-\infty}^{\infty} \left[\left(n + \frac{1}{2}\right)\pi, (n + 1)\pi\right].$$

See Fig. 2.94.

On Ω, $w = \sqrt{\tan z}$ has two branches:

$$w_1 = +\sqrt{\tan z} \left(\sqrt{\tan \frac{\pi}{4}} = \sqrt{1} = 1\right)$$

and

$$w_2 = -\sqrt{\tan z} \left(\sqrt{\tan \frac{\pi}{4}} = \sqrt{1} = -1\right).$$

Take two copies of Ω and paste them together crosswise along the branch cuts $\left[(n + \frac{1}{2})\pi, (n + 1)\pi\right]$, $n = 0, \pm 1, \dots$. The resulted Riemann surface for $w = \sqrt{\tan z}$ is shown in Fig. 2.95. There is an algebraic branch point of order 1 over each of the points $z = 0$, $\pm\frac{\pi}{2}$, $\pm\pi$, $\pm\frac{3\pi}{2}, \dots$, and ∞ is the limit point of these branch points.

Exercises A

(1) Try to use Example 3 in Sec. 2.7.3 to give a quick description of mapping properties and Riemann surface for $w = \cos^{-1} z$.

(2) Decompose $z = \cos w$ as $w \to \zeta = iw \to \eta = e^{\zeta} \to z = \frac{1}{2}(\eta + \frac{1}{\eta})$ and try to redo Section (1) and, eventually, formulate mapping properties similar to (2.7.4.5) for $w = \cos^{-1} z$.

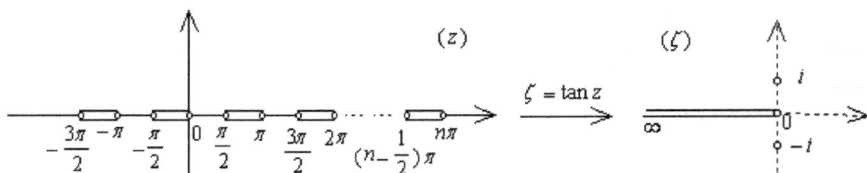

Fig. 2.94

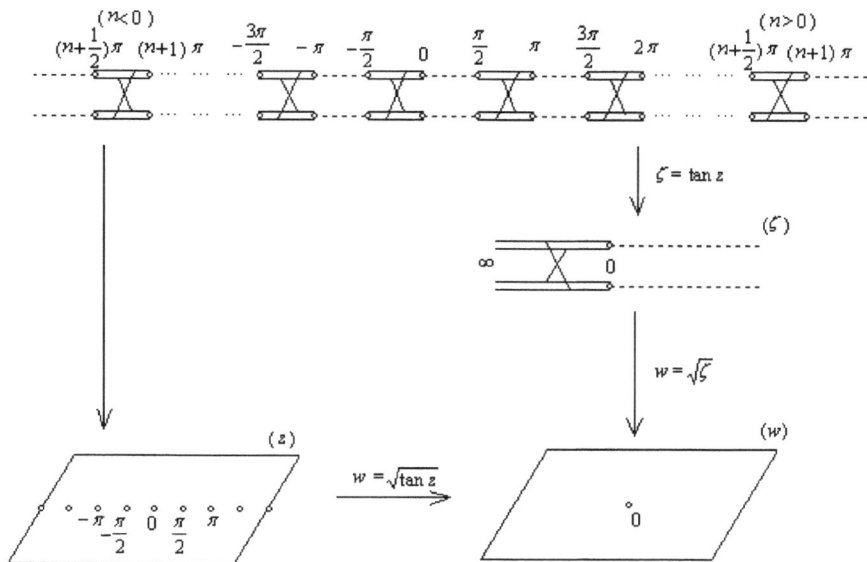

Fig. 2.95

(3) Decompose $z = \sin w$ as $w \to \zeta = iw \to \eta = e^{\zeta} \to \lambda = -i\eta \to z = \frac{1}{2}(\lambda + \frac{1}{\lambda})$ and try to find the image of $-\frac{\pi}{2} < \operatorname{Re} w < \frac{\pi}{2}$ stepwise under $z = \sin w$. Use this to reprove (2.7.4.5).

(4) Prove (2.7.4.7) in detail.

(5) Try to use (2.7.4.8) to formulate mapping properties from (2.7.4.7) for $w = \cot^{-1} z$. Then, use the decomposition $w \to \zeta = 2iw \to \eta = e^{\zeta} \to \lambda = \frac{\eta+1}{\eta-1} \to z = i\lambda$ to justify these results. Construct Riemann surface for $w = \cot^{-1} z$.

(6) Try to formulate $\sec^{-1} z$ and $\csc^{-1} z$ and construct their Riemann surfaces.

(7) Based on Examples 3 and 4 of Sec. 2.7.3, try to describe mapping properties and their Riemann surfaces for inverse hyperbolic functions (see (2.7.1)).

Exercises B

(1) For each of the following functions $w = w(z)$, do the following problems.

   (i) Find its single-valued domains and branches.
   (ii) Determine the ranges of branches.

(iii) Construct the Riemann surface for $w = w(z)$, including its branch points, orders, and branch cuts.

(a) $w = \sqrt{\cos z}$.   (b) $w = \cos(\sqrt{e^z - 1})$.   (c) $w = \sqrt{\sin z}$.

(d) $w = \csc\sqrt{z}$.   (e) $w = \sin^{-1}(\sqrt{1 - z^2})$.   (f) $w = z\sin^{-1} z^2$.

(g) $w = \frac{1}{\sqrt{z}}e^{\tan z}$.   (h) $w = e^{\tan^{-1}(i\sqrt{z})}$.

(i) $w = \sqrt{\cot z}$.   (j) $w = \cot^{-1}(\sqrt{z})$.

2.8.   Differentiation in Complex Notation

Suppose $f = f(x, y) : O$ (open in $\mathbf{R}^2$) $\to \mathbf{R}^2$ is a function. f is said to be *differentiable* at a point $(x_0, y_0) \in O$ if there exists a linear transformation $df_{(x_0, y_0)} : \mathbf{R}^2 \to \mathbf{R}^2$ so that, for all (h, k) with $|(h, k)| = \sqrt{h^2 + k^2} > 0$ sufficiently small,

$$f(x_0 + h, y_0 + k) = f(x_0, y_0) + df_{(x_0, y_0)}(h, k) + o((h, k)), \quad \text{and}$$

$$\lim_{|(h,k)| \to 0} \frac{o((h, k))}{|(h, k)|} = 0. \tag{2.8.1}$$

$df_{(x_0, y_0)}$ is called the *total differential* of f at (x_0, y_0). Suppose $f(x, y) = (u, v)$. In terms of the natural coordinate system for $\mathbf{R}^2$, $df_{(x_0, y_0)}$ has the matrix representation

$$[df_{(x_0, y_0)}] = \begin{bmatrix} \dfrac{\partial u}{\partial x} & \dfrac{\partial v}{\partial x} \\[2mm] \dfrac{\partial u}{\partial y} & \dfrac{\partial v}{\partial y} \end{bmatrix} (x_0, y_0), \quad \text{where} \dfrac{\partial u}{\partial x}(x_0, y_0) = u_x(x_0, y_0), \text{ etc.}$$

called the *Jacobian matrix* of f at (x_0, y_0).

In complex notation, let $z_0 = x_0 + iy_0, z = (x_0 + h) + i(y_0 + k)$, and $w = u + iv = f(z)$. Then, (2.8.1) can be rewritten in *complex notation* as

$$f(z) = f(z_0) + \left[\frac{1}{2}((z - z_0) + \overline{(z - z_0)}) \frac{1}{2i}((z - z_0) - \overline{(z - z_0)}) \right]$$

$$\begin{bmatrix} \dfrac{\partial u}{\partial x} & \dfrac{\partial v}{\partial x} \\[2mm] \dfrac{\partial u}{\partial y} & \dfrac{\partial v}{\partial y} \end{bmatrix} + o(z - z_0)$$

$$= f(z_0) + f_z(z_0)(z - z_0) + f_{\bar{z}}(z_0)(\bar{z} - \bar{z}_0) + o(z - z_0), \tag{2.8.2}$$

with the total differential

$$df_{z_0}(z - z_0) = f_z(z_0)(z - z_0) + f_{\bar{z}}(z_0)(\bar{z} - \bar{z}_0),$$

where

$$f_x(z_0) = u_x(z_0) + iv_x(z_0), f_y(z_0) = u_y(z_0) + iv_y(z_0),$$

and

$$f_z(z_0) = \frac{1}{2}[f_x(z_0) - if_y(z_0)],$$

also denoted as $\dfrac{\partial f}{\partial z}(z_0)$ or $\partial_z f(z_0)$ or $\partial f(z_0)$,

$$f_{\bar{z}}(z_0) = \frac{1}{2}[f_x(z_0) + if_y(z_0)],$$

also denoted as $\dfrac{\partial f}{\partial \bar{z}}(z_0)$ or $\partial_{\bar{z}} f(z_0)$ or $\bar{\partial} f(z_0)$,   (2.8.3)

are called the *complex derivatives* of f at z_0. These two complex derivatives can be easily and formally understood in the following sense: Treat $f(x, y) = f\left(\frac{1}{2}(z + \bar{z}), -\frac{i}{2}(z - \bar{z})\right)$ as a function of x and y, which, in turn, are functions of z and $\bar{z}$. Then, use chain rule for differentiation to obtain (2.8.3). See the Remark below.

The *directional derivatives* of f at z_0 in the *direction* θ is

$$D_\theta f(z_0) = \lim_{r \to 0} \frac{f(z_0 + re^{i\theta}) - f(z_0)}{re^{i\theta}}$$

(note that $e^{i\theta}$ appears in the denominator),

$$= e^{-i\theta}[f_x(z_0)\cos\theta + f_y(z_0)\sin\theta] \text{ (namely, } e^{-i\theta} df_{z_0}(\cos\theta, \sin\theta))$$

$$= e^{-i\theta}(\cos\theta \ \sin\theta)\begin{bmatrix} u_x(z_0) & v_x(z_0) \\ u_y(z_0) & v_y(z_0) \end{bmatrix}$$

$$= f_z(z_0) + f_{\bar{z}}(z_0)e^{-2i\theta}.$$ (2.8.4)

We give an example.

Example. Justify that

$$\frac{\partial}{\partial z}z = 1, \quad \frac{\partial}{\partial z}\bar{z} = 0 \quad \text{and} \quad \frac{\partial}{\partial \bar{z}}z = 0, \quad \frac{\partial}{\partial \bar{z}}\bar{z} = 1.$$

Then, compute $\frac{\partial f}{\partial z}$ and $\frac{\partial f}{\partial \bar{z}}$ and $D_\theta f(z)$ for each of the following functions $f(x, y)$ and determine if it is differentiable in case (3).

(1) $z^k \bar{z}^l$ (k and l are positive integers);
(2) $x^2 y - xy^2 + xyi$;
(3) $\dfrac{\bar{z}}{z}$ ($z \neq 0$).

Solution. Let $z = x + iy$. Then $z_x = 1$ and $z_y = i$. Hence

$$\frac{\partial}{\partial z} z = \frac{1}{2}(z_x - iz_y) = \frac{1}{2}(1 + 1) = 1 \quad \text{and}$$

$$\frac{\partial}{\partial \bar{z}} z = \frac{1}{2}(z_x + iz_y) = \frac{1}{2}(1 - 1) = 0.$$

While, for $\bar{z} = x - iy$, then $\bar{z}_x = 1$ and $\bar{z}_y = -i$ will induce that $\frac{\partial}{\partial z}\bar{z} = 0$ and $\frac{\partial}{\partial \bar{z}}\bar{z} = 1$.

(1) By computing f_x and f_y, it follows easily from (2.8.3) that

$$\frac{\partial}{\partial z} z^k \bar{z}^l = k z^{k-1} \bar{z}^l \quad \text{and} \quad \frac{\partial}{\partial \bar{z}} z^k \bar{z}^l = l z^k \bar{z}^{l-1}.$$

But it is advised to use directly the established differentiation rules mentioned below (refer to (2.8.6)).

(2) $f_x = 2xy - y^2 + yi, f_y = x^2 - 2xy + xi$. Hence

$$\frac{\partial f}{\partial z} = \frac{1}{2}[2xy - y^2 + yi - i(x^2 - 2xy + xi)]$$

$$= \frac{1}{2}[2xy - y^2 + x + i(-x^2 + 2xy + y)],$$

$$\frac{\partial f}{\partial \bar{z}} = \frac{1}{2}[2xy - y^2 + yi + i(x^2 - 2xy + xi)]$$

$$= \frac{1}{2}[2xy - y^2 - x + i(x^2 - 2xy + y)].$$

Readers are urged to express $f(x,y)$ as $f\left(\frac{1}{2}(z + \bar{z}), \frac{-i}{2}(z - \bar{z})\right)$ and then differentiate partially with respect to z and $\bar{z}$, respectively.

(3) $f_z(z) = -\frac{\bar{z}}{z^2}$ and $f_{\bar{z}}(z) = \frac{1}{z}$, where $z \neq 0$. Now, for $z_0 \neq 0$,

$$o(z - z_0) = f(z) - f(z_0) - f_z(z_0)(z - z_0) - f_{\bar{z}}(z_0)(\bar{z} - \bar{z}_0),$$

$$= \frac{\bar{z}}{z} - \frac{\bar{z}_0}{z_0} + \frac{\bar{z}_0}{z_0^2}(z - z_0) - \frac{1}{z_0}(\bar{z} - \bar{z}_0) = \frac{(z - z_0)(z\bar{z}_0 - \bar{z}z_0)}{zz_0^2},$$

$$\Rightarrow \lim_{z \to z_0} \frac{o(z - z_0)}{z - z_0} = \lim_{z \to z_0} \frac{z\bar{z}_0 - \bar{z}z_0}{zz_0^2} = 0.$$

So, $f(z) = \frac{\bar{z}}{z}$ is differentiable at $z_0 \neq 0$.

Remark. The complex differential operators $\frac{\partial}{\partial z}, \frac{\partial}{\partial \bar{z}}$ and the complex exterior derivations $\wedge, d$.

Let O be a nonempty open set in **C** (or **R**^{2}). Recall (from real analysis) that a function $u : O \to \mathbf{R}$ is called *k times continuously differentiable* on

O if all its partial derivatives up to and including order k exist and are continuous on O.

A complex-valued function $f = u + iv : O \rightarrow \mathbf{C}$ is called k *times contin-uously differentiable* on O if both u and v are such functions.

Let, for nonnegative integer k or ∞,

$$C^k(O) = \{k \text{ times continuously differentiable}$$
$$\text{complex-valued function on } \Omega \}. \qquad (2.8.5)$$

$C^k(O)$ is a *linear space* over $\mathbf{C}$ via usual additive and scalar multiplication of functions. In particular, a function in $C(O) = C^0(O)$ is just a continuous function.

It is quite straightforward to show the following properties for $\frac{\partial}{\partial z}$ and $\frac{\partial}{\partial \bar{z}}$: For $f, g \in C^1(O)$ and $a, b \in \mathbf{C}$,

(1) *Linearity*:

$$\frac{\partial}{\partial z}(af + bg) = a\frac{\partial f}{\partial z} + b\frac{\partial g}{\partial z}; \quad \frac{\partial}{\partial \bar{z}}(af + bg) = a\frac{\partial f}{\partial \bar{z}} + b\frac{\partial g}{\partial \bar{z}}.$$

(2) *Leibniz rules*:

$$\frac{\partial}{\partial z}(fg) = \frac{\partial f}{\partial z}g + f\frac{\partial g}{\partial z}; \quad \frac{\partial}{\partial \bar{z}}(fg) = \frac{\partial f}{\partial \bar{z}}g + f\frac{\partial g}{\partial \bar{z}}. \qquad (2.8.6)$$

Hence, we call, if $z = x + iy$ and $k \geq 1$,

$$\partial_z = \frac{\partial}{\partial z} = \frac{1}{2}\left(\frac{\partial}{\partial x} - i\frac{\partial}{\partial y}\right) : C^k(O) \rightarrow C^{k-1}(O);$$

$$\partial_{\bar{z}} = \frac{\partial}{\partial \bar{z}} = \frac{1}{2}\left(\frac{\partial}{\partial x} + i\frac{\partial}{\partial y}\right) : C^k(O) \rightarrow C^{k-1}(O); \qquad (2.8.7)$$

the *complex differential (linear) operators*. They will play a fundamental role in complex analysis from the standpoint of (2.8.2) and, in particular, in complex analysis of several complex variables. For further operational properties of ∂_z and $\partial_{\bar{z}}$, see Exercise B (1).

To give the differential $dz = dx + idy$, where $z = x + iy$, a possi-ble geometric interpretation, let us consider a differentiable curve γ with parametric representation $z(t) = x(t) + iy(t)$. Then, its tangent vector at t is

$z'(t) = x'(t) + iy'(t),$

$\Rightarrow$ (multiply both sides by dt, an increment in t)

$z'(t)dt = x'(t)dt + iy'(t)dt,$

$\Rightarrow$ (adopt differential notations for

$dx(t) = x'(t)dt$ and $dy(t) = y'(t)dt$)

$dz(t) \underset{\text{(def.)}}{=} z'(t)dt = dx(t) + idy(t),$

$\Leftrightarrow dz = dx + idy,$ and

$|dz| = \sqrt{dx^2 + dy^2},$ *the arc length element* (see (2.4.10)).

$$(2.8.8)$$

Similarly,

$$d\bar{z} = dx - idy = \overline{dz}. \tag{2.8.9}$$

As independent *real* variables x and y, we define the *exterior product* $dx \wedge dy$ of dx and dy according to the following rules:

(1) *Bilinearity:* for $a_1, a_2 \in \mathbf{R}$,

$$(a_1 dx_1 + a_2 dx_2) \wedge dy = a_1 dx_1 \wedge dy + a_2 dx_2 \wedge dy;$$

$$dx \wedge (a_1 dy_1 + a_2 dy_2) = a_1 dx \wedge dy_1 + a_2 dx \wedge dy_2.$$

(2) *Skew-symmetric or alternating:*

$$dx \wedge dy = -dy \wedge dx.$$

In particular,

$$dx \wedge dx = dy \wedge dy = 0. \tag{2.8.10}$$

Geometrically, $dx \wedge dy$ can be interpreted as the *oriented* area element in the xy-plane, as comparing to the conventional *absolute* area element $dxdy$ which always means positive in quantity. Some terminologies are as follows:

fundamental differential form of *order* 1: dx, dy;
fundamental differential form of *order* 2: $dx \wedge dy$; and
for *real-valued* functions $P, Q \in C^k(O), k \geq 0$.
differential form of *order* 0: $P, Q, \ldots$ (as function on O);
differential form of *order* 1: $Pdx + Qdy$;
differential form of *order* 2: $Pdx \wedge dy$.

$$(2.8.11)$$

They are subjected to the following operations:

(1) *Exterior product*: For $R, P_1, \ldots$ in $C^k(\mathrm{O})$,
$$R(Pdx + Qdy) = RPdx + RQdy;$$
$$(P_1 dx + Q_1 dy) \wedge (P_2 dx + Q_2 dy) = (P_1 Q_2 - P_2 Q_1)dx \wedge dy$$
(according to (2.8.10)).

(2) *Exterior differentiation*: For $P, Q, \ldots$ in $C^k(\mathrm{O}), k \geq 1$,
$$dP = P_x dx + P_y dy;$$
$$d(Pdx + Qdy) = dP \wedge dx + dQ \wedge dy = (Q_x - P_y)dx \wedge dy. \qquad (2.8.12)$$

Note that $d^2 = d \circ d = 0$ always holds.

Now, (2.8.10)–(2.8.12) can be extended correspondingly to complex variables z and $\bar{z}$. Let us go back to (2.8.8) and (2.8.9). We have

fundamental differential form of order 1: $dz, d\bar{z}$;

fundamental differential form of order 2: $dz \wedge d\bar{z}$, $\qquad (2.8.13)$

subject to the bilinearity and skew-symmetry as stated in (2.8.10) with $a_1, a_2 \in \mathbf{C}$. In particular,

$$dz \wedge dz = d\bar{z} \wedge d\bar{z} = 0;$$
$$\begin{aligned} dz \wedge d\bar{z} &= (dx + idy) \wedge (dx - idy) \\ &= dx \wedge dx + idy \wedge dx - idx \wedge dy + dy \wedge dy \text{ (see (2.8.12))} \\ &= -2idx \wedge dy \\ &= -d\bar{z} \wedge dz. \end{aligned} \qquad (2.8.14)$$

Now, for *complex-valued* $f, g \in C^k(\mathrm{O})$, (2.8.11) and exterior product in (2.8.12) are still valid in complex case. While, exterior differentiations in (2.8.12) are:

$$df = f_x dx + f_y dy \text{ (treat } f(z) \text{ as } f(x,y) \text{ and use (2.8.12)),}$$
$$= f_z dz + f_{\bar{z}} d\bar{z} \left(\text{treat } f(z) \text{ as } f\left(\frac{1}{2}(z + \bar{z}), -\frac{i}{2}(z - \bar{z}) \right) \right.$$
$$\text{and use (2.8.7)} \bigg),$$
$$= \frac{\partial f}{\partial z} dz + \frac{\partial f}{\partial \bar{z}} d\bar{z};$$
$$d(fdz + gd\bar{z}) = \left(\frac{\partial g}{\partial z} - \frac{\partial f}{\partial \bar{z}} \right) dz \wedge d\bar{z}. \qquad (2.8.15)$$

Exercises A

(1) Prove (2.8.6) in detail.

(2) Show that $p(z, \bar{z}) = \sum_{k,l=1}^{m,n} a_{kl} z^k \bar{z}^l$ is a polynomial in z only, namely, $a_{kl} = 0$ for all $l \geq 1$, if and only if $\frac{\partial p}{\partial \bar{z}} = 0$.

Note: If writing a complex polynomial $p(z) = \sum a_k z^k$ as a polynomial in x and y, namely, $p(x, y)$, then it follows that $\frac{\partial p}{\partial \bar{z}} = 0$ always holds. But it is no more true for any polynomial $p(x, y)$ in x and y with complex coefficients as the Example shows.

(3) Show that (2.8.2) is equivalent to the following expression.

$$\Delta w = f_z(z_0)\Delta z + f_{\bar{z}}(z_0)\overline{\Delta z} + \varepsilon_1 \Delta z + \varepsilon_2 \overline{\Delta z},$$

where $\Delta w = f(z) - f(z_0)$, $\Delta z = z - z_0$, $\overline{\Delta z} = \bar{z} - \bar{z}_0$ and $\varepsilon_1 = \frac{o(\Delta z)}{|\Delta z|^2}\overline{\Delta z}, \varepsilon_2 = \frac{o(\Delta z)}{|\Delta z|^2}\Delta z$ which approach zero if $\Delta z \to 0$.

(4) f is said to be *differentiable in the complex sense* at a point $z_0 \in O$ (open) if

$$\lim_{z \to z_0} \frac{f(z) - f(z_0)}{z - z_0} = \lim_{\Delta z \to 0} \frac{\Delta w}{\Delta z}$$

exists as a finite complex number, denoted as $\frac{df}{dz}(z_0)$ or $f'(z_0)$, and called the *derivative* of f at z_0. Let $f(z) = u(z) + iv(z)$. Show that

    (i) f is differentiable at z_0 in the complex sense (for details, see Sec. 3.2).

$\Leftrightarrow$ (ii) $\frac{\partial f}{\partial \bar{z}}(z_0) = \partial_{\bar{z}} f(z_0) = 0$ and $\frac{\partial f}{\partial z}(z_0) = \partial_z f(z_0) = f'(z_0)$.

$\Leftrightarrow$ (iii) (Cauchy–Riemann equations) $u_x(z_0) = v_y(z_0)$, $u_y(z_0) = -v_x(z_0)$, namely, $\partial_z u(z_0) = i\partial_z v(z_0)$ or $\partial_{\bar{z}} u(z_0) = -i\partial_{\bar{z}} v(z_0)$.

$\Rightarrow$ (iv) $\partial_{\bar{z}}\bar{f}(z_0) = \overline{f'(z_0)}$.

(5) A complex-valued function $f \in C^1(O)$ is called *analytic* on O if

$$\frac{\partial f}{\partial \bar{z}} = 0$$

at every point of O (for details, see Sec. 3.3) and then, $f'(z) = \frac{\partial f}{\partial z}(z)$, its *derivative* on O.

    (a) Show that e^z is analytic on $\mathbf{C}$ and $(e^z)' = e^z$.

    (b) Show that both $\cos z$ and $\sin z$ are analytic on $\mathbf{C}$ with $(\cos z)' = -\sin z$ and $(\sin z)' = \cos z$.

    (c) Show that $\tan z$ is analytic at points where $\cos z \neq 0$. What is $(\tan z)'$?

    (d) How about $\cosh z, \sinh z$ and $\tanh z$?

(6) Suppose $f(z) = u(r, \theta) + iv(r, \theta)$, where $z = re^{i\theta}$. Show that, in polar coordinate (r, θ),

$$\partial_z = \frac{\partial}{\partial z} = \frac{1}{2} e^{-i\theta} \left(\frac{\partial}{\partial r} - \frac{i}{r} \frac{\partial}{\partial \theta} \right),$$

$$\partial_{\bar{z}} = \frac{\partial}{\partial \bar{z}} = \frac{1}{2} e^{i\theta} \left(\frac{\partial}{\partial r} + \frac{i}{r} \frac{\partial}{\partial \theta} \right).$$

(7) Suppose $w = f(z)$ has the inverse function $z = f^{-1}(w)$. Show that

$$dz = \frac{\bar{w}_{\bar{z}}}{|w_z|^2 - |w_{\bar{z}}|^2} dw + \frac{-w_z}{|w_z|^2 + |w_{\bar{z}}|^2} d\bar{w}.$$

Exercises B

(1) *Fundamental properties* of ∂_z and $\partial_{\bar{z}}$. Except those stated in (2.8.6) and Exercises A (4), (6), ∂_z and $\partial_{\bar{z}}$ have the following operational properties.

(a) In case $g(z, \bar{z}) \neq 0$,

$$\partial_z \left(\frac{f}{g} \right) = \frac{g\partial_z f - f\partial_z g}{g^2}, \quad \partial_{\bar{z}} = \frac{g\partial_{\bar{z}} f - f\partial_{\bar{z}} g}{g^2}.$$

(b) *Composite operation with analytic function* (see Exercise A (5)). Let $\Phi(\zeta, \eta)$ be an analytic function of ζ and η (namely, for each fixed η, $\Phi(\cdot, \eta)$ is analytic in ζ and so is $\Phi(\zeta, \cdot)$) and $\zeta = f(z), \eta = g(z)$. Then,

$$\partial_z \Phi(\zeta, \eta) = \frac{\partial \Phi}{\partial \zeta} \partial_z \zeta + \frac{\partial \Phi}{\partial \eta} \partial_z \eta; \quad \partial_{\bar{z}} \Phi(\zeta, \eta) = \frac{\partial \Phi}{\partial \zeta} \partial_{\bar{z}} \zeta + \frac{\partial \Phi}{\partial \eta} \partial_{\bar{z}} \eta.$$

In particular, in case $\zeta = f(z) = z$ and $\eta = g(z) = \bar{z}$, then

$$\partial_z \Phi(z, \bar{z}) = \frac{\partial \Phi}{\partial z}; \quad \partial_{\bar{z}} \Phi(z, \bar{z}) = \frac{\partial \Phi}{\partial \bar{z}}.$$

(c) *Composite operation*. Let $w = f(z)$ and $\zeta = g(w)$. Then

$$\partial_z (g \circ f) = \partial_w g \cdot \partial_z f + \partial_{\bar{w}} g \cdot \partial_z \bar{f}; \quad \partial_{\bar{z}} (g \circ f) = \partial_w g \cdot \partial_{\bar{z}} f + \partial_{\bar{w}} g \cdot \partial_{\bar{z}} \bar{f}$$

where $\bar{f}(z) = \overline{f(z)}$. In case f is analytic (then $\partial_{\bar{z}} f = 0$ and $\partial_z f = f'$), then

$$\partial_z (g \circ f) = \partial_w g \cdot f'; \quad \partial_{\bar{z}} (g \circ f) = \partial_{\bar{w}} g \cdot \bar{f}'.$$

In case g is analytic (then $\partial_{\bar{w}} g = 0$ and $\partial_w g = g'$), thus

$$\partial_z (g \circ f) = g' \partial_z f; \quad \partial_{\bar{z}} (g \circ f) = g' \partial_{\bar{z}} f.$$

(d) Let $w = f(z) = u(z) + iv(z)$. Then

$$\partial_z f = \frac{1}{2}[(u_x + v_y) + i(-u_y + v_x)];$$

$$\partial_{\bar{z}} f = \frac{1}{2}[(u_x - v_y) + i(u_y + v_x)].$$

The *Jacobian determinant* of f at $z = x + iy$ is

$$J_f(z) = \det df_z(z) = \frac{\partial(u, v)}{\partial(x, y)} = \begin{vmatrix} u_x(z) & v_x(z) \\ u_y(z) & v_y(z) \end{vmatrix}$$

$$= |\partial_z f(z)|^2 - |\partial_{\bar{z}} f(z)|^2.$$

Also (see (2.8.4)),

$$\max_\theta |D_\theta f(z)| = |\partial_z f(z)| + |\partial_{\bar{z}} f(z)|;$$

$$\min_\theta |D_\theta f(z)| = ||\partial_z f(z)| - |\partial_{\bar{z}} f(z)||.$$

(e) The *Laplacian operator* is defined as

$$\Delta = \frac{\partial^2}{\partial x^2} + \frac{\partial^2}{\partial y^2}.$$

Suppose $f \in C^2(O)$. Then

$$\Delta f = \frac{\partial^2 f}{\partial x^2} + \frac{\partial^2 f}{\partial y^2} = 4\frac{\partial^2 f}{\partial z \partial \bar{z}} \underset{\text{(def.)}}{=} 4\partial^2_{z\bar{z}} f \underset{\text{(def.)}}{=} 4f_{z\bar{z}}.$$

In case $f(z) = u(z) + iv(z)$, then $f_{z\bar{z}} = \frac{1}{4}(\Delta u + i\Delta v)$.

(f) $\overline{\partial_z f} = \partial_{\bar{z}} \bar{f}; \overline{\partial_{\bar{z}} f} = \partial_z \bar{f}; \overline{(df)} = d\bar{f}$ (see (2.8.15)).

It is worthing mention that $\partial_z f$, $\partial_{\bar{z}} f$ and $J_f(z)$ can be used to study the following topics:

(1) the existence of single-valued inverse of $w = f(z)$;
(2) local maximum and minimum principles of $w = f(z)$;
(3) orientation-preserving of $w = f(z)$;
(4) influence on local behavior of the directional derivative of $w = f(z)$;
(5) the dilatation at a point of a nonconformal mapping $w = f(z)$;
(6) the rotation angle at a point of a nonconformal mapping $w = f(z)$;
(7) the angle-preserving at a point of a nonconformal mapping $w = f(z)$;

etc. They can all be used to study planar quasiconformal mappings (see Appendix C in Vol. 2). Readers may refer to Ref. [36] for details.

2.9. Integration in complex notation

It is supposed that readers are familiar with definition and basic properties of Riemann integral $\int_a^b f(x)dx$ which may be rewritten as $\int_\gamma f(x)dx$, where γ denotes the line segment $[a, b]$. We try to extend this form of integration to a complex-valued function $f(z)$ of a complex variable z along a plane curve γ and is similarly denoted as $\int_\gamma f(z)dz$.

From now on, γ will always denote a differentiable or piecewise differentiable curve with parametric equation $z(t) = x(t) + iy(t), a \le t \le b$ (see Sec. 2.4) and with the direction induced by increasing t as its positive direction, unless otherwise stated. In this case, γ is a rectifiable curve.

Section (1) Definition

Suppose $P(x, y)$ and $Q(x, y)$ are real-valued functions defined and continuous along γ, namely, $P(x(t), y(t))$ and $Q(x(t), y(t))$ are continuous on $[a, b]$. The *line integral* of P and Q *along* γ is defined as

$$\int_\gamma Pdx + Qdy = \int_a^b [P(x(t), y(t))x'(t) + Q(x(t), y(t))y'(t)]dt. \quad (2.9.1)$$

This definition also provides a way of computing the line integral.

Let $f(z) = u(z) + iv(z)$ be a complex-valued continuous function along γ, namely, $f(z(t))$ is continuous on $[a, b]$. Define the *complex integral* of f *along* γ as

$$\int_\gamma f(z)dz = \int_\gamma (udx - vdy) + i \int_\gamma (vdx + udy). \quad (2.9.2)$$

The right-hand side can be considered as a complex number formed by the line integrals along γ of the real and imaginary parts of the *differential form* $f(z)dz = (u(z) + iv(z))(dx + idy) = (udx - vdy) + i(vdx + udy)$ (see (2.8.11)). Note that (2.9.2) is computed according to (2.9.1).

We give two elementary examples.

Example 1. Let γ be a piecewise differentiable curve connecting z_1 to z_2 (refer to Remark 2 below). Then

(1) $\int_\gamma dz = z_2 - z_1$.

(2) $\int_\gamma zdz = \dfrac{1}{2}(z_2^2 - z_1^2)$.

In case $z_2 = z_1$ (i.e., γ is a closed curve), then both integrals have 0 as their values.

Solution. Let γ have a representation $z(t) = x(t) + iy(t), t \in [a, b]$, with $z(a) = z_1$ and $z(b) = z_2$. Then, via (2.9.1) and (2.9.2),

$$\int_\gamma dz = \int_a^b dx + i \int_a^b dy = x(b) - x(a) + i(y(b) - y(a))$$

$$= x(b) + iy(b) - [x(a) + iy(a)],$$

$$= z(b) - z(a) = z_2 - z_1;$$

$$\int_\gamma zdz = \int_\gamma (xdx - ydy) + i \int_\gamma (xdy + ydx)$$

$$= \int_a^b d\left(\frac{1}{2}(x^2 - y^2)\right) + i \int_a^b d(xy),$$

$$= \frac{1}{2}[(x(b)^2 - y(b)^2) - (x(a)^2 - y(a)^2)] + i[(x(b)y(b) - x(a)y(a)],$$

$$= \frac{1}{2}[x(b) + iy(b)]^2 - \frac{1}{2}[x(a) + iy(a)]^2 = \frac{1}{2}(z_2^2 - z_1^2).$$

Example 2. Suppose γ is the circle $z = a + re^{i\theta}, r > 0, 0 \le \theta \le 2\pi$, and n is an integer. Show that

$$\int_\gamma (z - a)^n dz = \begin{cases} 0, & n \ne -1 \\ 2\pi i, & n = -1 \end{cases}.$$

Solution. In case $n = -1$, since $dz = dx + idy = rie^{i\theta} d\theta$ (why?),

$$\int_\gamma \frac{1}{z - a} dz = \int_0^{2\pi} \frac{rie^{i\theta}}{re^{i\theta}} d\theta = i \int_0^{2\pi} d\theta = 2\pi i.$$

If $n \ne -1$, then (try to figure out the details)

$$\int_\gamma (z - a)^n dz = \int_0^{2\pi} r^n e^{in\theta} \cdot rie^{i\theta} d\theta = ir^{n+1} \int_0^{2\pi} e^{i(n+1)\theta} d\theta$$

$$= ir^{n+1} \frac{e^{i(n+1)\theta}}{i(n + 1)}\bigg|_0^{2\pi} = 0.$$

Four Remarks concerning complex integration are ready in the following.

Remark 1.   Equivalent of (2.9.2). Suppose $f(t) = u(t) + iv(t)$ is a complex-valued continuous function on the interval $[a, b]$. The *Riemann integral* of f on $[a, b]$ is defined as

$$\int_a^b f(t)dt = \int_a^b u(t)dt + i \int_a^b v(t)dt. \tag{2.9.3}$$

It enjoys most properties of Riemann integral of real-valued functions. In particular, if α is a complex number, then

$$\int_a^b \alpha f(t)dt = \alpha \int_a^b f(t)dt.$$

Now, let $f(z) = u(z) + iv(z)$ be continuous along a piecewise differentiable curve γ with equation $z(t) = x(t) + iy(t)$, $t \in [a,b]$. The *complex integral of f along γ* is defined as

$$\int_\gamma f(z)dz = \int_a^b f(z(t))z'(t)dt. \tag{2.9.4}$$

This new definition is equivalent to the old one (2.9.2): Starting from (2.9.2),

$$\int_\gamma f(z)dz = \int_\gamma [u(z)dx - v(z)dy] + i \int_\gamma [v(z)dx + u(z)dy],$$

$$= \int_a^b [u(x(t),y(t))x'(t) - v(x(t),y(t))y'(t)]dt$$

$$+ i \int_a^b [v(x(t),y(t))x'(t) + u(x(t),y(t))y'(t)]dt \quad \text{(see (2.9.1))},$$

$$= \int_a^b \{[u(z(t))x'(t) - v(z(t))y'(t)]$$

$$+ i[v(z(t))x'(t) + u(z(t))y'(t)]\}dt \quad \text{(see (2.9.3))},$$

$$= \int_a^b f(z(t))z'(t)dt.$$

According to well-known results about Riemann integral, once the set of discontinuity of $f(z(t))z'(t)$ on $[a,b]$ is a set of measure zero, then the integral on the right of (2.9.4) also exists as a finite complex number. In this case, we still use (2.9.4) as the definition for the *complex integral of f along γ*.

Remark 2. Complex integration along a rectifiable curve. Suppose $f(z) = u(z) + iv(z)$ is a complex-valued function along a rectifiable curve γ with representation $z(t) = x(t) + iy(t)$ on $[a,b]$ (see Sec. 2.4).

Take any partition $P : a = t_0 < t_1 < \cdots < t_{n-1} < t_n = b$ of $[a,b]$ and then, choose any fixed point $\zeta_j \in [t_j, t_{j+1}], 0 \le j \le n-1$. Formulate the sum

$$S(f, \gamma; P, \{\zeta_j\}) = \sum_{j=0}^{n-1} f(z(\zeta_j))(z(t_{j+1}) - z(t_j))$$

$$= \sum_{j=0}^{n-1} [u(z(\zeta_j))\Delta x_j - v(z(\zeta_j))\Delta y_j]$$

$$+ i \sum_{j=0}^{n-1} [v(z(\zeta_j))\Delta x_j + u(z(\zeta_j))\Delta y_j], \quad \text{where}$$

$$\Delta x_j = \text{Re}(z(t_{j+1}) - z(t_j)) \quad \text{and}$$

$$\Delta y_j = \text{Im}(z(t_{j+1}) - z(t_j)), \quad 0 \le j \le n - 1.$$

Letting $\delta(P) = \max_{0 \le j \le n-1} |t_{j+1} - t_j| \to 0$, if all such possible sums $S(f, \gamma; P, \{\zeta_j\})$ approach a finite complex number as their common limit, then f is said to be *integrable* along γ. The common limit is called the *complex integral* of f along γ and is denoted by

$$\int_\gamma f(z)dz = \lim_{\delta(P) \to 0} S(f, \gamma; P, \{\zeta_j\})$$

$$= \int_\gamma (udx - vdy) + i \int_\gamma (vdx + udy). \tag{2.9.5}$$

In case f is continuous along γ, then f is certainly integrable along γ; if, in addition, γ is piecewise differentiable, then this definition (2.9.5) will agree with that in (2.9.2) or (2.9.4). For beginners in complex analysis, (2.9.2) or (2.9.4) is good enough.

As a matter of fact, *Example 1 is still valid for rectifiable curve* γ. To see this, via (2.9.5),

$$\int_\gamma dz = \lim_{\delta(P) \to 0} \sum_{j=0}^{n-1} [z(t_{j+1}) - z(t_j)] = \lim_{\delta(P) \to 0} [z(t_n) - z(t_0)] = z(b) - z(a);$$

on the other hand, by choosing $\zeta_j = t_j$ and $\zeta_j = t_{j+1}$, respectively,

$$\int_\gamma zdz = \lim_{\delta(P) \to 0} \sum_{j=0}^{n-1} z(t_j)[z(t_{j+1}) - z(t_j)],$$

$$= \lim_{\delta(P) \to 0} \sum_{j=0}^{n-1} z(t_{j+1})[z(t_{j+1}) - z(t_j)],$$

$$= \frac{1}{2} \lim_{\delta(P)\to 0} \sum_{j=0}^{n-1} [z(t_{j+1}) + z(t_j)][z(t_{j+1}) - z(t_j)],$$

$$= \frac{1}{2} \lim_{\delta(P)\to 0} \sum_{j=0}^{n-1} [z(t_{j+1})^2 - z(t_j)^2],$$

$$= \frac{1}{2}(z(b)^2 - z(a)^2).$$

Remark 3.   Complex integration with respect to the conjugate complex variable $\bar{z}$. Suppose $f(z)$ is continuous along a piecewise differentiable or even a rectifiable curve γ. Define the *complex integral* of $f(z)$ *along γ with respect to $\bar{z}$* as

$$\int_\gamma f(z)d\bar{z} = \overline{\int_\gamma \overline{f(z)}dz} \quad \left(\text{or denoted as } \overline{\int_\gamma f(z)\overline{dz}} \right). \tag{2.9.6}$$

As a consequence, we are able to define the *complex integral* of $f(z)$ *along γ with respect to $x = \operatorname{Re} z$ or $y = \operatorname{Im} z$*, respectively, as

$$\int_\gamma f(z)dx = \frac{1}{2} \left(\int_\gamma f(z)dz + \int_\gamma f(z)d\bar{z} \right);$$

$$\int_\gamma f(z)dy = \frac{1}{2i} \left(\int_\gamma f(z)dz - \int_\gamma f(z)d\bar{z} \right). \tag{2.9.7}$$

Example 3.   Let γ be the circle $z = e^{i\theta}$, $0 \leq \theta \leq 2\pi$. Then (compare to Example 2),

$$\int_\gamma \frac{1}{z}d\bar{z} = 0,$$

$$\int_\gamma \frac{1}{z}dx = \pi i \quad \text{and} \quad \int_\gamma \frac{1}{z}dy = \pi.$$

Solution.   By definition,

$$\int_\gamma \frac{1}{z}d\bar{z} = \overline{\int_\gamma \frac{1}{\bar{z}}dz} = \overline{\int_0^{2\pi} \frac{1}{e^{-i\theta}} i e^{i\theta} d\theta} = \overline{\int_0^{2\pi} i e^{2i\theta} d\theta} = \overline{\left. \frac{e^{2i\theta}}{2} \right|_0^{2\pi}} = 0,$$

or, by direct computation,

$$\int_\gamma \frac{1}{z}d\bar{z} = \int_0^{2\pi} \frac{1}{e^{i\theta}} \overline{i e^{i\theta} d\theta} = \int_0^{2\pi} -i e^{-2i\theta} d\theta = \left. \frac{e^{-2i\theta}}{2} \right|_0^{2\pi} = 0.$$

Also, by Example 2 and by definition,

$$\int_\gamma \frac{1}{z} dx = \frac{1}{2}(2\pi i + 0) = \pi i; \quad \int_\gamma \frac{1}{z} dy = \frac{1}{2i}(2\pi i - 0) = \pi;$$

or, by direct computation,

$$\int_\gamma \frac{1}{z} dx = \int_0^{2\pi} \frac{1}{e^{i\theta}}(-\sin\theta)d\theta = \int_0^{2\pi}(\cos\theta - i\sin\theta)(-\sin\theta)d\theta$$

$$= i \int_0^{2\pi} \sin^2\theta d\theta = \pi i.$$

Similarly, $\int_\gamma \frac{1}{z} dy = \pi$.

Remark 4.   Complex integration with respect to arc length. Suppose $f(z)$ is a complex-valued continuous function along a rectifiable curve γ with a representation $z(t) = x(t) + iy(t)$, $t \in [a, b]$. The *complex integral of $f(z)$ along γ with respect to the arc length element* $|dz|$ (see (2.4.6) and (2.4.10)) is defined as

$$\int_\gamma f(z)|dz| = \lim_{\delta(P)\to 0} \sum_{j=0}^{n-1} f(z(\zeta_j))|z(t_{j+1}) - z(t_j)|, \qquad (2.9.8)$$

wherein the explanation of the limit process is exactly the same as in (2.9.5). Recall that, in the Euclidean plane, the arc length element is $ds = \sqrt{dx^2 + dy^2} = |dz|$.

In case γ is piecewise differentiable, it can be shown that

$$\int_\gamma f(z)|dz| = \int_a^b f(z(t))|z'(t)|dt. \qquad (2.9.9)$$

The integral on the right can be used as the definition for the left line integral with respect to arc length.

For example (see Ref. [1], p. 108),

$$\int_{|z|=r} |dz| = 2\pi r;$$

$$\int_{|z|=1} |z - 1||dz| = \int_0^{2\pi} \sqrt{(\cos\theta - 1)^2 + \sin^2\theta}\, d\theta = \int_0^{2\pi} 2\sin\frac{\theta}{2}\, d\theta = 8.$$

Section (2) Basic properties of integrations

Operational properties of complex integration. All curves concerned are oriented and piecewise differentiable (even, rectifiable), and $f(z)$ and $g(z)$ are complex-valued continuous functions along curves.

(1) Linear property. For $\alpha, \beta \in \mathbf{C}$,

$$\int_\gamma (\alpha f(z) + \beta g(z))dz = \alpha \int_\gamma f(z)dz + \beta \int_\gamma g(z)dz.$$

(2) Partition property. Suppose the terminal point of γ_1 is the initial point of γ_2 so that $\gamma_1 + \gamma_2$ is the resulting composite (or sum) curve. Then (see Fig. 2.96)

$$\int_{\gamma_1+\gamma_2} f(z)dz = \int_{\gamma_1} f(z)dz + \int_{\gamma_2} f(z)dz.$$

(3) Inverse curve property. Let $-\gamma$ denote the inverse directed curve of γ. Then

$$\int_{-\gamma} f(z)dz = - \int_\gamma f(z)dz.$$

(4) Invariance under change of variables. Suppose γ has a representation $z(t) = x(t) + iy(t)$, $t \in [a, b]$, and $t = t(\tau) : [c, d] \to [a, b]$ is an increasing piecewise differentiable function. Then γ has another representation

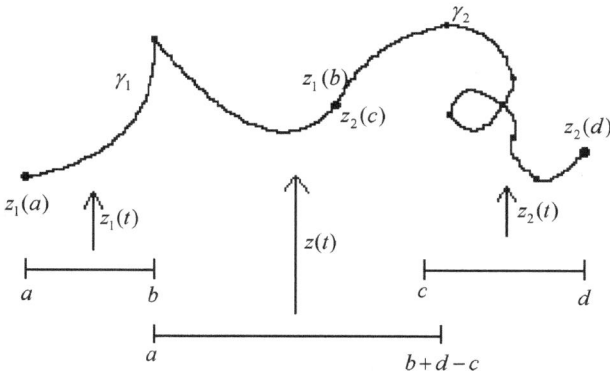

Fig. 2.96

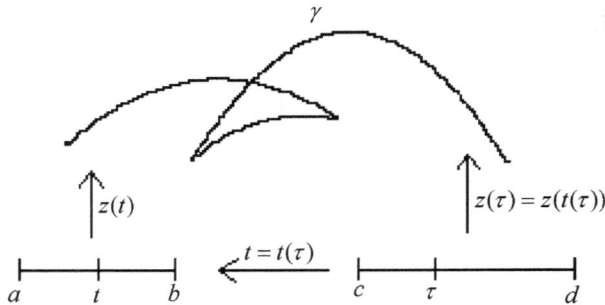

Fig. 2.97

$z(\tau) = z(t(\tau)) : [c, d] \to \mathbf{C}$ and

$$\int_{\gamma} f(z)dz = \int_{a}^{b} f(z(t))z'(t)dt = \int_{c}^{d} f(z(t(\tau)))[z(t(\tau))]'d\tau$$

(see Fig. 2.97). This means the definition of complex integration is independent of the parameter adopted.

(5) Invariance under the translation of the parameter. Suppose γ is a *closed* curve and $\tilde{\gamma}$ is another closed curve equivalent to γ (see Exercise B (1)(b) of Sec. 2.4). Then

$$\int_{\gamma} f(z)dz = \int_{\tilde{\gamma}} f(z)dz.$$

This means that, when dealing with integration along a closed curve, any point on the curve can be used as an initial and hence, terminal point.

(6) Integration by parts. Suppose a curve γ has a representation $z = z(t) :$ $[a, b] \to \Omega$ (a domain), and $f(z)$ and $g(z)$ are *analytic functions* on Ω (see Exercises A (4) and (5) of Sec. 2.8 and Sec. 3.3). Then

$$\int_{\gamma} f(z)g'(z)dz = f(z)g(z)|_{z=z(a)}^{z(b)} - \int_{\gamma} g(z)f'(z)dz,$$

where it is assumed that both $f'(z)$ and $g'(z)$ are continuous on Ω (see (3.4.2.6)).

(7) Termwise integration. Suppose each $f_n(z), n \geq 1$, is continuous along γ and the sequence $f_n(z)$ *converges uniformly* on γ to a limit function $f(z)$. Then

   (i) $\int_\gamma f(z)dz$ exists and
   (ii) $\lim_{n \to \infty} \int_\gamma f_n(z)dz = \int_\gamma f(z)dz$.   (2.9.10)

We *explain* some content of (2.9.10) and give a *proof* as follows.

(1) and (7) are obvious and are left to the readers.

What is $\gamma_1 + \gamma_2$? Suppose γ_1 has a representation $z_1(t) : [a, b] \to \mathbf{C}$, while γ_2 has a representation $z_2(t) : [c, d] \to \mathbf{C}$. Also, suppose $z_1(b) = z_2(c)$. See Fig. 2.96. Define a function $z = z(t) : [a, b + d - c] \to \mathbf{C}$ by

$$z(t) = \begin{cases} z_1(t), & t \in [a, b] \\ z_2(t - b + c), & t \in [b, b + d - c] \end{cases}.$$

Then, this $z(t)$ defines a piecewise differentiable curve, denoted as $\gamma_1 + \gamma_2$ and is called the *sum curve* of γ_1 and γ_2. Now,

$$\int_{\gamma_1 + \gamma_2} f(z)dz = \int_a^b f(z(t))z'(t)dt + \int_b^{b+d-c} f(z(t))z'(t)dt$$

(properties of Riemann integral),

$$= \int_a^b f(z_1(t))z_1'(t)dt + \int_b^{b+d-c} f(z_2(t - b + c))z_2'(t - b + c)dt$$

(by definition of $z(t)$),

$$= \int_{\gamma_1} f(z)dz + \int_c^d f(z(u))z'(u)du$$

(by change of variables $u = t - b + c$),

$$= \int_{\gamma_1} f(z)dz + \int_{\gamma_2} f(z)dz.$$

This is (2).

Suppose γ has a representation $z(t) = x(t) + iy(t), t \in [a, b]$. Then, the inverse curve $-\gamma$ has the representation $\lambda(t) = z(a + b - t), t \in [a, b]$ (see

(2.4.2)). Then,

$$\int_{-\gamma} f(z)dz = \int_a^b f(\lambda(t))\lambda'(t)dt \quad \text{(by definition (2.9.4))},$$

$$= \int_a^b f(z(a+b-t))z'(a+b-t)(-dt)$$

(substitute $\lambda(t)$ by $z(a+b-t)$),

$$= -\int_a^b f(z(a+b-t))z'(a+b-t)dt \quad \text{(by (1))},$$

$$= \int_b^a f(z(u))z'(u)du \quad \text{(by change of variables } u = a+b-t\text{)},$$

$$= -\int_a^b f(z(t))z'(t)dt \quad \text{(use } t \text{ to replace } u\text{)},$$

$$= -\int_\gamma f(z)dz \quad \text{(by (2.9.4))}.$$

This proves (3).

(4) See Fig. 2.97. Then

$$\int_\gamma f(z)dz = \int_a^b f(z(t))z'(t)dt \quad \text{(by (2.9.4))},$$

$$= \int_c^d f(z(t(\tau)))z'(t(\tau))t'(\tau)d\tau$$

(by change of variables $t = t(\tau)$),

$$= \int_c^d f(z(t(\tau)))[z(t(\tau))]'d\tau$$

(noting that $[z(t(\tau))]' = z'(t(\tau))t'(\tau)$).

This proves (4).

(5) is indeed a special case of (2). The initial point of the original representation of this closed curve and that of the new representation determine two subcurves γ_1 and γ_2 of γ. Then

$$\int_\gamma f(z)dz = \int_{\gamma_1+\gamma_2} f(z)dz = \int_{\gamma_2+\gamma_1} f(z)dz = \int_{\tilde{\gamma}} f(z)dz.$$

(6) Readers should make sure what analytic functions really mean before reading the following proof. According to (2.9.4),

$$\int_\gamma f(z)g'(z)dz = \int_a^b f(z(t))g'(z(t))z'(t)dt,$$

$$= \int_a^b f(z(t))[g(z(t))]'dt,$$

$$= f(z(t))g(z(t))|_{t=a}^b - \int_a^b g(z(t))[f(z(t))]'dt$$

(integration by parts of Riemann integral),

$$= f(z)g(z)|_{z=z(a)}^{z(b)} - \int_a^b g(z(t))f'(z(t))z'(t)dt,$$

$$= f(z)g(z)|_{z=z(a)}^{z(b)} - \int_\gamma g(z)f'(z)dz.$$

This proves (6).   □

Correspondingly, we also have the

Operational properties of complex integration with respect to arc length. Adopt the same assumption as in (2.9.10).

(1) Linear property.
(2) Partition property.
(3) Inverse curve property:

$$\int_{-\gamma} f(z)|dz| = \int_\gamma f(z)|dz|.$$

(4) Invariance under change of variables and translation of parameters (for closed curve).
(5) Integral inequality:

$$\left|\int_\gamma f(z)dz\right| \le \int_\gamma |f(z)||dz| \le \left(\sup_{z\in\gamma}|f(z)|\right) \cdot (\text{the length of } \gamma).$$

$$(2.9.11)$$

Proof of (5). Let θ be a real number. By linear property, we have

$$e^{-i\theta}\int_\gamma f(z)dz = \int_a^b e^{-i\theta}f(z(t))z'(t)dt,$$

$$\Rightarrow \text{Re}\left\{e^{-i\theta}\int_\gamma f(z)dz\right\}$$

$$= \int_a^b \mathrm{Re}\, e^{-i\theta} f(z(t)) z'(t)\, dt,$$

$$\le \int_a^b |f(z(t))||z'(t)|\, dt,$$

$$= \int_\gamma |f(z)||dz|.$$

In case $\int_\gamma f(z)dz = 0$, the required inequality does hold trivially. If $\int_\gamma f(z)dz \ne 0$, choose θ to be $\mathrm{Arg}(\int_\gamma f(z)dz)$, then the left side of the above expression is nothing but $|\int_\gamma f(z)dz|$. $\square$

Section (3) Some further examples (to illustrate (2.9.10) and (2.9.11))

Example 4. Let $\sqrt{z}$ be the branch determined by its value at the point z_0 and γ be the given curve with the initial point z_0. Compute

$$\int_\gamma \frac{dz}{\sqrt{z}}.$$

(1) $z_0 = 1$ and $\sqrt{1} = 1$. γ is the upper semicircle $|z| = 1, y \ge 0$ (see Figs. 2.98 and 2.99). What happens if $\sqrt{1} = -1$?

(2) $z_0 = 1$ and $\sqrt{1} = 1$. γ is the lower semicircle $|z| = 1$, $y \le 0$. What happens if $\sqrt{1} = -1$?

(3) $z_0 = 1$ and $\sqrt{1} = 1$. γ is the unit circle $|z| = 1$ (see Fig. 2.100).

(4) $z_0 = -1$ and $\sqrt{-1} = i$. γ is the unit circle $|z| = 1$ (see Fig. 2.101).

Solution. (1) The branch of $\sqrt{z}$ determined by $\sqrt{1} = 1$ is $|z|^{\frac{1}{2}} e^{i\frac{\theta}{2}}, -\pi < \mathrm{Arg}\, z = \theta < \pi$. This branch has $\mathbf{C} - [\infty, 0]$ as its domain of definition which contains γ except its terminal point -1. Observe that this branch is bounded and continuous on $[0, t], 0 < t < \pi$, and then can be extended to

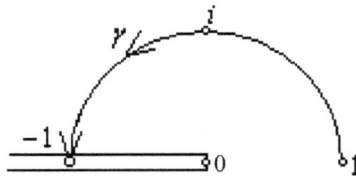

Fig. 2.98

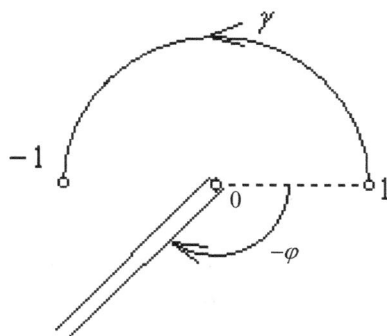

Fig. 2.99

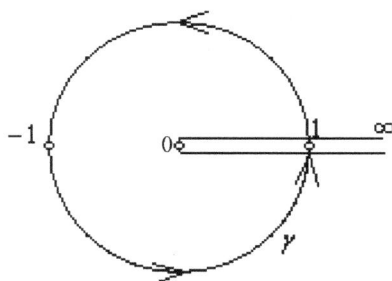

Fig. 2.100

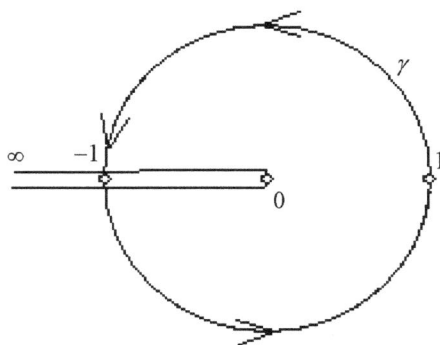

Fig. 2.101

be a continuous function on $[0, \pi]$ by assigning its value i at $z = -1$. See Fig. 2.98. Hence

$$\int_\gamma \frac{dz}{\sqrt{z}} = \lim_{t \to \pi^-} \int_0^t \frac{ie^{it}}{1 \cdot e^{i\frac{t}{2}}} dt = i \lim_{t \to \pi^-} \int_0^t e^{\frac{it}{2}} dt = i \lim_{t \to \pi^-} \left\{ \frac{2}{i} e^{\frac{it}{2}} \right\}_0^t,$$

$$= 2 \lim_{t \to \pi^-} \left\{ e^{\frac{it}{2}} - 1 \right\} = 2 \left\{ e^{\frac{\pi i}{2}} - 1 \right\} = 2(-1 + i).$$

The branch of $\sqrt{z}$ determined by $\sqrt{1} = -1$ on $\mathbf{C} - (\infty, 0]$ is $-|z|e^{\frac{i\theta}{2}}$, $-\pi < \text{Arg } z = \theta < \pi$. Similar computation shows that

$$\int_\gamma \frac{dz}{\sqrt{z}} = \lim_{t \to \pi^-} \int_0^t \frac{ie^{it}}{-1 \cdot e^{\frac{it}{2}}} dt = -2(-1 + i).$$

Note: Fix $0 < \varphi < \pi$. The branch of $\sqrt{z}$ determined by $\sqrt{1} = 1$ on the domain $\mathbf{C} - \{z \mid z = 0 \text{ or } \text{Arg } z = -\varphi\}$ is $|z|^{\frac{1}{2}} e^{\frac{i\theta}{2}}$, $-\varphi < \text{Arg } z < 2\pi - \varphi$. See Fig. 2.99. Direct computation shows that

$$\int_\gamma \frac{dz}{\sqrt{z}} = \int_0^\pi \frac{ie^{it} dt}{e^{\frac{it}{2}}} = 2e^{\frac{it}{2}} |_0^\pi = 2(-1 + i).$$

(2) Similar to (1),

$$\int_\gamma \frac{dz}{\sqrt{z}} = -2(1 + i) \quad \text{if } \sqrt{1} = 1; \quad \int_\gamma \frac{dz}{\sqrt{z}} = 2(1 + i) \quad \text{if } \sqrt{1} = -1.$$

(3) The branch of $\sqrt{z}$ determined by $\sqrt{1} = 1$ on $\mathbf{C} - [0, \infty)$ is $\sqrt{z} = |z|^{\frac{1}{2}} e^{i\frac{\theta}{2}}, 0 < \text{Arg } z < 2\pi$. Designate the value at $\theta = 0$ of this branch to be equal to 1 and the value at $\theta = 2\pi$ to be equal to $e^{i\frac{2\pi}{2}} = -1$. Then this branch is continuous on $[0, 2\pi]$. See Fig. 2.100. Then

$$\int_\gamma \frac{dz}{\sqrt{z}} = \int_0^{2\pi} \frac{ie^{it} dt}{e^{\frac{it}{2}}} = i \int_0^{2\pi} e^{\frac{it}{2}} dt = 2e^{\frac{it}{2}} |_{t=0}^{2\pi} = 2(e^{\pi i} - 1) = -4.$$

(4) The branch of $\sqrt{z}$ determined by $\sqrt{-1} = i$ on $\mathbf{C} - (\infty, 0]$ is $\sqrt{z} = |z|^{\frac{1}{2}} e^{\frac{i\theta}{2}}, -\pi < \text{Arg } z < \pi$. See Fig. 2.101. Then

$$\int_\gamma \frac{dz}{\sqrt{z}} = \int_{-\pi}^\pi \frac{ie^{it}}{e^{\frac{it}{2}}} dt = i \int_{-\pi}^\pi e^{\frac{it}{2}} dt = 2e^{\frac{it}{2}} |_{-\pi}^\pi = 4i.$$

Note that, when comparing (3) and (4), the integral value might depend on the initial point of the curves even though their loci are the same set $|z| = 1$. This does not violate (5) in (2.9.10) because the integrand does have a point of discontinuity on $|z| = 1$.

Example 5. Let $\log z$ be the branch determined by the given point z_0 and γ be the given curve with initial point z_0. Compute

$$\int_{\gamma} z^n \log z dz, \quad \text{where } n \text{ is an integer.}$$

(1) $\log 1 = 0.$ γ is the circle $|z| = 1.$
(2) $\log(-1) = \pi i.$ γ is the circle $|z| = 1.$

Solution. (1) The branch on $\mathbf{C} - [0, \infty)$ is $\log|z| + i \operatorname{Arg} z, 0 < \operatorname{Arg} z < 2\pi.$ On $|z| = 1$, this branch has values $i \operatorname{Arg} z$ which can be extended to be a continuous function on $[0, 2\pi]$. Hence (see Fig. 2.100)

$$\int_{|z|=1} z^n \log z dz = \int_0^{2\pi} e^{in\theta}(i\theta)ie^{i\theta}d\theta = -\int_0^{2\pi} \theta e^{i(n+1)\theta}d\theta,$$

which is equal to $-2\pi^2$ if $n = -1$. In case $n \neq -1$, by integration by parts,

$$-\int_0^{2\pi} \theta e^{i(n+1)\theta}d\theta = -\frac{e^{i(n+1)\theta}}{(n+1)i}\theta\Big|_{\theta=0}^{2\pi} + \int_0^{2\pi} \frac{e^{i(n+1)\theta}}{i(n+1)}d\theta = \frac{2\pi i}{n+1}.$$

(2) Since $\log(-1) = \pi i$ and $z_0 = -1$ is the initial point of the curve $|z| = 1$, the branch of $\log z$ determined by $\log(-1) = \pi i$ on $\mathbf{C} - (\infty, 0]$ is $\log|z| + i \arg z, \pi < \arg z < 3\pi.$ Then

$$\int_{|z|=1} z^n \log z dz = -\int_{\pi}^{3\pi} \theta e^{i(n+1)\theta}d\theta = \begin{cases} -4\pi^2, & \text{if } n = -1 \\ (-1)^{n+1}\dfrac{2\pi i}{n+1}, & \text{if } n \neq -1 \end{cases}.$$

Note: Which function has its derivative equal to $z^n \log z$?

It is well-known that any branch of $\log z$ has the derivative $\frac{1}{z}$ (see (3.3.1.7)). Hence, on any single-valued domain of $\log z$ (see (2.7.3.2)),

$$\frac{d}{dz}\frac{1}{2}(\log z)^2 = z^{-1}\log z (n = -1);$$

$$\frac{d}{dz}\left\{\frac{z^{n+1}}{n+1}\log z - \frac{z^{n+1}}{(n+1)^2}\right\} = z^n \log z (n \neq -1).$$

Using integration by parts (see (6) in (2.9.10)), one can evaluate the integrals. For (1): Even though the branch $\log|z| + i \operatorname{Arg} z (0 < \operatorname{Arg} z < 2\pi)$ is neither continuous nor analytic at $z = 1$, it is bounded and continuous on $|z| = 1$ except at $z = 1$. Then (in (6) of (2.9.10), let $f(z) = 1$ and $g'(z) = z^{-1} \log z$),

$$\int_{|z|=1} z^{-1} \log z\, dz = \frac{1}{2}(\log z)^2 \bigg|_{|z|=1} = \frac{1}{2}(2\pi i)^2 - \frac{1}{2}(0)^2 = -2\pi^2;$$

$$\int_{|z|=1} z^n \log z\, dz = \left\{\frac{z^{n+1}}{n+1} \log z - \frac{z^{n+1}}{(n+1)^2}\right\}\bigg|_{|z|=1} = \frac{1}{n+1}\{2\pi i - 0\}$$

$$= \frac{2\pi i}{n+1} \quad \text{if } n \neq -1.$$

(2) can be similarly evaluated.

Example 6. Let

$$I(r) = \int_{|z|=r} \frac{\log z}{z^2+1}dz, \quad r > 0 \quad \text{and} \quad r \neq 1.$$

Show that $\lim_{r\to 0} I(r) = \lim_{r\to\infty} I(r) = 0$.

Solution. Since 0 and ∞ should belong to the same component of $\mathbf{C}^* - \Omega$ if Ω is a single-valued domain Ω of $\log z$ (see (2.7.3.2)), no matter which branch of $\log z$ is chosen, there is at least one point of discontinuity on $|z| = r$. For example, the principal branch $\text{Log}\, z = \log|z| + i\,\text{Arg}\, z, -\pi < \text{Arg}\, z < \pi$ has its discontinuity occurring at $z = -r$ but it is continuous and bounded elsewhere on $|z| = r$. Hence

$$|I(r)| \leq \int_{|z|=r} \frac{|\log z|}{|z^2+1|}|dz| \leq \begin{cases} \dfrac{\log r + 2\pi}{1 - r^2} \cdot 2\pi r, & \text{if } 0 < r < 1 \\[2mm] \dfrac{\log r + 2\pi}{r^2 - 1} \cdot 2\pi r, & \text{if } r > 1 \end{cases}.$$

And the results follow.

Example 7. Suppose $f(z) = z^n + a_{n-1}z^{n-1} + \cdots + a_1 z + a_0$. Show that

$$\max_{|z|=1} |f(z)| \geq 1.$$

Solution. Let $a_k = |a_k|e^{i\theta_k}, 0 \leq k \leq n-1$, and $z = e^{i\theta}, 0 \leq \theta < 2\pi$. Then, on $|z| = 1$,

$$|f(z)| = |z|^n \left| 1 + \frac{a_{n-1}}{z} + \cdots + \frac{a_1}{z^{n-1}} + \frac{a_0}{z^n} \right|$$

$$= \left| 1 + |a_{n-1}| e^{i(\theta_n - \theta)} + \cdots + |a_1| e^{i(\theta_1 - (n-1)\theta)} + |a_0| e^{i(\theta_0 - n\theta)} \right|$$

$$\geq 1 + |a_{n-1}| \cos(\theta_n - \theta) + \cdots + |a_1| \cos(\theta_1 - (n-1)\theta)$$

$$+|a_0| \cos(\theta_0 - n\theta).$$

Now,

$$\int_0^{2\pi} \cos(\theta_k - (n-k)\theta) d\theta = \frac{-1}{n-k} \sin(\theta_k - (n-k)\theta)|_{\theta=0}^{2\pi} = 0,$$

$$0 \leq k \leq n-1,$$

$$\Rightarrow \int_0^{2\pi} |f(e^{i\theta})| d\theta \geq \int_0^{2\pi} 1 \cdot d\theta = 2\pi,$$

$$\Rightarrow 2\pi \leq \int_0^{2\pi} |f(e^{i\theta})| d\theta \leq \max_{|z|=1} |f(z)| \int_0^{2\pi} 1 \cdot d\theta = 2\pi \max_{|z|=1} |f(z)|,$$

$$\Rightarrow 1 \leq \max_{|z|=1} |f(z)|.$$

Example 8. Show that $|e^z - 1| < |z|$ if $\operatorname{Re} z < 0$.

Solution. Fix a point z_0 with $\operatorname{Re} z_0 < 0$. Let γ denote the line segment connecting 0 to z_0, namely, $z(t) = tz_0, 0 \leq t \leq 1$. Then, for $f(z) = e^z$,

$$\int_\gamma f(z)dz = \int_0^1 f(z(t))z'(t)dt = \int_0^1 e^{tz_0} z_0 dt = e^{z_0} - 1.$$

In case $\operatorname{Re} z_0 < 0$,

$$\max_{0 \leq t \leq 1} |f(z(t))| = \max_{0 \leq t \leq 1} |e^{tz_0}| = \max_{0 \leq t \leq 1} e^{t \operatorname{Re} z_0} = 1 \text{ occurs only at } t = 0.$$

$$\Rightarrow |e^{z_0} - 1| \leq \int_\gamma |f(z)||dz| < \int_\gamma 1 |dz| = |z_0|.$$

Example 9. (Jordan's lemma). Suppose $f(z)$ is continuous on the closed domain Ω : $|z| \geq R_0 > 0$, $\operatorname{Im} z \geq a$ (a is a fixed real number) and $\lim_{z \to \infty} f(z) = 0$ holds. Show that, for any positive number m,

$$\lim_{R \to \infty} \int_{\Gamma_R} e^{\mathrm{im}\, z} f(z)dz = 0,$$

where Γ_R denotes the part of the circle $|z| = R \geq R_0$ that lies in the domain Ω. See Fig. 2.102.

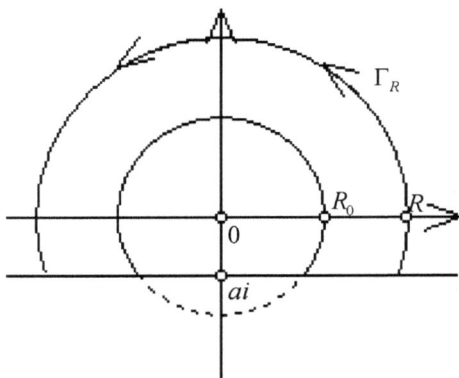

Fig. 2.102

Solution. In case $a < 0$, let γ_1 denote the subcurve of Γ_R that lies in the fourth quadrant with the representation $z(\theta) = \mathrm{R}e^{i\theta}, -\sin^{-1}\frac{|a|}{R} \leq \theta \leq 0$, while γ_3 that part of Γ_R that lies in the third quadrant with the representation $z(\theta) = \mathrm{R}e^{i\theta}, \pi \leq \theta \leq \pi + \sin^{-1}\frac{|a|}{R}$ (here, $0 < \sin^{-1}\frac{|a|}{R} \leq \frac{\pi}{2}$ is the principal value). γ_2 always denotes the part of Γ_R that lies in $\mathrm{Im}\, z \geq 0$.

Choose any fixed $\varepsilon > 0$. Since $\lim_{z\to\infty} f(z) = 0$, there exists a $\delta = \delta(\varepsilon) > 0$ so that $|f(z)| \leq \varepsilon$ on Ω with $|z| \geq \delta$. Note that

$$\int_{\Gamma_R} e^{\mathrm{im}\, z} f(z)dz = \int_{\gamma_1} \cdots + \int_{\gamma_2} \cdots + \int_{\gamma_3} \cdots .$$

To estimate $\int_{\gamma_2}$: On γ_2, $z = \mathrm{R}e^{i\theta}, 0 \leq \theta \leq \pi$. When choosing $R \geq \delta$, then

$$\int_{\gamma_2} e^{\mathrm{im}z} f(z)dz = \int_0^\pi e^{\mathrm{im}\mathrm{R}e^{i\theta}} f(\mathrm{R}e^{i\theta})\mathrm{R}ie^{i\theta}d\theta,$$

$$\Rightarrow \left|\int_{\gamma_2} e^{\mathrm{im}z} f(z)dz\right| \leq \int_0^\pi e^{-mR\sin\theta}|f(\mathrm{R}e^{i\theta})|\mathrm{R}d\theta$$

$$\leq 2\varepsilon\mathrm{R}\int_0^{\frac{\pi}{2}} e^{-m\mathrm{R}\sin\theta}d\theta,$$

$$\leq 2\varepsilon\mathrm{R}\int_0^{\frac{\pi}{2}} e^{-m\mathrm{R}\frac{2\theta}{\pi}}d\theta \left(\text{using } \theta \geq \sin\theta \geq \frac{2\theta}{\pi} \text{ if } 0 \leq \theta \leq \frac{\pi}{2}\right),$$

$$= \frac{\pi \varepsilon}{m}(1 - e^{-m\mathrm{R}}).$$

$$\Rightarrow \varlimsup_{R \to \infty} \left| \int_{\gamma_2} e^{\mathrm{i}mz} f(z) dz \right| \leq \frac{\pi \varepsilon}{m}$$

$$\Rightarrow \lim_{R \to \infty} \int_{\gamma_2} e^{\mathrm{i}mz} f(z) dz = 0.$$

If $a < 0$, we still need to estimate $\int_{\gamma_1}$ and $\int_{\gamma_3}$. Choose $R \geq \delta$. Then,

$$\int_{\gamma_1} e^{\mathrm{i}mz} f(z) dz = \int_{-\sin^{-1}\frac{|a|}{R}}^{0} e^{\mathrm{i}m\mathrm{R}e^{\mathrm{i}\theta}} f(\mathrm{R}e^{\mathrm{i}\theta}) R\mathrm{i}e^{\mathrm{i}\theta} d\theta,$$

$$\Rightarrow \left| \int_{\gamma_1} e^{\mathrm{i}mz} f(z) dz \right|$$

$$\leq \int_{-\sin^{-1}\frac{|a|}{R}}^{0} e^{-m\mathrm{R}\sin\theta} |f(\mathrm{R}e^{\mathrm{i}\theta})| R d\theta$$

$$\leq e^{-ma} R\varepsilon \sin^{-1}\frac{|a|}{R} \quad (\text{since } |R\sin\theta| \leq |a|).$$

$$\Rightarrow \left(\text{observing that } \lim_{R \to \infty} R\sin^{-1}\frac{|a|}{R} = |a| \right)$$

$$\varlimsup_{R \to \infty} \left| \int_{\gamma_1} e^{\mathrm{i}mz} f(z) dz \right| \leq e^{-ma} |a|\varepsilon,$$

$$\Rightarrow \lim_{R \to \infty} \int_{\gamma_1} e^{\mathrm{i}mz} f(z) dz = 0.$$

Similarly, $\lim_{R \to \infty} \int_{\gamma_3} e^{\mathrm{i}mz} f(z) dz = 0$. The result follows.

Example 10.   (See Ref. [1], p. 108). Let $P(z)$ be a polynomial and γ the circle $|z - a| = R$. Show that

(1) $\int_{\gamma} P(z) dz = 0$. Note that this result still holds even if γ is any closed rectifiable curve (see Remark 2 and (6) in (2.9.10)).
(2) $\int_{\gamma} P(\bar{z}) d\bar{z} = 0$.
(3) $\int_{\gamma} P(z) d\bar{z} = -2\pi i R^2 P'(a)$.

Solution.   Suppose $P(z) = a_n z^n + a_{n-1} z^{n-1} + \cdots + a_1 z + a_0$. Let γ be $z = a + \mathrm{R}e^{\mathrm{i}\theta}$, $0 \leq \theta \leq 2\pi$.

(1) By Example 2,

$$\int_\gamma P(z)dz = \sum_{k=0}^n a_k \int_\gamma z^k dz = \sum_{k=0}^n a_k \cdot 0 = 0.$$

Or, observing that $\left(\sum_{k=0}^n \frac{a_k}{k+1} z^{k+1}\right)' = P(z)$ and γ is a closed curve, by (6) in (2.9.10),

$$\int_\gamma P(z)dz = \left\{\sum_{k=0}^n \frac{a_k}{k+1} z^{k+1}\right\}\bigg|_\gamma = 0.$$

(2) Similar method such as (1) will work. As a matter of fact, this is also a consequence of (1) for

$$\int_\gamma P(\bar{z})d\bar{z} = \overline{\int_\gamma (\overline{a_n}z^n + \cdots + \overline{a_1}z + \overline{a_0})} = \bar{0} = 0.$$

(3) Just like (1), we have

$$\int_\gamma P(z)d\bar{z} = \sum_{k=0}^n a_k \int_\gamma z^k d\bar{z}.$$

In case $k = 0$, $\int_\gamma z^k d\bar{z} = \int_\gamma d\bar{z} = \bar{z}|_\gamma = 0$. While, for $1 \le k \le n$, using $\int_0^{2\pi} e^{ik\theta} d\theta = 0$, then

$$\int_\gamma z^k d\bar{z} = \int_0^{2\pi} (a + Re^{i\theta})^k Rie^{-i\theta}(-d\theta) = -Ri \int_0^{2\pi} (a + Re^{i\theta})^k e^{-i\theta} d\theta$$

$$= -Ri \int_0^{2\pi} ka^{k-1}Re^{i\theta} e^{-i\theta} d\theta = -2R^2 \pi i a^{k-1}k.$$

Therefore, the final result follows easily.

Example 11. (See Ref. [1], p. 108). Suppose $f(z)$ is analytic (see Exercise A (5) of Sec. 2.8 or Sec. 3.3) on a domain Ω and satisfies $|f(z) - 1| < 1$ for $z \in \Omega$. Then, for any closed rectifiable curve γ in Ω,

$$\int_\gamma \frac{f'(z)}{f(z)} dz = 0.$$

Here, we also assume $f'(z)$ is continuous on Ω (refer to (3.4.2.6) in Chap. 3).

Solution. $|w - 1| < 1$ is a single-valued domain of $\log w$ (see (2.7.3.2)). For any branch of $\log w$ on $|w - 1| < 1$, we always have (see (3.3.1.7))

$$\frac{d}{dw} \log w = \frac{1}{w}.$$

Since $|f(z) - 1| < 1$ for $z \in \Omega$, the corresponding branch for $\log f(z)$ then satisfies

$$\frac{d}{dz} \log f(z) = \frac{f'(z)}{f(z)}, \quad z \in \Omega$$

(Do you see why? Try to imitate (2.5.1.5) for polynomials to prove this or just refer to (4) in (3.1.6) or (3.3.1.8)). Then, using (6) in (2.9.10), we have

$$\int_\gamma \frac{f'(z)}{f(z)} dz = \log f(z)|_\gamma = 0.$$

Example 12. (See Ref. [1], p. 109). Under what circumstances that

$$\int_\gamma \log z \, dz = 0$$

is meaningful and true?

Solution. Suppose Ω is a domain in $\mathbf{C}$ so that 0 and ∞ belong to the same component of $\mathbf{C}^* - \Omega$. Then, Ω is a single-valued domain for $\log z$. For any branch of $\log z$ on Ω, we have

$$\frac{d}{dz} (z \log z - z) = \log z, \quad z \in \Omega.$$

If γ is any closed rectifiable curve in Ω, then by (6) in (2.9.10)

$$\int_\gamma \log z \, dz = \{z \log z - z\}|_\gamma = 0.$$

Section (4) Green or Stokes formula and generalized Cauchy integral formula

Let γ be a rectifiable Jordan closed curve endowed with positive orientation. Suppose $P(x, y)$ and $Q(x, y)$ are continuous real-valued functions on $\overline{\text{Int}}\, \gamma = \gamma \cup \text{Int}\, \gamma$, while both $\frac{\partial P}{\partial y}$ and $\frac{\partial Q}{\partial x}$ exist and are continuous on $\text{Int}\, \gamma$. Then

$$\int_\gamma P dx + Q dy = \iint_{\text{Int}\, \gamma} \left(\frac{\partial Q}{\partial x} - \frac{\partial P}{\partial y} \right) dx dy, \qquad (2.9.12)$$

called the *Green formula*. Adopt differential one-form $\omega = P dx + Q dy$ as in (2.8.11), then $d\omega = (Q_x - P_y) dx \wedge dy$ as in (2.8.12). Equation (2.9.12) can be rewritten in the form of the integration of differential as

$$\int_\gamma \omega = \iint_{\text{Int}\,\gamma} d\omega, \qquad (2.9.12)'$$

which is a special case of the *Stokes formula.* Furthermore, we still can write $(2.9.12)$ or $(2.9.12)'$ in complex differential form. Note that

$$Pdx + Qdy = P\left(\frac{1}{2}(z+\bar z), \frac{-i}{2}(z-\bar z)\right)\cdot \frac{1}{2}(dz+d\bar z)$$

$$+ Q\left(\frac{1}{2}(z+\bar z), \frac{-i}{2}(z-\bar z)\right)\cdot \frac{-i}{2}(dz-d\bar z),$$

$$= M(z,\bar z)dz + N(z,\bar z)d\bar z,$$

$$\Rightarrow d(M(z,\bar z)dz + N(z,\bar z)d\bar z) = \left(\frac{\partial M}{\partial z}dz + \frac{\partial M}{\partial \bar z}d\bar z\right)\wedge dz$$

$$+ \left(\frac{\partial N}{\partial z}dz + \frac{\partial N}{\partial \bar z}d\bar z\right)\wedge d\bar z$$

$$= \left(\frac{\partial M}{\partial \bar z} - \frac{\partial N}{\partial z}\right)d\bar z \wedge dz$$

$$= 2i\left(\frac{\partial M}{\partial \bar z} - \frac{\partial N}{\partial z}\right)dx \wedge dy \quad \text{(see (2.8.14))}.$$

Hence, the Green or Stokes formula in complex form is

$$\int_\gamma Mdz + Nd\bar z = \iint_{\text{Int}\gamma}\left(\frac{\partial M}{\partial \bar z} - \frac{\partial N}{\partial z}\right)d\bar z \wedge dz \quad \text{or}$$

$$\int_\gamma \omega = \iint_{\text{Int}\gamma} d\omega, \qquad (2.9.12)''$$

where $\omega = Mdz + Nd\bar z$.

Note that $(2.9.12)$–$(2.9.12)''$ are still valid if $\text{Int}\,\gamma$ is replaced by bounded open set D with its boundary ∂D consisting of a finite number of rectifiable Jordan closed curves.

As an application of $(2.9.12)''$, we have the generalized Cauchy or Pompeiu integral formula.

Generalized Cauchy or Pompeiu integral formula. Let γ be a rectifiable Jordan closed curve and let $f \in C^1(\overline{\text{Int}\,\gamma})$ be a complex-valued function.

Then, for any $z \in \operatorname{Int} \gamma$,

$$f(z) = \frac{1}{2\pi i} \left\{ \int_\gamma \frac{f(\zeta)}{(\zeta - z)} d\zeta + \iint_{\operatorname{Int} \gamma} \frac{\partial f(\zeta)}{\partial \bar\zeta} \cdot \frac{1}{\zeta - z} d\zeta \wedge d\bar\zeta \right\},$$

$$= -\frac{1}{2\pi i} \left\{ \int_\gamma \frac{f(\zeta)}{\bar\zeta - \bar z} d\bar\zeta - \iint_{\operatorname{Int} \gamma} \frac{\partial f(\zeta)}{\partial \zeta} \cdot \frac{1}{\bar\zeta - \bar z} d\zeta \wedge d\bar\zeta \right\}. \qquad (2.9.13)$$

In particular, if f is analytic on $\operatorname{Int} \gamma$ (see Exercise A (5) of Sec. 2.8 or Sec. 3.3) and is C^1 on $\overline{\operatorname{Int} \gamma}$, then

$$f(z) = \frac{1}{2\pi i} \int_\gamma \frac{f(\zeta)}{\zeta - z} d\zeta, \quad z \in \operatorname{Int} \gamma \qquad (2.9.14)$$

which is the classical *Cauchy integral formula* (see (3.4.2.4)).

Proof. Fix any point $z \in \operatorname{Int} \gamma$ and choose $\rho > 0$ so that the closed disk $|\zeta - z| \le \rho$ is contained in $\operatorname{Int} \gamma$. Then $D_\rho = \operatorname{Int} \gamma - \{|\zeta - z| \le \rho\}$ is a domain whose boundary $\partial D_\rho = \gamma - \{|\zeta - z| = \rho\}$, oriented so that D_ρ lies to the left of ∂D_ρ. Observe that

$$d\left(\frac{f(\zeta)}{\zeta - z} d\zeta \right) = \frac{\partial}{\partial \bar\zeta} \left(\frac{f(\zeta)}{\zeta - z} \right) d\bar\zeta \wedge d\zeta = \frac{\partial f(\zeta)}{\partial \bar\zeta} \cdot \frac{1}{\zeta - z} d\bar\zeta \wedge d\zeta.$$

Applying $(2.9.12)''$ to the above two-form on D_ρ, we have

$$\iint_{D_\rho} \frac{\partial f(\zeta)}{\partial \bar\zeta} \cdot \frac{1}{\zeta - z} d\bar\zeta \wedge d\zeta = \int_{\partial D_\rho} \frac{f(\zeta)}{\zeta - z} d\zeta$$

$$= \int_\gamma \frac{f(\zeta)}{\zeta - z} d\zeta - \int_{|\zeta - z| = \rho} \frac{f(\zeta)}{\zeta - z} d\zeta. \qquad (*)$$

Wherein, by writing $\zeta = z + \rho e^{i\theta}, 0 \le \theta \le 2\pi$, for $|\zeta - z| = \rho$ and using the continuity of f at z,

$$\int_{|\zeta - z| = \rho} \frac{f(\zeta)}{\zeta - z} d\zeta = \int_0^{2\pi} f(z + \rho e^{i\theta}) i d\theta \to 2\pi i f(z) \quad \text{as } \rho \to 0.$$

Since $\frac{1}{\zeta - z}$ is integrable over D_ρ (or $\frac{1}{\zeta - z} d\bar\zeta \wedge d\zeta$ is a bounded measure on the ζ-plane), by letting $\rho \to 0$ in $(*)$, we obtain the first identity. By taking the complex conjugate of this identity,

$$\bar{f}(z) = -\frac{1}{2\pi i}\left\{\int_\gamma \frac{\bar{f}(z)}{\bar{\zeta}-\bar{z}}d\bar{\zeta} + \iint_{\text{Int}\,\gamma} \overline{\frac{\partial f}{\partial \bar{\zeta}}}\cdot\frac{1}{\bar{\zeta}-\bar{z}}\overline{d\zeta\wedge d\bar{\zeta}}\right\},$$

$$= -\frac{1}{2\pi i}\left\{\int_\gamma \frac{\bar{f}(z)}{\bar{\zeta}-\bar{z}}d\bar{\zeta} + \iint_{\text{Int}\,\gamma} \frac{\partial\bar{f}}{\partial\bar{\zeta}}\cdot\frac{1}{\bar{\zeta}-\bar{z}}d\bar{\zeta}\wedge d\zeta\right\}$$

(using Exercise B (1) (f) of Sec. 2.8)

and then, by replacing $\bar{f}$ by f, the second identity follows.   □

Exercises A

(1) In Example 2, setting $a = 0$ and $z = re^{i\theta}, 0 \le \theta \le 2\pi$, then for integer n,

$$\int_0^{2\pi} e^{in\theta}d\theta = \begin{cases} 2\pi, & n = 0 \\ 0, & n \ne 0 \end{cases}.$$

(a) Try to show that, for $n \ge 0$,

$$\int_0^{2\pi} \cos^{2n}\theta d\theta = \frac{\pi}{2^{2n-1}}C_n^{2n}.$$

(b) For $n \ge 0$, evaluate $\int_0^{2\pi}\cos^n\theta d\theta$ and $\int_0^{2\pi}\sin^n\theta d\theta$.

(2) Evaluate

$$\int_{\gamma_j} xdz, \quad 1 \le j \le 7,$$

where γ_j is defined respectively as follows: γ_1 is the line segment $\overline{z_1 z_2}$ connecting z_1 to z_2; γ_2 the circle $|z| = r > 0$; γ_3 the boudary of the square $[0,1]\times[0,1]$; γ_4 the ellipse $\frac{x^2}{a^2} + \frac{y^2}{b^2} = 1$; γ_5 the polygonal curve $\overline{0, 2i + 2i, (4+2i)}$; γ_6 the polygonal curve $\overline{0,1 + 1,(1+i) + (1+i),0}$; and γ_7 the curve $z(t) = t + it^2, 0 \le t \le 1$.

(3) Let γ_1 be the curve $z(t) = e^{t+it}, 0 \le t \le 2\pi$ and γ_2 the curve $z(t) = (1-t)e^{2\pi} + t, 0 \le t \le 1$. Denote $\gamma = \gamma_1 + \gamma_2$. Evaluate the following integrals.

(a) $\int_\gamma \frac{1}{z}dz$. (b) $\int_\gamma \frac{1}{z}|dz|$. (c) $\int_\gamma \frac{1}{|z|}dz$. (d) $\int_\gamma \frac{1}{|z|}|dz|$.

(e) $\int_{\gamma_1} e^z dz$. (f) $\int_{\gamma_1} e^{|z|}|dz|$. (g) $\int_\gamma e^z dz$.

(4) Evaluate the following integrals:

(a) $\int_\gamma \bar{z}d\bar{z}$, where γ is $z(t) = e^{it}, 0 \le \theta \le 2\pi$.

(b) $\int_\gamma \frac{1}{z^2}dy$, where γ is as in (a).

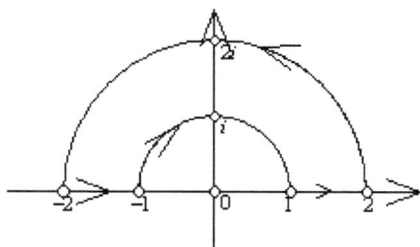

Fig. 2.103

(c) $\int_\gamma |z|\bar{z}d\bar{z}$, $\int_\gamma |z|\bar{z}dx$ and $\int_\gamma |z|\bar{z}dy$, where γ is the upper half-circle $e^{it}, 0 \le t \le \pi$, followed by the diameter $[-1, 1]$.

(d) $\int_\gamma \frac{1}{\sqrt{1+z^2}}|dz|$, where γ is $z(t) = te^{it}, 0 \le t \le \pi$.

(e) $\int_\gamma |2x + 3iy||dz|$, where γ is $z(t) = 3t^2 + 2t^3 i, 0 \le t \le 1$.

(f) $\int_\gamma xe^y |dz|$, where γ is $z(t) = (1+t^2)^{-1}(1 - t^2 + 2ti), -1 \le t \le 1$.

(g) $\int_\gamma (z^2 + iz)dz$, where γ is $z(t) = 3 - 3t^2 + i(t^3 - 3t + 1), -1 \le t \le 1$.

(h) $\int_\gamma \frac{1}{\sqrt{z}}dz$, where γ is the boundary of the square with vertices at ± 1 and $\pm i$.

(i) $\int_\gamma \frac{z}{\bar{z}}dz$, where γ is as in Fig. 2.103.

(5) Evaluate the following integrals.

(a) $\int_{|z|=1} \frac{1}{(z-2)^2}dz$.   (b) $\int_{|z|=1} \frac{1}{z^2-4}dz$.   (c) $\int_{|z|=1} \frac{1}{z^2+2z}dz$

(d) $\int_{|z|=1} \frac{1}{z^3+4z}dz$.   (e) $\int_{|z|=1} \frac{1}{\sqrt{4-z^2}}dz$.

(f) $\int_{|z|=1} \left(z + \frac{1}{z}\right)^n dz$ ($n > 0$ is an integer).

(6) Evaluate $\int_\gamma \log z dz$, according to

(a) $\text{Log } 1 = 0$ and γ is as in Fig. 2.104, and

(b) $\text{Log } i = \frac{\pi i}{2}$ and γ is as in Fig. 2.105.

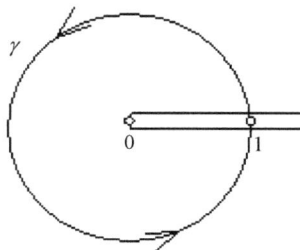

Fig. 2.104

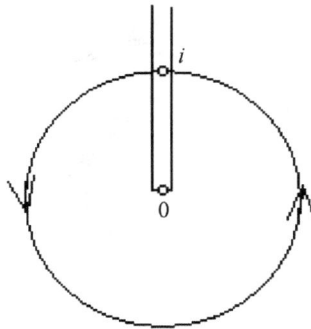

Fig. 2.105

(7) For $\alpha \in \mathbf{C}$, designate $1^\alpha = 1$. Show that

$$\int_{|z|=1} z^\alpha dz = \begin{cases} \dfrac{e^{2\alpha\pi i} - 1}{\alpha + 1}, & \text{if } \alpha \neq -1 \\ 2\pi i, & \text{if } \alpha = -1 \end{cases}.$$

(8) Show that $\int_{|z|=1} a^z dz = 0$.

(9) Let γ be the boundary of the sector $0 < |z| < R$, $0 < \operatorname{Arg} z < \frac{\pi}{4}$ as shown in Fig. 2.106. Evaluate $\int_\gamma (z^3 - i)dz$ according to (2.9.2) and (2.9.4), respectively.

(10) Graph the eight-figure curve γ:

$$z(t) = \begin{cases} 1 - e^{-it}, & 0 \leq t \leq 2\pi \\ 1 + e^{-it}, & 0 \leq t \leq 2\pi \end{cases}$$

and evaluate $\int_\gamma (z^2 - 1)dz$.

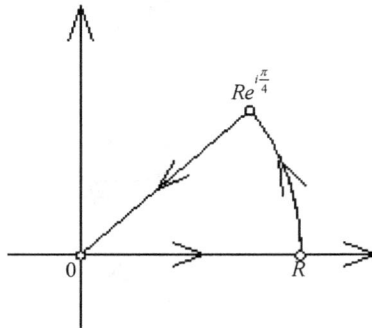

Fig. 2.106

(11) Evaluate

$$\int_{|z|=1} z^{m-1}\bar{z}^n dz$$

where m and n are integers. Then, deduce that

$$\frac{1}{2\pi}\int_0^{2\pi} e^{-imt}\cos ntdt = \begin{cases} 0, & |m| \neq |n| \\ \dfrac{1}{2}, & |m| = |n| > 0 \;; \\ 1, & m = n = 0 \end{cases}$$

$$\frac{1}{2\pi}\int_0^{2\pi} e^{-imt}\sin ntdt = \begin{cases} 0, & |m| \neq |n| \\ -\dfrac{i}{2}, & m = n \neq 0 \\ \dfrac{i}{2}, & m = -n \neq 0 \\ 0, & m = n = 0 \end{cases} .$$

(12) Suppose $P(z) = a_n z^n + a_{n-1}z^{n-1} + \cdots + a_1 z + a_0$ satisfies $\int_{|z|=1} P(z)\bar{z}^n dz = 0$ for $n = 0, 1, 2, \ldots$, then show that $P(z) \equiv 0$, namely, $a_n = a_{n-1} = \cdots = a_1 = a_0 = 0$.

(13) Let γ be the line segment $z(t) = te^{i\theta}$, $0 \leq t \leq 1$ and $0 \leq \theta < 2\pi$. Find the range for θ so that the integrals $\int_\gamma e^{-\frac{1}{z}}dz$ and $\int_\gamma e^{-\frac{1}{z^n}}dz$ $(n \geq 1)$ exist, respectively.

(14) (a) Evaluate $\int_{|z|=1} \frac{dz}{z+2}$.

(b) Use (a) to show that $\int_0^{2\pi} \frac{1+2\cos\theta}{5+4\cos\theta}d\theta = 0$; $\int_0^{2\pi} \frac{\sin\theta}{5+4\cos\theta}d\theta = 0$.

(15) Suppose $f(z)$ is continuous on $|z| \leq 1$ and, for any r, $0 < r < 1$,

$$\int_{|z|=r} f(z)dz = 0$$

always holds. Show that $\int_{|z|=1} f(z)dz = 0$.

(16) Suppose $f(z)$ is analytic (see Exercise A (5) of Sec. 2.8 or Sec. 3.3) on $0 < |z| < 1$ and, for any r, $0 < r < 1$,

$$\int_{|z|=r} f(z)dz = 0$$

always holds. Is it possible for $f(z)$ to be analytic at 0?

(17) Try to use (5) in (2.9.11) to show the following inequalities.

(a) $\left| \int_\gamma \frac{e^z}{z} dz \right| \le e^\pi$, where γ is $z(t) = e^{it}$, $0 \le t \le \pi$.

(b) $\left| \int_\gamma \frac{\log z}{z^2} dz \right| \le \frac{\pi}{R} \left(\log R + \frac{1}{2}\pi \right)$, where γ is $z(t) = Re^{it}$, $-\frac{\pi}{2} \le t \le \frac{\pi}{2}$ and $R > 1$.

(c) $\left| \int_\gamma \frac{e^{az}}{1+e^z} dz \right| \le \frac{2\pi e^{-aR}}{1-e^{-R}}$, where $a > 0$ and γ is $z(t) = -R + it$, $0 \le t \le 2\pi$, $R > 0$.

(d) $\left| \int_\gamma \frac{1}{\log z} dz \right| \le e \log(\pi + \sqrt{\pi^2 + 1})$, where γ is $z(t) = e^{1+it}$, $0 \le t \le \pi$.

(e) $\left| \int_\gamma \cos z^2 dz \right| \le \sqrt{a^2 + b^2} \cdot \frac{\sinh 2ab}{2ab}$, where $a > 0$, $b > 0$ and γ is the line segment from 0 to $a + bi$.

(f) $\left| \int_\gamma \frac{1}{1+z^3} dz \right| \le \frac{2\pi}{21}$, where γ is the circular arc $z(t) = 2e^{it}$, $-\frac{\pi}{6} \le t \le \frac{\pi}{6}$. Try to show that this estimate value can be improved to $\frac{2\pi}{3\sqrt{65}}$.

(g) $\left| \int_\gamma e^{iz^2} dz \right| \le \frac{\pi(1-e^{-R^2})}{4R}$, where γ is $z(t) = Re^{it}$, $0 \le t \le \frac{\pi}{4}$ and $R > 0$.

(h) $\left| \int_\gamma (e^{2z} + 2\bar{z}) dz \right| \le 108$, where γ is the triangle with vertices at 0, -3 and $4i$.

(i) $\left| \int_{|z-1|=2} \frac{z+1}{z-1} dz \right| \le 8\pi$.

(18) Let a and b be real numbers. If $s = \sigma + it$ with $\sigma > 0$, show that

$$|e^{bs} - e^{as}| \le |b - a||s|e^{\max\{a,b\}\sigma}.$$

(19) For $0 < r < \infty$, let

$$I(r) = \int_{\gamma_r} z^{-1} e^{iz} dz,$$

where γ_r is the circular arc $z(t) = re^{it}$, $0 \le t \le \pi$. Show that $\lim_{r \to \infty} I(r) = 0$ and $\lim_{r \to 0} I(r) = \pi i$.

20 (See Ref. [1], p. 108) Suppose $f(z)$ is analytic on a domain Ω (and assume $f'(z)$ is continuous there). For any rectifiable closed curve γ in Ω, show that $\int_\gamma \overline{f(z)} f'(z) dz$ is purely imaginary.

21 (a) Let γ be a rectifiable Jordan closed curve, positively oriented. Then

$$\text{the } \textit{area of } \text{Int } \gamma = \frac{1}{i} \int_\gamma (\text{Re } z) dz = -\int_\gamma (\text{Im } z) dz = \frac{1}{2i} \int_\gamma \bar{z} dz.$$

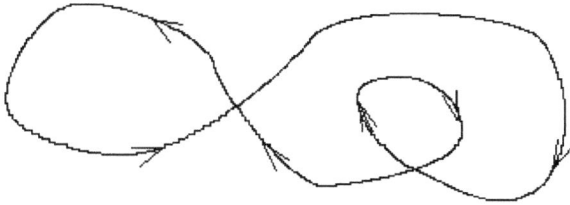

Fig. 2.107

(b) If γ is a rectifiable closed curve. Designate

$$\text{the } \textit{signed area} \text{ of the domain bounded by } \gamma = \frac{1}{2i}\int_\gamma \bar{z}\,dz.$$

Try to use the domain bounded by γ shown in Fig. 2.107 to explain this quantity.

(22) Adopt notations in (2.8.3) and (2.8.4). Suppose $f \in C^1(\Omega)$, where Ω is a domain.

(a) Show that, for $z_0 \in \Omega$,

$$f_z(z_0) = \frac{1}{2\pi}\int_0^{2\pi} D_\theta f(z_0)\,d\theta,$$

called the *average derivative* of f at z_0 (Kasner, 1928, 1936).

(b) Show that, for a rectifiable Jordan closed curve γ satisfying $\overline{\text{Int}\,\gamma} \subseteq \Omega$,

$$\int_\gamma f(z)\,dz = 2i\iint_{\text{Int}\,\gamma} f_{\bar{z}}(z)\,dx dy = \iint_{\text{Int}\,\gamma} f_{\bar{z}}(z)\,d\bar{z}\wedge dz.$$

In case f is analytic on $\overline{\text{Int}\,\gamma}$ (see Exercise A (5) of Sec. 2.8 or Sec. 3.3), then

$$\int_r f(z)\,dz = 0$$

which is called the *Cauchy integral theorem* (see (3.4.2.1)). If $f(z) = \bar{z}$, we get the area formula as in Exercise (21)(a).

(c) Suppose $z_0 \in \Omega$ and γ is a rectifiable Jordan closed curve so that $z_0 \in \text{Int}\,\gamma \subseteq \overline{\text{Int}\,\gamma} \subseteq \Omega$ holds. Use $|\text{Int}\,\gamma|$ to denote the area of Int

γ (see Exercise (21)(a)). Show that

$$f_{\bar{z}}(z_0) = \frac{1}{2\pi i} \lim_{|\mathrm{Int}\,\gamma| \to 0} \frac{1}{|\mathrm{Int}\,\gamma|} \int_\gamma f(z)dz,$$

called the *phase* or *area derivative* of f at z_0 (Pompeiu, 1912), which is a *derivation of f from analyticity at z_0* (Min, 1944). Here, $|\mathrm{Int}\,\gamma| \to 0$ means that, for any $\varepsilon > 0$, there corresponds a $\delta > 0$ so that for all rectifiable Jordan closed curves γ satisfying $\mathrm{Int}\,\gamma \subseteq \{|z - z_0| < \delta\}$, then $|\mathrm{Int}\,\gamma| < \varepsilon$ always holds.

(d) Let z_0 and γ be as in (c). Show that

$$f_z(z_0) = \lim_{|\mathrm{Int}\,\gamma| \to 0} \frac{i}{2|\mathrm{Int}\,\gamma|} \int_\gamma f(z)d\bar{z}.$$

(e) Suppose $\overline{f(z)}$ is analytic on $\overline{\mathrm{Int}\,\gamma}$. Show that, for a rectifiable Jordan closed curve γ with $\overline{\mathrm{Int}\,\gamma} \subseteq \Omega$,

$$\int_\gamma f(z)d\bar{z} = 0.$$

(23) Observe that $\int_0^{2\pi} e^{i\theta}d\theta = 0$ but there does not exist a θ_0, $0 \le \theta_0 \le 2\pi$ so that $\int_0^{2\pi} e^{i\theta}d\theta = e^{i\theta_0}(2\pi - 0)$. This means that a direct extension of the mean-value theorem for integral in real analysis to complex integral does not work. Suppose $f(z)$ is continuous on a domain. Let a and b be two points in Ω so that the segment $\overline{ab}$ is still in Ω. Show that there exist two points c_1 and c_2 on $\overline{ab}$ so that

$$\int_{\overline{ab}} f(z)dz = (\mathrm{Re}\, f(c_1) + i\mathrm{Im}f(c_2))(b - a).$$

This is the *mean-value theorem for complex integral.* Furthermore, suppose $f(z) \in C^1(\Omega)$ is analytic (see Exercise A (5) of Sec. 2.8 or Sec. 3.3). Then,

$$f(b) - f(a) = (\mathrm{Re}\, f'(c_1) + i\mathrm{Im}f'(c_2))(b - a),$$

called the *mean-value theorem for complex differentiation* (see (3.4.1.1)). Show that, there exist $\lambda, |\lambda| \le 1$, and a point c lying on $\overline{ab}$ so that

$$f(b) - f(a) = \lambda(b - a)f'(c).$$

(24) Suppose f is continuous in a neighborhood of a. Show that

$$\lim_{r \to 0} \int_{|z-a|=r} \frac{f(z)}{z - a}dz = 2\pi i f(a).$$

(25) Suppose $a < b$. For any real c, show that

$$\lim_{c \to \pm\infty} \int_{\gamma_c} e^{-z^2} dz = 0,$$

where γ_c is the line segment connecting $c + ia$ to $c + ib$.

(26) Suppose $f(z)$ is continuous on $|z - z_0| \geq r_0 > 0$. Let $M(r) = \max_{|z-z_0|=r>r_0} |f(z)|$. Suppose $\lim_{r \to \infty} rM(r) = 0$. Show that $\lim_{r \to \infty} \int_{|z-z_0|=r} f(z)dz = 0$.

(27) Suppose $f(z)$ is continuous on the closed half-plane $\operatorname{Re} z \geq \sigma$ (σ is a fixed real number) and $\lim_{z \to \infty} f(z) = 0$. Let Γ_R be the half-circle shown in Fig. 2.108. Show that

$$\lim_{R \to \infty} \int_{\Gamma_R} e^{zi} f(z)dz = 0.$$

(28) (a) Suppose $f(z)$ is continuous on the half-strip: $\operatorname{Re} z \geq \sigma, 0 \leq \operatorname{Im} z \leq h$ and $\lim_{x \to \infty} f(x+iy) = A$ (a constant independent of y). Let γ_x be the line segment $z(t) = x + it, 0 \leq t \leq h$, shown in Fig. 2.109. Show that

$$\lim_{x \to \infty} \int_{\gamma_x} f(z)dz = iAh.$$

(b) Suppose $f(z)$ is continuous on the sector: $0 < |z - a| \leq R, 0 \leq \operatorname{Arg}(z - a) \leq \alpha$ ($0 < \alpha \leq 2\pi$) and $\lim_{z \to a}(z - a)f(z) = A$. Let γ_r be the circular arc $z(t) = a + re^{it}, 0 < r < R, 0 \leq t \leq \alpha$, shown in Fig. 2.110. Show that

$$\lim_{r \to 0} \int_{\gamma_r} f(z)dz = iA\alpha.$$

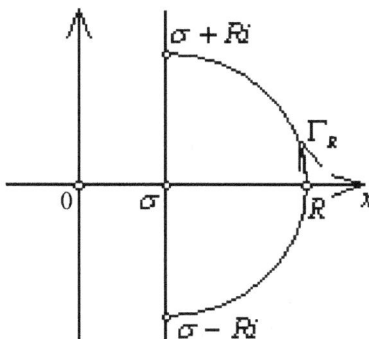

Fig. 2.108

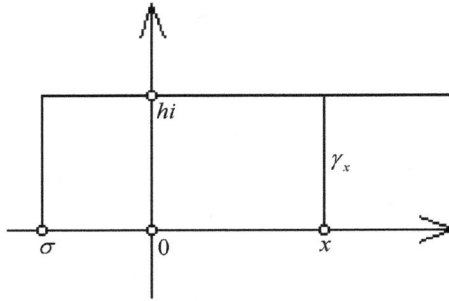

Fig. 2.109

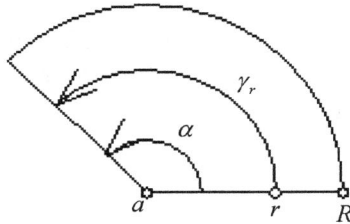

Fig. 2.110

(c) Suppose $f(z)$ is continuous on the sector: $R \le |z - a| < \infty$, $0 \le$ $\text{Arg}(z - a) \le \alpha$ $(0 < \alpha \le 2\pi)$ and $\lim_{z \to \infty}(z - a)f(z) = A$. Let Γ_r be the arc $z(t) = a + re^{it}$, $r > R$, $0 \le t \le \alpha$. Show that

$$\lim_{r \to \infty} \int_{\Gamma_r} f(z)\,dz = iA\alpha.$$

(29) Suppose $f \in C^1(\Omega)$ where Ω is an open set. Let $z_0 \in \Omega$. Show that f is differentiable in the complex sense (see Exercise A (4) of Sec. 2.8) if and only if

$$\lim_{r \to 0} \frac{1}{\pi r^2} \int_{|z - z_0| = r} f(z)\,dz = 0.$$

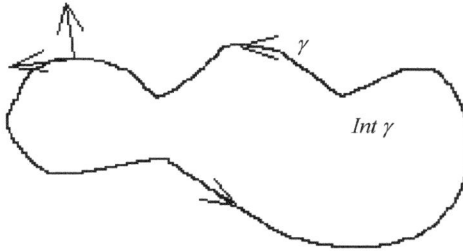

Fig. 2.111

Exercises B

(1) Suppose $G(x,y) \in C^1(\Omega)$, where Ω is an open set and $F(x,y) \in C^2(\Omega)$. In (2.9.12), letting $P = -GF_y$ and $Q = GF_x$, show that, for a rectifiable Jordan closed curve γ satisfying $\overline{\mathrm{Int}\,\gamma} \subseteq \Omega$,

$$\iint_{\mathrm{Int}\,\gamma} G\nabla^2 F\,dxdy + \iint_{\mathrm{Int}\,\gamma} <\nabla F, \nabla G>\,dxdy = \int_\gamma G\frac{\partial F}{\partial n}\,ds,$$

where $\nabla F = (F_x, F_y)$ is the gradient; $\nabla^2 F = \Delta F = F_{xx} + F_{yy}$ is the Laplacian of F; $\langle \nabla F, \nabla G \rangle$ is the natural inner product of ∇F and ∇G in $\mathbf{R}^2$, and $\frac{\partial F}{\partial n}$ is the directional derivative of F at a point along γ in the outward normal. See Fig. 2.111. In particular,

$$\iint_{\mathrm{Int}\,\gamma} \nabla^2 F\,dxdy = \int_\gamma \frac{\partial F}{\partial n}\,ds.$$

Hence, if F is harmonic on Ω, namely $\nabla^2 F = 0$ on Ω, then

$$\int_\gamma \frac{\partial F}{\partial n}\,ds = 0.$$

CHAPTER 3

Fundamental Theory: Differentiation, Integration, and Analytic Functions

Introduction

Given a complex-valued function $f(z)$ defined on a domain Ω (or open set), the following properties for $f(z)$ are *equivalent*:

(1) Differentiation. For each fixed point $z_0 \in \Omega$, the limit $\lim_{z \to z_0} \frac{f(z) - f(z_0)}{z - z_0}$ exists as a finite complex number, specifically denoted by $f'(z_0)$ (see (3.1.1), (3.2.2) and Sec. 3.3).

(2) Integration. For any triangle Δ such that Δ is contained in Ω, $\int_{\partial \Delta} f(z)dz = 0$ always holds (see (3.4.2.2) and Chap. 4). Here $f(z)$ is presumed to be continuous on Ω.

(3) Power series. For each fixed point $z_0 \in \Omega$, $f(z) = \sum_{n=0}^{\infty} a_n(z - z_0)^n$ converges in the disk $|z - z_0| < \text{dist}(z_0, \partial \Omega)$ (see (3.4.2.6) and Chap. 5).

Any one of these properties can be adopted as a definition for $f(z)$ being called an *analytic* (or *holomorphic*) *function* on Ω. For other equivalent conditions for analyticity, see (3.4.2.15). This kind of functions is the one that complex analysis studies.

The present chapter uses the differentiation process (1) as our starting point, and, in order to quickly acquire many important properties of analytic functions such as the continuity of $f'(z)$ and property (3), we lean on the integration process (2) in its simplified form.

Sketch of the Content

As a continuation of Sec. 2.8, Sec. 3.1 formally introduces the complex differentiation and its operational properties.

Section 3.2 focuses on the (complex) differentiability (see (3.2.2)): The *Cauchy–Riemann equations*, its origins and meanings from linearly algebraic, analytic, geometric, and physical viewpoints.

Formal definition for *analytic* (including *meromorphic*) *function* on a domain or at a point (or on a set) is given in Sec. 3.3. Besides their basic operational properties (see (3.3.1)) and *criterion for analyticity* (see (3.3.1.1)), we present a lot of *elementary analytic (meromerphic) functions*, such as e^z, $\cos z$, $\tan z$, $\sinh z$, $z^{\frac{1}{n}}$, $\log z$, z^{α}, and the ones defined through power series (see Sec. 3.3.2), and many others. Emphasis has been carefully paid to the *single-valued analytic branches* of a multiple-valued analytic function such as $\sqrt[n]{f(z)}$ (see (3.3.1.5)), $\log f(z)$ (see (3.3.1.8)) and $\log \log z$ (see Example 5 in Sec. 3.3.1). Crude (owing to intuitive and descriptive) idea about the Riemann surfaces on which multiple-valued functions turn out to be single-valued and analytic is clarified theoretically one step further in Sec. 3.3.3. However, Sec. 3.3.3 is better optional to a beginner or not-so-well-prepared students.

Sections 3.4 and 3.5 are the cores of this chapter (even of the whole elementary complex analysis).

Section 3.4 focuses on the fundamental *analytic properties* of analytic functions. The theoretical background lies mainly on the Cauchy–Riemann equations (see (2), (3), and (6) in (3.2.2)).

Section 3.4.1 uses a modified form of the *mean-valued theorem* to derive the most *direct* properties from the very definition of being analytic, such as the univalence on a convex domain, characteristic properties of a constant function, and a simple form of the symmetry principle.

The most striking property lies on the fact that an analytic function $f(z)$ is automatically *infinitely differentiable*. More precisely, it can be expressed locally by a *convergent* (Taylor) *power series* $\sum_{n=0}^{\infty} a_n(z - z_0)^n$ whenever it is analytic at a point z_0 (see (3.3.2.4) and (3.4.2.6)); equivalently this means that $f(z)$ can be locally expressed as $f(z) = f(z_0) + (z - z_0)f_1(z)$ where $f_1(z)$ is another analytic function at z_0 with $f_1(z_0) = f'(z_0)$ (see (3.4.2.9) and Exercise B (6) in Sec. 3.4.2). The easiest way to achieve this is via the complex integration. Section 3.4.2 introduces the *simple forms* of the *Cauchy integral theorem and formula* for these purposes. Some basic results, among all, are the *isolatedness of zeros* of a *nonconstant* analytic function and the *interior uniqueness theorem* (3.4.2.9), *Liouville's theorem* (3.4.2.10), and the *fundamental theorem of algebra* (3.4.2.12), *Morera's theorem* (3.4.2.13) and the *characteristic properties for a continuous function to be analytic* (3.4.2.15), and *the removable singularity* (3.4.2.17). The *Residue theorem* is previewed in Exercise A (2).

Both the real part Re $f(z) = u(z)$ and the imaginary part Im $f(z) = v(z)$ of an analytic $f(z) = u(z) + iv(z)$ are harmonic functions. A real-valued C^2 function $u(x, y)$ satisfying the *Laplacian equation* $\Delta u = u_{xx} + u_{yy} = 0$ in a domain Ω is called *harmonic*. A harmonic function $u(z)$ is locally expressible as the real part of an analytic function and hence enjoys the *mean-value property*

$$u(z_0) = \frac{1}{2\pi} \int_0^{2\pi} u(z_0 + re^{i\theta})d\theta, \quad r > 0,$$

which, in turn, can be used to characterize a *continuous* function to be harmonic (see (6.3.2.12)). Section 3.4.3 studies basic properties of a harmonic function and its interaction with analytic functions; in particular, the *max-min principle* for them (see (3.4.3.7)) is proved based on this mean-value property.

The max–min property is equivalent to the *open mapping property* (see (3.4.4.4)) of an analytic function. Section 3.4.4 presents some different proofs for them and extends the former to the unbounded domain, including the *Hadamard three-circle theorem*.

Section 3.4.5 singles out the *Schwarz lemma* as a meaningful application of the maximum principle. This lemma estimates the deviation of $|f(z)|$ from $|z|$ if an analytic function $f(z)$ is suitably constrained. Its generalization, the *Schwarz–Pick lemma*, and some *extremal problems* are included; the differential geometric aspect of it, reappeared as the *Schwarz–Ahlfors lemma*, will be probed in (5.8.4.14).

As a continuation of an easier version (3.4.1.5) of the symmetry principle, Sec. 3.4.6 uses the Morera theorem to prove the *Painleve's theorem* (3.4.6.1), which describes how to analytically extend an analytic function beyond the boundary curve. This will lead to the general *symmetry principles* (3.4.6.2) and (3.4.6.3) for analytic functions, and (3.4.6.4) for harmonic functions.

How to solve w as a multiple-valued function of z from the algebraic equation $z^3 + w^3 - zw = 0$? How to decide explicitly the single-valued branches of the resulted $w = w(z)$? Section 3.4.7 states, without giving proofs, the classical *inverse* and *implicit function theorems* in the style appeared in real analysis. Complex analysis treatment for them will be postponed to Secs. 3.5.1 and 3.5.6.

Section 3.5 will devote to the fundamental *geometric properties* of analytic functions, based on the primary concepts shown in (4) and (5) in (3.2.2).

Section 3.5.1 studies how $f'(z_0) \neq 0$ or $f'(z_0) = 0$ influences the *local behavior* of an analytic function $f(z)$ at a point z_0. At a point z_0 where $f'(z_0) \neq 0$, the mapping $w = f(z)$ is *conformal*, namely, preserving the angles and its senses between curves passing z_0, and acts as a *similarity motion infinitesimally* (see Fig. 3.45). While, at a point z_0 where

$f'(z_0) = \cdots = f^{(k-1)}(z_0) = 0$ but $f^{(k)}(z_0) \neq 0$ for $k \geq 2$, $w = f(z)$ *behaves exactly like* $w = z^k$ (see Sec. 2.5.2 and Fig. 3.53): It enlarges the angle between curves passing z_0 k times yet preserves its sense and assumes each value w sufficiently near $f(z_0)$ exactly k times in a neighborhood of z_0. In this case $w_0 = f(z_0)$ is called an *algebraic branch point* of order $k - 1$ of the inverse function $z = f^{-1}(w)$. As a global application, Sec. 3.5.1 also discusses the *branched covering properties* of a meromorphic function on $\mathbf{C}^*$ (i.e., a rational function).

Section 3.5.2 formalizes the concept of the *winding number* $n(\gamma; z_0)$, introduced in Sec. 2.7.1, by its integral representation and interprets its geometric meaning and properties.

As a generalization of the concept of the winding number, the *argument principle* and its *geometric interpretation* are the main theme in Sec. 3.5.3 (see (3.5.3.3) and (3.5.3.4)). A generalized version of it is in (4.11.3.4), as an application of the generalized Cauchy integral theorem (4.3.3.1) and the residue theorem (4.11.3.2).

Important applications of this principle are presented in Sec. 3.5.4 for *Rouché's theorem*, in Sec. 3.5.5 for *sufficient conditions of univalence*, and in Sec. 3.5.6 for *the inverse and the implicit function theorems*. There are complicated illustrative examples.

Finally, Sec. 3.5.7 gives concrete examples of constructing univalent conformal mappings between specified simply connected domains.

3.1.   (Complex) Differentiation

Suppose $A \subseteq \mathbf{C}$ is a point set and $z_0 \in A$ is a limit point of A. Let $f : A \to \mathbf{C}$ be a function. If the limit

$$\lim_{z \in A \to z_0} \frac{f(z) - f(z_0)}{z - z_0} \tag{3.1.1}$$

exists as a finite complex number, then f is said to be *differentiable* at z_0 *relative to A* and the finite limit is denoted as

$$f'_A(z_0) \quad \text{or} \quad \frac{d_A f}{dz}(z_0), \tag{3.1.2}$$

called the *derivative of f at z_0 relative to A.* In this case,

$$\lim_{z \in A \to z_0} (f(z) - f(z_0)) = \lim_{z \in A \to z_0} \frac{f(z) - f(z_0)}{z - z_0}(z - z_0) = f'_A(z_0) \cdot 0 = 0$$

shows that f is *continuous at z_0 relative to A.*

Unless otherwise stated, when dealing with differentiation in this text, we will always consider only the case that A *is an open set and* $z_0 \in A$ and simply denote (3.1.1) and (3.1.2) as

$$\lim_{z \to z_0} \frac{f(z) - f(z_0)}{z - z_0} = f'(z_0) \quad \text{or} \quad \frac{df}{dz}(z_0) \quad \text{(exists as a finite number).}$$

$$(3.1.3)$$

Note: For case $A \subseteq \mathbf{R}$ and f is a real-valued function on A, this definition coincides with the one we had in calculus. While, if $A \subseteq \mathbf{R}$ and f is complex-valued on A, then this definition is the one given in Sec. 2.4 for differentiable curve.

We illustrate some basic examples.

Example 1.   Let $f\colon \Omega(\text{open}) \to \mathbf{R}$ be a real-valued function. For any $z_0 \in \Omega$, then either f is not differentiable at z_0 or $f'(z_0) = 0$.

Solution.   Suppose f is differentiable. Then

$$\lim_{z \to z_0} \frac{f(z) - f(z_0)}{z - z_0}$$

$$= \begin{cases} \text{a real number,} & \text{if } z \to z_0 \text{ parallel to the real axis,} \\ \text{a pure imaginary,} & \text{if } z \to z_0 \text{ parallel to the imaginary axis.} \end{cases}$$

It follows that $f'(z_0) = 0$.

Example 2.   Show that

(1) $f(z) = \bar{z}$ is continuous everywhere on $\mathbf{C}$ but nowhere differentiable.
(2) $f(z) = z|z|$ is differentiable only at $z = 0$ and $f'(0) = 0$.
(3) $f(z) = |z|^2$ is differentiable only at $z = 0$ and $f'(0) = 0$.

Solution.

(1) Fix any $z_0 \in \mathbf{C}$. Let $z - z_0 = re^{i\theta}$, where r and θ depends on z. Then $\frac{\bar{z} - \bar{z_0}}{z - z_0} = e^{-2i\theta}$ does not approach a unique finite limit as $z \to z_0$ in all possible directions.
(2) Obviously f is differentiable at 0 and $f'(0) = 0$. For any $z_0 \in \mathbf{C}$, observe that

$$\frac{f(z) - f(z_0)}{z - z_0} = \frac{z|z| - z_0|z_0|}{z - z_0} = |z| + \frac{|z| - |z_0|}{z - z_0} z_0.$$

The problem reduces to the differentiability of $|z|$ at z_0. In what follows, we focus on the differentiability of $|z|$.

In case $z_0 = 0$: Let $z = |z|e^{i\theta}, -\pi < \theta \leq \pi$, and L_θ be the line $z(t) = te^{i\theta}$, $t \in \mathbf{R}$. Then

$$\lim_{z \in L_\theta \to 0} \frac{|z|}{z} = e^{-i\theta} \quad (\theta \text{ is kept fixed during } t \to 0^+).$$

Hence $|z|$ is not differentiable at 0. Note that the limit values of $\frac{|z|}{z}$ as $z \to 0$ fill up the whole circle $|z| = 1$.

In case $z_0 \neq 0$: Let $z - z_0 = |z - z_0|e^{i\theta}$, $-\pi < \theta \leq \pi$, and L_θ be the line $z(t) = z_0 + te^{i\theta}$, $t \in \mathbf{R}$. Then, along L_θ,

$$\frac{|z| - |z_0|}{z - z_0} = \frac{|z|^2 - |z_0|^2}{(|z| + |z_0|)(z - z_0)} = \frac{1}{|z| + |z_0|}\left\{\bar{z} + z_0 \frac{\bar{z} - \overline{z_0}}{z - z_0}\right\}$$

$$= \frac{1}{|z| + |z_0|}\{\bar{z} + z_0 e^{-2i\theta}\} \to \frac{1}{2|z_0|}\{\overline{z_0} + z_0 e^{-2i\theta}\}$$

$$= \frac{1}{2}e^{-i\theta_0} + \frac{1}{2}e^{i(\theta_0 - 2\theta)}$$

as $z \in L_\theta \to z_0$, where $\theta_0 = \text{Arg}\, z_0$. This shows that $|z|$ is not differentiable at z_0. Note that the limit values of $\frac{|z| - |z_0|}{z - z_0}$ as $z \to z_0$ fill up the circle with center $\frac{1}{2}e^{-i\theta_0}$ and radius $\frac{1}{2}$ (see (2.8.4)).

(3) is left to readers as an exercise.

Remark (About continuous everywhere and differentiable nowhere function). Treat $f(z) = \bar{z}$ as a function of two real variables x and y, where $z = x + iy$. Then $\text{Re}\, f(z) = u(x, y) = x$ and $\text{Im}\, f(z) = v(x, y) = -y$ are C^∞ functions on $\mathbf{R}^2$ and $f(x, y) = (x, -y)$ is differentiable everywhere in the "real sense". This striking contrast to (1) in Example 2 shows that real differentiation and complex differentiation differ vehemently by nature, even though their formal definitions appear seemly to be identical. For a detailed account, see Sec. 3.2.

Similar explanations apply to $\text{Arg}\, z = \tan^{-1}(\frac{y}{x})$, $|z|^2 = x^2 + y^2$, $\text{Re}\, z = x$, and $\text{Im}\, z = y$, etc. $\square$

Example 3. Let $n \geq 1$ be a positive integer. Then $f(z) = z^n$ is differentiable everywhere on $\mathbf{C}$ and

$$\frac{d}{dz}z^n = nz^{n-1}, \quad z \in \mathbf{C}. \tag{3.1.4}$$

Solution. (Just as in calculus) Using binomial expansion,

$$(z+h)^n = z^n + (nz^{n-1})h + h^2(C_2^n z^{n-2} + \cdots + C_n^n h^{n-2})$$

$$\Rightarrow \lim_{h\to 0} \frac{(z+h)^n - z^n}{h}$$

$$= \lim_{h\to 0}\{nz^{n-1} + h(C_2^n z^{n-2} + \cdots + C_n^n h^{n-2})\}$$

$$= nz^{n-1}.$$

Example 4. Let $n \geq 2$ be a positive integer. Fix any $z_0 \in \mathbf{C}$ with $z_0 \neq 0$. Then, any branch of $\sqrt[n]{z}$ on a neighborhood of z_0 (see (2.7.2.6)), still denoted by $\sqrt[n]{z}$, is differentiable at z_0 and

$$\frac{d}{dz}\sqrt[n]{z}\Big|_{z=z_0} = \frac{1}{n}z_0^{\frac{1}{n}-1} = \frac{\sqrt[n]{z_0}}{nz_0}. \qquad (3.1.5)$$

Solution. There are n distinct branches of $\sqrt[n]{z}$ on $|z-z_0| < |z_0|$. Let $\sqrt[n]{z}$ be any *one* of them, uniquely determined by its value $w_0 = \sqrt[n]{z_0}$. Then, letting $w = \sqrt[n]{z}$ on $|z-z_0| < |z_0|$,

$$\frac{\sqrt[n]{z} - \sqrt[n]{z_0}}{z - z_0} = \frac{w - w_0}{w^n - w_0^n} = \frac{1}{w^{n-1} + w^{n-2}w_0 + \cdots + w_0^{n-1}} \to \frac{1}{nw_0^{n-1}}$$

$$= \frac{1}{n}w_0^{-n+1} = \frac{1}{n}z_0^{-1+\frac{1}{n}}$$

as $z \to z_0$.

Finally, we list the
Operational properties of differentiation (refer to (2.5.1.4) and (2.5.1.5)).

(1) $\frac{dc}{dz} = 0$, where c represents a constant function.
(2) $\frac{dz}{dz} = 1$, where z represents the identity function on $\mathbf{C}$.
(3) Suppose f and g are differentiable at z and c is a constant.

  (i) Addition: $\frac{d}{dz}(f(z)+g(z)) = \frac{df(z)}{dz} + \frac{dg(z)}{dz}$ or $(f+g)'(z) = f'(z)+g'(z)$.

  (ii) Subtraction: $(f-g)'(z) = f'(z) - g'(z)$.

  (iii) Multiplication: $(cf)'(z) = cf'(z)$;
$$(fg)'(z) = f'(z)g(z) + f(z)g'(z).$$

  (iv) Division: $\left(\frac{f}{g}\right)'(z) = \frac{f'(z)g(z) - f(z)g'(z)}{g(z)^2}$, where $g(z) \neq 0$.

In particular, if $n \geq 1$ is an integer, then

$$\frac{d}{dz}(f(z))^n = n(f(z))^{n-1}\frac{df(z)}{dz}.$$

(4) Composite operation: Suppose $w = f(z)$ is differentiable at z_0 and $g(w)$ is differentiable at $w_0 = f(z_0)$. Then the composite function $g \circ f$ is differentiable at z_0 and

$$(g \circ f)'(z_0) = g'(f(z_0))f'(z_0),$$

called the *chain rule* for differentiation.

(5) Inverse function operation: Let $z = f^{-1}(w)$ be the inverse function of the one-to-one function $w = f(z)$. Suppose f is differentiable at z_0 and $f'(z_0) \neq 0$, then f^{-1} is differentiable at $w_0 = f(z_0)$ and

$$(f^{-1})'(w_0) = \frac{1}{f'(z_0)}. \tag{3.1.6}$$

Proof. Proof for each of (1)–(5) is almost exactly the same as the corresponding case in calculus. Hence, we prove only (4) and (5).

(4) Let $w = f(z)$ and $\ell = g(w)$. Also, let $\Delta z = z - z_0$, $\Delta w = f(z) - f(z_0)$ and $\Delta \ell = g(w) - g(w_0)$. Denote

$$\varphi(\Delta w) = \begin{cases} \dfrac{\Delta \ell}{\Delta w} & \text{if } \Delta w \neq 0, \\ g'(w_0) & \text{if } \Delta w = 0. \end{cases}$$

Note that in case $\Delta w = 0$ (namely, $f(z) - f(z_0) = 0$ for some z), the corresponding $\Delta \ell = g(f(z)) - g(f(z_0)) = 0$ holds too. Hence,

$$\frac{\Delta \ell}{\Delta z} = \varphi(\Delta w)\frac{\Delta w}{\Delta z}$$

holds for all z in a neighborhood of z_0. Letting $\Delta z \to 0$, the result follows.

(5) There exists a $\delta > 0$ so that, for $0 < |z - z_0| < \delta$,

$$\left| \frac{f(z) - f(z_0)}{z - z_0} - f'(z_0) \right| < \frac{1}{2}|f'(z_0)|$$

$$\Rightarrow \frac{1}{2}|f'(z_0)||z - z_0| < |f(z) - f(z_0)| < \frac{3}{2}|f'(z_0)||z - z_0|.$$

This suggests that $z = f^{-1}(w) \to z_0 = f^{-1}(w_0) \Leftrightarrow w = f(z) \to w_0 = f(z_0)$ holds.

Hence, in

$$\frac{f^{-1}(w) - f^{-1}(w_0)}{w - w_0} = \left(\frac{f(z) - f(z_0)}{z - z_0} \right)^{-1}, \quad 0 < |z - z_0| < \delta,$$

let $w \to w_0$ and the result follows. $\qquad\square$

As a consequence of (1)–(3) in (3.1.6),

$$\frac{d}{dz}(a_n z^n + a_{n-1} z^{n-1} + \cdots + a_1 z + a_0) = n a_n z^{n-1} + (n-1) a_{n-1} z^{n-2} + \cdots + a_1;$$

$$\frac{d}{dz}\frac{p(z)}{q(z)} = \frac{p'(z)q(z) - p(z)q'(z)}{q(z)^2}, \tag{3.1.7}$$

where $p(z)$ and $q(z)$ are polynomials, and $q(z) \neq 0$. These are what we obtained in (2.5.1.4) and (2.5.1.5).

Based on Example 4, in case m and $n > 0$ are integers so that $|m|$ and n are relatively prime, then

$$\frac{d}{dz} z^{\frac{m}{n}} = \frac{m}{n} z^{\frac{m}{n}-1}, \quad z \neq 0, \tag{3.1.8}$$

where $z^{\frac{m}{n}}$ represents any one of its local branches.

For further differentiation formulas for elementary functions, refer to Sec. 3.3.

Exercises A

(1) (a) Prove that $\operatorname{Re} z$ and $\operatorname{Im} z$ are continuous everywhere but differentiable nowhere.

    (b) Prove that $z \operatorname{Re} z$ and $z \operatorname{Im} z$ are differentiable only at $z = 0$.

(2) Is the function

$$f(z) = \begin{cases} 0, & z = 0, \\ |z|^2 (1+i)\operatorname{Im} z^2, & z \neq 0, \end{cases}$$

differentiable at $z = 0$? Let $f = u + iv$. Compute the partial derivatives $u_x(0)$, $u_y(0)$, $v_x(0)$ and $v_y(0)$.

(3) For each of the following functions $f(z) = u(z) + iv(z)$, do the following problems.

    (a) Find points of continuity and discontinuity, respectively.

    (b) Find points z of differentiability and then, compute $f'(z)$. In doing so, try the following methods (refer to Exercise A (4) of Sec. 2.8 or Sec. 3.2):

        (i) The definition (3.1.3).

        (ii) Compute the partial derivatives $u_x(z)$, $u_y(z)$, $v_x(z)$, and $v_y(z)$, where $z = x + iy$, and see if they satisfy $u_x(z) = v_y(z), u_y(z) = -v_x(z)$. If it is, then try to show that $f'(z) = u_x(z) + iv_x(z)$ holds.

        (iii) Compute the complex partial derivatives $\frac{\partial f}{\partial z}(z)$ and $\frac{\partial f}{\partial \bar{z}}(z)$, and see if $\frac{\partial f}{\partial \bar{z}}(z) = 0$ holds. If it is, why is $f'(z) = \frac{\partial f}{\partial z}(z)$?

(c) At point z_0 where differentiability fails, try to find all possible limit values of $\frac{f(z)-f(z_0)}{z-z_0}$ as $z \to z_0$.

(i) $f(z) = x^3 - 3xy^2 + i(3x^2y - y^3)$.

(ii) $f(z) = 3ze^{iz^2} + |z|^2 + 4$.

(iii) $f(z) = (\operatorname{Re} z)^2 + i(\operatorname{Im} z)^2$.

(iv) $f(z) = 2(\operatorname{Re} z)^3 + i(\operatorname{Im} z)^3$.

(v) $f(z) = \sqrt{|\operatorname{Re} z \cdot \operatorname{Im} z|}$.

(vi) $f(z) = (\operatorname{Re} z - \operatorname{Im} z)^2$.

(vii) $f(z) = \sin \bar{z}$.

(viii) $f(z) = \begin{cases} z^2, & \operatorname{Im} z \geq 0 \\ \bar{z}^2, & \operatorname{Im} z < 0 \end{cases}$.

(ix) $f(z) = \begin{cases} 0, & z = 0 \\ \dfrac{\operatorname{Re} z}{1 + |z|}, & z \neq 0 \end{cases}$.

(x) $f(z) = \begin{cases} 0, & z = 0 \\ \dfrac{\operatorname{Re} z}{z}, & z \neq 0 \end{cases}$.

(xi) $f(z) = \begin{cases} 0, & z = 0 \\ \dfrac{\operatorname{Re} z^2}{z^2}, & z \neq 0 \end{cases}$.

(xii) $f(z) = \begin{cases} 0, & z = 0 \\ \dfrac{(\operatorname{Re} z^2)^2}{z^2}, & z \neq 0 \end{cases}$.

(xiii) $f(z) = \begin{cases} 0, & z = 0 \\ \dfrac{x^3 - y^3 + i(x^3 + y^3)}{x^2 + y^2}, & z \neq 0 \end{cases}$.

(xiv) $f(z) = \begin{cases} 0, & z = 0 \\ \dfrac{xy^2(x + iy)}{x^2 + y^4}, & z \neq 0 \end{cases}$.

(xv) $f(z) = \begin{cases} 0, & z = 0 \\ \dfrac{x^3y(y - ix)}{x^6 + y^2}, & z \neq 0 \end{cases}$.

(xvi) $f(z) = \dfrac{z + 1}{\bar{z} - 1}$.

(xvii) $f(z) = x^4 y^5 + i x y^3$.

(xviii) $f(z) = y^2 \sin x + i y$.

(xx) $f(z) = \sin^2(x + y) + i \cos^2(x + y)$.

(xxi) $f(z) = -6(\cos x + i \sin x) + (2 - 2i)y^3 + 15(y^2 + 2y)$.

(4) Suppose $f(0) = 0$ and $f'(0) = 1$ hold. Compute the following limits.

(a) $\lim\limits_{z \to 0} \dfrac{f(az)}{z}$ ($a \neq 0$ is a constant).

(b) $\lim\limits_{z \to 0} \dfrac{f(z^{n+1} - z^n)}{z}$ ($n \geq 1$ is an integer).

(c) $\lim\limits_{z \to 0} \dfrac{f(z^n)}{z}$ ($n \geq 2$ is an integer).

(d) $\lim\limits_{z \to i} \dfrac{f(z^2 + 1)}{z - i}$.

(5) Try to use the methods shown in Exercise (3) to show that

$$\frac{d}{dz} e^z = e^z, \quad z \in \mathbf{C}$$

(refer to Exercise A (5) of Sec. 2.8). Then, use (5) in (3.1.6) to show that

$$\frac{d}{dz} \log z = \frac{1}{z}, \quad z \in \mathbf{C} - \{0\},$$

where $\log z$ represents any local branch in a neighborhood of a nonzero point.

(6) Let

$$f(z) = \begin{cases} 1, & z = \pm 1, \\ z^2 + (z - 1)(z^2 - 1)\mathrm{Log}(z^2 - 1), & z \neq \pm 1 \end{cases}.$$

Show that f is differentiable at 1 and $f'(1) = 2$, yet f is not differentiable at $z = -1$.

(7) Try to use Example 2 in Sec. 1.7 to show that, for $a \neq 0$,

$$\lim_{h \to 0} \frac{a^{z+h} - a^z}{h} = a^z \cdot \log a = \frac{d}{dz} a^z.$$

Discuss the case where $a = e$ (designate $\log e = 1$). Can we use Exercise (5) to prove this result?

3.2.   Differentiability: Cauchy–Riemann Equations, their Equivalents and Meanings

Definition (3.1.3) for *differentiability* of $f(z)$ can be rewritten as, when $f'(z)$ is supposed to be already defined,

$$\lim_{z \to z_0} \frac{f(z) - f(z_0) - f'(z_0)(z - z_0)}{z - z_0} = 0.$$

$\Leftrightarrow$ There exists a finite complex number ℓ satisfying that, for all sufficiently small $|z - z_0| > 0$,

$$f(z) = f(z_0) + \ell(z - z_0) + o(z - z_0) \quad \text{with} \quad \lim_{z \to z_0} \frac{o(z - z_0)}{z - z_0} = 0.$$

In this case, $\ell = f'(z_0)$ is uniquely defined. $\hspace{2em}$ (3.2.1)

Observe that the later relation can also be expressed as

$$f(z_0 + h) = f(z_0) + f'(z_0)h + o(h) \quad \text{with} \quad \lim_{h \to 0} \frac{o(h)}{h} = 0. \hspace{2em} (3.2.1)'$$

When comparing (3.2.1) to (2.8.2), it follows obviously that if $f(z)$ is differentiable in *the complex sense* (3.2.1), then definitely it is also differentiable in *the real sense* (2.8.2) and

$$f'(z_0) = f_z(z_0)$$
$$= f_x(z_0) = u_x(z_0) + iv_x(z_0)$$
$$= -if_y(z_0) = -i(u_y(z_0) + iv_y(z_0)) = v_y(z_0) - iu_y(z_0)$$

with $f_{\bar{z}}(z_0) = 0$.

Examples in Sec. 3.1 showed that the converse is not necessarily true, in general.

Hence, we figure out the following (one should refer to Sec. 2.8 for notations and Exercise A (4) there).

Differentiability in the complex sense. Suppose $w = f(z) = u(z) + iv(z)$: O (open in $\mathbf{C}$) $\to \mathbf{C}$ is differentiable at z_0 in the real sense (2.8.2). Then, the following are equivalent:

(1) f is differentiable at z_0 in the complex sense (3.2.1).
(2) The *total differential* (a linear operator) $df_{z_0} : \mathbf{R}^2 \to \mathbf{R}^2$ of f at z_0 is also a linear operator $df_{z_0} : \mathbf{C} \to \mathbf{C}$ on the complex vector space $\mathbf{C}$. In this case,

$$df_{z_0}(1) = f'(z_0).$$

(3) The total differential df_{z_0} of f at z_0 has the matrix representation, with respect to the natural basis for $\mathbf{R}^2$,

$$\begin{bmatrix} u_x(z_0) & v_x(z_0) \\ -v_x(z_0) & u_x(z_0) \end{bmatrix}.$$

Namely, the partial derivatives of $u(z)$ and $v(z)$ at z_0 satisfy the *Cauchy–Riemann differential equations*

$$u_x(z_0) = v_y(z_0), u_y(z_0) = -v_x(z_0).$$

In this case, the derivative

$$f'(z_0) = u_x(z_0) + iv_x(z_0) = -i(u_y(z_0) + iv_y(z_0)).$$

(4) Either $df_{z_0} = 0$ (zero linear operator on $\mathbf{R}^2$) or df_{z_0} is a *conformal mapping* on $\mathbf{R}^2$ and is of the form rP, where $r > 0$ and P is an orthogonal matrix of order 2 with $\det P = 1$.

(5) The *directional derivative* $e^{-i\theta} df_{z_0}(e^{i\theta})$ of f at z_0 in any direction $e^{i\theta}$ $(0 \le \theta < 2\pi)$ (or just in two distinct directions) is equal to a constant $f'(z_0)$, i.e.,

$$e^{-i\theta} df_{z_0}(e^{i\theta}) = f'(z_0), \quad 0 \le \theta < 2\pi.$$

In particular, $df_{z_0}(1) = -idf_{z_0}(i)$ is the *Cauchy–Riemann equations*

$$f_x(z_0) = -if_y(z_0) \quad (= f'(z_0) = f_z(z_0))$$

and $f_{\bar{z}}(z_0) = 0$.

(6) $f = f(x, y) = f(z, \bar{z})$ is independent of $\bar{z}$, namely, the *Cauchy–Riemann equations* in complex form

$$f_{\bar{z}}(z_0) = 0$$

holds. In this case,

$$f'(z_0) = f_z(z_0). \tag{3.2.2}$$

For other forms of Cauchy–Riemann equations, see Exercises A (5), (6) and (7) of Sec. 3.2.2 and Exercise A (4) 3 of Sec. 2.8.

Remark (Sufficient conditions for differentiability). Still denote $f(z) = u(z) + iv(z)$, defined on an open set or a domain Ω.

Even though u and v have partial derivatives at a point $z_0 \in \Omega$ and satisfy the Cauchy–Riemann equations there, it is not necessarily that u and v will be differentiable at z_0 and hence, f is differentiable at z_0 in the complex sense. For instance, $f(z) = \sqrt{|xy|}$ $(z = x + iy)$ is the case, where $u_x(0) = u_y(0) = v_x(0) = v_y(0) = 0$, yet if $z \to 0$ along the line $y = mx$, $\lim_{z \to 0} \frac{f(z) - f(0)}{z} = \lim_{x \to 0} \frac{\sqrt{|m|}x^2}{x + imx} = \pm \frac{\sqrt{|m|}}{1 + im}$, which indicates that the limits are not unique.

A Commonly cited

Sufficient condition for differentiability. Suppose

(1) $u(z)$ and $v(z)$ have partial derivatives in a neighborhood of z_0 and are continuous at the point z_0.
(2) The partial derivatives satisfy the Cauchy–Riemann equations at z_0, namely,

$$u_x(z_0) = v_y(z_0), \quad u_y(z_0) = -v_x(z_0). \tag{3.2.3}$$

Then, $f = u + iv$ is differentiable at z_0 in the complex sense.

This is because condition 1 that guarantees the differentiability of both u and v at z_0 and hence, the differentiability of f at z_0 in the real sense.

As a matter of fact, conditions in (3.2.3) can still be weakened to obtain the following.

Sufficient condition for differentiability on a domain. Suppose $f = u + iv$: Ω (domain) $\to$ **C** satisfying:

(1) The partial derivatives of u and v exist everywhere on Ω and satisfy the Cauchy–Riemann equations there.
(2) f is either continuous on Ω or locally bounded on Ω.   (3.2.4)

Then f is differentiable everywhere on Ω in the complex sense.
The former (continuity) is the Looman–Menchoff theorem (for a proof, see Ref. [61], pp. 5, 43–50), while the latter (local boundedness) was due to G. P. Tolstov (1942, 1943).

By the way, (3.2.2) provides clues to extend analytic functions (see Sec. 3.3) to other classes of functions, such as *analytic functions of several complex variables* (see (3.4.7.2)) and *quasiconformal mappings* (see Appendix C).   □

If the readers are familiar with the content of Sec. 2.8, it will turn out to be an easy job to prove the statements in (3.2.2). However, we divide this section into four subsections in order to probe into the appearance and meanings of the Cauchy–Riemann equations from different points of view — algebraic, analytic, geometric, and physical aspects.

3.2.1.   (*Linearly*) *Algebraic viewpoint*

Suppose Ω is an open set or a domain in $\mathbf{C}$ and $z_0 \in \Omega$ is a fixed point.

Let $f(z) = u(z) + iv(z) : \Omega \to \mathbf{C}$ be differentiable at z_0 in the real sense. Then (see (2.8.1)),

$$f(z) = f(z_0) + df_{z_0}(z - z_0) + o(z - z_0)$$

$$\Leftrightarrow u(z) = u(z_0) + du_{z_0}(z - z_0) + o(z - z_0), \quad \text{where } du_{z_0} = \operatorname{Re} df_{z_0};$$

$$v(z) = v(z_0) + dv_{z_0}(z - z_0) + o(z - z_0), \quad \text{where } dv_{z_0} = \operatorname{Im} df_{z_0}, \quad (3.2.1.1)$$

which means $\operatorname{Re} f = u$ and $\operatorname{Im} f = v$ are both differentiable at z_0 and *vice versa*. Note that $df_{z_0} : \mathbf{R}^2$ (or $\mathbf{C}$) $\to \mathbf{R}^2$ (or $\mathbf{C}$) is a linear operator while $\operatorname{Re} df_{z_0}$ and $\operatorname{Im} df_{z_0} : \mathbf{R}^2 \to \mathbf{R}$ are linear functions. Also, $\operatorname{Re} o(z - z_0) = o(z - z_0)$ and $\operatorname{Im} o(z - z_0) = o(z - z_0)$ hold.

In this case, the matrix representation of $df_{z_0} = du_{z_0} + idv_{z_0}$ with respect to the natural basis of $\mathbf{R}^2$ is

$$[df_{z_0}] = ([du_{z_0}][dv_{z_0}]) = \begin{bmatrix} u_x(z_0) & v_x(z_0) \\ u_y(z_0) & v_y(z_0) \end{bmatrix}.$$

Now, suppose f is also differentiable at z_0 in complex sense $(3.2.1)'$. Then $df_{z_0} : \mathbf{R}^2 \to \mathbf{R}^2$ acting on the vector $h = (s, t)$ as a linear operator is equivalent to the linear operator $f'(z_0) : \mathbf{C} \to \mathbf{C}$ acting on the complex number $h = s + it$ (recall Exercise C (1) of Sec. 1.3). Namely,

$$df_{z_0}(h) = (s \quad t) \begin{bmatrix} u_x(z_0) & v_x(z_0) \\ u_y(z_0) & v_y(z_0) \end{bmatrix} = f'(z_0)h \quad \text{for all } h = (s, t) = s + it.$$

Choosing $h = 1 = 1 + i0$ and $h = i = 0 + i \cdot 1$, respectively, then

$$f'(z_0) = (u_x(z_0), v_x(z_0)) = u_x(z_0) + iv_x(z_0), \quad \text{and}$$

$$if'(z_0) = (u_y(z_0), v_y(z_0)) = u_y(z_0) + iv_y(z_0) \quad \text{or}$$

$$f'(z_0) = v_y(z_0) - iu_y(z_0).$$

These two relations imply that the Cauchy–Riemann equations $u_x(z_0) = v_y(z_0)$ and $u_y(z_0) = -v_x(z_0)$ should hold at z_0. Moreover,

$$df_{z_0}(1, 0) = (u_x(z_0), v_x(z_0)), \quad \text{or}$$

$$df_{z_0}(1) = u_x(z_0) + iv_x(z_0) = f'(z_0) \quad \text{(as a complex linear operator)}.$$

$$(3.2.1.2)$$

Reverse process says that both u and v are differentiable at z_0, which, in turn, implies that f is differentiable at z_0 in the real sense.

This proves the equivalence among (1), (2), (3), and (4) in (3.2.2). Note that (4) is nothing but a geometric description of (3) in (3.2.2).

3.2.2.   *Analytic viewpoint*

Let $w = f(z) = u(z) + iv(z)$ be defined on an open set Ω. $z_0 \in \Omega$ is a fixed point.

The traditional classical way to treat the differentiability of f at z_0 is as follows.

Suppose f is differentiable at z_0. Then, if $z = x + iy$ and $z_0 = x_0 + iy_0$,

$$\lim_{z \to z_0} \frac{f(z) - f(z_0)}{z - z_0} = \lim_{z \to z_0} \left\{ \frac{u(z) - u(z_0)}{z - z_0} + i\frac{v(z) - v(z_0)}{z - z_0} \right\}$$

$$= \begin{cases} u_x(z_0) + iv_x(z_0), & \text{if } z - z_0 = x - x_0 \to 0 \\ & \text{along line parallel to } x\text{-axis,} \\ \dfrac{1}{i}(u_y(z_0) + iv_y(z_0)), & \text{if } z - z_0 = i(y - y_0) \to 0 \\ & \text{along line parallel to } y\text{-axis} \end{cases}$$

$$\Rightarrow f'(z_0) = u_x(z_0) + iv_x(z_0) = -i(u_y(z_0) + iv_y(z_0)).$$

$$(3.2.2.1)$$

And the Cauchy–Riemann equations follow.

Conversely, suppose $u(z)$ and $v(z)$ are differentiable at z_0 and $u_x(z_0) = v_y(z_0)$, $u_y(z_0) = -v_x(z_0)$ hold. Then, if $z = x + iy$, $z_0 = x_0 + iy_0$,

$$u(z) = u(z_0) + u_x(z_0)(x - x_0) + u_y(z_0)(y - y_0) + o(z - z_0),$$

$$\text{with } u_y(z_0) = -v_x(z_0),$$

$$v(z) = v(z_0) + v_x(z_0)(x - x_0) + v_y(z_0)(y - y_0) + o(z - z_0),$$

$$\text{with } v_y(z_0) = u_x(z_0),$$

$$\Rightarrow f(z) = u(z) + iv(z) = f(z_0) + (u_x(z_0) + iv_x(z_0))(z - z_0)$$

$$+ o(z - z_0),$$

which means that f is differentiable at z_0 in the complex sense with $f'(z_0) = u_x(z_0) + iv_x(z_0)$.

This proves the equivalence between (1) and (3) in (3.2.2).

For other forms of the Cauchy–Riemann equations, see Exercises A (5), (6), and (7).

Henceforth, *differentiability of functions always means in complex sense unless otherwise stated.*

We give three examples.

Example 1.   Determine these points z where $f(z) = 2\operatorname{Re} z \cdot \operatorname{Im} z + i|z|^2$ are differentiable and compute the corresponding derivatives $f'(z)$.

Solution. Let $z = x + iy$. Then $f(z) = 2xy + i(x^2 + y^2)$ with $u(z) = 2xy$ and $v(z) = x^2 + y^2$. Since $u_x(z) = 2y, u_y(z) = 2x, v_x(z) = 2x$, and $v_y(z) = 2y$ exist and are continuous everywhere on $\mathbf{C}$, $f(z)$ is differentiable at points z where $2y = 2y$ and $2x = -2x$ hold, namely, $x = 0$. Hence, by (3.2.2), $f(z)$ is differentiable only along the imaginary axis and, then, $f'(iy) = u_x(iy) + iv_x(iy) = 2y, y \in \mathbf{R}$.

Try to use the concept of directional derivative (see (2.8.4) and (5) in (3.2.2)). Fix any point $z = x + iy \in \mathbf{R}$ and $\theta, 0 \le \theta < 2\pi$. The directional derivative of f at z in the direction $e^{i\theta}$ is

$$\lim_{t \to 0+} \frac{f(z + te^{i\theta}) - f(z)}{te^{i\theta}}$$

$$= e^{-i\theta} \lim_{t \to 0+} \frac{1}{t}\{2(x + t\cos\theta)(y + t\sin\theta) + i[(x + t\cos\theta)^2$$

$$+ (y + t\sin\theta)^2] - 2xy - i(x^2 + y^2)\}$$

$$= e^{-i\theta}\{2x\sin\theta + 2y\cos\theta + i[2x\cos\theta + 2y\sin\theta]\}$$

$$= e^{-i\theta}\{2ye^{i\theta} + i \cdot 2xe^{-i\theta}\} = 2y + 2xe^{-2i\theta}i.$$

For these limits to be identical, i.e., independent of the choice of θ, $0 \le \theta < 2\pi$, it is necessary and sufficient that $x = 0$. Hence, $f(z)$ is differentiable along $x = 0$ and $f'(iy) = 2y, y \in \mathbf{R}$. By the way, observe (see (2.8.3)) that $f_x(z) = 2y + i2x, f_y(z) = 2x + i2y$ and

$$[df_z] = \begin{bmatrix} f_x(z) \\ f_y(z) \end{bmatrix} = \begin{bmatrix} 2y & 2x \\ 2x & 2y \end{bmatrix}$$

$$\Rightarrow df_z(e^{i\theta}) = df_z(\cos\theta, \sin\theta) = 2ye^{i\theta} + i2xe^{-i\theta}$$

$$\Rightarrow e^{-i\theta}df_z(e^{i\theta}) = 2y + 2xe^{-2i\theta}i.$$

The third way is to treat $f(z) = f(x, y)$ as a function of z and $\bar{z}$ (see (2.8.3)). Direct computation shows that

$$f_z(z) = \frac{1}{2}(f_x(z) - if_y(z)) = 2y \quad \text{and} \quad f_{\bar{z}}(z) = \frac{1}{2}(f_x(z) - if_y(z)) = 2xi.$$

Or, observe that

$$f(z) = 2 \cdot \frac{1}{2}(z + \bar{z}) \cdot \frac{1}{2i}(z - \bar{z}) + i|z|^2 = \frac{1}{2i}(z^2 - \bar{z}^2) + iz\bar{z},$$

$$\Rightarrow f_z(z) = \frac{1}{2i} \cdot 2z + i\bar{z} = -iz + i\bar{z} = i(\bar{z} - z) = 2y,$$

$$f_{\bar{z}}(z) = \frac{1}{2i} \cdot (-2\bar{z}) + iz = i\bar{z} + iz = 2xi.$$

Hence, f is differentiable at z if and only if $f_{\bar{z}}(z) = 2xi = 0$, or $x = 0$.

Example 2. Construct a polynomial $f(x,y)$ in x and y with complex coefficients, satisfying

(1) f is differentiable only at 0 and along the unit circle $x^2 + y^2 = 1$ and the hyperbola $x^2 - y^2 = 1$.
(2) f is not differentiable elsewhere.

Solution. According to (5) or (6) in (3.2.2), try to construct $f(x,y) = f(z)$ so that

$$f_{\bar{z}} = z(|z|^2 - 1) \cdot \frac{1}{2}(z^2 + \bar{z}^2) = \frac{1}{2}z(z\bar{z} - 1)(z^2 + \bar{z}^2)$$

$$= \frac{1}{2}(z^4\bar{z} + z^2\bar{z}^3 - z^3 - z\bar{z}^2).$$

Note that $\frac{1}{2}$ in the right might be omitted. A possible choice for $f(z)$ is $\frac{1}{4}z^4\bar{z}^2 + \frac{1}{8}z^2\bar{z}^4 - \frac{1}{2}z^3\bar{z} - \frac{1}{6}z\bar{z}^3$. Hence, choose a polynomial $p(z)$ in z only and set

$$f(z) = p(z) + \frac{1}{4}z^4\bar{z}^2 + \frac{1}{8}z^2\bar{z}^4 - \frac{1}{2}z^3\bar{z} - \frac{1}{6}z\bar{z}^3.$$

$$\Rightarrow f_z(z) = p'(z) + z^3\bar{z}^2 + \frac{1}{4}z\bar{z}^4 - \frac{3}{2}z^2\bar{z} - \frac{1}{6}\bar{z}^3,$$

$$f_{\bar{z}}(z) = \frac{1}{2}(z^4\bar{z} + z^2\bar{z}^3 - z^3 - z\bar{z}^2) = \frac{1}{2}z(|z|^2 - 1)(z^2 + \bar{z}^2).$$

Then, this $f(z)$ is one of the required polynomials.

Example 3. Fix a point $z_0 \in \mathbf{C}$ and a number $r > 0$. Try to construct a C^∞ function $f : \mathbf{C} \to \mathbf{C}$ satisfying

(1) f is differentiable on $|z - z_0| \le r$ and $f'(z) = 1$ there.
(2) f is not differentiable on $|z - z_0| > r$.

Solution. It is well known from calculus that the function

$$g(t) = \begin{cases} 0, & t \leq 0 \\ e^{-t^{-2}}, & t > 0 \end{cases}$$

is a C^∞ function. Consider the function $h : \mathbf{C} \to \mathbf{C}$ defined by

$$h(z) = g(|z - z_0|^2 - r^2) = \begin{cases} 0, & |z - z_0| \leq r \\ e^{-(|z-z_0|^2-r^2)^{-2}}, & |z - z_0| > r \end{cases}.$$

Since $|z - z_0|^2 - r^2$ is a C^∞ function on $\mathbf{C}$, so is h. Observe that

$$h_z(z) = \begin{cases} 0, & |z - z_0| \leq r, \\ e^{-(|z-z_0|^2-r^2)^{-2}} \cdot 2(|z - z_0|^2 - r^2)^{-3} \cdot (\bar{z} - \overline{z_0}), & |z - z_0| > r, \end{cases}$$

$$h_{\bar{z}}(z) = \begin{cases} 0, & |z - z_0| \leq r, \\ e^{-(|z-z_0|^2-r^2)^{-2}} \cdot 2(|z - z_0|^2 - r^2)^{-3} \cdot (z - z_0), & |z - z_0| > r. \end{cases}$$

Since $h_{\bar{z}}(z) \neq 0$ on $|z - z_0| > r$, $h(z)$ is not differentiable there but is differentiable on $|z - z_0| \leq r$. Finally, let

$$f(z) = z + h(z), \quad z \in \mathbf{C}.$$

Then $f'(z) = 1$ on $|z - z_0| \leq r$ and this f is a required one.

Exercises A

(1) Use the three methods indicated in Example 1 to do the following problems, where $f = u + iv$.

    (i) Compute u_x, u_y, v_x, and v_y.

    (ii) Compute the directional derivative $e^{-i\theta} df_z(e^{i\theta})$ (see (2.8.4)).

    (iii) Compute the complex derivatives $f_z(z)$ and $f_{\bar{z}}(z)$ (see (2.8.2) and (2.8.3)).

    (iv) Pinpoint these points z where f are differentiable and compute $f'(z)$.

      (a) $f(z) = 3z^2 + e^{\bar{z}} + z|z|^4$.

      (b) $f(z) = |z|^2(|z|^2 - 2)$.

      (c) $f(z) = \sin(|z|^2)$.

      (d) $f(z) = \text{Log}(1 - |z|^2)$.

      (e) $f(z) = z(z + \bar{z})^2$.

      (f) $f(z) = z \, Arc \, \sin|z|$ (principal branch for $\sin^{-1}$).

(g) $f(z) = \sqrt{\dfrac{\bar{z}}{z}}$.

(h) $f(z) = \begin{cases} |z|^{\alpha-1}z, & z \neq 0 \\ 0, & z = 0 \end{cases}$ ($\alpha > 0$ is a constant).

(2) Construct a C^1 function $f : \mathbf{C} \to \mathbf{C}$ so that the set of points of differentiability of f is each of the following given sets S. Then, compute $f'(z)$ at such points $z \in S$.

(a) $S = \{(x, y) \in \mathbf{R}^2 \mid y = x^2\}$.

(b) $S = \{(x, y) \in \mathbf{R}^2 \mid x^2 - y^2 = 1\}$.

(c) $S = \{z \in \mathbf{C} \mid |z|^4 - |z|^2 = 0\}$.

(d) $S = \{z \in \mathbf{C} \mid \bar{z}e^z - |z|^2 e^{\bar{z}} = 0\}$.

(3) Construct a C^1 function $f : \mathbf{C} \to \mathbf{C}$ satisfying: $z = x + iy$,

(i) f is differentiable on $|x| \leq 1$, $|y| \leq 1$ and $f'(z) = 1$ there.

(ii) f is not differentiable elsewhere.

(4) Construct a C^∞ function $f : \mathbf{C} \to \mathbf{C}$ satisfying

(i) f is differentiable on $|z| \leq 1$ with $f'(z) = 1$ and is differentiable on $|z| \geq 2$ with $f'(z) = 0$.

(ii) f is not differentiable on $1 < |z| < 2$.

(5) Suppose f is differentiable at z_0 in the real sense. Show that $f = u + iv$ is differentiable at z_0 if and only if, there are two perpendicular directions s and n through z_0 so that n is obtained by rotating s through $90°$ in the anticlockwise direction, and the directional derivative operations $\frac{\partial}{\partial s}$ and $\frac{\partial}{\partial n}$ of f at z_0 satisfy *the Cauchy–Riemann equations*

$$\frac{\partial u}{\partial s} = \frac{\partial v}{\partial n}, \quad \frac{\partial u}{\partial n} = -\frac{\partial v}{\partial s}.$$

What happens if s and n are just two linearly independent directions?

(6) Let $f(z) = u(x, y) + iv(x, y) = u(r, \theta) + iv(r, \theta)$, where (r, θ) is the polar coordinate.

(a) Show that $u(x, y)$ and $v(x, y)$ are differentiable at $z_0 = x_0 + iy_0 \neq 0$ (in the real sense) in the xy-coordinate system if and only if $u(r, \theta)$ and $v(r, \theta)$ are differentiable at z_0 in the $r\theta$-coordinate system. What are the relations between them?

(b) Show that $f(z) = u(r, \theta) + iv(r, \theta)$ is differentiable at $z_0 \neq 0$ if and only if $u(r, \theta)$ and $v(r, \theta)$ are differentiable in the real sense at z_0

and satisfy *the Cauchy–Riemann equations* at z_0 in *polar* form:

$$\frac{\partial u}{\partial r} = \frac{1}{r}\frac{\partial v}{\partial \theta}, \quad \frac{\partial u}{\partial \theta} = -r\frac{\partial v}{\partial r}.$$

In this case,

$$f'(z_0) = \frac{r}{z_0}\left(\frac{\partial u}{\partial r}(z_0) + i\frac{\partial v}{\partial r}(z_0)\right) = \frac{1}{z_0}\left(\frac{\partial v}{\partial \theta}(z_0) - i\frac{\partial u}{\partial \theta}(z_0)\right).$$

(7) Suppose $w = f(z)$ is differentiable at $z_0 = x_0 + iy_0$.

   (a) Let $w = f(z) = R(\cos\Phi + i\sin\Phi)$, where (R, Φ) is the polar coordinate in the w-plane. If $f(z_0) \neq 0$, show that *the Cauchy–Riemann equations* at z_0 are

$$\frac{\partial R}{\partial x} = R\frac{\partial \Phi}{\partial y}, \quad \frac{\partial R}{\partial y} = -R\frac{\partial \Phi}{\partial x}.$$

   In this case,

$$f'(z_0) = f(z_0)\left(\frac{1}{R}\frac{\partial R}{\partial x} + i\frac{\partial \Phi}{\partial x}\right) = f(z_0)\left(\frac{\partial \Phi}{\partial y} - i\frac{\partial R}{\partial y}\right).$$

   (b) In (a), let $z_0 = r(\cos\theta + i\sin\theta)$. Suppose $z_0 \neq 0$ and $f(z_0) \neq 0$. Show that *the Cauchy–Riemann equations* at z_0 are

$$\frac{\partial R}{\partial r} = \frac{R}{r}\frac{\partial \Phi}{\partial \theta}, \quad \frac{\partial R}{\partial \theta} = -rR\frac{\partial \Phi}{\partial r}.$$

   In this case,

$$\frac{z_0}{f(z_0)}f'(z_0) = r\left(\frac{1}{R}\frac{\partial R}{\partial r} + i\frac{\partial \Phi}{\partial r}\right) = \frac{\partial \Phi}{\partial \theta} - i\frac{1}{R}\frac{\partial R}{\partial \theta}.$$

3.2.3.   *Geometric viewpoint*

Suppose $w = f(z)$ is differentiable at z_0. Then

$$\lim_{z \to z_0}\frac{f(z) - f(z_0)}{z - z_0} = f'(z_0)$$

$$\Rightarrow \begin{cases} \displaystyle\lim_{z \to z_0}\frac{|f(z) - f(z_0)|}{|z - z_0|} = |f'(z_0)|, & (3.2.3.1) \\[2ex] \displaystyle\lim_{z \to z_0}\arg\frac{f(z) - f(z_0)}{z - z_0} = \arg f'(z_0) \text{ in case } f'(z_0) \neq 0. & (3.2.3.2) \end{cases}$$

One should refer to (1.7.3) for the meaning of the limit process concerning (3.2.3.2).

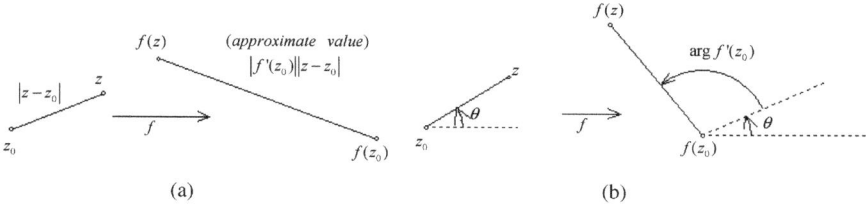

Fig. 3.1

Equation (3.2.3.1) means that if $|z - z_0| > 0$ is small enough, then the approximate estimate $|f(z) - f(z_0)| \cong |f'(z_0)||z - z_0|$ holds, namely, $|f(z) - f(z_0)|$ is $|f'(z_0)|$ multiple of $|z - z_0|$. Roughly speaking, since f is differentiable at z_0, then *the ratio of magnification of f at z_0 in any direction is the constant $|f'(z_0)|$.* See Fig. 3.1(a). While, (3.2.3.2) says that if $|z - z_0| > 0$ is small, then the approximate estimate $\arg(f(z) - f(z_0)) \cong \arg(z - z_0) + \arg f'(z_0)$ also holds. Hence, once $f'(z_0) \neq 0$, *the angle $\arg(f(z) - f(z_0))$ is obtained by rotating the angle $\arg(z - z_0)$ through a fixed constant angle $\arg f'(z_0)$.* See Fig. 3.1(b). For more accurate description about the geometric properties of $f'(z_0)$, see Sec. 3.6.1.

For the moment, we do care about the converse of (3.2.3.1) and (3.2.3.2).

The characterizations of differentiability via the ratio of magnification and rotation angle. Suppose $f(z) = u(z) + iv(z)$ is differentiable at z_0 in the real sense, i.e. both $u(z)$ and $v(z)$ are differentiable at z_0.

(1) If the limit

$$\lim_{z \to z_0} \left| \frac{f(z) - f(z_0)}{z - z_0} \right|$$

exists, then either $f(z)$ or $\overline{f(z)}$ is differentiable at z_0 (in the complex sense).

(2) If the limit

$$\lim_{z \to z_0} \text{Arg} \frac{f(z) - f(z_0)}{z - z_0} \tag{3.2.3.3}$$

exists, then $f(z)$ is differentiable at z_0.

Proof. Choose any differentiable (not necessarily continuously differentiable) curve $z(t) = x(t) + iy(t) : [0,1] \to \mathbf{C}$ passing z_0, namely, $z(0) = z_0$, and such that $z'(0) \neq 0$, the tangent vector of the curve at 0. Then

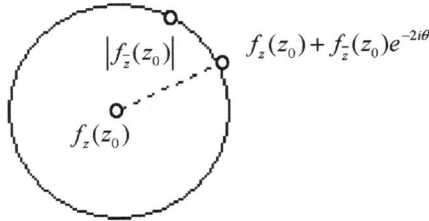

Fig. 3.2

$w(t) = f(z(t))$ is also a differentiable curve passing $w_0 = f(z(0)) = f(z_0)$.
The tangent vector of $w(t)$ at $t = 0$ is

$$w'(0) = \frac{\partial f}{\partial x}(z_0)x'(0) + \frac{\partial f}{\partial y}(z_0)y'(0)$$

$$= \frac{1}{2}\left\{\frac{\partial f}{\partial x}(z_0) - i\frac{\partial f}{\partial y}(z_0)\right\}z'(0) + \frac{1}{2}\left\{\frac{\partial f}{\partial x}(z_0) + i\frac{\partial f}{\partial y}(z_0)\right\}\overline{z'(0)}$$

$$= f_z(z_0)z'(0) + f_{\bar{z}}(z_0)\overline{z'(0)} \quad \text{(see (2.8.3))},$$

$$\Rightarrow \frac{w'(0)}{z'(0)} = f_z(z_0) + f_{\bar{z}}(z_0)e^{-2i\theta}, \tag{3.2.3.4}$$

where $z'(0) = |z'(0)|e^{i\theta}$ and $\frac{\overline{z'(0)}}{z'(0)} = e^{-2i\theta}$.

Note that (3.2.3.4) can also be derived directly from (2.8.4) by choosing there the direction θ of the tangent vector $z'(0)$ and hence, compute $\lim_{t\to 0+} \frac{w(t)-w(0)}{z(t)-z(0)}$, the direction derivative of f at z_0 along the curve $z(t)$. Observe that in (3.2.3.4), when $\theta = \text{Arg } z'(0)$ varies, $w = f_z(z_0) + f_{\bar{z}}(z_0)e^{-2i\theta}$ represents a circle with center at $f_z(z_0)$ and radius $|f_{\bar{z}}(z_0)|$. See Fig. 3.2.

(1) Now,

$$\lim_{z\to z_0}\left|\frac{f(z) - f(z_0)}{z - z_0}\right| \quad \text{exists}.$$

    ⇔ In (3.2.3.4), $\left|\frac{w'(0)}{z'(0)}\right|$ is a constant no matter how we choose the curve $z(t)$ passing 0.

    ⇔ The circle in Fig. 3.2 is a point circle, namely, $f_{\bar{z}}(z_0) = 0$, or its center is at 0, namely, $f_z(z_0) = 0$.

Then, $f_{\bar{z}}(z_0) = 0$ if and only if $f_x(z_0) = -if_y(z_0)$, and then f is differentiable at z_0; while $f_z(z_0) = 0$ if and only if $f_x(z_0) = if_y(z_0)$, i.e., $\bar{f}$ is differentiable at z_0.

(2) Similarly,

$$\lim_{z \to z_0} \text{Arg} \frac{f(z) - f(z_0)}{z - z_0} \quad \text{exists.}$$

$\Leftrightarrow$ In (3.2.3.4), $\text{Arg} \frac{w'(0)}{z'(0)}$ is a constant.
$\Leftrightarrow$ The circle in Fig. 3.2 is a point circle, i.e., $f_{\bar{z}}(z_0) = 0$.

Hence, f is differentiable at z_0. □

For another proof of (3.2.3.3), see Exercise A (1).

Exercises A

(1) For another proof of (3.2.3.3), note that

$$\frac{f(z) - f(z_0)}{z - z_0} = (u_x + iv_x)\frac{x - x_0}{z - z_0} + (u_y + iv_y)\frac{y - y_0}{z - z_0} + \frac{o(z - z_0)}{z - z_0},$$

where $u_x = u_x(z_0)$, etc.

(a) Let z approach z_0 along lines parallel to the x-axis, y-axis, and the lines $y - y_0 = \pm(x - x_0)$, respectively. Show that

$$|u_x + iv_x| = |u_y + iv_y| = \left| \frac{1}{1+i}[(u_x + iv_x) + (u_y + iv_y)] \right|$$

$$= \left| \frac{1}{1-i}[(u_x + iv_x) - (u_y + iv_y)] \right|.$$

And then, either $u_x = v_y, u_y = -v$, or $u_x = (-v)_y, u_y = -(-v)_x$ holds.

(b) Show that

$$\text{Arg} \frac{f(z) - f(z_0)}{z - z_0} = \tan^{-1}$$

$$\frac{-u_y(y - y_0)^2 + v_x(x - x_0)^2 + (-u_x + v_y)(x - x_0)(y - y_0) + o(z - z_0)}{u_x(x - x_0)^2 + v_y(y - y_0)^2 + (u_y + v_x)(x - x_0)(y - y_0) + o(z - z_0)},$$

where $\tan^{-1}$ takes values between $-\pi$ and π. Then, try to show that

$$\frac{v_x}{u_x} = \frac{-u_y}{v_y} = \frac{-u_x - u_y + v_x + v_y}{u_x + u_y + v_x + v_y}$$

and hence, equal to $\frac{-u_x + v_y}{u_y + v_x}$. Hence, $u_x = v_y$ and $u_y = -v_x$.

(2) Let $f(z) = u(z) + iv(z)$ and z_0 be a fixed point.

(a) If

$$\lim_{z \to z_0} \operatorname{Re} \frac{f(z) - f(z_0)}{z - z_0}$$

exists, show that both $u_x(z_0)$ and $v_y(z_0)$ exist and are equal.

(b) If

$$\lim_{z \to z_0} \operatorname{Im} \frac{f(z) - f(z_0)}{z - z_0}$$

exists, show that both $u_y(z_0)$ and $v_x(z_0)$ exist and $u_y(z_0) = -v_x(z_0)$.

(c) Suppose $u(z)$ and $v(z)$ are differentiable at z_0. If one of the limits in (a) and (b) exists, show that the other limit will also exist and then f is differentiable at z_0.

3.2.4. *Physical viewpoint*

Consider the motion of an incompressible fluid in a space domain Ω (a nonempty open connected set in $\mathbf{R}^3$). If, to each point $P \in \Omega$ and at any time t, there associates a vector function $\vec{v}(P, t)$ expressing the velocity of the fluid at point P and time t, then $\vec{v}$ is called a *velocity field* or *flow pattern* on Ω. In case $\vec{v}(P, t) = \vec{v}(P)$ is independent of time t, the flow is called *stationary*. It is called *plane-parallel* if $\vec{v}$ is the same on all planes parallel to a given plane Σ and has no component orthogonal to Σ.

Treat Σ as the complex plane. Then, the velocity field of a stationary plane-parallel flow can be and will be always expressed in this section as

$$w(z) = u(z) + iv(z), \quad z = x + iy,$$

where $u(z)$ and $v(z)$ are x- and y-components, respectively. In what follows, $u(z)$ and $v(z)$ are supposed to be continuously differentiable in a neighborhood of a point $z_0 = x_0 + iy_0$.

Section (1) Divergence (the flux per unit volume at a point)

Construct a square with center at z_0 and side length $2h\,(h > 0)$. We try to estimate the amount of the flux across the boundary of the square. See Fig. 3.3. More precisely, we have to consider a rectangular box with the square as base and having height equal to one unit. See Fig. 3.4. Then, "the amount of the flux across the boundary" of the square should be interpreted as "the amount of the flux across the four vertical faces" of the rectangular box.

The outward direction perpendicular to the boundary is considered as the *positive* direction of the normal to the boundary of the square.

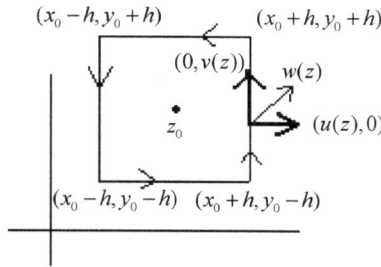

$(x_0 - h, y_0 + h)$ $(x_0 + h, y_0 + h)$

$(0, v(z))$ $w(z)$

z_0

$(u(z), 0)$

$(x_0 - h, y_0 - h)$ $(x_0 + h, y_0 - h)$

Fig. 3.3

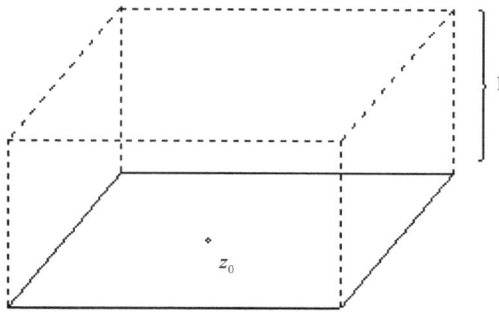

z_0

Fig. 3.4

The right vertical side of the square has its points $z = (x_0 + h, y_0 + s)$, $-h \le s \le h$, and arc length element ds, while the right face of the box has the corresponding area element $1 \cdot ds = ds$. On the other hand, the velocity at a point on the right face has its component $u(z)$ along the positive normal. See Fig. 3.5.

Now,

the flux across the area element $1 \cdot ds = u(z)ds$

$\Rightarrow$ the flux across the right face of the rectangular box (see Fig. 3.4)

$$= \int_{-h}^{h} u(z)ds = \int_{-h}^{h} u(x_0 + h, y_0 + s)ds.$$

Similarly,

the flux across the left face $= \int_{-h}^{h} -u(z)ds = -\int_{-h}^{h} u(x_0 - h, y_0 + s)ds;$

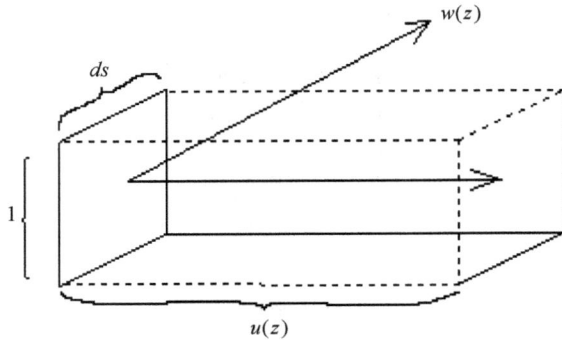

Fig. 3.5

the flux across the upper face $= \int_{-h}^{h} v(z)ds = \int_{-h}^{h} v(x_0 + s, y_0 + h)ds;$

the flux across the lower face $= \int_{-h}^{h} -v(z)ds = -\int_{-h}^{h} v(x_0 + s, y_0 - h)ds.$

Note that, on the left face, the component of $w(z)$ along the outward normal $(-1, 0)$ is $-u(z)$, instead of $u(z)$. Similar explanation holds for the lower face.

Combining together,

the total flux F across the square

$$= \int_{-h}^{h} \{[u(x_0 + h, y_0 + s) - u(x_0 - h, y_0 + s)]$$

$$+ [v(x_0 + s, y_0 + h) - v(x_0 + s, y_0 - h)]\}ds$$

$\Rightarrow$ the flux per unit volume (around z_0) $\dfrac{F}{4h^2}$

$$= \frac{1}{2h} \int_{-h}^{h} \left[\frac{u(x_0 + h, y_0 + s) - u(x_0 - h, y_0 + s)}{2h} \right.$$

$$+ \left. \frac{v(x_0 + s, y_0 + h) - v(x_0 + s, y_0 - h)}{2h} \right] ds$$

$$= \frac{1}{2h} \int_{-h}^{h} [u_x(x_0 + t_1, y_0 + s) + v_y(x_0 + s, y_0 + t_2)]ds$$

(by mean value theorem for differentiation),

where $-h \leq t_1$, $t_2 \leq h$. Then, by mean value theorem for integration, we have

$$\lim_{h \to 0} \frac{F}{4h^2} = u_x(z_0) + v_y(z_0),$$

which measures *the amount of flux per unit volume at* z_0. Specifically, call

$$\text{div } w = \frac{\partial u}{\partial x} + \frac{\partial v}{\partial y} \qquad (3.2.4.1)$$

the *divergence* (at a point) of the velocity field $w(z) = u(z) + iv(z)$ of the flow.

Section (2) Curl (the circulation around a point)

Orient the boundary of the square in Fig. 3.3 with anticlockwise direction. We try to calculate "the amount of flux (or circulation) around the boundary" of the square.

The velocity $w(z)$ at a point z on the right vertical side of the square has its component $v(z)$ along the boundary (see Fig. 3.3). Hence,

the flux along the arc length element $ds = v(z)ds$

$\Rightarrow$ the flux along the right vertical side

$$= \int_{-h}^{h} v(z)ds = \int_{-h}^{h} v(x_0 + h, y_0 + s)ds.$$

Similarly,

the flux along the left vertical side

$$= \int_{h}^{-h} v(z)ds = -\int_{-h}^{h} v(x_0 - h, y_0 + s)ds;$$

the flux along the upper horizontal side

$$= \int_{h}^{-h} u(z)ds = -\int_{-h}^{h} u(x_0 + s, y_0 + h)ds;$$

the flux along the lower horizontal side

$$= \int_{-h}^{h} u(z)ds = -\int_{-h}^{h} u(x_0 + s, y_0 - h)ds.$$

Combining together,

the circulation W along the boundary of the square

$$= \int_{-h}^{h} \{[v(x_0 + h, y_0 + s) - v(x_0 - h, y_0 + s)]$$
$$- [u(x_0 + s, y_0 + h) - u(x_0 + s, y_0 - h)]\}ds$$

$\Rightarrow$ the circulation per unit area (around z_0) $\frac{W}{4h^2}$

$$= \frac{1}{2h} \int_{-h}^{h} \left[\frac{v(x_0 + h, y_0 + s) - v(x_0 - h, y_0 + s)}{2h} \right.$$
$$\left. - \frac{u(x_0 + s, y_0 + h) - u(x_0 + s, y_0 - h)}{2h} \right] ds$$

$$= \frac{1}{2h} \int_{-h}^{h} [v_x(x_0 + t_1, y_0 + s) - u_y(x_0 + s, y_0 + t_2)]ds$$
$$(-h \le t_1, t_2 \le h)$$

$$\Rightarrow \lim_{h \to 0} \frac{W}{4h^2} = v_x(z_0) - u_y(z_0),$$

which measures *the circulation of the flow per unit area at z_0*. Call

$$\operatorname{curl} w = \frac{\partial v}{\partial x} - \frac{\partial u}{\partial y} \tag{3.2.4.2}$$

the *curl* (at a point) of the velocity field $w(z) = u(z) + iv(z)$ of the flow.

Section (3) The physical meaning of the Cauchy–Riemann equations

z_0 is called a *source* of the flow if div $w(z_0) = u_x(z_0) + v_y(z_0) > 0$, a *sink* if div $w(z_0) < 0$ and $|\text{div } w(z_0)|$ the *strength* of the source (sink). See Fig. 3.6.

z_0 is called a *vortex* of the flow if curl $w(z_0) = v_x(z_0) - u_y(z_0) \ne 0$, which is called the *strength* of the vortex. See Fig. 3.7.

If the divergence div w of the velocity field $w(z)$ of a fluid on a domain Ω is identically equal to zero, namely,

$$\text{div } w(z) = u_x(z) + v_y(z) = 0, \quad z \in \Omega,$$

then $w(z) = u(z) + iv(z)$ is said to be *solenoidal* on Ω; it is called *irrotational* if

$$\text{curl } w(z) = v_x(z) - u_y(z) = 0, \quad z \in \Omega.$$

Observe that $u_x(z) = -v_y(z) = (-v)_y(z)$ and $u_y(z) = v_x(z) = -(-v)_x(z)$.

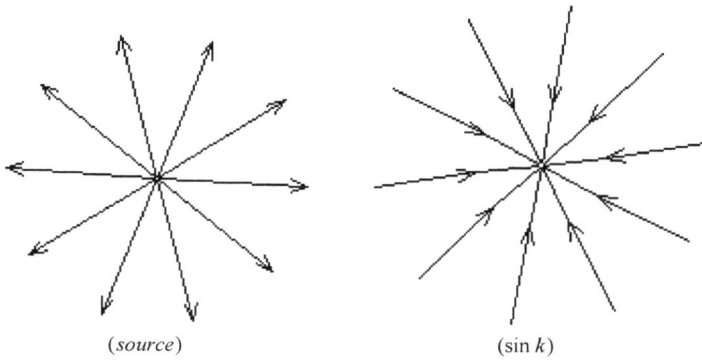

(*source*)   ($\sin k$)

Fig. 3.6

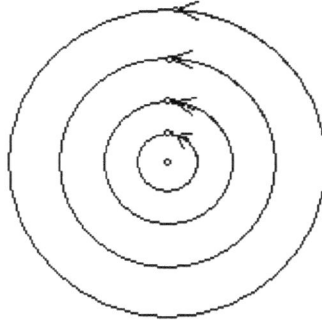

Fig. 3.7

The physical interpretation of differentiability. Suppose $u(z), v(z) : \Omega$ (simply connected domain) $\to \mathbf{R}$ are continuously differentiable in the real sense. Then

(1) $w(z) = u(z) + iv(z)$ is the velocity field of a solenoidal, irrotational flow on Ω.

$\Leftrightarrow$ (2) The conjugate function $\bar{w}(z) = u(z) - iv(z)$ is differentiable on Ω.

In particular, if $z_0 \in \Omega$ is neither a source (sink) nor a vortex of the velocity field $w(z)$, then $\bar{w}(z)$ is differentiable at z_0.   (3.2.4.3)

Remark (Integration approach).   Suppose $u(z)$ and $v(z)$ are C^1 functions on a subdomain D of the flow domain Ω. It is to allow that $u(z)$ and $v(z)$ may not be defined or continuously differentiable at *some* point of Ω.

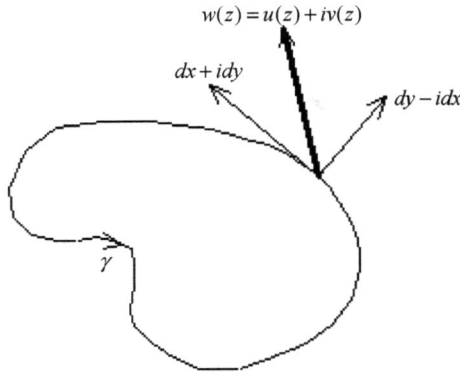

Fig. 3.8

Let γ be a piecewise smooth Jordan closed curve in D. Using differentials, the tangent vector at a point on γ is $dx + idy = (dx, dy)$, while the outward (external) normal to γ at the point is $dy - idx = (dy, -dx)$. See Fig. 3.8. Adopt the natural inner product $\langle\, , \rangle$ on $\mathbf{R}^2$. Then

$$\langle w(z), (dx, dy) \rangle = u(z)dx + v(z)dy,$$

$$\langle w(z), (dy, -dx) \rangle = -v(z)dx + u(z)dy,$$

$$\Rightarrow \int_\gamma w(z)dz = \int_\gamma \langle w(z), (dx, dy) \rangle + i \int_\gamma \langle w(z), (dy, -dx) \rangle, \qquad (3.2.4.4)$$

where, by using Green's formula (2.9.12),

$$\int_\gamma \langle w(z), (dx, dy) \rangle$$

$$= \iint_{\text{int } \gamma} (v_x(z) - u_y(z))dxdy \cdots \text{the } \textit{circulation} \text{ of the flow around } \gamma;$$

$$\int_\gamma \langle w(z), (dy, -dx) \rangle$$

$$= \iint_{\text{int } \gamma} (u_x(z) + v_y(z))dxdy \cdots \text{the } \textit{flux} \text{ of the flow through } \gamma.$$

$$(3.2.4.5)$$

If the circulation vanishes for every piecewise smooth Jordan closed curve γ in D, the flow $w(z) = u(z) + iv(z)$ is said to be *irrotational* in D; it is *solenoidal* in D if the flux vanishes for every such γ in D. In this case, (3.2.4.3) also follows.

Let $z_0 \in \Omega$, and $u(z)$ and $v(z)$ be C^1 in a deleted neighborhood $\tilde{D}_\delta(z_0) = \{0 < |z - z_0| < \delta\}$ of z_0. z_0 is called a *source* or a *sink* if

$$\int_\gamma \langle w(z), (dy, -dx) \rangle = \text{a constant} > 0 \quad \text{or} < 0,$$

for all $\gamma \subseteq \tilde{D}_\delta(z_0)$ with $z_0 \in \text{Int} \, \gamma$, respectively, and the absolute value of this constant is called the *strength* of this source (sin k). While, z_0 is called a *vortex* if

$$\int_\gamma \langle w(z), (dx, dy) \rangle = \text{a nonzero constant},$$

for all $\gamma \subseteq \tilde{D}_\delta(z_0)$ with $z_0 \in \text{Int} \, \gamma$ and this constant is called the *strength* of the vortex. $\qquad\qquad\qquad\qquad\qquad\qquad\qquad\qquad\qquad\qquad\qquad\qquad$ $\square$

Exercises A

(1) Try to use $w = \frac{1}{\bar{z}}$, $z \neq 0$, to justify the process and the result of this section.

3.3.   Analytic Functions

A function $f(z)$, which is differentiable at every point of an *open* set $O \subseteq \mathbf{C}$, simply called *differentiable on* O, is said to be *analytic* (or *holomorphic*) on O.

If $f(z)$ is analytic in a neighborhood of z_0, then f is said to be *analytic at* z_0. Under this restricted definition, $|z|^2$ and $z|z|$ are not analytic at 0 even though they both are differentiable at 0 (see Example 3 in Sec. 3.1).

To say that $f(z)$ is analytic on a nonempty set S, we mean that it is analytic on an open set containing S.

Remark (Analyticity at ∞ and at point $\mathbf{z_0}$ where $\mathbf{f(z_0)} = \infty$). Suppose $f(z)$ is defined on $r < |z| \le \infty$, including ∞, where $f(\infty) \in \mathbf{C}$. If the function $g(z) = f\left(\frac{1}{z}\right)$, $|z| < \frac{1}{r}$ is analytic at 0, then $f(z)$ is said to be *analytic* at ∞ with $f'(\infty) = g'(0)$. Note that, in general, $f'(\infty) \neq \lim_{z \to \infty} f'(z)$ which is equal to zero if f is analytic at ∞; but $f(\infty) = \lim_{z \to \infty} f(z) = \lim_{z \to 0} f\left(\frac{1}{z}\right)$ always exists as a finite complex number. For instance, $f(z) = \frac{1}{z}$ is analytic at ∞ with $f'(\infty) = 1$, so is $f(z) = \frac{1}{z^2}$ with $f'(\infty) = 0$.

In case $f(z)$ is defined on and $r < |z| \le \infty$ and $f(\infty) = \infty$, $f(z)$ is said to be *analytic at* ∞ if $g(z) = \left(f\left(\frac{1}{z}\right)\right)^{-1}$ is analytic at 0 and has $f'(\infty) = g'(0)$. For instance, $f(z) = z$ is analytic at ∞ with $f'(\infty) = 1$ and $f(z) = z^2$ is analytic at ∞ with $f'(\infty) = 0$.

346 Chap. 3. Differentiation, Integration, and Analytic Functions

If $f(z)$ is defined on $0 < |z - z_0| < r$ and $f(z_0) = \infty$, $f(z)$ is said to be *analytic at* z_0 if $g(z) = (f(z))^{-1}$ is analytic at z_0 and has $f'(z_0) = g'(z_0)$. For instance, $f(z) = \frac{1}{z}$ is analytic at 0 with $f'(0) = 1$, so is $f(z) = \frac{1}{z^2}$ with $f'(0) = 0$.

Note that, *in this text, we would not accept this generalized definition for analyticity unless otherwise specified.*

Later in (3.5.1.1), we will adopt these definitions for conformality at a point z_0 where $z_0 = \infty$ with $f'(\infty) \neq 0$ or $f(z_0) = \infty$ with $f'(z_0) \neq 0$. Refer to Section (1) in Sec. 2.5.4.

Based on (3), (4), and (5) of (3.1.6), it is easy to see the following

Basic operation properties of analytic functions.

(1) Suppose f and g are analytic at a point z. Then,

$$f + g, \quad f - g, \quad cf(c \in \mathbf{C}), \quad \frac{f}{g}\ (g(z) \neq 0)$$

are also analytic at z.

(2) Composite operation: Suppose f is analytic at z and g is analytic at $f(z)$, then

$$g \circ f$$

is also analytic at z.

(3) Inverse function: Suppose $w = f(z) : \mathrm{O(open)} \to f(\mathrm{O})(\mathrm{open})$ is a one-to-one analytic function, which implicitly implies that $f'(z) \neq 0$ for every $z \in \mathrm{O}$. Then, the inverse function $g = f^{-1}(w) : f(\mathrm{O}) \to \mathrm{O}$ is also an analytic function. \hfill (3.3.1)

As for (3), the inverse function theorem, see Secs. 3.4.7, 3.5.1, and 3.5.6 for detailed account.

As an easy consequence of (3.3.1), we have

the constant function $f(z) = c$;

the identity function $f(z) = z$, and

the polynomial $p(z) = a_n z^n + a_{z-1} z^{n-1} + \cdots + a_1 z + a_0$

are analytic on the entire plane $\mathbf{C}$; while

the rational function $\dfrac{p(z)}{q(z)}$

is analytic everywhere on $\mathbf{C}$ except at points z where $q(z) = 0$.

For further examples for analytic functions, see Secs. 3.3.1 and 3.3.2.

3.3.1.   *Basic examples*

According to (3.2.4), we have an easy

sufficient criterion on analyticity. Suppose both $u(z)$ and $v(z)$ are continuously differentiable functions on an open set O so that their partial derivatives satisfy the Cauchy–Riemann equations everywhere in O. Then $f(z) = u(z) + iv(z)$ is analytic on O and the derivative $f'(z) = u_x(z) + iv_x(z)$ is also continuous on O.   (3.3.1.1)

We present some illustrative examples.

Example 1. e^z is analytic on **C** and

$$\frac{d}{dz} e^z = e^z, \quad z \in \mathbf{C}. \tag{3.3.1.2}$$

Solution. Let $z = x + iy$. Now, both $u(z) = \operatorname{Re} e^z = e^x \cos y$ and $v(z) = \operatorname{Im} e^z = e^x \sin y$ are C^1-functions on **C**. Also, $u_x(z) = e^x \cos y = v_y(z)$ and $u_y(z) = -e^x \sin y = -v_x(z)$ hold everywhere on **C**. Hence, e^z is analytic on **C** and

$$\frac{d}{dz} e^z = u_x(z) + iv_x(z) = e^x \cos y + ie^x \sin y = e^z, \quad z \in \mathbf{C}.$$

Recall that $\cos z = \frac{1}{2}(e^{iz} + e^{-iz})$. Hence, $\cos z$ is analytic on **C** and

$$\frac{d}{dz} \cos z = \frac{1}{2}(ie^{iz} - ie^{-iz}) = \frac{-1}{2i}(e^{iz} - e^{-iz}) = -\sin z, \quad z \in \mathbf{C}.$$

Suppose it is known that $(\cos z)' = -\sin z$ and $(\sin z)' = \cos z$. In case $z \neq n\pi$ for $n = 0, \pm 1, \pm 2, \ldots$, then

$$\frac{d}{dz} \cot z = \frac{d}{dz} \frac{\cos z}{\sin z} = \frac{-\sin z \sin z - \cos z \cos z}{\sin^2 z} = \frac{-1}{\sin^2 z} = -\csc^2 z.$$

And so on. Therefore, we have the following
formulas for derivatives of trigonometric and hyperbolic functions.

(1) Trigonometric functions:

$$\frac{d}{dz} \cos z = -\sin z, \quad z \in \mathbf{C},$$

$$\frac{d}{dz} \sin z = \cos z, \quad z \in \mathbf{C},$$

$$\frac{d}{dz} \tan z = \sec^2 z, \quad z \neq \frac{\pi}{2} + n\pi, \quad n = 0, \pm 1, \pm 2, \ldots,$$

$$\frac{d}{dz} \cot z = -\csc^2 z, \quad z \neq n\pi, \quad n = 0, \pm 1, \pm 2, \ldots,$$

$$\frac{d}{dz} \sec z = \sec z \tan z, \quad z \neq \frac{\pi}{2} + n\pi, \quad n = 0, \pm 1, \pm 2, \dots,$$

$$\frac{d}{dz} \csc z = -\csc z \cot z, \quad z \neq n\pi, \quad n = 0, \pm 1, \pm 2, \dots.$$

(2) Hyperbolic functions:

$$\frac{d}{dz} \cosh z = \sinh z, \quad z \in \mathbf{C},$$

$$\frac{d}{dz} \sinh z = \cosh z, \quad z \in \mathbf{C},$$

$$\frac{d}{dz} \tanh z = \sec h^2 z, \quad z \neq \frac{\pi}{2} + n\pi i, \quad n = 0, \pm 1, \pm 2, \dots,$$

$$\frac{d}{dz} \coth z = -\csc h^2 z, \quad z \neq n\pi i, \quad n = 0, \pm 1, \pm 2, \dots,$$

$$\frac{d}{dz} \sec h z = -\sec h z \tanh z, \quad z \neq \frac{\pi i}{2} + n\pi i, \quad n = 0, \pm 1, \pm 2, \dots,$$

$$\frac{d}{dz} \csc h z = -\csc h z \coth z, \quad z \neq n\pi i, \quad n = 0, \pm 1, \pm 2, \dots.$$

$$(3.3.1.3)$$

Proofs are left to the readers.

Remark 1 (Entire and meromorphic functions). A function which is analytic in the whole plane $\mathbf{C}$ is called an *entire function*, such as polynomials, e^z, $\cos z$, $\sin z$, $\cosh z$, and $\sinh z$. This type of analytic functions inherits many algebraic and analytic properties of polynomials.

If $f(z)$ fails to be analytic at a point z_0 but is analytic at some point in any neighborhood of z_0, then such a point z_0 is called a *singular point* of $f(z)$. In case there exists a $r > 0$ so that $f(z)$ is analytic on $0 < |z - z_0| < r$, namely, $f(z)$ is analytic on $|z - z_0| < r$ except possibly at z_0, then z_0 is called an *isolated singularity* of $f(z)$.

Suppose z_0 is an isolated singularity of $f(z)$ so that

$$\lim_{z \to z_0} f(z) = \infty$$

holds. Such a point z_0 is called a *pole* of $f(z)$. A function $f(z)$, defined on an open set O in $\mathbf{C}$, is called a *meromorphic function* if $f(z)$ is analytic everywhere in O except poles. Such functions are rational functions, $\cot z, \tan z,$

etc. Meromorphic functions imitate many properties of rational functions. Observe that ∞ is a singular point of tan z but is not isolated.   □

When we are talking about the analyticity of a multi-valued function, we always mean the analyticity of its (single-valued continuous) *branches* defined on their respective single-valued domains. Under this circumstance, such a branch is called an *analytic branch* and the domain on which this analyticity occurs is called a *single-valued* or an *analytic domain*, for simplicity. Basic knowledge about multiple-valued functions are needed for better and clear understanding (see Sec. 2.7). For more rigorous description about analyticity of multiple-valued functions, see Sec. 3.3.3.

Example 2.   Show that, for positive integer $n \geq 2$,

$$\frac{d}{dz} z^{\frac{1}{n}} = \frac{1}{n} z^{\frac{1}{n}-1}, \quad z \neq 0.$$

This differentiation formula should be understood in two aspects (refer to Sec. 2.7.2):

(1) (local) $z^{\frac{1}{n}}$ represents any branch of the n-valued function $z^{\frac{1}{n}}$ in a neighborhood (not containing o) of any nonzero point in **C**. Hence, (such a branch) $z^{\frac{1}{n}}$ is analytic on that neighborhood with the derivative $\frac{1}{n} z^{\frac{1}{n}-1}$ (refer to Exercises A (16) and (17)).

(2) (global) Let Ω be a domain in **C** for which 0 and ∞ belong to the same component of $\mathbf{C}^* - \Omega$, such as $\mathbf{C}^* - [\infty, 0]$, etc. Then, any branch $z^{\frac{1}{n}}$ is analytic on Ω with the derivative $\frac{1}{n} z^{\frac{1}{n}-1}$.   (3.3.1.4)

Solution.   Fix a point $z_0 \in \mathbf{C}$, where $z_0 \neq 0$. Let θ_0 be any fixed value of $\arg z_0 = \operatorname{Arg} z_0 + 2n\pi$ $(n = 0, \pm 1, \pm 2, \ldots)$ and, for $0 < \delta < |z_0|$, $D_\delta(z_0) = \{|z - z_0| < \delta\}$ be a neighborhood of z_0 not containing 0. For $z \in D_\delta(z_0)$, choose the unique argument θ of z satisfying $|\theta - \theta_0| < \frac{\pi}{2}$ and thus, define a branch of $z^{\frac{1}{n}}$ on $D_z(z_0)$, still denoted as $z^{\frac{1}{n}} = u + iv$. Now, $z = re^{i\theta} \in D_\delta(z_0)$ and

$$u(z) = r^{\frac{1}{n}} \cos \frac{\theta}{n}, \quad v(z) = r^{\frac{1}{n}} \sin \frac{\theta}{n}$$

$\Rightarrow$ (by use of Exercise A (6) of Sec. 3.2.2)

$$\frac{\partial u}{\partial r} = \frac{1}{n} r^{\frac{1}{n}-1} \cos \frac{\theta}{n} = \frac{1}{r} \frac{\partial v}{\partial \theta}, \quad \frac{\partial u}{\partial \theta} = -\frac{1}{n} r^{\frac{1}{n}} \sin \frac{\theta}{n} = -r \frac{\partial v}{\partial r}$$

$\Rightarrow z^{\frac{1}{n}} = u + iv$ is differentiable at $z \in D_\delta(z_0)$, and

$$\frac{d}{dz} z^{\frac{1}{n}} = \frac{r}{z}\left(\frac{\partial u}{\partial r} + i\frac{\partial v}{\partial r}\right) = \frac{r}{z}\left(\frac{1}{n}r^{\frac{1}{n}-1}\cos\frac{\theta}{n} + i\,\frac{1}{n}r^{\frac{1}{n}-1}\sin\frac{\theta}{n}\right)$$

$$= \frac{1}{nz}\cdot z^{\frac{1}{n}} = \frac{1}{n}z^{\frac{1}{n}-1}.$$

Or, we consider the inverse function $z = w^n$. Choose $w_0 \neq 0$ so that $z_0 = w_0^n$. According to Sec. 2.5.2, choose a sector domain $\Omega : \varphi_0 < \arg w < \varphi_0 + \frac{2\pi}{n}$ so that $w_0 \in \Omega$ and then, $z = w^n$ maps Ω one-to-one and onto $\mathbf{C} - \{\arg z = n\varphi_0\}$ (including 0). On Ω, $\frac{dz}{dw} = nw^{n-1} \neq 0$. By using (3) in (3.3.1), the inverse function $w = z^{\frac{1}{n}} : \mathbf{C} - \{\arg z = n\varphi_0\} \to \Omega$ is analytic and

$$\frac{d}{dz} z^{\frac{1}{n}} = \frac{1}{\frac{dw^n}{dw}} = \frac{1}{nw^{n-1}} = \frac{1}{n}(z^{-\frac{1}{n}})^{n-1} = \frac{1}{n}z^{\frac{1}{n}-1},$$

$$z \in \mathbf{C} - \{\arg z = n\varphi_0\}.$$

Based on Example 2 and (2) in (3.3.1), we have the following important *Single-valued analytic branches of* $\sqrt[n]{f(z)}$. Suppose $f : \Omega$ (a domain in $\mathbf{C}$) $\to \mathbf{C}$ is analytic and $f(\Omega)$ is a domain (see (3.4.4.4), a superfluous assumption!).

(1) Suppose 0 and ∞ belong to the same component of the set $\mathbf{C}^* - f(\Omega)$. Then, there are exactly n distinct analytic branches $\sqrt[n]{f(z)}$ on Ω and, for each branch,

$$\frac{d}{dz}\sqrt[n]{f(z)} = \frac{1}{n}f(z)^{\frac{1}{n}-1}\cdot f'(z), \quad n \in \Omega.$$

(2) Suppose Ω is a simply connected domain on which $f(z) \neq 0$. Then Ω is a single-valued domain, namely, an analytic domain for $\sqrt[n]{f(z)}$ on which (1) holds.   (3.3.1.5)

For instance, if $|m|$ and n are positive integers, relatively prime, and Ω is as in (3.3.1.5), then $z^{\frac{m}{n}}$ is analytic on Ω and

$$\frac{d}{dz} z^{\frac{m}{n}} = \frac{m}{n}z^{\frac{m}{n}-1}, \quad z \in \Omega. \qquad (3.3.1.6)$$

Example 3. Determine domains on which each of the following multiple-valued functions has analytic branches and compute the corresponding derivatives.

(1) $f(z) = \sqrt{z^2 - 1}$.
(2) $f(z) = \sqrt{z+1} + \sqrt{z-1}$.
(3) $f(z) = \sqrt[3]{z^2 - 1}$.

(4) $f(z) = \sqrt[6]{(z-i)^2(z+i)^3}$.

(5) $f(z) = \sqrt[3]{\dfrac{z-1}{z^2}} + \sqrt{z-i}$.

(6) $f(z) = \sqrt{1+\sqrt{z}}$.

Solution. (1) (Refer to Example 2 in Sec. 2.7.2). For $z_0 \neq \pm 1$ and ∞, $f(z)$ has two analytic branches on $|z - z_0| < \min\{|z_0-1|, |z_0+1|\}$ on which each branch has the derivative

$$\frac{d}{dz}\sqrt{z^2-1} = \frac{1}{2}(z^2-1)^{\frac{1}{2}-1} \cdot (2z) = \frac{z}{\sqrt{z^2-1}}, \quad z \neq \pm 1.$$

For instance,

$$\frac{d}{dz}\sqrt{z^2-1}\Big|_{z=i} = \begin{cases} \dfrac{i}{\sqrt{2i}} = \dfrac{1}{\sqrt{2}} & \text{if } \sqrt{-2} = \sqrt{2}i \\[2mm] \dfrac{i}{-\sqrt{2i}} = -\dfrac{1}{\sqrt{2}} & \text{if } \sqrt{-2} = -\sqrt{2}i \end{cases},$$

while $\mathbf{C} - (\infty, -1] - [1, \infty)$ and $\mathbf{C} - [-1, 1]$ are global single-valued domains for $f(z)$. So is any domain Ω in $\mathbf{C}$, for which -1 and 1 lie on the same component of $\mathbf{C}^* - \Omega$.

(2) (Refer to Exercise A (7) in Sec. 2.7.2). Suppose $z_0 \neq \pm 1$ and ∞. $f(z)$ has four distinct analytic branches on $|z - z_0| < \min\{|z_0 + 1|, |z_0 - 1|\}$, wherein each branch has the derivative

$$\frac{d}{dz}(\sqrt{z+1} + \sqrt{z-1}) = \frac{1}{2}\left(\frac{1}{\sqrt{z+1}} + \frac{1}{\sqrt{z-1}}\right), \quad z \neq \pm 1.$$

For instance, $f'(0) = \frac{1}{2}(1-i)$, $\frac{1}{2}(1+i)$, $\frac{1}{2}(-1-i)$, $\frac{1}{2}(-1+i)$, depending on the branch determined by $\sqrt{1} = 1$ and $\sqrt{-1} = i$, $\sqrt{1} = 1$ and $\sqrt{-1} = -i$, $\sqrt{1} = -1$ and $\sqrt{-1} = i$, $\sqrt{1} = -1$ and $\sqrt{-1} = -i$, respectively. While any domain Ω in $\mathbf{C}$ is a global single-valued domain for analytic branches of $f(z)$ if ± 1 and ∞ lie on the same component of $\mathbf{C}^* - \Omega$, such as $\mathbf{C}^* - [\infty, -1] - [1, \infty]$.

(3) (Refer to Example 3 in Sec. 2.7.2). The same domains Ω as in (2) are global single-valued domains for analytic branches of $f(z)$. For $z_0 \neq \pm 1$ and ∞, $f(z)$ has three distinct analytic branches on $|z - z_0| < \min\{|z_0 + 1|, |z_0 - 1|\}$ and, for each branch,

$$\frac{d}{dz}\sqrt[3]{z^2-1} = \frac{1}{3}(z^2-1)^{\frac{1}{3}-1} \cdot (2z) = \frac{2z}{3(\sqrt[3]{z^2-1})^2}, \quad z \neq \pm 1.$$

Note that $f'(0) = 0$ for each branch. What are $f'(i)$?

(4) (Refer to Exercise A (8) in Sec. 2.7.2). Take any $z_0 \in \mathbf{C}$ but $z_0 \neq \pm i$. There will be six distinct branches of $f(z)$ on $|z - z_0| <$ $\min\{|z_0 - i|, |z_0 + i|\}$. Each such branch has the derivative

$$\frac{d}{dz} \sqrt[6]{(z-i)^2(z+i)^3} = \frac{1}{6}\{2(z-i)(z+i)^3 + 3(z-i)^2(z+i)^2\}$$

$$\times \{(z-i)^2(z+i)^3\}^{-\frac{5}{6}}, \quad z \neq \pm i.$$

What are $f'(0)$? Any domain in $\mathbf{C}$ is a global single-valued domain for analytic branches if $\pm i$ and ∞ belong to the same component of $\mathbf{C}^* - \Omega$, such as $\mathbf{C}^* - [i, i\infty] - [-i\infty, -i]$.

(5) (Refer to Exercise A (9) in Sec. 2.7.2). $f(z)$ has six distinct branches on any domain Ω where $0, 1, i$, and ∞ lie on the same component of the set $\mathbf{C}^* - \Omega$. For instance, let $\Omega = \mathbf{C}^* - [\infty, 0] - [1, \infty] - [i, i\infty]$. For $z \neq 0, 1, i, \infty$, let

$$z - 1 = r_1 e^{i\theta_1}, r_1 = |z - 1|, \theta_1 = \mathrm{Arg}(z-1); \quad z - i = r_2 e^{i\theta_2}, r_2 = |z - i|,$$

$$\theta_2 = \mathrm{Arg}(z-i); \quad z = r_3 e^{i\theta_3}, r_3 = |z|, \theta_3 = \mathrm{Arg}\, z.$$

Then the six branches of $f(z)$ on Ω are

$$f_1(z) = w_0 + \widetilde{w_0}, \quad \text{where } w_0 = \sqrt[3]{\frac{r_1}{r_3^2}} e^{i\frac{\theta_1 - 2\theta_3}{3}}, \quad \widetilde{w_0} = \sqrt{r_2}e^{\frac{i\theta_2}{2}};$$

$$f_2(z) = w_0\omega + \widetilde{w_0}, \quad \text{where } \omega = e^{\frac{2\pi i}{3}};$$

$$f_3(z) = w_0\omega^2 + \widetilde{w_0}; \quad f_4(z) = w_0 - \widetilde{w_0};$$

$$f_5(z) = w_0\omega - \widetilde{w_0}; \quad f_6(z) = w_0\omega^2 - \widetilde{w_0}.$$

They are all analytic. Each branch has the derivative

$$\frac{d}{dz}\left\{ \sqrt[3]{\frac{z-1}{z^2}} + \sqrt{z-i} \right\}$$

$$= \frac{1}{3} \cdot \frac{-z+2}{z^3}\left\{ \frac{z-1}{z^2} \right\}^{-\frac{2}{3}} + \frac{1}{2}\{z-i\}^{-\frac{1}{2}}, \quad z \neq 0, 1, i.$$

For instance, the branch that assumes the value $\frac{1}{\sqrt[3]{2}}e^{\frac{2\pi i}{3}} - 1$ at $1 + i$ is the one $f_5(z)$. In this case, $f_5(-i) = \sqrt[6]{2}e^{\frac{11\pi i}{12}} + \sqrt{2}e^{\frac{3\pi i}{4}}$ and

$$f_5'(-i) = \frac{1}{3} \cdot \frac{i+2}{(-i)^3}\{\sqrt[6]{2}e^{\frac{11\pi i}{12}}\}^{-2} + \frac{1}{2}\{\sqrt{2}e^{\frac{3\pi i}{4}}\}^{-1}$$

$$= \frac{1-2i}{3\sqrt[3]{2}}e^{-\frac{11\pi i}{6}} + \frac{1}{2\sqrt{2}}e^{-\frac{3\pi i}{4}}.$$

(6) (Refer to Exercise A (12) in Sec. 2.7.2). Let $\zeta = 1 + \sqrt{z}$, and then $w = \sqrt{\zeta}$. It is well known that w has two local analytic branches on a neighborhood of any nonzero point in the ζ-plane. So is ζ as a function of z. Therefore, $w = f(z) = \sqrt{1 + \sqrt{z}}$ has four distinct analytic branches on $|z - z_0| < \min\{|z_0|, |z_0 - 1|\}$ if $z_0 \neq 0, 1$, and has two distinct analytic branches on $|z - 1| < 1$ induced by the branch of $\sqrt{z}$ determined by $\sqrt{1} = 1$. Each branch has the derivative

$$\frac{d}{dz}\sqrt{1 + \sqrt{z}} = \frac{1}{2}(1 + \sqrt{z})^{-\frac{1}{2}} \cdot \frac{1}{2}(z)^{-\frac{1}{2}}$$

$$= \frac{1}{4\sqrt{z}\sqrt{1 + \sqrt{z}}}, \quad z \neq 0 \quad \text{and} \quad z = 1 \ (\sqrt{1} = 1).$$

For instance, to compute $f'(1)$, we need to consider the two branches $\sqrt{1 + |z|e^{i\frac{\theta}{2}}}, -\frac{\pi}{2} < \theta < \frac{\pi}{2}$ and $|z - 1| < 1$. Then $f'(1) = \pm\frac{1}{4\sqrt{2}}$ where $\sqrt{2} > 0$.

On the other hand, let $w = \sqrt{1 + \sqrt{z}}$, then $z = (w^2 - 1)^2$. Observe that $\frac{dz}{dw} = 4w(w^2 - 1) = 0$ if $w = 0, \pm 1$. For $w_0 \neq 0, \pm 1$, $\frac{dz}{dw} \neq 0$ on $|w - w_0| < \min\{|w_0|, |w_0 - 1|, |w_0 + 1|\}$. On a smaller disk $|w - w_0| < \delta$ if necessary, $z = z(w)$ will be one-to-one (see (3.4.1.2)) and maps the disk onto a domain O (see (3.4.4.4)). Hence, the inverse function $w = \sqrt{1 + \sqrt{z}}$: O $\to \{|w - w_0| < \delta\}$ exists as an analytic function and

$$\frac{dw}{dz} = \frac{1}{\frac{dz}{dw}} = \frac{1}{4w(w^2 - 1)} = \frac{1}{4\sqrt{1 + \sqrt{z}}(1 + \sqrt{z} - 1)} = \frac{1}{4\sqrt{z}\sqrt{1 + \sqrt{z}}}.$$

In general, any domain Ω for which $0, 1,$ and ∞ belong to the same component of $\mathbf{C}^* - \Omega$ is a global single-valued domain for the branches induced by $\sqrt{z} = \sqrt{1} = -1$; for instance, $\mathbf{C}^* - [\infty, 0] - [1, \infty]$ is such a domain. While a domain for which 0 and ∞ lie on the same component of $\mathbf{C}^* - \Omega$ is a single-valued domain for the two branches determined by $\sqrt{z} = \sqrt{1} = 1$.

Example 4.   Show that

$$\frac{d}{dz}\log z = \frac{1}{z}, \quad z \in C \quad \text{and} \quad z \neq 0.$$

(1) (local) $\log z$ has infinitely many analytic branches on $|z - z_0| < |z_0|$, where $z_0 \neq 0$ (see (1) in (2.7.3.2)).

(2) (global) A domain Ω in $\mathbf{C}^*$ is an analytic domain for $\log z$ if 0 and ∞ belong to the same component of $\mathbf{C}^* - \Omega$ (see (2) in (2.7.3.2)).

$$\text{(3.3.1.7)}$$

Recall that, a domain on which an (single-valued, continuous) analytic branch of a multiple-valued function can be defined is also called an analytic domain.

Solution. Consider $z = e^w$. For $z_0 \in \mathbf{C}$ and $z_0 \neq 0$, pick $w_0 \in \mathbf{C}$ so that $e^{w_0} = z_0$. There exists $\alpha \in \mathbf{R}$ so that w_0 lies in the parallel strip $\alpha < \operatorname{Im} w < \alpha + 2\pi$ on which $z = e^w$ is one-to-one and onto the z-plane with the ray $\arg z = \alpha$ (including $z = 0$) deleted. Since $\frac{d}{dw} e^w = e^w \neq 0$ everywhere, by (3) in (3.3.1), the inverse function $w = \log z : \mathbf{C} - \{\arg z = \alpha\} \to \{\alpha < \operatorname{Im} w < \alpha + 2\pi\}$ is analytic and

$$\frac{d}{dz} \log z = \frac{1}{\frac{d}{dw} e^w} = \frac{1}{e^w} = \frac{1}{z}.$$

Or, choose $\theta_0 < \operatorname{Arg} z_0 < \theta_0 + 2\pi$ and adopt the polar coordinates $r = |z|, \theta = \operatorname{Arg} z$ on the domain $\mathbf{C} - \{\operatorname{Arg} z = \theta_0 \text{ (including } 0)\}$. Consider the single-valued branch $w = \operatorname{Log} z$ on the domain $\mathbf{C} - \{\operatorname{Arg} z = \theta_0\}$. In this case, $u(r, \theta) = \log|z| = \log r$ and $v(r, \theta) = \operatorname{Im} \operatorname{Log} z = \theta$. Also

$$\frac{\partial u}{\partial r} = \frac{1}{r} = \frac{1}{r} \frac{\partial u}{\partial \theta}, \quad \frac{\partial u}{\partial \theta} = 0 = -r \frac{\partial v}{\partial r}$$

holds continuously on this domain. By Exercise A (6) of Sec. 3.2.2, $\operatorname{Log} z$ is analytic on the domain and

$$\frac{d}{dz} \operatorname{Log} z = \frac{r}{z} \left(\frac{\partial u}{\partial r}(z) + i \frac{\partial v}{\partial r}(z) \right) = \frac{r}{z} \cdot \frac{1}{r} = \frac{1}{z}.$$

Since all other branches of $\log z$ on the domain are of the forms $\operatorname{Log} z + 2n\pi i$ for integers n, the result follows.

Corresponding to (3.3.1.5), we have another important

Single-valued analytic branches of $\log f(z)$. Suppose $f : \Omega$ (domain) $\to \mathbf{C}$ is analytic so that $f(\Omega)$ is a domain (still a superfluous assumption).

(1) Suppose 0 and ∞ belong to the same component of $\mathbf{C}^* - f(\Omega)$. Then, there are countably infinitely many analytic branches of $\log f(z)$ on Ω, and for each branch,

$$\frac{d}{dz} \log f(z) = \frac{f'(z)}{f(z)}, \quad z \in \Omega.$$

In case $F(z)$ is any analytic branch of $\log f(z)$ on Ω, then all the analytic branches are of the form $F(z) + 2n\pi i, n = 0, \pm 1, \pm 2, \ldots$ (see Exercises A (19) and (20) or Exercise B (1) of Sec. 2.7.3).

(2) A simply connected domain Ω on which $f(z) \neq 0$ is an analytic domain for $\log f(z)$ on which (1) holds.   (3.3.1.8)

Example 5. Find analytic domains for each of the following multiple-valued functions $f(z)$ and compute $f'(z)$.

(1) $f(z) = \log(1 + z^3)$.

(2) $f(z) = \log(z + \sqrt{z^2 - 1})$.

(3) $f(z) = \log \dfrac{z-a}{z-b}$ $(a \neq b)$.

(4) (See Ref. [1], p. 72; also see Exercise B (1)(o) of Sec. 2.7.3) $f(z) = \log \log z$.

Solution. (1) Let $\eta = 1 + z^3$. By Example 4, we knew that $f(z) = \log \eta$ has infinitely many analytic branches on $\mathbf{C}^* - [\infty, 0]$. Hence, we try to determine a domain Ω in the z-plane so that its image under $\eta = 1 + z^3$ is $\mathbf{C}^* - [\infty, 0]$. This is equivalent to find these points z so that $1 + z^3$ are either 0 or negative real numbers. For any $x \leq 0, 1 + z^3 = x$ or $z^3 = x - 1$ has the solutions $z_k = |x - 1|^{\frac{1}{3}} e^{i(\pi + 2k\pi)/3}$, $k = 0, 1, 2$. Hence, the domain, obtained by deleting three rays from $\mathbf{C}$,

$$\Omega = \mathbf{C} - \{re^{\pi i/3} | r \geq 1\} - \{re^{\pi i} | r \geq 1\} - \{re^{-\pi i/3} | r \geq 1\}$$

is a required one. On Ω, each branch has the derivative

$$\frac{d}{dz} \log(1 + z^3) = \frac{3z^2}{1 + z^3}.$$

(2) (Refer to Example 3 of Sec. 2.7.3) Both $\mathbf{C}^* - [\infty, -1] - [1, \infty]$ and $\mathbf{C}^* - [\infty, 1]$ are analytic domains for $f(z)$ and on which

$$\frac{d}{dz} \log(z + \sqrt{z^2 - 1}) = \frac{1}{z + \sqrt{z^2 + 1}} \cdot \left(1 + \frac{1}{2}(z^2 - 1)^{-\frac{1}{2}} \cdot 2z\right) = \frac{1}{\sqrt{z^2 - 1}}.$$

(3) (Refer to Example 4 of Sec. 2.7.3) Let $\overset{\frown}{ab}$ denote a Jordan arc connecting a and b. The $\mathbf{C} - \overset{\frown}{ab}$ is an analytic domain for $f(z)$, and for each of the infinitely many branches,

$$\frac{d}{dz} \log \frac{z-a}{z-b} = \frac{1}{z-a} - \frac{1}{z-b}, \quad z \in \mathbf{C} - \overset{\frown}{ab}.$$

(4) Let $\zeta = \log z$ and $w = \log \zeta = \log \log z$.

$\zeta = \log z$ has analytic branches, for $n = 0, \pm 1, \pm 2, \ldots$

$\zeta_n = (\log z)_n = \operatorname{Log} z + 2n\pi i : \mathbf{C} - (\infty, 0] \to (2n - 1)\pi < \operatorname{Im} \zeta < (2n + 1)\pi$.

Now, adopt the branch cut $[0, \infty)$ in the ζ-plane. $w = \log \zeta$ has analytic branches, for $m = 0, \pm 1, \pm 2, \ldots$,

$w_m = (\log \zeta)_m = \operatorname{Log} \zeta + 2m\pi i : \mathbf{C} - [0, \infty) \to 2m\pi < \operatorname{Im} w < (2m + 2)\pi$.

Before we are able to composite these two families of analytic branches together, we need to figure out which parts of the z-plane that will be mapped onto the cut $[0, \infty)$ on the ζ-plane under $\zeta = \log z$. To see this, let $z = x + iy$ and $\zeta = s + it$. Then $z = e^\zeta$ implies that $x = e^s \cos t$ and $y = e^s \sin t$. Hence $z = e^\zeta$ maps the cut $[0, \infty) : t = 0, 0 \le s < \infty$ one-to-one and onto the cut $[1, \infty) : y = 0, 1 \le x < \infty$ in the z-plane. Thus, the branch $\zeta_n = (\log z)_n$ maps $[1, \infty)$ in the z-plane one-to-one and onto the horizontal line $t = 2n\pi, 0 \le s < \infty$, in the ζ-plane, for $n = 0, \pm 1, \pm 2, \ldots$. In conclusion, letting $\tilde{D}_n = \{(2n-1)\pi < \operatorname{Im} \zeta < (2n+1)\pi\}$, $n = 0, \pm 1, \pm 2, \ldots$,

$$\zeta_0 = \operatorname{Log} z : \mathbf{C} - (\infty, 0] - [1, \infty) \to \tilde{D}_0 - E_0, \quad \text{where}$$

$$E_0 = \{s + i0 \mid 0 \le s < \infty\} = [0, \infty);$$

$$\zeta_n = (\log z)_n : \mathbf{C} - (\infty, 0] \to \tilde{D}_n.$$

These mappings are both one-to-one and onto.

Finally, take any analytic branch $w_m = (\log \zeta)_m$, $m = 0, \pm 1, \pm 2, \ldots$ (recall that its domain of definition is $\mathbf{C} - [0, \infty)$, which always contains $\tilde{D}_0 - E_0$ and $\tilde{D}_n$ for $n = 0, \pm 1, \pm 2, \ldots$). Then, take another analytic branch $\zeta_n = (\log z)_n$, $n = 0, \pm 1, \pm 2, \ldots$ of $\zeta = \log z$. The composite functions

$$w_{mo} = (\log(\operatorname{Log} z))_m : \mathbf{C} - (\infty, 0] - [1, \infty) \to \mathbf{C}, \quad m = 0, \pm 1, \pm 2, \ldots;$$

$$w_{mn} = (\log(\log z)_n)_m : \mathbf{C} - (\infty, 0] \to \mathbf{C},$$
$$n = \pm 1, \pm 2, \ldots, m = 0, \pm 1, \pm 2, \ldots$$

are analytic branches of $f(z) = \log \log z$ on analytic domains $\mathbf{C} - (\infty, 0] - [1, \infty)$ and $\mathbf{C} - (\infty, 0]$, respectively, with derivatives

$$\frac{dw_{mn}}{dz} = \begin{cases} \dfrac{1}{\operatorname{Log} z} \cdot \dfrac{1}{z}, z \in \mathbf{C} - (-\infty, 0] - [1, \infty), & n = 0, m = 0, \pm 1, \pm 2, \ldots, \\[3mm] \dfrac{1}{\log z} \cdot \dfrac{1}{z}, z \in \mathbf{C} - (-\infty, 0], & \begin{aligned} n &= \pm 1, \pm 2, \ldots, \\ m &= 0, \pm 1, \pm 2, \ldots. \end{aligned} \end{cases}$$

As a whole, we use

$$\frac{d}{dz} \log \log z = \frac{1}{z \log z}, \quad z \ne 0 \quad \text{and} \quad \log z \in \mathbf{C} - [0, \infty)$$

to express each of the above derivative formulas.

Remark 2 (The derivatives for inverse trigonometric and hyperbolic functions).   Recalling the algebraic expressions for these inverse functions in (2.7.1) and using (2) and (3) in Example 5, we have the following:

$$\frac{d}{dz}\cos^{-1}z = \frac{-i}{\sqrt{z^2-1}}, \quad z \in \mathbf{C}^* - [\infty,-1] - [1,\infty] \quad \text{or} \quad \mathbf{C}^* - [\infty,1];$$

$$\frac{d}{dz}\sin^{-1}z = \frac{-1}{\sqrt{z^2-1}}, \quad z \in \mathbf{C}^* - [\infty,-1] - [1,\infty];$$

$$\frac{d}{dz}\tan^{-1}z = \frac{1}{1+z^2}, \quad z \in \mathbf{C}^* - [-i,i];$$

$$\frac{d}{dz}\cot^{-1}z = -\frac{1}{1+z^2}, \quad z \in \mathbf{C}^* - [-i,i];$$

$$\frac{d}{dz}\cosh^{-1}z = \frac{1}{\sqrt{z^2-1}}, \quad z \in \mathbf{C}^* - [\infty,-1] - [1,\infty];$$

$$\frac{d}{dz}\sinh^{-1}z = \frac{-i}{\sqrt{z^2-1}}, \quad z \in \mathbf{C}^* - [\infty,-1] - [1,\infty];$$

$$\frac{d}{dz}\tanh^{-1}z = \frac{1}{1-z^2}, \quad z \in \mathbf{C}^* - [-1,1];$$

$$\frac{d}{dz}\coth^{-1}z = \frac{1}{1-z^2}, \quad z \in \mathbf{C}^* - [-1,1]. \tag{3.3.1.9}$$

Example 6. For fixed $\alpha \in \mathbf{C}$,

$$\frac{d}{dz}z^\alpha = \alpha z^{\alpha-1}, \quad z \neq 0 \tag{3.3.1.10}$$

(see (2.7.3.4)). In case α is an irrational number or a complex number with $\text{Im}\,\alpha \neq 0$, z^α has exactly the same analytic domains as $\log z$ (see Example 4).

Example 7. For fixed $a \in \mathbf{C}$ with $a \neq 0$,

$$\frac{d}{dz}a^z = (\log a)a^z, \quad z \in \mathbf{C} \tag{3.3.1.11}$$

(see (2.7.3.6)). For each value of $\log a$, say $\text{Log}\,a + 2n\pi i$ $(n=0,\pm1,\pm2,\dots)$, then $e^{z(\text{Log}\,a+2n\pi i)}$ is an analytic function on $\mathbf{C}$.

Example 8. Find analytic domains for $f(z)=(z^2+1)^z$ and then compute $f'(z)$.

Solution. By definition, the analytic branches of $f(z)=e^{z\log(z^2+1)}$ originate from these of $\log(z^2+1)$ and have the same analytic domains. Hence,

$f(z)$ has infinitely many analytic branches on the domain $\mathbf{C}^* - [i, i\infty] - [-i\infty, -i]$. For each such branch on the domain,

$$\frac{d}{dz} \log(z^2 + 1)^z = \frac{d}{dz} e^{z \log(z^2+1)}$$

$$= e^{z \log(z^2+1)} \cdot \left\{ \log(z^2 + 1) + z \cdot \frac{1}{z^2 + 1} \cdot 2z \right\}$$

$$= (z^2 + 1)^z \cdot \left\{ \log(z^2 + 1) + \frac{2z^2}{z^2 + 1} \right\}.$$

Exercises A

(1) Suppose $f : \Omega(\text{domain}) \to \mathbf{C}$ is continuous and $f(\Omega) \subseteq \Omega'(\text{domain})$. If $g : \Omega' \to \mathbf{C}$ is an analytic function so that $g'(w) \neq 0$ for every $w \in \Omega'$ and the composite function $g \circ f : \Omega \to \mathbf{C}$ happens to be analytic, show that f is analytic on Ω and

$$f'(z) = \frac{(g \circ f)'(z)}{g'(f(z))}, \quad z \in \Omega.$$

(2) Prove differentiation formulas shown in (3.3.1.3) and (3.3.1.9) in detail.

(3) Use (3.3.1.3) to show the analyticity for each of the following functions $f(z)$, then use Cauchy–Riemann equations to find $f'(z)$, and then check $f'(z)$ by using (3.3.1) and (3.1.6).

(a) $f(z) = \dfrac{az + b}{cz + d}$ $(ad - bc \neq 0$, refer to Section (1) in Sec. 2.5.4).

(b) $f(z) = e^{z^2}$.

(c) $f(z) = e^{\sin z}$.

(d) $f(z) = \cos e^z$.

(e) $f(z) = z^n$ (n is a positive integer; what happens if n is negative).

(f) $f(z) = e^x(x \cos y - y \sin y) + ie^x(y \cos y + x \sin y)$.

(g) $f(z) = e^z \sin z$.

(4) Do the following problems for each function $f(z)$.

   (i) Figure out the point set of analyticity.

   (ii) Compute $f'(z)$ at points of analyticity.

   (iii) Find out all the points where $f(z)$ fail to be analytic. Are they singular points? isolated? poles?

   (a) $f(z) = \dfrac{4z^2 + 5z + 3}{2z^4 + z^3 - 13z^2 + z - 15}.$

(b) $f(z) = \left(z + \dfrac{1}{z}\right)^{-2}$.

(c) $f(z) = \dfrac{z}{z^n - 2}$ $(n \geq 1$ is an integer$)$.

(d) $f(z) = \dfrac{e^z}{z(z-1)(z-2)}$.

(e) $f(z) = \dfrac{1}{1 + e^z}$.

(f) $f(z) = \dfrac{1}{\cos z - 2}$.

(g) $f(z) = ze^{(z-1)^{-1}}$.

(h) $f(z) = \cot z^2$.

(i) $f(z) = \sin \dfrac{1}{z}$.

(j) $f(z) = \csc \dfrac{1}{z}$.

(k) $f(z) = \dfrac{1}{1 - e^{z^2}}$.

(5) Let $z = x + iy$. Show that $y + ix$, e^{x-iy}, and $e^y \cos x + ie^y \sin x$ are not analytic functions.

(6) Determine real constants a and b so that $x^2 + 2axy + by^2$ $(z = x + iy)$ is the real part of an entire function $f(z)$. Find all such $f(z)$.

(7) Show that $f(z) = |z^2 - z|$ is not analytic anywhere.

(8) Let $p(z) = (z - z_1) \cdots (z - z_k)$. Apply induction on n to show that

$$\frac{p'(z)}{p(z)} = \frac{1}{z - z_1} + \cdots + \frac{1}{z - z_n}, \quad z \neq z_1, \ldots, z_n$$

called the *logarithmic derivative* of $p(z)$.

(9) If $q(z)$ is a polynomial with distinct zeros $z_1, \ldots, z_n$, and $p(z)$ is a polynomial of degree less than n, show that

$$\frac{p(z)}{q(z)} = \sum_{k=1}^{n} \frac{p(z_k)}{q'(z_k)(z - z_k)}.$$

Use this to prove that there exists a unique polynomial $p(z)$ of degree less than n with given values α_k at the points z_k, $1 \leq k \leq n$ (*Lagrange's interpolation polynomial*, refer to Exercise A (10) of Sec. 2.5.1).

(10) Suppose f and g are analytic at z_0 and $f(z_0) = g(z_0) = 0$, but $g'(z_0) \neq 0$. Show that

$$\lim_{z \to z_0} \frac{f(z)}{g(z)} = \frac{f'(z_0)}{g'(z_0)}.$$

called the *L'hospital's rule*: Compute the following limits:

(a) $\lim_{z \to i} \dfrac{z^5 - i}{z^9 - i}$.

(b) $\lim_{z \to 1+i} \dfrac{z^3 + 3iz^2 - 2iz + 6}{z^4 + 4}$.

(c) $\lim_{z \to 0} \dfrac{e^z - 1}{z}$.

(d) $\lim_{z \to n\pi} \dfrac{\sin z}{z - n\pi}$ (n, integers).

(11) Suppose $f(z)$ is analytic on a domain Ω.

(a) Is $\overline{f(z)}$ analytic on Ω?
(b) On what domain will $\overline{f(\bar{z})}$ be analytic, too?

(12) Determine analytic function $f(z) = u(z) + iv(z)$ if $u(z) + v(z) = (x - y)(x^2 + 4xy + y^2)$, where $z = x + iy$.

(13) Let

$$f(z) = \begin{cases} 0, & z = 0 \\ e^{-z^{-4}}, & z \neq 0. \end{cases}$$

Show that f is analytic on $\mathbf{C} - \{0\}$ but not analytic at $z = 0$ even though it satisfies the Cauchy–Riemann equations at $z = 0$.

(14) Show that $f(z)$ in Exercise A (3) of Sec. 2.7.2 is an analytic branch of $w = z^{\frac{1}{3}}$ on $\mathbf{C} - [0, \infty)$. Find the other two analytic branches on $\mathbf{C} - [0, \infty)$.

(15) Show that the four branches of $w = z^{\frac{1}{4}}$ on $\mathbf{C} - [0, \infty]$ as required in Exercise A (4) of Sec. 2.7.2 are analytic.

(16) Show that it is impossible to define an analytic branch of $z^{\frac{1}{n}}$ ($n \geq 2$) on any open neighborhood of 0.

(17) Redo Exercise B (1) of Sec. 2.7.2.

(18) For each of the multiple-valued functions shown in Exercise B (5) of Sec. 2.7.2, do the following problems.

(i) (Local) At what points and on what neighborhoods of such points can $f(z)$ possess analytic branches? How many? Compute $f'(z)$ at such points.

(ii) (Global) Find (the possibly largest) analytic domains for $f(z)$, namely, domains on which analytic branches can be defined.

(19) Show that it is impossible to define an analytic branch of $\log z$ on any open neighborhood of 0.

(20) Redo Exercise B (1) of Sec. 2.7.3.

(21) Redo Exercises A (2), (3), (4), and (5) of Sec. 2.7.3 and show that branches obtained are analytic.

(22) For each of the multiple-valued functions shown in Exercise A (20) and Exercise B (2) of Sec. 2.7.3, and Exercise B (1) of Sec. 2.7.4, do problems as in Exercise (18).

Exercises B

One of the most prominent properties of an analytic function is that *its derivative turns out to be continuous automatically.* Yet its proof is far from being obvious and trivial. For an analytic proof via integration, see Sec. 3.4.2.

As a consequence of this remarkable fact, an analytic function $f(z)$ is then infinitely differentiable and its derivatives

$$f^{(0)}(z) = f(z);$$

$$f^{(n)}(z) = \frac{d}{dz}f^{(n-1)}(z) = (f^{(n-1)})'(z), \quad n \geq 1$$

are also analytic (see (3.4.2.6)). Moreover, its real parts $\operatorname{Re} f(z) = u(z)$ and imaginary parts $\operatorname{Im} f(z) = v(z)$ are infinitely differentiable, too; in particular, partial derivatives of all orders of both $u(z)$ and $v(z)$ exist and are continuous, and their mixed partial derivatives of the same type are equal.

Another prominent property of analytic functions is that *a nonconstant analytic function is open*, namely, it always map open sets onto open sets. Hence, it possesses *maximum modulus principal*. See Sec. 3.4.4 for detailed account.

In what follows, we take these properties for granted implicitly.

(1) Suppose $u(z) : \Omega(\text{domain}) \to \mathbf{R}^2$ is C^2 and satisfies the *Laplacian equation* (see Exercise B (1)(e) of Sec. 2.8):

$$\Delta u = \frac{\partial^2 u}{\partial x^2} + \frac{\partial^2 u}{\partial y^2} = 0$$

on Ω. Such a $u(z)$ is called *harmonic* on Ω (refer to Sec. 3.4.3 for details). Show that then $f(z) = \frac{\partial u}{\partial x}(z) - i\frac{\partial u}{\partial y}(z)$ is analytic on Ω.

(2) Suppose $f : \Omega(\text{domain}) \to \mathbf{C}$ is analytic and $f'(z_0) \neq 0$ for a point $z_0 \in \Omega$. Show that

$$\{\operatorname{Re} f(z) \mid z \in \Omega\}, \{\operatorname{Im} f(z) \mid z \in \Omega\} \quad \text{and} \quad \{|f(z)| \mid z \in \Omega\}$$

are open sets in $\mathbf{R}$. Under what conditions, will $\{\operatorname{Arg} f(z) \mid z \in \Omega\}$ be an open set in $\mathbf{R}$?

(3) Let $f(z) = u(z) + iv(z)$ be analytic on an open set Ω. Let Δu be as in Exercise (1), etc. Show that

(a) $\Delta u = \Delta v = \Delta f = 0$.

(b) $\Delta |f(z)|^p = p^2 |f(z)|^{p-2} |f'(z)|^2$, where p is a positive integer (p might be negative if $f(z) \neq 0$).

(c) $\Delta |u(z)|^p = p(p-1)|u(z)|^{p-2}|f'(z)|^2$, p is a positive integer.

(d) $\dfrac{\partial}{\partial z}|f(z)| = \dfrac{1}{2}|f(z)|\dfrac{f'(z)}{f(z)}$ if $f(z) \neq 0$.

(e) $\dfrac{\partial}{\partial z}u(z) = \dfrac{1}{2}f'(z)$ and $\dfrac{\partial}{\partial z}v(z) = \dfrac{1}{2i}f'(z)$.

(f) $\Delta e^{p|f(z)|} = p^2 e^{p|f(z)|}\left(p + \dfrac{1}{|f(z)|}\right)|f'(z)|^2$ if $f(z) \neq 0$, where $p \in \mathbf{C}$.

(g) $\Delta \log(1 + |f(z)|^2) = \dfrac{4|f'(z)|^2}{1 + |f(z)|^2}$.

3.3.2.   *The analyticity of functions defined by power series*

Call the series $\sum_{n=0}^{\infty} a_n(z - z_0)^n$ a *power series* with *center* z_0. The translation $w = z - z_0$ of variables changes it to a power series $\sum_{n=0}^{\infty} a_n w^n$ with center at 0. Henceforth, we focus our attention on $\sum_{n=0}^{\infty} a_n z^n$, often abbreviated as $\sum_0^{\infty} a_n z^n$ or simply as $\sum a_n z^n$.

$\sum a_n z^n$ always converges at $z = 0$. In what follows, we suppose the readers are familiar with basic facts about convergence of series of functions (see Sec. 2.3).

Suppose there exists a point $z_0 \neq 0$ so that $\sum a_n z^n$ converges. Hence, there exists a constant $M > 0$ so that $|a_n z_0^n| \leq M$ for $n \geq 0$. Fix any point $z, |z| < |z_0|$. Then,

$$\sum |a_n z^n| = \sum |a_n z_0^n| \left|\frac{z}{z_0}\right|^n \leq M \sum \left|\frac{z}{z_0}\right|^n, \quad \left|\frac{z}{z_0}\right| < 1.$$

This says that $\sum |a_n z^n|$ converges. Therefore $\sum a_n z^n$ converges absolutely on $|z| < |z_0|$.

As a matter of fact, we can obtain one further information. Choose $r, 0 < r < |z_0|$. On the closed disk $|z| \leq r$,

$$|a_n z^n| = |a_n z_0^n| \left| \frac{z}{z_0} \right|^n \leq M \left(\frac{r}{|z_0|} \right)^n.$$

Since $\sum \left(\frac{r}{|z_0|} \right)^n$ converges, by the Weierstrass test, $\sum a_n z^n$ converges uniformly on $|z| \leq r$ and hence, uniformly on every compact subset of $|z| < |z_0|$ (why?).

These results lead us to consider

$$r = \sup \left\{ |z| \mid \sum a_n z^n \text{ converges} \right\}. \tag{3.3.2.1}$$

In case $r = 0$, then $\sum a_n z^n$ converges only at $z = 0$. If $r > 0$, then $\sum a_n z^n$ converges absolutely and uniformly on every compact subset of $|z| < r$; yet $\sum a_n z^n$ diverges if $|z| > r$. Call r *the radius of convergence* and $|z| < r$ *the disk of convergence* for the power series $\sum a_n z^n$. In general, $0 \leq r \leq \infty$.

This r is completely determined by the coefficients a_n. To see this, observe that $\sum a_n z^n$ converges absolutely on $|z| < r$. Hence, series $\sum |a_n| |z|^n$ of positive terms can be used to determine r. According to root test (see Exercise B (3)(c) of Sec. 1.7), if

$$\varlimsup_{n \to \infty} \sqrt[n]{|a_n| |z|^n} = \left(\varlimsup_{n \to \infty} \sqrt[n]{|a_n|} \right) |z| < 1,$$

then $\sum a_n z^n$ converges absolutely; if the above quantity is greater than 1, a subsequence of $\{a_n z^n\}$ does not converge to zero and hence, $\sum a_n z^n$ diverges. Thus,

$$r = \frac{1}{\varlimsup \sqrt[n]{|a_n|}} = \varliminf \frac{1}{\sqrt[n]{|a_n|}} \left(= \lim \left| \frac{a_n}{a_{n+1}} \right|, \text{ if this limit exists} \right),$$

$$\tag{3.3.2.2}$$

called the *Hadamard formula* for the radius of convergence.

We summarize the above as

A basic property of a power series. Suppose $\sum a_n z^n$ has a *positive* radius r of convergence, where r is defined as in (3.3.2.1) and (3.3.2.2).

(1) $\sum a_n z^n$ converges absolutely and uniformly on every compact subset of $|z| < r$ to a continuous function $f(z)$.
(2) $\sum a_n z^n$ diverges on $|z| > r$.

We always use $f(z) = \sum a_n z^n$, $|z| < r$ to denote this fact. $\qquad$ (3.3.2.3)

What we really care is the analyticity of this $f(z)$ in (3.3.2.3). This affirmative result lies on the facts that each term $a_n z^n$ is analytic and that the series $\sum a_n z^n$ does converge uniformly on each compact subset of $|z| < r$ to $f(z)$.

Firstly, observe that $\overline{\lim} |n a_{n-1}|^{\frac{1}{n-1}} = \overline{\lim} |a_n|^{\frac{1}{n}} = \frac{1}{r}$, so $\sum_{n=1}^{\infty} n a_n z^{n-1}$ has the same radius r of convergence as the original series $\sum_{n=0}^{\infty} a_n z^n$. Similarly, the series, obtained by differentiating k-times the terms $a_n z^n$,

$$\sum_{n=k}^{\infty} a_n \cdot n(n-1)(n-2)\cdots(n-k+1)z^{n-k}, \quad k = 1,2,3,\ldots \qquad (*_1)$$

all have the same r as their radii of convergence. And hence, (3.3.2.3) is applicable to each of these derived series.

Fix a point z_0, $|z_0| < r$, and $\rho > 0$ so that the closed disk $|z - z_0| \le \rho$ lies in the disk $|z| < r$. See Fig. 3.9. On the open disk $|z - z_0| < \rho$,

$$\frac{f(z) - f(z_0)}{z - z_0} - \sum_{n=1}^{\infty} n a_n z_0^{n-1} = \sum_{n=1}^{\infty} a_n \left(\frac{z^n - z_0^n}{z - z_0} - n z_0^{n-1} \right). \qquad (*_2)$$

By applying binomial expansion to $z^n = (z - z_0 + z_0)^n$, for $n \ge 1$,

$$\left| \frac{z^n - z_0^n}{z - z_0} - n z_0^{n-1} \right| = |C_2^n z_0^{n-2}(z - z_0) + \cdots + C_k^n z_0^{n-k}(z - z_0)^{k-1}$$

$$+ \cdots + (z - z_0)^{n-1}|$$

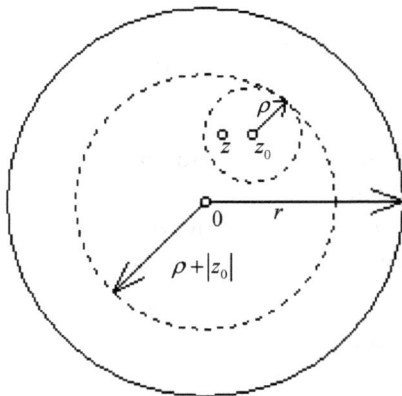

Fig. 3.9

$$\leq C_2^n |z_0|^{n-2}|z - z_0| + \cdots + C_k^n |z_0|^{n-k}|z - z_0|^{k-1}$$
$$+ \cdots + |z - z_0|^{n-1}$$
$$= \frac{(|z - z_0| + |z_0|)^n - |z_0|^n}{|z - z_0|} - n|z_0|^{n-1}.$$

Since $\sum a_n z^n$ converges uniformly on $|z| \leq \rho + |z_0|$, $|a_n|(\rho + |z_0|)^n \leq M$, $n \geq 0$, for some constant $M > 0$. Substitute these inequalities into $(*_2)$, we obtain

$$\left| \frac{f(z) - f(z_0)}{z - z_0} - \sum_{n=1}^{\infty} n a_n z_0^{n-1} \right|$$

$$\leq M \sum_{n=1}^{\infty} \frac{1}{(\rho + |z_0|)^n} \left\{ \frac{(|z - z_0| + |z_0|)^n - |z_0|^n}{|z - z_0|} - n|z_0|^{n-1} \right\}$$

$$= M \left\{ \frac{1}{|z - z_0|} \left(\frac{|z - z_0| + |z_0|}{\rho - |z - z_0|} - \frac{|z_0|}{\rho} \right) - \frac{\rho + |z_0|}{\rho^2} \right\}$$

$$= \frac{M(\rho + |z_0|)|z - z_0|}{\rho^2(\rho - |z - z_0|)}$$

$$\to 0 \quad \text{as } z \to z_0$$

$$\Rightarrow f'(z_0) = \sum_{n=1}^{\infty} n a_n z_0^{n-1}, \quad |z_0| < r.$$

Similarly, $f'(z)$ is also differentiable on $|z| < 1$ and so on. Hence, $(*_1)$ becomes

$$f^{(k)}(z) = \sum_{n=k}^{\infty} n(n-1) \cdots (n-k+1) a_n z^{n-k}, \quad |z| < r.$$

In particular, letting $z = 0$, we obtain the coefficient of the kth term in $\sum a_n z^n$ as

$$a_k = \frac{f^{(k)}(0)}{k!}, \quad k = 0, 1, 2, \ldots .$$

We summarize the above as the following important
Analyticity of a power series. Suppose $f(z) = \sum_0^{\infty} a_n(z - z_0)^n$ has a positive radius r of convergence.

(1) $f(z)$ is an analytic function on $|z - z_0| < r$ and

$$f'(z) = \sum_{n=1}^{\infty} n a_n(z - z_0)^n, \quad |z - z_0| < r.$$

(2) $f(z)$ is infinitely differentiable on $|z - z_0| < r$, i.e., each kth derivative $f^{(k)}(z)$ exists as an analytic function and

$$f^{(k)}(z) = \sum_{n=k}^{\infty} n(n-1)\cdots(n-k+1)a_n(z-z_0)^{n-k},$$
$$|z - z_0| < r, \quad k = 1, 2, \ldots.$$

(3) In particular,

$$a_n = \frac{f^{(n)}(z_0)}{n!}, \quad n \geq 0.$$

Hence, $f(z)$ can be expressed as

$$f(z) = \sum_{n=0}^{\infty} \frac{f^{(n)}(z_0)}{n!}(z-z_0)^n, \quad |z - z_0| \leq r,$$

and is called the *Taylor series expansion* of $f(z) = \sum_0^{\infty} a_n(z-z_0)^n$ at the center z_0 or $f(z)$ is *expandable as a Taylor series* at z_0.

(4) The power series may be integrated termwise along any rectifiable curve γ in $|z - z_0| < r$ to give

$$\int_{\gamma} f(z)dz = \sum_{n=0}^{\infty} \int_{\gamma} a_n(z-z_0)^n dz.$$

In particular, if z_0 is the initial point of γ and z is the terminal point, then

$$\int_{\gamma} f(z)dz = \sum_{n=0}^{\infty} \frac{a_n}{n+1}(z-z_0)^{n+1}, \quad |z - z_0| < r \qquad (3.3.2.4)$$

(see (7) in (2.9.10)).

Later in (3.4.2.6), we will show that *a function analytic at a point z_0 is expandable as a Taylor series at z_0 with positive radius of convergence.* This is the starting point of Weierstrass's theory for analytic functions (see Chap. 5).

We give three examples.

Example 1. Let $\log 1 = 0$. Then $\log(1+z)$ has the Taylor series expansion

$$\log(1+z) = \sum_{n=1}^{\infty} \frac{(-1)^{n-1}}{n}z^n, \quad |z| < 1.$$

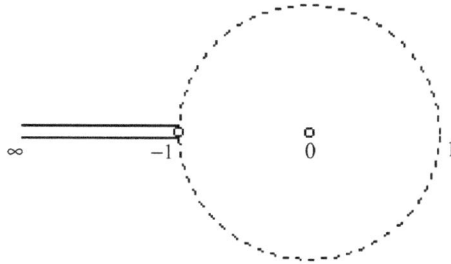

Fig. 3.10

Solution. The series on the right has radius of convergence equal to $\lim_{n\to\infty}\left|-\frac{n+1}{n}\right| = 1$. So, by (1) in (3.3.2.4), $f(z) = \sum_{n=1}^{\infty}\frac{(-1)^{n-1}}{n}z^n$ is an analytic function on $|z| < 1$. On the other hand, the branch of log $(1 + z)$ determined by log $1 = 0$ is $\mathrm{Log}\,(1 + z) = \mathrm{Log}\,|1 + z| + i\,\mathrm{Arg}(1 + z)$, $-\pi < \mathrm{Arg}(1 + z) < \pi$, which is analytic on $\mathbf{C} - (\infty, -1]$ and hence, on $|z| < 1$. See Fig. 3.10.

Thus, $f(z) - \mathrm{Log}(1 + z)$ is analytic on $|z| < 1$ and has its derivative equal to

$$f'(z) - \frac{1}{1+z} = \sum_{n=1}^{\infty}(-1)^{n-1}z^{n-1} - \frac{1}{1+z} = \frac{1}{1+z} - \frac{1}{1+z} = 0 \quad \text{on } |z| < 1.$$

Via the known result from calculus or just using (3.4.1.3), $f(z) - \mathrm{Log}(1 + z) = c$ is a constant on $|z| < 1$. Since $f(0) - \mathrm{Log}\,1 = 0$, $c = 0$ and the result follows.

Remark 1. This series representation of $\log(1 + z)$, $\log 1 = 0$, is valid only on $|z| < 1$. If the center 0 is moved to other point $z_0 \neq 0$, then the representation would be different. What is that? One might refer to Sec. 3.3.3 for preliminary examples.

By exactly the same argument, we have, when designating $\log 1 = 2k\pi i$, $k = 0, \pm 1, \ldots$,

$$\log(1 + z) = 2k\pi i + \sum_{n=1}^{\infty}\frac{(-1)^{n-1}}{n}z^n, \quad |z| < 1.$$

$$\Rightarrow \frac{1}{1+z} = \sum_{0}^{\infty}(-1)^n z^n; \quad \frac{1}{(1+z)^2} = \sum_{n=1}^{\infty}(-1)^{n-1}nz^{n-1}; \ldots;$$

$$\frac{1}{(1+z)^k} = \sum_{n=k-1}^{\infty}(-1)^{n-k+1}C_{k-1}^n z^{n-k+1}, \quad |z| < 1, \ k = 1, 2, 3, \ldots . \quad \square$$

Example 2. Let $\lambda \in \mathbf{C}$. Denote

$$
C_n^\lambda = \begin{cases} 1, & n = 0 \\ \dfrac{\lambda(\lambda - 1)\cdots(\lambda - n + 1)}{n!}, & \text{also denoted as } \begin{pmatrix} \lambda \\ n \end{pmatrix}, \quad n = 1, 2, \ldots \end{cases}
$$

which are called the *generalized binomial coefficients*. Define $1^\lambda = e^{2k\pi\lambda i}$ for any fixed k $(k = 0, \pm 1, \pm 2, \ldots)$. Then

$$
(1 + z)^\lambda = e^{2k\pi\lambda i} \sum_{n=0}^{\infty} C_n^\lambda z^n, \quad |z| < 1,
$$

called the *generalized binomial series*.

Solution. We just consider the principle branch of $(1 + z)^\lambda$ determined by $1^\lambda = 1$, namely, $k = 0$. Let $f(z)$ denote the power series on the right. Then, $f(z)$ is analytic on $|z| < 1$ and

$$
f'(z) = \sum_{n=1}^{\infty} C_n^\lambda n z^{n-1}, \quad |z| < 1,
$$

$$
\Rightarrow (1 + z)f'(z) = \sum_{n=1}^{\infty} C_n^\lambda n(z^{n-1} + z^n)
$$

$$
= \lambda \left\{ 1 + \sum_{n=2}^{\infty} [C_{n-1}^{\lambda-1} + C_{n-2}^{\lambda-1}] z^{n-1} \right\}
$$

$$
= \lambda \left\{ 1 + \sum_{n=2}^{\infty} C_{n-1}^{\lambda} z^{n-1} \right\} = \lambda f(z), \quad |z| < 1.
$$

On the other hand, the principle branch of $(1 + z)^\lambda$ is analytic on $|z| < 1$ (see Fig. 3.10). Let $\varphi(z) = \frac{f(z)}{(1+z)^\lambda}, |z| < 1$. $\varphi(z)$ is analytic on $|z| < 1$ and

$$
\varphi'(z) = \frac{1}{(1 + z)^{\lambda+1}}[(1 + z)f'(z) - \lambda f(z)] = 0 \quad \text{on } |z| < 1.
$$

So $\varphi(z)$ is a constant function on $|z| < 1$ (see (3.4.1.3)). Since $\varphi(0) = 1$ (or $e^{-2k\pi\lambda i}$), it follows that $\varphi(z) = 1$ on $|z| < 1$ and the result follows.

Example 3. Find the power series solution of the ordinary differential equation

$$
w'(z) = w(z), \quad w(0) = 1.
$$

Solution. Suppose $w(z) = \sum_0^\infty a_n z^n, |z| < r$ $(r > 0)$, satisfies the equation $w'(z) = w(z)$ with the initial condition $w(0) = 1$. Differentiate $w(z)$ termwise and compare coefficients of $w'(z) = w(z)$, we get

$$a_0 = 1; \quad (n+1)a_{n+1} = a_n, \quad n \geq 1$$

$$\Rightarrow a_n = \frac{1}{n!}, \quad n \geq 0$$

$$\Rightarrow w(z) = \sum_{n=0}^{\infty} \frac{1}{n!} z^n.$$

This series has radius of convergence equal to ∞. In case $z = x$ is real, $w(x) = e^x$ is the real exponential function of x.

Remark 2. Hence, one might define *the complex exponential function* as

$$e^z = \sum_{n=0}^{\infty} \frac{1}{n!} z^n, \quad z \in \mathbf{C}$$

(refer to Exercise B (1) of Sec. 2.6.1). Consequently,

$$\cos z = \frac{1}{2}(e^{iz} + e^{-iz}) = \sum_{n=0}^{\infty} \frac{(-1)^n}{(2n)!} z^{2n}, \quad z \in \mathbf{C};$$

$$\sin z = \frac{1}{2i}(e^{iz} - e^{-iz}) = \sum_{n=1}^{\infty} \frac{(-1)^{n-1}}{(2n-1)!} z^{2n-1}, \quad z \in \mathbf{C}.$$

By the way, using the equation $w' = w$, it is easy to show that

$$e^{z_1 + z_2} = e^{z_1} e^{z_2}, \quad z_1, z_2 \in \mathbf{C}.$$

In particular,

$$e^z \neq 0, \quad z \in \mathbf{C}.$$

To see this (due to the Caratheódory), let $f(z) = e^z e^{z_1 + z_2 - z}$. Then $f'(z) = 0$ on $\mathbf{C}$ and hence, $f(z) = f(z_1) = f(0)$ is a constant. □

Exercises A

(1) Determine the radius of convergence for each of the following power series.

(a) $\sum_1^\infty z^{n!}$.

(b) $\displaystyle\sum_{0}^{\infty} \frac{(2n)!}{(n!)^2} z^n$.

(c) $\displaystyle\sum_{1}^{\infty} \frac{n!}{n^n} z^n$.

(d) $\displaystyle\sum_{0}^{\infty} \frac{1}{2^n} z^{2^n}$.

(e) $\displaystyle\sum_{0}^{\infty} (n + a^n) z^n$.

(f) $\displaystyle\sum_{0}^{\infty} (\cos in) z^n$.

(g) $\displaystyle\sum_{0}^{\infty} 2^n z^{n!}$.

(2) The same as Exercise (1), do the following series.

(a) $\displaystyle\sum_{n=1}^{\infty} \frac{1 \cdot 3 \cdots (2n-1)}{2 \cdot 4 \cdots 2n} \left(1 + \frac{1}{2} + \cdots + \frac{1}{n} \right) z^n$.

(b) $\displaystyle\sum_{n=1}^{\infty} \frac{1}{n} (-1)^{[\sqrt{n}]} z^n$, where $[\sqrt{n}]$ denotes the largest integer$\leq \sqrt{n}$.

(c) $\displaystyle\sum_{n=0}^{\infty} (n!) z^{n^n}$.

(d) $\displaystyle\sum_{n=1}^{\infty} \frac{\varepsilon_n}{n \log n} z^n$, where $\varepsilon_n = \begin{cases} 1 & \text{if } 2^{2k} \leq n < 2^{2k+1} \\ -1 & \text{if } 2^{2k+1} \leq n < 2^{2k+2} \end{cases}$,
$k = 0, 1, 2, 3, \ldots$.

(3) Suppose $\sum a_n z^n$ and $\sum b_n z^n$ have r and ρ as its radius of convergence, respectively. Try to determine the radius of convergence for each of the following series.

(a) $\sum a_n z^{kn}$ (k, a positive integer).

(b) $\sum a_n z^{n^k}$ (k, a positive integer).

(c) $\sum \frac{a_n}{n!} z^n$.

(d) $\sum (1 + z_0^n) a_n z^n$.

(e) $\sum (a_n \pm b_n) z^n$.

(f) $\sum a_n b_n z^n$.

(g) $\sum \left(\dfrac{a_n}{b_n} \right) z^n$.

(h) $\sum (a_n b_0 + a_{n-1} b_1 + \cdots + a_0 b_n) z^n$.

(4) Suppose $f(z) = \sum_0^\infty a_n z^n$ has real values on the real interval $(-\delta, \delta)$, $\delta > 0$. Show that a_n is real for each $n \geq 0$.

(5) If the series $\sum_0^\infty a_n z^n$ has bounded coefficients a_n for $n \geq 0$, show that its radius r of convergence is ≥ 1. In case $r = 1$, it is possible that $\sum_0^\infty a_n z^n$ converges everywhere on $|z| = 1$ except at $z = 1$.

(6) Suppose $f(z) = \sum_0^\infty a_n z^n$. Find the sums of $\sum_0^\infty n^k a_n z^n$ for $k = 1, 2, 3$.

(7) Find the Taylor series expansion of each of the following functions $f(z)$ at the given point z_0 and determine its radius of convergence.

(a) $\dfrac{1}{(1-z)^k}$ (k, a positive integer), $z_0 = 0$.

(b) $\dfrac{2z+3}{z+1}$, $z_0 = 1$.

(c) $\dfrac{z^2}{(z+1)^2}$, $z_0 = 0$.

(d) $\dfrac{z}{z^2 - 4z + 13}$, $z_0 = 0$.

(e) $\cosh z$, $z_0 = 0$.

(f) $\sinh z$, $z_0 = 0$.

(g) $\log \dfrac{1+z}{1-z} (\log 1 = 0)$, $z_0 = 0$.

(8) Suppose $f(z) = \sum_0^\infty a_n (z - z_0)^n$ has positive radius r of convergence and $\sum_0^\infty f^{(n)}(z_0)$ converges. Show that $r = \infty$. The converse is not necessarily true and $e^z = \sum_0^\infty \dfrac{z^n}{n!}$ is such a series.

(9) Suppose $m \neq -1, -2, -3, \ldots$. Show that

$$w(z) = 1 + \sum_{n=1}^\infty \frac{1}{n! \Pi_{j=1}^n (m+j)} z^n$$

has radius of convergence equal to ∞ and satisfies $zw'' + (m+1)w' + w = 0$.

(10) Suppose the differential equation $w'' + w = 0$ has a series solution $w(z) = \sum_0^\infty a_n z^n$. Show that its general solution is $a \cos z + b \sin z$, where a and b are complex constants.

Exercises B

(1) Suppose the *hypergeometric equation*

$$z(1 - z)w'' + [\gamma - (\alpha + \beta + 1)z]w' - \alpha\beta w = 0,$$

where α, β, and γ are three real constants, which are not 0 or negative integers, has a power series solution $w(z) = \sum_0^\infty a_n z^n$ with $w(0) = 1$. Show that

$$w(z) = F(\alpha, \beta, \gamma, z)$$
$$\underset{\text{(def.)}}{=} 1 + \sum_{n=1}^\infty \frac{\alpha(\alpha + 1)\cdots(\alpha + n - 1)\beta(\beta + 1)\cdots(\beta + n - 1)}{n!\gamma(\gamma + 1)\cdots(\gamma + n - 1)} z^n$$

with 1 as its radius of convergence and is called the *Gauss hypergeometric series*. Show that

(a) (i) If $\gamma > \alpha + \beta$, the series converges uniformly and absolutely on $|z| \le 1$.

   (ii) If $\alpha + \beta - 1 < \gamma \le \alpha + \beta$, the series diverges at $z = 1$; conditionally converges on $|z| = 1$ except $z = 1$.

   (iii) If $\gamma \le \alpha + \beta - 1$, the series diverges everywhere on $|z| = 1$.

(b) $aF(\alpha, \beta, \gamma, z) + bz^{1-\gamma}F(\alpha + 1 - \gamma, \beta + 1 - \gamma, 2 - \gamma, z), a, b \in \mathbf{C}$, are also solutions.

(c) (i) $\log(1 + z) = zF(1, 1, 2, -z), \log 1 = 0$.

   (ii) $(\alpha - \beta)F(\alpha, \beta, \gamma, z) = \alpha F(\alpha + 1, \beta, \gamma, z) - \beta F(\alpha, \beta + 1, \gamma, z)$.

   (iii) $(\alpha - \gamma + 1)F(\alpha, \beta, \gamma, z) = \alpha F(\alpha, \beta - 1, \gamma, z) - (\gamma - 1)F(\alpha, \beta, \gamma - 1, z)$.

   (iv) $(1 - z)F(\alpha, \beta, \gamma, z) = F(\alpha, \beta - 1, \gamma, z) - \dfrac{\gamma - \alpha}{\gamma}zF(\alpha, \beta, \gamma + 1, z)$.

*3.3.3. *Analyticity of multiple-valued functions and the Riemann surfaces (revisited)*

Let $w = f(z)$ be a *multiple-valued* function defined on a domain $\Omega \subseteq \mathbf{C}$.

In this subsection, we try to connect together analytically all branches of $f(z)$ on Ω and then to define reasonably the analyticity of $f(z)$ on Ω. Meanwhile, $f(z)$ will turn out to be a *single-valued analytic function* on an ideal surface — its Riemann surface. We just sketch main ideas concerned and leave the detail to the readers. One might refer to Ref. [75], Chap. 3, for details. See also Example 4 in Sec. 3.4.6 and Chap. 7.

Firstly, we introduce some terminologies:

(1) Elements of $f(z)$. Let $N(\zeta)$ be a neighborhood of a point ζ in Ω, on which an analytic branch $g(z)$ of $f(z)$ can be defined. Then, $(N(\zeta), g(z))$ is called an *element* of $f(z)$.

(2) A branch of $f(z)$ along a curve γ. Let γ be a curve in Ω with representation $z = z(t) : [a, b] \to \Omega$ and $G(z) : [a, b] \to \mathbf{C}$ be a continuous function defined along γ, namely, $G(z(t))$ is continuous on $[a, b]$. Suppose that for each point $z(t)$ on γ, there exist a neighborhood $N(z(t))$ of $z(t)$ in Ω and a neighborhood $\widetilde{N}(z(t))$ of $z(t)$ along γ satisfying

 (a) $\widetilde{N}(z(t)) \subseteq N(z(t))$;
 (b) there exists an analytic branch $g(z)$ of $f(z)$ on $N(z(t))$; and
 (c) $G(z) = g(z)$ for all $z \in \widetilde{N}(z(t))$.

Then $G(z)$ is called *a branch of $f(z)$ along the curve γ*. See Fig. 3.11.

(3) Analytic continuation of an element along a curve γ. Let $(N(\zeta), g(z))$ and $(N(\eta), h(z))$ be two elements of $f(z)$. Let γ be a curve $z = z(t) :$ $[a, b] \to \Omega$ with its initial point $z(a) \in N(\zeta)$ and its terminal point $z(b) \in N(\eta)$. Suppose that $G(z)$ is a branch of $f(z)$ along the curve γ satisfying

 (a) $G(z) = g(z)$ on some neighborhood $\widetilde{N}(z(a)) \subseteq N(\zeta)$ of $z(a)$ along γ, and
 (b) $G(z) = h(z)$ on some neighborhood $\widetilde{N}(z(b)) \subseteq N(\eta)$ of $z(b)$ along γ.

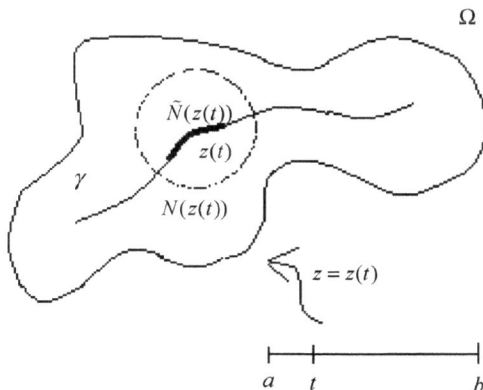

Fig. 3.11

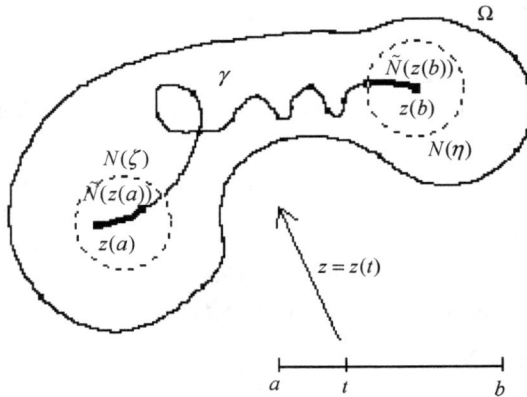

Fig. 3.12

Then, the element $(N(\eta), h(z))$ is said to be *an analytic continuation of the element* $(N(\zeta), g(z))$ *along the curve* γ. See Fig. 3.12.

Definition. Let $w = f(z)$ be a multiple-valued function defined on a domain Ω, having the following properties:

(1) For each point $\zeta \in \Omega$ and any one value w of the values of f at ζ, there exists an element $(N(\zeta), g(z))$ of $f(z)$ such that $g(\zeta) = w$.

(2) For any two elements $(N(\zeta), g(z))$ and $(N(\eta), h(z))$ of $f(z)$, $(N(\eta), h(z))$ is an analytic continuation of $(N(\zeta), g(z))$ along some curve γ in Ω (and vice versa).

  Then $w = f(z)$ is said to be *analytic* on Ω.   (3.3.3.1)

Contents of Sec. 2.7 and Examples of Sec. 3.3.1 provide information stated in condition 1 for elementary multiple-valued functions. What we have to do is to make certain the condition 2. The set

$$\mathbf{R}_{f(z)} = \{(\zeta, g(z)) | \zeta \in \Omega \text{ so that } (N(\zeta), g(z)) \text{ is an element of } f(z)\}$$
(3.3.3.2)

can be endowed with a Hausdorff topological structure and a complex structure so that it becomes a *Riemann surface* in the abstract sense (see Definition 2 (2.7.2.8) in Exercise B of Sec. 2.7.2 or Chap. 7). On $\mathbf{R}_{f(z)}$, $f(z)$ is then a *single-valued analytic function*. See Ref. [75].

  Some illustrative examples are much more convincing than words.

Example 1. $w = \sqrt{z}$ is analytic on $\Omega = \mathbf{C} - \{0\}$ and is single-valued on its Riemann surface $\mathbf{R}_{\sqrt{z}}$ (see Fig. 2.49) with 0 and ∞ as algebraic branch points of order 1 and $[\infty, 0]$ the branch cut.

See Example 4 of Sec. 3.4.6 for another treatment.

Explanation. Pick up the principle branch of $\arg z$ as $-\pi < \operatorname{Arg} z \leq \pi$. Denote, for simplicity,

$$(\sqrt{z})_+ = |z|^{\frac{1}{2}} e^{\frac{i \operatorname{Arg} z}{2}} \quad \text{and} \quad (\sqrt{z})_- = -(\sqrt{z})_+ = -|z|^{\frac{1}{2}} e^{\frac{i \operatorname{Arg} z}{2}} \quad \text{for } z \neq 0.$$

Then, $(\mathbf{C} - (\infty, 0], (\sqrt{z})_+)$ and $(\mathbf{C} - (\infty, 0], (\sqrt{z})_-)$ are two global elements of $\sqrt{z}$. For each point $z_0 \neq 0$, let $N(z_0)$, such as $|z - z_0| < |z_0|$, be a neighborhood of z_0 not containing 0. On $N(z_0)$, two elements of $\sqrt{z}$ can be defined, namely,

(1) If $N(z_0) \cap (\infty, 0] = \phi$, they are $(N(z_0), (\sqrt{z})_+)$ and $(N(z_0), (\sqrt{z})_-)$.
(2) If $N(z_0) \cap (\infty, 0] \neq \phi$, they are $(N(z_0), f_1(z))$ and $(N(z_0), f_2(z))$, where

$$f_1(z) = \begin{cases} (\sqrt{z})_+, & \text{if } \operatorname{Im} z \geq 0 \\ (\sqrt{z})_-, & \text{if } \operatorname{Im} z < 0 \end{cases} \quad \text{and}$$

$$f_2(z) = -f_1(z) = \begin{cases} (\sqrt{z})_-, & \text{if } \operatorname{Im} z \geq 0 \\ (\sqrt{z})_+, & \text{if } \operatorname{Im} z < 0 \end{cases}$$

(refer to Case 1 and Case 2 of local branches for $\arg z$ in Section (1) in Sec. 2.7.1).

The element $(N(z_0), (\sqrt{z})_-)$ is an analytic continuation of the element $(N(z_0), (\sqrt{z})_+)$ along a Jordan closed curve γ surrounding 0. The process is illustrated vividly both on the complex plane $\mathbf{C}$ and on the Riemann surface $\mathbf{R}_{\sqrt{z}}$ as shown in Fig. 3.13. The readers are urged to check all possible cases (refer to Remark after this example). According to Definition (3.3.3.1), $\sqrt{z}$ is a two-valued analytic function on $\mathbf{C} - \{0\}$ but it is single-valued on $\mathbf{R}_{\sqrt{z}}$.

Something more can be said about the Riemann surface $\mathbf{R}_{\sqrt{z}}$:

Points: $(\zeta, g(z))$ is considered as a point in $\mathbf{R}_{\sqrt{z}}$ (see (3.3.3.2)), where $\zeta \in \mathbf{C} - \{0\}$ and $g(z)$ is either $(\sqrt{z})_+, (\sqrt{z})_-$ or $f_1(z), f_2(z)$.

Open neighborhood of a point: Let $N(\zeta)$ be any neighborhood of the point ζ in $\mathbf{C} - \{0\}$ so that an analytic branch $g(z)$ of $\sqrt{z}$ can be defined on $N(\zeta)$. Then, the element $(N(\zeta), g(z))$ is considered as an open neighborhood of the point $(\zeta, g(z))$.

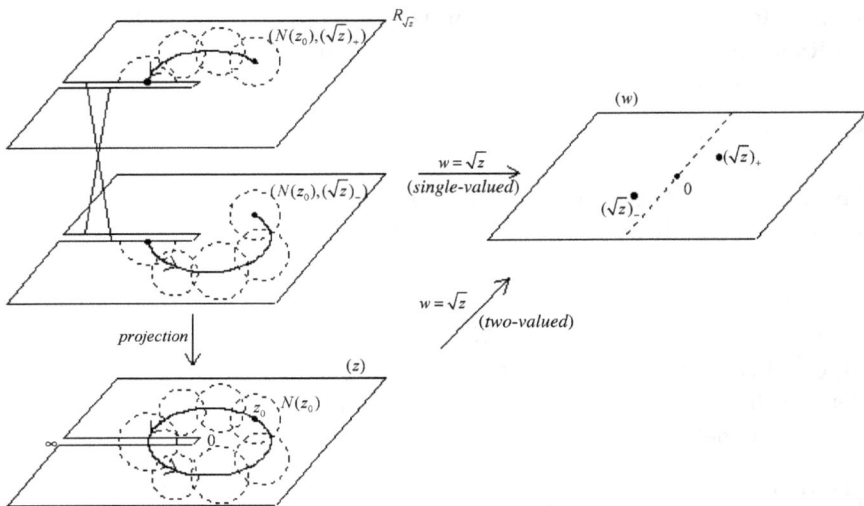

Fig. 3.13

Connected Hausdorff topological structure: $\mathbf{R}_{\sqrt{z}}$ is then a Hausdorff topological surface (see Exercise B (2) of Sec. 1.8). It is connected owing to condition 2 in (3.3.3.1). $\mathbf{R}_{\sqrt{z}}$, in Fig. 3.13, is obtained by pasting two copies of $\mathbf{C} - (\infty, 0]$ crosswise along the cut $(\infty, 0]$ and can be viewed as a geometrical model of $\mathbf{R}_{\sqrt{z}}$ defined in (3.3.3.2). Yet, another geometrical (or topological) model for $\mathbf{R}_{\sqrt{z}}$ is the Riemann sphere S (see Sec. 1.6) via stereographic projection.

Two-manifold structure (see (2.7.2.7)): Let $(N(\zeta), g(z))$ be a neighborhood of the point $(\zeta, g(z))$. Now, define $\varphi : (N(\zeta), g(z)) \to \mathbf{C}$ by

$$\varphi(t, g(z)) = t, \quad t \in N(\zeta).$$

This φ is a homeomorphism onto the open set $\varphi((N(\zeta), g(\zeta)) = N(\zeta)$ in the plane $\mathbf{C}$ and acts as a local coordinate system.

Complex (or Riemann surface) structure (see (2.7.2.8)): In case $U = (N(\zeta), g(z)) \cap (N(\eta), h(z)) \neq \phi$, i.e., $N(\zeta) \cap N(\eta) \neq \phi$, consider the local coordinate systems $\varphi : (N(\zeta), g(z)) \to \mathbf{C}$ and $\psi : (N(\eta), h(z)) \to \mathbf{C}$. Now,

$$\psi \circ \varphi^{-1}(t) = t : \varphi(U) \to \psi(U) \tag{*}$$

is a one-to-one and analytic function. Hence, $\mathbf{R}_{\sqrt{z}}$ as defined in (3.3.3.2) is indeed a Riemann surface in abstract sense.

Analyticity of $\sqrt{z}$ on $\mathbf{R}_{\sqrt{z}}$: At this moment, we do not think that 0 is a point in $\mathbf{R}_{\sqrt{z}}$. The value of $\sqrt{z}$ at a point $(\zeta, g(z)) \in \mathbf{R}_{\sqrt{z}}$ is defined as $g(\zeta)$. *More precisely*, for any local coordinate system $\varphi : (N(\zeta), g(z)) \to \mathbf{C}$, $\sqrt{z}$ is defined on $(N(\zeta), g(z))$ by

$$\left(\sqrt{z} \circ \varphi^{-1}\right)(t) = g(t) : \varphi(N(\zeta), g(z)) = N(\zeta) \to \mathbf{C}.$$

This definition is well defined owing to (*). Hence, $\sqrt{z}$ is a single-valued analytic function on $\mathbf{R}_{\sqrt{z}}$.

   $\mathbf{R}_{\sqrt{z}}$ *as a covering surface of* $\mathbf{C}^*$ (see (2.7.2.9)): Define a mapping $F : R_{\sqrt{z}} \to \mathbf{C}$ by

$$F(t, g(z)) = t, \quad t \in \mathbf{C} - \{0\}.$$

F is obviously continuous. Fix any point $(\zeta, g(z)) \in \mathbf{R}_{\sqrt{z}}$ and consider the local coordinate system $\varphi : (N(\zeta), g(z)) \to \mathbf{C}$ defined by

$$\varphi(t, g(z)) = t - \zeta, \quad t \in N(\zeta).$$

Then, $\varphi(\zeta, g(z)) = \zeta - \zeta = 0$. Also, define a mapping $\psi : \mathbf{C} \to \mathbf{C}$ by $\psi(t) = t - \zeta$. Then

$$(\psi \circ F \circ \varphi^{-1})(\varphi(t, g(z))) = (\psi \circ F \circ \varphi^{-1})(t - \zeta) = (\psi \circ F)(t, g(z))$$

$$= \psi(t) = t - \zeta, \quad t \in N(\zeta).$$

Combining together, this means that $\mathbf{R}_{\sqrt{z}}$ (without the points 0 and ∞) is a *smooth covering surface* of $\mathbf{C} - \{0\}$ and F is the *projection*.

   Finally, we try *to attach* 0 *and* ∞ *to* $\mathbf{R}_{\sqrt{z}}$, still denote as $\mathbf{R}_{\sqrt{z}}$, and to define open neighborhoods for 0 and ∞. Consider the open disk $D_r(0)$: $|z| < r$ in the z-plane. By cutting the radius $(-r, 0]$ from $D_r(0)$, we obtain the open set $D_r(0) - (-r, 0]$, which is mapped onto the right half of the open disk of $|w| < \sqrt{r}$ by $w = (\sqrt{z})_+$ and onto the left half of the disk of $|w| < \sqrt{r}$ by $w = (\sqrt{z})_-$. Observe that the radius $(-r, 0]$ is mapped, either by $(\sqrt{z})_+$ or by $(\sqrt{z})_-$, onto the diameter $(-\sqrt{r}i, \sqrt{r}i)$. See Fig. 3.14. Hence, denote the set

$$N_r(0) = \{0\} \cup \{(\zeta, g(z)) \mid \zeta \in \mathbf{C} - \{0\} \text{ and } |\zeta| < r\}$$

   = (geometrically) The figure obtained by pasting crosswise the radii
      $(-r, 0]$ of two copies of $D_r(0) - (-r, 0]$,

which might be called an *open disk* in $\mathbf{R}_{\sqrt{z}}$ with center 0 and radius r. More accurately, we just *define* $(N_r(0), \sqrt{z})$ as an *open disk* in $\mathbf{R}_{\sqrt{z}}$ with center 0 and radius r. Observe that $\sqrt{z}$ maps $(N_r(0), \sqrt{z})$ homeomorphically onto the open disk $|w| < \sqrt{r}$ and hence, $\sqrt{z}$ acts as a local coordinate.

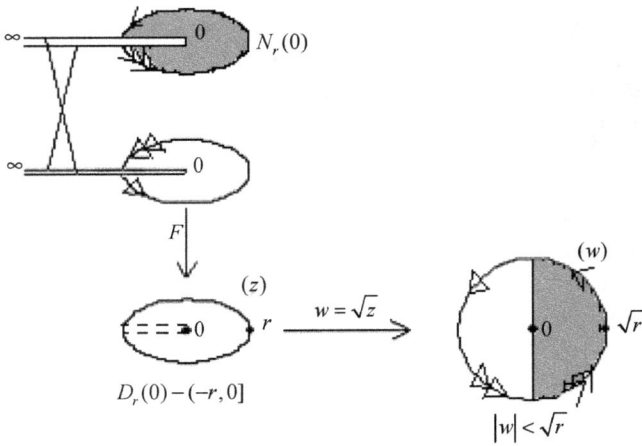

Fig. 3.14

Similarly, let $N_r(\infty) = \{\infty\} \cup \{(\zeta, g(z)) | \zeta \in \mathbf{C} - \{0\}$ and $|\zeta| > r\}$. Then $(N_r(\infty), \sqrt{z})$ is an *open neighborhood* of ∞ in $\mathbf{R}_{\sqrt{z}}$ with $\sqrt{z}$ acting as a local coordinate.

For simplicity, let $\varphi_0 : (N_r(0), \sqrt{z}) \to \mathbf{C}$ defined by $\varphi_0(0) = 0$ and $\varphi_0(\zeta, g(z)) = \sqrt{\zeta}$ be the local coordinate. Then, combining with the projection F,

$$(F \circ \varphi_0^{-1})(w) = (F \circ \varphi_0^{-1})(\sqrt{\zeta}) = F(\zeta, g(z)) = \zeta = w^2, \quad w \in D_r(0).$$

Hence, 0 in $\mathbf{R}_{\sqrt{z}}$ is called a *branch point of order* 1. So is ∞. And the ray $[\infty, 0]$, connecting 0 and ∞, is called a *branch cut* of $\mathbf{R}_{\sqrt{z}}$. In this sense, $\mathbf{R}_{\sqrt{z}}$ (together with 0 and ∞) is said to be a *branched covering surface* (of two sheets) of $\mathbf{C}^*$, with two branch points 0 and ∞ of order 1, and at the same time, $\mathbf{R}_{\sqrt{z}}$ is also an abstract Riemann surface in the sense of (2.7.2.8).

Remark. Choose any fixed point $z_0 \neq 0$. We try to expand the two branches of $\sqrt{z}$ around z_0 into the Taylor series as follows.

$$z = z_0 \left(1 + \frac{z - z_0}{z_0}\right)$$

$\Rightarrow$ (By Example 2 in Sec. 3.3.2) $\sqrt{z} = \sqrt{z_0}(1 + \frac{z-z_0}{z_0})^{\frac{1}{2}}$, where $\sqrt{z_0}$ is either $(\sqrt{z_0})_+$ or $(\sqrt{z_0})_-$, and $\sqrt{1} = 1$ has the power series representation

$$\sqrt{z} = \sqrt{z_0}\left\{1 + \sum_{n=1}^{\infty} C_n^{\frac{1}{2}}\left(\frac{z - z_0}{z_0}\right)^n\right\}, \quad |z - z_0| < |z_0|,$$

$$C_n^{\frac{1}{2}} = (-1)^{n-1}\frac{(2n)!}{2^{2n}(n!)^2(2n-1)}, \quad n \geq 1.$$

For simplicity, let $P(z - z_0)$ denote the power series inside the bracket $\{\ \}$ in the above expression. Then, (3.3.2.2) can be rewritten as

$$\mathbf{R}_{\sqrt{z}} = \{(z_0, (\sqrt{z_0})_+ P(z - z_0)), \quad (z_0, (\sqrt{z_0})_- P(z - z_0)) \mid z_0 \in \mathbf{C} - \{0\}\}.$$

$$(3.3.2.2')$$

If we continue $\sqrt{z}$ around the circle $\gamma : z = |z_0|e^{i(2\pi t + \text{Arg } z_0)}, 0 \leq t \leq 1$, then $P_0(z - z_0) = (\sqrt{z_0})_+ P(z - z_0)$ is changed continuously to

$$P_t(z - z_t) = e^{i\left(\pi t + \frac{\text{Arg } z_0}{2}\right)}\left\{1 + \sum_{n=1}^{\infty} C_n^{\frac{1}{2}}\left(\frac{z - z_t}{z_t}\right)^n\right\},$$

where

$$z_t = |z_0|e^{i(2\pi t + \text{Arg} z_0)} \quad \text{and} \quad |z - z_t| < |z_t|, \quad 0 \leq t \leq 1.$$

After one circulation around γ, $P_1(z - z_1) = -P_0(z - z_0) = (\sqrt{z_0})_- P(z - z_0)$. This amounts to say, on $\mathbf{R}_{\sqrt{z}}$, that the element $(z_0, (\sqrt{z})_+ P(z - z_0))$ moves continuously along the "circle" $(z_t, (\sqrt{z})_+ P_t(z - z_t))$ to $(z_0, (\sqrt{z_0})_- P(z - z_0))$, and after one more circulation, we will return to the original element $(z_0, (\sqrt{z_0})_+ P(z - z_0))$. This phenomenon leads to the geometrical model for $\mathbf{R}_{\sqrt{z}}$ as shown in the left figure in Figs. 3.13 and 3.14 and, at the same time, makes certain what we have said that 0 is a branch point of order 1. Similar explanation is still valid for ∞.

The advantage of using local power series to represent an element in a Riemann surface, such as in $\mathbf{R}_{\sqrt{z}}$, is universally beneficial in establishing abstract and general theory. This is because the only way of existence of a function analytic at a point is a power series with a positive radius of convergence. See Chap. 5. □

Example 2. $w = \log z$ is a multiple-valued analytic function on $\mathbf{C} - \{0\}$ but is a single-valued analytic function on its Riemann surface $R_{\log z}$ (see Fig. 2.73) with 0 and ∞ as logarithmic branch points and $[\infty, 0]$ as branch cut.

Explanation. Designate the principle branch of $\arg z$ as $-\pi < \operatorname{Arg} z \le \pi$. For simplicity, let

$$(\log z)_n = \operatorname{Log} z + 2n\pi i, \quad n = 0, \pm 1, \pm 2, \dots$$

with $\operatorname{Log} z = \log|z| + i\operatorname{Arg} z = (\log z)_0$. Then, $(\mathbf{C} - (\infty, 0], (\log z)_n)$, $n = 0, \pm 1, \pm 2, \dots$, are global elements of $\log z$. For $z_0 \ne 0$, let $N(z_0)$ be the disk $|z - z_0| < |z_0|$. There are countably infinitely many elements of $\log z$ on $N(z_0)$:

(1) If $N(z_0) \cap (\infty, 0] = \varphi$, they are $(N(z_0), (\log z)_n), n = 0, \pm 1, \pm 2, \dots$.
(2) If $N(z_0) \cap (\infty, 0] = \varphi$, they are $(N(z_0), f_n(z)), n = 0, \pm 1, \pm 2, \dots$, where

$$f_n(z) = \begin{cases} (\log z)_n & \text{if } \operatorname{Im} z \ge 0 \\ (\log z)_{n+1} & \text{if } \operatorname{Im} z < 0 \end{cases} \quad \text{for } n = 0, \pm 1, \pm 2, \dots.$$

For any two fixed elements $(N(z_0), (\log z)_m)$ and $(N(z_0), (\log z)_n)$ with $m < n$, let γ be Jordan closed curve passing z_0 and surrounding 0 as its interior point. Then $(N(z_0), (\log z)_n)$ is an analytic continuation of $(N(z_0), (\log z)_m)$ along γ circulating $n - m$ times in the anticlockwise direction. In case $(N(\zeta), (\log z)_n)$ and $(N(\eta), (\log z)_n)$ with $\zeta \ne \eta$, the path of analytic continuation can be chosen as a Jordan arc connecting ζ and η. Many other cases are left to the readers to think about.

According to Definition (3.3.1.1), $\log z$ is a multiple-valued analytic function on $\mathbf{C} - \{0\}$ but is single-valued on its Riemann surface (see (3.3.3.2))

$$R_{\log z} = \{(\zeta, g(z)) \,|\, \zeta \in \mathbf{C} - \{0\} \quad \text{so that } (N(\zeta), g(z))$$
$$\text{is an element of } \log z\}.$$

Observe that an analytic branch $g(z)$ on $N(\zeta)$ can be expressed as one of the following power series (refer to Example 1 in Sec. 3.3.2):

$$\log z = \log\left\{\zeta\left(1 + \frac{z-\zeta}{\zeta}\right)\right\} = \log\zeta + \operatorname{Log}\left(1 + \frac{z-\zeta}{\zeta}\right), \quad \operatorname{Log} 1 = 0$$

$$= \log\zeta + \sum_{n=1}^{\infty} \frac{(-1)^{n-1}}{n}\left(\frac{z-\zeta}{\zeta}\right)^n, \quad |z - \zeta| < |\zeta|.$$

Just like in Example 1, we can justify that $R_{\log z}$ is indeed a Riemann surface in abstract sense (see (2.7.2.8)) and is a smooth covering surface of $\mathbf{C} - \{0\}$ (see (2.7.2.9)). Recall that 0 and ∞ are not points belonging to $R_{\log z}$. The left figure in Fig. 2.73 is the familiar geometric model for $R_{\log z}$ which is topologically equivalent to the finite complex plane $\mathbf{C}$.

Exercises A

(1) Prove in detail that the Riemann surface $R_{\log z}$ in Example 2 is a Riemann surface in abstract sense and is a smooth (i.e., consisting of regular points only) covering surface of $\mathbf{C} - \{0\}$.
(2) Do as in Example 1 for the functions $w = z^{\frac{3}{2}}$ and $w = z^{\frac{2}{3}}$, respectively (see Example 1 in Sec. 2.7.2).

Exercises B

Try to model after Example 1 to justify that the multiple-valued functions presented in Sec. 2.7, especially $\cos^{-1} z, \sin^{-1} z, \tan^{-1} z$ etc. are analytic on their domains of definition and are single-valued, analytic on their Riemann surfaces, respectively.

3.4. Analytic Properties of Analytic Functions

As indicated implicitly in (2), (3), and (6) in (3.2.2), an analytic function $f(z) = u(z) + iv(z)$ is a rather special and restricted kind of functions $f(x,y) = (u(x,y), v(x,y))$ of two real variables x and y, even though in appearance it looks like a function of a real variable in many algebraic operations, less the ordering. In this section, it is supposed that the readers are better familiar with the basics of the theory of functions of one and two real variables and are able to compare them with the properties of analytic functions being introduced.

A total of seven subsections are divided.

3.4.1. *Elementary properties derived from definition*

Consider the entire function $f(z) = e^{2\pi i z}$. Observe that $f(1) = f(0)$ but, for any $0 < t < 1$, $f(1) - f(0) \neq f'(t) = 2\pi i e^{2\pi i t}$ holds. This indicates that the mean-value theorem in calculus is not more true for analytic functions. Anyway, we have the following modified

Mean-valued theorem. Let f be analytic on a domain Ω. Suppose a and b are two distinct points in Ω so that the line segment $\overline{ab}$ is still in Ω. Then, there exist two points c_1 and c_2 on $\overline{ab}$, other than a and b, so that

$$f(b) - f(a) = (\operatorname{Re} f'(c_1) + i \operatorname{Im} f'(c_2))(b - a). \tag{3.4.1.1}$$

Note that c_1 and c_2 might be distinct. See also Exercise A (23) of Sec. 2.9.

Proof. Note that $\overline{ab}$ has the parametric representation $z(t) = a + (b-a)t$, $0 \le t \le 1$. Let

$$\varphi(t) = \frac{1}{b-a}\, f(a + (b-a)t) = \alpha(t) + i\beta(t), \quad 0 \le t \le 1.$$

Now, both $\alpha(t)$ and $\beta(t)$ are differentiable on $0 \le t \le 1$. Hence there exist $t_1, t_2 \in (0,1)$ so that $\alpha(1) - \alpha(0) = \alpha'(t_1)$ and $\beta(1) - \beta(0) = \beta'(t_2)$. Let $c_1 = a + t_1(b-a)$ and $c_2 = a + t_2(b-a)$. Since $\alpha'(t) + i\beta'(t) = f'(a + (b-a)t)$, so

$$\frac{f(b) - f(a)}{b-a} = \varphi(1) - \varphi(0) = [\alpha(1) - \alpha(0)] + i[\beta(1) - \beta(0)]$$

$$= \alpha'(t_1) + \beta'(t_2)i = \operatorname{Re} f'(c_1) + i \operatorname{Im} f'(c_2). \qquad \square$$

Equation (3.4.1.1) can be used to prove

The univalence of analytic function on a convex domain (Noshiro–Warschawski). Suppose f is analytic on a *convex* domain Ω (namely, Ω contains the segment connecting any two of its points). Suppose $\operatorname{Re} f'(z) > 0$ holds on Ω (or $\operatorname{Re} f'(z) < 0$ or $\operatorname{Im} f'(z) > 0$ or $\operatorname{Im} f'(z) < 0$ on Ω). Then f is univalent (i.e., one-to-one) on Ω. (3.4.1.2)

For other versions of criterion for univalence, see Sec. 3.5.5. In this case, by Brower's theorem (2.4.13), $f(\Omega)$ would be a domain (see Sec. 3.4.4). For examples, $w = z^2$ is univalent either on $\operatorname{Re} z > 0$ or on $\operatorname{Im} z < 0$. Also, $w = z^3$ is univalent on $0 < |y| < x$ or $0 < |y| < -x$, where $z = x + iy$. While $w = e^z$ is univalent on each strip domain $(2n - \frac{1}{2})\pi < \operatorname{Im} z < (2n + \frac{1}{2})\pi, n = 0, \pm 1, \pm 2, \ldots$.

Another consequence of (3.4.1.1) is the following

Characteristic property of a constant analytic function. An analytic function f on a domain Ω is a constant if and only if

$$f'(z) = 0, \quad z \in \Omega. \tag{3.4.1.3}$$

Proof of Sufficiency. Fix a point $z_0 \in \Omega$. Then, by (1.8.11), any other point z in Ω can be joined to z_0 by a polygonal curve $\overline{z_0 z_1} + \overline{z_1 z_2} + \cdots + \overline{z_n z}$, consisting of line segments and lying entirely in Ω. Now, (3.4.1.1) guarantees that

$$f(z) = f(z_n) = f(z_{n-1}) = \cdots = f(z_1) = f(z_0), \quad z \in \Omega.$$

Or, observe that each segment $\overline{z_j z_{j+1}}$ can be chosen to be parallel to the coordinate axes. Since $u_x = u_y = v_x = v_y = 0$ everywhere on Ω, where $f = u + iv$, u and v are constants on each segment $\overline{z_j z_{j+1}}$. So is $f = u + iv$. $\square$

Remark 1 (A nonconstant analytic function has only isolated zeros, if any). It can be shown that *if there exists a sequence z_n in Ω having a limit point in Ω so that*

$$f'(z_n) = 0, \quad n = 1, 2, 3, \ldots,$$

then $f'(z) = 0$ for all $z \in \Omega$ and hence, f *is a constant.* Equivalently, this means that if f is not a constant, then $f'(z)$ can only have isolated zeros (namely, if $f'(z_0) = 0$, then there exists a $r > 0$ so that $f'(z) \neq 0$ on $0 < |z - z_0| < r$). This result is still valid for $f(z)$ itself. See (3.4.2.9).

Equation (3.4.1.3) can be used to prove the following

Sufficient conditions for constant analytic function. Suppose f is analytic on a domain Ω. Then, any one of the following conditions will imply that $f(z)$ is a constant.

(1) $\text{Re}\, f(z) = $ constant, $z \in \Omega$.
(2) $\text{Im}\, f(z) = $ constant, $z \in \Omega$.
(3) $|f(z)| = $ constant, $z \in \Omega$.
(4) $\arg f(z) = $ constant, $z \in \Omega$.
(5) $\overline{f(z)}$ is analytic on Ω.   (3.4.1.4)

Proof. Let $f = u + iv$.

(1) and (2): Suppose u or v is a constant function on Ω. Then $u_x = u_y = 0$ or $v_x = v_y = 0$ on Ω. By the Cauchy–Riemann equations, $f'(z) = u_x + iv_x = 0$ on Ω and hence, $f(z)$ is a constant.

(3) Let $|f(z)|^2 = u^2 + v^2 = c$ (a constant). If $c = 0$, then $f(z) \equiv 0$. In case $c \neq 0$, differentiate $u^2 + v^2 = c$ and we obtain $uu_x + vv_x = 0$ and $uu_y + vv_y = -uv_x + vu_x = 0$ on Ω. Solve these equations and hence, $u_x = u_y = v_x = v_y = 0$ on Ω. Thus, $f'(z) = 0$ and $f(z)$ is a constant.

(4) Suppose $\arg f(z) = \tan^{-1}\left(\frac{v}{u}\right)$ is a constant. If $u = 0$, then by (2), $f(z)$ is a constant. If $u \neq 0$, then $u = cv$ for some constant c. Since $\text{Re}(1 + ic)f = u - cv$ is now a constant, by (1), $(1 + ic)f(z)$ is constant and so is $f(z)$.

(5) Suppose $f(z) = u(z) + iv(z)$ and $\overline{f(z)} = u(z) - iv(z)$ are simultaneously analytic on the same domain. By the Cauchy–Riemann equations, both $u_x = v_y, u_y = -v_x$, and $u_x = -v_y, u_y = v_x$ hold on Ω. Hence $u_x = u_y = v_x = v_y = 0$ on Ω and $f(z)$ is a constant. Or, owing to the analyticity of $f(z)\overline{f(z)} = u^2 + v^2$ on Ω and Example 1 in Sec. 3.1, $f'(z) = 0$ is the only possibility.   □

Remark 2 (A nonconstant analytic function is an open mapping, i.e., mapping open sets onto open sets (see (3.4.4.4))). As to (5) in (3.4.1.4), we can compensate the following

Symmetry principle of analytic function. Let Ω be a domain. Call $\Omega^* = \{z \in \mathbf{C} \mid \bar{z} \in \Omega\}$ the *symmetric domain* of Ω *with respect to the real axis* and Ω itself a *symmetric domain* if $\Omega^* = \Omega$ holds (in this case, Ω has nonempty intersection with the real axis). See Fig. 3.15.

(1) Suppose f is a function on a domain Ω. Then

      (i) $f(z)$ is analytic on Ω.

$\Leftrightarrow$ (ii) $g(z) = \overline{f(\bar{z})}$ is analytic on Ω^* and

$$g'(z) = \overline{f'(\bar{z})}, \quad z \in \Omega^*.$$

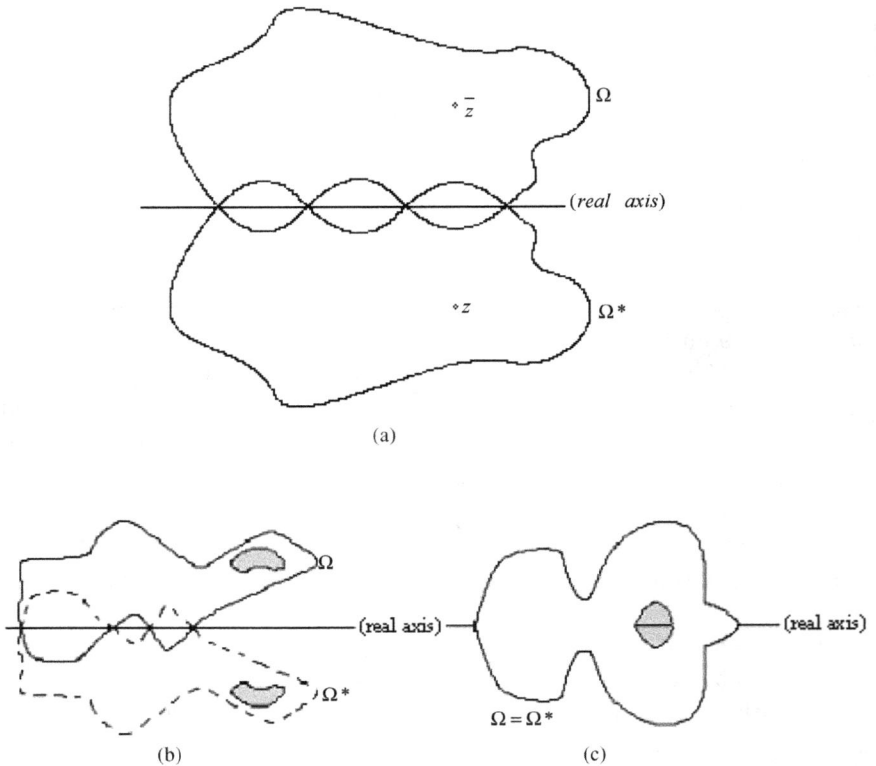

(a)

(b)   (c)

Fig. 3.15

(2) If f is analytic on a *symmetric* domain Ω, then f preserves symmetric points with respect to the real axis, namely,

$$f(\bar{z}) = \overline{f(z)}, \quad z \in \Omega,$$

if and only if, $f(z)$ takes real values on $\Omega \cap \{\operatorname{Im} z = 0\}$.   (3.4.1.5)

$\square$

Proof. (1) Fix a point $z_0 \in \Omega^*$. If $z \in \Omega^*$, then $\bar{z}$ and $\overline{z_0}$ are in Ω. Now, as $z \to z_0$,

$$\frac{g(z) - g(z_0)}{z - z_0} = \frac{\overline{f(\bar{z})} - \overline{f(\overline{z_0})}}{z - z_0} = \overline{\left(\frac{f(\bar{z}) - f(\overline{z_0})}{\bar{z} - \overline{z_0}} \right)} \to \overline{f'(\overline{z_0})}.$$

Or, set $z = x + iy$ and $f(z) = u(x,y) + iv(x,y)$. Then $\overline{f(\bar{z})} = U(x,y) + iV(x,y)$ where $U(x,y) = u(x,-y)$ and $V(x,y) = -v(x,-y)$ are differentiable on Ω^*. Known $u_x = v_y$ and $u_y = -v_x$, so

$$U_x(x,y) = u_x(x,-y) = v_y(x,-y) = -v_{(-y)}(x,-y) = V_{(-y)}(x,-y);$$

$$U_{(-y)}(x,y) = u_{(-y)}(x,-y) = -u_y(x,-y) = v_x(x,-y)$$

$$= -(-v)_x(x,-y) = -V_x(x,-y)$$

hold on Ω^*. Hence, $\overline{f(\bar{z})}$ is analytic on Ω^*. Conversely, if $\overline{f(\bar{z})}$ is analytic on Ω^*, then the known result shows that $\overline{\overline{f(\bar{z})}} = f(z)$ is analytic on $(\Omega^*)^* = \Omega$.
   (2) is an easy consequence of (1) above and (4) in (3.4.2.9). $\square$

Remark 3 (Analytic extension via symmetry principle).   Suppose f is analytic on a domain Ω, which lies on one side of the real axis and its boundary has a nonempty intersection with the real axis so that $\partial \Omega \cap \{\operatorname{Im} z = 0\} = \sigma$ is an open set. If f is continuous on $\Omega \cup \sigma$ and takes real values on σ, then the function $F(z)$ defined by

$$F(z) = \begin{cases} f(z), & z \in \Omega \cup \sigma, \\ \overline{f(\bar{z})}, & z \in \Omega^*, \end{cases}$$

is analytic on $\Omega \cup \sigma \cup \Omega^*$. For a proof, see (3.4.6.1) or (3.4.6.2).

Exercises A

(1) Prove (3.4.1.2) in detail.
(2) Suppose f is analytic on a domain Ω.

(a) If there exist real constants α and β, not all equal to zero, so that $\alpha \operatorname{Re} f(z) + \beta \operatorname{Im} f(z)$ is a constant on Ω. Show that f is a constant function. This is still valid for complex constants α and β.

(b) Suppose $F(u, v)$ is a real-valued function of two real variables u and v, and has continuous partial derivatives. Under what conditions that the relation $F(\operatorname{Re} f(z), \operatorname{Im} f(z)) = 0$ on Ω will induce that f is a constant? Justify it by using $F(u, v) = u^2 - v$.

(c) If $h : \mathbf{R} \to \mathbf{R}$ is a differentiable function so that $\operatorname{Re} f(z) = h(\operatorname{Im} f(z))$ on Ω, show that f is a constant.

(3) Suppose $f_1, \ldots, f_n$ are analytic on a domain Ω so that $|f_1(z)|^2 + \cdots + |f_n(z)|^2 = 1$ on Ω. Show that $f_1, \ldots, f_n$ are constant functions.

(4) Suppose f is analytic on a domain Ω so that $f(f(z)) = f(z)$ on Ω. Show that $f(z)$ is either a constant or $f(z) = z$. Here, the continuity of $f'(z)$ is implicitly assumed.

(5) Let f be analytic on a domain Ω and satisfy $f'(z) = \alpha f(z)$ on Ω for some constant α. Show that $f(z) = ce^{\alpha z}$.

(6) Suppose a function $f : \mathbf{C} - (\infty, 0] \to \mathbf{C}$ satisfying the following conditions: $(f(z))^2 = z$ and $\operatorname{Re} f(z) > 0$ for all $z \in \mathbf{C} - (\infty, 0]$. Show that f is continuous and differentiable on $\mathbf{C} - (\infty, 0]$ with $f'(z) = \frac{1}{2}(f(z))^{-1}$. What is f?

(7) Let $f(z) = z + e^z$.

(a) Show that f is univalent on $\Omega_0 : 0 < \operatorname{Im} z < \pi$ and justify that $f(\Omega_0) = \{w \in \mathbf{C} \,|\, \operatorname{Im} w > 0\} - \{w \in \mathbf{C} \,|\, \operatorname{Im} w = \pi, -\infty < \operatorname{Re} w \le -1\}$, denoted by D_0.

(b) Show that $f(\bar{z}) = \overline{f(z)}$ and hence, show that f maps Ω_0^* univalently onto D_0^*.

(c) Show that the restriction $f|_{\mathbf{R}}$ maps $\mathbf{R}$ onto itself univalently.

Now, let $\Omega = \Omega_0 \cup \mathbf{R} \cup \Omega_0^*$ and $D = D_0 \cup \mathbf{R} \cup D_0^*$. Then, f maps Ω univalently onto D. Try to construct the Riemann surface for the inverse function of $w = z + e^z$ (refer to Example 2 in Sec. 6.2.3).

(8) Let Ω be a domain.

(a) Suppose f and g are analytic on Ω so that $f'(z) = g'(z)$ on Ω. Show that $g(z) = f(z) + c$ for some constant c.

(b) Suppose f is analytic and n-times differentiable on $\Omega(f^{(k)}(z) = (f^{(k-1)})'(z), k \ge 2)$ and $f^{(n)}(z) = 0$ on Ω. Show that f is a polynomial of degree at most equal to $n - 1$ $(n \ge 1)$.

(9) Let $p(z) = a_{mn}x^m y^n + \cdots + a_{10}x + a_{01}y + a_{00}$ be a polynomial in x and y, where $z = x + iy$. In case $p(z)$ is also an analytic function of z,

show that $p(z)$ is a polynomial of z. Then, if $p(z) = a_n z^n + \cdots + a_1 z + a_0$, show that

$$a_k = \frac{1}{k!} \frac{\partial^k p}{\partial x^k}(0), \quad k = 0, 1, 2, \ldots, n.$$

(10) Show that $\overline{e^z} = e^{\bar{z}}, \overline{\cos z} = \cos \bar{z}$ and $\overline{\sin z} = \sin \bar{z}$ on $\mathbf{C}$.

3.4.2.   *Cauchy integral theorem and formula (simple forms) : The continuity (analyticity) of the derivative of an analytic function and its Taylor series representation*

Other meaningful properties of an analytic function, except those stated in Sec. 3.4.1, depend solely on the continuity of its derivative. The classical proofs for the continuity of the derivative use complex integration (see Sec. 2.9) as we will see in this section. For an integration-free proof but a difficult one than the classical proofs, see E. H. Connell and P. Porcelli's article on *Bull. Amer. Math. Soc.* 67, 1961, which leaned heavily on a topological theorem due to G. T. Whyburn, *Topological Analysis*, Princeton University Press, Princeton, NJ., 1958, 1964.

This section is divided into five subsections.

Section (1) Cauchy integral theorem (simple form)

A. L. Cauchy obtained in 1825 the following remarkable result, esteemed as

Cauchy integral theorem. Suppose f is analytic on a simply connected domain Ω. Then, for any rectifiable closed curve γ in Ω,

$$\int_\gamma f(z)dz = 0. \tag{3.4.2.1}$$

Cauchy proved this theorem under the additional assumption that the derivative $f'(z)$ is continuous and adopted Green's theorem (see (2.9.12)): Let $f(z) = u(z) + iv(z)$ and γ be a rectifiable Jordan closed curve, with positive orientation, in Ω. Then

$$\int_\gamma f(z)dz = \int_\gamma (udx - vdy) + i \int_\gamma (vdx + udy) \quad \text{(see (2.9.2) and (2.9.5))}$$

$$= \iint_{\overline{\text{Int } \gamma}} (-v_x - u_y)dxdy$$

$$+ i \iint_{\overline{\text{Int } \gamma}} (u_x - v_y) dx dy \quad \text{(by Green's theorem)}$$

$$= 0 \quad \text{(by Cauchy–Riemann equations)}. \tag{*}$$

Readers are strongly urged to review Fig. 3.8 (with $w(z)$ there replaced by $\overline{f(z)} = u(z) - iv(z)$) and (3.2.4.5) and then try to give (*) a vivid physical interpretation. On the other hand, if it is known that $f(z)$ can be locally represented by a power series $\sum_{n=0}^{\infty} a_n(z - z_0)^n$ with a positive radius r of convergence, then (4) in (3.3.2.4) guarantees that

$$\int_\gamma f(z) dz = \sum_0^\infty a_n \int_\gamma (z - z_0)^n dz = \sum_0^\infty a_n \cdot 0 = 0$$

for γ lying entirely in $|z - z_0| < r$. This local result will eventually lead to the global one (*) after a suitable readjustment.

However, it was E. Goursat who, in 1900, discovered that the classical hypothesis of a continuous $f'(z)$ in the Cauchy theorem is redundant and gave a proof that could hardly be simpler. In what follows, we follow a revised proof published by A. Pringsheim in 1901.

The crucial fundamental fact in this direction is the

Cauchy–Goursat theorem. Let f be analytic on a domain and Δ be any triangle (including its interior Int Δ and its bounding $\partial\Delta$) so that $\Delta =$ Int $\Delta \cup \partial\Delta \subseteq \Omega$ holds. Then,

$$\int_{\partial\Delta} f(z) \, dz = 0. \tag{3.4.2.2}$$

Proof. Let

$$I(\Delta) = \int_{\partial\Delta} f(z) dz.$$

Divide Δ into four congruent triangles $\Delta^1, \Delta^2, \Delta^3, \Delta^4$ as in Fig. 3.16. According to (2) and (3) in (2.9.10) and observing that the integrals over

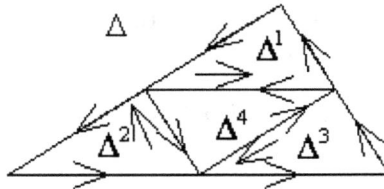

Fig. 3.16

the common sides cancel each other,

$$I(\Delta) = \sum_{j=1}^{4} \int_{\partial\Delta^j} f(z)dz = \sum_{j=1}^{4} I(\Delta^j),$$

where

$$I(\Delta^j) = \int_{\partial\Delta^j} f(z)dz, \quad 1 \leq j \leq 4, \quad \text{for simplicity.}$$

Since $|I(\Delta)| \leq \sum_{j=1}^{4} |I(\Delta^j)|$, at least one of the congruent triangles $\Delta^j, 1 \leq j \leq 4$, say Δ_1, must satisfy the conditions:

(1) $|I(\Delta)| \leq 4|I(\Delta_1)|$.

(2) $l(\partial\Delta_1) = \frac{1}{2}l(\partial\Delta)$ ($l(\partial\Delta)$ denotes the perimeter of $\partial\Delta$, etc.).

(3) $d(\Delta_1) = \frac{1}{2}d(\Delta)$ ($d(\Delta)$ denotes the diameter of Δ, etc.).

This process can be repeated infinitely, and we obtain a sequence of nested triangles

$$\Delta = \Delta_0 \supseteq \Delta_1 \supseteq \Delta_2 \supseteq \cdots \supseteq \Delta_n \supseteq \Delta_{n+1} \supseteq \cdots$$

satisfying the following properties:

(1) $|I(\Delta_{n-1})| \leq 4|I(\Delta_n)|, \quad n \geq 1$.

(2) $l(\partial\Delta_n) = \frac{1}{2}l(\partial\Delta_{n-1}) = \cdots = \frac{1}{2^n}l(\partial\Delta), \quad n \geq 1$.

(3) $d(\Delta_n) = \frac{1}{2}d(\Delta_{n-1}) = \cdots = \frac{1}{2^n}d(\Delta), \quad n \geq 1$.

Since Δ_n is compact for $n \geq 1$ and $\lim_{n\to\infty} d(\Delta_n) = 0$, by nested sets theorem (see (2) in (1.9.3)), there exists a unique point $z_0 \in \mathbf{C}$ so that $\bigcap_{n=1}^{\infty} \Delta_n = \{z_0\}$. Since Δ is closed and $\Delta \subseteq \Omega$, $z_0 \in \Delta$ and f is analytic at z_0.

For any fixed $\varepsilon > 0$, there exists a $\delta = \delta(z_0, \varepsilon) > 0$ so that, on $|z - z_0| < \delta$,

$$f(z) = f(z_0) + f'(z_0)(z - z_0) + o(z - z_0) \quad \text{with } |o(z - z_0)| \leq \varepsilon|z - z_0|.$$

For all sufficiently large n, Δ_n are contained in $|z - z_0| < \delta$ and hence, for such Δ_n,

$$\int_{\partial\Delta_n} f(z)dz$$

$$= f(z_0) \int_{\partial\Delta_n} dz + f'(z_0) \int_{\partial\Delta_n} (z - z_0)dz + \int_{\partial\Delta_n} o(z - z_0)dz$$

$$= \int_{\partial\Delta_n} o(z - z_0)dz \quad \text{(by using Examples 1 and 2 in Sec. 2.9)},$$

$\Rightarrow |I(\Delta_n)|$

$$= \left| \int_{\partial \Delta_n} o(z - z_0)dz \right| \le \int_{\partial \Delta_n} |o(z - z_0)||dz| \le \varepsilon \int_{\partial \Delta_n} |z - z_0||dz|$$

$$\le \varepsilon \cdot d(\Delta_n) \cdot l(\partial \Delta_n)$$

$$= \varepsilon \cdot \frac{d(\Delta)}{2^n} \cdot \frac{l(\partial \Delta)}{2^n} = \varepsilon \cdot \frac{d(\Delta) \cdot l(\partial \Delta)}{4^n}$$

$$\Rightarrow |I(\Delta)| \le 4|I(\Delta_1)| \le 4^2|I(\Delta_2)| \le \cdots \le 4^n|I(\Delta_n)| \le \varepsilon \cdot d(\Delta) \cdot l(\partial \Delta)$$

$$\Rightarrow I(\Delta) = \int_{\partial \Delta} f(z)dz = 0 \quad (\text{since } \varepsilon > 0 \text{ is arbitrary}).$$

The theorem is then proved.   □

A slight generalization of (3.4.2.2) is the following

Cauchy–Goursat theorem for closed polygonal curve. Suppose f is analytic on a domain Ω. Let D be a simply connected subdomain of Ω. Then

$$\int_\Gamma f(z)dz = 0,$$

where Γ is any closed polygonal curve contained in D.   (3.4.2.3)

Proof. Case 1. Suppose P is a convex polygon (including the interior Int P and its boundary ∂P) with n vertices $A_1, \ldots, A_n$ as shown in Fig. 3.17. The diagonals $A_1 A_3, \ldots, A_1 A_{n-1}$ divide P into $(n-2)$ triangles $\Delta_2, \ldots, \Delta_{n-1}$. Original orientation along ∂P induces orientation on each $\Delta_j, 2 \le j \le n-1$, so that the common sides of two triangles have opposite directions. Therefore, if $f(z)$ is analytic on P, then

$$\int_{\partial P} f(z)dz = \int_{\partial \Delta_2} f(z)dz + \cdots + \int_{\partial \Delta_{n-1}} f(z)dz = 0 + \cdots + 0 = 0.$$

Case 2. Next, suppose P is a Jordan polygon (including Int P and ∂P). In this case, ∂P is composed of non-self-intersecting line segments $A_1 A_2, \ldots, A_n A_1$ as shown in Fig. 3.18.

We may suppose that P is not convex so that at least one of $A_1 A_2, \ldots, A_n A_1$ has its extension lying in the interior Int P. Extend such a segment along one or two directions until it first meets a point on the boundary ∂P. This process will divide P into a finite number of convex

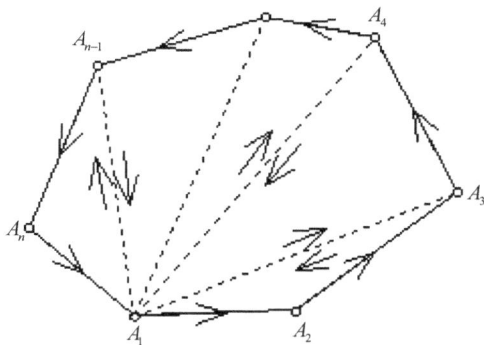

Fig. 3.17

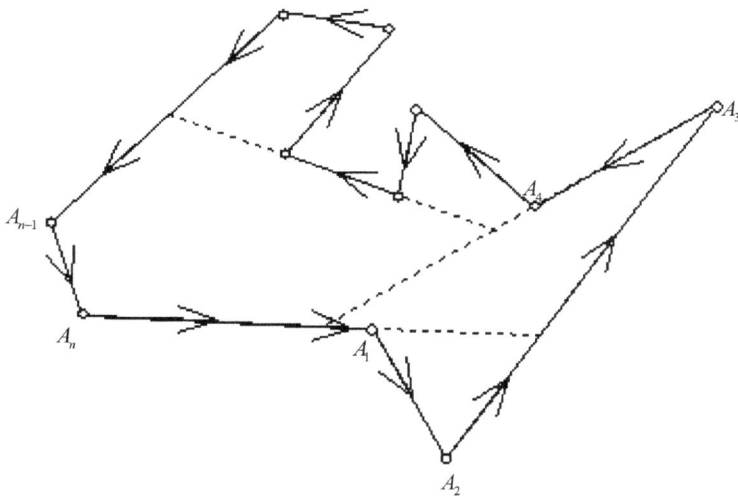

Fig. 3.18

polygons $P_1, \ldots, P_m$. If $f(z)$ is analytic on P, then

$$\int_{\partial P} f(z)dz = \int_{\partial P_1} f(z)dz + \cdots + \int_{\partial P_m} f(z)dz = 0 + \cdots + 0 = 0.$$

Case 3. Suppose P is an arbitrary polygon (including the domains it enclosed and its boundaries). When oriented the boundary ∂P, some of its boundary segments may be covered partially or wholly several times as

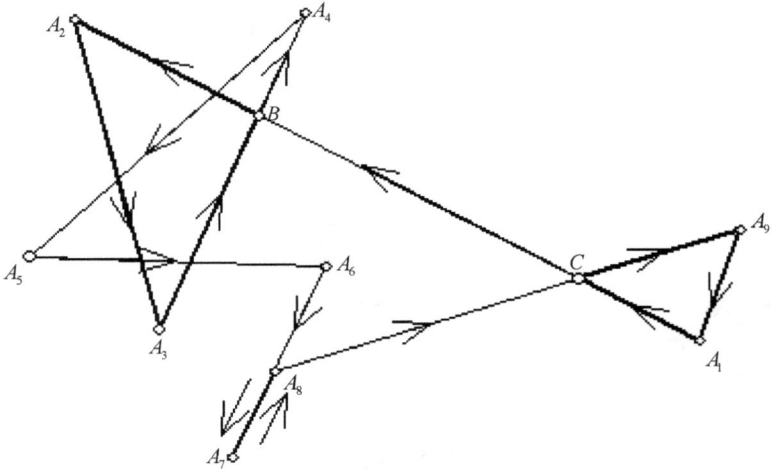

Fig. 3.19

shown in Fig. 3.19 (for simplicity, only nine vertices $A_1, \ldots, A_9$ are considered). Starting from A_1 and winding along ∂P once in a definite direction, there are two possibilities:

(1) A new segment might intersect a previously traversed segment for the first time. For instance, when going from A_3 to A_4, we meet the segment $A_1 A_2$ at B, and from A_8 to A_9, meet the segment at C. In these cases, by (3.4.2.2),

$$\int_{\partial \triangle B A_2 A_3} f(z)dz = 0; \quad \int_{\partial \triangle C A_9 A_1} f(z)dz = 0,$$

whenever f is analytic on both triangles.

(2) When going from a vertex (say A_6) to the next vertex (say A_7), we might meet a vertex never traversed before (say A_8). In this case, the segment $A_7 A_8$ is part of the segment $A_6 A_7$. Hence,

$$\int_{A_8 A_7 A_8} f(z)dz = \int_{A_8 A_7} f(z)dz + \int_{A_7 A_8} f(z)dz = 0.$$

These facts indicate that when removed $\partial \triangle B A_2 A_3, \partial \triangle C A_9 A_1$, and $A_7 A_8$ from ∂P and replaced it by the polygonal closed curve γ :

$CBA_4A_5A_6A_8C$, we still have

$$\int_{\partial P} f(z)dz = \int_{\gamma} f(z)dz = 0.$$

This finishes the proof. □

Proof of Cauchy integral theorem (3.4.2.1). Let γ be a rectifiable curve (not necessarily closed) in the simply connected domain Ω, with representation $z = z(t), a \le t \le b$. Since $\gamma = z([a, b])$ is compact, the distance $\rho_1 = \text{dist}(\gamma, \partial\Omega) > 0$ holds.

f is uniformly continuous on γ. For any $\varepsilon > 0$, there exists a $\rho_2 = \rho_2(\varepsilon) > 0$ so that, for any $z, z' \in \gamma$ with $|z - z'| < \rho_2, |f(z) - f(z')| < \frac{\varepsilon}{2l(\gamma)}$ always holds, where $l(\gamma)$ denotes the length of γ. On the other hand, $z = z(t)$ is also uniform continuous on $[a, b]$, and so there exists a $\delta_1 = \delta_1(\varepsilon) > 0$ such that whenever $t, t' \in [a, b]$ and $|t - t'| < \delta_1, |z(t) - z(t')| < \min(\rho_1, \rho_2)$ holds.

According to (2.9.5), for $\varepsilon > 0$, there exists a $\delta_2 = \delta_2(\varepsilon) > 0$ so that for any partition $\pi : a = t_0 < t_1 < \cdots < t_{n-1} < t_n = b$ with $\delta(\pi) = \max_{0 \le k \le n-1} |t_{k+1} - t_k| < \delta_2$, we always have

$$\left| \int_{\gamma} f(z)dz - \sum_{k=1}^{n} f(z_k)(z_k - z_{k-1}) \right| < \frac{\varepsilon}{2}, \quad z_k = z(t_k), \quad 0 \le k \le n. \quad (*_1)$$

Let $\delta = \min\{\delta_1, \delta_2\}$. Choose partition π so that $\delta(\pi) < \delta$. Construct the inscribed polygonal curve $\Gamma = \overline{z_0z_1} + \cdots + \overline{z_{n-1}z_n}$ of γ. Note that $z_n = z_0$ if γ is closed and thus, so is Γ. This Γ is contained in Ω. This is because, for $z \in \Gamma$, then $z \in \overline{z_{k-1}z_k}$ for some k $(1 \le k \le n)$ and hence, $|z - z_k| \le |z_{k-1} - z_k| < \min\{\rho_1, \rho_2\} \le \rho_1$. On the other hand, if $z \in \overline{z_{k-1}z_k}$, then

$$|z - z_k| \le |z_{k-1} - z_k| < \min\{\rho_1, \rho_2\} \le \rho_2,$$

$$\Rightarrow |f(z) - f(z_k)| < \frac{\varepsilon}{2l(\gamma)}, \quad z \in \overline{z_{k-1}z_k},$$

$$\Rightarrow \left| \int_{\overline{z_{k-1}z_k}} f(z)dz - f(z_k)(z_k - z_{k-1}) \right| = \left| \int_{\overline{z_{k-1}z_k}} [f(z) - f(z_k)]dz \right|$$

$$\le \int_{\overline{z_{k-1}z_k}} |f(z) - f(z_k)||dz|$$

$$\le \frac{\varepsilon}{2l(\gamma)} |z_k - z_{k-1}|,$$

$$\Rightarrow \left| \int_\Gamma f(z)dz - \sum_{k=1}^n f(z_k)(z_k - z_{k-1}) \right| \leq \sum_{k=1}^n \left| \int_{\overline{z_{k-1}z_k}} [f(z) - f(z_k)]dz \right|$$

$$\leq \frac{\varepsilon}{2l(\gamma)} \sum_{k=1}^n |z_k - z_{k-1}| \leq \frac{\varepsilon}{2}, \quad (*_2)$$

$$\Rightarrow \text{ (Combining } (*_1) \text{ and } (*_2)) \left| \int_\Gamma f(z)dz - \int_\gamma f(z)dz \right| < \varepsilon$$

$$\Rightarrow \left(\text{If } \gamma \text{ is closed, so is } \Gamma. \text{ Then, by use of (3.4.2.3), } \int_\Gamma f(z)dz = 0. \right)$$

$$\left| \int_\gamma f(z)dz \right| < \varepsilon,$$

$$\Rightarrow \int_\gamma f(z)dz = 0$$

and the theorem is proved.   □

For instance, see Examples 11 and 12 in Sec. 2.9; also, observe the following:

$$\int_{|z|=\frac{1}{2}} \frac{1}{1+z^2}\, dz = 0; \qquad \int_{|z|=1} \frac{z^3+1}{z^4-9}\, dz = 0;$$

$$\int_{|z|=2} \frac{\cos z + \sin h\left(\frac{z}{2}\right)}{(z^2-16)(z^2+25)}\, dz = 0; \qquad \int_{|z|=1} \frac{ze^z}{z^5 - 3z^4 - 16z + 48}\, dz = 0, \text{ etc.}$$

Section (2) Cauchy integral formula (simple form)

The most prominent direct application of the Cauchy integral theorem is that the value of an analytic function at any point interior to a Jordan closed curve can be expressed as a contour integral along the curve. This is the

Cauchy integral formula. Suppose f is analytic in an open set O. Let γ be a rectifiable Jordan closed curve so that $\overline{\text{Int}\,\gamma} = \gamma \cup \text{Int}\,\gamma$ is contained in O. Then

$$f(z) = \frac{1}{2\pi i} \int_\gamma \frac{f(\zeta)}{\zeta - z}d\zeta, \quad z \in \text{Int}\,\gamma,$$

where the integral along γ is always taken in the anticlockwise direction, unless otherwise stated.   (3.4.2.4)

See (3.5.2.5) for a generalization.

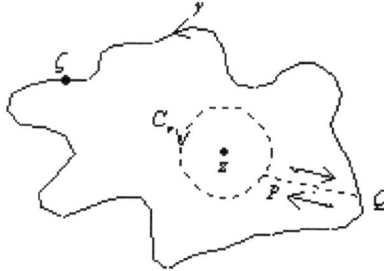

Fig. 3.20

Proof. Fix any $z \in \text{Int } \gamma$ and choose $r > 0$ so that the disk $|\zeta - z| \leq r$ lies completely in $\text{Int } \gamma$. Let C_r denote the circle $|\zeta - z| = r$ endowed with positive direction. Construct an auxiliary curve connecting a point P on C_r to a point Q on γ so that $\Gamma = \gamma \cup \overline{QP} \cup (-C_r) \cup \overline{PQ}$ is a closed curve Γ lying entirely within $\overline{\text{Int }} \gamma$. See Fig. 3.20. By the Cauchy integral theorem applying to $\frac{f(\zeta)}{\zeta - z}$ on $\overline{\text{Int }} \gamma - \text{Int } C_r$, then

$$\int_\Gamma \frac{f(\zeta)}{\zeta - z} d\zeta = \int_\gamma \frac{f(\zeta)}{\zeta - z} d\zeta + \int_{\overline{QP}} \frac{f(\zeta)}{\zeta - z} d\zeta + \int_{-C_r} \frac{f(\zeta)}{\zeta - z} d\zeta$$

$$+ \int_{\overline{PQ}} \frac{f(\zeta)}{\zeta - z} d\zeta = 0,$$

$$\Rightarrow \int_\gamma \frac{f(\zeta)}{\zeta - z} d\zeta = \int_{C_r} \frac{f(\zeta)}{\zeta - z} d\zeta.$$

On C_r, $\zeta = z + re^{i\theta}$, $0 \leq \theta \leq 2\pi$. Thus,

$$\int_{C_r} \frac{f(\zeta)}{\zeta - z} d\zeta = \int_0^{2\pi} f(z + re^{i\theta}) \cdot \frac{1}{re^{i\theta}} rie^{i\theta} d\theta = i \int_0^{2\pi} f(z + re^{i\theta}) d\theta$$

$$\to i \int_0^{2\pi} f(z) d\theta = 2\pi i f(z)$$

as $r \to 0$. In the last step, we have used the continuity of f at z. The formula is then proved. $\qquad\square$

Remark 1 (Mean-value property). In the process of this proof, we obtained a useful relation: For fixed $z \in \text{Int } \gamma$,

$$f(z) = \frac{1}{2\pi i} \int_\gamma \frac{f(\zeta)}{\zeta - z} d\zeta = \frac{1}{2\pi i} \int_{C_r} \frac{f(\zeta)}{\zeta - z} d\zeta = \frac{1}{2\pi} \int_0^{2\pi} f(z + re^{i\theta}) d\theta,$$

$$(3.4.2.5)$$

where $C_r : |\zeta - z| = r$ is any circle so that the disk $|\zeta - z| \leq r$ lies in $\operatorname{Int}\gamma$. The third identity is called *the mean-value property* of a function analytic at a point $z \cdot \frac{1}{\zeta - z}$ in (3.4.2.5) is called the *Cauchy (reproducing) kernel.*

By the way, the *Cauchy integral formula* would imply *Cauchy integral theorem* in a narrow sense: For any fixed point $z \in \operatorname{Int}\gamma$, apply the Cauchy integral formula to the analytic function $(\zeta - z)f(\zeta)$ and we obtain

$$\int_\gamma f(\zeta)d\zeta = \int_\gamma \frac{f(\zeta)}{\zeta - z}(\zeta - z)d\zeta = 2\pi i \cdot f(z) \cdot 0 = 0.$$

Also, the Cauchy integral formula can be used to evaluate some integrals. For instance,

$$\int_{|z|=1} \frac{2\sinh(z) + \cosh^2 3z}{z}dz = 2\pi i(2\sinh 0 + \cosh^2 0) = 2\pi i;$$

$$\int_{|z|=2} \frac{z^3}{z^4 - 1}dz = \int_{|z|=2} \frac{1}{4}\left\{\frac{1}{z - 1} + \frac{1}{z + 1} + \frac{1}{z - i} + \frac{1}{z + i}\right\}dz$$

$$= \frac{1}{4} \cdot 8\pi i = 2\pi i, \dots . \qquad \square$$

Remark 2 (Generalizations of the Cauchy integral theorem and formula).   Using uniform continuity of a continuous function on a compact set, it is not hard to extend (3.4.2.1) to the following

A generalized version of the Cauchy integral theorem. Let γ be a rectifiable Jordan closed curve. If f is continuous on $\overline{\operatorname{Int}\gamma}$ and analytic in $\operatorname{Int}\gamma$, then

$$\int_\gamma f(z)dz = 0. \qquad (3.4.2.1)'$$

See Exercise A (1). Applying this fact and the auxiliary techniques shown in Fig. 3.20, one can prove the following

Generalizations of the Cauchy integral theorem and formula. Let $\Gamma, \gamma_1, \dots, \gamma_n$ be $(n + 1)$ rectifiable Jordan closed curves, satisfying the following conditions:

(1) Each $\gamma_k, 1 \leq k \leq n$, lies entirely in $\operatorname{Int}\Gamma$, and
(2) each one of $\gamma_1, \dots, \gamma_n$ lies in the exterior of the remaining ones, namely,

$$\gamma_k \subseteq \operatorname{Ext}\gamma_j, \quad k \neq j, \quad 1 \leq j, \ k \leq n.$$

Orient $\Gamma, \gamma_1, \dots, \gamma_n$ with counterclockwise directions. Construct the $(n + 1)$ connected domain $\Omega = \operatorname{Int}\Gamma - \bigcup_{k=1}^n \overline{\operatorname{Int}\gamma_k}$ so that its boundary $\partial\Omega = \Gamma - \gamma_1 - \dots - \gamma_n$. Suppose $f(z)$ is continuous on $\overline{\Omega}$ and analytic in Ω. Then,

(1) $\int_{\partial\Omega} f(z)dz = 0$, namely,

$$\int_{\Gamma} f(z)dz = \int_{\gamma_1} f(z)dz + \cdots + \int_{\gamma_n} f(z)dz.$$

(2)

$$f(z) = \frac{1}{2\pi i}\int_{\partial\Omega}\frac{f(\zeta)}{\zeta - z}d\zeta = \frac{1}{2\pi i}\int_{\Gamma}\frac{f(\zeta)}{\zeta - z}d\zeta - \sum_{k=1}^{n}\frac{1}{2\pi i}\int_{\gamma_k}\frac{f(\zeta)}{\zeta - z}d\zeta,$$

$$z \in \Omega. \qquad (3.4.2.4)'$$

See Fig. 3.21.

For more detailed discussion about the Cauchy integral theorem and for-mula, see Chap. 4; in particular, Secs. 4.2.4 and 4.3.3.   □

Section (3) Power series representation of a function analytic at a point

There are many various important, yet direct, applications of the Cauchy integral formula. Here, we just single out the Taylor series expansion of an analytic function at a point for our immediate purpose and leave the others to Chap. 4 and in the Exercises.

The Taylor series representation of an analytic function. Suppose that f is analytic on an open set O.

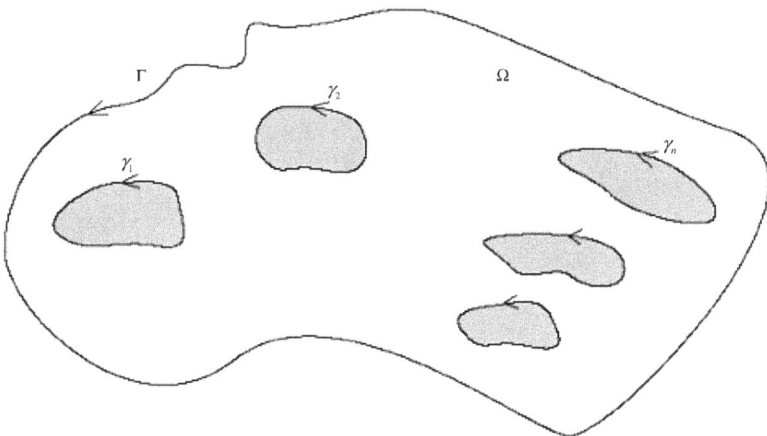

Fig. 3.21

(1) Let $z_0 \in O$ be any fixed point. Then

$$f(z) = \sum_{n=0}^{\infty} a_n(z - z_0)^n, \quad |z - z_0| < R = \text{dist}(z_0, \partial O),$$

where, for any $r > 0$ with $r < R$,

$$a_n = \frac{1}{2\pi i} \int_{|\zeta - z_0| = r} \frac{f(\zeta)}{(\zeta - z_0)^{n+1}} d\zeta, \quad n = 0, 1, 2, \ldots .$$

(2) Hence, as a consequence of (3.3.2.4), f is infinitely differentiable on O, namely, each derivative $f^{(n)}$ of f exists on O as an analytic function, and

$$f^{(n)}(z_0) = n! a_n = \frac{n!}{2\pi i} \int_{|\zeta - z_0| = r} \frac{f(\zeta)}{(\zeta - z_0)^{n+1}} d\zeta, \quad n \geq 0,$$

so that

$$f(z) = \sum_{n=0}^{\infty} \frac{f^{(n)}(z_0)}{n!} (z - z_0)^n, \quad |z - z_0| < R,$$

is the Taylor series expansion of f at z_0, uniquely defined in a neighborhood of z_0.   (3.4.2.6)

Proof. Let R denote the distance from z_0 to the boundary ∂O. Note that $R > 0$. Fix any point z so that $|z - z_0| < R$. Choose $r > 0$ so that $|z - z_0| < r < R$. By (3.4.2.4),

$$f(z) = \frac{1}{2\pi i} \int_{|\zeta - z_0| = r} \frac{f(\zeta)}{\zeta - z} d\zeta.$$

See Fig. 3.22. Expand the Cauchy kernel $\frac{1}{\zeta - z}$ at z_0 as follows:

$$\frac{1}{\zeta - z} = \frac{1}{\zeta - z_0} \cdot \frac{1}{1 - \frac{z - z_0}{\zeta - z_0}} = \frac{1}{\zeta - z_0} \sum_{n=0}^{\infty} \frac{(z - z_0)^n}{(\zeta - z_0)^n}, \quad |z - z_0| < |\zeta - z_0| = r$$

(see Remark 1 in Sec. 3.3.2). Remind that this power series converges uniformly on the circle $|\zeta - z_0| = r$. Hence, by using (7) in (2.9.10) or (4) in (3.3.2.4), termwise integration of it is permissible and

$$f(z) = \frac{1}{2\pi i} \int_{|\zeta - z_0| = r} \left(\sum_{n=0}^{\infty} \frac{(z - z_0)^n}{(\zeta - z_0)^{n+1}} f(\zeta) \right) d\zeta$$

$$= \sum_{n=0}^{\infty} \left(\frac{1}{2\pi i} \int_{|\zeta - z_0| = r} \frac{f(\zeta)}{(\zeta - z_0)^{n+1}} d\zeta \right) (z - z_0)^n = \sum_{0}^{\infty} a_n(z - z_0)^n.$$

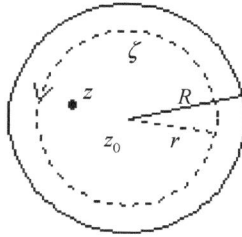

Fig. 3.22

Suppose $f(z) = \sum_0^\infty b_n(z - z_0)^n$ is another power series representation of $f(z)$ for $|z - z_0| < R'$. Then, on $|z - z_0| < \min(R, R')$, $\sum_{n=0}^\infty a_n(z - z_0)^n = \sum_{n=0}^\infty b_n(z - z_0)^n$ holds identically. Let $z = z_0$, we obtain $b_0 = a_0$ and hence, $\sum_{n=1}^\infty a_n(z - z_0)^{n-1} = \sum_{n=1}^\infty b_n(z - z_0)^{n-1}$. This will, in turn, induce that $b_1 = a_1$ and eventually, $b_n = a_n$ for $n \geq 2$. Hence, the power series expansion of f at z_0 is unique in a neighborhood of z_0 and this expansion is the Taylor series representation of f at z_0. □

Readers are asked to redo Examples 1, 2, and 3 in Sec. 3.3.2 by using results in (3.4.2.6) directly. For more complicated illustrative examples, see Sec. 4.8.

Section (4) Some consequences of (3.4.2.6)

Firstly, the integral expression in (3.4.2.6) can be used to prove

The Cauchy integral formula for higher derivatives of an analytic function. Suppose f is analytic on the set $\overline{\text{Int}\,\gamma} = \gamma \bigcup \text{Int}\,\gamma$, where γ is a rectifiable Jordan closed curve. Then, for $n \geq 0$,

$$f^{(n)}(z) = \frac{n!}{2\pi i} \int_\gamma \frac{f(\zeta)}{(\zeta - z)^{n+1}} d\zeta, \quad z \in \text{Int}\,\gamma. \tag{3.4.2.7}$$

Proof. Fix $z \in \text{Int}\,\gamma$. Adopt Fig. 3.20 and notations wherein. The function

$$F(\zeta) = \frac{f(\zeta)}{(\zeta - z)^{n+1}}$$

is analytic on $\overline{\text{Int}\,\gamma} - \text{Int}\,C_r$, where $C_r = \{|\zeta - z| = r\}$. Hence,

$$\int_\Gamma F(z)dz = \int_\gamma F(\zeta)d\zeta - \int_{C_r} F(\zeta)d\zeta = 0,$$

$$\Rightarrow \int_\gamma F(\zeta)d\zeta = \int_\gamma \frac{f(\zeta)}{(\zeta - z)^{n+1}} d\zeta = \int_{C_r} \frac{f(\zeta)}{(\zeta - z)^{n+1}} d\zeta$$

$$= \frac{2\pi i}{n!} f^{(n)}(z) \quad \text{(by using (2) in (3.4.2.6))},$$

$$\Rightarrow f^{(n)}(z) = \frac{n!}{2\pi i} \int_{\gamma} \frac{f(\zeta)}{(\zeta - z)^{n+1}} d\zeta, \quad z \in \text{Int } \gamma \quad \text{for } n \geq 0. \quad \square$$

Remark 3. Equation (3.4.2.5) has a counterpart for higher derivatives: For fixed $z \in \text{Int } \gamma$,

$$f^{(n)}(z) = \frac{n!}{2\pi i} \int_{\gamma} \frac{f(\zeta)}{(\zeta - z)^{n+1}} d\zeta = \frac{n!}{2\pi i} \int_{|\zeta - z| = r} \frac{f(\zeta)}{(\zeta - z)^{n+1}} d\zeta \quad (3.4.2.8)$$

where the disk $|\zeta - z| \leq r$ is contained in $\text{Int } \gamma$. A direct proof and a generalized version of (3.4.2.8) will be presented in (4.6.2) and (4.7.2). Similar modification can be applied to (3.4.2.4)' (see (4.7.3)). $\square$

Equation (3.4.2.8) can be used effectively to evaluate integrals.

Example 1. Evaluate the following integrals:

(1) $\int_{\gamma} \frac{e^{\pi z}}{(1+z^2)^2} dz$, where γ is the ellipse: $4x^2 + y^2 = 2y$.

(2) $\int_{\gamma} \frac{e^{3z} + 3\cosh z}{\left(z + \frac{\pi i}{2}\right)^4} dz$, where γ is any rectifiable Jordan closed curve with $-\frac{\pi i}{2} \in \text{Int } \gamma$.

(3) $\frac{1}{2\pi i} \int_{\gamma} \frac{dz}{(z^4 - 1)(z - 3)^2}$, where γ is the circle $|z| = 2$.

(4) $\int_{\gamma} \frac{dz}{z^3 (z+1)(z-2)}$, where γ is a rectifiable Jordan closed curve, not passing -1, 0, and 2.

Solution.

(1) Observe that $f(z) = \frac{e^{\pi z}}{(z+i)^2}$ is analytic on and inside the ellipse. Hence,

$$\int_{\gamma} \frac{e^{\pi z}}{(1+z^2)^2} dz = \int_{\gamma} \frac{f(z)}{(z-i)^2} dz = 2\pi i \cdot f'(i)$$

$$= 2\pi i \left\{ \frac{[\pi(z+i) - 2]e^{\pi z}}{(z+i)^3} \right\}_{z=i} = \frac{\pi(\pi i - 1)}{2}.$$

(2) Let $f(z) = e^{3z} + 3\cosh z$. Then

$$\int_{\gamma} \frac{e^{3z} + 3\cosh z}{(z + \frac{\pi i}{2})^4} dz = \frac{2\pi i}{3!} f^{(3)}\left(-\frac{\pi i}{2}\right) = \frac{2\pi i}{3!} \{27e^{3z} + 3\sinh z\}_{z=-\frac{\pi i}{2}}$$

$$= \frac{2\pi i}{3!} (27i - 3i) = -8\pi.$$

(3) Let $f(z) = \frac{1}{(z-3)^2}$. Note that

$$\frac{1}{z^4 - 1} = \frac{1}{2}\left(\frac{1}{z^2 - 1} - \frac{1}{z^2 + 1}\right) = \frac{1}{4}\left(\frac{1}{z - 1} - \frac{1}{z + 1}\right)$$
$$- \frac{1}{4i}\left(\frac{1}{z - i} - \frac{1}{z + i}\right),$$

$$\Rightarrow \int_\gamma \frac{dz}{(z^4 - 1)(z - 3)^2} = \frac{1}{4}\int_\gamma \frac{dz}{(z - 1)(z - 3)^2}$$
$$- \frac{1}{4}\int_\gamma \frac{dz}{(z + 1)(z - 3)^2} - \frac{1}{4i}\int_\gamma \frac{dz}{(z - i)(z - 3)^2}$$
$$+ \frac{1}{4i}\int_\gamma \frac{dz}{(z + i)(z - 3)^2}$$
$$= 2\pi i \cdot \frac{1}{4}\{f(1) - f(-1) + if(i) - if(-i)\} = \cdots = \frac{27}{1600}.$$

Or, by the reason used in Fig. 3.20 and in (3.4.2.5), choose $R > 3$ and note that

$$\frac{1}{2\pi i}\int_{|z|=R} \frac{dz}{(z^4 - 1)(z - 3)^2} - \frac{1}{2\pi i}\int_{|z|=2} \frac{dz}{(z^4 - 1)(z - 3)^2}$$
$$= \frac{d}{dz}\left(\frac{1}{z^4 - 1}\right)\Big|_{z=3} = \frac{-27}{1600},$$
$$\Rightarrow \frac{1}{2\pi i}\int_{|z|=2} \frac{dz}{(z^4 - 1)(z - 3)^2}$$
$$= \frac{27}{1600} + \frac{1}{2\pi i}\int_{|z|=R} \frac{dz}{(z^4 - 1)(z - 3)^2}$$
$$\to \frac{27}{1600} \quad \text{as } R \to \infty.$$

In the last step, we have used the estimate

$$\left|\int_{|z|=R} \frac{dz}{(z^4 - 1)(z - 3)^2}\right| \le \frac{2\pi R}{(R^4 - 1)(R - 3)^2} \to 0, \quad R \to \infty.$$

(4) Observe that

$$\frac{1}{z^3(z + 1)(z - 2)} = \frac{-\frac{3}{8}z^2 + \frac{1}{4}z - \frac{1}{2}}{z^3} + \frac{\frac{1}{3}}{z + 1} + \frac{\frac{1}{24}}{z - 2}.$$

In case $0, -1, 2 \notin \text{Int}\,\gamma$, then the integral is equal to zero. If only $0 \in \text{Int}\,\gamma$ (and hence, $-1, 2 \in \text{Ext}\,\gamma$): Choose $\rho > 0$ so small that the closed disk $|z| \le \rho$ lies in Int γ. Then, by (3.4.2.4)$'$,

$$
\int_\gamma \frac{dz}{z^3(z+1)(z-2)} = \int_{|z|=\rho} \frac{-\frac{3}{8}z^2 + \frac{1}{4}z - \frac{1}{2}}{z^3} dz
$$

$$
= -\frac{3}{8}\int_{|z|=\rho} \frac{dz}{z} + \frac{1}{4}\int_{|z|=\rho} \frac{dz}{z^2} - \frac{1}{2}\int_{|z|=\rho} \frac{dz}{z^3}
$$

$$
= -\frac{3}{8}\cdot 2\pi i + \frac{1}{4}\cdot 2\pi i \cdot 0 - \frac{1}{2}\cdot 2\pi i \cdot 0 = -\frac{3\pi}{4}i.
$$

If $0, -1 \in$ Int γ and $2 \in$ Ext γ: Choose $\rho_1 > 0$ and $\rho_2 > 0$ sufficiently small. Then,

$$
\int_\gamma \frac{dz}{z^3(z+1)(z-2)} = \int_{|z|=\rho_1} \frac{-\frac{3}{8}z^2 + \frac{1}{4}z - \frac{1}{2}}{z^3} dz
$$

$$
+ \int_{|z|=\rho_2} \frac{\frac{1}{3}}{z+1} dz = -\frac{3\pi i}{4} + \frac{2\pi i}{3} = -\frac{\pi i}{12}.
$$

If $0, -1, 2 \in$ Int γ: Choose $\rho_1 > 0, \rho_2 > 0, \rho_3 > 0$ sufficiently small. Then

$$
\int_\gamma \frac{dz}{z^3(z+1)(z-2)} = \int_{|z|=\rho_1} \frac{-\frac{3}{8}z^2 + \frac{1}{4}z - \frac{1}{2}}{z^3} dz
$$

$$
+ \int_{|z|=\rho_2} \frac{\frac{1}{3}}{z+1} dz + \int_{|z|=\rho_3} \frac{\frac{1}{24}}{z-2} dz
$$

$$
= -\frac{\pi i}{12} + \frac{1}{24}\cdot 2\pi i = 0.
$$

By the way, we have (why?)

$$
\int_{|z|=3} \frac{\sin z}{z^3(z+1)(z-2)} dz
$$

$$
= -\frac{3\pi i}{4}\sin 0 + \frac{1}{4}\cdot 2\pi i \cos 0 - \frac{1}{2}\cdot 2\pi i(-\sin 0)\cdot \frac{1}{2!}
$$

$$
+ \frac{2\pi i}{3}\sin(-1) + \frac{\pi i}{12}\sin 2
$$

$$
= \frac{-2\pi i}{3}\sin 1 + \frac{\pi i}{12}\sin 2 + \frac{\pi i}{2}.
$$

Example 2. Let γ be a piecewise differentiable Jordan closed curve such that $0 \in \text{Int } \gamma$. Suppose f is analytic on $\overline{\text{Int } \gamma}$. Then

$$\frac{1}{2\pi i} \int_\gamma f'(z)(\log z)_k dz = f(z_0) - f(0),$$

where z_0 is the initial (and terminal) point of γ and $(\log z)_k = \text{Log } z + 2\pi i k$ (k fixed) is a branch of $\log z$. Here, it is as usual, supposed that γ is positively oriented.

Proof. When z, starting from z_0, winds along γ once and back to z_0, the values of $(\log z)_k$ will change continuously from $(\log z_0)_k$ to $(\log z_0)_{k+1} = (\log z_0)_k + 2\pi i$. Using integration by parts (see (6) in (2.9.10)) and Example 4 in Sec. 3.3.1,

$$\int_\gamma f'(z)(\log z)_k dz = f(z)(\log z)_k |_\gamma - \int_\gamma \frac{f(z)}{z} dz$$
$$= f(z_0)[(\log z_0)_k + 2\pi i] - f(z_0) \cdot (\log z_0)_k - 2\pi i \cdot f(0)$$
$$= [f(z_0) - f(0)] \cdot 2\pi i. \qquad \square$$

Example 3.

(1) Let $r > 0$ and $r \neq 1$. Show that

$$\int_{|z|=r} \frac{dz}{\sqrt{z^2 + z + 1}}$$
$$= \begin{cases} 0, & \text{if } r < 1, \\ 2\pi i \text{ (in case } \sqrt{1} = 1), -2\pi i \text{ (in case } \sqrt{1} = -1), & \text{if } r > 1. \end{cases}$$

(2) Let γ be the parabola $y^2 = 2x$, endowed with the direction of increasing y. Designate $\sqrt{1 + x^2} > 0$ if $x > 0$. Show that

$$\int_\gamma \frac{dz}{(z^4 + 1)\sqrt{z^2 + 1}} = \frac{\pi i}{2} \sqrt{1 + \sqrt{2}}.$$

Solution. (1) Note that $z^2 + z + 1 = 0$ has solutions $w = e^{\frac{2\pi i}{3}}$ and w^2. Hence, $\sqrt{z^2 + z + 1}$ has two branches on $\Omega = \mathbf{C} - \{\text{ray Arg } z = \frac{2\pi}{3},$ $|z| \geq 1\} - \{\text{ray Arg } z = \frac{4\pi}{3}, |z| \geq 1\}$. See Fig. 3.23.

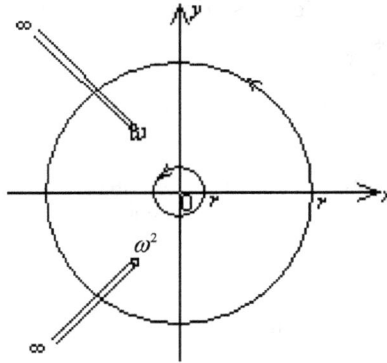

Fig. 3.23

In case $0 < r < 1$, no matter which branch of $\sqrt{z^2 + z + 1}$ has been chosen, we always have

$$\int_{|z|=r} \frac{dz}{\sqrt{z^2 + z + 1}} = 0.$$

Now, suppose $r > 1$. Each branch of $\sqrt{z^2 + z + 1}$ is discontinuous along the rays cut off and will change into another branch once z goes across the rays. Since $|z| = r$ has only two points of intersection with the rays, each branch is still integrable along $|z| = r$ (see the end of Remark 1 in Sec. 2.9). Invoking the change of variable $\zeta = \frac{1}{z}$, then (see (4) in (2.9.10))

$$\int_{|z|=r} \frac{dz}{\sqrt{z^2 + z + 1}}$$

$$= \int_{|\zeta|=\frac{1}{r}} \frac{-\frac{1}{\zeta^2} d\zeta}{\sqrt{\frac{1}{\zeta^2} + \frac{1}{\zeta} + 1}} \quad \left(|\zeta| = \frac{1}{r} \text{ is negatively oriented.} \right)$$

$$= \int_{|\zeta|=\frac{1}{r}} \frac{\left(\sqrt{\zeta^2 + \zeta + 1} \right)^{-1}}{\zeta} d\zeta \quad \left(|\zeta| = \frac{1}{r} \text{ is now positively oriented.} \right)$$

$$= 2\pi i \left(\sqrt{\zeta^2 + \zeta + 1} \right)^{-1} \Big|_{\zeta=0} = \pm 2\pi i.$$

(2) $\sqrt{z^2 + 1}$ has two branches on $\mathbf{C} - [i, +i\infty) - [-i, -i\infty)$. While $z^4 + 1 = 0$ has two roots $z_0 = e^{\frac{\pi i}{4}}$ and $\overline{z_0}$ lying in the domain $y^2 < 2x$. Choose $R > 1$ and construct the oriented closed curve Γ formed by $y^2 = 2x$ and the line $\operatorname{Re} z = R$ as shown in Fig. 3.24. Choose $r > 0$ so small that the closed disks

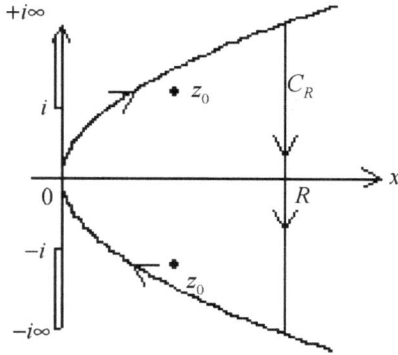

Fig. 3.24

$|z - z_0| \leq r$ and $|z - \overline{z_0}| \leq r$ all lying in Int Γ. Then,

$$\int_{\Gamma} \frac{dz}{(z^4 + 1)\sqrt{z^2 + 1}}$$

$$= \int_{|z-z_0|=r} \frac{dz}{(z^4 + 1)\sqrt{z^2 + 1}} + \int_{|z-\overline{z_0}|=r} \frac{dz}{(z^4 + 1)\sqrt{z^2 + 1}}$$

(both are negatively oriented)

$$= -2\pi i \left\{ \frac{\left(\sqrt{z_0^2 + 1}\right)^{-1}}{(z_0 - e^{\frac{3\pi i}{4}})(z_0 - e^{\frac{5\pi i}{4}})(z_0 - \overline{z_0})} \right.$$

$$\left. + \frac{\left(\sqrt{\overline{z_0}^2 + 1}\right)^{-1}}{(\overline{z_0} - e^{\frac{3\pi i}{4}})(\overline{z_0} - e^{\frac{5\pi i}{4}})(\overline{z_0} - z_0)} \right\}$$

(Why? See Exercise A (2).),

$$= -2\pi i \cdot 2 \operatorname{Re} \frac{1}{z\sqrt{2}(-1 + i)\sqrt{1 + i}},$$

$$= \frac{\pi i}{\sqrt{2}} \operatorname{Re}\sqrt{1 + i} = \frac{\pi i}{\sqrt{2}} \cos \frac{\pi}{8} = \frac{\pi i}{\sqrt{2}} \cdot \frac{\sqrt{2 + \sqrt{2}}}{2}$$

$$= \frac{\pi i}{2} \sqrt{1 + \sqrt{2}}.$$

Along $C_R : \operatorname{Re} z = R, -\sqrt{2R} \leq \operatorname{Im} z \leq \sqrt{2R}$, we have

$$|(z^4 + 1)\sqrt{z^2 + 1}| \geq (|z|^4 - 1)\sqrt{|z|^2 - 1} \geq (R^4 - 1)\sqrt{R^2 - 1}$$

holds. Hence,

$$\left| \int_{C_R} f(z)dz \right| \le \frac{2\sqrt{2}R}{(R^4 - 1)\sqrt{R^2 - 1}} \to 0 \quad \text{as } R \to \infty.$$

Then, letting $R \to +\infty$ in the above integral, the result follows.

Example 4. Suppose f is analytic on $|z| \le 1$ (or analytic in $|z| < 1$ and continuous on $|z| \le 1$). Show that

$$\frac{1}{2\pi i} \int_{|\zeta|=1} \frac{\overline{f(\zeta)}}{\zeta - z} d\zeta = \begin{cases} \overline{f(0)}, & |z| < 1, \\ \overline{f(0)} - \overline{f\left(\frac{1}{\overline{z}}\right)}, & |z| > 1. \end{cases}$$

Proof. If $|\zeta| = 1$, then $\zeta = e^{i\theta}$ ($0 \le \theta \le 2\pi$). Then $\bar{\zeta} = e^{-i\theta}$ and $d\zeta = ie^{i\theta}d\theta = -\frac{d\bar{\zeta}}{\bar{\zeta}^2}$. Owing to $\zeta\bar{\zeta} = 1$ if $|\zeta| = 1$, then

$$\frac{1}{2\pi i} \int_{|\zeta|=1} \frac{\overline{f(\zeta)}}{\zeta - z} d\zeta = \frac{1}{2\pi i} \int_{|\zeta|=1} \frac{\overline{f(\zeta)}}{\bar{\zeta} - z} \cdot \frac{-d\bar{\zeta}}{\bar{\zeta}^2} = \frac{1}{2\pi i} \int_{|\zeta|=1} \frac{\overline{f(\zeta)}}{\zeta(1 - \bar{z}\zeta)} d\zeta$$

$$= \frac{1}{2\pi i} \int_{|\zeta|=1} \left\{ \frac{1}{\zeta} - \frac{1}{\zeta - \frac{1}{\bar{z}}} \right\} \overline{f(\zeta)} d\zeta \quad \text{(in case } z \ne 0\text{)}.$$

Let $|z| < 1$. Then,

$$\frac{1}{2\pi i} \int_{|\zeta|=1} \frac{\overline{f(\zeta)}}{\zeta(1 - \bar{z}\zeta)} d\zeta = \frac{1}{2\pi i} \int_{|\zeta|=1} \frac{1}{\zeta} \cdot \frac{\overline{f(\zeta)}}{1 - \bar{z}\zeta} d\zeta = \left. \frac{\overline{f(\zeta)}}{1 - \bar{z}\zeta} \right|_{\zeta=0} = \overline{f(0)}.$$

If $|z| > 1$, then

$$\frac{1}{2\pi i} \int_{|\zeta|=1} \left\{ \frac{1}{\zeta} - \frac{1}{\zeta - \frac{1}{\bar{z}}} \right\} \overline{f(\zeta)} d\zeta = \overline{f(0)} - \overline{f\left(\frac{1}{\overline{z}}\right)}.$$

And the results follow obviously.

See Exercise B (4) for an application of this example. □

Secondly, we try to study the zeros of an analytic function f. A point z_0 where $f(z_0) = 0$ is called *a zero* of f. Suppose f is analytic on a domain Ω and $z_0 \in \Omega$. Then, by (2) in (3.4.2.6),

$$f(z) = \sum_0^\infty \frac{f^{(n)}(z_0)}{n!}(z - z_0)^n, \quad |z - z_0| < R = \text{dist}(z_0, \partial\Omega).$$

In case $f^{(n)}(z_0) = 0$ for all $n \ge 0$, then $f(z) \equiv 0$ on the disk $|z - z_0| < R$. Hence, the set $\{z \in \Omega \mid f^{(n)}(z) = 0 \text{ for all } n \ge 0\}$ is a nonempty open and closed subset of the connected set Ω and thus is equal to Ω itself. Therefore $f \equiv 0$ on Ω.

If this exists a smallest positive integer m such that $f(z_0) = f'(z_0) = \cdots = f^{(m-1)}(z_0) = 0$ but $f^{(m)}(z_0) \neq 0$ holds, then z_0 is said to be *a zero* of *order* m of f. In this case, on $|z - z_0| < R$,

$$f(z) = (z - z_0)^m \varphi(z), \quad \text{where } \varphi(z) = \sum_{n=m}^{\infty} \frac{f^{(n)}(z_0)}{n!}(z - z_0)^{n-m}.$$

Note that $\varphi(z)$ is analytic on $|z - z_0| < R$ and $\varphi(z_0) = \frac{1}{m!}f^{(m)}(z_0) \neq 0$. By continuity of φ at z_0, there exists a $\delta > 0$ so that $\varphi(z) \neq 0$ on $|z - z_0| < \delta$ and hence, $f(z) \neq 0$ on $0 < |z - z_0| < \delta$.

We summarize the above as

The zeros of an analytic function. Suppose f is analytic on a domain Ω.

(1) If there exists a point $z_0 \in \Omega$ so that $f^{(n)}(z_0) = 0$ for all $n \geq 0$, then $f(z) \equiv 0$ on Ω.

(2) $z_0 \in \Omega$ is a zero of order m of f if and only if there exists a $r > 0$ so that

$$f(z) = (z - z_0)^m \varphi(z), \quad |z - z_0| < r,$$

where $\varphi(z)$ is analytic on $|z - z_0| < r$ and $\varphi(z_0) \neq 0$. This r might be chosen smaller so that $\varphi(z) \neq 0$ on $|z - z_0| < r$ (compare with (2.5.1.3)).

(3) (Isolated-zero principle). If $f \not\equiv 0$ on Ω, then f has only *isolated* zeros, if any. Namely, if $f(z_0) = 0$, then there exists a $\delta > 0$ so that $f(z) \neq 0$ on the deleted disk $0 < |z - z_0| < \delta$.

(4) (Interior uniqueness theorem). Suppose $A \subseteq \Omega$ is a subset of Ω, which has a limit point in Ω, namely, $A' \cap \Omega \neq \phi$. If $f(z) = 0$ for all $z \in A$, then $f \equiv 0$ on Ω.   (3.4.2.9)

Comparison: Observe that $\text{Re } z$ is identically equal to zero on the imaginary axis and these zeros of $\text{Re } z$ are not isolated and are not of finite orders. Also, even though

$$f(z) = \begin{cases} 0, & |z| \leq 1, \\ e^{-(|z|^2-1)^{-1}}, & |z| > 1, \end{cases}$$

is C^∞ on $\mathbf{C}$ (see (2.8.5)), every point on $|z| \leq 1$ is a zero of infinite order of f. An interesting example cited quite often is the C^∞-function

$$\varphi(x) = \begin{cases} 0, & x = 0, \\ e^{-x^{-2}}, & x \neq 0. \end{cases}$$

It is not *real* analytic at $x = 0$, namely, $\varphi(x)$ cannot be expressed as a convergent power series in a nondegenerated interval with center at 0 because

$\varphi^{(n)}(0) = 0$ for $n \geq 0$. While $f(z) = \sin \frac{1}{z}$ is a nonconstant analytic function on $\mathbf{C} - \{0\}$ with isolated zeros $z_n = \frac{1}{n\pi}$ $(n = \pm 1, \pm 2, \ldots)$ converging to the boundary point 0 of $\mathbf{C} - \{0\}$.

Example 5. Determine the zeros and the corresponding orders for the following functions.

(1) $f(z) = 6 \sin z^3 + z^3(z^6 - 6)$ at $z = 0$.

(2) $f(z) = \dfrac{(z^2 - \pi^2)^2 \sin z}{z^7}$.

(3) $f(z) = (1 - \sqrt{2 - 2\cos z})^2$.

Solution. (1) Since $\sin z = \sum_0^\infty \frac{(-1)^n}{(2n+1)!} z^{2n+1}, |z| < \infty$,

$$\sin z^3 = \sum_0^\infty \frac{(-1)^n}{(2n+1)!} z^{6n+3} = z^3 \left(1 - \frac{1}{6} z^6\right) + \frac{1}{5!} z^{15} - \frac{1}{7!} z^{21} + \cdots,$$

$$\Rightarrow 6 \sin z^3 + z^3(z^6 - 6) = z^{15} \varphi(z),$$

where

$$\varphi(z) = \frac{6}{5!} - \frac{6}{7!} z^6 + \cdots, \quad |z| < \infty.$$

Note that $\varphi(0) \neq 0$. Hence $z = 0$ is a zero of order 15 of f.

(2) $f(z) = 0 \Leftrightarrow z = n\pi, n = \pm 1, \pm 2, \ldots$. Be observant that f is not defined at 0. Since

$$\sin z = -\sum_0^\infty \frac{(-1)^n}{(2n+1)!} (z - \pi)^{2n+1}, \quad |z - \pi| < \infty,$$

$$\Rightarrow f(z) = (z - \pi)^3 \varphi(z), \quad |z - \pi| < \infty,$$

where

$$\varphi(z) = \frac{(z + \pi)^2}{z^7} \left[-1 + \sum_{n=1}^\infty (-1)^{n-1} \frac{1}{(2n+1)!} (z - \pi)^{2n} \right].$$

Note that $\varphi(z)$ is analytic at $z = \pi$ and $\varphi(\pi) = \frac{-4}{\pi^5} \neq 0$. Thus, $z = \pi$ is a zero of order 3 of f. So is $z = -\pi$. While $z = n\pi (n = \pm 2, \pm 3, \ldots)$ are zeros of order 1.

(3) $f(z) = 0 \Leftrightarrow \sqrt{2 - 2\cos z} = 1 \Leftrightarrow z_n = 2n\pi \pm \frac{\pi}{3}, n = 0, \pm 1, \pm 2, \ldots$ (and $\sqrt{1} = 1$).

$\zeta = 2(1 - \cos z)$ maps the vertical strip $2n\pi < \operatorname{Re} z < (2n + 1)\pi, n = 0, \pm 1, \pm 2, \ldots$ univalently onto the domain $\mathbf{C} - (-\infty, 0] - [4, \infty)$ in the ζ-plane, which is contained in $\mathbf{C} - (-\infty, 0]$. Yet, on $\mathbf{C} - (-\infty, 0], \eta = \sqrt{\zeta}$

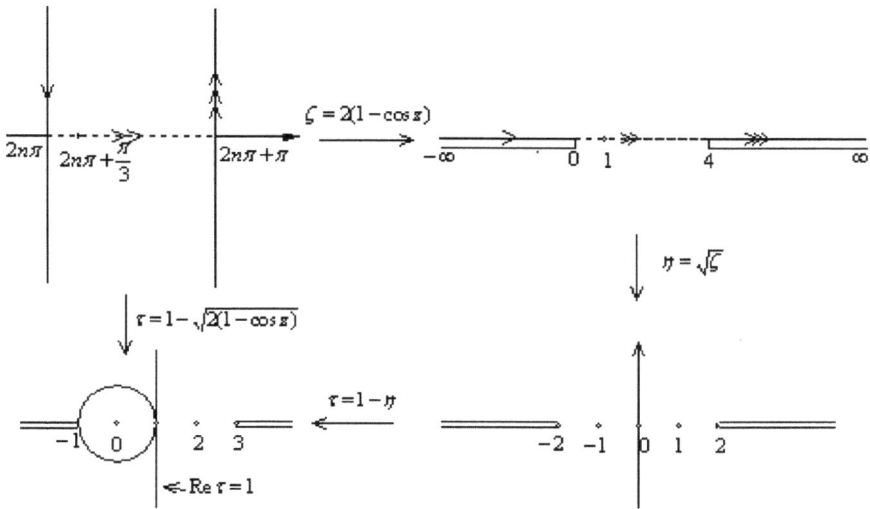

Fig. 3.25

has two branches η_+ (determined by $\sqrt{1} = 1$) and η_- (by $\sqrt{1} = -1$). See Fig. 3.25. Moreover, $\tau_+ = 1 - \eta_+$ maps $2n\pi < \mathrm{Re}\, z < (2n+1)\pi$ univalently onto the left half-plane $\mathrm{Re}\, \tau < 1$ with $(-\infty, -1]$ deleted (also, maps $\mathrm{Re}\, z = (2n+1)\pi$ onto $(-\infty, -1]$ in two-to-one manner; maps $\mathrm{Re}\, z = 2n\pi$ one-to-one and onto $\mathrm{Re}\, \tau = 1$). Note that τ_+ maps $2n\pi + \frac{\pi}{3}$ onto 0 for $n = 0, \pm 1, \pm 2, \ldots$.

This τ_+ has a zero of order 1 at $z_n = 2n\pi + \frac{\pi}{3}$, for $n = 0, \pm 1, \pm 2, \ldots$. To see this, for fixed n,

$$\cos z = \frac{1}{2} + \frac{1}{2}\left\{ -\sqrt{3}(z - z_n) - \frac{1}{2!}(z - z_n)^2 + \frac{\sqrt{3}}{3!}(z - z_n)^3 \right.$$

$$\left. + \frac{1}{4!}(z - z_n)^4 - \frac{\sqrt{3}}{5!}(z - z_n)^5 + \cdots \right\}, \quad |z - z_n| < \infty$$

$$\Rightarrow 2(1 - \cos z) = 1 + (z - z_n)\psi(z),$$

where

$$\psi(z) = \sqrt{3} + \frac{1}{2}(z - z_n) - \frac{\sqrt{3}}{3!}(z - z_n)^2 - \cdots, \quad |z - z_n| < \infty,$$

$\Rightarrow$ (by using Example 2 in Sec. 3.3.2)

$$\eta_+ = \sqrt{2(1 - \cos z)} = \sqrt{1 + (z - z_n)\psi(z)}$$

$$= 1 + \frac{1}{2}(z - z_n)\psi(z) - \frac{1}{8}(z - z_n)^2\psi(z)^2 + \cdots$$

$$= 1 + (z - z_n)\varphi(z),$$

where

$$\varphi(z) = \frac{1}{2}\psi(z) - \frac{1}{8}(z - z_n)\psi(z)^2 + \cdots, \quad |(z - z_n)\psi(z)| < 1.$$

Choose $r > 0$ so small such that $|(z - z_n)\psi(z)| < 1$ on $|z - z_n| < r$. In this case, $\varphi(z)$ is analytic on $|z - z_n| < r$ and $\varphi(z_n) = \frac{1}{2}\psi(z_n) = \frac{\sqrt{3}}{2} \neq 0$. Altogether $\tau_+ = 1 - \eta_+ = (z - z_n)(-\varphi(z)), \varphi(z_n) \neq 0$, has a zero of order 1 at z_n.

Therefore, $f(z) = \tau_+^2$ has $z_n = 2n\pi + \frac{\pi}{3}$ as its zeros, each of order 2, for $n = 0, \pm1, \pm2, \ldots$. Similarly, so are $z_n = 2n\pi - \frac{\pi}{3}, n = 0, \pm1, \pm2, \ldots$. We can reconfirm these facts by observing that

$$\tau_+'(z) = -\eta_+'(z) = -\frac{\sin}{\sqrt{2(1 - \cos z)}},$$

$$\Rightarrow \tau_+'(z_n) = \pm\frac{\sqrt{3}}{2} \neq 0, \quad n = 0, \pm1, \pm2, \ldots.$$

On the other hand, the branch $\tau_- = 1 - \eta_-$, determined by η_-, maps $2n\pi < \text{Re}\, z < (2n + 1)\pi$ onto $\text{Re}\, \tau > 1$ and hence has no zeros on this strip for $n = 0, \pm1, \pm2, \ldots$. As a whole, $2n\pi \pm \frac{\pi}{3}$ are not zeros of $f(z)$ in this case.

Thirdly, to the end, we prove the following famous theorems (also, refer to Exercise A (35)).

Liouville's Theorem. A bounded entire function is a constant.

$$(3.4.2.10)$$

Proof. Let f be an entire function, namely, f is analytic on the entire plane $\mathbf{C}$ (see Remark 1 in Sec. 3.3.1). Suppose $|f(z)| \leq M$ for all $z \in \mathbf{C}$ and some constant $M > 0$. For any $R > 0$, by the Cauchy integral formula for $f'(z)$ (see (3.4.2.7) for $n = 1$),

$$f'(z) = \frac{1}{2\pi i} \int_{|\zeta|=R} \frac{f(\zeta)}{(\zeta - z)^2} d\zeta, \quad |z| < R,$$

$$\Rightarrow |f'(z)| \leq \frac{1}{2\pi} \int_{|\zeta|=R} \frac{|f(\zeta)|}{|\zeta - z|^2} |d\zeta|$$

$$\leq \frac{M \cdot 2\pi R}{2\pi (R - |z|)^2} = \frac{MR}{(R - |z|)^2} \to 0 \quad \text{as } R \to \infty.$$

In the last limit process, we let $z \in \mathbf{C}$ being fixed and consider these R so that $|z| < R$ and then, let $R \to \infty$. Hence, $f'(z) = 0$ on $\mathbf{C}$ and thus, $f(z) = c$ is a constant by (3.4.1.3). $\qquad\square$

An alternative proof. Suppose $f(z)$ is analytic on $|z - z_0| \leq R$. Then, by (3.4.2.7), we have the following estimate for $f^{(n)}(z_0)$, called the *Cauchy inequality*,

$$|f^{(n)}(z_0)| \leq \frac{n!}{2\pi} \int_{|z-z_0|=R} \frac{|f(z)|}{|z - z_0|^{n+1}} |dz|$$

$$\leq \frac{n!}{2\pi} \cdot \frac{2\pi R}{R^{n+1}} \cdot \max_{|z-z_0|=R} |f(z)|$$

$$= \frac{n!}{R^n} \max_{|z-z_0|=R} |f(z)|, \quad n \geq 0. \qquad (3.4.2.11)$$

Since $|f(z)| \leq M$ on $\mathbf{C}$, this inequality implies that, for $n \geq 1$,

$$|f^{(n)}(0)| \leq \frac{n!M}{R^n} \to 0 \quad \text{as } R \to \infty$$

and thus, $f^{(n)}(0) = 0$ for $n \geq 1$. Since $f(z)$ is entire, it follows that $f(z) = \sum_0^\infty \frac{f^{(n)}(0)}{n!} z^n$ on $\mathbf{C}$ induces that $f(z) = f(0)$, $z \in \mathbf{C}$. $\qquad\square$

A third proof. Pick up any two distinct points a and b, and let $R > 0$ so that $|a|, |b| < R$. Then,

$$f(a) - f(b) = \frac{1}{2\pi i} \left[\int_{|z|=R} \frac{f(z)}{z - a} dz - \int_{|z|=R} \frac{f(z)}{z - b} dz \right]$$

$$= \frac{a - b}{2\pi i} \int_{|z|=R} \frac{f(z)}{(z - a)(z - b)} dz$$

$$\Rightarrow |f(a) - f(b)| \leq \frac{|a - b| RM}{(R - |a|)(R - |b|)} \to 0 \quad \text{as } R \to \infty.$$

Hence $f(a) = f(b)$, for any $a, b \in \mathbf{C}$. $\qquad\square$

In Exercise B (2) of Sec. 2.2 and in (2.5.1.1), we adopted basically a topological proof for

The fundamental theorem of algebra. Any polynomial of positive degree has at least one zero.   (3.4.2.12)

Here, we give an analytic **proof**, based on Liouville's theorem. For other proofs, see Exercise A (43).

Let $p(z) = a_n z^n + \cdots + a_1 z + a_0$ be a polynomial, where $a_n \neq 0$ and $n \geq 1$. If $p(z) \neq 0$ holds for all $z \in \mathbf{C}$, then $\frac{1}{p(z)}$ would be an entire function.

Since $p(z) = z^n \left(a_n + \frac{1_{n-1}}{z} + \cdots + \frac{a_0}{z^n} \right) \to \infty$ as $z \to \infty$, $|p(z)| \geq 1$ for $|z| \geq R$ and some $R > 0$. Hence, $\left| \frac{1}{p(z)} \right| \leq 1$ on $|z| \geq R$. On the other hand, $\frac{1}{p(z)}$ is continuous on the compact set $|z| \leq R$ and thus, $\left| \frac{1}{p(z)} \right| \leq M$ on $|z| \leq R$ for some $M > 0$. Combining together,

$$\left| \frac{1}{p(z)} \right| \leq \max\{1, M\}, \quad z \in \mathbf{C}.$$

As a bounded entire function, $\frac{1}{p(z)}$ is a constant and, in turn, $p(z)$ is a constant, which contradicts to the fact that $\lim_{z \to \infty} p(z) = \infty$. Consequently, there exists a $z_0 \in \mathbf{C}$ such that $p(z_0) = 0$.   □

Example 6.

(1) Let f be an entire function such that $|f(z)| \leq M|z|^k$ for sufficiently large $|z|$ where $M > 0$ is a constant and k is a positive integer. Show that f is a polynomial of degree less than or equal to k.

(2) If f is entire and $f(z) \to \infty$ as $z \to \infty$, then f is a polynomial.

(3) Suppose that f is analytic on a domain Ω. If for each $z \in \Omega$, there corresponds a nonnegative integer $n(z)$ so that $f^{(n(z))}(z) = 0$, then f is a polynomial.

Proof. (1) By Cauchy's inequality (3.4.2.11), for any $R > 0$ large enough,

$$|f^{(n)}(0)| \leq \frac{n!}{R^n} \cdot M \cdot R^k = \frac{n!M}{R^{n-k}} \to 0 \quad \text{as } R \to \infty$$

if $n > k$. Hence, $f^{(n)}(0) = 0$ for $n > k$ and

$$f(z) = f(0) + f'(0)z + \cdots + \frac{f^{(k)}(0)}{k!} z^k.$$

(2) There exists an $R > 0$ so that $|f(z)| \geq 1$ on $|z| \geq R$. Then f has only a finite number of zeros $z_1, \ldots, z_n$ in $|z| < R$ unless f is a constant. Consider $g(z) = \frac{(z - z_1) \cdots (z - z_n)}{f(z)}$, which is still entire (why?). Now, $|g(z)| \leq M + N|z|^n$

if $|z|$ is large enough. By (a), g is a polynomial without zeros (indeed, a constant) and so is f.

(3) Take any $r > 0$ so that the closed disk $|z - z_0| \le r$ lies in Ω. For any integer $n \ge 0$, let

$$A_n = \{z \,|\, |z - z_0| \le r \text{ and } f^{(n)}(z) = 0\}.$$

By assumption, it follows that $\{|z - z_0| \le r\} = \bigcup_{n=0}^\infty A_n$. Since $|z - z_0| \le r$ is an uncountable set, at least one of A_n, $n \ge 0$, is an infinite set (in fact, an uncountable set too).

Hence, by the Bolzano–Weierstrass theorem (see (3) in (1.9.3)), such an A_n has a limit point in $|z - z_0| \le r$ and thus, in Ω. By using the interior uniqueness theorem (see (4) in (3.4.2.9)), $f^{(n)}(z) = 0$ holds on Ω and hence, f is a polynomial.

Or, since $|z - z_0| \le r$ is a complete metric space as a closed set in $\mathbf{C}$, by the Baire category theorem (see Exercise B (4) of Sec. 1.9), at least one of A_n, $n \ge 0$, has a nonempty interior. And the result follows easily by (3.4.2.9).   □

Section (5) Morera theorem

The converse of the Cauchy integral theorem is

Morera's theorem. If f is continuous in a domain Ω, and if

$$\int_\gamma f(z)dz = 0$$

for every rectifiable closed curve γ in Ω, then f is analytic in Ω.

(3.4.2.13)

Proof. Fix a point $z_0 \in \Omega$. For any point $z \in \Omega$, choose two rectifiable curves γ_1 and γ_2, lying entirely in Ω and connecting z_0 to z. Then $\gamma_1 - \gamma_2$ is a closed curve and

$$\int_{\gamma_1 - \gamma_2} f(z)dz = \int_{\gamma_1} f(z)dz - \int_{\gamma_2} f(z)dz = 0,$$

$$\Rightarrow \int_{\gamma_1} f(z)dz = \int_{\gamma_2} f(z)dz,$$

which means that the integration from z_0 to z is independent of the choice of the rectifiable curve connecting z_0 and z. *This fact* is denoted by the following function:

$$F(z) = \int_{z_0}^{z} f(z)dz, \quad z \in \Omega, \tag{3.4.2.14}$$

which is well defined. There is a $\delta > 0$ so that $|\zeta - z| < \delta$ is contained in Ω. For $h \in \mathbf{C}$ and $h \neq 0$ such that $|h| < \delta$, we have

$$F(z) = \int_{z_0}^{z+h} f(\zeta)d\zeta + \int_{z+h}^{z} f(\zeta)d\zeta$$

$$= F(z+h) - \int_{z}^{z+h} f(\zeta)d\zeta,$$

$$\Rightarrow \frac{F(z+h) - F(z)}{h} = \frac{1}{h} \int_{z}^{z+h} f(\zeta)d\zeta,$$

or

$$\frac{F(z+h) - F(z)}{h} - f(z) = \frac{1}{h} \int_{z}^{z+h} [f(\zeta) - f(z)]d\zeta.$$

Since f is continuous at z, for any $\varepsilon > 0$, choose $\delta = \delta(z, \varepsilon) > 0$ so small that $|f(\zeta) - f(z)| < \varepsilon$ on $|\zeta - z| < \delta$. In this case,

$$\left| \frac{f(z+h) - F(z)}{h} - f(z) \right| \leq \frac{1}{|h|} \cdot \varepsilon \cdot \text{the length } |h| \text{ of the segment}$$

$$\text{connecting } z \text{ to } z + h$$

$$= \varepsilon.$$

Hence $F'(z) = f(z)$ everywhere on Ω, which, in turn, implies that $f'(z)$ exists on Ω. Hence f is analytic on Ω. □

Remark 4 (On hypothesis in Morera's theorem). Morera's theorem is still valid under the relaxed condition: For each disk D contained in Ω and each triangle or rectangle γ lying entirely in D, $\int_{\gamma} f(z)dz = 0$ always holds. Then $f(z)$ is analytic on Ω. This is because differentiability is a local property and $F(z)$ in the proof can be defined only in the disk to guarantee that $F'(z) = f(z)$ locally.

Therefore, we are eventually able to formulate the following

Characteristic properties for a continuous function to be analytic. Suppose f is continuous on a domain Ω. The following statements are equivalent:

(1) f is analytic on Ω (i.e., differentiable everywhere on Ω).
(2) $\int_{\partial\Delta} f(z)dz = 0$ for any triangle Δ contained in Ω.

(3) $\int_{\partial R} f(z)dz = 0$ for any rectangle R contained in Ω (in particular, only these rectangles whose sides are parallel to the axes are needed to be considered).

(4) f locally has a *primitive* everywhere in Ω, namely, for each point $z_0 \in \Omega$, there exist an open disk $|z - z_0| < \rho$ contained in Ω and a differentiable function $F(z)$ so that $F'(z) = f(z)$ holds on $|z - z_0| < \rho$. Such a $F(z)$ is of the form

$$\int_{z_0}^z f(z)dz + c, \quad |z - z_0| < \rho, \quad c \text{ is a constant,}$$

where the path of integration from z_0 to z is contained in $|z - z_0| < \rho$ (see (3.4.2.14) and Sec. 4.1).

(5) For each closed disk $|z - z_0| \le \rho$ $(\rho > 0)$ contained in Ω,

$$f(z) = \frac{1}{2\pi i} \int_{|\zeta - z_0| = \rho} \frac{f(\zeta)}{\zeta - z} d\zeta, \quad |z - z_0| < \rho.$$

Namely f is, locally, a *Cauchy-type integral* of itself (for details, see Sec. 4.7).

(6) f, locally, can be represented by a power series, namely, for each $z_0 \in \Omega$, there is a $\rho > 0$ so that

$$f(z) = \sum_0^\infty a_n(z - z_0)^n, \quad |z - z_0| < \rho$$

where $a_n = \frac{f^{(n)}(z_0)}{n!}, n \ge 0$. And thus, the power series is called the *Taylor* series of f at z_0 (for details, see Chap. 5). (3.4.2.15)

Note that (1) $\Rightarrow$ (2) $\Leftrightarrow$ (3) is the Cauchy–Goursat theorem (3.4.2.2); (2) $\Leftrightarrow$ (3) $\Rightarrow$ (1) is Morera's theorem (3.4.2.13); (3) $\Rightarrow$ (4) $\Rightarrow$ (1) is contained in the proof of (3.4.2.13); (6) $\Rightarrow$ (1) $\Rightarrow$ (5) $\Rightarrow$ (6) is contained in (3.3.2.4), (3.4.2.6) and its proof. Section 4.7 will give a detailed account for (5). □

Remark 5 *(A generalized version of Morera's theorem).* Suppose f is continuous on $|z - z_0| < r$ and analytic there except on a line segment L contained in $|z - z_0| < r$. After performing a rotation $z \to az$, we may suppose that L is parallel to the real axis. Let R be a rectangle in $|z - z_0| < r$ with sides parallel to the axes. Three cases are divided.

(1) If $R \cap L = \varphi$, the Cauchy–Goursat theorem implies that

$$\int_{\partial R} f(z)dz = 0.$$

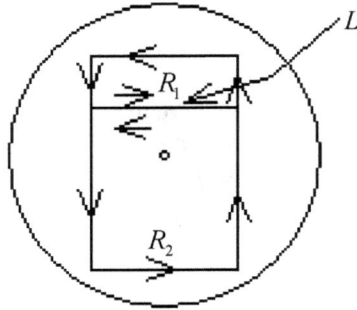

Fig. 3.26

(2) If L is a horizontal side of R, by $(3.4.2.1)'$,

$$\int_{\partial R} f(z)dz = 0.$$

(3) If L is inside the interior of R as indicated in Fig. 3.26, then L divided R into two smaller rectangles R_1 and R_2 with L as their common side. By 2,

$$\int_{\partial R} f(z)dz = \int_{\partial R_1} f(z)dz + \int_{\partial R_2} f(z)dz = 0 + 0 = 0.$$

Hence, f is analytic on $|z - z_0| < r$ by Morera's theorem.

We formulate as

A sufficient condition for a continuous function to be analytic. Suppose f is continuous on a domain Ω and analytic there, except on a line segment (or even a rectifiable Jordan curve) contained in Ω. Then f is analytic everywhere on Ω.   (3.4.2.16)

Sketch of another proof

We may suppose the exceptional segment L lying on the real axis. Take any fixed point $z_0 \in L$ and a $\rho > 0$ so that $|z - z_0| \leq \rho$ is contained in Ω. Try to show that, via $(3.4.2.1)'$ or $(3.4.2.4)'$,

$$f(z) = \frac{1}{2\pi i}\int_{\partial D}\frac{f(\zeta)}{\zeta - z}d\zeta, \quad z \in D = \{|z - z_0| < \rho\} \quad \text{and} \quad \text{Im}\, z \neq 0.$$

$$(*_3)$$

Choose any point $z^* \in D$ and $\text{Im}\, z^* = 0$. Let $d = \text{dist}(z^*, \partial D)$ and let $z \in D$, $\text{Im}\, z \neq 0$ with $|z - z^*| \leq \frac{d}{2}$. Then, if $\zeta \in \partial D$, $|\zeta - z^*| \geq d$, and

$|\zeta - z| \geq |\zeta - z^*| - |z^* - z| \geq \frac{d}{2}$ hold. Under these circumstance, by $(*_3)$

$$\left| f(z) - \frac{1}{2\pi i} \int_{\partial D} \frac{f(\zeta)}{\zeta - z^*} d\zeta \right| \leq \frac{|z - z^*|}{2\pi} \int_{\partial D} \frac{|f(\zeta)|}{|\zeta - z||\zeta - z^*|} |d\zeta|$$

$$\leq \frac{|z - z^*|}{2\pi} \cdot \max_{\zeta \in \partial D} |f(\zeta)| \frac{2\pi\rho}{d \cdot \frac{d}{2}} \to 0$$

as $z \to z^*$. Since $\lim_{z - z^*} f(z) = f(z^*)$, it follows that

$$f(z^*) = \frac{1}{2\pi i} \int_{\partial D} \frac{f(\zeta)}{\zeta - z^*} d\zeta, \quad z^* \in D \quad \text{and} \quad \operatorname{Im} z^* = 0 \qquad (*_4)$$

$$\Rightarrow (\text{Combining } (*_3) \text{ and } (*_4)) f(z) = \frac{1}{2\pi i} \int_{\partial D} \frac{f(\zeta)}{\zeta - z} d\zeta, \quad z \in D.$$

Hence, by (5) in (3.4.2.15), f is analytic on D and, in particular, analytic at points on the line L.   □

As an immediate application of Morera's theorem, we introduce the

Removable singularity of an analytic function. Suppose f is analytic in $0 < |z - z_0| < R$. Then, the following are equivalent.

(1) f can be defined at z_0 so that f is continuous at z_0, namely, $\lim_{z - z_0} f(z) = f(z_0)$.
(2) f is bounded in a deleted neighborhood $0 < |z - z_0| < R$ of z_0.
(3) $\lim_{z - z_0}(z - z_0)f(z) = 0$.
(4) f can be defined at z_0 so that f is analytic at z_0 and hence, analytic everywhere in $|z - z_0| < R$.

In this case, z_0 is said to be a *removable singularity* of f (see Sec. 4.10.1 for details).   (3.4.2.17)

Proof. (4) $\Rightarrow$ (1) $\Rightarrow$ (2) $\Rightarrow$ (3) are trivial.

To prove (1) $\Rightarrow$ (4): This is a consequence of (3.4.2.16). A direct proof is as follows. Let γ be any *triangle* in $|z - z_0| < R$. If $z_0 \in \operatorname{Ext} \gamma$, then by the Cauchy integral theorem (3.4.2.1) or just (3.4.2.2), $\int_\gamma f(z)dz = 0$ holds. If z_0 lies on γ, we can shrink γ to z_0 in the following sense: Since f is (uniformly) continuous in a neighborhood of z_0, which is contained in $|z - z_0| < R$, for fixed $\varepsilon > 0$, there is a $\delta = \delta(\varepsilon) > 0$ so that $|z - z_0| < \delta$

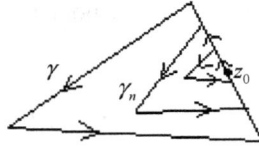

Fig. 3.27

implies that $|f(z) - f(z_0)| < \varepsilon$. Let γ_n, for each integer $n \geq 1$, be a triangle in $|z - z_0| < \delta$ satisfying the following conditions:

(1) γ_n lies in $\overline{\text{Int}\,\gamma}$ if n is large enough.
(2) $z_0 \in \gamma_n$ and the side of γ_n that contains z_0 is part of a side of γ as indicated in Fig. 3.27.
(3) The length $l(\gamma_n)$ of γ_n is less than $\frac{1}{n}$.

Then (reasoning as in (3.4.2.5)),

$$\int_\gamma f(z)dz = \int_{\gamma_n} f(z)dz,$$

$$\Rightarrow \left| \int_\gamma f(z)dz \right| \leq \int_{\gamma_n} |f(z)||dz| \leq (|f(z_0)| + \varepsilon)l(\gamma_n)$$

$$< \frac{1}{n}(|f(z_0)| + \varepsilon) \to 0 \quad \text{as } n \to \infty,$$

$$\Rightarrow \int_\gamma f(z)dz = 0.$$

If z_0 lies in Int γ, divide γ into two small triangles γ_1 and γ_2 as indicated in Fig. 3.28. Then,

$$\int_\gamma f(z)dz = \int_{\gamma_1} f(z)dz + \int_{\gamma_2} f(z)dz = 0 + 0.$$

Finally, by Morera's theorem, f is analytic on $|z - z_0| < R$.

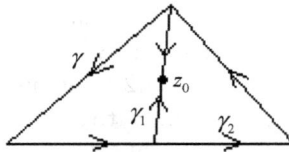

Fig. 3.28

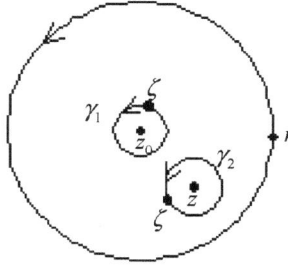

Fig. 3.29

To prove $(3) \Rightarrow (4)$: Fix any $0 < r < R$. Define

$$F(z) = \frac{1}{2\pi i} \int_{|\zeta - z_0| = r} \frac{f(\zeta)}{\zeta - z} d\zeta, \quad |z - z_0| < r.$$

It is not hard to show that F is analytic on $|z - z_0| < r$ (one may imitate the process leading to (3.4.2.6) or just refer to (3.4.2.15) or see (4.7.1)). We still need to show that $F(z) = f(z)$ on $0 < |z - z_0| < r$. To see this, fix any z in $0 < |z - z_0| < r$. Construct two *disjoint* circles γ_1 and γ_2 with center z_0 and z_1, respectively, as indicated in Fig. 3.29. For any $\varepsilon > 0$, choose $\delta = \delta(\varepsilon) > 0$ so that $|(\zeta - z_0)f(\zeta)| < \varepsilon$ if $0 < |\zeta - z_0| < \delta$, $|\zeta - z_0| < \delta$. Now, let the circle γ_1 be contained in $|\zeta - z_0| < \delta$. Then

$$F(z) = \frac{1}{2\pi i} \int_{\gamma_1} \frac{f(\zeta)}{\zeta - z} d\zeta + \frac{1}{2\pi i} \int_{\gamma_2} \frac{f(\zeta)}{\zeta - z} d\zeta = f(z) + \frac{1}{2\pi i} \int_{\gamma_1} \frac{f(\zeta)}{\zeta - z} d\zeta.$$

Now, on γ_1, $|\zeta - z| \geq |z - z_0| - |\zeta - z_0| = |z - z_0| - \rho$, where ρ is the radius of γ_1. Hence,

$$\left| \frac{1}{2\pi i} \int_{\gamma_1} \frac{f(\zeta)}{\zeta - z} d\zeta \right| \leq \frac{1}{2\pi} \int_{\gamma_1} \frac{|f(\zeta)|}{|\zeta - z|} |d\zeta|$$

$$\leq \frac{1}{2\pi} \frac{\varepsilon}{\rho} \cdot \frac{2\pi \rho}{|z - z_0| - \rho} = \frac{\varepsilon}{|z - z_0| - \rho},$$

$$\Rightarrow \left| \frac{1}{2\pi i} \int_{\gamma_1} \frac{f(\zeta)}{\zeta - z} d\zeta \right| \leq \frac{\varepsilon}{|z - z_0|} \quad \text{as } \rho \to 0,$$

$$\Rightarrow \frac{1}{2\pi i} \int_{\gamma_1} \frac{f(\zeta)}{\zeta - z} d\zeta = 0 \quad \text{as } \varepsilon \to 0,$$

$$\Rightarrow F(z) = f(z) \quad \text{on } 0 < |z - z_0| < r.$$

This proves (4).   □

For other applications of Morera's theorem, see Secs. 3.4.5 and 3.4.6.

Exercises A

(1) Suppose f is analytic on an open disk $|z - z_0| < R$ and is continuous on the closed disk $|z - z_0| \le R$. Show that

    (i) (Cauchy integral theorem)

$$\int_{|\zeta - z_0| = R} f(z)dz = 0, \quad \text{and}$$

    (ii) (Cauchy integral formula)

$$f^{(n)}(z) = \frac{n!}{2\pi i} \int_{|\zeta - z_0| = R} \frac{f(\zeta)}{(\zeta - z)^{n+1}} d\zeta, \quad |z - z_0| < R.$$

These results are still valid if $|z - z_0| = R$ is replaced by any rectifiable Jordan closed curve γ.

(2) (*Residue theorem*, for details see Sec. 4.11.3) Let γ be a rectifiable Jordan closed curve and $z_1, \ldots, z_n$ be distinct points in Int γ. Suppose f is analytic on $\overline{\text{Int}\,\gamma} - \{z_1, \ldots, z_n\}$, namely, f is analytic on $\overline{\text{Int}\,\gamma}$ except at isolated singularities $z_1, \ldots, z_n$. Then,

$$\int_\gamma f(z)dz = 2\pi i \sum_{k=1}^{n} \text{Re}\, s(f, z_k),$$

where, for $1 \le k \le n$,

$$\text{Re}\, s(f; z_k) = \frac{1}{2\pi i} \int_{|z - z_k| = r_k} f(z)dz$$

is called the *residue* of f at z_k. Note that $r_k > 0$, $1 \le k \le n$, are so small that $|\zeta - z_k| \le r_k$ are pairwise disjoint. For instance, let $f(z) = \frac{(z^2 - 1)^2}{z^2(z-a)(z-b)}$, where $ab = 1$ and $a \ne b$. Then, by the Cauchy integral formula,

$$\text{Re}\, s(f; 0) = a + b, \quad \text{Re}\, s(f; a) = a - b \quad \text{and} \quad \text{Re}\, s(f; b) = b - a.$$

(3) Let f be a C^1 function on a domain Ω. If

$$\int_\gamma f(z)dz = 0$$

for all circles γ in Ω satisfying $\overline{\text{Int}\,\gamma} \subseteq \Omega$, then f is analytic on Ω.

(4) Try to use the Cauchy integral theorem (3.4.2.1) or Cauchy integral formulae (3.4.2.4), (3.4.2.4)', (3.4.2.7) and (3.4.2.8) or Exercise (2) to

evaluate the following integrals:

(a) $\displaystyle\int_{|z|=1} \frac{dz}{z^5 + 2}$.

(b) $\displaystyle\int_{|z-1|=1} \frac{e^z}{z^2 + 1} dz$.

(c) $\displaystyle\int_{|z|=1} \frac{dz}{z^2 + 2z}$.

(d) $\displaystyle\int_{|z+i|=\frac{3}{2}} \frac{dz}{z^4 + z^2}$.

(e) $\displaystyle\int_{|z|=\frac{3}{2}} \frac{dz}{(z^2 + 1)(z^2 + 4)}$.

(f) $\displaystyle\int_{|z|=2} \frac{z^4 dz}{z^5 + 1}$.

(g) $\displaystyle\int_{|z|=2} \frac{dz}{z^3(z-1)^3(z-3)^5}$.

(h) $\displaystyle\int_{|z|=R} \frac{dz}{(z-a)^n(z-b)}$, where $|a|, |b| \neq R > 0$ and n is a positive integer.

(i) $\displaystyle\int_{|z|=1} \frac{\cos h^2 2z}{z^3} dz$.

(j) $\displaystyle\int_{|z|=2} \frac{e^{2z} \sin h3z}{(z-1)^4} dz$.

(k) $\displaystyle\int_{|z|=1} \frac{e^{az}}{z^{n+1}} dz$, $a \in \mathbf{C}$.

(5) Evaluate the following integrals (note that various $r > 0$ should be considered).

(a) $\displaystyle\int_{|z-i|=r} \frac{z^4 + z^2 + 1}{z(z^2 + 1)} dz$, where $|z-i| = r$ does not pass 0 and $\pm i$.

(b) $\displaystyle\int_{|z|=r} \frac{z^3}{(z-1)(z^2 + 2z + 3)} dz$.

(c) $\int_{|z|=r} \dfrac{1}{z^3(z+1)(z-2)}dz.$

(d) $\int_{|z|=r} \dfrac{dz}{z^4(z^2-1)^2}.$

(e) (See Ref. [1], p. 120) $\int_{|z|=r} \dfrac{|dz|}{|z-a|^2}$ $(|a| \neq r).$

(f) (See Ref. [1], p. 123) $\int_{|z|=r} \dfrac{|dz|}{|z-a|^4}$ $(|a| \neq r).$

(6) Evaluate the following integrals (note that branches on some domain should be defined).

(a) $\int_{|z|=1} \sqrt{9-z^2}dz.$

(b) $\int_{|z|=1} \log(z^2+1)dz.$

(c) $\int_{|z-i|=\frac{1}{2}} \dfrac{dz}{(1+z)z^a}$ $(0 < \alpha < 1).$

(d) $\int_{|z-1|=\frac{1}{3}} \dfrac{\log z}{\sqrt{z}(1+z)^2}dz.$

(e) $\int_{|z+i|=\frac{1}{2}} \dfrac{1}{\sqrt[3]{(1-z)(1+z)^2}}dz.$

(7) Find the Taylor series expansion of f at the indicated point z_0 and its radius of convergence.

(a) $f(z) = \dfrac{z-1}{z^2-z-1}$ at $z_0 = 0.$

(b) $f(z) = \dfrac{z^2}{(z+1)^2}$ at $z_0 = 1.$

(c) $f(z) = \dfrac{2z}{z^2-1}$ at $z_0 = i.$

(d) $f(z) = \sin(2z-z^2)$ at $z_0 = 1.$

(e) $f(z) = \sin^2 z$ at $z_0 = 0.$

(f) $f(z) = \int_0^z \dfrac{\sin \zeta}{\zeta}d\zeta$ at $z_0 = 0.$

(8) Find the Taylor series expansion for the branch of f, determined by the prescribed condition at the given point z_0, and its radius of convergence.

(a) $f(z) = \sqrt{z+i}$, $\quad f(0) = \sqrt{i} = \dfrac{1+i}{\sqrt{2}}$ $\quad$ at $z_0 = 0$.

(b) $f(z) = \sqrt{1 + \sqrt{1+z}}$, $\quad f(0) = \sqrt{2}$ $\quad$ at $z_0 = 0$.

(c) $f(z) = \text{Log}(1 + \sqrt{1+z^2})$, $\quad f(0) = \text{Log}(1 + \sqrt{1}) = \log 2$ $\quad$ at $z_0 = 0$.

(d) $f(z) = \dfrac{\text{Log } z}{z}$, $\quad \text{Log } 1 = 0$ at $z_0 = 1$.

(e) $f(z) = e^{z/(1-z)}$ $\quad$ at $z_0 = 0$.

(9) Find the zeros and their orders for each of the following functions $f(z)$.

(a) $f(z) = \dfrac{z^2 + 9}{z^4}$.

(b) $f(z) = z^2(e^{z^2} - 1)$.

(c) $f(z) = \dfrac{\sin^3 z}{z}$.

(d) $f(z) = \sin z^3$.

(e) $f(z) = (\sqrt{z} - 2)^3$.

(f) $f(z) = \sqrt{1 + \sqrt{z}}$.

(g) $f(z) = (\sqrt[3]{z + z^4} - 2)^2$.

(h) $f(z) = (\log \cos z)^2$ $\quad$ at $z_0 = 2\pi$.

(i) $f(z) = z^2 - \tan^{-1} z^2$ $\quad$ at $z_0 = 0$.

(j) $f(z) = 1 - \cos \sqrt{z^3}$ $\quad$ at $z_0 = (2\pi)^{2/3}$.

(k) $f(z) = z^{\frac{1}{2}} - z$ at $z_0 = 1$.

(10) (*L' Hospital's rule*) Suppose z_0 is a zero of order m and n of $f(z)$ and $g(z)$, respectively, where both f and g are analytic at z_0. Show that

$$\lim_{z \to z_0} \frac{f(z)}{g(z)} = \lim_{z \to z_0} \frac{f^{(n)}(z)}{g^{(n)}(z)} = \begin{cases} 0, & m > n, \\ \dfrac{f^{(n)}(z_0)}{g^{(n)}(z_0)}, & m = n, \\ \infty, & m < n. \end{cases}$$

Then, compute the following limits:

(a) $\lim\limits_{z \to 0} \dfrac{z^3 \cos z - 2z^4}{e^{z^4} - 1}$.

(b) $\lim\limits_{z \to 0} (1 + az)^{1/z}$, where a is a constant.

(c) $\lim\limits_{z \to 0} \dfrac{1 - \cos z^{3/2}}{z^2}$.

(d) $\lim\limits_{z \to 0} \dfrac{z^{1/z} - z}{(z - 1)^2}$.

(e) $\lim\limits_{z \to 2\pi} \dfrac{\log \cos z}{(1 - e^{iz})^2}$.

(f) $\lim\limits_{z \to 0} \left(\dfrac{1}{e^z - 1} - \dfrac{1}{z} \right)$.

(11) Determine if there exists a function f analytic at 0 and satisfying each of the following conditions. If certainty, find this f explicitly.

(a) $f\left(\frac{1}{n}\right) = \begin{cases} 0, & n = 1, 3, 5, \ldots \ . \\ \dfrac{1}{n}, & n = 2, 4, 6, \ldots \ . \end{cases}$

(b) $f\left(\dfrac{1}{n}\right) = \dfrac{n}{n+1}$, $n = 1, 2, 3, \ldots$.

(c) $f\left(\dfrac{1}{n}\right) = \dfrac{1}{1+n}$, $n = 1, 2, \ldots$.

(d) $\left| f\left(\dfrac{1}{n}\right) \right| \le e^{-n}$, $n \ge 1$ but f is not identically equal to zero.

(e) $|f^{(n)}(0)| \ge (n!)^2$, $n \ge 1$.

(12) Suppose that both f and g are analytic on a domain Ω and satisfy that on $f(z)g(z) = 0$ on Ω. Show that at least one of f and g is identically equal to zero on Ω.

(13) Suppose that f is analytic on $|z| < r$ and is an *even* function, namely, $f(-z) = f(z)$ for all z in $|z| < r$. Show that $g(z) = f(\sqrt{z})$ is an analytic function on $|z| < r$ and

$$g^{(n)}(0) = [(2n)(2n - 1) \cdots (n + 1)]^{-1} f^{(2n)}(0), \quad n \ge 0.$$

(14) Let $f = u + iv$ be an entire function satisfying $u_x(z)v_y(z) - u_y(z)v_x(z) = 1$, $z \in \mathbf{C}$. Show that $f(z) = az + b$, where a and b are constants with $|a| = 1$.

(15) Let $f = u + iv$ be an entire function satisfying $u_x(z) + v_y(z) = 1$, $z \in \mathbf{C}$. Show that $f(z) = az + b$, where a and b are constants with $\mathrm{Re}\, a = \frac{1}{2}$.

(16) Suppose f is an entire function, which assumes real values at $\frac{1}{n}$, $n = 1, 2, 3, \ldots$. Show that f assumes real values on the real axis and hence satisfies

$$f(\bar{z}) = \overline{f(z)}, \quad z \in \mathbf{C}.$$

(17) Let f be analytic on a domain Ω, which contains the closed disk $|z| \le 1$. If f assumes real values on $|z| = 1$, show that f is a constant function.

(18) Suppose f is analytic on the right half-plane $\mathrm{Re}\, z \ge 0$ and satisfies $f(z+1) = 2f(z)$ there. Show that there exists an entire function $F(z)$ such that $F(z) = f(z)$ on $\mathrm{Re}\, z \ge 0$.

(19) Show that

(a) $|e^z - 1| \le e^{|z|} - 1 \le |z|e^{|z|}$, $z \in \mathbf{C}$.

(b) $\frac{1}{4}|z| < |e^z - 1| \le \frac{7}{4}|z|$ on $0 < |z| < 1$.

(20) Let f be a nonconstant analytic function on a domain Ω and γ be a Jordan closed curve such that $\overline{\mathrm{Int}\, \gamma} \subseteq \Omega$ holds. For each w_0, show that $f(z_0) = w_0$ has only a finite number of solutions in $\mathrm{Int}\, \gamma$.

(21) Suppose that f is analytic on $|z| \le R$ so that $f(z) \ne 0$ on $|z| = R$. Show that $f(z)$ has only a finite number of zeros in $|z| < R$. If, instead, $|f(z)| = 1$ on $|z| = R$, show that f has at least one zero in $|z| < R$ except f is a constant.

(22) Suppose that f is analytic on $r_1 < |z| < r_2$ and is not identically equal to zero. Then, there exists at least one $r, r_1 < r < r_2$, so that $f(z) \ne 0$ on $|z| = r$.

(23) Let $f(z)$ be an entire function with only a finite number of zeros. Show that there is a polynomial $p(z)$ and an entire function $g(z)$ without zeros so that $f(z) = p(z)g(z)$.

(24) Suppose that f and g are analytic on a domain Ω so that, for some positive integer n, $f(z)^n = g(z)^n$ holds on Ω. Show that, there exists a constant λ, $|\lambda| = 1$, so that $f(z) = \lambda g(z)$.

(25) Let f and g be entire functions satisfying $f(g(x)) = x$ for all $x \in \mathbf{R}$. Show that both f and g be one-to-one mapping from $\mathbf{C}$ onto $\mathbf{C}$ so that either one is the inverse of the other.

(26) *Law of permanence*:
   (a) Try to deduce that $\cos^2 z + \sin^2 z = 1, z \in \mathbf{C}$, from the well-known fact that $\cos^2 x + \sin^2 x = 1$ for real variable x.
   (b) Try to show that $e^{z_1 + z_2} = e^{z_1 z_2}, z_1, z_2 \in \mathbf{C}$ from $e^{x_1 + x_2} = e^{x_1 x_2}, x_1, x_2 \in \mathbf{R}$.

(27) Let $f(z)$ be a complex-valued function defined on the disk $|z| < 1$. If both $(f(z))^2$ and $(f(z))^3$ are analytic there, then $f(z)$ itself is analytic.

(28) Suppose that $f(z) = \sum_0^\infty a_n z^n$, with $f(0) = a_0 \neq 0$, has radius $R > 0$ of convergence. Let $M(\rho) = \max_{|z|=\rho} |f(z)|$ for $0 < \rho < R$. If z_0 is the zero of f that is nearest to 0, then

$$|z_0| \geq \frac{\rho |a_0|}{M(\rho) + |a_0|}.$$

(29) An entire function, which maps a triangle onto a triangle, is necessary of the form $w = az + b, a \neq 0$.

(30) Let γ be analytic in $|z| < 1$ and satisfy $|f(z)| \leq 1 - |z|$ there. Show that $f(z) = 0, |z| < 1$.

(31) Let γ be a rectifiable Jordan closed curve. If f is continuous on $\overline{\operatorname{Int}\gamma}$ and analytic in $\operatorname{Int}\gamma$, show that

$$|f(z)| \leq \max_{\zeta \in \gamma} |f(\zeta)|, \quad z \in \operatorname{Int}\gamma.$$

This is a form of *maximum modulus principle* (see (2) in (3.4.4.2)).

(32) *Cauchy integral formula on the exterior of a Jordan domain*: Let γ be a rectifiable Jordan closed curve. Suppose f is continuous on $\overline{\operatorname{Ext}\gamma} = \gamma \bigcup \operatorname{Ext}\gamma$ and analytic in $\operatorname{Ext}\gamma$. If $\lim_{z\to\infty} f(z) = A$ (a finite complex number, see (3.4.2.17)), then

$$\frac{1}{2\pi i} \int_\gamma \frac{f(\zeta)}{\zeta - z} d\zeta = \begin{cases} -f(z) + A, & z \in \operatorname{Ext}\gamma, \\ A, & z \in \operatorname{Int}\gamma. \end{cases}$$

Note that γ is positively oriented with respect to $\operatorname{Int}\gamma$ and hence is negatively oriented with respect to $\operatorname{Ext}\gamma$. In case $0 \in \operatorname{Int}\gamma$, then

$$\frac{1}{2\pi i} \int_\gamma \frac{zf(\zeta)}{\zeta z - \zeta^2} d\zeta = \begin{cases} f(z), & z \in \operatorname{Ext}\gamma, \\ 0, & z \in \operatorname{Int}\gamma. \end{cases}$$

(33) *Integral mean-value theorem*: Suppose f is analytic in $|z - z_0| < R$. Then

$$f(z_0) = \frac{1}{2\pi} \int_0^{2\pi} f(z_0 + re^{i\theta}) d\theta, \quad 0 < r < R.$$

If, in addition, $f(z) \neq 0$ on $|z - z_0| < R$, then

$$\log |f(z_0)| = \frac{1}{2\pi} \int_0^{2\pi} \log |f(z_0 + re^{i\theta})|d\theta, \quad 0 < r < R.$$

Apply these identities to $f(z) = \cos z$ to show that

$$\int_0^{2\pi} \cos(\cos\theta)\cosh(\sin\theta)d\theta = 2\pi; \quad \int_0^{2\pi} \sin(\cos\theta)\sinh(\sin\theta)d\theta = 0.$$

What is $\int_0^{2\pi} \log[\cosh^2(\sin\theta) - \sin^2(\cos\theta)]d\theta = ?$

(34) Suppose f is analytic in $|z - z_0| \leq r$. Show that

(a) $f'(z_0) = \dfrac{1}{\pi r} \displaystyle\int_0^{2\pi} [\operatorname{Re} f(z_0 + re^{i\theta})]e^{-i\theta}d\theta, \quad$ and

(b) $|f'(z_0)| \leq \dfrac{2M}{r} \quad$ if $|\operatorname{Re} f(z)| \leq M$ holds.

(35) *Equivalent statements of Liouville's theorem*: The following are equivalent.

   (i) f is a bounded entire function.

   (ii) f is analytic on $\mathbf{C}^*$ (recall that f is said to be analytic at ∞ if $f\left(\frac{1}{z}\right)$ is analytic at 0. See Remark in Sec. 3.3).

   (iii) f is entire and has a removable singularity at ∞ (imitate (4) in (3.4.2.17)).

   (iv) f is a constant.

Try to prove the above claim and then try the following possible proofs of this theorem.

(a) Use Exercise (34).

(b) Use maximum modulus theorem as stated in Exercise (31).

(c) Use the uniqueness of the Taylor series expansions of f at 0 and ∞.

(d) Use Gutzmer's inequality (see Exercise B (2)).

(e) Use (a) in Example 6. In addition, try to show that (a) in Example 6 is equivalent to Liouville's theorem.

(f) Use Schwarz's lemma (see Sec. 3.4.5).

(g) Use integral mean-value theorem as stated in Exercise (33) (refer to E. Nelson: A proof of Liouville theorem, *Proc. Amer. Math. Soc.* **12** (1961), 995).

(36) Suppose that f is an entire function. Show that any one of the following conditions will guarantee that f is a constant.

   (i) $\displaystyle\lim_{|z| \to \infty} \frac{|f(z)|}{|z|} = 0.$

(ii) Let $g : (0, \infty) \to [0, \infty)$ be a function satisfying $\lim_{t \to \infty} g(t) = 0$. Then $|f(z)| \leq |z| g(|z|)$ holds on $|z| > 0$.

(iii) $f(0) = 0$ and $\lim_{z \to \infty} \operatorname{Re} f(z) = 0$ or $\lim_{z \to \infty} \operatorname{Im} f(z) = 0$.

(iv) $\operatorname{Re} f(z)$ is bounded if $|z| \to \infty$.

(v) One of the conditions that $\operatorname{Re} f(z) \leq 0$, $\operatorname{Re} f(z) \geq 0$, $\operatorname{Im} f(z) \leq 0$, and $\operatorname{Im} f(z) \geq 0$ hold throughout the plane $\mathbf{C}$.

(vi) There exists a $z_0 \in \mathbf{C}$ such that $|f(z)| \geq |f(z_0)| > 0$ on $\mathbf{C}$.

(vii) $\operatorname{Re} f(z)$ or $\operatorname{Im} f(z)$ does not have zero in $\mathbf{C}$.

(viii) f is bounded on $0 < |z| < \infty$.

(ix) $f(z) \neq 0$ on $\mathbf{C}$ and $\lim_{z \to \infty} f(z)$ exists as a nonzero complex number.

(x) $f(z)$ does not assume values on a nondegenerated line segment.

The most general result in this direction is the famous *Picard's little theorem: An entire function is a constant if it does not assume two distinct finite complex numbers.* See (4.10.3.3), (5.8.3.11) and Sec. 5.8.4.

(37) Let ω_1 and ω_2 be nonzero complex numbers so that $\frac{\omega_1}{\omega_2}$ is not a real number. Suppose that an entire function f satisfies $f(z) = f(z + \omega_1) = f(z + \omega_2)$, $z \in \mathbf{C}$. Then f is a constant. In short, *a double periodic entire function is necessarily a constant.* Observe that e^z, $\cos z$, and $\sin z$ are only simply periodic.

(38) Let f be an entire function so that $|f(z)| \geq M > 0$ on $|z| \geq R > 0$ for some constants R and M. Show that f is either a constant or has at least one zero in $|z| < R$.

(39) Suppose that an entire function f satisfies the property $|f(z)| \leq M e^{\alpha \operatorname{Re} z}$ on $\mathbf{C}$ for some constants $\alpha > 0$ and $M > 0$. Show that $f(z) = A e^{\alpha z}$, where A is a constant.

(40) Suppose that $f(z) = \sum_0^\infty a_n z^n$ is an entire function. Show that the series converges uniformly to $f(z)$ on the entire plane $\mathbf{C}$ if and only if $f(z)$ is a polynomial.

(41) Suppose that $f(z) = \sum_0^\infty a_n z^n$ is an entire function. If there exist constants $M > 0$, $\alpha \geq 0$ and a sequence r_n, $r_n > 0$, satisfying $\lim_{n \to \infty} r_n = \infty$ so that

$$|f(z)| \leq M r_n^\alpha, \quad |z| = r_n, \quad n \geq 1.$$

Show that $a_k = 0$ for $k > \alpha$ and hence, f is a polynomial. Then, try to prove the following:

(i) The Liouville theorem.

(ii) f is not a polynomial if $\sup_{a_k \neq 0} k = \infty$ holds.

(42) Let f be an entire function and $M(R, f) = \max_{|z|=R} |f(z)|$.

(a) If there exists a positive integer m so that $\lim_{R \to \infty} \frac{M(R,f)}{R^m} = 0$ holds, then f is a polynomial of degree not more than $m - 1$.

(b) Suppose f is not a polynomial. Show that $\lim_{R \to \infty} \frac{\log M(R,f)}{\log R} = \infty$.

(c) If there exist an entire function g and a positive integer k so that $|f(z)| \leq |z^k g(z)|$ for all sufficiently large $|z|$, then there is a rational function $Q(z)$ so that $f(z) = g(z)Q(z)$.

(43) *Proofs of the fundamental theorem of algebra*: The following are some proofs of this theorem among others: Let $p(z) = a_n z^n + \cdots + a_1 z + a_0$, $a_n \neq 0$ and $n \geq 1$.

(a) A topological proof in real analysis character, see Exercise B (2) of Sec. 2.2 and (2.5.1.1).

(b) A topological proof, see Exercise B (3) of Sec. 2.5.1.

(c) Use Liouville's theorem (see (3.4.2.12)).

(d) Show that

$$\frac{1}{2}|a_n| |z|^n \leq |p(z)| \leq \frac{3}{2}|a_n| |z|^n \quad \text{if } |z| \geq \max\left\{1, \frac{2}{|a_n|}\sum_{k=0}^{n-1}|a_k|\right\}$$

(refer to (2.5.1.7)). Write $p(z) = zQ(z) + a_0$ and observe that $\frac{1}{z} = \frac{p(z)}{zp(z)} = \frac{Q(z)}{p(z)} + \frac{a_0}{zp(z)}$ if $z \neq 0$ and $p(z) \neq 0$. Suppose $p(z) \neq 0$ on $\mathbf{C}$. Then show that

$$\int_{|z|=r} \frac{a_0}{zp(z)} dz = 2\pi i, \quad r > 0,$$

would lead to a contradiction.

(e) Use maximum modulus principle (see Exercise A(7) (b) of Sec. 3.4.4).

(f) Suppose $p(z) \neq 0$ on $\mathbf{C}$. Let $f(z) = \frac{1}{p(z)}$. By maximum modulus principle or integral mean-value theorem (see Exercise (33)), $|f(0)| \leq \max_{|z|=r} |f(z)|, r > 0$ holds. Since $\lim_{z \to \infty} f(z) = 0$, $f(0) = 0$ contradicting to $f(0) = \frac{1}{p(0)} \neq 0$.

(g) Use Exercise (38).

(h) Suppose $p(0) = a_0 \neq 0$. Choose $r > 0$ (for instance, $r > \sqrt[n]{2|a_0|/|a_n|}$) so that $|p(0)| < \frac{1}{2}|a_n|r^n$ holds. Via the left inequality in (d), choose r so large such that $|p(0)| < \min_{|z|=r} |p(z)|$. By minimum modulus principle (see (1) in (3.4.4.2)), $p(z)$ has at least one zero in $|z| < r$.

(44) Suppose f is analytic in a domain Ω and has a finite number of zeros, say $z_1, \ldots, z_n$, in Ω. Show that there exists an analytic function g on Ω such that $g(z) \neq 0$ on Ω and $f(z) = (z - z_1) \cdots (z - z_n)g(z)$ holds on Ω.

Exercises B

(1) Suppose f is a nonconstant entire function and $\lambda \neq 1$ is a constant so that $f(\lambda z) = f(z)$ for all $z \in \mathbf{C}$.

   (a) Show that there exists a positive integer m so that $\lambda^m = 1$.
   (b) Let m be the least positive integer such that $\lambda^m = 1$. Show that there exists an entire function g so that $f(z) = g(z^m), z \in \mathbf{C}$.

(2) Let $f(z) = \sum_0^\infty a_n(z - z_0)^n, |z - z_0| < R$, with $R > 0$.

   (a) For each fixed r, $0 < r < R$, prove the *Parseval identity*

$$\frac{1}{2\pi} \int_0^{2\pi} |f(z_0 + re^{i\theta})|^2 d\theta = \sum_0^\infty |a_n|^2 r^{2n}.$$

   (b) Set $M(r) = \max_{|z - z_0| = r} |f(z)|$, $0 < r < R$. Prove the *Gutzmers' inequality* (1888)

$$\sum_{n=0}^\infty |a_n|^2 r^{2n} \leq M(r)^2.$$

   If there exists an n_0 so that $|a_{n_0}|\gamma n_0 = M(r)$, then it is necessary that $f(z) = a_{n_0}(z - z_0)^{n_0}$; in particular, if $n_0 = 0$, then we obtain the *maximum modulus principle*: If $|f(z)|$ attains its maximum on $|z - z_0| \leq r < R$ at z_0, then f is a constant (see (3.4.4.2)).

   (c) Try to use (b) to prove the *Cauchy inequality*

$$|f^{(n)}(z_0)| \leq \frac{n!}{r^n} M(r), \quad n \geq 0.$$

   (d) Suppose f is bounded on $|z - z_0| < R$, say $|f(z)| \leq M$ (for instance, when f is continuous on $|z - z_0| \leq R$). Then (b) can be improved to be

$$\sum_{n=0}^\infty |a_n|^2 R^{2n} \leq M^2.$$

   This inequality will provide another proof of Liouville's theorem.

(e) Suppose f is continuous on $|z - z_0| \le R$. Then (a) can be improved to be

$$\frac{1}{2\pi} \int_0^{2\pi} |f(z_0 + Re^{i\theta})|^2 d\theta = \sum_{n=0}^{\infty} |a_n|^2 R^{2n}.$$

(f) Apply (a) to $f(z) = \frac{1}{1-z}$ and show that

$$\int_0^{2\pi} \frac{d\theta}{1 - 2r\cos\theta + r^2} = \frac{2\pi}{1 - r^2}, \quad 0 \le r \le 1.$$

(g) Apply (a) to

$$f(z) = \begin{cases} \dfrac{z^n - 1}{z - 1} = 1 + z + \cdots + z^{n-1}, & z \ne 1, \\ n, & z = 1, \end{cases}$$

to show that

$$\int_0^{2\pi} \left(\frac{\sin \frac{n\theta}{2}}{\sin \frac{\theta}{2}} \right)^2 d\theta = 2n\pi.$$

(3) (Extension of Exercise A (34)) Let f be analytic in $|z| < R$.

(a) In case $f(z) = \sum_0^{\infty} a_n z^n$, $|z| < R$, show that, for $0 < r < R$,

$$a_n = \frac{f^{(n)}(0)}{n!} = \frac{1}{\pi r^n} \int_0^{2\pi} e^{-in\theta} \operatorname{Re} f(re^{i\theta}) d\theta$$

$$= \frac{i}{\pi r^n} \int_0^{2\pi} e^{-in\theta} \operatorname{Im} f(re^{i\theta}) d\theta, \quad n = 1, 2, \ldots .$$

And then, deduce that letting $A(r) = \max_{|z|=r} \operatorname{Re} f(z)$,

$$|a_n| r^n \le \max\{4A(r), 0\} - 2 \operatorname{Re} f(0).$$

(b) Suppose $R = 1$. In case $f(0) = 1$ and $\operatorname{Re} f(z) > 0$ holds on $|z| < 1$, show that

(i) $|a_n| \le 2, \quad n = 1, 2, \ldots .$

(ii) $\frac{1-|z|}{1+|z|} \le |f(z)| \le \frac{1+|z|}{1-|z|}$ and equalities hold if $f(z) = \frac{1+z}{1-z}$ and z is real.

(4) In case $|z| < 1$ and $|\zeta| = 1$, apply Example 4 to the function

$$\frac{\operatorname{Re} f(\zeta)}{\zeta} \cdot \frac{\zeta + z}{\zeta - z} = \frac{2 \operatorname{Re} f(\zeta)}{\zeta - z} - \frac{\operatorname{Re} f(\zeta)}{\zeta} = \frac{f(\zeta) + \overline{f(\zeta)}}{\zeta - z} - \frac{1}{2} \cdot \frac{f(\zeta) + \overline{f(\zeta)}}{\zeta}$$

to show the following:

(a) *Schwarz integral formula on the unit disk:*

$$f(z) = \frac{1}{2\pi i} \int_{|\zeta|=1} \frac{\operatorname{Re} f(\zeta)}{\zeta} \frac{\zeta+z}{\zeta-z} d\zeta + i \operatorname{Im} f(0), \quad |z|<1$$

$$= \frac{1}{2\pi} \int_0^{2\pi} \frac{e^{i\theta}+z}{e^{i\theta}-z} \operatorname{Re} f(e^{i\theta}) d\theta + i \operatorname{Im} f(0),$$

$$\zeta = e^{i\theta}(0 \le \theta \le 2\pi), \quad |z|<1.$$

This result is still valid even if f is continuous on $|z| \le 1$ and analytic in $|z| < 1$.

(b) *Poisson integral for harmonic function:* $\operatorname{Re} f(z) = u(z)$ is harmonic on $|z|<1$ (see Sec. 3.4.3) and

$$u(z) = \frac{1}{2\pi} \int_0^{2\pi} \operatorname{Re} \frac{e^{i\theta}+z}{e^{i\theta}-z} u(e^{i\theta}) d\theta = \frac{1}{2\pi} \int_0^{2\pi} \frac{1-|z|^2}{|e^{i\theta}-z|^2} u(e^{i\theta}) d\theta$$

$$= \frac{1}{2\pi} \int_0^{2\pi} \frac{1-r^2}{1-2r\cos(\theta-\varphi)+r^2} u(e^{i\theta}) d\theta$$

$$z = re^{i\varphi}, \quad 0 \le r < 1.$$

This result is still valid if u is harmonic in $|z| < 1$ and continuous on $|z| \le 1$.

(c) In case $u(z) \ge 0$ on $|z| = 1$, show that

$$u(0)\frac{1-|z|}{1+|z|} \le u(z) \le \frac{1+|z|}{1-|z|}u(0), \quad |z|<1.$$

Try to restate (a), (b), and (c) if $f(z) = u(z) + iv(z)$ is analytic in $|z| < R$ and continuous on $|z| \le R$.

(5) Let $f : \Omega \to \Omega$ be an analytic function where Ω is a bounded domain containing 0. If $f(0) = 0$ and $f'(0) = 1$, show that $f(z) = z$ on Ω.

(6) Suppose that f is analytic in a domain Ω and $z_0 \in \Omega$ is any fixed point. For any integer $n \ge 1$, there exists an analytic function f_n on Ω so that

$$f(z) = \sum_{k=0}^{n-1} \frac{f^{(k)}(z)}{k!}(z-z_0)^k + f_n(z)(z-z_0)^n, \quad z \in \Omega$$

which is called the *Taylor formula with remainder* of f in Ω with center at z_0, where if $0 < \rho < \operatorname{dist}(z_0, \partial\Omega)$, then

$$f_n(z) = \frac{1}{2\pi i} \int_{|\zeta-z_0|=\rho} \frac{f(\zeta)}{(\zeta-z_0)^n(\zeta-z)} d\zeta, \quad |z-z_0| < \rho.$$

This is the most useful tool in the study of the local properties of analytic function. Try the following methods:

(i) In case Ω is the disk $|z - z_0| < R$, try to use the Taylor series expansion of f at z_0 (see (3.4.2.6)).

(ii) Repeated application of $f(z) = f(z_0) + \int_{z_0}^z f'(\zeta)d\zeta = f(z_o) - \int_{z_0}^z f'(\zeta)d(z - \zeta)$ (refer to (3.4.2.14)) will lead to

$$f(z) = \sum_{k=0}^{n-1} \frac{f^{(k)}(z)}{k!}(z - z_0)^k + R_n(z - z_0),$$

where

$$R_n(z - z_0)$$
$$= \frac{1}{(n-1)!} \int_{z_0}^z f^{(n)}(\zeta)(z - \zeta)^{n-1}d\zeta$$
$$= \left\{ \frac{1}{(n-1)!} \int_0^1 f^{(n)}(z_0 + t(z - z_0))(1 - t)^{n-1}dt \right\} (z - z_0)^n.$$

Could one use $f^{(n)}(z_0 + t(z - z_0)) = \frac{n!}{2\pi i} \int_{|\zeta - z_0|=\rho} \frac{f(\zeta)}{[\zeta - (z_0 + t(z - z_0))]^{n+1}}d\zeta$, to obtain the desired expression for $f_n(z)$?

(iii) Use the Cauchy integral formula for $f(z)$, $|z - z_0| \leq \rho < R$, and note that

$$\frac{1}{\zeta - z} = \frac{1}{\zeta - z_0} \left\{ \sum_{k=1}^{n-1} \left(\frac{z - z_0}{\zeta - z_0} \right)^k + \frac{1}{\zeta - z} \left(\frac{z - z_0}{\zeta - z_0} \right)^n \right\}.$$

(iv) In general situation, try to use (3.4.2.17) to show that

$$f_1(z) = \begin{cases} \dfrac{f(z) - f(z_0)}{z - z_0}, & z \neq z_0, z \in \Omega \\ f'(z_0), & z = z_0 \end{cases}$$

is analytic on Ω and hence, $f(z) = f(z_0) + f_1(z)(z - z_0)$. This procedure will eventually lead to the result. Then, for $|z - z_0| < |\zeta - z_0| = \rho < \text{dist}(z_0, \partial\Omega)$,

$$f_n(z) = \frac{1}{2\pi i} \int_{|\zeta - z_0|=\rho} \frac{f_n(\zeta)}{\zeta - z}d\zeta$$

$$= \frac{1}{2\pi i} \int_{|\zeta - z_0| = \rho} \frac{f(\zeta)}{(\zeta - z)(\zeta - z_0)^n} d\zeta$$

$$- \sum_{k=1}^{n-1} \frac{f^{(k)}(z_0)}{k!} \cdot \frac{1}{2\pi i} \int_{|\zeta - z_0| = \rho} \frac{1}{(\zeta - z)(\zeta - z_0)^{n-k}} d\zeta.$$

Finally, try to show that $\int_{|\zeta - z_0| = \rho} \frac{1}{(\zeta - z)(\zeta - z_0)^{n-k}} d\zeta = 0$ on $|z - z_0| < \rho$, for $1 \leq k \leq n-1$. (Why? Use (3.4.2.7) or (4.7.1)).

(7) (About the *Cauchy inequality* (3.4.2.11)) Suppose f is analytic in a domain Ω. Let $0 < R < \Delta_{z_0} = \text{dist}(z_0, \partial\Omega)$, where $z_0 \in \Omega$ is a fixed point. Then

$$|f^{(n)}(z_0)| \leq \frac{M(R)n!}{R^n}, \quad n \geq 0,$$

where $M(R) = \max_{|z - z_0| = R} |f(z)|$.

(a) Try to find the minimum of $\frac{M(R)n!}{R^n}$ on the interval $(0, \Delta_z)$ and one will obtain the optimal upper bound for $|f^{(n)}(z_0)|$. For instance, if f is analytic in $|z| < 1$ and $|f(z)| \leq \frac{1}{1 - |z|}$, then

$$|f^{(n)}(0)| \leq n! \left(\frac{1 + n}{n}\right)^n (n + 1) < e(n + 1)!.$$

(b) Show that it is not possible to find a nondecreasing sequence of positive real numbers $\alpha_n, n \geq 1$, satisfying $r\alpha_{n_0} > 1$ for some constant $r > 0$ and some integer n_0 so that $|f^{(n)}(z_0)| > n!\alpha_n^n$, $n \geq 1$. For instance, $|f^{(n)}(z_0)| > n!n^n$ for $n \geq 1$ is false.

(c) Show that

$$\varlimsup_{n \to \infty} \sqrt[n]{\frac{|f^{(n)}(z_0)|}{n}} \leq \frac{1}{\Delta_{z_0}},$$

which is called the *Cauchy–Hadamard inequality*. In particular, if f is an entire function, then $\lim_{n \to \infty} \sqrt[n]{\frac{|f^{(n)}(z_0)|}{n!}} = 0$ for any $z_0 \in \mathbf{C}$. Try to justify this claim by using $f(z) = e^z$.

(d) Suppose f is entire and $|f(z)| \leq Me^{|z|}$ on $\mathbf{C}$ for some constant $M > 0$. Show that

$$\frac{|f^{(n)}(0)|}{n!} \leq M \left(\frac{e}{n}\right)^n, \quad n \geq 1.$$

(e) Suppose f is an entire function. For any fixed $z_0 \in \mathbf{C}$, there exists a constant M dependent on z_0 but independent of $n \geq 1$ so that

$$|f^{(n)}(z_0)| \leq M \cdot n!, \quad n \geq 1.$$

(8) Adopt notations as in (3.4.2.4)′ and refer to Fig. 3.21. Suppose $f(z)$ is continuously differentiable in the real sense on $\bar{\Omega}$, namely C^1 on $\bar{\Omega}$ (see (2.8.2) and (2.8.5)). Then, for $\zeta = x + iy$,

$$f(z) = \frac{1}{2\pi i} \int_{\partial\Omega} \frac{f(\zeta)d\zeta}{\zeta - z} - \frac{1}{\pi} \iint_{\bar{\Omega}} \frac{f_{\bar{\zeta}}(\zeta)}{\zeta - z} dxdy$$
$$= \frac{1}{2\pi i} \left\{ \int_{\partial\Omega} \frac{f(\zeta)}{\zeta - z} + \iint_{\bar{\Omega}} \frac{f_{\bar{\zeta}}(\zeta)}{\zeta - z} d\zeta \wedge d\bar{\zeta} \right\}, \quad z \in \Omega.$$

This is the *Pompeiu formula* (1912, 1913), which is a generalization of (2.9.13). Note that in case f is analytic on $\bar{\Omega}$ and $f_{\bar{\zeta}} = 0$ is thus equal to zero, then the Pompeiu formula reduces to the Cauchy integral formula (3.4.2.4)′. Therefore, f can be approximated by an analytic function at z with the error

$$E(z) = -\frac{1}{\pi} \iint_{\bar{D}} \frac{f_{\bar{\zeta}}(\zeta)}{\zeta - z} dxdy,$$

where $\bar{D}$ is a closed disk $|z - z_0| \leq \rho$ contained in Ω. Let $\delta = \max_{\zeta \in \bar{D}} |f_{\bar{\zeta}}(\zeta)|$, called the *derivation* of f from analyticity at z. Show that $|E(z)| \leq 2\delta\rho$.

(9) Suppose $f \in C^1(\mathbf{C})$ with compact support, namely, $\overline{\{z | f(z) \neq 0\}} = \text{supp } f$ is a compact set. Show that

$$f(z) = \frac{1}{2\pi i} \iint_C \frac{f_{\bar{\zeta}}(\zeta)}{\zeta - z} d\zeta \wedge d\bar{\zeta}, \quad z \in \mathbf{C}.$$

(10) Let K be a compact subset of a domain (or an open set) Ω. Suppose f is analytic on Ω. Then, there is another compact set $A \subseteq \Omega$ so that $K \subseteq A$ and

$$f^{(n)}(z) = \frac{n!}{2\pi i} \iint_A f(\zeta)\varphi_{\bar{\zeta}}(\zeta) \frac{d\zeta \wedge d\bar{\zeta}}{(\zeta - z)^{n+1}}, \quad z \in K,$$

where $\varphi \in C^\infty(\Omega)$ has compact support with values identical equal to 1 in an open neighborhood of K and A may be chosen as supp $\varphi_{\bar{\zeta}}$.

3.4.3. The real and imaginary parts of an analytic function: Harmonic functions

In Sec. 3.4.2, we learned that the derivative function of an analytic function is also analytic and hence, all higher derivative functions exist and are analytic.

Hence, the real part $u = \operatorname{Re} f$ and the imaginary part $v = \operatorname{Im} f$ of an analytic function $f = u + iv$ are infinitely differentiable. In particular, partial derivatives $u_{xx}, u_{xy} = u_{yx}, u_{yy}, v_{xx}, v_{xy} = v_{yx}$, and v_{yy} all exist and are continuous. Now,

$$u_x = v_y \Rightarrow u_{xx} = v_{xy}, \quad v_{yy} = u_{xy};$$

$$u_y = -v_x \Rightarrow u_{yy} = -v_{yx} = -v_{xy}, \quad v_{xx} = -u_{yx} = -u_{xy};$$

$$\Rightarrow \Delta u = \frac{\partial^2 u}{\partial x^2} + \frac{\partial^2 u}{\partial y^2} = 0 \quad \text{and} \quad \Delta v = \frac{\partial^2 v}{\partial x^2} + \frac{\partial^2 v}{\partial y^2} = 0.$$

u and v are the so-called harmonic functions.

A real-valued function $u = u(z) = u(x,y)$ on a domain is said to be *harmonic* on Ω if u has continuous partial derivatives of order 2, namely, $u \in C^2(\Omega)$, and satisfies the *Laplaces equation*

$$\Delta u = \frac{\partial^2 u}{\partial x^2} + \frac{\partial^2 u}{\partial y^2} = 0 \quad \text{or} \quad r\frac{\partial}{\partial r}\left(r\frac{\partial u}{\partial r}\right) + \frac{\partial^2 u}{\partial \theta^2} = 0$$

$$\text{(in polar form) on } \Omega. \qquad (3.4.3.1)$$

If u and v are harmonic functions on Ω and satisfy the Cauchy–Riemann equations $u_x = v_y, u_y = -v_x$, then v is said to be a *conjugate harmonic function* or *harmonic conjugate* of u on Ω; in short, v is *conjugate harmonic* to u. *Such a v is unique up to an additive constant.* To see this, let $\tilde{v}$ be another conjugate harmonic function of u on Ω. Then both $u+iv$ and $u+i\tilde{v}$ are analytic on Ω. So is $(u + i\tilde{v}) - (u + iv) = i(\tilde{v} - v)$. By (2) in (3.4.1.4), it follows that $\tilde{v} - v = c$ is a constant and $\tilde{v} = v + c$ on Ω.

Suppose v is a conjugate harmonic function of u on Ω. *Then u is a conjugate harmonic function of $-v$ on Ω.* This follows easily from the fact that $i(u + iv) = -v + iu$ is analytic on Ω.

Summarize the above as

Analytic functions and harmonic functions (I): Let $f : \Omega(\text{domain}) \to \mathbf{C}$ be analytic.

(1) $\operatorname{Re} f = u$ and $\operatorname{Im} f = v$ are harmonic on Ω; v is a conjugate harmonic function of u on Ω, and u is a conjugate harmonic function of $-v$ on Ω (both are unique up to an additive constant).
(2) In case $f(z) \neq 0$ on Ω, then $\log |f(z)|$ is harmonic on Ω.
(3) If Ω is a single-valued domain for $\arg f(z)$, then each branch of $\arg f(z)$ is harmonic on Ω and is conjugate harmonic to $\log |f(z)|$. (3.4.3.2)

(2) and (3) are easy consequence of (3.3.1.8). Direct proofs of both are left as Exercise A (1).

A natural converse problem: Does a harmonic function u on a domain Ω have a conjugate harmonic function v (so that $u + iv$ is analytic on Ω)? The answer is not always in the affirmative. We present three elementary examples in what follows.

Example 1. Find a harmonic function $v(x,y)$, conjugate to $u(x,y) = x^2 - y^2$ on $\mathbf{C}$.

Solution. *Method* 1: Since $u_x = 2x$, $u_{xx} = 2$ and $u_y = -2y$, $u_{yy} = -2$, $\Delta u = 0$ and u indeed is harmonic on $\mathbf{C}$. Substitute $u_x = 2x$ and $u_y = -2y$ into the Cauchy–Riemann equations

$$v_x = -u_y = 2y, \quad v_y = u_x = 2x,$$

$$\Rightarrow v = 2xy + \varphi(y), \text{ where } \varphi(y) \text{ is a differentiable function of } y \text{ alone.}$$

$$\Rightarrow v_y = 2x + \varphi'(y) = 2x$$

$$\Rightarrow \varphi'(y) = 0 \text{ and hence, } \varphi(y) = c, \text{ a constant.}$$

Hence $v = 2xy + c$ is a required one and $u + iv = z^2 + ic$ is analytic on $\mathbf{C}$.

Method 2 (Some basic knowledge about line integration is needed (see Sec. 2.9)): Suppose $v(x,y)$ is a conjugate harmonic function of u on $\mathbf{C}$. Then,

$$dv = \frac{\partial v}{\partial x}dx + \frac{\partial v}{\partial y}dy = -\frac{\partial u}{\partial y}dx + \frac{\partial u}{\partial x}dy$$

$$= 2ydx + 2xdy = d(2xy), \quad \text{an exact form}$$

$$\Rightarrow v(x,y) = \int_{(0,0)}^{(x,y)} d(2xy) + c = 2xy + c, \quad \text{where } c \text{ is a constant.}$$

Since $2xy$ is C^1 on the simply connected domain $\mathbf{C}$, the path of integration can be chosen as any rectifiable curve connecting $(0,0)$ to (x,y); in particular, the polygonal path connecting $(0,0)$ to $(x,0)$ and then to (x,y). In this case,

$$v(x,y) = \int_{(0,0)}^{(x,y)} (2ydx + 2xdy) + c = \int_0^x 0dx + \int_0^y 2xdy + c$$

$$= \int_0^y 2xdy + c = 2xy + c.$$

Method 3: Suppose the existence of $v(x,y)$ is known so that $f(z) = u(z) + iv(z)$ is analytic on $\mathbf{C}$. Then

$$f'(z) = u_x(z) + iv_x(z) = u_x(z) - iu_y(z) = 2x + 2iy = 2z$$

implies that $f(z) = z^2 + c$ where c is a complex constant. Since $\operatorname{Re} f(z) = x^2 - y^2 + \operatorname{Re} c = u(x,y)$, $\operatorname{Re} c = 0$ and c is purely imaginary.

Remark 1 (Discussion of methods).   The essence of method 1 lies on the independence of $\varphi(y)$ on x, while that of method 2 is that the line integral $\int_{(0,0)}^{(x,y)} (2y\,dx + 2x\,dy)$ depends only on the end points $(0,\ 0)$ and (x,y) of the path of integration. Method 3 presumes the existence of a conjugate harmonic function. The main reason behind all these methods is that $u(x,y)$ is harmonic on a *simply connected* domain, such as **C**.

Method 3 can be formally reconstructed as follows. Still *suppose the existence of v, harmonic conjugate to u on Ω, so that $f(z) = u(z) + iv(z)$ is analytic on Ω. Let $z = x + iy$ and then $x = \frac{1}{2}(z + \bar{z})$ and $y = \frac{1}{2i}(z - \bar{z})$ so that both $u(x,y)$ and $v(x,y)$ can be treated as functions of z and $\bar{z}$ as we did in Sec. 2.8. Then the formal derivatives of $\overline{f(z)} = u(x,y) - iv(x,y)$ with respect to z is*

$$\frac{\partial \bar{f}}{\partial z} = \frac{1}{2}[(u_x - v_y) - i(u_y + v_x)] = 0 \quad \text{on } \Omega.$$

Hence, we treat $\bar{f}$ as a function of $\bar{z}$ only and rewrite it as $\bar{f}(\bar{z}) = u(z) - iv(z)$. We obtain the *formal* identity

$$u(x,y) = \frac{1}{2}[f(x + iy) + \bar{f}(x - iy)],$$

$$\Rightarrow \left(\text{letting } x = \frac{z}{2} \text{ and } y = \frac{z}{2i}\right) \ u\left(\frac{z}{2}, \frac{z}{2i}\right) = \frac{1}{2}[f(z) + \bar{f}(0)],$$

$$\Rightarrow f(z) = 2u\left(\frac{z}{2}, \frac{z}{2i}\right) - u(0,0) \qquad\qquad (3.4.3.3)$$

by noting that $f(z)$ *is uniquely determined by $u(z)$; up to a purely imaginary constant*, so that $\bar{f}(0)$ might be chosen as a real constant and then $\bar{f}(0) = f(0) = u(0,0)$. $\qquad\square$

Example 2. Determine the most general form of harmonic function $u(x,y) = ax^3 + bx^2 y + cxy^2 + dy^3$, where $a, b, c,$ and d are real constants, and its conjugate harmonic functions $v(x,y)$.

Solution. Three methods in Example 1 are still applicable.

$\Delta u = (6a + 2c)x + (2b + 6d)y = 0$ on **C** implies that $6a + 2c = 2b + 6d = 0$. So $c = -3a$ and $b = -3d$. Hence $u(x,y) = ax^3 - 3dx^2 y - 3axy^2 + dy^3$

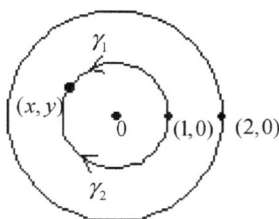

Fig. 3.30

is its most general form. The required

$$f(z) = 2\left[a\left(\frac{z}{2}\right)^3 - 3d\left(\frac{z}{2}\right)^2\left(\frac{z}{2i}\right) - 3a\left(\frac{z}{2}\right)\left(\frac{z}{2i}\right)^2 + d\left(\frac{z}{2i}\right)^3\right] = (a+di)z^3,$$

up to a purely imaginary constant (see (3.4.3.3)).

Example 3. The harmonic function $u(x,y) = \log|z| = \frac{1}{2}\log(x^2+y^2)$ does not have a *single-valued* conjugate harmonic function on $0 < |z| < 2$.

Solution. Since $u_x = \frac{x}{x^2+y^2}$ and $u_y = \frac{y}{x^2+y^2}$,

$$\int_{|z|=1} -u_y dx + u_x dy = \int_0^{2\pi} (\sin^2\theta + \cos^2\theta)d\theta = 2\pi$$

$$= \int_{\gamma_1} -u_y dx + u_x dy + \int_{-\gamma_2} -u_y dx + u_x dy$$

$$\Rightarrow \text{(see Fig. 3.30)} \quad \int_{\gamma_1} -u_y dx + u_x dy$$

$$- \int_{\gamma_2} -u_y dx + u_x dy = 2\pi. \tag{$*_1$}$$

If γ represents any rectifiable curve, in $0 < |z| < 2$, connecting $(1,0)$ to (x,y), then the value of the line integral

$$\int_\gamma -u_y dx + u_x dy$$

depends on γ chosen as indicated in $(*_1)$.

On the contrary, suppose $u(z)$ has a *single-valued* conjugate harmonic function $v(z)$ on $0 < |z| < 2$. Owing to $v_x = -u_y$ and $v_y = u_x$,

$$\int_\gamma -u_y dx + u_x dy = \int_\gamma v_x dx + v_y dy = \int_\gamma dv = v(x,y) - v(1,0)$$

shows that the value of the line integral depends only on the end points $(1, 0)$ and (x, y) of the path, independent of the path γ chosen. This contradiction says that such a v cannot exist.

We formulate the following general results, based on Examples 1–3.

Analytic function and harmonic function (II). Suppose $u(z) = u(x, y)$ is harmonic on a domain Ω.

(1) $f(z) = u_x(z) - iu_y(z)$ is analytic on Ω (this is the most natural way of passing a harmonic function to an analytic one).

(2) $u(z)$ has a conjugate harmonic function $v(z)$ on Ω if and only if, for any rectifiable (or piecewise differentiable) closed curve γ in Ω,

$$\int_\gamma -u_y dx + u_x dy = 0$$

always holds. In this case, fix any point $z_0 = x_0 + iy_0$ in Ω, the function $v(z)$ defined by

$$v(z) = \int_{z_0}^z -u_y dx + u_x dy, \quad z = x + iy \in \Omega$$

is well defined (i.e., independent of paths in Ω, connecting z_0 to z) and is harmonic on Ω, conjugate to $u(z)$.

(3) $u(z)$ always has a (single-valued) conjugate harmonic function $v(z)$ on Ω if Ω is simple-connected.   (3.4.3.4)

Proof. (1) is obvious.

(2) Suppose $u(z)$ has a conjugate harmonic function $v(z)$ on Ω. Then $v_x = -u_y$ and $v_y = u_x$ hold on Ω and hence, $dv = v_x dx + v_y dy = -u_y dx + u_x dy$ results in

$$\int_\gamma -u_y dx + u_x dy = \int_\gamma dv = v(\text{terminal point}) - v(\text{initial point}) = 0 \quad (*_2)$$

for any rectifiable closed curve γ in Ω.

Conversely, suppose $(*_2)$ holds. Choose two arbitrary rectifiable curves γ_1 and γ_2, connecting z_0 to z and lying entirely in Ω. Then $\gamma_1 - \gamma_2 = \gamma_1 + (-\gamma_2)$ is a closed curve in Ω and

$$\int_{\gamma_1 - \gamma_2} -u_y dx + u_x dy = \int_{\gamma_1} -u_y dx + u_x dy - \int_{\gamma_2} -u_y dx + u_x dy = 0,$$

$$\Rightarrow \int_{\gamma_1} -u_y dx + u_x dy = \int_{\gamma_2} -u_y dx + u_x dy.$$

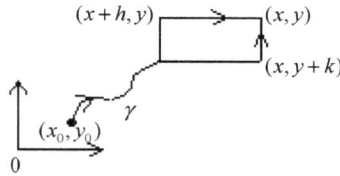

Fig. 3.31

This means that $v(z)$ is indeed well defined. Choose γ so that its last part near $z = x + iy$ is a horizontal segment connecting $(x + h, y)$ to (x, y) as shown in Fig. 3.31. In this case,

$$v(z) = \int_{z_0}^{z+h} -u_y dx + u_x dy + \int_{z+h}^{z} -u_y dx + u_x dy = v(z + h) - \int_{z+h}^{z} u_y dx$$

$$\Rightarrow \frac{v(z+h) - v(z)}{h} = -\frac{1}{h} \int_{z}^{z+h} u_y dx = -u_y(z + \theta h),$$

where $0 < \theta < 1$ and h is real.

$$\Rightarrow \lim_{h \to 0} \frac{v(z+h) - v(z)}{h} \text{ exists as } v_x(z) = -u_y(z).$$

Similarly, $v_y(z) = u_x(z)$ also holds. Recall that $u \in C^2(\Omega)$. Hence $v_{xx}(z) = -u_{xy}(z) = -u_{yx}(z) = -v_{yy}(z)$ implies that $\Delta v = 0$. This proves that $v(z)$ so defined is a conjugate harmonic function of $u(z)$ on Ω.

(3) Let $f(z) = u_x - iu_y$ be as in (1). Then, the differential (see Sec. 2.8)

$$f(z)dz = (u_x - iu_y)(dx + idy) = (u_x dx + u_y dy) + i(-u_y dx + u_x dy)$$
$$= du(x, y) + i(-u_y dx + u_x dy)$$

$\Rightarrow$ For any rectifiable (or piecewise differentiable) closed curve γ in Ω,

$$\int_{\gamma} f(z)dz = \int_{\gamma} du(x, y) + i \int_{\gamma} -u_y dx + u_x dy = i \int_{\gamma} -u_y dx + u_x dy$$

$$= 0, \quad \text{by observing that } \Omega \text{ is simply connected and using the}$$
$$\text{Cauchy integral theorem (see (3.4.2.1)).}$$

And the result follows by (2). $\qquad \square$

We list the following

Operational properties of harmonic functions. Let Ω be a domain.

(1) *Linear property:*

    (i) If $u(z)$ is harmonic on Ω, then so is $\alpha u(z)$ for any real constant α.
    (ii) If $u_1(z)$ and $u_2(z)$ are harmonic on Ω, then so is $u_1(z) + u_2(z)$.

(2) *Invariance under analytic composition*: Let $f : \Omega \to \mathbf{C}$ be analytic and $u : f(\Omega)$(a domain) $\to \mathbf{R}$ is harmonic. Then, the composite function $u \circ f$ is harmonic on Ω (refer to Exercises A (13) and (15)).

(3) Let $u(z)$ be harmonic on Ω. For any point $z_0 \in \Omega$ and every $r > 0$ so that the closed disk $|z - z_0| \leq r$ is contained in Ω, then

$$u(z_0) = \frac{1}{2\pi} \int_0^{2\pi} u(z_0 + re^{i\theta})d\theta,$$

called the *mean-value property* of harmonic function $u(z)$ at z_0.

$$(3.4.3.5)$$

Proof. (1) and (2) are left to the readers as Exercise A (2).

(3) According to (3) and then, (2) in (3.4.3.4), there exists a harmonic function $v(z)$ on $|z - z_0| < \rho$ so that $f(z) = u(z) + iv(z)$ is analytic there, where $\rho > r$, and $\{|z - z_0| < \rho\} \subseteq \Omega$ holds. By the Cauchy integral formula (see (3.4.2.4), (3.4.2.5), or Exercise A (33) of Sec. 3.4.2),

$$f(z_0) = \frac{1}{2\pi i} \int_{|\zeta - z_0| = r} \frac{f(\zeta)}{\zeta - z_0} d\zeta$$

$$= \frac{1}{2\pi} \int_0^{2\pi} f(z_0 + re^{i\theta})d\theta, \quad \text{by noting that } \zeta = z_0 + re^{i\theta},$$

$$0 \leq \theta \leq 2\pi.$$

After equating the real parts of both sides, the result follows.   □

Remark 2.   Results concerned with analytic functions are still valid for a pair of harmonic functions, conjugate to each other. But, when dealing with *multiple-valued* conjugate harmonic functions, it is not necessarily beneficial to adopt complex analysis method. Theory of harmonic functions can be singled out as an independent topic in mathematics and be extended to harmonic functions of n variables.

By the way, the definition of harmonicity can be essentially weakened to the following. A real-valued continuous function $u(z)$ on a domain is said to be *harmonic* if it enjoys the *mean-value property*: for any $z_0 \in \Omega$ and any closed disk $|z - z_0| \leq r$ contained in Ω,

$$u(z_0) = \frac{1}{2\pi} \int_0^{2\pi} u(z_0 + re^{i\theta})d\theta \qquad (3.4.3.6)$$

always holds. This new definition will be equivalent to the classical one in (3.4.3.1) (see (6.3.2.12)).   □

Finally, we have the important

Maximum–minimum principle for harmonic functions. Suppose $u(z)$ is a *nonconstant* harmonic function on a domain Ω. Then,

(1) $u(z)$ does not attain its *local* maximum or local minimum at any point in Ω.
(2) In case Ω is a bounded domain and $u(z)$ is continuous on $\bar{\Omega}$, then $u(z)$ will attain its *global* maximum and minimum on $\bar{\Omega}$ at the boundary $\partial\Omega$ of Ω.   (3.4.3.7)

This principle is equivalent to *the maximum–minimum principle for the modulus* and *the open mapping property* of an analytic function. For details, see Sec. 3.4.4. In (3.4.3.7), (1) is equivalent to the fact that a *harmonic function assuming its local maximum or local minimum at a point in its domain of definition should be a constant.* In Exercises B of Secs. 2.2, 2.6.1, and 2.6.2, we had arranged some elementary functions to practice this principle.

Proof. Suppose that there exist a point $z_0 \in \Omega$ and $r > 0$ so that $|z - z_0| \leq r$ is contained in Ω and

$$u(z_0) = \max_{|z-z_0|\leq r} u(z).$$

$$\Rightarrow \text{For any } 0 < \rho \leq r \quad \text{and} \quad z = z_0 + \rho e^{i\theta}, \quad 0 \leq \theta \leq 2\pi,$$

$$u(z) \leq u(z_0) \text{ always holds.}$$

We try to show that $u(z) = u(z_0)$ on $|z - z_0| = \rho$ for $0 < \rho \leq r$ and hence, $u(z) = u(z_0)$ on $|z - z_0| \leq r$.

If not, then there is a $\theta_1 (0 \leq \theta_1 \leq 2\pi)$ so that $u(z_0 + \rho e^{i\theta_1}) < u(z_0)$ holds for some ρ, $0 < \rho \leq r$. By continuity of $u(z)$ along $|z - z_0| = \rho$, there exists an arc $z = z_0 + \rho e^{i\theta}$, $\theta_1 - \delta < \theta < \theta_1 + \delta(\delta > 0)$ on which $u(z) < u(z_0)$ holds. Let γ be the remaining arc on $|z - z_0| = \rho$ other than the arc $\theta_1 - \delta < \theta < \theta_1 + \delta$. Using the mean-valued property in (3.4.3.5),

$$u(z_0) = \frac{1}{2\pi}\int_0^{2\pi} u(z_0 + \rho e^{i\theta})d\theta = \frac{1}{2\pi}\int_{\theta_1-\delta}^{\theta_1+\delta} u(z_0 + \rho e^{i\theta})d\theta$$

$$+ \frac{1}{2\pi}\int_\gamma u(z_0 + \rho e^{i\theta})d\theta$$

$$< \frac{2\delta}{2\pi}u(z_0) + \frac{2\pi - 2\delta}{2\pi}u(z_0) = u(z_0).$$

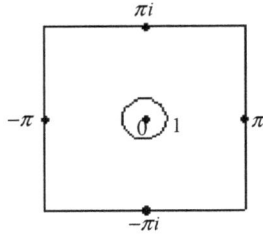

Fig. 3.32

This contradiction shows that $u(z) = u(z_0)$ on $|z - z_0| = \rho$ $(0 < \rho \le r)$ and hence, $u(z) = u(z_0)$ on the whole disk $|z - z_0| \le r$.

Since the set $\{z \in \Omega \,|\, u(z) = u(z_0)\}$ is a nonempty open and closed subset of the connected set Ω, it is necessary that $u(z) = u(z_0)$ on the whole domain Ω. This proves the maximum principle.

As for the minimum principle, apply known result to the harmonic function $-u(z)$ and the result follows.   □

Example 4. Let S denote the closed square domain $|x| \le \pi$, $|y| \le \pi$, where $z = x + iy$.

(1) Find the global maximum and minimum of $u(x, y) = \cos x \cosh y$ on S.
(2) Find the global maximum and minimum of $|f(z)|$, where $f(z) = \cos z$, on $|z| \le 1$ and on $\Omega = S - \{|z| < 1\}$, respectively. See Fig. 3.32.

Solution. (1) On $|x| < \pi$, $|y| < \pi$, $u_x = -\sin x \cosh y = 0$ and $u_y = \cos x \sinh y = 0$ do have only one common solution $x = y = 0$. While the Hessian matrix of $u(x, y)$ at $(0,0)$ is

$$\begin{bmatrix} u_{xx}(0,0) & u_{xy}(0,0) \\ u_{xy}(0,0) & u_{yy}(0,0) \end{bmatrix} = \begin{bmatrix} -1 & 0 \\ 0 & 1 \end{bmatrix}$$

with its determinant having value equal to -1. The origin is a saddle point of $u(x, y)$, and thus $u(0, 0)$ is neither a local maximum nor a local minimum of $u(x, y)$.

Along the boundary vertical sides $z = \pm\pi + iy$, $-\pi \le y \le \pi$: $u(\pm\pi, y) = -\cosh y$ has its maximum $u(\pm\pi, 0) = -1$ at $y = 0$ and its minimum $u(\pm\pi, \pm\pi) = -\cosh \pi$ at $y = \pm\pi$. While along the horizontal sides $z = x \pm i\pi$, $-\pi \le x \le \pi$: $u(x, \pm\pi) = \cosh \pi \cos x$ has its maximum $u(0, \pm\pi) = \cosh \pi$ at $x = 0$ and its minimum $u(\pm\pi, \pm\pi) = -\cosh \pi$ at

$x = \pm\pi$. Thus,

$$\max_{z \in S} u(z) = u(0, \pm\pi) = \cosh\pi \quad \text{and} \quad \min_{z \in S} u(z) = u(\pm\pi, \pm\pi) = -\cosh\pi.$$

(2) Recall that $|f(z)| = (\cos^2 x + \sinh^2 y)^{\frac{1}{2}}$ (see (2.6.2.6)). The extremes of $|f(z)|$ and of $|f(z)|^2$ will correspond, because $|f(z)|$ is nonnegative. Consider $g(z) = |f(z)|^2 = \cos^2 x + \sinh^2 y$ in what follows.

Observe that $g(z) \neq 0$ on $|z| < 1$. According to max–min principle (refer to (3.4.4.2), if necessary), $g(z)$ cannot attain its local extremes on $|z| < 1$. This claim can be checked as follows. As

$$g_x = -\sin 2x, \quad g_y = \sinh 2y$$

we see that $g_x = g_y = 0$ occurs only at $z = 0$ in $|z| < 1$. Now $g_{xx} = -2\cos 2x$, $g_{xy} = 0$ and $g_{yy} = 2\cosh 2y$, so the Hessian matrix of $g(z)$ at $z = 0$ is

$$\begin{bmatrix} g_{xx}(0) & g_{xy}(0) \\ g_{xy}(0) & g_{yy}(0) \end{bmatrix} = \begin{bmatrix} -2 & 0 \\ 0 & 2 \end{bmatrix}$$

whose determinant has value equal to -4. Thus, $z = 0$ is a saddle point of $g(z)$ and is not an extreme point of $g(z)$. Along the circle $|z|^2 = x^2 + y^2 = 1$, $g(x, y) = \cos^2 x + \sinh^2\sqrt{1 - x^2}$, $|x| \leq 1$. Both $\cos^2 x$ and $\sinh^2\sqrt{1 - x^2}$ are strictly decreasing on $0 \leq |x| \leq 1$, so the maximum of $g(z)$ occurs at $x = 0$, $y = \pm 1$ and the minimum at $x = \pm 1$, $y = 0$. Hence,

$$\max_{|z|=1} |f(z)| = |f(\pm i)| = \cosh 1 \quad \text{and} \quad \min_{|z| \leq 1} |f(z)| = |f(0)| = \cos 1.$$

On $\Omega = S - \{|z| < 1\}$, $f(z)$ has zeros at $z = \pm\frac{\pi}{2}$. So $|f(z)|$ has local minimum 0 at $z = \pm\frac{\pi}{2}$ and does not have local maximum on $\mathrm{Int}\,\Omega$ (why?). Along the line segments $z = \pm\pi + yi$, $-\pi \leq y \leq \pi$, $|f(z)| = (1 + \sinh^2 y)^{\frac{1}{2}}$ has minimum 1 at $y = 0$ and maximum $(1 + \sinh^2\pi)^{\frac{1}{2}}$ at $y = \pi$; while, along the line segments $z = x \pm \pi i$, $-\pi \leq x \leq \pi$, $|f(z)| = (\cos^2 x + \sinh^2\pi)^{\frac{1}{2}}$ has minimum $\sinh\pi$ at $x = \pm\frac{\pi}{2}$ and maximum $(1 + \sinh^2\pi)^{\frac{1}{2}}$ at $x = 0$. Note that the boundary $\partial\Omega$ is the disjoint union of the boundary of S and the unit circle $|z| = 1$. Thus,

$$\max_{z \in \Omega} = |f(z)| = |f(\pm\pi \pm \pi i)| = |f(\pm\pi i)| = (1 + \sinh^2\pi)^{\frac{1}{2}} = \cosh\pi;$$

$$\min_{z \in \Omega} |f(z)| = \left| f\left(\pm\frac{\pi}{2} \right) \right| = 0.$$

Exercises A

(1) Prove (2) and (3) of (3.4.3.2) directly by the very definition (3.4.3.1) of harmonic functions.
(2) Prove (1) and (2) of (3.4.3.5).
(3) Prove the polar form of Laplace's equation $\Delta u = 0$ as shown in (3.4.3.1). Note that this form cannot be used for $r = $ constant.

    (a) Show that both $\log r$ and θ (a branch on a domain) are harmonic functions.

    (b) Show that any harmonic function depending only on r is of the form $a \log r + b$, where a and b are real constants.

    (c) Show that, for any positive integer n, $r^n \cos n\theta$ and $r^n \sin n\theta$ are harmonic functions. Are they conjugate to each other?

    (d) Suppose R and φ are constants. Show that

$$\frac{R^2 - r^2}{R^2 - Rr\cos(\theta - \varphi) + r^2}$$

    is harmonic.

(4) Show that $ax + by$, where a and b are real constants, is the simplest nonconstant harmonic function.
(5) Determine analytic function $f(z) = u(z) + iv(z)$, where $u(z)$ or $v(z)$ is given below.

    (a) $u(z) = \dfrac{y}{x^2 + y^2}$, $f(2) = 0$.

    (b) $v(z) = (x - y)(x^2 + 4xy + y^2)$.

    (c) $u(z) = \dfrac{x^2 - y^2}{(x^2 + y^2)^2}$.

    (d) $u(z) = \dfrac{(1 + x^2 + y^2)x}{1 + 2(x^2 - y^2) + (x^2 + y^2)^2}$.

    (e) $v(z) = x(\sin x)\left(\dfrac{e^y + e^{-y}}{2}\right) - y(\cos x)\left(\dfrac{e^y + e^{-y}}{2}\right)$.

    (f) $u(z) = [(x^2 - y^2)\sin y - 2xy \cos y]e^x$.

    (g) $u(z) = e^x(x \cos y - y \sin x)$.

    (h) $u(z) = \dfrac{1}{2}\{\log(x^2 + y^2) - \log[(x - 1)^2 + y^2]\}$ on $\mathbf{C} - [0, 1]$.

    (i) $u(z) = \log(x^2 + y^2) - x^2 + y^2$ on $\mathbf{C} - (-\infty, 0]$.

(6) Let Ω^* be the symmetric domain of the domain Ω with respect to the real axis. Show that $u(z)$ is harmonic on Ω if and only if $u(\bar{z})$ is harmonic on Ω^*.

(7) Show that u^2 cannot be harmonic if $u = u(z)$ is a nonconstant harmonic function.

(8) Suppose f is analytic on a domain Ω so that $|f(z)|$ is harmonic on Ω. Show that f is a constant.

(9) Suppose $f(z) = u + iv$ and $(f(z))^2 = u^2 - v^2 + 2uvi$ are *complex* harmonic functions (namely, $f_{xx} + f_{yy} = 0$, etc.). Show that either f or $\bar{f}$ is analytic.

(10) Suppose $u = u(z)$ is harmonic on a simply connected domain Ω on which $u(z) \neq 0$. Show that there exist two harmonic functions u_1 and u_2 on Ω such that $u = u_1^2 - u_2^2$.

(11) Let $u = u(z)$ be harmonic. What kind of real-valued function φ of a real variable will guarantee that $\varphi \circ u$ is still harmonic? Does there exist a harmonic function of the following form? If it is, try to find a harmonic function conjugate to it.

(a) $u = \varphi\left(\dfrac{y}{x}\right)$.

(b) $u = \varphi(\log(x^2 + y^2))$.

(c) $u = \varphi\left(\dfrac{x}{x^2 + y^2}\right)$.

(d) $u = \varphi(x + \sqrt{x^2 + y^2})$.

(e) $u = \varphi(x^2 + y^2)$.

(f) $u = \varphi(x^2 + y)$.

(12) Suppose v is a conjugate harmonic function of u and u is also a conjugate harmonic function of v. Then both u and v are constants.

(13) Suppose $u = u(x, y)$ is harmonic. If $x = x(\zeta, \eta)$ and $y = y(\zeta, \eta)$ are harmonic functions so that $x + iy$ is analytic, then $u(x(\zeta, \eta), y(\zeta, \eta))$ is harmonic in ζ and η. See (2) in (3.4.3.5).

(14) Let u be harmonic on a domain Ω.

(a) Show that $\dfrac{\partial u}{\partial z} = \dfrac{1}{2}\left(\dfrac{\partial u}{\partial x} - i\dfrac{\partial u}{\partial y}\right)$ is harmonic on Ω (refer to Exercise (9)).

(b) Show that $\dfrac{\partial u}{\partial \bar{z}} = \dfrac{1}{2}\left(\dfrac{\partial u}{\partial x} + i\dfrac{\partial u}{\partial y}\right)$ has its conjugate function $\overline{\dfrac{\partial u}{\partial z}}$ harmonic on Ω.

(15) Suppose $f(z) = u(z) + iv(z)$ is analytic on a domain Ω and $\varphi = \varphi(x, y)$, $z = x + iy$, is sufficiently differentiable with equal mixed partial derivatives on Ω. Let $\psi = \varphi \circ f$. Show that

(i) The gradient $\nabla\psi = \psi_x + i\psi_y = (\varphi_u + i\varphi_v)\overline{f'(z)}$.

(ii) $\psi_x^2 + \psi_y^2 = (\varphi_u^2 + \varphi_v^2)|f'(z)|^2$.

(iii) The Laplacian $\Delta\psi = (\varphi_{uu} + \varphi_{vv})|f'(z)|^2$.

(16) Show that $u(x^2-y^2, 2xy)$ is harmonic if and only if $u(x,y)$ is harmonic.

(17) Let u be harmonic on a domain Ω and $u_y(x,y) \neq 0$ always holds. Show that $w = \tan^{-1}\frac{u_x}{u_y}$ is harmonic on Ω.

(18) Suppose u is harmonic on a domain Ω such that u is a homogeneous function of degree m, namely, $u(\lambda x, \lambda y) = \lambda^m u(x,y)$ holds. Show that $v = m^{-1}(yu_x - xu_y)$ is harmonic on Ω, conjugate to u.

(19) Suppose $f_j(z), 1 \le j \le n$, are analytic on a domain Ω.

   (a) If $\sum_{j=1}^{n} |f_j(z)|^2 = c$ (a constant) on Ω, then each $f_j(z), 1 \le j \le n$, is a constant function on Ω.

   (b) Show that $\sum_{j=1}^{n} |f_j(z)|^2$ is harmonic on Ω if and only if $f_j(z), 1 \le j \le n$, are constants.

(20) Suppose f is analytic on a domain Ω.

   (a) Let $w = |f(z)|^2$. Show that $w\Delta w = w_x^2 + w_y^2$ on Ω.

   (b) Show that both $u = \cos(\text{Im } f(z))$ and $v = \sin(\text{Im } f(z))$ are solutions of the equation $\Delta\varphi + |f'(z)|^2\varphi = 0$.

   (c) Suppose further that $f'(z) \neq 0$ and $|f(z)| < 1$ on Ω. Show that $w = \log\frac{|f'(z)|}{1-|f(z)|^2}$ satisfies $\Delta w = 4e^{2w}$ on Ω.

(21) Find the global maxima and minima of $\text{Re } f(z), \text{Im } f(z)$, and $|f(z)|$ for each of the following analytic functions $f(z)$ on the indicated domains Ω.

   (a) $f(z) = z^2 - 2z$ on $\Omega = \{z\,|\,0 \le \text{Re } z \le 1 \text{ and } 0 \le \text{Im } z \le 1\}$.

   (b) $f(z) = e^{z^2}$ on $\Omega : |z| < 1$.

   (c) $f(z) = \sin z$ on $\Omega : [0, 2\pi] \times [0, 2\pi]$.

   (d) $f(z) = e^z$ on $\Omega : |z - z_0| < 1$.

   (e) $f(z) = z^2 + 4$ on $\Omega : |z| < 1$.

   (f) $f(z) = (z+3)^2$ on the triangle (interior included) with vertices at 0, 1, and $-i$.

   (g) $f(z) = \dfrac{1}{z+4}$ on $\Omega : 1 \le |z| \le 2$.

(22) Find a harmonic conjugate $v(z)$ of $u(z) = \text{Arg}\frac{1+z}{1-z}$ on $|z| < 1$. Show that $v(z)$ is unbounded while $u(z)$ is bounded. What happens if

$$u(z) = \text{Arg}\frac{z-a}{1-\bar{a}z}, \quad |a| < 1, \quad |z| < 1?$$

Is max–min principle wrong in these cases?

(23) Let $x_0 \in \mathbf{R}$ be a fixed point. Suppose $u(z)$ is harmonic on $|z - x_0| < r$ so that $u(x, 0) = 0$ holds on $x_0 - r < x < x_0 + r$. Show that

$$u(z) = -u(\bar{z}), \quad |z - z_0| < r.$$

(24) Suppose that $u(z)$ is harmonic on a domain Ω. If there exists a sub-domain Ω_0 of Ω on which $u(z) = 0$, show that $u(z) = 0$ on Ω.

(25) A real-valued function $u(z)$ is said to be *harmonic* at point $z_0 \in \mathbf{C}$ if $u(z)$ is harmonic in a neighborhood z_0; in case $z_0 = \infty$, *harmonic* at ∞ if $u\left(\frac{1}{z}\right)$ is *harmonic* at 0.

    (a) Show that $u(z)$ is harmonic at ∞ if and only if $u\left(z_0 + \frac{a}{z}\right)$ is harmonic at $z = 0$, where $a \neq 0$ and z_0 are arbitrary constants.

    (b) Suppose that $u(z)$ is harmonic on Ext γ, where γ is a Jordan closed curve. Note that $\infty \in$ Ext γ. If $u(z)$ is continuous on $\gamma \cup$ Ext γ and is constant along γ, then $u(z)$ is a constant on Ext γ.

Exercises B

(1) Suppose $u(z) = u(x, y)$ is a bounded harmonic function on Im $z > 0$ and is continuous on Im $z \geq 0$ such that $u(x, 0) = 0$ for all $x \in \mathbf{R}$. Show that $u(z) = 0$ holds on Im $z \geq 0$. Is the boundedness of $u(z)$ a necessary condition?

(2) Suppose $u(z) = u(x, y)$ is harmonic on Im $z > 0$, continuous on Im $z \geq 0$ and assume constant value α along the boundary Im $z = 0$. In case $\lim_{z \to \infty} u(z) = \alpha$ holds, show that $u(z) = \alpha$ on Im $z \geq 0$.

(3) Suppose $u(z)$ is harmonic on $\mathbf{C}$ and bounded above, namely $u(z) \leq M, z \in \mathbf{C}$, for some constant M. Show that $u(z)$ is a constant.

(4) Suppose $u(z)$ is harmonic on $\mathbf{C}$ and $\lim_{z \to \infty} u(z) = 0$. Show that $u(z) \equiv 0$ on $\mathbf{C}$.

(5) Suppose $u(z)$ is harmonic on a domain Ω. Show that there exists an analytic function $f(z)$, not necessarily single-valued, on Ω so that

$$u(z) = \operatorname{Re} f(z).$$

Try to use $u(z) = \frac{1}{2} \log(x^2 + y^2)$, $z \in \mathbf{C} - \{0\}$, to justify this result and determine $f(z)$ explicitly.

(6) Suppose that $u(z)$ is harmonic on $\mathbf{C}$ and $u(z) \geq 0$ everywhere. Show that $u(z)$ is a constant.

3.4.4.   The maximum–minimum principle and the open mapping property

As important by themselves both in proofs and in applications, we single out this subsection to study these two peculiar properties of a nonconstant analytic function, even though they might be derived easily from the max–min principle for harmonic functions stated in (3.4.3.7). We had used elementary functions to preview them in Exercise B (2) of Sec. 2.2, and Exercise B of Secs. 2.5.1, 2.6.1, and 2.6.2.

Section (1) Basic theorems

Suppose f is analytic at a point z_0 and is not a constant function. According to (3.4.2.6) or (3.4.2.9), there exist a positive integer n and a $\rho > 0$ such that

$$f(z) = f(z_0) + (z - z_0)^n \{a_n + \varphi(z)\}, \quad |z - z_0| < \rho, \tag{$*_1$}$$

where $a_n = \dfrac{f^{(n)}(z_0)}{n!} \neq 0$ and $\varphi(z) = \sum_{k=1}^{\infty} \dfrac{1}{(n+k)!} f^{(n+k)}(z_0)(z - z_0)^k$ is analytic on $|z - z_0| < \rho$ with $\varphi(z_0) = 0$. This relation imitates closely that property of a polynomial (see (2.5.1.3)). Hence, the way we adopted in Exercise B (2) of Sec. 2.2 for polynomials can be modified slightly to prove the general case for analytic functions.

Now, $(*_1)$ implies that

$$|f(z)| \geq |f(z_0) + a_n(z - z_0)^n| - |\varphi(z)||z - z_0|^n, \quad |z - z_0| < \rho. \tag{$*_2$}$$

We may suppose $f(z_0) \neq 0$. Choose $0 < r < \rho$ so that $|\varphi(z)| < \dfrac{|a_n|}{2}$ on $|z - z_0| < r$. We try to pick up a point z_1 in $0 < |z - z_0| < r$ so that $|f(z_0) + a_n(z_1 - z_0)^n| = |f(z_0)| + |a_n||z_1 - z_0|^n$ holds. This could happen if and only if $\operatorname{Arg} f(z_0) = \operatorname{Arg}(a_n(z_1 - z_0)^n)$ (see (1.4.2.2)$'$). All we need to do is to choose z_1 satisfying $z_1 \neq z_0$ and $\operatorname{Arg}(z_1 - z_0) = \frac{1}{n}(\operatorname{Arg} f(z_0) - \operatorname{Arg} a_n)$. In this case, set $z = z_1$ in $(*_2)$ and we have

$$
\begin{aligned}
|f(z_1)| &\geq |f(z_0) + a_n(z_1 - z_0)^n| - |\varphi(z_1)||z_1 - z_0|^n \\
&= |f(z_0)| + |a_n||z_1 - z_0|^n - |\varphi(z_1)||z_1 - z_0|^n \\
&> |f(z_0)| + |a_n||z_1 - z_0|^n - \frac{1}{2}|a_n||z_1 - z_0|^n \\
&= |f(z_0)| + \frac{1}{2}|a_n||z_1 - z_0|^n \\
&> |f(z_0)| \quad (\text{since } a_n \neq 0 \text{ and } |z_1 - z_0| > 0). \tag{$*_3$}
\end{aligned}
$$

This indicates that $|f(z)|$ cannot attain its local maximum at z_0. In case $f(z_0) = 0$, since f is not identically equal to zero, there will exist a point z_1 near z_0 so that $f(z_1) \neq 0$ and $|f(z_1)| > 0 = |f(z_0)|$ still holds.

On the other hand, if $f(z_0) \neq 0$ *does hold*, then there exists z_2, $0 < |z_2 - z_0| < r$ so that

$$|f(z_0) + a_n(z_2 - z_0)^n| = |f(z_0)| - |a_n| \, |z_2 - z_0|^n$$

or

$$\text{Arg}(z_2 - z_0) = \frac{\pi}{n} + \frac{1}{n}(\text{Arg}\, f(z_0) - \text{Arg}\, a_n).$$

In this case, from $(*_1)$

$$
\begin{aligned}
|f(z_2)| &\leq |f(z_0) + a_n(z_2 - z_0)^n| + |\varphi(z_2)(z_2 - z_0)^n| \\
&= |f(z_0)| - |a_n| \, |z_2 - z_0|^n + |\varphi(z_2)| \, |z_2 - z_0|^n \\
&< f(z_0) - |a_n| \, |z_2 - z_0|^n + \frac{1}{2}|a_n| \, |z_2 - z_0|^n \\
&= |f(z_0)| - \frac{1}{2}|a_n| \, |z_2 - z_0|^n \\
&< |f(z_0)|.
\end{aligned}
\tag{$*_4$}
$$

Therefore, $|f(z)|$ does not attain its local minimum at z_0, too. In case $r > 0$ is so small that $f(z) \neq 0$ on $|z - z_0| < r$, we may apply known result from the last paragraph to $\frac{1}{f(z)}$ to obtain the same conclusion.

We summarize the above as the following important

Local maximum–minimum principle of analytic function. Suppose f is analytic at a point z_0 and is *not* a constant function.

(1) Maximum modulus principle. $|f(z)|$ cannot attain its local maximum at z_0.
(2) Minimum modulus principle. Suppose $f(z_0) \neq 0$, then $|f(z)|$ cannot attain its local minimum at z_0. $\hspace{2em}$ (3.4.4.1)

It is worthy to give three more *proofs*. We sketch as follows.

Proof 1. Since analytic function has the mean value property: $f(z_0) = \frac{1}{2\pi} \int_0^{2\pi} f(z_0 + re^{i\theta}) d\theta$, $0 \leq r < \rho$ (see (3.4.2.5)), the proof for the maximum principle of harmonic function (see (3.4.3.7)) is still applicable to $|f(z)|$. All we need to do is to replace $u(z)$ there by $|f(z)|$.

Proof 2. Suppose $f(z) \neq 0$ on $|z - z_0| < r$ so that $\log |f(z)|$ is harmonic there (see (3.4.2.2)). Applying (3.4.3.7) to $\log |f(z)|$, there exist z_1 and z_2 in $|z - z_0| < r$ so that

$$\log |f(z_2)| < \log |f(z_0)| < \log |f(z_1)|,$$
$$\Rightarrow |f(z_2)| < |f(z_0)| < |f(z_1)|.$$

And the result follows.

Proof 3. Use Gutzmer's inequality. See Exercise B (2) (b) of Sec. 3.4.2. □

By using of the interior uniqueness theorem (see (4) in (3.4.2.9)), we can rewrite (3.4.4.1) as follows.

Maximum–minimum modulus principle of analytic function. Suppose f is analytic on a domain Ω.

(1) (a) Maximum. $|f(z)|$ cannot attain local maximum at any point of Ω except f is a constant.
     (b) Minimum. Suppose $f(z) \neq 0$ on Ω. Then $|f(z)|$ cannot attain local minimum at any point of Ω except f is a constant.
(2) Suppose Ω is a *bounded* domain so that f is continuous on $\overline{\Omega}$ (and analytic in Ω).

     (a) Global maximum. $|f(z)|$ attains its global maximum on $\overline{\Omega}$ at point on the boundary $\partial\Omega$ of Ω, except f is a constant.
     (b) Global minimum. Suppose $f(z) \neq 0$ on Ω. Then $|f(z)|$ attains its global minimum on $\overline{\Omega}$ at point on the boundary $\partial\Omega$ of Ω, except f is a constant.   (3.4.4.2)

As an easy consequence of (2), we have the following fact: If $|f(z)|$ is identically equal to a *nonzero* constant on $\partial\Omega$, then it is necessary that

(i) either f is a constant in Ω, or
(ii) f has at least one zero in Ω.   (3.4.4.3)

For instance, $f(z) = z$ is such a case on $|z| < 1$.

As an important yet easy application of the maximum principle, we have the following remarkable **topological** property of a nonconstant function.

Open mapping theorem and its equivalents.

(1) Open mapping property. A nonconstant analytic function maps any (nonempty) open set onto an open set.
⇔ (2) The maximum–minimum principle of analytic function   ((3.4.4.2)).

$\Leftrightarrow$ (3) The maximum–minimum principle of harmonic function ((3.4.3.7)).

$$(3.4.4.4)$$

An independent proof of (1) can be easily seen by using (3.5.1.8). As a striking contrast, a real-valued C^∞ function of a real variable does not necessarily enjoy this open property. For instance, $f(x) = \frac{1}{1+x^2}$ maps $\mathbf{R}$ onto the set $(0, 1]$, which is not open.

Proof. (3) $\Rightarrow$ (2): See Proof 2 of (3.4.4.1).

(2) $\Rightarrow$ (1): Suppose f is analytic on an open set O. To prove $f(O)$ is open, we need to show that, for each $w_0 \in f(O)$, there associates a $\rho > 0$ so that the open disk $|w - w_0| < \rho$ is contained in $f(O)$.

Now, choose a point $z_0 \in O$ such that $f(z_0) = w_0$. Since f is not a constant, by (3) in (3.4.2.9), there exists a $r > 0$ so that $|z - z_0| \le r$ is contained in O and $f(z) \ne w_0$ on $0 < |z - z_0| \le r$. Let

$$2\rho = \min_{|z-z_0|=r} |f(z) - w_0|.$$

Then, for each $w, |w - w_0| < \rho$, we have on $|z - z_0| = r$,

$$|f(z) - w| \ge |f(z) - w_0| - |w - w_0| > 2\rho - \rho$$
$$= \rho > |w - w_0| = |f(z_0) - w|.$$

This means that $|f(z) - w|$ attains its local minimum on $|z - z_0| \le r$ in the interior $|z - z_0| < r$. Since $|f(z) - w|$ is not a constant, $f(z) - w$ has a zero in $|z - z_0| < r$. Hence $\{|w - w_0| < \rho\} \subseteq f(\{|z - z_0| < r\}) \subseteq f(O)$.

(1) $\Rightarrow$ (3): Suppose $u(z)$ is a nonconstant harmonic function on an open set O. Fix a point $z_0 \in O$ and choose $r > 0$ so that $|z - z_0| < r$ is contained in O. By (3.4.3.4), there exists a nonconstant analytic function $f(z) = u(z) + iv(z)$ in $|z - z_0| < r$.

Then, $e^{f(z)}$ is analytic in $|z - z_0| < r$ and its modulus function is $|e^{f(z)}| = e^{u(z)}$. By assumption, $g(\{|z - z_0| < r\}) = D$ is an open set containing $g(z_0)$, where $g(z) = e^{f(z)}$. Hence, there exists a point z_1 in $|z - z_0| < r$ so that $|g(z_0)| < |g(z_1)|$ and, by observing $g(z_0) \ne 0$, another point z_2 in $|z - z_0| < r$ so that $|g(z_2)| < |g(z_0)|$. Consequently,

$$e^{u(z_2)} < e^{u(z_0)} < e^{u(z_1)},$$

$$\Rightarrow u(z_2) < u(z_0) < u(z_1) \quad \text{for some } z_1, z_2 \text{ in } |z - z_0| < r.$$

This finishes the proof of (3.4.4.4). $\qquad\square$

Section (2) Examples

For numerical examples to illustrate max–min principle, see Example 4 in Sec. 3.4.3, and Exercise A (21) there for more practice. In what follows, we present more theoretical applications of this principle.

Example 1. Suppose both f and g are analytic in a domain Ω and continuous on $\overline{\Omega}$. Show that $|f(z)|+|g(z)|$ attains its maximum on the boundary $\partial\Omega$.

Proof. Fix any point $z_0 \in \Omega$. We may suppose $f(z_0)g(z_0) \neq 0$ and set $|f(z_0)| = e^{i\theta}f(z_0)$ and $|g(z_0)| = e^{i\varphi}g(z_0)$. Then the function $F(z) = f(z)e^{i\theta} + g(z)e^{i\varphi}$ is analytic in Ω and continuous on $\overline{\Omega}$. By (2) in (3.4.4.2), for any $z_0 \in \Omega$,

$$|f(z_0)| + |g(z_0)| = F(z_0) = |F(z_0)| \le \max_{\Omega}|F(z)|$$

$$= \max_{\partial\Omega}|F(z)| \le \max_{\partial\Omega}(|f(z)| + |g(z)|).$$

This proves the claim. □

Example 2. Suppose f is a nonconstant analytic function on $|z| \le 1$.

(1) If $|f(z)|$ attains its maximum on $|z| \le 1$ at the point z_0, where $|z_0| = 1$, then $f'(z_0) \neq 0$. This fact shows a surprising *contrast* to the well-known result for relative extreme point in calculus.

(2) And, $z_0\frac{f'(z_0)}{f(z_0)} > 0$ holds.

Observe that this result is not more true if z_0 is an interior point of a domain.

Solution. (1) A slight modification of $(*_1)$ to $(*_3)$ will justify the claim. To see this, we suppose $f'(z_0) = 0$ and then, there exists a positive integer $n \ge 2$ with $a_n \neq 0$ so that $(*_1)$ holds. Thus, in $(*_2)$, we need to consider a point z_1 in $0 < |z - z_0| < \rho$ so that $\operatorname{Arg} f(z_0) = \operatorname{Arg}(a_n(z_1 - z_0)^n)$ holds, namely, $\operatorname{Arg}(z_1 - z_0) = \pm\frac{1}{n}(\operatorname{Arg} f(z_0) - \operatorname{Arg} a_n)$ *unless these two directions are those of the tangent to the circle at* z_0. In these exceptional cases, we can switch z_1 slightly away from the tangent into the disk $|z| < 1$ so that

$$|f(z_0) + a_n(z_1 - z_0)^n| \ge |f(z_0)| + \frac{2}{3}|a_n||z_1 - z_0|^n$$

(note that for fixed $\alpha \neq 0$, $\beta \neq 0$ may be chosen so that $|\alpha + \beta| \ge |\alpha| + \frac{2}{3}|\beta|$). Whatsoever $(*_3)$ now becomes

$$|f(z_1)| \ge |f(z_0) + a_n(z_1 - z_0)^n| - |\varphi(z_1)||z_1 - z_0|^n$$

$$\ge |f(z_0)| + \frac{2}{3}|a_n||z_1 - z_0|^n - |\varphi(z_1)||z_1 - z_0|^n$$

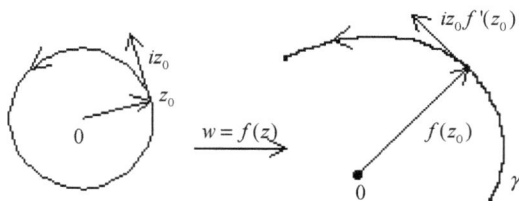

Fig. 3.33

$$> |f(z_0)| + \frac{2}{3}|a_n|\,|z_1 - z_0|^n - \frac{1}{2}|a_n|\,|z_1 - z_0|^n$$

$$= |f(z_0)| + \frac{1}{6}|a_n|\,|z_1 - z_0|^n > |f(z_0)|.$$

This contradiction shows that $f'(z_0) \neq 0$ is true. Observe that, in $(*_1)$ to $(*_3)$, $f'(z_0) = 0$ may hold and then the claim in (1) is no more true in a general domain Ω.

(2) The transformation $w = f(z)$ maps the circle $|z| = 1$ onto a differentiable curve γ in the w-plane on which $w_0 = f(z_0)$ has the largest distance from $w = 0$. γ has a tangent at w_0 because $f'(z_0) \neq 0$. This tangent vector $iz_0 f'(z_0)$ is perpendicular to the vector $f(z_0)$ since f is conformal at z_0 (see Secs. 3.5 or 3.2.3) and so $\frac{iz_0 f'(z_0)}{f(z_0)}$ is purely imaginary (see (1.4.3.4)). Also, by the conformality of f at z_0, in a neighborhood of w_0, the side of the image curve γ bended toward the origin corresponds to the side of the circle $|z| = 1$ bended toward the origin. It follows that $\frac{z_0 f'(z_0)}{f(z_0)} > 0$ holds. See Fig. 3.33.

Example 3. Suppose f is analytic in a bounded domain Ω and continuous on $\overline{\Omega}$. If $|f(z)| = 1$ for any $z \in \partial\Omega$, then either f is a constant or

$$f(\Omega) = \{w\,|\,|w| < 1\}.$$

Proof. By maximum principle, $|f(z)| \leq 1$ on Ω. In case there is a $z \in \Omega$ so that $|f(z)| = 1$, then f will be a constant; otherwise, $|f(z)| < 1$ holds for all $z \in \Omega$.

Now, suppose $|f(z)| < 1$ on Ω. Fix any point w_0, $|w_0| < 1$. We try to find a $z_0 \in \Omega$ so that $f(z_0) = w_0$. To see this, consider

$$g(z) = \frac{f(z) - w_0}{1 - \overline{w_0} f(z)}$$

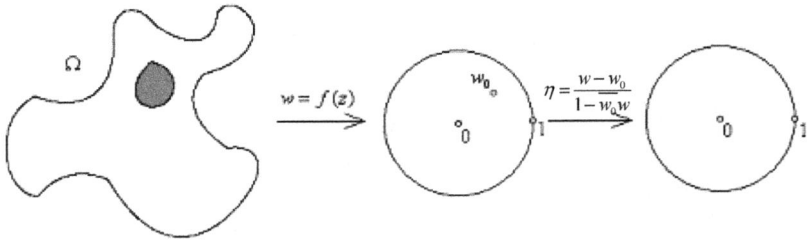

Fig. 3.34

and see Fig. 3.34 (note that (3.4.4.3) says that f takes the value zero in Ω). Observe that g is analytic on Ω and continuous on $\overline{\Omega}$. Also $|g(z)| = 1$ on $\partial\Omega$.

Since g is not a constant, by (3.4.4.3), $g(z)$ has a zero z_0 in Ω, which, in turn, implies that $f(z_0) = w_0$.   □

Example 4. Suppose f is a bounded analytic function on $|z| < 1$. If $|z| \to 1$ in the sector $\alpha \leq \text{Arg}\, z \leq \beta$ $(0 < \alpha < \beta)$, $f(z)$ converges uniformly to 0, show that $f(z) = 0$ on $|z| < 1$.

Proof. Choose a positive integer n so large such that $\beta - \alpha > \frac{2\pi}{n}$. Let $\omega = e^{\frac{2\pi i}{n}}$. Consider the function

$$g(z) = f(z)f(\omega z) \cdots f(\omega^{n-1} z), \quad |z| < 1.$$

Note that g is bounded and analytic on $|z| < 1$.

By assumption, for each $k, 0 \leq k \leq n - 1$, $f(\omega^k z)$ converges uniformly to 0 when $|z| \to 1$ in the sector $\alpha - \frac{2k\pi}{n} \leq \text{Arg}\, z \leq \beta - \frac{2k\pi}{n}$. Consequently, $g(z)$ converges uniformly to 0 as $|z| \to 1$ in $|z| < 1$. Thus, we may define $g(z) = 0$ on $|z| = 1$ and then g becomes continuous on $|z| \leq 1$. By max–min principle, $g(z) \equiv 0$ on $|z| \leq 1$. Therefore, $f(z) = 0$ for all z, $|z| < 1$ (see Exercise A (12) of Sec. 3.4.2).   □

Example 5. Suppose $p_n(z) = a_n z^n + a_{n-1} z^{n-1} + \cdots + a_1 z + a_0$ is bounded by M on $|z| \leq 1$. Show that $|p_n(z)| \leq M|z|^n$ on $1 \leq |z| < \infty$.

Solution. By (3.4.2.7), for $0 \leq k \leq n$,

$$a_k = \frac{p_n^{(k)}(0)}{k!} = \frac{1}{2\pi i} \int_{|z|=1} \frac{p_n(z)}{z^{k+1}} dz,$$

$$\Rightarrow |a_k| \leq \frac{1}{2\pi} \int_{|z|=1} \frac{|p_n(z)|}{|z|^{k+1}} |dz| \leq \frac{M}{2\pi} \cdot 2\pi = M, \quad 0 \leq k \leq n.$$

Now,

$$\frac{p_n(z)}{z^n} = a_n + \frac{a_{n-1}}{z} + \cdots + \frac{a_1}{z^{n-1}} + \frac{a_0}{z^n} \to a_n \quad \text{as } z \to \infty.$$

For any $\varepsilon > 0$, there exists an $R = R(\varepsilon) > 0$ so that $\left|\frac{p_n(z)}{z^n}\right| \leq |a_n| + \varepsilon \leq M + \varepsilon$ for all z, $|z| \geq R$. By maximum principle, $\left|\frac{p_n(z)}{z^n}\right| \leq M + \varepsilon$ on $1 \leq |z| < \infty$ for any $\varepsilon > 0$. Hence $|p_n(z)| \leq M|z|^n$ on $1 \leq |z| < \infty$. One might use Exercise A(6) (b) to give an alternative proof.

Example 6. **(Hadamard's three-circles theorem, 1896).** Suppose f is analytic on the ring domain $0 \leq R_1 < |z| < R_2 \leq \infty$. For $r, R_1 < r < R_2$, let

$$M(r) = \max_{|z|=r} |f(z)|.$$

Then, for arbitrary r_1 and $r_2, R_1 < r_1 < r_2 < R_2$, and any $r, r_1 < r < r_2$,

$$M(r) \leq M(r_1)^{1-\alpha} M(r_2)^{\alpha}, \quad \text{where } \alpha = \frac{\log r - \log r_1}{\log r_2 - \log r_1}$$

holds; or,

$$\log M(r) \leq \frac{\log r_2 - \log r}{\log r_2 - \log r_1} \log M(r_1) + \frac{\log r - \log r_1}{\log r_2 - \log r_1} \log M(r_2).$$

This means that $M(r)$ is a continuous function of r on (R_1, R_2) and $\log M(r)$ is a convex function of $\log r$. Also, the equality holds if and only if $f(z) = az^{\alpha}$, where a is a constant and α is a real constant.   (3.4.4.4)

Proof. Take a *real* constant α whose value is to be determined later on. Consider the multiple-valued function $z^{\alpha} f(z)$ on $r_1 \leq |z| \leq r_2$. Cut off the interval $[-r_2, -r_1]$ from $r_1 \leq |z| \leq r_2$. Then the principle branch $e^{\alpha \operatorname{Log} z} f(z)$ is analytic on the domain $\Omega = \{r_1 < |z| < r_2\} - [-r_2, -r_1]$. So does for any branch of $z^{\alpha} f(z)$.

Since α is a real constant, *any* branch of $z^{\alpha} f(z)$ has its modulus equal to $|z^{\alpha} f(z)| = |z|^{\alpha} |f(z)|$ at a point z. On the left half $\frac{\pi}{2} \leq \operatorname{Arg} z \leq \frac{3\pi}{2}$, $r_1 < |z| < r_2$, of the ring domain $r_1 < |z| < r_2$, we choose an arbitrary branch of $z^{\alpha} f(z)$ (say, $e^{\alpha \log z} f(z)$, where $0 < \arg z < 2\pi$). Then the modulus of this branch attains its maximum on the boundary: $|z| = r_1, |z| = r_2$ with $\frac{\pi}{2} \leq \operatorname{Arg} z \leq \frac{3\pi}{2}$, $[r_1 i, r_2 i]$ and $[-r_2 i, -r_1 i]$, but *not* on the segment $(-r_2, -r_1)$. See Fig. 3.35. Hence, the modulus $|z^{\alpha} f(z)|$ of $e^{\alpha \operatorname{Log} z} f(z)$ assumes its maximum on $|z| = r_1$ and $|z| = r_2$, but *not* on $(-r_2, -r_1)$.

According to (2) in (3.4.4.2),

$$r^{\alpha} M(r) = \max_{|z|=r} |z^{\alpha} f(z)| \leq \max\{r_1^{\alpha} M(r_1), r_2^{\alpha} M(r_2)\}, \quad r_1 \leq r \leq r_2.$$

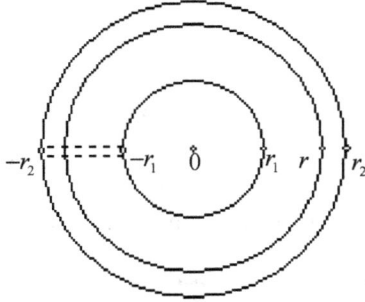

Fig. 3.35

Choose α so that $r_1^\alpha M(r_1) = r_2^\alpha M(r_2)$, namely, $\alpha = -\dfrac{\log(M(r_2)/M(r_1))}{\log(r_2/r_1)}$. Then $r^\alpha M(r) \leq r_1^\alpha M(r_1)$ induces that

$$M(r) \leq \left(\frac{r_1}{r}\right)^\alpha M(r_1)$$

$$\Rightarrow M(r)^{\log\left(\frac{r_2}{r_1}\right)} \leq \left(\frac{r_1}{r}\right)^{-\log\left(\frac{M(r_2)}{M(r_1)}\right)} M(r_1)^{\log\left(\frac{r_2}{r_1}\right)}$$

$$= \left(\frac{M(r_2)}{M(r_1)}\right)^{\log\left(\frac{r}{r_1}\right)} M(r_1)^{\log\left(\frac{r_2}{r_1}\right)}$$

$$= M(r_1)^{\log\left(\frac{r_2}{r}\right)} M(r_2)^{\log\left(\frac{r}{r_1}\right)}.$$

This proves the required inequality.   □

Rewrite this inequality as, in case $M(r_1) \leq M(r)$,

$$1 \leq \left(\frac{M(r)}{M(r_1)}\right)^{\log\left(\frac{r_2}{r_1}\right)} \leq \left(\frac{M(r_2)}{M(r_1)}\right)^{\log\left(\frac{r}{r_1}\right)}.$$

Letting $r \to r_1$, we see that $\frac{M(r)}{M(r_1)} \to 1$ or $M(r) \to M(r_1)$. Hence, $M(r)$ is a continuous function of r on $[r_1, r_2]$.

Section (3) Some extensions

Let us return to (2) in (3.4.4.2). The boundedness of the domain is a necessary assumption there. For instance, consider $f(z) = e^{-iz}$ on the open half-plane $\text{Im}\, z > 0$. Along the boundary $\text{Im}\, z = 0, |f(z)| = |e^{-ix}| = 1$ holds everywhere, where $z = x + iy$. Yet $f(z)$ neither is equal to a constant nor has a zero in $\text{Im}\, z > 0$. On the contrary, $f(z)$ is not even bounded since $|f(z)| = e^y \to +\infty$ as $y > 0$ and $y \to \infty$, not to say that $|f(z)|$

assumes its maximum on $\operatorname{Im} z = 0$. Another example is $g(z) = e^{e^z}$ on $\Omega = \left\{ z \mid |\operatorname{Im} z| < \frac{\pi}{2} \right\}$. Observe that g is analytic on $\bar{\Omega}$ (in $\mathbf{C}$, not in $\mathbf{C}^*$). Even though $|g(z)| = 1$ on $\partial\Omega$, yet $|g(x)| = e^{e^x} \to +\infty$ as $z = x \to \infty$.

Therefore, some restricted conditions should be imposed on the functions so that (3.4.4.2) can be suitably extended to unbounded domains.

The maximum–minimum modulus principle of analytic function on unbounded domain. Suppose that f is analytic on a domain Ω. Let $\partial_\infty \Omega$ denote the boundary of Ω in $\mathbf{C}^*$.

(1) Suppose $\overline{\lim}_{z \to \zeta} |f(z)| \le M$ for each $\zeta \in \partial_\infty \Omega$, where $M \ge 0$ is a constant. Then

$$|f(z)| \le M, \quad z \in \Omega.$$

(2) Suppose that $f(z) \ne 0$ on Ω and $\underline{\lim}_{z \to \zeta} |f(z)| \ge m$ for each $\zeta \in \partial_\infty \Omega$, where $m \ge 0$ is a constant. Then

$$|f(z)| \ge m, \quad z \in \Omega. \tag{3.4.4.5}$$

For some related problems, see Exercises B, (3) and (4).

Proof. (1) Fix any $\varepsilon > 0$. Let $A = \{z \in \Omega \mid |f(z)| > M + \varepsilon\}$. Try to show that A is the empty set. As a consequence, $|f(z)| \le M + \varepsilon$ for all $z \in \Omega$ and each $\varepsilon > 0$ and thus, $|f(z)| \le M$ on Ω.

A is open since $|f|$ is continuous on Ω. By assumption, for each $\zeta \in \partial_\infty \Omega$, there exists a disk $|z - \zeta| < r$ (in case $\zeta = \infty, |z| > r$) so that $|f(z)| < M + \varepsilon$ holds on $\Omega \cap \{|z - \zeta| < r\}$. This fact suggests that $\bar{A}$ is a bounded set in Ω and $\bar{A} \subseteq \{z \in \Omega \mid |f(z)| \ge M + \varepsilon\}$ holds. Also, $|f(z)| = M + \varepsilon$ if $z \in \partial A$.

Now, f is analytic on the bounded open set A and continuous on $\bar{A}$. According to (3.4.4.2), either f is a constant on A or $|f(z)| \le M + \varepsilon$ on A. In case f is not a constant, then $A = \phi$ should hold.

(2) Apply the method or the result of (1) to $\frac{1}{f(z)}$ and the claim follows. $\square$

As an application of (3.4.4.5), we have

Phragmén–Lindelöf theorem (1908). Suppose f is analytic on a *simply connected domain* Ω and g is a *bounded* analytic function on Ω such that $g(z) \ne 0$ on Ω. If the boundary $\partial_\infty \Omega$ of Ω in $\mathbf{C}^*$ can be expressed as $\partial_\infty \Omega = A \cup B$ satisfying:

(1) $\overline{\lim}_{z \to \zeta} |f(z)| \le M$ for each $\zeta \in A$ and some constant $M > 0$, and

(2) for any $\varepsilon > 0$, $\overline{\lim}_{z \to \zeta} |f(z)| |g(z)|^\varepsilon \leq M$ for each $\zeta \in B$.

Then, $|f(z)| \leq M, z \in \Omega$.   (3.4.4.6)

Exercises B(5) and (6) present some corollaries.

Proof. Suppose that $|g(z)| \leq N$ on Ω. According to (3.3.1.8), $\log g(z)$ has (single-valued) analytic branches on Ω; and hence, $h(z) = e^{\varepsilon \log g(z)}$ is a branch of $g(z)^\varepsilon$ on Ω for each such branch, where $\varepsilon > 0$. Observe that $|h(z)| = |g(z)|^\varepsilon \leq N^\varepsilon$ on Ω.

Consider the analytic function

$$F(z) = \frac{f(z) h(z)}{N^\varepsilon} \quad \text{on } \Omega.$$

Note that $|F(z)| \leq |f(z)|$ on Ω. Now, by assumption,

$$\overline{\lim_{z \to \zeta}} |F(z)| \leq \overline{\lim_{z \to \zeta}} |f(z)| \leq M, \quad \text{if } \zeta \in A;$$

$$\overline{\lim_{z \to \zeta}} |F(z)| = \overline{\lim_{z \to \zeta}} \frac{|f(z)| |h(z)|}{N^\varepsilon} \leq \frac{M}{N^\varepsilon}, \quad \text{if } \zeta \in B.$$

$\Rightarrow$ (by (3.4.4.5)) $|F(z)| \leq \max\{M, MN^{-\varepsilon}\}, \quad z \in \Omega \quad \text{and} \quad \varepsilon > 0,$

$\Rightarrow |f(z)| \leq |g(z)|^{-\varepsilon} N^\varepsilon \max\{M, MN^{-\varepsilon}\}, \quad z \in \Omega \quad \text{and} \quad \varepsilon > 0,$

$\Rightarrow$ (Let $\varepsilon \to 0$) $\ |f(z)| \leq M, \quad z \in \Omega.$ $\square$

Exercises A

(1) Suppose f is analytic at z_0 and, for all sufficiently small $r > 0$,

$$|f(z_0)| = \frac{1}{2\pi} \int_0^{2\pi} |f(z_0 + re^{i\theta})| d\theta$$

always holds. Show that f is a constant function.

(2) Let f be analytic in $|z| < R$ and continuous on $|z| \leq R$. Show that, for $n \geq 0$,

$$\max_{|z|=r} |f^{(n)}(z)| \leq \frac{Rn!}{(R-r)^{n+1}} \max_{|z|=R} |f(z)|, \quad 0 \leq r < R.$$

(3) Suppose that $p(z)$ is a polynomial of degree n. Let $M(r) = \max_{|z|=r} |p(z)|$. Show that, if $0 < r_1 < r_2$,

$$\frac{M(r_1)}{r_1^n} \geq \frac{M(r_2)}{r_2^n}$$

with equality if and only if $p(z) = az^n$, where a is a constant.

(4) (a) Suppose that f is analytic in $1 < |z| < 2$ and continuous on $1 \leq |z| \leq 2$. In case $|f(z)| \leq 1$ on $|z| = 1$ and $|f(z)| \leq 4$ on $|z| = 2$, show that $|f(z)| \leq |z|^2$ on $1 \leq |z| \leq 2$.

    (b) Show that an entire function f satisfying $|f(z)| = 1$ on $|z| = 1$ is of the form αz^n with $|\alpha| = 1$. Compare to Exercise B (2)(b) of Sec. 3.4.2.

(5) Let $\varphi(r)$ be a real-valued increasing function of r in $[0,1)$. Suppose that f is analytic in $|z| < 1, f(0) = 0$ and $|f(z)| \leq \varphi(|z|)$ holds on $|z| < 1$. Show that

$$|f(z)| \leq K|z|\varphi(|z|), \quad |z| < 1,$$

where K may be chosen as $\frac{2\phi(\frac{1}{2})}{\phi(0)}$ (which is ≥ 2).

(6) (a) Suppose that f is a nonconstant analytic function on $|z| < R$. Then

$$M(f;r) = \max_{|z|=r} |f(z)|$$

    is a strictly increasing continuous function of r in $(0, R)$.

    (b) Suppose that f is a nonconstant analytic function on $|z| > 1$, continuous on $|z| \geq 1$ and $\lim_{z\to\infty} f(z)$ exists as a finite complex number. Then $|f(z)|$ attains its maximum on $|z| = 1$, and

$$M(f;r) = \max_{|z|=r} |f(z)|$$

    is a strictly decreasing continuous function of r in $(1, \infty)$.

(7) Let $p_n(z) = a_n z^n + a_{n-1} z^{n-1} + \cdots + a_1 z + a_0, a_n \neq 0$ and $n \geq 1$, be a polynomial of degree n.

    (a) For each $\rho > 0$, show that

$$E_\rho = \{z \in \mathbf{C} \,|\, |p_n(z)| < \rho\}$$

    is a nonempty bounded open set having at most n pairwisely disjoint (open) components, each of them is a simply connected domain. Also, the boundary $\partial E_\rho = \{z \in \mathbf{C} \,|\, |p_n(z)| = \rho\}$ is composed of at most n compact connected closed curves, each of them is not degenerated to a single point. For $p_2(z) = z^2 - a^2$, see Fig. 1.23.

    (b) Try to use (a) to prove the fundamental theorem of algebra.

    (c) Show that each bounded component of $\mathbf{C} - \partial E_\rho$ is a subset of the set E_ρ, while its unbounded component is the set $\{z \in C \,|\, |p_n(z)| > \rho\}$.

(d) In case $\mathbf{C}-\partial E_\rho$ has exactly $(n+1)$ components, then $p_n(z)$ should be one-to-one on each of such components. Refer to (3.5.1.8) if necessary.

(8) Suppose that f is analytic in $|z| < 1$ and continuous on $|z| \leq 1$. If there exist constants $a > 0$ and $b > 0$ such that $|f(z)| \leq a$ on $|z| = 1$ and $\operatorname{Im} z \geq 0$; $|f(z)| \leq b$ on $|z| = 1$ and $\operatorname{Im} z \leq 0$, then $|f(0)| \leq \sqrt{ab}$.

(9) Let Q be an open square with center at 0 with its boundary $\partial\Omega$ composing of four sides $S_j, 1 \leq j \leq 4$. Suppose f is analytic in Q and continuous on $\bar{Q}$ so that $|f(z)| \leq M_j$ on S_j for $1 \leq j \leq 4$, where each $M_j > 0$. Show that $|f(0)|^4 \leq M_1 M_2 M_3 M_4$.

(10) Suppose that f is an entire function satisfying $|f(z)| \leq \frac{1}{\sqrt{|\operatorname{Im} z|}}$ on $|z| > R > 0$. Show that $f(z) \equiv 0$.

(11) Suppose that f is analytic in $\operatorname{Im} z > 0$, and bounded and continuous on $\operatorname{Im} z \geq 0$. If $|f(z)| \leq 1$ on $\operatorname{Im} z = 0$, show that $|f(z)| \leq 1$ on $\operatorname{Im} z > 0$. Give an example to show that the boundedness of f on $\operatorname{Im} z \geq 0$ is a necessary condition.

(12) Let f be analytic in a domain Ω and $\partial_\infty \Omega$ denote the boundary of Ω in $\mathbf{C}^*$. If there is a constant $M > 0$ so that, for each sequence $z_n \in \Omega$ converging to a point in $\partial_\infty \Omega$, $\overline{\lim}_{n\to\infty} |f(z_n)| \leq M$ always holds, then $|f(z)| < M$ in Ω except $|f(z)| = M$ for all $z \in \Omega$.

(13) Suppose that f is a bounded analytic function in $|z| < 1$. Also, suppose $f(\zeta) \to 0$ as $\zeta \to 1$ along the upper half-circle of $|\zeta - \frac{1}{2}| = \frac{1}{2}$. Show that $f(x) \to 0$ also holds if $x \to 1$ along the radius $[0, 1)$.

(14) Let $p(z)$ be a polynomial of degree n such that $|p(z)| \leq M$, $z = x \in [-1, 1]$. Show that $|p(z)| \leq M(a + b)^n$ in an ellipse of semiaxes a and b, and with ± 1 as its foci.

(15) Suppose that f and g are analytic in $|z| < R$ and $|f(z)| = |g(z)|$ holds on $|z| = R$. In case both f and g do not have zero point in $|z| < R$, show that there is a real θ so that $f(z) = e^{i\theta} g(z), |z| < R$.

(16) Suppose that f is analytic in $|z| < 1$ and satisfies $|f(z^2)| \geq |f(z)|$ there. Show that f is a constant.

(17) There does not exist an analytic function f in $|z| < 1$ satisfying that, $|f(z_n)| \to \infty$ as long as $|z_n| \to 1$.

(18) Suppose f and g are analytic in a domain Ω and are continuous on $\bar{\Omega}$ and both are bounded by 1 on $\bar{\Omega}$. Also, $g(\partial\Omega)$ is just the unit circle. In case $\frac{f(z)}{g(z)}$ is bounded on the set $\Omega - g^{-1}(0)$, show that $|f(z)| \leq |g(z)|, z \in \Omega$.

Exercises B

(1) Suppose that f is analytic in $|z| \leq R$ and $f(0) \neq 0$. Let $z_1, \ldots, z_n$ (counting multiplicity) be all zeros of f in $|z| < R$. Then

$$|f(0)| \leq \frac{M(R)}{R^n} |z_1 \cdots z_n|, \quad \text{where } M(R) = \max_{|z| \leq R} |f(z)|.$$

Let $v(r)$ denote the number of zeros of f on the closed disk $|z| \leq r < R$, then

$$\int_0^R \frac{v(r)}{r} dr \leq \log \frac{M(R)}{|f(0)|},$$

which is called the *Jensen inequality*, estimating $M(R)$ via the number of zeros in $|z| < R$.

(2) Hadamard's three-lines theorem. Suppose that f is analytic in the vertical strip domain $\Omega : a < \operatorname{Re} z < b$, and is bounded and continuous on $\bar{\Omega} : a \leq \operatorname{Re} z \leq b$. If there exist constants $M_1 > 0$ and $M_2 > 0$ so that

$$|f(a + iy)| \leq M_1, \ y \in \mathbf{R} \quad \text{and} \quad |f(b + iy)| \leq M_2, \ y \in \mathbf{R}$$

then

$$|f(z)| \leq M_1^{\frac{b-x}{b-a}} M_2^{\frac{x-a}{b-a}}, \quad z = x + iy \in \Omega.$$

Letting $M(x) = \sup_{y \in R} |f(x + iy)|$, $a < x < b$, the above inequality can be rewritten as

$$\log M(x) \leq \frac{b-x}{b-a} \log M(a) + \frac{x-a}{b-a} \log M(b), \quad a < x < b.$$

This shows that $\log M(x)$ is a convex function on (a, b).

(3) Suppose f is analytic in an *unbounded* domain Ω and continuous on $\bar{\Omega}$. If f is bounded on Ω and $|f(z)| \leq M$ on $\partial \Omega$, then $|f(z)| \leq M$ throughout the whole Ω. This is an easy consequence of (3.4.4.5). Yet a direct proof may be given as follows: Let $M = 1$.

   (i) Construct an auxiliary function g on Ω, which is analytic, and $g(z) \to 0$ as $z \to \infty$. For instance, $g(z) = \frac{f(z) - f(z_0)}{z - z_0}$, where $z_0 \in \Omega$ is a fixed point. Observe that g is also continuous on $\bar{\Omega}$ and hence is bounded on $\bar{\Omega}$.

   (ii) For any fixed positive integer n, choose $R > 0$ large enough so that $|f(z)|^n |g(z)| \leq \sup_\Omega |g(z)| = K < \infty$ on the boundary of $\{z \in \Omega | |z| \leq R\}$. By (3.4.4.2), $|f(z)|^n |g(z)| \leq K$ for any $z \in \Omega$. Then, try to let $n \to \infty$ in $|f(z)| \leq \left|\frac{K}{g(z)}\right|^{\frac{1}{n}}$ if $g(z) \neq 0$.

(4) A stronger form of the Liouville theorem. If f is a nonconstant entire function, then these exists a curve γ in $\mathbf{C}^*$ so that $\lim_{z \in \gamma \to \infty} f(z) = \infty$ (see (2.4.4)), namely, f has approximate value ∞ at ∞ along γ.

(5) Let $2a \geq 1$. Suppose that f is analytic in the angular domain Ω : $|\mathrm{Arg}\, z| < \frac{\pi}{2a}$ and satisfies:

  (i) for each $\zeta \in \partial\Omega, \overline{\lim}_{z \to \zeta}|f(z)| \leq M$ ($M > 0$ is a constant), and
  (ii) there exist positive numbers N and $b < a$ so that, for all sufficiently large $|z|$,

$$|f(z)| \leq Ne^{|z|^b}.$$

  Then, $|f(z)| \leq M$ on Ω.

(6) Let f be analytic in a bounded domain Ω. $z_0 \in \partial\Omega$ is a fixed point. Suppose that

  (i) for each $\zeta \in \partial\Omega - \{z_0\}$, $\overline{\lim}_{z \to \zeta} |f(z)| \leq M$, and
  (ii) for each $\varepsilon > 0$, $\lim_{z \to z_0}(z - z_0)^\varepsilon |f(z)| = 0$ holds.

  Then, $|f(z)| \leq M$ on Ω.

3.4.5. *Schwarz's lemma*

Maximum modulus principle can be used to give better estimates for the values of an analytic function in its domain of definition. A starting yet particular result in this direction was initiated by H. A. Schwarz in 1869–1870 and its importance in function theory was formally realized by C. Carathéodory in 1912. This is the

Schwarz's lemma. Let f be analytic in $|z| < 1$ and satisfy the conditions $|f(z)| \leq 1$ and $f(0) = 0$. Then,

(1) $|f(z)| \leq |z|$ in $|z| < 1$,   and
(2) $|f'(0)| \leq 1$.

If $|f(z_0)| = |z_0|$ for some $z_0 \neq 0$ or $|f'(0)| = 1$, then f is of the form $e^{i\theta}z$, where θ is a real constant.   (3.4.5.1)

Proof. The function g defined by

$$g(z) = \begin{cases} \dfrac{f(z)}{z}, & 0 < |z| < 1 \\ f'(0), & z = 0 \end{cases}$$

is analytic in $|z| < 1$ according to (3.4.2.17). For any fixed $0 < r < 1$,

$$|g(z)| = \frac{|f(z)|}{|z|} \leq \frac{1}{r} \quad \text{on } |z| = r$$

$\Rightarrow$ (by maximum modulus principle) $\quad |g(z)| \leq \dfrac{1}{r} \quad \text{on } |z| \leq r < 1$

$\Rightarrow$ (letting $r \to 1^-$) $\quad |g(z)| \leq 1 \quad \text{on } |z| < 1, \quad$ namely,

$$|f(z)| \leq |z| \quad \text{on } |z| < 1 \quad \text{and} \quad |f'(0)| \leq 1.$$

In case there is a $z_0 \neq 0$ so that $|f(z_0)| = |z_0|$, then $|g(z_0)| = 1$ and thus $|g(z)|$ obtains its maximum $|g(z_0)| = 1$ at a point z_0 in $|z| < 1$. Consequently, $g(z) = c$ is a constant in $|z| < 1$. Since $|g(z_0)| = |c| = 1$, $c = e^{i\theta}$ for some real θ and, in turn, $f(z) = e^{i\theta}z$ in $|z| < 1$. If $|f'(0)| = 1$, then $|g(0)| = 1$ will imply that $g(z) = e^{i\theta}$ for some real θ by the same reason and it follows that $f(z) = e^{i\theta}z$. $\qquad\square$

The rather specialized assumption in (3.4.5.1) is indeed a normalization of some general cases. A generalization of Schwarz lemma is as follows:

Schwarz–Pick's Lemma on the disk (1915). Suppose f is an analytic function from $|z| < r$ into $|w| < R$ and satisfies $f(z_0) = w_0$, where $|z_0| < r$ and $|w_0| < R$. Then,

(1) $\left| \dfrac{R(f(z)-f(z_0))}{R^2 - \overline{f(z_0)}f(z)} \right| \leq \left| \dfrac{r(z-z_0)}{r^2 - \overline{z}_0 z} \right|, \; |z| < r, \quad$ and

(2) $\dfrac{R|f'(z_0)|}{R^2 - |f(z_0)|^2} \leq \dfrac{r}{r^2 - |z_0|^2}, \quad |z_0| < r$

with equality in either (1) or (2) if and only if $w = f(z)$ is a linear fractional transformation (see Sec. 2.5.4). In case $r = R = 1$, (1) and (2) are usually restated as

(1)$'$ $\left| \dfrac{f(z_1)-f(z_2)}{1-\overline{f(z_2)}f(z_1)} \right| \leq \left| \dfrac{z_1-z_2}{1-\overline{z}_2 z_1} \right|, \; |z_1| < 1, |z_2| < 1, \text{ and } z_1 \neq z_2, \quad$ and

(2)$'$ $\dfrac{|f'(z)|}{1-|f(z)|^2} \leq \dfrac{1}{1-|z|^2}, |z| < 1$

with equality in either (1)$'$ or (2)$'$ if and only if $f(z) = e^{i\theta}\dfrac{z-a}{1-\overline{a}z}, \; |a| < 1$ and θ is a real number. $\qquad\qquad\qquad\qquad\qquad\qquad\qquad\qquad\qquad$ (3.4.5.2)

Proof. Consider two auxiliary linear fractional transformations

$$\eta = f_1(z) = \frac{r(z - z_0)}{r^2 - \overline{z}_0 z} : \{|z| < r\} \to \{|\eta| < 1\},$$

and

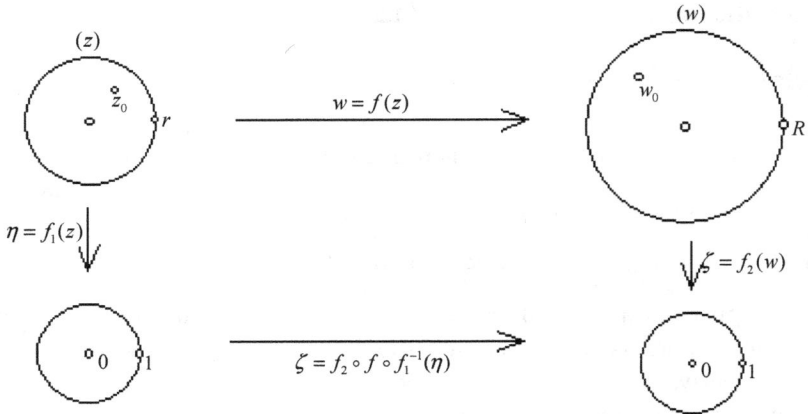

Fig. 3.36

$$\zeta = f_2(z) = \frac{R(w - w_0)}{R^2 - \overline{w_0}w} : \{|w| < R\} \to \{|\zeta| < 1\}, \quad \text{where } w_0 = f(z_0).$$

See Fig. 3.36.

Now, the composite function $f_2 \circ f \circ f_1^{-1}$ satisfies the conditions of the Schwarz lemma. Hence,

$$|f_2 \circ f \circ f_1^{-1}(\eta)| \leq |\eta|, \quad |\eta| < 1$$

$$\Rightarrow \text{(Note that } f_1^{-1}(\eta) = z \text{ and } \eta = f_1(z)) \quad |f_2 \circ f(z)| \leq |f_1(z)|, \quad |z| < r.$$

Equality $|f_2 \circ f \circ f_1^{-1}(\eta)| = |\eta|$ holds for some η, $0 < |\eta| < 1 \Leftrightarrow f_2 \circ f \circ f_1^{-1}(\eta) = e^{i\theta}\eta$ for some real θ, which, in turn, induces that $f(z) = f_2^{-1}(e^{i\theta}f_1(z))$, $|z| < 1$, is a linear fractional transformation. This is (1).

In case $z \neq z_0$, from (1) we have

$$\frac{R}{\left|R - \overline{f(z_0)}f(z)\right|} \cdot \frac{|f(z) - f(z_0)|}{|z - z_0|} \leq \frac{r}{|r^2 - \overline{z_0}z|}$$

$$\Rightarrow \text{(Letting } z \to z_0) \quad \frac{R|f'(z_0)|}{R^2 - |f(z_0)|^2} \leq \frac{r}{r^2 - |z_0|^2}, \quad |z_0| < r.$$

Equality holds if and only if $|(f_2 \circ f \circ f_1^{-1})'(0)| = 1$ holds, where

$$(f_2 \circ f \circ f_1^{-1})'(0) = f_2'(f(z_0))f'(z_0)(f_1^{-1})'(0)$$

$$= \frac{R}{R^2 - |f(z_0)|^2} \cdot f'(z_0) \cdot \frac{r^2 - |z_0|^2}{r}, \quad |z_0| < r$$

if and only if $(f_2 \circ f \circ f_1^{-1})(\eta) = e^{i\theta}\eta$ or f is then linear fractional. This is (2). □

We give two Remarks.

Remark 1. In $(1)'$, let $z_1 = z$ and $z_2 = 0$ and we obtain the inequality

$$\left| \frac{f(z) - f(0)}{1 - \overline{f(0)}f(z)} \right| \le |z|, \quad |z| < 1 \qquad (*)$$

with equality if and only if $f(z) = \frac{e^{i\theta}z + f(0)}{1 + \overline{f(0)}e^{i\theta}z}$, $|z| < 1$, for some real θ.

Note that

$$\left| \frac{w - f(0)}{1 - \overline{f(0)}w} \right| = |z| \quad \text{for a fixed } z, \ |z| < 1.$$

$\Leftrightarrow |w - \tilde{w}_0| = R$ is the circle with center $\tilde{w}_0 = \dfrac{f(0)(1 - |z|^2)}{1 - |z|^2|f(0)|^2}$ and radius

$$R = \frac{|z|(1 - |f(0)|^2)}{1 - |z|^2|f(0)|^2} \quad \text{(See Example 3 in Sec. 1.4.3)}.$$

$\Rightarrow |\tilde{w}_0| - R \le |w| \le |\tilde{w}_0| + R$ for all w on $|w - \tilde{w}_0| \le R$, no matter $|\tilde{w}_0| \ge R$ or $|\tilde{w}_0| \le R$.

See Fig. 3.37, where C is the circle $|w - \tilde{w}_0| = R$. Applying this geometric fact to $(*)$, we obtain the following

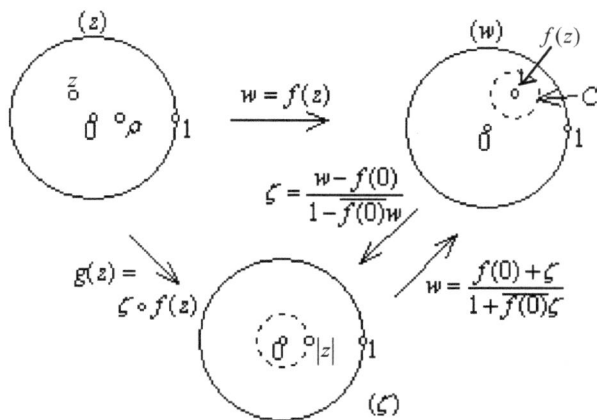

Fig. 3.37

Lower and upper bounded for the modulus. Suppose f is analytic in $|z| < 1$ and $|f(z)| \leq 1$ there. Then, if $f(0) \neq 0$,

$$\frac{|f(0)| - |z|}{1 - |f(0)|\,|z|} \leq |f(z)| \leq \frac{|f(0)| + |z|}{1 + |f(0)||z|}, \qquad |z| < 1;$$

if $f(0) = 0$, $\quad |f(z)| \leq |z|$, $\quad |z| < 1$ (this is the Schwarz's lemma). (3.4.5.3)

This figure suggests another proof for (3.4.5.3) and provides its geometric interpretations, including conditions when equalities hold. The inequalities $\frac{|a| - |b|}{1 - |a||b|} \leq \left|\frac{a - b}{1 - \bar{a}b}\right| \leq \frac{|a| + |b|}{1 + |a||b|} < 1$, $|a| < 1$, $|b| < 1$ (see Exercise A (4) of Sec. 1.4.2) might be helpful here. For detail, see Ref. [63], p. 167. $\qquad \square$

Remark 2. Fix a point z_0, $|z_0| < 1$. We pose two *problems*: Fix another point ζ, $|\zeta| < 1$.

(1) What is the possible maximum value $|f(\zeta)|$ among all functions f, which are analytic and bounded by 1 in the unit disk and satisfies $f(z_0) = 0$? When?

(2) What is the possible maximum value $|f'(z_0)|$ among the same set of analytic functions? When?

To answer these questions, we let $z_1 = z$ and $z_2 = z_0$ in $(1)'$ and $(2)'$ of (3.4.5.2). Then

$$|f(z)| \leq \left|\frac{z - z_0}{1 - \bar{z}_0 z}\right|, \qquad |z| < 1,$$

says that the maximum value for $|f(\zeta)|$ is $\left|\frac{\zeta - z_0}{1 - \bar{z}_0 \zeta}\right|$, which is obtained only when $f(z) = e^{i\theta}\frac{z - z_0}{1 - \bar{z}_0 z}$. On the other hand, in case $f(z_0) \neq 0$, consider

$$g(z) = \frac{f(z) - f(z_0)}{1 - \overline{f(z_0)}f(z)}, \qquad |z| < 1.$$

Then $g(z_0) = 0$ and $|g(z)| < 1$ on $|z| < 1$. Observe that

$$g'(z) = \frac{f'(z)(1 - |f(z_0)|^2)}{(1 - \overline{f(z_0)}f(z))^2},$$

$$\Rightarrow |g'(z_0)| = \frac{|f'(z_0)|}{1 - |f(z_0)|^2} \leq \frac{1}{1 - |z_0|^2} \quad \text{(by } (2)'\text{)},$$

$$\Rightarrow |g'(z_0)| > |f'(z_0)| \quad \text{in case } f(z_0) \neq 0.$$

This fact says that $|f'(z_0)|$ will attain its maximum $\frac{1}{1 - |z_0|^2}$ only when $f(z_0) = 0$ and $f(z) = e^{i\theta}\frac{z - z_0}{1 - \bar{z}_0 z}$. We summarize the above as

An extremal problem. Let z_0, $|z_0| < 1$, be a fixed point. Let $\Im$ be the family of analytic functions f on $|z| < 1$ satisfying $|f(z)| \le 1$ there. Then

(1) For any fixed point ζ, $|\zeta| < 1$,

$$\max_{\substack{f \in \Im \\ f(z_0)=0}} |f(\zeta)| = \left| \frac{\zeta - z_0}{1 - \overline{z_0}\zeta} \right|,$$

which is obtained if and only if $f(z) = e^{i\theta} \frac{z-z_0}{1-\overline{z_0}z}$ for real θ.

(2) $\max_{f \in \Im} |f'(z_0)| = \frac{1}{1-|z_0|^2}$, which is obtained if and only if $f(z_0) = 0$, and in this case, $f(z) = e^{i\theta} \frac{z-z_0}{1-\overline{z_0}z}$ for real θ.   (3.4.5.4)

By the way, we state another

Extremal problem. Let $\Im = \{f \mid f$ is analytic in $|z| < 1$, $f(0) = 0$ and $|f(z)| \le 1$ there$\}$. Then

$$\sup_{f \in \Im} |f'(0)| \le 1$$

always holds and $f(z) = e^{i\theta}z$, where θ is any real constant, are the *extremal functions*, namely, the functions in $\Im$ that obtain the supreme 1.   (3.4.5.5)

This deeper meaning of Schwarz lemma provides a fundamental idea in the existence proof of the *Riemann mapping theorem* (see Sec. 6.1.1 in Chap. 6).   □

It is well known that $\operatorname{Im} z > 0$ is conformally equivalent to $|z| < 1$ via linear fractional transformations (see Example 1 in Sec. 2.5.4). Hence, (3.4.5.2) can be restated as the

Schwarz–Picks' lemma on the upper half plane. Suppose f is analytic in $\operatorname{Im} z > 0$ and satisfies $\operatorname{Im} f(z) > 0$ there. Then,

(1) $\left| \frac{f(z_1)-f(z_2)}{f(z_1)-\overline{f(z_2)}} \right| \le \left| \frac{z_1-z_2}{z_1-\overline{z_2}} \right|$, $\operatorname{Im} z_1 > 0$, $\operatorname{Im} z_2 > 0$,   and $z_1 \ne z_2$;

(2) $\frac{|f'(z)|}{\operatorname{Im} f(z)} \le \frac{1}{\operatorname{Im} z}$, $\operatorname{Im} z > 0$,

with equality in (1) holds for some z_1 and z_2 or in (2) holds for a z, if and only if f is a linear fractional transformation.   (3.4.5.6)

Proof is left as Exercise A (2).

Both (3.4.5.2) and (3.4.5.6) can be used to study the classical non-Euclidean hyperbolic geometry via *Poincará's metric or hyperbolic metric* $\lambda(z) = \frac{1}{1-|z|^2}$, $|z| < 1$, in which length and area are conformal invariants. See Appendix B for details. It was L. V. Ahlfors (1938) who used the curvature to describe the Schwarz lemma and initiated the study of geometric aspects of analytic functions by means of differential geometry. See Refs. [2, 5, 51] and we will come back to this topic in (5.8.4.14). The Schwarz Lemma can also be extended to analytic functions of several complex variables and to hyperbolic manifolds. See Refs. [24, 49, 68].

In what follows, we present some examples.

Example 1. (Ahlfors, Ref. [1], p. 136). Let f be a one-to-one analytic function mapping $|z| < 1$ onto itself. Then f is given by a linear transformation of the form

$$e^{i\theta} \frac{z - z_0}{1 - \overline{z_0}z}, \quad |z_0| < 1 \quad \text{and } \theta \text{ is a real number.}$$

Proof. Since f is onto, there is a unique z_0 with $|z_0| < 1$ so that $f(z_0) = 0$. Consider

$$\zeta = g(z) = \frac{z - z_0}{1 - \overline{z_0}z} : \{|z| < 1\} \to \{|\zeta| < 1\}$$

and the composite function $f \circ g^{-1} : \{|\zeta| < 1\} \to \{|w| < 1\}$, which is one-to-one and maps 0 into 0. By the Schwarz lemma,

$$|f \circ g^{-1}(\zeta)| \leq |\zeta|, \quad |\zeta| < 1.$$

$$\Rightarrow (\text{Letting } z = g^{-1}(\zeta), |\zeta| < 1) \quad |f(z)| \leq |g(z)|, \quad |z| < 1.$$

Similarly, applying Schwarz lemma to $(f \circ g^{-1})^{-1} = g \circ f^{-1}$ will result in

$$|g \circ f^{-1}(w)| \leq |w|, \quad |w| < 1.$$

$$\Rightarrow (\text{Letting } z = f^{-1}(w), |w| < 1) \quad g(z)| \leq |f(z)|, \quad |z| < 1.$$

Combining together, $|f(z)| = |g(z)|$, $|z| < 1$, which amounts to $|(f \circ g^{-1})(\zeta)| = |\zeta|$ on $|\zeta| < 1$. Hence there is a real θ so that $(f \circ g^{-1})(\zeta) = e^{i\theta}\zeta$ or $f(z) = e^{i\theta}g(z)$, $|z| < 1$ and the result follows.   □

Example 2 (Revisited to Example 2 in Sec. 3.4.4).

(1) Suppose f is analytic in $|z| \leq 1$ and there is a z_0, $|z_0| = 1$, so that $|f(z_0)| = \max_{|z| \leq 1} |f(z)|$. Then $f'(z_0) \neq 0$ except f is a constant.
(2) Suppose f is analytic in $|z| < 1$, $f(0) = 0$ and $|f(z)| < 1$ always holds. In case f is also analytic at 1 and $f(1) = 1$, then $|f'(1)| \geq 1$.

Proof. (1) Suppose f is not a constant. We may suppose $z_0 = 1$ by using $z_0 e^{-i\mathrm{Arg}z_0}$ to replace z_0. By maximum modulus principle, $f(1) \neq 0$ should hold and then, we may suppose $f(1) = 1$ by using $\frac{f(z)}{f(1)}$ to replace $f(z)$.

Now, let $z_0 = 1$ and $f(1) = 1$. Hence, $|f(z)| \le f(1) = 1$ holds on $|z| \le 1$.

In case $f(0) = 0$, by Schwarz's lemma, $|f(z)| \le |z|$ holds on $|z| \le 1$. In particular, choose $z = r$ where $0 \le r < 1$, then $|f(r)| \le r < 1$ implies that

$$\left| \frac{f(r) - f(1)}{r - 1} \right| = \left| \frac{f(r) - 1}{r - 1} \right| \ge \frac{1 - |f(r)|}{1 - r} \ge \frac{1 - r}{1 - r} = 1.$$

$$\Rightarrow |f'(1)| = \lim_{r \to 1^-} \left| \frac{f(r) - f(1)}{r - 1} \right| \ge 1.$$

This shows that $f'(1) \ne 0$.

If $f(0) = w_0 \ne 0$, applying the result of the last paragraph to $g(z) = \frac{f(z) - w_0}{1 - \overline{w_0} f(z)}$, $|z| \le 1$, to obtain $|g'(1)| \ge 1$. Now

$$g'(z) = \frac{[1 - |w_0|^2] f'(z)}{(1 - \overline{w_0} f(z))^2}.$$

$$\Rightarrow |f'(1)| = \left| \frac{g'(1)[1 - \overline{w_0} f(1)]^2}{1 - |w_0|^2} \right| \ge \frac{|g'(1)|(1 - |w_0|)^2}{1 - |w_0|^2} \ge \frac{1 - |w_0|}{1 + |w_0|} > 0.$$

Note that (2) is shown in the proof of (1). □

Example 3. (Hadamard–Borel–Carathéodory theorem) Suppose f is analytic on $|z| \le R$. Let $0 < r \le R$ and $M(r) = \max_{|z|=r} |f(z)|$, $A(r) = \max_{|z|=r} \operatorname{Re} f(z)$. Then,

$$M(r) \le \frac{R + r}{R - r} |f(0)| + \frac{2r}{R - r} A(R);$$

$$A(r) \le \frac{R - r}{R + r} \operatorname{Re} f(0) + \frac{2r}{R + r} A(R).$$

This last inequality indicates how the supremum of the real or imaginary part of an analytic function on a larger disk can be used to estimate the modulus of that function on a smaller disk.

Proof. We may suppose that f is not a constant.

In case $f(0) = 0$, then it is necessary that $A(R) > \operatorname{Re} f(0) = 0$ by using (3.4.3.7). Consider

$$F(z) = \frac{f(z)}{2A(R) - f(z)}.$$

Since $\operatorname{Re}(2A(R) - f(z)) = 2A(R) - \operatorname{Re} f(z) \ge A(R) > 0$, F is analytic on $|z| \le R$. Also, $F(0) = 0$. Since $-2A(R) + \operatorname{Re} f(z) \le \operatorname{Re} f(z) \le 2A(R) -$

$\operatorname{Re} f(z)$, it follows that $|F(z)|^2 \le 1$ and hence $|F(z)| \le 1$ on $|z| \le R$. By Schwarz's lemma or (3.4.5.2),

$$|F(z)| \le \frac{r}{R}, \quad |z| = r < R.$$

$$\Rightarrow |f(z)| = \left| \frac{2A(R)F(z)}{1 + F(z)} \right| \le \frac{2A(R)r}{R - r}.$$

In case $f(0) \ne 0$, applying known result to $f(z) - f(0)$, we have

$$|f(z) - f(0)| \le \frac{2r}{R - r} \max_{|z| = R} \operatorname{Re}[f(z) - f(0)]$$

$$\le \frac{2r}{R - r}(A(R) + |f(0)|), \quad |z| \le r.$$

$$\Rightarrow M(r) \le \frac{2r}{R - r}A(R) + \frac{R + r}{R - r}|f(0)|$$

$$\le \frac{R + r}{R - r}(A(R) + |f(0)|) \quad \text{in case } A(R) \ge 0.$$

Other method. May suppose $A(R) > \operatorname{Re} f(0)$.

The univalent function

$$g(z) = f(0) - [A(R) - \operatorname{Re} f(0)]\frac{2z}{1 - z}$$

maps $|z| < 1$ onto $\operatorname{Re} w < A(R)$. By the subordinate principle (observe that $|z| < r < R \Leftrightarrow \frac{|z|}{R} < \frac{1}{R} < 1$ and see Exercise B (1)), it follows that

$$f(\{|z| < r\}) \subseteq g\left(\left\{|z| < \frac{r}{R}\right\}\right), \quad |z| < r \quad \text{and} \quad 0 \le r < R. \qquad (**)$$

Then,

$$M(r) \le \sup_{|z| \le \frac{r}{R}} |g(z)| \le |f(0)| + [A(R) - \operatorname{Re} f(0)]\frac{2\frac{r}{R}}{1 - \frac{r}{R}}$$

and the first inequality follows.

Observe that $\operatorname{Re}\left(\frac{-z}{1 - z}\right) \le \frac{|z|}{1 + |z|} \le \frac{\rho}{1 + \rho}$, $|z| \le \rho < 1$. Applying this inequality to $(**)$, we have

$$A(r) \le \operatorname{Re} f(0) + [A(R) - \operatorname{Re} f(0)]\frac{2r}{R + r}$$

$$= \frac{R - r}{R + r} \operatorname{Re} f(0) + \frac{2r}{R + r}A(R). \qquad \square$$

Application. Let f be an entire function. If there exist constants $\alpha > 0$ and $M > 0$ such that $\operatorname{Re} f(z) \le M|z|^\alpha$ for all sufficiently large $|z|$, then f is necessarily a polynomial of degree not greater than α (refer to Exercise A (36) of Sec. 3.4.2). $\hspace{2cm}$ (3.4.5.7)

Proof. There exists $r > 0$ so that $\operatorname{Re} f(z) \leq M|z|^\alpha$ on $|z| \geq r$. Choose $R = 2r > r$ in Example 3. Then $A(2r) \leq M(2r)^\alpha$ and

$$\max_{|z|=r} |f(z)| \leq 3|f(0)| + 2A(2r) \leq 3|f(0)| + 2M(2r)^\alpha$$
$$= 3|f(0)| + 2^{\alpha+1}M|z|^\alpha.$$

According to Example 6(1) in Sec. 3.4.2, the result follows obviously. Or, try to use Exercise B (3) of Sec. 3.4.2 to give another proof. $\hspace{1cm}\square$

Example 4. Let f be analytic in $|z| < 1$ and $|f(z)| < 1$ holds everywhere. In case f has two distinct *fixed* points a and b in $|z| < 1$, namely $f(a) = a$ and $f(b) = b$, then $f(z) = z$ on $|z| < 1$.

Note that, if in addition that f is *continuous* on $|z| \leq 1$ is imposed, then Rouché theorem (3.4.5.1) can be used to show that f has a unique fixed point in $|z| < 1$ unless f is a constant. This fact is a special case of the *Brouwer fixed point theorem*.

Proof. We may suppose that $a \neq 0$. Let $\varphi(z) = \frac{z-a}{1-\bar{a}z}$ and $F(z) = \varphi \circ f \circ \varphi^{-1}(z)$. Then F is analytic in $|z| < 1$, $|F(z)| < 1$ and $F(0) = \varphi \circ f \circ \varphi^{-1}(0) = \varphi \circ f(a) = \varphi(a) = 0$. Yet F has a fixed point $z_0 = \varphi(b)$ in $|z| < 1$. Schwarz lemma says that $F(z) = z$ holds everywhere in $|z| < 1$, which implies that $f(z) = z$ in $|z| < 1$. $\hspace{1cm}\square$

Example 5. Suppose f is analytic in $|z| < 1$, $f(0) = 0$ and $|f(z)| \leq 1$ there. Show that

(1) If there exist z_1, z_2 with $z_1 \neq z_2$ and $|z_1| = |z_2| = \rho < 1$ such that $f(z_1) = f(z_2) = \beta$ with $|\beta| < 1$, then

$$\frac{f(z)-\beta}{1-\bar{\beta}f(z)} = \left(\frac{z-z_1}{1-\bar{z_1}z}\right)\left(\frac{z-z_2}{1-\bar{z_2}z}\right) h(z), \quad |z| < 1,$$

where h is analytic in $|z| < 1$ and $|h(z)| \leq 1$. Also, $|\beta| \leq \rho^2$ holds.

(2) In case $|f'(0)| = \alpha$, then $\rho(\alpha - \rho) \leq (1 - \alpha\rho)|f(\rho e^{i\theta})|$, $0 \leq \theta \leq 2\pi$ and f is univalent in the open disk

$$|z| < \rho_0 = \frac{\alpha}{1 + \sqrt{1-\alpha^2}}.$$

Proof. (1) The function

$$\frac{f(z) - \beta}{1 - \bar{\beta} f(z)}$$

has two distinct zeros at z_1 and z_2. Therefore, the function

$$h(z) = \frac{f(z) - \beta}{1 - \bar{\beta} f(z)} \cdot \frac{1 - \overline{z_1} z}{z - z_1} \cdot \frac{1 - \overline{z_2} z}{z - z_2}$$

has removable singularities at z_1 and z_2, and hence is analytic in $|z| < 1$ (see (3.4.2.17) for analyticity at z_1 and z_2) and also $|h(z)| \le 1$ everywhere. And the result follows. Letting $z = 0$, then $-\beta = z_1 z_2 h(0)$ implies that $|\beta| = |z_1||z_2||h(0)| \le \rho^2$.

(2) Apply the left inequality in (3.4.5.3) to the analytic function

$$F(z) = \begin{cases} \dfrac{f(z)}{z}, & 0 < |z| < 1, \\ f'(0), & z = 0, \end{cases}$$

and we get

$$\frac{|f'(0)| - |z|}{1 - |f'(0)||z|} \le \frac{|f(z)|}{|z|}, \quad 0 < |z| < 1,$$

$$\Rightarrow \text{(Letting } z = \rho e^{i\theta}, 0 \le \theta \le 2\pi)$$

$$\rho(\alpha - \rho) \le (1 - \alpha\rho)|f(\rho e^{i\theta})|, \quad 0 \le \theta \le 2\pi.$$

In case the assumption in (1) holds, then we have $\rho(\alpha - \rho) \le (1 - \alpha\rho)|\beta| \le \rho^2(1 - \alpha\rho)$, which implies $\rho^2 - \frac{2}{\alpha}\rho + 1 \le 0$ (note that $\alpha \le 1$ by the Schwarz's lemma). Hence,

$$\frac{\alpha}{1 + \sqrt{1 - \alpha^2}} \le \rho \le \frac{1 + \sqrt{1 - \alpha^2}}{\alpha}.$$

This means that it is not possible to have z_1, z_2, $z_1 \ne z_2$ and $|z_1| = |z_2| = \rho < \rho_0 = \frac{\alpha}{1 + \sqrt{1 - \alpha^2}}$ so that $f(z_1) = f(z_2)$ holds. As a consequence, f is univalent in $|z| < \rho_0$ (see (3.5.5.1) if necessary). □

Example 6. Suppose f is analytic in $|z| < 1$, $f(0) = 0$, and $|f(z)| < 1$ everywhere. Let $\omega = e^{\frac{2\pi i}{n}}$ where $n \ge 1$ is an integer. Then $\left|\sum_{k=1}^{n} f(\omega^k z)\right| \le n|z|^n$ in $|z| < 1$ with equality at a point z_0, $0 < |z_0| < 1$, if and only if $f(z) = e^{i\theta} z^n$ for some real θ.

Proof. We can imitate the proof of the Schwarz's lemma and consider

$$
g(z) = \begin{cases} \dfrac{1}{nz^{n-1}} \displaystyle\sum_{j=1}^{n} f(\omega^j z), & 0 < |z| < 1, \\[2mm] 0, & z = 0. \end{cases}
$$

g is analytic at 0 (use Exercise A (14)(b) of Sec. 1.5 and Exercise A (10) of Sec. 3.3.1 and see (3.4.2.17)) and thus, analytic everywhere in $|z| < 1$ and $|g(z)| < 1$ holds.

By the Schwarz's lemma, $|g(z)| \leq |z|$, $|z| < 1$. Hence the inequality holds. In case there is a z_0, $0 < |z_0| < 1$ so that $|g(z_0)| = |z_0|$, then $g(z) = e^{i\theta} z$ for real θ. To show that $f(z) = e^{i\theta} z^n$ holds identically in $|z| < 1$ is equivalent to show that $h(z) = f(z) - e^{i\theta} z^n$ is identically equal to zero in $|z| < 1$. Now, $g(z) = e^{i\theta} z$ means that $h(\omega z) + \cdots + h(\omega^n z) = 0$ in $|z| < 1$. For $1 \leq j \leq n$,

$$
h(\omega^j z) + e^{i\theta} \omega^{jn} z^n = f(\omega^j z) \quad \text{and} \quad |f(\omega^j z)| < 1,
$$

$$
\Rightarrow |e^{i\theta} z^n|^2 + 2\,\mathrm{Re}\{e^{i\theta} z^n \overline{h(\omega^j z)}\} + |h(\omega^j z)|^2 < 1, \quad 1 \leq j \leq n,
$$

$$
\Rightarrow \Bigg(\text{Adding the above inequalities side by side for } 1 \leq j \leq n,
$$

$$
\text{and using } \sum_{j=1}^{n} h(\omega^j z) = 0 \Bigg)
$$

$$
n|z|^{2n} + |h(z)|^2 \leq n|z|^2 + \sum_{j=1}^{n} |h(\omega^j z)|^2 < n
$$

$$
\Rightarrow |h(z)|^2 \leq n(1 - |z|^{2n}), \quad |z| < 1.
$$

This fact indicates that $h(z)$ converges to zero uniformly as $|z| \to 1$. By maximum modulus principle (see (3.4.4.5)), $h(z) = 0$ holds everywhere in $|z| < 1$. Thus $f(z) = e^{i\theta} z^n$, $|z| < 1$. □

Exercises A

(1) Suppose f is analytic in $|z| < 1$ and $|f(z)| \leq 1$ there. In case $f(0) = 0$ and 0 is a zero of order n of f, then $|f(z)| \leq |z|^n$ on $|z| < 1$ and $|f^{(n)}(0)| \leq n!$. If there is a z_0, $0 < |z_0| < 1$, so that $|f(z_0)| = |z_0|^n$ or $|f^{(n)}(0)| = n!$ then $f(z) = e^{i\theta} z^n$ for some real θ.

(2) Prove (3.4.5.6) in detail.

(3) Let $w = f(z)$ be an analytic function from $|z| < 1$ into $|w| < 1$.

(a) Suppose that $f\left(\frac{2}{3}\right) = 0$. Find the possible largest value for $\left|f\left(\frac{1}{2}\right)\right|$ and, in this case, find such function f.

(b) Show that $\left|f'\left(\frac{2}{3}\right)\right|$ will assume its largest value only when $f\left(\frac{2}{3}\right) = 0$. Find this value and the corresponding f.

(4) Suppose that f is analytic in $|z| < 1$ and $|f(z)| \leq 1$ there. Show that

$$|f'(z)| \leq \frac{1 - |f(z)|^2}{1 - |z|^2}, \quad |z| < 1$$

by using the following three methods:

(i) (3.4.5.2).

(ii) Fix any point z_0, $|z_0| < 1$. Consider (say reason why!)

$$\left(\frac{f(z) - f(z_0)}{1 - \overline{f(z_0)}f(z)}\right) \Big/ \left(\frac{z - z_0}{1 - \overline{z_0}z}\right), \quad |z| < 1$$

and then, let $z \to z_0$.

(iii) Fix any point z_0, $|z_0| < 1$. Consider (say reason why!)

$$\left[f\left(\frac{z + z_0}{1 + \overline{z_0}z}\right) - f(z_0)\right] \Big/ \left[1 - \overline{f(z_0)}f\left(\frac{z + z_0}{1 + \overline{z_0}z}\right)\right], \quad |z| < 1.$$

Then, try to use $f(z_2) - f(z_1) = \int_{z_1}^{z_2} f'(z)dz$, $|z_1| < 1$, $|z_2| < 1$, where the path of integration is the line segment connecting z_1 to z_2, to show that

$$\left|\frac{f(z_2) - f(z_1)}{z_2 - z_1}\right| \leq \frac{1}{1 - r^2}, \quad |z_1| < r, \quad |z_2| < r \quad (0 < r < 1).$$

(5) (a) Does there exist an analytic function f in $|z| < 1$ satisfying $|f(z)| \leq 1$ everywhere,

$$f(0) = \frac{1}{2} \quad \text{and} \quad f'(0) = \frac{3}{4}?$$

(b) Does there exist an analytic function f in $|z| < 1$ satisfying $|f(z)| \leq 1$ everywhere,

$$f\left(\frac{1}{2}\right) = \frac{3}{4} \quad \text{and} \quad f'\left(\frac{1}{2}\right) = \frac{2}{3}?$$

(6) Let f be analytic in $|z| < 1$ and $|f(z)| < 1$ everywhere. Try to find bound for $\left|f\left(\frac{3}{4}\right)\right|$ if $f\left(\frac{1}{2}\right) = 0$. Is there a function to attain this bound?

(7) Show that $|\sin z| \leq \left(\frac{1}{2}\cosh 2\right)|z|$ on $|z| \leq 2$.

(8) Suppose f is analytic in $|z| < 1$, $f(0) = f'(0) = 0$ and, for each z in $|z| < 1$, $|f'(z)| \leq 1$ holds. Show that $|f(z)| \leq \frac{|z|^2}{2}$, $|z| < 1$. When does equality hold?

(9) Suppose f is analytic in $|z| < 1$, $f(0) = 0$ and $|\operatorname{Re} f(z)| < 1$ there. Show that

   (i) $|\operatorname{Re} f(z)| \leq \frac{4}{\pi} \operatorname{Arc\,tan} |z|$ (principle branch), $|z| < 1$, and
   (ii) $|\operatorname{Im} f(z)| \leq \frac{2}{\pi} \log \frac{1+|z|}{1-|z|}$, $|z| < 1$.

(10) If f is analytic in $|z| < 1$ and $\operatorname{Re} f(z) < 0$ everywhere, then $|f'(0)| \leq 2|\operatorname{Re} f(0)|$.

(11) Suppose $f(z) = \sum_{n=1}^{\infty} a_n z^n$ is analytic in $|z| < 1$ and $|f(z)| < 1$ everywhere. Show that $|a_2| \leq 1 - |a_1|^2$.

(12) Suppose $f(z)$ is analytic in $|z| < 1$ and $f(0) = 0$. Show that $\sum_{n=1}^{\infty} f(z^n)$ converges uniformly on compact subsets of $|z| < 1$.

Exercises B

(1) (Subordinate principle) Let f and g be analytic in $|z| < 1$, $f(0) = g(0) = 0$ and $g(\{|z| < 1\}) \subseteq f(\{|z| < 1\})$. In case f is univalent, then

$$g(\{|z| < r\}) \subseteq f(\{|z| < r\}), \quad 0 \leq r < 1.$$

In this case, g is said to be *subordinate* to f. If only $f(0) = g(0)$ is required, try to show that $|g'(0)| \leq |f'(0)|$ with equality only if $g(z) = f(e^{i\theta} z)$ for some real θ. Let $f(z) = z$, then this is the Schwarz's lemma.

(2) Let f be analytic in $|z| < 1$ and $\operatorname{Re} f(z) \geq 0$ holds everywhere. Show that

   (i) $\left| \dfrac{f(z_1) - f(z_2)}{f(z_1) + \overline{f(z_2)}} \right| \leq \left| \dfrac{z_1 - z_2}{1 - \overline{z_2} z_1} \right|$, $|z_1| < 1$, $|z_2| < 1$, $z_1 \neq z_2$, and

   (ii) $|f'(z)| \leq \dfrac{2 \operatorname{Re} f(z)}{1 - |z|^2}$, $|z| < 1$ with equality in 1 or 2 only if f is a linear fractional transformation.

   (iii) If, in addition, $f(0) = 1$, then

$$\frac{1-|z|}{1+|z|} \leq \operatorname{Re} f(z) \leq \frac{1+|z|}{1-|z|}, \quad |\operatorname{Im} f(z)| \leq \frac{2|z|}{1-|z|^2}, \quad \text{and}$$

$$\frac{1-|z|}{1+|z|} \leq |f(z)| \leq \frac{1+|z|}{1-|z|}.$$

   Observe that $f_0(z) = \frac{1+z}{1-z} = 1 + 2 \sum_{n=1}^{\infty} z^n$ is analytic in $|z| < 1$, $f(0) = 0$ and $\operatorname{Re} f(z) > 0$ holds; when choosing $z = -|z|$ and

$z = |z|$, respectively, then the equalities in the third formula in 3 will hold. This suggests the following *extremal problem*: Let $\mathfrak{F} = \{f(z) = 1 + \sum_{n=1}^{\infty} a_n z^n$ is analytic in $|z| < 1$ and $\mathrm{Re}\, f(z) \geq 0\}$. Then

$$\max_{f \in \mathfrak{F}} |a_n| = 2, \quad n \geq 1$$

and either $f_0(z) = \frac{1+z}{1-z}$ or $\frac{1-z}{1+z}$ is *the extremal function* which attains the maximum 2.

(3) Let f be analytic in $|z| < 1$ and $|f(z)| \leq M$ holds in $|z| < 1$. If f has a finite number of zeros $z_1, \ldots, z_n$ in $|z| < 1$, show that

   (i) $|f(z)| \leq M \prod_{k=1}^{n} \left| \frac{z - z_k}{1 - \overline{z_k} z} \right|, \quad |z| < 1$, and

   (ii) $|f(0)| \leq M \prod_{k=1}^{n} |z_k|.$

   In case f has infinitely many zeros $z_1, z_2, \ldots, z_n, \ldots$ in $0 < |z| < 1$ and these zeros do not have any limit point in $|z| < 1$, then

   (iii) $\sum_{k=1}^{\infty} (1 - |z_k|)$ converges and hence $\sum_{k=1}^{\infty} \log |z_k|$ converges, too.

(4) (E. Study) Let f be an analytic function on $|z| < 1$, which maps $|z| < 1$ univalently onto a domain Ω. Let $\Omega_r = f(\{|z| < r\})$, $0 < r < 1$.

   (a) If Ω is a convex domain, then each Ω_r is a convex domain, too.
   (b) If Ω is a *starlike* domain with center at $f(0)$ (namely, for each point w in Ω, the line segment connecting $f(0)$ to w is also contained in Ω), then each Ω_r is also a starlike domain with center at $f(0)$.

(5) The *non-Euclidean distance* between z_1 and z_2 in $|z| < 1$ is defined by

$$d(z_1, z_2) = \inf_{\gamma} \int_{\gamma} \frac{|dz|}{1 - |z|^2},$$

where γ is any piecewise differentiable curve in $|z| < 1$, connecting z_1 and z_2. Show that $d(z_1, z_2)$ is obtained by the circular arc, connecting z_1 to z_2, which is orthogonal to the unit circle $|z| = 1$ and, in this case,

$$d(z_1, z_2) = \frac{1}{2} \log \frac{1 + \left| \frac{z_1 - z_2}{1 - \overline{z_1} z_2} \right|}{1 - \left| \frac{z_1 - z_2}{1 - \overline{z_1} z_2} \right|}.$$

Define and write out the explicit formula for the non-Euclidean distance in the upper half-plane, see Appendix B if necessary.

Exercises C

Read Appendix B in Volume 2.

3.4.6.   *The symmetry (or reflection) principle*

Morera's theorem (3.4.2.13) can be used to extend the analyticity of a function beyond its original domain of definition in certain cases.

As a general result in this direction, we have the

Painleve's theorem. Let Ω_1 and Ω_2 be two *disjoint* domains such that their boundaries $\partial\Omega_1$ and $\partial\Omega_2$ contain a *common* rectifiable Jordan curve (regardless of endpoints) γ. Suppose that

(1) The function f_k is analytic in Ω_k and is continuous on $\Omega_k \cup \gamma$, for $k = 1, 2$.
(2) $f_1(z) = f_2(z)$ for all $z \in \gamma$.

Then the function f defined by

$$f(z) = \begin{cases} f_1(z), & z \in \Omega_1 \\ f_1(z) = f_2(z), & z \in \gamma \\ f_2(z), & z \in \Omega_2 \end{cases}$$

is analytic on the domain $\Omega_1 \cup \gamma \cup \Omega_2$. In this case, f_1 and f_2 are said to be the *analytic continuation* of each other across the boundary curve γ.

$$(3.4.6.1)$$

Proof. All we need to do is to show the analyticity of f along γ. Fix any point $z_0 \in \gamma$ and construct an open disk $D : |z - z_0| < \rho$, contained in $\Omega_1 \cup \gamma \cup \Omega_2$. Let Γ be any rectifiable Jordan closed curve in D. If Γ lies either in $\Omega_1 \cup \gamma$ or $\Omega_2 \cup \gamma$, then by the Cauchy integral theorem (also, refer to $(3.4.2.4)'$),

$$\int_\Gamma f(z)dz = \int_\Gamma f_1(z)dz \quad \text{or} \quad \int_\Gamma f_2(z)dz = 0.$$

If $\Gamma \cap \Omega_1 \neq \phi$ and $\Gamma \cap \Omega_2 \neq \phi$ hold, let $\Gamma_1 = \Gamma \cap \Omega_1, \Gamma_2 = \Gamma \cap \Omega_2$ and $\gamma_0 = \gamma \cap \text{Int}\Gamma$. Then $\Gamma = (\Gamma_1 + \gamma_0) + (-\gamma_0 + \Gamma_2)$ as indicated in Fig. 3.38, and

$$\int_\Gamma f(z)dz = \int_{\Gamma_1 + \gamma_0} f(z)dz + \int_{-\gamma_0 + \Gamma_2} f(z)dz = 0 + 0 = 0.$$

By Morera's theorem, f is analytic in D; in particular, f is analytic at z_0. The theorem is thus proved. $\square$

As a particular case and important application of (3.4.6.1), we have the extension of (3.4.1.5):

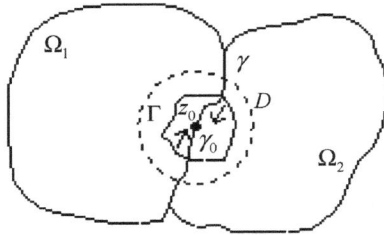

Fig. 3.38

Symmetry principle. Let Ω be a domain, lying entirely in an open half plane determined by a line L (or in the interior or exterior of a circle S) such that its boundary $\partial\Omega$ contains an *open* segment γ of L (or an *open* arc γ of S). Suppose a function f satisfies the conditions:

(1) f is analytic in Ω and is continuous on $\Omega \cup \gamma$; and
(2) for each $z \in \gamma$, $f(z)$ always lies on a line L' (or a circle S').

Then, the function F defined by

$$F(z) = \begin{cases} f(z), & z \in \Omega \cup \gamma, \\ f(z^*)^*, & z \in \Omega^* = \{z \in \mathbf{C} \mid \text{the symmetric point } z^* \text{ of} \\ & \quad z \text{ w.r.t. } L \text{ (or } S) \text{ lies in } \Omega, \end{cases}$$

is analytic on $\Omega \cup \gamma \cup \Omega^*$ (see Fig. 3.39), where $f(z^*)^*$ is the symmetric point of $f(z^*)$ w.r.t. L' (or S'), and satisfies

$$F(z^*) = F(z)^*, \quad z \in \Omega \cup \gamma \cup \Omega^*.$$

In this case, F is called *the analytic continuation of f to Ω^* across the boundary curve γ.*   (3.4.6.2)

Proof. After performing a suitable bilinear transformation (see Figs. 3.39(b) and 3.39(c)), a line or a circle can always be mapped univalently onto the real axis (see Sec. 2.5.4). Therefore, without losing its generality, we may just consider the case that both L and L' are real axes. In this case, Ω^* is the symmetric domain of Ω w.r.t. the real axis (see Fig. 3.15 or Fig. 3.39(a) in which $\Omega = \Omega^+$ and $\Omega^* = \Omega^-$) and thus, $z^* = \bar{z}$ and $f(z)^* = \overline{f(z)}$ for $z \in \Omega \cup \gamma \cup \Omega^*$. We already showed in (3.4.1.5) that $\overline{f(\bar{z})}$ is analytic in Ω^* and $f(\bar{z}) = f(z)$ is real for any $z \in \gamma$ (an open segment on the real axis). According to Painleve's theorem, the function

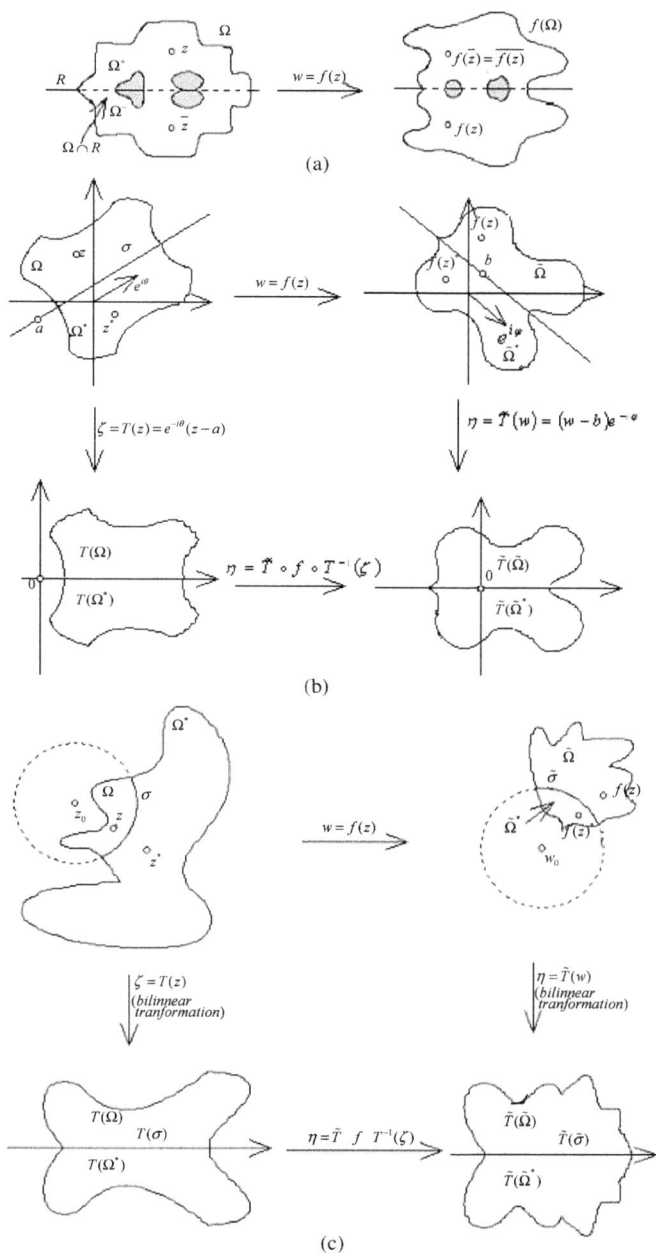

(a)

(b)

(c)

Fig. 3.39

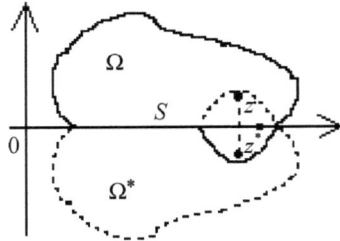

Fig. 3.40

$$F(z) = \begin{cases} f(z), & z \in \Omega \cup \gamma, \\ \overline{f(\bar{z})}, & z \in \Omega^*, \end{cases}$$

is analytic on $\Omega \cup \gamma \cup \Omega^*$ and $F(\bar{z}) = \overline{F(z)}$ there.   □

Remark 1 (On hypotheses in symmetry principle). The assumption that Ω lies on one side of a line L (or a circle S) guarantees that the image domain $f(\Omega)$ under f also lies on one side of the line L' (or the circle S'), because $f(L) \subseteq L'$ and the continuity of f preserves connectedness of a set. Consequently, the extended function F will be *single-valued* on $\Omega \cup \gamma \cup \Omega^*$. On the contrary, if Ω lies on both sides of L (namely, $\Omega \cap L \neq \phi$), there would be a point $z \in \Omega$ so that its symmetric point z^* with respect to L is still in Ω. (See Fig. 3.40.) In case $f(z) \neq f(z^*)^*$, then the extended F will be *double-valued* at z, because $F(z) = f(z)$ and $F(z) = F(z^*)^* = f(z^*)^* \neq f(z)$.

In (3.4.6.2), if f is univalent on $\Omega \cup \gamma$ so that $f(\gamma)$ is an open segment on L' or an open arc on S' and that $f(\Omega)$ lies on one side of L' (or S'), then F will be *univalent* on $\Omega \cup \gamma \cup \Omega^*$. In case f is univalent on $\Omega \cup \gamma$ but $f(\Omega)$ does *not* lies on one side of L' (or S'), the resulted F will be *multivalent*, namely, not univalent on $\Omega \cup \gamma \cup \Omega^*$. This is because there exists a point $w \in f(\Omega)$ whose symmetric point w^* w.r.t. L' still lies in $f(\Omega)$. Let z and z' be points in Ω so that $f(z) = w$ and $f(z') = w^*$. Let z^* be the symmetric point of z w.r.t. L. Then $z^* \in \Omega^*$. Now, the extended function F satisfies $F(z^*) = f(z)^* = w^* = f(z') = F(z')$. See Fig. 3.41.

A real or complex-valued function $\varphi : (a, b)$ (real interval) $\to \mathbf{C}$ of a real variable t is said to be *real analytic* if, for each $t_0 \in (a, b)$, $\varphi(t)$ can be expressed as a convergent Taylor series $\sum_{n=0}^{\infty} \frac{\varphi^{(n)}(t_0)}{n!}(t - t_0)^n$ in some interval $(t_0 - \rho, t_0 + \rho)$, $\rho > 0$. By (3.3.2.3) and (3.3.2.4), $\varphi(z)$ can be defined

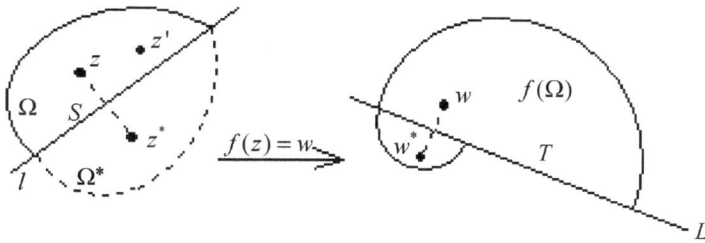

Fig. 3.41

as an analytic function in $|z - t_0| < \rho$ and, as a whole, $\varphi(z)$ can be defined as an analytic function in a domain Ω, symmetric to the real axis, which contains the interval (a, b). In these circumstances, φ is said to determine an *analytic arc*. It is *regular* if $\varphi'(t) \neq 0$ for each $t \in (a, b)$, and is *simple* if $\varphi(t_1) = \varphi(t_2)$ only when $t_1 = t_2$.

An arc γ of the boundary $\partial\Omega$ of a domain Ω is said to a *free boundary arc* if every point of γ has an open neighborhood disk O whose intersection with the whole boundary $\partial\Omega$ is the same as its intersection with γ. Let this intersection part be $\sigma = O \cap \partial\Omega = O \cap \gamma$. Then σ divides O into two parts (open connected subsets). If only one is inside Ω, then we call the point a *one-sided boundary point*. If every point of γ is a one-sided boundary point of Ω, γ is said to be a *free one-sided boundary arc*.

Then, (3.4.6.2) can be extended to

A generalization of the symmetry principle. Let f be analytic in a domain Ω, in $\mathbf{C}^*$, whose boundary $\partial\Omega$ contains a *regular, simple, free one-sided boundary analytic arc* γ. Suppose f is continuous on $\Omega \cup \gamma$ in *the generalized sense* (see (2.2.2)) so that $\tilde{\gamma} = f(\gamma)$ is a regular, simple analytic arc. Then f can be analytically extended to $\Omega \cup \gamma$ and beyond.   (3.4.6.3)

The detail is left as Exercise B (3) or just refer to (6.1.2.3). That γ can be mapped onto an arc of the unit circle if Ω is simply connected is a consequence of the Riemann mapping theorem (see (6.1.2)). For much more detailed account about symmetry principle, see Ref. [58], Vol. III, Chap. 9.

Some illustrated examples are as follows.

Example 1. Let f be analytic in $|z| < 1$ and continuous on $|z| \leq 1$.

(1) If $f(z) \neq 0$ in $|z| < 1$ and $|f(z)| = 1$ on $|z| = 1$, then f is a constant.
(2) If f maps $|z| < 1$ univalently onto $|w| < 1$, then f is a linear fractional transformation (compare with Example 1 of Sec. 3.4.5).

(3) (See Ref. [1], p. 174) If $|f(z)| = 1$ on $|z| = 1$, then f is a rational function (refer to Exercise B (1) of Sec. 2.5.3).

Proof.

(1) By the symmetry principle (3.4.6.2) and the condition that $f(z) \neq 0$ in $|z| < 1$, f can be extended to an entire function, still denoted by f; in particular, $f(z^*) = f(z)^*$ on $|z| \leq 1$, where z^* is the symmetric point of z with respect to the unit circle. Since $0 < m \leq |f(z)| \leq 1$ on $|z| \leq 1$, by maximum–minimum modulus principle (3.4.4.2), it follows that f is a bounded entire function (why?). Liouville theorem says that this f should be a (nonzero) constant.

(2) Since f is the continuous on $|z| \leq 1$, the condition that $|f(z)| < 1$ on $|z| < 1$ implies that $|f(z)| \leq 1$ on $|z| \leq 1$. Furthermore, $|f(z)| = 1$ on $|z| = 1$ should hold. Otherwise, there is a point ζ_0, $|\zeta_0| = 1$, so that $|f(\zeta_0)| < 1$. By univalence of f from $|z| < 1$ onto $|w| < 1$, there is a unique point z_0, $|z_0| < 1$, so that $f(z_0) = f(\zeta_0)$ holds. Choose $\delta > 0$ small enough so that the disks $D_\delta(z_0) : |z - z_0| < \delta$ and $D_\delta(\zeta_0) : |z - \zeta_0| < \delta$ are disjoint. Observe that $V = f(D_\delta(z_0)) \cap f(D_\delta(\zeta_0))$ is nonempty and open, and $f^{-1}(V) \subseteq D_\delta(z_0)$ and $f^{-1}(V) \subseteq D_\delta(\zeta_0)$ hold simultaneously. This would contradict to the univalence of f in $|z| < 1$. By symmetry principle, f can be extended analytically from $|z| < 1$ into $|z| \geq 1$. There is a unique point z_0, $|z_0| < 1$, so that $f(z_0) = 0$. Then the function

$$g(z) = f(z) \cdot \frac{1 - \overline{z_0} z}{z - z_0}$$

is analytic in $|z| < 1$, continuous on $|z| \leq 1$ and $|g(z)| = 1$ on $|z| = 1$ and $g(z) \neq 0$ in $|z| < 1$. By (a), this g is a nonzero constant $e^{i\theta}$ (θ real) and $f(z) = e^{i\theta} \frac{z - z_0}{1 - \overline{z_0} z}$.

(3) Suppose f has a zero of order α at 0 and $a_1, \ldots, a_k$ are all distinct zeros of f in $|z| < 1$, with respective order $\alpha_1, \ldots, \alpha_k$. Say reason why f has only finitely many zeros in $|z| < 1$. Then the function

$$F(z) = \frac{f(z)}{z^\alpha \prod_{j=1}^k \left(\frac{z - a_j}{1 - \overline{a}_j z} \right)^{\alpha_j}}, \quad |a_j| < 1, \quad 1 \leq j \leq k$$

is analytic in $|z| < 1$ and satisfies $|F(z)| = 1$ on $|z| = 1$. Observe that $F(z)$ does not have zero in $|z| < 1$. By what we obtained in (a) above, $F(z) = e^{i\theta}$ is a constant for some $\theta \in \mathbf{R}$ and the claim follows. $\qquad\square$

Example 2. Show that

$$f(z) = \sum_{n=0}^{\infty} a_n z^n, \quad |z| < r,$$

takes real values on the diameter $(-r, r)$ if and only if a_n, $n \geq 0$, are reals (refer to Exercise A (4) of Sec. 3.3.2).

Solution. Only the necessity is needed to be proved.

Note that $f(z)$ is analytic on $|z| < r$, $\operatorname{Im} z > 0$ and is continuous on $|z| < r$, $\operatorname{Im} z \geq 0$. Also, $f(x)$ is real for each $x \in (-r, r)$. By symmetry principle, $f(\bar{z}) = \overline{f(z)}$ on $|z| < 1$ implies that

$$\sum_{n=0}^{\infty} a_n \bar{z}^n = \sum_{n=0}^{\infty} \overline{a_n} \bar{z}^n, \quad |z| < 1.$$

Let $z \to 0$ in the above identity, we have $\overline{a_0} = a_0$. This fact, in turn, results in

$$\sum_{n=1}^{\infty} a_n \bar{z}^{n-1} = \sum_{n=1}^{\infty} \overline{a_n} \bar{z}^{n-1}.$$

Again, $z \to 0$ induces that $\overline{a_1} = a_1$. Inductively, $\overline{a_n} = a_n$, $n \geq 0$, hold and then, a_n, $n \geq 0$ are reals.

For instance, let $f(z) = \sum_{n=0}^{\infty} \frac{1}{n!} z^n$, $|z| < \infty$. We know already that e^z is an entire function with the property that

$$f(x) = \sum_{n=0}^{\infty} \frac{1}{n!} x^n = e^x, \quad x \in \mathbf{R}.$$

By interior uniqueness theorem (see (4) in (3.4.2.9)), it follows that

$$e^z = \sum_{n=0}^{\infty} \frac{1}{n!} z^n, \quad |z| < \infty.$$

Similarly, since $\cos x = \sum_{n=0}^{\infty} \frac{(-1)^n}{(2n)!} x^n$, $\sin x = \sum_{n=0}^{\infty} \frac{(-1)^n}{(2n+1)!} x^n$, $x \in \mathbf{R}$, and $\log(1+x) = \sum_{n=1}^{\infty} \frac{(-1)^{n-1}}{n} x^n$, $|x| < 1$, we have, respectively, the following series representations:

$$\cos z = \sum_{n=0}^{\infty} \frac{(-1)^n}{(2n)!} z^n; \quad \sin z = \sum_{n=0}^{\infty} \frac{(-1)^n}{(2n+1)!} z^n, \quad |z| < \mathbf{R}, \quad \text{and}$$

$$\operatorname{Log}(1+z) = \sum_{n=1}^{\infty} \frac{(-1)^{n-1}}{n} z^n, \quad |z| < 1.$$

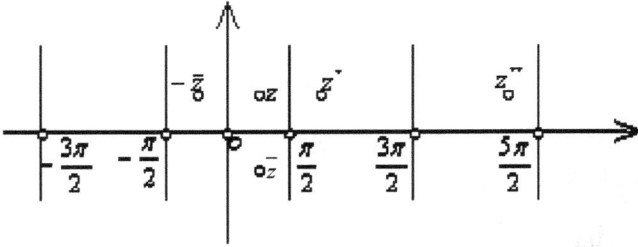

Fig. 3.42

Example 3. Suppose f is analytic in the strip $-\frac{\pi}{2} < \operatorname{Re} z < \frac{\pi}{2}$ and satisfies the following conditions:

(1) f is continuous on $-\frac{\pi}{2} \le \operatorname{Re} z \le \frac{\pi}{2}$ and assumes real values along the boundary lines $\operatorname{Re} z = \pm\frac{\pi}{2}$;
(2) f also assumes real values on the real interval $-\frac{\pi}{2} < \operatorname{Re} z < \frac{\pi}{2}$, $\operatorname{Im} z = 0$.
(3) $f(\bar{z}) = \overline{f(z)}$ in $-\frac{\pi}{2} < \operatorname{Re} z < \frac{\pi}{2}$.

Then f can be analytically extended to an entire function satisfying

$$f(\pi - z) = f(z), \quad f(z + 2\pi) = f(z), \quad z \in \mathbf{C}.$$

For instance, $\sin z$ is such a function.

Proof. See Fig. 3.42.

Let Ω_n denote the strip $n\pi - \frac{\pi}{2} < \operatorname{Re} z < n\pi + \frac{\pi}{2}$, $n = 0, \pm 1, \pm 2, \dots$. Then the translation $z \to z + n\pi$ maps Ω_0 univalently onto Ω_n.

The symmetric point of $z \in \Omega_0$ with respect to the line $\operatorname{Re} z = \frac{\pi}{2}$ is $z^* = -\bar{z} + \pi$. Since $f(z)$ are real along $\operatorname{Re} z = \frac{\pi}{2}$,

$$f(z^*) = f(\pi - \bar{z}) = f(z)^* \quad \text{(by symmetry principle)}$$

$$= \overline{f(z)} \quad \text{(the symmetric point of } f(z) \text{ with respect to}$$

$$\text{the real axis)}$$

$$= f(\bar{z}) \quad \text{(by assumption 3).}$$

Hence, $f(z^*) = f(\bar{z})$, where $-\frac{\pi}{2} < \operatorname{Re} z < \frac{\pi}{2}$ and $\frac{\pi}{2} < \operatorname{Re} z^* < \frac{3\pi}{2}$, indicates that f can be analytically extended from Ω_0 onto Ω_1 across the common boundary $\operatorname{Re} z = \frac{\pi}{2}$, and $f(\pi - z) = f(z)$ throughout Ω_0.

Similarly, the symmetric point of $z^* \in \Omega_1$ with respect to the line $\operatorname{Re} z = \frac{3\pi}{2}$ is $z^{**} = 3\pi - \overline{z^*} = 3\pi - \overline{(\pi - \bar{z})} = 2\pi + z$. Note that, by continuity of f along $\operatorname{Re} z = -\frac{\pi}{2}$ from Ω_0 and $f(\pi - z) = f(z)$, it follows that the extended f takes real values along $\operatorname{Re} z = \frac{3\pi}{2}$ and $f(\pi - z) = f(z)$ holds along $\operatorname{Re} z = -\frac{\pi}{2}$. Hence,

$$f(z^{**}) = f(2\pi + z) = f(z^*)^* \quad \text{(by symmetry principle)}$$

$$= \overline{f(z^*)} \quad \text{(the symmetric point of } f(z^*) \text{ with respect to}$$

$$\text{the real axis)}$$

$$= f(z) \quad \text{(it is known before that as } f(z^*) = \overline{f(z)}), \quad z \in \Omega_0.$$

This fact indicates that f can be analytically extended from Ω_1 to Ω_2 across the common boundary $\operatorname{Re} z = \frac{3\pi}{2}$. Furthermore,

$$f(z + 2\pi) = f(z), \quad z \in \Omega_0,$$

$$f(\pi - z) = f(z), \quad z \in \Omega_0 \quad \text{and} \quad f(\pi - z^*) = \overline{f(z)} = f(z^*), \quad z^* \in \Omega_1.$$

Repeating the above process, f can then be extended analytically from Ω_0 onto the whole plane $\mathbf{C}$ and satisfies $f(\pi - z) = f(z)$ and $f(z + 2\pi) = f(z)$, $z \in \mathbf{C}$. $\qquad \square$

*Example 4. $w = \sqrt[n]{z}(n \geq 2)$ is a single-valued univalent analytic function on its Riemann surface $R_{\sqrt[n]{z}}$.

Explanation. We have already confirmed this statement for $n = 2$ in Example 1 of Sec. 3.3.3.

In what follows, we adopt notations used in (2.7.2.3) with $\varphi = -\frac{\pi}{n}$ and $\mathbf{C}_{-\frac{\pi}{n}} = \mathbf{C} - (\infty, 0]$, $\Omega_k : \frac{(2k-1)\pi}{n} < \operatorname{Arg} w < \frac{(2k+1)\pi}{n}$, $k = 0, 1, 2, \ldots, n - 1$.

For fixed k, $0 \leq k \leq n - 1$,

$$w_k = |z|^{\frac{1}{n}} e^{i\frac{\operatorname{Arg} z + 2k\pi}{n}} : \mathbf{C} - (\infty, 0] = \{z | (2k - 1)\pi < \arg z < (2k + 1)\pi\},$$

$$\text{denoted by } D_k \to \Omega_k,$$

is a univalent analytic function, and is continuous from D_k up to its upper side $\arg z = (2k - 1)\pi$ and to its lower side $\arg z = (2k + 1)\pi$. Also, $w = w_k(z)$ maps $\arg z = (2k - 1)\pi$ and $\arg z = (2k + 1)\pi$ univalently onto the upper edge $\operatorname{Arg} w = \frac{(2k-1)\pi}{n}$ and the lower edge $\operatorname{Arg} w = \frac{(2k+1)\pi}{n}$ of Ω_k, respectively. Note that $w_k(z) = w_{k+1}(z)$ along the common side $\operatorname{Arg} z = (2k + 1)\pi$. By Painleve's theorem, the function $F(z)$ defined by

$$F(z) = \begin{cases} w_k(z), & (2k - 1)\pi < \arg z \leq (2k + 1)\pi \\ w_{k+1}(z), & (2k + 1)\pi \leq \arg z < (2k + 3)\pi \end{cases}$$

is single-valued, univalent, and analytic on $(2k - 1)\pi < \arg z < (2k + 3)\pi$. Hence, $w_k(z)$ is analytically continued into $w_{k+1}(z)$ across the ray $\arg z =$

$(2k+1)\pi$ for $k = 0, 1, 2, \ldots, n-1$. Observe that, when $k = n-1$, it should be understood that $w_n(z) = w_0(z)$.

Now, paste D_0 and D_1 along $\arg z = \pi$, D_1 and D_2 along $\arg z = 3\pi, \ldots, D_{n-2}$ and D_{n-1} along $\arg z = (2n-3)\pi$, and finally, D_{n-1} and D_0 along $\arg z = (2n-1)\pi$. The resulted n-sheet "surface"

$$R_{\sqrt[n]{z}} = \bigcup_{k=0}^{n-1} D_k \cup \left(\bigcup_{k=0}^{n-1} \{\arg z = (2k+1)\pi\} \right)$$

is the Riemann surface we described in Fig. 2.49. On $R_{\sqrt[n]{z}}$, $w = \sqrt[n]{z}$ is single-valued, univalent, and analytic.

It should be mentioned that all the Riemann surfaces for multiple-valued functions we described in Sec. 2.7 can be reconfirmed, via Painleve's theorem, just like the manner we just did for $w = \sqrt[n]{z}$.   □

Based on (3) in (3.4.3.4), it is easy to obtain the

Symmetry (or reflection) principle for harmonic functions. Let Ω be a symmetric domain (see (3.4.1.5) and Fig. 3.15 or Fig. 3.39(a)) and $\Omega^+ = \Omega \cap \{\operatorname{Im} z > 0\}$, the upper part of Ω; $\sigma = \Omega \cap \{\operatorname{Im} z = 0\}$, the part of the real axis in Ω. Suppose

(1) $v(z)$ is harmonic in Ω^+.
(2) $v(z)$ is continuous on $\Omega^+ \cup \sigma$.
(3) $v(z) = 0$ along σ.

The $v(z)$ has a harmonic extension to Ω, which satisfies the symmetric relation

$$v(\bar{z}) = -v(z). \tag{3.4.6.4}$$

Note that if σ is replaced by a circular arc and, in this case, $v(z)$ has the constant value 0 along σ, then the result is still valid with $v(z^*) = -v(z)$ where z^* is the symmetric point of z with respect to the circle. See (6.3.2.13) for a proof.

Example 4. (Ahlfors, Ref. [1], p. 174) If $u(z)$ is harmonic and $0 \le u(z) \le Ky$ for $y = \operatorname{Im} z > 0$, prove that $u(z) = ky$ with $0 \le k \le K$.

Proof. $0 \le u(z) \le Ky$ for $y > 0$ suggests that $\lim_{y \to 0^+} u(z) = 0$ uniformly. Hence, we define the boundary value of $u(z)$ along the real axis $y = 0$ to be identically equal to zero. By (3.4.6.4), $u(z)$ can be harmonically extended to the lower half-plane across $y = 0$ so that $u(\bar{z}) = -u(z)$. We still use $u(z)$

to denote the extended function on **C**. Note that for z with $\operatorname{Im} z = y < 0$, $0 \le -u(z) = u(\bar{z}) \le K(-y)$ holds, which, in turn, implies that $0 \ge u(z) \ge Ky$. Hence, $|u(z)| \le K|y|$ for all $z \in \mathbf{C}$.

By (3) in (3.4.3.4), there is an entire function g so that $g(z) = -v(z) + iu(z)$, $z \in \mathbf{C}$. Rewrite $g(z)$ by $-ig(z) = f(z) = u(z) + iv(z)$. This f is still entire and $\operatorname{Re} f(z) = u(z)$.

Applying the Schwarz integral formula to f on the disk $|z| < R$ (see Exercise B (4) of Sec. 3.4.2 or (6.3.2.3)), we have

$$f(z) = \frac{1}{2\pi i} \int_{|\zeta|=R} \frac{\zeta + z}{\zeta - z} u(\zeta) \frac{1}{\zeta} d\zeta + i \operatorname{Im} f(0), \quad |z| < R,$$

$$\Rightarrow f'(z) = \frac{-z}{\pi i} \int_{|\zeta|=R} \frac{u(\zeta)}{(\zeta - z)^2} \frac{1}{\zeta} d\zeta, \quad |z| < R.$$

When considering z with $|z| \le \frac{R}{2}$, then $|\zeta - z| \ge |\zeta| - |z| \ge R - \frac{R}{2} = \frac{R}{2}$. Observe that, on $|\zeta| = R$, $|u(\zeta)| \le K|\operatorname{Im} \zeta| \le K|\zeta| = KR$. Therefore,

$$|f'(z)| \le \frac{R}{2\pi} \int_{|\zeta|=R} \frac{KR}{\left(\frac{R}{2}\right)^2} \frac{1}{R} |d\zeta| = \frac{R}{2\pi} \cdot \frac{4K}{R^2} \cdot 2\pi R = 4K \quad \text{on } |z| \le \frac{R}{2}.$$

Thus $|f'(z)| \le 4K$ on **C**, and, by the Liouville theorem, $f'(z) = a$ is a constant. Hence, $f(z) = az + b$ where b is a constant. Since $f = u + iv$, so $u(z) = a_1 x - a_2 y + b_1$ where $a = a_1 + ia_2$ and $b = b_1 + ib_2$. When restricting to $y = \operatorname{Im} z > 0$, by assumption, $0 \le a_1 x - a_2 y + b_1 \le Ky$ always holds. This inequality forces both a_1 and b_1 to be equal to zero, and $-a_2 = k \ge 0$ and $k \le K$. The resulted $u(z) = ky$, $0 \le k \le K$, then follows. $\qquad\square$

Before proceeding to further applications, we stop for the moment and introduce a reference lemma of general interest (refer to Remark 3 in Sec. 2.4).

A topological lemma. Let f be a homeomorphism (i.e., topological mapping) from a domain Ω_1 onto a domain Ω_2.

(1) If $z_n \in \Omega_1$ *tends to the boundary* of Ω_1 (namely, for each compact subset K of Ω_1, there is an n_0 so that $z_n \notin K$ holds for $n \ge n_0$), then $f(z_n)$ tends to the boundary of Ω_2.

(2) In case both Ω_1 and Ω_2 are *Jordan domains* (i.e., the interior domains of the Jordan closed curves) and f is any univalent analytic function mapping Ω_1 onto Ω_2, then f can be extended to a homeomorphism from $\overline{\Omega_1}$ onto $\overline{\Omega_2}$ so that f maps $\partial\Omega_1$ homeomorphically onto $\partial\Omega_2$.

$$(3.4.6.5)$$

In (2), the existence of such f between Ω_1 and Ω_2 is provided by the Riemann mapping theorem (see (6.1.1)). For a detailed proof, see Ref. [58], Vol. III, p. 70 or see (6.1.2) and its proof in Sec. 6.1.2.

Example 5. (Continued from Example 4 of Sec. 2.5.4) There exists a univalent analytic function f mapping the circular ring domain $\Omega_1 : 0 < r_1 < |z| < r_2 < \infty$ onto another one $\Omega_2 : 0 < R_1 < |w| < R_2 < \infty$ if and only if

$$\frac{r_2}{r_1} = \frac{R_2}{R_1}.$$

In this case, $f(z) = az\left(|a| = \frac{R_1}{r_1}\right)$ or $\frac{a}{z}(|a| = r_1 R_2)$.

The ratio $\frac{r_2}{r_1}$ of radii or $\log \frac{r_2}{r_1}$ is called the *modulus* of the ring $r_1 < |z| < r_2$. It is a *conformal invariant* and can be used to classify the ring domains (i.e., a domain whose complement in $\mathbf{C}^*$ has two nondegenerated components).

Proof. Only the necessity is needed to be proved. Using $\frac{z}{r_1}$ to replace z, and $\frac{w}{R_1}$ to replace w if necessary, we may suppose that $r_1 = R_1 = 1$. Applying (3.4.6.5) to the simply connected domain $\Omega_1 - \{z \,|\, 1 \le \operatorname{Re} z \le r_2$ and $\operatorname{Im} z = 0\}$, we may suppose that $|f(z)| = 1$ on $|z| = 1$. Hence, $|f(z)| = R_2 = R$ on $|z| = r_2 = r$.

By symmetry principle (3.4.6.2), f can be extended analytically to $\overline{\Omega_1}$: $1 \le |z| \le r$ and maps $\overline{\Omega_1}$ univalently onto $\overline{\Omega_2} : 1 \le |w| \le R$. Consider the harmonic function (see (3.4.3.2))

$$u(z) = \log |f(z)| - \frac{\log R}{\log r} \log |z|, \quad z \in \Omega,$$

whose boundary values along $|z| = 1$ and $|z| = r$ are all equal to zero. By maximum principle for harmonic functions (see (3.4.3.7)), it follows that $u(z) \equiv 0$ on Ω. Consequently (see (3.4.1.4)), there is a real constant α so that

$$F(z) = \log f(z) - \frac{\log R}{\log r} \log z = i\alpha, \quad z \in \Omega.$$

Choose any fixed $0 < \rho < r$. Let $\Gamma_\rho = f\{|z| = \rho\}$, the image curve of $|z| = \rho$ under f. Since Γ_ρ is a Jordan closed curve containing 0 in its interior, thus (see (2.7.1.12) or (5) in (3.5.2.4)),

$$1 = \frac{1}{2\pi i} \int_{\Gamma_\rho} \frac{dw}{w} = \frac{1}{2\pi i} \int_{|z|=\rho} \frac{f'(z)}{f(z)} dz.$$

This indicates that, if z winds around $|z| = \rho$ once so that $\operatorname{Arg} z$ increases 2π, $w = f(z)$ will wind around Γ_ρ once so that $\operatorname{Arg} f(z)$ increases 2π, too (see also (3.5.3.4)). Under this circumstance, the amount increased by $\operatorname{Im} F(z) = \alpha$ (a constant) is $2\pi\left(1 - \frac{\log R}{\log r}\right)$, which is necessarily equal to zero. It follows that $\frac{\log R}{\log r} = 1$ and $R = r$ holds.

For *another proof*, suppose $|f(z)| \to R_j$ as $|z| \to r_j$ for $j = 1, 2$.

By symmetry principle (3.4.6.2), f can be analytically extended to $r_2 < |z| < \frac{r_2^2}{r_1}$ across the boundary $|z| = r_2$, and maps $r_2 < |z| < \frac{r_2^2}{r_1}$ univalently onto $R_2 < |w| < \frac{R_2^1}{R_2}$. Similarly, f can also be analytically extended to $\frac{r_1^2}{r_2} < |z| < r_1$ across $|z| = r_1$ and maps it univalently on $\frac{R_1^2}{R_2} < |w| < R_1$. Proceed in this way, $w = f(z)$ eventually can be extended to an analytic function, mapping $0 < |z| < \infty$ univalently onto $0 < |w| < \infty$.

0 is a removable singularity of the extended f and $f(0) = 0$ holds (see (3.4.2.17)).

$z = \infty$ is an isolated singularity of f. Owing to the univalence of f on $|z| < 1$, ∞ can not be a removable or an essential singularity but only a simple pole (see (4.10.2.1)). Let az, $a \neq 0$, be the singular part of f at ∞. Then it follows that $f(z) = az + b$, where b is a constant. Since $f(0) = 0$, it follows that $b = 0$. Also, $|f(z)| = |a||z| = |a|r_1 = R_1$ on $|z| = r_1$ and so $|a| = \frac{R_1}{r_1}$ and $\frac{R_1}{r_1} = \frac{R_2}{r_2}$ follows.

In case $|f(z)| \to R_2$ as $|z| \to r_1$, and $|f(z)| \to R_1$ as $|z| \to r_2$. Apply the known result to $\frac{1}{f(z)}$ and the result follows.

Yet for *another proof*, see Ref. [58], Vol III, p. 4.   □

We will present more applications of the symmetry principle in suitable places as we proceed after introducing more concepts or tools needed.   □

Exercises A

(1) If f is an entire function and real on the real axis, purely imaginary on the imaginary axis, show that f is odd, i.e., $f(-z) = -f(z)$.

(2) Show that every function, which is analytic in a symmetric domain Ω, can be expressed as $f_1 + if_2$ where f_1, f_2 are analytic in Ω and real on the real axis.

(3) Does there exist a nonconstant analytic function in $|z| \leq 1$, which is real on $|z| = 1$?

(4) Suppose f is bounded and analytic in $\operatorname{Im} z \geq 0$ and real on the real axis. Show that f is a constant.

(5) Suppose f is analytic in the open square $0 < \operatorname{Re} z < 1$, $0 < \operatorname{Im} z < 1$, and continuous on the closed square $0 \leq \operatorname{Re} z \leq 1$,

$0 \leq \operatorname{Im} z \leq 1$ except at four vertices where it assumes ∞. If f also assumes real values on the boundary (except at the four vertices) of the square, try to describe how to extend f analytically beyond the square.

(6) There exists a univalent analytic function f mapping a rectangle Ω_1 onto a rectangle Ω_2, with continuous extension to $\overline{\Omega_1}$ and $\overline{\Omega_2}$, so that the four vertices correspond to each other, if and only if Ω_1 and Ω_2 are similar, namely, they have equal ratio of length to width. Refer to Exercise (3) in Appendix B.

(7) Let f be analytic in $|z| \leq 1$, $\operatorname{Im} z > 0$ and real on $|z| = 1$, $\operatorname{Im} z > 0$. Show that f can be analytically extended to the upper half-plane. In what manner?

(8) Suppose f is an analytic function mapping $0 < \rho < |z| < 1$ onto $|w| < 1$ minus the real segment $-\alpha \leq \operatorname{Re} w \leq \alpha$, $\operatorname{Im} w = 0$ $(0 < \alpha < 1)$.

   (a) Show that $w = f(z)$ can be analytically extended so that it maps $\rho < |z| < \frac{1}{\rho}$ onto the whole plane $\mathbf{C}^*$ minus the real segments $-\alpha \leq \operatorname{Re} w \leq \alpha$ $(\operatorname{Im} w = 0)$ and rays $-\infty \leq \operatorname{Re} w \leq -\frac{1}{\alpha}$, $\frac{1}{\alpha} \leq \operatorname{Re} w \leq \infty$ $(\operatorname{Im} w = 0)$.

   (b) Show that f can be analytically extended across $|z| = 1$ according to the relation $f(z)\overline{f(\frac{1}{\bar{z}})} = 1$; across $|z| = \rho$ according to $f(z) = \overline{f(\frac{\rho^2}{\bar{z}})}$.

(9) Suppose that an analytic function f maps the interior of a rectangle, with vertices $-a$, a, $a + bi$ and $-a + bi$ $(a > 0, b > 0)$, univalently onto $\operatorname{Im} w > 0$ and satisfies $f(-a) = \alpha$, $f(a) = \beta$ with $\alpha < \beta$. Show that f can be analytically extended so that it maps the interior of the rectangle, with vertices $-a - bi$, $a - bi$, $a + bi$ and $-a + bi$, univalently onto the whole plane $\mathbf{C}$, minus the rays $-\infty < w \leq \alpha$ and $\beta \leq w < \infty$. In case $\alpha > \beta$, then f can be so analytically extended that it maps the interior of a larger rectangle onto $C - \{w \mid \beta \leq \operatorname{Re} w \leq \alpha$ and $\operatorname{Im} w = 0\}$.

(10) Let f be an analytic function mapping $\operatorname{Im} z > 0$ univalently onto the interior of the ellipse $\frac{u^2}{a^2} + \frac{v^2}{b^2} = 1$, where $w = f(z)$, $z = x + iy$ and $w = u + iv$. In case $w = f(z)$ maps the real axis $\operatorname{Im} z = 0$ univalently onto the ellipse, then it necessarily satisfies

$$\frac{b^2}{4}[f(z) + \overline{f(\bar{z})}]^2 - \frac{a^2}{4}[f(z) - \overline{f(\bar{z})}]^2 = a^2 b^2.$$

Try to find such a mapping $w = f(z)$.

Exercises B

(1) Let f be a univalent analytic function mapping $|z| < 1$ onto a domain Ω, which is symmetric with respect to a line L. If $f(0) = 0$, then there is a diameter γ of $|z| < 1$ so that $f(\gamma) = \Omega \cap L$.

(2) Let $f : \Omega = \{z \mid 0 < \operatorname{Re} z < 1 \text{ and } \operatorname{Im} z > 0\} \to \mathbf{C}$ be analytic, bounded, and continuous on $\overline{\Omega}$, assume real values on $(0, 1)$, and

$$|f(iy)| \le M_1, \quad |f(1 + iy)| \le M_2, \quad y \ge 0,$$

where M_1, M_2 are constants. Show that $|f(z)| \le M_1^{1 - \operatorname{Re} z} M_2^{\operatorname{Re} z}$, $z \in \Omega$.

(3) Prove (3.4.6.3) and (3.4.6.4) in detail.

*3.4.7.   The inverse and implicit function theorems

In this section, we state these two fundamental theorems without proofs. They all can be proved by the well-known methods in the so-called *Advanced Calculus* course. For other treatment in the content of complex analysis, refer to Secs. 3.5.1 and 3.5.6.

Section (1) Inverse function theorem

Suppose that f is analytic at z_0 and $f'(z_0) \ne 0$. There is an $r > 0$ so that $|f'(z) - f'(z_0)| < \frac{1}{2}|f'(z_0)|$ if $|z - z_0| < r$, by the continuity of f' at z_0 (see (2) in (3.4.2.6)).

Let

$$R(z, z_0) = f(z) - f(z_0) - (z - z_0)f'(z_0), \quad |z - z_0| < r, \qquad (*_1)$$

which we denoted as $o(z - z_0)$ in (3.2.1). Choose z_1 and z_2 so that $|z_1 - z_0| < r$ and $|z_2 - z_0| < r$. Then,

$$R(z_2, z_0) - R(z_1, z_0) = \int_{z_1}^{z_2} [f'(z) - f'(z_0)]dz$$

$$\Rightarrow |R(z_2, z_0) - R(z_1, z_0)| \le \frac{1}{2}|f'(z_0)||z_1 - z_2|,$$

$$|z_1 - z_0| < r, \quad |z_2 - z_0| < r, \qquad (*_2)$$

which can also be proved by using the mean-value theorem (3.4.1.1).

Rewrite $(*_1)$ as: Letting $w_0 = f(z_0)$ and $w = f(z)$,

$$z = z_0 + \frac{1}{f'(z_0)}(w - w_0) - \frac{1}{f'(z_0)}R(z, z_0), \quad |z - z_0| < r.$$

With the crucial estimate $(*_2)$ in mind and adopting successive approximation method, one can prove the following

Local inverse function theorem. Suppose f is analytic at z_0 and $f'(z_0) \neq 0$. Let $w = f(z)$ and $w_0 = f(z_0)$.

(1) Existence and uniqueness. There exist $r > 0$ and $R > 0$ so that the functional equation

$$f(g(w)) = w$$

has a unique analytic solution $z = g(w)$ on $|w - w_0| < R$, satisfying

(i) $g(w_0) = z_0$.
(ii) $|g(w) - z_0| < r$ on $|w - w_0| < R$.

In this case, $f'(z) \neq 0$ holds on $|z - z_0| < r$ and

$$g'(w) = \frac{1}{f'(z)}, \quad z = g(w), \quad |w - w_0| < R.$$

Let $\Omega = g(\{|w - w_0| < R\})$. Then Ω is a domain containing z_0 so that $f : \Omega \to \{|w - w_0| < R\} = D$ is univalently analytic and has an *analytic inverse function* $g : D \to \Omega$, satisfying

(i) $f(g(w)) = w, \quad w \in D$.
(ii) $g(f(z)) = z, \quad z \in \Omega$.

This g is a *local (single-valued, analytic) branch* of the "multiple-valued" inverse function $z = f^{-1}(w)$ at w_0. As a metter of fact, g is the only such local branch owing to $f'(z_0) = 0$ (See (3.5.1.4)).

(2) Recursive approximation sequence. Let

$$g_0(w) = z_0,$$

$$g_n(w) = g_{n-1}(w) - \frac{1}{f'(z_0)}[f(g_{n-1}(w)) - w], \quad n \geq 1, \quad |w - w_0| < R.$$

Then the sequence $g_n(w)$ of functions defined on $|w - w_0| < R$ has the properties:

(i) $g_n(w)$ is analytic;
(ii) $|g_n(w) - z_0| \leq \left(1 - \frac{1}{2^n}\right) r, \quad n \geq 0$;
(iii) $|g_n(w) - g_{n-1}(w)| < \frac{1}{2^n} r, \quad n \geq 1, \quad$ and $\quad |g_n(w) - g(w)| < \frac{1}{2^{n-2}} r,$
     $n \geq 2$.

Therefore, $g_n(w)$ converges uniformly on $|w - w_0| < R$ to the inverse function $z = g(w)$. (3.4.7.1)

R in (1) usually is the shortest distance from w_0 to the *branch points* of $z^{-1} = f(w)$, namely, these points w given by these points z such that

$f(z) = w$ and $f'(z) = 0$ (see (3.5.1.8)). For concrete illustrative examples, see Sec. 3.5.1.

Remark 1 (Open mapping property and max-min modulus principle). Suppose f is analytic on a domain Ω and $f'(z) \neq 0$ everywhere (an unnecessary assumption). Then (3.4.7.1) shows easily that $f(\Omega)$ is a domain too, and $|f|$ does not assume any local maximum in Ω; and, if in addition that $f(z) \neq 0$ in Ω, $|f|$ does not even assume any local minimum in Ω.   ☐

Section (2) Implicit function theorem

Under what conditions can one solve w as a single-valued analytic function of z from the equation $z^3 + w^3 - 3zw = 0$? In what neighborhood of z? How many such solutions are there? To formulate a general result in this direction, we give some preparatory explanations as follows.

Let $\mathbf{C}^2 = \{(z, w) \,|\, z, \ w \ \in \ \mathbf{C}\}$ be the two-dimensional complex vector space over $\mathbf{C}$. $\mathbf{C}^2$, endowed with the natural inner product $\langle (z, w), (z', w') \rangle = z\bar{z}' + w\bar{w}'$, is an inner product space. The *length* of the vector (z, w) is $|(z, w)| = \langle (z, w), (z, w) \rangle^{\frac{1}{2}} = \sqrt{|z|^2 + |w|^2}$. The point set $B_r(z, w) = \{(z', w') \in \mathbf{C}^2 \,|\, |z - z'|^2 + |w - w'|^2 < r^2\}$ is called an *open disk* with center (z, w) and radius $r > 0$. A set O in $\mathbf{C}^2$ is said to be *open* if, for each point $(z, w) \in O$, there is some $r > 0$ so that $B_r(z, w) \subseteq O$ holds. Thus, $\mathbf{C}^2$ is a *topological space* (see Exercise B (2) of Sec. 1.8) which is equivalent to the *product topology* $\mathbf{C} \times \mathbf{C}$ with $U \times V = \{(z, w) \,|\, z \in U \text{ and } w \in V\}$ as its basic open sets, where U and V are nonempty open sets in $\mathbf{C}$. Note that, when viewed $\mathbf{C}^2$ as a real vector space with $z = (x, y) = x + iy$ and $w = (u, v) = u + iv$, it is linearly isomorphic to $\mathbf{R}^4 = \{(x, y, u, v) \,|\, x, y, u, v \in \mathbf{R}\}$ and has the same topological structure as $\mathbf{R}^4$.

As a generalized counterpart of (3.2.2) and (3.4.2.15) for $\mathbf{C}^2$, we have the

Characteristic properties for analytic function of two complex variables. Let $F = F(z, w) : \Omega \to \mathbf{C}$ be a complex valued function of two complex variables z and w, where $\Omega \subseteq \mathbf{C}^2$ is a nonempty open set. Then, the following are equivalent.

(1) For each fixed w, $F(\cdot, w)$ is an analytic function of z; and, for each fixed z, $F(z, \cdot)$ is an analytic function of w.
(2) For each point $(z_0, w_0) \in \Omega$, there are $r > 0$ and $R > 0$ so that, on $|z - z_0| < r$ and $|w - w_0| < R$, $F(z, w)$ can be expressed as a convergent

double series

$$F(z,w) = \sum_{j,k=0}^{\infty} a_{jk}(z-z_0)^j (w-w_0)^k.$$

(3) $F(z,w)$ is continuously differentiable in the real sense with respect to (x,y,u,v) and satisfies the *Cauchy–Riemann equations* in Ω:

$$\frac{\partial F}{\partial \bar{z}} = \frac{1}{2}\left(\frac{\partial F}{\partial x} + i\frac{\partial F}{\partial y}\right) = 0;$$

$$\frac{\partial F}{\partial \bar{w}} = \frac{1}{2}\left(\frac{\partial F}{\partial u} + i\frac{\partial F}{\partial v}\right) = 0.$$

In any of these cases, $F(z,w)$ is said to be *analytic* in Ω.   (3.4.7.2)

(1) $\Rightarrow$ (2) is *Hartog's theorem*, see Ref. [11], Chap. 7; if continuity of $F(z,w)$ is imposed, then (1) $\Rightarrow$ (2) is *Osgood's theorem*, which can be proved by the classical method of the Cauchy integral formula. (2) $\Rightarrow$ (1) is proved by imitating Sec. 3.3.2, while (1) $\Leftrightarrow$ (3) is similar to (1) $\Leftrightarrow$ (6) in (3.2.2), but the continuity of the derivative is needed here.

As usual in calculus, the *partial differentiation (derivative)* of $F(z,w)$ with respect to either z or w is defined as

$$\frac{\partial F}{\partial z}(z,w) = F_z(z,w) = \lim_{h\to 0}\frac{F(z+h,w)-F(z,w)}{h}$$

$$= \frac{1}{2}\left(\frac{\partial F}{\partial x} - i\frac{\partial F}{\partial y}\right)(z,w);$$

$$\frac{\partial F}{\partial w}(z,w) = F_w(z,w) = \lim_{k\to 0}\frac{F(z,w+k)-F(z,w)}{k}$$

$$= \frac{1}{2}\left(\frac{\partial F}{\partial u} - i\frac{\partial F}{\partial v}\right)(z,w).$$

They also enjoy the well-known operational properties (see Sec. 2.8).

Suppose $F(z,w)$ is analytic and $F(z_0,w_0)=0$, $F_w(z_0,w_0)\neq 0$. Applying Banach fixed point theorem to $\varphi(z,\cdot)$ defined by $\varphi(z,w)=w-\frac{F(z,w)}{F_w(z_0,w_0)}$, we can prove the following

Implicit function theorem of two complex variables. Let $U,V \subseteq \mathbf{C}$ be nonempty open sets and $F=F(z,w):U\times V \to \mathbf{C}$ be analytic. Suppose a point $(z_0,w_0)\in U\times V$ satisfies:

(1) $F(z_0, w_0) = 0$, and

(2) $\dfrac{\partial F}{\partial w}(z_0, w_0) \neq 0$.

Then, there exist an open neighborhood I of z_0 in U, and an open neighborhood J of w_0 in V and a *unique* analytic function $w = f(z) : I \to J$ satisfying:

(1) $w_0 = f(z_0)$; and, if $z \in I$, then $w = f(z)$ satisfies the equation $F(z, f(z)) = 0$;

(2) $\dfrac{\partial F}{\partial w}(z, w) \neq 0$ holds on $z \in I$, and

$$\frac{dw}{dz} = f'(z) = -\frac{\frac{\partial F}{\partial z}(z, f(z))}{\frac{\partial F}{\partial w}(z, f(z))}, \quad z \in I.$$

Furthermore, the sequence $f_n : I \to \mathbf{C}$ defined by

$$f_0(z) = w_0,$$

$$f_{n+1}(z) = f_n(z) - \frac{F(z, f_n(z))}{\frac{\partial F}{\partial w}(z_0, w_0)}, \quad n \geq 0, z \in I,$$

converges uniformly to f on I, namely, $\{f_n\}$ is a successive approximation sequence of f on I. \hfill (3.4.7.3)

For other proofs, see Ref. [58], Vol. III, pp. 105–110; [39]. pp. 13–16. Sections 3.5.1 and 3.5.6 will present some illustrative examples.

Remark 2. The inverse and implicit function theorems formulated in (3.4.7.1) and (3.4.7.3) are mainly the forms we usually encountered in Advanced Calculus textbooks. They are single-valued theorems by nature. It is the very characteristic of an analytic function to be representable locally by a convergent power series, or equivalent, if f is a nonconstant function analytic at a point z_0, then $f(z) = f(z_0) + (z - z_0)^n \varphi(z)$ where φ is analytic at z_0 and $\varphi(z_0) \neq 0$ for some positive integer n, which enables us to study the *multi-valued inverse and implicit function theorems*, namely, the case that $f'(z_0) = \cdots = f^{(k-1)}(z_0) = 0$ but $f^{(k)}(z_0) \neq 0$ for $k \geq 2$ in (3.4.7.1), and the case that $\dfrac{\partial F}{\partial w}(z_0, w_0) = \cdots = \dfrac{\partial^{k-1} F}{\partial w^{k-1}}(z_0, w_0) = 0$ but $\dfrac{\partial^k F}{\partial w^k}(z_0, w_0) \neq 0$ in (3.4.7.3). We will do these in Secs.3.5.1 and 3.5.6 by purely complex analytic methods. \hfill $\square$

3.5.   Geometric Properties of Analytic Functions

Recall that a nonzero complex number is not just a planar vector but also represents a rotation followed by a one-way stretch (see Sec. 1.1). This section is devoted to emphasize the (infinitesimally and globally) geometric properties of analytic functions influenced by (4) and (5) in (3.2.2). Before starting, readers are asked to review the content of Sec. 3.2.3.

Still seven subsections are divided.

3.5.1.   *Local behavior of an analytic function at a point:*
   Conformality, etc.

The geometric behavior of an analytic function at a point originated primarily from the geometric characterization of the differentiability at that point as shown in Sec. 3.2.3, in particular, the result (3.2.3.3). Even as far as in Sec. 2.5.1, we have shown the conformality of a polynomial at points where its first derivatives are nonzero (see (2.5.1.6)) and, henceforth, applied this concept to rational functions (Sec. 2.5.3) including linear transformations (Sec. 2.5.4) and the Joukowski function (Sec. 2.5.5), and elementary transcendental functions (Sec. 2.6) in some graphic illustrations (e.g., see Figs. 2.13–2.15, 2.20, 2.21, 2.28, etc.).

Here we formally introduce these related concepts to analytic functions in a general setting.

Two section are divided. It is always supposed that *f is analytic at the point z_0.*

Section (1) The case that $f'(z_0) \neq 0$

Firstly, we try to review (2.5.1.6) with the polynomial $p(z)$ there replaced by an analytic function $f(z)$.

Let $z(t) : [-1, 1] \to \mathbf{C}$ be a differentiable curve passing z_0, say $z(0) = z_0$, and $z'(0) \neq 0$. Then the image curve $w(t) = f(z(t)) : [-1, 1] \to \mathbf{C}$ is also differentiable and passes $w(0) = f(z(0)) = f(z_0)$, and $w'(0) = f'(z_0)z'(0) \neq 0$. This implies that

$$\operatorname{Arg} w'(0) = \operatorname{Arg} f'(z_0) + \operatorname{Arg} z'(0),$$
$$\Rightarrow \operatorname{Arg} w'(0) - \operatorname{Arg} z'(0) = \operatorname{Arg} f'(z_0). \qquad (3.5.1.1)$$

Hence, the *difference* between the argument of the directed tangent vector at $f(z_0)$ of the image curve $f(z(t))$ and that of the tangent vector at z_0 of *any* differentiable curve $z(t)$ passing z_0 is always equal to a constant, namely, $\operatorname{Arg} f'(z_0)$, which is called the *rotation angle* of $w = f(z)$ at z_0.

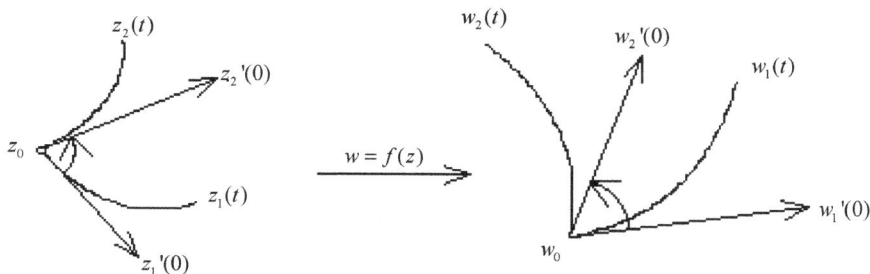

Fig. 3.43

Choose any two differentiable curves $z_1(t)$ and $z_2(t)$ with $z_1(0) = z_2(0) = z_0$ and $z_1'(0) \neq 0$, $z_2'(0) \neq 0$. Their image curves $w_1(t)$ and $w_2(t)$ under $w = f(z)$ all pass $f(z_0)$ and

$$\operatorname{Arg} w_j'(0) = \operatorname{Arg} f'(z_0) + \operatorname{Arg} z_j'(0), \quad j = 1, 2,$$

$$\Rightarrow \operatorname{Arg} w_2'(0) - \operatorname{Arg} w_1'(0) = \operatorname{Arg} z_2'(0) - \operatorname{Arg} z_1'(0). \quad (3.5.1.2)$$

Consequently, $w = f(z)$, as a mapping, *preserves angle and its sense at the point z_0*. See Fig. 3.43. And thus, the mapping $w = f(z)$ is said to be *conformal* at a point z_0 with $f'(z_0) \neq 0$.

Definition. Let $w = f(z) : O(\text{open set}) \to \mathbf{C}$ be a continuous mapping. Let $z_0 \in O$ be a fixed point. If $w = f(z)$ has the following properties at z_0:

(1) the image curve $w(t) = f(z(t))$, under $w = f(z)$, of any differentiable curve $z(t)$ with $z(0) = z_0$ and $z'(0) \neq 0$ is also differentiable, $w(0) = f(z_0)$ and $w'(0) \neq 0$, and

(2) two differentiable curves that form an angle at z_0 (such as $\operatorname{Arg} z_2'(0) - \operatorname{Arg} z_1'(0)$ in (3.5.1.2)) are mapping, via $w = f(z)$, upon curves forming the same angle at $f(z_0)$, in sense as well as in size,

then $w = f(z)$ is said to be *conformal* at z_0. $w = f(z)$ is said to be a *conformal mapping* on O if it is conformal at every point in O. In case $w = f(z)$ preserves angles at z_0 *but* reverses the sense, then $w = f(z)$ is called *sense-reversing conformal* at z_0. For instance, $f(z) = \bar{z}$ is sense-reversing conformal on $\mathbf{C}$. The *conformality* of f at points z_0 where $z_0 = \infty$ or $f(z_0) = \infty$ is defined as in the Remark in Sec. 3.3 (see also Section (1) in Sec. 2.5.4).   (3.5.1.3)

Even though $f(z) = z^2$ is conformal on $\mathbf{C} - \{0\}$, it is not univalent on $\mathbf{C} - \{0\}$ but only locally univalent in a neighborhood of any point in $\mathbf{C} - \{0\}$.

In general, the continuity of f' at z_0 (recall that $f'(z_0) \neq 0$) implies that $f'(z) \neq 0$ in an (even connected and convex) open neighborhood O of z_0, say an open disk $|z - z_0| < r$. Then (3.4.1.2) guarantees that $w = f(z)$ will be univalent on O. By (5) in (3.1.6) or using the inverse function theorem (3.4.7.1), plus the open mapping property (3.4.4.4), $w = f(z) : O \to f(O)$ (open or a domain) has an analytic inverse $z = f^{-1}(w) : f(O) \to O$, which is still conformal because $(f^{-1})'(w) = \frac{1}{f'(z)}$ with $z \in O$ and $w = f(z)$.

We summarize what we obtained so far in the following

Local behavior of an analytic function f at a point z_0 where $f'(z_0) \neq 0$. As a mapping $w = f(z)$:

(1) (Refer to (3.2.3.1) and (3.2.3.2), and (3.2.3.3)) the *dilatation* of f at z_0 in *any* direction: $|f'(z_0)| = \lim_{z \to z_0} \frac{|f(z) - f(z_0)|}{|z - z_0|}$; the *rotation angle* of f at z_0 : $\operatorname{Arg} f'(z_0) = \lim_{z \to z_0} \operatorname{Arg} \frac{f(z) - f(z_0)}{z - z_0}$.

(2) f is conformal at z_0.

(3) (*Local single-valued inverse function theorem*, refer to (3.4.7.1)). There exists an open disk $|z - z_0| < r$ (or an open neighborhood of z_0) so that the analytic function $w = f(z)$ maps $|z - z_0| < r$ *topologically* and *conformally* onto a domain (in this case, $f'(z) \neq 0$ on $|z - z_0| < r$ and the inverse function $z = f^{-1}(w)$ is also conformal).

(4) Let $z_0 = x_0 + iy_0$.

    (i) $w = f(z)$ maps each pair of the following orthogonal curves:

        (a) line segments $x = x_0$ and $y = y_0$, and

        (b) (in case $z_0 \neq 0$) the circle $|z| = |z_0|$ and the ray $\operatorname{Arg} z = \operatorname{Arg} z_0$, onto a pair of orthogonal curves at $f(z_0)$.

    (ii) Each pair of the following *level curves*

        (a) The *equipotential line* $u(z) = u(x, y) = \operatorname{Re} f(z_0)$ and the *streamline line* $v(z) = v(x, y) = \operatorname{Im} f(z_0)$, and

        (b) (in case $f(z_0) \neq 0$) $|w| = |f(z_0)|$ and $\operatorname{Arg} w = \operatorname{Arg} f(z_0)$

    intersects orthogonally at z_0, where $f = u + iv$ is viewed as the *complex potential* of a flow (see Sec. 3.2.4).   (3.5.1.4)

Readers are asked to give a direct proof for (4) without recourse to (2) and (3). See Fig. 3.44.

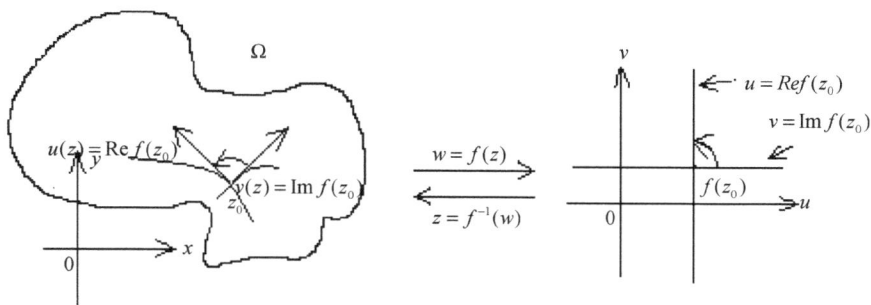

Fig. 3.44

Second, we try to describe the *infinitesimally* geometrical mapping behavior of f at z_0 in case $f'(z_0) \neq 0$.

Recall (see (3.2.1)) that the differentiability of f at z_0 can be expressed as

$$f(z) = f(z_0) + f'(z_0)(z - z_0) + o(z - z_0) \quad \text{with} \quad \lim_{z \to z_0} \frac{o(z - z_0)}{z - z_0} = 0,$$

where $w = f'(z_0)(z - z_0)$ is called the *linear part* of f at z_0 while $w = f(z_0) + f'(z_0)(z - z_0)$ the *affine part* of f at z_0 and $o(z - z_0)$ the *error estimate* (function) of f at z_0 from its affine part. Note that the affine transformation $w = f(z_0) + f'(z_0)(z - z_0)$ is the composite of the following motions:

$z \to |f'(z_0)|(z - z_0)$, a stretching with center z_0 and dilatation $|f'(z_0)|$

$\to |f'(z_0)|e^{i \operatorname{Arg} f'(z_0)}(z - z_0) = f'(z_0)(z - z_0)$, a rotation with center z_0 and angle $\operatorname{Arg} f'(z_0)$

$\to f(z_0) + f'(z_0)(z - z_0) = w$, a translation along the vector $f(z_0)$.

Namely, $w = f(z_0) + f'(z_0)(z - z_0)$ is a similarity mapping, preserving angle and its sense. It maps a triangle with a vertex at z_0 onto a similar triangle with a vertex at $f(z_0)$. See Fig. 3.45. The side $\overline{z_0 z_1} : z = z_0 + te^{i\alpha}$, $0 \leq t \leq |z_1 - z_0|$, $\alpha = \operatorname{Arg}(z_1 - z_0)$, has its images, under $w = f(z_0) + f'(z_0)(z - z_0)$ and $w = f(z)$, respectively, the line segment $\overline{w_0 w_1} : w = f(z_0) + tf'(z_0)e^{i\alpha}$ and the curve $w = f(z_0 + te^{i\alpha}) = f(z_0) + tf'(z_0)e^{i\alpha} + o(te^{i\alpha})$. They all pass $w_0 = f(z_0)$ and tangent to each other at w_0, and the distance between $f(z_0 + te^{i\alpha})$ and the corresponding point on $\overline{w_0 w_1}$ is $|o(te^{i\alpha})| = |o(z - z_0)|$. As a consequence, the triangle $\Delta w_0 w_1 w_2$ and the curvilinear triangle $f(z_0) f(z_1) f(z_2)$ have the same angle and sense at the vertex w_0. On the

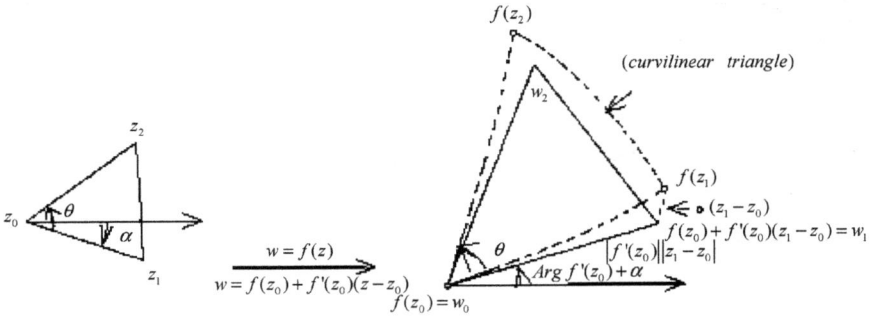

Fig. 3.45

other hand, the side $\overline{w_0 w_1}$ has its length equal to $|f'(z_0)||z_1 - z_0|$, while that of the curvilinear side $\overline{f(z_0)f(z_1)}$ is approximately equal to $|f'(z_0)||z_1 - z_0|$, too, if $|z_1 - z_0| > 0$ is sufficiently small.

Readers are urged to describe how a square with a vertex at z_0 or a circle with center at z_0 is mapped infinitesimally by $w = f(z_0) + f'(z_0)(z - z_0)$ and $w = f(z)$, respectively.

Third, we have formulas for

Arc length and area under a univalent analytic function $w = f(z) : \Omega \to \mathbf{C}$:

(1) Suppose $z(t) : [a, b] \to \Omega$ is a continuously differentiable curve γ. Then the image curve $f(z(t)) : [a, b] \to \mathbf{C}$ is also continuously differentiable and has its length

$$l(f(\gamma)) = \int_a^b |f'(z(t))||z'(t)|dt \quad \text{or simply as} \quad \int_\gamma |f'(z)||dz|.$$

(2) Suppose $A \subseteq \Omega$ has planar area (for instance, A is a domain with a compact closure $\overline{A} \subseteq \Omega$ so that the boundary ∂A is a set of measure zero). Then, $f(A)$ has the planar area

$$\iint_A |f'(z)|^2 dx dy. \tag{3.5.1.5}$$

Proofs are left as Exercise A (2).

Some illustrative examples are as follows. For $f(z) = z^2$, refer to the Example in Sec. 2.5.2 and Figs. 2.14 and 2.15 there.

Example 1. (Ahlfors, Ref. [1], p. 91) Try to use $w = f(z) = z^3$ to explain (3.5.1.4) and (3.5.1.5).

Solution. $f'(z) = 3z^2 = 0 \Leftrightarrow z = 0$. Hence, $f(z)$ is conformal everywhere in $\mathbf{C} - \{0\}$, yet it is three-to-one there and hence is not univalent.

Let $z = x + iy$ and $w = u + iv$. Then, $w = z^3$ implies that $u(x, y) = x^3 - 3xy^2$ and $v(x, y) = 3x^2y - y^3$.

The affine part of f at $z_0 = 1$ is $w = f(z_0) + f'(z_0)(z - z_0) = 1 + 3(z - 1) = 3z - 2$ and the error estimate is $o(z - 1) = z^3 - (1 + 3(z - 1)) = z^3 - 3z + 2 = (z - 1)^2(z + 2)$ if $|z - 1| > 0$ is sufficiently small. $w = z^3$ maps the segment $0 < x < 2$ on the real axis univalently onto the segment $0 < u < 8$, and maps the segment $z = 1 + iy$, $|y| \leq 1$, univalently onto the curve $u = 1 - 3y^2$, $v = 3y - y^3$, which intersects the segment $0 < u < 8$ at $w_0 = 1$ orthogonally (see Fig. 3.50).

To determine the equipotential line $u(x, y) = u_0$, namely, $x^3 - 3xy^2 = u_0$: In case $u_0 = 0$, then $x(x^2 - 3y^2) = 0$. $w = z^3$ maps each of the rays:

$$x = 0 \text{ and } y \geq 0; \quad x = -\sqrt{3}y \text{ and } y \leq 0; \quad \text{and } x = \sqrt{3}y \text{ and } y \leq 0,$$

univalently onto the lower imaginary axis $u = 0$ and $v \leq 0$ in the w-plane. See Fig. 3.46. These three rays divide the plane $\mathbf{C}$ into three sector domains of angle $\frac{2}{3}\pi$. Each of them is mapped by $w = z^3$ univalently onto the domain $\mathbf{C} - \{w \mid u = 0 \text{ and } v \leq 0\}$. Similarly, $w = z^3$ maps each of the rays:

$$x = 0 \text{ and } y \leq 0; \quad x = -\sqrt{3}y \text{ and } y \geq 0;$$
$$x = \sqrt{3}y \text{ and } y \geq 0,$$

univalently onto the upper imaginary axis $u = 0$ and $v \geq 0$, and the three sector domains divided by these rays univalently onto the domain $\mathbf{C} - \{w \mid u = 0 \text{ and } v \geq 0\}$. See Fig. 3.47.

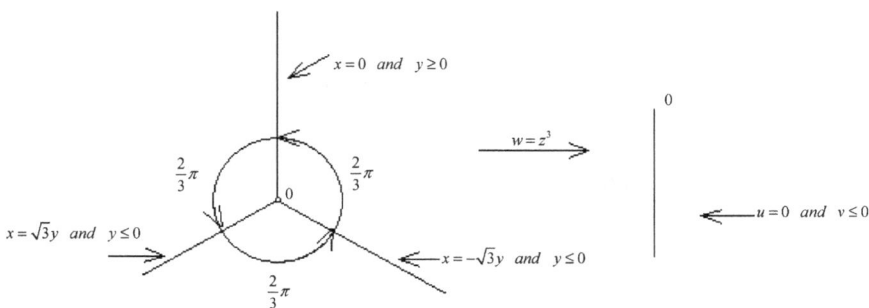

Fig. 3.46

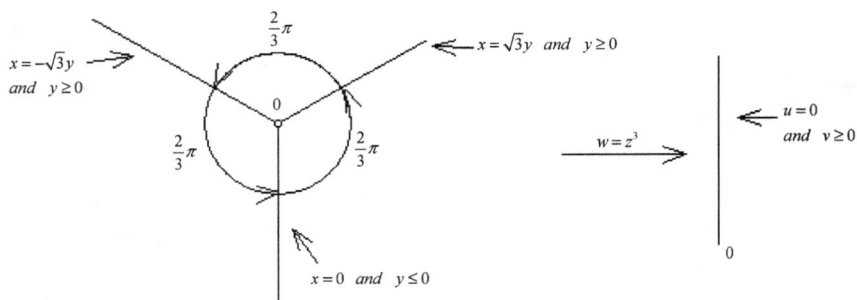

Fig. 3.47

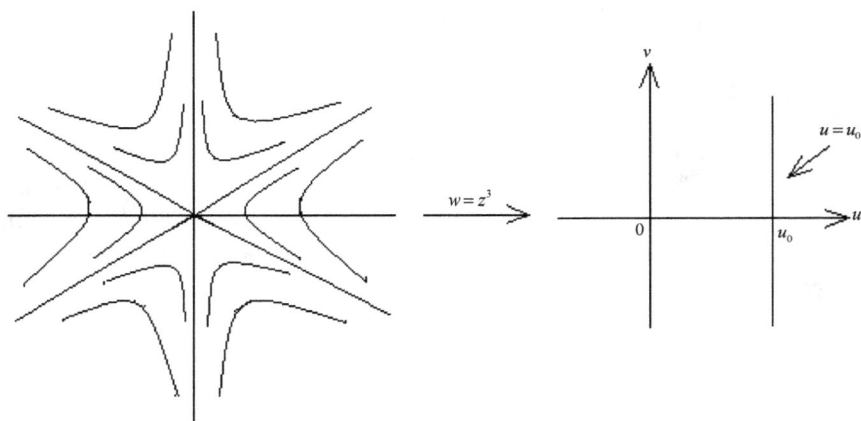

Fig. 3.48

If $u_0 \neq 0$, then $x^3 - 3xy^2 = r^3 \cos 3\theta = u_0$ $(x = r \cos \theta, y = r \sin \theta)$ is sketched as in Fig. 3.48.

Observe that $(-x)((-x)^2 - 3y^2) = -u_0$, which indicates that the graph of $x(x^2 - 3y^2) = u_0 < 0$ is obtainable from that of $x(x^2 - 3y^2) = u_0 > 0$ by reflecting the latter with respect to the y-axis. Hence, both graphs coincide for any $u_0 \neq 0$.

The streamlines $v(x,y) = 3x^2y - y^3 = r^3 \sin 3\theta = v_0$ is sketched similarly. Or, replacing z by iz in $w = z^3$, we have $w = -iz^3$. This means that the streamlines are obtained by rotating the equipotential lines through $-90°$ with center at 0. See Fig. 3.49, where the black curves are streamlines while the dotted ones are equipotential lines. These two families of curves intersect orthogonally wherever they meet each other.

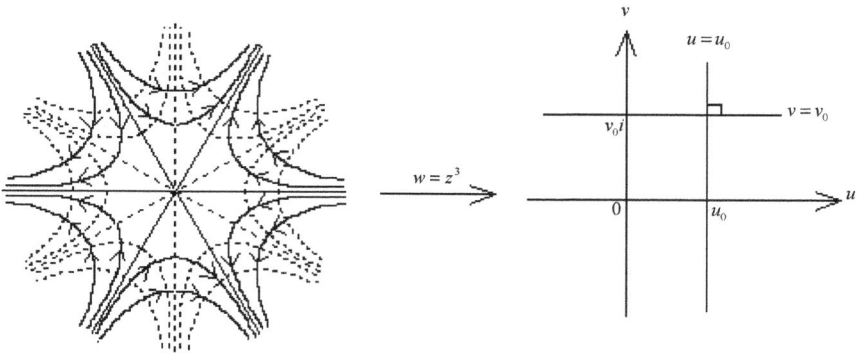

Fig. 3.49

The image curve of the vertical line $x = x_0$ under $w = z^3$ is given parametrically by $u = x_0^3 - 3x_0 y^2$, $v = 3x_0^2 y - y^3$ with $y \in \mathbf{R}$. In particular, the imaginary axis $x_0 = 0$ is mapped univalently onto the imaginary axis in the w-plane. In case $x_0 \neq 0$, observing that $u(x, -y) = u(x, y)$ and $v(x, -y) = -v(x, y)$, the image curve is symmetric with respect to the real axis by the symmetry principle (3.4.6.2). See Fig. 3.50 in case $x_0 > 0$. Please pay attention to the fact that the continuous change of $\arg w = 3 \arg z$, as y increases from $y = 0$ to $y = \infty$ along the line $z = x_0 + iy$, will sketch the loop shape of the image curve. It is a folium of Descartes without asymptote. Since $u(-x_0, y) = -u(x_0, y)$ and $v(-x_0, y) = v(x_0, y)$, the image curve in case $x_0 < 0$ is obtained by reflecting the image curve in case $x_0 > 0$ with respect to the imaginary axis. Also, $u(x, y) = -v(y, x)$ indicates that the image curves of $y = y_0$ are obtained by these of $x = x_0$ by reflecting them with respect to $u = v$ followed by reflecting the resulted curves with respect to the u-axis. These two families of image curves intersect orthogonally, too. By the way, the loop γ in Fig. 3.50 has the length

$$\int_\gamma |dw| = 2 \int_0^{\sqrt{3}x_0} 3|z|^2 |dz| = 2 \int_0^{\sqrt{3}x_0} 3(x_0^2 + y^2) dy = 12\sqrt{3}x_0^3;$$

while, by using Exercise A (21)(a) of Sec. 2.9 instead of (2) in (3.5.1.5), the domain $\operatorname{Int} \gamma$ has the area

$$\frac{1}{i} \int_\gamma (\operatorname{Re} w) dw = \int_{-\sqrt{3}x_0}^{\sqrt{3}x_0} (x_0^3 - 3x_0 y^2)(3x_0^2 - 3y^2) dy = \frac{36\sqrt{3}}{5} x_0^6.$$

The unit circle $|z| = 1$ is mapped onto $|w| = 1$, covered three times, by $w = z^3$. If $|z| = 1$ is distorted into an ellipse $\frac{x^2}{a^2} + \frac{y^2}{b^2} = 1$ $(a > 0, b > 0)$, what

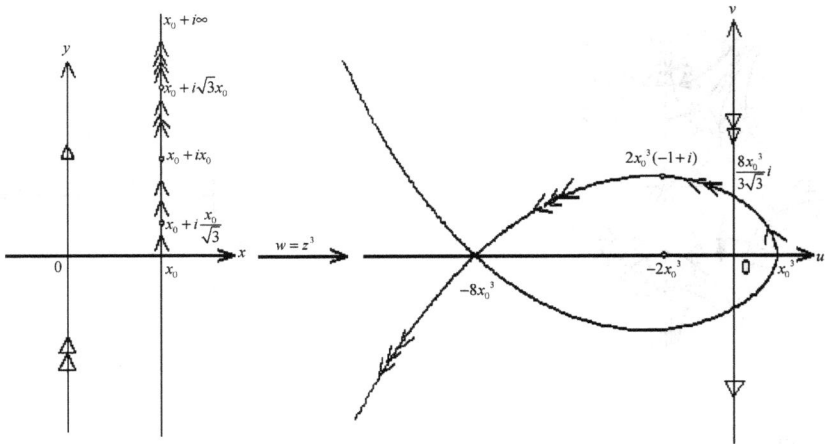

Fig. 3.50

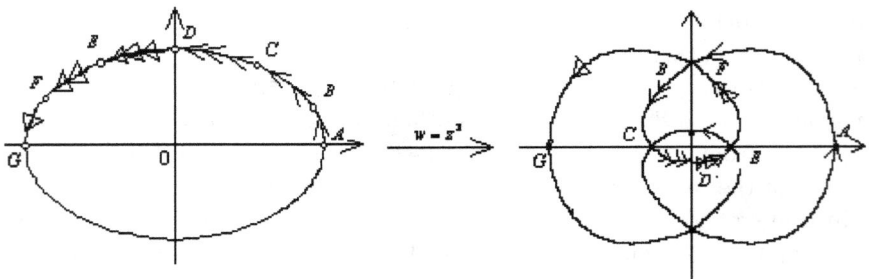

Fig. 3.51

is the image curve of this ellipse under $w = z^3$? Since $\arg w = 3 \arg z = 0$, $\frac{\pi}{2}, \pi, \frac{3\pi}{2}, 2\pi, \frac{5\pi}{2}, 3\pi$ if and only if $\arg z = 0, \frac{\pi}{6}, \frac{\pi}{3}, \frac{\pi}{2}, \frac{2\pi}{3}, \frac{5\pi}{6}, \pi$, respectively, by continuity of $\arg z$ along the upper half of the ellipse (see (2.7.1.4)), we can sketch the image of the upper half of the ellipse as in Fig. 3.51, while the image of the lower half is obtained by reflecting the former with respect to the real axis. Note that the image curve still winds around the origin o three times. Readers are asked to sketch the image curve of the square $|x| \leq 1, |y| \leq 1$ under $w = z^3$ by using Fig. 3.50. Can you guess how many times this new image curve will wind around 0? Why?

Example 2. (Nevanlinna, Ref. [64], p. 293) Try to describe the infinitesimal mapping behavior of $f(z) = ze^{-z}$ at $z = 0$.

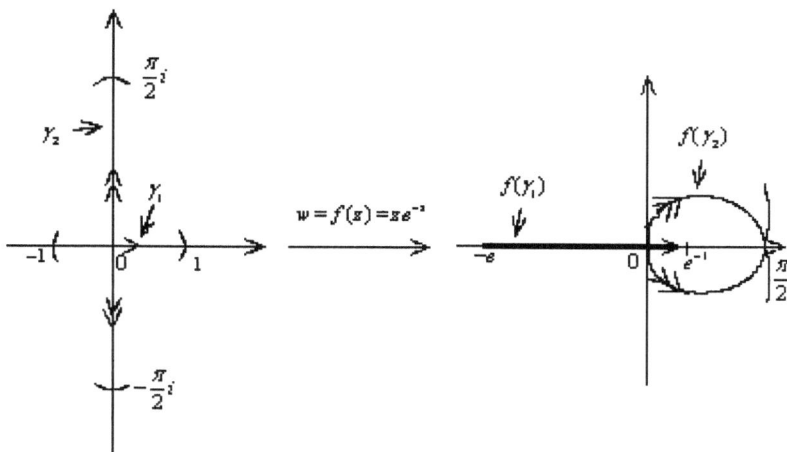

Fig. 3.52

Solution. $f'(z) = (1-z)e^{-z} = 0 \Leftrightarrow z = 1$. Hence f is conformal at points $z \neq 1$.

The affine part of f at $z = 0$ is $w = f(0) + f'(0)z = z$ with the error estimate $o(z - 0) = o(z) = ze^{-z} - z = z(e^{-z} - 1)$, where $|z| > 0$ is sufficiently small. The orthogonal line segments $\gamma_1 : z_1(t) = t$, $-1 \leq t \leq 1$, and $\gamma_2 : z_2(t) = it$, $-\frac{\pi}{2} < t < \frac{\pi}{2}$, are kept fixed under $w = z$ but are mapped onto the line segment $f(\gamma_1) : w_1(t) = te^{-t}$, $-1 \leq t \leq 1$ and an arc of the spiral $f(\gamma_2) : w_2(t) = ite^{-it}$, $-\infty < t < \infty$, respectively, under $w = f(z)$. See Fig. 3.52. Note that the whole spiral $f(\gamma_2)$ winds around the point e^{-1} countably infinitely many times.

We will continue the discussion of $w = ze^{-z}$ in the following Example 4.

Section (2) The case that $f'(z_0) = \cdots = f^{(k-1)}(z_0) = 0$ but $f^{(k)}(z_0) \neq 0$ for $k \geq 2$

According to (2) in (3.4.2.9), there exist a $\rho > 0$ and an analytic function $g(z)$ such that $g(z_0) = \frac{f^{(k)}(z_0)}{k!} \neq 0$ and $f(z) = f(z_0) + (z - z_0)^k g(z)$ on $|z - z_0| < \rho$.

Choose a differentiable curve γ with parametric equation $z(t) : [0, 1] \to$ **C** so that $z(0) = z_0$ and $z'(0) \neq 0$. Then

$$f(z(t)) = f(z_0) + (z(t) - z_0)^k g(z(t)), \quad \text{a differentiable curve passing } f(z_0)$$

$$\Rightarrow \arg(f(z(t)) - f(z_0)) = k\arg(z(t) - z_0) + \arg g(z(t)),$$

$$\Rightarrow \arg \frac{(f(z(t)) - f(z_0))}{t - 0} = k \arg \frac{z(t) - z_0}{t - 0} + \arg g(z(t)), \quad t > 0,$$

$\Rightarrow$ (taking limit to both sides as $t \to 0$),

$$\lim_{t \to 0^+} \arg \frac{(f(z(t)) - f(0))}{t - 0} = k \arg z'(0) + \arg g(z_0). \qquad (3.5.1.6)$$

Compare to (3.5.1.1). This fact shows that $w = f(z)$ *enlarges any angle between two curves passing z_0 k times but still preserves the sense.*

Furthermore, choose $\rho > 0$ small enough so that $|g(z) - g(z_0)| < |g(z_0)|$ for $|z - z_0| < \rho$. Since then, $g(z) \neq 0$ on $|z - z_0| < \rho$, it is possible to define a (single-valued analytic) branch of $\sqrt[k]{g(z)}$ on $|z - z_0| < \rho$ (see (2) in (3.3.1.5)), which is denoted by $\varphi(z)$. Then,

$$f(z) - f(z_0) = \zeta(z)^k, \quad \zeta(z) = (z - z_0)\varphi(z). \qquad (3.5.1.7)$$

Now $\zeta'(z_0) = \varphi(z_0) \neq 0$ and we still can choose $\rho > 0$ smaller so that $\zeta'(z) \neq 0$ on a neighborhood O of z_0 so that the univalent analytic function $\zeta = \zeta(z)$ maps O topologically onto the disk $|\zeta| < r^{\frac{1}{k}}$ (see (3) in (3.5.1.4)). For each w, $0 < |w - f(z_0)| < r$, the mapping $w = f(z_0) + \zeta^k$ determines k equally spaced values ζ in $|\zeta| < r^{\frac{1}{k}}$, namely, $w = f(z_0) + \zeta^k$ has k distinct roots in $|\zeta| < r^{\frac{1}{k}}$ (see Sec. 2.5.2). By performing the mapping $w = f(z)$ in two steps: $z \to \zeta = (z - z_0)\varphi(\zeta) \to w = f(z_0) + \zeta^k$, we obtain a very clear picture of the local correspondence. See Fig. 3.53, which shows the inverse image O of a small disk $|w - f(z_0)| < r$ and the k arcs, emanating from z_0, that are mapped onto the negative radius, and the Riemann surface R_w over $|w - f(z_0)| < r$ for the k-valued inverse function $z = f^{-1}(w)$.

We summarize the above as the

Local behavior of an analytic function f at a point z_0 where $f'(z_0) = \cdots = f^{(k-1)}(z_0) = 0$ but $f^{(k)}(z_0) \neq 0$ for $k \geq 2$. As a mapping: $w_0 = f(z_0)$.

(1) $w = f(z)$ enlarges any angle between two curves passing z_0 k times but preserves the sense.

(2) Choose $\rho > 0$ so that $f(z) \neq w_0$ and $f'(z) \neq 0$ on $0 < |z - z_0| \leq \rho$. Let $0 < R_\rho = \inf_{|z - z_0| = \rho} |f(z) - w_0|$. Then, for each point w, $0 < |w - w_0| < R_\rho$, the equation $w = f(z)$ has exactly k distinct solutions on $0 < |z - z_0| < \rho$. Hence, the inverse function $z = f^{-1}(w)$: $\{|w - w_0| < R_\rho\} \to \{|z - z_0| < \rho\}$ is k-valued. Letting $O = f^{-1}\{|w - w_0| < R_\rho\}$, then $w = f(z)$ maps O onto $|w - w_0| < R_\rho$, covered k-times. See Fig. 3.53.

(3) (Local multiple-valued inverse function theorem). Let R_ρ and O be as in (2). For each point w, $0 < |w - w_0| < R_\rho$, let $f^{-1}(w) =$

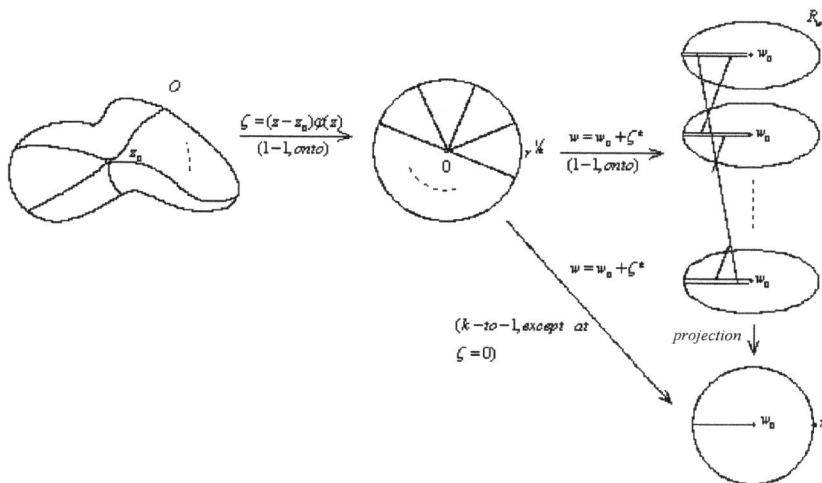

Fig. 3.53

$\{z_1(w), \ldots, z_k(w)\}$ where $z_j(w) \in O$ for $1 \leq j \leq k$. Then the inverse function $z = f^{-1}(w)$ has k *single-valued analytic* branches in a neighborhood of w. Namely, there exist a neighborhood U of w, and a neighborhood $O_j \subseteq O$ of $z_j(w)$ and a single-valued analytic function $z = g_j(\zeta)$ mapping U univalently onto O_j for $1 \leq j \leq k$ so that

$$f(g_j(w)) = w \quad \text{and} \quad f(g_j(\zeta)) = \zeta, \ \zeta \in U \quad \text{for } 1 \leq j \leq k.$$

The point $w_0 = f(z_0)$ is called an *algebraic branch point of order* $k - 1$ of the multiple-valued function $z = f^{-1}(w)$.   (3.5.1.8)

We have mentioned the concept of algebraic branch point for rational functions in the Note beneath Exercise A (4) of Sec. 2.5.3 and used them throughout the examples appear in Sec. 2.7 when we constructed descriptive the Riemann surfaces for elementary functions.

As an easy consequence of (3.5.1.4) and (3.5.1.8), we have

The univalence and conformality. Let $f : \Omega(\text{domain}) \to \mathbf{C}$ be a univalent analytic function. Then,

(1) $f'(z) \neq 0$ on Ω and hence, $w = f(z)$ is conformal everywhere on Ω.
(2) The inverse function $z = f^{-1}(w) : f(\Omega) \to \Omega$ is also univalent and conformal on the domain $f(\Omega)$.   (3.5.1.9)

We give two more examples.

Example 3. Try to use $f(z) = 1 + z^2 + z^4$ to illustrate the local inverse function theorem (3.4.7.1) and (3.5.1.8).

Solution. $f'(z) = 2z + 4z^3 = 0 \Leftrightarrow z = 0$ and $z = \pm\frac{i}{\sqrt{2}}$. Therefore, $f(0) = 1$ and $f\left(\pm\frac{i}{\sqrt{2}}\right) = \frac{3}{4}$ are branch points of the inverse function $z = f^{-1}(w)$.

Take, for instance, $z_0 = 1$. Then $f'(z_0) = f'(1) = 6 \neq 0$. Let $w_0 = f(z_0) = 3$. Define the recursive sequence of functions g_n:

$$g_0(w) = 1;$$

$$g_1(w) = g_0(w) - \frac{1}{f'(z_0)}[f(g_0(w)) - w] = 1 - \frac{1}{6}(3 - w) = 1 + \frac{1}{6}(w - 3);$$

$$g_2(w) = g_1(w) - \frac{1}{f'(z_0)}[f(g_1(w)) - w] = 1 + \frac{1}{6}(w - 3)$$

$$- \frac{1}{6}\left\{1 + \left[1 + \frac{1}{6}(w - 3)\right]^2 + \left[1 + \frac{1}{6}(w - 3)\right]^4 - w\right\}$$

$$= 1 + \frac{1}{6}(w - 3) - \frac{7}{216}(w - 3)^2 - \frac{1}{324}(w - 3)^3 - \frac{1}{7776}(w - 3)^4; \ldots$$

$\ldots$

The nth approximation function $g_n(w)$, $n \geq 1$, is a polynomial in $(w - 3)$ of degree 4^{n-1}. The theorem (3.4.7.1) says that there is a local branch $z = g(w)$ of $z = f^{-1}(w)$ on $|w - 3| < R$, which is inverse to $w = f(z)$ in a neighborhood of $z_0 = 1$. As a matter of fact, the largest value for R is the shorter distance from $w_0 = 3$ to 1 and $\frac{3}{4}$, namely, $3 - 1 = 2$.

The inverse function of $w = f(z) = 1 + z^2 + z^4$ is the four-valued function

$$z = \sqrt{\frac{1}{2}(-1 + \sqrt{4w - 3})} = f^{-1}(w).$$

Note that $\sqrt{4w - 3}$ has two branches on $\mathbf{C} - \left(\infty, \frac{3}{4}\right]$, namely, the one $(\sqrt{4w - 3})_+$ determined by $\sqrt{1} = 1$ and the other one $(\sqrt{4w - 3})_-$ determined by $\sqrt{1} = -1$. Hence, the double-valued function $f_+^{-1}(w) = \sqrt{\frac{1}{2}[-1 + (\sqrt{4w - 3})_+]}$ takes value 0 at $w = 1$, and $w = 1$ and $w = \infty$ are its algebraic branch points of order 1, and has, in turn, two branches on $\mathbf{C} - \left(\infty, \frac{3}{4}\right] - [1, \infty)$. The Riemann surface of $z = f_+^{-1}(w)$ is shown in Fig. 3.54. Within Fig. 3.54, $z = f_+^{-1}(w)$ maps the domain $\mathbf{C} - \left(\infty, \frac{3}{4}\right]$ in the w-plane onto the domain between the two branches of the hyperbola $x^2 - y^2 = -\frac{1}{\sqrt{2}}$ ($z = x + iy$) in one-to-two manner; while it maps its Riemann

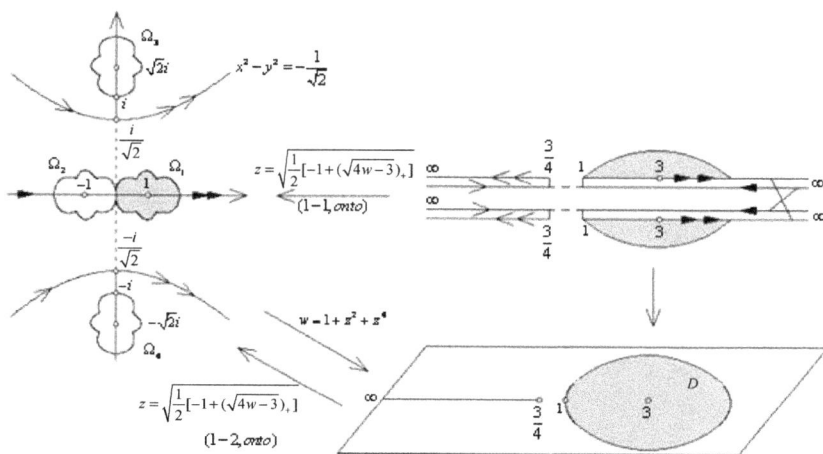

Fig. 3.54

surface onto the same domain univalently. The branch of $z = f_+^{-1}(w)$ that assumes the value 1 at $w = 3$ defines $z = g_1(w) : D = \{|w - 3| < 2\} \to \Omega_1$ (containing 1) as a local inverse of $w = f(z)$ at 1. This is what we obtained above according to (3.4.7.1).

Solve $1 + z^2 + z^4 = f(z) = 3$ and we get $z = \pm 1, \pm\sqrt{2}i$. By (3.5.1.8) this means that, on D, we can define four local branches $g_j(w)$, $1 \leq j \leq 4$, of $z = f^{-1}(w)$. Except $g_1(w)$ mentioned above, $z = g_2(w) : D \to \Omega_2$ is determined by $f_+^{-1}(3) = -1$; $z = g_3(w) : D \to \Omega_3$ by $f_-^{-1}(3) = \sqrt{2}i$ where

$$f_-^{-1}(w) = \sqrt{\frac{1}{2}[-1 + (\sqrt{4w - 3})_-]};$$

$z = g_4(w) : D \to \Omega_4$ by $f_-^{-1}(3) = -\sqrt{2}i$. Furthermore, $w = f(z)$ is univalent when restricting to Ω_j for $1 \leq j \leq 4$ and has $z = g_j(w)$ as its local inverse. This phenomenon is universally true for any pair of points z_0 and $w_0 = f(z_0)$, where $z_0 \neq 0, \pm\frac{i}{\sqrt{2}}$ and $w_0 \neq 1, \frac{3}{4}$. The *Riemann surface* of $z = f^{-1}(w)$ can be described as in Fig. 3.55 and its line complex in Fig. 3.56. There is an algebraic branch point of order 1 over $w = 1$; two algebraic branch points of order 1 over $w = \frac{3}{4}$ and an algebraic point of order 3 over ∞.

Since $f(0) = 1$ is an algebraic branch point of order 1, the image curve Γ of the circle $|z| = r < \frac{1}{\sqrt{2}}$ will wind around $w = 1$ twice. Do you know how to graph Γ on the Riemann surface for $z = f^{-1}(w)$? This fact can be

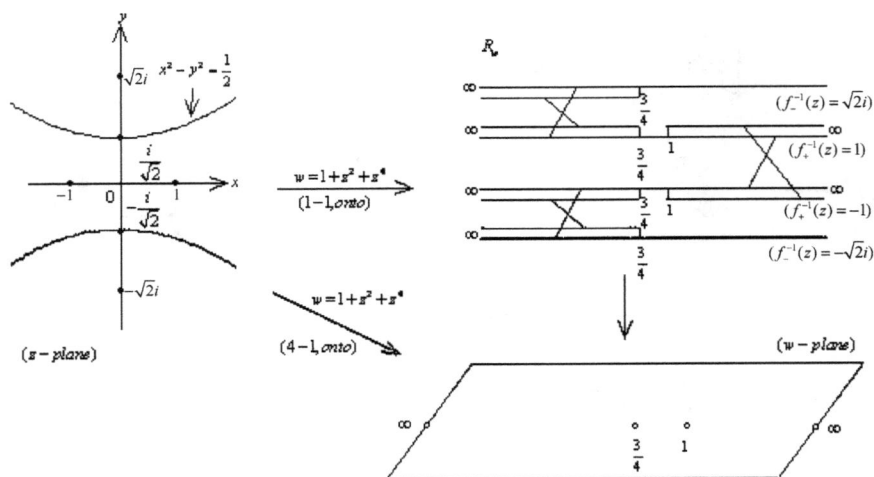

Fig. 3.55

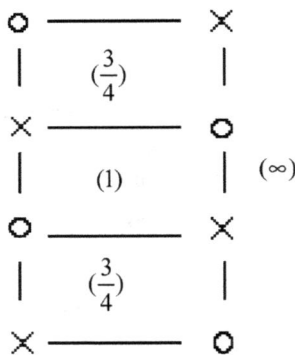

Fig. 3.56

expressed as, owing to reason to be clarified in Sec. 3.5.2 (see also (2.7.1.4) and Exercise B of Sec. 2.7.1),

$$\frac{1}{2\pi i}\int_{\Gamma}\frac{dw}{w-1}=\frac{1}{2\pi i}\int_{|z|=r}\frac{f'(z)dz}{f(z)-1}=\frac{1}{2\pi i}\int_{|z|=r}\frac{2z+4z^3}{z^2+z^4}dz$$

$$=\frac{2}{2\pi i}\int_{|z|=r}\frac{1}{z}\frac{1+2z^2}{1+z^2}dz=2\cdot\left\{\frac{1+2z^2}{1+z^2}\right\}_{z=0}=2.$$

This number 2 is called the *winding number* of Γ around $w=1$.

Example 4. (Continued from Example 2) Try to use $f(z) = ze^{-z}$ to illustrate (3.4.7.1) and (3.5.1.8).

Solution. Since $f'(1) = 0$ and $f''(1) = -e^{-1} \neq 0$, $f(1) = e^{-1}$ is an algebraic branch point of order 1 of $z = f^{-1}(w)$, where $w = f(z) = ze^{-z}$.

The local behavior of f at $z = 1$ is better and more convincingly illustrated by sketching the Riemann surface for $z = f^{-1}(w)$. To do this, let $z = x + iy$ and $w = u + iv$. Then $u(z) = e^{-x}(x \cos y + y \sin y)$ and $v(z) = e^{-x}(y \cos y - x \sin y)$. The curves $e^{-x}(y \cos y - x \sin y) = 0$, namely, $y = 0$ and $x = y \cot y$, are mapped onto parts of the u-axis. Let

$\Omega_0 = $ the simply connected domain bounded by the x-axis and the curve $x = y \cot y$, $\pi < y < 2\pi$;

$\Omega_{2n+1} = $ the simply connected domain bounded by the curve $x = y \cot y$, $(2n+1)\pi < y < (2n+2)\pi$ and the curve $x = y \cot y$, $(2n+3)\pi < y < (2n+4)\pi$, $n \geq 0$; and

$\Omega_0^* = $ the symmetric domain of Ω_0 with respect to the x-axis;

$\Omega_{2n+1}^* = $ the symmetric domain of Ω_{2n+1} with respect to the x-axis, $n \geq 0$.

These are fundamental domains (see (2.7.4)) for f: f maps Ω_0 univalently onto $\mathbf{C}^* - [-\infty, e^{-1}] = \mathbf{C} - (\infty, e^{-1}]$ and Ω_{2n+1} univalently onto $\mathbf{C}^* - [\infty, 0] = \mathbf{C} - (\infty, 0]$. The details are left to the readers (see Sec. 3.5.5 for univalence). The sketch of the Riemann surface R_w for $z = f^{-1}(w)$ is in Fig. 3.57. It has an algebraic branch point of order 1 over $w = e^{-1}$; there is a logarithmic branch point over each of $w = 0, \infty$. Note that 0 lies only on the sheets Ω_0 and Ω_0^* of R_w but ∞ does not lie on R_w since f does not assume ∞ on $\mathbf{C}$.

How many points z are mapped onto the point e^{-1}? Solve $ze^{-z} = e^{-1}$, which is equivalent to the equation $e^{-y \cot y} \frac{y}{\sin y} = e^{-1}$. This last equation has a unique solution $y = 0$ in $[-\pi, \pi]$ and a unique solution in each interval $[(2n+2)\pi, (2n+3)\pi], [-(2n+3)\pi, -(2n+2)\pi]$, for $n = 0, 1, 2, \ldots$. Going back to the z-plane in the left side of Fig. 3.57, this means that $ze^{-z} = e^{-1}$ has a unique solution $z = 1$ in the horizontal strip $-\pi < y < \pi$, which is mapped onto the algebraic branch point e^{-1}; and has a unique solution z_{2n+1} in $(2n+2)\pi < y < (2n+3)\pi$ and a unique solution $\bar{z}_{2n+1} = z_{2n+1}^*$ in $-(2n+3)\pi < y < -(2n+2)\pi$ for $n = 0, 1, 2, \ldots$, which are mapped onto the ordinary points e^{-1} on R_w, lying on Ω_{2n+1} and Ω_{2n+1}^*, respectively. The line complex (see the Explanation in Sec. 2.7.2) is as in Fig. 3.58.

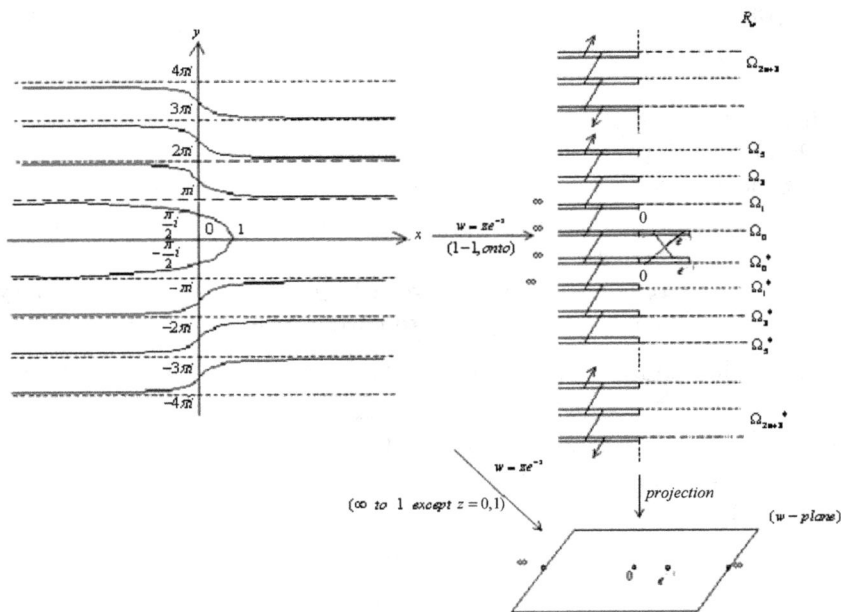

Fig. 3.57

Explanation.

(1) The four-sided polygon means one circulation around the algebraic branch point e^{-1} of order one.

(2) Each of the two-sided polygons means one circulation around an ordinary point e^{-1}, corresponding to z_{2n+1} or z_{2n+1}^*.

(3) The two infinitely-sided polygons surround separately the logarithmic branch points 0 and ∞.

Choose $\rho > 0$ so that $f(z) \neq e^{-1}$ and $f'(z) \neq 0$ hold on $0 < |z - 1| < \rho$. Let $R_\rho = \inf_{|z-1|=\rho} |ze^{-z} - e^{-1}|$. Then $R_\rho > 0$. The inverse function $z = F^{-1}(w)$ on $|w - e^{-1}| < R_\rho$ can be expressed as the Lagrange series (see (3.5.6.6)):

$$z = 1 + \sum_{n=1}^{\infty} \frac{1}{n!} \left\{ \frac{d^{n-1}}{dz^{n-1}} \left(\frac{z-1}{G(z)} \right)^n \right\} \Bigg|_{z=1} (w - e^{-1})^{\frac{n}{2}}$$

$$= 1 - \sqrt{2}ei(w - e^{-1})^{\frac{1}{2}} - \frac{2}{3}e(w - e^{-1})$$

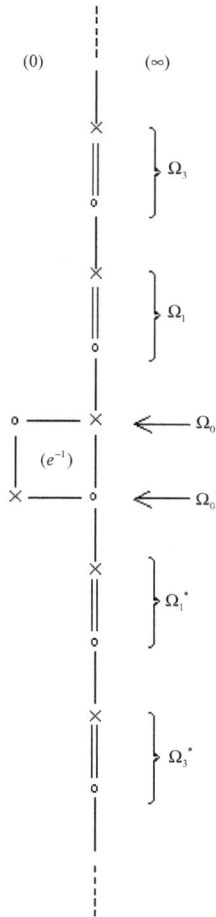

Fig. 3.58

$$+ \sum_{n=3}^{\infty} \frac{1}{n!} \left\{ \frac{d^{n-1}}{dz^{n-1}} \left(\frac{z-1}{G(z)} \right)^n \right\} \bigg|_{z=1} (w - e^{-1})^{\frac{n}{2}},$$

$$|w - e^{-1}| < R_\rho, \tag{$*$}$$

where $G(z)$ is any fixed branch of the two-valued function $\sqrt{ze^{-z} - e^{-1}}$ on the disk $|z - 1| < \rho$. On the other hand, $w = f(z)$ maps the circle $z = 1 + \rho e^{i\theta}$ ($0 \le \theta \le 2\pi$) onto the curve $\Gamma_\rho : w(\theta) = (1 + \rho e^{i\theta})e^{-(1+\rho e^{i\theta})}$, winding around e^{-1} twice. Equivalently, this means that the series $(*)$ is

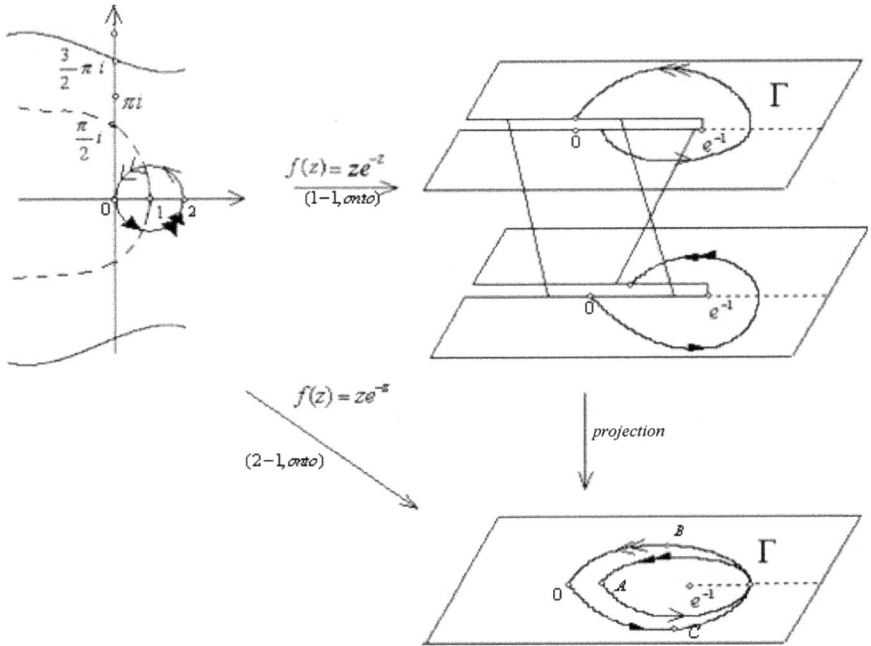

Fig. 3.59(a)

double-valued in $|w - e^{-1}| < R_\rho$. See Fig. 3.59(a) in case $\rho = 1$, where $\Gamma = \Gamma_1$. On the Riemann surface R_w, Γ_ρ (or $\Gamma = \Gamma_1$) is a Jordan closed curve whose interior Int Γ_ρ is a simply connected domain on which $z = f^{-1}(w)$ turns out to be single-valued. And this means that the series $(*)$ is singe-valued on Int Γ_ρ. This is the monodromy theorem (see (5.2.2.1) and (5.2.2.2)).

In Fig. 3.59:

(a) $\rho = 1$; $\Gamma_\rho = \Gamma$:

A: $w(0) = \dfrac{2}{e^2} = 0.270\ldots$;

B: $w(\dfrac{\pi}{2}) = (1 + i)e^{-(1+i)} = 0.374\cdots + 0.367\cdots i$;

C: $w\left(\dfrac{3\pi}{2}\right) = (1 - i)e^{-(1-i)} = \overline{w\left(\dfrac{\pi}{2}\right)}$.

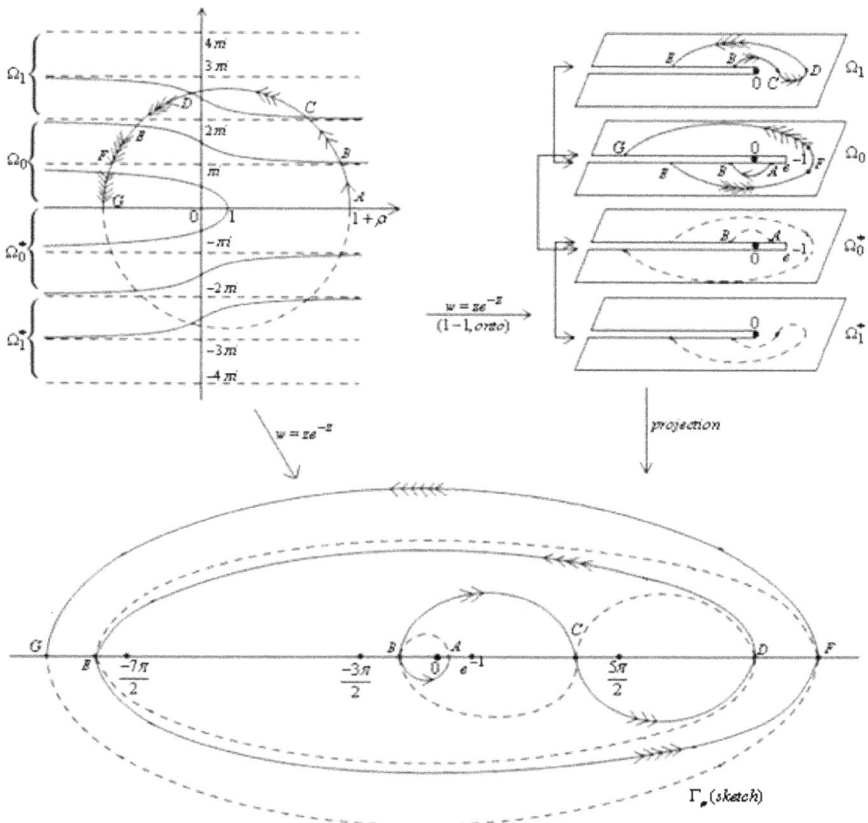

Fig. 3.59(b)

(b) For general ρ:

$$w(1 + \rho) \to 0 \quad \text{as } \rho \to +\infty;$$
$$w\left(\frac{\pi}{2}\right) = (1 + \rho i)e^{-(1+\rho i)} \to \infty \text{ (in } \mathbf{C}^*) \quad \text{as } \rho \to +\infty;$$
$$w(\pi) = (1 - \rho)e^{-(1+\rho)} \to -\infty \quad \text{as } \rho \to +\infty.$$

Hence, $0 < R_\rho < e^{-1}$ for any ρ, $0 < \rho < +\infty$. In case ρ is large enough, the circular arc BCD on $|z - 1| = \rho$ plays no role on the determination of R_ρ, because its image on the Riemann surface does not lie on Ω_0 and Ω_0*. As a consequence, only the curve $ABEFGF'E'B'A'$ as shown in (c) is good enough to find out R_ρ.

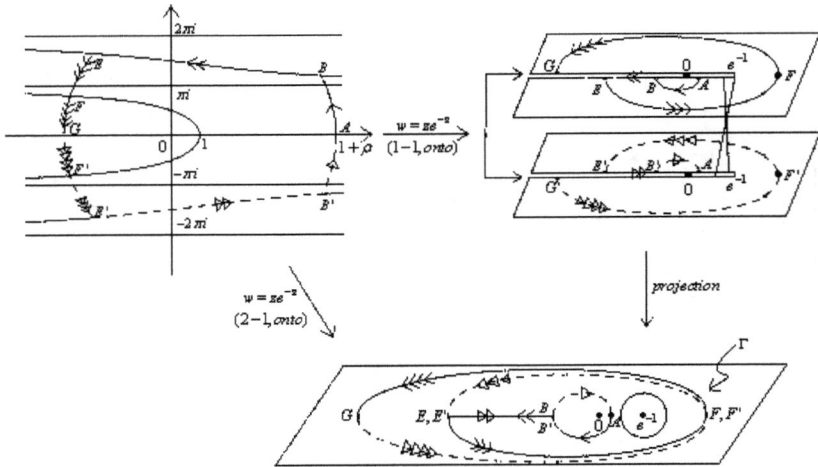

Fig. 3.59(c)

Remark (The graphs of an analytic function f as a mapping w = f(z)). Suppose $w = f(z) : O$ (open set in $\mathbf{C}$) $\to \mathbf{C}$ is an analytic or any complex-valued function. Its *graph* $G(f) = \{(z,w) \mid z \in O$ and $w = f(z) \in \mathbf{C}\}$ is a point set in $\mathbf{C}^2$, which can be viewed linearly and topologically as $\mathbf{R}^4$. Hence, we cannot picture $G(f)$ physically. In complex analysis, one usually takes two copies of the complex plane, one acting as the z-plane (or xy-plane) and the other as the w-plane (or uv-plane), to describe symbolically the action of the mapping $w = f(z)$ as we did all the way down.

To compensate this shortcoming in its own right and to study geometric mapping properties more accurately, three methods are adopted to graph $w = f(z) : O \to \mathbf{C}$, $z = x + iy$ and $w = u + iv$.

Method 1: *Try to determine what $f(O)$ might look like.* For instance, what is the image domain of the sector $0 < \operatorname{Arg} z < \frac{\pi}{n}$ under $w = z^n$? See Secs. 2.5–2.7 for more examples.

Try to use a net (such as $x = $ const., $y = $ const. or $r = $ const., $\theta = $ const.) *to cover O (locally)* and then, try to consider the image of the net under $w = f(z)$. See Figs. 2.14, 2.15, etc.

Try to describe the image curve $f(\gamma)$ if γ is a curve in D. For instance, what is the image curve of $|z| = 1$ under $w = \frac{1}{2}(z + \frac{1}{z})$? See Secs. 2.5–2.7 for more examples.

In particular, even the concept of how many times a closed curve winding around a specified point (namely, the *winding number* or *topological index*) will sometimes play a decisive role in many situations. See Secs. 2.7.1 and 3.5.2.

Of course, a *sketch of the Riemann surface*, even a descriptive one, for the inverse function $z = f^{-1}(w)$ is illustrative and convincing. See Sec. 2.7.

By the way, the geometric properties of iterative mappings $f^n = f \circ f^{n-1}$, $n \geq 2$, are fascinating and provide fruitful contribution to *fractal geometry* and *complex dynamics*, both of current research interest. See, for instance, Secs. 5.3.4 and 5.8.5, and for details, Ref. [15].

Method 2 : *Try to determine the level curves* $u(z) = c$, $v(z) = c$ (see (4) in (3.5.1.4)), especially in the theory of flow.

Method 3 : *Try to graph the modular surface*

$$\{(z, \mu) \mid z \in O \text{ and } |f(z)| = \mu\}$$

in $\mathbf{R}^3$, such as the one appeared in Ref. [58], Vol. I, p. 149.

No matter which method is adopted, computer work will help a lot in the process.   □

Section (3) Branched covering properties of meromorphic functions

Recall that z_0 is an *isolated singularity* of f if f is analytic in a deleted disk $0 < |z - z_0| < r$ for some $r > 0$. In this case, z_0 is said to be *a pole of order k of f* if z_0 is a zero of order of k of $\frac{1}{f(z)}$. Equivalently, there is an analytic function $\varphi(z)$ in $|z - z_0| < r$ so that $\varphi(z_0) \neq 0$ and

$$f(z) = \frac{\varphi(z)}{(z - z_0)^k}, \quad 0 < |z - z_0| < r. \tag{3.5.1.10}$$

This follows easily from (2) in (3.4.2.9). In case $z_0 = \infty$, one should replace $0 < |z - z_0| < \rho$ by $r < |z| \leq \infty$ and then, $f(z) = z^k \varphi(z)$ in $r < |z| < \infty$ with φ analytic in $r < |z| \leq \infty$ and $\varphi(\infty) \neq 0$ (see Remark in Sec. 3.3 and Remark 1 in Sec. 3.3.1, including the definition for *meromorphic function*).

Applying (3.5.1.8) to $\frac{1}{f(z)}$, we have the following

Local branched (or ramified) covering property of a meromorphic function at a pole. Let $z_0 \in \mathbf{C}^*$ be a pole of order k of f:

(1) In case $z_0 \in \mathbf{C}$: For sufficiently small $\rho > 0$, there is a $\delta = \delta(z_0, \rho) > 0$ such that for each $w^*, |w^*| > \delta$, the equation $f(z) = w^*$ has exactly k solutions in $|z - z_0| < \rho$. Moreover, ρ may be so chosen that if $\delta < |w^*| < \infty$, $f(z) = w^*$ has k *distinct* solutions (roots) in $0 < |z - z_0| < \rho$.

(2) In case $z_0 = \infty$: There are $\rho > 0$ and $\delta > 0$ so that for each $w^*, |w^*| > \delta$, $f(z) = w^*$ has exactly k solutions in $|z| > \rho$; and if $\delta < |w^*| < \infty$, $f(z) = w^*$, can have k *distinct* solutions in $|z| > \rho$.

The point $w_0 = f(z_0)$ is then called an *algebraic branch point* of order $k - 1$ of $z = f^{-1}(w)$ if $k \geq 2$.   (3.5.1.11)

Consequently, it follows

Some properties of nonconstant meromorphic functions. Let $f : \Omega$ (a domain in $\mathbf{C}^*$) $\to \mathbf{C}^*$ be a *nonconstant* meromorphic function.

(1) $w = f(z)$ maps open sets onto open sets in $\mathbf{C}^*$; in particular, $f(\Omega)$ is a domain in $\mathbf{C}^*$.
(2) In case $w = f(z)$ is univalent, then its inverse $z = f^{-1}(w) : f(\Omega) \to \Omega$ is also a univalent meromorphic function.
(3) For each fixed point $w_0 \in f(\Omega)$, the w_0-set $f^{-1}(w_0) = \{z \in \Omega \,|\, f(z) = w_0\}$ is a *discrete set*, namely, each of its points is an isolated point.   (3.5.1.12)

Note that, here, $\mathbf{C}^*$ is endowed with the spherical chord metric $d(,)$ defined in (1.6.8) and thus, is a compact, complete metric space (see Remark in Sec. 1.9). Meanwhile, (3) is equivalent to the fact that if $f^{-1}(w_0)$ has a limit point in Ω, then $f(z)$ would be equal to the constant w_0 throughout Ω.

In what follows, suppose that f is meromorphic on the entire extended complex plane $\mathbf{C}^*$ and is not a constant.

Since $f(\mathbf{C}^*)$ is open as well as closed (compact) in $\mathbf{C}^*$, as a nonempty connected set, it follows that $f(\mathbf{C}^*) = \mathbf{C}^*$ holds. Hence, f maps $\mathbf{C}^*$ *onto* $\mathbf{C}^*$.

We *claim* that $w = f(z)$ *assumes each point in* $\mathbf{C}^*$ (of course, including the point ∞) *the same number of times.* To see this, we introduce

$$b_f(z_0) = k - 1, \quad k \geq 1, \qquad (3.5.1.13)$$

and call it the *ramified number* of $w = f(z)$ at the point z_0, where, z_0 is either a zero of order k of $f(z) - f(z_0)$ as in (3.5.1.8) or z_0 is pole of order k of f as in (3.5.1.11). In case $k \geq 2$, this $b_f(z_0)$ is the order of the algebraic branch point $w_0 = f(z_0)$ of $z = f^{-1}(w)$.

For each integer $n \geq 1$, let

$$A_n = \left\{ w \in \mathbf{C}^* \,\middle|\, \sum_{z \in f^{-1}(w)} [b_f(z) + 1] \geq n \right\}.$$

According to either (3.5.1.8) or (3.5.1.11), A_n is an *open* set in $\mathbf{C}^*$ for each $n \geq 1$. To show that each A_n is also *closed* in $\mathbf{C}^*$, choose $w_k \in A_n, k \geq 1$, and suppose $\lim_{k \to \infty} w_k = w$. We have to show that $w \in A_n$. For $n \geq 2$, $z = f^{-1}(w)$ can have only a finite number of algebraic branch points of order $n-1$ in $\mathbf{C}^*$; otherwise, $w = f(z)$ would be identical with a constant or ∞ throughout $\mathbf{C}^*$. Hence, without loss of generality, we may suppose that for all $k \geq 1$ and for such point $z \in f^{-1}(w_k)$, $b_f(z) = 0$ always holds. Under this circumstance, each $f^{-1}(w_k)$ contains at least n distinct points, say

$$z_{k1}, z_{k2}, \ldots, z_{kn}, \quad k \geq 1.$$

Since $\mathbf{C}^*$ is compact, for each j $(1 \leq j \leq n)$, the sequence $\{z_{kj}\}_{k=1}^{\infty}$ has a subsequence, say itself, that converges to a point z_j in $\mathbf{C}^*$. Of course, $z_1, \ldots, z_n$ might not be distinct. Still for each j $(1 \leq j \leq n)$, since $f(z_{kj}) = w_k, k \geq 1$, as $k \to \infty$, we get

$$f(z_j) = w.$$

This indicates that $z_j \in f^{-1}(w)$ for $1 \leq j \leq n$ and then $\sum_{z \in f^{-1}(w)} [b_f(z) + 1] \geq n$. This proves that $w \in A_n$ and A_n is closed.

Now, each such A_n is open as well as closed in $\mathbf{C}^*$, which is a connected set. Therefore, each A_n is either empty or equal to $\mathbf{C}^*$.

For any fixed point $w_0 \in \mathbf{C}^*$, let $m = \sum_{z \in f^{-1}(w_0)} [b_f(z) + 1]$. Since f is not a constant, it follows that $0 < m < \infty$ should hold. Since $w_0 \in A_m$, $A_m = \mathbf{C}^*$ holds; and $w_0 \notin A_{m+1}$, $A_{m+1} = \phi$.

We summarize the above as

The covering property of a meromorphic function on $\mathbf{C}^*$. Let $w = f(z) :$ $\mathbf{C}^* \to \mathbf{C}^*$ be meromorphic.

(1) Then either f is a constant or f is *onto* the whole extended plane $\mathbf{C}^*$.
(2) In case f is *not* a constant, then there exists a (unique) positive integer m so that $w = f(z)$ assumes each point in $\mathbf{C}^*$ the same number of times m (counting multiplicities), namely, for each $w \in \mathbf{C}^*$,

$$\sum_{z \in f^{-1}(w)} [b_f(z) + 1] = m.$$

In particular, f has the same number of zeros and poles in $\mathbf{C}^*$.

$$(3.5.1.14)$$

(2) says that $w = f(z)$ maps $\mathbf{C}^*$ univalently onto an ideal "surface" R_w, which covers $\mathbf{C}^*$ exactly m times; or, the Riemann surface R_w of its inverse function $= f^{-1}(w)$ is m-sheeted and covers the w-plane $\mathbf{C}^*$ exactly m times. We saw lots of such examples in Sec. 2.7.2 and Exercises there. As a matter of fact, meromorphic functions on $\mathbf{C}^*$ are of the form $\frac{p(z)}{q(z)}$, where $p(z)$ and

$q(z)$ are polynomials, relatively prime (see Example 6 in Sec. 4.10.2). In this case, we already claimed the validity of (3.5.1.14) in (2.5.3.2). Moreover, (3.5.1.14) and its proof are still valid for analytic mappings between compact Riemann surfaces (see (7.2.16)).

Exercises A

(1) Prove (4) in (3.5.1.4) directly by using the Cauchy–Riemann equations.

(2) Prove (3.5.1.5) in detail.

(3) The function $f(z) = z|z|$ is differentiable at $z = 0$ and $f'(0) = 0$ but is not analytic at 0. Show that f is conformal at 0.

(4) Let $w = f(z) = z^2 + 2z$, where $z = x + iy$ and $w = u + iv$.

   (a) Describe the infinitesimal mapping behavior of f at $z = 0$. Determine the images of a small square with a vertex at 0 under the affine part of f and $w = f(z)$, respectively.

   (b) Determine these point sets where $|f'(z)| < 1$, $= 1$, and > 1, respectively. Find the image of the point set $\{z \| f'(z)| = 1\}$ under $w = f(z)$ and give a suitable geometric interpretation.

   (c) Determine the fundamental domains of f; in particular, these domains on which $w = f(z)$ is univalent and conformal.

   (d) Find the algebraic branch points for $z = f^{-1}(w)$. What are the orders?

   (e) Graph the image curves of $\operatorname{Re} z = x_0$ and $\operatorname{Im} z = y_0$ under $w = f(z)$, where x_0 and y_0 are constants.

   (f) Graph the level lines $u(z) = u_0$ and $v(z) = v_0$.

   (g) Find the family of curves in the z-plane whose images under $w = f(z)$ are concentric circles $|w| = \lambda \geq 0$? Also, determine another family of curves, which are orthogonal to the former.

   (h) Find the length of the perimeter and the area of the image domain of the domain $0 < x + 1 < 1$ and $0 < y + 1 < 1$ under $w = f(z)$.

   (i) Find the length of the image curve of the line segment $\operatorname{Re} z = 0$, $-1 \leq \operatorname{Im} z \leq 1$, under $w = f(z)$.

   (j) Find the area of the image domain of the domain $1 \leq |z + 1| \leq 2$, $-\frac{\pi}{4} \leq \operatorname{Arg}(z + 1) \leq \frac{\pi}{4}$.

(5) Let $w = f(z) = az + bz^2$ where a and b are nonzero complex constants. Find the image domain of the circle $|z| < R$, where $0 < R \leq \frac{|a|}{2|b|}$, and interpret its geometric meaning. What happens if $R > \frac{|a|}{2|b|}$?

(6) Let $w = f(z) = 3z + z^3$.

    (a) Show that f is univalent (and hence, conformal) on $|z| < 1$.

    (b) Determine the algebraic branch points of $z = f^{-1}(w)$? What are their orders?

    (c) Rewrite $z^3 + 3z = 2i + (z - i)^2(z + 2i)$ and then, try to interpret Fig. 3.53 in this case.

    (d) Rewrite $z^3 + 3z = -2i + (z + i)^2(z - 2i)$ and do the same problem as in (c).

(7) Do the same problem as in Exercise (5) to the function $f(z) = az + bz^3$, $a, b \in \mathbf{C}$.

(8) Do Example 1 to $f(z) = z^4$.

(9) Model after Examples 3 and 4 and use $f(z) = z^3 - 3z$ to study (3.4.7.1) and (3.5.1.8).

(10) Apply the representation $f(z) = f(z_0) + \zeta(z)^n$ to $\cos z$ with $z_0 = 0$. Determine $\zeta(z)$ explicitly.

(11) If f is analytic at 0 and $f'(0) \neq 0$, prove the existence of an analytic $g(z)$ satisfying $f(z^n) = f(0) + g(z)^n$ in a neighborhood of 0.

(12) Suppose $w = f(z) : \Omega$ (a bounded domain) $\to D$ (a bounded domain) is analytic. If f is continuous on $\bar{\Omega}$ and satisfies $f(\Omega) \subseteq D$ and $f(\partial\Omega) \subseteq \partial D$, show that

$$f(\Omega) = D; \quad f(\partial\Omega) = \partial D; \quad f(\bar{\Omega}) = \bar{D}.$$

(13) Suppose f is analytic in $|z| < 1$ and is univalent in $0 < |z| < 1$. Show that f is also univalent in $|z| < 1$.

(14) Suppose f is a univalent meromorphic function on a domain Ω. Show that f can only have a simple pole (a pole of order 1) in Ω.

Exercises B

(1) Local multiple-valued implicit function theorem. Adopt assumptions and notations in (3.4.7.3). Suppose that $F(z_0, w_0) = 0$, $F(z_0, w) \not\equiv 0$ and

$$\frac{\partial F}{\partial w}(z_0, w_0) = \cdots = \frac{\partial^{k-1} F}{\partial w^{k-1}}(z_0, w_0) = 0 \quad \text{but} \quad \frac{\partial^k F}{\partial w^k}(z_0, w_0) \neq 0 \quad (k \geq 2).$$

Then, there exist an $r > 0$ and an $R > 0$ so that

    (i) for each z, $0 < |z - z_0| < r$, the equation $F(z, w) = 0$ has k distinct solutions $w_1(z), \ldots, w_k(z)$ in $0 < |w - w_0| < R$; and

(ii) each of the solution functions $w_1(\zeta), \ldots, w_k(\zeta)$ in an open neigh-
borhood of a point z, $0 < |z - z_0| < r$, is a single-valued analytic
function of ζ satisfying:
the value of $w_j(\zeta)$ at $\zeta = z$ is $w_j(z)$ as shown in 1, for $1 \leq j \leq k$.

(2) Try to use $F(z, w) = z^3 + w^3 - 3zw$ to justify Exercise (1) and try
to construct the Riemann surface for $w = f(z)$ implicitly defined by
$F(z, w) = 0$. Refer to Example 3 in Sec. 3.5.6 if necessary.

3.5.2.   *The winding number: Its integral representation and geometric meaning*

Let γ be a continuous closed curve in $\mathbf{C}$ and z_0 be a point in $\mathbf{C}$, not on
γ. Recall that, in Sec. 2.7.1, the *winding number* $n(\gamma; z_0)$ of γ around z_0 is
defined to be the integer $\frac{1}{2\pi}\Delta_\gamma \arg(z - z_0)$, where $\Delta_\gamma \arg(z - z_0)$ denotes the
variation of any preassigned continuous branch of the $\arg(z - z_0)$ along γ.

Henceforth, γ will always represent a *piecewise differential closed curve*
(see (2.4.9)) unless otherwise stated, and the point $z_0 \notin \gamma$. In this case, we
try to use complex line integral of $\frac{1}{z - z_0}$ along γ to represent $n(\gamma; z_0)$ and
then, formulate its basic properties.

According to (1) in (2.7.1.4), we may choose a (continuous) *branch* of
$\arg(z - z_0)$ along γ and then, fix a (continuous) *branch* of $\log(z - z_0) =$
$\log|z - z_0| + i\arg(z - z_0)$. Under this circumstance,

$$\frac{1}{2\pi i} \int_\gamma \frac{dz}{z - z_0} = \frac{1}{2\pi i} \int_\gamma d\log(z - z_0)$$

$$= \frac{1}{2\pi i}\{\log|z - z_0||_\gamma + i\arg(z - z_0)|_\gamma\}$$

$$= \frac{1}{2\pi}\arg(z - z_0)|_\gamma = \frac{1}{2\pi}\Delta_\gamma \arg(z - z_0) = n(\gamma; z_0),$$

$$(3.5.2.1)$$

where $\arg(z - z_0)|_\gamma = \Delta_\gamma \arg(z - z_0)$ means the difference of the terminal
value and the initial value of the branch $\arg(z - z_0)$ as z winds along γ
once.

That $n(\gamma; z_0)$ is an integer, which can also be proved analytically as
follows. Let $z = z(t)$, $a \leq t \leq b$, be a *parametric representation of* γ.
Consider

$$F(t) = \frac{1}{2\pi i} \int_a^t \frac{z'(t)}{z(t) - z_0} dt, \quad a \leq t \leq b.$$

Then F is continuous on $[a, b]$ and is differentiable except at finitely many points where $z'(t)$ ceases to be continuous. Hence,

$$F'(t) = \frac{1}{2\pi i} \frac{z'(t)}{z(t) - z_0},$$

$$\Rightarrow z'(t) - 2\pi i F'(t)[z(t) - z_0] = 0,$$

$$\Rightarrow \frac{d}{dt} e^{-2\pi i F(t)} [z(t) - z_0] = e^{-2\pi i F(t)} \{z'(t) - 2\pi i F'(t)[z(t) - z_0]\} = 0,$$

$$\Rightarrow \text{(by continuity of } F) e^{-2\pi i F(t)} (z(t) - z_0)$$

$$= e^{-2\pi i F(a)} (z(a) - z_0), \quad \text{a constant not equal to } 0,$$

$$\Rightarrow e^{2\pi i F(t)} = \frac{z(t) - z_0}{z(a) - z_0}, \quad a \le t \le b$$

$$\Rightarrow \text{(since } z(b) = z(a)) \ e^{2\pi i F(b)} = 1,$$

$$\Rightarrow 2\pi i F(b) = 2\pi i n, \quad \text{where } n \text{ is an integer,}$$

$$\Rightarrow F(b) = n.$$

We summarize this fact as

The winding number. Let γ be a piecewise differentiable closed curve in $\mathbf{C}$ and z_0 be a point in $\mathbf{C}$, not on γ. The *integer*

$$n(\gamma; z_0) = \frac{1}{2\pi i} \int_\gamma \frac{dz}{z - z_0}$$

is called the *winding number* of γ around z_0 or the *topological index* of z_0 with respect to γ. (3.5.2.2)

Geometric interpretation: As a point set, $\gamma = z([a, b])$ is a compact and connected set in $\mathbf{C}$ and hence, z_0 is in a component of the open set $\mathbf{C}^* - \gamma$. Choose $\rho > 0$ so that the closed disk $|z - z_0| \le \rho$ has empty intersection with the curve γ. For each t, $a \le t \le b$, the line segment $\overline{z_0 z(t)}$, where $z(t) \in \gamma$, has the unique intersection point

$$\tilde{z}(t) = z_0 + \rho \frac{z(t) - z_0}{|z(t) - z_0|}, \quad a \le t \le b,$$

with the circle $\Gamma : |z - z_0| = \rho$. While as t increases from a to b, $z(t)$ will move from $z(a)$ along γ back to $z(b) = z(a)$ and the pointer $\overrightarrow{v}(t) = \rho \frac{z(t) - z_0}{|z(t) - z_0|}$ (considered as a plane vector with z_0 as the initial point) will then move along the circle Γ. See Fig 3.60. On Γ, designate $\overrightarrow{v}(a)$ as the *reference vector* with the initial point z_0 and the terminal point $\tilde{z}(a) = \tilde{z}(b)$.

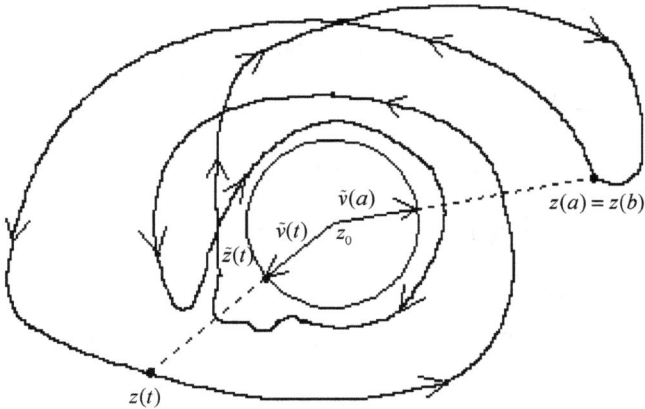

Fig. 3.60 $\tilde{z}(a) = \tilde{z}(b)$

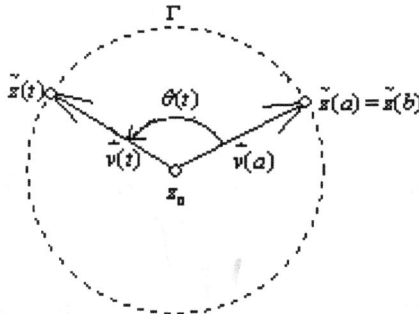

Fig. 3.61

As $\overrightarrow{v}(a)$ rotates an angle $\theta(t)$ in the counterclockwise direction to reach the vector $\overrightarrow{v}(t)$, then $0 \leq \theta(t) \leq 2\pi$ and $\theta(t)$ may be considered as the *reference angle* of $\overrightarrow{v}(t)$ with respect to $\overrightarrow{v}(a)$. See Fig. 3.61. In case there exist $c, d, a \leq c < d \leq b$, so that

(1) $\tilde{z}(c) = \tilde{z}(d) = \tilde{z}(a)$;
(2) $\tilde{z}([c, d]) = \Gamma$; and
(3) $\tilde{z}(t) \neq \tilde{z}(a)$ for all $t, c < t < d$.

Then we say $\overrightarrow{v}(t)$ performs a *complete revolution* of Γ on $[c, d]$. Observe that $\tilde{z}(t)$ may not be one-to-one on $[c, d]$, yet $\theta(t)$ is continuous on (c, d) with $\theta((c, d)) = (0, 2\pi)$. Under this circumstance, we call $\theta(t)$ or $\tilde{z}(t)$

or $\overrightarrow{v}(t)$ performs a *positive* revolution along Γ if $\lim_{t \to c^+} \theta(t) = 0$ and $\lim_{t \to d^-} \theta(t) = 2\pi$, and a *negative* revolution along Γ if $\lim_{t \to c^+} \theta(t) = 2\pi$ and $\lim_{t \to d^-} \theta(t) = 0$. Correspondingly, $z(t)$ is said to perform a *positive or negative* revolution along γ on $[c, d]$. By uniform continuity of $\widetilde{z}(t)$ on $[a, b]$, there is a partition $a \le c_1 < d_1 \le c_2 < d_2 \le \cdots < d_{n-1} \le c_n < d_n \le b$ so that $z(t)$ performs a revolution on each subinterval $[c_j, d_j]$, $1 \le j \le n$, but *not* on all other subintervals. This means that the total number of revolutions of $z(t)$ along γ as t varies from a to b is *finite* and, above all, the winding number (compare with (3.5.2.1))

$n(\gamma; z_0) = $ the number of positive revolution of $z(t)$ along Γ
$\qquad\qquad - $ the number of negative revolution of $z(t)$ along Γ.   (3.5.2.3)

For a detailed and neat treatment about winding numbers, refer to Ref. [66], pp. 153–160 and Appendix B within; for wonderful applications of winding numbers in complex analysis, see Ref. [8].

Now, we turn to the proof of the basic properties of winding numbers mentioned in Exercise B (2) of Sec. 2.7.1.

There is an $r > 0$ so that $|z - z_0| < r$ is contained in $\mathbf{C}^* - \gamma$, where $\gamma = z([a, b])$, a compact set. Choose z, $|z - z_0| < \frac{r}{2}$. Then

$$n(\gamma; z) - n(\gamma; z_0) = \frac{1}{2\pi i} \int_\gamma \left[\frac{1}{\zeta - z} - \frac{1}{\zeta - z_0} \right] d\zeta$$

$$= \frac{z - z_0}{2\pi i} \int_\gamma \frac{d\zeta}{(\zeta - z)(\zeta - z_0)},$$

$$\Rightarrow \quad \left(\text{Since } |\zeta - z_0| \ge r \text{ and } |\zeta - z| \right.$$

$$\left. \ge |\zeta - z_0| - |z_0 - z| \ge r - \frac{r}{2} = \frac{r}{2} \right)$$

$$|n(\gamma; z) - n(\gamma; z_0)| \le \frac{|z - z_0|}{2\pi} \cdot \frac{l(\gamma)}{r \cdot \frac{r}{2}}$$

$$= \frac{l(\gamma)}{\pi r^2} |z - z_0|, \quad \text{where } l(\gamma) \text{ is the length of } \gamma.$$

This shows that $n(\gamma; z)$ is a continuous function of z.

As a continuous function, the integer-valued function $n(\gamma; z)$ should be a constant on each component of $\mathbf{C}^* - \gamma$. For a *direct* proof: Choose two points z_1, z_2 in a component Ω of $\mathbf{C}^* - \gamma$. Owing to (2) in (1.8.11), we may suppose without losing its generality that the segment $\overline{z_1 z_2}$ lies in Ω. Observe that the function $g(z) = \frac{z - z_1}{z - z_2}$ maps z_1 and z_2 into 0 and ∞, respectively, and assume negative real values if and only if $z \in \overline{z_1 z_2}$ (but $z \neq z_1, z_2$). Hence, $\eta(z) = \mathrm{Log} \frac{z - z_1}{z - z_2}$ is a single-valued analytic function on $\Omega - \overline{z_1 z_2}$. Since $\gamma \subseteq \mathbf{C} - \overline{z_1 z_2}$ is a closed curve, then

$$\int_\gamma g'(z) dz = \int_\gamma \left[\frac{1}{z - z_1} - \frac{1}{z - z_2} \right] dz = 2\pi i [n(\gamma; z_1) - n(\gamma; z_2)] = 0$$

and the required result follows.

Let Ω_∞ be the unique unbounded component of $\mathbf{C}^* - \gamma$ that contains the point at infinity ∞. Choose any $z \in \Omega_\infty$ and $z \neq \infty$, then

$$|n(\gamma; z)| = \left| \frac{1}{2\pi i} \int_\gamma \frac{d\zeta}{\zeta - z} \right|$$

$$\leq \frac{1}{2\pi} \int_\gamma \frac{|d\zeta|}{|\zeta - z|} \leq \frac{1}{2\pi} \cdot \frac{l(\gamma)}{\mathrm{dist}(z, \gamma)} \to 0 \quad \text{as } z \to \infty.$$

It follows that $n(\gamma; z) = 0$ on Ω_∞.

We summarize these results again in the following (see also Exercise B of Sec. 2.7.1).

Basic properties of the winding numbers. Let γ be a piecewise differentiable closed curve in $\mathbf{C}$, parametrised by $z(t)$, $a \leq t \leq b$, and $z \notin \gamma$.

(1) Denote by $\bar{\gamma}$ the conjugate curve of γ (defined by $\overline{z(t)}$, $a \leq t \leq b$), then

$$n(\bar{\gamma}; \bar{z}) = -n(\gamma; z).$$

(2) Let $\alpha \gamma + \beta$ be the curve defined by $\alpha z(t) + \beta$, $a \leq t \leq b$, where α and β are complex constants. Then $n(\alpha \gamma + \beta; \alpha z + \beta) = n(\gamma; z)$.

(3) $n(-\gamma; z) = -n(\gamma; z)$.

(4) $n(\gamma; z)$ is an integer-valued constant function of z on each component of $\mathbf{C}^* - \gamma$. In particular, if Ω_∞ is the unbounded component of $\mathbf{C}^* - \gamma$ that contains ∞, then $n(\gamma; z) = 0$, $z \in \Omega_\infty$.

*(5) In case γ is a Jordan closed curve, then either $n(\gamma; z) = 1$ for all $z \in \mathrm{Int}\,\gamma$ or $n(\gamma; z) = -1$ for all $z \in \mathrm{Int}\,\gamma$ (see Ref. [66], Appendix B).

If both γ_1 and γ_2 are piecewise differentiable closed curves so that the sum curve $\gamma_1 + \gamma_2$ (see Fig. 2.96) is a closed curve, then the *addition property* says: for z not on $\gamma_1 + \gamma_2$,

$$n(\gamma_1 + \gamma_2; z) = n(\gamma_1; z) + n(\gamma_2; z). \tag{3.5.2.4}$$

Note that the addition property can be *formally* extended to and defined by

$$n\left(\sum_{k=1}^{m} a_k \gamma_k; z\right) = \sum_{k=1}^{m} a_k n(\gamma_k; z), \tag{3.5.2.5}$$

where the a_k are integers and the γ_k are piecewise differentiable closed curves, and z is not on any of the curves γ_k. Such a sum curve $\sum_{k=1}^{m} a_k \gamma_k$ is called a *closed chain* (see (4.3.2)).

Based on the concept of winding numbers, (3.4.2.4) can be extended to

A generalized Cauchy integral formula. Suppose f is analytic in a simply connected domain Ω and γ is a piecewise differentiable closed curve in Ω. Then, for any $z \in \mathbf{C} - \gamma$,

$$\frac{1}{2\pi i} \int_{\gamma} \frac{f(\zeta)}{\zeta - z} d\zeta = n(\gamma; z) f(z).$$

In particular, if γ is indeed a Jordan closed curve endowed with counterclockwise direction, then

$$\frac{1}{2\pi i} \int_{\gamma} \frac{f(\zeta)}{\zeta - z} d\zeta = f(z), \quad z \in \mathrm{Int}\, \gamma. \tag{3.5.2.6}$$

The **proof** is easy: Apply the fundamental Cauchy integral theorem (3.4.2.1) to the function $F(\zeta)$ defined by

$$F(\zeta) = \begin{cases} \dfrac{f(\zeta) - f(z)}{\zeta - z}, & \zeta \in \Omega - \{z\}, \\ f'(z), & \zeta = z, \end{cases}$$

which has a removable singularity at z (see (3.4.2.17)) and the result follows.

Exercises A

(1) Suppose the curves γ_1, γ_2, and γ_3 have, respectively, the equation $z_1(t) = e^{it}$, $0 \le t \le 4\pi$; $z_2(t) = -1 + 2e^{-2it}$, $0 \le t \le 2\pi$, and $z_3(t) = 1 - i + e^{it}$, $\frac{\pi}{2} \le t \le \frac{9\pi}{2}$. Sketch the sum curve $\gamma = \gamma_1 + \gamma_2 + \gamma_3$ and evaluate all possible values of $n(\gamma; z)$.

(2) Let γ_1 be the circle $|z| = \rho_1$ and γ_2 be the circle $|z| = \rho_2$, where $\rho_1 < \rho_2$. Evaluate $n(\gamma; z_0)$, where $\gamma = \gamma_1 + \gamma_2$ or $\gamma_1 - \gamma_2$ and $|z_0| < \rho_1$, $\rho_1 < |z_0| < \rho_2$ or $|z_0| > \rho_2$, respectively.

(3) Suppose γ is a closed chain so that $a, b \notin \gamma$ and $n(\gamma; a) = n(\gamma; b)$.

(a) Show that

$$\int_\gamma \frac{dz}{(z-a)(z-b)} = 0$$

and thus, deduce that, for positive integers m and n,

$$\int_\gamma \frac{dz}{(z-a)^m(z-b)^n} = 0.$$

(b) If $p_n(z)$ is a polynomial of degree at most equal to n and $|a| < R$. Show that

$$\int_{|z|=R} \frac{p_n(z)}{z^{n+1}(z-a)} dz = 0.$$

(4) Suppose that $|a| < \rho < |b|$.

(a) Show that $\int_{|z|=\rho} \frac{1}{(z-a)(z-b)} dz = \frac{2\pi i}{a-b}$.

(b) Evaluate $\int_{|z|=\rho} \frac{1}{(z-a)^m(z-b)^n} dz$, where m and n are positive integers.

(5) Let γ be a closed chain.

(a) If $p(z)$ is a polynomial, show that $\int_\gamma p(z) dz = 0$.

(b) In case $z_0 \notin \gamma$, show that $\int_\gamma \frac{1}{(z-z_0)^n} dz = 0$ for any integer $n \geq 2$.

(c) If the rational function $R(z)$ has the partial fraction expansion $R(z) = \frac{A_1}{z-a_1} + \cdots + \frac{A_n}{z-a_n} +$ higher-order terms of $(z-a_j)^{-1}$ for $1 \leq j \leq n$, and γ does not pass $a_1, \ldots, a_n$, then

$$\int_\gamma R(z) dz = 2\pi i \sum_{j=1}^n A_j n(\gamma; a_j).$$

Exercises B

(1) Let γ be a piecewise differentiable curve, not necessarily closed. Suppose g and h are complex-valued continuous functions along γ. Define $f : \mathbf{C} - h(\gamma) \to \mathbf{C}$ by

$$f(z) = \int_\gamma \frac{g(\zeta)d\zeta}{h(\zeta) - z}, \quad z \in \mathbf{C} - h(\gamma).$$

Show that for each $z_0 \in \mathbf{C} - h(\gamma)$, $f(z)$ can be expressed as

$$f(z) = \sum_{n=0}^\infty a_n(z_0)(z - z_0)^n, \quad |z - z_0| < \text{dist}(z_0, h(\gamma)),$$

where

$$a_n(z_0) = \frac{f^{(n)}(z_0)}{n!} = \int_\gamma \frac{g(\zeta)}{[h(\zeta) - z_0]^{n+1}} d\zeta, \quad n \geq 0.$$

This fact shows that f is analytic on each component of $\mathbf{C} - h(\gamma)$. In case $g(z) = h(z) = z$, then $f(z) = n(\gamma; z)$ is the winding number of γ around z and thus, $n(\gamma; z)$ is necessarily a continuous function and then a constant function on each component of $\mathbf{C} - \gamma$.

3.5.3.   *The argument principle*

Suppose f is analytic at z_0. Let γ be a closed curve around z_0 and $w_0 = f(z_0)$. What is interesting to us is the number of times that the image curve $f(\gamma)$ will surround the point w_0. This situation reduces to the concept of winding numbers detailed in Sec. 3.5.2 in case $f(z)$ is the function $z - z_0$.

For a little more general setting, suppose f is a nonconstant meromorphic function in a domain Ω (see Remark (1) in Sec. 3.3.1).

Let $a \in \Omega$ be *a zero of order* n of f. Then, there is an $r > 0$ so that $f(z) = (z - a)^n \varphi(z)$ on $|z - a| \leq r$ where φ is analytic and $\varphi(z) \neq 0$ on $|z - a| \leq r$ (see (2) in (3.4.2.9)). Now,

$$f'(z) = n(z - a)^{n-1}\varphi(z) + (z - a)^n \varphi'(z),$$

$$|z - a| \leq r \quad \text{(which is contained in } \Omega, \text{ supposedly)}.$$

$$\Rightarrow \frac{f'(z)}{f(z)} = \frac{n}{z - a} + \frac{\varphi'(z)}{\varphi(z)},$$

$$\Rightarrow \frac{1}{2\pi i} \int_{|z-a|=r} \frac{f'(z)}{f(z)} dz = \frac{1}{2\pi i} \int_{|z-a|=r} \frac{n}{z - a} dz = n.$$

In case $g(z)$ is another function analytic in Ω, then

$$\frac{1}{2\pi i} \int_{|z-a|=r} \frac{f'(z)}{f(z)} g(z) dz = \frac{n}{2\pi i} \int_{|z-a|=r} \frac{g(z)}{z - a} dz = ng(a). \qquad (3.5.3.1)$$

On the other hand, let $b \in \Omega$ be *a pole of order* m of f (refer to (3.5.1.10)). There is a $\rho > 0$ so that $f(z) = \frac{\psi(z)}{(z-b)^m}$ on $|z - b| \leq \rho$ where ψ is analytic and $\psi(z) \neq 0$ on $|z - b| \leq \rho$. In this case,

$$f'(z) = -m(z - b)^{-m-1}\psi(z) + (z - b)^{-m}\psi'(z),$$

$$|z - b| \leq \rho \quad \text{(which is contained in } \Omega, \text{ too)}.$$

$$\Rightarrow \frac{f'(z)}{f(z)} = \frac{-m}{z - b} + \frac{\psi'(z)}{\psi(z)},$$

$$\Rightarrow \frac{1}{2\pi i} \int_{|z-b|=\rho} \frac{f'(z)}{f(z)} dz = \frac{-m}{2\pi i} \int_{|z-b|=\rho} \frac{1}{z - b} dz = -m,$$

or

$$\frac{1}{2\pi i} \int_{|z-b|=\rho} \frac{f'(z)}{f(z)} g(z) dz = -mg(b) \quad \text{if } g \text{ is analytic in } \Omega. \qquad (3.5.3.2)$$

For any rectifiable Jordan closed curve γ in Ω, which does not pass any zeros or any poles, there are only a finite number of zeros and poles in the Int γ (why? see (4) in (3.4.2.9) and be sure to give precise reasons). Applying (1) in (3.4.2.4)′ to $\frac{f'(z)}{f(z)}$ and $\frac{f'(z)}{f(z)} g(z)$ or using directly Exercise A (2), the residue theorem of Sec. 3.4.2, we obtain the following theorem.

Theorem. *Let $f(z)$ be nonconstant and meromorphic (see Remark 1 in Sec. 3.3.1) in a domain Ω and γ be a rectifiable Jordan closed curve in Ω, not passing zeros and poles of f, so that its interior Intγ is still contained in Ω. Let $a_1, \ldots, a_k$ be zeros of f in Intγ with orders $n_1, \ldots, n_k$, respectively, and $b_1, \ldots, b_\ell$ be poles of f in Intγ with orders $m_l, \ldots, m_l$, respectively.*

(1) *In case g is an analytic function in Ω, then*

$$\frac{1}{2\pi i} \int_\gamma g(z) \frac{f'(z)}{f(z)} dz = \sum_{j=1}^{k} n_j g(a_j) - \sum_{p=1}^{\ell} m_p g(b_p).$$

(2) *If $g(z) \equiv 1$, then the argument principle says*

$$\frac{1}{2\pi i} \int_\gamma \frac{f'(z)}{f(z)} dz = \sum_{j=1}^{k} n_j - \sum_{p=1}^{\ell} m_p = N - P,$$

where $N = \sum_{j=1}^{k} n_j$ is the total number of zeros (counting multiplicities) of f in Intγ, and $P = \sum_{p=1}^{\ell} m_p$ is the total number of poles (counting multiplicities) of f in Intγ.

(3) *In case f is analytic in Ω and $g(z) \equiv 1$, then*

$$\frac{1}{2\pi i} \int_\gamma \frac{f'(z)}{f(z)} dz = \text{the total number of zeros of } f \text{ in Int } \gamma. \qquad (3.5.3.3)$$

The results in (1) and (2) can be extended one step further to the case that γ is a closed chain, which is homologous to zero in Ω (namely, $n(\gamma; z) = 0$ for all $z \in \mathbf{C} - \Omega$). See (4.11.3.4) for details. This depends essentially on the Cauchy integral theorem in its general setting (see (4.3.3.1)).

Geometric interpretation: To interpret (3.5.3.1) and (3.5.3.2) geometrically, let us suppose for simplicity that a and b are, respectively, the *only* zero of order n and the *only* pole of order m of f in Int γ. γ can be deformed

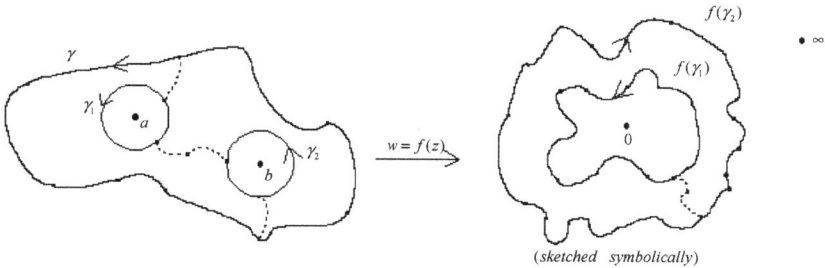

Fig. 3.62

continuously into a circle $\gamma_1 : |z - a| = r$ and a circle $\gamma_2 : |z - b| = \rho$ by auxiliary arcs as shown in Fig. 3.62. According to (2) in (3.5.1.8) and (3.5.1.11), and imitating Fig. 3.51, the circle γ_1 is mapped via $w = f(z)$ onto a curve $f(\gamma_1)$, which will wind around the origin n times in the counterclockwise direction. Recall that b is a zero of order m of $\frac{1}{f(z)}$. Hence, the interior of the circle γ_2 is mapped onto an open neighborhood of ∞ and γ_2 itself is onto a curve $f(\gamma_2)$, surrounding ∞ m times in the counterclockwise sense as viewed from ∞, which is in the clockwise direction when viewed from the origin 0. This is the reason why the zero a of order n contributes n and the pole b of order m contributes $-m$ to the integral

$$\frac{1}{2\pi i} \int_\gamma \frac{f'(z)}{f(z)} dz = n - m = \frac{1}{2\pi i} \int_{f(\gamma)} \frac{dw}{w},$$

which is equal to the winding number $n(f(\gamma); 0)$, the net number of times the image curve $f(\gamma)$ surrounding the origin 0. Apply this result to the function $f(z) - w_0$ and we will have the following

Geometric meaning of argument principle: Let f be a meromorphic function in the interior Int γ of a rectifiable Jordan closed curve γ and be continuous on $\overline{\text{Int}\, \gamma}$ except poles in Int γ. For any fixed point $w_0 \in \mathbf{C}$, if γ does not pass any zero or pole of the function $f(z) - w_0$, then

$$\frac{1}{2\pi i} \int_{f(\gamma)} \frac{dw}{w - w_0} = \frac{1}{2\pi i} \int_\gamma \frac{f'(z)dz}{f(z) - w_0}, \quad \text{where } w = f(z)$$
$$= n(\Gamma; 0) = \frac{1}{2\pi} \Delta_\Gamma \arg(f(z) - w_0), \text{ the number of times the}$$
curve $\Gamma = f(\gamma) - w_0$ surrounding the origin 0 in the w-plane

$= n(f(\gamma); w_0) = \frac{1}{2\pi}\Delta_{f(\gamma)} \arg(f(z) - w_0)$, the number of times the curve $f(\gamma)$ surrounding the point w_0

$= N_{w_0} - P$,   where N_{w_0} is the number of zeros (counting multiplicities) of $f(z) - w_0$ in Int γ and P is the number of poles (counting multiplicities) of $f(z)$ in Int γ. (3.5.3.4)

In case γ is the circle $|z| = r$ and $w_0 = 0$, (3.5.3.4) can be rewritten as

$$\frac{1}{2\pi}\int_0^{2\pi} F(re^{i\theta})d\theta = N - P, \quad \text{where } N = N_0 \text{ and } F(z) = z\frac{f'(z)}{f(z)}.$$

Petrovitch (1908) and Montel (1938) extended this form of the argument principle to $\int_0^{2\pi} g(t)\,\text{Re}\,F(t)dt = (N - P)\int_0^{2\pi} g(t)dt$, where $g(t)$ is continuous on $[0, 2\pi]$ and $\int_0^{2\pi} g(t)\cos nt\,dt = 0$ for $n \geq 1$.

Take, for instance, the function $f(z) = \frac{z-1}{z}$ and γ the circle $|z| = 2$. Since this f has a simple zero at $z = 1$ and a simple pole at $z = 0$ in $|z| < 2$, by (3.5.3.4),

$$\frac{1}{2\pi i}\int_{|z|=2} \frac{f'(z)}{f(z)}dz = \frac{1}{2\pi i}\int_{|z|=2} \frac{dz}{z(z-1)} = 1 - 1 = 0.$$

Actual computation will show that this is true indeed and the image curve $f(\gamma)$ is the circle $|w - 1| = \frac{1}{2}$, which excludes 0 and ∞ from $|w - 1| \leq \frac{1}{2}$. As another example, let $f(z) = \frac{1}{2}(z + \frac{1}{z})$, which has a simple pole at 0 and two simple zeros at $\pm i$. Then, by (3.5.3.4), and referring to (2.5.5.4) and Fig. 2.26,

$$\frac{1}{2\pi i}\int_{|z|=2} \frac{f'(z)}{f(z)}dz = 2 - 1 = 1.$$

The argument principle has a variety of applications. Some classical ones will be given in Sec. 3.5.4 for the Rouché's theorem, Sec. 3.5.5 for sufficient conditions for univalent functions, and Sec. 3.5.6 for inverse and implicit function theorems. For further interesting development, see Ref. [8].

In what follows, we illustrate some examples.

Example 1.   Show that the equation $z^4 + z^3 + 5z^2 + 2z + 4 = 0$ does not have pure imaginary, positive real roots and complex roots with positive real parts.

Solution.   Let $p(z) = z^4 + z^3 + 5z^2 + 2z + 4$.

If $z = iy(y \in \mathbf{R})$, then $p(iy) = (y^2 - 1)(y^2 - 4) - iy(y^2 - 2)$. Since $(y^2 - 1)(y^2 - 4) = 0$ and $y(y^2 - 2) = 0$ do not have common solution,

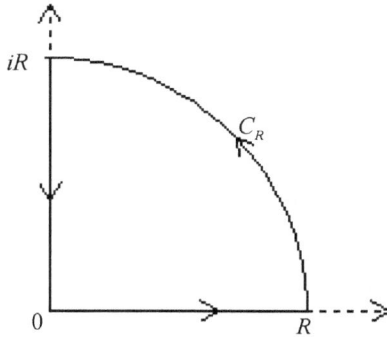

Fig. 3.63

thus $p(iy) \neq 0$ for all $y \in \mathbf{R}$ and $p(z) = 0$ does not have pure imaginary roots.

In case $z = x > 0$, $p(x) > 0$ always holds. So $p(z) = 0$ does not have positive real roots.

Since $p(z)$ is a polynomial with real coefficients, the complex roots of $p(z) = 0$ will appear in conjugate pair. It is thus sufficient to show that $p(z) = 0$ does not have roots in the first quadrant. To do this, consider the Jordan closed curve γ composed of a quarter C_R of the circle $|z| = R$ and its two radii as shown in Fig. 3.63.

Along $z = x$, $0 \leq x \leq R : p(z) = p(x) > 0$ holds. Then

$$\arg p(z) = 0,$$

$$\Rightarrow \frac{1}{2\pi} \Delta_{[0,R]} \arg p(z) = 0 \quad \text{for } R > 0.$$

Along the circle arc $C_R : z = Re^{i\theta}$, $0 \leq \theta \leq \frac{\pi}{2}$: Note that

$$p(z) = z^4 \left(1 + \frac{1}{z} + \frac{5}{z^2} + \frac{2}{z^3} + \frac{4}{z^4} \right).$$

Since

$$1 + \frac{1}{z} + \frac{5}{z^2} + \frac{2}{z^3} + \frac{4}{z^4} \to 1 \quad \text{as } z \to \infty,$$

it follows that

$$\arg \left(1 + \frac{1}{z} + \frac{5}{z^2} + \frac{2}{z^3} + \frac{4}{z^4} \right) \to 0 \quad \text{as } z \to \infty.$$

Now

$$\arg p(\mathrm{Re}^{i\theta})$$

$$= \arg R^4 e^{i4\theta} + \arg\left(1 + \frac{1}{z} + \frac{5}{z^2} + \frac{2}{z^3} + \frac{4}{z^4}\right) = 4\theta + \arg(\cdots).$$

$$\Rightarrow \Delta_{C_R} \arg p(\mathrm{Re}^{i\theta}) = 4\left(\frac{\pi}{2} - 0\right) + \arg(\cdots) = 2\pi + \arg(\cdots),$$

$$\Rightarrow \lim_{R\to\infty} \frac{1}{2\pi} \Delta_{C_R} \arg p(\mathrm{Re}^{i\theta}) = \frac{1}{2\pi} \cdot 2\pi = 1.$$

Along the radius $z = iy$, $0 \le y \le R$: In this case,

$$\arg p(iy) = \tan^{-1} \frac{-(y+\sqrt{2})y(y-\sqrt{2})}{(y+1)(y+2)(y-1)(y-2)}.$$

Observe how $\arg p(iy)$ varies as y changes from $R > 2$ down to 0 in the following table:

y	∞	$\downarrow$	2	$\downarrow$	$\sqrt{2}$	$\downarrow$	1	$\downarrow$	0
$\tan \arg p(iy)$	0	< 0	$-\infty$	> 0	0	< 0	$-\infty$	> 0	0
$\arg p(iy)$	0	$-\frac{\pi}{2} < $ $\cdots < 0$	$-\frac{\pi}{2}$	$-\pi <$ $\cdots < -\frac{\pi}{2}$	$-\pi$	$-\frac{3\pi}{2} <$ $\cdots < -\pi$	$-\frac{3\pi}{2}$	$-2\pi <$ $\cdots < -\frac{3\pi}{2}$	-2π

This shows that

$$\lim_{R\to\infty} \Delta_{[0,iR]} \arg p(iy) = \frac{1}{2\pi}(-2\pi - 0) = -1.$$

Combining together, we have the fact that

$$\lim_{R\to\infty} \Delta_{p(\gamma)} \arg p(z) = 0 + 1 - 1 = 0.$$

By (3.5.3.4), $p(z) = 0$ does not have roots in the first quadrant.

Example 2.   Suppose that $\alpha, \beta \ne 0$ are real numbers and n is a positive integer. Show that the equation $z^{2n} + \alpha z^{2n-1} + \beta^2 = 0$ has n roots in the right half-plane if n is even, and has $n-1$ or $n+1$ roots there if n is odd, according to $\alpha > 0$ or $\alpha < 0$, respectively.

Solution.   Let $p(z) = z^{2n} + \alpha z^{2n-1} + \beta^2$. In case $y \in \mathbf{R}$, $p(iy) = (-1)^n y^{2n} + \beta^2 - i(-1)^n \alpha y^{2n-1} = 0$ does not have solution. Hence, $p(z) = 0$ does not have pure imaginary roots.

Let γ be the Jordan closed curve shown in Fig. 3.64, where C_R is the semicircle $z = \mathrm{Re}^{i\theta}$, $-\frac{\pi}{2} \le \theta \le \frac{\pi}{2}$.

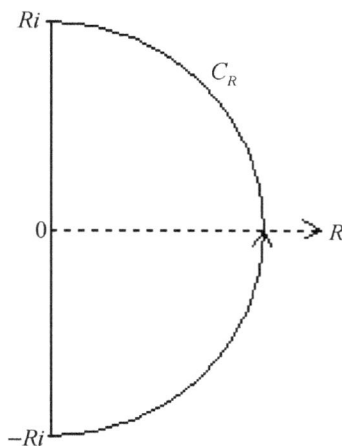

Fig. 3.64

Along C_R:

$$p(z) = z^{2n}\left(1 + \frac{\alpha}{z} + \frac{\beta^2}{z^{2n}}\right),$$

$$\Rightarrow \arg p(Re^{i\theta}) = 2n\theta + \arg\left(1 + \frac{\alpha}{Re^{i\theta}} + \frac{\beta^2}{R^{2n}e^{2ni\theta}}\right),$$

$$\Rightarrow \lim_{R\to\infty} \Delta_{C_R} \arg p(Re^{i\theta}) = 2n\left(\frac{\pi}{2} - \left(-\frac{\pi}{2}\right)\right) = 2n\pi.$$

Along the line segment $[-Ri, Ri]$ on the imaginary axis: *In case n is even,*

$$p(iy) = y^{2n} + \beta^2 - i\alpha y^{2n-1}, \quad y \in \mathbf{R} \text{ and } -R \le y \le R,$$

$$\Rightarrow \arg p(iy) = \tan^{-1} \frac{-\alpha y^{2n-1}}{y^{2n} + \beta^2},$$

where $y^{2n} + \beta^2 > 0$ always holds. Observing the following table:

y	∞	$\downarrow$	0	$\downarrow$	$-\infty$
$\tan \arg p(iy)$	0	$(\alpha > 0)$ $(\alpha < 0)$ < 0 or > 0	0	$(\alpha > 0)$ $(\alpha < 0)$ > 0 or < 0	0
$\arg p(iy)$	0	$-\dfrac{\pi}{2} < \cdots < 0$ or $0 < \cdots < \dfrac{\pi}{2}$	0	$0 < \cdots < \dfrac{\pi}{2}$ or $-\dfrac{\pi}{2} < \cdots < 0$	0

therefore

$$\lim_{R\to\infty} \Delta_{y\in[-R,R]} \arg p(iy) = 0.$$

Combining together,

$$\lim_{R\to\infty} \frac{1}{2\pi} \Delta_{p(\gamma)} \arg p(z) = \frac{1}{2\pi}(2n\pi + 0) = n.$$

Thus $p(z) = 0$ has n roots in $\operatorname{Re} z > 0$ if n is even. *In case n is odd,*

$$p(iy) = -y^{2n} + \beta^2 + i\alpha y^{2n-1}, \quad y \in \mathbf{R} \text{ and } -R \le y \le R,$$

$$\Rightarrow \arg p(iy) = \tan^{-1} \frac{-\alpha y^{2n-1}}{y^{2n} - \beta^2},$$

where $y^{2n} - \beta^2 = 0$ has only two real roots $y = \pm\beta^{\frac{1}{n}}$ (we may suppose that $\beta > 0$ in the sequel). Observing how $\arg p(iy)$ varies in the following table,

y	∞	$\downarrow$	$\beta^{\frac{1}{n}}$	$\downarrow$
$\tan\arg p(iy)$	0	$(\alpha>0)\ (\alpha<0)$ <0 or >0	∞	$(\alpha>0)\ (\alpha<0)$ >0 or <0
$\arg p(iy)$	0	$-\dfrac{\pi}{2} <\cdots< 0$ or $0 <\cdots< \dfrac{\pi}{2}$	$(\alpha>0)\ (\alpha<0)$ $-\dfrac{\pi}{2}$ or $\dfrac{\pi}{2}$	$-\pi <\cdots< -\dfrac{\pi}{2}$ or $\dfrac{\pi}{2} <\cdots< \pi$

y	0	$\downarrow$	$-\beta^{\frac{1}{n}}$
$\tan\arg p(iy)$	0	$(\alpha>0)\ (\alpha<0)$ <0 or >0	∞
$\arg p(iy)$	$(\alpha>0)\ (\alpha<0)$ $-\pi$ or π	$-\dfrac{3\pi}{2} <\cdots< -\pi$ or $\pi <\cdots< \dfrac{3\pi}{2}$	$(\alpha>0)\ (\alpha<0)$ $-\dfrac{3\pi}{2}$ or $\dfrac{3\pi}{2}$

y	$\downarrow$	$-\infty$
$\tan\arg p(iy)$	$(\alpha>0)\ (\alpha<0) >0$ or <0	0
$\arg p(iy)$	$(\alpha>0)\ (\alpha<0)$ $-2\pi <\cdots< -\dfrac{3\pi}{2}$ or $\dfrac{3\pi}{2} <\cdots< 2\pi$	$(\alpha>0)\ (\alpha<0)$ -2π or 2π

we obtain

$$\lim_{R\to\infty} \Delta_{y\in[-R,R]} \arg p(iy) = \begin{cases} -2\pi - 0 = -2\pi, & \text{if } \alpha > 0, \\ 2\pi - 0 = 2\pi, & \text{if } \alpha < 0. \end{cases}$$

Combining together,

$$\lim_{R\to\infty} \frac{1}{2\pi} \Delta_{p(\gamma)} \arg p(z) = \begin{cases} \dfrac{1}{2\pi}(2n\pi - 2\pi) = n - 1, & \text{if } \alpha > 0, \\ \dfrac{1}{2\pi}(2n\pi + 2\pi) = n + 1, & \text{if } \alpha < 0, \end{cases}$$

and the claimed results follow.

Example 3. (Continuation of the Eneström–Kakeya theorem (2.5.1.8)). Suppose $0 < a_0 < a_1 < \cdots < a_n$. It is well known that the zeros of the polynomial $p(z) = a_0 + a_1 z + \cdots + a_n z^n$ all lie in $|z| < 1$. Try to deduce that the trigonometric polynomial

$$a_0 + a_1 \cos \theta + a_2 \cos 2\theta + \cdots + a_n \cos n\theta$$

has $2n$ zeros in the interval $(0, 2\pi)$ and does not have pure imaginary roots.

Solution. Let $z = e^{i\theta}$, $0 \le \theta \le 2\pi$. Then,

$$\arg p(e^{i\theta}) = \tan^{-1} \frac{a_1 \sin \theta + \cdots + a_n \sin n\theta}{a_0 + a_1 \cos \theta + \cdots + a_n \cos n\theta}$$

$\Rightarrow$ (by assumption and (3.5.3.4)),

$$\frac{1}{2\pi} \Delta_{\theta\in[0,2\pi]} \arg p(e^{i\theta}) = n.$$

This indicates that the image curve Γ of the unit circle $|z| = 1$, under $w = p(z)$, winds around the origin 0 exactly n times in the counterclockwise direction.

As long as $w = p(e^{i\theta})$ winds once along Γ and back to its starting point, the values assumed by $\arg p(e^{i\theta})$ will change continuously from a certain φ_0 (suppose $a_0 + a_1 \cos \varphi_0 + \cdots + a_n \cos n\varphi_0 \ne 0$) to $\varphi_0 + 2\pi$ (refer to (2.7.1.4)). During the process, $\arg p(e^{i\theta})$ should take the values $\varphi_0 + \frac{\pi}{2}$ and $\varphi_0 + \frac{3\pi}{2}$ on each of which $a_0 + a_1 \cos \theta + \cdots + a_n \cos n\theta$ assumes the value zero.

In other words, if z winds along $\gamma: |z| = 1$ once, the corresponding point $w = p(z)$ will wind along $\Gamma = p(\gamma)$ exactly n times; and if Γ performs one revolution, it will intersect the imaginary axis at least twice. This means that there are at least $2n$ distinct θ in $(0, 2\pi)$ ensuring that $p(e^{i\theta})$ lies on the imaginary axis. Equivalently speaking, this means that $a_0 + a_1 \cos \theta + \cdots + a_n \cos n\theta = 0$ has at least $2n$ distinct roots in $(0, 2\pi)$.

To prove that $a_0 + a_1 \cos \theta + \cdots + a_n \cos n\theta = 0$ has exactly $2n$ distinct roots in $(0, 2\pi)$, let $\zeta = e^{i\theta}$. Observing that $\cos k\theta = \frac{1}{2}(\zeta^k + \zeta^{-k})$ for $1 \le k \le n$; then

$$a_0 + a_1 \cos \theta + \cdots + a_n \cos n\theta = \frac{1}{2} \zeta^{-n} q(\zeta),$$

where $q(\zeta) = a_n + a_{n-1}\zeta + \cdots + a_1\zeta^{n-1} + 2a_0\zeta^n + a_1\zeta^{n+1} + \cdots + a_{n-1}\zeta^{2n-1} + a_n\zeta^{2n}$. Now, $q(\zeta) = 0$ has $2n$ roots $\zeta_1, \ldots, \zeta_{2n}$, and any possible root θ_0 of $\operatorname{Re} p(e^{i\theta}) = 0$ should be of the form $e^{i\theta_0} = \zeta_k$ where k is one of $1, 2, \ldots, 2n$. This fact shows that $\operatorname{Re} p(e^{i\theta}) = 0$ has at most $2n$ roots in $(0, 2\pi)$, and hence, has exactly $2n$ distinct roots in $(0, 2\pi)$.

Since these $2n$ roots of $\operatorname{Re} p(e^{i\theta}) = 0$ are all distinct, $\zeta_1, \ldots, \zeta_{2n}$ should be all distinct, too. Yet $|\zeta_k| = 1$, $1 \le k \le 2n$, say that $\operatorname{Re} p(e^{i\theta}) = 0$ does not have pure imaginary roots.

Note: A general setting. Suppose $f(z) = u(z) + iv(z)$ is analytic on $\overline{\operatorname{Int} \gamma}$ where γ is a rectifiable Jordan closed curve. If $\operatorname{Re} f(z) = u(z)$ has $2n$ distinct zeros on γ, then $f(z)$ will have at most n zeros in $\operatorname{Int} \gamma$.

Proof. Arguments for this is similar to that used in Example 3. Suppose $f(z) = 0$ has k roots in $\operatorname{Int} \gamma$, then

$$k = \frac{1}{2\pi} \Delta_{f(\gamma)} \arg f(z), \quad \text{where } \arg f(z) = \tan^{-1} \frac{v(z)}{u(z)}.$$

Fix $z_0 \in \gamma$ so that $u(z_0) \ne 0$. We may suppose $\arg f(z_0) \in \left(-\frac{\pi}{2}, \frac{\pi}{2}\right)$. Let z, starting from z_0, move along γ. Once $u(z)$ assumes a zero value, then the value of $\arg f(z)$ will move out of the restricted interval $\left(-\frac{\pi}{2}, \frac{\pi}{2}\right)$ and go into the interval $\left(\frac{\pi}{2}, \frac{3\pi}{2}\right)$ or $\left(-\frac{3\pi}{2}, -\frac{\pi}{2}\right)$. Only when $u(z)$ assumes a zero value again, $\arg f(z)$ will then assume value in $\left(\frac{3\pi}{2}, \frac{5\pi}{2}\right)$ or $\left(-\frac{5\pi}{2}, -\frac{3\pi}{2}\right)$, and so on. If z winds one revolution along γ and back to z_0 and thus $u(z)$ has two distinct zeros, then it is necessary that $\Delta_{f(\gamma)} \arg f(z) \le 2\pi$ or $k \le 1$ holds. The general case follows by a similar argument. $\square$

Let $f(z)$ be an analytic function. A point z_0 on a level curve $|f(z)| = c$ (a nonzero constant) is said to be a *double point* if the level curve crosses itself once at z_0. *In this case,* $f'(z_0) = 0$ *and vice versa.* To see this, let

$f(z) = u(z) + iv(z)$. Since $|f(z)| = c$ and $|f(z_0)| = c$, by invoking Taylor's formula to $|f(z)|^2 - |f(z_0)|^2 = u(z)^2 - u(z_0)^2 + v(z)^2 - v(z_0)^2$, we obtain that z_0 is a double point if and only if

$$u(z_0)u_x(z_0) + v(z_0)v_x(z_0) = 0 \quad \text{and} \quad u(z_0)u_y(z_0) + v(z_0)v_y(z_0) = 0$$

hold. By using the Cauchy–Riemann equations, these two equations will ensure that $|f'(z_0)|^2|f(z_0)| = 0$ and hence $f'(z_0) = 0$ follows.

Example 4.   (Titchmarsh, Ref. [76], pp. 121–123). Let the *Jordan* closed curve γ be a level curve of a nonconstant analytic function $f(z)$ so that $f(z)$ is analytic on $\overline{\operatorname{Int} \gamma} = \gamma \cup \operatorname{Int} \gamma$. Then,

(1) $f(z)$ has at least one zero in $\operatorname{Int} \gamma$; and
(2) in case $f(z)$ has n zeros in $\operatorname{Int} \gamma$, the derivative $f'(z)$ has $(n - 1)$ zeros in $\operatorname{Int} \gamma$.

Proof. (1) is an easy consequence of (3.4.4.3). Yet the following proof is available for the sake of (2). Suppose that γ is given by $|f(z)| = c$, where c is a positive constant and $f(z) = u(z) + iv(z)$.

On γ, $f(z) = ce^{i\theta}$ by noting that $\theta = \theta(z)$ is a differentiable function of z along γ. Let $s = s(z)$ be the length function of γ measured from a certain point on it. Then $ds^2 = dx^2 + dy^2$ is the arc length element along γ. Now,

$$u(z)^2 + v(z)^2 = c^2 \Rightarrow \frac{dc}{ds} = \frac{1}{c}\left(u\frac{du}{ds} + v\frac{dv}{ds}\right) = 0, \quad \text{and}$$

$$\tan^{-1}\frac{v(z)}{u(z)} = \theta \Rightarrow \frac{d\theta}{ds} = \frac{1}{c^2}\left(u\frac{dv}{ds} - v\frac{du}{ds}\right).$$

We try to show that $\frac{d\theta}{ds} \neq 0$ *along* γ. In case there is a z_0 on γ so that $|f(z_0)| = c$ and $\frac{d\theta}{ds}(z_0) = 0$, then at the point z_0, the sum of the squares of the above two equations results in

$$(u^2 + v^2)(z_0)\left\{\left(\frac{du}{ds}(z_0)\right)^2 + \left(\frac{dv}{ds}(z_0)\right)^2\right\} = 0,$$

$$\Rightarrow (\text{since } (u^2 + v^2)(z_0) = c^2 \neq 0) \quad \frac{du}{ds}(z_0) = \frac{dv}{ds}(z_0) = 0.$$

On the other hand, by using the Cauchy–Riemann equations,

$$\frac{du}{ds} = u_x \frac{dx}{ds} + u_y \frac{dy}{ds},$$

$$\frac{dv}{ds} = v_x \frac{dx}{ds} + v_y \frac{dy}{ds} = -u_y \frac{dx}{ds} + u_x \frac{dy}{ds} \quad \text{(both evaluated at } z_0\text{)},$$

$\Rightarrow$ (squaring both sides and then adding side by side)   At z_0,

$$(u_x^2 + u_y^2)\left\{ \left(\frac{dx}{ds}\right)^2 + \left(\frac{dy}{ds}\right)^2 \right\} = u_x^2 + u_y^2$$

$$= |f'(z_0)|^2 = \left(\frac{du}{ds}\right)^2 + \left(\frac{dv}{ds}\right)^2 = 0.$$

Hence $f'(z_0) = 0$ and z_0 is a double point of γ, which contradicts to the fact that γ is a Jordan curve. This proves the claim.

As a continuous function of z, $\frac{d\theta}{ds}$ should be always positive or negative along γ, and then $\theta = \theta(z)$ is a strictly increasing or decreasing function along γ. This fact suggests that

$$\Delta_{z\in\gamma} \arg f(z) = \Delta_{z\in\gamma}\theta(z) \neq 0$$

holds. The argument principle (3.5.3.4) then implies that $f(z)$ has at least one zero in Int γ.

As for (2): Along γ, recalling that $f(z) = ce^{i\theta}$,

$$f'(z) = cie^{i\theta}\frac{d\theta}{dz},$$

$$\Rightarrow \arg f'(z) = \frac{\pi}{2} + \theta + \arg\frac{d\theta}{dz},$$

$$\Rightarrow \Delta_{z\in\gamma} \arg f'(z) = \Delta_{z\in\gamma} \arg f(z) + \Delta_{z\in\gamma} \arg\frac{d\theta}{dz}. \qquad (*)$$

Since $\frac{d\theta}{dz} = \frac{d\theta}{ds}\frac{ds}{dz}$ and $\frac{d\theta}{ds} > 0$ or < 0 holds along γ, $\Delta_{z\in\gamma} \arg\frac{d\theta}{dz} = \Delta_{z\in\gamma} \arg\frac{ds}{dz}$ is true. Denote by φ the angle between the tangent of γ at z and the positive real axis. Then,

$$\frac{dz}{ds} = \frac{dx}{ds} + i\frac{dy}{ds} = \cos\varphi + i\sin\varphi = e^{i\varphi},$$

$$\Rightarrow \Delta_{z\in\gamma} \arg\frac{ds}{dz} = -\Delta_{z\in\gamma} \arg\frac{dz}{ds} = -\Delta_{z\in\gamma}\varphi = -2\pi$$

(see (5) in (3.5.2.4)),

$\Rightarrow$ (substituting into (*)) $\dfrac{1}{2\pi}\Delta_{z\in\gamma}\arg f'(z) = -1 + \dfrac{1}{2\pi}\Delta_{z\in\gamma}\arg f(z)$

and (2) follows.   □

We turn to *a general problem* concerning Examples 1 and 2:

Suppose a polynomial $p(z) = z^n + a_{n-1}z^{n-1} + \cdots + a_1 z + a_0$ has *each* of its coefficients $a_{n-1}, \ldots, a_0$ a real-valued continuous function of two real variables α and β. Since zeros of a polynomial depend continuously on its coefficients (see Example 4 in Sec. 3.5.4), it follows that the zeros also depend continuously on α and β in this case. This kind of dependence can be described more accurately as follows:

(1) Construct the curve $p(it) = 0$, where t is a real parameter, in the $\alpha\beta$-plane. Points (α, β) on this curve will ensure that $p(z) = 0$ has zero roots or pure imaginary roots.
(2) $p(z) = 0$ can have only real roots or conjugate complex roots with nonzero imaginary parts on each component Ω of $\mathbf{C}^* - \{p(it) = 0\}$. Via arguments similar to the proof of (3.5.1.14), the number of roots, with positive real parts, of $p(z) = 0$ on such a Ω is a *constant*.
(3) How to determine this constant? Choose a representative point (α, β) in Ω and try to find the number of roots of the resulted polynomial equation $p(z) = 0$ in the right half-plane $\operatorname{Re} z > 0$:

(i) As $y = \operatorname{Im} z$ moves from $+i\infty$ along the imaginary axis down to $-i\infty$, evaluate the variation of $\arg p(iy)$, say $\Delta_{y\in(-\infty,\infty)}\arg p(iy) = k\pi$, $|k| \le n$.
(ii) Evaluate the variation of $\arg p(z)$ along the right half-circle $C_R : z = Re^{i\theta}$, $-\frac{\pi}{2} \le \theta \le \frac{\pi}{2}$, say $\lim_{R\to\infty}\Delta_{C_R}\arg p(z) = m\pi$.

Then, the required number, where $\gamma_R = [-iR, iR] \cup C_R$ with counterclockwise direction, is

$$\lim_{R\to\infty}\frac{1}{2\pi}\Delta_{\gamma_R}\arg p(z) = \frac{1}{2}(k+m), \quad \text{a non-negative integer.} \qquad (3.5.3.5)$$

We illustrate the following example.

Example 5.   Let α and β be real parameters. Try to find domains on the $\alpha\beta$-plane on each of which the number of roots, with positive real parts, of the equation $p(z) = z^3 + \alpha z^2 + \beta z + 1$ is equal to a constant. Determine these constants.

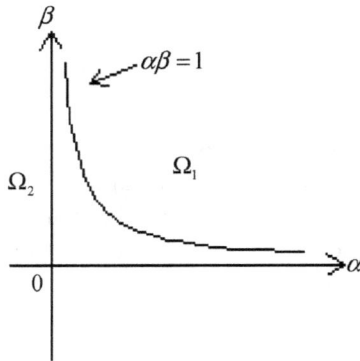

Fig. 3.65

Solution. Let t be a real parameter. Then $p(it) = -it^3 - \alpha t^2 + \beta it + 1 = (-\alpha t^2 + 1) + i(-t^3 + \beta t) = 0$, which implies that $-\alpha t^2 + 1 = -t^3 + \beta t = 0$. After eliminating $t \neq 0$ from both equations, we obtain the plane curve $p(it) = 0$, namely, the hyperbola $\alpha\beta = 1$, in the $\alpha\beta$-plane.

Along $\alpha\beta = 1$, $p(z) = z^3 + \alpha z^2 + \frac{1}{\alpha}z + 1 = (z + \alpha)\left(z^2 + \frac{1}{\alpha}\right) = 0$ has roots $z = -\alpha$, $\pm\frac{i}{\sqrt{\alpha}}$ (if $\alpha > 0$) or $\pm\frac{1}{\sqrt{-\alpha}}$ (if $\alpha < 0$).

The curve $\alpha\beta = 1$ divides the plane into two domains:

$$\Omega_1 = \left\{ (\alpha, \beta) \mid \alpha > 0 \text{ and } \beta > \frac{1}{\alpha} \right\}, \quad \text{and}$$

$$\Omega_2 = \left\{ (\alpha, \beta) \mid \alpha \leq 0 \text{ or } \alpha > 0 \text{ and } \beta < \frac{1}{\alpha} \right\}.$$

See Fig. 3.65.

On Ω_1: Pick a representative point, say $\alpha = 2$ and $\beta = 2$. In this case, $p(z) = z^3 + 2z^2 + 2z + 1$. Let $z = x + iy$. Now $p(iy) = (-2y^2 + 1) - i(y^3 - 2y) \neq 0$ holds for all $y \in \mathbf{R}$, and

$$\arg p(iy) = \tan^{-1}\frac{y^3 - 2y}{2y^2 - 1} = \tan^{-1}\frac{(y + \sqrt{2})y(y - \sqrt{2})}{2\left(y + \frac{1}{\sqrt{2}}\right)\left(y - \frac{1}{\sqrt{2}}\right)}.$$

Observing the variation of $\arg p(iy)$ as y decreases from $+\infty$ down to $-\infty$ in the following table:

y	∞	$\downarrow$	$\sqrt{2}$	$\downarrow$	$\dfrac{1}{\sqrt{2}}$
$\tan \arg p(iy)$	∞	>0	0	<0	$-\infty$
$\arg p(iy)$	$\dfrac{\pi}{2}$	$0<\cdots<\dfrac{\pi}{2}$	0	$-\dfrac{\pi}{2}<\cdots<0$	$-\dfrac{\pi}{2}$
y		$\downarrow$	0	$\downarrow$	$-\dfrac{1}{\sqrt{2}}$
$\tan \arg p(iy)$		>0	0	<0	$-\infty$
$\arg p(iy)$		$-\pi<\cdots<-\dfrac{\pi}{2}$	$-\pi$	$-\dfrac{3\pi}{2}<\cdots<-\pi$	$-\dfrac{3\pi}{2}$
y		$\downarrow$	$-\sqrt{2}$	$\downarrow$	$-\infty$
$\tan \arg p(iy)$		>0	0	<0	$-\infty$
$\arg p(iy)$		$-2\pi<\cdots<-\dfrac{3\pi}{2}$	-2π	$-\dfrac{5\pi}{2}<\cdots<-2\pi$	$-\dfrac{5\pi}{2}$

it follows that

$$\Delta_{y\in(-\infty,\infty)} \arg p(iy) = -\frac{5\pi}{2} - \frac{\pi}{2} = -3\pi.$$

Let C_R be the half-circle $z = \mathrm{Re}^{i\theta}$, $-\frac{\pi}{2} \le \theta \le \frac{\pi}{2}$, and γ be as in Fig. 3.64. Then, as in Example 2,

$$\lim_{R\to\infty} \Delta_{z\in C_R} \arg p(z) = 3\left(\frac{\pi}{2} - \left(-\frac{\pi}{2}\right)\right) = 3\pi$$

$$\Rightarrow \text{(combining together)} \quad \lim_{R\to\infty} \frac{1}{2\pi}\Delta_{p(\gamma)} \arg p(z) = \frac{1}{2\pi}(-3\pi + 3\pi)$$
$$= 0.$$

This means that $p(z) = 0$ does not have roots with positive real parts if $(\alpha, \beta) \in \Omega_1$.

On Ω_2: Pick $\alpha = -1$ and $\beta = -1$ as representative. In this case, $p(z) = z^3 - z^2 - z + 1$ and $p(iy) = (y^2 + 1) - i(y^3 + y) \neq 0$ also holds for any $y \in \mathbf{R}$. Then

$$\arg p(iy) = \tan^{-1} \frac{-y(y^2 + 1)}{y^2 + 1} = \tan^{-1}(-y),$$

$$\Rightarrow \Delta_{y\in(-\infty,\infty)} \arg p(iy) = \frac{\pi}{2} - \left(-\frac{\pi}{2}\right) = \pi.$$

On the other hand,

$$\lim_{R \to \infty} \Delta_{z \in C_R} \arg p(z) = 3\pi,$$

$\Rightarrow$ (combining together) $\lim\limits_{R \to \infty} \dfrac{1}{2\pi} \Delta_{p(\gamma)} \arg p(z) = \dfrac{1}{2\pi}(\pi + 3\pi)$

$$= 2.$$

Therefore, if $(\alpha, \beta) \in \Omega_2$, two of the three roots of $p(z) = 0$ will have positive real parts.

Exercises A

(1) Try to use (3.5.3.3), in particular, the argument principle, to evaluate the following integrals.

(a) $\frac{1}{2\pi i} \int_\gamma \frac{f'(z)}{f(z)} dz$, where γ is the circle $|z - 1 - i| = 2$ and $f(z)$ is one of the functions $\frac{z-2}{z(z-1)}$, $\frac{z-2}{z(z-1)^2}$, and $\frac{z^2-9}{z^2+1}$.

(b) $\int_{|z|=\pi} \cot \pi z \, dz$.

(c) $\int_{|z|=1} \dfrac{dz}{\sin z}$.

(d) $\int_{|z|=3\pi} \dfrac{e^z}{e^z - 1} dz$.

(e) $\int_{|z-1|=2} \tanh z \, dz$.

(f) $\int_{|z|=2.5} (z^2 + z - 1) \dfrac{f'(z)}{f(z)} dz$,   where $f(z) = \dfrac{(z^2+1)^2}{z^3 - 6z^2 + 11z - 6}$.

(g) $\int_{|z|=2} z \dfrac{f'(z)}{f(z)} dz$,   where $f(z) = \dfrac{z^n - 1}{z^n + 1}$.

(h) $\int_{|z|=3} z \dfrac{f'(z)}{f(z)} dz$,   where $f(z) = \dfrac{(z^2+4)^2}{(z^2 + z + 1)^2}$.

(2) Suppose γ is a rectifiable Jordan closed curve whose interior $\text{Int}\,\gamma$ contains all the zeros of the polynomial $p(z) = a_0 + a_1 z + \cdots + a_{n-1} z^{n-1} + z^n$. Show that

(a) $\int_\gamma z \dfrac{p'(z)}{p(z)} dz = -2\pi i a_{n-1}$,   and

(b) $\int_\gamma z^2 \dfrac{p'(z)}{p(z)}\,dz = 2\pi i(a_{n-1}^2 - 2a_{n-2})$.

What is $\int_\gamma z^3 \frac{p'(z)}{p(z)}\,dz$?

(3) Applying the argument principle to $e^z = \sum_{n=0}^\infty \frac{1}{n!}z^n$, $|z| < \infty$, to show that $e^z \neq 0$ on $\mathbf{C}$.

(4) Do each of the following problems.

(a) Show that $z^8 + az^3 + bz + c$ always has two zeros in the first quadrant where a, b, and c are any positive real constants.

(b) Show that $z^4 + 3z + 3$ has only one zero in the horizontal strip $0 < \operatorname{Im} z < 1$.

(c) How many roots of the equation $z^4 + 8z^3 + 3z^2 + 8z + 3 = 0$ lie in the right half-plane?

(d) Same as (c) for the equation $z^6 + z^5 + 6z^4 + 5z^3 + 8z^2 + 4z + 1 = 0$.

(e) Same as (c) for the equation $z^4 + 2z^3 + 3z^2 + z + 2 = 0$. In the first quadrant?

(f) How many roots of the equation $2z^4 - 3z^3 + 3z^2 - z + 1 = 0$ lie in each quadrant?

(g) Where are the roots of the equation $z^4 + z^3 + 4z^2 + 2z + 3 = 0$ situated?

(h) Show that the equation $z^8 + 3z^3 + 7z + 5 = 0$ has exactly two roots in the first quadrant. Where are the other roots?

(i) Show that $z^4 + 2z^3 - 2z + 10 = 0$ has exactly one root in each quadrant.

(5) Show that $\tan z - z = 0$ has only real roots and each of the intervals $\left((n - \tfrac{1}{2})\pi, (n + \tfrac{1}{2})\pi\right)$, $n = 0, \pm 1, \pm 2, \ldots$ contains exactly only one root x_n of it. Try to find the sum

$$\sum_{\substack{n=-\infty \\ n \neq 0}}^{\infty} \frac{1}{x_n^2}$$

by evaluating the integral $\int_\gamma \left[\frac{1}{z - \tan z} - \frac{1}{z}\right]\,dz$, where γ is the square with vertices at $n\pi(\pm 1 \pm i)$, as shown in Fig. 3.66.

(6) Show that the equation $(z + 1)e^{-z} = z + 2$ does not have roots in the right half-plane.

(7) Suppose f is analytic in the ring domain $r < |z| < R$ and $f(z) \neq w_0$ holds there. Let $\Gamma_t = f(\{|z| = t\})$, $r < t < R$. Show that $n(\Gamma_t : w_0)$ is a constant for $r < t < R$.

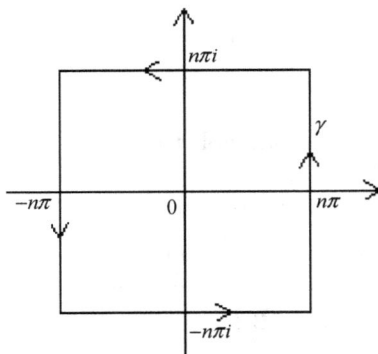

Fig. 3.66

(8) Suppose f is analytic in $|z| \leq R$ and $f(z) \neq 0$ on $|z| = R$. Show that the maximum value of $\mathrm{Re}\left(z\frac{f'(z)}{f(z)}\right)$ on $|z| = R$ is at least equal to the number of zeros of $f(z)$ in $|z| < R$.

(9) In the theory of automatic control, the stability of the solutions of the differential equation $a_n w^{(n)}(z) + \cdots + a_1 w'(z) + a_0 = f(z)$, appeared in some physical problems, is reduced to require that the roots of its characteristic equation $p(z) = a_n z^n + \cdots + a_1 z + a_0 = 0$ all lie in the left plane. Show that this is the case if and only if

$$\Delta_{y \in (-\infty, \infty)} \arg p(iy) = n\pi,$$

where y increases from $-\infty$ to $+\infty$.

(10) Imitate Example 5 and do the following problems.

   (a) $p(z) = z^3 + (\alpha + \beta)z^2 + (\alpha - \beta)z + \alpha$.
   (b) $p(z) = z^3 + \alpha z^2 + \alpha z + \beta$.

Exercises B

Finally, we raise *another type of problem*:

   Let $p_n(z)$ and $q_m(z)$ be polynomials of respective degrees n and m, relatively prime, and $n > m$. Also, suppose that

(1) The function $f(z) = p_n(z) + q_m(z)e^{-tz}$ $(t > 0)$ does not have zeros along the imaginary axis.

(2) $p_n(z)$ has N zeros, in total, in the right half-plane.

Then,

   (1) $f(z)$ does not have zeros in the right half-plane.

⇔ (2) When $z = iy$ moves from $+i\infty$ downward to $-i\infty$ along the imaginary axis, the points $w = -\dfrac{q_m(z)}{p_n(z)}e^{-tz}$ wind around the point 1 exactly N times in the clockwise direction, namely,

$$\frac{1}{2\pi}\Delta_{y\in(-\infty,\infty)}\arg\left(\frac{q_m(z)}{p_n(z)}e^{-tz}+1\right) = -N. \qquad (3.5.3.6)$$

Note that in case $p_n(z)$ having zeros on the imaginary axis, a dented small half-circle toward the right half-plane with center at such a zero point should be invoked as z moves downward along the imaginary axis.

Sketch of Proof

Let γ be as in Fig. 3.64. It is well known that $\lim_{R\to\infty}\Delta_{z\in C_R}\arg p_n(z) = n\left(\frac{\pi}{2}-\left(-\frac{\pi}{2}\right)\right) = n\pi$. By Condition 2,

$$\lim_{R\to\infty}\frac{1}{2\pi}\Delta_{z\in\gamma}\arg p_n(z) = \frac{1}{2\pi}\Delta_{y\in(-\infty,\infty)}\arg p_n(iy)$$

$$+ \lim_{R\to\infty}\frac{1}{2\pi}\Delta_{z\in C_R}\arg p_n(z) = N,$$

$$\Rightarrow \frac{1}{2\pi}\Delta_{y\in(-\infty,\infty)}\arg p_n(iy) = N - \frac{1}{2}n. \qquad (*_1)$$

Recall Condition 1 and consider, as R large enough,

$$\Delta_{z\in\gamma}\arg f(z) = \Delta_{y\in[-R,R]}\arg p_n(iy) + \Delta_{y\in[-R,R]}\arg\left\{1 + \frac{q_m(iy)}{p_n(iy)}e^{-ity}\right\}$$

$$+ \Delta_{z\in C_R}\arg p_n(z)\left[1 + \frac{q_m(z)}{p_n(z)}e^{-tz}\right]. \qquad (*_2)$$

Since $n > m$ and $\operatorname{Re} z > 0$ along C_R,

$$\lim_{R\to\infty}\Delta_{z\in C_R}\arg p_n(z)\left[1 + \frac{q_m(z)}{p_n(z)}e^{-tz}\right] = \lim_{R\to\infty}\Delta_{z\in C_R}\arg p_n(z) = n\pi,$$

$$\Rightarrow \text{(by } (*_1) \text{ and } (*_2))$$

$$\lim_{R\to\infty}\frac{1}{2\pi}\Delta_{z\in\gamma}\arg f(z) = N - \frac{1}{2}n + \frac{1}{2}n$$

$$+ \frac{1}{2\pi}\Delta_{y\in(-\infty,\infty)}\arg\left\{1 + \frac{q_m(iy)}{p_n(iy)}e^{-ity}\right\}.$$

And the result follows obviously.

We illustrate the following problem.

(1) Try to find a domain Ω in the ab-plane on which each of the following functions does not have zeros in the right half-plane.

(a) $f(z) = z + a + be^{-tz}, \quad t > 0.$
(b) $f(z) = z^2 + az + be^{-tz}, \quad t > 0.$
(c) $f(z) = z^2 + (az + b)e^{-tz}, \quad t > 0.$

3.5.4.   The Rouché's theorem

One of the standard applications of the argument principle is the following

Rouché's theorem. Suppose that f and g are analytic in a domain Ω. Let γ be a rectifiable Jordan closed curve in Ω such that $\text{Int}\,\gamma$ is contained in Ω. If

$$|f(z) - g(z)| < |g(z)|$$

holds along γ, then f and g have the same number of zeros in $\text{Int}\,\gamma$.

(3.5.4.1)

This result is still valid for a closed chain γ in Ω which is homologous to zero in Ω (namely, $n(\gamma; z) = 0$ for all z not in Ω) and $n(\gamma; z) = 0$ or 1 for all $z \in \Omega - \gamma$. For an extension to meromorphic functions, see Exercise B (2).

Proof. The condition that $|f(z) - g(z)| < |g(z)|$, $z \in \gamma$, suggests that both $f(z)$ and $g(z)$ do not assume zero values along γ. Consequently, $\frac{f(z)}{g(z)}$ is analytic on γ and

$$\left| \frac{f(z)}{g(z)} - 1 \right| < 1, \quad z \in \gamma.$$

This means that the image curve $F(\gamma)$ of γ under $F(z) = \frac{f(z)}{g(z)}$ lies in the disk $|w - 1| < 1$ and thus

$$\Delta_{z \in \gamma} \arg F(z) = 0.$$

On the other hand, by the argument principle (3.5.3.4),

$$\frac{1}{2\pi i} \int_\gamma \frac{F'(z)}{F(z)} dz = 0$$

$$= \{\text{the number of zeros of } F \text{ or } f \text{ in Int}\,\gamma\}$$

$$- \{\text{the number of poles of } F \text{ or the number of zeros of } g \text{ in Int}\,\gamma\}$$

and the result follows.   □

Another proof. Each of the functions defined by $h_t(z) = g(z) + t(f(z) - g(z))$, $0 \leq t \leq 1$, is analytic in Ω. By assumption, $|h_t(z)| \geq |g(z)| - t|f(z) - g(z)| \geq |g(z)| - |f(z) - g(z)| > 0$ on γ, for all t, $0 \leq t \leq 1$. By argument principle,

$$\frac{1}{2\pi i} \int_\gamma \frac{h_t'(z)}{h_t(z)} dz = \text{the number of zeros of} h_t(z) \text{inInt } \gamma, \quad 0 \leq t \leq 1.$$

But the left integral in the above relation is a continuous function of $t \in [0, 1]$ and assumes only integer values. It is necessarily a constant (see Remark 2 in Sec. 2.2). In particular, $h_0(z) = g(z)$ and $h_1(z) = f(z)$. Assume the same number of zeros in Int γ. $\square$

Some concrete examples are as follows.

Example 1.

(a) How many roots does the equation $z^4 - 6z + 3 = 0$ have in the annulus domain $1 < |z| < 2$?
(b) Let $\lambda > 1$, show that $f(z) = \lambda - z - e^{-z}$ has only one zero in Re $z \geq 0$, which is a real one.

Solution.

(a) On $|z| = 1$, $|z^4 + 3| \leq 1 + 3 = 4 < 6 = |-6z|$, so $z^4 - 6z + 3 = 0$ has only one root in $|z| < 1$. Observe that $z^4 - 6z + 3 \neq 0$ on $|z| = 1$. On the other hand, on $|z| = 2$, $|-6z + 3| \leq 12 + 3 = 15 < 16 = |z^4|$, so $z^4 - 6z + 3 = 0$ has all its four roots in $|z| < 2$. Therefore, the equation has three roots in $1 < |z| < 2$.
(b) Let γ be the closed half-circle $[-iR, iR] \cup C_R$ where $R > 0$ and C_R is $z = Re^{i\theta}$, $-\frac{\pi}{2} \leq \theta \leq \frac{\pi}{2}$ (see Fig. 3.64). Let $g(z) = \lambda - z$.
   Along the segment $[-iR, iR]$: Let $z = x + iy$. Then $|g(iy)| = |\lambda - iy| \geq \lambda > 1$ and $|f(iy) - g(iy)| = |e^{-iy}| = 1 < |g(iy)|$. While, along C_R ($R > \lambda + 1$ is tacitly assumed),

$$|f(z) - g(z)| = |e^{-z}| = e^{-x} \leq 1 < R - \lambda \leq |z| - \lambda \leq |z - \lambda| = |g(z)|.$$

By the Rouché's theorem, $f(z)$ and $g(z)$ have the same number of zero in Int γ. Since $g(z)$ has only one zero in Int γ, so does $f(z)$.
In case $R > \lambda + 1$, $|f(z)| > R - \lambda - e^{-\text{Re } z} \geq R - \lambda - 1 > 0$ along C_R. So $f(z) \neq 0$ on C_R. Since $R > 0$ is arbitrary, $f(z)$ has only one zero in Re $z \geq 0$.
If there is a z_0 with Re $z_0 \geq 0$ and Im $z_0 \neq 0$ so that $f(z_0) = 0$, since $\overline{f(z)} = f(\bar{z})$ holds on Re $z \geq 0$, it follows that $f(\overline{z_0}) = 0$ is true.

This contradicts to the fact that $f(z)$ has only one zero in $\operatorname{Re} z \geq 0$. Hence $\operatorname{Im} z_0 = 0$ should hold and the only zero of $f(z)$ in $\operatorname{Re} z \geq 0$ should be a real one.

Example 2.

(a) If n is large enough, show that the polynomial $\sum_{k=0}^{n} \frac{z^k}{k!}$ does not have zero in any bounded domain Ω. Equivalently, the function $\sum_{k=0}^{n} \frac{1}{k!} \frac{1}{z^k}$ has all its zeros lying in $|z| < \rho$ no matter how small $\rho > 0$ is if n is sufficiently large.

(b) In case $0 < \rho < 1$, show that the polynomial $\sum_{k=1}^{n} kz^{k-1} \neq 0$ in $|z| < \rho$ if n is large enough.

Proof.

(a) The sequence of polynomials $p_n(z) = \sum_{k=0}^{n} \frac{z^k}{k!}$ converges uniformly on Ω to e^z. Let $m = \min_{\partial\Omega} |e^z|$. Then $m > 0$. Choosing $0 < \varepsilon < m$, there is a positive integer N so that $|p_n(z) - e^z| < \varepsilon$ holds on $\overline{\Omega}$ if $n \geq N$. Then, along $\partial\Omega$,

$$|p_n(z) - e^z| < \varepsilon < m \leq |e^z|. \tag{$*$}$$

So $p_n(z)$ and e^z have the same number of zeros in Ω (here, we may suppose that Ω is the interior $\operatorname{Int} \gamma$ of a rectifiable Jordan closed curve γ). Since $e^z \neq 0$ everywhere, $p_n(z) \neq 0$ on Ω. Or, by the minimum modulus principle (see (2)(b) in (3.4.4.2)), $m = \min_{\overline{\Omega}} |e^z|$ holds, too. If $p_n(z_0) = 0$ for some $z_0 \in \Omega$, then $(*)$ says that $|p_n(z_0) - e^{z_0}| = |e^{z_0}| < m$, a contradiction. Hence $p_n(z) \neq 0$ for any $z \in \Omega$.

(b) It is well known that $p_n(z) = \sum_{k=1}^{n} kz^{k-1}$ converges uniformly to $\frac{1}{(1-z)^2}$ on every compact subset of $|z| < 1$. Along $|z| = \rho$,

$$\left| p_n(z) - \frac{1}{(1-z)^2} \right| = \left| -\sum_{k=n+1}^{\infty} kz^{k-1} \right| \leq \sum_{k=n+1}^{\infty} k\rho^{k-1}$$

$$= \frac{n\rho^{n-1}}{1-\rho} + \frac{\rho^n}{(1-\rho)^2}.$$

Let $m = \min_{|z|=\rho} \left| \frac{1}{(1-z)^2} \right|$. Choose n large enough so that $\frac{n\rho^{n-1}}{1-\rho} + \frac{\rho^n}{(1-\rho)^2} < m$. For all such n, then $\left| p_n(z) - \frac{1}{(1-z)^2} \right| < \frac{1}{|1-z|^2}$ holds along $|z| = \rho$. Since $\frac{1}{(1-z)^2} \neq 0$ on $|z| < \rho$, the result follows. □

Example 3. Try to use the Rouché's theorem and the argument principle, respectively, to prove the fundamental theorem of algebra (see (3.4.2.12)).

Proof. Let $p(z) = a_n z^n + \cdots + a_1 z + a_0$, where $n \geq 1$ and $a_n \neq 0$, be the polynomial.

By Rouché's theorem: Choose $g(z) = a_n z^n$. Then,

$$\lim_{z \to \infty} \frac{p(z) - g(z)}{g(z)} = \lim_{z \to \infty} \frac{1}{a_n} \left\{ \frac{a_{n-1}}{z} + \cdots + \frac{a_1}{z^{n-1}} + \frac{a_0}{z^n} \right\} = 0.$$

$\Rightarrow$ There is an $R > 0$ so that

$$\frac{|p(z) - g(z)|}{|g(z)|} < 1 \quad \text{on } |z| \geq R.$$

$$\Rightarrow |p(z) - g(z)| < |g(z)| \quad \text{on } |z| = R$$

(and implicitly, $p(z) \neq 0$ on $|z| \geq R$).

Hence, $p(z)$ and $g(z)$ has the same number of zeros in $|z| < R$. Since $g(z)$ has a zero of order n at $z = 0$, $p(z)$ has exactly n zeros in $|z| < R$.

By the argument principle: Since $\lim_{z \to \infty} p(z) = \infty$, there is an $R > 0$ such that $|p(z)| \geq 1$ on $|z| \geq R$. Then,

$$\frac{1}{2\pi i} \int_{|z| = R} \frac{p'(z)}{p(z)} dz = \text{the number of zeros of } p(z) \text{ in } |z| < R$$

$$= -\mathrm{Re}\, s \left(\frac{p'(z)}{p(z)}; \infty \right) \quad \text{(see (4.11.1.3))}$$

$$= \mathrm{Re}\, s \left(\frac{p'(z^{-1})}{p(z^{-1})} \cdot \frac{1}{z^2}; 0 \right) = n.$$

Note: The determination of R so that $p(z)$ has its n zeros in $|z| < R$.
Let $g(z) = a_n z^n$. Along $|z| = R$,

$$|p(z) - g(z)| \leq |a_{n-1}| R^{n-1} + \cdots + |a_1| R + |a_0|$$

$$< (|a_{n-1}| + \cdots + |a_1| + |a_0|) R^{n-1}, \quad \text{if } R > 1$$

$$< |a_n| R^n, \quad \text{if } R > \frac{1}{|a_n|}(|a_{n-1}| + \cdots + |a_1| + |a_0|)$$

$$= |g(z)|.$$

Therefore, choose

$$R > \max \left\{ 1, \frac{1}{|a_n|}(|a_{n-1}| + \cdots + |a_1| + |a_0|) \right\}.$$

Rouché's theorem guarantees that $p(z)$ has all its n zeros lying in $|z| < R$.

The other way is to choose (here, suppose $a_n = 1$)

$$R = 2 \max_{0 \leq k \leq n-1} |a_k|^{\frac{1}{n-k}}.$$

Then $|a_k| \leq \left(\frac{R}{2}\right)^{n-k}$ for $0 \leq k \leq n-1$. Along $|z| = R$,

$$|a_{n-1}z^{n-1} + \cdots + a_1 z + a_0| \leq R^n \left(\frac{1}{2} + \frac{1}{2^2} + \cdots + \frac{1}{2^n}\right)$$

$$= \left(1 - \frac{1}{2^n}\right) R^n < R^n = |z|^n.$$

By Rouché's theorem, $p(z)$ has all its n zeros in $|z| < R$. What happens if the condition $a_n = 1$ is dropped?

By the way, in case $z_1, \ldots, z_n$ are the zeros of $p(z)$, then $f(z) = \frac{p(z)}{(z-z_1)\cdots(z-z_n)}$ is entire (see (3.4.2.17)). Since $\lim_{z \to \infty} f(z) = a_n$, $f(z)$ is bounded and, by Liouville's theorem, $f(z)$ is a constant, namely $f(z) \equiv a_n$. Hence, it follows the factorization $p(z) = a_n(z - z_1)\cdots(z - z_n)$. □

Example 4.   (The continuous dependence of zeros of a polynomial on its coefficients). Suppose the polynomial $p(z) = z^n + a_{n-1}z^{n-1} + \cdots + a_1 z + a_0$ has k distinct zeros $z_1, \ldots, z_k$ with multiplicities $m_1, \ldots, m_k$, respectively, where $m_1 \geq 1, \ldots, m_k \geq 1$ and $m_1 + \cdots + m_k = n$. Let

$$\rho = \min_{1 \leq j < l \leq k} |z_j - z_l|.$$

Then, for each ε, $0 < \varepsilon < \rho$, there corresponds a $\delta > 0$ so that, for any polynomial $q(z) = z^n + b_{n-1}z^{n-1} + \cdots + b_1 z + b_0$ with its coefficients satisfying $|b_j - a_j| < \delta$ for $0 \leq j \leq n-1$, each open disk

$$|z - z_l| < \varepsilon, \quad 1 \leq l \leq k,$$

contains exactly m_j zeros of $q(z)$.

Proof.   Note that $p(z) = (z - z_1)^{m_1} \cdots (z - z_k)^{m_k}$.

For each fixed l, $1 \leq l \leq k$, along $|z - z_l| = \varepsilon$,

$$|p(z)| = |z - z_1|^{m_1} \cdots |z - z_k|^{m_k} \leq \varepsilon^{m_l}(\rho + \varepsilon)^{n - m_l}.$$

On the other hand, on $|z - z_l| < \varepsilon$,

$$|q(z) - p(z)| = |(b_{n-1} - a_{n-1})z^{n-1} + \cdots + (b_1 - a_1)z + (b_0 - a_0)|$$
$$\leq |b_{n-1} - a_{n-1}||z|^{n-1} + \cdots + |b_1 - a_1||z| + |b_0 - a_0|$$
$$< \delta\mu_l, \quad \text{where } \mu_l = \sum_{j=0}^{n-1}(|z_l| + \varepsilon)^j, \quad \text{if } |b_j - a_j| < \delta$$

for $0 \leq j \leq n - 1$.

Now, choose

$$0 < \delta < \min_{1 \leq l \leq k} \mu_l^{-1}\varepsilon^{m_l}(\rho + \varepsilon)^{n - m_l}.$$

Then, by what we have established above, $|q(z) - p(z)| < |p(z)|$ along $|z - z_l| = \varepsilon$. By Rouché's theorem, $q(z)$ and $p(z)$ have the same number of zeros in each disk $|z - z_l| < \varepsilon$, $1 \leq l \leq k$. □

Exercises A

(1) For each of the following polynomials, determine the number of its zeros in the indicated domain.

(a) $z^7 - z^3 + 12$ in $1 < |z| < 2$.

(b) $z^4 + 7z + 1$ in $\dfrac{3}{2} < |z| < 2$.

(c) $2z^5 - z^3 + 3z^2 - z + 8$ in $|z| < 1$.

(d) $z^8 - 4z^5 + z^2 - 1$ in $|z| < 1$.

(e) $z^5 - z + 16$ in $1 < |z| < 2$. Also, show that it has two zeros in the right half-plane.

(f) $z^4 + iz^3 + 1$ in $|z| < \frac{3}{2}$. Show that it has exactly one zero in the first quadrant.

(2) For each real number $\lambda > 1$, show that $f(z) = ze^{\lambda - z} - 1$ has a unique positive real zero in $|z| < 1$.

(3) Show that $e^z - az^n = 0$ has n roots in $|z| < 1$ if $|a| > e$.

(4) Redo Exercise A (6) of Sec. 3.5.3 by using the Rouché's theorem.

(5) Suppose that $0 < |a| < 1$ and p is a positive integer. Show that $(z - 1)^p - ae^{-z}$ has exactly p simple zeros with positive real parts and lying in $|z - 1| < 1$.

(6) For any complex constant a, show that $az^n + z + 1$ $(n \geq 2)$ has at least one zero in $|z| < 2$.

(7) Show that the equation $\sin z = 2z^4 - 7z + 1$ has exactly one root in $|z| < 1$.

(8) Let w_0 be a complex number and n be a positive integer such that $\left(n + \frac{1}{2}\right)\pi > |w_0|$. Show that $f(z) = z\sin z$ assumes the value w_0 exactly $2n + 2$ times in the strip $|\operatorname{Re} z| < \left(n + \frac{1}{2}\right)\pi$.

(9) (A continuation of Example 7 of Sec. 2.9) Let $p(z) = a_n z^n + a_{n-1}z^{n-1} + \cdots + a_1 z + a_0$, where $n \geq 1$ and $a_n \neq 0$. For each $\rho > 0$, show that there is a point z_ρ, $|z_\rho| = \rho$, so that $|p(z_\rho)| \geq |a_n|\rho^n$. Hence,

$$\max_{|z|=\rho} |p(z)| \geq |a_n|\rho^n.$$

(10) Consider the transcondental equation $\tan z = \alpha z$ (α is a complex constant).

   (a) If $|\alpha n\pi\sqrt{2}| > \frac{1+e^{-2n\pi}}{1-e^{-2n\pi}}$ holds for a positive integer n, show that $\tan z = \alpha z$ has exactly $2n + 1$ roots in the square γ with vertices at $n\pi (\pm 1 \pm i)$.

   (b) In case $\alpha > 0$ is a real number, then $\tan z = \alpha z$ has infinitely many roots. More precisely, all the roots are real if $\alpha \geq 1$; and there are only two pure imaginary roots if $0 < \alpha < 1$.

(11) Show that $f(z) = (z+a)e^z + (z-a)e^{-z}$, $a > 0$, has only one real root and infinitely many pure imaginary roots.

(12) Let f be analytic on $\overline{\operatorname{Int}\gamma}$, where γ is a rectifiable Jordan closed curve. Show that it is *not* possible to find a point z_0 in $\operatorname{Int}\gamma$ so that $|f(z)| < |f(z_0)|$ on γ.

(13) Let $|a_k| < 1$ for $1 \leq k \leq n$. Let

$$f(z) = \prod_{k=1}^{n} \frac{z - a_k}{1 - \overline{a_k}z}.$$

Show that the equation $f(z) = b$ has exactly n roots in $|z| < 1$ if $|b| < 1$, while $f(z) = b$ has exactly n roots in $|z| > 1$ if $|b| > 1$.

(14) Suppose that f is analytic in $|z| < R$ and continuous on $|z| \leq R$. Also, suppose

   (i) $|f(0)| < m$ ($m > 0$ is a constant); and

   (ii) $|f(z)| \geq m$ along $|z| = R$.

   Show that f has at least one zero in $|z| < R$ by the following two methods:

   (a) Use Rouché's theorem.

   (b) Use the Cauchy integral formula via a contradictory argument.

(15) (Compare with (3.4.4.3)) Let f be analytic on $|z| \leq R$ and not constant. Assume that $|f(z)|$ is constant on $|z| = R$. Use Rouché's theorem to show that f has at least one zero in $|z| < R$.

(16) Suppose f is analytic on $|z| \leq 1$ and $|f(z)| < 1$ on $|z| = 1$. Show that f has a unique fixed point in $|z| < 1$, namely, there is a unique point z_0 in $|z| < 1$ such that $f(z_0) = z_0$. Give an example to illustrate that the condition "$|f(z)| < 1$ on $|z| = 1$" is necessary. If this condition is replaced by "$|f(z)| \leq 1$ on $|z| = 1$", show that f has at least one fixed point in $|z| \leq 1$ and that if f fixes no point of the circle $|z| = 1$, it has exactly one fixed point in $|z| \leq 1$.

(17) (Compare with Example 3 in Sec. 3.4.4) Let f be analytic on $|z| \leq 1$ and not constant. Assume that $|f(z)| = 1$ if $|z| = 1$. Show that, by Rouché's theorem, the image $f(\{|z| < 1\})$ contains $|w| < 1$.

(18) Let f be analytic on $|z| \leq 1$. If there is a point z_0, $|z_0| < 1$, so that $|f(z_0)| < 1$, and $|f(z)| \geq 1$ on $|z| = 1$, show that the image $f(\{|z| < 1\})$ contains $|w| < 1$.

(19) Let f be analytic on $|z| \leq 1$ and $|f(z) - z| < |z|$ on $|z| = 1$. Show that $|f'(\frac{1}{2})| \leq 8$ and f has exactly one zero in $|z| < 1$.

(20) Let f and g be analytic on $|z| \leq R$, and assume that $f(z) \neq 0$ on $|z| = R$. Show that there is an $\varepsilon > 0$ so that $f(z)$ and $f(z) + \varepsilon g(z)$ have the same number of zeros in $|z| < R$.

(21) Let f be analytic on $|z| \leq R$. Let $z_1, \ldots, z_n$ be distinct points in $|z| < R$ and $g(z) = (z - z_1) \cdots (z - z_n)$. Show that

$$p(z) = \frac{1}{2\pi i} \int_{|\zeta| = R} f(\zeta) \frac{1 - \frac{g(z)}{g(\zeta)}}{\zeta - z} d\zeta$$

is a polynomial of degree $n - 1$ such that $p(z_j) = f(z_j)$ for $1 \leq j \leq n$.

Exercises B

(1) (A version of Rouché's theorem) Let f, g, Ω, and γ be as in (3.5.4.1). If

$$|f(z) - g(z)| < |f(z)| + |g(z)|, \quad z \in \gamma,$$

then f and g have the same number of zeros in Int γ.

Note: If f and g are analytic in a bounded, simply connected domain Ω and continuous on $\overline{\Omega}$ and if $|f(z) - g(z)| < |f(z)| + |g(z)|$ holds along $\partial\Omega$, then f and g have the same number of zero in Ω.

The proof of this generalized version is based on the Riemann mapping theorem, to be proved in Chap. 6, which insures the existence of a univalent analytic function φ mapping $|z| < 1$ onto Ω. Then, use a uniform continuity argument to show that Exercise (1) is applicable to

f and g in the domain $\Omega_r = \varphi(\{|z| < r\})$ for all r sufficiently close to 1.

(2) (A generalized version of Rouché's theorem, Estermann (1962) and Glicksberg (1976)) Let f and g be meromorphic in $\operatorname{Int}\gamma$, where γ is a rectifiable Jordan closed curve, and analytic on $\overline{\operatorname{Int}\gamma}$ except poles. In case,

$$|f(z) + g(z)| < |f(z)| + |g(z)|, \quad z \in \gamma,$$

then the difference of the number of zeros of f and g in $\operatorname{Int}\gamma$ is the same as the difference of the number of poles of f and g in $\operatorname{Int}\gamma$.

3.5.5.   *Some sufficient conditions for analytic functions to be univalent*

It is well known that an analytic function f is locally univalent at points z where $f'(z) \neq 0$ (see (3.4.1.2), (3.4.7.1), and (3.5.1.4)). What we do care in this section is the global univalence about analytic functions. The argument principle will help a lot in this topic.

The most basic one among all is the following

Univalence in the interior and the exterior of a rectifiable Jordan closed curve γ.

(1) If f is analytic on $\overline{\operatorname{Int}\gamma}$ and is one-to-one along γ, then f is univalent in $\operatorname{Int}\gamma$ and maps $\operatorname{Int}\gamma$ onto $\operatorname{Int}f(\gamma)$, namely, $f(\operatorname{Int}\gamma) = \operatorname{Int}f(\gamma)$.
(2) If f is analytic on $\overline{\operatorname{Ext}\gamma} = \gamma \cup \operatorname{Ext}\gamma$ (including ∞) and is one-to-one along γ, then f is univalent in $\operatorname{Ext}\gamma$ and maps $\operatorname{Ext}\gamma$ onto $\operatorname{Int}f(\gamma)$, namely, $f(\operatorname{Ext}\gamma) = \operatorname{Int}f(\gamma)$.   (3.5.5.1)

Proof.

(1) Since f is a one-to-one continuous function on γ, it follows that $f(\gamma)$ is a Jordan closed curve. $f(\gamma)$ is also rectifiable owing to f being analytic. Fix any $w_0 \in \operatorname{Int}f(\gamma)$. By argument principle (3.5.3.4),

$$n(f(\gamma); w_0) = \text{the number of times that } f(z) \text{ assumes } w_0 \text{ in } \operatorname{Int}\gamma.$$

Since $n(f(\gamma); w_0) = 1$ or -1 (see (5) in (3.5.2.4)) and the number of zeros of $f(z) - w_0$ is a nonnegative integer, hence $n(f(\gamma); w_0) = 1$ should hold. This, in turn, implies that

(i) $f(z)$ assumes each $w_0 \in \operatorname{Int}f(r)$ exactly once in $\operatorname{Int}\gamma$ and then $\operatorname{Int}f(\gamma) \subseteq f(\operatorname{Int}\gamma)$; and

(ii) $w = f(z)$ preserves the orientation of γ.

To prove the converse inclusion $f(\operatorname{Int}\gamma) \subseteq \operatorname{Int} f(\gamma)$, observe that, for any fixed $w_0 \in \operatorname{Ext} f(\gamma)$,

$$n(f(\gamma); w_0) = 0 = \text{the number of zeros of } f(z) - w_0 \text{ in } \operatorname{Int}\gamma.$$

This implies that

(iii) $f(z)$ does not assume w_0 in $\operatorname{Int}\gamma$, i.e., $\operatorname{Ext} f(\gamma) \cap f(\operatorname{Int}\gamma) = \phi$ which indicates that $f(\operatorname{Int}\gamma) \subseteq C^* - \operatorname{Ext} f(\gamma) = f(\gamma) \cup \operatorname{Int} f(\gamma)$; and

(iv) also, $f(\gamma) \cap f(\operatorname{Int}\gamma) = \phi$ holds.

Consequently, $f(\operatorname{Int}\gamma) \subseteq \operatorname{Int} f(\gamma)$.

Combining together, $f(\operatorname{Int}\gamma) = \operatorname{Int} f(\gamma)$ and f is univalent on $\operatorname{Int}\gamma$.

(2) Take a point $z_0 \in \operatorname{Int}\gamma$. The bilinear transformation $\zeta = \frac{1}{z-z_0}$ maps γ one-to-one and onto a rectifiable Jordan closed curve γ^* in the ζ-plane, and maps $\operatorname{Ext}\gamma$ univalently and conformally onto the Jordan domain $\operatorname{Int}\gamma^*$. See Fig. 3.67. The composite maps $f^*(\zeta) = f\left(z_0 + \frac{1}{\zeta}\right)$ are thus analytic in $\overline{\operatorname{Int}\gamma^*}$ and one-to-one on γ^*. By established result in (1), $f^*(\zeta)$ maps $\operatorname{Int}\gamma^*$ univalently onto $\operatorname{Int} f^*(\gamma^*) = f^*(\operatorname{Int}\gamma^*) = \operatorname{Int} f(\gamma)$. This means that $f(z)$ maps $\operatorname{Ext}\gamma$ univalently onto $\operatorname{Int} f(\gamma)$. $\square$

Remark 1 (Generalized versions of (3.5.5.1)).

(1) Suppose f is analytic on $\overline{\operatorname{Int}\gamma}$ except at a *simple* pole $z_0 \in \operatorname{Int}\gamma$ and is one-to-one along γ. Then f maps $\operatorname{Int}\gamma$ univalently onto $\operatorname{Ext} f(\gamma)$.

$$(3.5.5.2)$$

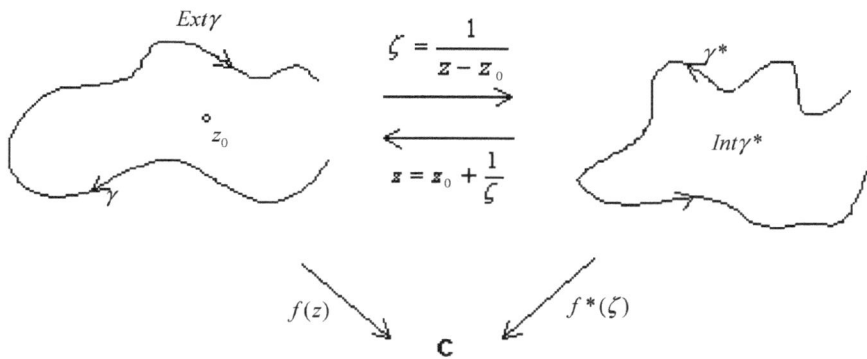

Fig. 3.67

Applying the argument principle and the method in proving (3.5.5.1), the claim follows easily. In case $f(z) \neq 0$ in Int γ, one can always apply the known result in (3.5.1.1) to $f^*(z) = \frac{1}{f(z)}$; in case $f(z)$ has a zero in Int γ, choose any fixed w_0 in Int $f(\gamma)$ other than 0 and apply to $f^*(z) = \frac{1}{f(z)+w_0}$.

(2) Suppose f is analytic on $\overline{\text{Int}\,\gamma}$ except at a finite number of points on γ but is still continuous on $\overline{\text{Int}\,\gamma}$. If f is one-to-one along γ, then f maps Int γ univalently onto Int $f(\gamma)$. Similar result holds on $\overline{\text{Ext}\,\gamma}$ as (2) of (3.5.5.1).   (3.5.5.3)

At a point $z_0 \in \gamma$ where f ceases to be analytic, replace a small arc γ_{z_0} of γ that contains z_0 by a small circular arc σ_{z_0} with center at z_0 and lying entirely in Int γ and with the same end points as γ_{z_0}. This process will produce a rectifiable Jordan closed curve γ' such that f is analytic on $\overline{\text{Int}\,\gamma'}$. Then, the image curves $f(\gamma_{z_0})$ and $f(\sigma_{z_0})$ will lie in a circle $|w - w_0| = R$, where $w_0 = f(z_0)$, with the same end points. If w^* is a point in $|w - w_0| > R$, then (4) in (3.5.2.4) guarantees that $\Delta_{f(\sigma_{z_0})} \arg(w - w^*) = \Delta_{f(\gamma_{z_0})} \arg(w - w^*)$. Finally, apply uniform continuity argument and repeat the method in proving (3.5.5.1). For details, see Ref. [58], Vol. II, pp. 120– 122.

(3) Suppose f is analytic on $\overline{\text{Int}\,\gamma}$ except at a finite number of points on γ and f assumes ∞ at one of these exceptional points. If f is continuous on $\overline{\text{Int}\,\gamma}$ in the generalized sense (see (2.2.2)) and one-to-one along γ, and if *there is a point* $w_0 \notin f(\overline{\text{Int}\,\gamma})$, then f maps Int γ univalently onto one of the components of $\mathbf{C}^* - f(\gamma)$ (note that $f(\gamma)$ passes ∞). See Exercise A (2) for a generalized version.   (3.5.5.4)

Observe that $g(z) = \frac{1}{f(z)-w_0}$ is continuous in the Euclidean sense on $\overline{\text{Int}\,\gamma}$ and is analytic on $\overline{\text{Int}\,\gamma}$ except for a finite number of points on γ and one-to-one along γ. Also, $g(\gamma)$ is a bounded curve in $\mathbf{C}$ and thus, is rectifiable. (3.5.5.3) is applicable to this g and the result follows.   □

Let γ be a curve in $\mathbf{C}^*$ that passes ∞, with the representation $z(t) = x(t) + iy(t) : [a, b] \to \mathbf{C}^*$ (see (2.4.3) and Remark 1 in Sec. 2.4). If $z(a) = z(b)$, γ is said to be *closed*; if $z(t)$ is one-to-one on $[a, b]$, it is called a *Jordan curve*. In case there is a $t_0 \in [a, b]$ so that $\lim_{t \to t_0} z(t) = \infty$ and, on each closed subinterval $[\alpha, \beta]$ of $[a, b] - \{t_0\}$, $z([\alpha, \beta])$ is rectifiable, then γ is said to be *locally rectifiable*.

Another generalization of (3.5.5.1) is the following

Univalence on a domain bounded by a locally rectifiable Jordan closed curve γ passing ∞.

Let Ω_1 and Ω_2 be the two simply connected domains (unbounded) so that $\mathbf{C}^* - \gamma = \Omega_1 \cup \Omega_2$ and $\Omega_1 \cap \Omega_2 = \phi$. Suppose f is analytic on $\overline{\Omega_1}$ and is one-to-one along $\partial\Omega_1 = \gamma$. If the image curve γ^* under $\zeta = \frac{1}{z-z_0}$ of γ, where z_0 is an arbitrarily fixed point of Ω_2, is a *rectifiable* Jordan closed curve, then f maps Ω_1 univalently onto Int $f(\gamma)$. $\qquad$ (3.5.5.5)

Sketch of Proof.

In case $w_0 \in \Omega_2$ such that $w_0 \neq z_0$, let γ^{**} be the image under $\eta = \frac{1}{z-w_0}$ of γ. Since $\eta = \frac{\zeta}{(z_0-w_0)\zeta+1}$ is univalently conformal (see (2) in (2.5.4.13)), it follows that γ^* is rectifiable if and only if γ^{**} is.

Applying (3.5.5.1) to $f^*(\zeta) = f(z_0 + 1/\zeta)$ on $\zeta(\Omega_1) = $ Int γ^*, then f^* maps Int γ^* univalently onto $f^*(\text{Int } \gamma^*) = $ Int $f(\gamma)$ and so f does on Ω_1.

Remark 2 (A generalized version of (3.5.5.5)).

Suppose f is analytic on $\overline{\Omega_1} = \Omega_1 \cup \gamma$ except at a finite number of points on γ and is continuous on $\overline{\Omega_1}$. If f is also one-to-one along γ and γ^* in (3.5.5.5) is rectifiable, then f maps Ω_1 univalently onto Int $f(\gamma)$. (3.5.5.6)

The method in proving (3.5.5.3) is applicable in this case. $\qquad\qquad$ □

We had used ideas in this section, even though implicitly and not so rigorously, in the construction of fundamental domains for elementary multiple-valued functions throughout Sec. 2.7. Later in Sec. 6.2, we will apply results here to the Schwarz–Christoffel transformation, mapping polygonal domain univalently onto a half-plane or a disk. Some preliminary examples for univalent conformal mappings will be presented in Sec. 3.5.7.

At the moment, we should mention that the converse of either (3.5.5.1) or (3.5.5.5) is not necessarily true. For instance, $f(z) = z^2$ maps $\Omega = \{\text{Im } z > 0\}$ univalently onto the domain $\mathbf{C} - [0, \infty)$ yet it maps $\partial\Omega$ onto the nonnegative real axis $[0, \infty)$ in two-to-one manner except at $z = 0$. While the Joukowski function $f(z) = \frac{1}{2}\left(z + \frac{1}{z}\right)$ maps any circle passing ± 1 in two-to-one manner onto a circular arc connecting ± 1 except at $z = \pm 1$ (for details, see Sec. 2.5.5).

To emphasize the necessity of the condition "$w_0 \notin f(\overline{\text{Int } \gamma})$" in (3.5.5.4), we give the following example.

Example 1. Investigate the univalence of

$$f_n(z) = \left(\frac{z-1}{z+1}\right)^n$$

on $|z| = 1$ and $|z| < 1$, respectively, for $n = 1, 2, 3$.

Solution. Observe that $w = f_1(z)$ has a simple pole of order 1 at $z = -1$. f_1 maps $|z| = 1$ univalently onto the imaginary axis $\operatorname{Re} w = 0$ and $|z| < 1$ univalently onto the left plane $\operatorname{Re} w < 0$. Hence $f_1(|z| \leq 1) = \{w \mid \operatorname{Re} w \leq 0\}$. Theoretically, (3.5.5.4) then guarantees that f_1 is indeed univalent on $|z| < 1$.

f_2 has a pole of order 2 at $z = -1$. f_2 maps $|z| = 1$ onto the nonpositive real axis $[\infty, 0]$ in two-to-one manner except at $z = 1$ and $z = -1$, while it maps $|z| < 1$ univalently onto $\mathbf{C}^* - [\infty, 0]$. Hence $f_2(|z| \leq 1) = \mathbf{C}^*$ holds.

f_3 has a pole of order 3 at $z = -1$. In case $z = e^{i\theta}$, $-\pi < \theta \leq \pi$,

$$f_3(e^{i\theta}) = -i \tan^3 \frac{\theta}{2}$$

is the composite of the following one-to-one mappings: $(-\pi, \pi) \xrightarrow{\zeta = \tan \frac{\theta}{2}}$ $\mathbf{R} \xrightarrow{\eta = \zeta^3} \mathbf{R} \xrightarrow{w = -i\eta}$ the imaginary axis. Hence, $w = f_3(z)$ maps $|z| = 1$ onto the imaginary axis univalently.

On the other hand, letting $z = x + iy$ and $w = u + iv$, simple computation shows that $w = f_3(z)$ maps each of the following circles:

$$\gamma_1 : x^2 + y^2 = 1; \quad \gamma_2 : x^2 + y^2 - 2\sqrt{3}y - 1 = 0;$$
$$\gamma_3 : x^2 + y^2 + 2\sqrt{3}y - 1 = 0$$

univalently onto the imaginary axis. It maps, according to (3.5.5.4), each of the domains:

(I) $\operatorname{Int} \gamma_1 \cap \operatorname{Int} \gamma_2 \cap \operatorname{Ext} \gamma_3$; (II) $\operatorname{Int} \gamma_1 \cap \operatorname{Int} \gamma_3 \cap \operatorname{Ext} \gamma_2$; (III) $\operatorname{Ext} \gamma_1 \cap \operatorname{Ext} \gamma_2 \cap \operatorname{Ext} \gamma_3$

univalently onto the right half-plane $\operatorname{Re} w = u > 0$. And, it maps each of the domains:

(IV) $\operatorname{Int} \gamma_1 \cap \operatorname{Int} \gamma_2 \cap \operatorname{Int} \gamma_3$; (V) $\operatorname{Ext} \gamma_1 \cap \operatorname{Int} \gamma_2 \cap \operatorname{Ext} \gamma_3$; (VI) $\operatorname{Ext} \gamma_1 \cap \operatorname{Ext} \gamma_2 \cap \operatorname{Int} \gamma_3$

univalently onto the left half-plane $u < 0$. See Fig. 3.68.

Consequently $w = f_3(z)$ maps $|z| < 1$ onto $\mathbf{C}^* - \{0, \infty\}$ but not univalently even though it is univalent on $|z| = 1$. Observe that $f_3(|z| \leq 1) = \mathbf{C}^*$ holds. The other way to do this is to investigate the mapping behavior of the mappings $z \to \zeta = \frac{z-1}{z+1} \to w = f_3(z) = \zeta^3$ on $|z| < 1$ or to try to construct the Riemann surface for the inverse function of $w = f_3(z)$.   □

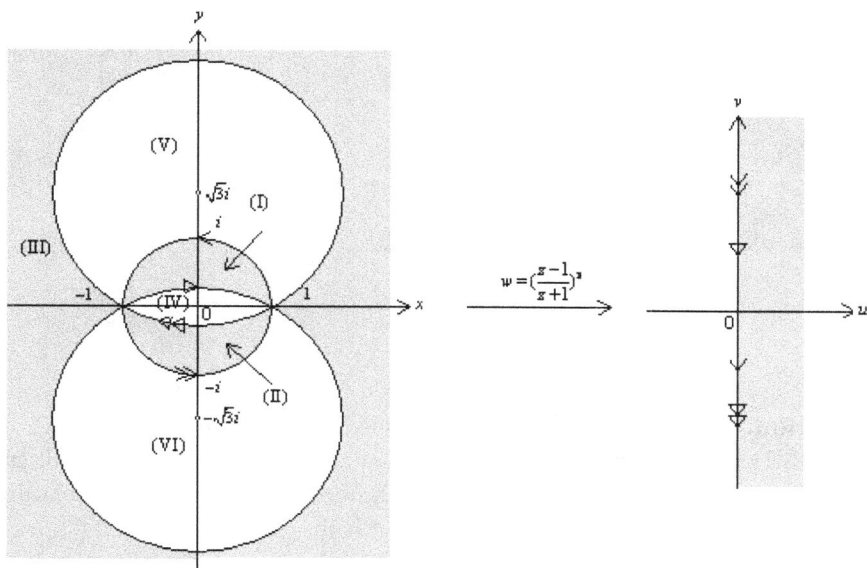

Fig. 3.68

As meaningful applications, we present the following two examples.

Example 2. Let the function f be defined by

$$f(z) = \int_0^z \frac{d\zeta}{\sqrt{1 - \zeta^2}}, \quad z = x + iy \quad \text{with } x \geq 0 \text{ and } y \geq 0,$$

where the path of integration is any rectifiable curve γ connecting 0 to z, lying entirely in the first quadrant (including its boundary) yet missing the point 1. Show that $w = f(z)$ maps

(1) the nonnegative imaginary axis $[0, +i\infty]$ univalently onto $[0, +i\infty]$;
(2) the segment $[0, 1]$ on the real axis univalently onto $\left[0, \frac{\pi}{2}\right]$; and
(3) the segment $[1, \infty]$ on the real axis univalently onto the segment $[0, +i\infty]$ on the vertical line $\operatorname{Re} w = \frac{\pi}{2}$.

Hence (by (3.5.5.4) or Exercise A (2)), $w = f(z)$ maps the first quadrant (boundary excluded) univalently onto the simple connected domain

$$\left\{ w \mid 0 < \operatorname{Re} w < \frac{\pi}{2} \text{ and } \operatorname{Im} w > 0 \right\}, \quad \text{a half-strip. See Fig. 3.69.}$$

See the *Note* after this Example.

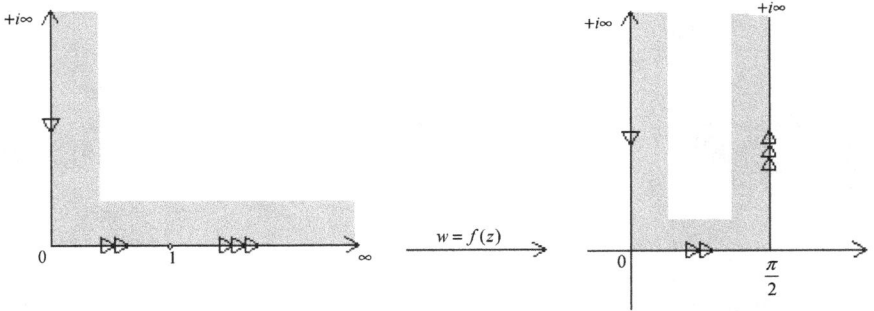

Fig. 3.69

Solution. $\sqrt{1-\zeta^2}$ has two (simple-valued analytic) branches on $\mathbf{C}$ – $[-1,1]$. Here we *fix* the branch, which assumes the value 1 at $\zeta = 0$. In this case, a path of integration in the definition of f can at most contain part of the upper side of the branch cut $[-1,1]$.

By the Cauchy integral theorem, $f(0)$ is well defined and $f(0) = 0$ holds.

Along $[0,+i\infty]$: $z = iy$, $0 \le y < \infty$. Still by the Cauchy integral theorem, the path of integration may be chosen as the segment connecting 0 to iy and thus

$$f(iy) = \int_0^{iy} \frac{d\zeta}{\sqrt{1-\zeta^2}}(\zeta = it, 0 \le t \le y) = \int_0^y \frac{id\zeta}{\sqrt{\zeta^2+1}}(0 \le \zeta \le y)$$

$$= i\log(y + \sqrt{y^2+1}).$$

It follows that $w = f(z)$ maps $[0,+i\infty]$ univalently onto $[0,+i\infty]$.

Along $[0,1]$: $z = x$, $0 \le x \le 1$. In this case, the path of integration may be chosen as the upper side of the cut $[-1,1]$ on which $\sqrt{1-\zeta^2}$ is strictly decreasing. Hence,

$$f(x) = \int_0^x \frac{d\zeta}{\sqrt{1-\zeta^2}} \quad (0 \le x \le 1 \text{ and } 0 \le \zeta \le x)$$

is a strictly increasing function of x until x reaches 1, where

$$f(1) = \int_0^1 \frac{d\zeta}{\sqrt{1-\zeta^2}} = \frac{\pi}{2}.$$

Hence, $w = f(z)$ maps $[0,1]$ univalently onto $\left[0, \frac{\pi}{2}\right]$.

Along $[1,\infty)$: $z = x$, $1 \le x < \infty$. Since the integrand $\frac{1}{\sqrt{1-\zeta^2}}$ ceases to be analytic at $\zeta = 1$, we select a dented path of integration at 1, namely,

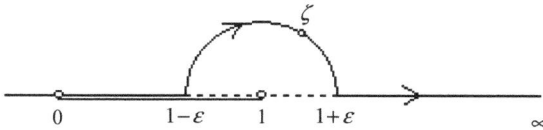

Fig. 3.70

the upper half C_ε of the circle $|\zeta - 1| = \varepsilon$ or $\zeta = 1 + \varepsilon e^{i\theta}$, $0 \le \theta \le \pi$, oriented in the decreasing of θ for sufficiently small $\varepsilon > 0$. See Fig. 3.70. Observe that

$$\Delta_{C_\varepsilon} \arg(1 - \zeta) = -\pi - 0 = -\pi,$$

$$\Rightarrow \Delta_{C_\varepsilon} \arg(1 - \zeta)^{\frac{1}{2}} = -\frac{\pi}{2} - 0 = -\frac{\pi}{2}.$$

Then, as $1 + \varepsilon < x$,

$$f(x) = \int_0^x \frac{d\zeta}{\sqrt{1 - \zeta^2}} = \int_0^{1-\varepsilon} \frac{d\zeta}{\sqrt{1 - \zeta^2}} + \int_{1-\varepsilon}^{1+\varepsilon} \frac{d\zeta}{\sqrt{1 - \zeta^2}} + \int_{1+\varepsilon}^x \frac{d\zeta}{\sqrt{1 - \zeta^2}},$$

$$\rightarrow \int_0^1 \frac{d\zeta}{\sqrt{1 - \zeta^2}} + 0 + \int_1^x \frac{d\zeta}{\sqrt{1 + \zeta}\sqrt{\zeta - 1}e^{\frac{-\pi i}{2}}}$$

$$= \frac{\pi}{2} + i \int_1^x \frac{d\zeta}{\sqrt{\zeta^2 - 1}} \qquad \text{as } \varepsilon \to 0.$$

$$\Rightarrow f(x) = \frac{\pi}{2} + i \int_1^x \frac{d\zeta}{\sqrt{\zeta^2 - 1}}, \qquad 1 \le x < \infty.$$

This shows that $w = f(z)$ maps $[1, \infty)$ univalently onto the segment $w = \frac{\pi}{2} + iv$, $0 \le v < \infty$.

According to the last statement in (3.5.5.4), $w = f(z)$ maps the first quadrant univalently onto the desired half-strip. See also Exercise A (2).

Note: The analytic continuation of $f(z)$ to the entire plane via the symmetry principle (3.4.6.2).

Based on Fig. 3.69, reflect the first quadrant with respect to $(0, +i\infty)$. The symmetry principle says that $f(z)$ can be analytically continued from the first quadrant into the secord quadrant such that the resulted $f(z)$ maps the half-plane $\operatorname{Im} z > 0$ univalently onto the vertical half-strip: $-\frac{\pi}{2} < \operatorname{Re} w < \frac{\pi}{2}$ and $\operatorname{Im} w > 0$, and satisfies the relation

$$f(-\bar{z}) = -\overline{f(z)}, \qquad \operatorname{Im} z > 0.$$

In this case, $w = f(z)$ maps $[-1, 1]$ univalently on the segment $[-\frac{\pi}{2}, \frac{\pi}{2}]$ in the w-plane.

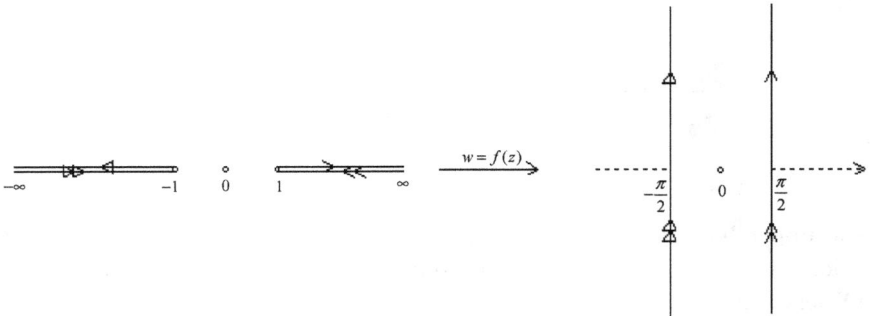

Fig. 3.71

Reflect the upper half-plane Im $z > 0$ into the lower half-plane Im $z < 0$ across the open interval $(-1, 1)$. Then $f(z)$ is analytically continued into the lower half-plane Im $z < 0$ across $(-1, 1)$ and the resulted $f(z)$ maps Im $z < 0$ univalently onto the lower half-strip: $-\frac{\pi}{2} < \text{Re}\, w < \frac{\pi}{2}$ and Im $w < 0$, and satisfies the relation

$$f(\bar{z}) = \overline{f(z)}, \quad z \in \mathbf{C} - (\infty, -1] - [1, \infty).$$

Combining together, the resulted $w = f(z)$ maps $\mathbf{C} - (\infty, -1] - [1, \infty)$ univalently and hence conformally (see (3.5.1.9)) onto the strip $-\frac{\pi}{2} < \text{Re}\, w < \frac{\pi}{2}$. See Fig. 3.71.

Based on Fig. 3.71, reflect the domain $\mathbf{C} - (\infty, -1] - [1, \infty)$ across the upper and the lower sides of the cut $(1, \infty)$, respectively, and across the upper and the lower sides of the cut $(\infty, -1)$, respectively. Correspondingly, this means that the range of $w = f(z)$ goes from $-\frac{\pi}{2} < \text{Re}\, w < \frac{\pi}{2}$ into $\frac{\pi}{2} < \text{Re}\, w < \frac{3\pi}{2}$ across the vertical line $\text{Re}\, w = \frac{\pi}{2}$, and into $-\frac{3\pi}{2} < \text{Re}\, w < -\frac{\pi}{2}$ across the line $\text{Re}\, w = -\frac{\pi}{2}$. The resulted $w = f(z)$ maps $\mathbf{C} - \{-1, 1\}$ onto $-\frac{3\pi}{2} < \text{Re}\, w < \frac{3\pi}{2}$ with the two points $\pm\frac{\pi}{2}$ deleted, while $f(1) = \frac{\pi}{2}$ and $f(-1) = -\frac{\pi}{2}$. See Fig. 3.72. There is a branch point of order 1 over each of the points $z = \pm 1$. Note that f is single-valued on $\mathbf{C} - (\infty, -1] - [1, \infty)$ with range $-\frac{\pi}{2} < \text{Re}\, z < \frac{\pi}{2}$ yet *multi-valued* on $\mathbf{C} - \{-1, 1\}$ with range $-\frac{3\pi}{2} < \text{Re}\, z < \frac{3\pi}{2}$, but it becomes *single-valued* on the three-sheets surface over $\mathbf{C} - \{-1, 1\}$.

This process can be repeated infinitely until the range of $w = f(z)$ covers the entire w-plane with its domain of definition on $\mathbf{C} - \{-1, 1\}$. Thus, $w = f(z)$: $\mathbf{C} - \{-1, 1\} \to \mathbf{C}$ is infinitely valued but is single-valued on the infinite-sheets surface R_f (refer to Figs. 2.85 and 2.86). There are

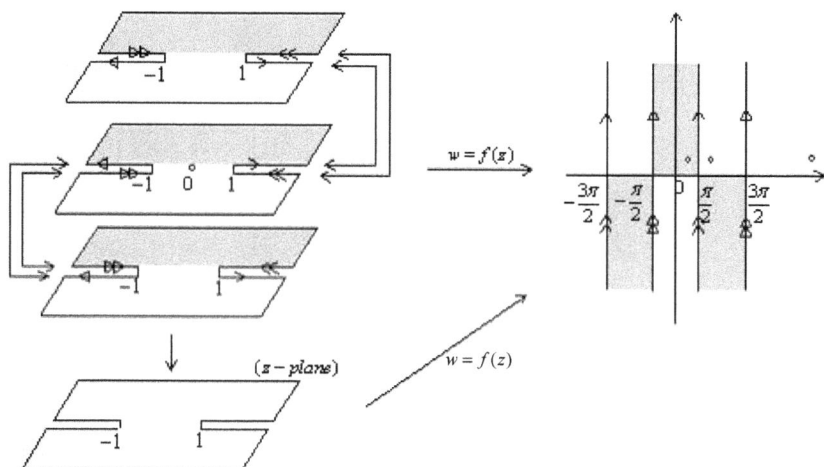

Fig. 3.72

infinitely many algebraic branch points of order 1 over $z = \pm 1$ and two logarithmic branch points over ∞.

This final $w = f(z)$ seems to be familiar to us. It is the inverse function $w = \sin^{-1} z$ of the entire function $z = \sin w$. To see this, recall that $w = f(z)$: $\mathbf{C} - (\infty, -1] - [1, \infty) \rightarrow \{ w \mid -\frac{\pi}{2} < \operatorname{Re} w < \frac{\pi}{2} \}$ is a single-valued analytic function. When restricted to the open interval $(-1, 1)$ within, it is, by calculus,

$$f(x) = \int_0^x \frac{d\zeta}{\sqrt{1 - \zeta^2}} = \operatorname{Arc\,sin}^{-1} x \quad \text{(principle branch of } \sin^{-1} x \text{)}.$$

By the interior uniqueness theorem (see (4) in (3.4.2.9)), $f(z) = \operatorname{Arc\,sin}^{-1} z$ on $\mathbf{C} - (\infty, -1] - [1, \infty)$ and the $f(z) = \sin^{-1} z$ on $\mathbf{C} - \{-1, 1\}$.

For fixed $z \in \mathbf{C} - \{-1, 1\}$, suppose the corresponding $w = f(z)$ lies in $-\frac{\pi}{2} < \operatorname{Re} w < \frac{\pi}{2}$. The point z is changed to $\bar{z}$ by reflecting it with respect to the upper side of the cut $[1, \infty)$ and then back to itself $\bar{\bar{z}} = z$ by reflecting $\bar{z}$ with respect to the lower side of the cut $(\infty, -1]$. The corresponding $w = f(z)$ is changed to its symmetric point $\pi - w$ with respect to the line $\operatorname{Re} w = \frac{\pi}{2}$ and then to the symmetric point $w + 2\pi$ of $\pi - w$ with respect to the line $\operatorname{Re} w = \frac{3\pi}{2}$. Hence, the inverse function $z = \sin w$ of $w = \sin^{-1} z$ satisfies

$$\sin w = \sin(w + 2\pi), \quad w \in \mathbf{C}.$$

Consequently, $z = \sin w$ is a *simple periodic function* with period 2π.

Example 3. Let

$$f(z) = \int_0^z \frac{d\zeta}{\sqrt{(1 - \zeta^2)(1 - k^2\zeta^2)}} \quad (0 < k < 1),$$

called the *elliptic integral of the first kind*,

where the path of integration is any rectifiable curve connecting 0 and z. Show that $w = f(z)$ maps

(1) the point $-\infty, -\frac{1}{k}, -1, 0, 1, \frac{1}{k}, +\infty$ on the real axis onto the points

$$iK', -K + iK', -K, 0, K, K + iK', iK',$$

respectively, where

$$K = \int_0^1 \frac{d\zeta}{\sqrt{(1 - \zeta^2)(1 - k^2\zeta^2)}}, \quad \text{and}$$

$$K' = \int_0^{\frac{1}{k}} \frac{d\zeta}{\sqrt{(\zeta^2 - 1)(1 - k^2\zeta^2)}} = \int_0^1 \frac{dt}{\sqrt{(1 - t^2)[1 - (1 - k^2)t^2]}}$$

$$\left(\zeta = \frac{1}{\sqrt{1 - (1 - k^2)t^2}} \right);$$

(2) the intervals $[-\infty, -\frac{1}{k}]$, $[-\frac{1}{k}, -1]$, $[-1, 0]$, $[0, 1]$, $[1, \frac{1}{k}]$, $[\frac{1}{k}, +\infty]$ univalently onto the sides of the rectangle R, respectively, as shown in Fig. 3.73.

Hence, $w = f(z)$ maps the real axis univalently onto the rectangle R with vertices at $\pm K$, $\pm K + iK'$ and the upper half-plane $\text{Im } z > 0$ univalently onto the interior $\text{Int } R$.

See *Notes* after this example.

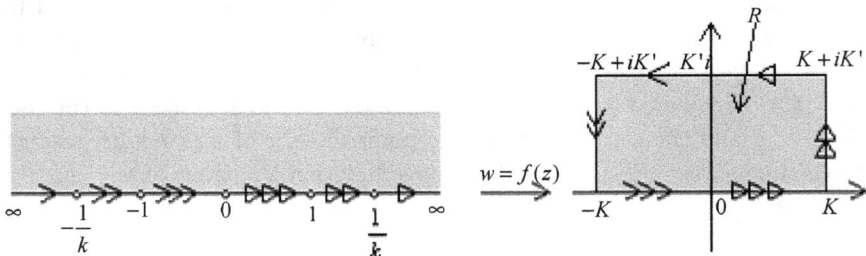

Fig. 3.73

Before we start to handle this problem, note that $\sqrt{(1-\zeta^2)(1-k^2\zeta^2)}$ is double-valued except at $\zeta = \pm 1, \pm\frac{1}{k}$ and it has two branches on $\mathbf{C}^* - \left[-\frac{1}{k}, -1\right] - \left[1, \frac{1}{k}\right]$. Its Riemann surface has an algebraic branch point of order 1 over each of the points $\pm 1, \pm\frac{1}{k}$ and has a node point over ∞ (refer to Exercise B (3)(a) of Sec. 2.7.2). As a consequence, the function $f(z)$ defined by the integral is double-valued, too.

We *claim* that both branches of f are analytic at ∞. To see this, on $|z| > \frac{1}{k}$, the integrand has the Taylor expansion (see Example 2 in Sec. 3.3.2)

$$\frac{1}{\sqrt{(1-\zeta^2)(1-k^2\zeta^2)}} = \frac{1}{k\zeta^2}(1-\zeta^{-2})^{-\frac{1}{2}}(1-k^{-2}\zeta^{-2})^{-\frac{1}{2}}$$

$$= \frac{\pm 1}{k\zeta^2}\left(1 + \frac{1}{2\zeta^2} + \cdots\right)\left(1 + \frac{1}{2k^2\zeta^2} + \cdots\right)$$

$$= \pm\left[\frac{1}{k\zeta^2} + \frac{1}{2}\left(\frac{1}{k} + \frac{1}{k^3}\right)\frac{1}{\zeta^4} + \cdots\right],$$

$$\Rightarrow w = f(z) = c \mp \left\{\frac{1}{k\zeta} + \frac{1}{6}\left(\frac{1}{k} + \frac{1}{k^3}\right)\frac{1}{\zeta^3} + \cdots\right\}, \quad |z| > \frac{1}{k},$$

where c is a constant. And the claim follows.

Hence, each branch of $w = f(z)$ is analytic on $\{\mathrm{Im}\, z \geq 0\}\setminus\{\pm\frac{1}{k}, \pm 1\}$, including the infinite point ∞, and continuous on $\mathrm{Im}\, z \geq 0$. In what follows, we choose the branch of $f(z)$ determined by the branch $\sqrt{(1-\zeta^2)(1-k^2\zeta^2)}$ subject to the condition $\sqrt{1} = 1$. For (3.5.5.6), in this case, choose γ to be the real axis, Ω_1 the upper half-plane $\mathrm{Im}\, z > 0$, and Ω_2 the lower half-plane $\mathrm{Im}\, z < 0$. For any point $z_0 \in \Omega_2$, the mapping $z \to \frac{1}{z-z_0}$ maps γ onto a circle, which is certainly a rectifiable Jordan closed curve.

Solution. What we do care is if $w = f(z)$ is one-to-one along the real axis γ.

Along the segment $[0, 1] : z = x, 0 \leq x \leq 1$. The function

$$f(x) = \int_0^x \frac{d\zeta}{\sqrt{(1-\zeta^2)(1-k^2\zeta^2)}}$$

(the path of integration is the segment $[0, x]$) is positive and increases from 0 to K (K is defined since the improper integral, if $x = 1$, is convergent) as x increases from 0 to 1.

Along the segment $\left[1, \frac{1}{k}\right] : z = x, 1 \leq x \leq \frac{1}{k}$. Attention should be paid to the change of sign of $1 - \zeta$ as ζ goes across the point 1.

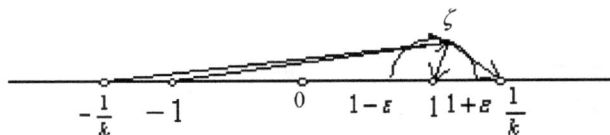

Fig. 3.74

Choose $0 < \varepsilon < \min\{1, \frac{1}{k} - 1\}$. Construct the half-circle $C_\varepsilon : z = 1 + \varepsilon e^{i\theta}$, $0 \le \theta \le \pi$, oriented in the clockwise direction. See Fig. 3.74. Then, along C_ε,

$$\Delta_{C_\varepsilon} \arg\left(\zeta + \frac{1}{k}\right) = 0 - 0 = 0;$$

$$\Delta_{C_\varepsilon} \arg(\zeta + 1) = 0 - 0 = 0;$$

$$\Delta_{C_\varepsilon} \arg(1 - \zeta) = -\pi - 0 = -\pi; \quad \text{and}$$

$$\Delta_{C_\varepsilon} \arg\left(\frac{1}{k} - \zeta\right) = 0 - 0 = 0.$$

$$\Rightarrow \Delta_{C_\varepsilon} \arg \sqrt{(1 - \zeta^2)(1 - k^2\zeta^2)} = -\frac{\pi}{2}.$$

Therefore, if $1 + \varepsilon \le x \le \frac{1}{k}$,

$$f(x) = \int_0^x \frac{d\zeta}{\sqrt{(1 - \zeta^2)(1 - k^2\zeta^2)}}$$

$$= \int_0^{1-\varepsilon} \cdots + \int_{1-\varepsilon}^{1+\varepsilon} \cdots + \int_{1+\varepsilon}^x \frac{d\zeta}{\sqrt{(\zeta^2 - 1)(1 - k^2\zeta^2)}e^{\frac{-\pi i}{2}}}$$

$$\to K + i \int_1^x \frac{d\zeta}{\sqrt{(\zeta^2 - 1)(1 - k^2\zeta^2)}} \quad \text{as } \varepsilon \to 0.$$

This indicates that the function

$$f(x) = K + i \int_1^x \frac{d\zeta}{\sqrt{(\zeta^2 - 1)(1 - k^2\zeta^2)}}$$

(the path of integration is the segment $[1, x]$) increases from K, along the line $\operatorname{Re} w = K$, to the point $K + iK'$.

Along the segment $[\frac{1}{k}, \infty] : z = x$, $\frac{1}{k} \le x \le \infty$. Choose $0 < \varepsilon < \frac{1}{k} - 1$ and construct the half-circle $C_\varepsilon : z = \frac{1}{k} + \varepsilon e^{i\theta}$, $0 \le \theta \le \pi$, oriented in the

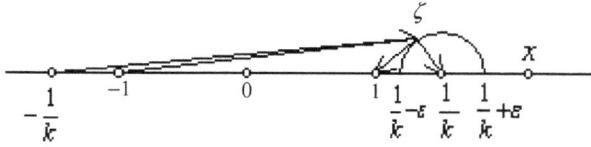

Fig. 3.75

clockwise direction. See Fig. 3.75. Then, along C_ε,

$$\Delta_{C_\varepsilon} \arg\left(\frac{1}{k} + \zeta\right) = 0 - 0 = 0; \quad \Delta_{C_\varepsilon} \arg(1 + \zeta) = 0 - 0 = 0;$$

$$\Delta_{C_\varepsilon} \arg(1 - \zeta) = (-\pi) - (-\pi) = 0 \quad \text{and}$$

$$\Delta_{C_\varepsilon} \arg\left(\frac{1}{k} - \zeta\right) = (-\pi) - 0 = -\pi.$$

Therefore, as ζ moves from $\frac{1}{k} - \varepsilon$, along C_ε, to $\frac{1}{k} + \varepsilon$, $\sqrt{1 - \zeta}$ assumes the initial value $\sqrt{\left|1 - \left(\frac{1}{k} - \varepsilon\right)\right|} e^{\frac{-\pi i}{2}}$ and changes continuously to the terminal value $\sqrt{\left|1 - \left(\frac{1}{k} + \varepsilon\right)\right|} e^{\frac{-\pi i}{2}}$; while $\sqrt{\frac{1}{k} - \zeta}$ assumes the value $\sqrt{\varepsilon} e^{\frac{0i}{2}} = \sqrt{\varepsilon}$ and changes continuously to the value $\sqrt{\varepsilon} e^{0i/2 + i\Delta_{C_\varepsilon} \arg\left(\frac{1}{k} - \zeta\right)/2} = \sqrt{\varepsilon} e^{\frac{-\pi i}{2}}$ (refer to Examples 2 and 3 in Sec. 2.7.2 for more practice). Now, if $\frac{1}{k} < x < \infty$,

$$f(x) = \int_0^x \frac{d\zeta}{\sqrt{(1 - \zeta^2)(1 - k^2\zeta^2)}} \to K + iK'$$

$$+ \int_{\frac{1}{k}}^x \frac{d\zeta}{\sqrt{\zeta^2 - 1} e^{\frac{-\pi i}{2}} \cdot \sqrt{k^2\zeta^2 - 1} e^{\frac{-\pi i}{2}}}$$

$$= K + iK' - \int_{\frac{1}{k}}^x \frac{d\zeta}{\sqrt{(\zeta^2 - 1)(k^2\zeta^2 - 1)}}$$

$$\left(\text{the path of integration is } \left[\frac{1}{k}, x\right]\right) \quad \text{as } \varepsilon \to 0,$$

where, if $x = \infty$,

$$\int_{\frac{1}{k}}^\infty \frac{d\zeta}{\sqrt{(\zeta^2 - 1)(k^2\zeta^2 - 1)}} = \int_0^1 \frac{dt}{\sqrt{(1 - t^2)(1 - k^2t^2)}} \left(\zeta = \frac{1}{kt}\right) = K.$$

Hence, as x moves from $\frac{1}{k}$ to ∞, $f(x)$ moves from the point $K + iK'$, along the line $\operatorname{Im} w = K'$, to the point $K + iK' - K = iK'$. See Fig. 3.73.

In case $x < 0$,

$$f(x) = \int_0^x \frac{d\zeta}{\sqrt{(1-\zeta^2)(1-k^2\zeta^2)}}$$

$$= -\int_0^{-x} \frac{d\zeta}{\sqrt{(1-\zeta^2)(1-k^2\zeta^2)}} \quad \text{(replace } \zeta \text{ by } -\zeta\text{)}.$$

This shows that $f(x)$ assumes consecutively the values from 0 to $-K$, then to $-K + iK'$ (along $\mathrm{Re}\, w = -K$) and then to iK' (along $\mathrm{Im}\, w = K'$) as x decreases from 0 to -1, then to $-\frac{1}{k}$, and then to $-\infty$.

Combining together, $w = f(z)$ maps the real axis univalently onto the side segments of R as indicated in Fig. 3.73 and thus, by (3.5.5.6), maps $\mathrm{Im}\, z > 0$ univalently onto the interior $\mathrm{Int}\, R$.

Note 1: The univalent conformal mapping from the upper half-plane onto the interior of a presumed rectangle.

As k increases from 0 to 1, K increases from $\int_0^1 \frac{d\zeta}{\sqrt{(1-\zeta^2)}} = \frac{\pi}{2}$ to $\int_0^1 \frac{d\zeta}{1-\zeta^2}$
$= \log \frac{1+\zeta}{1-\zeta}\Big|_0^1 = \infty$; while K' decreases from $\int_0^1 \frac{d\zeta}{1-\zeta^2} = \infty$ to $\int_0^1 \frac{d\zeta}{\sqrt{1-\zeta^2}} = \frac{\pi}{2}$. The ratio of the altitude of the rectangle R shown in Fig. 3.73 to its base

$$\lambda(k) = \frac{K'}{2K}$$

is thus a strictly decreasing function of k on $(0, 1)$ with range from ∞ downward to 0.

Given a rectangle with base $2b$ and altitude a, there is a unique $k \in (0, 1)$ so that the resulted K and K' satisfy

$$\frac{a}{2b} = \frac{K'}{2K}. \qquad (*)$$

The mapping, corresponding to this k,

$$w = f(z) = \int_0^z \frac{d\zeta}{\sqrt{(1-\zeta^2)(1-k^2\zeta^2)}},$$

then maps the upper half-plane $\mathrm{Im}\, z > 0$ univalently and conformally onto the interior of a rectangle with vertices at $\pm K$ and $\pm K + iK'$, and is similar to the given one. By $(*)$, choose $\mu = \frac{a}{K'} = \frac{b}{K}$, then $w = \mu f(z)$ maps $\mathrm{Im}\, z > 0$ univalently onto the interior of the rectangle with vertices at $\pm b$

and $\pm b + ia$. Finally, the mapping

$$w = \alpha + \beta f(z)$$

maps $\operatorname{Im} z > 0$ univalently onto the interior of the presumed rectangle, provided the constants α and β are suitably chosen.

Two rectangles R and R', with base $2b$ and altitude a, and $2b'$ and a', respectively, are called *similar* if $\frac{a}{2b} = \frac{a'}{2b'}$ holds. In this case, a similarity transformation $w = \alpha z + \beta$ will map one onto the other. Conversely, every univalent conformal mapping between two similar rectangles is given by a similarity transformation. This follows from the symmetry principle and the interior uniqueness theorem (refer to Exercise A (29) of Sec. 3.4.2). Consequently the ratio $\lambda(k) = \frac{a}{2b}$ is a ***conformal invariant*** and is called the *conformal modulus* of a rectangle. This concept is the core of the geometric definition for the planar *quasiconformal mappings*. See Ref. [54], p. 16.

Note 2: The simplest nonconstant elliptic function. Based on Fig. 3.73, reflect $\operatorname{Im} z > 0$ with respect to the open interval $(-1, 1)$. Then $w = f(z)$ is analytically continued into the rectangle R_1 with vertices at $\pm K$ and $\pm K - iK'$ across the lower side $(-K, K)$ of R and satisfies

$$f(\bar{z}) = \overline{f(z)}, \quad z \in \mathbf{C}^* - [\infty, -1] - [1, \infty].$$

See Fig. 3.76. The extended $w = f(z)$ maps $\mathbf{C}^* - [\infty, -1] - [1, \infty]$ univalently and thus conformally onto the domain $\operatorname{Int} R \cup (-K, K) \cup \operatorname{Int} R_1$.

Now, based on Fig. 3.76, reflect $\operatorname{Im} z > 0$ and $\operatorname{Im} z > 0$, respectively, with respect to the upper side and the lower side of the cut $\left(1, \frac{1}{k}\right)$. This amounts to the fact that the extended $w = f(z)$ is, respectively, analytically continued into the rectangle R_2 with vertices at K, $3K$, $3K + iK'$, and $K + iK'$ across the interval $(0, iK')$ on the line $\operatorname{Re} w = K$; and into the rectangle R_3 with vertices at K, $3K$, $3K - iK'$, and $K - iK'$ across the

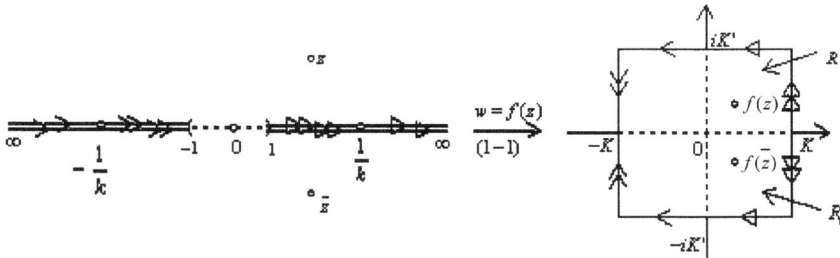

Fig. 3.76

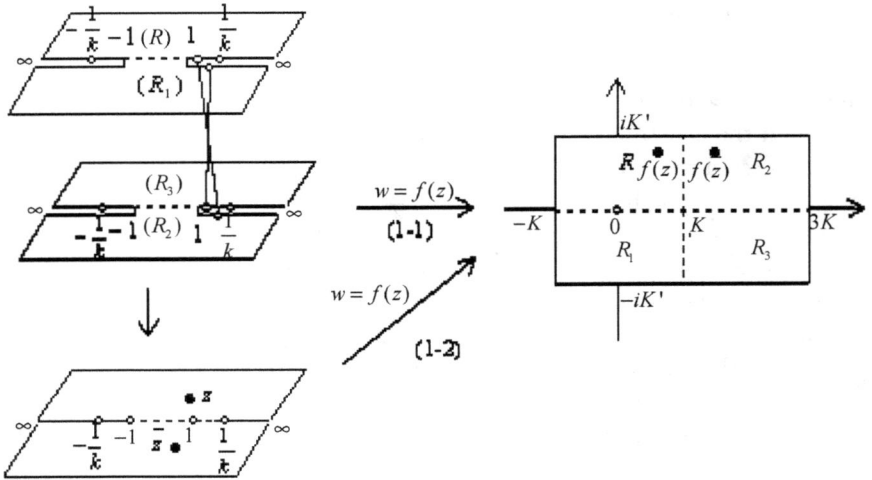

Fig. 3.77

interval $(-iK', 0)$ on the line $\operatorname{Re} w = K$. See Fig. 3.77. This newly extended function $w = f(z)$ maps $\mathbf{C}^* - [\infty, -1] - \left[\frac{1}{k}, \infty\right]$ onto the interior of a larger rectangle with vertices at $-K \pm iK'$ and $3K \pm iK'$ in one-to-two manner except at the point $z = 1$ where $f(z)$ assumes the value K twice; but it is one-to-one on the two-sheets surface over $\mathbf{C}^* - [\infty, -1] - \left[\frac{1}{k}, \infty\right]$. Moreover,

$$f(\bar{z}) = 2K - f(z), \quad \mathbf{C}^* - [\infty, -1] - \left[\frac{1}{k}, \infty\right].$$

If, in Fig. 3.76, we reflect $\operatorname{Im} z > 0$ and $\operatorname{Im} z < 0$ with respect to the upper side and the lower side of the cut $\left(\frac{1}{k}, \infty\right] \cup [\infty, -\frac{1}{k})$ (connected at ∞), then $w = f(z)$ will be analytically continued, respectively, into R_4 across the segment $(-K, K)$ on the line $\operatorname{Im} w = K'$; and into R_5 across the segment $(-K, K)$ on the line $\operatorname{Re} w = -K'$. See Fig. 3.78. This extended $w = f(z)$ maps $\mathbf{C}^* - \left[-\frac{1}{k}, -1\right] - \left[1, \frac{1}{k}\right]$ onto the interior of the rectangle with vertices at $\pm K \pm 2iK'$ in one-to-two manner; in particular, $f(\infty) = \pm iK'$. Moreover,

$$f(\bar{z}) = 2iK' - f(z), \ z \in R, \quad \text{and} \quad f(\bar{z}) = -2iK' - f(z), \ z \in R_1.$$

Still based on Fig. 3.76, reflect $\operatorname{Im} z > 0$ and $\operatorname{Im} z > 0$, respectively, with respect to the upper side and the lower side of the cut $\left(-\frac{1}{k}, -1\right)$. Correspondingly, $w = f(z)$ is analytically continued, respectively, into the rectangle R_6 across the segment $(0, iK')$ on the segment $\operatorname{Re} w = -K$; and into the rectangle R_7 across the segment $(-iK', 0)$, as shown in Fig. 3.79.

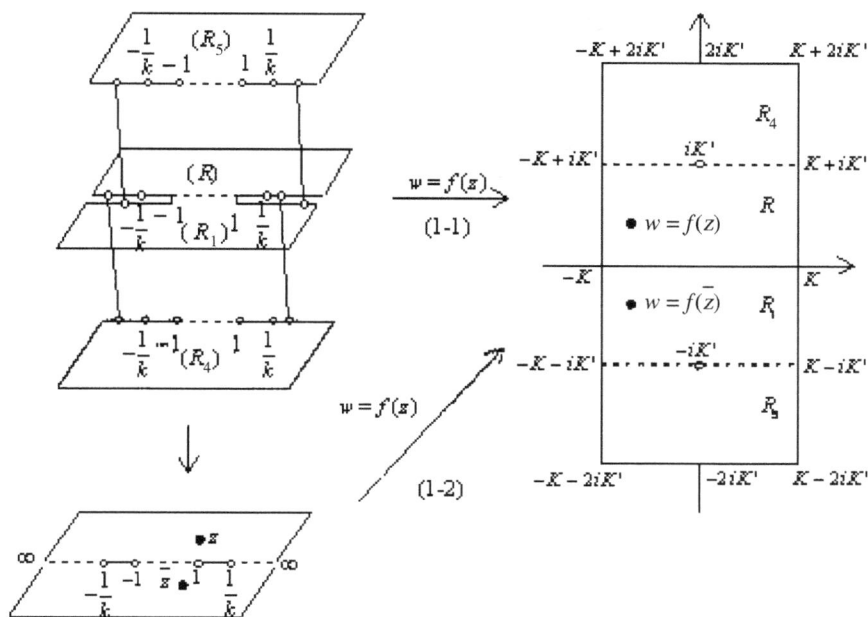

Fig. 3.78

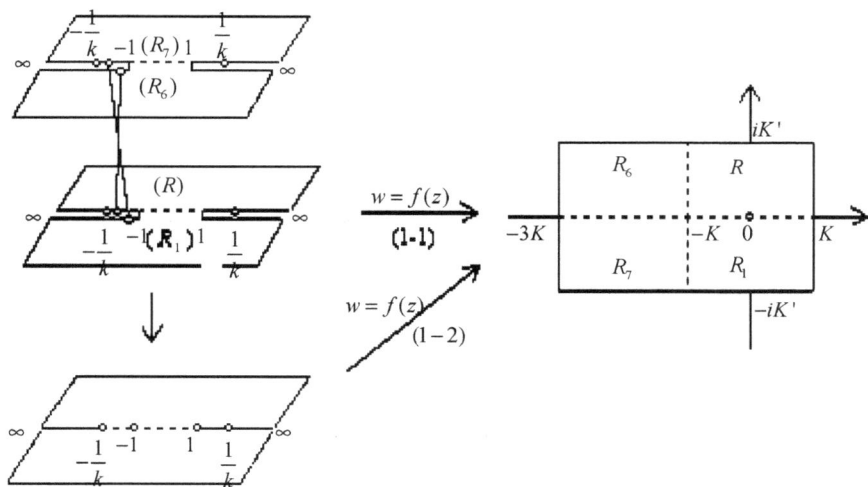

Fig. 3.79

Then, $w = f(z)$ maps $\mathbf{C}^* - [1, \infty] - \left[\infty, -\frac{1}{k}\right]$ onto the interior of the rectangle with vertices at $K \pm iK'$ and $-3K \pm iK'$ in one-to-two manner except at the point $z = -1$ where $f(z)$ assumes the value $-K$ twice. Observe that, in this case,

$$f(\bar{z}) = -2K - f(z), \quad z \in \mathbf{C}^* - [1, \infty] - \left[\infty, -\frac{1}{k}\right].$$

Combining Figs. 3.74 to 3.79, we obtain the following Fig. 3.80. The extended $w = f(z)$ assumes $K + K'i$, the common vertex of R, R_2, R_9, and

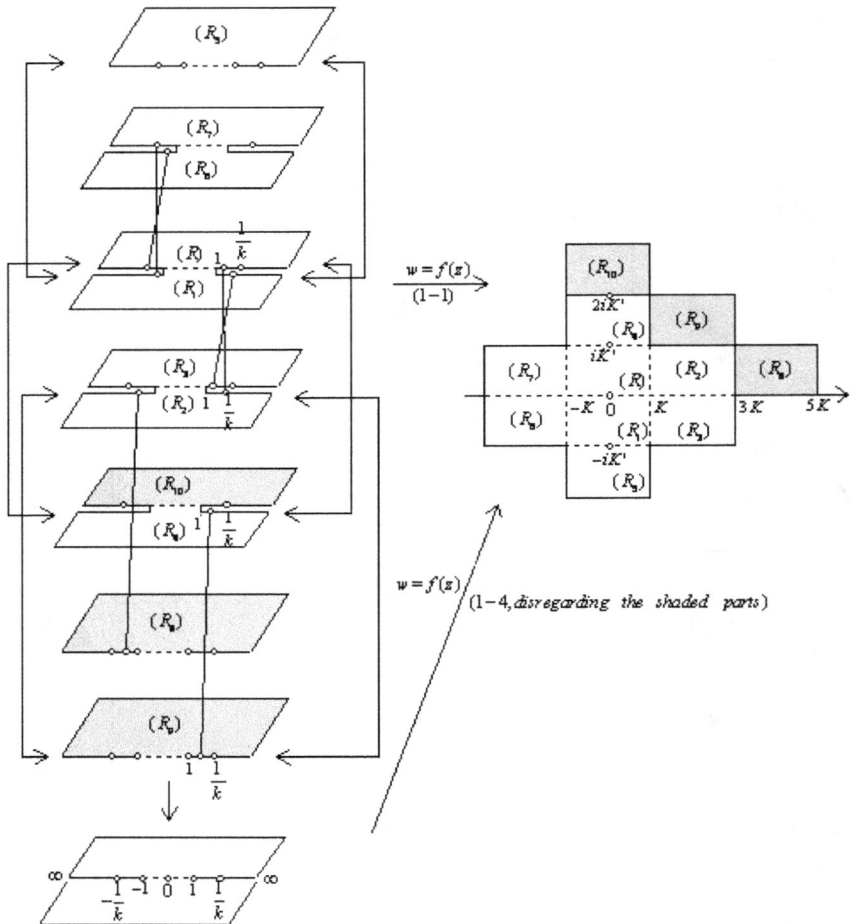

Fig. 3.80

R_4, twice at the point $z = \frac{1}{k}$, and K at $z = 1$ twice, and $-K$ at $z = -1$ twice, too. Repeat the above process. Increase the number of the upper half-planes $\operatorname{Im} z > 0$ and the lower half-plane $\operatorname{Im} z > 0$, and paste them together properly along the branch cuts $[-1, 1]$, $\left[1, \frac{1}{k}\right]$, $\left[\frac{1}{k}, \infty\right] \cup \left[\infty, -\frac{1}{k}\right]$, and $\left[-\frac{1}{k}, -1\right]$. Correspondingly, increase the number of the associated rectangles, and paste them properly along their common sides so that they tessellate the whole w-plane. The resulted surface $R_{f(z)}$, composed of infinitely many sheets (or layers) of the planes pasting properly together along cuts, is the Riemann surface of the multiple-valued function $w = f(z)$. There are infinitely many algebraic branch points, each of order 2, in $R_{f(z)}$ over each of the points ± 1, $\pm \frac{1}{k}$. Observe that $w = f(z)$ is single-valued and univalent on $R_{f(z)}$, yet it is infinitely valued in the classical z-plane with range the whole w-plane. The line complex of $R_{f(z)}$ is shown in Fig. 3.81.

The understanding of this extended $w = f(z)$ is better tried from its inverse function $z = f^{-1}(w)$:

domain: analytic everywhere in $\mathbf{C}^*$ except isolated poles;
zeros: $2mK + 2niK'$, $m, n = 0, \pm 1, \pm 2, \ldots$;
poles: $2mK + (2n + 1)iK'$, $m, n = 0, \pm 1, \pm 2, \ldots$; and
double periods $4K$ and $2iK'$: $z(w + 4K) = z(w)$, $z(w + 2iK') = z(w)$, $w \in \mathbf{C}^*$.

Hence, this $z = f^{-1}(w)$ is a *double periodic meromorphic function*. It is the simplest *nonconstant elliptic function* with the rectangle with vertices at $-K \pm iK'$ and $3K \pm iK'$ as its *fundamental domain of period* (see the right rectangle in Fig. 3.77).

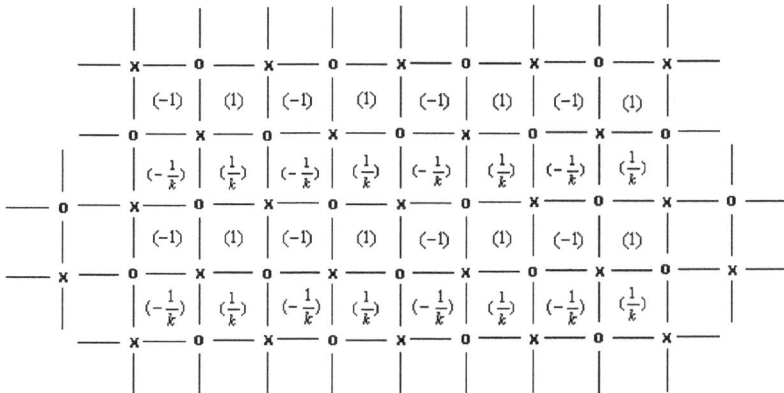

Fig. 3.81

Exercises A

(1) Try to prove (3.5.5.3) in detail.

(2) Let γ, Ω_1, Ω_2, and γ^* be as in (3.5.5.5). Suppose f is analytic on $\overline{\Omega_1}$ except a finite number of points on γ and f assumes ∞ at one of these exceptional points. If f is continuous on $\overline{\Omega_1}$ in the generalized sense and one-to-one along γ and if there is a point $w_0 \notin f(\overline{\Omega_1})$, then f maps Ω_1 univalently onto one of the components of $\mathbf{C}^* - f(\gamma)$.

(3) (Compare with (3.5.5.3)) Let γ be a rectifiable Jordan closed curve. Suppose f is analytic on $\overline{\mathrm{Int}\,\gamma}$ except at a point $z_0 \in \gamma$ where $f'(z) = o(|z - z_0|^\alpha)$, $\alpha > -1$, as $z \in \overline{\mathrm{Int}\,\gamma} - \{z_0\} \to z_0$ and is continuous on $\overline{\mathrm{Int}\,\gamma}$. If f is one-to-one along γ, show that f maps $\mathrm{Int}\,\gamma$ univalently onto $\mathrm{Int}\,f(\gamma)$.

(4) (Compare with (3.5.5.4)) Let γ be a rectifiable Jordan closed curve. Suppose f is analytic on $\overline{\mathrm{Int}\,\gamma}$ except at a point $z_0 \in \gamma$ where f has a simple pole. Suppose that, via a certain change of variable, one may choose $z_0 = 0$ and $\mathrm{Re}\,z \geq 0$ holds for all $z \in \mathrm{Int}\,\gamma$. If f is continuous on $\overline{\mathrm{Int}\,\gamma}$ in the generalized sense and is one-to-one along γ, then f maps $\mathrm{Int}\,\gamma$ univalently onto a component of $\mathbf{C}^* - f(\gamma)$.

(5) Show that each branch f of $\frac{z}{\sqrt{1-z}}$ is univalent on $|z| < 1$. Determine $f(|z| < 1)$ for each f.

(6) Show that $f(z) = \frac{z}{(1-z)^3}$ is univalent in $|z| < \frac{1}{2}$ but not in any larger concentric disk. Determine $f\left(|z| < \frac{1}{2}\right)$. What is the Riemann surface for $z = f^{-1}(w)$?

(7) Let $\rho > 1$. Define

$$f(z) = \int_0^z \frac{d\zeta}{\sqrt{\zeta(\zeta - 1)(\zeta - \rho)}},$$

where the path of integration is a rectifiable curve connecting 0 to z, and the values of $\sqrt{\zeta}$, $\sqrt{\zeta - 1}$, and $\sqrt{\zeta - \rho}$ are designated to lie in the first quadrant. Show that $w = f(z)$ maps

(i) The points $0, 1, \rho, \infty$ (in this ordering) on the real axis onto the points $0, -K, -K - iK', -iK'$, respectively, where

$$K = \int_0^1 \frac{d\zeta}{\sqrt{\zeta(\zeta - 1)(\zeta - \rho)}} \quad \text{and}$$

$$K' = \int_1^\rho \frac{d\zeta}{\sqrt{\zeta(\zeta - 1)(\zeta - \rho)}}, \quad \text{and}$$

(ii) The intervals $[0,1], [1,\rho], [\rho,\infty], [\infty,0]$ univalently onto the sides of the rectangle R with vertices $0, -K, -K - iK', -iK'$ (in this ordering).

Hence, $w = f(z)$ maps the upper plane $\text{Im }z > 0$ univalently onto the interior $\text{Int }R$. Imitate Note 2 in Example 3 to extend this f analytically into an infinitely valued function with range the whole w-plane $\mathbf{C}^*$. Construct the Riemann surface for $z = f^{-1}(w)$.

(8) Show that $f(z) = z + \frac{1}{n}z^n$ is univalent in $|z| < 1$.

(9) Let $p(z) = a_n z^n + a_{n-1}z^{n-1} + \cdots + a_1 z + a_0$, where $n \geq 2$ and $a_n \neq 0$. In case $\sum_{k=2}^{n} k|a_k| \leq |a_1|$, show that $p(z)$ is univalent in $|z| < 1$.

(10) Suppose $f(z) = \sum_{n=0}^{\infty} a_n z^n$ has a positive radius R of convergence. If, for some $r(0 < r \leq R), |a_1| \geq \sum_{n=2}^{\infty} n|a_n|r^{n-1}$ holds, show that f is univalent in $|z| < 1$.

(11) Suppose $f(z) = z + \sum_{n=2}^{\infty} a_n z^n$ converges in $|z| < R$ and $|f(z)| \leq M$ holds there. Show that $f(z) \neq 0$ on $0 < |z| < \frac{R^2}{M+R}$.

(12) Suppose f is analytic at 0. Define

$$\rho = \sup\{r > 0 \,|\, f \text{ is univalent on } |z| < r\}.$$

Find the corresponding ρ for each of the following functions $f(z)$:

(a) $f(z) = \dfrac{z}{1-z}$.

(b) $f(z) = z + az^2$ (a is a real constant).

(c) $f(z) = \dfrac{z}{(1-z)^2}$.

(d) $f(z) = e^z - 1$.

*3.5.6.   *The inverse and implicit function theorems (revisited)*

Even though we stated the inverse and implicit function theorems in Sec. 3.4.7 in the realm of real analysis and proved the former in Sec. 3.5.1 via the factorization $f(z) = f(z_0) + (z - z_0)^k g(z)$, it is still worthy to mention that the integral formula in (1) of (3.5.3.3) and, in particular, the argument principle, provides a powerful tool in the study of these two theorems. And, in turn, the implicit theorem can be used to solve the algebraic equation $P(z,w) = p_n(z)w^n + \cdots + p_1(z)w + p_0(z) = 0$ (see (2.9)) completely.

We will do these in this section for the purpose of completion and will use the Lagrange series to represent the inverse function obtained. Anyway, the material is optional and may be skipped without handicapping the understanding of the remaining parts of the book.

As a beginning, we try to use the Rouché's theorem to give (2) in (3.5.1.8) another yet concise proof. We restate it in the following.

Let f be analytic at z_0 and not constant. Let z_0 be a zero of order k of $f(z) - f(z_0)$.

(1) If $\rho > 0$ so that $f(z) - f(z_0) \neq 0$ on $0 < |z - z_0| \leq \rho$, then there is an $R > 0$ so that for each w in $|w - f(z_0)| < R$, the equation $f(z) = w$ has exactly k solutions in $|z - z_0| < \rho$.

(2) Let $\rho > 0$ be so chosen that, in addition, $f'(z) \neq 0$ on $0 < |z - z_0| < \rho$, then $f(z) = w$ has k distinct solutions in $0 < |z - z_0| < \rho$ for each w, $0 < |w - f(z_0)| < R$.   (3.5.6.1)

Proof. Along $|z - z_0| = \rho$, $|f(z) - f(z_0)| \geq \min |f(z) - f(z_0)| = R > 0$. For w, $|w - f(z_0)| < R$, observe the inequality $|w - f(z_0)| < R \leq |f(z) - f(z_0)|$ on $|z - z_0| = \rho$. Rouché's theorem says that $f(z) - f(z_0)$ and $(f(z) - f(z_0)) - (w - f(z_0)) = f(z) - w$ have the same number of zeros in $|z - z_0| < \rho$. In case $f'(z) \neq 0$ on $0 < |z - z_0| < \rho$, each zero of $f(z) - w$ is of order 1 for each w, $0 < |w - f(z_0)| < R$. This finishes the proof. $\square$

In what follows, three sections are divided.

Section (1) The single-valued inverse function and its Lagrange series representation

In case $k = 1$ in (3.5.6.1), then the inverse function $z = f^{-1}(w):\{|w - w_0| < R\} \to \{|z - z_0| < \rho\}$ is single-valued and hence, univalently analytic there; where $w_0 = f(z_0)$. According to (1) in (3.5.3.3), $f^{-1}(w)$ can be represented explicitly by

$$f^{-1}(w) = \frac{1}{2\pi i} \int_{|z-z_0|=\rho} \frac{f'(z)}{f(z) - w} z\, dz, \quad |w - w_0| < R. \qquad (3.5.6.2)$$

To put in a much more general setting, suppose g is analytic on $|z - z_0| \leq \rho$. Still by using (1) in (3.5.3.3), we have the corresponding integral representation of g in terms of w as

$$(g \circ f^{-1})(w) = \frac{1}{2\pi i} \int_{|z-z_0|=\rho} \frac{f'(z)}{f(z) - w} g(z)\, dz, \quad |w - w_0| < R. \qquad (3.5.6.3)$$

Observe that $|f(z) - w_0| \geq R$ should hold on $|z - z_0| = \rho$. Also, R can be chosen as the distance from w_0 to the image curve $f(\{|z - z_0| = \rho\})$.

Therefore, in case $|w - w_0| < R$ and $|z - z_0| = \rho$,

$$\frac{f'(z)}{f(z) - w} = \frac{f'(z)}{f(z) - w_0} \cdot \frac{1}{1 - \frac{w - w_0}{f(z) - w_0}}$$

$$= \frac{f'(z)}{f(z) - w_0} \cdot \sum_{n=0}^{\infty} \frac{1}{(f(z) - w_0)^n}(w - w_0)^n. \qquad (*_1)$$

Let $M = \max_{|z - z_0| = \rho} |f'(z)|$. Then, for $n \geq 0$,

$$\left| \frac{f'(z)}{f(z) - w_0} \right| \cdot \left| \frac{w - w_0}{f(z) - w_0} \right|^n \leq \frac{M}{R}\left(\frac{|w - w_0|}{R} \right)^n,$$
$$\text{if } |z - z_0| = \rho \quad \text{(and hence } |f(z) - w_0| \geq R).$$

$\Rightarrow$ The series in $(*_1)$ converges uniformly on $|z - z_0| = \rho$ for any fixed w with $|w - w_0| < R$.

$\Rightarrow$ The series in $(*_1)$ can be integrated termwise so that $(3.5.6.3)$ becomes

$$(g \circ f^{-1})(w) = \sum_{n=0}^{\infty} \left\{ \frac{1}{2\pi i} \int_{|z-z_0|=\rho} g(z) \frac{f'(z)}{(f(z) - w_0)^{n+1}} dz \right\}$$
$$\times (w - w_0)^n, \quad |w - w_0| < R. \qquad (*_2)$$

What remains is to try to show that

$$\frac{1}{2\pi i} \int_{|z-z_0|=\rho} g(z) \frac{f'(z)}{(f(z) - w_0)^{n+1}} dz$$
$$= \frac{1}{n!} \left\{ \frac{d^{n-1}}{dz^{n-1}} g'(z) \left[\frac{z - z_0}{f(z) - w_0} \right]^n \right\} \Bigg|_{z=z_0}, \quad n \geq 1. \qquad (*_3)$$

In case $n = 0$, by $(3.5.6.3)$,

$$(g \circ f^{-1})(w_0) = g(z_0) = \frac{1}{2\pi i} \int_{|z-z_0|=\rho} g(z) \frac{f'(z)}{f(z) - w_0} dz.$$

For $n \geq 1$, by integration by parts (see (6) in $(2.9.10)$),

$$\frac{1}{2\pi i} \int_{|z-z_0|=\rho} g(z) \frac{f'(z)}{(f(z) - w_0)^{n+1}} dz$$
$$= \frac{-1}{2n\pi i} \int_{|z-z_0|=\rho} g(z) d\left[\frac{1}{(f(z) - w_0)^n} \right]$$

$$= \frac{1}{2n\pi i} \int_{|z-z_0|=\rho} \frac{g'(z)}{(f(z)-w_0)^n} dz$$

$$= \frac{2\pi i}{2n\pi i} \cdot \text{the residue of } \frac{g'(z)}{(f(z)-w_0)^n} \text{ at the point } z_0 \quad (\text{see } (4.11.1.1))$$

$$= \frac{1}{n} \cdot \frac{1}{(n-1)!} \left\{ \frac{d^{n-1}}{dz^{n-1}} \frac{(z-z_0)^n g'(z)}{(f(z)-w_0)^n} \right\} \Bigg|_{z=z_0} \quad (\text{see } (4.11.2.1)).$$

This finishes the proof for $(*_3)$.

We summarize the above as

The single-valued inverse function and its representations (local version).
Let f be analytic at z_0 and $f'(z_0) \neq 0$. Denote $w_0 = f(z_0)$.

(1) There exist a $\rho > 0$ and an $R > 0$ so that the function $w = f(z)$ has a univalent inverse function $z = f^{-1}(w)$ mapping $|w - w_0| < R$ into $|z - z_0| < \rho$.

(2) In case $g(z)$ is analytic on $|z - z_0| \leq \rho$. Then $g \circ f^{-1}(w)$ is analytic on $|w - w_0| < R$ and

$$(g \circ f^{-1})(w) = \frac{1}{2\pi i} \int_{|z-z_0|=\rho} g(z) \frac{f'(z)}{f(z)-w} dz, \quad |w - w_0| < R$$

$$= g(z_0) + \sum_{n=1}^{\infty} \frac{1}{n!} \left\{ \frac{d^{n-1}}{dz^{n-1}} g'(z) \left[\frac{z-z_0}{f(z)-w_0} \right]^n \right\} \Bigg|_{z=z_0}$$

$$\times (w - w_0)^n, \quad |w - w_0| < R,$$

where the latter series is called the *Lagrange series* of $g \circ f^{-1}$ at w_0. In particular, in case $g(z) = z$, then

$$z = f^{-1}(w) = z_0 + \sum_{n=1}^{\infty} \frac{1}{n!} \left\{ \frac{d^{n-1}}{dz^{n-1}} \left[\frac{z-z_0}{f(z)-w_0} \right]^n \right\} \Bigg|_{z=z_0}$$

$$\times (w - w_0)^n, \quad |w - w_0| < R. \tag{3.5.6.4}$$

How to determine the radius R of convergence for the Lagrange series? Try the following steps:

Step 1. Choose $r > 0$ so that $f(z) \neq w_0$ and $f'(z) \neq 0$ on $0 < |z - z_0| < r$.

Step 2. Let $\delta(r) = \min_{|z-z_0|=r} |f(z) - w_0|$, the distance from the point $w_0 = f(z_0)$ to the image curve $f(|z - z_0| = r)$. Choose

$$R = \sup_r \delta(r), \tag{3.5.6.5}$$

where r runs through all such possible values as indicated in Step 1.

   Since the Lagrange series converges on $|w - w_0| < \delta(r)$, its radius R of convergence satisfies $R \geq \delta(r)$, where r is as in Step 1. Hence $R \geq \delta_0 = \sup \delta(r)$. These w satisfying $w = f(z)$ and $f'(z) = 0$ cannot lie in $|w - w_0| < R$ because $z = f^{-1}(w)$ is univalent in $|w - w_0| < R$. In case there exists w satisfying $|w - w_0| = \delta_0$ and $f(z) = w$ and $f'(z) = 0$, then it is necessary that $\delta_0 \geq R$. Hence $R = \delta_0$ holds and this R is the required one.

Example 1. Try to use $w = f(z) = ze^{-az}$ $(a \neq 0)$ and $g(z) = e^{bz}$ $(b \neq 0)$ to illustrate (3.6.5.4) at the points $z_0 = 0$ and $w_0 = f(0) = 0$.

Solution. $f'(z) = (1 - az)e^{-az} \Rightarrow f'(0) = 1 \neq 0$, and $f'(z) = 0 \Leftrightarrow z = \frac{1}{a}$.
   The image of the circle $|z| = \frac{1}{|a|}$ or $z = \frac{1}{|a|}e^{i\theta}$, $0 \leq \theta \leq 2\pi$, under $w = f(z)$ is the curve $w = \frac{1}{|a|}e^{i\theta}e^{-\frac{a}{|a|}e^{i\theta}}$. Since $|w| = \frac{1}{|a|}e^{-\cos(\alpha+\theta)}$, where $\frac{a}{|a|} = e^{i\alpha}$, the nearest point on this curve to $w_0 = 0$ occurs at $\theta = -\alpha$ and then, the distance from w_0 to the curve is $\frac{1}{|a|e}$ which will be the radius of convergence for the Lagrange series of $g \circ f^{-1}(w)$ as expected from (3.5.6.5).
   Since $g(z)$ is entire,

$$(g \circ f^{-1})(w) = e^{bf^{-1}(w)} = 1 + \sum_{n=1}^{\infty} \frac{1}{n!} \left\{ \frac{d^{n-1}}{dz^{n-1}} g'(z) \left[\frac{z}{f(z)} \right]^n \right\} \bigg|_{z=0} w^n$$

is valid on $|w| < \frac{1}{|a|e}$. Actual computation shows that, for $n \geq 1$,

$$\frac{d^{n-1}}{dz^{n-1}} g'(z) \left[\frac{z}{f(z)} \right]^n$$

$$= \frac{d^{n-1}}{dz^{n-1}} be^{bz} \cdot e^{anz} = b(b + an)^{n-1}e^{(b+an)z},$$

$$\Rightarrow (g \circ f^{-1})(w) = e^{bf^{-1}(w)} = 1 + b\sum_{n=1}^{\infty} \frac{(b + an)^{n-1}}{n!} w^n, \quad |w| < \frac{1}{|a|e}.$$

Since $\lim_{n \to \infty} \sqrt[n]{\frac{n!}{n^n}} = \lim_{n \to \infty} \frac{n^n}{(n+1)^n} = \frac{1}{e}$, the right series has its radius of convergence $\left(\overline{\lim}_{n \to \infty} \sqrt[n]{\frac{|b+an|^{n-1}}{n!}} \right)^{-1} = \frac{1}{e|a|}$, as we claimed above by using (3.5.6.5).
   Note: In case $a = b = 1$, $g(z) = e^z = \frac{z}{f(z)}$ and then $(g \circ f^{-1})(w) = \frac{f^{-1}(w)}{w}$. Therefore, $w = f(z) = ze^{-z}: \{|z| < 1\} \to \{|w| < e^{-1}\}$ has the

univalent inverse function

$$z = f^{-1}(w) = \sum_{n=0}^{\infty} \frac{(n+1)^{n-1}}{n!} w^{n+1}, \qquad |w| < e^{-1}. \qquad (*_4)$$

This series *converges absolutely* on $|w| = e^{-1}$: Let $a_n = \frac{(n+1)^{n-1}}{n!} \cdot \frac{1}{e^n}$. Then

$$\frac{a_{n+1}}{a_n} = \left(\frac{n+2}{n+1}\right)^n e^{-1} \leq \frac{n+1}{n+2} < 1,$$

$$\Rightarrow \lim_{n\to\infty} n\left(1 - \frac{a_{n+1}}{a_n}\right) = \lim_{n\to\infty} n\left[1 - \left(1 + \frac{1}{n+1}\right)^n e^{-1}\right]$$

$$= \lim_{n\to\infty} \frac{-e^{-1}\left(1 + \frac{1}{n+1}\right)^n \left[\log\left(1 + \frac{1}{n+1}\right) - \frac{n}{(n+1)(n+2)}\right]}{-\frac{1}{n^2}} = 2.$$

By the Raabe test, the series $\sum a_n$ converges. Since $\left|\frac{(n+1)^{n-1}}{n!} w^n\right| = |a_n|$ if $|w| = e^{-1}$, the series in $(*_4)$ does converge absolutely on $|w| = e^{-1}$. On the other hand, $a_{n+1} < a_n$ and $\lim_{n\to\infty} a_n = 0$, so $\sum_{n=0}^{\infty} a_n e^{in\theta}$ is convergent for $0 < \theta < 2\pi$ and thus, the series in $(*_4)$ converges uniformly on $0 < \delta \leq \theta \leq 2\pi - \delta$.

The point $w = e^{-1}$ is the only *singular point* of the series $(*_4)$ on the circle $|w| = e^{-1}$, namely, the series $(*_4)$ *cannot* be analytically continued beyond the disk $|w| < e^{-1}$ along the radius segment $[0, e^{-1})$. For details, see (5.1.3.6) in Chap. 5. But, it can be analytically continued along any rectifiable curve, *not* passing the point e^{-1}, to the whole plane $\mathbf{C}$ except e^{-1}. The resulted multiple-valued function is the inverse function $z = f^{-1}(w)$, which turns out to be single-valued on its Riemann surface. Refer to Example 4 in Sec. 3.5.1 and Fig. 3.55 there. Monodromy theorem (see (5.2.2.2)) guarantees that $z = f^{-1}(w)$ is single-valued on every simply connected subdomain of $\mathbf{C} - \{e^{-1}\}$.

The disk $|w| < e^{-1}$ of convergence for $(*_4)$ is shown on the Riemann surface in Fig. 3.82.

Section (2) The multiple-valued inverse function and its Lagrange series representation

In case $k \geq 2$ in (3.5.6.1), for a given w, $0 < |w - w_0| < R$ where $w_0 = f(z_0)$, how can one find k distinct points $z_1, \ldots, z_k$ in $0 < |z - z_0| < \rho$ so that $f(z_j) = w$ holds for $1 \leq j \leq k$?

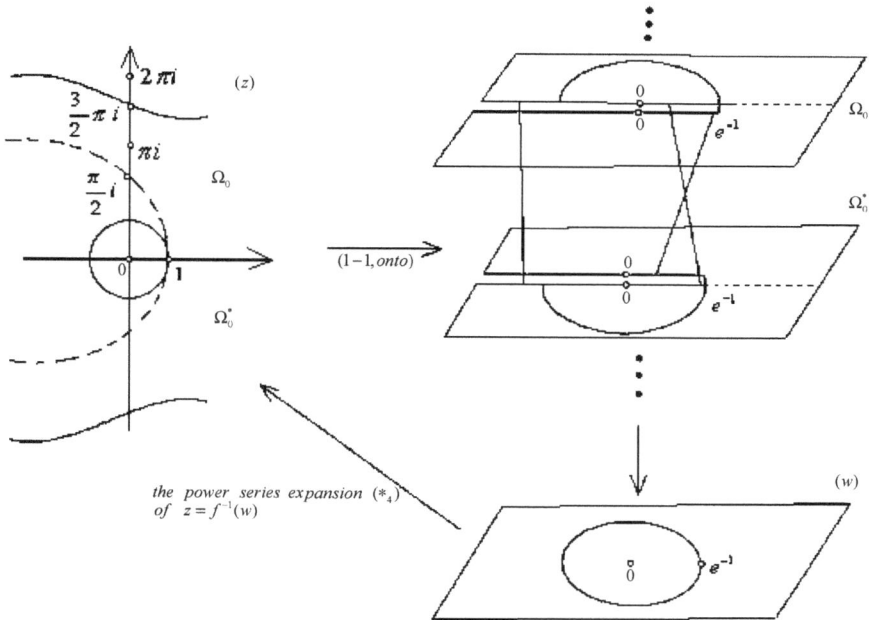

Fig. 3.82

Let us go back to the situation stated in (2) of (3.5.1.8) and watch Fig. 3.53. Recall that $w = f(z) = w_0 + (z - z_0)^k g(z)$, where g is analytic on $|z - z_0| < \rho$ and $g(z_0) \neq 0$. Note that $\rho > 0$ is so chosen that $g(z) \neq 0$ on $|z - z_0| < \rho$. Let $\varphi(z)$ be any *fixed* branch of $\sqrt[k]{g(z)}$ on $|z - z_0| < \rho$ (see (2) in (3.3.1.5); any other branch is of the form $\varphi(z)\omega^m$, $1 \leq m \leq k - 1$, where ω is a primitive kth root of 1). Under this circumstance, $w = f(z)$ can be decomposed as

$$z \to \zeta = (z - z_0)\varphi(z) \to w = f(z) = w_0 + \zeta^k,$$

where $\zeta = (z - z_0)\varphi(z)$ is univalent from the domain $O = \zeta^{-1}(|\zeta| < R^{\frac{1}{k}})$ onto $|\zeta| < R^{\frac{1}{k}}$. Observe that O is an open subset of $|z - z_0| < \rho$, containing z_0.

Now, let w, $0 < |w - w_0| < R$, be given. Let ζ be any one of the k distinct points $\zeta_1, \ldots, \zeta_k$ in $|\zeta| < R^{\frac{1}{k}}$ so that $w = w_0 + \zeta^k$ (recall that we may designate $\zeta_j = \zeta\omega^j$, $1 \leq j \leq k$). Then, let z be one of the k-distinct points $z_1, \ldots, z_k$ in O so that $\zeta = (z - z_0)\varphi(z)$ (also, recall that we may designate $z_j = z\omega^j$, $1 \leq j \leq k$). Then $f(z\omega^j) = w_0 + (\zeta(z\omega^j))^k = w_0 + \zeta_j^k = w$ holds for $1 \leq j \leq k$.

According to (3.5.6.4), we have

$$z = \zeta^{-1}(\zeta(z)) = z_0 + \sum_{n=1}^{\infty} \frac{1}{n!} \left\{ \frac{d^{n-1}}{dz^{n-1}} \left(\frac{z - z_0}{G(z)} \right)^n \right\} \Bigg|_{z=z_0} \zeta^n, \quad |\zeta| < R^{\frac{1}{k}},$$

where $G(z)$ is a fixed branch of $\sqrt[k]{f(z) - w_0} = (z - z_0) \sqrt[k]{g(z)} = (z - z_0)\varphi(z)$ on $|z - z_0| < \rho$ and ζ within can be one of $\zeta_1, \dots, \zeta_k$ and the corresponding z can be one of $z_1, \dots, z_k$. In the above formula, if we replace ζ by $(w - w_0)^{\frac{1}{k}}$, $w_0 = f(z_0)$, then $z = \zeta^{-1}(\zeta(z))$ should be changed to $z = f^{-1}(w)$.

We summarize the above as

the multiple-valued inverse function and its Lagrange series representation *(local version).* Let f be analytic at z_0 and

$$f'(z_0) = \cdots = f^{(k-1)}(z_0) = 0, \quad f^{(k)}(z_0) \neq 0 \quad \text{for } k \geq 2.$$

Denote $w_0 = f(z_0)$.

(1) Choose $\rho > 0$ so that $f(z) \neq w_0$ and $f'(z) \neq 0$ on $0 < |z - z_0| \leq \rho$. Let $R = \inf_{|z-z_0|=\rho} |f(z) - w_0|$. Then for each w, $0 < |w - w_0| < R$, the equation $f(z) = w$ has exactly k distinct solutions in $0 < |z - z_0| < \rho$. Moreover, the k-valued inverse function $z = f^{-1}(w)$ is analytic (see Sec. 3.3.3) on $|w - w_0| < R$ and can be represented as the *Lagrange series*

$$z = f^{-1}(w) = z_0$$
$$+ \sum_{n=1}^{\infty} \frac{1}{n!} \left\{ \frac{d^{n-1}}{dz^{n-1}} \left(\frac{z - z_0}{G(z)} \right)^n \right\} \Bigg|_{z=z_0} (w - w_0)^{\frac{n}{k}}, \quad |w - w_0| < R,$$

where $G(z)$ is any fixed branch of the k-valued function $\sqrt[k]{f(z) - w_0}$ on $|z - z_0| < \rho$ (namely, $(z - z_0)\sqrt[k]{g(z)}$, where $f(z) = w_0 + (z - z_0)^k g(z)$ and $g(z) \neq 0$ on $|z - z_0| < \rho$). Hence, $w = w_0$ is an *algebraic branch point* of order $k - 1$ (see Fig. 3.53).

(2) Let $O = f^{-1}(|w - w_0| < R)$. Then O is a domain in $|z - z_0| < \rho$ and $w = f(z) : O \to |w - w_0| < R$ has a k-valued inverse function $z = f^{-1}(w)$ mapping $|w - w_0| < R$ onto O. \hfill (3.5.6.6)

It should be mentioned that both (3.5.6.4) and (3.5.6.6) are still valid in case $z_0 = \infty$ or $w_0 = \infty$ (namely, f has a pole of order k at z_0). Details are left to the readers.

One should refer to Example 4 in Sec. 3.5.1 for an illustrative example. Yet another one is the following.

Example 2.   (Ahlfors, Ref. [1], p. 99) Let $w = f(z) = z^3 - 3z$. Try to find local representations of the inverse function $z = f^{-1}(w)$ on the classical w-plane and on its Riemann surface.

Solution. For the construction of the Riemann surface R_w of the inverse function $z = f^{-1}(w)$, refer to Example 4 in Sec. 2.7.2; in particular, see Fig. 2.66 for the descriptive R_w. Since $w = f(z)$ sets up a one-to-one correspondence between $\mathbf{C}^*$ and R_w, one can view $\mathbf{C}^*$ or the Riemann sphere (see Sec. 1.6) as a topological model for R_w. Note that R_w has an algebraic branch point of order 1 over each of 2 and -2, and an algebraic branch point of order 2 over ∞.

As a beginning, observe that $f'(z) = 3(z^2 - 1)$ and $f'(z) = 0$ only if $z = \pm 1$. In what follows, we will choose different $z_0 \in \mathbf{C}$ to justify either (3.5.6.4) or (3.5.6.6).

$z_0 = 0 : f(0) = 0$ yet $f'(0) = -3 \neq 0$. Since $f(z) \neq 0$ and $f'(z) \neq 0$ on $0 < |z| < 1$ and

$$\inf_{|z|=1} |f(z) - f(0)| = \inf_{0 \le \theta \le 2\pi} |f(e^{i\theta})| = \inf_{0 \le \theta \le 2\pi} (10 - 6\cos 2\theta)^{\frac{1}{2}} = 2,$$

$w = f(z)$ has a univalent inverse function $z = f^{-1}(w) : \{|w| < 2\} \to \{|z| < 1\}$. Its Lagrange series expansion is

$$z = -\frac{1}{3}w - \frac{1}{81}w^3 + \sum_{n=4}^{\infty} \frac{1}{n!} \left\{ \frac{d^{n-1}}{dz^{n-1}} \left(\frac{1}{z^2 - 3} \right)^n \right\}\Bigg|_{z=0} w^n, \quad |w| < 2.$$

If we adopt the recursive approximation method in (2) of (3.4.7.1), the approximation functions are

$$g_0(w) = z_0 = 0; \quad g_1(w) = -\frac{1}{3}w; \quad g_2(w) = -\frac{1}{3}w - \frac{1}{81}w^3;$$

$$g_3(w) = g_2(w) - \frac{1}{f'(z_0)}[f(g_2(w)) - w]$$

$$= -\frac{1}{3}w - \frac{1}{81}w^3 - \frac{1}{729}w^5 - \frac{1}{19,693}w^7 - \frac{1}{3 \times 81^3}w^9; \dots .$$

See Fig. 3.83. The right figure in Fig. 3.83 should be put in the 2nd sheet (II) if we lift it to R_w (see Fig. 2.66).

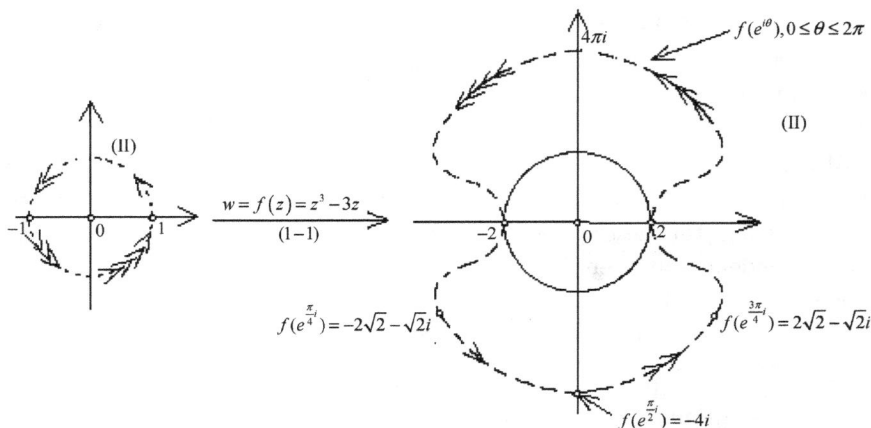

Fig. 3.83

$z_0 = -\sqrt{3} : f(z_0) = w_0 = 0$ and $f'(z_0) = 6 \neq 0$. On $0 < |z + \sqrt{3}| < (-1) - (-\sqrt{3}) = \sqrt{3} - 1$, $f(z) \neq 0$, and $f'(z) \neq 0$ hold. Since

$$\inf_{|z+\sqrt{3}|=\sqrt{3}-1} |f(z) - w_0|$$
$$= \inf_{0\leq\theta\leq 2\pi} |[-\sqrt{3} + (\sqrt{3} - 1)e^{i\theta}]^3 - 3[-\sqrt{3} + (\sqrt{3} - 1)e^{i\theta}]| = 2,$$

the inverse function $z = f^{-1}(w)$ is univalent on $|w| < 2$ and, observing $\frac{z-z_0}{f(z)-w_0} = \frac{1}{z(z-\sqrt{3})}$, we have

$$z = f^{-1}(w) = z_0 + \sum_{n=1}^{\infty} \frac{1}{n!} \left\{ \frac{d^{n-1}}{dz^{n-1}} \left(\frac{z - z_0}{f(z) - w_0} \right)^n \right\} \bigg|_{z=z_0} (w - 0)^n,$$

$$= -\sqrt{3} + \frac{1}{6}w + \frac{\sqrt{3}}{72}w^2 - \frac{48}{6^5}w^3 + \cdots, \quad |w| < 2.$$

The image curve of the circle $z = -\sqrt{3} + (\sqrt{3} - 1)e^{i\theta}$, $0 \leq \theta \leq 2\pi$, under $w = f(z)$ is given by $f(-\sqrt{3} + (\sqrt{3} - 1)e^{i\theta}) = e^{i\theta}[6(\sqrt{3} - 1) - 3\sqrt{3}(4 - 2\sqrt{3})e^{i\theta} + (-10+6\sqrt{3})e^{2i\theta}]$, $0 \leq \theta \leq 2\pi$. This can be sketched approximately as in Fig. 3.84, where $f(\sqrt{-3}) = 0$; $f(-1) = 2$; $A = f(-\sqrt{3}+(\sqrt{3}-1)e^{\frac{\pi i}{4}}) = 6.211\cdots + i3.427\ldots$; $B = f(-\sqrt{3} + (\sqrt{3} - 1)e^{\frac{\pi i}{2}}) = 2.784\cdots + 4i$; $C = f(-\sqrt{3} + (\sqrt{3} - 1)e^{\pi i}) = -7.569\ldots$. The lift of the right figure to R_w should be on the 3rd sheet (III).

$z_0 = \sqrt{3}$: This case is left to the readers but note that $f(-z) = -f(z)$.

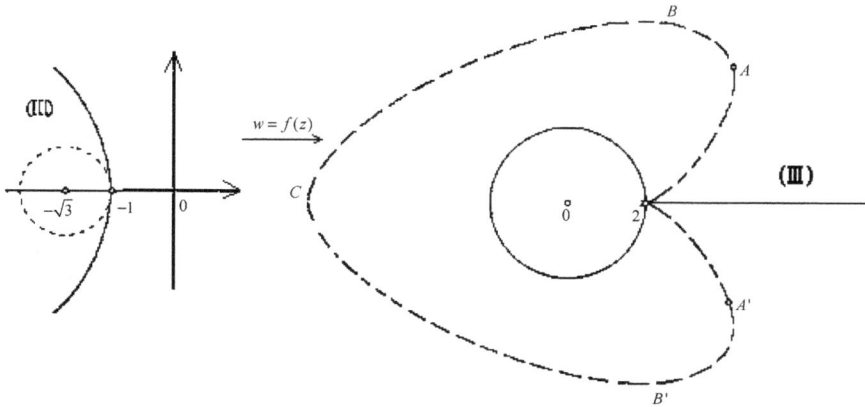

Fig. 3.84

As a whole, $z = f^{-1}(w)$ has three single-valued analytic branches on $|w| < 2$ with range in $|z - 0| < 1$, $|z + \sqrt{3}| < \sqrt{3} - 1$, and $|z - \sqrt{3}| < \sqrt{3} - 1$, respectively.

$z_0 = 2$: The equation $f(z) = z^3 - 3z = 2$ has solution -1, -1, and 2. $f(z) \neq 2$ on $0 < |z - 2| < |-1 - 2| = 3$ yet $f'(1) = 0$. So, we choose $\rho = 1$ and thus, $f(z) \neq 2$ and $f'(z) \neq 0$ on $0 < |z - 2| < 1$. The inverse function of $w = f(z)$ on $|z - 2| < 1$ is univalent on $|w - 2| < \inf_{|z-2|=1} |f(z) - 2| = 4$ and is given by

$$z = f^{-1}(w) = 2 + \sum_{n=1}^{\infty} \frac{1}{n!} \left\{ \frac{d^{n-1}}{dz^{n-1}} \left(\frac{z - 2}{z^3 - 3z - 2} \right)^n \right\} \Bigg|_{z=2} (w - 2)^n$$

$$= 2 + \sum_{n=1}^{\infty} \frac{(-1)^{n-1}(3n - 2)!}{n!(2n - 1)!} \cdot \frac{1}{3^{3n-1}} (w - 2)^n, \quad |w - 2| < 4.$$

The right series has its radius of convergence equal to 4. Indeed, $\lim_{n \to \infty} \left| \frac{a_n}{a_{n+1}} \right| = 4$ does hold. See Fig. 3.85 for a sketch of $|z - 2| < 1$, $|w - 2| < 4$, and the image curve $f(2 + e^{i\theta})$, $0 \leq \theta \leq 2\pi$. Wherein $f(2) = 2$; $f(1) = -2$; $f(3) = 18$; $A = f(2 + e^{\frac{\pi i}{4}}) = 7.656 \cdots + i13.071 \ldots$; $B = f(2 + e^{\frac{\pi i}{2}}) = -4 + 8i$; $C = f(2 + e^{\frac{3\pi i}{4}}) = -3.656 \cdots + i1.0710 \ldots$. The lift of the right figure to R_w is on the 1st sheet (I).

$z_0 = -2$: This case is similar to $z_0 = 2$ and is left to the readers.

In case z_0 is a fixed point in $\mathbf{C}^*$ other than $\pm 1, \pm 2, \pm \sqrt{3}, 0$ and ∞: Solve $f(z) = f(z_0)$, namely, $z^3 - 3z = z_0^3 - 3z_0$ and the solutions are $z = z_0$,

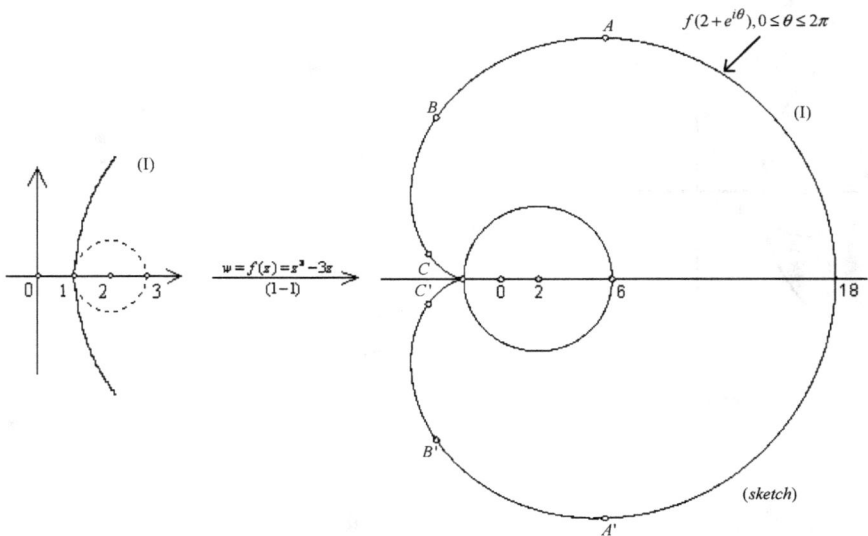

Fig. 3.85

$\frac{-z_0 \pm \sqrt{12 - 3z_0^2}}{2}$. Choose

$$\rho = \min \left\{ \left| z_0 + \frac{z_0 + \sqrt{12 - 3z_0^2}}{2} \right|, \right.$$

$$\left. \left| z_0 - \frac{z_0 - \sqrt{12 - 3z_0^2}}{2} \right|, |z_0 + 1|, |z_0 - 1| \right\}.$$

Then $f(z) \neq f(z_0)$ and $f'(z) \neq 0$ on $0 < |z - z_0| < \rho$. Let $R = \inf_{|z-z_0|=\rho} |f(z) - f(z_0)|$. Thus, the inverse function $z = f^{-1}(w)$ is univalent on $|w - w_0| < R$ with its range a domain in $|z - z_0| < \rho$. Take, for instance, $z = i$. The solutions of $f(z) = f(i) = -4i$ are i and $\frac{-i \pm \sqrt{15}}{2}$. Choose $\rho = |i + 1| = \sqrt{2}$. The corresponding $R = \inf_{0 \le \theta \le 2\pi} |f(i + \sqrt{2}e^{i\theta}) - (-4i)| = |-2 + 4i| = 2\sqrt{5}$. The restriction of $z = f^{-1}(w)$ to the disk $|w + 4i| < 2\sqrt{5}$ has the Lagrange series expansion

$$z = i + \sum_{n=1}^{\infty} \frac{1}{n!} \left\{ \frac{d^{n-1}}{dz^{n-1}} \left(\frac{1}{z^2 + iz - 4} \right)^n \right\} \bigg|_{z=i} (w + 4i)^n$$

$$= i - \frac{1}{6}(w + 4i) + \frac{i}{72}(w + 4i)^2 + \frac{1}{648}(w + 4i)^3 + \cdots, \quad |w + 4i| < 2\sqrt{5}.$$

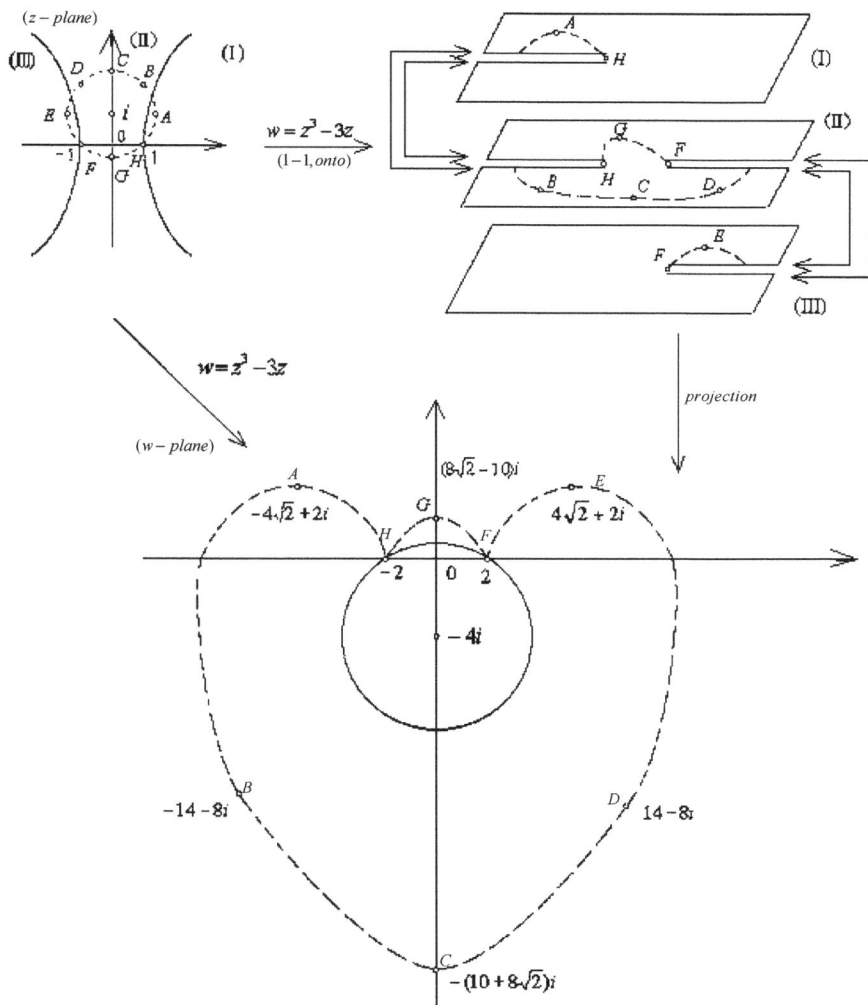

Fig. 3.86

See Fig. 3.86 for a sketch of related curves. Wherein the points A, B, C, D, E, H, G, and F are the image points, under $w = f(z)$, of the points $z = i + \sqrt{2}e^{i\theta}$ with $\theta = 0, \frac{\pi}{4}, \frac{\pi}{2}, \frac{3\pi}{4}, \pi, -\frac{\pi}{4}, -\frac{\pi}{2}$, and $-\frac{3\pi}{4}$, respectively.

$z_0 = 1$: $f(z) = z^3 - 3z = f(z_0) = -2$ has solutions 1, 1, and -2. Choose $\rho = 2$ and then $f(z) \neq -2$ and $f'(z) \neq 0$ on $0 < |z - 1| < 2$. In this case, $R = \inf_{0 \leq \theta \leq 2\pi} |f(1 + 2e^{i\theta}) + 2| = |f(1 + 2e^{\pi i}) + 2| = 4$. Since $f'(1) = 0$

while $f''(1) \neq 0$, the inverse function $z = f^{-1}(w)$ is double-valued on $|w + 2| < 4$ and its Lagrange series expansion is

$$z = 1 + \sum_{n=1}^{\infty} \frac{1}{n!} \left\{ \frac{d^{n-1}}{dz^{n-1}} \left(\frac{z-1}{G(z)} \right)^n \right\} \bigg|_{z=1} (w+2)^{\frac{n}{2}}, \quad |w+2| < 4, \quad (*)$$

where $G(z)$ is a fixed branch of $\sqrt{f(z) - w_0} = (z-1)\sqrt{z+2}$ on $|z-1| < 2$. Choose, for instance, the branch in $|z-1| < 2$ determined by $\sqrt{2} > 0$. Then,

$$\left(\frac{z-1}{G(z)} \right)^n = (z+2)^{-\frac{n}{2}} \quad \text{with } \sqrt{2} > 0,$$

$$\Rightarrow \frac{z-1}{G(z)} \bigg|_{z=1} = \frac{1}{\sqrt{3}} \quad (\sqrt{3} > 0);$$

$$\frac{d^{n-1}}{dz^{n-1}} (z+2)^{-\frac{n}{2}} = (-1)^{n-1} \cdot \frac{n(n+2)(n+4) \cdots (n+2n-4)}{2^{n-1}}$$

$$\times (z+2)^{-\frac{n+2n-2}{2}};$$

$$\Rightarrow \left\{ \frac{d^{n-1}}{dz^{n-1}} (z+2)^{-\frac{n}{2}} \right\} \bigg|_{z=1}$$

$$= (-1)^{n-1} \cdot \frac{n(n+2)(n+4) \cdots (3n-4)}{2^{n-1}} 3^{-\frac{3n-2}{2}}, \quad n \geq 2.$$

Explicitly the Lagrange series $(*)$ is then

$$z = 1 + \frac{1}{\sqrt{3}}(w+2)^{\frac{1}{2}} + \sum_{n=2}^{\infty} (-1)^{n-1} \cdot \frac{n(n+2)(n+4) \cdots (3n-4)}{2^{n-1} \cdot n!}$$

$$\times 3^{-\frac{3n-2}{2}} (w+2)^{\frac{n}{2}}, \quad |w+2| < 4.$$

The image curve $f(1+2e^{i\theta}) = -2+12e^{2i\theta}+8e^{3i\theta}$, $0 \leq \theta \leq 2\pi$, winds around the point $f(1) = -2$ twice. Its lift in the Riemann surface R_w is a Jordan closed curve. See Fig. 3.87. The points A, B, C, D, E, H, G, and F are, respectively, the points $f(1+2e^{i\theta})$ for $\theta = 0, \frac{\pi}{4}, \frac{\pi}{2}, \frac{3\pi}{4}, \pi, -\frac{\pi}{4}, -\frac{\pi}{2}$, and $-\frac{3\pi}{4}$, respectively. Note that $z = f^{-1}(w)$ becomes single-valued on $|w+2| < 4$ in R_w but double-valued on $|w+2| < 4$ in the classical $\mathbf{C}^*$.

$z_0 = -1$: This case is similar to $z_0 = 1$.

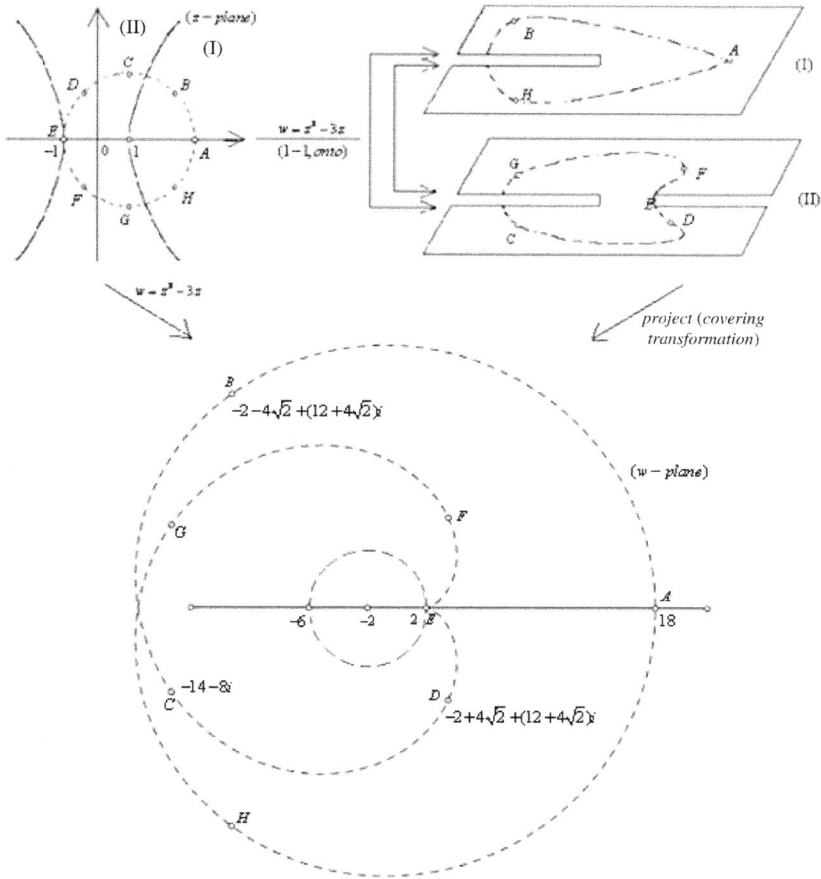

Fig. 3.87

$z_0 = \infty$: Since $f(\infty) = \infty = w_0$, the change of variables $\zeta = \frac{1}{z}$ and $\eta = \frac{1}{w}$ change $w = f(z) = z^3 - 3z$ into

$$\eta = g(\zeta) = \frac{\zeta^3}{1 - 3\zeta^2}.$$

$\eta = g(\zeta)$ is analytic on $|\zeta| < \frac{1}{\sqrt{3}}$ and $g'(0) = g''(0) = 0$ but $g'''(0) = 6 \neq 0$. Observe that $g(\zeta) \neq 0$ and $g'(\zeta) \neq 0$ on $0 < |\zeta| < \frac{1}{\sqrt{3}}$. On the other hand,

$$\inf_{0 \leq \theta \leq 2\pi} \left| g\left(\frac{1}{\sqrt{3}} e^{i\theta}\right) - g(0) \right| = \frac{1}{3\sqrt{3}} \inf_{0 \leq \theta \leq 2\pi} \left| \frac{e^{3i\theta}}{1 - e^{2i\theta}} \right| = \frac{1}{6\sqrt{3}}.$$

Hence, the inverse function $\zeta = g^{-1}(\eta)$ is triple-valued on $|\eta| < \frac{1}{6\sqrt{3}}$ and has the Lagrange series expansion

$$\zeta = 0 + \sum_{n=1}^{\infty} \frac{1}{n!} \left\{ \frac{d^{n-1}}{d\zeta^{n-1}} \left(\frac{\zeta - 0}{G(\zeta)} \right)^n \right\}\bigg|_{\zeta=0} \eta^{\frac{n}{3}}, \quad |\eta| < \frac{1}{6\sqrt{3}}, \qquad (**)$$

where $G(\zeta)$ is a branch of $\sqrt[3]{g(\zeta) - 0} = \frac{\zeta}{\sqrt[3]{1-3\zeta^2}}$ in $|\zeta| < \frac{1}{\sqrt{3}}$. Take, for instance, the branch in $|\zeta| < \frac{1}{\sqrt{3}}$ determined by $\sqrt[3]{1} = 1$. Then $(**)$ is

$$\zeta = \eta^{\frac{1}{3}} + \frac{-6}{3!}\eta^{\frac{3}{3}} + \cdots = \eta^{\frac{1}{3}} - \eta + \cdots, \quad |\eta| < \frac{1}{6\sqrt{3}},$$

$$\Rightarrow z = \frac{1}{w^{-\frac{1}{3}} - w^{-1} + \cdots} = w^{\frac{1}{3}} + w^{-\frac{1}{3}} + \cdots, \quad |w| > 6\sqrt{3}.$$

This shows that $z = f^{-1}(w)$ is triple-valued on $|w| > 6\sqrt{3}$ with range in $|z| > \sqrt{3}$. Also the image curve $f(\sqrt{3}e^{i\theta}) = 3\sqrt{3}e^{i\theta}(e^{2i\theta} - 1)$, $0 \le \theta \le 2\pi$, winds around ∞ three times. The following table illustrates some correspondence of points between curves $z = \sqrt{3}e^{i\theta}$, $0 \le \theta \le 2\pi$, and $f(\sqrt{3}e^{i\theta})$, $0 \le \theta \le 2\pi$. Note that $f(\bar{z}) = \overline{f(z)}$ holds. See Fig. 3.88 for a sketch. $z = f^{-1}(w)$ is single-valued on $|w| > 6\sqrt{3}$ in the Riemann surface R_w.

θ	$0(A)$	$\frac{\pi}{6}(B)$	$\frac{\pi}{4}(C)$	$\frac{\pi}{3}(D)$	$\frac{\pi}{2}(E)$	$\frac{2\pi}{3}(F)$
$f(\sqrt{3}e^{i\theta})$	0	$\frac{3\sqrt{3}}{2}(-\sqrt{3}+i)$	$-3\sqrt{6}$	$\frac{9}{2}(-\sqrt{3}-i)$	$-6\sqrt{3}i$	$\frac{9}{2}(\sqrt{3}-i)$
$\frac{3\pi}{4}(G)$	$\frac{5\pi}{6}(H)$			$\pi(I)$		
$3\sqrt{6}$	$\frac{3\sqrt{3}}{2}(\sqrt{3}+i)$			0		

Section (3) Implicit function theorem for two complex variables

It is supposed that readers are familiar with basic knowledge about analytic functions of two complex variables as presented in Sec. 3.4.7.1.

Recall that the key point to the inverse function theorem (3.5.6.6) is the factorization property $f(z) = f(z_0) + (z - z_0)^k g(z)$, where g is analytic at z_0 and $g(z_0) \ne 0$. As a counterpart, we have the

Weierstrass factorization theorem. Suppose $F(z, w)$ is analytic at (z_0, w_0), namely, in a neighborhood of (z_0, w_0), and satisfies

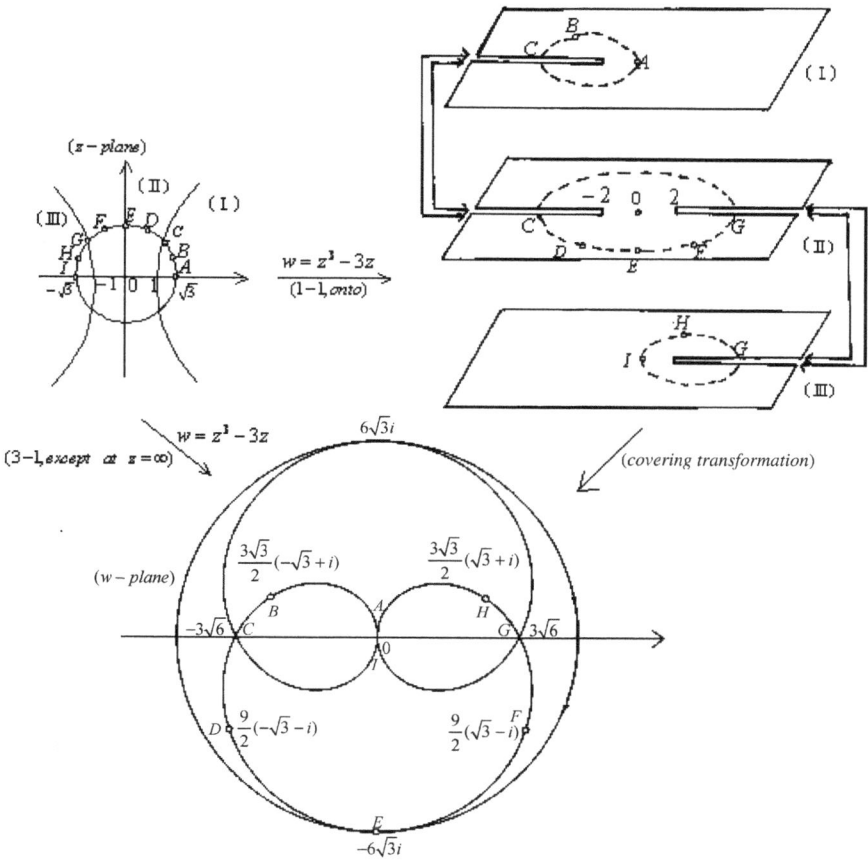

Fig. 3.88

(1) $F(z_0, w_0) = 0$,   and

(2) $F(z_0, w) \not\equiv 0$.

Then there exist $r > 0$ and $R > 0$ such that, on $|z - z_0| < r$ and $|w - w_0| < R$, $F(z, w)$ can be expressed as

$$F(z, w) = [w^k + A_{k-1}(z)w^{k-1} + \cdots + A_1(z)w + A_0(z)]G(z, w),$$

where $k \geq 1$, is such that

$$\frac{\partial F}{\partial w}(z_0, w_0) = \cdots = \frac{\partial^{k-1} F}{\partial w^{k-1}}(z_0, w_0) = 0, \quad \frac{\partial^k F}{\partial w^k}(z_0, w_0) \neq 0;$$

moreover,

(a) the coefficient functions $A_{k-1}(z), \ldots, A_1(z)$, $A_0(z)$ are analytic on $|z - z_0| < r$, and

(b) $G(z, w)$ is analytic and nonzero on $|z - z_0| < r$, $|w - w_0| < R$.

$$(3.5.6.7)$$

For an elementary proof of this result, see Ref. [58], Vol. II, pp. 105–109. The **proof** for the following (3.5.6.9) is taken from pp. 109–110 in this book.

Let $F(z, w)$ be as in (3.5.6.7) in what follows.

In case $k = 1$: Then $F(z_0, w_0) = 0$ and $\frac{\partial F}{\partial w}(z_0, w_0) \neq 0$, and

$$F(z, w) = (w + A_0(z))G(z, w), \quad |z - z_0| < r, \quad |w - w_0| < R$$

holds. By (b) in (3.5.6.7), $G(z, w) \neq 0$ there and hence

$$F(z, w) = 0, \quad |z - z_0| < r, \quad |w - w_0| < R,$$
$$\Leftrightarrow w + A_0(z) = 0, \quad \text{namely,} \quad w = -A_0(z) \text{ on}$$
$$|z - z_0| < r, \quad |w - w_0| < R.$$

Since $F(z_0, w_0) = 0$ while $G(z_0, w_0) \neq 0$, $w_0 = -A_0(z_0)$ holds. This means that, for each z, $|z - z_0| < r$, $F(z, w) = 0$ has a unique single-valued analytic solution $w = -A_0(z)$ satisfying $|w - w_0| < R$ and $w_0 = -A_0(z_0)$. Compare with (3.4.7.3).

In case $k \geq 2$: Then,

$$F(z, w) = 0, \quad |z - z_0| < r, \quad |w - w_0| < R,$$
$$\Leftrightarrow H(z, w) = w^k + A_{k-1}(z)w^{k-1} + \cdots + A_1(z)w + A_0(z) = 0 \text{ on}$$
$$|z - z_0| < r, \quad |w - w_0| < R.$$

For each z, $|z - z_0| < r$, $H(z, w) = 0$ has k roots $w_1(z), \ldots, w_k(z)$ (counting multiplicity). But, for a point z, $|z - z_0| < r$, so that

$$H(z, w) = 0 \text{ has a multiplicity root } w.$$
$$\Leftrightarrow \text{For some } w_j(z), w = w_j(z) \text{ holds so that}$$
$$H(z, w) = \frac{\partial H}{\partial w}(z, w) = 0 \text{ holds.}$$

Such points z occur as the *zeros* of the *discriminant* for $H(z, w) = 0$:

$$D(z) = \prod_{1 \le j < l \le k} (w_j(z) - w_l(z))^2$$

$$= \begin{vmatrix} k & s_1(z) & s_2(z) & \cdots & s_{k-1}(z) \\ s_1(z) & s_2(z) & s_3(z) & \cdots & s_k(z) \\ \vdots & \vdots & \vdots & & \vdots \\ s_{k-1}(z) & s_k(z) & s_{k+1}(z) & \cdots & s_{2k-2}(z) \end{vmatrix}, \quad \text{where}$$

$$s_m(z) = \sum_{j=1}^{k} w_j(z)^m \text{ for } 1 \le m \le 2k - 2. \tag{3.5.6.8}$$

On the other hand, by (1) in (3.5.3.3),

$$s_m(z) = \frac{1}{2\pi i} \int_{|w-w_0|=R} w^m \frac{\frac{\partial F}{\partial w}(z, w)}{F(z, w)} dw, \quad 1 \le m \le 2k - 2,$$

show that $s_m(z)$ are analytic in $|z - z_0| < r$ for $1 \le m \le 2k - 2$ (see, for instance, Ref. [58], Vol. I, p. 418, Theorem 17.19 or refer to (4.7.4) and, more accurately, to Exercise B (1) of Sec. 5.3.3). Consequently, $D(z)$ is analytic in $|z - z_0| < r$.

By assumption, $F(z_0, w)$ has a zero of order k at w_0, and so is $H(z_0, w)$. Hence, $D(z_0) = 0$.

Assume $D(z) \not\equiv 0$ on $|z - z_0| < r$. Therefore, $D(z)$ has only isolated zeros in $|z - z_0| < r$. And thus there exists $0 < \rho < r$ so that $D(z) \neq 0$ on $0 < |z - z_0| < \rho$. Fix any z, $0 < |z - z_0| < \rho$, $D(z) \neq 0$ shows that $H(z, w) = 0$ has k *distinct* roots $w_1(z), \ldots, w_k(z)$ satisfying

$$\frac{\partial H}{\partial w}(z, w_j(z)) \neq 0, \quad 1 \le j \le k.$$

Applying the result for the case $k = 1$ above to such z and each of the corresponding $w_j(z)$, $1 \le j \le k$, there exist an open disk $I_z \subseteq \{0 < |z - z_0| < \rho\}$, with center at z, and an open disk $J_j \subseteq \{0 < |w - w_0| < R\}$, with center at $w_j(z)$, such that

(1) to each $\zeta \in I_z$, $F(\zeta, w) = 0$ has a unique solution $\widetilde{w_j} = \widetilde{w_j}(\zeta)$ in J_j, and

(2) $\widetilde{w_j} = \widetilde{w_j}(\zeta)$ is a single-valued analytic function of ζ in I_z, satisfying $\widetilde{w_j}(z) = w_j(z)$.

Summarize the above result as

The implicit function theorem for analytic function of two complex variables. Suppose $F(z, w)$ is analytic at (z_0, w_0) and satisfies

$$F(z_0, w_0) = 0 \quad \text{and} \quad F(z, w) \not\equiv 0.$$

Let $k \geq 1$ be a positive integer such that

$$\frac{\partial F}{\partial w}(z_0, w_0) = \cdots = \frac{\partial^{k-1} F}{\partial w^{k-1}}(z_0, w_0) = 0 \quad \text{but} \quad \frac{\partial^k F}{\partial w^k}(z_0, w_0) \neq 0.$$

(1) *The single-valued case* ($k = 1$): There exist $r > 0$ and $R > 0$ such that

  (i) to each z, $|z - z_0| < r$, the equation $F(z, w) = 0$ has a unique solution $w = w(z)$ in $|w - w_0| < R$ satisfying $w(z_0) = w_0$; and

  (ii) the solution $w = w(z)$ is single-valued and analytic in $|z - z_0| < r$.

(2) *The multiple-valued case* ($k \geq 2$): There exist $r > 0$ and $R > 0$ such that

  (i) to each z, $0 < |z - z_0| < r$, the equation $F(z, w) = 0$ has exactly k distinct solutions $w_1(z), \ldots, w_k(z)$ in $|w - w_0| < R$; and

  (ii) in a neighborhood of the point z, each solution function $w_j(\zeta)$, $1 \leq j \leq k$, of $F(z, w) = 0$ is a single-valued analytic function satisfying the condition that $w_j(\zeta)$ assumes the value $w_j(z)$ at $\zeta = z$.

  In short, for each z, $0 < |z - z_0| < r$, the multiple-valued solution $w = w(z)$ of the equation $F(z, w) = 0$ has k distinct single-valued analytic branches in an open neighborhood of z that assumes the values $w_1(z), \ldots, w_k(z)$ at the point z.   (3.5.6.9)

Note that this result and its proof are not more true for a point z where $D(z) = 0$ (see (3.5.6.8)), not to say the case that $D(z) \equiv 0$. Also, it is well known that the inverse function theorem is equivalent to the implicit function theorem. Južakov (1974, 1975) extended the Lagrange series to implicit function of several complex variables and, then, deduced the inverse function theorem of several complex variables.

One of the elementary and useful applications of (3.5.6.9) is that it can be used to solve completely an algebraic equation (see (2.9)), namely,

The existence of algebraic functions. The solution of the algebraic equation of order n

$$P(z, w) = p_n(z)w^n + p_{n-1}(z)w^{n-1} + \cdots + p_1(z)w + p_0(z) = 0$$

is an n-valued function $w = f(z)$, defined on $\mathbf{C}^*$. Let $\{z_1, \ldots, z_r\}$ be the total of the distinct roots of $p_n(z) = 0$ and the discriminant $D(z) = 0$

(see (3.5.6.8) with k replaced by n). Then $w = f(z)$ has the following *local representations* at a point $z_0 \in \mathbf{C}^*$:

(1) In case $z_0 \neq z_1, \ldots, z_r$ and ∞: in a neighborhood of z_0, $w = f(z)$ can be expressed by n *distinct* power series

$$w_j(z) = \sum_{l=0}^{\infty} a_l^{(j)}(z - z_0)^l, \quad j = 1, 2, \ldots, n.$$

(2) In case z_0 is one of $z_1, \ldots, z_r$: in a neighborhood of z_0, $w = f(z)$ can be expressed as

$$\sum_{l=m}^{\infty} a_l(z - z_0)^{\frac{l}{k}} \quad (1 \leq k \leq n, m \leq 0).$$

(3) In case $z_0 = \infty$: in a neighborhood of ∞, $w = f(z)$ can be expressed as

$$\sum_{l=m}^{\infty} a_l z^{-\frac{l}{k}} \quad (1 \leq k \leq n, m \leq 0). \tag{3.5.6.10}$$

Case 1 is an easy consequence of (3.5.6.9). For a complete proof, see Ref. [58], Vol. III, p. 303. $w_j(z)$ in Case 1 are called *regular elements* of $f(z)$ and can be analytically continued to any point in $\mathbf{C} - \{z_1, \ldots, z_r\}$ along a rectifiable curve within. $z_1, \ldots, z_r$ and ∞ could be a *singular point* (see Remark 1 in Sec. 3.3.1) or a *pole* (if $k = 1$ and $m < 0$) or an *algebraic branch point*. As a consequence, the Riemann surface R_w of $w = f(z)$ is an n-sheeted "surface" over the z-plane $\mathbf{C}^*$, obtained by pasting together properly and crosswise the branch cuts connecting $z_1, \ldots, z_r$ and/or ∞. Its *topological model* is a sphere with handles as shown in Fig. 3.89. One may refer to Sec. 2.7.2 for elementary examples. For more information about

Fig. 3.89

algebraic functions, see Refs. [1], pp. 300–306; [58], Vol. III, pp. 297–308; for much more detailed account, see Refs. [34], Vol. II, Chap. 1; [44], Vol. II, Chap. 12.

Example 3. Construct the Riemann surface R_w for $z = z(w)$ and the Riemann surface R_z for $w = w(z)$, where $z = z(w)$ and $w = w(z)$ are implicit functions defined by

$$F(z, w) = z^3 + w^3 - 3zw = 0.$$

And then, try to use $F(z, w) = 0$ to justify statements in (3.5.6.9) and (3.5.6.10). This problem appeared in the first edition (1953) of Ahlfors's book [1].

Solution. $F(z, w) = 0$ defines both $w = w(z)$ and $z = z(w)$ as triple-valued functions.

Adopting complex variable t as a parameter and setting $w = tz$, $F(z, w) = 0$ can be expressed as

$$z = \frac{3t}{1 + t^3} = z(w), \quad w = \frac{3t^2}{1 + t^3} = w(z).$$

This indicates that, for a fixed z, $z = \frac{3t}{1+t^3}$ defines t as a triple-valued function of z (counting multiplicities). When these resulted values of t are substituted into $\frac{3t^2}{1+t^3}$, we obtain $w = w(t) = w(t(z))$ as a triple-valued function of z. Therefore, the Riemann surface of the inverse function $t = t(z)$ of $z = \frac{3t}{1+t^3}$ is the Riemann surface R_z of $w = w(z)$ over the z-plane. Similar explanation is applied to $w = \frac{3t^2}{1+t^3}$ and the Riemann surface R_w.

To construct R_z: Consider $z = \frac{3t}{1+t^3}$ between t-plane and z-plane.

Observe that $t = \infty$ is a zero of order 2 of $z = z(t)$. So $z(\infty) = 0$ is an algebraic branch point of order 1 (see (3.5.1.8) and (3.5.1.11)).

On the other hand,

$$z' = \frac{-3(2t^3 - 1)}{(1 + t^3)^2}, \quad z'' = \frac{-18t^2(2 - t^3)}{(1 + t^3)^3}$$

$$\Rightarrow z' = 0 \text{ if and only if } t_k = \frac{1}{\sqrt[3]{2}} e^{\frac{2k\pi i}{3}}, \quad k = 0, 1, 2; \text{ but } z''(t_k) \neq 0.$$

Hence, $z_k = z(t_k) = \sqrt[3]{4} e^{\frac{2k\pi i}{3}}$, $k = 0, 1, 2$, are algebraic branch points of order 1.

Where are the fundamental domains for $z = z(t)$ (see (2.7.4))? For this object, first try to determine the images of the rays $\operatorname{Arg} t = \frac{2k\pi}{3}$, $k = 0, 1, 2$,

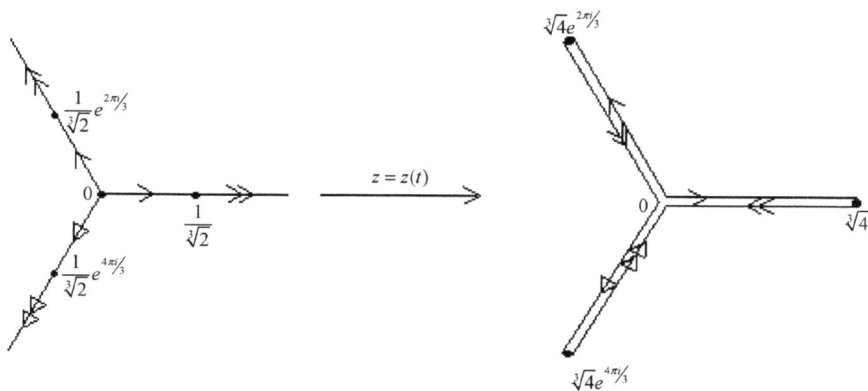

Fig. 3.90

under $z = z(t)$.

(1) $\operatorname{Arg} t = 0$, namely, the nonnegative real axis: In this case, $z = z(t)$ maps the segments $[0, \frac{1}{\sqrt[3]{2}}]$ and $[\frac{1}{\sqrt[3]{2}}, \infty]$ univalently onto $[0, \sqrt[3]{4}]$, respectively. See Fig. 3.90.

(2) $\operatorname{Arg} t = \frac{2\pi}{3}$: This ray can be expressed as $t = |t| e^{\frac{2\pi i}{3}}, 0 \le |t| \le \infty$. Then

$$z(|t| e^{\frac{2\pi i}{3}}) = \frac{3|t| e^{\frac{2\pi i}{3}}}{1 + |t|^3}.$$

This indicated that $z = z(t)$ maps the segments $[0, \frac{1}{\sqrt[3]{2}} e^{\frac{2\pi i}{3}}]$ and $[\frac{1}{\sqrt[3]{2}} e^{\frac{2\pi i}{3}}, \infty]$ on the ray $\operatorname{Arg} t = \frac{2\pi}{3}$ univalently onto the segment $[0, \sqrt[3]{4} e^{\frac{2\pi i}{3}}]$ on the original ray, respectively. See Fig. 3.90.

(3) $\operatorname{Arg} t = \frac{4\pi}{3}$: Just like Cases 1 and 2, $z = z(t)$ maps, respectively, the segments $[0, \frac{1}{\sqrt[3]{2}} e^{\frac{4\pi i}{3}}]$ and $[\frac{1}{\sqrt[3]{2}} e^{\frac{4\pi i}{3}}, \infty]$ on the ray $\operatorname{Arg} t = \frac{4\pi}{3}$ univalently onto the segment $[0, \sqrt[3]{4} e^{\frac{4\pi i}{3}}]$ on the same ray. See Fig. 3.90.

Since $z = z(t)$ maps the t-plane onto the z-plane in three-to-one manner, what remains to do is to justify the three sector domains of angle $\frac{2\pi}{3}$ each in the left figure of Fig. 3.90 are fundamental domains for $z = z(t)$.

Take, for instance, the sector domain $0 < \operatorname{Arg} z < \frac{2\pi}{3}$ for consideration. We try to show that $z = z(t)$ maps this domain univalently onto the domain $\mathbf{C}^* - [0, \sqrt[3]{4}] - [0, \sqrt[3]{4} e^{\frac{2\pi i}{3}}]$ in the z-plane. To start with a little more general setting, consider the circle $t = re^{i\theta}, 0 \le \theta \le 2\pi$, where $0 < r \le \frac{1}{\sqrt[3]{2}}$: If there are θ_1, θ_2 with $0 \le \theta_1$, $\theta_2 \le 2\pi$ so that

$$z(re^{i\theta_1}) = z(re^{i\theta_2}),$$

$$\Rightarrow |z(re^{i\theta_1})| = |z(re^{i\theta_2})| \quad \text{or} \quad |1 + r^3 e^{3i\theta_1}| = |1 + r^3 e^{3i\theta_2}|.$$

$$\Rightarrow \cos 3\theta_1 = \cos 3\theta_2 \quad \text{and hence,} \quad \theta_2 = \theta_1 + \frac{2k\pi}{3}, \ k = 0, 1, 2.$$

This indicates that if $z = \frac{3t}{1+t^3}$ has three distinct solutions $t_1(z), t_2(z)$, and $t_3(z)$ for a presumed z, then they should be equally spaced on a circle. In particular, $z = z(t)$ is univalent on the circular arc $t = \frac{1}{\sqrt[3]{2}} e^{i\theta}$, $0 < \theta < \frac{2\pi}{3}$. Combining with Case 1, and using (3.5.5.1) and (3.5.5.5), $z = z(t)$ is univalent on the domain bounded by the ray segments $\left[0, \frac{1}{\sqrt[3]{2}}\right]$, $\left[0, \frac{1}{\sqrt[3]{2}} e^{\frac{2\pi i}{3}}\right]$ and the circular arc, and on the domain bounded by the ray segments $\left[\frac{1}{\sqrt[3]{2}}, \infty\right]$, $\left[\frac{1}{\sqrt[3]{2}} e^{2\pi i/3}, \infty\right]$ and the circular arc. See Fig. 3.91. Therefore, $z = z(t)$ maps the sector domain univalently onto $\mathbf{C}^* - [0, \sqrt[3]{4}] - [0, \sqrt[3]{4} e^{\frac{2\pi i}{3}}]$. Similar argument holds for the other two sector domains.

As a whole, $z = z(t)$ maps the sector domains (I), (II), (III) in the t-plane univalently onto the domains (I), (II), (III) in the z-plane, respectively:

t-plane		z-plane
(I) $0 < \operatorname{Arg} t < \frac{2\pi}{3}$		(I) $\mathbf{C}^* - \left[0, \sqrt[3]{4}\right] - \left[0, \sqrt[3]{4} e^{\frac{2\pi i}{3}}\right]$
(II) $\frac{2\pi}{3} < \operatorname{Arg} t < \frac{4\pi}{3}$	$\xrightarrow{z=z(t)}$	(II) $\mathbf{C}^* - \left[0, \sqrt[3]{4} e^{\frac{2\pi i}{3}}\right] - \left[0, \sqrt[3]{4} e^{\frac{4\pi i}{3}}\right]$
(III) $\frac{4\pi}{3} < \operatorname{Arg} t < 2\pi$		(III) $\mathbf{C}^* - \left[0, \sqrt[3]{4} e^{\frac{4\pi i}{3}}\right] - [0, \sqrt[3]{4}].$

Paste (I), (II), and (III) together crosswise along the branch cuts over the z-plane just like the manner how the sector domains (I), (II), and (III) are related to each other in the t-plane. Observe that $z = z(t)$ is conformal at

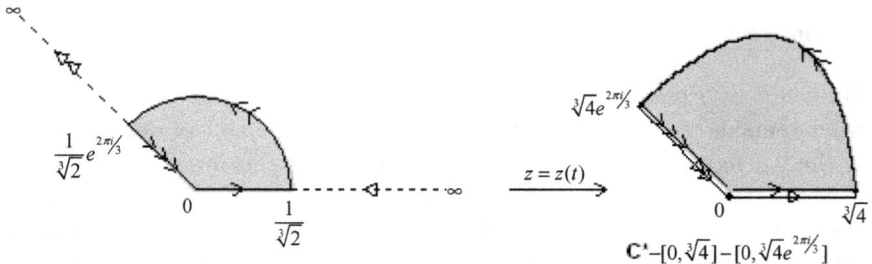

Fig. 3.91

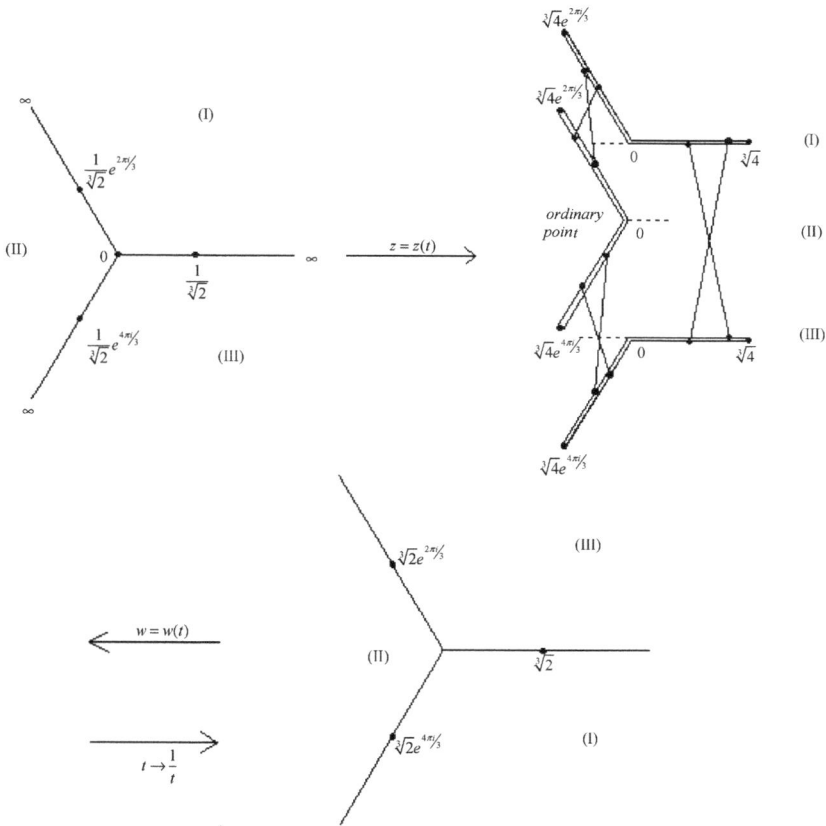

Fig. 3.92

$t = 0$ and is univalent on $|t| < \frac{1}{\sqrt[3]{2}}$. The resulted three-sheets surface R_z is the Riemann surface for $t = t(z)$ or $w = w(t) = w(t(z)) = w(z)$. See Fig. 3.92. Recall that R_z has an algebraic branch point of order 1 over each of $z = 0$, $\sqrt[3]{4}e^{\frac{2k\pi i}{3}}$ for $0 \le k \le 2$. Its topological model is $\mathbf{C}^*$ or the Riemann sphere.

Owing to the fact that $w = tz$ or the change of variable $t \to \frac{1}{t}$, $z = \frac{3t}{1+t^3}$ is changed to $w = \frac{3t^2}{1+t^3}$. Therefore, the Riemann surface R_w for $z = z(w)$ is essentially the same as R_z, even though the sector domains (I), (II), (III) *now* are situated as in the last figure in Fig. 3.92. *Remind* that $z = z(t)$ is univalent in $|t| < \frac{1}{\sqrt[3]{2}}$ while $w = w(t)$ is univalent in $|t| > \sqrt[3]{2}$. Also, *note* that, in R_z, there is an *ordinary point* over the point 0 and, at the same time, there is an algebraic *branch point* of order 1 over 0 (originated from $t = \infty$ in $z = z(t)$ and $t = 0$ in $w = w(t)$).

We turn to (3.5.6.9) *and* (3.5.6.10):

Since $\frac{\partial F}{\partial w}(z, w) = 3(w^2 - z)$, $\frac{\partial^2 F}{\partial w^2}(z, w) = 6w$, and $\frac{\partial^3 F}{\partial w^3}(z, w) = 6$, hence

$$F(z, w) = 0,$$

$$\frac{\partial F}{\partial w}(z, w) = 0.$$

$$\Leftrightarrow (z, w) = (0, 0), (\sqrt[3]{4}, \sqrt[3]{2}), (\sqrt[3]{4}e^{\frac{4\pi i}{3}}, \sqrt[3]{2}e^{\frac{2\pi i}{3}}) \text{ and}$$

$$(\sqrt[3]{4}e^{\frac{2\pi i}{3}}, \sqrt[3]{2}e^{\frac{4\pi i}{3}});$$

$$F(z, w) = 0,$$

$$\frac{\partial F}{\partial w}(z, w) = \frac{\partial^2 F}{\partial w^2}(z, w) = 0 \Leftrightarrow (z, w) = (0, 0).$$

On the other hand, for each fixed w, the discriminant (see (3.5.6.8)) of $F(z, w) = 0$ is $D(z) = 27z^3(4 - z^3)$. The solutions of $D(z) = 0$ are $z = 0$, $\sqrt[3]{4}e^{\frac{2k\pi i}{3}}$ ($k = 0, 1, 2$), which are just the first coordinates of the above four exceptional points.

Fix a point (z_0, w_0).

Let $(z_0, w_0) = (0, 0)$ be the ordinary point in R_z, which lies over $z = 0$: This point is obtained from $z = z(t)$ by setting $t = 0$ and $w = w(t)$ by setting $t = \infty$. Via (3.5.6.4), the inverse $t = t(z)$ of $z = z(t)$ is

$$t = t(z) = 0 + \sum_{n=1}^{\infty} \frac{1}{n!} \left\{ \frac{d^{n-1}}{dt^{n-1}} \left(\frac{t}{\left(\frac{3t}{1+t^3} \right)} \right)^n \right\} \Bigg|_{t=0} z^n = \frac{z}{3} + \frac{z^4}{3^4} + \cdots,$$

$$|z| < \sqrt[3]{4}, \quad |t| < \frac{1}{\sqrt[3]{2}}.$$

Readers can also justify this expansion by long division or by using recursive approximation sequence shown in (2) of (3.4.7.1). Treat this expression as expansion for $\frac{1}{t}$ and substitute into

$$w = \frac{3t^2}{1+t^3} = \frac{3}{t} \cdot \frac{1}{1 + \frac{1}{t^3}} = \frac{3}{t} \left\{ 1 - \frac{1}{t^3} + \frac{1}{t^6} - \cdots \right\}$$

$$= 3 \left(\frac{z}{3} + \frac{z^4}{3^4} + \cdots \right) \left\{ 1 - \left(\frac{z}{3} + \frac{z^4}{3^4} + \cdots \right)^3 + \left(\frac{z}{3} + \frac{z^4}{3^4} + \cdots \right)^6 - \cdots \right\}$$

$$= z + a_5 z^5 + \cdots, \quad |z| < \sqrt[3]{4}, \quad |t| > \sqrt[3]{2}.$$

This is the local expansion of the univalent function $w = w(z)$ in $|z| < \sqrt[3]{4}$.

Let $(z_0, w_0) = (0,0)$ be the branch point of order 1 in R_z: This point is obtained from $z = z(t)$ by setting $t = \infty$ and $w = w(t)$ by setting $t = 0$. Set $\zeta = \frac{1}{t}$ temporarily in $z = z(t)$ to obtain $z = z\left(\frac{1}{\zeta}\right) = \frac{3\zeta^2}{1+\zeta^3}$. Via (3.5.6.6),

$$\zeta = \frac{1}{t} = 0 + \sum_{n=1}^{\infty} \frac{1}{n!} \left\{ \frac{d^{n-1}}{d\zeta^{n-1}} \left(\frac{\zeta - 0}{G(\zeta)} \right)^n \right\}\Bigg|_{\zeta=0} (z - 0)^{\frac{n}{2}},$$

$$|z| < \sqrt[3]{4} \quad \text{and} \quad |\zeta| < \frac{1}{\sqrt[3]{2}},$$

where $G(\zeta)$ is a branch of $\sqrt{z(\zeta) - 0} = \frac{\sqrt{3}\zeta}{\sqrt{1+\zeta^3}}$, say the one determined by $\sqrt{1} = 1$. Now

$$\frac{\zeta}{G(\zeta)} = \frac{\sqrt{1+\zeta^3}}{\sqrt{3}} \Rightarrow \frac{\zeta}{G(\zeta)}\bigg|_{\zeta=0} = \frac{1}{\sqrt{3}};$$

$$\frac{d}{d\zeta} \left(\frac{\sqrt{1+\zeta^3}}{\sqrt{3}} \right)^2 = \zeta^2 \Rightarrow \frac{d}{d\zeta} \left(\frac{\sqrt{1+\zeta^3}}{\sqrt{3}} \right)\bigg|_{\zeta=0} = 0;$$

$$\frac{d^2}{d\zeta^2} \left(\frac{\sqrt{1+\zeta^3}}{\sqrt{3}} \right)^3 = \frac{1}{3\sqrt{3}} \left[9\zeta(1+\zeta^3)^{\frac{1}{2}} + \frac{27}{4}\zeta^4(1+\zeta^3)^{-\frac{1}{2}} \right]$$

$$\Rightarrow \frac{d^2}{d\zeta^2} \left(\frac{\sqrt{1+\zeta^3}}{\sqrt{3}} \right)^3 \Bigg|_{\zeta=0} = 0;$$

$$\frac{d^3}{d\zeta^3} \left(\frac{\sqrt{1+\zeta^3}}{\sqrt{3}} \right)^4 = \frac{1}{9}[12(1+\zeta^3) + 108\zeta^3]$$

$$\Rightarrow \frac{d^3}{d\zeta^3} \left(\frac{\sqrt{1+\zeta^3}}{\sqrt{3}} \right)^4 \Bigg|_{\zeta=0} = \frac{4}{3}; \dots$$

$$\Rightarrow \zeta = \frac{1}{t} = \frac{1}{\sqrt{3}}z^{\frac{1}{2}} + \frac{1}{18}z^2$$

$$+ \sum_{n=5}^{\infty} \frac{1}{n!} \left\{ \frac{d^{n-1}}{d\zeta^{n-1}} \left(\frac{\zeta}{G(\zeta)} \right)^n \right\}\Bigg|_{\zeta=0} z^{\frac{n}{2}},$$

$$|z| < \sqrt[3]{4} \quad \text{and} \quad |\zeta| < \frac{1}{\sqrt[3]{2}}, \text{or}$$

$$t = \sqrt{3}z^{-\frac{1}{2}} - \frac{1}{6}z + \cdots, |z| < \sqrt[3]{4} \quad \text{and} \quad |t| > \sqrt[3]{2}.$$

Substitute the above expression of $\frac{1}{t}$ into

$$
\begin{aligned}
w = \frac{3}{t} \cdot \frac{1}{1 + \frac{1}{t^3}} &= \frac{3}{t} \sum_{n=0}^{\infty} (-1)^n \frac{1}{t^{3n}} = 3 \left\{ \left[\frac{1}{\sqrt{3}} z^{\frac{1}{2}} + \frac{1}{18} z^2 + \cdots \right] \right. \\
&\left. - \left[\frac{1}{\sqrt{3}} z^{\frac{1}{2}} + \frac{1}{18} z^2 + \cdots \right]^4 + \cdots \right\} \\
&= \sqrt{3} z^{\frac{1}{2}} - \frac{1}{6} z^2 + \cdots, \quad |z| < \sqrt[3]{4} \quad \text{and} \quad |t| > \sqrt[3]{2}.
\end{aligned}
$$

In case $(z_0, w_0) = (\sqrt[3]{4}, \sqrt[3]{2})$: Then $F(z_0, w_0) = 0$ and $F(z_0, w) = w^3 - 3\sqrt[3]{4}w + 4 = (w - \sqrt[3]{2})^2 (w + 2\sqrt[3]{2}) = 0$ has $w_0 = \sqrt[3]{2}$ as a root of order 2. Choose R such that $0 < R < \sqrt[3]{2} - (-2\sqrt[3]{2}) = 3\sqrt[3]{2}$. Then $F(z_0, w) \neq 0$ on $0 < |w - \sqrt[3]{2}| < R$ and the corresponding r can be chosen as $0 < r < \sqrt[3]{4}$. For any z, $|z - \sqrt[3]{4}| < r$, the equation $F(z, w) = 0$ implicitly defines $w = w(z)$ as a triple-valued function.

What is the local expression of $w = w(z)$ at z_0? Solve $\frac{3t}{1+t^3} = z_0 = \sqrt[3]{4}$ and we obtain the roots $\frac{\sqrt[3]{4}}{2}$ (of multiplicity 2) and $-\sqrt[3]{4}$. Also, $w\left(\frac{\sqrt[3]{4}}{2}\right) = \sqrt[3]{2} = w_0$. Altogether, setting $t_0 = \frac{\sqrt[3]{4}}{2} = \frac{1}{\sqrt[3]{2}}$, then $z(t_0) = z_0$, $z'(t_0) = 0$, and $z''(t_0) \neq 0$ show that t_0 is a root of multiplicity 2 of $z(t) = z_0$. Now, $z(t) \neq 0$ and $z'(t) \neq 0$ on $0 < |t - t_0| < t_0 = \rho$. Choose δ so that

$$
0 < \delta \leq \inf_{|t - t_0| = t_0} |z(t) - z(t_0)|.
$$

Then the inverse function $t = t(z) : \{|z - z_0| < \delta\} \to \{|t - t_0| < \rho\}$ is double-valued. Let $G(t)$ be a branch of $\sqrt{z(t) - z_0} = \sqrt{\frac{-\sqrt[3]{4}(t + \sqrt[3]{4})}{1 + t^3}}(t - t_0)$, say the one determined by $\sqrt{-(\sqrt[3]{4})^2} = \sqrt[3]{4}i$. Then the Lagrange series expansion of this local $t = t(z)$ is

$$
\begin{aligned}
t &= \frac{\sqrt[3]{4}}{2} + \sum_{n=1}^{\infty} \frac{1}{n!} \left\{ \frac{d^{n-1}}{dt^{n-1}} \left(\frac{t - t_0}{G(t)} \right)^n \right\} \bigg|_{t=t_0} (z - z_0)^{\frac{n}{2}} \\
&= \frac{\sqrt[3]{4}}{2} - \frac{i}{\sqrt[3]{4}} (z - \sqrt[3]{4})^{\frac{1}{2}} - \frac{1}{6} (z - \sqrt[3]{4}) \\
&\quad + \sum_{n=3}^{\infty} a_n (z - \sqrt[3]{4})^{\frac{n}{2}}, \quad |z - z_0| < \delta,
\end{aligned}
$$

$$\Rightarrow w = tz = t(z - \sqrt[3]{4}) + \sqrt[3]{4}t$$

$$= \sqrt[3]{2} - i(z - \sqrt[3]{4})^{\frac{1}{2}} + \frac{\sqrt[3]{4}}{3}(z - \sqrt[3]{4}) - \frac{i}{\sqrt[3]{4}}(z - \sqrt[3]{4})^{\frac{3}{2}}$$

$$- \frac{1}{6}(z - \sqrt[3]{4})^2 + \cdots, \qquad |z - z_0| < \delta. \qquad (\Delta)$$

In case $(z_0, w_0) = \left(\sqrt[3]{4}e^{\frac{4\pi i}{3}}, \sqrt[3]{2}e^{\frac{2\pi i}{3}} \right)$: Substitute $\zeta = e^{\frac{2\pi i}{3}}z$ and $\eta = e^{\frac{4\pi i}{3}}w$ into $w^3 + z^3 - 3zw = \eta^3 + \zeta^3 - 3\zeta\eta$. Then (Δ) holds for η and ζ in place of w and z, respectively. Therefore, the local expression of $w = w(z)$ at z_0 is

$$w = w_0 - i(z - z_0)^{\frac{1}{2}} + \frac{\sqrt[3]{4}}{3}(z - z_0) - \cdots, \qquad |z - z_0| < \delta.$$

In case $(z_0, w_0) = \left(\sqrt[3]{4}e^{\frac{2\pi i}{3}}, \sqrt[3]{2}e^{\frac{4\pi i}{3}} \right)$: Similar expression as above holds, too.

Readers are asked to find the local expression for $w = w(z)$ at $z = \infty$.

Note: Južakov showed that the branch of $w^3 + z^3 - 3zw = 0$ that tangents to $z = 0$ is given by $z = \sum_{n=0}^{\infty} \frac{(3n)!}{3^{3n+1}n!(2n+1)!} w^{3n+2}$.

Exercises A

(1) Expand e^z in powers of w subject to the equation $(z + 1)^2 w = z$ and determine its radius of convergence.

(2) Expand e^{-z} in powers of w, given that $zw = z - a$, and determine its radius of convergence.

(3) To the inverse function $z = f^{-1}(w)$ of each of the following functions $w = f(z)$, do the following problems (imitating Example 2).

 (i) Construct the Riemann surface R_w of $z = f^{-1}(w)$, over the w-plane.

 (ii) Find the local Lagrange series representation of $z = f^{-1}(w)$ on the classical w-plane and on R_w, respectively. Determine the radius of convergence.

 (iii) Endow R_w a complex structure so that R_w is indeed an abstract Riemann surface (see (2.7.2.8) and Sec. 3.3.3).

 (iv) Show that R_w is a covering surface of the w-plane (see (2.7.2.9)). How many branches points? orders?

 (v) Show that both $z = f^{-1}(w)$ and the projection are single-valued analytic functions on R_w.

 *(vi) These Riemann surfaces R_w are *compact* (usually called *closed surfaces* in the literature). In case R_w has n *sheets* and the sum

of the orders of its branched points is α, then R_w is topologically equivalent to a sphere with g *handles* (called *genus*), where

$$\alpha = 2(n + g - 1).$$

For a proof, see Refs. [75], p. 275; [32], p. 18 and III.4.12 on p. 75; or (7.2.22) in Chap. 7. Determine g for each R_w.

(a) $w = \dfrac{z^2 - z + 1}{z^2 + z + 1}$ (see Exercise B (6) of Sec. 2.7.2 and Fig. 2.70).

(b) $w = \dfrac{z^4 - z^2 + 1}{z^4 + z^2 + 1}$ (see Exercise B (7) of Sec. 2.7.2 and Fig. 2.71).

(c) $w = (z^2 - 1)^2$.

(d) $w = \dfrac{1}{2}\left(z^n + \dfrac{1}{z^n}\right)$.

(e) $w = \dfrac{z}{(1 + z^n)^2}$.

(f) $w = z - \dfrac{z^n}{n}$.

(g) $w = \dfrac{1}{z} + \dfrac{z^n}{n}$.

(4) To the inverse function $z = f^{-1}(w)$ of each of the following functions $w = f(z)$, do Problems 1 to 5 in Exercise (3). Note that R_w in these cases are not compact (usually called *open surfaces* in the literature).

(a) $z = \log(1 + w^3)$ (see Exercise B (2)(*l*) for $n = 3$ in Sec. 2.7.3).

(b) $z = \log\left(w - \frac{1}{w}\right)$.

(c) $z = \log\log w$ (see Example 5(4) of Sec. 3.3.1).

(d) $w = z + e^z$.

(e) $w = e^{z^2}$.

(f) $w = e^{\sqrt{z}}$.

(g) $z = \log(e^w - 1)$.

(5) Model after Example 3 and do each of the following equations $F(z, w) = 0$.

(a) $F(z, w) = w^2(1 - z) - z^3$.

(b) $F(z, w) = z^2(1 + z) - w^2(1 - z)$.

(c) $F(z, w) = w^n(1 + z^n)^2 - z^n$.

(d) $F(z, w) = w^2 - z^2 - (z^2 + w^2)^2$.

(e) $F(z, w) = w^2 - 2z(z^2 + w^2) - (z^2 + w^2)^2$.

(f) $F(z, w) = w^2 - z^4 - w^4$.

(g) $F(z, w) = w^3 z^2 - wz + 1$.

(h) $F(z, w) = z^3 w^4 + w^3 - 8z^7$.

(i) (See Ref. [1], p. 306) $F(z, w) = w^3 - 3wz + 2z^3$.

3.5.7.   Examples of (univalently) conformal mappings

A univalent analytic function will always be abbreviated as a *conformal mapping* when we emphasize the geometric mapping properties of the function (see (3.5.1.9)).

Given two domains Ω and D in $\mathbf{C}^*$. We pose the following *geometric mapping problems*:

(1) Does there exist a conformal mapping between Ω and D?

(2) If it does, how can one determine such a mapping explicitly?

(3) Is it unique if it does exist?

These kinds of problems also include the one: How to find the image domain of a domain under an analytic function?

The most fundamental theory in this direction is the *Riemann mapping theorem* (Sec. 6.1): *If both Ω and D are simply connected domains whose boundaries contain at least two distinct points, then there exists a conformal mapping between Ω and D.* Such a mapping is *unique* under some additional assumption, and can be *continuously extended* to a homeomorphism between $\bar{\Omega}$ and $\bar{D}$ if both Ω and D are the Jordan domains. In case $\bar{\Omega}$ is a Jordan domain whose boundary $\partial\Omega$ is a nonself-intersecting closed polygonal curve, then the *Schwarz–Christoffel formula* (Sec. 6.2) provides an explicit conformal mapping which maps Ω onto the upper half-plane or the unit disk. See Example 3 in Sec. 3.5.5 for a preliminary result. We will come back to these topics in Chap. 6.

In this section, we try to apply the well-known elementary functions to realize the aforementioned geometric mapping problems when the domains Ω and D are particularly chosen. To start with, one should be familiar with the geometric mapping properties of the following functions:

(1) The power functions $w = z^\alpha$ (see Secs. 2.5.2, 2.7.2, and 2.7.3.4).

(2) The linear fractional transformation $w = \frac{az+b}{cz+d}$ (see Sec. 2.5.4).

(3) The Joukowski function $w = \frac{1}{2}\left(z + \frac{1}{z}\right)$ (see Sec. 2.5.5).

(4) The exponential function $w = e^z$ and the logarithm $w = \log z$ (see Secs. 2.6.1 and 2.7.3).

(5) The trigonometric functions $w = \sin z$, etc. and the hyperbolic functions $w = \sinh z$, etc. (see Secs. 2.6.2 and 2.7.4).

We had touched this topic a little in Secs. 2.5.2 and 2.5.4. And we can indeed arrange and elaborate more examples in Secs. 2.5, 2.6, and 2.7.

But we do not do this until now because we need more solid theoretical foundations, such as the symmetry principle (Sec. 3.4.6) and criteria for univalence (Sec. 3.5.5), in some particular problems.

In what follows, we list a series of examples. *All mappings are required to be conformal (namely, univalent and analytic) unless otherwise stated.*

Example 1.

(1) Map the complement of the ray $\mathrm{Arg}\left(z - \frac{1}{4}e^{i(\pi-\theta)}\right) = \pi - \theta$ $(0 \leq \theta \leq \pi)$ in $\mathbf{C}^*$ onto the unit disk so that the point ∞ corresponds to the point 1.

(2) (See Ref. [1], p. 97) Map the complement of the circular arc $|z| = 1$, $\mathrm{Im}\, z \geq 0$ onto $|w| > 1$ so that $w(\infty) = \infty$.

Solution. (1) See Fig. 3.93. The composite map

$$w = \frac{\sqrt{e^{i\theta}z + \frac{1}{4}} - \frac{1}{2}}{\sqrt{e^{i\theta}z + \frac{1}{4}} + \frac{1}{2}}, \quad \sqrt{1} = 1, \quad \text{or equivalently } z = \frac{e^{-i\theta}w}{(w-1)^2}$$

is a required one. If we replace w by $e^{i\theta}w$, then $z = \frac{w}{(1 - e^{i\theta}w)^2}$, which is the composite of the mappings

$$w \to w_1 = -e^{i\theta}w \to w_2 = \frac{1}{2}\left(w_1 + \frac{1}{w_1}\right) \to w_3$$

$$= -2e^{i\theta}(w_2 + 1) \to z = \frac{1}{w_3}.$$

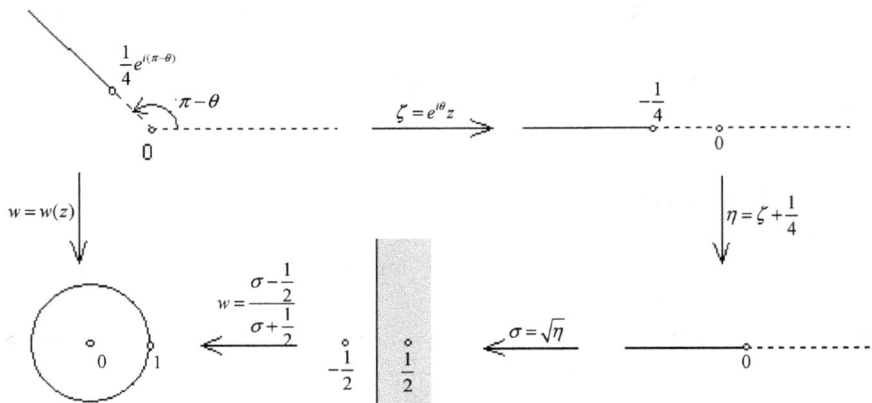

Fig. 3.93

(2) Similar method will lead to the required mapping

$$w = \frac{\sqrt{-i\frac{z-1}{z+1}} + e^{\frac{\pi}{4}i}}{\sqrt{-i\frac{z-1}{z+1}} - e^{-\frac{\pi}{4}i}} \quad \text{or} \quad z = \frac{1+i}{2}\left\{\left(w - \frac{1+i}{2}\right) + \frac{-\frac{i}{2}}{w - \frac{1+i}{2}}\right\}.$$

Remark. The conformal mapping $w = f(z)$ that maps $|z| < 1$ onto the slit domain $\mathbf{C}^* - \left[-\infty, -\frac{1}{4}\right]$ and satisfies $f(0) = 0$ and $f'(0) > 0$ is given by and is called *Koebe's extremal function*

$$f_0(z) = \frac{z}{(1-z)^2} = \sum_{n=1}^{\infty} nz^n, \quad |z| < 1. \tag{3.5.7.1}$$

This function or $e^{-i\theta} f_0(e^{i\theta} z)$, $0 \le \theta \le 2\pi$, solves the *extremal problem*: Let $\mathcal{S} = \{f \mid f(z) \text{ is univalent and analytic on } |z| < 1, \text{ and satisfies } f(0) = 0 \text{ and } f'(0) = 1\}$.

Any such $f(z)$ has a Taylor expansion

$$f(z) = z + a_2 z^2 + a_3 z^3 + \cdots + a_n z^n + \cdots, \quad |z| < 1.$$

Then, we have the following:

(1) Bieberbach conjecture. For any $f \in \mathcal{S}$, $|a_n| \le n$ holds for all $n \ge 1$ with equalities hold if and only if f is given by f_0 in (3.5.7.1).

2. For any $f \in \mathcal{S}$, the image domain $f(|z| < 1)$ contains the disk $|w| < \frac{1}{4}$. This result is sharp, namely, the constant $\frac{1}{4}$ cannot be replaced by any larger number as shown by f_0 in (3.5.7.1).

Bieberbach conjecture (1935) is a longstanding one in function theory until 1985, when completely solved by de Branges [21]: A proof of the Bieberbach conjecture, *Acta Math. 154* (1985) 137–152. Refer to Ref. [28] for a detailed account about univalent functions. For some results in this direction, refer to Exercise B (2)–(6) of Sec. 4.8.

Sketch of Proof of (3.5.7.1)

Since $w = f(z)$ maps $|z| = 1$ onto $\left[-\infty, -\frac{1}{4}\right]$, by symmetry principle (3.4.6.2), $f(z) = \overline{f\left(\frac{1}{\bar{z}}\right)}$ holds for $|z| \ge 1$. We still use $w = f(z)$ to denote the extended function. Now, $w = f(z)$ maps $|z| < 1$ and $1 < |z| \le \infty$ univalently onto $\mathbf{C}^* - \left[-\infty, -\frac{1}{4}\right]$, respectively. Since $f(z) = \overline{f(z)}$ along $|z| = 1$, $f(z)$ is real on $|z| = 1$ and the segment $\left[-\infty, -\frac{1}{4}\right]$ is covered twice by $|z| = 1$ under $w = f(z)$. Moreover, f has two distinct simple poles or a pole of order 2 on $|z| = 1$ (see (3) in (3.5.1.8), (3.5.1.11), and (3.5.1.14)) and is analytic elsewhere. Consequently, f is a rational function of order 2 (see Example 6 in Sec. 4.10.2, if needed).

Suppose ζ is a pole of f on $|z| = 1$. Since $f(\bar{\zeta}) = f(\zeta) = \infty$, $\bar{\zeta}$ is another pole. Observe that $\zeta = \bar{\zeta}$ if and only if $\zeta = \pm 1$.

In case $\zeta \neq \bar{\zeta}$: Then (why?)

$$f(z) = \frac{a}{z - \zeta} + \frac{b}{z - \bar{\zeta}} + c \quad \text{where } a \neq 0, \ b \neq 0, \text{ and } c \text{ is a constant.}$$

$\Rightarrow c = 0$ since $f(\infty) = 0$, and $a\bar{\zeta} + b\zeta = 0$ since $f(0) = 0$. Moreover,

$$f'(0) = -a\bar{\zeta}^2 - b\zeta^2 > 0.$$

$\Rightarrow a + b > 0$.

Then, $f(1) = \frac{a}{1-\zeta} + \frac{b}{1-\bar{\zeta}} = \frac{a+b}{|1-\zeta|^2} \leq -\frac{1}{4}$ implies $a + b < 0$, a contradiction. And $\zeta = \pm 1$ should hold and either 1 or -1 is a pole of order 2 of f.

In case $\zeta = -1$: Since $f(\infty) = 0$,

$$f(z) = \frac{a}{(z+1)^2} + \frac{b}{z+1}, \quad a \neq 0 \text{ (why?)}$$

$$\Rightarrow a + b = 0 \quad \text{since } f(0) = 0.$$

$$\Rightarrow f(z) = \frac{a}{(z+1)^2} - \frac{a}{z+1} = \frac{-az}{(z+1)^2}.$$

$$\Rightarrow f'(z) = \frac{a(z-1)}{(z+1)^3} \quad \text{and thus, } f'(0) = -a > 0 \text{ and then } a < 0.$$

On the other hand, $f(i) = -\frac{a}{2} \leq -\frac{1}{4}$ and thus, $a \geq \frac{1}{2}$, a contradiction.

Therefore, 1 is a pole of order 2 of f. Since $f(\infty) = f(0) = 0$, it follows that

$$f(z) = \frac{a}{(z-1)^2} + \frac{a}{z-1} = \frac{az}{(z-1)^2}, \quad a \neq 0.$$

Since $f'(0) = a$, $a > 0$ holds. On the other hand, $f(-1) = \frac{-a}{4} \leq -\frac{1}{4}$ so that $a \geq 1$. The question is that which point on $|z| = 1$ will be mapped onto $-\frac{1}{4}$. Solve $f(z) = \frac{az}{(z-1)^2} = -\frac{1}{4}$ and the solutions are $(1-2a) \pm 2\sqrt{a(a-1)}$. For these two points to lie on $|z| = 1$, it is necessary that $a = 1$. Hence $f(z) = f_0(z)$ as claimed in (3.5.7.1). □

Example 2.

(1) (See Ref. [1], p. 97) Map the outside of the ellipse $\frac{x^2}{a^2} + \frac{y^2}{b^2} = 1$ onto $|w| < 1$ with the preservation of symmetries (namely, $w(\infty) = 0$ and $w'(\infty) > 0$).

(2) Map the domain, obtained from the upper half-plane $\operatorname{Im} z > 0$ by deleting the set $\frac{x^2}{a^2} + \frac{y^2}{b^2} \leq 1$, $y \geq 0$, onto the upper half-plane $\operatorname{Im} z > 0$.

Solution. (1) One may suppose that $a > b$, otherwise the mapping $w = \frac{a}{z}$ will work if $a = b$. The ellipse $\frac{x^2}{a^2} + \frac{y^2}{b^2} = 1$ has its foci at $\pm\sqrt{a^2 - b^2}$. The mapping $\zeta = \frac{1}{\sqrt{a^2-b^2}} z$ transforms the ellipse into ellipse $\frac{(\operatorname{Re}\zeta)^2}{\frac{a^2}{(a^2-b^2)}} + \frac{(\operatorname{Im}\zeta)^2}{\frac{b^2}{(a^2-b^2)}} = 1$ with foci at ± 1. See Fig. 3.94. The composite map is

$$w = \sqrt{\frac{a-b}{a+b}} \cdot \frac{1}{\frac{z}{\sqrt{a^2-b^2}}} + \sqrt{\frac{z^2}{(a^2-b^2)} - 1}$$

$$= \frac{a-b}{z + \sqrt{z^2 - (a^2 - b^2)}} \quad (\sqrt{1} = 1).$$

Let $g(z) = w(\frac{1}{z})$. Then $g'(0) = w'(\infty) = a - b$ and thus, $\operatorname{Arg} w'(\infty) = 0$. So $w = w(z)$ is the required one.

(2) In the last step shown in Fig. 3.94, replace $w = \sqrt{\frac{a-b}{a+b}} \cdot 1\eta$ by $\sigma = \sqrt{\frac{a-b}{a+b}}\eta$. Then,

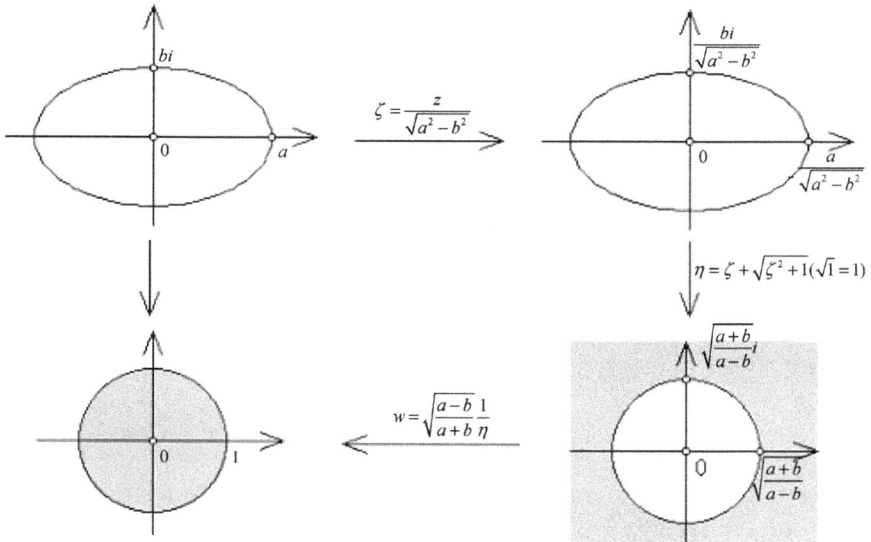

Fig. 3.94

$$\sigma = \sqrt{\frac{a-b}{a+b}}\left(\frac{z}{\sqrt{a^2-b^2}} + \sqrt{\frac{z^2}{a^2-b^2}-1}\right)$$

$$= \frac{1}{a+b}(z + \sqrt{z^2-(a^2-b^2)}) \quad (\sqrt{1}=1)$$

maps the given domain onto $\{\operatorname{Im}\sigma > 0\} - \{|\sigma| \le 1 \text{ and } \operatorname{Im}\sigma > 0\}$, which, in turn, is mapped onto $\operatorname{Im} w > 0$ by $w = \frac{1}{2}\left(\sigma + \frac{1}{\sigma}\right)$ (see (4)(b) in (2.5.5.4)). Hence, the composite map

$$w = \frac{1}{2}\left(\sigma(z) + \frac{1}{\sigma(z)}\right) = \frac{az - b\sqrt{z^2-(a^2-b^2)}}{a^2-b^2} \quad (\sqrt{1}=1)$$

is the required one.

Example 3. Map the interior (the part containing 1) and the exterior of the right-hand branch of the hyperbola $\frac{x^2}{\cos^2\alpha} - \frac{y^2}{\sin^2\alpha} = 1$ onto the upper half-plane. In case $\alpha = 45°$, see Ref. [1], p. 97.

Solution. Before we start, note that the Joukowski function $z = \frac{1}{2}\left(\zeta + \frac{1}{\zeta}\right)$ *alone* cannot solve this problem completely. The reasons why are as follows: Even though the sector $-\alpha < \operatorname{Arg}\zeta < \alpha$ is mapped onto the interior of the right-hand branch of the hyperbola, yet it maps the segments $[0,1]$ and $[1,\infty)$ both onto the segment $[1,\infty)$ in the z-plane. Hence, the inverse function $\zeta = z + \sqrt{z^2-1}$ does *not* map the interior univalently onto $-\alpha < \operatorname{Arg}\zeta < \alpha$ (see Fig. 2.28(b)). Similarly, it maps $|\zeta| > 1$, $-\alpha < \operatorname{Arg}\zeta < \alpha$, univalently onto the interior of the right branch with the segment $[\cos\alpha, 1]$ deleted and thus, maps $|\zeta| \ge 1$, $-\alpha < \operatorname{Arg}\zeta < \alpha$, onto the interior but it is not one-to-one on the circular arc $|\zeta| = 1$, $-\alpha < \operatorname{Arg}\zeta < \alpha$.

So, just look at the part of the interior that lies in the first quadrant. See Fig. 3.95. The composite map

$$\delta = \delta(z) = \left\{\frac{1}{2}\left[(z + \sqrt{z^2-1})^{\frac{\pi}{\alpha}} + (z + \sqrt{z^2-1})^{-\frac{\pi}{\alpha}}\right] + 1\right\}^{\frac{1}{2}}$$

$$= \frac{1}{\sqrt{2}}\left[(z + \sqrt{z^2-1})^{\frac{\pi}{2\alpha}} + (z + \sqrt{z^2-1})^{-\frac{\pi}{2\alpha}}\right] \quad (\sqrt{1}=1)$$

maps that part univalently onto the first quadrant in the δ-plane: the quarter branch of the hyperbola in the first quadrant is mapped univalently onto the positive imaginary axis in the δ-plane, the segment $[\cos\alpha, 1]$ univalently onto $[0, \sqrt{2}]$ and $[1, \infty)$ univalently onto $[\sqrt{2}, \infty)$. By symmetry principle

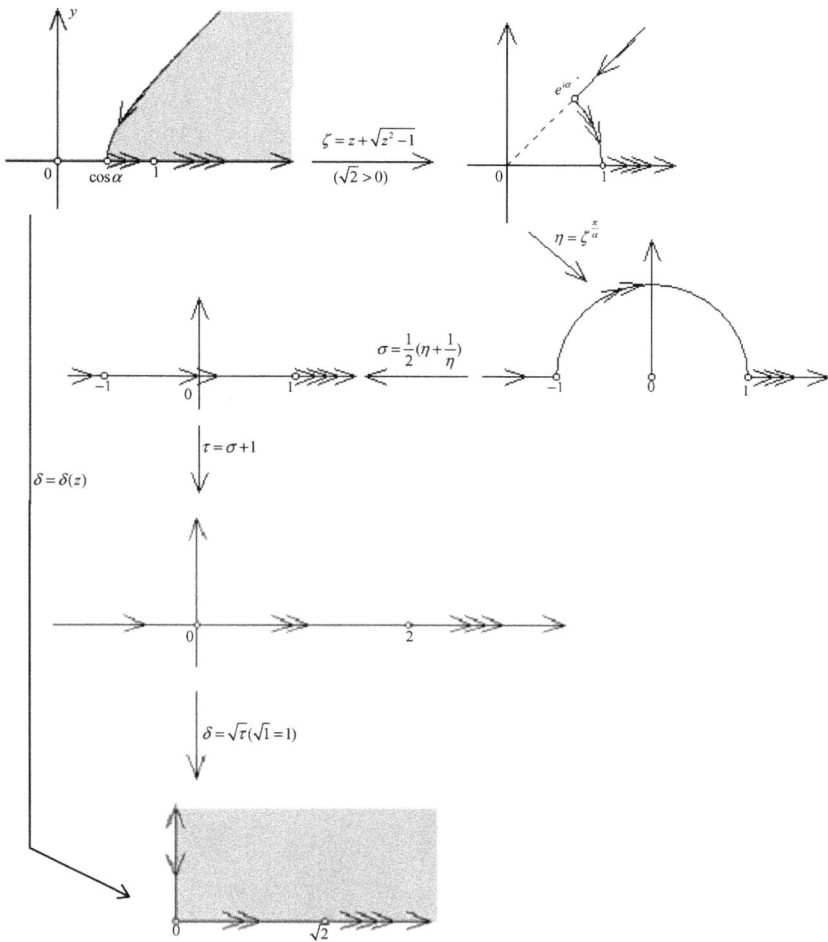

Fig. 3.95

(3.4.6.2), this $\delta = \delta(z)$ maps the whole interior of the right branch univalently onto the right half-plane $\operatorname{Re}\delta > 0$; then, a rotation of $90°$ will, in turn, map the latter onto the upper half-plane. Hence,

$$w = i\delta(z) = \frac{i}{\sqrt{2}}\left[(z + \sqrt{z^2 - 1})^{\frac{\pi}{2\alpha}} + (z + \sqrt{z^2 - 1})^{-\frac{\pi}{2\alpha}}\right], \quad \sqrt{1} = 1,$$

will work. In particular, if $\alpha = 45° = \frac{\pi}{4}$, then $w = \frac{i}{\sqrt{2}}[(z + \sqrt{z^2 - 1})^2 + (z + \sqrt{z^2 - 1})^{-2}]$ will map the interior of the right-hand branch onto the upper half-plane.

Similar process will lead to the mapping

$$w = [e^{-i\alpha}(z + \sqrt{z^2 - 1})]^{\frac{\pi}{2\beta}} - [e^{-i\alpha}(z + \sqrt{z^2 - 1})]^{-\frac{\pi}{2\beta}},$$

$$\sqrt{1} = 1 \quad \text{and} \quad \beta = \pi - \alpha$$

that maps the exterior of the right-hand branch onto the upper half-plane.

Note (another solution for the case $\alpha = 45°$): Map the exterior of the right-hand branch of $x^2 - y^2 = 1$ onto a half-plane (Lindelöf). How about the interior onto $|w| < 1$ so that the focus and the vertex are mapped into $w = 0$ and -1, respectively?

Solution. Let $w = w(z)$ be the mapping that maps the upper half of the indicated domain onto the fourth quadrant. By symmetry principle (3.4.6.2), this $w = w(z)$ will map the exterior onto the right half-plane. See Fig. 3.96. The composite map $w = w(z)$ defined by $z^2 = 1 + \frac{1}{2}(w^3 - 3w)$ is a required one.

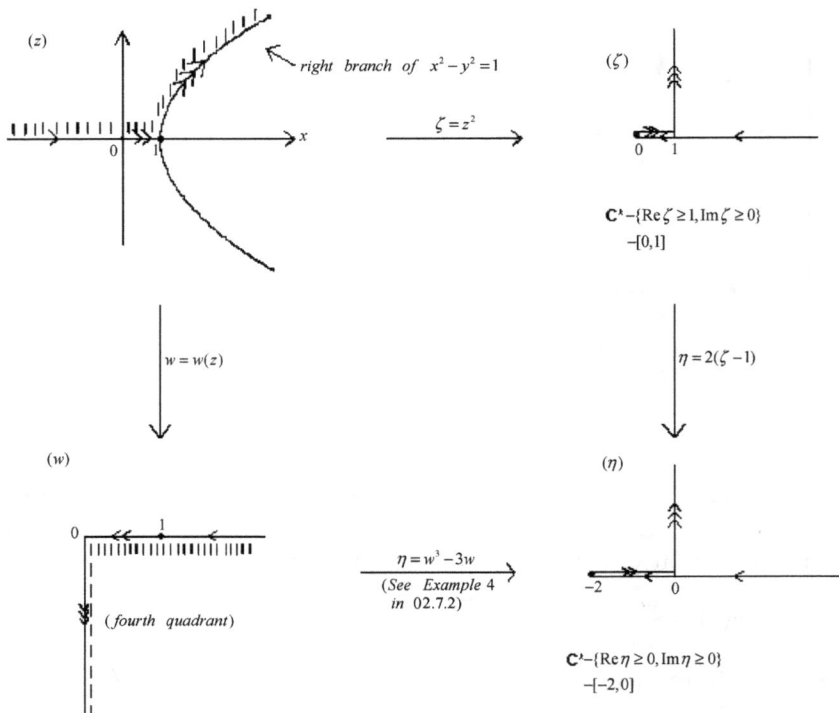

Fig. 3.96

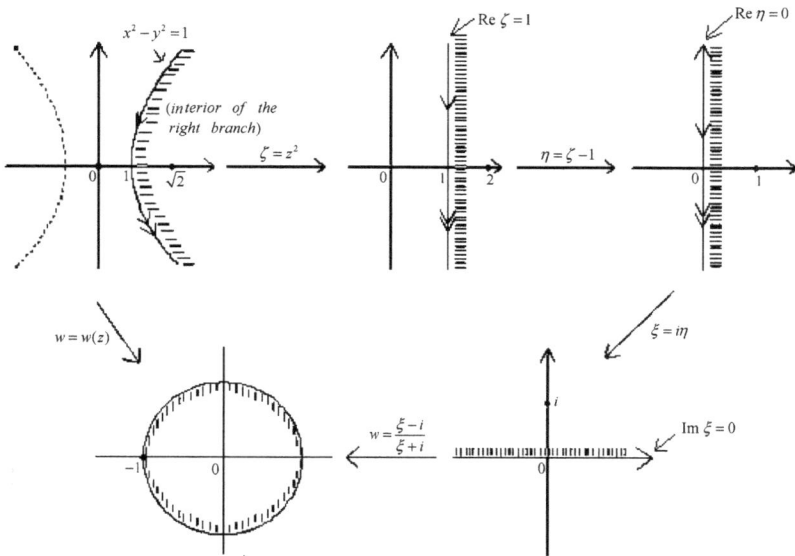

Fig. 3.97

Observe that $\zeta = z^2$ maps the interior of the right-hand branch onto $\operatorname{Re} \zeta > 1$. See Fig. 3.97. The composite map $w = \frac{i(z^2-1)-i}{i(z^2-1)+i} = 1 - \frac{2}{z^2}$ is a required one. Refer to Exercise A (2) of Sec. 2.5.2.

Example 4. (See Ref. [1], p. 97) Map the inside of the lemniscate $|z^2 - a^2| = \rho^2$ $(\rho > a)$ onto the disk $|w| < 1$ so that the symmetries are preserved (Lindelöf).

Solution. Remind that we learned the lemniscate in Example 4 of Sec. 1.4.3. See Fig. 3.98. $\zeta = z^2$ map the domain $|z^2 - a^2| < \rho^2$ univalently onto a two-sheets disk $|\zeta - a^2| < \rho^2$, pasted together crosswise along the segment $[0, a^2 + \rho^2]$, over the ζ-plane. Try to map $|\zeta - a^2| < \rho^2$ onto the two-sheets unit disk $|\eta| < 1$ univalently and preserve the symmetries: $0 \to 0$, the symmetric point $a^2 - \frac{\rho^4}{a^2} \to \infty$. Hence $\eta = k \cdot \frac{\zeta - 0}{\zeta - (\frac{a^2 - \rho^4}{a^2})} = \frac{ka^2 \zeta}{\rho^4 + a^2(\zeta - a^2)}$ will work and for $\eta(a^2 + \rho^2) = 1$, it follows that $ka^2 = \rho^2$. The final map is $w = \sqrt{\eta}$, which maps the two-sheets unit disk univalently onto the disk $|w| < 1$. The composite map

$$w = \frac{\rho z}{\sqrt{\rho^4 + a^2(z^2 - 1)}}, \quad \sqrt{1} = 1,$$

is the required one.

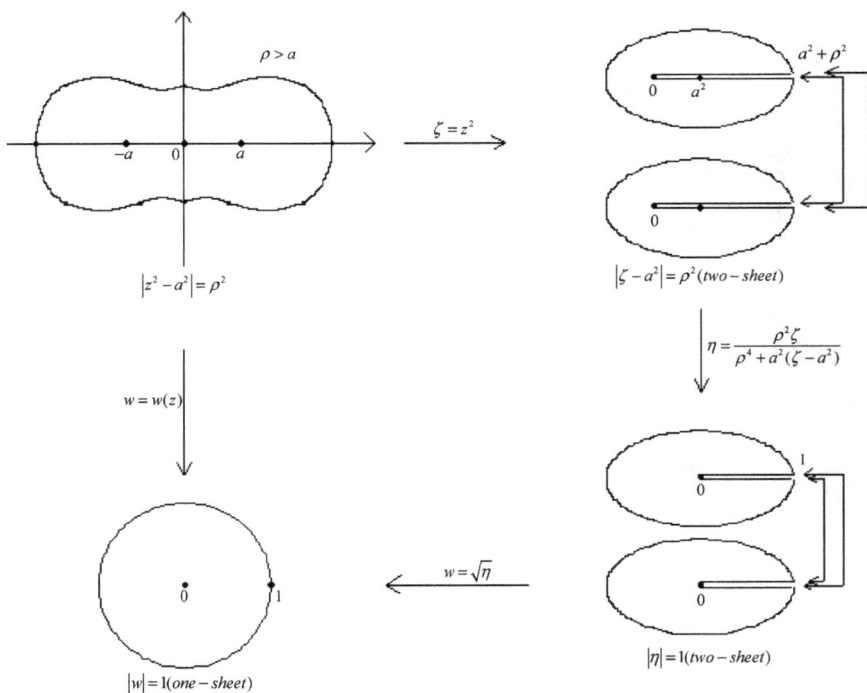

Fig. 3.98

Example 5. Map the open disk $|z| < 1$, with the segment $[a, 1)$, $0 \le a < 1$, deleted univalently onto the upper half-plane.

Solution. See Fig. 3.99. A required one is the following composite map

$$w = \left(\frac{1 + \sqrt{\frac{z-a}{1-az}}}{1 - \sqrt{\frac{z-a}{1-az}}} \right)^2 , \quad 0 \le a < 1 \text{ and } \sqrt{-1} = i. \qquad (*_1)$$

Or, see Fig. 3.100. Another one is the composite map

$$w = i \sqrt{\left(\frac{1-z}{1+z} \right)^2 - \left(\frac{1-a}{1+a} \right)^2} , \quad 0 \le a < 1 \text{ and } \sqrt{1} = 1. \qquad (*_2)$$

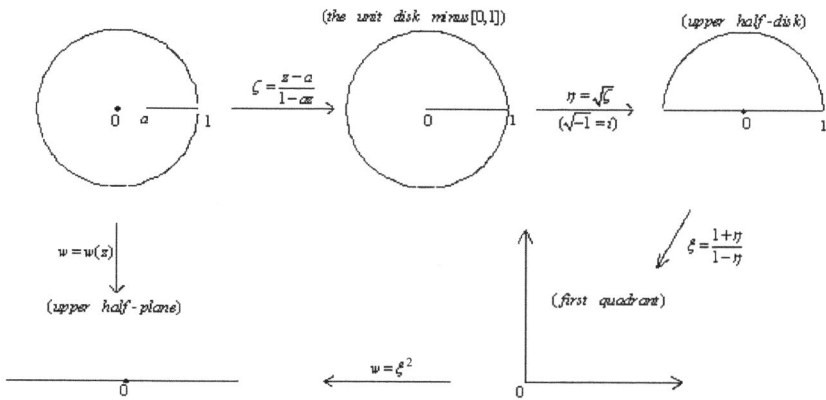

Fig. 3.99

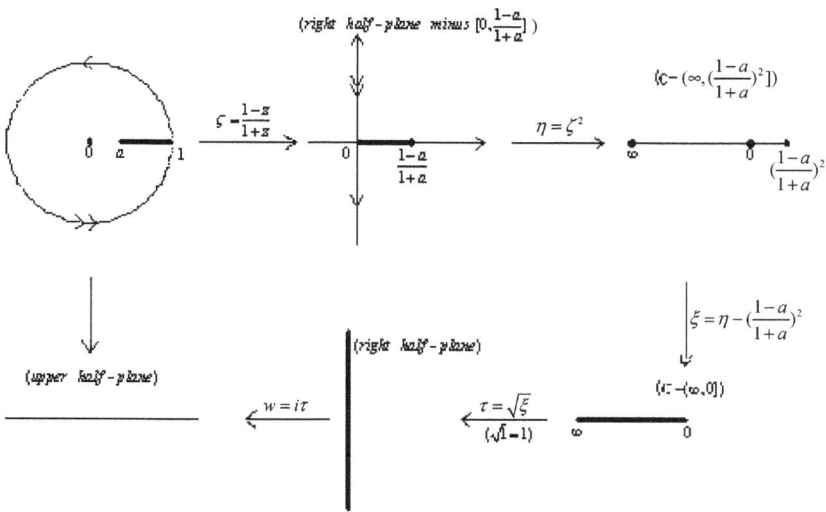

Fig. 3.100

Note that the linear fractional transformation mapping the points $1, \infty, -1$ onto $0, \frac{1-a}{1+a}, \infty$, respectively, is given by

$$w_2 = \frac{1-a}{1+a} \cdot \frac{w_1 - 1}{w_1 + 1},$$

which maps the upper half-plane onto the upper half-plane. Treat w in $(*_1)$ as w_1 and substitute it into the above equation, and the resulted w_2 is the one given by $(*_2)$.

Example 6.

(1) Map the sector domain $|z| < 1$, $0 < \operatorname{Arg} z < \frac{\pi}{3}$, onto the upper half-plane and the unit disk, respectively.

(2) Map the curvilinear triangle with vertices at $0, \frac{1}{2}, \omega = \frac{1+\sqrt{3}i}{2}$ of respective angle $\frac{\pi}{2}, \frac{\pi}{2}, \frac{\pi}{3}$ onto the upper half-plane (see Fig. 3.102)

Solution.

(1) See Fig. 3.101. The composite maps

$$\zeta = \left(\frac{z^3+1}{z^3-1}\right)^2 \quad \text{and} \quad w = \frac{(1+z^3)^2 - i(1-z^3)^2}{(1+z^3)^2 + i(1-z^3)^2}$$

are the required ones, respectively.

(2) Observe that $\omega^2 = -\bar{\omega} = \omega - 1$.

$$\eta = \frac{0+\omega^2}{0-\omega} \cdot \frac{z-\omega}{z+\omega^2} = -\omega \frac{z-\omega}{z+\omega^2}$$

maps the given curvilinear triangle (see Fig. 3.102) onto the sector domain given in (1) (see the first figure in Fig. 3.101). Hence, the composite map

$$w = \left(\frac{\eta^3+1}{\eta^3-1}\right)^2 \quad \text{where} \quad \eta = -\omega \frac{z-\omega}{z+\omega^2}$$

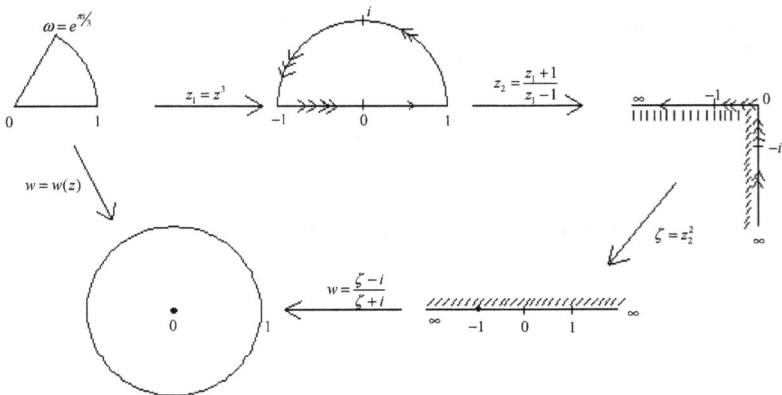

Fig. 3.101

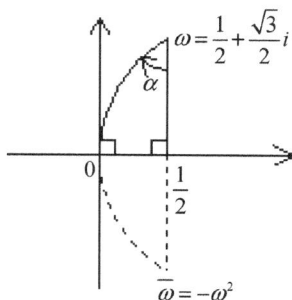

Fig. 3.102

is the required one. Note, after simplification,

$$w = 1 - \frac{4}{27} \cdot \frac{(z^2 - z + 1)^3}{(z^2 - z)^2},$$

which maps $0, \frac{1}{2}, \omega$ onto $\infty, 0, 1$, respectively.

Note: The Riemann surfaces for $z = z^{-1}(\zeta)$ in (1) and for $z = z^{-1}(w)$ in (2).

$\zeta = \zeta(z)$ in (1) maps $0, 1, \omega$ onto $1, \infty, 0$, respectively; maps the boundary segment $[0, 1]$, circular arc $\widehat{1\omega}$, and segment $[\omega, 0]$ onto $[1, \infty]$, $[\infty, 0]$, and $[0, 1]$, respectively. By the use of the symmetry principle (3.4.6.2), $\zeta = \zeta(z)$ can be analytically extended to $\mathbf{C}^*$ except poles. The extended $\zeta = \zeta(z)$ maps the sector domains $\Omega_k : \frac{(k-1)\pi}{3} < \operatorname{Arg} z < \frac{k\pi}{3}$ $(1 \le k \le 6)$ onto the slit domain $\mathbf{C}^* - [0, 1] - [1, \infty]$ in the ζ-plane. The Riemann surface for the inverse function $z = z^{-1}(\zeta)$ is shown in right figure in Fig. 3.103 and its line complex in Fig. 3.104. There arc two algebraic branch points of order 2 over $\zeta = 1$, three algebraic branch points of order 1 over $\zeta = 0$, and three algebraic branch points of order 1 over $\zeta = \infty$.

$w = w(z)$ in (2) maps the boundary segment $\left[\frac{1}{2}, \omega\right]$, circular arc $\widehat{\omega 0}$, and segment $\left[0, \frac{1}{2}\right]$ onto $[0, 1], [1, \infty]$, and $[\infty, 0]$ respectively. Extend $w = w(z)$, via the symmetry principle, analytically to $\mathbf{C}^*$ except poles. The extended $w = w(z)$ maps domains marked by ① and ② in Fig. 3.105, respectively, onto the upper and the lower half-planes. Take six copies of the plane $\mathbf{C}^*$ with branch cuts along $[\infty, 0]$, $[0, 1]$, and $[1, \infty]$ and paste them crosswise according to how ① and ② are situated in the z-plane. The line complex for the Riemann surface is shown in Fig. 3.106, where there are two algebraic branch points of order 2 over $w = 1$, and three algebraic branch points of order 1 over each of $w = 0$ and ∞.

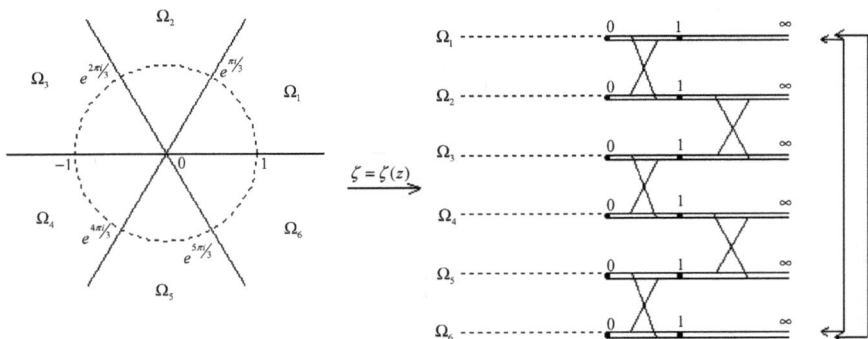

Fig. 3.103

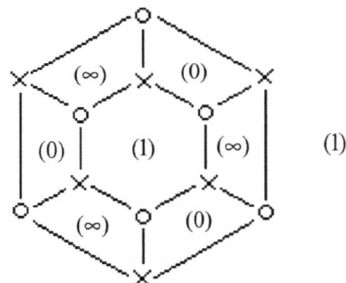

Fig. 3.104

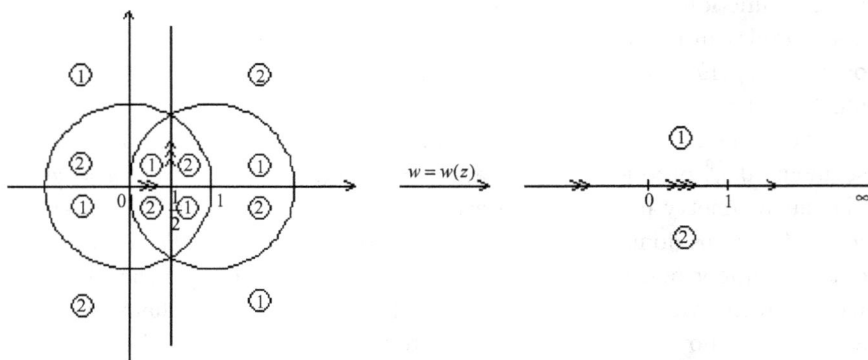

Fig. 3.105

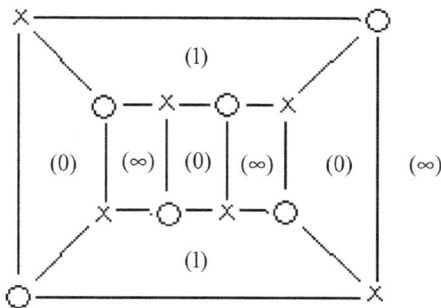

Fig. 3.106

Example 7. Map the half-plane $\operatorname{Im} z < 1$, with the unit disk $|z| \leq 1$ deleted, onto the unit disk $|w| < 1$ with the requirement that $w(-3i) = \frac{-1+i}{2}$ and $\operatorname{Arg} w'(-3i) = \frac{\pi}{2}$.

Solution. The circle $|z| = 1$ is tangent to the line $\operatorname{Im} z = 1$ at $z = i$. The bilinear transformation $\zeta = \frac{z+i}{z-i}$ maps the indicated domain onto the vertical strip and $\zeta(-3i) = \frac{1}{2}$. See Fig. 3.107, where, in the last transformation, a constant k appears in order to adjust to the condition that $\operatorname{Arg} w'(-3i) = \frac{\pi}{2}$. Now, compute as follows:

$$\zeta'(z) = \frac{-2i}{(z-i)^2} \Rightarrow \zeta'(-3i) = \frac{\pi i}{8};$$

$$\eta'(\zeta) = \pi i \Rightarrow \eta'\left(\frac{1}{2}\right) = \pi i;$$

$$\tau'(\eta) = e^\eta \Rightarrow \tau'\left(\frac{\pi i}{2}\right) = i;$$

$$\frac{w'_\tau(\tau)\left[\frac{-1+i}{2} - (-1+i)\right]}{[w - (-1+i)]^2} = k \cdot \frac{2i}{(\tau+i)^2} \Rightarrow w'_\tau(i) = \frac{k}{2i} \cdot \frac{1-i}{2},$$

where w'_τ denotes the differentiation of $w = w(\tau)$ with respect to τ. By chain rule,

$$w'(-3i) = w'_\tau(i) \cdot \tau'\left(\frac{\pi i}{2}\right) \cdot \eta'\left(\frac{1}{2}\right) \cdot \zeta'(-3i) = \frac{-k\pi}{32}(1-i),$$

$$\Rightarrow \operatorname{Arg} w'(-3i) = \pi + \operatorname{Arg} k + \operatorname{Arg}(1-i) = \frac{\pi}{2},$$

$$\Rightarrow \operatorname{Arg} k = -\pi + \frac{\pi}{2} + \frac{\pi}{4} = -\frac{\pi}{4}.$$

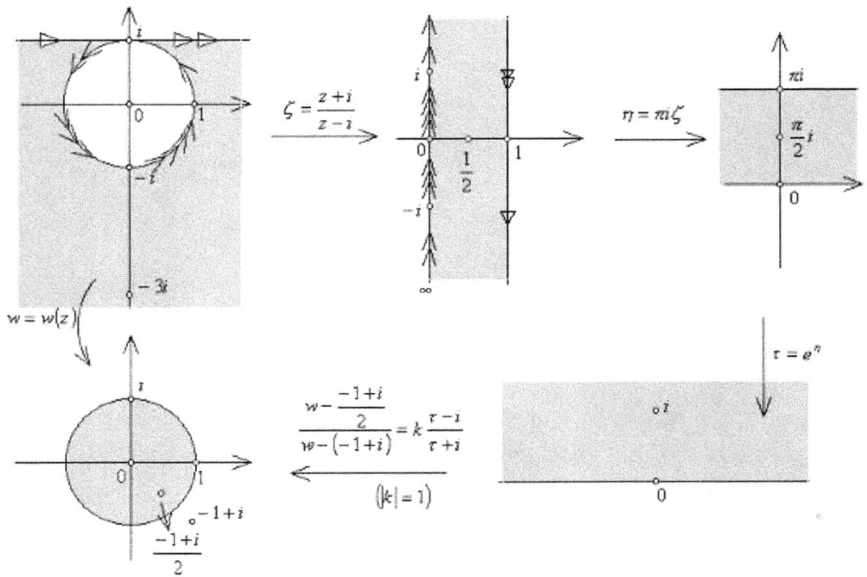

Fig. 3.107

Since $|k| = 1$, $k = \frac{1-i}{\sqrt{2}}$. Then, the required map is

$$w = \frac{e^{\frac{\pi i(z+i)}{(z-i)} + 2 - i}}{e^{\frac{\pi i(z+i)}{(z-i)}} + 2 + i}.$$

Example 8. Map the domain $\{|z| < 1\} - [0,1)$ or $\{|z| > 1\} - (1,\infty]$ onto the domains $\{0 < \operatorname{Im} w < 2\pi\} - \{w = x + \pi i \,|\, -\infty \le x \le 0\}$ and $\{0 < \operatorname{Im} w < \pi\}$, respectively.

Solution. Note that $z \to \frac{1}{z}$ interchanges $\{|z| < 1\} - [0,1)$ and $\{|z| > 1\} - (1,\infty]$ univalently and conformally. So we consider only the former domain.

Figure 3.108 shows that $w = \log \frac{1}{2}(z + \frac{1}{z})$ maps the domain onto the slit horizontal strip, while Fig. 3.109 shows that $w = \frac{1}{2}(\sqrt{z} + \frac{1}{\sqrt{z}})$ maps the domain onto the horizontal strip.

Note: The Riemann surfaces for $w = \log \frac{1}{2}(\zeta + \frac{1}{\zeta})$ and $w = \log \frac{1}{2}(\sqrt{z} + \frac{1}{\sqrt{z}})$.

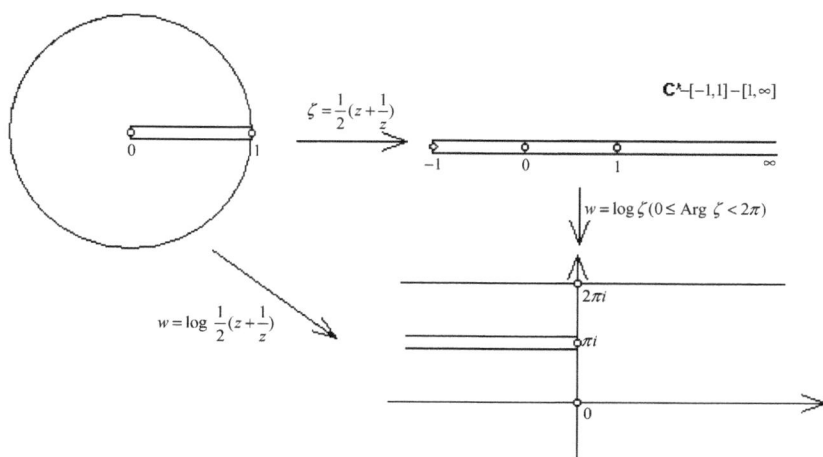

Fig. 3.108

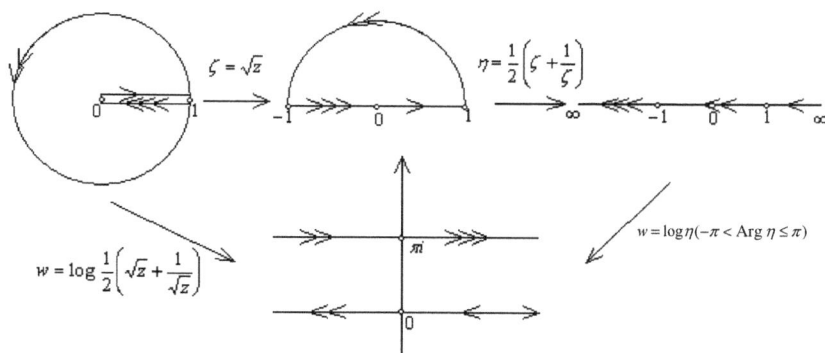

Fig. 3.109

In Fig. 3.110, R_ζ is the Riemann surface for $w = \log \frac{1}{2}(\zeta + \frac{1}{\zeta})$ while R_z is the one for $w = \log \frac{1}{2}(\sqrt{z} + \frac{1}{\sqrt{z}})$. Readers are urged to determine the branch points and their orders, and branch cuts on both surfaces. Refer to Exercise B $(2)(q)$ of Sec. 2.7.3.

Example 9. Map the upper half-strip $0 < \operatorname{Re} z < \pi, \operatorname{Im} z > 0$, with the ray $\operatorname{Re} z = \frac{\pi}{2}, h \leq \operatorname{Im} z < \infty (h > 0)$ deleted, onto the upper half-plane.

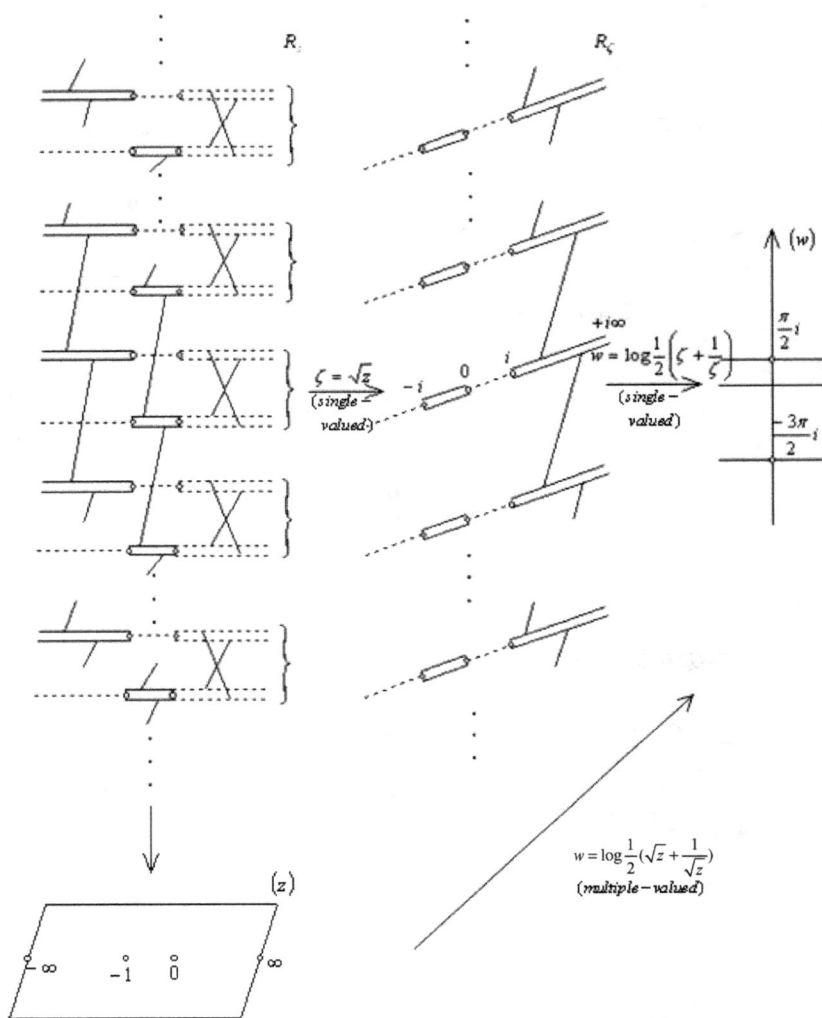

Fig. 3.110

Solution. Recall that $w = \cos z$ maps the half-strip $0 < \operatorname{Re} z < \pi$ univalently onto the slit domain $\mathbf{C} - (\infty, -1] - [1, \infty)$, the upper half of it onto the lower half-plane (see (2)(a) in (2.6.2.6)). See Fig. 3.111. The required map is the composite

$$w = \sqrt{\frac{\cos 2z + \cosh 2h}{\cos 2z + 1}}.$$

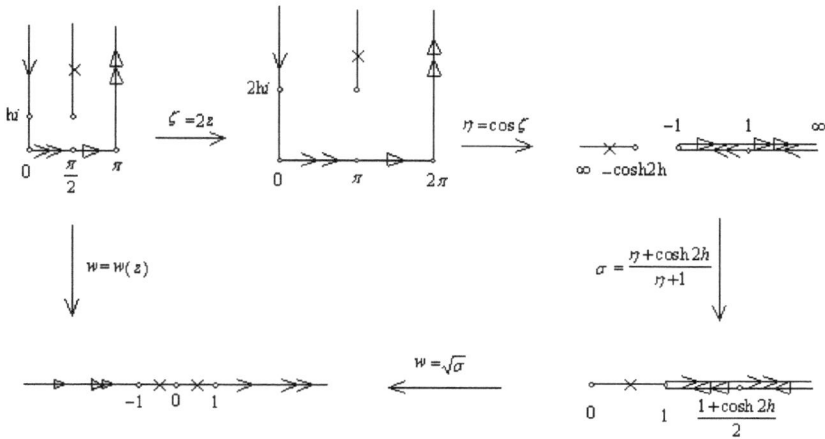

Fig. 3.111

Example 10.

(1) Map the domain, obtained by deleting the segments $\mathrm{Re}\,z = \left(n+\frac{1}{2}\right)\pi, 0 \le \mathrm{Im}\,z \le a(n = 0, \pm 1, \pm 2, \ldots)$ from the upper half-plane (see Fig. 3.112), onto the upper half-plane.

(2) Map the domain, obtained by deleting the horizontal segments $-a \le \mathrm{Re}\,z \le a, \mathrm{Im}\,z = \frac{\pi}{2} + n\pi(n = 0, \pm 1, \pm 2, \ldots)$ from the plane, onto the domain $\mathbf{C}^* - [\mathbf{R} - \cup_{n=-\infty}^{\infty}[n\pi - b, n\pi + b]]$, where $0 < b < \frac{\pi}{2}$, and $[n\pi - b, n\pi + b]$ are intervals on the real axis.

Solution. (1) Recall that $\zeta = \sin z$ maps the half-strip $-\frac{\pi}{2} < \mathrm{Re}\,z < \frac{\pi}{2}$, $\mathrm{Im}\,z > 0$ onto the upper half-plane with its boundaries $\mathrm{Re}\,z = -\frac{\pi}{2}$ and $\mathrm{Im}\,z \ge 0, -\frac{\pi}{2} \le \mathrm{Re}\,z \le \frac{\pi}{2}$ and $\mathrm{Im}\,z = 0, \mathrm{Re}\,z = \frac{\pi}{2}$ and $\mathrm{Im}\,z \ge 0$ onto the segments $(-\infty, -1], [-1, 1]$ and $[1, \infty)$ on the real axis, respectively (see (2.6.2.6)). Observe that, at the same time, it maps the half-rays $\mathrm{Re}\,z = -\frac{\pi}{2}$, $a \le \mathrm{Im}\,z < \infty$ and $\mathrm{Re}\,z = \frac{\pi}{2}$, $a \le \mathrm{Im}\,z < \infty$ onto the segments $(-\infty, -\cosh a]$ and $[\cosh a, +\infty)$, respectively. The composite map $w = \sin^{-1}\frac{\sin z}{\cosh a}$ in Fig. 3.112 then maps the upper half-strip $-\frac{\pi}{2} < \mathrm{Re}\,z < \frac{\pi}{2}, \mathrm{Im}\,z > 0$ onto itself in the Ed:variable-plane, wherein $w = \sin^{-1}\eta$ is the negative principle branch of $z = \sin w$ (see (2.7.4.5) with $n = 0$ and $\sqrt{-1} = -i$).

Apply the symmetry principle (3.4.6.2) to $w = \sin^{-1}\frac{\sin z}{\cosh a}$ on $-\frac{\pi}{2} < \mathrm{Re}\,z < \frac{\pi}{2}$ with respect to right half-ray $\mathrm{Re}\,z = \frac{\pi}{2}$, $a \le \mathrm{Im}\,z < \infty$. Consequently, $w = \sin^{-1}\frac{\sin z}{\cosh a}$ is analytically extended to $\frac{\pi}{2} < \mathrm{Re}\,z < \frac{3\pi}{2}, \mathrm{Im}\,z >$

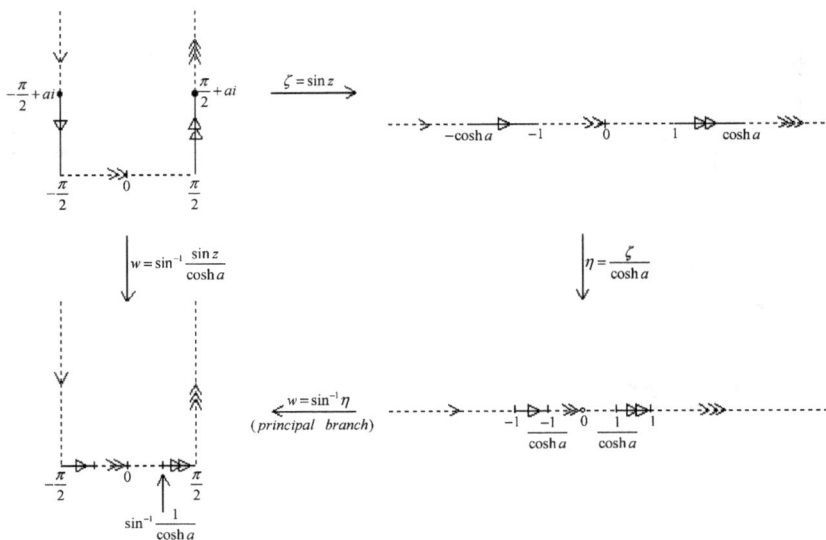

Fig. 3.112

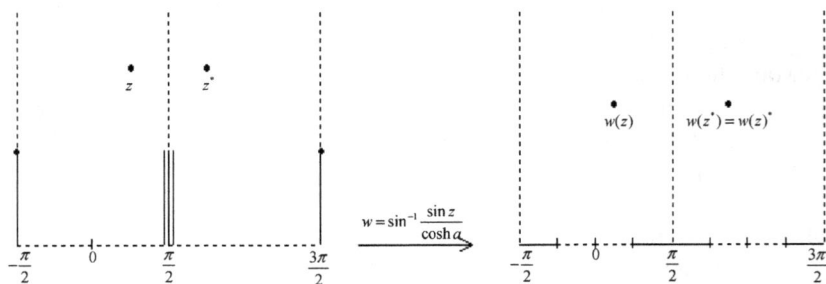

Fig. 3.113

0 with its corresponding range extended to $\frac{\pi}{2} < \operatorname{Re} w < \frac{3\pi}{2}, \operatorname{Im} w > 0$, yet it is not defined on the segment $\operatorname{Re} z = \frac{\pi}{2}, 0 \le \operatorname{Im} z \le a$. See Fig. 3.113. Repeat this process successively to the whole upper half-plane and the result $w = \sin^{-1} \frac{\sin z}{\cosh a}$ is thus required. See Fig. 3.114.

(2) Based on both the method and the result of (1), see Fig. 3.115. The required map is the composite

$$ w = \frac{b \sin^{-1}\left(\frac{\sin iz}{\cosh a}\right)}{\sin^{-1}\left(\frac{1}{\cosh a}\right)}. $$

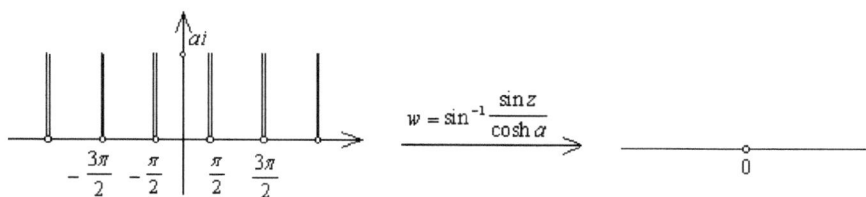

$$w = \sin^{-1}\frac{\sin z}{\cosh a}$$

Fig. 3.114

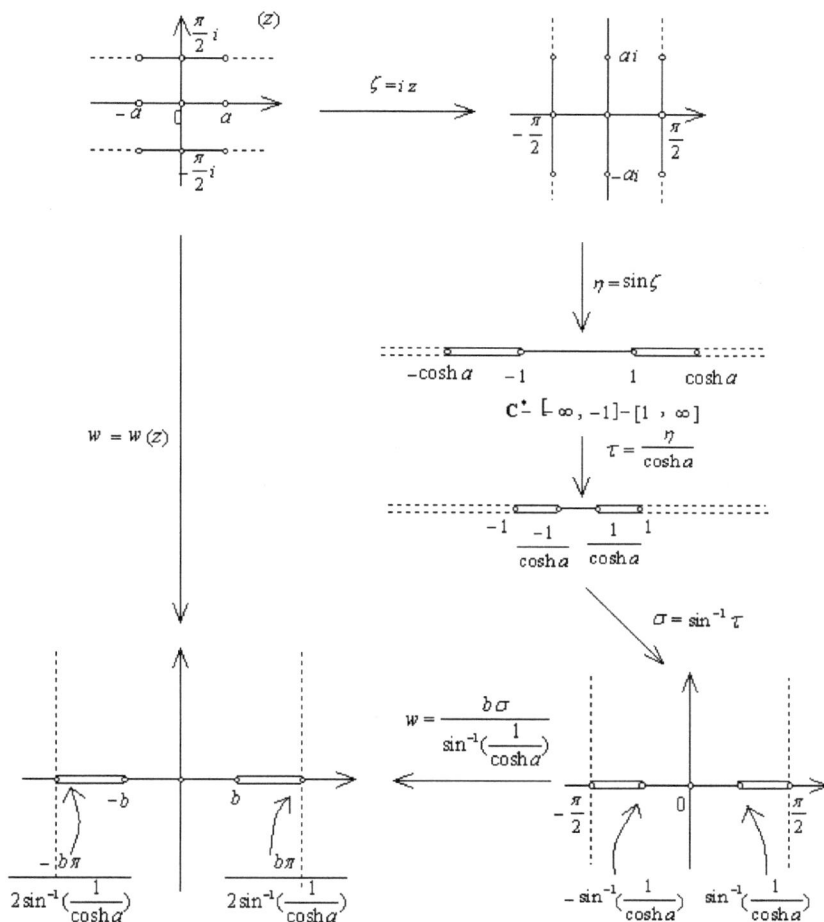

Fig. 3.115

Example 11.

(1) Map the domain, obtained by deleting the rays $\left(-\infty, -\frac{\pi}{2}\right]$ and $\left[\frac{\pi}{2}, \infty\right)$ along the real axis and the vertical segments $\mathrm{Re}\, z = \frac{\pi}{2} - n\pi$ $(n = 0, \pm 1, \pm 2, \ldots), |\mathrm{Im}\, z| \leq a$ from the plane, onto the outside of the unit disk (see the left figure in Fig. 3.117).

(2) Map the domain

$$\mathbf{C}^* - (-\infty, p] - [q, +\infty) - \cup_{n=-\infty}^{\infty} \left\{ z \,\middle|\, \mathrm{Re}\, z = \left(n + \frac{1}{2}\right)\pi, |\mathrm{Im}\, z| \leq a \right\}$$

onto the upper half-plane (see the left figure in Fig. 3.118).

Solution. (1) In Fig. 3.113, $w = \sin^{-1}\left(\frac{\sin z}{\cosh a}\right)$ maps the half-strip $\frac{\pi}{2} < \mathrm{Re}\, z < \frac{3\pi}{2}, 0 < \mathrm{Im}\, z < \infty$, onto itself in the w-plane:

the half-ray $\mathrm{Re}\, z = \frac{\pi}{2}, a < \mathrm{Im}\, z < \infty \to \mathrm{Re}\, w = \frac{\pi}{2}, 0 < \mathrm{Im}\, w < \infty$ (univalently);

the half-ray $\mathrm{Re}\, z = \frac{3\pi}{2}, a < \mathrm{Im}\, z < \infty \to \mathrm{Re}\, w = \frac{3\pi}{2}, 0 < \mathrm{Im}\, w < \infty$;

the segment $\mathrm{Re}\, z = \frac{\pi}{2}, 0 \leq \mathrm{Im}\, z < a \to$ the segment $\mathrm{Im}\, w = 0, \frac{\pi}{2} \leq \mathrm{Re}\, w \leq \frac{\pi}{2} + \sin^{-1}\frac{1}{\cosh a}$;

the segment $\mathrm{Re}\, z = \frac{3\pi}{2}, 0 \leq \mathrm{Im}\, z \leq a \to$ the segment $\mathrm{Im}\, w = 0, \frac{3\pi}{2} - \sin^{-1}\frac{1}{\cosh a} \leq \mathrm{Re}\, w \leq \frac{3\pi}{2}$;

the segment $\mathrm{Im}\, w = 0, \frac{\pi}{2} \leq \mathrm{Re}\, z \leq \frac{3\pi}{2} \to$ the segment $\mathrm{Im}\, w = 0, \frac{\pi}{2} + \sin^{-1}\frac{1}{\cosh a} \leq \mathrm{Re}\, w \leq \frac{3\pi}{2} - \sin^{-1}\frac{1}{\cosh a}$.

See Fig. 3.116. Repeat this process to the half-strips $\left(n - \frac{1}{2}\right) < \mathrm{Re}\, z < \left(n + \frac{1}{2}\right)\pi$, $n = \pm 1, \pm 2, \ldots$ (except $n = 0$). Consequently $w = \sin^{-1}\left(\frac{\sin z}{\cosh a}\right)$ maps the upper half-plane plus the rays $\left(-\infty, -\frac{\pi}{2}\right]$ and $\left[\frac{\pi}{2}, +\infty\right)$ along the real axis onto the upper half plus the rays $\left(-\infty, -\sin^{-1}\mathrm{sech}\, a\right]$ and $\left[\sin^{-1}\mathrm{sech}\, a, +\infty\right)$ along the real axis in the w-plane. Then, apply the symmetry principle (3.4.6.2) to $w = \sin^{-1}\left(\frac{\sin z}{\cosh a}\right)$ on the upper half-plane with respect to the open interval $\left(-\frac{\pi}{2}, \frac{\pi}{2}\right)$. The extended function

Fig. 3.116

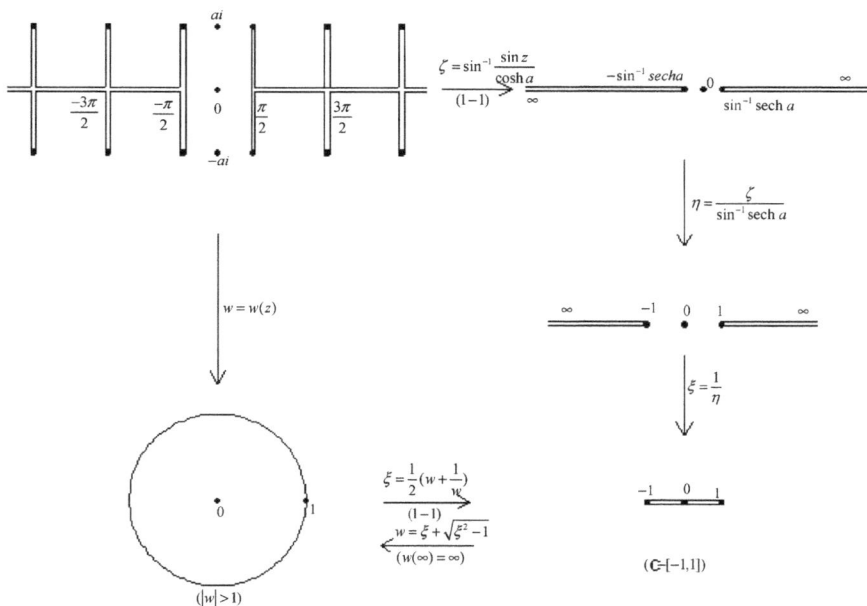

$$\zeta = \sin^{-1}\frac{\sin z}{\cosh a}$$

$$-\sin^{-1}\mathrm{sech}\,a \qquad \bullet\, 0 \qquad\qquad \infty$$

$$\sin^{-1}\mathrm{sech}\,a$$

$$\eta = \frac{\zeta}{\sin^{-1}\mathrm{sech}\,a}$$

$$w = w(z)$$

$$\xi = \frac{1}{\eta}$$

$$\xi = \frac{1}{2}\left(w+\frac{1}{w}\right)$$
$$w = \xi + \sqrt{\xi^2 - 1}$$
$$(w(\infty)=\infty)$$

$$(\mathbf{C}\text{-}[-1,1])$$

$$(|w|>1)$$

Fig. 3.117

maps the indented domain onto the slit domain $\mathbf{C}^* - [\infty, -\sin^{-1}\mathrm{sech}\,a] - [\sin^{-1}\mathrm{sech}\,a, \infty]$. See Fig. 3.117. The required map is the composite

$$w = w(z) = \frac{\sin^{-1}\frac{1}{\cosh a} + \sqrt{\left(\sin^{-1}\frac{1}{\cosh a}\right)^2 - \left(\sin^{-1}\frac{\sin z}{\cosh a}\right)^2}}{\sin^{-1}\frac{\sin z}{\cosh a}}.$$

(2) Model after (1) and try to use result there. The composite map in Fig. 3.118

$$w = \sqrt{\frac{\sin^{-1}\left(\frac{\sin z}{\cosh a}\right) - \sin^{-1}\left(\frac{\sin p}{\cosh a}\right)}{\sin^{-1}\left(\frac{\sin q}{\cosh a}\right) - \sin^{-1}\left(\frac{\sin z}{\cosh a}\right)}}$$

is the required one.

Exercises A

All mappings are required to be univalently conformal unless otherwise stated.

(1) (See Ref. [1], p. 96) Map the common part of the disk $|z| < 1$ and $|z - 1| < 1$ onto the unit disk so that the symmetries are preserved.

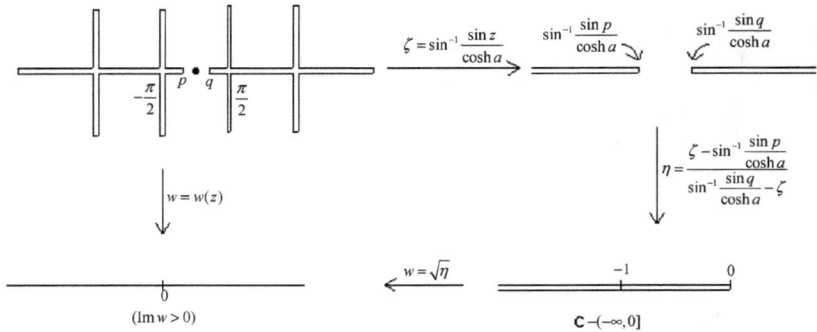

$$\zeta = \sin^{-1}\frac{\sin z}{\cosh a}$$

$$\sin^{-1}\frac{\sin p}{\cosh a}$$

$$\sin^{-1}\frac{\sin q}{\cosh a}$$

$$w = w(z)$$

$$\eta = \frac{\zeta - \sin^{-1}\dfrac{\sin p}{\cosh a}}{\sin^{-1}\dfrac{\sin q}{\cosh a} - \zeta}$$

$$w = \sqrt{\eta}$$

(Im $w > 0$)

$\mathbf{C} - (-\infty, 0]$

Fig. 3.118

In addition, try to map each of the domains $|z| < 1$ and $|z - 1| > 1$; $|z| > 1$ and $|z - 1| < 1$; $|z| > 1$ and $|z - 1| > 1$ onto $|w| < 1$.

(2) Try to map each of the following domains onto the upper half-plane:

(a) $|z - i| < 2$ and $\operatorname{Im} z > 0$.

(b) $|z + i| > \sqrt{2}$ and $|z - i| < \sqrt{2}$.

(c) $|z| < 2$ and $|z - 1| > 1$.

(3) Map the slit domain $\{|z| < 1\} - [0, 1]$ onto $|w| < 1$ so that 1 in the upper side of the cut $[0, 1]$ is mapped onto 1, while 1 in the lower side onto -1.

(4) Map the upper half-plane $\operatorname{Im} z > 0$, with the segment $\operatorname{Re} z = a$ ($a > 0$), $0 \leq \operatorname{Im} z \leq h$ deleted, onto $\operatorname{Im} w > 0$.

(5) Map the sector domain $-\frac{\pi}{4} < \operatorname{Arg} < \frac{\pi}{2}$ onto $\operatorname{Im} w > 0$ so that the points $1 - i, i$, and 0 are mapped onto $2, -1$, and 0, respectively.

(6) Map the domain $\Omega = \mathbf{C}^* - [-a, b] - [-ic, ic]$, where $a > 0$, $b > 0$ and $c > 0$, onto $\operatorname{Im} w > 0$.

(7) Map the domain common to the exteriors of two tangent circles $|z - r_1| = r_1$ and $|z + r_2| = r_2$ ($r_1 > 0, r_2 > 0$) onto the domain bounded by two parabolas (with foci at 0 and the vertices on the negative real axis), and onto the domain bounded by two coaxial hyperbolas (with asymptotes of the bisectors of the rectangular coordinate axes).

(8) Map the domain bounded by the two branches of the hyperbola $\frac{x^2}{a^2} - \frac{y^2}{b^2} = 1$ ($a > b > 0$) onto the upper half-plane (see Sec. 2.5.5).

(9) Map the domain bounded by the two cofocal ellipses $\frac{x^2}{a^2} + \frac{y^2}{b^2} = 1$ and $\frac{x^2}{a^2+k^2} + \frac{y^2}{a^2+k^2} = 1$ ($a > b > 0, k > 0$) onto a domain bounded by concentric circles with center at 0 (see Sec. 2.5.5).

(10) Try to use the Joukowski function (see Sec. 2.5.5) to map each of the following domains onto the upper half-plane.

(a) The unit disk $|z| < 1$ with the segments $\left[\frac{1}{2}, 1\right)$ deleted.

(b) The unit disk $|z| < 1$ with the segments $(-1, 0]$ and $[a, 1)$, $0 < a < 1$, deleted.

(c) The upper half-disk $|z| < 1, \operatorname{Im} z > 0$ with the segment $[\alpha i, i)$, $0 < \alpha < 1$ deleted. What happens if, instead, the segment $[0, \alpha i]$ is deleted?

(d) The domain obtained by deleting the segments $(i, bi]$, $[-bi, -i)$, $(1, a]$ and $[-a, 1)$, where $a > 1$ and $b > 1$, from $|z| > 1$.

(e) The domain obtained by deleting the segments $[a, 1)$, and $(-1, -b]$, where $0 < a < 1$ and $0 < b < 1$, from $|z| < 1$. Firstly, map it onto $|\zeta| < 1$ so that $\zeta(0) = 0$ and $\zeta'(0) > 0$ and then onto the upper half-plane $\operatorname{Im} w > 0$.

(11) (a) Find the image of the sector domain $0 < \operatorname{Arg} z < \frac{\pi}{n}$ under $w = \frac{1}{2}\left(z^n + \frac{1}{z^n}\right)$.

(b) Map the sector domain $|z| < 1, 0 < \operatorname{Arg} z < \frac{\pi}{n}$ onto itself so that the radial segments $|z| \leq \alpha, \operatorname{Arg} z = 0$ and $|z| \leq \alpha, \operatorname{Arg} z = \frac{\pi}{n}$ $(0 < \alpha < 1)$ are mapped, respectively, onto the radius $|z| \leq 1$, $\operatorname{Arg} z = 0$ and $|z| \leq 1, \operatorname{Arg} z = \frac{\pi}{n}$.

(c) Map the domain $\{|z| > 1\} - \cup_{k=0}^{n-1}\{1 \leq |z| \leq \alpha, \operatorname{Arg} z = \frac{2k\pi}{n}\}$ onto $|w| > 1$.

(12) Let

$$w = \frac{z}{(1 + z^n)^{\frac{2}{n}}} \quad (\sqrt{1} = 1).$$

(a) Find the image of the sector domain $|z| < 1$, $-\frac{\pi}{n} < \operatorname{Arg} z < \frac{\pi}{n}$ under $w = w(z)$.

(b) Find the image of $|z| < 1$ under $w = w(z)$.

(c) Map $|z| < 1$ onto the exterior of a "starlike" domain, namely,

$$\mathbf{C}^* - \left\{ w \,\middle|\, |w| \leq 1, \operatorname{Arg} w = \frac{2k\pi}{n} (0 \leq k \leq n - 1) \right\}.$$

(13) Let Ω be the domain bounded by the line $\operatorname{Im} z = 1$ and the circle $|z| = 1$ (see the left figure in Fig. 3.107). Determine the conformal mapping $w = w(z)$ to map Ω onto the upper half-plane subject to $w(-3i) = 1 + i$ and $w'(-3i) = \pi$, and another $w = w(z)$ onto $|w| < 1$ subject to $w(-3i) = 0$ and $w'(-3i) = \frac{\pi}{3}$.

(14) Use mainly the exponential function (see Sec. 2.6) to map each of the following domains onto the upper half-plane in (a)–(d) and onto the unit disk in (e). Let $z = x + iy$.

(a) The domain between the parabolas $y^2 = 4(x + 1)$ and $y^2 = 8(x + 2)$.

(b) The domain between $y = x$ and $y = x + h$.

(c) The lune domain between $|z| = 2$ and $|z - 1| = 1$.

(d) The domain bounded by $|z| = 2$ and $|z - 3| = 1$, namely, $\mathbf{C}^* - \{|z| \leq 2\} - \{|z - 3| \leq 1\}$.

(e) The horizontal strip $-\frac{\pi}{2} < \operatorname{Im} z < \frac{\pi}{2}$.

(15) Map each of the following domains onto the upper half-plane.

(a) The vertical strip $0 < \operatorname{Re} z < 1$ with the ray $\operatorname{Re} z = \frac{1}{2}, h \leq \operatorname{Im} z < \infty$ deleted.

(b) The vertical strip $0 < \operatorname{Re} z < 1$, with the rays $\operatorname{Re} z = \frac{1}{2}, h_1 \leq \operatorname{Im} z < \infty$, and $\operatorname{Re} z = \frac{1}{2}, -\infty < \operatorname{Im} z \leq h_2$ $(h_2 < h_1)$ deleted.

(c) The domain bounded by $|z - 1| = 1$ and $|z + 1| = 1$, namely, $\mathbf{C}^* - \{|z - 1| \leq 1\} - \{|z + 1| \leq 1\}$, with the segment $\operatorname{Re} z = 0, -\alpha \leq \operatorname{Im} z \leq \beta$ $(\alpha \geq 0, \beta \geq 0)$ deleted.

(16) Map the half-strip $\operatorname{Re} z < 1, 0 \leq \operatorname{Im} z \leq h$, onto the upper half-plane.

(17) Map the upper half-plane $\operatorname{Im} z > 0$, with two closed disks $|z - 1| \leq 1$ and $|z + 1| \leq 1$ deleted, onto the upper half-plane.

(18) Map each of the following domains onto the upper half-plane.

(a) $\{0 < \operatorname{Re} z < 1\} - \{0 < \operatorname{Re} z < h \text{ and } \operatorname{Im} z = 0\}$, where $0 < h < 1$.

(b) $\{0 < \operatorname{Re} z < 1\} - \{0 < \operatorname{Re} z < h_1 \text{ and } \operatorname{Im} z = 0\} - \{1 - h_2 < \operatorname{Re} z < 1 \text{ and } \operatorname{Im} z = 0\}$, where $h_1 + h_2 < 1$.

(c) $\{0 < \operatorname{Re} z < \pi \text{ and } \operatorname{Im} z > 0\} - \{\operatorname{Re} z = \frac{\pi}{2} \text{ and } 0 < \operatorname{Im} z \leq h\}$.

(d) $\{0 < \operatorname{Re} z < \pi \text{ and } \operatorname{Im} z > 0\} - \{\operatorname{Re} z = \frac{\pi}{2} \text{ and } 0 < \operatorname{Im} z \leq h_1\} - \{\operatorname{Re} z = \frac{\pi}{2} \text{ and } h_2 \leq \operatorname{Im} z < \infty\}$, where $h_1 < h_2$.

(e) $\{|z - 1| < 1\} \cap \{|z + 1| < 1\} - \{2 \leq \operatorname{Re} z < \infty \text{ and } \operatorname{Im} z = 0\}$.

(f) $\{|z - 1| < 1\} \cap \{|z - 2| < 2\} - \{2 \leq \operatorname{Re} z \leq a \text{ and } \operatorname{Im} z = 0\}$, where $a < 4$.

(g) $\{|z - 1| < 1\} \cap \{|z - 2| < 2\} - \{2 \leq \operatorname{Re} z \leq a \text{ and } \operatorname{Im} z = 0\} - \{b \leq \operatorname{Re} z \leq 4 \text{ and } \operatorname{Im} z = 0\}$, where $a < b$.

(h) The domain bounded by $\operatorname{Re} z = 0$ and $|z - 1| = 1$, with the segments $2 \leq \operatorname{Re} z \leq a, \operatorname{Im} z = 0$, and $b \leq \operatorname{Re} z \leq 4, \operatorname{Im} z = 0$ $(a < b)$ deleted.

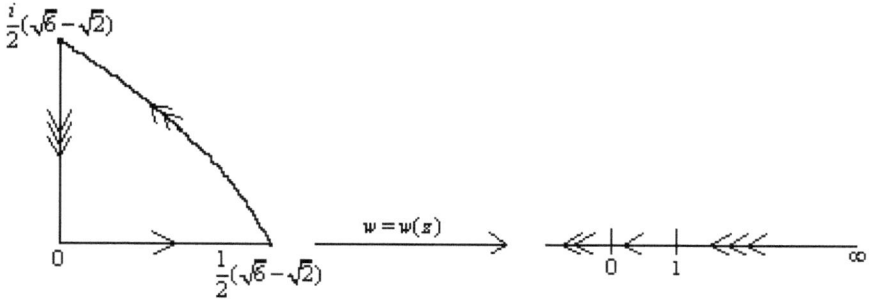

Fig. 3.119

(i) $\{|z - 1| > 1\} \cap \{|z + 1| > 1\} \cap \{\operatorname{Im} z > 0\} - \{\operatorname{Re} z = 0 \text{ and } 0 \leq \operatorname{Im} z \leq h\}$.

(19) Map the interior of the parabola $y^2 = 4\alpha^2(x + \alpha^2)$ onto $\operatorname{Im} w > 0$ and $|w| < 1$, respectively.

Exercises B

(1) Show that

$$ w = \left(\frac{z^4 + 2\sqrt{3}z^2 - 1}{z^4 - 2\sqrt{3}z^2 - 1} \right)^3 $$

maps the curvilinear triangle $\left(\frac{\pi}{2}, \frac{\pi}{3}, \frac{\pi}{3} \right)$ onto the lower half-plane as shown in Fig. 3.119. Also, do the following problems.

(a) The singular points (points where $w'(z) = 0$ and poles of order $n \geq 2$) of $w = w(z)$: $0, \frac{\varepsilon}{2}(\sqrt{6} - \sqrt{2}), \frac{\varepsilon}{2}(\sqrt{6} + \sqrt{2}), \varepsilon e^{\frac{\pi i}{4}}, \infty$, where $\varepsilon = \pm 1, \pm i$.

(b) $0, \infty, \pm e^{\frac{\pi i}{4}}$, and $\pm e^{-\frac{\pi i}{4}}$ are mapped into 1.

(c) $\pm \frac{1}{2}(\sqrt{6} - \sqrt{2})$ and $\pm \frac{i}{2}(\sqrt{6} + \sqrt{2})$ are mapped into 0.

(d) $\pm \frac{1}{2}(\sqrt{6} + \sqrt{2})$ and $\pm \frac{i}{2}(\sqrt{6} - \sqrt{2})$ are mapped into ∞.

(e) The lines $x = 0, y = 0$ and the circles $|z \pm e^{\frac{\pi i}{4}}| = \sqrt{2}, |z \pm e^{-\frac{\pi i}{4}}| = \sqrt{2}$ are mapped onto the real axis. Determine which parts of them are mapped onto $[\infty, 0], [0, 1]$ and $[1, \infty]$, respectively.

(f) The domains ① and ② in Fig. 3.120 are mapped onto the lower and the upper half-planes, respectively.

(g) Figure 3.121 shows the line complex of the Riemann surface R_w for $z = z^{-1}(w)$. Try to pinpoint all its branch points and determine their orders.

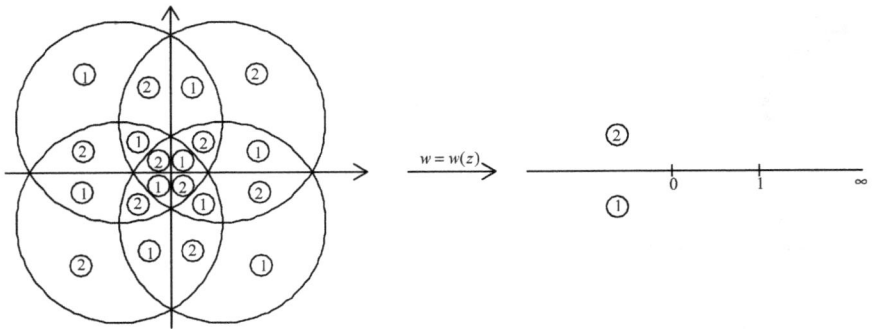

Fig. 3.120

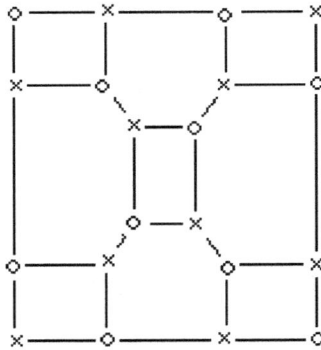

Fig. 3.121

(2) Show that

$$w = \frac{(z^4 + z^{-4} + 14)^3}{(z^4 + z^{-4} - 2)^2}$$

maps the curvilinear triangle $\left(\frac{\pi}{2}, \frac{\pi}{3}, \frac{\pi}{4}\right)$ onto the lower half-plane as shown in Fig. 3.122. Do the following problems.

(a) Singular points: $0, \varepsilon(\sqrt{2} \pm 1), \varepsilon, \varepsilon e^{\frac{\pi i}{4}}, \frac{\varepsilon}{2}(\sqrt{6} - \sqrt{2})e^{\frac{\pi i}{4}}, \frac{\varepsilon}{2}(\sqrt{6} + \sqrt{2})e^{\frac{\pi i}{4}}, \infty$ where $\varepsilon = \pm 1, \pm i$.

(b) $\varepsilon z, \frac{\varepsilon}{z}, \varepsilon\frac{z+1}{z-1}, \varepsilon\frac{z-1}{z+1}, z\frac{z+i}{z-i}, \varepsilon\frac{z-i}{z+i}$ → (mapped into the same point) $w(z)$.

(c) $0, \varepsilon, \infty \to \infty$.

(d) $\varepsilon(\sqrt{2} \pm 1), \varepsilon e^{\frac{\pi i}{4}} \to 108$.

(e) $\frac{\varepsilon}{2}(\sqrt{6} \pm \sqrt{2})e^{\frac{\pi i}{4}} \to 0$.

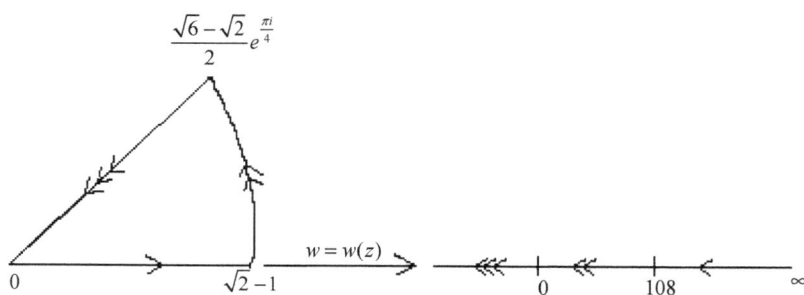

Fig. 3.122

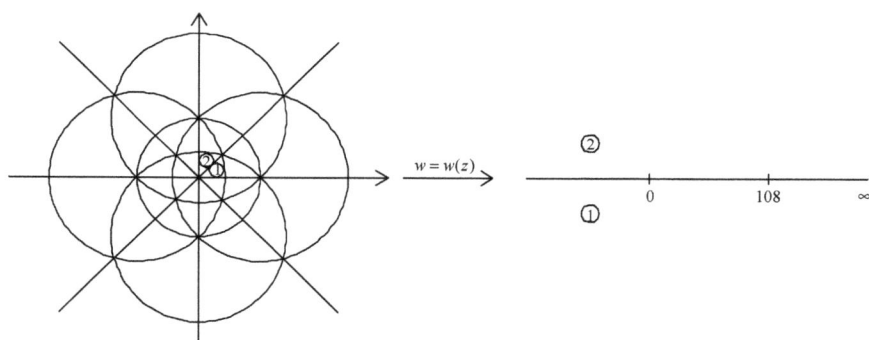

Fig. 3.123

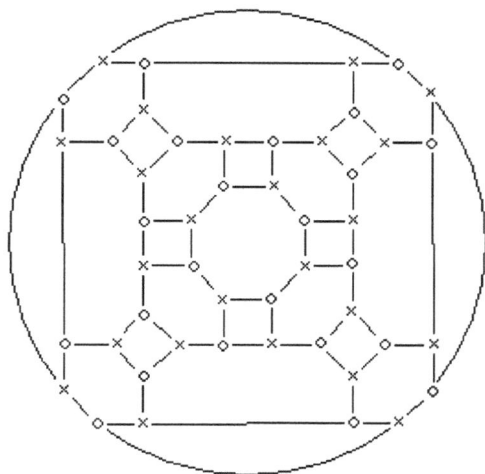

Fig. 3.124

(f) The lines $x = 0, y = 0$; the rays $\operatorname{Arg} z = \pm \frac{\pi}{4}, \operatorname{Arg} z = \pm \frac{3\pi}{4}$ and the circles $|z| = 1, |z \pm 1| = \sqrt{2}, |z \pm i| = \sqrt{2}$ are mapped onto the real axis. Which parts of them are mapped onto $[\infty, 0], [0, 108]$, and $[108, \infty]$, respectively.

(g) The domains ① and ② in Fig. 3.123 are mapped onto the lower and the upper half-planes, respectively.

(h) Figure 3.124 shows the line complex of the Riemann surface R_w for $z = z^{-1}(w)$. Try to locate all its branch points and determine their orders.

CHAPTER 4

Fundamental Theory: Integration (Advanced)

Introduction

A complex-valued *continuous* function $f(z)$ in a domain (or open set) Ω is said to be *analytic* in Ω if, for each triangle Δ contained in Ω, $\int_{\partial\Delta} f(z)dz = 0$ always holds (see (3.4.2.2), the Cauchy–Goursat theorem). This definition of analyticity will be equivalent to the one defined through differentiability as adopted in Chap. 3 by observing the following process:

$\int_\gamma f(z)dz = 0$ for any closed rectifiable curve γ in a *simply connected domain* Ω (see (3.4.2.1), the *Cauchy integral theorem* in simpler form).

$\Rightarrow f(z) = \frac{1}{2\pi i} \int_{|\zeta - z_0| = \rho} \frac{f(\zeta)}{\zeta - z} d\zeta, |z - z_0| < \rho < \text{dist}(z_0, \partial\Omega)$ for each fixed point z_0 in Ω (see (3.4.2.4), the *Cauchy integral formula* in simpler form).

$\Rightarrow f(z) = \sum_{n=0}^{\infty} a_n(z - z_0)^n, |z - z_0| < \text{dist}(z_0, \partial\Omega)$ for each fixed point z_0 in Ω (see (3.4.2.6), the *Taylor series expansion* of $f(z)$ at $z_0 \in \Omega$).

$\Rightarrow f(z)$ is *infinitely differentiable* in the complex sense (see (3.3.2.4)).

In particular, the first derivative $f'(z)$ will result in the Cauchy–Riemann equations. Refer to (3.4.2.15) for other equivalent conditions. There is no essential difference between these two definitions.

Sketch of the Content

After a quick description about the existence of a single-valued *primitive function* in Sec. 4.1, we turn immediately to the theoretical investigation of when $\int_\gamma f(z)dz = 0$ will hold for any analytic function f in Ω. This relies on the topological structure of the domain Ω itself and the shape of the path γ of the integration relative to the domain.

Section 4.2 emphasizes the stronger geometric flavor of the concept of homotopy concerning continuous deformation of a continuous curve in a domain; and proves the *homotopic invariance of the winding numbers* (see Sec. 3.5.2) and the *Cauchy integral theorem in homotopic form.*

While Sec. 4.3 takes advantageously the algebraic linear property of the line integral and introduces the *homology* between two curves, a concept combining the closed chains of curves and the winding numbers. We adopt Artin's proof to prove the *Cauchy integral theorem in homologous form*; in particular, in a $(n+1)$-connected domain in terms of a *homology basis* and the concept of *periods* (see (4.3.3.2)), and a byproduct is the *residue theorem*.

Section 4.4, as a review, uses *topological, analytic*, and *algebraic* properties to characterize a domain to be simply connected. And above all, any analytic function in a simply connected domain always has a single-valued primitive.

As a contrast, primitive functions in a multiply connected domain are, in general, multiple-valued. Section 4.5 tells us how to determine all possible *single-valued branches* of the primitive in a simply-connected subdomain.

Based on Sec. 4.3, Sec. 4.6 gives a general form of the Cauchy integral formula (3.4.2.4) for an analytic function and its higher derivatives.

Integrals of Cauchy type and the concept of Cauchy principal value are introduced in Sec. 4.7. Section 4.8 presents complicated examples in Taylor series expansions, especially, of the branches of multiple-valued functions. The Laurent series expansion of an analytic function at an isolated singularity is in Sec. 4.9. The classification of isolated singularities into removable one, pole and essential one, and their characteristic properties are the main theme of Sec. 4.10.

The coefficient a_{-1} in the Laurent series expansion $f(z) = \sum_{n=-\infty}^{\infty} a_n(z-z_0)^n$, $0 < |z - z_0| < R$, plays a crucial role in determining whether $\int_\gamma f(z)dz = 0$ is true or false for any closed rectifiable curve γ in $0 < |z - z_0| < R$. a_{-1} is specifically called the *residue* $\text{Res}(f(z); z_0)$ of $f(z)$ at its isolated singularity z_0. Section 4.11 formally introduces this important concept and ways of computing it. The *residue theorem* (4.3.3.6) is reformulated as (4.11.3.2).

What remains in this Chapter is devoted to the various *applications* of the residue theorem: The evaluation of the integrals (Sec. 4.12); the Fourier and the Laplace transforms (Sec. 4.13); the asymptotic expansion (Sec. 4.14) and the summation of the series (Sec. 4.15).

4.1.   Complex Integration Independent of Paths: Primitive Functions

Let Ω be a domain in the complex plane $\mathbf{C}$.

Suppose both $P(z)$ and $Q(z)$ are real- or complex-valued continuous functions in Ω. Let γ be a rectifiable or piecewise differentiable curve, varying in Ω. Then, the values of the line integral (see Sec. 2.9)

$$\int_\gamma P dx + Q dy$$

depend on γ and can be considered as a function of *the path γ of integration*.

In particular, we raise a *question*: If the initial point z_0 of γ is kept fixed and let its terminal points z vary throughout Ω, under what circumstance the seemingly multiple-valued function $F(z) = \int_\gamma P dx + Q dy$ will turn out to be single-valued? This is not always the case as Example 3 in Sec. 3.4.3 indicated (see Fig. 3.30):

$$F(z) = \int_\gamma \frac{1}{\zeta} d\zeta \quad \text{(with } z_0 = 1 \text{ and varying in } \mathbf{C} - \{0\})$$

$$= \log |z| + i \arg z = \log z, \tag{4.1.1}$$

is indeed a multiple-valued function of z in $\mathbf{C} - \{0\}$. This result can be easily seen using the concept of winding number (see Sec. 3.5.2) and the details are left to the readers.

The proof of the validity of (2) in (3.4.3.4) is universal and can be modified a little to give the following

Characteristic properties of an exact differential $P dx + Q dy$. Let Ω, P, Q, and γ be as above.

(1) The following are equivalent:

(i) The line integral $\int_\gamma P dx + Q dy$ is independent of γ and depends only on the initial and the terminal points of γ (see Fig. 4.1):

$$\int_{\gamma_1} P dx + Q dy = \int_{\gamma_2} P dx + Q dy,$$

simply denoted as $\int_a^b P dx + Q dy$.

(ii) For any rectifiable closed curve γ in Ω,

$$\int_\gamma P dx + Q dy = 0.$$

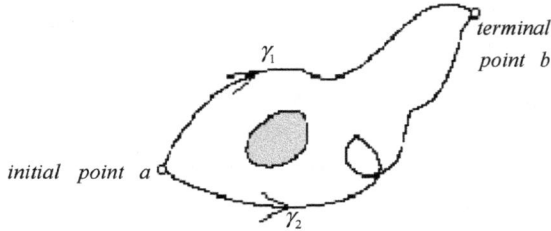

Fig. 4.1

(iii) There exists a continuously differentiable function $u(z) = u(x,y)$ on Ω so that, on Ω,

$$\frac{\partial u}{\partial x} = P \quad \text{and} \quad \frac{\partial u}{\partial y} = Q.$$

In this case, call $Pdx + Qdy = \frac{\partial u}{\partial x}dx + \frac{\partial u}{\partial y}dy = du$ an *exact differential form* on Ω.

(2) Let $f : \Omega \to \mathbf{C}$ be a complex-valued continuous function. Then

    (i) $\int_\gamma f(z)dz$ is independent of the path γ of integration.

$\Leftrightarrow$ (ii) There is an analytic function F on Ω such that, on Ω,

$$F'(z) = f(z).$$

Such a F is called a *primitive function* of f on Ω and any primitive function of f on Ω is of the form $F(z) + c$, where c is an arbitrary constant (see (3.4.1.3)).   (4.1.2)

Proofs are left to the readers.

    Take, for instance,

Example 1. Let γ be any rectifiable closed curve in $\mathbf{C} - \{0\}$. Then

$$\int_\gamma (z-a)^n dz = 0 \quad \text{where } n \neq -1 \text{ is any integer.}$$

While, in case γ is the circle $z = a + re^{i\theta}$ $(r > 0, 0 \leq \theta \leq 2\pi)$,

$$\int_\gamma \frac{dz}{z-a} = \int_{|z-a|=r} \frac{dz}{z-a} = 2\pi i.$$

This shows that it is not possible to define any single-valued branch of $\log(z-a)$ in the domain $0 \leq r_1 < |z-a| < r_2$.

Example 2. (Without recourse to Cauchy integral theorem in Sec. 3.4.2). For any rectifiable closed curve γ in $\mathbf{C}$,

$$\int_\gamma \cos z\, dz = 0; \quad \int_\gamma \sin z\, dz = 0; \quad \int_\gamma \cosh z\, dz = 0 \quad \text{and} \quad \int_\gamma \sinh z\, dz = 0.$$

Moreover,

$$\int_\gamma e^{-z^2}\, dz = 0 \quad \text{and} \quad \int_\gamma \frac{\sin z}{z}\, dz = 0.$$

For the validity of the last two integrals, one may refer to (4.1.3).

As a consequence of (2) in (4.1.2), we have

The integral of the derivative (supposed to be continuous) of an analytic function. Let $f : \Omega(\text{domain}) \to \mathbf{C}$ be analytic and γ be any rectifiable curve with a parametric representation $z(t), a \le t \le b$. Then

$$\int_\gamma f'(z)\, dz = f(z(b)) - f(z(a));$$

in particular, if γ is closed, then

$$\int_\gamma f'(z)\, dz = 0.$$

Hence, in case F is a primitive function of f in Ω, then

$$\int_\gamma f(z)\, dz = F(z(b)) - F(z(a)).$$

Therefore, one can deduce the following facts:

(1) If $f(z) = \sum_0^\infty a_n z^n$ has a positive radius r of convergence, then for any rectifiable closed curve γ in $|z| < r$, $\int_\gamma f(z)\, dz = 0$.
(2) In case $f'(z) = 0$ on Ω, then f is a constant on Ω.
(3) Local univalence. If f is analytic at z_0 and $f'(z_0) \ne 0$, then there is an $r > 0$ so that f is univalent in $|z - z_0| < r$.   (4.1.3)

Proofs for 2 and 3.   Once $|z - z_0| < r$ lies in Ω, we have

$$\int_{z_0}^z 0\, d\zeta = f(z) - f(z_0) = 0 \Rightarrow f(z) = f(z_0), \quad |z - z_0| \le r.$$

Then, a routine argument will show that $f(z) = f(z_0)$ on Ω (see the proof of (3.4.3.7)).

Since $f'(z_0) \ne 0$ and f' is continuous at z_0, there is an $r > 0$ so that f is analytic on $|z - z_0| \le r$ and $|f'(z) - f'(z_0)| < |f'(z_0)|$ there. Observe that $F(z) = f(z) - f'(z_0)z$ is a primitive of $f'(z) - f'(z_0)$ in $|z - z_0| \le r$.

Hence, for z_1, z_2 with $|z_j - z_0| < r, j = 1, 2,$

$$\int_{z_1}^{z_2} [f'(z) - f'(z_0)]dz = F(z_2) - F(z_1)$$

$$= f(z_2) - f(z_1) - f'(z_0)(z_2 - z_1),$$

$$\Rightarrow |f(z_2) - f(z_1) - f'(z_0)(z_2 - z_1)|$$

$$\leq \sup_{|z-z_0|\leq r} |f'(z) - f'(z_0)| \cdot |z_2 - z_1|,$$

$$\Rightarrow (\text{in case } z_1 \neq z_2) \quad \left| \frac{f(z_2) - f(z_1)}{z_2 - z_1} - f'(z_0) \right|$$

$$\leq \sup_{|z-z_0|\leq r} |f'(z) - f'(z_0)| < |f'(z_0)|,$$

$$\Rightarrow (\text{in case } z_1 \neq z_2) \quad f(z_1) \neq f(z_2).$$

This shows the univalence of f on $|z - z_0| < r$.

Exercises A

(1) Show that the following integrals all have zero value without actual computation.

(a) $\int_{|z|=1} \dfrac{dz}{\cos^2 z}$.

(b) $\int_{|z|=1} \dfrac{dz}{z^2 + 2z + 2}$.

(c) $\int_{|z|=1} z \cos z^2 dz$.

(2) Evaluate the following intervals, where γ is the segment connecting z_1 to z_2.

(a) $\int_i^{-4} \dfrac{1}{\sqrt{z}} dz$.

(b) $\int_{-1}^i \dfrac{1}{z \log z} dz$.

(c) $\int_1^{\sqrt{2i}} \log(1 + z^2) dz$.

(d) $\int_1^{2i} \tan^{-1} z \, dz$.

(e) $\int_{-i}^i \sqrt{z} dz$.

Note: $\int_\gamma f(z)dz = F(z(b)) - F(z(a))$ is still valid under the weaker conditions:

  (i) $f(z)$ and $F(z)$ are continuous along γ; and
  (ii) $F(z)$ is differentiable and $F'(z) = f(z)$ for all $z \in \gamma - z(S)$, where S is a finite set in $[a, b]$.

(3) Show that $f(z) = \bar{z}$ does not have any primitive on $\mathbf{C} - \{\mathrm{Im}\, z = 0\}$. How about on $\mathbf{C}$?

(4) Suppose $P(z)$ and $Q(z)$: Ω (simply connected domain) $\to \mathbf{R}$ are continuous and have continuous partial derivatives $P_y(x,y)$ and $Q_x(x,y)$. Then

(i) For any rectifiable closed curve γ in Ω, $\int_\gamma P dx + Q dy = 0$.
$\Leftrightarrow$ (ii) $P_y = Q_x$ on Ω.

In this case, $P dx + Q dy$ is an exact form.

(5) Prove (4.1.1) in details.
(6) Prove (4.1.2) in details.

Exercises B

(1) (Extension of (4) in (2.9.10)) Suppose $g : \Omega$ (domain in $\mathbf{C}$) $\to \mathbf{C}$ is a nonconstant analytic function and γ is a rectifiable curve in Ω with a representation $z(t), a \le t \le b$. Then

1. the image curve $g(\gamma)$, with representation $g(z(t))$, $a \le t \le b$, is rectifiable; and
2. if $f(w)$ is continuous along $g(\gamma)$, then

$$\int_{g(\gamma)} f(w) dw = \int_\gamma f(g(z)) g'(z) dz.$$

4.2. The General Form of Cauchy Integral Theorem: Homotopy

The *Cauchy integral theorem* we proved in Sec. 3.4.2 states: In case Ω is a simply connected domain and γ is any rectifiable closed curve in Ω, as long as f is analytic in Ω, then

$$\int_\gamma f(z) dz = 0$$

always holds (see (3.4.2.1)).

In the present section, Ω is just required to be a domain and γ is a closed rectifiable (even a continuous) curve. We try to find conditions under which $\int_\gamma f(z) dz = 0$ will hold for any analytic function f in Ω. The complete solution to this problem relies on two aspects: the topological structure of the domain itself, e.g., the simple connectedness, and the shape of the path of integration relative to the domain, e.g., the homotopic (Sec. 4.2.2) and the homologous (Sec. 4.3.2) properties.

Based on Cauchy–Goursat theorem (see (3.4.2.2)), Sec. 4.2.1 introduces the line integral of an *analytic* function along a *continuous* curve as a preparatory work in establishing a general form of Cauchy's theorem in (4.2.4.1).

When considering continuous deformation of a continuous curve in a domain, Sec. 4.2.2 introduces the concept of homotopy which has a stronger geometric flavor than homology to be given in Sec. 4.3.2, but less algebraic operation properties (group) than the latter (connection with the line integration).

Section 4.2.3 proves the homotopic invariance of the winding number. Exactly the same proof then establishes *the Cauchy integral theorem in homotopic form* in Sec. 4.2.4.

4.2.1. *The line integral of an analytic function along a*
  continuous curve

Here f is always supposed to be *analytic* in its domain Ω of definition. Also, γ always represents a *continuous* curve $z = z(t) : [a, b] \to \Omega$, the domain of f.

Three sections are divided. The method we adopted is completely similar to that used in Sec. 2.7.1.

Section (1) Local definition

In case $\Omega = D$ is a disk, say $|z - z_0| < R$. Choose two arbitrary points z_1 and z_2 in D and an arbitrary curve γ in D, connecting $z_1 = z(a)$ to $z_2 = z(b)$. See Fig. 4.2.

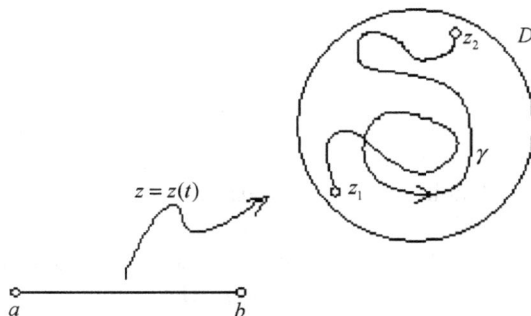

Fig. 4.2

Applying Cauchy integral theorem (3.4.2.1) to f in D and result in (4.1.2), f has a primitive F in D. Therefore, we define *the line integral of f along a continuous curve γ in D* by

$$\int_\gamma f(z)dz = F(z(b)) - F(z(a)) = F(z_2) - F(z_1). \qquad (4.2.1.1)$$

This definition is well defined in the sense that it is independent of either the choice of the primitive function F or the continuous curve connecting z_1 to z_2. It depends solely on the analyticity of the function f itself and the points z_1 and z_2 pre-summed.

In case γ is piecewise differentiable, it is worthy mentioned that (4.2.1.1) agrees with that given in Sec. 2.9.

Section (2) Global definition

In case Ω is a domain in general character, even though f is not necessarily to possess a primitive in Ω, but it does own local primitive in an open disk neighborhood of each point in Ω. Combining with the fact that the point set $\gamma = z([a, b])$ is *compact* and $z(t)$ is *uniformly continuous* on $[a, b]$, it is still possible to define $\int_\gamma f(z)dz$, where γ is any continuous curve lying entirely in Ω.

The details are as follows:

The distance $d = \text{dist}(\gamma; \mathbf{C} - \Omega)$ of γ to $\mathbf{C} - \Omega$ is positive (see Exercise A (2) of Sec. 1.9). The uniform continuity of $z = z(t)$ on $[a, b]$ says that, for each $0 < \varepsilon < d$, there is a $\delta = \delta(\varepsilon) > 0$ so that, whenever t, $t' \in [a, b]$ with $|t - t'| < \delta, |z(t) - z(t')| < \varepsilon$ always holds. Hence, the disk $D: |z - z(t)| < \varepsilon$, with center at $z(t)$, lies entirely in Ω and, in particular, the subarc $z([t, t']), a \le t < t' \le b$ with $|t - t'| < \delta$, also lies in D and hence, in Ω.

Construct a partition $a = t_0 < t_1 < \cdots < t_{n-1} < t_n = b$ so that $\max_{0 \le j \le n-1}(t_{j+1} - t_j) < \delta$. Denote $z_j = z(t_j)$ for $0 \le j \le n$. Then, the disk $D_j: |z - z_j| < \varepsilon$ lies in Ω and, moreover, the segment $\overline{z_j z_{j+1}}$ lies entirely in D_j for $0 \le j \le n-1$. Since $z_{j+1} \in D_j \cap D_{j+1}$, so $D_j \cap D_{j+1} \ne \phi$ for $0 \le j \le n-1$. See Fig. 4.3.

Let $F_j(z)$ be a primitive of $f(z)$ in D_j, for $0 \le j \le n$. Since $D_j \cap D_{j+1} \ne \phi$, both $F_j(z)$ and $F_{j+1}(z)$ are primitives of f on $D_j \cap D_{j+1}$ and so $F_j(z) = F_{j+1}(z) + c_{j+1}$ for some constant c_{j+1}, where $z \in D_j \cap D_{j+1}$. In this case, we readjust the primitive of f in D_{j+1} as $F_{j+1}(z) + c_{j+1}$ instead of $F_{j+1}(z)$ (this change is harmless to the definition in (4.2.1.1)). This guarantees that $F_j(z) = F_{j+1}(z)$ on $D_j \cap D_{j+1}$ for $0 \le j \le n-1$. Consequently, we are able to define a *continuous function* $F(t): [a, b] \to \Omega$ by

$$F(t) = F_j(z(t)), \quad t \in [t_j, t_{j+1}], \quad 0 \le j \le n-1. \qquad (4.2.1.2)$$

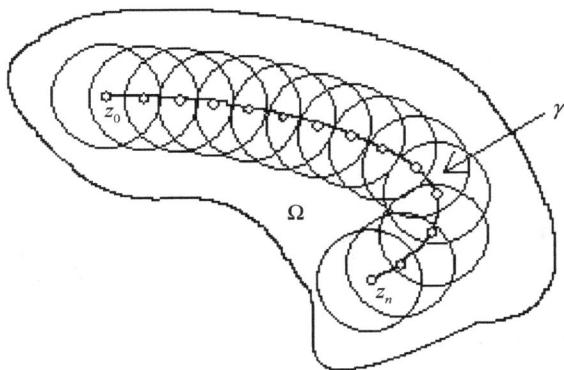

Fig. 4.3

$F(t)$ is well defined and is called a *primitive function* (for reason see (4.2.1.3) below) of *the analytic function* $f(z)$ along *the continuous curve* γ. It is *unique* up to a constant. As a matter of fact, the function $\tilde{F}(z): \cup_{j=0}^{n-1} D_j \to$ **C** defined by $\tilde{F}(z) = F_j(z), z \in D_j, 0 \le j \le n-1$, is well defined and is a multiple-valued analytic function in general. Yet its restriction to γ is the single-valued continuous function defined in (4.2.1.2).

Let us go back to Fig. 4.3. Let γ_j denote the subcurve of $z = z(t)$ restricting to the subinterval $[t_j, t_{j+1}]$ for $0 \le j \le n-1$. Then the *line integral of the analytic function f* along the *continuous curve* γ is defined as

$$\int_\gamma f(z)dz = \sum_{j=0}^{n-1} \int_{\gamma_j} f(z)dz \tag{$*_1$}$$

$$= \sum_{j=0}^{n-1} [F_j(z_{j+1}) - F_j(z_j)] \tag{$*_2$}$$

$$= F_{n-1}(z_n) - F_0(z_0) = F_{n-1}(z(b)) - F_0(z(a))$$

$$= F(b) - F(a) \quad (F \text{ is defined in } (4.2.1.2)). \tag{4.2.1.3}$$

We still need to show that the equalities in $(*_1)$ and $(*_2)$ are *independent* of the specified partition $a = t_0 < t_1 < \cdots < t_{n-1} < t_n = b$ of $[a, b]$.

The *proof* of independence of partition in $(*_1)$ and $(*_2)$:

Firstly suppose the jth subinterval $[t_j, t_{j+1}]$ has an additional partitioned point t_j^*, i.e., $t_j < t_j^* < t_{j+1}$ for $0 \le j \le n-1$. Let D_j^* be the disk $|z - z_j^*| < \varepsilon$, where $z_j^* = z(t_j^*)$, and $F_j^*(z)$ be a primitive of $f(z)$ in D_j^*. Observe that $F_j(z) - F_j^*(z)$ is a constant in $D_j \cap D_j^*$, and

$z_j, z_{j+1} \in D_j \cap D_j^*$. Thus,

$$F_j(z_j) - F_j^*(z_j) = F_j(z_{j+1}) - F_j^*(z_{j+1}), \quad 0 \le j \le n-1$$
$$\Rightarrow F_j(z_{j+1}) - F_j(z_j) = F_j^*(z_{j+1}) - F_j^*(z_j), \quad 0 \le j \le n-1$$
$$\Rightarrow [F_j^*(z_j^*) - F_j^*(z_j)] + [F_j^*(z_{j+1}) - F_j^*(z_j^*)]$$
$$= F_j(z_{j+1}) - F_j(z_j), \quad 0 \le j \le n-1.$$

This shows that the equality in $(*_2)$ remains unchanged if one additional partition point is added in $\lfloor t_j, t_{j+1} \rfloor$. And hence, it remains unchanged, too, for any refinement P' of the partition $P : a = t_0 < t_1 < \cdots < t_{n-1} < t_n = b$ (namely, all partition points of P are the ones of P').

Let P_1 and P_2 be two partitions of $[a, b]$ and P be the refinement of both P and P', obtained by rearranging all the partition points of P and P' according to the ascending order. By the result in the last paragraph, the quantities in $(*_2)$, according to P_1 and P_2, are equal to the same quantity according to P and hence, both are equal to each other. This finishes the proof. □

Section (3) Main result

Well, definition in (4.2.1.3) is good in establishing the general form of Cauchy's theorem, yet it is not good in computing the line integral.

To compensate such a shortcoming, we can find a piecewise differentiable curve Γ with the same end points of γ, so close to γ that the integrals of f along γ and Γ have the same value.

Let $\gamma_1 : z_1(t), a \le t \le b$ and $\gamma_2 : z_2(t), a \le t \le b$ be two continuous curve in Ω. If the common domain $[a, b]$ has a partition $a = t_0 < t_1 < \cdots < t_{n-1} < t_n = b$ so that, for each $0 \le j \le n-1$, there is an open disk $D_j \subseteq \Omega$ satisfying

$$z_k(\lfloor t_j, t_{j+1} \rfloor) \subseteq D_j, \quad k = 1, 2 \tag{4.2.1.4}$$

(see Fig. 4.4), then γ_1 and γ_2 are said to be *close to each other* in Ω.

For a continuous curve γ in Ω, still denote $\delta = \mathrm{dis}(\gamma, \mathbf{C} - \Omega) > 0$ and choose $0 < \varepsilon < \delta$. Let $a = t_0 < t_1 < \cdots < t_{n-1} < t_n = b$ be a partition of $[a, b]$ so that, if $\max_{0 \le j \le n-1}(t_{j+1} - t_j) < \delta$, then $|z(t_{j+1}) - z(t_j)| < \varepsilon$. As in Section (2), the disk $D_j : |z - z_j| < \varepsilon$ lies in Ω, where $z_j = z(t_j)$ for $0 \le j \le n$. Use segments, parallel to the coordinate axes, to connect z_j and z_{j+1} as in Fig. 4.5: this polygonal curve Γ_j consists of one horizontal and one vertical segment. Beginning at $z_0 = z(a)$ and ending at $z_n = z(b)$, *the composed*

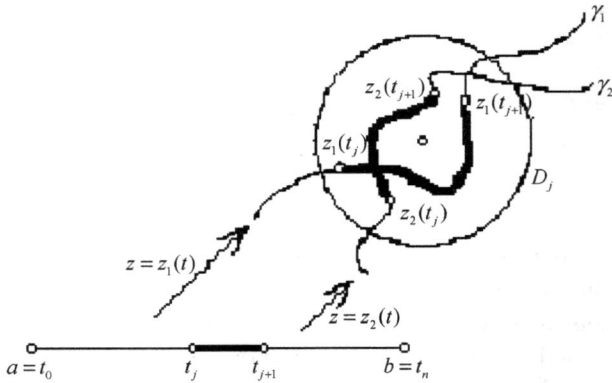

Fig. 4.4

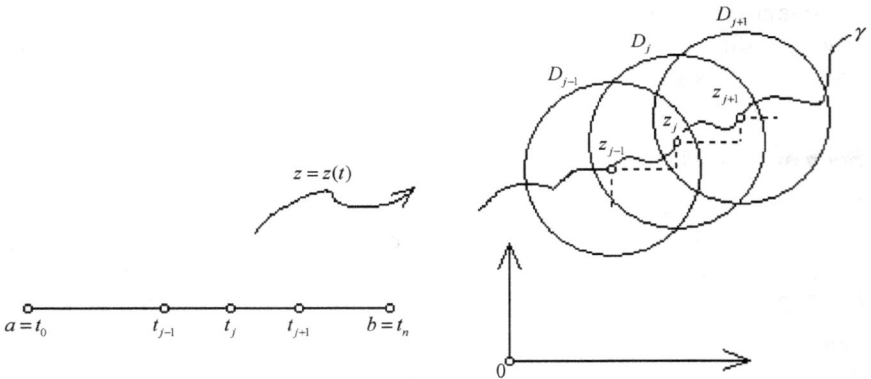

Fig. 4.5

polygonal curve $\Gamma = \Gamma_0 + \Gamma_1 + \cdots + \Gamma_{n-1}$ *is a piecewise differentiable curve, having the same end points as* γ *and close to* γ *according to* (4.2.1.4).

Combining the concepts in (4.2.1.3) and (4.2.1.4) together, suppose γ_1 and γ_2 are two continuous curves in Ω, close to each other. Then

$$\int_{\gamma_1} f(z)dz - \int_{\gamma_2} f(z)dz$$

$$= [F_{n-1}(z_1(b)) - F_0(z_1(a))] - [F_{n-1}(z_2(b)) - F_0(z_2(a))]$$

$$= [F_{n-1}(z_1(b)) - F_{n-1}(z_2(b))] - [F_0(z_1(a)) - F_0(z_2(a))]$$

$= 0$, if γ_1 and γ_2 have the same end points $z_1(a) = z_2(a)$ and

$z_1(b) = z_2(b)$, and

$= 0$, if γ_1 and γ_2 are closed curves, namely, $z_1(a) = z_1(b)$ and

$z_2(a) = z_2(b)$.

Under these stated conditions, $\int_{\gamma_1} f(z)dz = \int_{\gamma_2} f(z)dz$ holds.

We summarize the above as

The main result. Suppose f is analytic in Ω, and γ_1 and γ_2 are two continuous curves in Ω, close to each other (see (4.2.1.4)). In case

(1) either γ_1 and γ_2 are closed curves;
(2) or γ_1 and γ_2 has the same end points,

then it follows that

$$\int_{\gamma_1} f(z)dz = \int_{\gamma_2} f(z)dz. \qquad (4.2.1.5)$$

Consequently, *the path of integration* in the line integral of an analytic function along a *continuous curve* (4.2.1.3) can be replaced by *a piecewise differentiable curve.*

4.2.2.   *Homotopy of curves*

In this section, Ω is an *open* set in $\mathbf{C}$. Let $\gamma_1 : z_1(t)$, $a \leq t \leq b$, and $\gamma_2 : z_2(t)$, $a \leq t \leq b$, be two continuous curves *in Ω* with the *same* parametric interval $[a, b]$.

If $H = H(t, s) : [a, b] \times [0, 1] \to \Omega$ is a continuous mapping that satisfies the condition

$$H(t, 0) = z_1(t), \quad H(t, 1) = z_2(t), \quad a \leq t \leq b, \qquad (4.2.2.1)$$

then γ_1 is said to be *homotopic* to γ_2 in Ω, and H is the *homotopy* between γ_1 and γ_2. In this case, for each fixed s, $0 \leq s \leq 1$, $H_s(t) = H(t, s)$: $[a, b] \to \Omega$ is a continuous curve and, if s increases from 0 to 1, the curves H_s vary continuously from γ_1 to γ_2 in Ω. See Fig. 4.6. It is noticed that γ_2 is homotopic to γ_1 via the homotopy $\tilde{H}$ defined by $\tilde{H}(t, s) = H(t, 1-s)$, $(t, s) \in [a, b] \times [0, 1]$.

Three special kinds of homotopy are as follows.

(1) In case $H_s(t)$, $0 \leq s \leq 1$, are closed curves (in this case, γ_1 and γ_2 are presumed to be closed curves), namely, $H_s(a) = H_s(b)$ for $0 \leq s \leq 1$, then γ_1 is said to be *freely homotopic* to γ_2 in Ω. See Fig. 4.7.

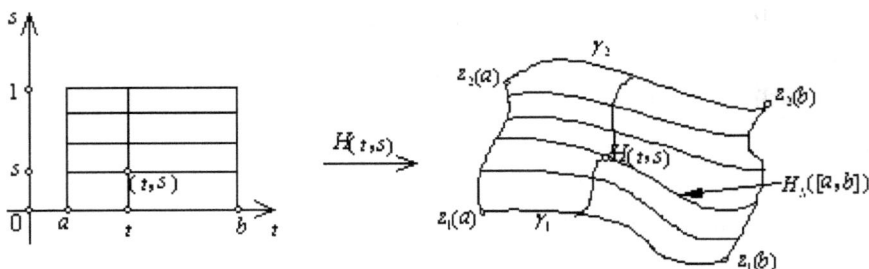

Fig. 4.6

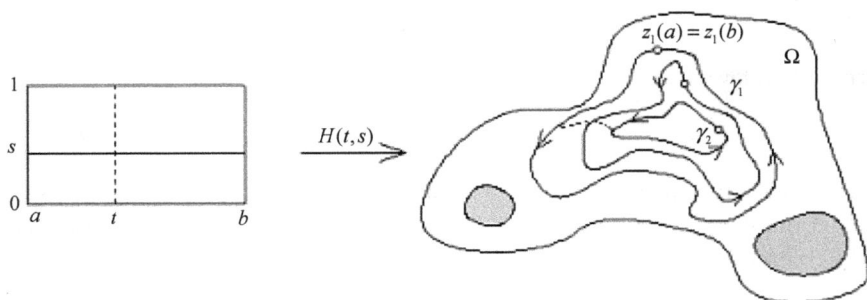

Fig. 4.7

(2) In case all the curves $H_s(t)$, $0 \le s \le 1$, have the same initial and terminal points (in this case, γ_1 and γ_2 are assumed to satisfy $z_1(a) = z_2(a)$ and $z_1(b) = z_2(b)$), namely, $H_s(a) = z_1(a)$ and $H_s(b) = z_1(b)$ for $0 \le s \le 1$, then γ_1 and γ_2 are said to be *homotopic with fixed end points in* Ω. See Fig. 4.8.

(3) In case γ_2 is a point z_0, namely, a constant curve and the closed curve γ_1 is freely homotopic to z_0 in Ω, then γ_1 is said to be *contractible to the point* z_0 in Ω and the associated H a *null homotopy*. See Fig. 4.9.

For primary properties of homotopy, see Exercises A (1) and (2); for details, see Refs. [57] or [74], and refer to Sec. 7.4 for a concise introduction to fundamental group.

We present three basic examples. Each curve mentioned is supposed to be continuous.

Example 1. The ellipse $\gamma_1 : z(t) = a\cos t + ib\sin t$, $0 \le t \le 2\pi$, $0 < a \le b$, and $\gamma_2 : z(t) = e^{it}$, $0 \le t \le 2\pi$, are freely homotopic in $\mathbf{C}$ and in $\mathbf{C} - \{0\}$.

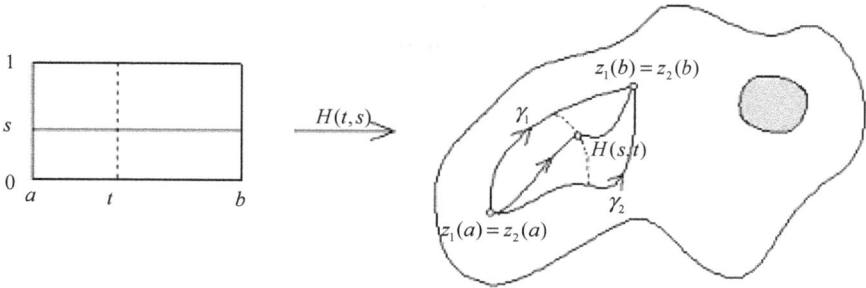

Fig. 4.8

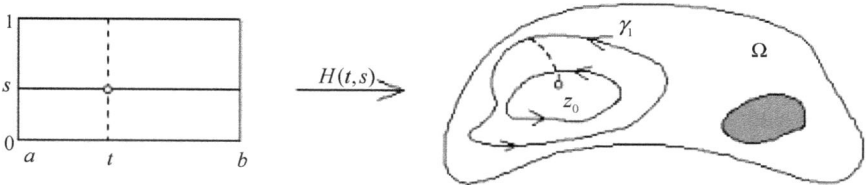

Fig. 4.9

Solution. Define $H(t, s) : [0, 2\pi] \times [0, 1] \to \mathbf{C}$ by

$$H(t, s) = (1 - s)(a \cos t + ib \sin t) + s(\cos t + i \sin t).$$

Then H is a free homotopy between γ_1 and γ_2 either in $\mathbf{C}$ or in $\mathbf{C} - \{0\}$.

Similarly, the circle $z(t) = e^{it}$, $0 \le t \le 2\pi$, and the circle $z(t) = 2i + 4e^{it}$, $0 \le t \le 2\pi$, are freely homotopic in $\mathbf{C} - \{0\}$.

Example 2.

(1) Suppose Ω is a *starlike domain* with *center* z_* (namely, a domain Ω such that there is a point $z_* \in \Omega$ with the property that every segment $\overline{zz_*}$ always lies in Ω for each point $z \in \Omega$. Refer to Fig. 4.11). Then each curve (especially, closed one) is homotopic to z_*.

(2) Suppose Ω is a convex domain. Then any two curves in Ω are homotopic.

Solution.

(1) Let $\gamma : z(t)$, $a \le t \le b$, be any curve in Ω. Define $H(t, s) : [a, b] \times [0, 1] \to \Omega$ by

$$H(t, s) = (1 - s)z(t) + sz_*.$$

Remind that the image of $H(t,s)$ is indeed in Ω because Ω is star-like. This H shows that γ is homotopic to z_*. In case γ is closed, i.e., $z(a) = z(b)$, so is $H_s(t)$ for each $0 \leq s \leq 1$ for $H_s(a) = H(a,s) = (1-s)z(a) + sz_* = (1-s)z(b) + sz_* = H(b,s) = H_s(b)$. Then, γ is freely homotopic to z_* in Ω.

(2) Let $\gamma_k : z_k(t)$, $a \leq t \leq b$, for $k = 1,2$, be two curves in Ω. The mapping $H(t,s)$ defined by $H(t,s) = (1-s)z_1(t) + sz_2(t)$, $(t,s) \in [a,b] \times [0,1]$, indeed is a homotopy between γ_1 and γ_2 in Ω.

In case γ_1 and γ_2 are closed curves, so is each $H_s(t)$ for $0 \leq s \leq 1$. And γ_1 and γ_2 are thus freely homotopic in Ω. If γ_1 and γ_2 have the same end points, namely, $z_1(a) = z_2(a)$ and $z_1(b) = z_2(b)$, then so is $H_s(t)$ for $0 \leq s \leq 1$. In this case, γ_1 and γ_2 are homotopic with fixed end points in Ω.

Example 3. Let Ω be a domain in $\mathbf{C}$. Then:

(1) Ω is a simply connected domain (namely, $\mathbf{C}^* - \Omega$ is connected).

$\Leftrightarrow$ (2) Any closed curve in Ω is homotopic to a point in Ω (in this case, Ω is said to be *contractible to a point* in it).

Sketch of Proof

(1) $\Rightarrow$ (2): In case $\Omega = \mathbf{C}$, the mapping $w = \frac{z}{1+|z|}$ realizes a homeomorphism between Ω and the unit disk $|w| < 1$. In case $\Omega \subsetneq C$. The *Riemann mapping theorem* (see (6.1.1) in Chap. 6) says that there is a univalent analytic function mapping Ω onto $|w| < 1$ and this function is homeomorphic (see (3.5.1.9)). In either case, let $w = f(z) : \Omega \to \{|w| < 1\}$ be such a homeomorphism.

Let $\gamma : z(t)$, $a \leq t \leq b$, be any closed curve in Ω. Define $H(t,s) : [a,b] \times [0,1] \to \Omega$ by

$$H(t,s) = f^{-1}(sf(z(t))).$$

Observe that $H(t,0) = f^{-1}(0)$, $a \leq t \leq b$, is a point in Ω and $H(t,1) = f^{-1}(f(z(t))) = z(t)$, $a \leq t \leq b$, is the given curve γ. Hence γ is homotopic to $f^{-1}(0)$ in Ω.

(2) $\Rightarrow$ (1). Suppose $\mathbf{C}^* - \Omega$ is not connected. Then there are nonempty closed subsets Ω_1 and Ω_2 in $\mathbf{C}^*$ so that $\mathbf{C}^* - \Omega = \Omega_1 \cup \Omega_2$ and $\Omega_1 \cap \Omega_2 = \phi$. We may suppose $\infty \in \Omega_2$ and hence, Ω_1 is a bounded set. Exactly as in the construction of the rectangular Jordan closed curve in Fig. 4.21, there is a curve γ in Ω so that the winding number

$$n(\gamma, z) = 1, \quad z \in \Omega_1.$$

On the other hand, by the homotopic invariance of the winding number (see (4.2.3.1)) and the condition (2) above, $n(\gamma, z) = 0$ should hold for any closed curve γ in Ω and for each $z \in \Omega_1$ (observe that $z \notin \Omega$). This contradiction shows that $\mathbf{C}^* - \Omega$ is necessarily connected and hence, Ω is simply connected.

Exercises A

(1) Use $\gamma_1 \simeq \gamma_2(\Omega)$ to denote the fact that γ_1 is homotopic to γ_2 in Ω, for simplicity. Show that

   (a) $\gamma \simeq \gamma$.
   (b) $\gamma_1 \simeq \gamma_2 \Rightarrow \gamma_2 \simeq \gamma_1$.
   (c) $\gamma_1 \simeq \gamma_2$ and $\gamma_2 \simeq \gamma_3 \Rightarrow \gamma_1 \simeq \gamma_3$.

   Note: These properties show that homotopy among curves is an equivalent relation. This fact will eventually lead to the concept of fundamental group for equivalent classes concerned (see Sec. 7.4).

(2) Prove the following statements.

   (a) If γ_1 is either freely homotopic or homotopic with fixed end points to γ_2 in Ω, then so is $-\gamma_1$ to $-\gamma_2$.
   (b) Suppose $\gamma_1 \simeq \gamma_2(\Omega)$ and $\tau_1 \simeq \tau_2(\Omega)$ with fixed end points, and $\gamma_1 + \tau_1$ and $\gamma_2 + \tau_2$ are well defined. Then $(\gamma_1 + \tau_1) \simeq (\gamma_2 + \tau_2)(\Omega)$.
   (c) Let $h(t) : [a, b] \to [a, b]$ is a nondecreasing continuous function satisfying $h(a) = a$ and $h(b) = b$. Let γ be the curve $z(t)$, $a \le t \le b$, and $\tilde{\gamma}$ the curve $z = z(h(t))$, $a \le t \le b$. Then $\tilde{\gamma} \simeq \gamma$ freely or with fixed end points.
   (d) Suppose $\gamma_1 \simeq \gamma_2(\Omega)$ freely or with fixed end points and $f : \Omega \to \Omega$ is a continuous mapping. Then $f \circ \gamma_1 \simeq f \circ \gamma_2(\Omega)$ holds.

(3) Is each of the following pairs of curves homotopic in the indicated domain Ω?

   (a) $\gamma_1 : z(t) = e^{2it}$; $\gamma_2 : z(t) = -1 + 2e^{it}$, $0 \le t \le 2\pi$; $\Omega = \mathbf{C} - \{0\}$.
   (b) $\gamma_1 : z(t) = e^{it}$; $\gamma_2 : z(t) = i$, $0 \le t \le 2\pi$; $\Omega = \mathbf{C} - \{i/2\}$.
   (c) $\gamma_1 : z(t) = 2ie^{\frac{it}{4}}$, $0 \le t \le 4\pi$; $\gamma_2 = \alpha_1 + \alpha_2$ where $\alpha_1 : z(t) = -1 + e^{it}$ and $\alpha_2 : z(t) = 1 - e^{it}$, $0 \le t \le 2\pi$; $\Omega = \mathbf{C}$ and $\Omega = \mathbf{C} - \{\pm 1\}$, respectively.

4.2.3.   *Homotopic invariance of the winding numbers*

Ω will always be a nonempty open set and z_0 a fixed point in Ω in the sequel.

   Let γ_1 and γ_2 be two closed curves in $\Omega - \{z_0\}$, freely homotopic to each other via the homotopy $H(t, s) : [a, b] \times [0, 1] \to \Omega - \{z_0\}$.

Since $K = H([a, b] \times [0, 1])$ is compact in $\Omega - \{z_0\}$, the distance d between K and the closed set $\mathbf{C} - (\Omega - \{z_0\}) = (\mathbf{C} - \Omega) \cup \{z_0\}$ is positive.

Choose $0 < \varepsilon < d$. Since $H(t, s)$ is uniformly continuous on $[a, b] \times [0, 1]$, there is a $\delta = \delta(\varepsilon) > 0$ such that, whenever (t, s), $(t', s') \in [a, b] \times [0, 1]$ satisfying $|t - t'| \le \delta$ and $|s - s'| \le \delta$,

$$|H(t, s) - H(t', s')| < \varepsilon$$

always holds. Let $z_{ts} = H(t, s)$ for $(t, s) \in [a, b] \times [0, 1]$. Then, the disk $|z - z_{ts}| < \varepsilon$ is contained in $\Omega - \{z_0\}$ and contains the image of the rectangle $|t - t'| \le \delta$, $|s - s'| \le \delta$ under $H(t, s)$. See Fig. 4.10.

Choose a partition $P : a = t_0 < t_1 < \cdots < t_{n-1} < t_n = b$ of $[a, b]$ such that $\max_{0 \le j \le n-1}(t_{j+1} - t_j) \le \delta$, and another partition $P' : 0 = s_0 < s_1 < \cdots < s_{m-1} < s_m = 1$ of $[0, 1]$ such that $\max_{0 \le k \le m-1}(s_{k+1} - s_k) \le \delta$. Denote

$$R_{jk} = [t_j - \delta, t_j + \delta] \times [s_k - \delta, s_k + \delta];$$

$$z_{jk} = H(t_j, s_k), \quad \text{and}$$

$$D_{jk} = \{|z - z_{jk}| < \varepsilon\}, \quad 0 \le j \le n-1, \quad 0 \le k \le m-1.$$

$$\Rightarrow H(R_{jk}) \subseteq D_{jk} \subseteq \Omega - \{z_0\}, \quad 0 \le j \le n-1, \quad 0 \le k \le m-1.$$

In particular, the midway curves $H_{s_k}(t) = H_k(t)$ and $H_{s_{k+1}}(t) = H_{k+1}(t)$ satisfy

$$H_k([t_j, t_{j+1}]), H_{k+1}([t_j, t_{j+1}]) \subseteq D_{jk}, \quad 0 \le j \le n-1, \quad 0 \le k \le m-1.$$

Hence, the curves $\tilde{\gamma}_k$ defined by $H_k(t)$ and $\tilde{\gamma}_{k+1}$ by $H_{k+1}(t)$ are close to each other in $\Omega - \{z_0\}$ (see (4.2.1.4)) for $0 \le k \le m-1$. Note that $\tilde{\gamma}_0 = \gamma_1$ and $\tilde{\gamma}_m = \gamma_2$.

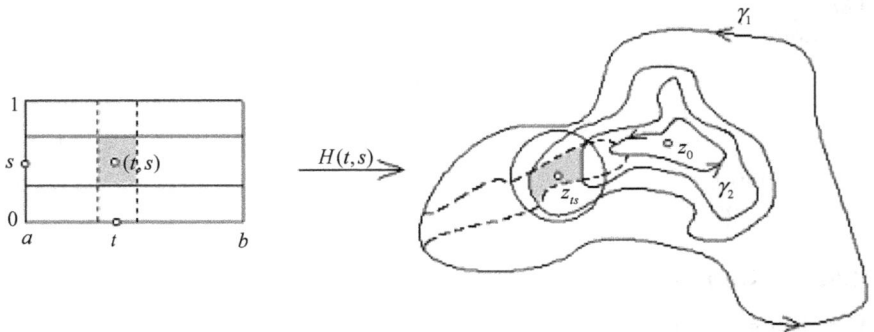

Fig. 4.10

Now the function $f(z) = \frac{1}{z-z_0}$ is analytic in $\Omega - \{z_0\}$. According to (4.2.1.5), we have

$$\int_{\gamma_1} \frac{dz}{z-z_0} = \int_{\tilde{\gamma}_0} \frac{dz}{z-z_0} = \int_{\tilde{\gamma}_1} \frac{dz}{z-z_0} = \cdots = \int_{\tilde{\gamma}_{m-1}} \frac{dz}{z-z_0}$$

$$= \int_{\tilde{\gamma}_m} \frac{dz}{z-z_0} = \int_{\gamma_2} \frac{dz}{z-z_0} \qquad\qquad (*)$$

$$\Rightarrow n(\gamma_1; z_0) = n(\gamma_2; z_0).$$

We summarize the above as

The homotopic invariance of the winding numbers. Let Ω be a domain or a nonempty open set in $\mathbf{C}$ and $z_0 \notin \Omega$ be a fixed point. Suppose two closed curves γ_1 and γ_2 are freely homotopic in Ω, then the winding numbers

$$n(\gamma_1; z_0) = n(\gamma_2; z_0). \qquad\qquad (4.2.3.1)$$

Roughly speaking, this means that when γ_1 is deformed continuously in Ω to γ_2 (so that z_0 is never touched during the process of deformation), then the winding number $n(\gamma; z_0)$ remains unchanged. Readers are strongly urged to use the definition (2.7.1.11) to give yet another proof of (4.2.3.1) (see Exercise B (2)(d) of Sec. 2.7.1).

4.2.4. *Homotopic form of Cauchy integral theorem*

In the last step $(*)$ of the proof for (4.2.3.1), use $f(z)$, analytic in Ω, to replace $\frac{1}{z-z_0}$ and we obtain the following

Homotopic form of the Cauchy integral theorem. Let f be analytic in a domain Ω. Suppose two (continuous) curves γ_1 and γ_2 in Ω satisfy the following condition:

(1) either γ_1 and γ_2 are freely homotopic in Ω;
(2) or γ_1 and γ_2 are homotopic with fixed end points in Ω.

Then

$$\int_{\gamma_1} f(z)dz = \int_{\gamma_2} f(z)dz.$$

In particular, if a closed curve γ in Ω is homotopic to a point in Ω, then

$$\int_{\gamma} f(z)dz = 0. \qquad\qquad (4.2.4.1)$$

Eberlein (1975) presented another proof using Green's formula of discrete type.

Combining (4.2.4.1) and Example 3 in Sec. 4.2.3, we obtain

The Cauchy integral theorem in a simply connected domain. Let f be analytic in a *simply connected* domain Ω.

(1) For any two (continuous) curves γ_1 and γ_2 with the same end points in Ω,

$$\int_{\gamma_1} f(z)dz = \int_{\gamma_2} f(z)dz.$$

(2) For any closed (continuous) curve γ in Ω,

$$\int_{\gamma} f(z)dz = 0. \tag{4.2.4.2}$$

We illustrate three examples.

Example 1. Show that $\gamma_1 : z(t) = e^{it}$, $0 \le t \le 2\pi$ and $\gamma_2 : z(t) = e^{-it}$, $0 \le t \le 2\pi$ are freely homotopic in $\mathbf{C}$ but not in $\mathbf{C} - \{0\}$.

Solution. As a point set, both γ_1 and γ_2 represent the unit circle $|z| = 1$ but in opposite directions. $H(t,s) : [0, 2\pi] \times [0,1] \to \mathbf{C}$ defined by $H(t,s) = e^{it} - 2is\sin t$ is a homotopic between γ_1 and γ_2 in $\mathbf{C}$.

On the other hand, $n(\gamma_1; 0) = 1$ while $n(\gamma_2; 0) = -1$. By (4.2.3.1), γ_1 and γ_2 are never homotopic in $\mathbf{C} - \{0\}$.

Example 2. Let Ω be a starlike domain (see Example 2 in Sec. 4.2.2) with center z_*.

(1) If f is analytic in Ω and γ is any rectifiable closed curve in Ω, then

$$\int_{\gamma} f(z)dz = 0.$$

Hence f has a primitive in Ω (see (4.1.2)).

Therefore, any analytic function in a domain has a primitive *locally* (namely, in a neighborhood of each point in the domain).

(2) If f is continuous on Ω and analytic in $\Omega - \{z_*\}$, then f has a primitive in Ω. Consequently if f is continuous on a domain D and analytic in $D - \{z_0\}$, where z_0 is a fixed point in D, then f definitely has a primitive in D.

Before engaging in the proof, observe that:

(1) A convex domain is starlike, with each of its points as center.
(2) The slit domain $\mathbf{C} - (\infty, 0]$ is starlike, with each point in the positive real axis as center.

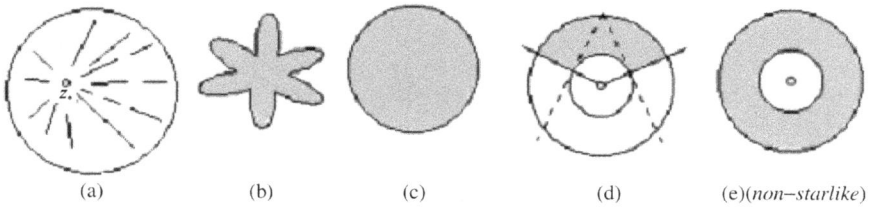

(a)   (b)   (c)   (d)   (e)(*non–starlike*)

Fig. 4.11

(3) The domain obtained by deleting finitely many segments in $|z| < 1$, all their extended lines passing through the same point z_* with $|z_*| < 1$, and having the end points on $|z| = 1$, is starlike with center z_*. See Fig. 4.11(a).

(4) The shaded ringlike sector domain in Fig. 4.11(d) is starlike.

(5) Yet $\mathbf{C} - \{0\}$ and $r < |z| < R$ are not starlike.

Sketch of Proof

(1) According to (1) in Example 2 of Sec. 4.2.2 and (4.2.4.1), it follows easily that $\int_\gamma f(z)dz = 0$ should hold.

Let

$$F(z) = \int_{z_*}^z f(z)dz, \quad z \in \Omega,$$

where the path of integration is the segment $\overline{z_* z}$. The usual argument will lead to $F'(z) = f(z)$.

(2) All we need to do is to show that the Cauchy–Goursat theorem (3.4.2.2) is still valid for each triangle $\Delta_{z_* z_1 w_1}$, where z_1 and w_1 are any two points in Ω, so that the triangle is still contained in Ω. Then $\int_\Delta f(z)dz = 0$ for any triangle Δ in Ω and the argument as in the proof of (3.4.3.4) will lead the existence of a primitive.

Fix such a triangle $\Delta_{z_* z_1 w_1}$ and construct a sequence of decreasing triangles $\Delta_{z_* z_n w_{n+1}}$ with its perimeter and diameter approaching zero as $n \to \infty$. See Fig. 4.12. Then

$$\int_{\partial \Delta_{z_* z_1 w_1}} f(z)dz = \int_{\partial \Delta_{z_* z_1 w_2}} f(z)dz = \int_{\partial \Delta_{z_* w_2 z_2}} f(z)dz = \cdots$$

$$= \int_{\partial \Delta_{z_* z_n w_{n+1}}} f(z)dz \to 0$$

as $n \to \infty$ by continuity of f at z_* (compare to the proof of (3.4.2.17) and Fig. 3.27).

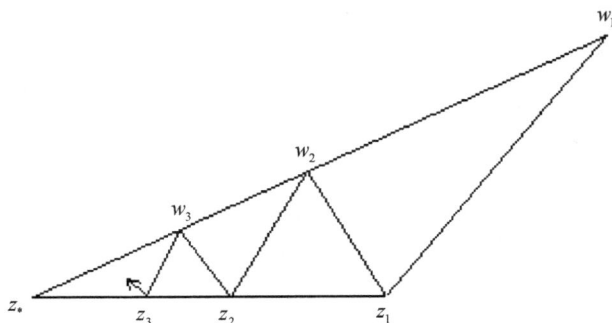

Fig. 4.12

Note: Take, for instance, $f(z) = \frac{1}{z}$ in $\mathbf{C} - (\infty, 0]$. Then

$$F(z) = \int_1^z \frac{1}{\zeta} d\zeta = \log|z| + i \operatorname{Arg} z = \operatorname{Log} z, \quad z \in \mathbf{C} - (\infty, 0].$$

Example 3. Let γ be a rectifiable Jordan closed curve. Suppose f is analytic in $\operatorname{Int} \gamma$ and continuous on $\overline{\operatorname{Int} \gamma}$. If there is a point $z_* \in \mathbf{C}$ such that each ray starting from z_* will intersect γ at exactly one point, then

$$\int_\gamma f(z) dz = 0.$$

Note that the existence of such a z_* is not necessarily, and an argument of compactness and uniform continuity still guarantees that $\int_\gamma f(z) dz = 0$ (see $(3.4.2.1)'$ and Exercise A (1) there).

Sketch of Proof

After performing a translation $z \to z - z_*$, we may suppose that $z_* = 0$. Treat $z_* = 0$ as the polar point, γ can be put in polar form $z(t) = r(t)e^{it}$, $0 \le t \le 2\pi$ and $z(0) = z(2\pi)$. For any fixed ρ, $0 < \rho < 1$, the curve $\gamma_\rho : z(t) = \rho r(t)e^{it}$, $0 \le t \le 2\pi$, lies in $\operatorname{Int} \gamma$ and is similar to γ itself. Then, by $(4.2.4.2)$,

$$\int_{\gamma_\rho} f(\zeta) d\zeta = 0$$

$$\Rightarrow \int_\gamma f(\rho\zeta) d(\rho\zeta) = \rho \int_\gamma f(\rho\zeta) d\zeta = 0$$

$$\Rightarrow \int_\gamma f(\rho\zeta) d\zeta = 0 \quad \text{for } 0 < \rho < 1.$$

Now, consider $\int_\gamma f(\zeta)d\zeta = \int_\gamma [f(\zeta) - f(\rho\zeta)]d\zeta$. By uniform continuity of f on $\overline{\text{Int}\,\gamma}$, for each $\varepsilon > 0$, there is a $\delta = \delta(\varepsilon) > 0$ so that for any two points z_1, z_2 in $\overline{\text{Int}\,\gamma}$ with $|z_1 - z_2| < \delta$, $|f(z_1) - f(z_2)| < \varepsilon$ always holds. In order to have, for a given $\zeta = r(t)e^{it} \in \gamma$, $|\zeta - \rho\zeta| = (1 - \rho)|r(t)| < \delta$, we only need to choose $\rho > 1 - \frac{\delta}{M}$, where $M = \max_{0 \le t \le 2\pi} |r(t)|$. Choose such a ρ. Then $|f(\zeta) - f(\rho\zeta)| < \varepsilon$ holds for any $\zeta \in \gamma$. This leads to

$$\left| \int_\gamma f(\zeta)d\zeta \right| \le \int_\gamma |f(\zeta) - f(\rho\zeta)||d\zeta| \le \varepsilon \cdot \text{ the length of } \gamma$$

$$\Rightarrow \int_\gamma f(\zeta)d\zeta = 0.$$

4.3.   The General Form of Cauchy Integral Theorem: Homology

According to (4.2.1.5), we may restrict *the path γ of integration* in $\int_\gamma f(z)dz$ to *rectifiable* or *piecewise differentiable curve* in case the integrand f is known to be *analytic*.

From now on, all curves concerned are supposed to be rectifiable or piecewise differential, unless stated specifically.

Let a curve γ be divided into subcurves $\gamma_1, \ldots, \gamma_n$ so that $\gamma = \gamma_1 + \cdots + \gamma_n$ holds (see (2) in (2.9.10)). In case f is continuous along γ, then so is f along each γ_j and

$$\int_{\gamma_1 + \cdots + \gamma_n} f(z)dz = \int_{\gamma_1} f(z)dz + \cdots + \int_{\gamma_n} f(z)dz. \tag{4.3.1}$$

Conversely, for any finitely many curves $\gamma_1, \ldots, \gamma_n$ regardless of whether the terminal point of a curve is the initial point of another one, as long as f is continuous along each γ_j for $1 \le j \le n$, $\int_{\gamma_j} f(z)dz$ is well defined and the right sum in (4.3.1) is meaningful. Hence, we formulate the *formal sum* $\gamma_1 + \cdots + \gamma_n$ and *define the integral of f along it* by (4.3.1).

Call the *formal sum*

$$\sum_{j=1}^{n} \gamma_j = \gamma_1 + \cdots + \gamma_n, \tag{4.3.2}$$

a *chain* composed by curves $\gamma_1, \ldots, \gamma_n$. The key point in the introduction of chains lies on the benefit of the linear property of line integration as shown in (4.3.1).

Four sections are divided.

Section 4.3.1 develops preliminary explanations about chains and *closed* chains or *cycles*, composed of closed curves. Combining closed chains and the concept of winding numbers (see Exercise B of Sec. 2.7.1 and Sec. 3.5.2), we introduce the homology between two curves and, in particular, the concept of a cycle being *homologous to zero* in a domain.

A cycle being homologous to zero can be used to characterize the *simple connectedness* of a domain. Section 4.3.2 also introduces a *homology basis* for a finitely many connected domain.

Section 4.3.3 states homologous forms of Cauchy integral theorem in a simply or multiply connected domain.

Section 4.3.4 gives Emil Artin's proof of this general Cauchy integral theorem. Artin's method uses only Cauchy–Goursat theorem (3.2.4.2) and topological properties of curves. Beardon (See Ref. [8], pp. 163–165) gave another proof. One might also refer to Ref. [1], pp. 142–144; 66, pp. 188–191. Yet a third proof was given by Dixon (see Ref. [26], A brief proof of Cauchy's integral theorem, *Proc. Amer. Math. Soc.* (1971) **29**, 625–626). Dixon proved this theorem based on properties of analytic function without recourse to topological considerations.

4.3.1.   *Cycles and homology of two cycles*

In (4.3.1), there are many different ways of dividing a given curve γ into subcurves $\gamma_1, \ldots, \gamma_n$ but the right-hand sum is always equal to $\int_\gamma f(z)dz$. Similarly, the same function f might have equal integral values when integrating along different chains. Hence, we designate two chains *equal*, namely,

$$\gamma_1 + \cdots + \gamma_n = \tilde{\gamma}_1 + \cdots + \tilde{\gamma}_m$$

$\Leftrightarrow \int_{\gamma_1 + \cdots + \gamma_n} f(z)dz = \int_{\tilde{\gamma}_1 + \cdots + \tilde{\gamma}_m} f(z)dz$ for all continuous functions f on
γ_j $(1 \leq j \leq n)$ and $\tilde{\gamma}_k$ $(1 \leq k \leq m)$.   (4.3.1.1)

According to properties of line integrals shown in (2.9.10), we can realize that the following "operations" cannot change the equality of two chains defined by (4.3.1.1):

(1) The interchange of two curves γ_1 and γ_2 in the sum: $\gamma_1 + \gamma_2 = \gamma_2 + \gamma_1$.
(2) The different division of a given curve γ into subcurves $\gamma_1, \ldots, \gamma_n$.
(3) The combination of several subcurves into a "larger" curve.

    Take, for instance, in case 2 above, $\gamma_1 + \cdots + \gamma_n$ represents γ and thus, we can use γ to replace $\gamma_1 + \cdots + \gamma_n$. In case $\gamma_1, \ldots, \gamma_n$ are all equal to the same curve γ, since

$$\int_{\gamma_1} f(z)dz = \cdots = \int_{\gamma_n} f(z)dz = \int_{\gamma} f(z)dz,$$

we denote $\gamma_1 + \cdots + \gamma_n$ by $n\gamma$ specifically so that

$$\int_{\gamma_1} f(z)dz + \cdots + \int_{\gamma_n} f(z)dz = \int_{\gamma_1 + \cdots + \gamma_n} f(z)dz$$

$$= \int_{n\gamma} f(z)dz = n\int_{\gamma} f(z)dz$$

is meaningful.

(4) The change of the parameters of a curve.

(5) Two curves γ and $-\gamma$, with same loci but with opposite directions, can be cancelled out in the sum, namely $\gamma + (-\gamma) = 0$ (0 denotes the point curve determined by the initial point of γ). For any integer n, it is understood that

$$n\gamma = -n(-\gamma).$$

Also $0\gamma = 0$.

Consequently, we *redefine* and *denote* a chain as follows: For arbitrarily given finitely many distinct curves $\gamma_1, \ldots, \gamma_n$ and any integers $m_1, \ldots, m_n$, call the *formal sum*

$$\gamma = m_1\gamma_1 + \cdots + m_n\gamma_n = \sum_{j=1}^{n} m_j\gamma_j, \qquad (4.3.1.2)$$

a *chain* and $\gamma_1, \ldots, \gamma_n$ its *component curves*, and $m_1, \ldots, m_n$ the corresponding *coefficients*.

The term with zero coefficient (for example $0 \cdot \gamma_{n+1}$) can be added freely to a sum, so that two chains have the same number of component curves. Therefore, it is possible to introduce the *addition operation* among chains. For instance,

$$\gamma = m_1\gamma_1 + \cdots + m_n\gamma_n = m_1\gamma_1 + \cdots + m_n\gamma_n + 0\tilde{\gamma}_1 + \cdots + 0\tilde{\gamma}_k;$$

$$\tilde{\gamma} = n_1\tilde{\gamma}_1 + \cdots + n_k\tilde{\gamma}_k = 0\gamma_1 + \cdots + 0\gamma_n + n_1\tilde{\gamma}_1 + \cdots + n_k\tilde{\gamma}_k$$

$$\Rightarrow \gamma + \tilde{\gamma} = (m_1 + 0)\gamma_1 + \cdots + (m_n + 0)\gamma_n$$

$$+ (0 + n_1)\tilde{\gamma}_1 + \cdots + (0 + n_k)\tilde{\gamma}_k$$

$$= m_1\gamma_1 + \cdots + m_n\gamma_n + n_1\tilde{\gamma}_1 + \cdots + n_k\tilde{\gamma}_k, \quad (4.3.1.3)$$

call the *sum* (chain) of γ and $\tilde{\gamma}$. *Zero chain* 0 means a chain with all its coefficients equal to zero or the empty set.

If the component curves $\gamma_1, \ldots, \gamma_n$ of a chain $\gamma = \sum_{j=1}^{n} m_j \gamma_j$ all lie in a domain or an open set Ω, then we call γ *a chain in* Ω.

A chain is called a *cycle* if it can be represented as a sum of closed curves. In this case, in any representation the initial and the terminal points of each component curve are identical in pairs.

Let $\gamma = \sum_{j=1}^{n} m_j \gamma_j$ be a chain in a domain Ω and f be a complex-valued function in Ω. If each integral $\int_{\gamma_j} f(z) dz$ exists for $1 \le j \le n$, the *line integral* of f *along* γ is defined linearly by

$$\int_{\gamma} f(z) dz = \sum_{j=1}^{n} m_j \int_{\gamma_j} f(z) dz. \qquad (4.3.1.4)$$

In particular, if γ is a *continuous* chain (namely, each component curve γ_j is continuous), then f is required to be *analytic* in Ω; if γ is a *rectifiable* or *piecewise differentiable* chain, then f is supposed to be *continuous* in Ω.

If $\gamma = \sum_{j=1}^{n} m_j \gamma_j$ is a cycle and z_0 is a fixed point which does not lie on any one of γ_j for $1 \le j \le n$, the *winding number* of γ about z_0 is defined and denoted by

$$n(\gamma; z_0) = \frac{1}{2\pi i} \int_{\gamma} \frac{dz}{z - z_0} = \sum_{j=1}^{n} m_j n(\gamma_j; z_0). \qquad (4.3.1.5)$$

In case γ and $\tilde{\gamma}$ are cycles in Ω and $z_0 \notin \gamma \cup \tilde{\gamma}$, we have the *addition property*:

$$n(\gamma + \tilde{\gamma}; z_0) = n(\gamma; z_0) + n(\tilde{\gamma}; z_0). \qquad (4.3.1.6)$$

Suppose Ω is an open set or a domain in $\mathbf{C}$. A cycle γ in Ω is said to be *homologous to zero* in Ω if

$$n(\gamma; z) = 0 \quad \text{for all } z \in \mathbf{C} - \Omega \qquad (4.3.1.7)$$

and is denoted as $\gamma \sim 0 \pmod{\Omega}$ or simply as $\gamma \sim 0$. If O is an open set containing Ω, then $\gamma \sim 0 \pmod{\Omega}$ implies that $\gamma \sim 0 \pmod{O}$.

Two cycles γ_1 and γ_2 in Ω are said to be *homologous to each other* in Ω if

$$n(\gamma_1; z) = n(\gamma_2; z), \quad z \in \mathbf{C} - \Omega$$

$$\Leftrightarrow n(\gamma_1 - \gamma_2; z) = 0, \quad z \in \mathbf{C} - \Omega, \text{ namely,}$$

$$\gamma_1 - \gamma_2 \text{ is homologous to zero in } \Omega, \qquad (4.3.1.8)$$

and is denoted as $\gamma_1 \sim \gamma_2 \pmod{\Omega}$ or simply as $\gamma_1 \sim \gamma_2$.

Let γ_1 and γ_2 be two curves in Ω but not closed, yet they have the same
end points. In case the closed curve $\gamma_1 + (-\gamma_2) = \gamma_1 - \gamma_2$ is homologous to
zero in Ω, then γ_1 and γ_2 are said to be *homologous to each other* in Ω and
is denoted as $\gamma_1 \sim \gamma_2 \pmod{\Omega}$ or $\gamma_1 \sim \gamma_2$.

For instance, let z_1, z_2, and z_3 be three distinct points in $\mathbf{C}$ and γ_j be a
circle with center at z_j and endowed with the counterclockwise direction for
$j = 1, 2, 3$. Then the cycle γ in Fig. 4.13 is homologous to $-2\gamma_1 + 2\gamma_2 - \gamma_3$ in
$\mathbf{C} - \{z_1, z_2, z_3\}$. While in Fig. 4.14, $\gamma \sim \gamma_1 + \gamma_2$ in Ω yet γ is not homologous
to either γ_1 or γ_2; neither is α homologous to β, too.

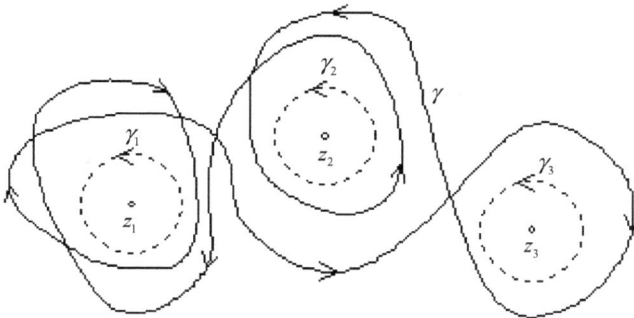

Fig. 4.13

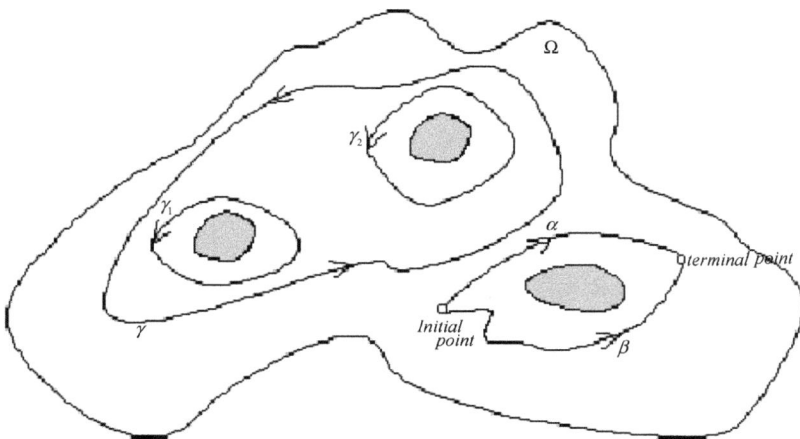

Fig. 4.14

Exercises A

(1) In Fig. 4.15, find out the cycles which are homologous to γ in Ω.

(2) Let α be a curve with the initial point 1 and terminal point i in $\mathbf{C} - \{0, -1\}$ as shown in Fig. 4.16. Try to find another curve β in $\mathbf{C} - \{0, 1\}$ with the same end points as α such that $\alpha \sim \beta$ in Ω.

(3) Let $\gamma_1 : z(t) = e^{it}$, $0 \le t \le 2\pi$; $\gamma_2 : z(t) = \frac{5}{3} + e^{it}$, $0 \le t \le 2\pi$, and $\gamma_3 : z(t) = -1 + 2e^{it}$, $0 \le t \le 2\pi$. Determine if the following cycles are homologous to zero in Ω.

  (a) $\gamma = 2\gamma_1 + \gamma_2 - \gamma_3$ in $\Omega = \{0 < |z| < 5\}$.
  (b) $\gamma = \gamma_1 - \gamma_2 + \gamma_3$ in $\mathbf{C} - \{\sqrt{2}i\}$.
  (c) $\gamma = \gamma_1 + \gamma_3 - 2\gamma_2$ in $\mathbf{C} - [-\frac{1}{2}, \frac{1}{2}]$.

4.3.2. Simply and finitely connected domains: Homology basis

Recall that a domain Ω in $\mathbf{C}$ is called *simply connected* if $\mathbf{C}^* - \Omega$ is connected in $\mathbf{C}^*$; *n-connected* if $\mathbf{C}^* - \Omega$ has n components in $\mathbf{C}^*$ (see Section (4) in Sec. 2.4).

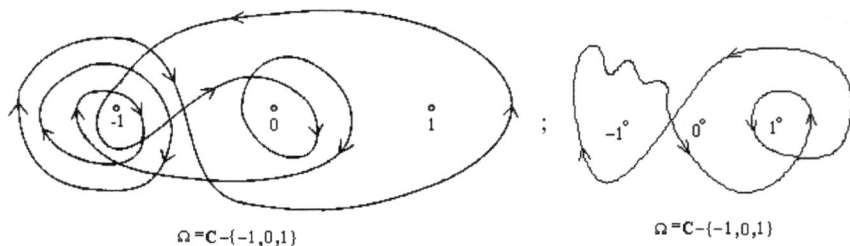

$\Omega = \mathbf{C} - \{-1, 0, 1\}$ $\Omega = \mathbf{C} - \{-1, 0, 1\}$

Fig. 4.15

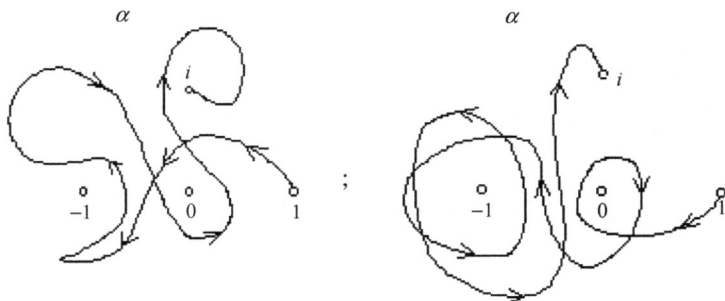

Fig. 4.16

Here, we try to use homology to characterize a simply connected domain and then, find a homology basis for a $(n+1)$-connected domain.

Fix a domain Ω in $\mathbf{C}$ whose complement $\mathbf{C}^* - \Omega$ in $\mathbf{C}^*$ has two components Ω_1 and Ω_2. Suppose Ω_1 is bounded while Ω_2 is unbound and thus, $\infty \in \Omega_2$. Let

$$d = \text{dist}(\Omega_1, \Omega_2),$$

the distance between Ω_1 and Ω_2. Note that $d > 0$ holds.

Choose a point $z_0 \in \Omega_1$. Use a *net* of squares, each side of which is less than $\frac{d}{\sqrt{2}}$, to cover the entire plane $\mathbf{C}$ such that z_0 is the center of one of these squares. See Fig. 4.17. A general square in the net is denoted by S (consisting of ∂S and Int S) and endow ∂S with the counterclockwise direction. Consider the cycle

$$\gamma = \sum_{\Omega_1 \cap S_j \neq \phi} \partial S_j. \tag{4.3.2.1}$$

This is a finite sum since Ω_1 is compact. Such a net can be refined so that no vertices of the squares S_j in (4.3.2.1) lie on the boundary $\partial \Omega_1$ of Ω_1. Since z_0 lies exactly in only one of such squares,

$$n(\gamma; z_0) = 1.$$

Each square S has its diameter less than d. Thus γ has no point of intersection with Ω_2. To a square S_j, appearing in the definition of γ, which possesses one side with nonempty intersection with Ω_1, then this side should be the common edges of two neighboring such squares. Hence, $\gamma \cap \Omega_1 = \phi$ holds and γ is a *rectangular Jordan closed curve in* Ω. Observe that $\mathbf{C}^* - \gamma$

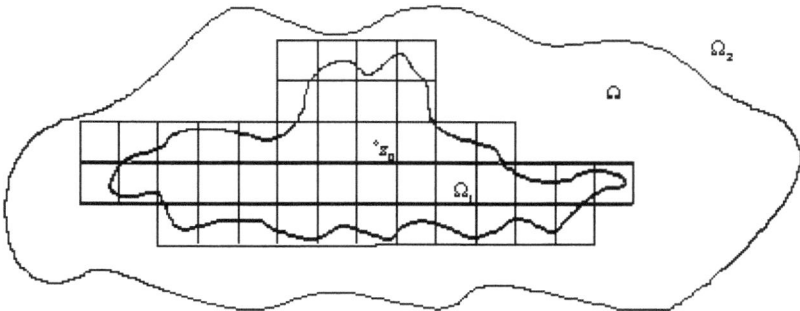

Fig. 4.17

has two components. According to (4) in (3.5.2.4),

$$n(\gamma; z) = n(\gamma; z_0) = 1, \quad z \in \Omega_1, \quad \text{and}$$
$$n(\gamma; z) = 0, \quad z \in \Omega_2,$$

(4.3.2.2)

since Ω_2 lies in the unbounded component of $\mathbf{C}^* - \gamma$ or any point $z \in \Omega_2$ never appears in the interior $\text{Int}\, S_j$ of each S_j in the definition of γ.

As a first consequence of (4.3.2.2), we have (compare to Example 3 in Sec. 4.2.2)

A characteristic property of simply connected domains. Let Ω be a domain in $\mathbf{C}$. Then:

(1) Ω is a simply connected domain (namely, $\mathbf{C}^* - \Omega$ is connected in $\mathbf{C}^*$).

$\Leftrightarrow$ (2) Any cycle γ in Ω is homologous to zero in Ω, namely,

$$n(\gamma; z) = 0, \quad z \in \mathbf{C}^* - \Omega. \tag{4.3.2.3}$$

Owing to the process of construction leading to (4.3.2.2), we only need to **prove** (1) $\Rightarrow$ (2): Since $\mathbf{C}^* - \Omega$ is connected, it should lie in a component of $\mathbf{C}^* - \gamma$; since $\infty \in \mathbf{C}^* - \Omega$, it lies in the unbounded component of $\mathbf{C}^* - \gamma$. By (4) in (3.5.2.4), it follows that $n(\gamma; z) = 0$ for $z \in \mathbf{C}^* - \Omega$.

As a second application of (4.3.2.2), let Ω be a $(n+1)$-connected domain in $\mathbf{C}$. Suppose $\Omega_1, \ldots, \Omega_n$, Ω_∞ are the components of $\mathbf{C}^* - \Omega$ with $\infty \in \Omega_\infty$, the unbounded one.

Apply the process of construction leading to (4.3.2.2) to each Ω_j, $1 \le j \le n$, and Ω_∞. There is a Jordan closed curve γ_j in Ω satisfying

$$n(\gamma_j; z) = \begin{cases} 1, z \in \Omega_j, \\ 0, z \in \Omega_k, \quad k \ne j \text{ (including } k = \infty). \end{cases} \tag{4.3.2.4}$$

To each cycle γ in Ω, consider the cycle

$$\Gamma = \gamma - \sum_{j=1}^{n} n(\gamma; z_j)\gamma_j, \quad z_j \in \Omega_j, \quad 1 \le j \le n.$$

Recall that $n(\gamma; z_j)$ is a constant for all $z_j \in \Omega_j, 1 \le j \le n$. In case $z_k \in \Omega_k$, $1 \le k \le n$,

$$n(\Gamma; z_k) = n(\gamma; z_k) - \sum_{j=1}^{n} n(\gamma; z_j)n(\gamma_j; z_k) = n(\gamma; z_k) - n(\gamma; z_k)$$

(by using (4.3.2.4))

$$= 0;$$

if $z \in \Omega_\infty$,

$$n(\Gamma; z) = n(\gamma; z) - \sum_{j=1}^{n} n(\gamma; z_j)n(\gamma_j; z) = 0 - 0 = 0.$$

Combining together, these facts show that Γ is homologous to zero in Ω.

Summarize the above as the

Homology basis for a $(n+1)$-connected domain Ω in $\mathbf{C}$: Let $\Omega_1, \ldots, \Omega_n$ and Ω_∞ be the $(n+1)$ components of $\mathbf{C}^* - \Omega$ such that $\infty \in \Omega_\infty$, the unbounded component. Then there are n polygonal Jordan closed curves $\gamma_1, \ldots, \gamma_n$ in Ω, satisfying: For $1 \le j \le n$,

$$n(\gamma_j; z) = \begin{cases} 1, & z \in \Omega_j, \\ 0, & z \in \Omega_1, \ldots, \Omega_{j-1}, \Omega_{j+1}, \ldots, \Omega_n, \Omega_\infty \end{cases}$$

and, for any cycle γ in Ω, there are unique constants $c_j = n(\gamma; z)$, $z \in \Omega_j$ for $1 \le j \le n$, so that γ is homologous to $\sum_{j=1}^{n} c_j \gamma_j$ in Ω, i.e.,

$$\gamma \sim \sum_{j=1}^{n} c_j \gamma_j \quad (\bmod \Omega).$$

In this case, $\{\gamma_1, \ldots, \gamma_n\}$ is called a *homology basis* for Ω.   (4.3.2.5)

For a detailed account about homology basis, see Ref. [6], Chap. 1, and [75], Chap. 5.

4.3.3.   *Homologous form of Cauchy integral theorem*

Cauchy integral theorem in simple form (3.4.2.1) says that, for any rectifiable closed curve γ in a simply connected domain Ω and any function f analytic in Ω,

$$\int_\gamma f(z)dz = 0$$

always holds. In case Ω is not simply connected, then exist a point $z_0 \notin \Omega$ and a cycle γ in Ω so that $n(\gamma; z_0) \ne 0$ (see (4.3.2.3)). Even though $f(z) = \frac{1}{z-z_0}$ is analytic in Ω, yet

$$\int_\gamma f(z)dz = \int_\gamma \frac{dz}{z - z_0} = 2\pi i n(\gamma; z_0) \ne 0.$$

On the other hand, suppose there is a cycle γ in Ω such that

$$\frac{1}{2\pi i} \int_\gamma \frac{dz}{z - a} = 0 \quad \text{for all } a \notin \Omega.$$ $(*_1)$

Then, for any function f analytic in Ω,

$$\int_\gamma f(z)dz = 0, \qquad\qquad (*_2)$$

always holds. As a matter of fact, $(*_1)$ and $(*_2)$ *are equivalent.* This is the content of the following remarkable fact (4.3.3.1). For the moment, to realize how $(*_1) \Rightarrow (*_2)$, we consider a *simpler* case that f is analytic in a deleted disk $0 < |z - z_0| < R$. Later in Sec. 4.9, we are going to show that f can be represented by a Laurent series $\sum_{n=-\infty}^{\infty} a_n (z - z_0)^n$ which converges to $f(z)$ absolutely in $0 < |z - z_0| < R$ and uniformly on every compact set of $0 < |z - z_0| < R$. Consequently, for a cycle γ in $0 < |z - z_0| < R$ which is homologous to zero,

$$\int_\gamma f(z)dz = \int_\gamma \left[\sum_{n=-\infty}^{\infty} a_n (z - z_0)^n \right] dz$$

$$= \sum_{n=-\infty}^{\infty} a_n \int_\gamma (z - z_0)^n dz \quad \text{(by uniform convergence)}$$

$$= a_{-1} \int_\gamma \frac{dz}{z - z_0} \quad \text{(using results in Example 1 of Sec. 4.1)}$$

and the claim follows. The coefficient $a_{-1} = \int_{|z-z_0|=\rho} f(z)dz$ $(0 < \rho < R)$ is called the *residue* of f at z_0, to be explored fully in Sec. 4.11.

Consequently it is reasonable to expect the following

Homologous form of Cauchy integral theorem. Let γ be a cycle in a domain Ω. If γ is homologous to zero in Ω, then for analytic functions f in Ω,

$$\int_\gamma f(z)dz = 0$$

always holds. In particular, for any two cycles γ_1 and γ_2 in Ω, or any two continuous curves γ_1 and γ_2 in Ω with the same end points, if γ_1 is homologous to γ_2 in Ω, then

$$\int_{\gamma_1} f(z)dz = \int_{\gamma_2} f(z)dz. \qquad\qquad (4.3.3.1)$$

The proof is postponed to Sec. 4.3.4.

Applying (4.3.3.1) to $\gamma \sim \sum_{j=1}^{n} c_j \gamma_j \pmod{\Omega}$ in (4.3.2.5), we have (compare to (1) in $(3.4.2.4)'$)

The Cauchy integral theorem in a $(n+1)$-connected domain Ω. Adopt notations and assumptions in (4.3.2.5). If f is analytic in Ω, then

$$\int_\gamma f(z)dz = \sum_{j=1}^{n} c_j \int_{\gamma_j} f(z)dz$$

where, each integral value

$$P_j = \int_{\gamma_j} f(z)dz, \quad 1 \leq j \leq n$$

is dependent only on f but independent of γ_j, and is called the *period* of the *indefinite integral* $\int_\gamma f(z)dz$ or f around the *component* Ω_j of $\mathbf{C}^* - \Omega$.

(4.3.3.2)

Consequently, the integral

$$\int_\gamma f(z)dz = \sum_{j=1}^{n} c_j P_j, \qquad (4.3.3.3)$$

is a linear combination of these periods P_j with integer coefficients c_j. Fix a point $z_0 \in \Omega$ and let z be a varying point in Ω. If Γ_1 and Γ_2 are two curves in Ω, connecting z_0 to z. Then $\Gamma_1 - \Gamma_2$ is a cycle and is homologous

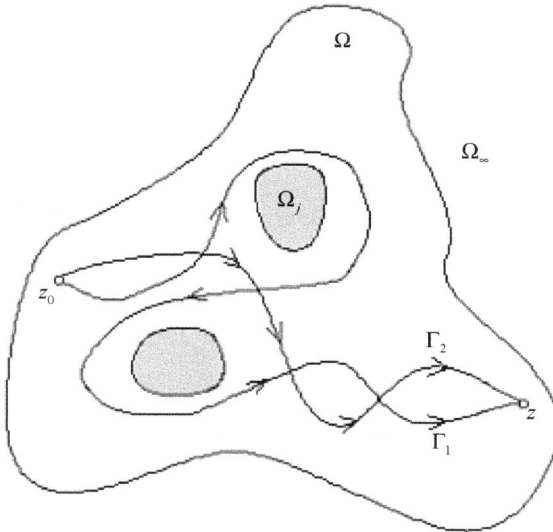

Fig. 4.18

to some $\sum_{j=1}^{n} c_j \gamma_j$. See Fig. 4.18. In this case,

$$\int_{\Gamma_1} f(z)dz = \int_{\Gamma_2} f(z)dz + \int_{\Gamma_1 - \Gamma_2} f(z)dz$$

$$= \int_{\Gamma_2} f(z)dz + \sum_{j=1}^{n} c_j P_j \qquad (4.3.3.4)$$

where c_j is the winding number of $\Gamma_1 - \Gamma_2$ surrounding the component Ω_j, $1 \le j \le n$. Therefore,

$$P_j = 0, \quad 1 \le j \le n, \qquad (4.3.3.5)$$

is a necessary and sufficient condition for the existence of a *single-valued indefinite integral* $\int_r f(z)dz$ in Ω, or the existence of a single-valued *primitive function* of $f(z)$ in Ω (see Sec. 4.1).

As an important application, we have (compare to Exercise A (2) of Sec. 3.4.2),

The residue theorem. Let $z_1, \ldots, z_n$ be distinct points in a domain Ω. Suppose that

(1) f is analytic in $\Omega - \{z_1, \ldots, z_n\}$; and
(2) γ is a cycle in $\Omega - \{z_1, \ldots, z_n\}$ and is homologous to zero in Ω.

Choose closed disks $D_j : |z - z_j| \le r_j$, $1 \le j \le n$, satisfying:

(1) $D_j \subseteq \Omega$, $\quad 1 \le j \le n$;
(2) $D_j \cap D_k = \phi$, $\quad 1 \le j \ne k \le n$; and
(3) the boundary $\partial D_j : |z - z_0| = r_j$ is endowed with the counterclockwise direction.

Let $c_j = n(\gamma; z_j)$, $1 \le j \le n$. Then γ is homologous to $\sum_{j=1}^{n} c_j \partial D_j$ in $\Omega - \{z_1, \ldots, z_n\}$ and

$$\int_{\gamma} f(z)dz = \sum_{j=1}^{n} c_j \int_{\partial D_j} f(z)dz.$$

See Fig. 4.19. In particular, if γ is a rectifiable Jordan closed curve such that $z_1, \ldots, z_n$ lie in the interior Int γ, then

$$\int_{\gamma} f(z)dz = \sum_{j=1}^{n} \int_{\partial D_j} f(z)dz. \qquad (4.3.3.6)$$

For applications of residue theorem, see Secs. 3.5.3, 3.5.4, and 4.12.

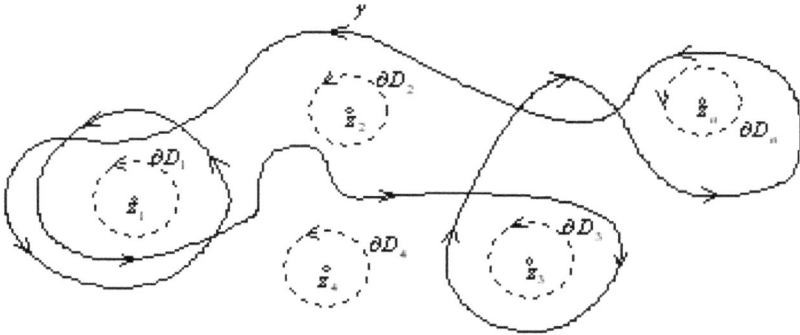

Fig. 4.19

4.3.4.   *Artin's proof*

Artin proved

An extension of Cauchy integral theorem. Let $P(z)$ and $Q(z)$ be complex-valued continuous functions on a domain Ω in $\mathbf{C}$. Then

    (1) $Pdx + Qdy$ is *locally exact* (namely, it is exact in some neighborhood of each point in Ω (see (1) in (4.1.2)).

$\Leftrightarrow$ (2) For each cycle γ in Ω which is homologous to zero in Ω,

$$\int_\gamma Pdx + Qdy = 0.$$

In particular, if f is analytic on Ω, then

$$\int_\gamma f(z)dz = 0$$

for any cycle γ in Ω which is homologous to zero in Ω.   (4.3.4.1)

According to (4.1.2), it is obvious that $Pdx + Qdy$ is locally exact if and only if

$$\int_{\partial R} Pdx + Qdy = 0 \qquad\qquad\qquad (4.3.4.2)$$

for every rectangle R contained in Ω (i.e., $\overline{\text{Int } R} \subseteq \Omega$). In case f is analytic in Ω, then Cauchy–Goursat theorem (3.4.2.2) shows that $f(z)dz$ is definitely locally exact in Ω.

The **proof** of (1) $\Rightarrow$ (2) is separated into three steps.

Step 1. *Try to use a rectangular cycle to approximate a cycle* γ.

We state this formally in the following:

Lemma 1. *Let $\gamma = \sum_{j=1}^{n} m_j \gamma_j$ be a cycle, homologous to zero, in Ω. Then, there is a rectangular cycle (composed of segments parallel to the coordinate axes) Γ such that, for all locally exact forms $Pdx + Qdy$ in Ω,*

$$\int_{\gamma} Pdx + Qdy = \int_{\Gamma} Pdx + Qdy.$$

In this case, Γ is also homologous to zero in Ω. (4.3.4.3)

Proof.   That such a Γ is homologous to zero in Ω is easily seen as follows. For any $z_0 \notin \Omega$, $\frac{1}{z-z_0}$ is analytic in Ω and hence, $\frac{1}{z-z_0} dz$ is locally exact in Ω. Then

$$n(\Gamma; z_0) = \frac{1}{2\pi i} \int_{\Gamma} \frac{dz}{z - z_0} = \frac{1}{2\pi i} \int_{\gamma} \frac{dz}{z - z_0} = n(\gamma; z_0) = 0$$

and Γ is homologous to zero in Ω.

Mainly owing to the linear properties concerned, without loss of generality, we may prove the validity of (2) by assuming, at the outset, that γ is a continuous curve such that $\int_{\gamma} Pdx + Qdy$ exists.

Let γ be parametrized by $z(t)$, $a \leq t \leq b$. Since $z([a,b])$ is compact in Ω and $\mathbf{C} - \Omega$ is closed, the *distance* $d = \text{dist}(\gamma, \mathbf{C} - \Omega) > 0$ holds. Here, in case $\Omega = \mathbf{C}$, the simple connectedness of $\mathbf{C}$ will ensure that the local exactness of $Pdx + Qdy$ on $\mathbf{C}$ is indeed a global one on $\mathbf{C}$. Namely, the result holds by using (1) in (4.1.2) and imitating the proof of the Cauchy integral theorem (3.4.2.1). Henceforth we just suppose Ω is a *proper* domain in $\mathbf{C}$.

Since $z(t)$ is uniformly continuous in $[a,b]$, for $0 < \varepsilon < d$, there is a $\delta = \delta(\varepsilon) > 0$ so that, wherever $t, t' \in [a,b]$ with $|t - t'| < \delta$, $|z(t) - z(t')| < \varepsilon$ always holds. Corresponding to this δ, choose a partition $a = t_0 < t_1 < \cdots < t_{n-1} < t_n = b$ satisfying $\max_{0 \leq j \leq n-1}(t_{j+1} - t_j) < \delta$.

Let γ_j denote the subcurve of γ, obtained by restricting $z(t)$ to the subinterval $[t_j, t_{j+1}]$. Fix an arbitrary point $t_j^* \in [t_j, t_{j+1}]$. Construct the disk $D_j : |z - z(t_j^*)| < \varepsilon$, $0 \leq j \leq n-1$. See Fig. 4.20: $z_j = z(t_j)$, $0 \leq j \leq n$, and $z_j^* = z(t_j^*)$, $0 \leq j \leq n - 1$. It follows obviously that

(1) $\gamma_j([t_j, t_{j+1}]) \subseteq D_j \subseteq \Omega$, $0 \leq j \leq n - 1$, and
(2) $z_j, z_{j+1} \in D_j$, $0 \leq j \leq n - 1$.

In D_j, use a polygonal curve Γ_j, composed of a horizontal segment and a vertical segment, to connected z_j to z_{j+1} for $0 \leq j \leq n-1$. Then as j goes from 0 to $n-1$, $\Gamma_0, \Gamma_1, \ldots, \Gamma_{n-1}$, in this ordering, form a *polygonal curve* Γ,

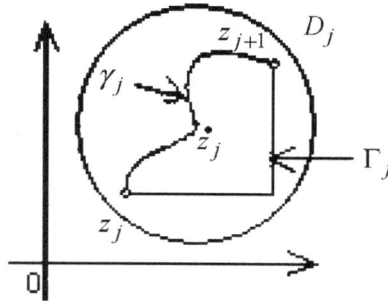

Fig. 4.20

having its segments parallel to the coordinate axes and with the same end points as γ.

Since each D_j is simply connected, $Pdx + Qdy$ is exact in D_j for $0 \le j \le n - 1$. By (4.1.2),

$$\int_{\gamma_j} Pdx + Qdy = \int_{\Gamma_j} Pdx + Qdy, \quad 0 \le j \le n - 1$$

$$\Rightarrow \left(\text{since } \gamma = \sum_{j=0}^{n-1} \gamma_j \text{ and } \Gamma = \sum_{j=0}^{n-1} \Gamma_j \right) \quad \int_{\gamma} Pdx + Qdy = \int_{\Gamma} Pdx + Qdy.$$

Note that, in case $f(z)dz = Pdx + Qdy$, where f is analytic in Ω, then this is a direct consequence of (4.2.1.5).

Step 2. *To express the rectangular cycle in terms of the boundaries of finitely many rectangles.*

Lemma 2. *If Γ is a rectangular cycle, homologous to zero, in Ω, then there are finitely many rectangles $R_1, \ldots, R_m$ in Ω so that*

$$\Gamma = \sum_{j=1}^{m} c_j \partial R_j$$

where each ∂R_j is the boundary of R_j in the counterclockwise direction and c_j is an integer.   (4.3.4.4)

Proof. Extend each component segment of Γ into a line. These horizontal and vertical lines divide the whole plane $\mathbf{C}$ into rectangles and unbounded rectangular domains. See Fig. 4.21. Pick up a fixed point a_j in the interior

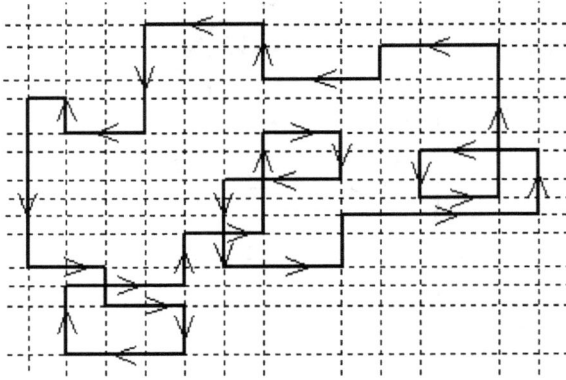

Fig. 4.21

of each *finite* (or bounded) rectangle R_j. Construct the cycle

$$\Gamma_0 = \sum_j n(\Gamma; a_j) \partial R_j,$$

where the sum runs over all finite rectangles. Γ_0 is well defined since each a_j is not on Γ. Recall again that ∂R_j is the boundary of R_j in the counterclockwise direction. In what follows, we try to show that

(a) whenever $n(\Gamma; a_j) \neq 0$, then R_j is contained in Ω (i.e., $\overline{\operatorname{Int} R_j} \subseteq \Omega$); and

(b) $\Gamma = \sum_j n(\Gamma; a_j) \partial R_j = \Gamma_0$.

For (a): For such a R_j, $n(\Gamma; a_j) \neq 0$ is a constant on $\operatorname{Int} R_j$. Since $\Gamma \sim 0$ in Ω, $a_j \in \Omega$. This shows that $\operatorname{Int} R_j \subseteq \Omega$.

In case $\partial R_j \cap \Gamma \neq \phi$, then $\partial R_j \cap \Gamma \subseteq \Omega$ holds since $\Gamma \subseteq \Omega$. If R_j has a boundary point $a \notin \Gamma$, then $n(\Gamma; a)$ is meaningful and, by continuity of the winding numbers, $n(\Gamma; a) \neq 0$ holds. Hence $a \in \Omega$. This proves that $\partial R_j \subseteq \Omega$.

Combining together, $R_j = \operatorname{Int} R_j \cup \partial R_j \subseteq \Omega$ holds.

For (b): Let the cycle

$$\sigma = \Gamma - \Gamma_0 = \Gamma - \sum_j n(\Gamma; a_j) \partial R_j.$$

Try to show that $n(\sigma; a) = 0$ for all $a \notin \sigma$ and then, to show that $\sigma = 0$. Thus $\Gamma = \Gamma_0$.

In case $a \notin \sigma$, four cases are divided.

Case 1. If a lies in an unbounded rectangular domain, then $n(\Gamma; a) = 0$ and $n(\partial R_j; a) = 0$ for all finite rectangles. It follows that $n(\sigma; a) = 0$.

Case 2. If $a \in \text{Int } R_k$ and $n(\Gamma; a_k) \neq 0$, then $n(\partial R_k; a) = 1$ yet $n(\partial R_j; a) = 0$ for $j \neq k$. Then $n(\sigma; a) = n(\Gamma; a) - \sum_j n(\Gamma; a_j) n(\partial R_j; a) = n(\Gamma; a) - n(\Gamma; a_k) = n(\Gamma; a_k) - n(\Gamma; a_k) = 0$.

Case 3. If $a \in \text{Int } R_k$ and $n(\Gamma; a_k) = 0$, then $n(\Gamma; a) = 0$ in this case yet $n(\partial R_k; a) = 1$ and $n(\partial R_j; a) = 0$ for $j \neq k$. Then $n(\sigma; a) = n(\Gamma; a) - \sum_j n(\Gamma; a_j) n(\partial R_j; a) = n(\Gamma; a) - n(\Gamma; a_k) n(\partial R_k; a) = 0 - 0 \cdot 1 = 0$.

Case 4. If a lies on the boundary of an unbounded rectangular domain or the boundary of a finite rectangle, then $n(\sigma; a) = 0$ by continuity of the winding numbers.

Finally, we turn to prove $\sigma = 0$ in the sense that $\Gamma = \Gamma_0$ holds regardless of these segments possessing reciprocal (namely, back and forth) directions. Note that such segments can be canceled out in the formal sum for cycles.

Let σ_{jk} be the common edge of two neighboring rectangles R_j and R_k. See Fig. 4.22. Suppose $\sigma = \Gamma - \Gamma_0$ is in its simplest form disregarding terms with zero coefficient and σ contains a term $c\sigma_{jk}$, where c is a integer. Then the cycle $\sigma - c\partial R_j$ does not contains σ_{jk}. Then

$$n(\sigma - c\partial R_j; a_j) = n(\sigma - c\partial R_j; a_k).$$

While $n(\sigma - c\partial R_j; a_j) = n(\sigma; a_j) - cn(\partial R_j; a_j) = 0 - c \cdot 1 = -c$ and $n(\sigma - c\partial R_j; a_k) = n(\sigma; a_k) - cn(\partial R_j; a_k) = 0 - c \cdot 0 = 0$. It follows that $c = 0$.

Let σ_{jk} be the common edge of a rectangle R_j and an unbounded rectangular domain R'_k. See Fig. 4.23. If $\sigma = \Gamma - \Gamma_0$, in its simplest form, contains a term $c\sigma_{jk}$. Then the cycle $\sigma - c\partial R_j$ (passing ∞) contains σ_{jk} no more. For any $a'_k \in \text{Int } R'_k$, $n(\sigma - c\partial R_j; a_j) = 0 - c \cdot 1 = -c$ while $n(\sigma - c\partial R_j; a'_k) = 0 - c \cdot 0 = 0$. Thus $c = 0$.

Combining together, the coefficient of each side of a finite rectangle R_j in $\sigma = \Gamma - \Gamma_0$, in its simplest form, is zero. Hence $\sigma = 0$ and $\Gamma = \Gamma_0$.

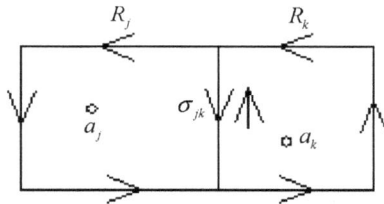

Fig. 4.22

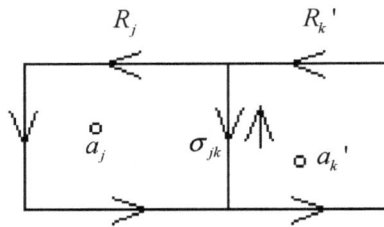

Fig. 4.23

Step 3. *What remains is easy now!*

By assumption, $\int_{\partial R_j} P dx + Q dy = 0$ for all finite rectangles $R_j \subseteq \Omega$. By (4.3.4.3) and (4.3.4.4), then

$$\int_\gamma P dx + Q dy = \int_\Gamma P dx + Q dy = \sum_j n(\Gamma; a_j) \int_{\partial R_j} P dx + Q dy = 0.$$

This finishes the proof.   □

4.4.   Characteristic Properties of Simply Connected Domains (a Review): The Single-Valuedness of a Primitive Function

In this section Ω will always denote a simply connected domain in **C**.

By Cauchy–Goursat theorem (3.4.2.2), needless of the Cauchy integral theorem in its general forms such as in (4.2.4.2) and (4.3.3.1), it follows easily from (4.1.2) the following:

The single-valuedness of a primitive function on a simply connected domain Ω. Let f be analytic in Ω. Then:

(1) f has a single-valued primitive function $F(z) = \int_{z_0}^z f(\zeta) d\zeta$ in Ω.
(2) Any other primitive function of f in Ω is of the form $F(z) + c$, where c is a constant.
(3) For any primitive $G(z)$ of f in Ω, $\int_{z_0}^z f(\zeta) d\zeta = G(z) - G(z_0)$.   (4.4.1)

As an important application, we try to *reprove* (3.3.1.5) *and* (3.3.1.8) *in a simply connected domain* Ω.

Let f be analytic in Ω such that f' is known supposedly to be analytic, too (see (3.4.2.6)). In case $f(z) \neq 0$ throughout Ω, then $\frac{f'(z)}{f(z)}$ is analytic in

Ω and, by (4.4.1),

$$F(z) = \int_{z_0}^{z} \frac{f'(\zeta)}{f(\zeta)} d\zeta, \quad z \in \Omega$$

is a primitive of $\frac{f'(z)}{f(z)}$ in Ω, satisfying $F(z_0) = 0$. Then $\varphi(z) = \frac{f(z)}{e^{F(z)}}$ is analytic in Ω and

$$\varphi'(z) = e^{-F(z)} \left[f'(z) - f(z) \frac{f'(z)}{f(z)} \right] = 0, \quad z \in \Omega.$$

Therefore $\varphi(z) = c$, a constant. It follows that

$$\varphi(z) = \varphi(z_0) = \frac{f(z_0)}{e^{F(z_0)}} = f(z_0)$$

$$\Rightarrow f(z) = f(z_0)e^{F(z)} = e^{F(z)+\log f(z_0)}$$

$$\Rightarrow F(z) + \log f(z_0) = \int_{z_0}^{z} \frac{f'(\zeta)}{f(\zeta)} d\zeta + \log f(z_0), \quad z \in \Omega \qquad (*)$$

are single-valued analytic branches of $\log f(z)$ for any particular value of $\log f(z_0)$.

(*Recall*: A continuous function $g : \Omega(\text{open set}) \to \mathbf{C}$ is a *single-valued branch* of $\log f$ if $e^{g(z)} = f(z)$ holds throughout Ω. Refer to (2.2.5)).

Conversely, if $g(z)$ is any single-valued analytic branch of $\log f(z)$, then $e^{g(z)} = f(z)$ induces that

$$g'(z) = \frac{f'(z)}{f(z)}, \quad z \in \Omega$$

$\Rightarrow$ (since $F'(z) = \frac{f'(z)}{f(z)}$, $z \in \Omega$, is known) $g(z) - F(z) = c$, a constant, $z \in \Omega$.

$\Rightarrow$ (since $F(z_0) = 0$ and hence $c = g(z_0)$) $g(z) = F(z) + g(z_0)$, $z \in \Omega$.

Since $e^{g(z_0)} = f(z_0)$, $g(z_0)$ is one of the values of $\log f(z_0)$. Hence, any single-valued analytic branch of $\log f(z)$ should be of the form $(*)$.

Remark 1. Any single-valued continuous branch of the inverse function $z = f^{-1}(w)$ of a nonconstant analytic function $w = f(z)$ is necessarily analytic. $\square$

Sketch of Proof

Let $w = f(z) : \Omega$ (open) $\to \mathbf{C}$ be analytic and $D \subseteq f(O)$ be a domain so that $z = g(w) : D \to \mathbf{C}$ is a single-valued continuous branch of $z = f^{-1}(w)$

on D. We need to show that g is differentiable everywhere in D. Refer to (5) in (3.1.6).

Let O be a component of Ω that contains $g(D)$. For any two distinct points w_1 and w_2 in D, $f(g(w_1)) = w_1 \neq w_2 = f(g(w_2))$ shows that f is nonconstant on O. Hence f' has only isolated zeros in O (see (3) in (3.4.2.9)).

Fix a point $w_0 \in D$. Let $z_0 = g(w_0)$.

In case $f'(z_0) \neq 0$: By continuity of g at w_0 and the differentiability of f at z_0,

$$\frac{g(w) - g(w_0)}{w - w_0} = \frac{1}{\frac{f(z)-f(z_0)}{z-z_0}} \rightarrow \frac{1}{f'(z_0)}$$

as $z \to z_0$. Therefore, g is differentiable at w_0 and $g'(w_0) = \frac{1}{f'(z_0)}$.

If $f'(z_0) = 0$, then there is a $\rho > 0$ so that $f'(z) \neq 0$ on $0 < |z - z_0| < \rho$. By continuity of g at w_0, there is a $r > 0$ so that $g(\{|w - w_0| < r\}) \subseteq \{|z - z_0| < \rho\}$. Since g is one-to-one on D and $g(w_0) = z_0$, it follows that $g(\{0 < |w - w_0| < r\}) \subseteq \{0 < |z - z_0| < \rho\}$ holds. In particular, $f'(z) \neq 0$ for any point $z = g(w)$, where $0 < |w - w_0| < r$. The result obtained in the last paragraph says that g is analytic on $0 < |w - w_0| < r$. Since g is continuous at z_0, by (3.4.2.17), w_0 is a removable singularity of g. Thus, g is analytic on $|w - w_0| < r$ and, in particular, analytic at w_0.

This finishes the proof.

Consequently, any single-valued continuous branch of $\log z$ is necessarily analytic. So is any single-valued continuous branch of $\log f(z)$ if $f(z)$ is analytic on a simply connected domain and $f(z) \neq 0$ there.

As we mentioned in Sec. 3.3.1, a (*analytic*) *branch of the multiple-valued analytic function* (see (3.3.3)) $z = f^{-1}(w)$ *will always assume to be single-valued and analytic* unless otherwise stated.

We summarize the above as

The existence of branches of $\log f(z)$, $f(z)^\lambda$, *and* $\sqrt[n]{f(z)}$ *on a simply connected domain* Ω *where* $f(z)$ *is analytic and* $f(z) \neq 0$ *throughout.*

(1) $\log f(z)$. Fix a point $z_0 \in \Omega$. Let

$$F(z) = \int_{z_0}^{z} \frac{f'(\zeta)}{f(\zeta)} d\zeta, \quad z \in \Omega.$$

    (i) Choose w_0 so that $e^{w_0} = f(z_0)$, then $g(z) = F(z) + w_0$ is the unique branch of $\log f(z)$ in Ω that satisfies $g(z_0) = w_0$.

(ii) Any branch of $\log f(z)$ in Ω is of the form (fix a value of $\log f(z_0)$)

$$F(z) + \log f(z_0) = \int_{z_0}^{z} \frac{f'(\zeta)}{f(\zeta)} d\zeta + \log |f(z_0)| + 2k\pi i,$$

$$k = 0, \pm 1, \pm 2, \ldots.$$

Refer to Exercises A (4) and (5) of Sec. 4.5.

(2) $f(z)^\lambda$ (λ is a nonzero complex constant): Fix any branch $g(z)$ of $\log f(z)$ in Ω. Then

$$e^{\lambda g(z)} \quad \text{or} \quad \exp(\lambda g(z))$$

is a branch of $f(z)^\lambda$ in Ω corresponding to $g(z)$. In case λ is an integer, $e^{\lambda g(z)}$ is always the λ-power function $f(z)^\lambda$ no matter which branch $g(z)$ of $\log f(z)$ is chosen.

(3) $\sqrt[n]{f(z)}$ ($n \geq 2$). The branches of $\sqrt[n]{f(z)}$ on Ω are of the form

$$\exp \frac{1}{n} \left\{ \int_{z_0}^{z} \frac{f'(\zeta)}{f(\zeta)} d\zeta + \log |f(z_0)| + 2k\pi i \right\}, \quad k = 0, \pm 1, \pm 2, \ldots$$

where z_0 is any fixed point in Ω.   (4.4.2)

What should be mentioned is that, as long as $\log f(z)$ has a branch $g(z)$ on an *open* set O, then so is $\sqrt[n]{f(z)}$ ($n \geq 2$), namely, $\exp(\frac{1}{n}g(z))$ is a branch of $\sqrt[n]{f(z)}$ on O. But conversely, this is not necessarily true. Take, for instance, $\log \frac{z-a}{z-b}$ ($a, b \in \mathbf{C}$ and $a \neq b$) has infinitely many branches on $\mathbf{C} - [a, b]$ (see Example 4 in Sec. 2.7.3) and $\sqrt[n]{\frac{z-a}{z-b}}$ has only n distinct branches there. On the other hand, $\log(z^2 - 1)$ has branches on $\mathbf{C} - [\infty, -1] - [1, \infty]$ (refer to Exercise A (17) of Sec. 2.7.3) but not on $\mathbf{C} - [-1, 1]$, while $\sqrt{z^2 - 1}$ has two branches on $\mathbf{C} - [-1, 1]$ (remind that ∞ is a node point of $\sqrt{z^2 - 1}$). Furthermore, a branch of $\log z$ in a domain Ω is not necessarily always of the form $\log |z| + i\theta(z)$, where $\theta_0 < \theta(z) \leq \theta_0 + 2\pi$. For example, designate the principle branch of $\log z$ as $\mathrm{Log}\, z = \log |z| + i \, \mathrm{Arg}\, z$, $-\pi < \mathrm{Arg}\, z \leq \pi$. Then a branch of $\log z$ on the left plane $\mathrm{Re}\, z < 0$ is given by

$$g(z) = \begin{cases} \mathrm{Log}\, z, & \mathrm{Im}\, z \geq 0 \text{ and } \mathrm{Re}\, z < 0, \\ \mathrm{Log}\, z + 2\pi i, & \mathrm{Im}\, z < 0 \text{ and } \mathrm{Re}\, z < 0. \end{cases}$$

For further explanation, see Exercises A (4)–(7) of Sec. 4.5.

However, the equivalence of (1) and (3) in (4.4.1) can be used to characterize the simple connectedness of a domain. More precisely, we have, besides (2.4.12),

The characteristic properties of a simply connected domain. Let Ω be a domain in $\mathbf{C}$. Then the following are equivalent:

(1) Ω is a simply connected domain (see (2.4.12)).
(2) Any rectifiable closed curve γ in Ω is homologous to zero in Ω (see (4.3.2.3)).
(3) For any function f analytic in Ω and any continuous closed curve γ in Ω,

$$\int_\gamma f(z)dz = 0$$

   always holds (see (4.2.4.2)).
(4) Any analytic function in Ω has a single-valued primitive (see (4.1.2) and (4.4.1)).
(5) For any function f, analytic in Ω and $f(z) \neq 0$ there, $\log f(z)$ always has an analytic branch in Ω (see (4.4.2)).
(6) For any function f, analytic in Ω and $f(z) \neq 0$ there, then $\sqrt[n]{f(z)}$ ($n \geq 2$) always has an analytic branch in Ω (see (4.4.2)).
(7) Ω is homeomorphic to the unit disk $|z| < 1$ (see the proof of Example 3 in Sec. 4.2.2).
(8) Any continuous closed curve γ in Ω is homotopic to a point in Ω (see Example 3 in Sec. 4.2.2). $\qquad (4.4.3)$

Among these equivalent conditions, (1), (2), (7), and (8) are concerned with the *topological properties* of the domain Ω; (3), (4), and (5) are with the *analytic properties*, while (6) is, basically, with the *algebraic property*. We have proved all of them and their equivalence, except the step (6) $\Rightarrow$ (7), to which one might refer to the proof of the Riemann mapping theorem (see Sec. 6.1.1) and (3.5.1.9).

Remark 2. For curves in a general *open* set, homotopy is a stronger concept than that of homology. A curve could be homologous to zero without contracting to a point. For an example, see Ref. [70], p. 262. However, in a simply connected domain, they are equivalent as shown in (2) and (8) of (4.4.3).

We had tacitly used these general results (5) and (6) in (4.4.3) when differentiating or integrating some particular functions. See, for instance, Example 3 in Sec. 3.4.2 and Exercise A (6) there. We will encounter more applications such as these in Sec. 4.12. As another basic illustration, since $\log z$ has the principle branch $\operatorname{Log} z$ in $\mathbf{C} - (\infty, 0]$, we have, for $\operatorname{Re} \lambda \geq 1$,

$$\lim_{R \to \infty} \int_{r-iR}^{r+iR} \frac{1}{z^\lambda \operatorname{Log} z} dz = \begin{cases} 0, & \text{if } r > 1, \\ 2\pi i, & \text{if } 0 < r < 1. \end{cases}$$

To justify this, choose the path γ of integration in the segment $\operatorname{Re} z = r$, $-R < \operatorname{Im} z < R$ plus the half-circle $z(t) = r + Re^{it}$, $-\frac{\pi}{2} \leq t \leq \frac{\pi}{2}$, and then, let $R \to \infty$.   □

4.5.   The Branches of a Multiple-Valued Primitive Function on a Multiple-Connected Domain

Suppose Ω is a multiple-connected domain in $\mathbf{C}$ throughout the whole section. Let f be analytic in Ω.

If there is a rectifiable closed curve γ_0 so that

$$\int_{\gamma_0} f(z)dz \neq 0,$$

then, according to (4.1.2), the function

$$F(z) = \int_{\gamma_z} f(\zeta)d\zeta \quad (z_0 \text{ is a fixed point in } \Omega)$$

is multiple-valued and its values depend on the paths γ_z of integration connecting z_0 to z. For instance, if γ is the curve $z(t) = e^{int}$, $0 \leq t \leq 2\pi$ and $n = \pm 1, \pm 2, \ldots$, then $\int_\gamma \frac{1}{z} dz = 2n\pi i \neq 0$ and the function $F(z) = \int_{\gamma_z} \frac{1}{\zeta} d\zeta = \log z$ is multiple-valued in $\mathbf{C} - \{0\}$, where γ_z is a curve connecting 1 to z in $\mathbf{C} - \{0\}$.

Let D be a simply connected subdomain of Ω and z_0 a fixed point in Ω. Choose a fixed point z_* in D. Let $z \in D$ be a varying point. Then

$$F(z) = \int_{\gamma_z} f(\zeta)d\zeta = \int_{\gamma_{z_*}} f(\zeta)d\zeta + \int_{z_*}^{z} f(\zeta)d\zeta, \quad z \in D$$

where $\varphi(z) = \int_{z_*}^{z} f(\zeta)d\zeta$ has its paths of integration lying entirely in D. See Fig. 4.24. According to (4.4.1), φ is single-valued in D, while

$$F(z_*) = \int_{\gamma_{z_*}} f(\zeta)d\zeta$$

depends on the choice of the curve γ_{z_*} and thus, is multiple-valued. Pick a particular value c of $F(z_*)$, then $F_c(z) = c + \varphi(z)$ is a single-valued analytic function in D and $F_c'(z) = f(z)$ holds in D. We call such a $F_c(z)$ a *branch* of the *multiple-valued primitive function* $F(z)$ of $f(z)$ in D.

If Γ is an arbitrary curve in Ω, connecting z_0 to z, then

$$\int_\Gamma f(\zeta)d\zeta = \int_\Gamma f(\zeta)d\zeta + \int_z^{z_*} f(\zeta)d\zeta + \int_{z_*}^z f(\zeta)d\zeta$$

(the paths of integration of the last two integrals lie in D)

$$= \int_{\Gamma + \widehat{zz_*}} f(\zeta)d\zeta \quad \text{(where } \widehat{zz_*} \text{ is the curve in } D \text{ connecting}$$

$$z \text{ to } z_*) + \varphi(z), \quad z \in D.$$

See Fig. 4.24. Hence, $\int_\Gamma f(\zeta)d\zeta = F_c(z) = c + \varphi(z)$, where $c = \int_{\Gamma + \widehat{zz_*}} f(\zeta)d\zeta$ is one of the values of $F(z_*)$.

We summarize the above as (compare to (4.3.3.5))

The multiple-valuedness of a primitive function on a multiple-connected domain and its branches on a simply connected subdomain. Suppose f is analytic in a multiple-connected domain Ω such that there is a rectifiable closed curve γ_0 in Ω satisfying $\int_{\gamma_0} f(z)dz \neq 0$. Fix a point $z_0 \in \Omega$ and a simply connected subdomain D of Ω. Let

$$F(z) = \int_{\gamma_z} f(\zeta)d\zeta, \quad \text{where } \gamma_z \text{ is any rectifiable curve in } \Omega \text{ connecting } z_0$$

to z, $z \in \Omega$.

(1) $F(z)$ is multiple-valued in Ω and is still called a *primitive function* of f in Ω.

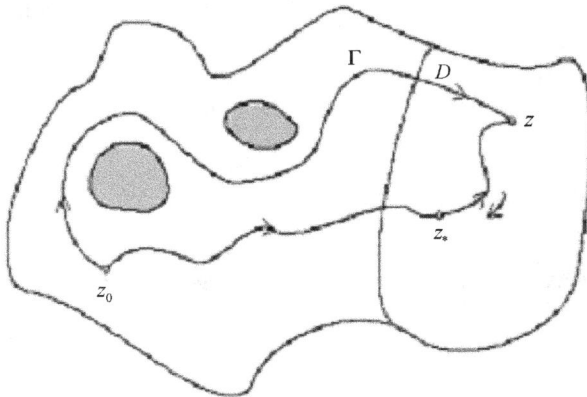

Fig. 4.24

(2) *Fix* a point $z_* \in D$ and choose a *particular* rectifiable curve γ^* in Ω connecting z_0 to z_*. Then

$$F_{\gamma^*}(z) = \int_{\gamma^*} f(\zeta)d\zeta + \int_{z_*}^{z} f(\zeta)d\zeta \quad \text{(path of integration lying in } D)$$

is single-valued and analytic in D, and $F'_{\gamma^*}(z) = f(z)$ holds in D. Call $F_{\gamma^*}(z)$ a *branch* of $F(z)$ in D.

(3) All branches of $F(z)$ on D are of the forms $\int_{\gamma^*} f(\zeta)d\zeta + \int_{z_*}^{z} f(\zeta)d\zeta$ as γ^* runs through all possible rectifiable curves in Ω connecting z_0 to z_*.

(4) Any two branches of $F(z)$ on D differ by a constant, determined by *one* of the many values of $\int_{\widehat{z_*z_*}} f(\zeta)d\zeta$, where $\widehat{z_*z_*}$ denotes an arbitrarily closed curve in Ω passing z_*. See Fig. 4.25:

$$F_{\gamma}(z) = F_{\gamma^*}(z) + \int_{-\gamma^*+\gamma+\widehat{zz_*}} f(\zeta)d\zeta$$

where γ is a curve in Ω connecting z_0 to $z \in D$ and $\widehat{zz_*}$ is a curve in D connecting z to z_*. In this case $\widehat{z_*z_*} = -\gamma_* + \gamma + \widehat{zz_*}$. $\qquad$ (4.5.1)

Applying (4.5.1) to (4.3.2.5) and (4.3.3.3), we obtain

The line integral in a $(n+1)$-connected domain Ω. Adopt notations and results in (4.3.2.5). Let D be a simply connected subdomain of Ω, and $z_0 \in \Omega$ and $z_* \in D$ be fixed points.

(1) Then the branches of the multiple-valued primitive function $F(z)$ of $f(z)$ in D are of the forms

$$F_{\gamma^*}(z) + \sum_{j=1}^{n} n(\gamma; z_j)\omega_j, \quad \text{with}$$

$$\omega_j = \int_{\gamma_j} f(z)dz \quad \text{the period of } f \text{ around } \Omega_j \text{ and } z_j \in \Omega_j, \quad 1 \leq j \leq n,$$

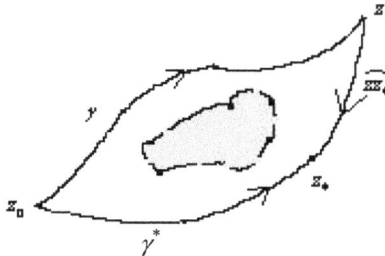

Fig. 4.25

where $F_{\gamma^*}(z)$ is the one in (2) of (4.5.1) and γ^* is any particularly chosen rectifiable curve in Ω connecting z_0 to z_* and γ is an arbitrary rectifiable closed curve in Ω passing through z_*.

(2) Then $F(z)$ is single-valued on every simply connected subdomain of Ω and hence, on the whole Ω, if and only if $w_1 = \cdots = w_n = 0$.   (4.5.2)

We present some examples.

Example 1. Let $\Omega = \mathbf{C} - \{0\}$ and $D = \mathbf{C} - (\infty, 0]$, a simply connected subdomain of Ω. Show that

$$\log z = \int_1^z \frac{1}{\zeta} d\zeta \quad \text{(path of integration lying in } \Omega\text{)}$$
$$= \operatorname{Log} z + 2n\pi i, \quad n = 0, \pm 1, \pm 2, \ldots$$

where the principle branch

$$\operatorname{Log} z = \int_1^z \frac{1}{\zeta} d\zeta \quad \text{(path of integration lying in } D\text{)}$$
$$= \log|z| + i \operatorname{Arg} z, \quad -\pi < \operatorname{Arg} z < \pi, \quad z \in D.$$

Solution. Take $z_0 = z_* = 1$. Let γ^* be any rectifiable closed curve in Ω, starting at $z_0 = 1$. Then,

$$\int_{\gamma^*} \frac{1}{\zeta} d\zeta = 2n\pi i, \quad \text{where } n = n(\gamma^*; 0) = 0, \pm 1, \pm 2, \ldots.$$

According to Cauchy integral theorem, a path of integration Γ from z_* to z in D can be chosen as the one composed of the line segment $[1, |z|]$ and a circular arc along $|\zeta| = |z|$. See Fig. 4.26. In this case,

$$\varphi(z) = \int_\Gamma \frac{1}{\zeta} d\zeta$$
$$= \int_1^{|z|} \frac{1}{x} dx + \int_0^{\operatorname{Arg} z} \frac{ie^{i\theta}}{e^{i\theta}} d\theta$$
$$= \log|z| + i \operatorname{Arg} z = \operatorname{Log} z$$

which is the principle branch of $\log z$ in D.

On the other hand, let γ be any rectifiable curve in Ω connecting $z_0 = 1$ to a point $z \in D$. See Fig. 4.27 and compare to Fig. 4.26. Use $\tilde{\gamma}$ to denote

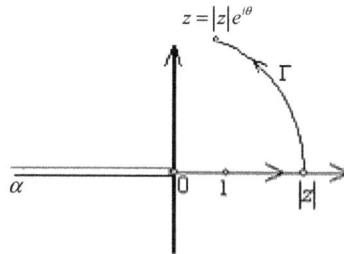

Fig. 4.26

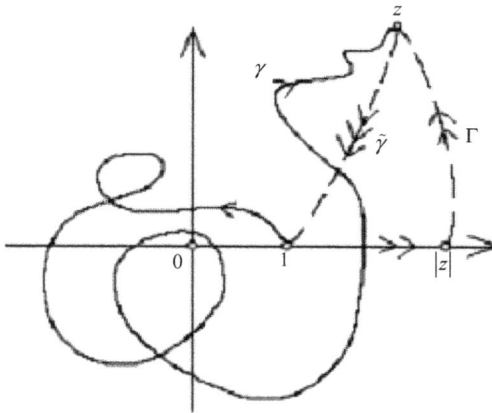

Fig. 4.27

any rectifiable curve in D, connecting z to $z_* = 1$. Then

$$\int_\gamma \frac{1}{\zeta} d\zeta = \int_{\gamma + \tilde{\gamma} - \tilde{\gamma}} \frac{1}{\zeta} d\zeta$$

$$= \int_{\gamma + \tilde{\gamma}} \frac{1}{\zeta} d\zeta + \int_{-\tilde{\gamma}} \frac{1}{\zeta} d\zeta = \int_{\gamma_*} \frac{1}{\zeta} d\zeta + \int_\Gamma \frac{1}{\zeta} d\zeta$$

$$= \operatorname{Log} z + 2n\pi i.$$

This justifies (4) in (4.5.1).

Conversely, *fix* any value $\operatorname{Log} z + 2n\pi i$ of $\log z$, where $z \in D$. Construct a curve γ in Ω, composed of $\gamma_1 : z(t) = e^{\mathrm{i}nt}$, $0 \le t \le 2\pi$, and Γ. Then

$$\int_\gamma \frac{1}{\zeta} d\zeta = \int_{\gamma_1} \frac{1}{\zeta} d\zeta + \int_\Gamma \frac{1}{\zeta} d\zeta = 2n\pi i + \operatorname{Log} z.$$

Combining together, we show that $\log z = \int_1^z \frac{1}{\zeta} d\zeta$, $z \in D$.

Example 2. Let $\Omega = \mathbf{C} - \{\pm i\}$ and $D = \mathbf{C} - (-i\infty, -i] - [i, i\infty)$, where $[i, i\infty)$ is the ray from i to $i\infty$ along the imaginary axis, etc. Then

$$\tan^{-1} z = \int_0^z \frac{1}{1 + \zeta^2} d\zeta \quad \text{(path of integration lying in } \Omega\text{)}$$

$$= \operatorname{Arc} \tan z + n\pi, \quad n = 0, \pm 1, \pm 2, \ldots$$

where the principle branch

$$\operatorname{Arc} \tan z = \int_0^z \frac{1}{1 + \zeta^2} d\zeta \quad \text{(path of integration lying in } D\text{)}$$

$$= \frac{-i}{2} \operatorname{Log} \frac{i - z}{i + z}, \quad z \in D.$$

Solution. Choose $z_0 = z_* = 0$. Let γ^* and γ have the corresponding meanings as in Example 1. Then

$$\int_{\gamma^*} \frac{1}{1 + \zeta^2} d\zeta = \frac{1}{2i} \left\{ \int_{\gamma^*} \frac{1}{\zeta - i} d\zeta - \int_{\gamma^*} \frac{1}{\zeta + i} d\zeta \right\}$$

$$= \frac{1}{2i} \{ 2\pi i n(\gamma^*; i) - 2\pi i n(\gamma^*; -i) \}$$

$$= \{ n(\gamma^*; i) - n(\gamma^*; -i) \} \pi$$

$$= n\pi, \quad n = 0, \pm 1, \pm 2, \ldots .$$

While, as shown in Fig. 4.28,

$$\int_{\gamma} \frac{1}{1 + \zeta^2} d\zeta = \int_{\gamma + \tilde{\gamma}} \frac{1}{1 + \zeta^2} d\zeta + \int_{-\tilde{\gamma}} \frac{1}{1 + \zeta^2} d\zeta$$

$$= n\pi + \int_{-\tilde{\gamma}} \frac{1}{1 + \zeta^2} d\zeta, \quad n = 0, \pm 1, \pm 2, \ldots \qquad (*_1)$$

in which

$$\varphi(z) = \int_{-\tilde{\gamma}} \frac{1}{1 + \zeta^2} d\zeta = \frac{1}{2i} \left\{ \int_{-\tilde{\gamma}} \frac{d\zeta}{\zeta - i} - \int_{-\tilde{\gamma}} \frac{d\zeta}{\zeta + i} \right\} \qquad (*_2)$$

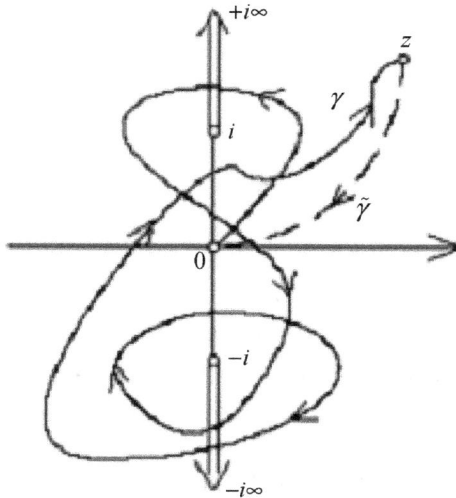

Fig. 4.28

where, if $z \in D$,

$$\int_{-\tilde{\gamma}} \frac{d\zeta}{\zeta - i} = \int_0^z \frac{d\zeta}{\zeta - i} \quad \text{(path of integration lying in } \mathbf{C} - [i, i\infty))$$

$$= \operatorname{Log}(z - i) - \operatorname{Log}(-i) = \operatorname{Log}(z - i) - \left(-\frac{\pi}{2}i\right),$$

$$-\frac{3\pi}{2} < \operatorname{Arg} z < \frac{\pi}{2} \quad \text{and} \quad z \in \mathbf{C} - [i, i\infty);$$

$$\int_{-\tilde{\gamma}} \frac{d\zeta}{\zeta + i} = \int_0^z \frac{d\zeta}{\zeta + i} \quad \text{(path of integration lying in } \mathbf{C} - (-i\infty, -i])$$

$$= \operatorname{Log}(z + i) - \operatorname{Log} i = \operatorname{Log}(z + i) - \frac{\pi}{2}i,$$

$$-\frac{\pi}{2} < \operatorname{Arg} z < \frac{3\pi}{2} \quad \text{and} \quad z \in \mathbf{C} - (-i\infty, -i].$$

Substituting these results into $(*_2)$, we obtain

$$\varphi(z) = \frac{1}{2i} \left\{ \operatorname{Log}(z - i) + \frac{\pi}{2}i - \operatorname{Log}(z + i) + \frac{\pi}{2}i \right\}$$

$$= \frac{-i}{2} \left\{ \operatorname{Log} \frac{z - i}{z + i} + \pi i \right\} = \frac{-i}{2} \operatorname{Log} \frac{-z + i}{z + i} = \frac{-i}{2} \operatorname{Log} \frac{i - z}{i + z} = \operatorname{Arc} \tan z;$$

and then, into $(*_1)$, the final result is

$$\int_\gamma \frac{d\zeta}{1+\zeta^2} = \frac{-i}{2}\mathrm{Log}\frac{i-z}{i+z} + n\pi = \tan^{-1}z, \quad z \in D$$

where $n = 0, \pm1, \pm2, \dots$.

Example 3. Show that a domain Ω in $\mathbf{C}$ is a *single-valued domain* of $\log\frac{z^2+1}{z^2-1}$ (namely, a domain on which $\log\frac{z^2+1}{z^2-1}$ can be defined a single-valued analytic branch. Refer to (1) in (2.7.1.6) and (3.3.1.8)) if and only if, for any rectifiable closed curve γ in Ω,

$$n(\gamma; i) + n(\gamma; -i) = n(\gamma; 1) + n(\gamma; -1).$$

Which of the following domains is a single-valued domain? See Fig. 4.29. Also, is Ω such a domain if ±1 and $\pm i$ belong to the same component of $\mathbf{C} - \Omega$? How about a domain if two of ±1 and $\pm i$ lie in a component of $\mathbf{C}-\Omega$ and the other two in another component of $\mathbf{C}-\Omega$? Take, for instance, in Fig. 4.29(a), try to determine the branch $g(z)$ satisfying the condition $g(0) = -\pi i$ and compute $g(-2+i)$.

Solution. Suppose g is a branch of $F(z) = \log\frac{z^2+1}{z^2-1}$ in Ω. Then

$$g'(z) = \frac{2z}{z^2+1} - \frac{2z}{z^2-1} = \frac{-4z}{z^4-1}$$

$$= \frac{1}{z-i} + \frac{1}{z+i} - \left(\frac{1}{z-1} + \frac{1}{z+1}\right), \quad z \in \Omega$$

$\Rightarrow$ (By using (2) in (4.1.2))   For any such a curve γ in Ω,

$$\int_\gamma g'(z)dz = 0 = 2\pi i\{n(\gamma; i) + n(\gamma; -i) - n(\gamma; 1) - n(\gamma; -1)\} = 0.$$

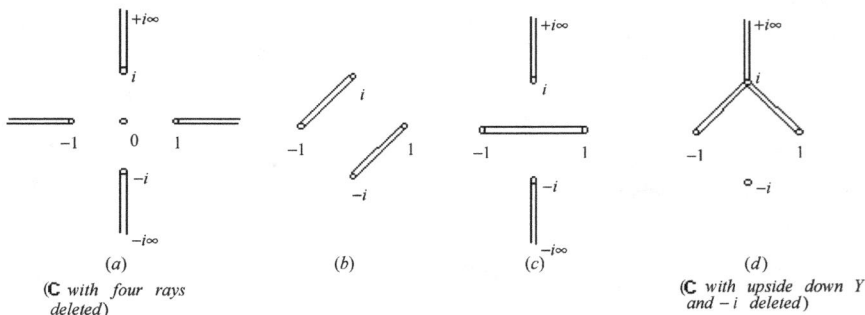

(a)	(b)	(c)	(d)
(**C** with four rays deleted)			(**C** with upside down Y and $-i$ deleted)

Fig. 4.29

Conversely, if the stated equality holds for all such γ, by argument similar to that used in Section (2) of Sec. 2.7.2 (including Figs. 2.51 and 2.52), one can realize that it is indeed possible to define branches on Ω. Try out the detail!

Domains in Fig. 4.29(a) and (b) are single-valued domains; while, in (c) and (d), are not.

If ± 1 and $\pm i$ belong to the same component of $\mathbf{C} - \Omega$, such as Fig. 4.29(a), then Ω always is such a domain. Otherwise, it depends on how ± 1 and $\pm i$ are related to each other.

The branch determined by $g(0) = -\pi i$ is

$$g(z) = g(0) + \int_0^z g'(\zeta)d\zeta = -\pi i + \int_0^z \frac{4\zeta}{1-\zeta^4}d\zeta,$$

$$z \in \Omega = \mathbf{C} - (-\infty, -1] - [1, \infty) - [i, i\infty) - (-i\infty, -i].$$

Since Ω is a starlike domain with center at 0 (see Example 2 in Sec. 4.2.2), (4.2.4.1) says that the path of integration in Ω may be chosen as the line segment $\overline{oz}$. Therefore,

$$\int_0^{-2+i} \frac{4\zeta}{1-\zeta^4}d\zeta = \operatorname{Log}\frac{z^2+1}{z^2-1}\bigg|_0^{-2+i} = \operatorname{Log}\frac{(-2+i)^2+1}{(-2+i)^2-1} - \operatorname{Log}(-1)$$

$$= \operatorname{Log}\frac{2(3+i)}{5} + \pi i$$

$$\Rightarrow g(-2+i) = \operatorname{Log}\frac{2(3+i)}{5}, \quad \text{where}$$

$$\operatorname{Log}w = \log|w| + i\operatorname{Arg}w, -\pi < \operatorname{Arg}w < \pi.$$

Example 4. Find the necessary and sufficient condition for a domain Ω in $\mathbf{C}$ to be a single-valued domain of $\sqrt{z^2-1}$. Determine which domain in Fig. 4.30 is such a domain? Find the branch $g(z)$ defined by $g(0) = -i$ in Fig. 4.30(a) and the other branch on the same domain.

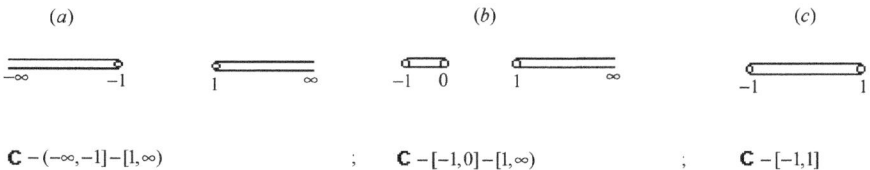

(a)   (b)   (c)

$\mathbf{C} - (-\infty, -1] - [1, \infty)$; $\mathbf{C} - [-1, 0] - [1, \infty)$; $\mathbf{C} - [-1, 1]$

Fig. 4.30

Solution. Let $g(z)$ be such a branch in a single-valued domain Ω of $\sqrt{z^2 - 1}$. Then $g(z)^2 = z^2 - 1 \Rightarrow 2g(z)g'(z) = 2z \Rightarrow \frac{g'(z)}{g(z)} = \frac{z}{z^2-1}$ on Ω. For any rectifiable closed curve γ in Ω (not passing ± 1),

$$\int_\gamma \frac{g'(z)}{g(z)} dz = \int_{g(\gamma)} \frac{dw}{w} = n(g(\gamma); 0) \cdot 2\pi i$$

$$= \int_\gamma \frac{z}{z^2 - 1} dz = \frac{1}{2} \int_\gamma \left(\frac{1}{z-1} + \frac{1}{z+1} \right) dz$$

$$= \frac{1}{2} \cdot 2\pi i [n(\gamma; 1) + n(\gamma; -1)]$$

$$\Rightarrow n(\gamma; 1) + n(\gamma; -1) = 2n(g(\gamma); 0).$$

This means that $n(\gamma; 1) + n(\gamma; -1)$ should be an integral multiple of 2. And this is a necessary condition. The argument in Section (2) of Sec. 2.7.2 shows that, for this *particular* function $\sqrt{z^2 - 1}$, it is also a sufficient condition.

(a) In case γ is a closed curve in $\Omega = \mathbf{C} - (-\infty, -1] - [1, \infty)$, $n(\gamma; -1) = n(\gamma; 1) = 0$ holds. Hence, $\log(z^2 - 1)$ has a branch $h(z)$ in Ω (refer to Example 3 and, for a general setting, see Exercise A (4)). Consequently $e^{\frac{1}{2}h(z)} = g(z)$ is a branch of $\sqrt{z^2 - 1}$ in Ω (see (4.4.2) and the explanation beneath it). If $h(0) = -\pi i$ is imposed, then

$$h(z) = -\pi i + \int_0^z \frac{2\zeta}{\zeta^2 - 1} d\zeta = -\pi i + \mathrm{Log}(z^2 - 1)$$

$$- \mathrm{Log}(-1) = -\pi i + \mathrm{Log}(1 - z^2)$$

$$\Rightarrow g(z) = e^{\frac{1}{2}[-\pi i + \mathrm{Log}(1-z^2)]} = -i\sqrt{1 - z^2} \quad (\sqrt{1} = 1)$$

$$= \sqrt{z^2 - 1} \quad (\sqrt{-1} = -i).$$

The other branch of $\log(z^2 - 1)$ is

$$\tilde{h}(z) = 2\pi i + h(z) = \pi i + \log(1 - z^2)$$

$$\Rightarrow \tilde{g}(z) = e^{\frac{1}{2}\tilde{h}(z)} = i\sqrt{1 - z^2} \quad (\sqrt{1} = 1) = \sqrt{z^2 - 1} \quad (\sqrt{-1} = i)$$

which is the other branch of $\sqrt{z^2 - 1}$ on Ω.

(b) Suppose $\sqrt{z^2 - 1}$ has a branch $g(z)$ on $\Omega = \mathbf{C} - [-1, 0] - [1, \infty)$. Then $g(z)^2 = z^2 - 1 \Rightarrow 2g(z)g'(z) = 2z \Rightarrow$ Both $g(z) \neq 0$ and $g'(z) \neq 0$ on Ω. Let γ be the circle $z(t) = -\frac{1}{2} + re^{i\theta}$, $0 \leq \theta \leq 2\pi$ and $\frac{1}{2} < r < \frac{3}{2}$. Then $g(\gamma)$ is a closed curve and

$$n(g(\gamma); 0) = \frac{1}{2\pi i} \int_{g(\gamma)} \frac{dw}{w} = \frac{1}{2\pi i} \int_\gamma \frac{g'(z)}{g(z)} dz$$

$$= \frac{1}{4\pi i} \int_\gamma \frac{2z}{z^2 - 1} dz = \frac{2\pi i}{4\pi i} = \frac{1}{2}.$$

This is impossible since $n(g(\gamma); 0)$ is an integer. Consequently Ω is not a single-valued domain of $\sqrt{z^2 - 1}$. The argument in Example 2 of Sec. 2.7.2 will work, too.

(c) According to Exercise A (14) or Example 4 in Sec. 2.7.3, $\log(z^2 - 1)$ does not have branches on $\Omega = \mathbf{C} - [-1, 1]$ (why?), but $\log \frac{z+1}{z-1}$ does have one, say $h(z)$. Let $g(z) = (z - 1)e^{\frac{1}{2}h(z)}$. Then $g(z)$ is single-valued and analytic in Ω. Since $g(z)^2 = (z - 1)^2 e^{h(z)} = z^2 - 1$, $g(z)$ indeed is a branch of $\sqrt{z^2 - 1}$ in Ω.

Exercises A

(1) Imitate Examples 1 and 2, and find explicitly each of the following functions:

(a) $F(z) = \int_0^z \frac{d\zeta}{\sqrt{\zeta^2 - 1}}$, $z \in \mathbf{C} - \{\pm 1\}$.

(b) $F(z) = \int_0^z \frac{3\zeta^2}{1 + \zeta^3} d\zeta$, $z \in \mathbf{C} - \{e^{\frac{\pi i}{3}}, e^{\pi i}, e^{-\frac{\pi i}{3}}\}$.

(c) $F(z) = \int_{z_0}^z \frac{d\zeta}{(\zeta - a)(\zeta - b)}$, where a and b are distinct points in $\mathbf{C}$, $z_0 \neq a, b$, and $z \in \mathbf{C} - \{a, b\}$.

(2) Try to use line integrals (indefinite integrals) to express each of the following multiple-valued functions.

(a) $\cos^{-1} z$.
(b) $\sin^{-1} z$.
(c) $\cot^{-1} z$.
(d) $\tanh^{-1} z$.

(3) Compute

$$F(z) = \int_{z_0}^z \frac{d\zeta}{(\zeta - z_1)(\zeta - z_2) \cdots (\zeta - z_n)}$$

where $z_1, \ldots, z_n$ are distinct points, $z_0 \neq z_j$ for $1 \leq j \leq n$ and $z \in \mathbf{C} - \{z_1, \ldots, z_n\}$.

(4) (Extension of (4.4.2)) Let f be analytic in a domain Ω (not necessarily simply connected) and $f(z) \neq 0$ throughout Ω.

    (a) Show that $\log f$ has a branch in Ω if and only if $\int_\gamma \frac{f'(\zeta)}{f(\zeta)} d\zeta = 0$ for any rectifiable closed curve γ in Ω.

    (b) In case g is a branch of $\log f$ in Ω, then all branches of $\log f$ are of the forms

$$g(z) + 2n\pi i, \quad n = 0, \pm 1, \pm 2, \ldots.$$

(5) (a) Suppose f is analytic in a domain Ω and $\log w$ has a branch in a domain D so that $f(\Omega) \subseteq D$ holds. Show that $\log f$ has a branch in Ω, and $g(f(z))$ is a branch of $\log f$ in Ω if g is a branch of $\log w$ in D.

    (b) Give an example to show that (a) fails if $f(\Omega) \subseteq D$ does not hold. *Note*: In (4.4.2) and Exercise (4) above, a branch of $\log f$ should be considered as a branch of the logarithmic function $\log f$ rather than as the composite of a branch of $\log w$ followed by the single-valued function f.

(6) Let the rational function

$$R(z) = a(z - z_1)^{m_1} \cdots (z - z_k)^{m_k}$$

where $z_1, \ldots, z_k$ are distinct points, $a \neq 0$ and $m_1, \ldots, m_k$ are nonzero integers. Let Ω be a domain in $\mathbf{C}$.

    (a) Show that $\log R(z)$ has a branch in the domain $\Omega - \{z_1, \ldots, z_k\}$ if and only if, for any rectifiable closed γ in $\Omega - \{z_1, \ldots, z_k\}$,

$$m_1 n(\gamma; z_1) + \cdots + m_k n(\gamma; z_k) = 0.$$

    (b) Let $n \geq 2$. Show that if $\sqrt[n]{R(z)}$ has a branch in the domain $\Omega - \{z_1, \ldots, z_k\}$, then it is necessary that, for any rectifiable closed curve γ in $\Omega \backslash \{z_1, \ldots, z_k\}$,

$$m_1 n(\gamma; z_1) + \cdots + m_k n(\gamma; z_k)$$

is an integral multiple of n. Is this a sufficient condition?

(7) (a) Give an example to show that it is possible for $\sqrt[n]{f(z)}$ to have a branch in a domain Ω while $\log f(z)$ is not.

    (b) Show that $\log z$ has a branch in a domain Ω if and only if $n(\gamma; 0) = 0$ for each rectifiable closed curve γ in Ω. Refer to (1) in (2.7.1.6) and Sec. 2.7.3.

(c) Show that the conclusion in (b) is still valid for $\sqrt[n]{z}$. Refer to (1) in (2.7.1.6) and Sec. 2.7.2.

(8) Let Ω be a domain in $\mathbf{C}$. Show that "$n(\gamma; -2) + n(\gamma; 1) + n(\gamma; 2)$ is an integral multiple of 2" is a sufficient condition for $\Omega - \{-2, 1, 2\}$ to be a single-valued domain of $\sqrt{(z+2)(z-1)(z-2)}$ (see Exercise (6)). Choose $\Omega = \mathbf{C} - (-\infty, -2] - [1, 2]$. Let g be the branch in Ω determined by $g(0) = 2$. Find $g(z)$ explicitly.

(9) Let $z_1, \ldots, z_n$ be distinct points in $\mathbf{C}$. Find the necessary and sufficient conditions for a domain Ω to be a single-valued domain for $\sqrt{(z - z_1) \cdots (z - z_n)}$ (refer to Exercise B (3) of Sec. 2.7.2). Also, find a domain that does not contain $z_1, \ldots, z_n$ and is not a single-valued domain.

(10) Let $f(z) = az^2 + bz + c$, and z_1 and z_2 be two roots of $f(z) = 0$.

   (a) In case $z_1 \neq z_2$, show that a domain Ω is a single-valued domain for $\sqrt{f(z)}$ if and only if, for each rectifiable closed curve γ in Ω, $n(\gamma; z_1) = n(\gamma; z_2)$.

   (b) In case $z_1 \neq z_2$, try to show, without recourse to (a), that if Ω is a single-valued domain for $\sqrt{f(z)}$, then neither z_1 nor z_2 is in Ω.

   (c) In case $z_1 = z_2$, does the result in (b) still hold?

(11) Determine if each of the following functions $f(z)$ has a nth root function (i.e., a branch of $\sqrt[n]{f(z)}$) in the indicated domain Ω. If yes, find all branches explicitly; if not, try to readjust the domain Ω so that it will turn out to be so.

   (a) $f(z) = z(z-1)(z+1) : n = 3, \quad \Omega = \mathbf{C} - (-\infty, -1] - [0, 1]$.

   (b) $f(z) = z^3 + z : n = 4, \quad \Omega = \mathbf{C} - (-\infty, 0] - [1, \infty) - \{e^{i\theta} | |\theta| \leq \frac{\pi}{2}\}$.

   (c) $f(z) = z^4 - 1 : n = 2, \quad \Omega = \mathbf{C} - \{e^{i\theta} | -\frac{\pi}{2} \leq \theta \leq 0 \text{ or } \frac{\pi}{2} \leq \theta \leq \pi\}$.

   (d) $f(z) = z^3 + z^2 + z + 1 : n = 3, \quad \Omega = \mathbf{C} - \{e^{i\theta} | \frac{\pi}{2} \leq \theta \leq \frac{3\pi}{2}\}$.

(12) Show that $\log \frac{z+1}{z-1}$ has a branch in a domain Ω if and only if, for each rectifiable closed curve γ in Ω, $n(r; 1) = n(r; -1)$ always holds. In this case, fix a point $z_0 \in \Omega$ and choose a w_0 of $\log \frac{z_0+1}{z_0-1}$. Then the so-determined branch is given by

$$g(z) = w_0 + \int_{z_0}^{z} \frac{-2}{\zeta^2 - 1} d\zeta$$

where the path of integration is any rectifiable curve in Ω, connecting z_0 to z. Can this condition be replaced by the requirement that ± 1 belongs to the same component of $\mathbf{C} - \Omega$? Take

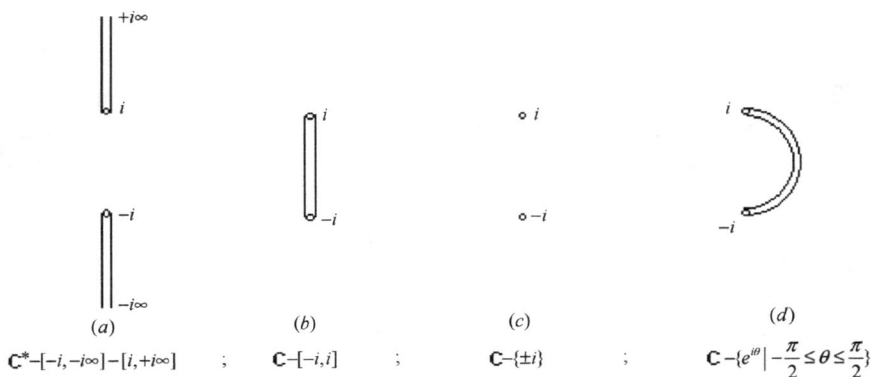

(a) $\mathbf{C}^* - [-i, -i\infty] - [i, +i\infty]$; (b) $\mathbf{C} - [-i, i]$; (c) $\mathbf{C} - \{\pm i\}$; (d) $\mathbf{C} - \{e^{i\theta} \mid -\dfrac{\pi}{2} \le \theta \le \dfrac{\pi}{2}\}$

Fig. 4.31

$\Omega = \mathbf{C} - \{e^{i\theta} \mid \pi \le \theta \le 2\pi\}$ now, and let g be the branch determined by $g(0) = \pi i$. Find out g explicitly and compute $g(i)$.

(13) (Compare to Example 2) In Exercise (6), let $R(z) = (z - i)(z + i)^{-1}$. Then a domain Ω is a single-valued domain of $\log \frac{z-i}{z+i}$ if and only if $n(\gamma; i) = n(\gamma; -i)$ holds for each rectifiable closed curve γ in Ω. Which domain in Fig. 4.31 is such a domain? Can this condition be replaced by the one that $\pm i$ belongs to the same component of $\mathbf{C} - \Omega$? Take Ω as in Fig. 4.31(d). Find the values at $\pm\sqrt{3}$ of the branch $g(z)$ determined by $g(0) = \pi$. By the way, try to show directly that this condition is still valid for $\tan^{-1} z$ to have a branch in Ω.

(14) Find conditions by which Ω is a single-valued domain of $\log(1 - z^{-2})$ and determine if the following domains Ω are single-valued domains.

(a) $\mathbf{C} - (-\infty, 0] - [1, \infty)$.

(b) $\mathbf{C} - [-1, 0] - [1, \infty)$.

(c) $\mathbf{C} - \{te^{it} \mid -1 \le t \le 1\}$.

(d) $\mathbf{C} - \{e^{it} \mid 0 \le t \le \pi\} - \{it \mid -\infty < t < \infty\}$.

(15) Same as Exercise (14) for $\log z^{-2}(z + 1)^{-1}(z^2 + 1)$.

(a) $\mathbf{C} - \{(-\infty, 1] \cup [0, \infty) \cup \{ti \mid t \in \mathbf{R} \text{ and } |t| \ge 1\}\}$.

(b) $\mathbf{C} - (-\infty, -1] - \{ti \mid -1 \le t \le 1\}$.

(c) $\mathbf{C} - (-\infty, 0] - \{e^{ti} \mid -\dfrac{\pi}{2} \le t \le \dfrac{\pi}{2}\}$.

(d) $\mathbf{C} - \{e^{it} \mid \dfrac{\pi}{2} \le t \le \pi\} - \{ti \mid -\infty < t < \infty\}$.

Exercises B

(1) Let f be analytic on a domain Ω where $f(z) \neq 0$.

   (a) Fix a point $z_0 \in \Omega$. Let γ be any rectifiable curve in Ω, connecting z_0 to a varying point z in Ω. Show that

   $$f(z) = f(z_0) \exp\left\{ \int_\gamma \frac{f'(\zeta)}{f(\zeta)} d\zeta \right\}, \quad z \in \Omega.$$

   (b) Let γ be any rectifiable closed curve in Ω. Then, there is an integral n so that

   $$\int_\gamma \frac{f'(\zeta)}{f(\zeta)} d\zeta = 2n\pi i.$$

   In particular, this shows that the winding number $\frac{1}{2\pi i} \int_\gamma \frac{1}{\zeta-z} d\zeta$ is an integer for any $z \notin \gamma$.

(2) (a) Suppose $f(z)$ is analytic in a domain Ω and is not identically equal to zero. If there is a sequence of positive integers $n_1 < n_2 < \cdots < n_k < \cdots (\lim_{k\to\infty} n_k = \infty)$ such that, for each k, $f(z)^{\frac{1}{n_k}}$ has a branch on Ω, then $\log f(z)$ has a branch on Ω.

   (b) Suppose, for *every* function $f(z)$ analytic in a domain Ω on which $f(z) \neq 0$ holds, $\sqrt{f(z)}$ always has a branch on Ω. Show that, for each such a function $f(z)$,

   (i) $\log f(z)$ has a branch in Ω; and
   (ii) for each integer $n \geq 2$, $\sqrt[n]{f(z)}$ has a branch in Ω.

4.6.   The General Form of Cauchy Integral Formula

Based on (4.3.3.1), the Cauchy integral formula (3.4.2.4) can be extended as follows.

Cauchy integral formula on a domain. Let f be analytic in a domain Ω. Suppose γ is a cycle in Ω and is homologous to zero in Ω. Then

$$\frac{1}{2\pi i} \int_\gamma \frac{f(\zeta)}{\zeta - z} d\zeta = n(\gamma; z) f(z), \quad z \in \Omega - \gamma.$$

In case γ is a Jordan closed curve with $\overline{\text{Int}\,\gamma} \subseteq \Omega$, then

$$\frac{1}{2\pi i} \int_\gamma \frac{f(\zeta)}{\zeta - z} d\zeta = f(z), \quad z \in \text{Int}\,\gamma. \tag{4.6.1}$$

Proof. Consider the function

$$F(\zeta) = \frac{f(\zeta) - f(z)}{\zeta - z}, \zeta \in \Omega - \{z\} \quad \text{where} z \text{ is a fixed point in } \Omega.$$

F is analytic in $\Omega - \{z\}$. Since $\lim_{\zeta \to z}(\zeta - z)F(\zeta) = 0$, by (3.4.2.17), z is a removable singularity of F. Now, apply (4.3.3.1) to F and we have

$$\int_\gamma F(\zeta)d\zeta = \int_\gamma \frac{f(\zeta) - f(z)}{\zeta - z}d\zeta = 0$$

$$\Rightarrow \int_\gamma \frac{f(\zeta)}{\zeta - z}d\zeta = \int_\gamma \frac{f(z)}{\zeta - z}d\zeta = f(z)\int_\gamma \frac{1}{\zeta - z}d\zeta = f(z) \cdot 2\pi i \cdot n(\gamma; z)$$

This finished the proof.   □

Based on (4.7.1) to be formally stated and proved in Sec. 4.7, (3.4.2.7) can be similarly generalized to

Cauchy integral formula for higher derivatives on a domain. Adopt assumptions and notations as in (4.6.1). Then f is infinitely differentiable on Ω and

$$\frac{n!}{2\pi i}\int_\gamma \frac{f(\zeta)}{(\zeta - z)^{n+1}}d\zeta = n(\gamma; z)f^{(n)}(z), \quad z \in \Omega - \gamma.$$

In particular, if γ is a Jordan closed curve with $\overline{\text{Int }\gamma} \subseteq \Omega$, then

$$\frac{n!}{2\pi i}\int_\gamma \frac{f(\zeta)}{(\zeta - z)^{n+1}}d\zeta = f^{(n)}(z), \quad z \in \text{Int }\gamma. \tag{4.6.2}$$

We present two examples.

Example 1. Let γ be the curve as shown in the first figure of Fig. 4.15 and f be analytic in $\mathbf{C}$.

(1) Evaluate $\int_\gamma \frac{f(z)}{z}dz$, $\int_\gamma \frac{f(z)}{z-1}dz$, $\int_\gamma \frac{f(z)}{z+1}dz$, and $\int_\gamma \frac{f(z)}{z(z-1)^2}dz$.

(2) Evaluate $\int_\gamma \frac{1}{z^m(z-1)^n(z+1)^p}dz$, where m, n, and p are positive integers.

Solution. Choose $0 < r < \frac{1}{2}$ and construct the circles $\gamma_1 : |z + 1| < r$, $\gamma_2 : |z| < r$, and $\gamma_3 : |z - 1| < r$. Then γ is homologous to $(-3 + 1)\gamma_1 + (-2 + 1)\gamma_2 + \gamma_3 = -2\gamma_1 - \gamma_2 + \gamma_3$ in $\Omega - \{-1, 0, 1\}$.

(1) According to (4.3.3.1), (4.3.1.4), and (4.6.1), (4.6.2),

$$\int_\gamma \frac{f(z)}{z}dz = -2\int_{\gamma_1} \frac{f(z)}{z}dz - \int_{\gamma_2} \frac{f(z)}{z}dz$$

$$+ \int_{\gamma_3} \frac{f(z)}{z}dz = -1 \cdot 2\pi i \cdot f(0) = -2\pi i f(0).$$

$$\int_\gamma \frac{f(z)}{z-1}dz = \int_{\gamma_3}\frac{f(z)}{z-1}dz = 2\pi i f(1);$$

$$\int_\gamma \frac{f(z)}{z+1}dz = -2\int_{\gamma_1}\frac{f(z)}{z+1}dz = -4\pi i f(-1).$$

$$\int_\gamma \frac{f(z)}{z(z-1)^2}dz = -\int_{\gamma_2}\frac{f(z)}{z(z-1)^2}dz + \int_{\gamma_3}\frac{f(z)}{z(z-1)^2}dz$$

$$= -2\pi i \cdot \frac{f(0)}{(-1)^2} + 2\pi i \frac{d}{dz}\left.\frac{f(z)}{z}\right|_{z=1}$$

$$= -2\pi i f(0) + 2\pi i(f'(1) - f(1)).$$

(2)

$$\int_\gamma \frac{1}{z^m(z-1)^n(z+1)^p}dz = -2\int_{\gamma_1}\cdots - \int_{\gamma_2}\cdots + \int_{\gamma_3}\cdots.$$

$$= -2\frac{2\pi i}{(p-1)!}\frac{d^{p-1}}{dz^{p-1}}\left\{\frac{1}{z^m(z-1)^n}\right\}\Bigg|_{z=-1}$$

$$-\frac{2\pi i}{(m-1)!}\frac{d^{m-1}}{dz^{m-1}}\left\{\frac{1}{(z-1)^n(z+1)^p}\right\}\Bigg|_{z=0}$$

$$+\frac{2\pi i}{(n-1)!}\frac{d^{n-1}}{dz^{n-1}}\left\{\frac{1}{z^m(z+1)^p}\right\}\Bigg|_{z=1}.$$

Example 2. Let γ be the curve as shown in the second figure of Fig. 4.15. Evaluate $\int_\gamma \frac{\operatorname{Arc\,tan} z}{z^3-z}dz$, where $\operatorname{Arc\,tan} z$ is the principal branch of $\tan^{-1}z$ (see (2.7.4.7)).

Solution. Observe that γ is homologous to $-\gamma_1 + \gamma_2 + 2\gamma_3$ in $\mathbf{C}-\{-1,0,1\}$, where γ_1, γ_2, and γ_3 are as in Example 1. Also, $\tan^{-1}z$ has the principal branch $\operatorname{Arc\,tan} z$ in $\mathbf{C} - (\infty i, -i] - [i, i\infty)$ with range $0 < \operatorname{Re} w < \pi$. Now

$$\frac{1}{z^3-z} = \frac{1}{2}\left(\frac{1}{z-1} + \frac{1}{z+1}\right) - \frac{1}{z}$$

$$\Rightarrow \int_\gamma \frac{\operatorname{Arc\,tan} z}{z^3-z}dz = -\frac{1}{2}\int_{\gamma_1}\frac{\operatorname{Arc\,tan} z}{z+1}dz$$

$$+\frac{2}{2}\int_{\gamma_3}\frac{\operatorname{Arc\,tan} z}{z-1}dz - \frac{1}{2}\int_{\gamma_2}\frac{\operatorname{Arc\,tan} z}{z}dz$$

$$= -\frac{1}{2} \cdot 2\pi i \cdot \text{Arc}\tan(-1) + 2\pi i \cdot \text{Arc}\tan(1) - \frac{1}{2} \cdot 2\pi i \cdot \text{Arc}\tan(0)$$

$$= -\frac{1}{2} \cdot 2\pi i \cdot \frac{3\pi}{4} + 2\pi i \cdot \frac{\pi}{4} = -\frac{\pi^2 i}{4}.$$

Exercises A

(1) Let γ be the closed curve shown in Fig. 4.32 and f be analytic in $\mathbf{C}$. Evaluate the following integrals.

(a) $\displaystyle\int_{\gamma} \frac{f(z)}{z+1}\,dz.$

(b) $\displaystyle\int_{\gamma} \frac{f(z)}{z^2-1}\,dz.$

(c) $\displaystyle\int_{\gamma} \frac{f(z)}{z(z+1)}\,dz.$

(d) $\displaystyle\int_{\gamma} \frac{f(z)}{z(z-1)^2}\,dz.$

(2) Evaluate the following integrals.

(a) $\displaystyle\int_{\gamma} \frac{z^4+z^2+1}{z^3+z^2}\,dz,$ where γ is as in the first figure of Fig. 4.16.

(b) $\displaystyle\int_{\gamma} \frac{\log(z-2)}{z^3-z}\,dz,$ where γ is as in the second figure of Fig. 4.16.

(3) Let $\gamma = [i, -1-i] + [-1-i, -\frac{1}{2}] + \{\frac{1}{2}e^{it} \mid 0 \le t \le \pi\} + [\frac{1}{2}, 1-i] + [1-i, i]$, where $[z_1, z_2]$ denotes the segment joining z_1 to z_2. Evaluate $\int_{\gamma} \frac{\text{Arc}\sin z}{z^4+2z^2}\,dz$, where $\text{Arc}\sin z$ is the principal branch of $\sin^{-1} z$ (see (2.7.4.5)).

(4) Let $\gamma_1 : z(t) = 3\cos t + i\sin t$, $-\frac{\pi}{2} \le t \le \frac{\pi}{2}$, and $\gamma_2 : z(t) = e^{-it}$, $-\frac{\pi}{2} \le t \le \frac{\pi}{2}$. Evaluate $\int_{\gamma_1+\gamma_2} \frac{\log z}{z(z-2)^2(z-4)}\,dz.$

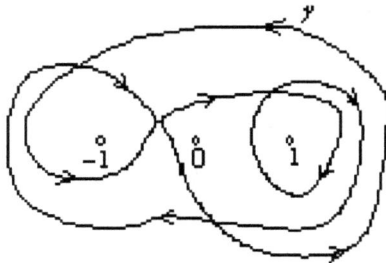

Fig. 4.32

4.7.   Integrals of Cauchy Type and Cauchy Principal Value

Let $\gamma : z(t), a \le t \le b$, be a rectifiable curve in $\mathbf{C}$. Suppose $\varphi(z)$ is a complex-valued continuous function along γ, namely, $\varphi(z(t)) : [a,b] \to \mathbf{C}$ is continuous. As long as $z \notin \gamma$, $\frac{\varphi(\zeta)}{\zeta - z}$ is a continuous function of $\zeta \in \gamma$; therefore, the line integral

$$F(z) = \int_\gamma \frac{\varphi(\zeta)}{\zeta - z} d\zeta \quad \text{or} \quad \frac{1}{2\pi i} \int_\gamma \frac{\varphi(\zeta)}{\zeta - z} d\zeta, \quad z \in \mathbf{C} - \gamma, \qquad (*)$$

exists and defines a single-valued function $F(z)$ on *each* component of $\mathbf{C} - \gamma$. Call such integral an *integral of Cauchy type* and $\varphi(z)$ the corresponding *density function*.

Usually, $F(z)$ represents different functions in different components of $\mathbf{C} - \gamma$. The Cauchy's integral formula (4.6.1) provides an important illustrative example.

Two sections are divided.

Section (1) The differentiability (or analyticity) of the functions $F(z)$

Our main result is the following

Analytic functions defined by an integral of Cauchy type. Let $\varphi(z)$ be a complex-valued function, continuous along a rectifiable curve γ in $\mathbf{C}$. Define

$$F_n(z) = \int_\gamma \frac{\varphi(\zeta)}{(\zeta - z)^n} d\zeta, \quad n = 1, 2, \dots, z \notin \gamma.$$

Then,

(1) $F_n(z)$ is analytic in each component of $\mathbf{C} - \gamma$; and
(2) $F_n'(z) = n F_{n+1}(z), \quad n = 1, 2, 3, \dots$.   (4.7.1)

Even though $\varphi(\zeta) d\zeta$ is a complex measure with compact support in $\mathbf{C}$, these results are still valid. For details, see Refs. [9], p. 98; [45], p. 3. For integrals of Cauchy type with parameters, see (4.7.4) below and for related examples, see Sec. 4.13.

The **proof** is based on induction with respective to n. For *another proof*, see the proof of (3.4.2.6) and Exercise B (1) of Sec. 3.5.2.

In case $n = 1$:

Choose any fixed point $z_0 \notin \gamma$. These is a $\delta > 0$ so that the disk $|z - z_0| < \delta$ lies in a component of $\mathbf{C} - \gamma$. Let z be such that $|z - z_0| < \frac{\delta}{2}$.

Then

$$F_1(z) - F_1(z_0) = (z - z_0) \int_\gamma \frac{\varphi(\zeta)}{(\zeta - z)(\zeta - z_0)} d\zeta. \tag{$*_1$}$$

To show the continuity of F_1 at z_0: If $\zeta \in \gamma$, then

$$|\zeta - z| \geq \frac{\delta}{2}, \quad |\zeta - z_0| \geq \frac{\delta}{2}$$

$$\Rightarrow |F_1(z) - F_1(z_0)| \leq \frac{4|z - z_0|}{\delta^2} \int_\gamma |\varphi(\zeta)||d\zeta|$$

$$\leq |z - z_0| \cdot \frac{4}{\delta^2} \max_{\zeta \in \gamma} |\varphi(\zeta)| \cdot \ell(\gamma) \to 0 \quad \text{as } z \to z_0,$$

where $\ell(\gamma)$ denotes the length of γ.

To show the differentiability of F_1 at z_0 and $F_1'(z_0) = F_2(z_0)$: By $(*_1)$,

$$\frac{F_1(z) - F_1(z_0)}{z - z_0} = \int_\gamma \frac{\varphi(\zeta)}{(\zeta - z)(\zeta - z_0)} d\zeta = \int_\gamma \frac{\varphi_1(\zeta)}{\zeta - z} d\zeta, \quad z, z_0 \notin \gamma$$

where $\varphi_1(\zeta) = \frac{\varphi(\zeta)}{\zeta - z_0}, \zeta \in \gamma$, is continuous along γ. By the known result proved in the last paragraph, as $z \to z_0$,

$$\int_\gamma \frac{\varphi_1(\zeta)}{\zeta - z} d\zeta \to \int_\gamma \frac{\varphi_1(\zeta)}{\zeta - z_0} d\zeta = \int_\gamma \frac{\varphi(\zeta)}{(\zeta - z_0)^2} d\zeta = F_2(z_0).$$

This shows that F_1 is differentiable at z_0 and $F_1'(z_0) = F_2(z_0)$. Therefore, F_1 is analytic in each component of $\mathbf{C} - \gamma$.

Assume the conclusion in (4.7.1) holds for $n-1$: Namely, F_{n-1} is analytic in each component of $\mathbf{C} - \gamma$ and $F_{n-1}'(z) = (n-1)F_n(z), z \in \gamma$, holds.

The general case n: Let z_0 and z be as in the case of $n = 1$.

Write $F_n(z) - F_n(z_0)$ as

$$F_n(z) - F_n(z_0) = \int_\gamma \left\{ \frac{1}{(\zeta - z)^n} - \frac{1}{(\zeta - z_0)^n} \right\} \varphi(\zeta)d\zeta = I_1 + I_2, \tag{$*_2$}$$

where

$$I_1 = \int_\gamma \left\{ \frac{1}{(\zeta - z)^{n-1}(\zeta - z_0)} - \frac{1}{(\zeta - z_0)^n} \right\} \varphi(\zeta)d\zeta \quad \text{and}$$

$$I_2 = \int_\gamma \left\{ \frac{1}{(\zeta - z)^n} - \frac{1}{(\zeta - z)^{n-1}(\zeta - z_0)} \right\} \varphi(\zeta)d\zeta.$$

To show the continuity of F_n at z_0: In the integral I_1, let $\varphi_1(\zeta) = \frac{\varphi(\zeta)}{\zeta - z_0}$. Observe that $\varphi_1(\zeta)$ is continuous along γ. Apply the inductive assumption to $\tilde{F}_{n-1}(z)$ defined by $\varphi_1(\zeta)$ (see (*3)' below). Then

$$I_1 = \tilde{F}_{n-1}(z) - \tilde{F}_{n-1}(z_0) \to 0 \quad \text{as } z \to z_0.$$

Also, note that, as $z \to z_0$,

$$|I_2| \leq |z - z_0| \int_\gamma \frac{|\varphi(\zeta)|}{|\zeta - z|^n |\zeta - z_0|} |d\zeta|$$

$$\leq |z - z_0| \left(\frac{2}{\delta}\right)^{n+1} \cdot \max_{\zeta \in \gamma} |\varphi(\zeta)| \ell(\gamma) \to 0.$$

Combining together, the continuity of F_n at z_0 follows.

To show the differentiability of F_n at z_0 and $F_n'(z_0) = nF_{n+1}(z_0)$:
By (*2),

$$\frac{F_n(z) - F_n(z_0)}{z - z_0} = \frac{1}{z - z_0} I_1 + \int_\gamma \frac{\varphi(\zeta)}{(\zeta - z)^n (\zeta - z_0)} d\zeta. \qquad (*3)$$

Within (*3),

$$\frac{1}{z - z_0} I_1 = \frac{\tilde{F}_{n-1}(z) - \tilde{F}_{n-1}(z_0)}{z - z_0}, \quad \text{where}$$

$$\tilde{F}_{n-1}(z) = \int_\gamma \frac{\varphi_1(\zeta)}{(\zeta - z)^{n-1}} d\zeta \quad \text{with } \varphi_1(\zeta) = \frac{\varphi(\zeta)}{\zeta - z_0}. \qquad (*3)'$$

Apply the inductive assumption to $\tilde{F}_{n-1}(z)$ and we have $\frac{1}{z - z_0} I_1 \to$
$\tilde{F}_{n-1}'(z_0) = (n-1)\tilde{F}_n(z_0)$ as $z \to z_0$. On the other hand, using the established result in the last paragraph to $\tilde{F}_{n-1}(z)$, we have

$$\int_\gamma \frac{\varphi(\zeta)}{(\zeta - z)^n (\zeta - z_0)} d\zeta = \int_\gamma \frac{\varphi_1(\zeta)}{(\zeta - z)^n} d\zeta = \tilde{F}_n(z) \to \tilde{F}_n(z_0) \quad \text{as } z \to z_0.$$

Substituting these two new results into (*3),

$$\lim_{z \to z_0} \frac{F_n(z) - F_n(z_0)}{z - z_0} = F'_{n-1}(z_0) + \tilde{F}_n(z_0)$$

$$= (n-1)\tilde{F}_n(z_0) + \tilde{F}_n(z_0) = n\tilde{F}_n(z_0) = nF_{n+1}(z_0).$$

This finishes the proof.   □

The **main consequences** of (4.7.1) are as follows:

(1) Let f be analytic in a domain Ω. Applying (4.7.1) to a circle which is contained in Ω with its interior, then f is infinitely differentiable and

each $f^{(n)}(z)$ is also analytic in Ω. Moreover,

$$f^{(n)}(z) = \frac{n!}{2\pi i} \int_{|\zeta - z| = \rho} \frac{f(\zeta)}{(\zeta - z)^{n+1}} d\zeta, \quad 0 < \rho < \text{dist}(z, \partial\Omega)$$

$$\text{for } n = 1, 2, \ldots . \qquad (4.7.2)$$

This is (3.4.2.7) obtained by using Taylor series expansion of f at z. In case γ is a cycle homologous to zero in Ω, the replacement of the circle $|\zeta - z| = \rho$ in (4.7.2) by γ results in (4.6.2).

(2) Let f be analytic in a $(n + 1)$-connected domain Ω and continuous on $\bar{\Omega}$. Adopting notations introduced in (3.4.2.4)', then

$$f^{(n)}(z) = \frac{n!}{2\pi i} \int_{\partial\Omega} \frac{f(\zeta)}{(\zeta - z)^{n+1}} d\zeta, \quad z \in \Omega \qquad (4.7.3)$$

where $\partial\Omega = \Gamma - \gamma_1 - \cdots - \gamma_n$.

As a generalization of (4.7.1), we have

The analytic function defined by integral of Cauchy type with a parameter. Let Ω be a domain in the z-plane and γ a rectifiable curve in the w-plane. Let $\varphi(z, w)$ be a function of $(z, w) \in \Omega \times \gamma$ satisfying:

(1) $\varphi(z, w)$ is a continuous function of $z \in \Omega$ and $w \in \gamma$.
(2) For each w, $\varphi(z, w)$ is analytic in $z \in \Omega$.

Then

$$F(z) = \int_\gamma \varphi(z, w) dw$$

is analytic in $z \in \Omega$ and

$$F'(z) = \int \frac{\partial\varphi(z, w)}{\partial z} dw, \quad z \in \Omega. \qquad (4.7.4)$$

For detailed accounts, refer to Ref. [58], Vol. I, pp. 417–422. See also Exercise B (2) of Sec. 5.3.1 and Exercise B of Sec. 5.3.3 for general settings.

Sketch of Proof

Fix a point $z_0 \in \Omega$ and choose $\rho > 0$ so that $|z - z_0| \leq \rho$ is contained in Ω. Then, for each fixed $w \in \gamma$,

$$\varphi(z, w) = \frac{1}{2\pi i} \int_{|\zeta - z_0| = \rho} \frac{\varphi(\zeta, w)}{\zeta - z} d\zeta$$

$$\text{(by Cauchy integral formula)}, \quad |z - z_0| < \rho$$

$$\Rightarrow F(z) = \frac{1}{2\pi i} \int_\gamma \left[\int_{|\zeta - z_0| = \rho} \frac{\varphi(\zeta, w)}{\zeta - z} d\zeta \right] dw$$

$$= \frac{1}{2\pi i} \int_{|\zeta - z_0| = \rho} \frac{1}{\zeta - z} \left[\int_\gamma \varphi(\zeta, w) dw \right] d\zeta \quad \text{(by Fubini's theorem)}$$

$$= \frac{1}{2\pi i} \int_{|\zeta - z_0| = \rho} \frac{F(\zeta)}{\zeta - z} d\zeta, \quad |z - z_0| < \rho.$$

By (4.7.1), $F(z)$ is analytic in $|z - z_0| < \rho$ and hence, at z_0, in particular. Moreover,

$$F'(z) = \frac{1}{2\pi i} \int_{|\zeta - z_0| = \rho} \frac{F(\zeta)}{(\zeta - z)^2} d\zeta = \frac{1}{2\pi i} \int_{|\zeta - z_0| = \rho} \left[\int_\gamma \frac{\varphi(\zeta, w)}{(\zeta - z)^2} dw \right] d\zeta$$

$$= \int_\gamma \left[\frac{1}{2\pi i} \int_{|\zeta - z_0| = \rho} \frac{\varphi(\zeta, w)}{(\zeta - z)^2} d\zeta \right] dw = \int_\gamma \frac{\partial \varphi(z, w)}{\partial z} dw, \quad |z - z_0| < \rho.$$

We give two more examples than those in Sec. 3.4.2.

Example 1. Evaluate the following integrals.

(1) $\frac{1}{2\pi i} \int_\gamma \frac{e^z}{z(1-z)^3} dz$, where γ is a rectifiable Jordan closed curve, not passing 0 and 1.

(2) $\int_\gamma \frac{\text{Log}(1+z)}{(2z-1)^3}$, where γ is given by $z(t) = (2\cos - 1)e^{it}, \quad 0 \le t \le 2\pi.$

(3) (Ahlfors [1], p. 123) $\int_{|z| = \rho} \frac{|dz|}{|z - a|^4}, \quad |a| \ne \rho.$

Solution.

(1) In case 0 and 1 are in Ext γ, the integral is equal to zero. While

the integral

$$= \begin{cases} \left. \dfrac{e^z}{(1-z)^3} \right|_{z=0} = 1, & \text{if } 0 \in \text{Int } \gamma \text{ and } 1 \in \text{Ext } \gamma, \\[2em] \left. -\dfrac{1}{2!} \dfrac{d^2}{dz^2} \left\{ \dfrac{e^z}{z} \right\} \right|_{z=1} = -\dfrac{1}{2} e, & \text{if } 1 \in \text{Int } \gamma \text{ and } 0 \in \text{Ext } \gamma. \end{cases}$$

If 0 and 1 are in Int γ, choose $\rho > 0$ sufficiently small so that $|z| \le \rho$ and $|z - 1| \le \rho$ are disjoint and all lie in Int γ. Then

$$\text{the integral} = \frac{1}{2\pi i} \int_{|z| = \rho} \frac{\left[\frac{e^z}{(1-z)^3} \right]}{z} dz + \frac{1}{2\pi i} \int_{|z-1| = \rho} \frac{\left[\frac{e^z}{z} \right]}{(1-z)^3} dz = 1 - \frac{1}{2} e.$$

(2) Observe that $\mathrm{Log}(1+z)$ is single-valued and analytic in $\mathbf{C} - (\infty, -1] = \Omega$ and γ is homologous to $2\gamma_0$ in Ω, where γ_0 is the circle $\left|z - \frac{1}{2}\right| = \rho$ for sufficiently small $\rho > 0$. In other words, $n\left(\gamma; \frac{1}{2}\right) = 2$ and it follows from (4.7.2) that

$$\text{the integral} = 2\pi i \cdot \frac{1}{2^3} \cdot n\left(\gamma; \frac{1}{2}\right) \cdot \frac{1}{2!}\frac{d^2}{dz^2}\mathrm{Log}(1+z)|_{z=\frac{1}{2}}$$

$$= 2\pi i \cdot \frac{1}{2^3} \cdot \frac{-1}{\left(1 + \frac{1}{2}\right)^2} = -\frac{\pi i}{9}.$$

(3) Write $|z| = \rho$ as $z = \rho e^{i\theta}, 0 \le \theta \le 2\pi$. Now $dz = \rho i e^{i\theta}d\theta$ implies that $|dz| = \rho d\theta = \frac{-i\rho dz}{z}$. On the other hand, $z\bar{z} = \rho^2$ helps to reduce $|z - a|^4$ to $(z - a)^2(\bar{z} - \bar{a})^2 = (z - a)^2\left[\frac{\rho^2}{z} - \bar{a}\right]^2$. Hence,

$$\text{the integral} = \int_{|z|=\rho}\frac{1}{(z - a)^2\left[\frac{\rho^2}{z} - \bar{a}\right]^2}\frac{-i\rho}{z}dz$$

$$= \frac{-i\rho}{\bar{a}^2}\int_{|z|=\rho}\frac{zdz}{(z - a)^2\left[z - \frac{\rho}{\bar{a}}\right]^2}.$$

Note that a and $\frac{\rho}{\bar{a}}$ (in case $a \ne 0$) are symmetric with respect to $|z| = \rho$. If $a \ne 0$, then

the integral

$$= \begin{cases} \dfrac{-i\rho}{\bar{a}^2} \cdot 2\pi i \cdot \dfrac{d}{dz}\left\{\dfrac{z}{(z - a)^2}\right\}\Big|_{z=\frac{\rho^2}{\bar{a}}} = 2\pi\rho \cdot \dfrac{|a|^2 + \rho^2}{(|a|^2 - \rho^2)^3}, & \text{if } 0 < \rho \le |a| \\[3mm] \dfrac{-i\rho}{\bar{a}^2} \cdot 2\pi i \cdot \dfrac{d}{dz}\left\{\dfrac{z}{\left(z - \frac{\rho^2}{\bar{a}}\right)^2}\right\}\Bigg|_{z=a} = 2\pi\rho \cdot \dfrac{|a|^2 + \rho^2}{(\rho^2 - |a|^2)^3}, & \text{if } |a| < \rho \end{cases}$$

If $a = 0$, then

$$\text{the integral} = \int_{|z|=\rho}\frac{|dz|}{|z|^4} = \int_0^{2\pi}\frac{\rho d\theta}{\rho^4} = \frac{2\pi}{\rho^3}.$$

Example 2. Evaluate the following integral

$$\int_{|z|=1}\frac{\bar{z}^k p(z)}{z - z_0}dz, \quad \text{where } |z_0| \ne 1, 0 \le k \le n, \quad \text{and}$$

$$p(z) = a_n z^n + \cdots + a_1 z + a_0.$$

In particular,

$$\int_{|\zeta|=1} \frac{\bar{\zeta}}{\zeta - z} d\zeta = \begin{cases} 0, & |z| < 1, \\ -\dfrac{1}{z}, & |z| > 1. \end{cases}$$

Solution. Observe that, on $|z| = 1$, $z\bar{z} = 1$, and thus $\bar{z}^k = (z^{-1})^k = z^{-k}$, $0 \le k \le n$. In case $|z_0| < 1$: if $z_0 = 0$, then

$$\text{the integral} = \int_{|z|=1} \frac{p(z)}{z^{k+1}} dz = \frac{2\pi i}{k!} \frac{d^k}{dz^k} p(z)\big|_{z=0} = 2\pi i a_k, \quad 0 \le k \le n.$$

Suppose $z_0 \ne 0$. Then

$$\text{the integral} = \int_{|z|=1} \frac{p(z)}{z^k(z - z_0)} dz = 2\pi i \left\{ \frac{1}{k!} \frac{d^k}{dz^k} \left[\frac{p(z)}{z - z_0} \right]\bigg|_{z=0} + \frac{p(z_0)}{z_0^k} \right\}.$$

In case $|z_0| > 1$: then

$$\text{the integral} = \int_{|z|=1} \frac{p(z)}{z^k(z - z_0)} dz = 2\pi i \cdot \frac{1}{k!} \frac{d^k}{dz^k} \left[\frac{p(z)}{z - z_0} \right]\bigg|_{z=0}.$$

Section (2) Cauchy principal value

We relax the condition imposed in $(*)$ by granting a particular point z_0 lying on the path of integration. More precisely, if $z_0 \in \gamma$, we are going to consider whether the integral

$$\int_\gamma \frac{\varphi(\zeta)}{\zeta - z_0} d\zeta \quad \text{or} \quad \frac{1}{2\pi i} \int_\gamma \frac{\varphi(\zeta)}{\zeta - z_0} d\zeta, \tag{4.7.5}$$

has a meaning. This depends on some additional restriction on the curve γ and how we define the limit process concerned.

Suppose $\gamma : z(t), a \le t \le b$, is a piecewise differentiable curve.

Let $z_0 = z(t_0) \in \gamma$ be a *regular point*, namely, z_0 is not end points of γ and $z'(t_0)$ exists as a *nonzero* number. In this case, the curve γ has a tangent $z = z_0 + z'(t_0)t, t \in \mathbf{R}$, at the point z_0. See Fig. 4.33. And we designate the *interior angle* of γ at z_0 as

$$\theta(z_0) = \pi. \tag{4.7.6}$$

Let $z_0 = z(t_0) \in \gamma$ be a *turning point*, namely, $a < t_0 < b$, and

(1) $z(t)$ has the left derivative $z'_-(t_0)$ at t_0 and is equal to

$$\lim_{t \to t_0^-} z'(t) = z'(t_0^-) \ne 0;$$

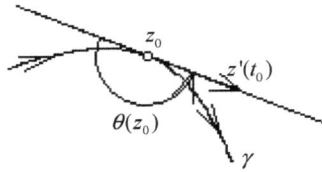

Fig. 4.33

(2) $z(t)$ has the right derivative $z'_+(t_0)$ at t_0 and is equal to

$$\lim_{t \to t_0^+} z'(t) = z'(t_0^+) \neq 0; \quad \text{and}$$

(3) $z'_-(t_0) \neq z'_+(t_0)$.

Refer to (2.4.9) and Fig. 4.34. In this case, designate the *interior angle* of γ at z_0 as

$$\theta(z_0) = \pi - (\operatorname{Arg} z'_-(t_0) - \operatorname{Arg} z'_+(t_0)) = \pi + \operatorname{Arg} z'_+(t_0) - \operatorname{Arg} z'_-(t_0).$$
$$(4.7.7)$$

Fix $z_0 \in \gamma$ (a regular or turning point). Construct a circle $|z - z_0| = \rho$ with center z_0 as in Fig. 4.35. If the following limit exists as *finite* number,

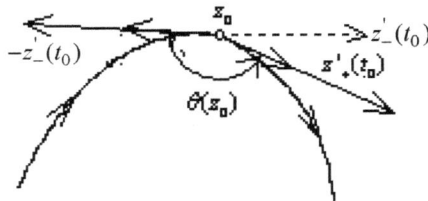

Fig. 4.34

Fig. 4.35

we call it the *Cauchy principal value* of $\int_\gamma \frac{\varphi(\zeta)}{\zeta - z_0} d\zeta$ and is denoted as

$$\text{P.V.} \int_\gamma \frac{\varphi(\zeta)}{\zeta - z_0} d\zeta = \lim_{\rho \to 0} \int_{\substack{\zeta \in \gamma \\ |\zeta - z_0| \ge \rho}} \frac{\varphi(\zeta)}{\zeta - z_0} d\zeta. \qquad (4.7.8)$$

Note that the original integral $\int_\gamma \frac{\varphi(\zeta)}{\zeta - z_0} d\zeta$ might diverge.

Example 3. Evaluate

$$\text{P.V.} \int_{|z|=1} \frac{z^2}{z^3 - 1} dz.$$

Solution. Decompose the integrand into partial fraction as

$$\frac{z^2}{z^3 - 1} = \frac{1}{3} \left(\frac{1}{z - 1} + \frac{1}{z - \omega} + \frac{1}{z - \omega^2} \right), \quad \omega = e^{\frac{2\pi i}{3}}.$$

In Fig. 4.36, consider the circle $\gamma : z = 1 + \rho e^{i\theta}, 0 \le \theta \le 2\pi$. This circle intersects $|z| = 1$ at two points $1 + \rho e^{i\alpha}$ and $1 + \rho e^{i(2\pi - \alpha)}$. Let γ_ρ denote the part of $|z| = 1$ outside γ, while $\tilde{\gamma}_\rho$ the part of γ inside $|z| = 1$. By Cauchy integral theorem,

$$\int_{\gamma_\rho - \tilde{\gamma}_\rho} \frac{1}{z - 1} dz = 0$$

$$\Rightarrow \int_{\gamma_\rho} \frac{1}{z - 1} dz = \int_{\tilde{\gamma}_\rho} \frac{1}{z - 1} dz = \int_\alpha^{2\pi - \alpha} \frac{\rho i e^{i\theta}}{\rho e^{i\theta}} d\theta$$

$$= i(2\pi - 2\alpha) \to i \left(2\pi - 2 \cdot \frac{\pi}{2} \right) = \pi i, \quad \text{as } \rho \to 0$$

$$\Rightarrow \text{P.V.} \int_{|z|=1} \frac{1}{z - 1} dz = \pi i.$$

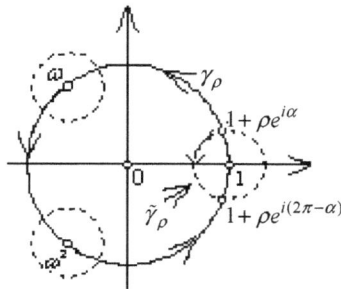

Fig. 4.36

But the line integral $\int_{|z|=1} \frac{1}{z-1} dz$ does not exist as a finite number. Similarly,

$$\text{P.V.} \int_{|z|=1} \frac{1}{z-\omega} dz = \pi i \quad \text{and} \quad \text{P.V.} \int_{|z|=1} \frac{1}{z-\omega^2} dz = \pi i.$$

Finally,

$$\text{P.V.} \int_{|z|=1} \frac{z^2}{z^3-1} dz = \frac{1}{3}(\pi i + \pi i + \pi i) = \pi i.$$

In general setting, we have

Integral with Cauchy principal value. Let γ be a piecewise differentiable Jordan closed curve and $z_0 \in \gamma$ be a fixed point which is either a regular point or a turning point. Suppose f is analytic on $\overline{\text{Int}\,\gamma}$. Then

$$\text{P.V.} \int_\gamma \frac{f(\zeta)}{\zeta - z_0} d\zeta = i\theta(z_0)f(z_0),$$

where $\theta(z_0)$ is the interior angle of γ at z_0 (see (4.7.6) and (4.7.7)).

$$(4.7.9)$$

Proof. Choose $\rho > 0$ so small that the circle $\tilde{\gamma}: z = z_0 + \rho e^{i\theta}$, $0 \le \theta \le 2\pi$, will intersect γ at exactly two points $z_1(\theta_1)$ and $z_2(\theta_2)$ as shown in Fig. 4.37. Let γ_ρ denote the part of γ that lies outside $\tilde{\gamma}$, and $\tilde{\gamma}_\rho$ the circular arc of $\tilde{\gamma}$ that lies in $\text{Int}\,\gamma$ and connects z_1 to z_2 so that the sum curve $\gamma_\rho + \tilde{\gamma}_\rho$ is a piecewise differentiable Jordan closed curve having z_0 in the unbounded component of $\mathbf{C} - \{\gamma_\rho + \tilde{\gamma}_\rho\}$. Therefore,

$$\int_{\gamma_\rho + \tilde{\gamma}_\rho} \frac{f(\zeta)}{\zeta - z_0} d\zeta = 0$$

$$\Rightarrow \int_{\gamma_\rho} \frac{f(\zeta)}{\zeta - z_0} d\zeta = -\int_{\tilde{\gamma}_\rho} \frac{f(\zeta)}{\zeta - z_0} d\zeta$$

$$= -f(z_0) \int_{\tilde{\gamma}_\rho} \frac{1}{\zeta - z_0} d\zeta - \int_{\tilde{\gamma}_\rho} \frac{f(\zeta) - f(z_0)}{\zeta - z_0} d\zeta. \quad (*4)$$

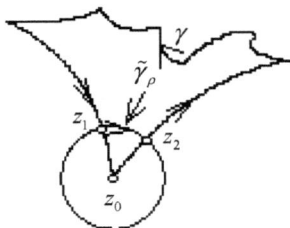

Fig. 4.37

Since f is continuous at z_0, for each $\varepsilon > 0$, there is a $\delta > 0$ so that $|f(\zeta) - f(z_0)| < \varepsilon$ whenever $|\zeta - z_0| < \delta$. Choose $\rho < \delta$ in $(*_4)$. It follows that

$$\left| \int_{\tilde{\gamma}_\rho} \frac{f(\zeta) - f(z_0)}{\zeta - z_0} d\zeta \right| \leq \frac{\varepsilon}{\rho} \cdot \text{the length of } \tilde{\gamma}_\rho \leq \frac{\varepsilon}{\rho} \cdot 2\pi\rho = 2\pi\varepsilon$$

$$\Rightarrow \lim_{\rho \to 0} \int_{\tilde{\gamma}_\rho} \frac{f(\zeta) - f(z_0)}{\zeta - z_0} d\zeta = 0.$$

On the other hand, letting $z_1 = z_0 + \rho e^{i\theta_1}$ and $z_2 = z_0 + \rho e^{i\theta_2}$, we have

$$-\int_{\tilde{\gamma}_\rho} \frac{1}{\zeta - z_0} d\zeta = -\int_{\theta_1}^{\theta_2} \frac{\rho i e^{i\theta}}{\rho e^{i\theta}} d\theta = -i(\theta_2 - \theta_1) = i(\theta_1 - \theta_2).$$

Substituting these two results into $(*_4)$, then

$$\lim_{\rho \to 0} \int_{\gamma_\rho} \frac{f(\zeta)}{\zeta - z_0} d\zeta = 0 + \text{ if}(z_0) \lim_{\rho \to 0} (\theta_1 - \theta_2) = i\theta(z_0)f(z_0).$$

This finishes the proof. □

Let γ be a curve passing the point ∞ (see Remark (1) in Sec. 2.4), namely, $z = z(t)$, $-\infty < t < \infty$ and $\lim_{t \to \infty} z(t) = \lim_{t \to -\infty} z(t) = \infty$. In case $z_0 \notin \gamma$, the *Cauchy principal value* of the line integral $\int_\gamma \frac{\varphi(\zeta)}{\zeta - z_0} d\zeta$ is defined and denoted as

$$\text{P.V.} \int_\gamma \frac{\varphi(\zeta)}{\zeta - z_0} d\zeta = \lim_{\rho \to \infty} \int_{\substack{\zeta \in \gamma \\ |\zeta| \leq \rho}} \frac{\varphi(\zeta)}{\zeta - z_0} d\zeta, \qquad (4.7.10)$$

provided the limit on the right exists as a finite complex number.

A main result in this direction is the following

Cauchy integral formula in a half-plane. Suppose f is analytic in the half-plane $\text{Re } z > \alpha$. Also there are constants $A > 0$ and $m > 0$ so that $|f(z)| \leq \frac{A}{|z|^m}$ for all sufficiently large $|z|$. In case $a > \alpha$, then

$$f(z) = \text{P.V.} \left\{ -\frac{1}{2\pi i} \int_{a-i\infty}^{a+i\infty} \frac{f(\zeta)}{\zeta - z} d\zeta \right\}, \qquad \text{Re } z > a$$

$$= -\frac{1}{2\pi i} \lim_{R \to \infty} \int_{a-iR}^{a+iR} \frac{f(\zeta)}{\zeta - z} d\zeta.$$

Letting $\zeta = a + iy$, the above formula can be rewritten as

$$f(z) = -\frac{1}{2\pi} \int_{-\infty}^{\infty} \frac{f(a + iy)}{a + iy - z} dy, \qquad \text{Re } z > a. \qquad (4.7.11)$$

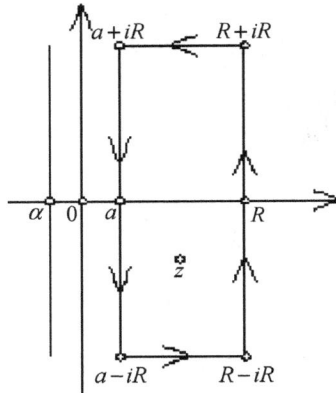

Fig. 4.38

Proof. Fix any z, $\text{Re}\,z > a$. Choose $R > \max\{|\text{Re}\,z|, |\text{Im}\,z|\}$ so that z lies in the rectangle $[a, R] \times [-R, R]$ as shown in Fig. 4.38. Let γ denote the boundary of this rectangle, endowed with counterclockwise direction, and L denote the left vertical boundary: $z(t) = a + it$, $-R \le t \le R$. Then

$$f(z) = \frac{1}{2\pi i} \int_\gamma \frac{f(\zeta)}{\zeta - z} d\zeta = -\frac{1}{2\pi i} \int_{a-iR}^{a+iR} \frac{f(\zeta)}{\zeta - z} d\zeta + \frac{1}{2\pi i} \int_{\gamma - L} \frac{f(\zeta)}{\zeta - z} d\zeta. \qquad (*_5)$$

Choose R large, say $R \ge 2|z|$. Then, if $\zeta \in \gamma - L$, $|\zeta| \ge R$ holds and so does $\left|\frac{z}{\zeta}\right| \le \frac{1}{2}$. By assumption,

$$\left| \frac{f(\zeta)}{\zeta - z} \right| = \frac{|f(\zeta)|}{|\zeta|\left|1 - \frac{z}{\zeta}\right|} \le \frac{A}{|\zeta|^{m+1}\left(1 - \left|\frac{z}{\zeta}\right|\right)} \le \frac{2A}{R^{m+1}}$$

$$\Rightarrow \left| \int_{\gamma - L} \frac{f(\zeta)}{\zeta - z} d\zeta \right| \le \frac{2A}{R^{m+1}}(4R - 2a) \to 0 \quad \text{as } R \to \infty.$$

Now, in $(*_5)$, let $R \to \infty$ and the first half of the claim follows.
    Let $\zeta = a + iy$. Then the integrand has its absolute value

$$\left| \frac{f(a+iy)}{a+iy - z} \right| \le \frac{M}{|y|^{m+1}} \quad (M > 0, \text{ a constant}) \text{ if } |y| \text{ is large enough.}$$

⇒ The improper integral

$$\int_{-\infty}^{\infty} \frac{f(a+iy)}{a+iy-z} dy = \lim_{\substack{R \to \infty \\ R' \to \infty}} \int_{-R'}^{R} \frac{f(a+iy)}{a+iy-z} dy, \quad \operatorname{Re} z > a$$

exists as a finite complex number and is equal to its Cauchy principal value (namely, the case $R' = R$). □

We give an

Example 4. Let $c > 0$ and t be a real parameter. Then

$$\text{P.V.} \ \frac{1}{2\pi i} \int_{c-i\infty}^{c+i\infty} \frac{e^{zt}}{z} dz = \begin{cases} 1, & t > 0, \\ \dfrac{1}{2}, & t = 0, \\ 0, & t < 0. \end{cases}$$

which is called the *Heaviside function*.

Solution. Integrate $f(z) = \frac{e^{tz}}{z}$ along the path of integration γ shown in Fig. 4.39 (in case $t > 0$, adopt (a); $t < 0$, adopt (b)).
   In case $t > 0$: By Cauchy integral formula,

$$\int_{\gamma} f(z)dz = \int_{c-iR}^{c+iR} f(z)dz + \int_{C_R} f(z)dz = 2\pi i e^0 = 2\pi i, \qquad (*6)$$

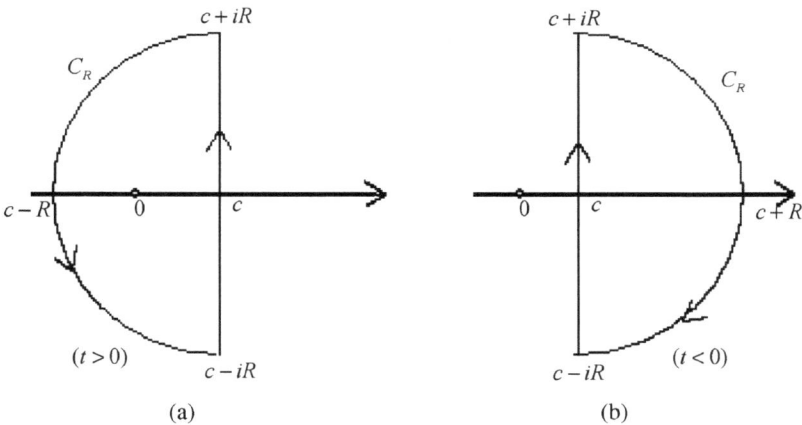

(a)   (b)

Fig. 4.39

where $C_R : z = c + Re^{i\theta}, \frac{\pi}{2} \le \theta \le \frac{3\pi}{2}$ and $R > c$. Along C_R,

$$\left| \int_{\frac{\pi}{2}}^{\frac{3\pi}{2}} \frac{e^{tz}}{z} dz \right| \le \int_{\frac{\pi}{2}}^{\frac{3\pi}{2}} \frac{|e^{tz}|}{|z|} |dz| \le \int_{\frac{\pi}{2}}^{\frac{3}{\pi/2}} \frac{e^{tc} \cdot e^{tR\cos\theta}}{R - c} R d\theta$$

$$= \frac{Re^{tc}}{R - c} \int_0^\pi e^{-tR\sin\theta} d\theta$$

$$= \frac{2e^{tc} R}{R - c} \int_0^{\frac{\pi}{2}} e^{-tR\sin\theta} d\theta$$

$$\le \frac{2e^{tc} R}{R - c} \int_0^{\frac{\pi}{2}} e^{\frac{-2tR\theta}{\pi}} d\theta \quad \left(\text{using } \frac{2}{\pi} \le \frac{\sin\theta}{\theta} \text{ if } 0 \le \theta \le \frac{\pi}{2} \right)$$

$$= \frac{e^{ct}}{t(R - c)} \pi (1 - e^{-tR}) \to 0 \quad \text{as } R \to \infty.$$

Therefore, by $(*_6)$, letting $R \to \infty$,

$$\text{P.V. } \frac{1}{2\pi i} \int_{c-i\infty}^{c+i\infty} \frac{e^{tz}}{z} dz = \frac{2\pi i}{2\pi i} = 1.$$

In case $t < 0$: By Cauchy integral theorem,

$$\int_\gamma f(z) dz = \int_{c-iR}^{c+iR} f(z) dz + \int_{C_R} f(z) dz = 0, \qquad (*_7)$$

where $C_R : z = c + Re^{i\theta}, -\frac{\pi}{2} \le \theta \le \frac{\pi}{2}$. Along C_R,

$$\left| \int_{C_R} \frac{e^{t^2}}{z} dz \right| \le \int_{-\frac{\pi}{2}}^{\frac{\pi}{2}} \frac{e^{tc} e^{tR\cos\theta}}{R} R d\theta = 2e^{tc} \int_0^{\frac{\pi}{2}} e^{tR\cos\theta} d\theta$$

$$= 2e^{tc} \int_0^{\frac{\pi}{2}} e^{tR\sin\theta} d\theta$$

$$\le 2e^{tc} \int_0^{\frac{\pi}{2}} e^{2tR\theta/\pi} d\theta = \frac{\pi}{tR} e^{tc} (e^{tR} - 1)$$

$$< \frac{\pi}{-tR} e^{tc} \to 0 \quad \text{as } R \to \infty.$$

Substituting into $(*_7)$ and letting $R \to \infty$, then

$$\text{P.V. } \frac{1}{2\pi i} \int_{c-i\infty}^{c+i\infty} \frac{e^{tz}}{z} dz = 0.$$

In case $t = 0$: Observe that $\text{Log}\, z$ is single-valued in $\mathbf{C} - (\infty, -0]$. Then

$$\frac{1}{2\pi i} \int_{c-iR}^{c+iR} \frac{1}{z}\, dz = \frac{1}{2\pi i} [\text{Log}(c + iR) - \text{Log}(c - iR)]$$

$$= \frac{1}{\pi} \tan^{-1} \frac{R}{c} \to \frac{1}{\pi} \cdot \frac{\pi}{2} = \frac{1}{2} \quad \text{as } R \to \infty.$$

$$\Rightarrow \text{P.V.} \ \frac{1}{2\pi i} \int_{c-i\infty}^{c+i\infty} \frac{1}{z}\, dz = \frac{1}{2}.$$

Remark (Boundary values of a function defined by integral of the Cauchy type).   Let γ be a piecewise differentiable Jordan closed curve with parameter equation $z(t) : a \leq t \leq b$. Let $z_0 = z(t_0)$ be a regular point of γ, i.e., $z'(t_0) \neq 0$, where $a \leq t_0 \leq b$. *Suppose $\varphi(z)$ is analytic along γ.* Choose $\rho > 0$ so small that the circle $\Gamma : |z - z_0| = \rho$ intersects γ at exactly two points z_1 and z_2 as shown in Fig. 4.40 (see also Fig. 4.37). Denote by σ the subcurve of γ that connects z_1 to z_2, σ_1 the one connecting $z(a)$ to z_1 and σ_2 the one connecting z_2 to $z(b)$, while Γ_r the circular arc of Γ that lies on the right of γ and Γ_ℓ the one on the left of γ. Then, by (4.7.2),

$$\int_{\sigma_1 + \Gamma_r + \sigma_2} \frac{\varphi(\zeta)}{\zeta - z}\, d\zeta = F(z) \text{ is analytic in } \text{Int}(\sigma - \Gamma_\ell). \qquad (*_8)$$

Just like the proof after $(*_4)$,

$$\frac{1}{2\pi i} \int_{\Gamma_r} \frac{\varphi(\zeta)}{\zeta - z}\, d\zeta = \frac{1}{2\pi i} \int_{\Gamma_r} \frac{\varphi(\zeta) - \varphi(z_0)}{\zeta - z_0}\, d\zeta + \frac{\varphi(z_0)}{2\pi i} \int_{\Gamma_r} \frac{d\zeta}{\zeta - z_0}$$

$$\to 0 + \frac{\varphi(z_0)}{2\pi i} \cdot \pi i = \frac{\varphi(z_0)}{2} \quad \text{as } \rho \to 0.$$

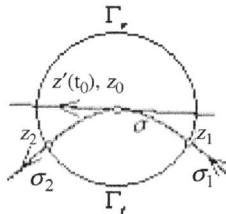

Fig. 4.40

Substituting into ($*_8$), we have

$$\lim_{\rho \to 0} \int_{\sigma_1 + \Gamma_r + \sigma_1} \frac{\varphi(\zeta)}{\zeta - z} d\zeta = \text{P.V.} \frac{1}{2\pi i} \int_{\gamma} \frac{\varphi(\zeta)}{\zeta - z_0} d\zeta + \frac{\varphi(z_0)}{2}$$

$$\underset{\text{(def.)}}{=} \lim_{\substack{z \in \text{Int}(\sigma - \Gamma_\ell) \\ z \to z_0}} F(z) \underset{\text{(def.)}}{=} F_l(z_0), \qquad (4.7.12)$$

called the *left boundary value* of F at z_0. Similarly, the *right boundary value* of F at z_0 is

$$\lim_{\rho \to 0} \int_{\sigma_1 + \Gamma_\ell + \sigma_2} \frac{\varphi(\zeta)}{\zeta - z} d\zeta = \text{P.V.} \frac{1}{2\pi i} \int_{\gamma} \frac{\varphi(\zeta)}{\zeta - z_0} d\zeta - \frac{\varphi(z_0)}{2}$$

$$\underset{\text{(def.)}}{=} \lim_{\substack{z \in \text{Int}(\sigma - \Gamma_r) \\ z \to z_0}} F(z) \underset{\text{(def.)}}{=} F_r(z_0). \qquad (4.7.13)$$

This is a rather special case of *Plemelj (or Sokhotski) formula*. For details, see Ref. [58], Vol. I, pp. 306–316.   □

Exercises A

(1) In (4.7.1), let Ω_∞ be the unbounded component of $\mathbf{C} - \gamma$ that contains the point ∞. Show that $F_1^{(n)}(z) \to 0$ as $z \in \Omega_\infty$ and $z \to \infty$, for $n = 0, 1, 2, \ldots$.

(2) Evaluate the following integrals.

(a) $\text{P.V.} \displaystyle\int_{|z|=a} \frac{dz}{z - a}$.

(b) $\text{P.V.} \displaystyle\int_{|z|=a} \frac{dz}{z^2 + a^2}$.

(c) $\text{P.V.} \displaystyle\int_{|z|=1} \frac{dz}{z^4 + 1}$.

(d) $\text{P.V.} \displaystyle\int_{-\infty}^{\infty} \frac{dx}{x^2 - 1}$.

(3) Suppose f is analytic in $\text{Im } z \geq 0$ and $|f(z)| \leq \frac{A}{|z|^m}$ ($A > 0, m > 0$ are constants) if $|z|$ is large. Show that

$$f(z) = \frac{1}{2\pi i} \int_{-\infty}^{\infty} \frac{f(x)dx}{x - z}, \quad \text{Im } z > 0.$$

(4) Let γ be the half-circle $|\zeta| = R$, $\text{Im } \zeta \geq 0$, and $\varphi(\zeta) \equiv 1$. Let $F(z)$ be $F_1(z)$ in (4.7.1) (see also ($*_8$)).

(a) Evaluate $F(z)$, $F_\ell(\zeta)$, and $F_r(\zeta)$.

(b) In particular, find $F(0)$, $F_\ell(iR)$, and $F_r(iR)$.

(5) Let $\varphi(\zeta) = \zeta^{-n}$ ($n \geq 1$ is an integer). Evaluate the integral $F(z) = F_1(z)$ of the Cauchy type (see (4.7.1)) in each case.

   (a) γ is the boundary of the ring domain $0 < r < |z| < R$.

   (b) γ is the half-circle $|\zeta| = R$, $\operatorname{Im} \zeta \geq 0$. Also find $F(0)$, $F_\ell(\zeta)$ and $F_r(\zeta)$.

Exercises B

(1) Let γ be a piecewise differentiable Jordan closed curve and $\varphi(\zeta)$ be continuous along γ except at most finitely many points. Let

$$F(z) = \frac{1}{2\pi i} \int_\gamma \frac{\varphi(\zeta)}{\zeta - z} d\zeta = \begin{cases} F_1(z), & z \in \operatorname{Int} \gamma, \\ F_2(z), & z \in \operatorname{Ext} \gamma. \end{cases}$$

Designate $F_1^+(\zeta) = \lim\limits_{\substack{z \in \operatorname{Int}\gamma \\ z \to \zeta \in \gamma}} F_1(z)$ and $F_2^-(\zeta) = \lim\limits_{\substack{z \in \operatorname{Ext}\gamma \\ z \to \zeta \in \gamma}} F_2(z)$.

Show that

   (a) $\varphi(\zeta) = F_1^+(\zeta) + F_2^-(\zeta)$, $\zeta \in \gamma$,   and

   (b) $F_\ell(\zeta) = F_1^+(\zeta)$, $F_r(\zeta) = -F_2^-(\zeta) + F_2^-(\infty)$.

Note: In case $F_2^-(\zeta)$ or $F_1^+(\zeta)$ is identically equal to 0, then the result reduces, respectively, to Cauchy integral on $\operatorname{Int}\gamma$ or $\operatorname{Ext}\gamma$ (see Exercise A (32) of Sec. 3.4.2).

(2) Let γ be a piecewise differentiable Jordan closed curve. Let a and b be two distinct points so that the segment $[a, b]$ is still contained in $\operatorname{Int}\gamma$. If $\varphi(\zeta)$ is the branch in $\mathbf{C} - [a, b]$, try to determine $F_\ell(\zeta)$ and $F_r(\zeta)$.

   (a) $\varphi(\zeta) = \operatorname{Log}\frac{\zeta-a}{\zeta-b}$ determined by $\varphi(\infty) = \operatorname{Log} 1 = 0$.

   (b) $\varphi(\zeta) = \sqrt{\frac{\zeta-a}{\zeta-b}}$ determined by $\varphi(\infty) = 1$.

(3) Let γ be as in Exercise (2). Let $a \in \operatorname{Int}\gamma$ and $b \in \operatorname{Ext}\gamma$. Suppose $\varphi(\zeta) = \operatorname{Log}\frac{\zeta-a}{\zeta-b}$ is the branch determined by $\operatorname{Log} 1 = 0$ in $\mathbf{C} - [a, b]$. Determine the corresponding $F(z)$, $F_\ell(\zeta)$ and $F_r(\zeta)$. Here, it is supposed that $[a, b]$ has only one point of intersection with γ.

(4) Let $0 < \lambda < 1$. Show that

   (a) $\displaystyle\int_0^1 \left(\frac{t}{1-t}\right)^\lambda \frac{dt}{t-z} = \frac{\pi}{\sin\lambda\pi}\left[1 - \left(\frac{z}{z-1}\right)^\lambda\right]$, $z \notin [0, 1]$.

   (b) $\displaystyle\int_0^1 \left(\frac{t}{1-t}\right)^\lambda \frac{dt}{t-z} = \frac{\pi}{\sin\lambda\pi}\left[1 - \left(\frac{z}{z-1}\right)^\lambda \cos\pi z\right]$, $z \in (0, 1)$.

(5) Let $\varphi(x) = \frac{x^2}{x^2+1}$, $x \in \mathbf{R}$. The *Cauchy principal value* of the improper integral of the Cauchy type is defined as

$$f(z) = \text{P.V.} \frac{1}{2\pi i} \int_{-\infty}^{\infty} \frac{\varphi(x)}{x - z} dz = \lim_{R \to \infty} \frac{1}{2\pi i} \int_{-R}^{R} \frac{\varphi(x)}{x - z} dx, \quad \text{Im } z \neq 0.$$

Show that

$$f(z) = \begin{cases} \dfrac{z}{2(z + i)}, & \text{if Im } z > 0, \\[3mm] -\dfrac{z}{2(z - i)}, & \text{if Im } z < 0. \end{cases}$$

(6) Let γ denote the segment $[-1, 1]$ on the real axis, oriented according to the increasing of x. Define

$$F(z) = \frac{1}{2\pi i} \int_{\gamma} \frac{d\zeta}{\zeta - z}, \quad z \in \mathbf{C} - [-1, 1].$$

Show that, if $-1 < x < 1$,

$$F_\ell(x) = \frac{1}{2\pi i} \log \frac{1 - x}{1 + x} + \frac{1}{2} \quad \text{and} \quad F_r(x) = \frac{1}{2\pi i} \log \frac{1 - x}{1 + x} - \frac{1}{2}.$$

This shows that $F(z)$ cannot be extended continuously to $\mathbf{C}$.

4.8. Taylor Series (Complicated Examples)

We have introduced Taylor series before, mainly in two spots. One is in Sec. 3.3.2, where $\log(1 + z), (1 + z)^\lambda, e^z, \cos z$ and $\sin z$ are represented, respectively, by their Taylor series. The other is in (3.4.2.6), where a function is expressed as its Taylor series at a point where it is analytic.

Here in this section, more complicated examples will be illustrated.

For reference, we give a general *comment* on how to expand an analytic function into its Taylor series at a point. Suppose

$$f(z) = \sum_{n=0}^{\infty} a_n(z - z_0)^n, \quad |z - z_0| < R \quad \text{(the radius of convergence)},$$

where

$$a_n = \frac{f^{(n)}(z_0)}{n!} = \frac{1}{2\pi i} \int_{|z-z_0|=r} \frac{f(z)}{(z - z_0)^{n+1}} dz, \quad n = 0, 1, 2, \ldots, 0 < r < R.$$

Try the following *methods*:

(1) Direct computation of $f^{(n)}(z_0)$, $n = 0, 1, 2, \ldots$.
(2) Use integral formula to compute a_n, $n = 0, 1, 2, \ldots$.

(3) Apply termwise differentiation or integration (owing to uniform convergence) to well-known Taylor series (see (3.3.2.4) for theoretical backgrounds), such as the geometric series $(1-z)^{-1} = \sum_{n=0}^{\infty} z^n$, $|z| < 1$.

(4) To expand a proper rational function $\frac{p(z)}{q(z)}$ in a Taylor series, it suffices to represent this function as a sun of partial fractions and represent each of these fractions by a Taylor series, using (3) in most cases.

(5) Apply algebraic operations or composition operation to well-known Taylor series (see Sec. 5.1 for details). For instance, let

$$f(z) = \sum_{0}^{\infty} a_n(z - z_0)^n, \quad |z - z_0| < r_1 \quad \text{and}$$

$$g(z) = \sum_{0}^{\infty} b_n(z - z_0)^n, \quad |z - z_0| < r_2.$$

Then

(i) $f(z) \pm g(z) = \sum_{0}^{\infty} (a_n \pm b_n)(z - z_0) < \min\{r_1, r_2\}$.

(ii) *Cauchy product* of f and g (owing to absolute convergence):

$$f(z)g(z) = \sum_{0}^{\infty} (a_n b_0 + a_1 b_{n-1} + \cdots + a_0 b_n)(z - z_0)^n,$$

$$|z - z_0| < \min\{r_1, r_2\}.$$

(iii) If $b_0 = g(z_0) \neq 0$, $\frac{f(z)}{g(z)} = \sum_{0}^{\infty} c_n(z - z_0)^n$, $|z - z_0| < r$ with $r > 0$.

(iv) If $f'(z_0) = a_1 \neq 0$, $z = f^{-1}(w) = \sum_{0}^{\infty} d_n(w - w_0)^n$, $|w - w_0| < \rho$ with $\rho > 0$.

In addition, let $h(w) = \sum_{0}^{\infty} c_n(w - w_0)^n$, $|w - w_0| < \delta$.

(v) If $w_0 = f(z_0) = a_0$, then $h \circ f = \sum_{0}^{\infty} a_n(z - z_0)$, $|z - z_0| < \varepsilon$ with $\varepsilon > 0$, where $w = f(z)$.   (4.8.1)

Owing to the uniqueness of the Taylor series expansion (if existed), the power series obtained by any of these methods should be the required one. Also, remember that the radius of convergence of the Taylor series of an analytic function at a point z_0 is the distance from z_0 to the nearest singular point of that function.

Now, we come to a series of examples.

Example 1. Find the Taylor series expansion of $f(z) = \frac{z^2}{(z+1)^2}$ at 0 and 1, respectively.

Solution. At $z_0 = 0$:

$$\frac{1}{z+1} = \sum_0^\infty (-1)^n z^n, \quad |z| < 1$$

$\Rightarrow$ (by termwise differentiation) $\quad \dfrac{-1}{(z+1)^2} = \sum_1^\infty (-1)^n \cdot n z^{n-1}, \quad |z| < 1$

$\Rightarrow$ (multiplied by z^2 and changing n to $n-1$)

$$\frac{z^2}{(z+1)^2} = \sum_{n=2}^\infty (-1)^n (n-1) z^n, \quad |z| < 1.$$

At $z_0 = 1$: $f(z)$ fails to be analytic at $z = -1$. So the radius of convergence of the expected Taylor series is $1 - (-1) = 2$. Now,

$$\frac{1}{1+z} = \frac{1}{2\left(1 + \frac{z-1}{2}\right)} = \frac{1}{2} \sum_0^\infty (-1)^n \frac{(z-1)^n}{2^n}$$

$$= \sum_0^\infty (-1)^n \frac{(z-1)^n}{2^{n+1}}, \quad |z - 1| < 2.$$

Since $z^2 = (z - 1 + 1)^2 = (z-1)^2 + 2(z-1) + 1$, therefore

$$\frac{z^2}{(1+z)^2} = [1 + 2(z-1) + (z-1)^2] \left[\sum_{n=0}^\infty (-1)^n \frac{n+1}{2^{n+2}} (z-1)^n \right]$$

$$= \frac{1}{4} + \sum_{n=0}^\infty (-1)^n \frac{(n-3)}{2^{n+2}} (z-1)^n, \quad |z-1| < 2.$$

Example 2. Find the Taylor series expansion for each of the following functions at the indicated point z_0.

(1) $f(z) = \sqrt{z+i}, \quad f(0) = \sqrt{i} = \frac{1+i}{\sqrt{2}}, \quad$ at $z_0 = 0$.

(2) $f(z) = \sqrt{1 + \sqrt{1+z}}, \quad f(0) = \sqrt{2} > 0, \quad$ at $z_0 = 0$.

(3) $f(z) = \mathrm{Log}(1 + \sqrt{1+z^2}), \quad f(0) = \mathrm{Log}(1 + \sqrt{1}) = \log 2, \quad$ at $z_0 = 0$.

Solution.

(1) The condition $\sqrt{i} = \frac{1+i}{\sqrt{2}}$ determines a unique branch of $\sqrt{z+i}$ in the slit domain $\mathbf{C} - [-i, -i\infty)$. The radius of convergence is thus equal to $|0 - (-i)| = 1$. Actual differentiation shows that, for $n \geq 1$,

$$f^{(n)}(z) = \frac{1}{2} \left(\frac{1}{2} - 1 \right) \left(\frac{1}{2} - 2 \right) \cdots \left(\frac{1}{2} - n + 1 \right) (z+i)^{\frac{1}{2} - n}$$

$$\Rightarrow f^{(n)}(0) = \frac{(-1)^{n-1} 1 \cdot 3 \cdot 5 \cdots (2n-3)}{2^n} i^{\frac{1}{2}} i^{-n}.$$

And the Taylor series expansion at $z_0 = 0$ is

$$\sqrt{z+i} = \frac{1+i}{\sqrt{2}}\left\{1 + \frac{1}{2}\cdot\frac{z}{i} + \sum_{n=2}^{\infty}(-1)^{n-1}\frac{1\cdot3\cdot5\cdots(2n-3)}{2^n\cdot n!}\left(\frac{z}{i}\right)^n\right\},$$

$$|z| < 1.$$

(2) Choose the branch of $\sqrt{1+z}$, uniquely determined by $\sqrt{1} = 1$, in $\mathbf{C} - (-\infty, -1]$. Then the condition $f(0) = \sqrt{2}$ uniquely determine the branch $f(z) = \sqrt{1+\sqrt{1+z}}$ in $\mathbf{C} - (-\infty, -1]$. The radius of convergence is $0 - (-1) = 1$.

For any real number t, we have (see Exercise B (1) of Sec. 1.3)

$$\sqrt{1+it} = \sqrt{\frac{1+\sqrt{1+t^2}}{2}} + (\mathrm{sgn}\, t)\sqrt{\frac{-1+\sqrt{1+t^2}}{2}}$$

$\Rightarrow$ (using expansion for $(1+z)^\lambda$ with $z = it$ and $\lambda = \frac{1}{2}$, see Example 2 of Sec. 3.3.2)

$$\sqrt{1+\sqrt{1+t^2}}$$

$$= \sqrt{2}\,\mathrm{Re}\sqrt{1+it}$$

$$= \sqrt{2}\,\mathrm{Re}\left\{1 + \frac{1}{2}it + \sum_{n=2}^{\infty}(-1)^{n-1}\frac{1\cdot3\cdot5\cdots(2n-3)}{2^n\cdot(2n)!}(it)^n\right\}, \quad |t| < 1$$

$$= \sqrt{2}\left\{1 + \sum_{n=1}^{\infty}\frac{1\cdot3\cdot5\cdots(4n-3)}{2^{2n}(4n)!}(-1)^{n-1}(t)^{2n}\right\}, \quad |t| < 1. \qquad (*)$$

In case t in the right series in $(*)$ is a complex variable, the radius of convergence of this new series is still equal to 1. On the other hand, the complex function $\sqrt{1+\sqrt{1+t^2}} = g(t)$ has a unique branch, determined by $g(0) = \sqrt{2}$, in $\mathbf{C} - [i, +\infty) - [-i, -i\infty) - (-\infty, -1]$. By the interior uniqueness theorem (see (3.4.2.9)), $(*)$ still holds for complex variable t whenever $|t| < 1$. Now, using z to replace t^2 in $(*)$, the required series is

$$\sqrt{1+\sqrt{1+z}} = \sqrt{2}\left\{1 + \sum_{n=1}^{\infty}(-1)^{n-1}\frac{1\cdot3\cdot5\cdots(4n-3)}{2^{2n}(4n)!}z^n\right\}, \quad |z| < 1.$$

(3) The branch of $\zeta = \sqrt{z^2+1}$, determined by $\sqrt{1} = 1$, maps $\mathbf{C} - [i, +\infty) - [-i, -i\infty)$ univalently onto the right half-plane $\mathrm{Re}\,\zeta > 0$; while $w = \mathrm{Log}(1 + \zeta)$ is single-valued in $\mathbf{C} - (-\infty, -1]$. See Fig. 4.41. Consequently the composite function $w = \mathrm{Log}(1+\sqrt{1+z^2})$ is analytic in $|z| < 1$ and its Taylor series expansion at $z_0 = 0$ will have the radius of convergence equal to 1.

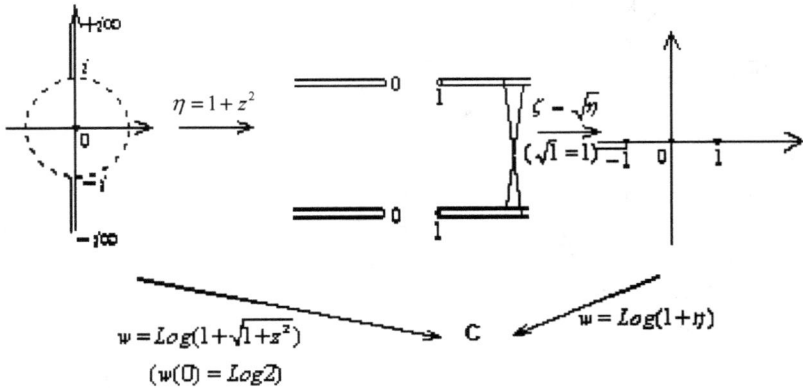

Fig. 4.41

Expand $\text{Log}(1+\eta)$ into Taylor series at $\eta = 1$ as (refer to Examples 1 and 2 of Sec. 3.3.2)

$$\text{Log}(1+\eta) = \text{Log}\,2 + \text{Log}\left(1 + \frac{\eta - 1}{2}\right)$$

$$= \text{Log}\,2 + \sum_{n=1}^{\infty} \frac{(-1)^{n-1}}{n}\left(\frac{\eta - 1}{2}\right)^n, \quad |\eta - 1| < 2.$$

Then, substituting $\eta = \sqrt{1+z^2} = 1 + \sum_{n=1}^{\infty}(-1)^{n-1}\frac{1 \cdot 3 \cdots (2n-3)}{2^n \cdot n!} z^{2n}$, $|z| < 1$, into $\text{Log}(1+\eta)$, the required expansion is

$$\text{Log}(1+\eta) = \text{Log}\,2 - \sum_{n=1}^{\infty} \frac{1}{2n} C_n^{\frac{-1}{2}} z^{2n}, |z| < 1$$

$$\left(\text{recall}\;\; \binom{-\frac{1}{2}}{n} = C_n^{-\frac{1}{2}} \text{ as defined in Example 2 of Sec. 3.3.2}\right).$$

An easy way to do this problem is to use differentiation and then integration as follows.

$$f'(z) = \frac{1 - (1+z^2)^{-\frac{1}{2}}}{z} = -\sum_{n=1}^{\infty} C_n^{-\frac{1}{2}} z^{2n}, \quad |z| < 1.$$

$\Rightarrow$ (integrating from $z_0 = 0$ to z along the radius of $|z| < 1$)

$$f(z) = f(0) + \int_0^z f'(\zeta)d\zeta = \text{Log}\,2 - \sum_{n=1}^{\infty} \frac{1}{2n} C_n^{-\frac{1}{2}} z^{2n} z^{2n}, \quad |z| < 1.$$

Example 3. Show that

$$\sin^2 z = \sum_{n=1}^{\infty}(-1)^{n+1}\frac{2^{2n-1}}{(2n)!}z^{2n}, \quad |z| < \infty.$$

Solution. We present four methods in the sequel.

(1) Let $f(z) = \sin^2 z$. By differentiation of f,

$$f'(z) = 2\sin z \cos z = \sin 2z$$

$$f''(z) = 2\cos 2z = 2\sin\left(2z + \frac{\pi}{2}\right)$$

$$f'''(z) = 2^2 \cos\left(2z + \frac{\pi}{2}\right) = 2^2 \sin\left(2z + \frac{2\pi}{2}\right)$$

$$\vdots$$

$$f^{(n)}(z) = 2^{n-1}\sin\left(2z + \frac{(n-1)\pi}{2}\right), \quad n \geq 1$$

$$\Rightarrow f^{(n)}(0) = \begin{cases} 2^{n-1}\sin\dfrac{(n-1)\pi}{2} = 0, & \text{if } n \text{ is odd} \\[2mm] 2^{n-1}\sin\dfrac{(n-1)\pi}{2} = (-1)^{\frac{n}{2}+1}\cdot 2^{n-1}, & \text{if } n \text{ is even} \end{cases}.$$

And the result follows.

(2) Using $\sin^2 z = \frac{1}{2} - \frac{1}{2}\cos 2z$ and $\cos 2z = \sum_{n=0}^{\infty}\frac{(-1)^n}{(2n)!}(2z)^n, |z| < \infty$, the result follows.

(3) Using differentiation and then integration, we have

$$f'(z) = \sin 2z = \sum_{n=1}^{\infty}(-1)^{n-1}\frac{(2z)^{2n-1}}{(2n-1)!}, \quad |z| < \infty,$$

$$\Rightarrow f(z) = \sin^2 z = \int_0^z \sin 2\zeta\, d\zeta$$

$$= \sum_{n=1}^{\infty}(-1)^{n-1}\frac{2^{2n-1}}{(2n-1)!}\int_0^z \zeta^{2n-1}d\zeta$$

$$= \sum_{n=1}^{\infty}(-1)^{n-1}\frac{2^{2n-1}}{(2n)!}z^{2n}, \quad |z| < \infty.$$

(4) Using Cauchy product of $\sin z = \sum_{n=1}^{\infty}(-1)^{n-1}\frac{1}{(2n-1)!}z^{2n-1}$, $|z| < \infty$, with itself, then

$$\sin^2 z = \left[\sum_{n=1}^{\infty}(-1)^{n-1}\frac{1}{(2n-1)!}z^{2n-1}\right]^2$$

$$= z^2 - \left(\frac{1}{3!} + \frac{1}{3!}\right)z^4 + \cdots + (-1)^{-1}\left[\frac{1}{(2n-1)!} + \frac{1}{3!(2n-3)!}\right.$$

$$\left. + \frac{1}{5!(2n-5)!} + \cdots + \frac{1}{(2n-1)!}\right]z^{2n} + \cdots$$

$$= \sum_{n=1}^{\infty}(-1)^{n-1}\frac{2^{2n-1}}{(2n)!}z^{2n}, \quad |z| < \infty.$$

Example 4.

(1) Let $\operatorname{Arc}\tan z$ denote the principal branch of $\tan^{-1}z$, determined by $\tan^{-1}0 = 0$, in $\mathbf{C} - [i, +i\infty) - [-i, -i\infty)$. Show that $f(z) = \sqrt{\frac{z}{1-z}}\operatorname{Arc}\tan\sqrt{\frac{z}{1-z}}$ is analytic at $z_0 = 0$ and find the radius of convergence of its Taylor series at $z_0 = 0$.

(2) Show that $f(z)$ satisfies the differential equation $2z(1-z)f'(z) = f(z) + z$, $f(0) = 0$, and then deduce that

$$f(z) = z + \sum_{n=1}^{\infty}\frac{2\cdot 4\cdots 2n}{3\cdot 5\cdots(2n+1)}z^{n+1}, \quad |z| < 1.$$

Solution.

(1) Since $\frac{d}{dw}\operatorname{Arc}\tan w = \frac{1}{1+w^2}$, $w \in \mathbf{C} - [i, +i\infty) - [-i, -i\infty)$,

$$\frac{1}{1+w^2} = \sum_{n=0}^{\infty}(-1)^n w^{2n}, \quad |w| < 1$$

$$\Rightarrow \operatorname{Arc}\tan w = \int_0^w \frac{d\zeta}{1+\zeta^2} = \sum_{n=0}^{\infty}(-1)^n\frac{w^{2n+1}}{2n+1}, \quad |w| < 1,$$

$$\Rightarrow w\operatorname{Arc}\tan w = \sum_{n=0}^{\infty}(-1)^n\frac{w^{2n+2}}{2n+1}, \quad |w| < 1.$$

Note that $w\operatorname{Arc}\tan w$ is a single-valued analytic function in $|w| < 1$ and is an even one. So

$$\sqrt{w}\operatorname{Arc}\tan\sqrt{w} = \sum_{n=0}^{\infty}(-1)^n\frac{1}{2n+1}w^{n+1}, \quad |w| < 1$$

is single-valued and analytic, too.

$\zeta = \frac{z}{1-z}$ maps the disk $|z| < 1$ onto the half-plane $\text{Re}\,\zeta > -\frac{1}{2}$ which, in turn, is mapped, under $w = \sqrt{\zeta}$, onto the domain between the two branches of the hyperbola $u^2 - v^2 = -\frac{1}{2}$, where $w = u + iv$. This domain is contained in $\mathbf{C} - [i, +i\infty) - [-i, -i\infty)$. Hence, $f(z)$ is single-valued on $|z| < 1$ and 1 is the radius of convergence of its Taylor series at $z_0 = 0$, since $|i - 0| = 1$.

(2) If $|z| < 1$, differentiate $f(z)$ and get

$$f'(z) = \frac{1}{2}\left(\frac{z}{1-z}\right)^{-\frac{1}{2}} \frac{1}{(1-z)^2} \text{Arc}\tan\left(\frac{z}{1-z}\right)^{\frac{1}{2}}$$

$$+ \left(\frac{z}{1-z}\right)^{\frac{1}{2}} \cdot \frac{\frac{1}{2}\left(\frac{z}{1-z}\right)^{-\frac{1}{2}} \cdot \frac{1}{(1-z)^2}}{1 + \frac{z}{1-z}}$$

$$\Rightarrow 2z(1-z)f'(z) = \left(\frac{z}{1-z}\right)^{\frac{1}{2}} \text{Arc}\tan\left(\frac{z}{1-z}\right)^{\frac{1}{2}} + z = f(z) + z.$$

Since $f(0) = 0$, let $f(z) = \sum_{n=1}^{\infty} a_n z^n$, $|z| < 1$, be its Taylor series expansion. Substituting into the differential equation, we get

$$2z(1-z)\left(\sum_{n=1}^{\infty} n a_n z^{n-1}\right) = \sum_{n=1}^{\infty} a_n z^n + z$$

$$\Rightarrow a_1 = 1 \quad \text{and} \quad (2n+1)a_{n+1} = 2n a_n, \quad n \geq 1$$

$$\Rightarrow a_{n+1} = \frac{2 \cdot 4 \cdots (2n)}{3 \cdot 5 \cdots (2n+1)}, \quad n \geq 1.$$

Example 5. Show that

$$\frac{z}{e^z - 1} = 1 - \frac{z}{2} + \sum_{n=1}^{\infty} \frac{B_{2n}}{(2n)!} z^{2n}, \quad |z| < 2\pi$$

has its radius of convergence equal to 2π, where the coefficients B_n, $n \geq 0$, are called *Bernoulli's numbers* and satisfy:

(1) $B_0 = 1$, $B_1 = -\frac{1}{2}$.
(2) $B_0 C_0^{n+1} + B_1 C_1^{n+1} + \cdots + B_n C_n^{n+1} = 0, n \geq 1$ (C_k^n denotes the binomial coefficients, $0 \leq k \leq n$).
(3) $B_{2n+1} = 0$ for $n \geq 1$ and each B_n is a rational number.
(4) For each $\varepsilon > 0$, there are infinitely many n so that $|B_{2n}| > \frac{(2n)!}{(2\pi+\varepsilon)^{2n}}$.

In particular, $B_2 = \frac{1}{6}$, $B_4 = -\frac{1}{30}$, $B_6 = \frac{1}{42}$, $B_8 = -\frac{1}{30}, \ldots, B_{20} = -\frac{174611}{330}, \ldots$.

Solution. It is known that $e^z - 1 = \sum_{n=1}^{\infty} \frac{1}{n!} z^n, |z| < \infty$. Also $e^z - 1 = 0$ if and only if $z = 2n\pi i$, $n = 0, \pm 1, \pm 2 \ldots$. Since $\lim_{z \to 0} \frac{z}{e^z - 1} = \lim_{z \to 0} \frac{1}{e^z} = 1$, $f(z) = \frac{z}{e^z - 1}$ is analytic at $z_0 = 0$ (see (3.4.2.15)). Therefore, $f(z)$ is analytic in $|z| < 2\pi$ and 2π will be the radius of convergence for the Taylor series expansion at 0.

Let

$$f(z) = \frac{z}{e^z - 1} = \sum_{n=0}^{\infty} c_n z^n, \quad |z| < 1,$$

$$\Rightarrow z = \left(\sum_{n=0}^{\infty} c_n z^n \right) (e^z - 1)$$

$$= \left(\sum_{n=0}^{\infty} c_n z^n \right) \left(\sum_{n=1}^{\infty} \frac{1}{n!} z^n \right), \quad \text{where } c_0 = 1. \tag{$*$}$$

$$\Rightarrow c_n = \begin{vmatrix} 1 & 0 & 0 & \cdots & 1 \\ \dfrac{1}{2!} & 1 & 0 & \cdots & 0 \\ \dfrac{1}{3!} & \dfrac{1}{2!} & 1 & \cdots & 0 \\ \vdots & \vdots & \vdots & & \vdots \\ \dfrac{1}{(n+1)!} & \dfrac{1}{n!} & \dfrac{1}{(n-1)!} & \cdots & 0 \end{vmatrix}$$

$$= (-1)^n \begin{vmatrix} \dfrac{1}{2!} & 1 & 0 & \cdots & 0 \\ \dfrac{1}{3!} & \dfrac{1}{2!} & 1 & \cdots & 0 \\ \vdots & \vdots & \vdots & & \vdots \\ \dfrac{1}{(n+1)!} & \dfrac{1}{n!} & \dfrac{1}{(n-1)!} & \cdots & \dfrac{1}{2!} \end{vmatrix}.$$

Designate $B_n = n! c_n$ and is called the *Bernoulli's number*, for $n \geq 0$.

However, an effective way to compute B_n is to set up a recursive relation. From $(*)$, it follows that $c_0 = 1$ and, by Cauchy product,

$$c_0 \frac{1}{(n+1)!} + c_1 \frac{1}{n!} + \cdots + c_{n-2} \frac{1}{2!} + c_n = 0,$$

$\Rightarrow \Bigg($multiply both sides by $(n+1)!$, and recall that

$$B_n = n!c_n \text{ and } C_k^{n+1} = \frac{(n+1)!}{k!(n+1-k)!}\Bigg)$$

$$B_0 C_0^{n+1} + B_1 C_1^{n+1} + \cdots + B_n C_n^{n+1} = 0, \quad n \geq 1.$$

From this recursive relation, we can evaluate successively the B_n's.
   To show that $B_1 = -\frac{1}{2}$ and $B_{2n+1} = 0$ for $n \geq 1$: Observe that

$$f(z) = \frac{z}{e^z - 1} = \sum_{n=0}^{\infty} \frac{B_n}{n!} z^n \quad \text{and}$$

$$f(-z) = \frac{ze^z}{e^z - 1} = \sum_{n=0}^{\infty} (-1)^n \frac{B_n}{n!} z^n, \quad |z| < 2\pi,$$

$$\Rightarrow f(z) - f(-z) = -z = 2 \cdot \frac{B_1}{1!} z + 2 \cdot \frac{B_3}{3!} z^3 + \cdots$$

$$+ 2 \cdot \frac{B_{2n+1}}{(2n+1)!} z^{2n+1} + \cdots,$$

$$\Rightarrow B_1 = -\frac{1}{2} \quad \text{and} \quad B_{2n+1} = 0 \quad \text{for } n \geq 1.$$

Therefore, we get the final expansion

$$\frac{z}{e^z - 1} = 1 - \frac{z}{2} + \sum_{n=1}^{\infty} \frac{B_{2n}}{(2n)!} z^{2n}, \quad |z| < 2\pi.$$

   According to Hadamard's formula for the radius of convergence (see (3.3.2.2)),

$$\varlimsup_{n \to \infty} \sqrt[2n]{\frac{|B_{2n}|}{(2n)!}} = \frac{1}{2\pi},$$

$$\Rightarrow \varlimsup_{n \to \infty} \sqrt[2n]{|B_{2n}|} = \infty \quad \text{or} \quad \varlimsup_{n \to \infty} |B_{2n}| = \infty.$$

In particular, for a given $\varepsilon > 0$, there are infinitely many n such that $|B_{2n}| > \frac{(2n)!}{(2\pi+\varepsilon)^{2n}}$.

Example 6. Show that

$$\sec z = \sum_{n=0}^{\infty} (-1)^n \frac{E_{2n}}{(2n)!} z^{2n}, \quad |z| < \frac{\pi}{2}$$

and its radius of convergence is equal to $\frac{\pi}{2}$, where $E_{2n}, n \geq 0$, are called the *Euler's numbers* and satisfy:

(1) $E_0 = 1$, $E_2 = -1$, $E_4 = 5$, $E_6 = -61$, $E_8 = 1385$, $E_{10} = -50,521,\ldots$.

(2) $E_0 + E_2 C_2^{2n} + E_4 C_4^{2n} + \cdots + E_{2n-2} C_{2n-2}^{2n} + E_{2n} = 0$, $n \geq 1$.

(3) Each E_{2n} is an integer.

Solution. $\cos z = 0$ if and only if $z = n\pi + \frac{\pi}{2}, n = 0, \pm 1, \pm 2, \ldots$. Therefore, $\sec z = \frac{1}{\cos z}$ is analytic at $z_0 = 0$ and can be expanded into Taylor series there with $\frac{\pi}{2}$ as its radius of convergence. Since $\sec(-z) = \sec z$, its Taylor series will lack terms of odd powers.

Let

$$\sec z = \frac{1}{\cos z} = \frac{1}{\sum_{n=0}^{\infty} (-1)^n \frac{1}{(2n)!} z^{2n}} = \sum_{n=0}^{\infty} c_{2n} z^{2n}$$

$$= \sum_{n=0}^{\infty} (-1)^n \frac{E_{2n}}{(2n)!} z^{2n}, \quad |z| < \frac{\pi}{2}$$

where $c_{2n} = (-1)^n \frac{E_{2n}}{(2n)!}$, $n \geq 0$, and E_{2n} are called the *Euler's numbers*.

To determine E_{2n}: By Cauchy product,

$$\left(\sum_{n=0}^{\infty} (-1)^n \frac{1}{(2n)!} z^{2n} \right) \left(\sum_{n=0}^{\infty} (-1)^n \frac{E_{2n}}{(2n)!} z^{2n} \right) = 1,$$

$$\Rightarrow E_0 = 1, \frac{E_0}{2!} + \frac{E_2}{2!} = 0, \quad \frac{E_0}{4!} + \frac{E_2}{(2!)^2} + \frac{E_4}{4!} = 0, \ldots.$$

The details are left to the readers.

Exercises A

(1) Find the Taylor series expansion for each of the following functions $f(z)$ at the indicated point z_0 and determine its radius of convergence.

(a) $f(z) = \dfrac{z-1}{z^2 - z - 1}$ at $z_0 = 0$.

(b) $f(z) = \dfrac{1}{(z^2 + z - 2)^2}$ at $z_0 = 0$.

(c) $f(z) = \dfrac{2z}{z^2 - 1}$ at $z_0 = i$.

(d) $f(z) = \dfrac{z}{z^2 - 2z + 5}$ at $z_0 = 1$.

(e) $f(z) = \sin(2z - z^2)$ at $z_0 = 1$.

(f) $f(z) = \dfrac{2z+1}{(z^2+1)(z+1)^2}$ at $z_0 = 0$.

(g) $f(z) = \displaystyle\int_0^z e^{\zeta^2}\,d\zeta$ at $z_0 = 0$.

(h) $f(z) = \displaystyle\int_0^z \dfrac{\sin\zeta}{\zeta}\,d\zeta$ at $z_0 = 0$.

(2) Find the Taylor series expansion of the branch at z_0 for each of the following multiple-valued functions $f(z)$ and determine its radius of convergence.

(a) $f(z) = \sinh^{-1} z,\ \sinh^{-1} 0 = 0,$ at $z_0 = 0$.

(b) $f(z) = \log(z^2 - 2z + 2),\ \log 2 > 0,$ at $z_0 = 0$.

(c) $f(z) = \dfrac{1}{2}\left(\log\dfrac{1}{1-z}\right)^2,\ \log 1 = 0,$ at $z_0 = 0$.

(d) $f(z) = (\tan^{-1} z)^2,\ \tan^{-1} 0 = 0,$ at $z_0 = 0$.

(e) $f(z) = (\tan^{-1} z)\log(1+z^2),\ \tan^{-1} 0 = 0$ and $\log 1 = 0$,   at $z_0 = 0$.

(f) $f(z) = (\sin^{-1} z)^2,\ \sin^{-1} 0 = 0,$ at $z_0 = 0$.

(g) $f(z) = \log(\sqrt{1+z} + \sqrt{1-z}),\ \sqrt{1} = 1$ and $\log 1 = 0,$ at $z_0 = 0$.

(h) $f(z) = (1+\sqrt{1+z})^{-\frac{1}{2}},\ f(0) = 2^{-\frac{1}{2}} > 0,$ at $z_0 = 0$.

(i) $f(z) = \dfrac{\log z}{z},\ \log 1 = 0,$ at $z_0 = 0$.

(j) $f(z) = (\log z)^2,\ \log 1 = 0,$ at $z_0 = 0$.

(k) $f(z) = (\log(1-z))^2,\ \log 1 = 2\pi i,$ at $z_0 = 0$.

(l) $f(z) = e^{\frac{z}{(1-z)}}$ at $z_0 = 0$.

(3) Find the Taylor series expansion, up to z^4 or z^5, of each of the following functions $f(z)$ at z_0.

(a) $f(z) = e^{\cos z}$.

(b) $f(z) = \sin\left(\dfrac{z}{1-z}\right)$.

(c) $f(z) = \dfrac{\sin z}{\cos 2z}$.

(d) $f(z) = \sin(\sin z)$.

(e) $f(z) = \dfrac{\sin \pi z^2}{\sin \pi z}$.

(f) $f(z) = \sqrt{\cos z}$ ($\sqrt{\cos 0} = 1$).

(g) $f(z) = e^{z\sin z}$.

(4) (a) Show that

$$e^z \cos z = \sum_{n=0}^{\infty} \frac{1}{n!} \cdot 2^{\frac{n}{2}} \cos \frac{n\pi}{4} z^n$$

$$= 1 + z - \frac{2}{3!} z^3 - \frac{2^2}{4!} z^4 - \frac{2^2}{5!} z^5 + \cdots, \quad |z| < \infty.$$

(b) Use integral formula for $f^{(n)}(0)$, where $f(z) = e^z \cos z$, to show that

$$e^z \cos z = \sum_{n=0}^{\infty} \left\{ \sum_{0 \le k \le \frac{n}{2}} (-1)^k \frac{1}{(2k)!(n-2k)!} \right\} z^n, \quad |z| < \infty.$$

(5) Based on Example 5, try to show the following expansions.

(a) $z \cot z = 1 + \displaystyle\sum_{n=1}^{\infty} (-1)^n \frac{2^{2n} B_{2n}}{(2n)!} z^{2n}, \quad |z| < \pi.$

(b) $\tan z = \displaystyle\sum_{n=1}^{\infty} (-1)^{n-1} \frac{2^{2n}(2^{2n}-1)}{(2n)!} B_{2n} z^{2n-1}, \quad |z| < \frac{\pi}{2}.$

(c) $z \csc z = 1 + \displaystyle\sum_{n=1}^{\infty} (-1)^{n-1} \frac{(2^{2n}-2)B_{2n}}{(2n)!} z^{2n}, \quad |z| < \pi.$

(d) $\mathrm{Log} \dfrac{\sin z}{z} = \displaystyle\sum_{n=0}^{\infty} (-1)^n \frac{2^{2n} B_{2n}}{2n(2n)!} z^{2n}, \quad |z| < \pi.$

(e) $\mathrm{Log} \cos z = \displaystyle\sum_{n=0}^{\infty} (-1)^n \frac{2^{2n}(2^n-1)B_{2n}}{2n(2n)!} z^{2n}, \quad |z| < \frac{\pi}{2}.$

(f) $\dfrac{z}{\sin z} = \displaystyle\sum_{n=0}^{\infty} (-1)^{n-1} \frac{(4^n-2)B_{2n}}{(2n)!} z^{2n}, \quad |z| < \pi.$

(g) $\dfrac{1}{\cos^2 z} = \displaystyle\sum_{n=0}^{\infty} (-1)^n \frac{2^{2n+1}(2^{n+2}-1)B_{2n+2}}{(n+1)(2n)!} z^{2n}, \quad |z| < \frac{\pi}{2}.$

(h) $\mathrm{Log} \dfrac{\tan z}{z} = \displaystyle\sum_{n=1}^{\infty} (-1)^{n-1} \frac{2^{2n}(2^{n-1}-1)B_{2n+2}}{n(2n)!} z^{2n}, \quad |z| < \frac{1}{2}\pi.$

(i) $\dfrac{1}{2} z \coth \dfrac{z}{2} = 1 + \displaystyle\sum_{n=0}^{\infty} \frac{B_{2n}}{(2n)!} z^{2n-1}, \quad |z| < 2\pi.$

(j) $\mathrm{Log} \cosh z = \displaystyle\sum_{n=1}^{\infty} \frac{2^{2n}(2^{2n}-1)B_{2n+2}}{2n(2n)!} z^{2n}, \quad |z| < \frac{1}{2}\pi.$

(6) Suppose $f(z) = \sum_{n=0}^{\infty} a_n(z - z_0)^n$, $|z - z_0| < r$, where $r > 0$. If f is unbounded on $|z - z_0| < r$, show that r is the radius of convergence of the series.

(7) If f is analytic in $|z - z_0| < R$, show that

$$f(z) = f(z_0) + 2\sum_{n=1}^{\infty} \frac{1}{2^n \cdot n!} f^{(n)}\left(\frac{z + z_0}{2}\right)(z - z_0)^n, \quad |z - z_0| < R.$$

(8) Let $f(z) = \sum_{n=0}^{\infty} a_n z^n$ be analytic in $|z| \leq r$. Show that

$$\varphi(z) = \sum_{n=0}^{\infty} \frac{a_n}{n!} z^n$$

is entire, and there is a constant $M > 0$ such that $|\varphi(z)| \leq M e^{\frac{|z|}{r}}$ and $|\varphi^{(n)}(z)| \leq \frac{M}{r^n} e^{\frac{|z|}{r}}$, $n \geq 1$.

(9) Show that $f(z) = \sum_{n=1}^{\infty} \frac{1}{n!} z^{n!}$ is analytic in $|z| < 1$ and continuous on $|z| \leq 1$. In case Ω is a domain containing $|z| \leq 1$, then it is not possible to have a function F, analytic in Ω, so that $F(z) = f(z)$ on $|z| \leq 1$.

(10) Suppose $f(z) = \sum_{n=0}^{\infty} a_n z^n$ and $g(z) = \sum_{n=0}^{\infty} b_n z^n$ have, respectively, the radius of convergence R_1 and R_2, where $0 < R_1, R_2 < \infty$. Show that if $|z| < R_1 R_2$, then

$$\sum_{n=0}^{\infty} a_n b_n z^n = \frac{1}{2\pi i} \int_{|\zeta|=\rho} f(\zeta) g\left(\frac{z}{\zeta}\right) \frac{d\zeta}{\zeta},$$

where $\frac{|z|}{R_2} < \rho < R_1$.

(11) Suppose f is analytic in $|z| < r$, where $r > 1$. Show that

$$\frac{1}{\pi} \int_0^{2\pi} f(e^{i\theta}) \cos^2 \frac{\theta}{2} d\theta = f(0) + \frac{1}{2} f'(0);$$

$$\frac{1}{\pi} \int_0^{2\pi} f(e^{i\theta}) \sin^2 \frac{\theta}{2} d\theta = f(0) - \frac{1}{2} f'(0).$$

(12) Suppose both f and g are analytic in a domain which contains the closed disk $|z| \leq r$. Suppose there is a point $z_0, |z_0| = r$, satisfying the conditions:

(i) $g(z_0) = 0, g'(z_0) \neq 0$, and z_0 is the only zero of $g(z)$ on $|z| \leq r$, and
(ii) $f(z_0) \neq 0$.

If $\sum_0^{\infty} a_n z^n$ is the Taylor series expansion of $\frac{f}{g}$ at 0, show that $\lim_{n \to \infty} \frac{a_n}{a_{n+1}} = z_0$. So $|z_0| = r$ is the radius for convergence of the series. For an extension, see Example 4 of Sec. 4.10.2.

(13) Let z_0 be a point in a domain Ω. Suppose f is analytic in $\Omega - \{z_0\}$ and there is an analytic function $G(z)$ on Ω such that $G'(z) = f(z)$ on $\Omega - \{z_0\}$. Show that there is an analytic function $F(z)$ on Ω so that $F(z) = f(z)$ on $\Omega - \{z_0\}$.

(14) Determine all entire functions f satisfying $f(z^2) = (f(z))^2, z \in \mathbf{C}$.

(15) Let f and g be entire functions without common zeros in $\mathbf{C} - \{0\}$. Let z_0 be the zero of $g(z)$ in $\mathbf{C} - \{0\}$ having the smallest absolute value. Suppose $\frac{f}{g}$ is analytic in $0 < |z| < |z_0|$ and there is an analytic function $F(z)$ in $|z| < |z_0|$ satisfying $F(z) = \frac{f}{g}$ in $0 < |z| < |z_0|$. Show that the Taylor series of $\frac{f}{g}$ at 0 has its radius of convergence equal to $|z_0|$.

(16) Let Ω be a domain containing the closed disk $|z| \leq R$ and z_0 be a point with $|z_0| = R$. Suppose f is analytic in $\Omega - \{z_0\}$ and $f(z) = \sum_{n=0}^{\infty} a_n z^n$ is its Taylor series. If, in a neighborhood of z_0, $(z - z_0)f(z)$ is bounded, then for all sufficiently small $r > 0$ and integers $k \geq 0$,

$$\int_{|\zeta - z_0| = r} \frac{f(\zeta)}{(\zeta - z_0)^k} d\zeta = 2\pi i \sum_{n=k}^{\infty} C_k^n (a_{n-1} - a_n z_0) z_0^{n-k},$$

where $a_{-1} = 0$.

Exercises B

(1) (*Area theorem*) Suppose $f(z) = \sum_{n=0}^{\infty} a_n z^n$ is univalent in $|z| < 1$. Show that the area of the image $f(\{|z| < r\})$ of $|z| < r < 1$ under $w = f(z)$ is

$$\pi \sum_{n=1}^{\infty} n|a_n|^2 r^{2n}.$$

(2) (*Area theorem*) Suppose $f(z) = \frac{1}{z} + \sum_{n=0}^{\infty} a_n z^n$ is univalent and analytic in $0 < |z| < 1$ (refer to Sec. 4.9). Show that

$$\sum_{n=1}^{\infty} n|a_n|^2 \leq 1.$$

Note that this is the area of the *complement* of the image $f(\{|z| < r\})$ with $f(0) = \infty$. Also, the equality holds if and only if $|a_1| = 1$ and $a_n = 0$ for $n \geq 2$. In this case, $f(z) = \frac{1}{z} + a_0 + a_1 z$ with $|a_1| = 1$ or $f_0(z) = \frac{1}{z} + e^{2i\theta} z$ (after a parallel shift) is the *extremal function*. This latter function $w = f_0(z)$ maps $|z| < 1$ onto $\mathbf{C}^*$ with a slit of length 4 and inclination angle θ. Compare to Example 1 of Sec. 3.5.7 and the Remark after it.

(3) Let $f(z) = z + \sum_{n=2}^{\infty} a_n z^n$ be univalent in $|z| < 1$.

    (a) Show that $|a_2^2 - a_3| \le 1$ by considering $\frac{1}{f(z)}$ on $0 < |z| < 1$.

    (b) Show that $g(z) = \sqrt{f(z^2)}$ is analytic and univalent in $|z| < 1$. Use (a) to show that

$$|a_2| \le 2$$

    with equality if and only if $f(z) = e^{-i\theta} f_0(e^{i\theta} z)$, where $f_0(z) = \frac{z}{(1-z)^2} = \sum_{n=1}^{\infty} n z^n$, $|z| < 1$ (see (3.5.7.1)).

(4) (*Koebe's quarter theorem*, see the Remark after Example 1 in Sec. 3.5.7). Let $f(z) = z + \sum_{n=2}^{\infty} a_n z^n$ be univalent in $|z| < 1$. Note that $f(0) = 0$ and $f'(0) = 1$. Show that the image $f(\{|z| < 1\})$ contains the open disk $|w| < \frac{1}{4}$, where $w = f(z)$. Note that $\frac{1}{4}$ cannot be replaced by any larger number as shown by $f_0(z) = \frac{z}{(1-z)^2}$.

(5) Suppose f is analytic in $|z| < 1$ and satisfies:

    (i) $f(0) = 0$ and $f'(0) = 1$; and

    (ii) $M = \sup_{|z|<1} |f'(z)| < \infty$.

    Then, for each $w \in \mathbf{C} - f(\{|z| < 1\})$, $|w| \ge \frac{1}{4M}$ always holds. Namely, the image domain $f(\{|z| < 1\})$ contains the open disk $|w| < \frac{1}{4M}$.

(6) (A. Bloch–E. Landan) Suppose f is analytic in $|z| < 1$ and satisfies:

    (i) $|f'(0)| = 1$; and

    (ii) $f'(z)$ is bounded in $|z| < 1$.

    Show that the image domain $f(\{|z| < 1\})$ contains an open disk of radius $\frac{1}{16}$, its center not necessarily at $f(0)$.

4.9.   Laurent Series

The prerequisite for this section is the result stated in (3.3.2.4) for power series $\sum_0^{\infty} a_n (z - z_0)^n$ and the Hadamard formula for the radius of convergence: $r = \overline{\lim} \sqrt[n]{|a_n|})^{-1}$.

Given a series of negative powers in $z - z_0$ as

$$\sum_{n=1}^{\infty} \frac{a_{-n}}{(z - z_0)^n} = \sum_{n=1}^{\infty} a_{-n}(z - z_0)^{-n} = \sum_{n=-\infty}^{-1} a_n (z - z_0)^n, \quad z \ne z_0. \quad (4.9.1)$$

The mapping $z \to w = (z - z_0)^{-1}$ transforms it into $\sum_{n=1}^{\infty} a_{-n} w^n$. In case $r = (\overline{\lim} \sqrt[n]{|a_{-n}|})^{-1} > 0$, then $\sum a_{-n} w^n$ converges absolutely and uniform

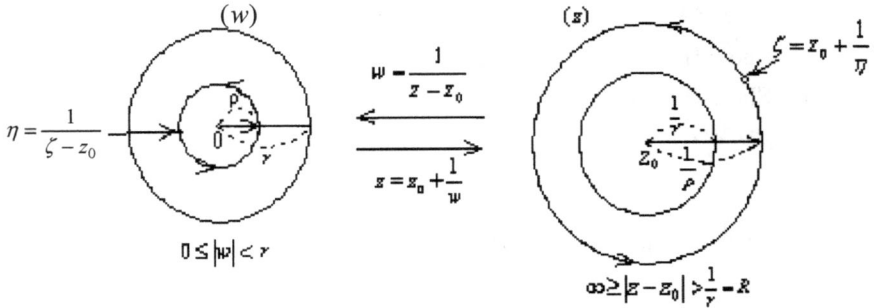

Fig. 4.42

on every compact subset of $|w| < r$ to an analytic function $g(w)$, and

$$a_{-n} = \frac{1}{2\pi i} \int_{|\eta|=\rho} \frac{g(\eta)}{\eta^{n+1}} d\eta \quad \text{(in counterclockwise direction)},$$

$$0 < \rho < r, \quad n \geq 1.$$

See Fig. 4.42. Let $\eta = (\zeta - z_0)^{-1}$. Then $|\zeta - z_0| = \frac{1}{|\eta|} = \frac{1}{\rho}$ on $|\eta| = \rho$. And, in this case,

$$a_{-n} = \frac{1}{2\pi i} \int_{|\zeta-z_0|=\frac{1}{\rho}} \frac{g\left(\frac{1}{\zeta-z_0}\right)}{(\zeta-z_0)^{-n-1}} \cdot \frac{-1}{(\zeta-z_0)^2} d\zeta \quad \text{(in clockwise direction)}$$

$$= \frac{1}{2\pi i} \int_{|\zeta-z_0|=\frac{1}{\rho}} \frac{g\left(\frac{1}{\zeta-z_0}\right)}{(\zeta-z_0)^{-n+1}} d\zeta \quad \text{(in counterclockwise direction)}.$$

Setting $\varphi(z) = g\left(\frac{1}{z-z_0}\right)$, then this $\varphi(z)$ is analytic in $\frac{1}{r} < |z - z_0| \leq \infty$ (including the point ∞).

We summarize the above as

The analytic function defined a power series of negative powers in $(z - z_0)$. Given a power series $\sum_{n=-\infty}^{-1} a_n(z - z_0)^n$ as in (4.9.1). Let

$$r = \varlimsup_{n \to \infty} \sqrt[n]{|a_{-n}|}.$$

Then $0 \leq r \leq \infty$. In case $0 \leq r < \infty$, the power series in $(z - z_0)^{-1}$ converges absolutely on $r < |z - z_0| \leq \infty$ and uniformly on every compact subset of $r < |z - z_0| \leq \infty$ (in $\mathbf{C}^*$ with spherical chord metric) to an analytic function $\varphi(z)$, and

$$a_{-n} = \frac{1}{2\pi i} \int_{|\zeta - z_0| = \rho > r} \frac{\varphi(\zeta)}{(\zeta - z_0)^{-n+1}} d\zeta$$

$$\text{(in counterclockwise direction)}, \quad n \geq 1. \tag{4.9.2}$$

r is called its *radius of convergence* and $|z - z_0| > r$ the *"disk"* of convergence.
The *formal* sum

$$\sum_{-\infty}^{\infty} a_n (z - z_0)^n = \sum_{n=-\infty}^{-1} a_n (z - z_0)^n + \sum_{n=0}^{\infty} a_n (z - z_0)^n \tag{4.9.3}$$

is called a *Laurent series* at $z_0 \in \mathbf{C}$ with $\sum_{-\infty}^{-1} a_n (z - z_0)^n$ as its *essential
or singular part* and $\sum_{n=0}^{\infty} a_n (z - z_0)^n$ its *analytic or regular part*. It is said
to *converge at a point* ζ if its singular and analytic parts both converge at
the point ζ; otherwise, *diverges at a point* ζ.

Suppose the singular part has the radius of convergence r, $0 \leq r < \infty$,
while that for the analytic part is R, $o < R \leq \infty$.

In case $r > R$: The Laurent series diverges everywhere.

In case $r = R$: The Laurent series diverges except at possible points on
$|z - z_0| = r$. For instance, $\sum_{-\infty}^{\infty} \frac{z^n}{n^2} (n \neq 0)$ converges everywhere on $|z| = 1$;
$\sum_{-\infty}^{\infty} z^n$ diverges everywhere on $|z| = 1$, while $\sum_{-\infty}^{\infty} \frac{z^n}{n} (n \neq 0)$ converges
everywhere on $|z| = 1$ except at $z = 1$.

In case $r < R$: The Laurent series converges in $r < |z - z_0| < R$, called
the (circular) *ring domain of convergence* (note that, in case $r = 0$ and
$R = \infty$, $0 < |z - z_0| < \infty$ is the punctured domain $\mathbf{C} - \{0\}$; while, if $r \geq 0$
and $R = \infty$, $r < |z - z_0| < \infty$ is a generalized ring domain). In this case,
we summarize as

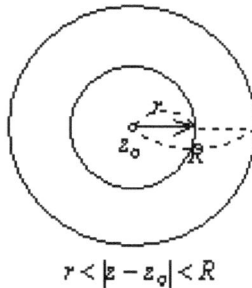

$$r < |z - z_0| < R$$

Fig. 4.43

The analytic function defined by a Laurent series on a ring domain. Let $f(z) = \sum_{-\infty}^{\infty} a_n(z - z_0)^n$ converge in the ring domain $0 \leq r < |z - z_0| < R \leq \infty$. Then $f(z)$ is analytic in $r < |z - z_0| < R$ and

$$a_n = \frac{1}{2\pi i} \int_{|\zeta - z_0| = \rho} \frac{f(\zeta)}{(\zeta - z_0)^{n+1}} d\zeta, \quad r < \rho < R, \ n = 0, \pm 1, \pm 2, \ldots.$$

Also, the series

(1) converges absolutely to $f(z)$ in $r < |z - z_0| < R$; and
(2) converges uniformly to $f(z)$ on every compact subset of $r < |z - z_0| < R$.

The series diverges in $|z| < r$ and $|z| > R$. $\hspace{2em}$ (4.9.4)

In this case, $f(z)$ is said to have a *Laurent series expansion* or *representation* in $r < |z - z_0| < R$ or at the isolated singularity z_0, for simplicity. This representation, if existed, is unique in the following sense.

The uniqueness of the Laurent series expansion of an analytic function in a ring domain. Give two Laurent series

$$f(z) = \sum_{-\infty}^{\infty} a_n(z - z_0)^n, \quad z \in \Omega_1 \quad \text{(a ring domain with center at } z_0\text{)},$$

$$g(z) = \sum_{-\infty}^{\infty} b_n(z - z_0)^n, \quad z \in \Omega_2 \quad \text{(a ring domain with center at } z_0\text{)}$$

If there is a circle $|z - z_0| = \rho$, containing in $\Omega_1 \cap \Omega_2 (\neq \emptyset)$, so that $f(z) = g(z)$ holds along $|z - z_0| = \rho$, then

$$a_n = b_n, \quad n = 0, \pm 1, \pm 2 \ldots.$$

Hence, $f(x) = g(x)$ holds in the ring domain $\Omega_1 \cap \Omega_2$. $\hspace{2em}$ (4.9.5)

This follows easily by applying the coefficient integral shown in (4.9.4) to both a_n, and b_n.

Section 4.9.1 will prove the converse of (4.9.4), namely, an analytic function in a ring domain $r < |z - z_0| < R$ is always representable as a Laurent series at z_0. Concrete examples are given in Sec. 4.9.2.

4.9.1.   *The Laurent series expansion of an analytic function in a (circular) ring domain*

Suppose f is analytic in $0 \leq r < |z - z_0| < R \leq \infty$.

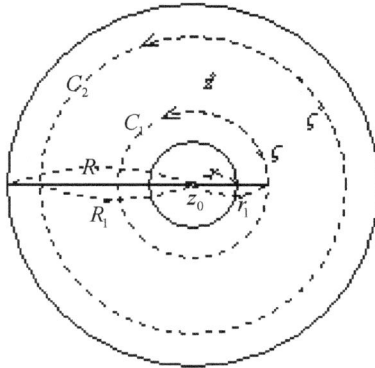

Fig. 4.44

Fix a point z, $r < |z - z_0| < R$. Choose r_1 and R_1 so that $r < r_1 < |z - z_0| < R_1 < R$. See Fig. 4.44. Let $C_1 : |\zeta - z_0| = r_1$ and $C_2 : |\zeta - z_0| = R_1$. Then (see (3.4.2.4'))

$$f(z) = \frac{1}{2\pi i} \int_{C_2} \frac{f(\zeta)}{\zeta - z} d\zeta - \frac{1}{2\pi i} \int_{C_1} \frac{f(\zeta)}{\zeta - z} d\zeta$$

where, as in the proof of (3.4.2.6),

$$f_1(z) = \frac{1}{2\pi i} \int_{C_2} \frac{f(\zeta)}{\zeta - z} d\zeta = \sum_{n=0}^{\infty} a_n (z - z_0)^n, \quad |z - z_0| < R_1 \quad \text{with}$$

$$a_n = \frac{1}{2\pi i} \int_{|\zeta - z_0| = \rho \leq R_1} \frac{f(\zeta)}{(\zeta - z_0)^{n+1}} d\zeta, \quad n \geq 0. \tag{$*_1$}$$

On the other hand, we want to show that

$$f_2(z) = -\frac{1}{2\pi i} \int_{C_1} \frac{f(\zeta)}{\zeta - z} d\zeta = \sum_{-\infty}^{-1} a_n (z - z_0)^n, \quad r_1 < |z - z_0| \quad \text{with}$$

$$a_n = \frac{1}{2\pi i} \int_{|\zeta - z_0| = \rho \geq r_1} \frac{f(\zeta)}{(\zeta - z_0)^{n+1}} d\zeta, \quad n = -1, -2 \ldots. \tag{$*_2$}$$

To do this, observe that $|z - z_0| > |\zeta - z_0| = r_1$. Hence

$$\frac{-1}{\zeta - z} = \frac{1}{z - \zeta} = \frac{1}{z - z_0} \cdot \frac{1}{\frac{\zeta - z_0}{z - z_0}} = \frac{1}{z - z_0} \sum_{n=0}^{\infty} \left(\frac{\zeta - z_0}{z - z_0} \right)^n$$

$$= \sum_{n=0}^{\infty} \frac{(\zeta - z_0)^n}{(z - z_0)^{n+1}}, \quad |\zeta - z_0| < |z - z_0|.$$

Recall that z is a fixed point. Note that the geometric series $\sum_{n=0}^{\infty} \left(\frac{\zeta - z_0}{z - z_0} \right)^n$ converges uniformly on the circle $|\zeta - z_0| = r_1$, owing to Weierstrass M-test. Therefore, termwise integration is permissible and

$$f_2(z) = \frac{1}{2\pi i} \int_{C_1} \left(\sum_{n=0}^{\infty} \frac{(\zeta - z_0)^n}{(z - z_0)^{n+1}} \right) f(\zeta) d\zeta$$

$$= \sum_{n=0}^{\infty} \left\{ \frac{1}{2\pi i} \int_{C_1} f(\zeta)(\zeta - z_0)^n d\zeta \right\} \frac{1}{(z - z_0)^{n+1}}.$$

This is $(*_2)$.

Summarize the above as

The Laurent series expansion of an analytic function in a ring domain. Let f be analytic in $0 \leq r < |z - z_0| < R \leq \infty$. Then f can be expressed as

$$f(z) = \sum_{-\infty}^{\infty} a_n (z - z_0)^n, \quad r < |z - z_0| < R$$

where

$$a_n = \frac{1}{2\pi i} \int_{|\zeta - z_0| = \rho} \frac{f(\zeta)}{(\zeta - z_0)^{n+1}} d\zeta, \quad r < \rho < R, \quad n = 0, \pm 1, \pm 2, \dots .$$

$$(4.9.1.1)$$

As in the case of power series, a Laurent series $\sum_{-\infty}^{\infty} a_n (z - z_0)^n$, $r < |z - z_0| < R$, is the Laurent series expansion of its sum function $f(z)$ in $r < |z - z_0| < R$.

Concrete examples are referred to Sec. 4.9.2.

We give two Remarks.

Remark 1 (The Laurent series expansion at ∞). Suppose f is analytic in $0 \leq R < |z| < \infty$ so that ∞ is an isolated singularity of f. Let $w = \frac{1}{z}$. Then the function $g(w) = f(\frac{1}{w})$ is analytic in $0 < |w| < \frac{1}{R}$ and has the Laurent series expansion at $w = 0$ as $g(w) = \sum_{-\infty}^{\infty} a_n w^n$, $0 < |w| < \frac{1}{R}$. Consequently, we designate the *Laurent series expansion* of $f(z)$ at ∞ as

$$f(z) = \sum_{-\infty}^{\infty} \frac{a_n}{z^n}, \quad R < |z| < \infty,$$

with its *singular* or *essential part:* $\sum_{n=-\infty}^{-1} \frac{a_n}{z^n} = \sum_{n=1}^{\infty} a_{-n} z^n$ and *regular part:* $\sum_{n=0}^{\infty} \frac{a_n}{z^n}$, where *the coefficient integrals* are

$$a_n = \frac{1}{2\pi i} \int_{|\zeta|=\rho} \frac{f(\zeta)}{\zeta^{n+1}} d\zeta \quad \text{(in counterclockwise direction)}, \quad \rho > R,$$

$$n = 0, \pm 1, \pm 2, \ldots . \tag{4.9.1.2}$$

$\square$

Remark 2 (An extension of the proof of (4.9.1.1)). We can deduce from the proof of (4.9.1.1) the following

Laurent representation of an analytic function in a (circular) ring domain. Let f be analytic in $0 \leq r < |z - z_0| < R \leq \infty$.

(1) There exist an analytic function $f^+(z)$ in $|z - z_0| < R$ and an analytic function $f^-(z)$ in $r < |z - z_0| < \infty$, satisfying:

   (i) $f(z) = f^+(z) + f^-(z), \quad r < |z - z_0| < R;$ and
   (ii) $\lim_{z \to \infty} f^-(z) = 0$.

(2) Conditions 1 and 2 in (1) uniquely define $f^+(z)$ and $f^-(z)$ in the following sense: For each ρ, $r < \rho < R$,

$$f^+(z) = \frac{1}{2\pi i} \int_{|\zeta - z_0| = \rho} \frac{f(\zeta)}{\zeta - z} d\zeta, \quad |z - z_0| < \rho;$$

$$f^-(z) = \frac{-1}{2\pi i} \int_{|\zeta - z_0| = \rho} \frac{f(\zeta)}{\zeta - z} d\zeta, \quad \rho < |z - z_0| < \infty. \tag{4.9.1.3}$$

For further extension, see Exercise A (3).

Sketch of Proof

For each ρ, $r < \rho < R$, define $f_\rho^+(z)$ as

$$f_\rho^+(z) = \frac{1}{2\pi i} \int_{|\zeta - z_0| = \rho} \frac{f(\zeta)}{\zeta - z} d\zeta, \quad |z - z_0| < \rho.$$

$f_\rho^+(z)$ is analytic in $|z - z_0| < \rho$. For $r < \rho < \sigma < R$, the Cauchy integral theorem says that $f_\rho^+(z) = f_\sigma^+(z)$ in $|z - z_0| < \rho$. Hence, the function $f^+(z)$ defined by

$$f^+(z) = f_\rho^+(z), \quad |z - z_0| < \rho < R$$

is well defined and analytic in $|z - z_0| < R$. Similarly, the one defined by

$$f^-(z) = f_\sigma^-(z) = \frac{-1}{2\pi i} \int_{|\zeta - z_0| = \sigma} \frac{f(\zeta)}{\zeta - z} d\zeta, \quad |z - z_0| > \sigma > r$$

is well defined and analytic in $|z - z_0| > r$.

Fixed any z, $r < |z - z_0| < R$, there are ρ and σ such that $r < \sigma < |z - z_0| < \rho < R$. Then, by Cauchy integral formula (3.4.2.4'),

$$f(z) = \frac{1}{2\pi i} \int_{|\zeta - z_0| = \rho} \frac{f(\zeta)}{\zeta - z} d\zeta - \frac{1}{2\pi i} \int_{|\zeta - z_0| = \sigma} \frac{f(\zeta)}{\zeta - z} d\zeta$$
$$= f_\rho^+(z) + f_\sigma^-(z) = f^+(z) + f^-(z).$$

On the other hand, if $|z - z_0| > \sigma$,

$$|f^-(z)| = |f_\sigma^-(z)| \le \frac{2\pi\sigma}{2\pi} \cdot \max_{|\zeta - z_0| = \sigma} \left| \frac{f(\zeta)}{\zeta - z} \right|$$
$$\le \frac{\sigma}{|z - z_0| - \sigma} \max_{|\zeta - z_0| = \sigma} |f(\zeta)| \to 0 \quad \text{as } z \to \infty.$$

Let $g^+(z)$ and $g^-(z)$ be analytic functions in $|z - z_0| < R$ and $|z - z_0| > r$, respectively, and satisfy the conditions 1 and 2 in (1), namely $f(z) = g^+(z) + g^-(z)$ in $r < |z - z_0| < R$ and $\lim_{z \to \infty} g^-(z) = 0$. Then $f^+(z) - g^+(z) = g^-(z) - f^-(z)$ holds in $r < |z - z_0| < R$. This guarantees that the function $h(z)$ defined by

$$h(z) = \begin{cases} f^+(z) - g^+(z), & |z - z_0| < R, \\ g^-(z) - f^-(z), & |z - z_0| > r, \end{cases}$$

is entire and satisfies $\lim_{z \to \infty} h(z) = 0$. By Liouville's theorem, $h(z)$ is a constant function and $h(z) \equiv 0$ holds. Hence $f^+(z) = g^+(z)$ and $f^-(z) = g^-(z)$ holds in **C**.   □

Exercises A

(1) Do the following problems.

  (a) Find the ring domain of convergence for $\sum_{n=-\infty}^{\infty}(-1)^n n \cdot 2^{-|n|}z^n$ and the sum of the series.

  (b) Determine these complex numbers λ so that $\sum_{-\infty}^{\infty}\lambda^{|n|}z^n$ has a nonempty ring domain of convergence. Find also the sum of such series.

  (c) In case $f(z) = \sum_{0}^{\infty} a_n z^n$ has the radius r of convergence, find the ring domain of convergence for $\sum_{-\infty}^{\infty} a_{|n|} z^n$ and its sum.

(2) Given two Laurent series $f(z) = \sum_{-\infty}^{\infty} a_n(z - z_0)^n$ and $g(z) = \sum_{-\infty}^{\infty} b_n(z - z_0)$ in $0 \le r \le |z - z_0| < R \le \infty$.

  (a) Show that the series $c_n = \sum_{k=-\infty}^{\infty} a_k b_{n-k}$ converge absolutely for $n = 0, \pm 1, \pm 2, \ldots$.

  (b) Show that $f(z)g(z)$ has the Laurent series expansion $\sum_{-\infty}^{\infty} c_n(z - z_0)^n$ in $r < |z - z_0| < R$.

(3) Let γ_1 and γ_2 be rectifiable Jordan closed curves such that $\gamma_1 \subseteq \text{Int}\,\gamma_2$. If f is analytic in $\Omega = \text{Int}\,\gamma_2 - \overline{\text{Int}\,\gamma_1}$, show that there are an analytic function f_2 in $\text{Int}\,\gamma_2$ and an analytic function f_1 in $\text{Ext}\,\gamma_1$ (including ∞) so that $f = f_1 + f_2$ in Ω. Also, f_1 and f_2 are uniquely defined up to a constant.

(4) Suppose $f(z) = \sum_{-\infty}^{\infty} a_n(z - z_0)^n$, $0 \le r < |z - z_0| < R \le \infty$.

  (a) For each ρ, $r < \rho < R$, prove the *Gutzmer's inequality*: Letting $M(\rho) = \max_{|z-z_0|=\rho} |f(z)|$,

  $$\sum_{-\infty}^{\infty} |a_n|^2 \rho^{2n} = \frac{1}{2\pi} \int_0^{2\pi} |f(z_0) + \rho e^{i\theta})|^2 d\theta \le M(\rho)^2;$$

  in particular, the *Cauchy's inequality*:

  $$|a_n| \le \frac{M(\rho)}{\rho^n}, \quad n = 0, \pm 1, \pm 2, \ldots.$$

  (b) Let $r < \rho < \sigma < R$. Show that the area of the image domain of $\rho \le |z - z_0| \le \sigma$, under $w = f(z)$, is given by

  $$\pi \sum_{-\infty}^{\infty} n|a_n|^2(\sigma^{2n} - \rho^{2n}).$$

(5) Suppose f is analytic in $r < |z| < R$ where $r > 0$ and $R < \infty$. Also suppose $\lim_{n \to \infty} f(z_n) = 0$ for any sequence z_n in $r < |z| < R$ satisfying $\lim_{n \to \infty} |z_n| = r$ or R. Show that $f(z) = 0$ in $r < |z| < R$. What happens if either $r = 0$ or $R = \infty$?

(6) Let $f(z) = \sum_{-\infty}^{\infty} a_n z^n$ be analytic in $0 < |z| < \infty$ and satisfy $\overline{f(\bar{z})} = f\left(\frac{1}{z}\right)$ everywhere there. Show that $\overline{a_n} = a_{-n}$, $n \geq 1$, and a_0 is a real constant. Calculate $f(z)$ if $|z| = 1$.

Exercises B

(1) Let f be analytic in $r < |z| < R$ and $f(z) \neq 0$ everywhere there.

  (a) Suppose $\frac{f'(z)}{f(z)} = \sum_{-\infty}^{\infty} a_n z^n$ is the Laurent series expansion. Show that:

    (i) The coefficient a_{-1} is an integer α, which is equal to the winding number $n(f(\gamma_\rho); 0)$ where γ_ρ is the circle $|z| = \rho$, $r < \rho < R$.

    (ii) $\log f(z) z^{-\alpha}$ has a single-valued branch in $r < |z| < R$.

    (iii) $\log f(z)$ has a branch in $r < |z| < R$ if and only if $\alpha = 0$.

    (iv) $\sqrt[n]{f(z)}$ has a branch in $r < |z| < R$ if and only if n divides α.

    Try to use $f(z) = \prod_{k=1}^{m} (z - a_k)^{\alpha_k}$ to justify these statements, where $a_1, \ldots, a_m$ are not in $r < |z| < R$ and $\alpha_1, \ldots, \alpha_m$ are integers.

  (b) Show that there is a function $g(z)$, analytic in $r < |z| < R$, so that $f(z) = z^\alpha e^{g(z)}$, where $\alpha = a_{-1}$ is as in (a). In case there are an integer β and an analytic function $h(z)$ in $r < |z| < R$ such that $f(z) = z^\beta e^{h(z)}$, then it is necessary that $\beta = \alpha$.

  (c) There are functions $h(z)$, analytic in $|z| < R$, and $H(z)$, analytic in $|z| > r$, such that $h(z) \neq 0$ in $|z| < R$, $H(z) \neq 0$ in $|z| > r$, $\lim_{z \to \infty} H(z) z^{-\alpha} = 1$ and $f(z) = h(z) H(z)$ in $r < |z| < R$. Under what conditions will $h(z)$ and $H(z)$ be unique?

(2) Let f be analytic in $0 < |z| < \infty$ and $f(z) \neq 0$ there. Show that there are entire functions g and h, both without zeros, such that

$$f(z) = z^n g(z) h\left(\frac{1}{z}\right), \quad 0 < |z| < \infty$$

where n is an integer.

(3) Let γ, g, h and f be as in Exercise B (1) of Sec. 3.5.2, where

$$f(z) = \int_\gamma \frac{g(\zeta)}{h(\zeta) - z} d\zeta, \quad z \in \mathbf{C} - h(\gamma)$$

is analytic on each component of $\mathbf{C} - h(\gamma)$. Suppose there is a $R > 0$ so that $h(\gamma)$ is contained in $|z| < R$. Therefore, f is analytic in $R < |z| < \infty$. Show that f has its Laurent series at ∞ as

$$\sum_{n=1}^{\infty} \frac{a_n}{z^n}, \quad \text{where } a_n = \int_\gamma g(\zeta)(h(\zeta))^{n-1}d\zeta, \quad n \geq 1.$$

In particular, f has a simple zero at ∞.

(4) Suppose $f(z)$ is analytic on the circle $|z| = 1$, one-to-one, and $f'(z) \neq 0$ holds there.

   (a) Show that there are r_1 and r_2, $0 < r_1 < 1 < r_2$, such that f is univalent and analytic in $r_1 < |z| < r_2$.

   (b) Let $z = g(w)$ be the inverse of $w = f(z)$ in $r_1 < |z| < r_2$. In case $w = f(z)$ maps $|z| = 1$ onto $|w| = 1$. Show the kth coefficient of the Laurent series expansion of $g(w)$ in $\rho_1 < |w| < \rho_2$ $(0 < \rho_1 < 1 < \rho_2)$ is given by

$$\frac{n(f(|z| = 1); 0)}{2\pi i} \int_{|z|=1} \frac{z f'(z)}{[f(z)]^{k+1}} dz$$

   where $n(f(|z| = 1); 0)$ is the winding number of the curve $f(|z| = 1)$ around 0.

   (c) Show that $f(z) = \frac{3z+z^3}{1+3z^2}$ is analytic on $|z| = 1$, univalent, and maps $|z| = 1$ onto $|w| = 1$, yet $f'(1) = 0$.

(5) Suppose $f(z) = \sum_{-\infty}^{\infty} a_n(z - z_0)^n$, $r < |z - z_0| < \infty$. Let $f(z) = \sum_{-n}^{\infty} b_n z^n$, $|z| > R$, be the Laurent series expansion of f at ∞.

   (a) Show that $\sup_{b_k \neq 0} k = \sup_{a_k \neq 0} k$.

   (b) Show that $a_{-1} = b_{-1}$, and, in case $k > 0$, try to use $b_{-1}, \ldots, b_{-k}$ to express a_{-k}.

   (c) If ∞ is a pole of f, find out corresponding formulas for a_k, where $k \geq 0$.

4.9.2.   Examples

We give a series of examples in the following.

Example 1. Find the Laurent series expansions of $f(z) = \frac{z^2+1}{z(z^2-3z+2)}$ at $z_0 = -1$ in terms of $z + 1$.

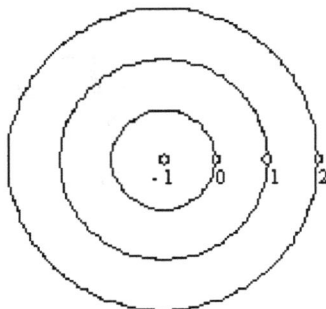

Fig. 4.45

Solution. Factorize $f(z)$ into partial fraction as

$$f(z) = \frac{1}{2} \cdot \frac{1}{z} - \frac{2}{z-1} + \frac{5}{2} \cdot \frac{1}{z-2}$$

and try to expand it into Laurent series in the domains $|z+1| < 1$, $1 < |z+1| < 2$, $2 < |z+1| < 3$ and $|z+1| > 3$, respectively. See Fig. 4.45.

Imitating the geometric series, we have the following individual expansions:

$$\frac{1}{z} = \begin{cases} \dfrac{-1}{1-(z+1)} = -\displaystyle\sum_{0}^{\infty}(z+1)^n, & \text{if } |z+1| < 1, \\[4mm] \dfrac{1}{(z+1)\left(1 - \frac{1}{z+1}\right)} = \displaystyle\sum_{0}^{\infty}\dfrac{1}{(z+1)^{n+1}}, & \text{if } |z+1| > 1 \end{cases}$$

$$\frac{1}{z-1} = \begin{cases} \dfrac{1}{-2\left(1 - \frac{(z+1)}{2}\right)} = -\dfrac{1}{2}\displaystyle\sum_{0}^{\infty}\dfrac{1}{2^n}(z+1)^n, & \text{if } |z+1| < 2, \\[4mm] \dfrac{1}{(z+1)\left(1 - \frac{2}{z+1}\right)} = \displaystyle\sum_{0}^{\infty}\dfrac{2^n}{(z+1)^{n+1}}, & \text{if } |z+1| > 2, \end{cases}$$

and

$$\frac{1}{z-2} = \begin{cases} \dfrac{1}{-3\left(1 - \frac{(z+1)}{3}\right)} = -\dfrac{1}{3}\displaystyle\sum_{0}^{\infty}\dfrac{1}{3^n}(z+1)^n, & \text{if } |z+1| < 3, \\[4mm] \dfrac{1}{(z+1)\left(1 - \frac{3}{z+1}\right)} = \displaystyle\sum_{0}^{\infty}\dfrac{3^n}{(z+1)^{n+1}}, & \text{if } |z+1| > 3. \end{cases}$$

Combining together, the required Laurent series are

$$
f(z) = \begin{cases}
\dfrac{1}{2}\displaystyle\sum_{0}^{\infty}\left\{-1+\dfrac{1}{2^{n-1}}-\dfrac{5}{3^{n+1}}\right\}(z+1)^n, & |z+1|<1, \\[2.5ex]
\dfrac{1}{2}\displaystyle\sum_{1}^{\infty}\dfrac{1}{(z+1)^n}+\dfrac{1}{2}\displaystyle\sum_{0}^{\infty}\left\{\dfrac{1}{2^{n+1}}-\dfrac{5}{3^{n+1}}\right\}(n+1)^n, & 1<|z+1|<2, \\[2.5ex]
\displaystyle\sum_{1}^{\infty}\left\{\dfrac{1}{2}-2^n\right\}\dfrac{1}{(z+1)^n}-\dfrac{5}{6}\displaystyle\sum_{0}^{\infty}\dfrac{1}{3^n}(z+1)^n, & 2<|z+1|<3, \\[2.5ex]
\dfrac{1}{2}\displaystyle\sum_{1}^{\infty}\{1-2^{n+1}+5\cdot3^{n-1}\}\dfrac{1}{(z+1)^n}, & |z+1|>3.
\end{cases}
$$

Example 2. Find the Laurent series expansions of $f(z) = \frac{z^2-2z+5}{(z-2)(z^2+1)}$ in $1<|z|<2$ and $2<|z|<\infty$, respectively.

Solution. Observe that $f(z) = \frac{1}{z-2} - \frac{2}{z^2+1}$. In case $1<|z|<2$,

$$
f(z) = -\frac{1}{2}\cdot\frac{1}{1-\frac{z}{2}} - \frac{2}{z^2}\cdot\frac{1}{1+\frac{1}{z^2}} = -\frac{1}{2}\sum_{0}^{\infty}\left(\frac{z}{2}\right)^n - \frac{2}{z^2}\sum_{0}^{\infty}\left(\frac{-1}{z^2}\right)^n
$$

$$
= 2\sum_{1}^{\infty}\frac{(-1)^n}{z^{2n}} - \sum_{0}^{\infty}\frac{1}{2^{n+1}}z^n.
$$

In case $2<|z|<\infty$,

$$
f(z) = \frac{1}{z}\cdot\frac{1}{1-\frac{2}{z}} - \frac{2}{z^2}\cdot\frac{1}{1+\frac{1}{z^2}} = \sum_{1}^{\infty}\frac{2^{n-1}}{z^n} + 2\sum_{1}^{\infty}\frac{(-1)^n}{z^{2n}} = \sum_{1}^{\infty}\frac{a_{-n}}{z^n}
$$

where

$$
a_{-n} = \begin{cases} 2^{2m}, & \text{if } n=2m+1 \\ 2^{2m-1}+2(-1)^m, & n=2m \end{cases}, \quad m=1,2,3,\dots.
$$

As a matter of fact, in the case $|z|>2$, the coefficient $a_n, n = 0, \pm1, \pm2, \dots$, can be computed directly by using the original coefficient integrals

$$
a_n = \frac{1}{2\pi i}\int_{|z|=\rho>2}\frac{z^2-2z+5}{z^{n+1}(z-2)(z^2+1)}dz.
$$

If $n \geq 0$, there is a constant $M > 0$ so that, for $\rho > 2$ large enough,

$$
|a_n| \leq \frac{1}{2\pi}\int_{|z|=\rho}\frac{\rho^2 M}{\rho^{n+1}\cdot\rho\cdot\rho^2}|dz|
$$

$$
= \frac{1}{2\pi}\cdot M\rho^2\cdot\frac{2\pi\rho}{\rho^{n+4}} = \frac{M}{\rho^{n+1}} \to 0 \quad \text{as } \rho \to \infty.
$$

So $a_n = 0$ for $n \geq 0$. If $n \leq -1$, using $-n$ to replace n, then, for $n \geq 1$,

$$a_{-n} = \frac{1}{2\pi i} \int_{|z|=\rho} \frac{z^{n-1}(z^2 - 2z + 5)}{(z-2)(z^2+1)} dz = \frac{1}{2\pi i} \int_{|z|=\rho} \left\{ \frac{z^{n-1}}{z-2} - \frac{2z^{n-1}}{z^2+1} \right\} dz$$

$$= 2^{n-1} - \frac{2}{2\pi i} \int_{|z|=\rho} \frac{z^{n-1}}{z^2+1} dz.$$

Recall that $\frac{z^{n-1}}{z^2+1} = z^{n-3} \sum_0^\infty \frac{(-1)^k}{z^{2k}}$, $|z| > 2$. By uniform convergence of this series along $|z| = \rho > 2$, in case $n = 2m+1$,

$$\int_{|z|=\rho} \frac{z^{n-1}}{z^2+1} dz = \sum_0^\infty (-1)^k \int_{|z|=\rho} \frac{z^{2m-2}}{z^{2k}} dz = 0;$$

in case $n = 2m$,

$$\int_{|z|=\rho} \frac{z^{n-1}}{z^2+1} dz = \sum_0^\infty (-1)^k \int_{|z|=\rho} \frac{z^{2m-3}}{z^{2k}} dz = (-1)^{m-1} \int_{|z|=\rho} \frac{z^{2m-3}}{z^{2m-2}} dz$$

$$= (-1)^{m-1} \int_{|z|=\rho} \frac{1}{z} dz = 2\pi i (-1)^{m-1}.$$

And the results follow.

Example 3. The Laurent series expansion of $f(\zeta) = e^{\frac{z}{2}(\zeta - \zeta^{-1})}$ at $\zeta = 0$ is given by $\sum_{-\infty}^\infty J_n(z)\zeta^n$, $0 < |\zeta| < \infty$, whose coefficients $J_n(z)$ are called the *Bessel functions of the first kind*. Show that, for $n = 0, 1, 2, \ldots$,

$$J_n(z) = (-1)^n J_{-n}(z)$$

$$= \frac{1}{2\pi} \int_0^{2\pi} \cos(n\theta - z\sin\theta) d\theta = \frac{1}{\pi} \int_0^\pi \cos(n\theta - z\sin\theta) d\theta$$

$$= \sum_{k=0}^\infty \frac{(-1)^k}{k!(n+k)!} \left(\frac{z}{2}\right)^{n+2k},$$

and $J_n(z)$ satisfies the *Bessel equation* $z^2 w''(z) + z w'(z) + (z^2 - n^2)w(z) = 0$.

Solution. For $n = 0, \pm 1, \pm 2, \ldots$

$$J_n(z) = \frac{1}{2\pi i} \int_{|\zeta|=1} \frac{e^{\frac{z}{2}(\zeta - \zeta^{-1})}}{\zeta^{n+1}} d\zeta$$

$$= \frac{1}{2\pi} \int_{-\pi}^\pi e^{iz\sin\theta} \cdot e^{-in\theta} d\theta \qquad (\zeta = e^{i\theta}, -\pi \leq \theta \leq \pi) \qquad (*)$$

$$= \frac{1}{2\pi} \int_{-\pi}^{\pi} \cos(n\theta - z\sin\theta)d\theta - \frac{i}{2\pi} \int_{-\pi}^{\pi} \sin(n\theta - z\sin\theta)d\theta$$

$$= \frac{1}{2\pi} \int_{-\pi}^{\pi} \cos(n\theta - z\sin\theta)d\theta$$

(since $\sin(n\theta - z\sin\theta)$ is odd in $[-\pi, \pi]$)

$$= \frac{1}{\pi} \int_{0}^{\pi} \cos(n\theta - z\sin\theta)d\theta$$

(since $\cos(n\theta - z\sin\theta)$ is even in $[-\pi, \pi]$).

On the other hand,

$$J_{-n}(z)$$

$$= \frac{1}{\pi} \int_{0}^{\pi} \cos(n\theta + z\sin\theta)d\theta = \frac{1}{\pi} \int_{\pi}^{0} \cos[n(\pi - \theta) + z\sin(\pi - \theta)](-d\theta)$$

$$= (-1)^n \cdot \frac{1}{\pi} \int_{0}^{\pi} \cos(n\theta - z\sin\theta)d\theta = (-1)^n J_n(z), \quad n = 0, 1, 2, \ldots.$$

Observe that

$$e^{\frac{z}{2}(\zeta - \zeta^{-1})} = e^{\frac{z}{2}\zeta} \cdot e^{-\frac{z}{2}\zeta^{-1}}$$

$$= \left[\sum_{k=0}^{\infty} \frac{1}{k!} \left(\frac{z}{2}\right)^k \zeta^k \right] \left[\sum_{\ell=0}^{\infty} \frac{1}{\ell!} \left(-\frac{z}{2}\right)^\ell \zeta^{-\ell} \right], \quad 0 < |\zeta| < \infty.$$

The absolute convergence of both series in the right permits us to multiply the series out by Cauchy product and the uniform convergence of the resulted series along $|\zeta| = 1$ guarantees the termwise integration in (∗). And the final series expression for $J_n(z)$ follows. Direct computation will show that $J_n(z)$ satisfies the differential equation.

Example 4. Find the Laurent series expansions of $f(z) = \frac{\text{Log}(2-z)}{z(z-1)}$ in $0 < |z| < 1$ and $0 < |z - 1| < 1$, respectively.

Solution. $\text{Log}(2 - z)$ is the principal branch of $\log(2 - z)$ in $\mathbf{C} - [2, \infty)$. See Fig. 4.46. Note that

$$\frac{1}{z - 1} = -\sum_{0}^{\infty} z^n, \quad |z| < 1;$$

$$\frac{1}{z} = \sum_{0}^{\infty} (-1)^n (z - 1)^n, \quad |z - 1| < 1, \quad \text{and}$$

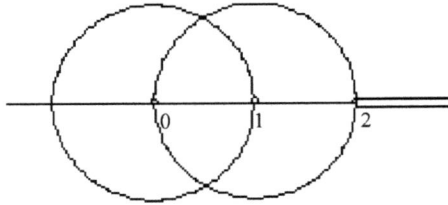

Fig. 4.46

$$\text{Log}(2-z) = \begin{cases} \text{Log}\,2 + \text{Log}\left(1 - \dfrac{z}{2}\right) = \text{Log}\,2 - \displaystyle\sum_{n=1}^{\infty} \dfrac{z^n}{n \cdot 2^n}, & |z| < 1, \\[4mm] \text{Log}(1 - (z-1)) = -\displaystyle\sum_{1}^{\infty} \dfrac{(z-1)^n}{n}, & |z-1| < 1. \end{cases}$$

Using the method of Cauchy product, it follows that

$$f(z) = \begin{cases} -\dfrac{\text{Log}\,2}{z} + \displaystyle\sum_{n=0}^{\infty} \left\{ \sum_{k=1}^{n+1} \dfrac{1}{k \cdot 2^k} \right\} z^n, & 0 < |z| < 1, \\[5mm] -\displaystyle\sum_{n=0}^{\infty} \left\{ \sum_{k=0}^{n} \dfrac{(-1)^{n-k}}{k+1} \right\} (z-1)^n, & 0 < |z-1| < 1. \end{cases}$$

Example 5. Find the Laurent series expansions for each of the following functions in the indicated domains.

(1) The branch of $f(z) = \sqrt{\frac{z}{(z-1)(z-2)}}$ determined by $\text{Im}\, f\left(\frac{3}{2}\right) > 0$ in $1 < |z| < 2$.

(2) The branch of $f(z) = \sqrt{z(z-1)}$ determined by $f(2) = \sqrt{2} > 0$ in $1 < |z| < \infty$ and $1 < |z-1| < \infty$, respectively.

(3) The two branches of $f(z) = \sqrt{(z-a)(z-b)}, |a| \le |b|$, at $z_0 = \infty$.

Solution.

(1) $f(z)$ has branches in $\mathbf{C}^* - [0,1] - [2,\infty]$. See Fig. 4.47. The branch we want is the one determined by $f\left(\frac{3}{2}\right) = \sqrt{-6} = \sqrt{6}i$. Then, for this branch in $1 < |z| < 2$,

$$f(z) = \sqrt{\dfrac{1}{-2(1 - \frac{1}{z})(1 - \frac{z}{2})}} = \sqrt{-\dfrac{1}{2}} \left(1 - \dfrac{1}{z}\right)^{-\frac{1}{2}} \left(1 - \dfrac{z}{2}\right)^{-\frac{1}{2}}$$

$$= \dfrac{i}{\sqrt{2}} \left(1 - \dfrac{1}{z}\right)^{-\frac{1}{2}} \left(1 - \dfrac{z}{2}\right)^{-\frac{1}{2}}$$

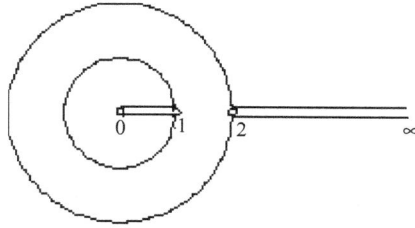

Fig. 4.47

$$= \frac{i}{\sqrt{2}} \left\{ 1 + \sum_{n=1}^{\infty} (-1)^n \begin{pmatrix} -\frac{1}{2} \\ n \end{pmatrix} \frac{1}{z^n} \right\} \left\{ 1 + \sum_{m=1}^{\infty} (-1)^m \begin{pmatrix} -\frac{1}{2} \\ m \end{pmatrix} \left(\frac{z}{2}\right)^m \right\}$$

$$= \sum_{n=-\infty}^{\infty} a_n z^n,$$

where $a_n = 2^{-n} a_{-n}, n = 1, 2, 3, \ldots$, and

$$a_0 = \frac{i}{\sqrt{2}} \left\{ 1 + \sum_{m=1}^{\infty} \begin{pmatrix} -\frac{1}{2} \\ m \end{pmatrix} 2^{-m} \right\};$$

$$a_{-n} = \frac{(-1)^n i}{\sqrt{2}} \left\{ \begin{pmatrix} -\frac{1}{2} \\ n \end{pmatrix} + \sum_{m=1}^{\infty} \begin{pmatrix} -\frac{1}{2} \\ m+n \end{pmatrix} \begin{pmatrix} -\frac{1}{2} \\ m \end{pmatrix} 2^{-m} \right\}.$$

(2) $f(z)$ has branches in $\mathbf{C}^* - [0,1]$. See Fig. 4.48. For the branch determined by $f(2) = \sqrt{2} > 0$, in $1 < |z| < \infty$,

$$f(z) = z\sqrt{1 - \frac{1}{z}} \quad \left(\text{wherein } \sqrt{1 - \frac{1}{z}} \text{ assumes positive values if } z = r > 1 \right)$$

$$= z \left\{ 1 + \sum_{n=1}^{\infty} (-1)^n \begin{pmatrix} \frac{1}{2} \\ n \end{pmatrix} \frac{1}{z^n} \right\} = z - \frac{1}{2} + \sum_{n=2}^{\infty} (-1)^n \begin{pmatrix} \frac{1}{2} \\ n \end{pmatrix} \frac{1}{z^{n-1}};$$

while, in $1 < |z - 1| < \infty$,

$$f(z) = \sqrt{(z-1)(z-1+1)} = (z-1)\sqrt{1 + \frac{1}{z-1}}$$

$$= (z-1) \left\{ 1 + \sum_{n=1}^{\infty} \begin{pmatrix} \frac{1}{2} \\ n \end{pmatrix} \frac{1}{z^n} \right\} = (z-1) + \frac{1}{2} + \sum_{n=2}^{\infty} \begin{pmatrix} \frac{1}{2} \\ n \end{pmatrix} \frac{1}{(z-1)^{n-1}}.$$

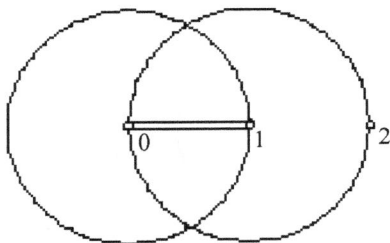

Fig. 4.48

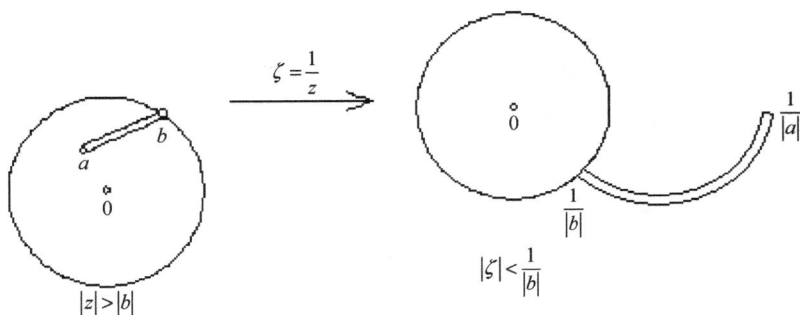

Fig. 4.49

(3) $f(z)$ has two branches in $\mathbf{C} - [a, b]$ (refer to Example 2 in Sec. 2.7.2). See Fig. 4.49. Let $\zeta = \frac{1}{z}$ and consider the function $g(\zeta) = f(\frac{1}{\zeta}) = \frac{1}{\zeta}\sqrt{(1 - a\zeta)(1 - b\zeta)}$. In case $|\zeta| < \frac{1}{|b|} \le \frac{1}{|a|}$,

$$\sqrt{1 - a\zeta} = \pm\left\{1 + \sum_{n=1}^{\infty} \binom{\frac{1}{2}}{n}(-a\zeta)^n\right\};$$

$$\sqrt{1 - b\zeta} = \pm\left\{1 + \sum_{m=1}^{\infty} \binom{\frac{1}{2}}{m}(-b\zeta)^m\right\}.$$

Substituting these two power series into $g(\zeta) = f(\frac{1}{\zeta})$ and using the Cauchy product and then replacing $\frac{1}{\zeta}$ by z, we have

$$f(z) = \pm\left\{z - \frac{1}{2}(a + b) + \sum_{n=1}^{\infty} \frac{c_{-n}}{z^n}\right\}, \qquad \text{where, for } n = 2, 3, 4, \ldots,$$

$$c_{-(n-1)} = (-1)^n \left[\binom{\frac{1}{2}}{n} b^n + \binom{\frac{1}{2}}{n-1} \binom{\frac{1}{2}}{1} b^{n-1} a \right.$$

$$\left. + \binom{\frac{1}{2}}{n-2} \binom{\frac{1}{2}}{2} b^{n-2} a^2 + \cdots + \binom{\frac{1}{2}}{n} a^n \right].$$

Example 6. Determine if the following functions can be expanded into Laurent series at the indicated points z_0.

(1) $\cos \dfrac{1}{z}$, $z_0 = 0, \infty$;

(2) $\sec \dfrac{1}{z-1}$, $z_0 = 1$;

(3) $\mathrm{Log}\, z$, $z_0 = 0$;

(4) $\mathrm{Log}\, \dfrac{z-i}{z+i}$, $z_0 = \infty$;

(5) $\log(z-1)(z-2)$, $z_0 = \infty$.

Solution.

(1) $\cos \frac{1}{z}$ is analytic in $0 < |z| \leq \infty$. So $z_0 = 0$ is an isolated singularity and it can be expanded as $1 + \sum_{n=1}^{\infty} (-1)^n \frac{1}{(2n)!} \cdot \frac{1}{z^{2n}}$, $0 < |z| \leq \infty$.

(2) Observe that $\cos \frac{1}{z-1} = 0$ if and only if $\frac{1}{z-1} = n\pi + \frac{\pi}{2}$ or $z_n = 1 + \left(n\pi + \frac{\pi}{2}\right)^{-1}$, $n = 0, \pm 1, \pm 2, \ldots$. Therefore $z_0 = 1$, as the limit of the isolated singularities z_n, is *not* an isolated singularity of $\sec \frac{1}{z-1}$. Thus, $\sec \frac{1}{z-1}$ cannot be expanded into the Laurent series at $z_0 = 1$.

(3) $\mathrm{Log}\, z$ is the principal branch of $\log z$ in $\mathbf{C} - (\infty, 0]$. $\mathrm{Log}\, z$ is not defined on $(\infty, 0]$, not to say the continuity, and in fact, every point of $(\infty, 0]$ is a singular point of $\mathrm{Log}\, z$ (see (2.2.5)). $z_0 = 0$ is thus not an isolated singularity of $\mathrm{Log}\, z$. No such Laurent series exists.

(4) $\mathrm{Log}\, \frac{z-i}{z+i}$ is the principal branch of $\log \frac{z-i}{z+i}$ in $\mathbf{C} - [-i, i]$ and is analytic at $z_0 = \infty$ (refer to Example 4 in Sec. 2.7.3). The Taylor series expansion of it at $z_0 = \infty$ is given by:

$$\mathrm{Log}\, \frac{1-iz}{1+iz} = \sum_{n=1}^{\infty} (-1)^{n-1} \frac{(-i)^n}{n} z^n - \sum_{n=1}^{\infty} (-1)^{n-1} \frac{i^n}{n} z^n$$

$$= \sum_{n=1}^{\infty} (-1)^{n-1} \frac{(-i)^n - i^n}{n} z^n, \quad |z| < 1$$

$$\Rightarrow \mathrm{Log}\, \frac{z-i}{z+i} = \sum_{n=1}^{\infty} (-1)^{n-1} \frac{(-i)^n - i^n}{n} \frac{1}{z^n}, \quad |z| > 1.$$

(5) It is $\mathbf{C}^* - [\infty, 1] - [2, \infty]$, not $\mathbf{C}^* - [1, 2]$, that is a single-valued domain of $\log(z-1)(z-2)$. The Riemann surface for $\log(z-1)(z-2)$ has a logarithmic branch point over each of the points $z = 1$ and 2, two logarithmic branch points over $z = \infty$ (refer to Exercise A (17) of Sec. 2.7.3). No branch can exist in a neighborhood of ∞, not to say the Laurent series expansion at ∞.

Example 7. To each of the following functions, do the same problem as Example 6.

(1) $\sqrt{z}, \quad z_0 = 0$.

(2) $\sqrt[3]{(z-1)(z-2)(z-3)}, \quad z_0 = \infty$.

(3) $\sqrt{1 + \sqrt{z}}, \quad z_0 = 1, 0$.

(4) $\sqrt{1 + \sqrt[3]{\dfrac{z}{z+1}}}, \quad z_0 = \infty$.

Solution.

(1) No branch can be defined in a neighborhood $z_0 = 0$ (see (2.7.2.6)). The answer is in the negative.

(2) $\sqrt[3]{(z-1)(z-2)(z-3)}$ has three branches in $\mathbf{C} - [1, 2] - [2, 3]$ (refer to Exercise B (4) of Sec. 2.7.2). If $f(z)$ is one of them, then the other two are given by $\omega f(z)$ and $\omega^2 f(z)$, where $\omega = e^{2\pi i/3}$. Take, for instance,

$$f(z) = z\left(1 - \frac{1}{z}\right)^{\frac{1}{3}}\left(1 - \frac{2}{z}\right)^{\frac{1}{3}}\left(1 - \frac{3}{z}\right)^{\frac{1}{3}}$$

where, if $z = r > 3$, $\left(1 - \frac{1}{r}\right)^{\frac{1}{3}}$, $\left(1 - \frac{2}{r}\right)^{\frac{1}{3}}$ and $\left(1 - \frac{3}{r}\right)^{\frac{1}{3}}$ should assume positive values. See Fig. 4.50. Note that $z_0 = \infty$ is a simple pole (see Sec. 10.4.2, if necessarily) for each of these branches. For $f(z)$

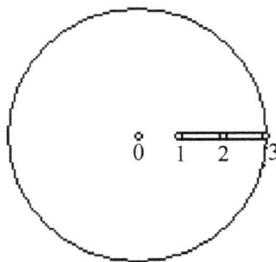

Fig. 4.50

above, if $|z| > 3$, then

$$\left(1 - \frac{1}{z}\right)^{\frac{1}{3}} = 1 + \sum_{n=1}^{\infty} (-1)^n \binom{\frac{1}{3}}{n} \frac{1}{z^n},$$

$$\left(1 - \frac{2}{z}\right)^{\frac{1}{3}} = 1 + \sum_{n=1}^{\infty} (-1)^n \binom{\frac{1}{3}}{n} \frac{2^n}{z^n}, \quad \text{and}$$

$$\left(1 - \frac{3}{z}\right)^{\frac{1}{3}} = 1 + \sum_{n=1}^{\infty} (-1)^n \binom{\frac{1}{3}}{n} \frac{3^n}{z^n}$$

$\Rightarrow$ (by using the Cauchy product) $\quad f(z) = z + a_0 + \sum_{n=1}^{\infty} a_n \frac{1}{z^n}$,
$$|z| > 1.$$

(3) $w = \sqrt{1 + \sqrt{z}}$ assumes four distinct values except at $z = 0$ and 1.

At $z_0 = 0$: $w = w(z)$ cannot have single-valued branches in a neighborhood of 0 because so is $\sqrt{z}$.

At $z_0 = 1$: Since $\sqrt{1} = \pm 1$ in $\mathbf{C}$, in case $\sqrt{1} = -1$, $w(1) = \pm\sqrt{1 + \sqrt{1}} = \pm\sqrt{1 - 1} = \pm\sqrt{0} = 0$; in case $\sqrt{1} = 1$, $w(1) = \pm\sqrt{1 + \sqrt{1}} = \pm\sqrt{1 + 1} = \pm\sqrt{2}$. Thus, the two branches of $w = w(z)$, determined by $w(1) = \pm\sqrt{2}$, in $\mathbf{C}^* - [\infty, 0]$ are analytic and can be expanded into Taylor series at $z_0 = 1$. To find these Taylor series, substitute $\sqrt{z} = 1 + \sum_{n=1}^{\infty} \binom{\frac{1}{2}}{n} (z - 1)^n, |z - 1| < 1$, into $\sqrt{1 + \sqrt{z}}$ and proceed according to the methods suggests in (4.8.1).

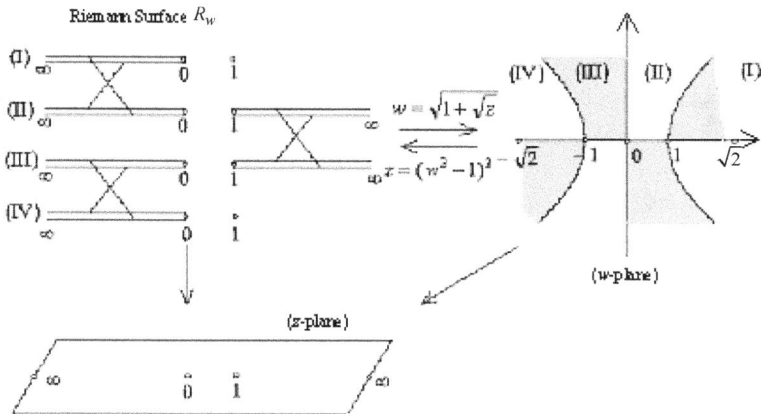

Fig. 4.51

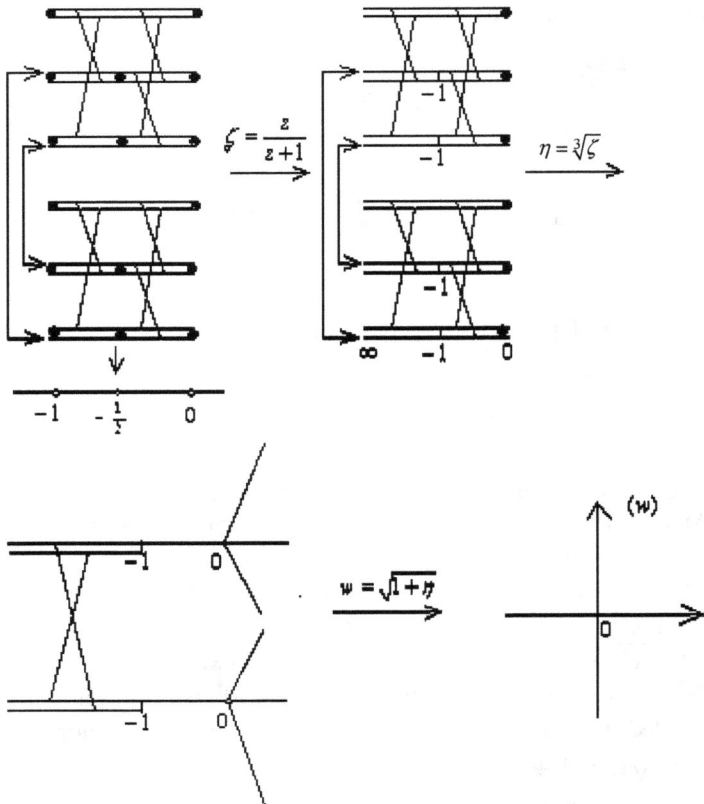

Fig. 4.52

By the way, note that $w = w(z)$ is the inverse function of $z = (w^2 - 1)^2$. Its Riemann surface can be described as in Fig. 4.51. The four branches are determined by $w(1) = \pm\sqrt{2}$ and $w(\frac{1}{4}) = \pm\sqrt{\frac{1}{2}}$, respectively. The two branches we mentioned in the last paragraph are (I) and (IV).

(4) Construct successively the Riemann surfaces for $z \to \zeta = \frac{z}{1+z} \to \eta = \sqrt[3]{\zeta} \to w = \sqrt{1+\eta}$ and we will obtain the descriptive Riemann surface for $w = \sqrt{1 + \sqrt[3]{\frac{z}{1+z}}}$. See Fig. 4.52. There are two algebraic branch points of order 2 over $z = 0$, two branch points of order 1 over $z = -\frac{1}{2}$ and one branch point of order 5 over $z = -1$. There are six distinct branches in a neighborhood of ∞, determined by $w(\infty) = \pm\sqrt{2}, \pm\sqrt{1+\omega}, \pm\sqrt{1+\omega^2}(\omega = e^{2\pi i/3})$, respectively. They all can be expanded into Taylor series. Try them out!

Example 8. Show that the function f defined by the series

$$f(z) = \sum_{n=0}^{\infty} \frac{1}{n!} \frac{1}{1 + a^{2n} z^2} \quad (a > 1)$$

can be expanded into the Laurent series in $|z| > 1$ as

$$f(z) = \sum_{n=0}^{\infty} (-1)^{n+1} e^{\frac{1}{a^{2n}}} \frac{1}{z^{2n}}.$$

But $f(z)$ cannot be expressed as a Laurent series in any neighborhood of $z_0 = 0$.

Solution. Each $f_n(z) = \frac{1}{1+a^{2n} z^2}$ is analytic in $\mathbf{C}^*$ except at $\pm \frac{1}{a^n}$; in particular, it is analytic in $1 < |z| \le \infty$. So $f_n(z)$ has the Taylor series expansion at ∞ as

$$f_n(z) = \frac{1}{a^{2n} z^2 (1 + \frac{1}{a^{2n} z^2})} = \sum_{k=1}^{\infty} (-1)^{k-1} \frac{1}{a^{2nk}} \frac{1}{z^{2k}}, \quad |z| > 1.$$

Recall that $f_n^{(2k)}(\infty) = (-1)^{k-1} \frac{(2k)!}{a^{2nk}}, k \ge 1$, which is equal to $g_n^{(2k)}(0)$ where $g_n(z) = f_n(\frac{1}{z})$ (see Remark in Sec. 3.3).

By Weierstrass M-test (see (1) in (2.3.5)), $\sum_{n=0}^{\infty} \frac{1}{n!} f_n(z)$ converges uniformly on every compact subset of $|z| > 1$ to the continuous function $f(z)$. By another Weierstrass theorem (see (5.3.1.1)), this $f(z)$ is also analytic on $1 < |z| \le \infty$ and even more, $f^{(k)}(z) = \sum_{n=0}^{\infty} \frac{1}{n!} f_n^{(k)}(z)$ uniformly on each compact subset of $1 < |z| \le \infty$ for $k = 1, 2, 3, \ldots$. In particular, $f^{(k)}(\infty) = \sum_{n=0}^{\infty} \frac{1}{n!} f_n^{(k)}(\infty)$ holds. Since f is analytic in $1 < |z| \le \infty$ and is even there, f has the Taylor series expansion at ∞ as

$$f(z) = \sum_{k=0}^{\infty} \frac{f^{(k)}(\infty)}{k!} \frac{1}{z^k} = \sum_{k=0}^{\infty} \frac{f^{(2k)}(\infty)}{(2k)!} \frac{1}{z^{2k}} = \sum_{k=0}^{\infty} \left\{ \sum_{n=0}^{\infty} \frac{(-1)^{k-1}}{a^{2nk}} \cdot \frac{1}{n!} \right\} \frac{1}{z^{2k}}$$

$$= \sum_{k=0}^{\infty} (-1)^{k-1} e^{\frac{1}{a^{2k}}} \frac{1}{z^{2k}}, \quad |z| > 1.$$

The process leading to this final result is just a special case of the so-called Weierstrass double series theorem (see (5.3.1.2)).

Exercises A

(1) Find the Laurent series expansions of each of the following functions $f(z)$ at the indicated points (and pinpoint its ring domain of

convergence) or domains.

(a) $f(z) = \dfrac{1}{(z-a)(z-b)}$ $(0 < |a| < |b|)$, $z_0 = 0, a, \infty$,   and
$|a| < |z| < |b|$.

(b) $\dfrac{1}{(z^2+1)^2}$, $z_0 = i, \infty$.

(c) $\dfrac{z^2-1}{(z+2)(z+3)}$, $2 < |z| < 3$ and $|z| > 3$.

(d) $\dfrac{1}{2-3z+z^2}$, $1 < |z| < 2$ and $\sqrt{2} < |z+i| < \sqrt{5}$.

(e) $\dfrac{z}{(z-1)(z-3)(z-5)}$, $z_0 = 1, 3, 5$.

(f) $\dfrac{1-2z}{(z^2-z)^2}$, $0 < |z| < 1$; $|z| > 1$; $0 < |z-1| < 1$; $|z-1| > 1$;
$1 < |z+i| < \sqrt{2}$.

(g) $\dfrac{4z-z^2}{(z^2-4)(z+1)}$, $1 < |z| < 2$; $|z| > 2$; $0 < |z+1| < 1$.

(h) $\dfrac{1}{(z+2)^2(z^2-9)}$, $2 < |z| < 3$; $|z| > 3$; $0 < |z+2| < 1$.

(2) Do the same problems as Exercise (1).

(a) $e^{(1-z)^{-1}}$, $z_0 = 1, \infty$.

(b) $e^{z+z^{-1}}$, $0 < |z| < \infty$.

(c) $\sin z \sin \dfrac{1}{z}$, $0 < |z| < \infty$.

(d) $\sin \dfrac{z}{1-z}$, $z_0 = 1, \infty$ (in case $z_0 = \infty$, find the coefficients
up to z^{-3}).

(e) $\cot z$, $\pi < |z| < 2\pi$.

(f) $\operatorname{Log} \dfrac{z-a}{z-b}$, $z_0 = \infty$.

(g) $\dfrac{1}{z-2} \operatorname{Log} \dfrac{z-i}{z+i}$, $z_0 = \infty$ and $1 < |z| < 2$.

(h) $(b-a)\left[\operatorname{Log} \dfrac{z-a}{a-b}\right]^{-1}$, $\left|z - \dfrac{1}{2}(a+b)\right| > \dfrac{1}{2}|b-a|$.

(i) $\dfrac{e^z}{(z-i)^3(z+2)^2}$, $z_0 = i, -2$.

(j) $\csc z$, $z_0 = \pi$.

(k) $\sec z$, $z_0 = \dfrac{\pi}{2}$.

(l) $z^6 \cos^2 \dfrac{1}{z^2}$, $|z| > 0$.

(3) Show that $\cosh(z + z^{-1}) = a_0 + \sum_{n=1}^{\infty} a_n(z + z^{-1})^n, |z| > 0$, where

$$a_n = \frac{1}{2\pi} \int_0^{2\pi} \cos n\theta \cosh(2\cos\theta)d\theta, \quad n \geq 0.$$

State and prove a similar result for $\sinh(z + z^{-1}), |z| > 0$.

(4) Find the Laurent series expansion of $e^{z^{-1}}$ in $|z| > 0$ and show that

$$\frac{1}{\pi} \int_0^{\pi} e^{\cos\theta} \cos(\sin\theta - n\theta)d\theta = \frac{1}{n!}, \quad n \geq 1.$$

(5) Determine if each of the following functions $f(z)$ can be expressed by a Laurent series at the indicated point z_0. Give precise reasons.

(a) $f(z) = \cot z$, $z_0 = \infty$.

(b) $\tanh z^{-1}$, $z_0 = 0$.

(c) $z^2 \csc z^{-1}$, $z_0 = 0$.

(d) $\mathrm{Log}\,\dfrac{1}{z-1}$, $z_0 = \infty$.

(e) $z^\alpha, z_0 = 0$.

(6) Determine if each of the following multiple-valued functions has analytic branches in a neighborhood of the indicated point z_0 so that they can be expanded into the Laurent series at z_0.

(a) $f(z) = \sqrt{z(z-2)}$, $z_0 = 1$.

(b) $\sqrt[4]{z(z-1)^2}$, $z_0 = \infty$.

(c) $\sqrt{z + \sqrt{z^2 - 1}}$, $z_0 = \infty, 1$.

(d) $\log \dfrac{(z-\alpha)(z-\beta)}{(z-\gamma)(z-\delta)}$, $z_0 = \infty$,   where α, β, γ, and δ are distinct complex numbers.

(e) $\sin^{-1} z$, $z_0 = 0$.

(f) $\tan^{-1}(1 + z)$, $z_0 = 0$.

(g) $\sinh^{-1}(z + i)$, $z_0 = 0$.

(h) $\sqrt{\dfrac{\pi}{2} - \sin^{-1} z}$, $z_0 = 1$.

(i) $\sqrt{\dfrac{\pi}{4} - \sin^{-1} z}$, $z_0 = \dfrac{\sqrt{2}}{2}$.

Exercises B. Complex Fourier Series

Suppose f is analytic in a domain Ω except possible isolated singularities. If there is a *nonzero* complex number ω such that, for all $z \in \Omega$, $z + \omega$ is also in Ω and $f(z + \omega) = f(z)$ holds, then f is said to be a *periodic function* with *period* ω. For instance, e^z has period $2\pi i$; $\cos z$ and $\sin z$ have period 2π, while $\tan z$ and $\cot z$ have period π.

Let S be a *parallel strip* defined by $z = a + \omega t$ and $z = b + \omega t$, $t \in \mathbf{R}$, where $\operatorname{Re} a < \operatorname{Re} b$ and $\operatorname{Im} \frac{b-a}{\omega} < 0$. See Fig. 4.53, where $z \to \eta = \frac{z}{\omega} \to \zeta = e^{2\pi i \eta}$ maps S univalently onto a circular ring domain $0 \le r < |\zeta| < R \le \infty$. Let f be analytic in S with period ω. The inverse function $z = z(\zeta) = \frac{\omega}{2\pi i} \log \zeta$ of $\zeta = \zeta(z) = e^{\frac{2\pi i z}{\omega}}$ makes the composite function $F(\zeta) = f \circ z(\zeta)$ *single-valued* everywhere in $r < |z| < R$. According to (4.9.1.1), F has the Laurent series expansion

$$F(\zeta) = \sum_{-\infty}^{\infty} a_n \zeta^n, \quad r < |\zeta| < \mathbf{R}, \quad \text{where}$$

$$a_n = \frac{1}{2\pi i} \int_{|\zeta|=\rho} \frac{F(\zeta)}{\zeta^{n+1}} d\zeta, \quad r < \rho < R, \quad n = 0, \pm 1, \pm 2, \dots.$$

Substituting $\zeta = \zeta(z)$ into the above expression, we obtain the *complex Fourier series* expansion of f in S:

$$f(z) = \sum_{-\infty}^{\infty} a_n e^{\frac{2n\pi i z}{\omega}}, \quad z \in S, \quad \text{where}$$

$$a_n = \frac{1}{\omega} \int_{\sigma}^{\sigma+\omega} f(z) e^{-\frac{2n\pi i z}{\omega}} dz, \quad \sigma \in S \text{ (a fixed point)}, \quad n = 0, \pm 1, \pm 2, \dots.$$

$$(4.9.2.1)$$

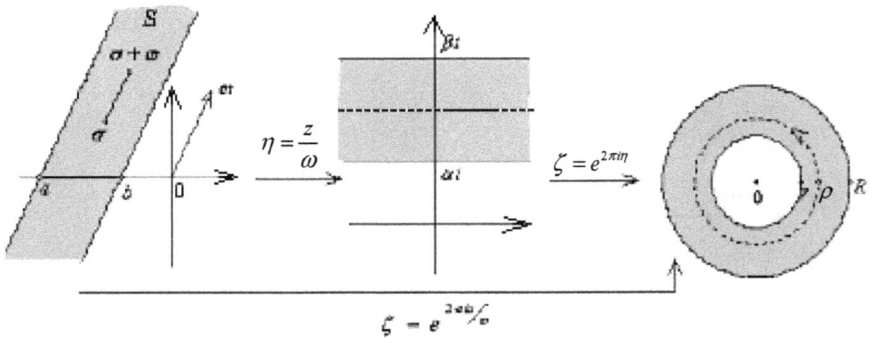

Fig. 4.53

The series *converges absolutely* in S and *uniformly on every compact subset* of S to $f(z)$. Rewriting $a_n e^{2n\pi iz} + a_{-n} e^{-2n\pi iz} = A_n \cos 2n\pi z + B_n \sin 2n\pi z, n \geq 0$, (4.9.2.1) can be put in the *Fourier series in trigono- metric form* as

$$f(z) = \frac{1}{2} A_0 + \sum_{n=1}^{\infty} \left(A_n \cos \frac{2n\pi z}{\omega} + B_n \sin \frac{2n\pi z}{\omega} \right), \quad z \in S, \quad \text{where}$$

$$A_0 = 2a_0 = \frac{2}{\omega} \int_{\sigma}^{\sigma+\omega} f(z)dz,$$

$$A_n = a_n + a_{-n} = \frac{2}{\omega} \int_{\sigma}^{\sigma+\omega} f(z) \cos \frac{2n\pi z}{\omega} dz,$$

$$B_n = i(a_n - a_{-n}) = \frac{2}{\omega} \int_{\sigma}^{\sigma+\omega} f(z) \sin \frac{2n\pi z}{\omega} dz, \quad n \geq 1. \qquad (4.9.2.2)$$

Do the following problems.

(1) Show that

$$\tan \pi z = \begin{cases} i\left(1 + 2\sum_{1}^{\infty}(-1)^n e^{2n\pi iz}\right), & \text{Im } z > 0, \\ -i\left(1 + 2\sum_{1}^{\infty}(-1)^n e^{-2n\pi iz}\right), & \text{Im } z < 0 \end{cases}.$$

(2) Find the Fourier series expansions for each of the following functions in the indicated domains.

    (a) $f(z) = \cot \pi z$, $\text{Im } z > 0$ and $\text{Im } z < 0$.

    (b) $f(z) = \sec 2\pi z$, $\text{Im } z > 0$ and $\text{Im } z < 0$.

    (c) $f(z) = \sec^2 \pi z$, $\text{Im } z > 0$ and $\text{Im } z < 0$.

    (d) $f(z) = \cos 2\pi z, \mathbf{C}$.

(3) Let f be an entire function with period 1. If there are constants $M > 0$ and $\alpha > 0$ so that $|f(z)| \leq Me^{\alpha|z|}$, show that there is an integer $N > 0$ so that

$$f(z) = \sum_{n=-N}^{N} a_n e^{2n\pi iz}.$$

(4) Suppose f is analytic in the strip $-\alpha < \text{Im } z < \alpha$ ($\alpha > 0$). Also suppose the series $F(z) = \sum_{n=-\infty}^{\infty} f(z+n)$ converges uniformly on every compact subset of $-\alpha < \text{Im } z < \alpha$ that contains the real segment

[0, 1]. Show that

$$F(z) = \sum_{n=-\infty}^{\infty} a_n e^{2n\pi i z}, \quad -\alpha < \operatorname{Im} z < \alpha, \quad \text{where}$$

$$a_n = \int_{-\infty}^{\infty} f(x) e^{-2n\pi i x} dx, \quad n = 0, \pm 1, \pm 2, \ldots.$$

Deduce the *Poisson summation formula*

$$\sum_{-\infty}^{\infty} f(n) = \sum_{-\infty}^{\infty} \int_{-\infty}^{\infty} f(x) e^{-2n\pi i x} dx.$$

4.10.   Classification and Characteristic Properties of Isolated Singularities of an Analytic Function

If f is analytic in $0 < |z-z_0| < R$, then z_0 is said to be an *isolated singularity* (of single-valued character) of f (refer to Remark (1) in Sec. 3.3.1).

In what follows, we always assume that z_0 is an isolated singularity of f. If there is an analytic function F in $|z - z_0| < R$ such that

$$F(z) = f(z), \quad 0 < |z - z_0| < R \tag{4.10.1}$$

holds, then z_0 is called a *removable singularity* of f and F *the analytic continuation* (or *extension*) of f in $|z - z_0| < R$. Note that such a F is unique if it exists. If we redefine the value of f at z_0 as $F(z_0)$, then f turns out to be analytic in $|z - z_0| < R$. We had introduced this concept in (3.4.2.17) and used it in many situations henceforth.

If

$$\lim_{z \to z_0} f(z) = \infty \quad \left(\text{or } \lim_{z \to \infty} |f(z)| = \infty\right), \tag{4.10.2}$$

then z_0 is called a *pole* of f. While, if z_0 is neither a removable singularity nor a pole of f, z_0 is called an *essential singularity* of f.

In case f is analytic in $0 \le r < |z| < \infty$, then ∞ is called an *isolated singularity* of f. ∞ is said to be a *removable singularity*, a *pole* or an *essential singularity*, respectively, according to 0 is the same type of singularity of the function

$$g(w) = f\left(\frac{1}{w}\right), \quad 0 < |w| < \frac{1}{r}. \tag{4.10.3}$$

Take, for instance, the function $z + \frac{1}{z^2}$ has a pole at $= 0$ and the function $e^{\frac{1}{z}}$ has an essential singularity at $z = 0$.

Basic properties for these three types of isolated singularities are developed in Secs. 4.10.1, 4.10.2, and 4.10.3, respectively.

4.10.1.   *Removable singularity*

We repeat (3.4.2.17) and enlarge its content to the following

Characteristic properties of a removable singularity. Let $z_0 \in \mathbf{C}$ be an isolated singularity of f. Then the following are equivalent.

(1) z_0 is a removable singularity of f.
(2) $\lim_{z \to z_0} f(z)$ exists as a finite complex number (and hence, f can be defined at z_0 so that it is continuous at z_0).
(3) There are constants $M > 0$ and $R > 0$ so that $|f(z)| \leq M$ in $0 < |z - z_0| < R$.
(4) $\lim_{z \to z_0} (z - z_0) f(z) = 0$.
(5) The Laurent series expansion of f at z_0 has only analytic (or regular) part, i.e.,

$$f(z) = \sum_0^\infty a_n (z - z_0)^n, \quad 0 < |z - z_0| < R. \qquad (4.10.1.1)$$

Proof. (1) $\Rightarrow$ (2) $\Rightarrow$ (3) $\Rightarrow$ (4) are trivial and refer to the proof of (3.4.2.17).

(4) $\Rightarrow$ (5). Fix any $\varepsilon > 0$. There is a $\rho_0 > 0$ so that $|(z - z_0) f(z)| < \varepsilon$ in $0 < |z - z_0| < \rho_0$. Choose arbitrarily fixed $0 < \rho < \rho_0$. Then, according to the coefficient integrals in the Laurent series expansion (4.9.1.1), if $n \leq -1$,

$$|a_n| \leq \frac{1}{2\pi} \int_{|z - z_0| = \rho} \frac{\varepsilon}{\rho^{n+2}} |dz| = \frac{\varepsilon}{\rho^{n+1}} = \varepsilon \rho^{-n-1} \to 0 \quad \text{as } \varepsilon \to 0.$$

Hence, $a_n = 0$ for $n \leq -1$.
  (5) $\Rightarrow$ (1). The series $\sum_0^\infty a_n (z - z_0)^n$ has radius of convergence $\sigma \geq R$ and represents an analytic function $F(z)$ in $|z - z_0| < \sigma$. Obviously $F(z) = f(z)$ in $0 < |z - z_0| < R \leq \sigma$. $\square$

Using (4.10.3), (4.10.1.1) can be transformed to the

Characteristic properties for ∞ being a removable singularity. Let ∞ be an isolated singularity of f. Then the following are equivalent.

(1) ∞ is a removable singularity of f.
(2) $\lim_{z \to \infty} f(z)$ exists as a finite complex number.
(3) $f(z)$ is bounded in a neighborhood of ∞.
(4) $\lim_{z \to \infty} \dfrac{f(z)}{z} = 0$.

(5) The Laurent series expansion of f at ∞ has only regular part, namely,

$$f(z) = \sum_0^\infty \frac{a_{-n}}{z^n}, \quad r < |z| < \infty. \qquad (4.10.1.2)$$

The details are left to the readers.

Remark (The value assumed at a removable singularity). According to (2) in (4.10.1.1), the following could happen:

(1) f is not defined at z_0;
(2) f is defined at z_0 yet $f(z_0) \neq \lim_{z \to z_0} f(z)$; and
(3) f is defined at z_0 and $f(z_0) = \lim_{z \to z_0} f(z)$ holds.

As a removable singularity, this means that we could redefine the value of f at z_0 in cases 1 and 2 so that case 3 holds and this will turn f to be analytic at z_0.

In general setting, let f be analytic in a domain Ω and z_0 be a fixed point in Ω. Define

$$F(z) = \begin{cases} \dfrac{f(z) - f(z_0)}{z - z_0}, & z \in \Omega - \{z_0\} \\ f'(z_0), & z = z_0 \end{cases}.$$

Then F will have a removable singularity at z_0 and thus is still analytic everywhere in Ω. This technique had been used in quite a few places; in particular, in the proof of the Schwarz's lemma (see (3.4.5.1)). For instance $\frac{z^2-1}{z-1}$ $(z \neq 1)$, $\frac{z}{e^z-1}$ $(z \neq 0)$, $\frac{\sin z}{z}$ $(z \neq 0)$, and $\frac{2(1-\cos z)}{z^2}$ $(z \neq 0)$ have, respectively, a removable singularity at $z = 1, 0, 0,$ and 0.

If both f and g have a zero of order n at z_0, let $f(z) = (z - z_0)^n \varphi(z)$ and $g(z) = (z - z_0)^n \Psi(z)$, where φ and Ψ are analytic at z_0 and $\varphi(z_0) \neq 0$, $\Psi(z_0) \neq 0$. Then $\frac{f}{g}$ has a removable singularity at z_0 and

$$\lim_{z \to z_0} \frac{f(z)}{g(z)} = \frac{\varphi(z_0)}{\Psi(z_0)}. \qquad \square$$

We give some examples.

Example 1. Find the removable singularity z_0 for each of the following functions f and its Laurent series expansion at z_0.

(1) $f(z) = \dfrac{1}{(z-1)(z-2)}$.

(2) $\log \dfrac{z-a}{z-b}$ $(a \neq b)$.

(3) $e^{\frac{z}{(z+2)}}$.

(4) $\dfrac{\tan(z-1)}{z-1}$.

Solution

(1) $z_0 = \infty$ is a removable singularity and $f(\infty) = 0$. In case $|z| > 2$,

$$f(z) = \frac{1}{z-2} - \frac{1}{z-1} = \sum_0^\infty \frac{2^n - 1}{z^{n+1}}.$$

(2) f has infinitely many branches in $\mathbf{C}^* - [a, b]$, each of which having ∞ as a removable singularity. Observe that $f(\infty) = \log 1 = 2k\pi i$, $k = 0, \pm 1, \pm 2, \dots$. Let $f_k(z) = \operatorname{Log} \frac{z-a}{z-b} + 2k\pi i$, $z \in \mathbf{C}^* - [a, b]$ and $k = 0, \pm 1, \pm 2, \dots$. The branch, for fixed k,

$$f_k(z) = 2k\pi i + \sum_{n=1}^\infty \frac{b^n - a^n}{n} \cdot \frac{1}{z^n}, \quad |z| > \max\{|a|, |b|\}.$$

(3) Let $w = \frac{1}{z} \cdot g(w) = f(\frac{1}{w}) = e^{(1+2w)^{-1}}$ is analytic at $w = 0$. So $z_0 = \infty$ is a removable singularity of f. Direct computation shows that $g'(0) = -2e$, $g''(0) = 12e$. Hence $g(w) = e(1 - 2w + 6w^2 + \cdots)$, $|w| < \frac{1}{2}$ implies that

$$f(z) = e\left(1 - \frac{2}{z} + \frac{6}{z^2} + \cdots\right), \quad 2 < |z| \le \infty.$$

(4) Since

$$\lim_{z \to 1} \frac{\tan(z-1)}{z-1} = \lim_{z \to 1} \frac{\sin(z-1)}{z-1} \cdot \frac{1}{\cos(z-1)} = 1,$$

$z_0 = 1$ is a removable singularity of f. By Exercise A (5)(b) of Sec. 4.8, it follows that

$$\frac{\tan(z-1)}{(z-1)} = \sum_{n=1}^\infty (-1)^{n-1} \frac{2^{2n}(2^{2n} - 1)B_{2n}}{(2n)!}(z-1)^{2n-2}, \quad |z-1| < \frac{\pi}{2}.$$

Example 2. Suppose f is analytic in $0 < |z - z_0| < R$ and there is a constant $M > 0$ so that either $\operatorname{Re} f(z) < M$ or $\operatorname{Re} f(z) > M$ always holds there. Show that z_0 is a removable singularity of f.

Proof. Suppose $\operatorname{Re} f(z) < M$ holds everywhere in $0 < |z - z_0| < R$. Via a bilinear transformation $\zeta = T(w)$ mapping $\operatorname{Re} w < M$ onto the unit disk $|\zeta| < 1$, the composite function $T \circ f$ is analytic in $0 < |z - z_0| < R$ and bounded there, so z_0 is a removable singularity of $T \circ f$. Since $\zeta = T(w)$ is univalent and analytic in $\operatorname{Re} w < M$, z_0 is a removable singularity of $f(z) = T^{-1} \circ T \circ f(z)$.

For another proof, consider the function $g(z) = e^{f(z)}$. Now $g(z)$ is analytic in $0 < |z - z_0| < R$ and $|g(z)| = e^{\operatorname{Re} f(z)} \leq e^M$ there. This shows that z_0 is a removable singularity of $g(z)$. Define $g(z_0) = \lim_{z \to z_0} g(z)$. Then $g(z)$ is analytic $|z - z_0| < R$ and $g(z) \neq 0$ there. Consequently, $\log g(z)$ has branches in $|z - z_0| < R$ (see (3.3.1.8)). Fix z_1, $0 < |z_1 - z_0| < R$. Choose $F(z)$ the branch of $\log g(z)$ determined by $\log g(z_1) = f(z_1)$. Then F is analytic in $|z - z_0| < R$ and $F(z) = f(z)$ in $0 < |z - z_0| < R$. This shows that z_0 is a removable singularity of f.

Example 3. Let ∞ be an isolated singularity of f. Show that ∞ is a removable singularity of f if and only if

$$\lim_{z \to \infty} f'(z) = 0.$$

Note that $f'(\infty)$ is not necessarily equal to zero (see Remark in Sec. 3.3).

Proof. *The necessity:* According to (4.10.3), $g(z) = f(\frac{1}{z})$ is analytic at 0. Hence

$$g'(z) = f'\left(\frac{1}{z}\right) \cdot \frac{-1}{z^2}, \quad |z| < R, \quad \text{and } g'(0) \text{ exists.}$$

$$\Rightarrow f'(z) = -\frac{1}{z^2} g'\left(\frac{1}{z}\right), \quad \frac{1}{R} < |z| < \infty.$$

$$\Rightarrow \lim_{z \to \infty} f'(z) = \lim_{z \to \infty} \frac{-1}{z^2} g'\left(\frac{1}{z}\right) = 0.$$

Or, by (5) in (4.10.1.2), $f(z) = \sum_{n=0}^{\infty} \frac{a_{-n}}{z^n}$, $\frac{1}{R} < |z| < \infty$ implies that (see (3.3.2.4)) $f'(z) = \sum_{n=1}^{\infty} \frac{-n a_{-n}}{z^{n+1}}$, $\frac{1}{R} < |z| < \infty$. This implies that $\lim_{z \to \infty} f'(z) = 0$.

The sufficiency: Since f' is analytic in $r < |z| < \infty$,

$$f'(z) = \sum_{1}^{\infty} b_n z^n + \sum_{n=0}^{\infty} \frac{b_{-n}}{z^n}, \quad r < |z| < \infty.$$

In case there is an $m \geq 1$ so that $b_n = 0$ for $n > m$ and $b_m \neq 0$, then $\lim_{z \to \infty} f'(z) = \infty$ and ∞ is a pole of f'; in case there are infinitely many $b_n \neq 0$ $(n > 0)$, then $\lim_{z \to \infty} f'(z)$ does not exist and ∞ is an essential singularity of f' (see Sec. 4.10.3). According to the assumption, $b_n = 0$ should hold for $n \geq 0$. Thus,

$$f'(z) = \sum_{1}^{\infty} \frac{b_{-n}}{z^n}, \quad r < |z| < \infty,$$

and ∞ is a removable singularity of f'. Any primitive function of $f'(z)$ in $r < |z| < \infty$ differs from the *single-valued* analytic function $f(z)$ by a constant (see (4.1.2)). It is thus necessary that $b_{-1} = 0$. Consequently, up to a constant,

$$f(z) = \sum_0^\infty \frac{a_{-n}}{z^n}, \quad r < |z| < \infty,$$

where $a_{-n} = -\frac{1}{n}b_{-n-1}$, $n \geq 1$, and a_0 is a constant. This shows that ∞ is a removable singularity of f.   □

Example 4. Suppose f is analytic in $0 < |z| \leq R$. For $0 < p \leq 2$, let

$$I(p) = \lim_{\varepsilon \to 0} \iint_{\varepsilon \leq |z| \leq R} |f(z)|^p dx dy.$$

Show that

(1) 0 is a removable singularity of $f \Leftrightarrow I(2) < \infty$.
(2) If 0 is a pole of order m of f (see Sec. 4.10.2), then $I(p) < \infty$ if $0 < p < \frac{2}{m}$ and $I(p) = \infty$ if $\frac{2}{m} \leq p \leq 2$.
(3) If $I(p) = \infty$ for all $0 < p \leq 2$, then 0 is an essential singularity of f (see Sec. 4.10.3).

Proof. Let $f(z) = \sum_{-\infty}^\infty a_n z^n$, $0 < |z| \leq R$. For each $0 < \varepsilon < R$, we have (see Exercise A (4) of Sec. 4.9.1)

$$\iint_{\varepsilon \leq |z| \leq R} |f(z)|^2 dx dy = 2\pi \sum_{-\infty}^\infty |a_n|^2 \int_\varepsilon^R r^{2n+1} dr. \qquad (*)$$

(1) If 0 is a removable singularity of f, then

$$I(2) = \lim_{\varepsilon \to 0} \iint_{\varepsilon \leq |z| \leq R} |f(z)|^2 dx dy \leq \pi R^2 \cdot \max_{|z| \leq R} |f(z)|^2 < \infty.$$

Conversely, if $I(2) < \infty$ holds and $a_{-1} \neq 0$, then by $(*)$

$$\iint_{\varepsilon \leq |z| \leq R} |f(z)|^2 dx dy \geq 2\pi |a_{-1}|^2 \int_\varepsilon^R r^{-1} dr$$

$$= 2\pi |a_{-1}|^2 \log \frac{R}{\varepsilon} \to \infty \quad \text{as } \varepsilon \to 0$$

contradicting to $I(2) < \infty$; while if $a_n \neq 0$ for some $n \leq -2$, then

$$\iint_{\varepsilon \leq |z| \leq R} |f(z)|^2 dx dy \geq 2\pi |a_n|^2 \int_\varepsilon^R r^{2n+1} dr$$

$$= 2\pi |a_n|^2 \frac{R^{2n+2} - \varepsilon^{2n+2}}{2n+2} \to \infty \quad \text{as } \varepsilon \to 0$$

a contradiction, too. Thus $a_n = 0$ for all $n \leq -1$ and $f(z) = \sum_0^\infty a_n z^n$ shows that 0 is a removable singularity of f.

(2) There is a ρ, $0 < \rho \leq R$, so that $f(z) = \frac{\varphi(z)}{z^m}$, where $\varphi(z)$ is analytic and nonzero in $|z| \leq \rho$. Let $m' = \min_{|z|\leq\rho} |\varphi(z)|$ and $M = \max_{|z|\leq\rho} |\varphi(z)|$. Then $M \geq m' > 0$. In case $0 < \varepsilon < \rho$,

$$\iint_{\varepsilon \leq |z| \leq \rho} |f(z)|^p dz = \iint_{\varepsilon \leq |z| \leq \rho} \frac{|\varphi(z)|^p}{|z|^{mp}} dxdy$$

$$\leq 2\pi M^p \int_\varepsilon^\rho \frac{1}{r^{mp-1}} dr \quad \text{and} \quad \geq 2\pi m'^p \int_\varepsilon^\rho \frac{1}{r^{mp-1}} dr.$$

Since $\int_0^\rho \frac{dr}{r^{mp-1}}$ converges if $mp - 1 < 1$ and diverges if $mp - 1 \geq 1$, the result follows.

(3) is an immediate consequence of (1) and (2).   □

Exercises A

(1) Prove (4.10.1.2) in details.
(2) Suppose f is analytic in $0 < |z| < \infty$ and there is a constant $M > 0$ such that $|f(z)| \leq M|z||\log z|$ there. Show that $f(z) = 0$ in $0 < |z| < \infty$.
(3) Suppose f and g are analytic in a domain Ω except isolated singularities. Let $E_f = \{z \in \Omega \mid z$ is a singularity of f but is not removable$\}$ and E_g, similarly. In case, there is a nonempty open subset A of $\Omega - (E_f \cup E_g)$ on which $f(z) = g(z)$ holds and A has a limit point in Ω, then $E_f = E_g$ and $f(z) = g(z)$ for all z in $\Omega - E_f$.
(4) Let f and g have a removable singularity at z_0. Show that $f(z) + g(z)$ and $f(z)g(z)$ have a removable singularity at z_0, too. What happens to $\frac{f(z)}{g(z)}$?
(5) Suppose f and g are analytic in a domain Ω. If the condition $f(z_1) = f(z_2)$ always implies that $g(z_1) = g(z_2)$ holds, too, then there is analytic function h in $f(\Omega)$ so that $g(z) = h(f(z))$, $z \in \Omega$.

Exercises B

(1) Let f be analytic in $0 < |z - z_0| < R$ and $M(r) = \max_{|z-z_0|=r} |f(z)|$, $0 < r < R$.

(a) In case z_0 is not removable, show that $\lim_{r\to 0} rM(r)$ exists as a positive number (could be ∞).
(b) Use (a) to prove (4) $\Rightarrow$ (1) in (4.10.1.1) in the following sense: If there is a decreasing sequence of r_n $(0 < r_n < R)$ satisfying $\lim_{n\to\infty} r_n = 0$ and $\lim_{n\to\infty} r_n M(r_n) = 0$, then z_0 is a removable singularity of f.

(2) Suppose f is analytic in $0 \leq r < |z| < \infty$. Let $M(\rho) = \max_{|z|=\rho} |f(z)|$, $r < \rho < \infty$ and $a^+ = \max\{a, 0\}$. Show that the limit

$$\lim_{\rho \to \infty} \frac{\log M(\rho)^+}{\log \rho} = \alpha$$

exists in the range $[0, \infty]$, and

(i) $\alpha = 0 \Leftrightarrow \infty$ is a removable singularity of f;
(ii) $\alpha = m$ (a positive integer) $\Leftrightarrow \infty$ is a pole of order m (see Sec. 4.10.2); and
(iii) $\alpha = \infty \Leftrightarrow \infty$ is an essential singularity of f (see Sec. 4.10.3).

4.10.2. Pole

Our main result is the following

Characteristic properties of a pole. Let $z_0 \in \mathbf{C}$ be an isolated singularity of an analytic function f. Then the following are equivalent.

(1) z_0 is a *pole of order n* of f, namely, $\lim_{z \to z_0} f(z) = \infty$ and $\lim_{z \to z_0} (z - z_0)^n f(z)$ exists as a *nonzero* finite number while $\lim_{z \to z_0} (z - z_0)^k f(z) = 0$ for $k \geq n + 1$, where $n \geq 1$ is an integer.
(2) z_0 is a removable singularity of $\frac{1}{f(z)}$ so that z_0 is a zero of order n of $\frac{1}{f(z)}$.
(3) There are a $R > 0$ and an analytic function $\varphi(z)$ in $|z - z_0| < R$, $\varphi(z_0) \neq 0$, such that

$$f(z) = \frac{\varphi(z)}{(z - z_0)^n}, \quad 0 < |z - z_0| < R.$$

(4) There are $R > 0$ and positive constants m and M so that

$$\frac{m}{|z - z_0|^n} \leq |f(z)| \leq \frac{M}{|z - z_0|^n}, \quad 0 < |z - z_0| < R.$$

(5) There is a polynomial $p(z)$ of degree n, with $p(0) = 0$, such that $f(z) - p\left(\frac{1}{z - z_0}\right)$ has a removable singularity at z_0.
(6) If $\sum_{-\infty}^{\infty} a_k(z - z_0)^k$ is the Laurent series expansion of f at z_0, then $a_{-n} \neq 0$ but $a_{-k} = 0$ for all $k > n$, namely,

$$f(z) = \frac{a_{-n}}{(z - z_0)^n} + \cdots + \frac{a_{-1}}{z - z_0} + \sum_{k=0}^{\infty} a_k(z - z_0)^k, \quad 0 < |z - z_0| < R.$$

The polynomial $p(z)$ in (5) is the one in (6), namely,

$$p(z) = a_{-n}z^n + a_{-n+1}z^{n-1} + \cdots + a_{-1}z, \quad a_{-n} \neq 0,$$

and is called the *singular* or *principal part* of f at its pole z_0 of order n. In case $n = 1$, a pole of order 1 is simply called a *simple pole* and if $n = 2$, a *double pole*.   (4.10.2.1)

The **proof** adopts the following ordering:

$$(1) \Rightarrow (2) \Rightarrow (3) \Rightarrow (4) \searrow$$
$$\Downarrow \qquad\qquad\qquad (1)$$
$$(5) \Rightarrow (6) \nearrow$$

(1) $\Rightarrow$ (2): By assumption, $\lim_{z \to z_0} \frac{1}{f(z)} = 0$. So z_0 is a removable singularity of $g(z) = \frac{1}{f(z)}$ and we may redefine $g(z_0) = 0$ so that z_0 is a zero of $g(z)$. Since $f(z)$ is not a constant, so is $g(z)$ and thus z_0 is an isolated zero of $g(z)$, say $g(z) = (z - z_0)^m \Psi(z)$, where $\Psi(z)$ is analytic at z_0 and $\Psi(z_0) \neq 0$. The condition that $\lim_{z \to z_0}(z - z_0)^n f(z) = \lim_{z \to z_0} \frac{(z-z_0)^{n-m}}{\Psi(z)}$ exists as a nonzero finite number suggests that $m = n$ should hold.

(2) $\Rightarrow$ (3): There is a $R > 0$ and there is an analytic function $g(z)$ on $|z - z_0| < R$ such that $g(z) \neq 0$ there and $\frac{1}{f(z)} = (z - z_0)^n g(z)$ on $z - z_0 < R$. Therefore,

$$f(z) = \frac{\varphi(z)}{(z - z_0)^n}, \quad 0 < |z - z_0| < R$$

where $\varphi(z) = \frac{1}{g(z)}$ is analytic in $|z - z_0| < R$ and $\varphi(z_0) = \frac{1}{g(z_0)} \neq 0$.

(3) $\Rightarrow$ (4): Take $R > 0$ smaller if necessarily so that $\varphi(z) \neq 0$ on $|z - z_0| \leq R$. Let $m = \min_{|z-z_0| \leq R} |\varphi(z)|$ and $M = \max_{|z-z_0| \leq R} |\varphi(z)|$. And the result follows.

(4) $\Rightarrow$ (1) and (6) $\Rightarrow$ (1): These two cases are obvious.

(3) $\Rightarrow$ (5): Let $\varphi(z) = \sum_{k=0}^{\infty} b_k (z - z_0)^k$, $|z - z_0| < R$, where $b_0 = \varphi(z_0) \neq 0$. Then, in $0 < |z - z_0| < R$,

$$f(z) = \frac{\varphi(z)}{(z - z_0)^n}$$

$$= \frac{b_0}{(z - z_0)^n} + \frac{b_1}{(z - z_0)^{n-1}} + \cdots + \frac{b_{n-1}}{z - z_0} + \sum_{k=0}^{\infty} b_{n+k}(z - z_0)^k.$$

Choose the polynomial $p(z) = b_0 z^n + b_1 z^{n-1} + \cdots + b_1 z, b_0 \neq 0$. Then

$$f(z) - p\left(\frac{1}{z - z_0}\right) = \sum_{k=0}^{\infty} b_{n+k}(z - z_0)^k, \quad 0 < |z - z_0| < R$$

has z_0 as a removable singularity.

(5) $\Rightarrow$ (6): By assumption, there is a $R > 0$ such that $f(z) - p\left(\frac{1}{z-z_0}\right) = \sum_{k=0}^{\infty} a_k(z - z_0)^k$ in $0 < |z - z_0| < R$. Let $p(z) = a_{-n}z^n + a_{-n+1}z^{n-1} + \cdots + a_{-1}z$ (the constant term can be absorbed into a_0), where $a_{-n} \neq 0$. Therefore,

$$f(z) = p\left(\frac{1}{z-z_0}\right) - \sum_{k=0}^{\infty} a_k(z - z_0)^k$$

$$= \frac{a_{-n}}{(z-z_0)^n} + \cdots + \frac{a_{-1}}{z-z_0} + \sum_{k=0}^{\infty} a_k(z - z_0)^k, \quad 0 < |z - z_0| < R.$$

By the uniqueness of the Laurent series expansion, the above expression is just the Laurent series expansion of f at z_0.

Via (4.10.3), (4.10.2.1) can be transformed to the

Characteristic properties for ∞ being a pole. Let ∞ be an isolated singularity of f. Then the following are equivalent.

(1) ∞ is a pole of order n of f, namely, $\lim_{z\to\infty} f(z) = \infty$, $\lim_{z\to\infty} \frac{f(z)}{z^n} \neq 0$, but $\lim_{z\to\infty} \frac{f(z)}{z^{n+1}} = 0$, where $n \geq 1$ is an integer.
(2) ∞ is a removable singularity of $\frac{1}{f(z)}$ and, indeed, is a zero of order n of $\frac{1}{f(z)}$.
(3) There are a $r > 0$ and an analytic function $\varphi(z)$ in $r < |z| \leq \infty$, $\varphi(\infty) \neq 0$, and $f(z) = z^n\varphi(z)$, $r < |z| < \infty$.
(4) There are a $r > 0$ and positive constants m and M so that

$$m|z|^n \leq |f(z)| \leq M|z|^n, \quad r < |z| < \infty.$$

(5) There is a polynomial $p(z)$ of degree n, $p(0) = 0$, such that ∞ is a removable singularity of $f(z) - p(z)$.
(6) If $\sum_{-\infty}^{\infty} a_k z^k$ is the Laurent series expansion of f at ∞, then $a_n \neq 0$ but $a_k = 0$ for all $k > n$, namely,

$$f(z) = a_n z^n + a_{n-1}z^{n-1} + \cdots + a_1 z + \sum_{k=0}^{\infty} \frac{a_{-k}}{z^k}, \quad r < |z| < \infty.$$

The polynomial in (5) is just the $p(z) = a_n z^n + \cdots + a_1 z$ in (6), called the *singular* or *essential* part of f at ∞. In case $n = 1$, ∞ is called a *simple pole*.

$$(4.10.2.2)$$

The details are left to the readers.

Remark (The behavior of $f(z)$ as $z \to z_0$ in terms of $(z - z_0)^\alpha$ if z_0 is an isolated singularity of f).   If there exists a real number α such that $\lim_{z \to z_0} |z - z_0|^\alpha |f(z)|$ exists as a *nonzero* finite real number, then

$$\lim_{z \to z_0} |f(z)| = \lim_{z \to z_0} \frac{|z - z_0|^\alpha |f(z)|}{|z - z_0|^\alpha} = \infty. \tag{4.10.2.3}$$

In this case, f is said to *approach* ∞ in a neighborhood of z_0 to the αth power of $(z - z_0)^{-1}$.

(4.10.1.1) and (4.10.2.1) tell us the fact that, at a removable singularity or a pole, $f(z)$ cannot approach ∞ to *any* positive *fractional* power of $(z - z_0)^{-1}$. Conversely, suppose there is a positive fraction α (not an integer) such that $\lim_{z \to z_0} |(z - z_0)|^\alpha |f(z)| \neq 0$. In case $0 < \alpha < 1$, then

$$\lim_{z \to z_0} |z - z_0| \, |f(z)| = \lim_{z \to z_0} |z - z_0|^{1-\alpha} \cdot \lim_{z \to z_0} |z - z_0|^\alpha |f(z)| = 0$$

and thus, z_0 is removable; in case $\alpha > 1$, letting $[\alpha]$ the largest integer not larger than α, then

$$\lim_{z \to z_0} |z - z_0|^{[\alpha]+1} |f(z)| = \lim_{z \to z_0} |z - z_0|^{[\alpha]+1-\alpha} \cdot \lim_{z \to z_0} |z - z_0|^\alpha |f(z)| = 0$$

and thus, z_0 is pole of f of order $\leq [\alpha]$.

As a matter of fact, only the following three cases could happen:

Case 1. For all real α,

$$\lim_{z \to z_0} |z - z_0|^\alpha |f(z)| = 0 \tag{4.10.2.4}$$

always holds. Then $f(z) \equiv 0$. To see this, apply Gutzmer's identity (see Exercise A (4) of Sec. 4.9.1) to $f(z) = \sum_{-\infty}^{\infty} a_n (z - z_0)^n$, then, for $0 < \rho < R$ and $M(\rho) = \max_{|z - z_0| = \rho} |f(z)|$,

$$\sum_{-\infty}^{\infty} |a_n|^{2n} \rho^{2n} = \frac{1}{2\pi} \int_0^{2\pi} |f(z_0 + \rho e^{i\theta})|^2 d\theta \leq M(\rho)^2$$

$$\Rightarrow M(\rho)^2 \geq |a_n|^{2n} \rho^{2n}, \quad n = 0, \pm 1, \pm 2, \ldots$$

$$\Rightarrow \rho^{-n} M(\rho) = |z - z_0|^{-n} M(\rho) \geq |a_n|^n, \quad n = 0, \pm 1, \pm 2, \ldots$$

$$\Rightarrow \text{(by assumption)} \quad a_n = 0, \quad n = 0, \pm 1, \pm 2, \ldots.$$

It follows that $f(z) \equiv 0$.

Case 2. There is an integer h such that

$$\lim_{z \to z_0} |z - z_0|^\alpha |f(z)| = 0, \text{ if } \alpha > h, \quad \text{and}$$

$$\lim_{z \to z_0} |z - z_0|^\alpha |f(z)| = \infty, \text{ if } \alpha < h, \qquad (4.10.2.5)$$

hold. In this case, call h the *algebraic order* of f at z_0: if $h > 0$, z_0 is a pole of order h; if $h < 0$, z_0 is a zero of order-h.

Case 3. For all real α, both

$$\lim_{z \to z_0} |z - z_0|^\alpha |f(z)| = 0, \quad \text{and} \quad \lim_{z \to z_0} |z - z_0|^\alpha |f(z)| = \infty \qquad (4.10.2.6)$$

do not hold. In Sec. 4.10.3, we will show that z_0 is an essential singularity in this case.

In conclusion, we obtain the fact: In $0 < |z - z_0| < R$, there is *not* any single-valued nonzero analytic function that approaches 0 or ∞, as $z \to z_0$, in the manner as $|z - z_0|^\alpha$, where α is a fraction which is not an integer.

□

We list a series of examples or applications.

Example 1. Find the singularities of each of the following functions $f(z)$ and the Laurent series expansions at isolated ones. If it is a pole, also determine its order.

(1) $f(z) = \dfrac{z^5}{(1 - z)^2}$.

(2) $e^{z - z^{-1}}$.

(3) $\dfrac{\cot z}{z^2}$.

(4) $\sin\left(\dfrac{1}{\cos\dfrac{1}{z}}\right)$.

(5) $\dfrac{1}{(1 - \sqrt{2 - 2\cos z})^2}$.

(6) $\dfrac{1}{e^z - 1} - \dfrac{1}{z - 2\pi i}$.

(7) $\dfrac{2\cos(\pi z/2a)}{(z - a)(z^2 + b^2)^7 \sin^5 z}$ ($a, b \in \mathbf{R}$ and $a \neq b \neq 0$).

(8) $\dfrac{(1 - z^n)e^z}{z^m}$ (m, n integers and $m > n > 0$).

Solution.

(1) Since $\lim_{z \to 1}(z - 1)^2 f(z) = 1$, $z = 1$ is a pole of order 2; while $\lim_{z \to \infty} \frac{f(z)}{z^3} = 1$, $z = \infty$ is a pole of order 3.

The binomial expansion shows that

$$z^5 = (z - 1 + 1)^5 = (z - 1)^5 + 5(z - 1)^4 + 10(z - 1)^3 + 10(z - 1)^2$$
$$+ 5(z - 1) + 1$$

The Laurent series expansion of f at 1 is thus

$$f(z) = \frac{1}{(z - 1)^2} + \frac{5}{z - 1} + 10 + 10(z - 1)$$
$$+ 5(z - 1)^2 + (z - 1)^3, \quad 0 < |z - 1| < \infty.$$

On the other hand, $\frac{1}{1-z} = -\sum_0^\infty \frac{1}{z^{n+1}}$, $|z| > 1$, implies that $\frac{1}{(1-z)^2} = \sum_0^\infty \frac{n+1}{z^{n+2}}$, $|z| > 1$. Therefore, the Laurent series expansion of f at ∞ is

$$f(z) = z^3 + 2z^2 + 3z + \sum_0^\infty \frac{n+4}{z^n}, \quad |z| > 1.$$

(2) It is known that e^z has an essential singularity at ∞, while $e^{-z^{-1}}$ has a removable one at ∞. Thus $f(z)$ has an essential singularity at ∞ and at $z = 0$, too, by the same reason. One can use (4.10.2.6) to test that both 0 and ∞ are essential singularities.

Applying the Cauchy product to $e^z = \sum_0^\infty \frac{1}{n!} z^n$, $|z| < \infty$, and $e^{-z^{-1}} = \sum_0^\infty \frac{(-1)^n}{n!} z^{-n}$, $|z| > 0$, we obtain the Laurent series expansion (see Example 3 of Sec. 4.9.2), at $z = 0$ and $z = \infty$,

$$e^{z - z^{-1}} = \sum_{-\infty}^\infty J_n(2) z^n, \quad 0 < |z| < \infty, \quad \text{where}$$

$$J_n(2) = (-1)^n J_{-n}(2) = \sum_{k=0}^\infty \frac{(-1)^k}{k!(n - k)}, \quad n = 0, 1, 2, \ldots.$$

(3) Observe that $f(z) = \frac{\cos z}{z^2} \cdot \frac{1}{\sin z}$, and

$$\lim_{z \to 0} z^3 f(z) = 1 \cdot 1 = 1 \quad \text{and} \quad \lim_{z \to n\pi} (z - n\pi) f(z) = \frac{1}{n^2 \pi^2}$$
$$\text{for } n = \pm 1, \pm 2, \ldots.$$

It follows that $z = 0$ is a pole of order 3; $n\pi$ ($n = \pm 1, \pm 2, \ldots$) are simple poles and ∞, as a limit point of these simple poles, is not an isolated singularity.

By Exercise A (5)(a) of Sec. 4.8, we have

$$\cot z = \frac{1}{z} + \sum_{1}^{\infty} (-1)^n \frac{2^{2n} B_{2n}}{(2n)!} z^{2n-1}, \quad |z| < \pi \qquad (*_1)$$

$$\Rightarrow f(z) = \frac{\cot z}{z^2} = \frac{1}{z^3} - 2B_2 \cdot \frac{1}{z}$$

$$+ \sum_{2}^{\infty} (-1)^n \frac{2^{2n} B_{2n}}{(2n)!} z^{2n-3}, \quad 0 < |z| < \pi.$$

This is the Laurent series expansion of f at $z = 0$. For fixed k, $k = \pm 1, \pm 2, \ldots$, by $(*_1)$ we obtain

$$\cot z = \frac{1}{z - k\pi} + \sum_{n=1}^{\infty} (-1)^n \frac{2^{2n} B_{2n}}{(2n)!} (z - k\pi)^{2n-1}, \quad |z - k\pi| < \pi. \ (*_2)$$

On the other hand, by geometric series expansion,

$$\frac{1}{z} = \frac{1}{k\pi(1 + \frac{z - k\pi}{k\pi})} = \frac{1}{k\pi} \sum_{n=0}^{\infty} \frac{(-1)^n}{(k\pi)^n} (z - k\pi)^n, \quad |z - k\pi| < \pi (\leq |k|\pi)$$

$$\Rightarrow \frac{1}{z^2} = \sum_{n=1}^{\infty} \frac{(-1)^n n}{(k\pi)^{n+1}} (z - k\pi)^{n-1}, \quad |z - k\pi| < \pi. \qquad (*_3)$$

Applying the Cauchy product to $(*_2)$ and $(*_3)$, the Laurent series expansion of f at $z = k\pi$ is

$$f(z) = \frac{1}{k^2\pi^2} \cdot \frac{1}{z - k\pi} + \sum_{n=0}^{\infty} a_n (z - k\pi)^n, \quad |z - k\pi| < \pi,$$

$$k = \pm 1, \pm 2, \ldots .$$

(4) $f(z)$ ceases to be analytic at points z where $\cos \frac{1}{z} = 0$ because $\sin w$ is not analytic at $w = \infty$. Now

$$\cos \frac{1}{z} = 0 \Leftrightarrow \frac{1}{z} = k\pi + \frac{\pi}{2} \quad \text{or} \quad z_k = \frac{2}{(2k + 1)\pi}, \quad k = 0, \pm 1, \pm 2, \ldots .$$

Since

$$\sin \frac{1}{w} = \sum_{n=0}^{\infty} (-1)^n \frac{1}{(2n + 1)!} w^{-(2n+1)}, \quad |w| > 0,$$

$\sin \frac{1}{w}$ has an essential singularity at $w = 0$. Consequently z_k ($k = 0, \pm 1, \pm 2, \ldots$) are essential singularities of $f(z)$. $z = 0$, as the limit point of z_k, is a singularity of $f(z)$, but it is not isolated. On the other hand, $z = \infty$ is a removable singularity of $\cos \frac{1}{z}$, and $\cos \frac{1}{\infty} = \cos 0 = 1$. Hence, $f(z)$ is analytic at $z = \infty$.

The Laurent series of f at each z_k is complicated and one can use the methods suggested in (4.8.1) to find it.

(5) By Example 5(3) of Sec. 3.4.2, $(1 - \sqrt{2 - 2\cos z})^2$ has a zero of order 2 at each of the points $z_n = 2n\pi + \frac{\pi}{3}$, $n = 0, \pm1, \pm2, \dots$. Hence $f(z)$ has a pole of order 2 at each z_n, $n = 0, \pm1, \pm2, \dots$. Note that ∞ is a limit point of these poles z_n and hence, ∞ is a nonisolated singularity.

Try to use the methods indicated in (4.8.1) to find the Laurent series at z_k.

(6) Since

$$\lim_{z \to 2\pi i}(z - 2\pi i)f(z) = \lim_{z \to 2\pi i}\left\{\frac{1}{e^z} - 1\right\} = 0,$$

$z = 2\pi i$ is a removable singularity of f. By Example 5 of Sec. 4.8,

$$\frac{1}{e^z - 1} = \frac{1}{e^{z-2\pi i} - 1}$$

$$= \frac{1}{z - 2\pi i} - \frac{1}{2} + \sum_{n=1}^{\infty}\frac{B_{2n}}{(2n)!}(z - 2\pi i)^{2n-1}, \quad |z - 2\pi i| < 2\pi$$

$$\Rightarrow f(z) = -\frac{1}{2} + \sum_{n=1}^{\infty}\frac{B_{2n}}{(2n)!}(z - 2\pi i)^{2n-1}, \quad |z - 2\pi i| < 2\pi.$$

This is the Taylor series expansion of f at $2\pi i$.

Since $e^z - 1 = 0 \Leftrightarrow z_k = 2k\pi i$, $k = 0, \pm1, \pm2, \dots$, each z_k, except $z_1 = 2\pi i$, is a simple pole of f. At each fixed z_k $(k \neq 1)$,

$$\frac{1}{e^z - 1} = \frac{1}{e^{z-z_k} - 1}$$

$$= \frac{1}{z - z_k} - \frac{1}{2} + \sum_{n=1}^{\infty}\frac{B_{2n}}{(2n)!}(z - z_k)^{2n-1}, \quad |z - z_k| < 2\pi;$$

$$\frac{1}{z - 2\pi i} = \frac{1}{z - z_k + (z_k - 2\pi i)}$$

$$= \sum_{n=0}^{\infty}(-1)^n\frac{(z - z_k)^n}{(z_k - 2\pi i)^{n+1}}, \quad |z - z_k| < |z_k - 2\pi i|.$$

Then the Laurent series expansion of f at z_k follows obviously.

(7) $f(z)$ has a pole of order 7 at each of $\pm ib$; a pole of order 5 at each of $n\pi$ $(n = 0, \pm1, \pm2, \dots)$, and a removable singularity at a. ∞, as a limit point of poles $n\pi$, is not isolated. Readers are urged to find corresponding Laurent series.

(8) 0 is a pole of order m and ∞ is an essential singularity. It is not hard to find the associated Laurent series expansions.

Example 2. Let f be a meromorphic function in a domain Ω and $z_1, \ldots, z_n$ be all distinct poles of f in Ω, with respective orders $r_1, \ldots, r_n$. Then there exists an analytic function g in Ω such that $g(z_k) \neq 0$ for $1 \leq k \leq n$ and

$$f(z) = \frac{g(z)}{(z - z_1)^{r_1} \cdots (z - z_n)^{r_n}}, \quad z \in \Omega - \{z_1, \ldots, z_n\}.$$

Proof. For each fixed k, $1 \leq k \leq n$, $\lim_{z \to z_k} (z - z_k)^{r_k} f(z)$ exists as a nonzero finite number. This indicates that z_k is a removable singularity of $(z - z_k)^{r_k} f(z)$.

Consequently, $(z - z_1)^{r_1} f(z)$ is analytic in $\Omega - \{z_2, \ldots, z_n\}$. Since $z_2 \neq z_1$ and $(z - z_1)^{r_1}$ is analytic in $\mathbb{C}$, $(z - z_1)^{r_1} f(z)$ has a pole of order r_2 at z_2. In turn, $(z - z_1)^{r_1} (z - z_2)^{r_2} f(z)$ has a removable singularity at z_2 (and, of course, at z_1 too), and thus is analytic in $\Omega - \{z_3, \ldots, z_n\}$ and does not assume zero values at z_1 and z_2. The result follows by induction on n.

Example 3.

(1) Let $f(z) = \sum_{-\infty}^{\infty} a_n (z - z_0)^n$ be analytic in $0 \leq r < |z - z_0| < R \leq \infty$. Then f has a single-valued primitive in $r < |z - z_0| < R$ if and only if $a_{-1} = 0$ (refer to Exercise B (1) of Sec. 4.9.1).
(2) Suppose f is analytic in $0 < |z - z_0| < R$. Then f has a pole of order n at z_0 if and only if f' has a pole of order $(n+1)$ at z_0, and the Laurent series expansion $f'(z) = \sum_{k=-(n+1)}^{\infty} b_k (z - z_0)^k$, $0 < |z - z_0| < R$, has its coefficient $b_{-1} = 0$.

Proof.

(1) Suppose f has a single-valued primitive F in $r < |z - z_0| < R$. Then F can be expanded as

$$F(z) = \sum_{-\infty}^{\infty} A_n (z - z_0)^n, \quad r < |z - z_0| < R$$

$$\Rightarrow F'(z) = \sum_{n=-\infty}^{-1} n A_n (z - z_0)^{n-1} + \sum_{n=1}^{\infty} n A_n (z - z_0)^{n-1} = f(z)$$

$$= \sum_{n=-\infty}^{\infty} a_n (z - z_0)^n, \quad r < |z - z_0| < R.$$

By the uniqueness of the Laurent series expansion, the corresponding coefficients of both sides should be equal in pairs; in particular, $a_{-1} = 0$ holds.

Conversely, if $f(z) = \sum_{-\infty}^{\infty} a_k(z - z_0)^n$ with $a_{-1} = 0$, in $r < |z - z_0| < R$, consider

$$F(z) = \sum_{-\infty}^{-2} \frac{a_n}{n+1}(z - z_0)^{n+1} + \sum_{n=0}^{\infty} \frac{a_n}{n+1}(z - z_0)^{n+1}, \quad r < |z - z_0| < R$$

which is single-valued and analytic in $r < |z - z_0| < R$. Since $F'(z) = f(z)$ throughout $r < |z - z_0| < R$, thus $F'(z)$ is a primitive of $f(z)$.

(2) By assumption,

$$f(z) = \frac{a_{-n}}{(z - z_0)^n} + \cdots + \frac{a_{-1}}{z - z_0} + \sum_{k=0}^{\infty} a_k(z - z_k)^k,$$

$$a_{-n} \neq 0, \quad 0 < |z - z_0| < R$$

$$\Rightarrow f'(z) = \frac{-na_{-n}}{(z - z_0)^{n+1}} + \cdots + \frac{-a_{-1}}{(z - z_0)^2} + \sum_{k=1}^{\infty} ka_k(z - z_k)^{k-1},$$

$$0 < |z - z_0| < R.$$

Since $-na_{-n} \neq 0$, so f' has a pole of order $n + 1$ at z_0 and the coefficient b_{-1} of $(z - z_0)^{-1}$ term is equal to zero. The converse follows easily from (1).

Example 4. Suppose $f(z) = \sum_0^{\infty} a_n z_n$ has the radius of convergence $r > 0$. Also suppose $f(z)$ is analytic everywhere on $|z| = r$ except at z_0 ($|z_0| = r$) which is a pole of order m. Then

$$\lim_{n \to \infty} \frac{a_n}{a_{n+1}} = z_0$$

and hence, $\lim_{n \to \infty} \left|\frac{a_n}{a_{n+1}}\right| = |z_0| = r$. In case $m = 1$, this is Exercise A (12) of Sec. 4.8. Refer to Exercise B (1).

Proof. Let $\frac{a_{-1}}{z - z_0} + \cdots + \frac{a_{-m}}{(z-z_0)^m}$, $a_{-m} \neq 0$, be the singular part of f at z_0. Then, there is a $\sigma > 0$ such that the function

$$g(z) = f(z) - \left\{ \frac{a_{-1}}{z - z_0} + \cdots + \frac{a_{-m}}{(z - z_0)^m} \right\}$$

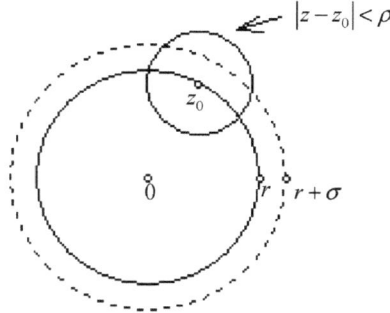

Fig. 4.54

is analytic in $|z| < r + \sigma$. See Fig. 4.54. Suppose $g(z) = \sum_0^\infty b_n z^n$ in $|z| < r + \sigma$, where

$$b_n = \frac{g^{(n)}(0)}{n!} = a_n + \sum_{k=1}^m (-1)^{k+1} C_n^{n+k-1} \frac{a_{-k}}{z_0^{n+k}},$$

where

$$C_n^{n+k-1} = \binom{n+k-1}{n} = \frac{(n+k-1)!}{n!(k-1)!}, \quad 1 \leq k \leq m.$$

Therefore,

$$\frac{a_n}{a_{n+1}} = z_0 \cdot \frac{\frac{b_n z_0^n}{C_{n+1}^{n+m}} - \left\{ \sum_{k=1}^{m-1} (-1)^{k+1} \frac{a_{-k}}{z_0^k} \cdot \frac{C_n^{n+k-1}}{C_{n+1}^{n+m}} \right\} - (-1)^{m+1} \cdot \frac{a_{-m}}{z_0^m} \cdot \frac{C_n^{n+m-1}}{C_{n+1}^{n+m}}}{\frac{b_{n+1} z_0^{n+1}}{C_{n+1}^{n+m}} - \left\{ \sum_{k=1}^{m-1} (-1)^{k+1} \frac{a_{-k}}{z_0^k} \cdot \frac{C_{n+1}^{n+k}}{C_{n+1}^{n+m}} \right\} - (-1)^{m+1} \cdot \frac{a_{-m}}{z_0^m}}.$$

$$(*)$$

By *Stirling's formula* $n! = \sqrt{2\pi n} \cdot n^n \cdot e^{-n + \frac{\theta_n}{12n}}$, $0 < \theta_n < 1$ (see Secs. 4.14 and 5.6.3), we have, as $n \to \infty$,

$$C_{n+1}^{n+m} = \frac{1}{(m-1)!} \cdot \frac{(n+m)!}{(n+1)!} = \frac{1}{(m-1)!} \sqrt{\frac{n+m}{n+1}}$$

$$\times \left(1 + \frac{m-1}{n+1} \right)^{n+1} \cdot (n+m)^{m-1} \cdot e^{-m+1} \cdot e^{\frac{\theta_{n+m}}{12(n+m)} - \frac{\theta_{n+1}}{12(n+1)}}$$

$$\to \begin{cases} 1, & \text{if } m = 1 \\ \infty, & \text{if } m \geq 2 \end{cases};$$

$$\frac{C_{n+1}^{n+k}}{C_{n+1}^{n+m}} = \frac{(m-1)!}{(k-1)!} \cdot \frac{1}{(n+m)\cdots(n+k+1)} \to 0, \quad 1 \le k \le m-1,$$

$$\frac{C_{n}^{n+k-1}}{C_{n+1}^{n+m}} = \frac{(m-1)!}{(k-1)!} \cdot \frac{n+1}{(n+m)\cdots(n+k)} \to 0, \quad 1 \le k \le m-1, \quad \text{and}$$

$$\frac{C_{n}^{n+m-1}}{C_{n+1}^{n+m}} = \frac{(m-1)!}{(m-1)!} \cdot \frac{n+1}{n+m} \to 1.$$

Combining with $\lim_{n\to\infty} b_n z_0^n = 0$, and substituting these results into $(*)$, we obtain

$$\lim_{n\to\infty} \frac{a_n}{a_{n+1}} = z_0.$$

Note that $a_{-m} \neq 0$ has been used in this last limit process. Also, in case $m = 1$, the process can be simplified significantly.

Example 5. Suppose f and g are meromorphic functions in $\mathbf{C}$ satisfying $|f(z)| \le |g(z)|$, $z \in \mathbf{C}$. Then $f(z) = cg(z)$ for some constant c.

Proof.   Fix any point $z_0 \in \mathbf{C}$.

In case f and g are analytic at z_0: If $g(z_0) \neq 0$, then $\frac{f}{g}$ is analytic at z_0, too. If z_0 is a zero of order n of g, then $g(z) = (z - z_0)^n \varphi(z)$ where $\varphi(z)$ is analytic in some $|z - z_0| < r$ and $\varphi(z_0) \neq 0$. In this case, $|f(z)| \le |g(z)| = |z - z_0|^n |\varphi(z)|$ shows that z_0 is a zero of f, of order at least equal to n. Hence, $\frac{f}{g}$ has a removable singularity at z_0.

In case f has a pole of order n at z_0: Since $\lim_{z\to z_0} g(z) = \infty$, z_0 is a pole of order m of g. In a deleted neighborhood of z_0, $|f(z)| \le |g(z)|$ will imply that $n \le m$. Hence $\frac{f}{g}$ has a zero of order $(m - n)$ at z_0.

In case g has a pole of order m at z_0: It is possible that f is analytic at z_0 and thus, $\frac{f}{g}$ has a zero at z_0, of order at least equal to m. While, if f has a zero of order n at z_0, just like the last paragraph showed, $n \le m$ holds and $\frac{f}{g}$ is analytic at z_0.

Anyway, the function $\frac{f}{g}$ is an entire function which is bounded by 1. By Liouville's theorem, $\frac{f}{g}$ is a constant c and $f = cg$ holds.

Example 6. A function meromorphic in the extended plane $\mathbf{C}^*$ is necessarily a rational function $\frac{p(z)}{q(z)}$, where $p(z)$ and $q(z)$ are polynomials, relatively prime. So is the converse.

Proof.   Let f be a meromorphic function in $\mathbf{C}^*$ which is not a constant.

Since $\mathbf{C}^*$ is compact, f can only have a finite number of poles $z_1, \ldots, z_n$ in $\mathbf{C}$ and ∞ may be a removable singularity or a pole of f.

Suppose z_k is a pole of order m_k, $1 \le k \le n$. Then the singular part of f at z_k is

$$p_k(z) = \frac{a_{-1}^{(k)}}{z - z_k} + \cdots + \frac{a_{-m_k}^{(k)}}{(z - z_k)^{m_k}}, \quad a_{-m_k}^{(k)} \ne 0, 1 \le k \le n.$$

If ∞ is pole of order m, then the singular part of f at ∞ is

$$p_\infty(z) = a_1 z + \cdot + a_m z^m, \quad a_m \ne 0;$$

while if ∞ is a removable singularity of f, then $p_\infty(z) = 0$.

According to (5) in (4.10.2.1) and (4.10.2.2), $f(z) - p_k(z)$ has a removable singularity at z_k for $1 \le k \le n$ while $p_j(z)$, $j \ne k$, $1 \le j \le n$, are analytic at z_k. Similarly, $f(z) - p_\infty(z)$ has a removable singularity at ∞ while $p_k(z)$, $1 \le k \le n$, are analytic at ∞. Hence, the function

$$F(z) = f(z) - p_\infty(z) - \sum_{k=1}^{n} p_k(z)$$

is analytic on $\mathbf{C}^*$. By the Liouville's theorem (see (3.4.2.10) and Exercise A (35) of Sec. 3.4.2), $F(z) = c$ is a constant. Therefore,

$$f(z) = c + p_\infty(z) + \sum_{k=1}^{n} p_k(z)$$

is a rational function.

For *another proof*: By Example 2, the function $f(z)(z - z_1)^{m_1} \cdots (z - z_n)^{m_n}$ is analytic on $\mathbf{C}$ and has a pole of order m at ∞ or a removable singularity at ∞. Let $a_1 z + \cdots + a_m z^m$, $a_m \ne 0$, be its singular part at ∞. Then

$$F(z) = f(z)(z - z_1)^{m_1} \cdots (z - z_n)^{m_n} - (a_1 z + \cdots + a_m z^m)$$

is analytic on $\mathbf{C}^*$ and hence, is a constant a_0. It follows that

$$f(z) = \frac{a_0 + a_1 z + \cdots + a_m z^m}{(z - z_1)^{m_1} \cdots (z - z_1)^{m_n}}, \quad z \in \mathbf{C}.$$

Example 7. Prove the following statements.

(1) Let $p(z)$ be a nonconstant polynomial and f be analytic in $0 < |z - z_0| < R$ (in case $z_0 = \infty$, in $r < |z| < \infty$). Then f has a removable singularity, a pole or an essential singularity at z_0, if and only if $p(f(z))$ has the same type of singularity at z_0, respectively.

(2) Let f be analytic at z_0 or have a pole at z_0 (z_0 could be ∞), and $w_0 = f(z_0)$ be a removable singularity, a pole or an essential singularity of $g(w)$. What kind of singularity of $g(f(z))$ at z_0 is?

Proof.

(1) Let $p(z) = a_m z^m + \cdots + a_1 z + a_0$, where $a_m \neq 0$ and $m \geq 1$. Then $p(f(z))$ is also analytic in $0 < |z - z_0| < R$, namely, z_0 is an isolated singularity of $p(f(z))$.

   By characteristic properties shown in (4.10.1.1) and (4.10.2.1), it is easy to see that $p(f(z))$ has a removable singularity or a pole at z_0 if $f(z)$ does.

   Now, suppose z_0 is an essential singularity of f. According to (4.10.3.1), for any given $w_0 \in \mathbf{C}^*$, there corresponds a sequence $z_n \neq z_0$ satisfying

$$\lim_{n \to \infty} z_n = z_0 \quad \text{and} \quad \lim_{n \to \infty} f(z_n) = w_0.$$

Fix any $\zeta_0 \in \mathbf{C}^*$. In case $\zeta_0 \in \mathbf{C}$: By the fundamental theorem of algebra (see (3.4.2.12)), there is at least one $w_0 \in \mathbf{C}$ so that $p(w_0) = \zeta_0$. In this case, choose z_n as in the last paragraph. Then, we have

$$\lim_{n \to \infty} p(f(z_n)) = a_0 + \sum_{k=1}^{m} a_k \lim_{n \to \infty} (f(z_n))^k = p(w_0) = \zeta_0.$$

In case $\zeta_0 = \infty$: For any $K > 0$, there corresponds a $R > 0$ so that, if $|w| \geq R$,

$$|p(w)| \geq \frac{1}{2}|a_m|\,|w|^m \geq K$$

holds. Since there is a sequence $z_n \neq z_0 \to z_0$ such that $\lim_{n \to \infty} f(z_n) = \infty$, there exists a positive integer n_k so that $|f(z_n)| \geq R$ if $n \geq n_k$. Consequently, if $n \geq n_k$,

$$|p(f(z_n))| \geq \frac{1}{2}|a_m|\,|f(z_n)|^m \geq K.$$

This shows that $\lim_{n \to \infty} p(f(z_n)) = \infty$ holds and thus, z_0 is an essential singularity of $p(f(z))$.

   Combining what we have proved, the sufficient statement also holds.

(2) *If $w_0 = f(z_0)$ is a removable singularity of $g(w)$*: Since g is bounded in $0 < |w - w_0| < R$, so is $g(f(z))$ in $0 < |z - z_0| < r$ and then z_0 is a removable singularity of $g(f(z))$.

In case w_0 is a pole of order n of $g(w)$: Suppose f is analytic at z_0. Then $g(w) = \frac{\varphi(w)}{(w-w_0)^n}$ where φ is analytic at w_0 and $\varphi(w_0) \neq 0$. Note that $\varphi(f(z))$ is also analytic at z_0. Let z_0 be a zero of order m of $f(z_0) - w_0$ and hence, $f(z) = w_0 + (z - z_0)^m \psi(z)$ where ψ is analytic at z_0 and $\psi(z_0) \neq 0$. Therefore,

$$g(f(z)) = \frac{\varphi(f(z))}{(f(z) - w_0)^m} = \frac{\varphi(f(z))}{(z - z_0)^{mn}\psi(z)^m}, \quad 0 < |z - z_0| < r.$$

Since $\frac{\varphi(f(z_0))}{\psi(z_0)^m} = \frac{\varphi(w_0)}{\psi(z_0)^m} \neq 0$, z_0 is a pole of order mn of $g(f(z))$.

On the other hand, if z_0 is a pole of order m of f, then $f(z_0) = \infty = w_0$ and $f(z) = \frac{\psi(z)}{(z-z_0)^m}$ where ψ is analytic at z_0 and $\psi(z_0) \neq 0$. In this case, $g(w) = w^n \varphi(w)$, where φ is analytic at ∞ and $\varphi(\infty) \neq 0$. Then

$$g(f(z)) = f(z)^n \varphi(f(z)) = \frac{\psi(z)^n}{(z - z_0)^{mn}}\varphi(f(z)), \quad r < |z| < \infty.$$

Since $\varphi(f(z_0))\psi(z_0)^n = \varphi(\infty)\psi(z_0)^n \neq 0$, so z_0 is also a pole of order mn of $g(f(z))$.

In case w_0 is an essential singularity of $g(w)$: For each fixed $\zeta_0 \in \mathbf{C}^*$, by (4.10.3.1), there is a sequence $w_n \neq w_0 \to w_0$ such that $g(w_n) \to \zeta_0$ as $n \to \infty$. If f is analytic at z_0, there is a sequence $z_n \neq z_0 \to z_0$ so that $f(z_n) = w_n \to f(z_0) = w_0$. In this case, $g(f(z_n)) \to \zeta_0$ as $n \to \infty$ and z_0 is an essential singularity of $g(f(z))$. If z_0 is a pole of f, then $w_0 = f(z_0) = \infty$. Similarly, there is a sequence $z_n \neq z_0 \to z_0$ so that $f(z_n) = w_n \to \infty$ which, in turn, implies that $g(f(z_n)) \to \zeta_0$ as $n \to \infty$. This proves that z_0 is also an essential singularity of $g(f(z))$.

Exercises A

(1) Find the isolated and nonisolated singularities of each of the following functions. If it is a pole, determine its order and the singular part; if it is an essential singularity, find its Laurent series expansion if possible.

(a) $\dfrac{z^4}{(z^4 + 16)^2}$.

(b) $\dfrac{z^2}{(z - bi)^4(z + c)^3}$ (b and c are real numbers, distinct from 0).

(c) $\dfrac{1}{\cos z - 1}$.

(d) $\dfrac{e^{(z-1)^{-1}}}{e^z - 1}$.

(e) $\cot z - \dfrac{1}{z}$.

(f) $\dfrac{1}{\cos z + \cos a}$.

(g) $\cot \dfrac{1}{z}$.

(h) $e^{-z} \cos \dfrac{1}{z}$.

(i) $e^{\cot z^{-1}}$.

(j) $e^{\tan z^{-1}}$.

(k) $\sin\left(\dfrac{1}{\sin z^{-1}}\right)$.

(l) $\dfrac{z^2 \sin z}{(z^2 - \pi^2)^2}$.

(m) $(e^{z-1} - 1)^{-1} e^{(z-1)^{-1}}$.

(n) $\dfrac{\sin^2 z}{(z - \pi)^6}$.

(o) $\dfrac{\sin z}{\cos z^3 - 1}$.

(p) $\dfrac{\sin z^3}{(1 - \cos z)^3}$.

(q) $\dfrac{z}{(z + 1) \sin z^n}$.

(r) $(e^z - 1)^{-1} e^{(1-z)^{-1}}$.

(s) $\dfrac{(z - 1)^2 (z + 2)}{1 - \sin(\pi z/2)}$.

(2) For each branch of the following functions, do the same problems as in Exercise (1).

(a) $(\tan^{-1} z)^2$.

(b) $(z - 1)^{-5} (\log z)^2$.

(c) $z(\log(1 + z))^{-3}$.

(d) $(z^2 - \tan^{-1} z^2)^{-1}$.

(e) $\sqrt{1 + \sqrt[3]{\dfrac{z}{1 + z}}}$ (see Example 7(4) of Sec. 4.9.2).

(3) Suppose f is analytic in $|z| < r$ $(r > 1)$ except a simple pole at $z_0 = 1$. In case

$$f(z) = \frac{a_{-1}}{z - 1} + \sum_{n=0}^{\infty} a_n (z - 1)^n, \quad 0 < |z - 1| < \rho,$$

show that

$$\lim_{n\to\infty} \frac{1}{n!}f^{(n)}(0) = -a_{-1},$$

where $a_{-1} = \operatorname{Re} s(f;1)$ is called the *residue* of f at $z_0 = 1$ (see Sec. 4.11).

(4) Suppose f is analytic in $0<|z-z_0|<r$ with a simple pole at z_0. For $0 < \rho < r$, let γ_ρ be the circular arc $z(t) = z_0 + \rho e^{it}$, $a \le t \le b$. Show that

$$\lim_{\rho\to 0} \int_{\gamma_\rho} f(z)dz = i(b-a)a_{-1},$$

where $a_{-1} = \operatorname{Re} s(f;z_0)$ is the coefficient of $(z-z_0)^{-1}$ in the Laurent series expansion of f at z_0.

(5) Let f be a univalent entire function. Show that $f(z) = az + b$ with $a \ne 0$.

(6) Suppose f is analytic in $\mathbf{C}^*$ except a simple pole. Show that

$$f(z) = \frac{az+b}{cz+d}, \quad ad - bc \ne 0.$$

In other words, any univalent meromorphic function from $\mathbf{C}^*$ onto $\mathbf{C}^*$ (see Section (1) in Sec. 2.5.4) is necessarily a bilinear fractional transformation.

(7) Suppose an entire function f has the property that, $f(z)$ is real if and only if $z = x$ is real.

(a) Show that $f'(z) \ne 0$ if $z = x$ is real and then, deduce that $f(z)$ is one-to-one along the real axis.

(b) Show that there are real constants $a \ne 0$ and b so that $f(z) = az + b$, $z \in \mathbf{C}$.

(8) Let f be a nonconstant meromorphic function satisfying $f(\lambda z) = f(z)$, $z \in \mathbf{C}$, for some constant λ. Show that there is a positive integer p such that $\lambda^p = 1$.

(9) Let f be meromorphic in $|z-z_0| < r$ and the analytic function g have an isolated singularity at $w_0 = f(z_0)$. Show that $g(f(z))$ has, at z_0, the same type of singularities as $g(w)$ has at w_0. Note that z_0 or w_0 could be ∞.

(10) Let f be meromorphic in $0 < |z - z_0| < r$. If its set of poles has z_0 as a limit point, show that $\mathbf{C}^* - f(\{0 < |z - z_0| < r\})$ does not have interior points.

(11) (a) Suppose f is analytic in $|z - z_0| < R$ and is not identically equal to zero. Show that $\frac{1}{f}$ has a removable singularity or pole at z_0.

(b) Suppose f is analytic in $0 < |z-z_0| < R$ and does not assume values in $|w-w_0| < \delta$. Then z_0 is a removable singularity or pole of f.

(12) Find the most general meromorphic function in $\mathbf{C}^*$ satisfying each of the following conditions.

   (a) There is only one simple pole.
   (b) There is only a pole of order m.
   (c) There is a pole of order 2 at 0 with its singular part $\frac{1}{z^2}$.
   (d) There are a pole of order n at $z = 0$ and a pole of order m at $z = \infty$.
   (e) There are only n simple poles.

(13) Let z_0 be an isolated singularity of f. If there is a positive integer n so that $(z - z_0)^n f(z)$ is bounded in $0 < |z - z_0| < R$, show that z_0 is a pole of order not greater than n or a removable singularity of f.

(14) Let f be a meromorphic function in $\mathbf{C}$. If there are a positive integer n, constants $M > 0$ and $R > 0$ such that $|f(z)| \leq M|z|^n$ for $|z| > R$, show that f is a rational function.

(15) Suppose f is analytic in $0 < |z - z_0| < R$ and z_0 is a pole of order n ($n > 1$) of f. Discuss the singularity of $F(z) = \int_a^z f(\zeta)d\zeta$ at z_0, where $0 < |z - z_0| < R$, and $a \neq z_0$ and the path of integration lies in $0 < |z - z_0| < R$.

(16) (a) Suppose F is analytic in $|z - z_0| < R$. Show that

$$|F(z_0)|^2 \leq \frac{1}{\pi r^2} \iint_{|z-z_0|\leq r} |F(z)|^2 dxdy, \quad 0 < r < R.$$

   (b) In case f is analytic in $0 < |z - z_0| < R$ and z_0 is a pole of f, then

$$\iint_{0<|z-z_0|\leq r} |f(z)|^2 dxdy = \infty, \quad 0 < r < R;$$

   while if $\iint_{0<|z-z_0|<r} |f(z)|^2 dxdy < \infty$, $0 < r < R$, holds, then z_0 is a removable singularity of f but never an essential singularity.

(17) Let f be analytic in a domain Ω except poles. Show that the logarithmic derivative $\frac{f'(z)}{f(z)-w_0}$ of $\log(f(z) - w_0)$ has its simple poles at poles of $f(z)$ and zeros of $f(z) - w_0$, and nowhere else.

(18) Let f and g be analytic in $|z - z_0| < R$. Suppose $f(z_0) \neq 0$, $g(z_0) = 0$ but $g(z) \neq 0$ in $0 < |z - z_0| < R$. Let z_0 be a zero of order m of g such that $g(z) = (z - z_0)^m \varphi(z)$, where φ is analytic at z_0 and $\varphi(z_0) \neq 0$. Show that $\frac{f}{g}$ has a pole of order m at z_0 and the coefficients a_k of the Laurent series expansion of $\frac{f}{g}$ at z_0 are given by

$$a_k = \frac{1}{(m+k)!} \frac{d^{m+k}}{dz^{m+k}} \left(\frac{f(z)}{\varphi(z)} \right) \Big|_{z=z_0}, \quad k \geq -n.$$

Try to find a_{-2} and a_{-1} in the Laurent series expansion of $(z-1)^{-2}(z-2)^{-2}(z-3)^{-3}$ at $z_0 = 2$.

(19) (a) Suppose f is a nonconstant meromorphic function in $\mathbf{C}^*$ and satisfies $|f(z)| = 1$ if $|z| = 1$. Show that f is of the form

$$e^{i\theta} \prod_{k=1}^{n} \left(\frac{z - a_k}{1 - \overline{a_k} z} \right)^{p_k} \prod_{l=1}^{m} \left(\frac{1 - \overline{b_l} z}{z - b_l} \right)^{q_l}$$

where θ is a real constant; $a_1, \ldots, a_n$ are distinct zeros of respective orders $p_1, \ldots, p_n$ of f in $|z| < 1$ and $b_1, \ldots, b_m$ are distinct poles of respective orders $q_1, \ldots, q_m$ of f in $|z| < 1$. Refer to Exercise B (1) of Sec. 2.5.3 and Example 1(3) of Sec. 3.4.6.

(b) Let f be meromorphic in $\mathbf{C}^*$. If there exist two circles C_1 and C_2 in $\mathbf{C}$ such that the image $f(C_1)$ is a subset of C_2, show that f is a rational function. Find this f explicitly.

(c) In (b), replace C_1 and C_2 by two lines L_1 and L_2 so that $f(L_1)$ is a subset of L_2. Show that the result is still valid.

(20) Suppose an analytic function $w = f(z)$ maps a circular arc (or a line segment) γ onto a circular arc (or a line segment) γ^*. If z_0 (or ∞) is either a pole of order m or an essential singularity of f and z_0^* is the symmetric point of z_0 with respect to γ, try to discuss the analyticity or singularity of f at z_0^*.

Exercises B

(1) Suppose $f(z) = \sum_0^\infty a_n z^n$ has the radius of convergence $R > 0$. If f has one and only one pole on $|z| = R$, show that the series $\sum_0^\infty a_n z^n$ diverges everywhere on $|z| = R$. For instance, $\frac{1}{1-z} = \sum_0^\infty z^n$, $|z| < 1$, is such a case.

4.10.3.   Essential singularity

The behavior of an analytic function f at its essential singularity z_0 is rather complicated when comparing to that of removable singularity or pole.

Since z_0 is not a removable singularity, for each $\rho > 0$, $f(z)$ is not bounded in $0 < |z - z_0| < \rho$. This means that, there is a sequence $z_n \neq z_0 \to z_0$ so that $f(z_n) \to \infty$ as $n \to \infty$. Of course, neither $\lim_{z \to z_0} f(z)$ exists. Indeed, for any given $w_0 \in \mathbf{C}$, $(f(z) - w_0)^{-1}$ does not have a removable singularity or a pole at z_0 (why?) and then, there is a sequence $z_n \neq z_0 \to z_0$ so that $(f(z_n) - w_0)^{-1} \to \infty$ which, in turn, implies that $f(z_n) \to w_0$ as $n \to \infty$. This means that, in a neighborhood of z_0, $f(z)$ comes arbitrarily close to any presumed $w_0 \in \mathbf{C}^*$.

Since z_0 is not a pole of f, for each integer $n \geq 1$, $(z - z_0)^n f(z)$ is not bounded in $0 < |z - z_0| < \rho$ for arbitrary $\rho > 0$. Otherwise, if $|z - z_0|^n |f(z)| \leq M$ (a constant) in $0 < |z - z_0| < \rho$ for same positive integer n, then z_0 will be a removable singularity or a pole of order not larger than n, a contradiction.

Precisely, we have the

Characteristic properties of an essential singularity. Let $z_0 \in \mathbf{C}$ be an isolated singularity of an analytic function f. Then the following are equivalent.

(1) z_0 is an essential singularity of f, namely, z_0 is neither a removable singularity nor a pole (or, neither $f(z)$ nor $(z - z_0)^n f(z)$ (for any integer $n \geq 1$) is bounded in a neighborhood of z_0).

(2) For any fixed real number α, neither

$$\lim_{z \to z_0} |z - z_0|^\alpha |f(z)| = 0 \quad \text{nor} \quad \lim_{z \to z_0} |z - z_0|^\alpha |f(z)| = \infty$$

holds.

(3) (*Casorati–Weierstrass*) For any fixed $w_0 \in \mathbf{C}^*$, there is a sequence $z_n \neq z_0 \to z_0$ so that

$$\lim_{n \to \infty} f(z_n) = w_0.$$

Equivalently, for each $\rho > 0$ so that $f(z)$ is analytic in $0 < |z - z_0| < \rho$, the image set $S = f(\{0 < |z - z_0| < \rho\})$ is dense in $\mathbf{C}^*$, namely, $\bar{S} = \mathbf{C}^*$ holds.

(4) There is a sequence $z_n \neq z_0 \to z_0$ such that the sequence $f(z_n)$ does not converge in $\mathbf{C}^*$. Hence, $\lim_{z \to z_0} f(z)$ does not exist.

(5) If $\sum_{-\infty}^{\infty} a_n (z - z_0)^n$, $0 < |z - z_0| < R$, is the Laurent series expansion of f at z_0, there are infinitely many coefficients a_{-n} ($n \geq 1$) not equal to zero.   (4.10.3.1)

Proof. (1) $\Rightarrow$ (2): Suppose there is a real α so that $\lim_{z \to z_0} |z - z_0|^\alpha |f(z)| = 0$ holds. Then for any integer $m \geq \alpha$, $\lim_{z \to z_0} |z - z_0|^m |f(z)| = 0$ holds; therefore, z_0 is a removable singularity of $(z - z_0)^m f(z)$. Let z_0 be a zero of order k of $(z - z_0)^m f(z)$. In this case, there is a function φ analytic at z_0 and $\varphi(z_0) \neq 0$ so that $(z - z_0)^m f(z) = (z - z_0)^k \varphi(z)$ in $0 < |z - z_0| < \rho$ for some $\rho > 0$. In case $k > m$, z_0 is a zero of $k - m$ of f; if $k = m$, z_0 is a removable singularity of f; if $k < m$, z_0 is a pole of order $m - k$. All these are contradicting to our assumption. Note that, in this case,

(1) if $\alpha > k - m$, then $\lim_{z \to z_0} |z - z_0|^{-\alpha} |f(z)| = \infty$; and
(2) if $\alpha < k - m$, then $\lim_{z \to z_0} |z - z_0|^{-\alpha} |f(z)| = 0$.

Next, suppose there is a real α so that $\lim_{z \to z_0} |z - z_0|^\alpha |f(z)| = \infty$ holds. Then for any integer $n \le \alpha$, $\lim_{z \to z_0} |z - z_0|^n |f(z)| = \infty$ holds and then, z_0 is a pole of order l of $(z - z_0)^n f(z)$. Let φ be a function analytic at z_0 and $\varphi(z_0) \ne 0$ so that $(z - z_0)^n f(z) = \frac{\varphi(z)}{(z - z_0)^l}$, $0 < |z - z_0| < \rho$, which, in turn, implies that $f(z) = \frac{\varphi(z)}{(z - z_0)^{n+l}}$, $0 < |z - z_0| < \rho$. If $n < -l$, z_0 is a zero of order $-(n + l)$ of f; if $n = -l$, z_0 is a removable singularity; if $n > -l$, z_0 is a pole of order $n + l$. These cases all contradict to our assumption. Note that, in this case,

(1) if $\alpha > l + n$, then $\lim_{z \to z_0} |z - z_0|^\alpha |f(z)| = 0$; and
(2) if $\alpha < l + n$, then $\lim_{z \to z_0} |z - z_0|^\alpha |f(z)| = \infty$.

Compare to (4.10.2.4), (4.10.2.5) and (4.10.2.6).

(2) $\Rightarrow$ (1): According to (4.10.1.1) and (4.10.2.1), this is obvious.

(1) $\Rightarrow$ (3): Refer to the second paragraph in the beginning of this section.

If $w_0 = \infty$: Since f is not bounded in $0 < |z - z_0| < \rho$, for each integer $n \ge 1$, there is a point z_n satisfying both $0 < |z_n - z_0| < \frac{1}{n}$ and $|f(z_n)| \ge n$. Hence, $z_n \ne z_0 \to z_0$ and $\lim_{n \to \infty} f(z_n) \to w_0 = \infty$ as $n \to \infty$.

If $w_0 \in \mathbf{C}$: Suppose $\lim_{z \to z_0} g(z) = L \in \mathbf{C}^*$ exists, where $g(z) = (f(z) - w_0)^{-1}$. This implies that

$$\lim_{z \to z_0} f(z) = \lim_{z \to z_0} \left\{ \frac{1}{g(z)} + w_0 \right\} = \frac{1}{L} + w_0$$

holds in $\mathbf{C}^*$: if $L = 0$, z_0 is a pole of f; if $L \ne 0$ (L could be ∞), z_0 is a removable singularity of f. Both contradict to the original assumption. This shows that z_0 is still an essential singularity of g. And a sequence $z_n \ne z_0 \to z_0$ so that $g(z_n) \to \infty$ implies that $f(z_n) \to w_0$.

Note: That $g(z) = (f(z) - w_0)^{-1}$ has an essential singularity at z_0 can also be proved as follows. If not, since $f(z)$ is not bounded near z_0, z_0 is a zero, say, of order m, of g. Let $g(z) = (z - z_0)^m \varphi(z)$ where φ is analytic at z_0 and $\varphi(z_0) \ne 0$. Then $f(z) = w_0 + \frac{1}{(z - z_0)^m \varphi(z)}$, $0 < |z - z_0| < \rho$, indicates that z_0 is either a removable singularity or a pole of order m of f, a contradiction.

(3) $\Rightarrow$ (4): Choose $w_1, w_2 \in \mathbf{C}^*$ but $w_1 \ne w_2$. There are a sequence $z_n^{(1)} \ne z_0 \to z_0$ so that $f(z_n^{(1)}) \to w_1$ and another sequence $z_n^{(2)} \ne z_0 \to z_0$ so that $f(z_n^{(2)}) \to w_2$. Define a sequence z_n as

$$z_n = \begin{cases} z_n^{(1)}, & \text{if } n = 2m + 1 \quad (m \ge 0), \\ z_n^{(2)}, & \text{if } n = 2m \quad (m \ge 1). \end{cases}$$

Then $z_n \neq z_0 \to z_0$ but $f(z_n)$ does not converge in $\mathbf{C}^*$.

(4) $\Rightarrow$ (1): This follows easily from (4.10.1.1) and (4.10.2.1).

(1) $\Leftrightarrow$ (5): This follows easily from (4.10.1.1) and (4.10.2.1), too. $\square$

Based on (4.10.3) and (4.10.3.1), we have the following

Characteristic properties for ∞ *being an essential singularity.* Let ∞ be an isolated singularity of an analytic function f. Then the following are equivalent.

(1) ∞ is an essential singularity of f.

(2) For any fixed real number α, neither $\lim_{z \to \infty} |z|^\alpha |f(z)| = 0$ nor $\lim_{z \to \infty} |z|^\alpha |f(z)| = \infty$ holds.

(3) (*Casorati–Weierstrass*) For any fixed $w_0 \in \mathbf{C}^*$, there is a sequence z_n in $\mathbf{C}$ so that $z_n \to \infty$ and $f(z_n) \to w_0$ as $n \to \infty$.

(4) There is a sequence $z_n \in \mathbf{C} \to \infty$ but the sequence $f(z_n)$ does not converge in $\mathbf{C}^*$.

(5) If $\sum_{-\infty}^{\infty} a_n z^n$, $r < |z| < \infty$, is the Laurent series expansion of f at ∞, then there are infinitely many coefficients $a_n \neq 0$ ($n \geq 1$).   (4.10.3.2)

The details are left to the readers.

Before going further, we give an

Example 1. Use $f(z) = e^{z^{-1}}$, $0 < |z| < \infty$, to illustrate (4.10.3.2). How about $g(z) = \sin \frac{1}{z}$, $0 < |z| < \infty$?

Solution. The Laurent series expansion of f at $z = 0$ or $z = \infty$ is $\sum_0^\infty \frac{1}{n!} \cdot \frac{1}{z^n}$. Hence, $z = 0$ is an essential singularity of f while $z = \infty$ is a removable singularity. Observe that $e^{z^{-1}} \neq 0$ in $0 < |z| < \infty$.

If $w_0 = 0$ or ∞: By $f(z) = e^{\frac{\bar{z}}{|z|^2}} = e^{\frac{x}{(x^2+y^2)} - \frac{iy}{(x^2+y^2)}}$, $z = x + iy$, it follows that, when choosing $z = x + ix$ and $x \neq 0$,

$$|f(z)| = |e^{\frac{1}{2x}}||e^{\frac{-i}{2x}}| = e^{\frac{1}{2x}} \to \begin{cases} \infty, & \text{if } x > 0, \\ 0, & \text{if } x < 0, \end{cases}$$

as $z \to 0$.

If $w_0 \in \mathbf{C}$ and $w_0 \neq 0$: Let $r_0 = |w_0|$ and $\theta_0 = \text{Arg} \, w_0$. Solve $e^{z^{-1}} = w_0$ and we get $\frac{1}{z} = \log w_0 = \log r_0 + i(\theta_0 + 2n\pi)$. Let $z_n = (\log r_0 + i(\theta_0 + 2n\pi))^{-1}$, $n = 0, \pm 1, \pm 2, \ldots$. Then $z_n \to 0$ and $f(z_n) = e^{\frac{1}{z_n}} = w_0$ indeed holds, not to say $f(z_n) \to w_0$. This shows that $f(z) = e^{z^{-1}}$ assumes each $w_0 \in \mathbf{C}$ ($w_0 \neq 0$) infinitely many times in $0 < |z| < \infty$.

As for $g(z) = \sin\frac{1}{z}$, $0 < |z| < \infty$, g is analytic at $z = \infty$. In case $w_0 \neq 0, \infty$: Solve $\sin z^{-1} = w_0$ and we get $z = \frac{i}{\log(iw_0+\sqrt{1-w_0^2})}$. Let

$$z_n = \frac{i}{\log\left|iw_0 + \sqrt{1 - w_0^2}\right| + i\left[\mathrm{Arg}\left(iw_0 + \sqrt{1 - w_0^2}\right)\right] + 2n\pi i},$$

$$n = 0, \pm1, \pm2, \dots .$$

Then $z_n \to 0$ as $n \to \infty$ and $g(z_n) \to w_0$. In fact, $g(z_n) = w_0$, $n = 0, \pm1, \pm2, \dots$. Therefore, 0 is an essential singularity of g.

The property that both $e^{z^{-1}}$ and $\sin z^{-1}$ showed in Example 1 has its generality.

Picard's Theorem

(1) Little theorem or the first theorem. Every nonconstant entire function assumes each finite complex number with at most one possible exception.

(2) Great theorem or the second theorem. In any deleted neighborhood of an essential singularity, an analytic function assumes each finite complex number with at most one possible exception.   (4.10.3.3)

For proofs, see Ref. [58], Vol. III, pp. 339–345, or Secs. 5.8.3 and 5.8.4. Recall that a polynomial of degree $n \geq 1$ assumes every (nonzero) finite complex number exactly n times (counting multiplicities), while e^z assumes every *nonzero* finite complex number infinitely many times.

Let f be an entire function. Then

(1) $f(z) = \sum_0^\infty a_n z^n$, $|z| < \infty$, has infinitely many coefficient $a_n \neq 0$;
⟺ (2) ∞ is an essential singularity of $f(z)$;
⟺ (3) 0 is an essential singularity of $f\left(\frac{1}{z}\right)$.   (4.10.3.4)

In this case, f is said to be a *transcendental (entire) function*. An entire function which is *not* a polynomial is transcendental.

Based on (2) in (4.10.3.3), it is easy to deduce

The number of times a complex number assumed by a meromorphic function defined in **C**.

(1) A transcendental entire function assumes every finite complex number, with at most one possible exception, infinitely many times.

(2) A nonconstant meromorphic function, defined on **C**, assumes every finite complex number, with at most two possible exceptions.

(3) A nonconstant meromorphic function, defined on $\mathbf{C}$ and having ∞ as its essential singularity, assumes every finite complex number, with at most two possible exceptions, infinitely many times.   (4.10.3.5)

Sketch of Proof

Let f be the function concerned.

(1) If f assumes two distinct finite complex numbers a and b only a finite number of times in $\mathbf{C}$, then there is a $R > 0$ so that $f(z) \neq a$ and $f(z) \neq b$ in $R < |z| < \infty$. This contradicts to (2) in (4.10.3.3) because ∞ is an essential singularity of f.

(2) In case f does not assume three distinct finite complex numbers a, b, and c, where $c \neq 0$, then the entire function $g(z) = \frac{1}{f(z)-c}$ does not assume two distinct finite complex numbers $\frac{1}{a-c}$ and $\frac{1}{b-c}$ in $\mathbf{C}$, contradicting to (1).

(3) This follows easily by combining the proofs of (1) and (2).

In what follows, we present some illustrative examples.

Example 2. Let z_0 be an isolated singularity of an analytic function f. Then z_0 cannot be a pole of $e^{f(z)}$, but is a removable singularity or an essential singularity (in case z_0 is not a removable one of f).

Solution. If z_0 is a removable singularity of f, then so is for $e^{f(z)}$.
Suppose z_0 is a pole of order m of f. Let $f(z) = \frac{\varphi(z)}{(z-z_0)^m}$, $0 < |z-z_0| < \rho$, where φ is analytic at z_0 and $\varphi(z_0) \neq 0$. Let $z = z_0 + re^{i\theta}$, $0 < r < \rho$ and $0 \leq \theta \leq 2\pi$. Then

$$\frac{\varphi(z)}{(z-z_0)^m} = \frac{\varphi(z_0 + re^{i\theta})}{r^m e^{im\theta}} = \frac{1}{r^m}\varphi(z_0 + re^{i\theta})e^{-im\theta}$$

$$\Rightarrow \operatorname{Re}\frac{\varphi(z)}{(z-z_0)^m} = \frac{1}{r^m} \times \{\operatorname{Re}\varphi(z_0 + re^{i\theta})\cos m\theta + \operatorname{Im}\varphi(z_0 + re^{i\theta})\sin m\theta\}.$$

Since $\varphi(z_0) \neq 0$, we may suppose $\operatorname{Re}\varphi(z_0) > 0$ (and leave the remaining cases $\operatorname{Re}\varphi(z_0) < 0$ or $\operatorname{Im}\varphi(z_0) > 0$, <0 to the readers). Choose θ such that $m\theta = 0$, i.e., $\theta = 0$, then, as $r \to 0$,

$$|e^{f(z)}| = e^{\operatorname{Re}\frac{\varphi(z)}{(z-z_0)^m}} \to e^{\infty} = \infty;$$

choose θ such that $m\theta = \pi$, i.e., $\theta = \frac{\pi}{m}$, then, as $r \to \infty$,

$$|e^{f(z)}| = e^{\operatorname{Re} \frac{\varphi(z)}{(z-z_0)^m}} \to e^{-\infty} = 0.$$

This shows that z_0 is neither removable nor a pole for $e^{f(z)}$. And z_0 should be an essential singularity of $e^{f(z)}$.

In case z_0 is an essential singularity of f, fix any point $w_0 \in \mathbf{C}^*$. Let $F(z) = e^{f(z)}$.

If $w_0 = \infty$: There is a sequence $z_n \neq z_0 \to z_0$ so that $\operatorname{Re} f(z_n) \to +\infty$ (why?). Hence, $|F(z_n)| = e^{\operatorname{Re} f(z_n)} \to +\infty$ as $n \to \infty$ and consequently, $F(z_n) \to \infty$ as $n \to \infty$.

If $w_0 = 0$: There is a sequence $z_n \neq z_0 \to z_0$ so that $\operatorname{Re} f(z_n) \to -\infty$ which, in turn, implies that $F(z_n) \to 0$ as $n \to \infty$.

If $w_0 \neq 0, \infty$: There is a sequence $z_n \neq z_0 \to z_0$ so that $f(z_n) \to \operatorname{Log} w_0$. This implies that $F(z_n) \to e^{\operatorname{Log} w_0} = w_0$.

These facts show that z_0 is an essential singularity of $e^{f(z)}$.

Example 3. Suppose f is analytic in $0 < |z - z_0| < \rho$ and z_0 is an essential singularity of f. Then, for each positive real number α, there are infinitely many points z in $0 < |z - z_0| < \rho$ so that $|f(z)| = \alpha$. In case $z_0 = \infty$, then these z are in $\rho < |z - z_0| < \infty$.

Solution. Fix $\alpha > 0$.

By Casorati–Weierstrass's theorem, there is a z_1 in $0 < |z - z_0| < \rho$ such that $|f(z_1)| = |f(z_1) - 0| < \alpha$ (in this case, $w_0 = 0$ in (3) of (4.10.3.1)). Since f is unbounded in $0 < |z - z_0| < \rho$, there is a point z_2, $0 < |z_2 - z_0| < \rho$, such that $|f(z_2)| > \alpha$.

In $0 < |z - z_0| < \rho$, let $\gamma : z = z(t)$, $a \le t \le b$, be a continuous curve connecting $z_1 = z(a)$ to $z_2 = z(b)$. Then, the continuous curve $f(\gamma) :$ $w = f(z(t))$, $a \le t \le b$, connects the point $f(z_1)$ in $|w| < \alpha$ to the point $f(z_2)$ in $|w| > \alpha$. Applying intermediate value theorem to $|f(z(t))| = \varphi(t)$ on $[a, b]$, there is a point t^*, $a < t^* < b$, so that $\varphi(t^*) = \alpha$ (recall that $\varphi(a) < \alpha < \varphi(b)$). Let $z^* = z(t^*)$. Then $0 < |z^* - z_0| < \rho$ and $|f(z^*)| = \alpha$. Continue this process to $f(z)$ in $0 < |z - z_0| < |z^* - z_0|$, and we will get a sequence of distinct points $z_1^*, z_2^*, \ldots, z_n^*, \ldots$ so that $|f(z_n^*)| = \alpha$, $n \ge 1$.

Example 4. Let f be a transcendental function. Then there is a sequence z_n satisfying

(1) $\lim_{n \to \infty} z_n = \infty$;   and
(2) $\lim_{n \to \infty} f(z_n) p(z_n) = 0$ for each polynomial $p(z)$.

Proof. If it happens that $f(z_n) = 0$ for $n \geq 1$, then 2 holds. Consequently, we may suppose f has only finitely many zeros in $\mathbf{C}$. In this case, there is a $R > 0$ so that $f(z) \neq 0$ on $|z| \geq R$.

Suppose there is a sequence z_n having the property that, $|z_n| \geq n$ and $|f(z_n)| \leq |z|^{-n}$ hold for all sufficiently large n. Let $p(z)$ be any polynomial, say, of degree m. Then there is a constant $M(p)$, depending on $p(z)$, so that $|p(z)| \leq M(p)|z|^m$ if $|z|$ is large enough. For such large $|z|$ and n,

$$|f(z_n)||p_n(z)| \leq |z_n|^{-n} \cdot M(p)|z|^m = M(p)|z|^{m-n} \to 0$$

as $n \to \infty$.

In case there is no such a sequence z_n as stated in the last paragraph, then there is an integer $n_0 \geq 1$ so that $|f(z)| \geq |z|^{-n_0}$ if $|z| \geq n_0$. This means that $|z^{n_0} f(z)| \geq 1$ if $|z| \geq n_0$ which, in turn, shows that the transcendental function $z^{n_0} f(z)$ does not assume values in $|w| < 1$ if $|z| \geq n_0$, contradicting to Picard's great theorem.   □

Example 5. Suppose f is an entire function and $f(z) \neq 0$ for all $z \in \mathbf{C}$. Then f should be of the form ae^z ($a \neq 0$, a constant) or $f(z) + e^z$ has infinitely many zeros.

Proof.   Since $f(z) \neq 0$ on $\mathbf{C}$, so $\log f(z)$ has branches in $\mathbf{C}$. Let $g(z)$ be one of these branches. Then g is entire and $e^{g(z)} = f(z)$, $z \in \mathbf{C}$.

By Example 2, $e^{g(z)-z}$ is a transcendental function except it is a constant function.

In case $e^{g(z)-z}$ is a constant, then $e^{g(z)-z} = a \neq 0$ on $\mathbf{C}$ and it follows that $f(z) = ae^z$.

If $e^{g(z)-z}$ is not a constant, then $f(z) + e^z = e^{g(z)} + e^z = e^z(e^{g(z)-z} + 1)$ holds. Since $e^{g(z)-z}$ is transcendental and does not assumes the value 0, it must assume the value -1. According to (1) in (4.10.3.5), $e^{g(z)-z}$ assumes -1 at infinitely many points and, correspondingly, $f(z) + e^z$ has infinitely many zeros.   □

Example 6. Suppose an entire function f does not assume two *distinct* complex numbers a and b at *finitely* many points. Then f is a polynomial.

Proof.   If f is not a polynomial, then f is a transcendental and has ∞ as its essential singularity. Since $f^{-1}(a) \neq \phi$ is a finite set, (1) in (4.10.3.5) says that $f^{-1}(b)$ should be an infinite set, a contradiction.

This shows that ∞ is a pole of order $m \geq 1$ of f and f is thus a polynomial.   □

Example 7. A nonconstant entire function mapping a given line L_1 into another given line L_2 is of the form $az + b$, where $a \neq 0$ and b are constants. Refer to Exercise A (29) of Sec. 3.4.2.

Proof. Let f be such a entire function that $f(L_1) \subseteq L_2$ holds. f, when restricted to L_1, cannot be constant, otherwise f will be a constant function by the interior uniqueness theorem (see (4) in (3.4.2.9)). Let z_1 and z_2 be two distinct points in L_1 so that $f(z_1) \neq f(z_2)$. Since $f(z_1)$ and $f(z_2)$ are two distinct points, $w = f(z)$ maps the segment $\overline{z_1 z_2}$ lying in L_1 onto the segment $\overline{f(z_1)f(z_2)}$ lying in L_2 owing to the facts that $f(L_1) \subseteq L_2$ and f, as a continuous function, maps connected sets onto connected sets. Then there is a unique bilinear transformation $w = az + b, a \neq 0$, maps $\overline{z_1 z_2}$ onto $\overline{f(z_1)f(z_2)}$. Since $f(z) = az + b$ holds everywhere on the segment $\overline{z_1 z_2}$, still by the interior uniqueness theorem, $f(z) = az + b$ holds throughout $\mathbf{C}$.

Remark. If the condition $f(L_1) \subseteq L_2$ is replaced by $L_1 \supseteq f^{-1}(L_2)$, the result is still valid. To see this, let $A_j, B_j, j = 1, 2$, be the two open half-planes into which $\mathbf{C}$ is divided by L_j. We try to show that

(1) $f(A_1) = A_2$ and $f(B_1) = B_2$ after a readjustment of the indexes;
(2) $f(L_1) = L_2$; and
(3) $f(z) = az + b, a \neq 0$.

Sketch of Proof

(1) In case $f(A_1) \cap L_2 \neq \phi$ holds, then $A_1 \cap f^{-1}(L_2) \neq \phi$ which, in turn, implies that $A_1 \cap L_1 \neq \phi$, a contradiction. Hence $f(A_1) \cap L_2 = \phi$ and $f(A_1)$, as a connected set, should lie either in A_2 or in B_2. Take for instance, $f(A_1) \subseteq A_2$. Similarly, $f(B_1)$ lies in A_2 or in B_2, too. If $f(B_1) \subseteq A_2$ holds, then $f(\mathbf{C}) \subseteq \overline{A_2}$ holds, which contradicts to the Casorati–Weierstrass's theorem or Picard's great theorem. Hence $f(B_1) \subseteq B_2$ holds if $f(A_1) \subseteq A_2$.

As a matter of fact, both $f(A_1) = A_2$ and $f(B_1) = B_2$ hold. For, if $f(A_1) \subsetneq A_2$ (a proper subset), then $f(B_1) \subsetneq B_2$ also holds. By the symmetry principle (see (3.4.6.1)) that says that f preserves symmetric points with respect to L_1, $f(\mathbf{C})$ does not contain at least two distinct finite complex numbers, still contradicting to the Picard's great theorem.

(2) $L_1 \supseteq f^{-1}(L_2)$ implies that $f(L_1) \supseteq L_2$. If $f(L_1) \not\supseteq L_2$, then there is a point $z_0 \in L_1$ so that $f(z_0) \notin L_2$ and thus, $f(z_0)$ is in A_2 or in B_2.

Suppose $f(z_0) \in A_2$. By continuity of f at z_0, there is a $r > 0$ so that $f(\{|z - z_0| < r\}) \subseteq A_2$ (open) holds. On the other hand, since $z_0 \in L_1$, $|z - z_0| < r$ has a nonempty intersection with B_1 and thus, $f(B_1 \cap \{|z - z_0| < r\}) \subseteq f(B_1) \subseteq B_2$. Since $A_2 \cap B_2 = \varphi$, this leads to a contradiction. It follows that $f(L_1) = L_2$.

(3) What remains is the same as the argument shown in the Example 7. Here, we present *another proof*, based on real analysis and of its own general interest. Performing suitable similarity motions $z \to az + b$ $(a \neq 0)$ and $w \to cz + d$ $(c \neq 0)$, we may suppose that both L_1 and L_2 are the real axes, and A_1 and A_2 are the upper half-planes, and B_1 and B_2 are the lower half-planes. So $w = f(z)$ maps $\mathbf{R}$ onto $\mathbf{R}$, the upper and the lower half-planes onto themselves, respectively.

Let $f(z) = u(z) + iv(z)$, $z = x + iy$.

Fix any point $z = x + i0$. Then $v(x, 0) = 0$ and $v(x, h) > 0$ for $h > 0$. This shows that

$$v_y(x, 0) = \lim_{h \to 0^+} \frac{v(x, h) - v(x, 0)}{h} \geq 0.$$

By the Cauchy–Riemann equations, $u_x(x, 0) = v_y(x, 0) \geq 0$ holds on $\mathbf{R}$. Hence, $u(x, 0)$ is monotone increasing in $x \in \mathbf{R}$, and, indeed, is strictly increasing, for otherwise the constancy of $u(x, 0)$ on a nondegenerated interval would lead to the constancy of f on $\mathbf{C}$.

In case $\underline{\lim}_{x \to -\infty} u(x, 0) > -\infty$ or $\overline{\lim}_{x \to +\infty} u(x, 0) < \infty$ happens, $f(z)$ will not assume infinitely many real numbers, contradicting to the Picard's great theorem. Hence $u(x, 0)$ maps $\mathbf{R}$ *onto* itself.

Combining together, then $w = f(z)$ assumes all the real numbers and each real number is assumed exactly once in the real axis (and in $\mathbf{C}$, too, referred to (3.5.5.5)).

If $f(z)$ is a transcendental entire function, then $f(z)$ will assume every finite real number, with at most one exception, infinitely many times (see (1) in (4.10.3.5)), contradicting to the results obtained in the last paragraph. Consequently, $f(z) = \sum_{k=0}^{n} a_k z^n$, $a_k \neq 0$, is a polynomial with real coefficients. Since $\operatorname{Re} f(x, 0) = u(x, 0)$ is one-to-one, so n must be an odd integer.

In case $n \geq 3$, then $f'(x, 0)$, as a polynomial in x, has only a finite number of zeros; and, there exists a $\alpha \in \mathbf{R}$ and a point $x_0 \in \mathbf{R}$ such that $f(x_0, 0) = \alpha$ and $f'(x_0, 0) \neq 0$, otherwise $f'(x, 0) \equiv 0$ will lead to $f'(z) \equiv 0$ and thus, $f(z) = a_0 + a_1 z$. This means that x_0 is the only real zero of $f(x,0) = u(x,0) = \alpha$ and the other zeros of $f(z) = \alpha$ are nonreal complex conjugate numbers in pairs. These pairs of conjugate complex numbers are mapped, under $w = f(z)$, into the upper and lower half-planes, but not into the real number α. Hence $n \geq 3$ does not hold.

Then $n = 1$ is the only possibility and $f(z) = a_0 + a_1 z$, $z \in \mathbf{C}$.

Example 8. Let A be a *discrete* set in a domain Ω (namely, for each $z \in A$, there is a $r > 0$ so that $0 < |z - z_0| < r$ contains no point of A) and do not have limit point in Ω. Suppose $f : \Omega - A \to \mathbf{C}$ is a univalent analytic function. Then

(1) each point of A is not an essential singularity of f;
(2) if $z_0 \in A$ is a pole of f, it is a simple pole; and
(3) if each point of A is a removable singularity of f, then f can be analytically extended to a univalent analytic function in Ω.

Proof.

(1) Fix $z_0 \in A$. Let $r > 0$ be such that $0 < |z - z_0| < r$ does not contain points in A and the set $O = \Omega - (A \cup \{|z - z_0| < r\})$ is nonempty. By the open mapping theorem (see (3.4.4.4)), it follows that $f(O)$ is a nonempty open set. By the univalence of f, $f(\{0 < |z - z_0| < r\})$ has empty intersection with $f(O)$ and hence, is not dense in $\mathbf{C}^*$. By Casorati–Weierstrass's theorem, z_0 is not an essential singularity of f.
(2) If z_0 is a pole of order n of f, then z_0 is a zero of order n of $\frac{1}{f}$ and vice versa. According to (3.5.1.8) and (or) (3.5.1.11) and the univalence of f, $n = 1$ should hold.
(3) By assumption, there is a function analytic in Ω such that $F(z) = f(z)$ in $\Omega - A$. Suppose F is not univalent in Ω. Then there are two distinct points z_1, z_2 in Ω so that $F(z_1) = F(z_2) = w_0$. Choose $r_1 > 0$, $r_2 > 0$ so that $|z - z_1| < r_1$ and $|z - z_2| < r_2$ are disjoint, and both $0 < |z - z_1| < r_1$ and $0 < |z - z_2| < r_2$ lie in $\Omega - A$. Since $F(\{|z - z_1| < r_1\}) \cap F(\{|z - z_2| < r_2\})$ is an open neighborhood of w_0, there are a point z_1^*, $0 < |z_1^* - z_1| < r_1$ and a point z_2^*, $0 < |z_2^* - z_2| < r_2$ so that $F(z_1^*) = F(z_2^*)$. Since z_1^*, $z_2^* \in \Omega - A$ and $z_1^* \neq z_2^*$, then $f(z_1^*) = f(z_2^*)$ will contradict to the univalence of f in $\Omega - A$. This proves that the extended F is univalent in Ω. $\square$

Exercises A

(1) True or false: if true, prove it; if false, give a counterexample. All functions concerned are analytic except at singular points.

   (a) Let z_0 be an essential singularity of f and $f(z) \neq 0$ in $0 < |z - z_0| < \rho$. Then z_0 is also an essential singularity of $\frac{1}{f}$. Is it possible that z_0 is a limit point of poles of $\frac{1}{f}$?

   (b) Suppose z_0 is an essential singularity of f and z_0 is a removable singularity or a pole of g. If g is not identically equal to zero, then $f(z) + g(z)$, $f(z)g(z)$ and $f(z)g(z)^{-1}$ all have z_0 as an essential singularity.

   (c) Let z_0 be an essential singularity of f and $p(z)$ a nonconstant polynomial. Then z_0 is an essential singularity of $p(f(z))$.

   (d) Suppose 0 is an essential singularity of f and $m \geq 1$ is an integer. Then 0 is an essential singularity of $f(z^m)$.

   (e) Suppose z_0 is an essential singularity of f and $f(z) \neq 0$ in $0 < |z - z_0| < \rho$. What kind of singularities does f^2 have at z_0?

   (f) Suppose z_0 is an essential singularity of f and g. What kind of singularities do $f(z) + g(z)$, $f(z)g(z)$ and $\frac{f(z)}{g(z)}$ have at z_0?

(2) Show that 0 is an essential singularity of each of the following functions. Based on Picard's great theorem, justify if $f(\{0 < |z| < 1\}) = \mathbf{C}$ holds; if not, find the exceptional value.

   (a) $f(z) = (z - 1)e^{z-1}$.

   (b) $f(z) = \cos z^{-1}$.

   (c) $f(z) = z \cos z^{-1}$.

   (d) $f(z) = z e^{z-2}$.

   (e) $f(z) = z^{-1} \sin z^{-1}$.

   (f) $f(z) = e^{\frac{1}{z}} - e^{-\frac{1}{z}}$.

   (g) $f(z) = z^3 e^{z^{-1}} \sin z^{-1}$.

(3) Suppose f is analytic in $0 < |z| < r$ and is an odd function (i.e., $f(-z) = -f(z)$). If 0 is an essential singularity of f, use Picard's great theorem to show that $f(\{0 < |z| < r\}) = \mathbf{C}$ or $\mathbf{C} - \{0\}$.

(4) Suppose f is analytic in $0 < |z| < r$ with 0 as its essential singularity. If $f(z)$ assumes real values on the subset $(-r, r) - \{0\}$ of the real axis and $\mathbf{R} \subseteq f(\{0 < |z| < r\})$ holds, show that $f(\{0 < |z| < r\}) = \mathbf{C}$.

(5) If z_0 is a limit point of poles of f (but is not a limit point of essential singularities), then Casorati–Weierstrass theorem is still valid at $\dot z_0$.

(6) Show that f is a transcendental function if and only if there is a sequence $z_n \in \mathbf{C}$ satisfying $z_n \to \infty$ as $n \to \infty$ and $\sup_{n\geq1}|f(z_n)|<\infty$.

(7) Prove or give a counterexample. $f(z)$ is an entire function.

(a) Is it possible that $f(\mathbf{C}) = \mathbf{C}$ for some transcendental function f?
(b) Is it possible that $f(\mathbf{C})$ is the half-plane $\operatorname{Re} w > 0$?
(c) In case f is univalent, is it possible that $f(\mathbf{C}) \subsetneq \mathbf{C}$?

(8) Suppose $f(z)$ and $g(z)$ are entire functions, and $p(z)$ and $q(z)$ are polynomials. In case $e^{f(z)} + p(z) = e^{g(z)} + q(z)$ holds for all $z \in \mathbf{C}$, show that $p(z) \equiv q(z)$.

(9) If f is a nonconstant entire function, then $[f^3 + f^2](\mathbf{C}) = \mathbf{C}$ holds.

(10) Let f be a nonconstant entire function. To each $0 < \sigma < \infty$, let $E_\sigma = \{z \in \mathbf{C}|\,|f(z)| < \sigma\}$. Show that

(i) $\overline{E_\sigma} = \{z \in \mathbf{C}|\,|f(z)| \leq \sigma\}$;   and
(ii) each bounded component of E_σ contain a zero of $f(z)$.

Try to use $f(z) = e^z$ to illustrate 1 and 2.

(11) Let $m \geq 1$ be an integer. Define

$$f(z) = \int_0^z e^{\zeta^{2m}} d\zeta.$$

Then $f(z)$ is entire but not univalent. Also $f(\mathbf{C}) = \mathbf{C}$ holds.

(12) Let f and g be entire functions satisfying $e^{f(z)} + e^{g(z)} = 1$, $z \in \mathbf{C}$. Then both f and g are constant functions.

4.11.   Residues and Residue Theorem

Suppose f is analytic in $|z| < R$. Then, by Cauchy's integral theorem, Morera's theorem and (4.1.2),

$$\int_\gamma f(z)dz = 0, \quad \text{for any rectifiable closed curve } \gamma \text{ in} |z| < R,$$

$$\Leftrightarrow F(z) = \int_{z_0}^z f(\zeta)d\zeta \text{ is a (single-valued) primitive of } f \text{ in } |z| < R.$$

$$(4.11.1)$$

While, suppose f is analytic in $0 < |z| < R$ so that 0 is an isolated singularity of f. By Sec. 4.9.1, the Laurent series expansion of f at 0 is

$$f(z) = \sum_{n=-\infty}^{\infty} a_n z^n, \quad 0 < |z| < R,$$

$\Rightarrow$ (by the uniform convergence of the series on compact sets in $0 < |z| < R$)

$$\int_\gamma f(z)dz = \sum_{n=-\infty}^{\infty} a_n \int_\gamma z^n dz = 2\pi i \cdot a_{-1}, \quad \text{for any rectifiable closed curve}$$

$$\gamma \text{ in } 0 < |z| < R. \tag{4.11.2}$$

Consequently,

The coefficient of z^{-1} in the Laurent series expansion of f at 0 is

$$a_{-1} = 0,$$

$\Leftrightarrow f(z)$ has a (single-valued) primitive F in $0 < |z| < R$, given by

$$F(z) = \sum_{-\infty}^{-2} \frac{a_n}{n+1} z^{n+1} + \sum_{0}^{\infty} \frac{a_n}{n+1} z^{n+1}, \quad 0 < |z| < R,$$

$$\Leftrightarrow \int_\gamma f(z)dz = 0, \quad \text{for any rectifiable closed curve } \gamma \text{ in } 0 < |z| < R.$$

$$\tag{4.11.3}$$

We then realize that whether f satisfies the Cauchy's integral theorem or has a primitive in $0 < |z| < R$ depends mainly on whether a_{-1} is equal to zero or not. Also, refer to Secs. 4.4 and 4.5. This particular coefficient a_{-1} is called the *residue* of f at the isolated singularity 0.

Section 4.11.1 will give a formal definition of residues of an analytic function at a finite isolated singularity and at the point ∞.

The methods of how to compute a residue is given in Sec. 4.11.2. So are illustrative examples.

The important residue theorem is stated and proved in Sec. 4.11.3. An immediate application of this theorem is used to evaluate some *complex* line integrals. Further applications will be postponed to Sec. 4.12.

4.11.1.   *Definition of residues*

Suppose f is analytic in $0 < |z - z_0| < R$ and $\sum_{-\infty}^{\infty} a_n(z - z_0)^n$ is its Laurent series expansion (see Sec. 4.9.1). For $0 < r < R$, the series converges uniformly on $|z - z_0| = r$ and termwise integration of it is permissible. This leads to

$$\int_{|z-z_0|=r} f(z)dz = \sum_{-\infty}^{\infty} a_n \int_{|z-z_0|=r} (z - z_0)^n dz = 2\pi i \cdot a_{-1}, \quad 0 < r < R.$$

This is equivalent to the fact that a_{-1} is the *unique* constant such that

$$\int_{|z-z_0|=r} \left\{ f(z) - \frac{a_{-1}}{z - z_0} \right\} dz = 0, \quad 0 < r < R.$$

In this case,

$$f(z) - \frac{a_{-1}}{z - z_0} = \sum_{-\infty}^{-2} a_n(z - z_0)^n + \sum_0^{\infty} a_n(z - z_0)^n, \quad 0 < |z - z_0| < R,$$

$$\Rightarrow F(z) = \sum_{-\infty}^{-2} \frac{a_n}{n+1}(z - z_0)^{n+1} + \sum_0^{\infty} \frac{a_n}{n+1}(z - z_0)^{n+1}, \quad 0 < |z - z_0| < R$$

is a primitive of $f(z) - \frac{a_{-1}}{z-z_0}$ in $0 < |z - z_0| < R$.

We summarize the above as the

Definition of residue at a point $z_0 \in \mathbf{C}$. Let f be analytic in $0 < |z - z_0| < R$. Then

(1) a_{-1} is the coefficient of the term $(z - z_0)^{-1}$ in the Laurent series expansion of f at z_0.

$\Leftrightarrow$ (2) For any r, $0 < r < R$, $\frac{1}{2\pi i} \int_{|z-z_0|=r} f(z)dz = a_{-1}$, where $|z - z_0| = r$ is oriented in the *counterclockwise* direction.

$\Leftrightarrow$ (3) a_{-1} is the unique constant such that

$$f(z) - \frac{a_{-1}}{z - z_0}$$

has a (single-valued) primitive in $0 < |z - z_0| < R$.

Such a number a_{-1} is called the *residue* of f at the isolated singularity z_0 and is denoted as

$$\mathrm{Re}\, s(f; z_0) \quad \text{or} \quad \mathrm{Re}\, s(f(z); z_0) \quad \text{or} \quad \mathrm{Re}\, s_{z=z_0} f(z) \qquad (4.11.1.1)$$

On the other hand, suppose f is analytic in $0 \le r < |z| < \infty$ so that ∞ is an isolated singularity of f. For ρ, $r < \rho < \infty$, the positive orientation along $|z| = \rho$, *when viewed from the point* ∞, is clockwise which is opposite to usual counterclockwise one, when viewed from the point 0

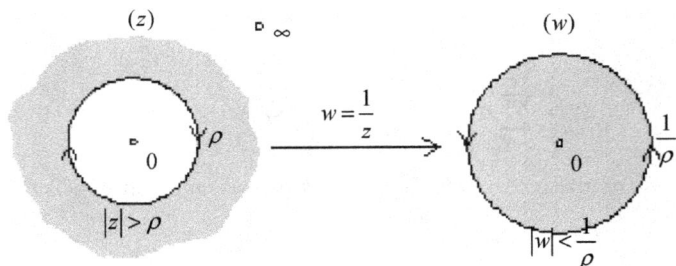

Fig. 4.55

or the interior of the circle $|z| = \rho$. This is because $w = \frac{1}{z}$ will imply that $\operatorname{Arg} w = -\operatorname{Arg} z$. See Fig. 4.55. Recall that the function $g(w) = f\left(\frac{1}{w}\right)$ is analytic in $0 < |w| < \frac{1}{r}$. Hence (refer to (4.9.1.2)),

$\int_{|z|=\rho} f(z)dz$ (integrated along the *counterclockwise* direction of $|z| = \rho$)

$$= \int_{|w|=\frac{1}{\rho}} f\left(\frac{1}{w}\right) \frac{-1}{w^2} dw \quad \left(\text{along the } \textit{clockwise} \text{ direction of } |w| = \frac{1}{\rho}\right)$$

$$= \int_{|w|=\frac{1}{\rho}} f\left(\frac{1}{w}\right) \frac{1}{w^2} dw = \int_{|w|=\frac{1}{\rho}} g(w) \frac{dw}{w^2}$$

$$\left(\text{both in the } \textit{counterclockwise} \text{ direction of } |w| = \frac{1}{\rho}\right) \qquad (4.11.1.2)$$

Based on this fact and (4.11.1.1), we give the

Definition of residue at ∞. Let f be analytic in $0 \leq r < |z| < \infty$. Then

(1) $-a_{-1}$ is the *negative* of the coefficient a_{-1} of the term z^{-1} in the Laurent series expansion of f at ∞.

$\Leftrightarrow$ (2) For any ρ, $r < \rho < \infty$, then $\frac{1}{2\pi i} \int_{|z|=\rho} f(z)dz$ (in clockwise direction) $= -a_{-1}$ or $-\frac{1}{2\pi i} \int_{|z|=\rho} f(z)dz = -a_{-1}$ ($|z| = \rho$ is oriented in the counterclockwise direction).

$\Leftrightarrow$ (3) $-a_{-1}$ is the unique constant so that $f(z) - \frac{a_{-1}}{z}$ has a (single-valued) primitive in $r < |z| < \infty$.

The number $-a_{-1}$ is called the *residue* of f at the isolated singularity ∞ and is denoted as

$$\operatorname{Re} s(f; \infty) \quad \text{or} \quad -\operatorname{Re} s(f(z); \infty) \quad \text{or} \quad \operatorname{Re} s_{z=\infty} f(z).$$

Note that (see (4.11.1.2))

$$\mathrm{Re}\,s(f(z);\infty) = -\mathrm{Re}\,s\left[\frac{1}{z^2}f\left(\frac{1}{z}\right);0\right]. \qquad (4.11.1.3)$$

Remark (The curiosity of residue at ∞ if ∞ is a removable singularity). In case z_0 is a *finite* point and is a removable singularity of f, then $\mathrm{Re}\,s(f;z_0) = 0$ *always* holds. This is not necessarily true if ∞ is a removable singularity. For instance,

$$\mathrm{Re}\,s\left[\frac{1}{z};\infty\right] = -1; \quad \mathrm{Re}\,s\left[\frac{z^3 - z^2 + 1}{z^3};\infty\right] = 1 \quad \text{and} \quad \mathrm{Re}\,s\left[\frac{1}{z^2};\infty\right] = 0.$$

$\square$

Residues have an important property, namely,

The conformal invariance of a residue. Let $g(z)$ be analytic at a point $z_0 \in \mathbf{C}^*$ with $g'(z_0) \neq 0$ or have a simple pole at z_0. Let $w_0 = g(z_0)$ be an isolated singularity of $f(w)$. Then,

(1) the inverse function $z = g^{-1}(w)$ from a neighborhood U of w_0 onto a neighborhood O of z_0 is either analytic at w_0 or has a simple pole at w_0; and

(2) $$\mathrm{Re}\,s(f(g(z)); z_0) = \mathrm{Re}\,s(f(w)(g^{-1})'(w); w_0) \quad \text{or}$$

$$\mathrm{Re}\,s(f(g(z))g'(z); z_0) = \mathrm{Re}\,s(f(w); w_0) \qquad (4.11.1.4)$$

Explanation. According to (3) in (3.5.1.4), (3.5.1.11), and , by assumption, the mapping $w = g(z) : O \to U$ is univalently analytic or meromorphic and so is the inverse mapping $z = g^{-1}(w) : U \to O$. Therefore, $w = g(z)$ represents a *change of variable* between O and U. The *differential form* $f(w)dw$ (see Sec. 2.8) is subject to the invariant rule:

$$f(w)dw = f(g(z))g'(z)dz, \quad w \in U, \quad z = g^{-1}(w) \in O. \qquad (4.11.1.5)$$

This means that $f(w)dw$ is invariant no matter in the w-coordinate or in the z-coordinate. Consequently, we can define *the residue* of the differential $f(w)dw$ at w_0 by

$$\mathrm{Re}\,s(f(w)dw; w_0) = \mathrm{Re}\,s(f; w_0). \qquad (4.11.1.6)$$

Now, (4.11.1.4) tells us the fact that *the residue of the differential $f(w)dw$ is a conformal invariant*. It is right this property of the residue that enables us to define the residue of a differential 1-form on an abstract Riemann surface. Refer to Ref. [9], and other books about Riemann surfaces, such as Refs. [75, 32, and 33].

Proof. Firstly, let z_0 be a finite point and $w = g(z)$ be analytic at z_0:

Choose $\rho > 0$ so that $f(w)$ is analytic in $0 < |w - w_0| \le \rho$ and $z = g^{-1}(w)$ exists as a univalent function in $|w - w_0| \le \rho$. Then

$$\operatorname{Re} s(f(w); w_0) = \frac{1}{2\pi i} \int_{|w-w_0|=\rho} f(w)\,dw$$

$$= \frac{1}{2\pi i} \int_\gamma f(g(z))g'(z)\,dz = \operatorname{Re} s(f(g(z))g'(z); z_0)$$

where $\gamma = g^{-1}(\{|w - w_0| = \rho\})$ is a rectifiable Jordan closed curve having z_0 in its interior $\operatorname{Int}\gamma$.

Or, denote $\operatorname{Re} s(f(w); w_0) = b_{-1}$ temporarily as in (4.11.1.1). Let $F(w)$ be a primitive of $f(w) - \frac{b_{-1}}{w-w_0}$ in $0 < |w - w_0| < \rho$. Then

$$f(g(z))g'(z) = F'(g(z))g'(z) + \frac{b_{-1}g'(z)}{g(z) - g(z_0)}, \quad 0 < |z - z_0| < R. \quad (*_1)$$

Since $F'(g(z))g'(z) = (F \circ g)'(z)$,

$$\operatorname{Re} s((F \circ g)'(z); z_0) = \frac{1}{2\pi i} \int_{|z-z_0|=r} (F \circ g)'(z)\,dz = 0;$$

since $g'(z_0) \ne 0$,

$$\operatorname{Re} s\left[\frac{g'(z)}{g(z) - g(z_0)}; z_0\right] = \frac{1}{2\pi i} \int_{|z-z_0|=r} \frac{g'(z)}{g(z) - g(z_0)}\,dz$$

$$= \frac{1}{2\pi i} \int_\Gamma \frac{dw}{w - w_0} = 1$$

where $\Gamma = g(\{|z - z_0| = r\})$ is a rectifiable Jordan closed curve. Substituting these results into $(*_1)$, we obtain $\operatorname{Re} s(f(g(z))g'(z); z_0) = b_{-1} = \operatorname{Re} s(f(w); w_0)$.

In case z_0 is a simple pole of $w = g(z)$: Let $g(z) = \frac{\varphi(z)}{z-z_0}$, $0 < |z - z_0| < R$, where φ is analytic and $\varphi(z) \ne 0$ throughout

$|z - z_0| < R$. Note $g(z_0) = w_0 = \infty$ in this case. Let $F(w)$ be a primitive of $f(w) - \frac{b_{-1}}{w}$ in $r < |w| < \infty$. Then

$$f(w) = F'(w) + \frac{b_{-1}}{w}, \quad r < |w| < \infty$$

$$\Rightarrow \left(\text{since } g'(z) = \frac{\varphi'(z)}{z - z_0} - \frac{\varphi(z)}{(z - z_0)^2}, \ 0 < |z - z_0| < R \right)$$

$$f(g(z))g'(z) = F'(g(z))g'(z) + \frac{b_{-1}}{g(z)} g'(z)$$

$$= F'(g(z))g'(z) + b_{-1} \left\{ \frac{\varphi'(z)}{\varphi(z)} - \frac{1}{z - z_0} \right\},$$

$$0 < |z - z_0| < R. \qquad (*_2)$$

Since $F'(g(z))g'(z) = (F \circ g)'(z)$ and $\frac{\varphi'(z)}{\varphi(z)}$ is analytic in $|z - z_0| < R$, hence, for $0 < \rho < R$,

$$\int_{|z - z_0| = \rho} F'(g(z))g'(z)dz = 0; \quad \int_{|z - z_0| = \rho} \frac{\varphi'(z)}{\varphi(z)} dz = 0,$$

and

$$\frac{1}{2\pi i} \int_{|z - z_0| = \rho} \frac{1}{z - z_0} dz = 1.$$

Substituting these results into $(*_2)$, it follows that $\operatorname{Re} s(f(g(z))g'(z); z_0) = -b_{-1} = \operatorname{Re} s(f(w); \infty)$.

Secondly, let $z_0 = \infty$ and ∞ be a simple pole of $w = g(z)$ (the case that $g(z)$ is analytic at ∞ is left to the readers):

Then $g(z) = z\varphi(z), r < |z| < \infty$, where $\varphi(z)$ is analytic and $\varphi(z) \neq 0, \infty$ throughout $r < |z| \leq \infty$. Then

$$f(w) = F'(w) + \frac{b_{-1}}{w}, \quad r < |w| < \infty$$

$$\Rightarrow f(g(z))g'(z) = F'(g(z))g'(z) + \frac{b_{-1}}{g(z)} g'(z)$$

$$= (F \circ g)'(z) + b_{-1} \left\{ \frac{\varphi'(z)}{\varphi(z)} + \frac{1}{z} \right\}, \quad r < |z| < \infty. \qquad (*_3)$$

For $r < \rho < \infty$, we have

$$\int_{|z| = \rho} (F \circ g)'(z)dz = 0; \quad \int_{|z| = \rho} \frac{\varphi'(z)}{\varphi(z)} dz = 0, \quad \text{and} \quad \frac{1}{2\pi i} \int_{|z| = \rho} \frac{dz}{z} = 1.$$

Substituting these results into $(*_3)$, we have (recall where the minus sign appears (see (2) in (4.11.1.3)) $\operatorname{Re} s(f(g(z))g'(z); \infty) = -b_{-1} = \operatorname{Re} s(f(w); \infty)$. □

Suppose f is analytic in $\mathbf{C}^*$ except a finite number of isolated singularity $z_1, \ldots, z_n$ in $\mathbf{C}$ and, probably, the point ∞. Choose $R > 0$ large enough so that the points $z_1, \ldots, z_n$ all lie in $|z| < R$. By using (1) in $(3.4.2.4)'$ or Exercise A (2) of Sec. 3.4.2 or Sec. 4.11.3,

$$\int_{|z|=R} f(z)dz = \sum_{k=1}^{n} \int_{\gamma_k} f(z)dz = 2\pi i \sum_{k=1}^{n} \operatorname{Re} s(f; z_k)$$

where each $\gamma_k : |z - z_k| = r_k$ is a circle, $1 \le k \le n$, and the disks $|z - z_k| \le r_k$ are pairwise disjoint. But

$$-\int_{|z|=R} f(z)dz = 2\pi i \operatorname{Re} s(f; \infty).$$

Hence, we obtain the following

Sum of residues of an analytic function in $\mathbf{C}^*$*, having a finite number of isolated singularities.* Let f be analytic in $\mathbf{C}^*$, having a finite number of isolated singularities at $z_1, \ldots, z_n$ and probably at ∞. Then

$$\sum_{k=1}^{n} \operatorname{Re} s(f; z_k) + \operatorname{Re} s(f; \infty) = 0,$$

namely, the sum of the residues of f at $z_1, \ldots, z_n$ and ∞ is equal to zero.

$$(4.11.1.7)$$

This identity can be used to evaluate the residue of f at an isolated singularity (usually, a harder one to compute!) if the remaining n residues of f are known already.

Exercises A

(1) Prove, in detail, the case that $z_0 = \infty$ and $w = g(z)$ is analytic at ∞ in (4.11.1.4).

(2) In (4.11.1.7), show that f has a single-valued primitive in $\mathbf{C}^*$ if and only if $\operatorname{Re} s(f; z_k) = 0, 1 \le k \le n$.

Exercises B

Here, the readers are asked to be familiar with the contents in Sec. 2.8, (2.9.13), Exercise A (22) of Sec. 2.9 and Exercise B (8) of Sec. 3.4.2. For the sake of reference, we repeat some of these results in the

Generalized Cauchy's integral theorem and formula. Let γ be a rectifiable Jordan closed curve in $\mathbf{C}$. Suppose f is C^1 in $\overline{\text{Int}\,\gamma}$ (i.e., the complex partial derivatives of f exist and are continuous on $\overline{\text{Int}\,\gamma}$). Then

(1) *Cauchy's integral theorem:* $z = x + iy$

    (i) $\displaystyle\int_\gamma f(z)dz = 2i \iint_{\text{Int}\,\gamma} \frac{\partial f}{\partial \bar{z}} dxdy = \iint_{\text{Int}\,\gamma} f_{\bar z}(z)d\bar z \wedge dz;$

    (ii) $\displaystyle\int_\gamma f(z)d\bar z = -2i \iint_{\text{Int}\,\gamma} \frac{\partial f}{\partial z} dxdy = -\iint_{\text{Int}\,\gamma} f_z(z)d\bar z \wedge dz.$

(2) *Cauchy's integral formula* (or *Cauchy–Pompieu–Green formula*):

    (i) $\displaystyle f(z) = \frac{1}{2\pi i} \int_\gamma \frac{f(\zeta)}{\zeta - z} d\zeta - \frac{1}{\pi} \iint_{\text{Int}\,\gamma} \frac{\partial f}{\partial \bar z}(\zeta) \frac{1}{\zeta - z} dxdy$ $(2idxdy = d\bar z \wedge dz);$

    (ii) $\displaystyle f(z) = -\frac{1}{2\pi i} \int_\gamma \frac{f(\zeta)}{\bar z - \bar z} d\bar\zeta - \frac{1}{\pi} \iint_{\text{Int}\,\gamma} \frac{\partial f}{\partial z}(\zeta) \frac{1}{\bar\zeta - \bar z} dxdy.$

Note that $\text{Int}\,\gamma$ can be replaced by a finitely connected domain Ω.

$$(4.11.1.8)$$

Recall again that, if f is analytic in $\text{Int}\,\gamma$ (i.e., $f_{\bar z}(z) = 0$ in $\text{Int}\,\gamma$) and continuous on $\overline{\text{Int}\,\gamma}$, then (1) turns out to be the classical Cauchy's integral theorem and (2) the classical Cauchy's integral formula. By the way, these formulas are important and basic in the study of singular integral operators and the Beltrami equations, both leading to the study of planar quasiconformal mappings. See Ref. [3].

Suppose f is continuous in $0 < |z - z_0| < R$. In this case, we also call z_0 an *isolated singularity* of f. V. C. Poor, in 1930, introduced *the residue of f at z_0* as

$$\mathrm{Re}\,s(f; z_0) = \lim_{r \to 0} \frac{1}{2\pi i} \int_{|z - z_0| = r} f(z)dz; \qquad (4.11.1.9)$$

and, in 1953, introduced the *partial residues of f* on *the simply connected domain* $\text{Int}\,\gamma$ as

$$R_1(f; \text{Int}\,\gamma) = \frac{1}{2\pi i} \int_\gamma f(z)dz - \frac{1}{\pi} \iint_{\text{Int}\,\gamma} \frac{\partial f}{\partial \bar z} dxdy;$$

$$R_2(f; \text{Int}\,\gamma) = \frac{-1}{2\pi i} \int_\gamma f(z)d\bar z - \frac{1}{\pi} \iint_{\text{Int}\,\gamma} \frac{\partial f}{\partial z} dxdy, \qquad (4.11.1.10)$$

and the *total residue of f* on $\text{Int}\,\gamma$ as

$$R(f; \text{Int}\,\gamma) = R_1(f; \text{Int}\,\gamma) + R_2(f; \text{Int}\,\gamma). \qquad (4.11.1.11)$$

In (4.11.1.10) and (4.11.1.11), $\operatorname{Re} f(z) = u(z)$ and $\operatorname{Im} f(z) = v(z)$ are supposed to have partial derivatives with respect to x and y so that the integrals in the right will exist. Also, $\operatorname{Int} \gamma$ can be replaced by a bounded domain Ω.

Try to do the following problems.

(1) Let f be as in (4.11.1.8). Show that $R_1(f; \operatorname{Int} \gamma) = R_2(f; \operatorname{Int} \gamma) = 0$.

(2) Suppose f is differentiable (in the real sense) in a bounded domain Ω and continuous on $\bar{\Omega}$. Fix a point $z_0 \in \Omega$. Let $g(z) = \frac{f(z)}{z - z_0}$ in $\Omega - \{z_0\}$. Show that

$$R_1(g; \Omega - \{z_0\}) = f(z_0) \quad \text{and} \quad R_2(g; \Omega - \{z_0\}) = 0.$$

Extend these results to the function $g(z) = \frac{f(z)}{(z - z_1) \cdots (z - z_n)}$ in $\Omega -$ $\{z_1, \ldots, z_n\}$, where $z_1, \ldots, z_n$ are distinct points in Ω.

(3) Let $f(z) = \frac{1}{az + b\bar{z}}$ and Ω, the punctured disk $0 < |z| < 1$. Compute $R_1(f; \Omega)$ and $R_2(f; \Omega)$ according to $|b| \neq |a|$ and $|b| = |a|$, respectively.

4.11.2.   *The computation of residues and examples*

How to compute the residue of an analytic function at an isolated singularity? In general, we have to use the methods provided by the definitions (4.11.1.1) and (4.11.1.3): the Laurent series expansion and the integral representation of its coefficient a_{-1}. For residue at poles, there are some formulated and applicable formulas; for residues at essential singularities, it could turn out to be quite complicated in computation and, in most cases, it is evaluated via the Laurent series expansion or using (4.11.1.7) if available.

Firstly, we have the

Residue at pole of order k ($k \geq 1$). Let $z_0 \in \mathbf{C}^*$ be a pole of order k.

(1) In case $z_0 \in \mathbf{C}$:

$$\operatorname{Re} s(f; z_0) = \frac{1}{(k-1)!} \lim_{z \to z_0} \frac{d^{k-1}}{dz^{k-1}} [(z - z_0)^k f(z)]$$

$$= \frac{1}{(k-1)!} \frac{d^{k-1}}{dz^{k-1}} [(z - z_0)^k f(z)]|_{z = z_0}.$$

In particular,

(i) If z_0 is a simple pole, then $\operatorname{Re} s(f; z_0) = \lim_{z \to z_0} (z - z_0) f(z)$. Conversely, if $\lim_{z \to z_0} (z - z_0) f(z) = \alpha \neq \infty$ exists, then z_0 is a removable singularity or a simple pole and

$$\operatorname{Re} s(f; z_0) = \alpha.$$

(ii) If $f(z) = \frac{p(z)}{q(z)}$, where $p(z)$ and $q(z)$ are analytic at z_0, $p(z_0) \neq 0$ and z_0 is a simple zero of $q(z)$, then $\operatorname{Re} s(f; z_0) = \frac{p(z_0)}{q'(z_0)}$.

(2) In case $z_0 = \infty$: Then $\operatorname{Re} s(f; \infty) = -\operatorname{Re} s\left[\frac{1}{z^2} f\left(\frac{1}{z}\right); 0\right]$ and compute the latter by using (1). In particular, if $\lim_{z \to \infty} z f(z) = \alpha \neq \infty$ exists, then $\operatorname{Re} s(f; \infty) = -\alpha$.

Note: In case ∞ is a zero of order k of f, then $f(z) = O(z^{-k})$, namely, $|f(z)| \leq M|z|^{-k}$ for constant $M > 0$ if $|z|$ is large enough. Hence

$$\operatorname{Re} s(f; \infty) = \begin{cases} -a_{-1}, & \text{if } k = 1, \\ 0, & \text{if } k \geq 2. \end{cases} \tag{4.11.2.1}$$

Proof. (1) Let $f(z) = \frac{\varphi(z)}{(z - z_0)^k}$, $0 < |z - z_0| < R$, where φ is analytic at z_0 and $\varphi(z_0) \neq 0$. Fix $0 < r < R$, then

$$\operatorname{Re} s(f; z_0) = \frac{1}{2\pi i} \int_{|z - z_0| = r} \frac{\varphi(z)}{(z - z_0)^k} dz = (k - 1)! \left. \frac{d^{k-1}}{dz^{k-1}} \varphi(z) \right|_{z = z_0}$$

$$= (k - 1)! \varphi^{(k-1)}(z_0).$$

This is the required formula. Or, by using the Laurent series expansion of f at z_0,

$$f(z) = \frac{a_{-k}}{(z - z_0)^k} + \cdots + \frac{a_{-1}}{z - z_0} + a_0 + \psi(z), \quad 0 < |z - z_0| < R,$$

where ψ is analytic at z_0.

$$\Rightarrow (z - z_0)^k f(z) = a_{-k} + a_{-k+1}(z - z_0) + \cdots + a_{-1}(z - z_0)^{k-1}$$

$$+ a_0(z - z_0)^k + \psi(z)(z - z_0)^k, \quad 0 < |z - z_0| < R.$$

Differentiate both sides of the above identity $(k - 1)$ times and then, put $z = z_0$.

In case z_0 is a simple pole, let $k = 1$ and the result follows.

In case $f(z) = \frac{p(z)}{q(z)}$, then z_0 is a simple pole of f by assumption and

$$\operatorname{Re} s(f; z_0) = \lim_{z \to z_0} (z - z_0) f(z) = \lim_{z \to z_0} \frac{p(z)}{\frac{q(z) - q(z_0)}{z - z_0}} = \frac{p(z_0)}{q'(z_0)}.$$

(2) By assumption, $\lim_{z \to \infty} z f(z) = \alpha \neq \infty$ if and only if

$$\lim_{z \to 0} \frac{1}{z} f\left(\frac{1}{z}\right) = \alpha = \lim_{z \to 0} z \left[\frac{1}{z^2} f\left(\frac{1}{z}\right)\right]$$

$$= \operatorname{Re} s\left(\frac{1}{z^2} f\left(\frac{1}{z}\right); 0\right) = -\operatorname{Re} s(f; \infty). \qquad \square$$

We give a series of examples.

Example 1. Find the residues of each of the following functions $f(z)$ at their isolated singularities.

(1) $\dfrac{\sin z}{z^4}$.

(2) $e^{\frac{1}{z^2}}$.

(3) $\dfrac{(1+z^5)\sinh z}{z^6}$.

(4) $\dfrac{\cosh z}{z^2(z+\pi i)^3}$.

(5) $\tan z$.

(6) $\dfrac{e^z}{\sinh z}$.

Solution.

(1) Since $\lim_{z\to 0}\frac{\sin z}{z}=1$, $z=0$ is a pole of order 3 of f. It is known that $\sin z = \sum_0^\infty (-1)^n \frac{z^{2n+1}}{(2n+1)!}$, therefore $f(z)=\frac{1}{z^3}-\frac{1}{3!}\cdot\frac{1}{z}+\frac{1}{5!}\cdot z+\cdots$, $0< |z|<\infty$. It follows that
$$\operatorname{Re}s(f;0)=-\frac{1}{6}.$$
$\Rightarrow$ (∞ is an essential singularity. Use (4.11.1.7).) $\operatorname{Re}s(f;\infty)= -\left(-\frac{1}{6}\right)=\frac{1}{6}$.

(2) $f(z)=\sum_0^\infty \frac{1}{n!}\frac{1}{z^{2n}}, 0<|z|<\infty$. $z=0$ is an essential singularity and $z=\infty$ is a removable one. Hence $\operatorname{Re}s(f;0)=\operatorname{Re}s(f;\infty)=0$. Note that f is an even function.

(3) Since $\lim_{z\to 0}\frac{\sinh z}{z}=1$, 0 is a pole of order 5 of f; while ∞ is an essential singularity. It is known that $\sinh z=\sum_0^\infty \frac{z^{2n+1}}{(2n+1)!}$, $|z|<\infty$. Therefore,
$$f(z)=\frac{(1+z)^5}{z^6}\left\{z+\frac{1}{3!}z^5+\frac{1}{5!}z^5+\cdots\right\}$$
$$=\frac{1}{z^5}+\frac{1}{3!}\cdot\frac{1}{z^3}+\frac{1}{5!}\cdot\frac{1}{z}+1+\frac{1}{7!}\cdot z+\cdots,\quad 0<|z|<\infty,$$
$$\Rightarrow \operatorname{Re}s(f;0)=\frac{1}{5!}\quad\text{and}\quad \operatorname{Re}s(f;\infty)=-\frac{1}{5!}.$$

(4) f has a pole of order 2 at 0, a pole of order 3 at $-\pi i$ and an essential singularity at ∞. Then
$$\operatorname{Re}s(f;0)=\frac{1}{1!}\lim_{z\to 0}\frac{d}{dz}\{z^2 f(z)\}$$

$$= \lim_{z \to 0} \frac{(z + \pi i) \sinh z - 3 \cosh z}{(z + \pi i)^4}$$

$$= -\frac{3}{\pi^4};$$

$$\operatorname{Re} s(f; -\pi i) = \frac{1}{2!} \lim_{z \to -\pi i} \frac{d^2}{dz^2} \{(z + \pi i)^3 f(z)\}$$

$$= \frac{1}{2!} \lim_{z \to -\pi i} \frac{-4z \sinh z + (z^2 + 6) \cosh z}{z^4}$$

$$= \frac{\pi^2 - 6}{2\pi^4};$$

$$\operatorname{Re} s(f; \infty) = -[\operatorname{Re} s(f; 0) + \operatorname{Re} s(f; -\pi i)] = -\frac{\pi^2 - 12}{2\pi^4}.$$

(5) f has simple pole at each of the points $n\pi + \frac{\pi}{2} (n = 0, \pm 1, \pm 2, \ldots)$, while ∞ is not an isolated singularity. Then $\operatorname{Re} s \left(f; n\pi + \frac{\pi}{2}\right) = -1, n = 0, \pm 1, \pm 2, \ldots$.

(6) f has simple pole at each of the points $n\pi i \ (n = 0, \pm 1, \pm 2, \ldots)$, Then

$$\operatorname{Re} s(f; n\pi i) = \frac{e^{n\pi i}}{\cosh(n\pi i)} = \frac{(-1)^n}{(-1)^n} = 1, \quad n = 0, \pm 1, \ldots .$$

Note that ∞ is not an isolated singularity.

Example 2. Find the residues for $f(z)$ at isolated singularities.

(1) $\dfrac{(z^2 - 1)^2}{z^2(z - a)(z - b)}$, $ab = 1$ and $a \neq b$.

(2) $\dfrac{e^{z^{-1}}}{1 - z}$.

(3) $\dfrac{e^{z^{-1}} z^n}{1 + z}$, $n = 1, 2, 3, \ldots$.

Solution.

(1) f has a pole of order 2 at 0, simple pole at a and b, and a removable singularity at ∞. It is easy to find

$$\operatorname{Re} s(f; a) = \frac{(z^2 - 1)^2}{z^2(z - b)}\bigg|_{z=a} = \frac{(a^2 - 1)^2}{a^2(a - b)} = \frac{a^2(a - b)^2}{a^2(a - b)} = a - b;$$

$$\operatorname{Re} s(f; b) = b - a.$$

To find the residue at 0, by (4.11.2.1),

$$\operatorname{Re} s(f;0) = \frac{1}{1!}\frac{d}{dz}\{z^2 f(z)\}|_{z=0} = \frac{a+b}{(ab)^2} = a+b;$$

or, by the Laurent series expansion of f at 0,

$$\frac{(z^2-1)^2}{z^2(z-a)(z-b)} = \frac{(z^2-1)^2}{z^2(a-b)}\left\{\frac{1}{z-a} - \frac{1}{z-b}\right\}$$

$$= \frac{(z^2-1)^2}{z^2(a-b)}\left\{\frac{1}{b} - \frac{1}{a} + \sum_1^\infty\left(\frac{1}{b^{n+1}} - \frac{1}{a^{n+1}}\right)z^n\right\},$$

$$\Rightarrow \operatorname{Re} s(f;0) = \frac{1}{a-b} = \left(\frac{1}{b^2} - \frac{1}{a^2}\right) = a+b.$$

Consequently, $\operatorname{Re} s(f;\infty) = -[\operatorname{Re} s(f;a) + \operatorname{Re} s(f;b) + \operatorname{Re} s(f;0)] = -(a+b)$.

(2) 1 is a simple pole of f, 0 is an essential singularity and ∞ is a zero of order 1. If $1 < |z| < \infty$,

$$f(z) = -\frac{1}{z}\cdot\frac{1}{1-\frac{1}{z}}e^{\frac{1}{z}} = -\frac{1}{z}\left(\sum_0^\infty\frac{1}{z^n}\right)\left(\sum_0^\infty\frac{1}{n!}\frac{1}{z^n}\right) = -\frac{1}{z} + \cdots,$$

$$\Rightarrow \operatorname{Re} s(f;\infty) = 1.$$

Since $\operatorname{Re} s(f;1) = -e$, $\operatorname{Re} s(f;0) = -[\operatorname{Re} s(f;1) + \operatorname{Re} s(f;\infty)] = e-1$.

(3) f has a simple pole at -1, an essential singularity at 0 and a pole of order $n-1$ at ∞ $\left(\text{observe that } f(\frac{1}{z}) = \frac{e^z}{z^{n-1}(1+z)}\right)$. According to (2) in (4.11.2.1),

$$\operatorname{Re} s(f;\infty) = -\operatorname{Re} s\left(\frac{1}{z^2}f\left(\frac{1}{z}\right);0\right)$$

$$= -\operatorname{Re} s\left(\frac{e^z}{z^{n+1}(1+z)};0\right) = \frac{-1}{n!}\lim_{z\to0}\frac{d^n}{dz^n}\left\{\frac{e^z}{1+z}\right\}$$

$$= \frac{-1}{n!}\lim_{z\to0}\sum_{k=0}^n C_k^n\frac{(-1)^k k! e^z}{(1+z)^{k+1}} = \sum_{k=0}^n\frac{(-1)^{k+1}}{(n-k)!}.$$

Since $\operatorname{Re} s(f;-1) = (-1)^n e^{-1}$,

$$\operatorname{Re} s(f;0) = -\operatorname{Re} s(f;-1) - \operatorname{Re} s(f;\infty)$$

$$= (-1)^{n+1}e^{-1} - \sum_{k=0}^n\frac{(-1)^{k+1}}{(n-k)!} = \sum_{k=1}^\infty\frac{(-1)^{k+1}}{(n+k)!}.$$

Or, by using the Laurent series expansions

$$e^{z^{-1}} = \sum_{k=0}^{\infty} \frac{1}{k!z^k}, \quad 0 < |z| < \infty, \quad \text{and} \quad \frac{1}{1+z} = \sum_{k=0}^{\infty} (-1)^k z^k, \quad |z| < 1,$$

the Cauchy product process of these two series will enable us to find the coefficient $\operatorname{Re} s(f;0)$ of z^{-1}.

Example 3. Find the residues for $f(z)$ at isolated singularities.

(1) $\dfrac{\sin \pi z}{(z-1)^3}$.

(2) $\dfrac{\sin z}{\cos z^3 - 1}$.

(3) $z \cos \dfrac{1}{z-1}$.

(4) $\exp\left(z + \dfrac{1}{z}\right)$.

Solution.

(1) $\sin \pi z$ has a simple zero at $z = 1$ and an essential singularity at ∞. Hence f has a pole of order 2 at $z = 1$ and an essential singularity at ∞ (recall that $f\left(\frac{1}{z}\right) = z^3(1-z)^{-3}\sin\frac{\pi}{z}$ has an essential singularity at 0). Using the Laurent series expansion

$$f(z) = \frac{1}{(z-1)^3}\sum_{n=0}^{\infty}(-1)^{n+1}\frac{\pi^{2n+1}}{(2n+1)!}(z-1)^{2n+1}$$

$$= -\frac{\pi}{(z-1)^2} + \frac{\pi^3}{3!} - \frac{\pi^5}{5!}(z-1)^2 + \cdots, \quad 0 < |z-1| < \infty,$$

$$\Rightarrow \operatorname{Re} s(f;1) = 0 \quad \text{and} \quad \operatorname{Re} s(f;\infty) = 0.$$

We can also obtain these two results directly by noting that $f(z) = -\frac{\sin\pi(z-1)}{(z-1)^3}$ is an even function in $z-1$ so that all the terms of the odd powers of $z - z_0$ disappear. Alternatively,

$$\operatorname{Re} s(f;1) = \lim_{z\to 1}\frac{d}{dz}\{(z-1)^2 f(z)\} = \lim_{z\to 1}\frac{\pi(z-1)\cos\pi z - \sin\pi z}{(z-1)^2}$$

$$= \lim_{z\to 1}\frac{-\pi^2(z-1)\sin\pi z}{2(z-1)} = -\frac{\pi^2}{2}\cdot 0 = 0$$

and $\operatorname{Re} s(f;\infty) = -\operatorname{Re} s(f;1) = 0$.

(2) Note that $\sin z$ has a simple zero at $z = 0$. Also, $\cos z^3 - 1 = 0$ if and only if $z^3 = 2n\pi$ $(n = 0, \pm 1, \pm 2, \ldots)$. Then $z_{nk} = \sqrt[3]{|2n\pi|}e^{\frac{i(2k\pi + \operatorname{Arg} n)}{3}}$,

$k = 0, 1, 2$ and $n = \pm 1, \pm 2, \ldots$, and $z_{01} = z_{02} = z_{03} = 0$ are zeros of $\cos z^3 - 1$.

Let $g(z) = \cos z^3 - 1$ temporarily. Then $g(0) = g'(0) = \cdots = g^{(5)}(0) = 0$ yet $g^{(6)}(0) = -360 \neq 0$, and $z = 0$ is a zero of order 6 of $g(z)$. Or, by using the Taylor series expansion, $\cos z^3 - 1 = \sum_1^\infty (-1)^n \frac{1}{(2n)!} z^{6n} = z^6\{-\frac{1}{2!} + \frac{1}{4!}z^6 - \cdots\}$ shows that 0 is a zero of order 6 of $g(z)$. Using the algorithm of long division (see Sec. 5.1 or (4.8.1)),

$$f(z) = \frac{z \sum_0^\infty (-1)^n \frac{z^{2n}}{(2n+1)!}}{z^6 \sum_1^\infty (-1)^n \frac{z^{6n-6}}{(2n)!}}$$

$$= \frac{1}{z^5}\left\{-2 + \frac{z^2}{3} - \frac{z^4}{60} + O(z^5)\right\}, \quad 0 < |z| < 1$$

$$\Rightarrow \operatorname{Re} s(f;0) = -\frac{1}{60}.$$

What is the order of z_{nk} ($k = 0, 1, 2$) and $n = \pm 1, \pm 2, \ldots$ as a zero of g? Since $g(z_{nk}) = g'(z_{nk}) = 0$ and $g''(z_{nk}) \neq 0$, z_{nk} is a zero of order 2 of g and hence, is a pole of order 2 of f because $\sin z_{nk} \neq 0$. According to (4.11.2.2) below,

$$\operatorname{Re} s(f; z_{nk}) = \frac{1}{1!}\frac{d}{dz}\{(z - z_{nk})^2 f(z)\}|_{z=z_{nk}}$$

$$= 2\frac{h'(z_{nk})}{g''(z_{nk})} - \frac{2}{3}\frac{h(z_{nk})g'''(z_{nk})}{[g''(z_{nk})]^2},$$

where $h(z) = \sin z$ and $g(z) = \cos z^3 - 1$: $h(z_{nk}) = \sin z_{nk}$, $h'(z_{nk}) = -\cos z_{nk}$, $g''(z_{nk}) = -9z_{nk}^4$ and $g'''(z_{nk}) = -18z_{nk}^4 - 36z_{nk}^3 = -18z_{nk}^4 - 72n\pi$, for $k = 0, 1, 2$ and $n = \pm 1, \pm 2, \ldots$.

(3) Since $\cos w$ has an essential singularity at $w = \infty$, f has an essential singularity at $z = 1$, and, by noting $f(\frac{1}{z}) = \frac{1}{z}\cos\frac{z}{1-z}$, f has a simple pole at $z = \infty$. Using

$$\cos w = \sum_{n=0}^\infty (-1)^n \frac{w^{2n}}{(2n)!}, \quad |w| < \infty,$$

it follows that

$$\cos \frac{1}{z-1} = \sum_{n=0}^\infty (-1)^n \frac{1}{(2n)!}(z-1)^{-2n}, \quad 0 < |z-1| < \infty,$$

$$\Rightarrow f(z) = (z-1)\cos\frac{1}{z-1} + \cos\frac{1}{z-1}$$

$$= (z-1) + 1 - \frac{1}{2!}\cdot\frac{1}{z-1} - \frac{1}{2!}\cdot\frac{1}{(z-1)^2}$$

$$+ \sum_{2}^{\infty} (-1)^n \frac{1}{(2n)!} \left[\frac{1}{(z-1)^{2n-1}} - \frac{1}{(z-1)^{2n}} \right],$$

$$0 < |z-1| < \infty,$$

$$\Rightarrow \operatorname{Re} s(f;1) = -\frac{1}{2},$$

$$\Rightarrow \operatorname{Re} s(f;\infty) = -\operatorname{Re} s(f;1) = \frac{1}{2}.$$

(4) Since $f\left(\frac{1}{z}\right) = f(z)$, both 0 and ∞ are essential singularities of f. By the Cauchy product of the series $e^z = \sum_0^\infty \frac{1}{n!} z^n$, $|z| < \infty$, and $e^{-z} = \sum_0^\infty \frac{1}{n!} z^{-n}$, $|z| > 0$, we obtain the Laurent series expansion of f at 0 and at ∞

$$f(z) = \sum_{n=-\infty}^{\infty} \left\{ \sum_{k=\max\{0,n\}}^{\infty} \frac{1}{k!(k-n)!} \right\} z^n, \quad 0 < |z| < \infty.$$

Or, the coefficient a_n in $f(z) = \sum_{-\infty}^{\infty} a_n z^n, 0 < |z| < \infty$, is given by

$$a_n = \frac{1}{2\pi i} \int_{|z|=1} \frac{f(z)}{z^{n+1}} dz = \frac{1}{2\pi i} \sum_{k=0}^{\infty} \frac{1}{k!} \int_{|z|=1} z^{k-n-1} e^{z^{-1}} dz,$$

where

$$\int_{|z|=1} z^{k-n-1} e^{z^{-1}} dz = \sum_{l=0}^{\infty} \frac{1}{l!} \int_{|z|=1} z^{k-n-1} \frac{1}{z^l} dz$$

$$= \begin{cases} \dfrac{2\pi i}{(k-n)!}, & k-n \geq 0, \\ 0, & k-n < 0. \end{cases}$$

Consequently

$$\operatorname{Re} s(f;0) = a_{-1} = \sum_{k=0}^{\infty} \frac{1}{k!(k+1)!} \quad \text{and} \quad \operatorname{Re} s(f;\infty) = -a_{-1}.$$

Example 4. Find the residues, at the isolated singularities, for the branches of the following multiple-valued functions $f(z)$.

(1) $\dfrac{(\log z)^2}{(1+z)^2}$.

(2) $\dfrac{1}{\sqrt[3]{(1-z)(1+z)^2}}$.

(3) $\dfrac{1}{z^2 - \tan^{-1} z^2}$ at $z = 0$.

(4) $\dfrac{\log z}{z^2 + a^2}$ $(a > 0)$, where $\log z = \log|z| + i\,\text{Arg}\,z,\ z \in \mathbf{C} - \{re^{i\alpha} \mid r \geq 0\}$
with α a fixed real number and $\alpha - 2\pi < \text{Arg}\,z < \alpha$.

Solution.

(1) In $\mathbf{C}^* - [0, \infty]$, $\log z$ has infinitely many branches $(\log z)_k = \log|z| + i(\text{Arg}\,z + 2k\pi)$, $0 < \text{Arg}\,z < 2\pi$ and $k = 0, \pm 1, \pm 2, \ldots$, each completely determined by $\log(-1) = (2k+1)\pi i$. For each fixed k, by Example 1 in Sec. 3.3.2,

$$(\log z)_k = \log[-1 + (z + 1)]$$

$$= (2k+1)\pi i - \sum_{n=1}^{\infty} \frac{1}{n}(z+1)^n, \quad |z + 1| < 1,$$

$$\Rightarrow ((\log z)_k)^2 = -(2k+1)^2\pi^2 - 2(2k+1)\pi i(z+1) + \cdots,$$

$$|z + 1| < 1,$$

$$\Rightarrow \text{Re}\,s\left(\frac{((\log z)_k)^2}{(1+z)^2}; -1\right) = -2(2k+1)\pi i, \quad k = 0, \pm 1, \pm 2, \ldots.$$

Alternatively, this residue can be evaluated by

$$\frac{1}{1!} \lim_{z \to -1} \frac{d}{dz}\left\{(1+z)^2 \cdot \frac{((\log z)_k)^2}{(1+z)^2}\right\} = 2 \cdot \frac{1}{z}(\log z)_k\Big|_{z=-1}$$

$$= -2(\log(-1))_k = -2(2k+1)\pi i.$$

(2) The function f has three branches in $\mathbf{C}^* - [-1, 1]$ (refer to Exercise B (4) of Sec. 2.7.2):

$$f_1(z) = \begin{cases} |(1-z)(1+z)^2|^{-\frac{1}{3}} e^{-\frac{i\text{Arg}(1-z)(1+z)^2}{3}}, & \text{Im}\,z \geq 0 \\ |(1-z)(1+z)^2|^{-\frac{1}{3}} e^{-\frac{i\text{Arg}(1-z)(1+z)^2}{3} + \frac{2\pi i}{3}}, & \text{Im}\,z < 0 \end{cases};$$

$$f_2(z) = f_1(z)e^{\frac{2\pi i}{3}}; \quad \text{and} \quad f_3(z) = f_1(z)e^{\frac{4\pi i}{3}},$$

where $f_1(z)$ is determined by the condition that $\text{Arg}\,f(x) = 0$ on the upper side of $-1 < x < 1$. They all are analytic at ∞.

To find $\text{Re}\,s(f_1; \infty)$: In case $R > 1$, then $\text{Arg}\,f_1(R) = -\frac{\pi}{3}$. Consider

$$f_1(z) = e^{-\frac{\pi i}{3}} \cdot \frac{1}{z} \cdot \left(1 - \frac{1}{z}\right)^{-\frac{1}{3}} \left(1 + \frac{1}{z}\right)^{-\frac{2}{3}}$$

where both $\left(1 - \frac{1}{z}\right)^{-\frac{1}{3}}$ and $\left(1 + \frac{1}{z}\right)^{-\frac{2}{3}}$ are the branches, in $\mathbf{C}^* - [-1, 1]$, determined uniquely by having their respective principal argument equal to zero at $z = R$. The Laurent series expansion of each of these two branches at ∞ has its coefficient $a_0 = 1$. This results in

$$\operatorname{Re} s(f_1; \infty) = -e^{-\frac{\pi i}{3}},$$

$$\Rightarrow \operatorname{Re} s(f_2; \infty) = -e^{\frac{\pi i}{3}} \quad \text{and} \quad \operatorname{Re} s(f_3; \infty) = -e^{\pi i} = 1.$$

(3) $\tan^{-1} w$ is multiple-valued (see (2.7.4.4) or Example 2 of Sec. 4.5) and is given by $\tan^{-1} w = n\pi + \operatorname{Arc} \tan w, n = 0, \pm 1, \pm 2, \ldots$, where

$$\operatorname{Arc} \tan w = -\frac{i}{2} \operatorname{Log} \frac{1 + iw}{1 - iw} : \ \mathbf{C}^* - [i, i\infty] - [-i, -i\infty]$$

$$\to \{0 < \operatorname{Re} z < \pi\}$$

is the principal branch. Only the branch $\operatorname{Arc} \tan w$ assumes the value 0 at $w = 0$. Since $\operatorname{Arc} \tan w = \sum_0^\infty (-1)^n \frac{w^{2n+1}}{2n+1}$, $|w| < 1$, we have

$$z^2 - \operatorname{Arc} \tan z^2 = z^6 \left\{ \frac{1}{3} - \frac{1}{5} z^4 + \frac{1}{7} z^8 - \cdots \right\},$$

$$\Rightarrow f(z) = \frac{1}{z^2 - \operatorname{Arc} \tan z^2} = \frac{1}{z^6} \varphi(z), \quad 0 < |z| < 1$$

where φ is analytic at 0 and $\varphi(0) = 3$. Hence, $z = 0$ is a pole of order 6 of f. Since $\varphi(z) = 3 + \frac{9}{5} z^4 + \cdots$ is an even function (i.e., $\varphi(-z) = \varphi(z)$), so

$$\operatorname{Re} s(f; 0) = 0.$$

Refer to Exercise A (4)(b).

(4) In case $\alpha \neq \frac{\pi}{2}$: $z = \pm ai$ are simple poles of f and

$$\operatorname{Re} s(f; ai) = \lim_{z \to ai} (z - ai) f(z) = \frac{\log ai}{2ai} = \frac{1}{2ai} \left(\log a + \frac{\pi}{2} i \right);$$

$$\operatorname{Re} s(f; -ai) = \frac{-1}{2ai} \left(\log a + \frac{3\pi}{2} i \right).$$

In case $\alpha = \frac{\pi}{2}$: f has a simple pole at $-ai$ with the residue as above.
In case $\alpha = \frac{3\pi}{2}$: f has a simple pole at ai and the residue is as above.

Example 5. Try to use the method of residues to reprove (1) the Liouville's theorem. (2) The fundamental theorem of algebra.

Proof.

(1) Let f be a bounded entire function. For any two distinct fixed points a and b, the function $F(z) = \frac{f(z)}{(z-a)(z-b)}$, $z \in \mathbf{C} - \{a, b\}$, has isolated singularities at a, b and possibly, at ∞. According to (4.11.1.7),

$\operatorname{Re} s(F;a) + \operatorname{Re} s(F;b) + \operatorname{Re} s(f;\infty) = 0$. Since f is bounded and thus $\lim_{z\to\infty} zF(z) = 0$, by (2) in (4.11.2.1),

$$\operatorname{Re} s(f;\infty) = -0 = 0,$$

$$\Rightarrow \operatorname{Re} s(F;a) + \operatorname{Re} s(F;b) = 0 \quad \text{or} \quad \frac{f(a)}{a-b} + \frac{f(b)}{b-a} = 0,$$

$$\Rightarrow (\text{since } b \neq a) \quad f(a) = f(b).$$

This shows that f is a constant.

(2) Let $p(z) = a_n z^n + \cdots + a_1 z + a_0$, where $n \geq 1$ and $a_n \neq 0$. ∞ is a pole of order n of p. Choose $R > 0$ so that $|p(z)| > 0$ in $|z| \geq R$. Then

$$\frac{1}{2\pi i}\int_{|z|=R} \frac{p'(z)}{p(z)} dz = -\operatorname{Re} s\left(\frac{p'(z)}{p(z)};\infty\right) = \operatorname{Re} s\left(\frac{1}{z^2}p'\left(\frac{1}{z}\right)\cdot\frac{1}{p\left(\frac{1}{z}\right)};0\right)$$

$$= \lim_{z\to 0} z \cdot \frac{1}{z^2}p'\left(\frac{1}{z}\right)\cdot\frac{1}{p\left(\frac{1}{z}\right)} = \frac{na_n}{a_n} = n.$$

By the argument principle (3.5.3.3), this shows that $p(z)$ has n zeros in $|z| < R$. Or, by observing that, if R is large enough,

$$\frac{p'(z)}{p(z)} = \frac{n}{z} + \sum_{k\geq 2} b_k \cdot \frac{1}{z^k}, \quad R < |z| < \infty,$$

$$\Rightarrow \frac{1}{2\pi i}\int_{|z|=R} \frac{p'(z)}{p(z)} dz = \frac{1}{2\pi i}\int_{|z|=R} \frac{n}{z} dz = n.$$

Example 6. (Titchmarch, Ref. [76], p. 108). Try to use the definition of residue as a line integral to show that

$$\operatorname{Re} s\left(\tan^{2k-1}\pi z; \frac{1}{2}\right) = \frac{(-1)^k}{\pi}, \quad k = 1, 2, 3, \ldots.$$

Solution. $f(z) = \tan^{2k-1}\pi z$ has a pole of order $(2k-1)$ at $z = \frac{1}{2}$. Choose $R > 0$ and consider a rectangle γ with vertices $-iR, 1-iR, 1+iR$, and iR. See Fig. 4.56. Then

$$\operatorname{Re} s\left(f; \frac{1}{2}\right) = \frac{1}{2\pi i}\int_\gamma f(z)dz$$

$$= \frac{1}{2\pi i}\left\{\int_{-iR}^{1-iR} + \int_{1-iR}^{1+iR} + \int_{1+iR}^{iR} + \int_{iR}^{-iR}\right\} f(z)dz.$$

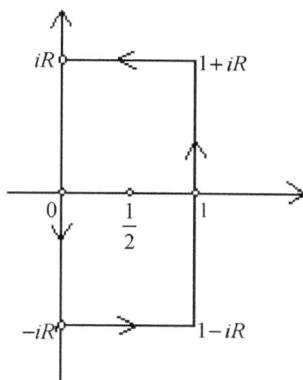

Fig. 4.56

Since $\tan \pi z$ has the primitive period 1, so

$$\int_{1-iR}^{1+iR} f(z)dz = \int_{-iR}^{iR} f(z)dz = -\int_{iR}^{-iR} f(z)dz,$$

$$\Rightarrow \operatorname{Re} s\left(f; \frac{1}{2}\right) = \frac{1}{2\pi i}\left\{\int_{-iR}^{1-iR} + \int_{1+iR}^{iR}\right\} f(z)dz.$$

But $\lim_{\operatorname{Im} z \to \infty} \tan \pi z = -\frac{1}{i}$ and $\lim_{\operatorname{Im} z \to -\infty} \tan \pi z = \frac{1}{i}$ (see (c) in (2.6.2.8)), thus

$$\lim_{R \to \infty} \int_{-iR}^{1-iR} f(z)dz = \left(\frac{1}{i}\right)^{2k-1}$$

and

$$\lim_{R \to \infty} \int_{1+iR}^{iR} f(z)dz = -\left(-\frac{1}{i}\right)^{2k-1} = \left(\frac{1}{i}\right)^{2k-1}.$$

Putting these results in $(*)$, we obtain the claimed residue of f at $\frac{1}{2}$.

To the end, we give some additional formulas to compute the residues and leave the details to the readers.

Suppose f is analytic at z_0 and $f(z_0) \neq 0$, while g has a zero of order k at z_0. Then z_0 is a pole of order k of $\frac{f}{g}$, and then

$$\frac{f(z)}{g(z)} = \frac{\sum_{n=0}^{\infty} a_n(z-z_0)^n}{\sum_{n=k}^{\infty} b_n(z-z_0)^n} = \frac{1}{(z-z_0)^k} \cdot \frac{\sum_{n=0}^{\infty} a_n(z-z_0)^n}{\sum_{n=k}^{\infty} b_n(z-z_0)^{n-k}},$$

$$a_0 \neq 0, \quad b_k \neq 0, \quad 0 < |z-z_0| < R.$$

Via the algorithm of long division (see (5)3 in (4.8.1) or Example 5 in Sec. 4.8), the division of the two power series in the right above can expressed as $\sum_0^\infty c_n(z-z_0)^n$, $|z-z_0| < R$. The coefficient c_{k-1} is then the residue of $\frac{f}{g}$ at z_0. We summarize in the

Formula for residue at a pole of order k. Let f and g be analytic at z_0. If $f(z_0) \neq 0$ and $g(z_0) = g'(z_0) = \cdots = g^{(k-1)}(z_0) = 0$, but $g^{(k)}(z_0) \neq 0$ ($k \geq 1$), then z_0 is a pole of order k of $\frac{f}{g}$ and

$$\operatorname{Re}s\left(\frac{f}{g}; z_0\right) = \frac{1}{b_k^k}\begin{vmatrix} b_k & 0 & 0 & \cdots & 0 & a_0 \\ b_{k+1} & b_k & 0 & \cdots & 0 & a_1 \\ b_{k+2} & b_{k+1} & b_k & \cdots & 0 & a_2 \\ \vdots & \vdots & \vdots & & \vdots & \vdots \\ b_{2k-2} & b_{2k-3} & b_{2k-4} & \cdots & b_k & a_{k-2} \\ b_{2k-1} & b_{2k-2} & b_{2k-3} & \cdots & b_{k+1} & a_{k-1} \end{vmatrix},$$

where

$$a_p = \frac{f^{(p)}(z_0)}{p!}, \quad 0 \leq p \leq k-1, \quad \text{and} \quad b_q = \frac{g^{(q)}(z_0)}{q!}, \quad k \leq q \leq 2k-1.$$

In particular,

(1) if $k = 1$, z_0 is a simple pole of $\frac{f}{g}$ and $\operatorname{Re}s\left(\frac{f}{g}; z_0\right) = \frac{f(z_0)}{g'(z_0)}$; and

(2) if $k = 2$, z_0 is a pole of order 2 of $\frac{f}{g}$ and

$$\operatorname{Re}s\left(\frac{f}{g}; z_0\right) = 2\frac{f'(z_0)}{g''(z_0)} - \frac{2}{3}\frac{f(z_0)g'''(z_0)}{[g''(z_0)]^2}. \tag{4.11.2.2}$$

Furthermore, we still have some established

Formulas for residues at poles. Let f and g be analytic at z_0.

(1) If z_0 is a zero of *same* order of both f and g, then z_0 is a removable singularity of $\frac{f}{g}$ and

$$\operatorname{Re}s\left(\frac{f}{g}; z_0\right) = 0.$$

(2) In case $f(z_0) \neq 0$ and $g(z) = (z-z_0)^2$, then $\operatorname{Re}s\left(\frac{f}{g}; z_0\right) = f'(z_0)$.

(3) In case $f(z_0) = 0$, $f'(z_0) \neq 0$ and $g(z_0) = g'(z_0) = g''(z_0) = 0$, $g'''(z_0) \neq 0$, then z_0 is a pole of order 2 of $\frac{f}{g}$ and

$$\operatorname{Re} s\left(\frac{f}{g}; z_0\right) = 3\frac{f''(z_0)}{g'''(z_0)} - \frac{3}{2}\frac{f'(z_0)g^{(4)}(z_0)}{[g'''(z_0)]^2}.$$

(4) If z_0 is a zero of order k of f and z_0 is a zero of order $k + l$ ($l \geq 1$) of g, then z_0 is a pole of order l of $\frac{f}{g}$ and

$$\operatorname{Re} s\left(\frac{f}{g}; z_0\right) = \frac{1}{(l-1)!}\lim_{z \to z_0}\frac{d^{l-1}}{dz^{l-1}}\left\{(z - z_0)^l\frac{f(z)}{g(z)}\right\}. \qquad (4.11.2.3)$$

As a final drop in this section, we briefly remark that residues can be used to expand a *proper* rational function into its partial fraction form. Let us go back to Example 6 in Sec. 4.10.2 and adopt notations there. Let the proper rational function $\frac{p(z)}{q(z)}$ be expanded as

$$\frac{p(z)}{q(z)} = \sum_{k=1}^{n} p_k(z), \quad \text{where } p_k(z) = \sum_{l=1}^{m_k}\frac{a_{-l}^{(k)}}{(z - z_k)^l}, \quad 1 \leq k \leq n.$$

Fix k ($1 \leq k \leq n$). Then

$$(z - z_k)^{m_k}\frac{p(z)}{q(z)} = \sum_{l=1}^{m_k}a_{-l}^{(k)}(z - z_k)^{m_k - l} + (z - z_k)^{m_k}\sum_{\substack{r=1 \\ r \neq k}}^{n}p_r(z).$$

To determine $a_{-l}^{(k)}$ ($1 \leq l \leq m_k$), we differentiate the left above ($m_k - l$) times and then let $z \to z_k$. Then,

$$a_{-l}^{(k)} = \frac{1}{(m_k - l)!}\lim_{z \to z_k}\frac{d^{m_k - l}}{dz^{m_k - l}}\left\{(z - z_k)^{m_k}\frac{p(z)}{q(z)}\right\}$$

$$= \operatorname{Re} s\left(\frac{1}{(z - z_k)^{-l+1}} \cdot \frac{p(z)}{q(z)}; z_k\right), \quad 1 \leq k \leq n \quad \text{and} \quad 1 \leq l \leq m_k.$$

$$(4.11.2.4)$$

This is especially useful when computing only one particular coefficient among $a_{-l}^{(k)}$, $1 \leq k \leq n$ and $1 \leq l \leq m_k$.

Exercises A

(1) Prove (4.11.2.2) in detail.

(2) Prove (4.11.2.3) in detail.

(3) Do the following problems:

    (i) $\operatorname{Re} s(af + bg; z_0) = a \operatorname{Re} s(f; z_0) + b \operatorname{Re} s(g; z_0)$.

    (ii) Suppose $f(z_0) \neq 0$ and g has a simple pole at z_0. Then $\operatorname{Re} s(fg; z_0) = f(z_0) \operatorname{Re} s(g; z_0)$.

    (iii) Suppose $f(z_0) \neq 0$ and g has a pole of order 2 at z_0. Then

$$\operatorname{Re} s(fg; z_0) = f'(z_0) \operatorname{Re} s((z - z_0)g(z); z_0) + f(z_0) \operatorname{Re} s(g; z_0).$$

    (iv) In case both f and g have a simple pole at z_0, what is $\operatorname{Re} s(fg; z_0)$?

    (v) If f is analytic at z_0 and z_0 is either a zero of order n or a pole of order n of g, what is

$$\operatorname{Re} s\left(f(z)\frac{g'(z)}{g(z)}; z_0\right)?$$

    (vi) If f is analytic at ∞, what is $\operatorname{Re} s(f(z)^2; \infty)$?

(4) (a) If f is an odd function, show that $\operatorname{Re} s(f; z_0) = \operatorname{Re} s(f; -z_0)$.

    (b) If f is an even function, show that $\operatorname{Re} s(f; z_0) = -\operatorname{Re} s(f; -z_0)$.

(5) Let f be a rational function whose denominator has degree at least 2 more than the numerator. Show that $\sum_{z_0 \in \mathbf{C}} \operatorname{Re} s(f; z_0) = 0$ where z_0's are poles of f.

(6) If f is an entire function, show that $\operatorname{Re} s(f; \infty) = 0$.

(7) Some special cases in computing $\operatorname{Re} s(f; \infty)$:

    (a) In case $f(z) = a_0 + \frac{a_{-1}}{z} + \frac{a_{-2}}{z^2} + \cdots$ is analytic at ∞:

        (i) If $f(\infty) \neq 0$, then $\operatorname{Re} s(f; \infty) = \lim_{z\to\infty} z^2 f'(z) = -a_{-1}$.

        (ii) If f has a simple zero at ∞, then $\operatorname{Re} s(f; \infty) = -\lim_{z\to\infty} z f(z) = -a_{-1}$.

        (iii) If z_0 is a zero of order m $(m \geq 2)$ of f, then $\operatorname{Re} s(f; \infty) = 0$.

    (b) In case $f(z) = a_1 z + a_0 + \frac{a_{-1}}{z} + \frac{a_{-2}}{z^2} + \cdots$ has a simple pole at ∞:

$$\operatorname{Re} s(f; \infty) = -\frac{1}{2} \lim_{z\to\infty} z^3 f'''(z) = -a_{-1}.$$

(8) Show that $\operatorname{Re} s\left(\frac{(1+z)^n}{z^{k+1}}; 0\right) = C_k^n,\ 0 \leq k \leq n$, where $n \geq 1$ is an integer. Then, show that

$$\operatorname{Re} s\left(\frac{(1+z)^n}{z^{k+1}} z^k \left(1 + \frac{1}{z}\right)^n; 0\right) = \sum_{k=0}^{n}(C_k^n)^2 = C_n^{2n}.$$

(9) Suppose f has only one isolated singularity z_0 $(z_0 \neq \infty)$ which is essential.

   (a) Let $g(z) = f(z_0 + \frac{1}{z})$. Show that $\operatorname{Re} s(f; z_0) = g'(0)$. Try to compute $\operatorname{Re} s(e^{\frac{1}{z}}; 0)$.

   (b) Let m be an integer and $F(z) = \frac{f(z)}{(z-z_0)^m}$. Then z_0 is still an essential singularity of F. Let $G(z) = F(z_0 + \frac{1}{z})$. Show that, if $m < 0$, $\operatorname{Re} s(f; z_0) = \frac{1}{(-m-1)!}G^{(-m-1)}(0)$. Try to deduce that $\operatorname{Re} s(z^3 e^{\frac{1}{z}}; 0) = -1$. In case $m \geq 0$, then $\operatorname{Re} s(f; z_0) = 0$.

(10) Suppose ∞ is a removable singularity or a pole of f. Let $g(z) = f(\frac{1}{z})$. Show that

$$\operatorname{Re} s\left(\frac{f'}{f}; \infty\right) = \operatorname{Re} s\left(\frac{g'}{g}; 0\right).$$

(11) If z_0 is an isolated singularity of f, then $\operatorname{Re} s(f'; z_0) = 0$ always holds.

(12) Suppose f is analytic at 0. Let $g_k(z) = z^{-k} f(z)$, $k = 0, 1, 2, \ldots$. In case $\operatorname{Re} s(g_k; 0) = 0$ for all k, then $f \equiv 0$.

(13) Suppose f is meromorphic in Ω and $z_0 \in \Omega$ is a pole of f.

   (a) If Ω is symmetric with respect to the real axis, namely, $\Omega^* = \{\bar{z} \mid z \in \Omega\} = \Omega$ holds, and $f(z)$ assumes real values if $z \in \Omega \cap \mathbf{R}$, then $\operatorname{Re} s(f; \overline{z_0}) = \overline{\operatorname{Re} s(f; z_0)}$.

   (b) Let the boundary $\partial\Omega$ contain a circular arc γ of the circle $|z - a| = R$ and Ω be symmetric with respect to γ. Suppose f assumes real values on $\bar{\Omega} \cap \gamma$. Let z_0^* denote the symmetric point of z_0 with respect to γ. In case f has the principal part $\sum_{k=1}^{n} a_k(z - z_0)^{-k}$ at z_0, then

$$\operatorname{Re} s(f; z_0^*) = \sum_{k=1}^{n}(-1)^k \frac{k\overline{a_k}(z_0^* - a)^{k+1}}{R^{2k}}.$$

   (c) In (b), suppose $|f(z)| = r$ instead if $z \in \bar{\Omega} \cap \gamma$, where $r > 0$ is a constant. Find $\operatorname{Re} s(f; z_0^*)$.

(14) Let f be analytic at 0 and n be a positive integer. In case $z_0 \neq 0$, then

$$\operatorname{Re} s\left(\frac{f(z)}{z^n(z - z_0)}; 0\right) = \sum_{k=0}^{n-1} \frac{f^{(k)}(0)}{k!} z_0^{k-n}.$$

(15) The function $f(z) = (1 + z + z^2 + \cdots + z^{n-1})^{-1}$, $n \geq 2$, has simple poles at $z_k = e^{\frac{2k\pi i}{n}}$, $1 \leq k \leq n - 1$. Show that

$$\operatorname{Re} s(f; z_k) = 2i\frac{e^{\frac{2k\pi i}{n}}}{n} \sin\frac{k\pi}{n}.$$

(16) Let $p(z) = z^4 + 16z^2 + 13$. Find the residues of $\frac{z^2}{p(z)}$ at zeros of $p(z)$ in the upper half-plane $\text{Im } z > 0$ and then, deduce its residues at zeros of $p(z)$ in the lower half-plane. What is $\text{Re } s\left(\frac{z^2}{p(z)}; 0\right)$?

(17) Suppose $\Gamma(z)$ is analytic in $\mathbf{C} - \{0, -1, -2, \ldots, -n, \ldots\}$ and has simple poles at $0, -1, -2, \ldots, -n, \ldots$. Suppose $\Gamma(1) = 1$, and $\Gamma(z+1) = \Gamma(z)$ at points z where f are analytic. Show that

$$\text{Re } s(\Gamma(z); -n) = \frac{(-1)^n}{n!}, \quad n = 0, 1, 2, \ldots.$$

(18) $f(z) = \frac{1}{z^4 + a^n}$ $(a \neq 0)$ has four simple poles z_k, $1 \leq k \leq n$. Show that

$$\text{Re } s(f; z_k) = -\frac{z_k}{4a^4}, \quad 1 \leq k \leq 4.$$

(19) Let z_0 be any simple pole of $f(z) = \frac{z^{n-1}}{z^n + a^n}$ $(a \neq 0)$. Show that $\text{Re } s(f; z_0) = \frac{1}{n}$ always holds.

(20) Let $f(z) = \frac{1}{z} \sum_{k=0}^n \frac{1}{(z+a)^k}$. Show that

$$\text{Re } s(f; -a) = \begin{cases} 1, & a = 0, \\ -\dfrac{1 - a^n}{a^n(1 - a)}, & a \neq 1, \\ -n, & a = 1. \end{cases}$$

(21) Suppose f is analytic in a domain Ω which contains the real axis. Let $z_k = \left(k + \frac{1}{2}\right)\pi$, $k = 0, \pm 1, \pm 2, \ldots$. Show that $\text{Re } s\left(\frac{f(z)}{\cos^2 z}; z_k\right) = f'(z_k)$, $k = 0, \pm 1, \pm 2, \ldots$.

(22) Imitate Example 6 to compute $\text{Re } s(\cot^{2n+1} \pi z; 0)$, $n = 0, \pm 1, \pm 2, \ldots$.

(23) Let $p(z)$ and $q(z)$ be polynomials, relatively prime, of degrees m and n, respectively, where $0 \leq m \leq n$. Set $f(z) = \frac{p(z)}{q(z)}$. Suppose $f(z) - z = 0$ and $f'(z) = 1$ do not have common solutions. Let $E = \{z \mid f(z) - z = 0\}$. Show that $\sum_{z_0 \in E} \frac{1}{1 - f'(z_0)} = 1$.

(24) Let $p(z)$ be a polynomial of positive degree.

   (a) Find the residues of $\frac{p'(z)}{p(z)}$ at poles.
   (b) Suppose $p(z)$ has only simple zeros and $f(z)$ is analytic. Find the residues of $\frac{f(z)}{p(z)}$ at poles.

(25) Find the residues of the following functions $f(z)$ at their isolated singularities, including ∞ if ∞ is such a point.

   (a) $\dfrac{z^2}{(z^2 + 1)^2}$.

   (b) $\dfrac{z^{2n}}{(1 + z)^n}$ (n is a positive integer).

(c) $\dfrac{z^m}{1 + z^{2n}}$ (m and n are positive integers).

(d) $\dfrac{1}{(1 + z^2)^n}$ (n is a positive integer).

(e) $\dfrac{z^2}{(z - 1)^3(z + 1)}$.

(f) $\dfrac{4z}{(z^2 + 2pz + 1)^2}$ ($p > 1$).

(g) $\dfrac{z^{2m}}{(z - 1)^m}$ (m is a positive integer).

(h) $\dfrac{z}{(z + 3i)^3(z + 2)^2}$.

(i) $\dfrac{1}{(z^3 - 1)(z + 1)^2}$.

(26) Same as Exercise (25).

(a) $\cot^2 z$.

(b) $\cot^3 z$.

(c) $z^3 \cos \dfrac{1}{z - 2}$.

(d) $\sin z \sin \dfrac{1}{z}$.

(e) $\cos \dfrac{z^2 + 4z - 1}{z + 3}$.

(f) $\dfrac{1}{z(1 - e^{-hz})}$ ($h \neq 0$).

(g) $z^n \sin \dfrac{1}{z}$ (n is an integer).

(h) $\left(\sin \dfrac{1}{z} \right)^{-1}$.

(i) $\dfrac{\sqrt{z}}{\sin \sqrt{z}}$.

(j) $\dfrac{\tan z}{z^n}$ (n is a positive integer).

(k) $\dfrac{\cos\left(\frac{\pi z}{2}\right)}{(z - 1)^3}$.

(l) $\dfrac{\sin^2 z}{(z - \pi)^6}$.

(m) $(z + 1)^4 \sin \dfrac{1}{\pi(z + 1)}$.

(n) $(z^2 + z) \cos \dfrac{1}{z}$.

(o) $\dfrac{\sin z^3}{(1 - \cos z)^3}$.

(p) $\dfrac{z^2 + 1}{(e^{\pi z} + 1)^4}$.

(q) $\dfrac{\cos \sqrt{z}}{z^3}$.

(r) $e^{z^{-1}} - e^{-z^{-1}}$.

(s) $z^3 e^{z^{-1}} \sin \dfrac{1}{z}$.

(t) $\dfrac{e^{\alpha z}}{z^{n+1}}$ ($\alpha \in \mathbf{C}$ and $n \geq$ is an integer).

(u) $\dfrac{\sin \alpha z}{z(z^2 + 1)}$ ($\alpha \in \mathbf{C}$).

(v) $\dfrac{e^{\alpha z}}{1 + e^z}$ ($\alpha \in \mathbf{C}$).

(w) $\pi \dfrac{\cot \pi z}{z^2}$.

(x) $\dfrac{\sin 3z - 3 \sin z}{(\sin z - z) \sin z}$.

(y) $\dfrac{\cot \pi z}{(z + a)^2}$ ($a \in \mathbf{R}$ but a is not an integer).

(z) $\dfrac{\csc \pi z}{z^2 + a^2}$ ($a > 0$).

(a)$'$ $\dfrac{e^{imz}}{(z^2 + a^2)^2}$ ($a \neq 0$).

(b)$'$ $\dfrac{1}{(1 + z^2) \cosh \dfrac{1}{2} \pi z}$.

(c)$'$ $\dfrac{\sin z}{z^3(z - 2)(z + 1)}$.

(d)′ $\dfrac{\cot z}{z^2(z+i)^2}$.

(e)′ $\dfrac{\sin \alpha z}{z^3 \sin \beta z}$ $(\alpha \neq \beta$ and $\beta \neq 0)$.

(f)′ $\dfrac{e^z}{\sin^3 z}$.

(g)′ $\dfrac{e^{z^2}}{(z-1)^2}$.

(h)′ $\dfrac{(z-1)^2}{(e^3-1)^3}$.

(i)′ $\dfrac{z^2}{(z-2)^2(\cos z - 1)^3}$.

(j)′ $\sin\left(1+\dfrac{1}{z}\right)\cos\left(1+\dfrac{1}{z^2}\right)$.

(27) Find the residues of the branches of the following multiple-valued functions $f(z)$ at the indicated isolated singularity z_0.

(a) $\dfrac{\sqrt{z}}{1-z}$, $z_0 = 1$.

(b) $\dfrac{1}{\sqrt{2-z}+1}$, $z_0 = 1$.

(c) $\dfrac{z^\alpha}{1-\sqrt{z}}$ $(\alpha \in \mathbf{C}$ but $\alpha \neq 0)$, $z_0 = 1$.

(d) $\sqrt{(z-a)(z-b)}\,(a \neq b)$, $z_0 = \infty$.

(e) $\log \dfrac{z-a}{z-b}(a \neq b)$, $z_0 = \infty$.

(f) $e^z \log \dfrac{z-a}{z-b}(a \neq b)$, $z_0 = \infty$.

(g) $(\log z)\sin \dfrac{1}{z-1}$, $z_0 = 1$.

(h) $(\log z)\cos \dfrac{1}{z-1}$, $z_0 = 1$.

(i) $\dfrac{\tan^{-1} z}{z}$, $z_0 = 0$ and ∞.

(j) $z^n \log \dfrac{z-a}{z-b}$ $(a \neq 0,\ b \neq 0,$ and n is an integer$)$, $z_0 = 0$ and ∞.

(k) $\dfrac{(\log z)^2}{(z-1)^5}$, $z_0 = 1$.

(l) $\dfrac{1}{(z^2-1)^3}$, $z_0 = 1$.

(m) $\dfrac{z}{[\log(1+z)]^3}$, $z_0 = 0$.

(n) $\dfrac{z^a}{(1+z^2)^2}$ $(a \in \mathbf{R})$, $z_0 = i$.

(o) $\dfrac{1}{z^2}e^{z^{-1}}\log\dfrac{1-az}{1-bz}$ $(a \neq b)$, $z_0 = 0$.

(28) Try to use (4.11.2.4) to expand $\dfrac{z^3+z^2+2}{z(z^2-1)^2}$ into its partial fraction form.

Exercises B

(1) Let f be analytic and $g(z) = f(z^n)$, where n is a positive integer.

    (a) If z_0 is a pole of g, then $z_k = z_0 w^k$ $(w = e^{\frac{2\pi i}{n}})$, $1 \leq k \leq n-1$, are also poles of g, and, of course, are of the same order as that of z_0.

    (b) Show that $\operatorname{Re} s(g; z_k) = w^k \operatorname{Re} s(f; z_0)$, $1 \leq k \leq n-1$.

(2) Let $f(z) = g(z) - h(z)$, where

$$g(z) = \left[\left(2 - \frac{1}{z^2}\right)e^{z^2} + \frac{1}{z^2}\right](\operatorname{Arc\,tan} z)^2, \qquad h(z) = \frac{1}{z^2}(\operatorname{Arc\,tan} z)^2$$

and $\operatorname{Arc\,tan} z$ is the principal branch of $\tan^{-1} z$ (note that $\lim_{z\to\infty} \operatorname{Arc\,tan} z = \frac{\pi}{2}$). Integrate f along γ as shown in Fig. 4.57 to show that $\operatorname{Re} s(f; \infty) = \pi\left(1 - \frac{1}{e}\right)$ by letting $r \to 0$ and $\delta \to 0$.

(3) Let $n \geq 1$ be an integer and $p(t) = t^n - nt^{n-1} + n(n-1)t^{n-2} + \cdots + (-1)^n n!$. Show that

$$\operatorname{Re} s\left(z^n e^z \operatorname{Log}\frac{z-a}{z-b}; \infty\right) = e^a p(a) - e^b p(b).$$

4.11.3.   *The residue theorem*

We have purposely arranged the *classical form* (4.11.3.2) of this theorem in Exercise A (2) of Sec. 3.4.2 for a preview. As a matter of fact, we have tacitly used this theorem in evaluating the complex line integrals so far, such as in Examples 1, 3 and Exercises A (4), (5) of Sec. 3.4.2, and Examples and Exercise A of Sec. 4.6. Also, we stated a *general form* of this theorem in (4.3.3.6).

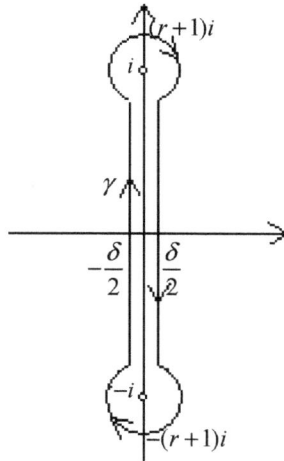

Fig. 4.57

Now, we formally introduce this important theorem.

Suppose f is analytic in a domain Ω except isolated singularities.

Let γ be a cycle in Ω, not passing the isolated singularities of f and homologous to zero in Ω (see (4.3.1.7)). We *claim* that only finitely many isolated singularities z_0 of f in Ω satisfy $n(\gamma; z_0) \neq 0$. To *prove* this, let the set $D = \{z \in \mathbf{C} \mid n(\gamma; z) = 0\}$. By property (4) in (3.5.2.4), this set D is nonempty and open. Since γ, as a planar point set, is compact, hence there is a $R > 0$ so that γ lies entirely in $|z| < R$. In this case, $n(\gamma; z) = 0$ on $|z| > R$, and hence, D contains the set $|z| > R$. Consequently, $\mathbf{C} - D$ is compact and thus, contains only finitely many isolated singularities of f owing to the interior uniqueness theorem (see (4) in (3.4.2.9)).

Let $z_1, \ldots, z_n$ be all the isolated singularities of f in Ω such that $n(\gamma; z_k) \neq 0$ for $1 \leq k \leq n$. Construct *pairwise disjoint* disks $D_k : |z - z_k| \leq r_k$, $1 \leq k \leq n$, all contained in Ω and oriented in the counterclockwise direction. Then γ is homologous to

$$\sum_{k=1}^{n} n(\gamma; z_k)\partial D_k$$

in Ω (see (4.3.2.5)). By using (4.3.3.1) and (4.3.3.2), it follows that

$$\int_{\gamma} f(z)dz = \sum_{k=1}^{n} n(\gamma; z_k) \int_{\partial D_k} f(z)dz = 2\pi i \sum_{k=1}^{n} n(\gamma; z_k)\mathrm{Re}\, s(f; z_k).$$

$$(4.11.3.1)$$

We formally summarize the above as (see also (4.3.3.6))

The residue theorem (general form). Suppose f is analytic in a domain Ω except isolated singularities. For any cycle γ in Ω which does not pass the isolated singularities of f and is homologous to zero in Ω,

$$\int_\gamma f(z)dz = 2\pi i \sum_{z_k} n(\gamma; z_k)\mathrm{Re}\, s(f; z_k)$$

where, in the summation $\sum_{z_k}$, only finitely many terms are not equal to zero.   (4.11.3.2)

We try to narrow the content of (4.11.3.2) so that it becomes much more useful in practical application. We say that a cycle γ *bounds a domain* or *has an interior domain* if either $n(\gamma; z) = 0$ or $n(\gamma; z) = 1$ for all $z \notin \gamma$. In this case, we call the set

$$\mathrm{Int}\,\gamma = \{z \in \mathbf{C} \,|\, n(\gamma; z) = 1\}$$

the *interior domain* of γ or the *domain bounded by* γ. By basic properties (3.5.2.4) of winding numbers, $\mathrm{Int}\,\gamma$ is open and hence, each of its components is a *domain*. Moreover, γ *should contain the boundary of its interior* $\mathrm{Int}\,\gamma$, when viewed as a point set, namely,

$$\partial(\mathrm{Int}\,\gamma) \subseteq \gamma,$$

but not conversely (see Fig. 4.58). To see this, let $z_0 \in \partial(\mathrm{Int}\,\gamma)$. If there is a $\rho > 0$ such that $|z - z_0| < \rho$ has an empty intersection with γ, then $n(\gamma; z)$ is well-defined in $|z - z_0| < \rho$ and is equal to the constant 0 or 1. This contradicts to the fact that $|z - z_0| < \rho$ has nonempty intersection with $\mathrm{Int}\,\gamma$ and $\mathbf{C} - \mathrm{Int}\,\gamma$. This shows that any open neighborhood of z_0 contains points of γ which is a closed set. Consequently, $z_0 \in \gamma$ holds.

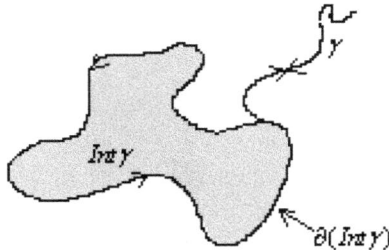

Fig. 4.58

After this preliminary, we are able to formulate (see also Exercise A (2) of Sec. 3.4.2).

The residue theorem (classical form). Let a cycle γ in $\mathbf{C}$ has the interior domain $\mathrm{Int}\,\gamma$. Suppose f is analytic in $\mathrm{Int}\,\gamma$ except isolated singularities $z_1, \ldots, z_n$ in $\mathrm{Int}\,\gamma$ and is continuous on $\overline{\mathrm{Int}\,\gamma} - \{z_1, \ldots, z_n\}$. Then

$$\int_\gamma f(z)dz = 2\pi i \sum_{k=1}^n \mathrm{Re}\, s(f; z_k).$$

In particular, this is the case when γ is a rectifiable Jordan closed curve.

(4.11.3.3)

What we should emphasize once again is that the cycle γ or the rectifiable Jordan closed curve γ cannot be allowed to pass through any isolated singularity of f. For an exceptional case, see Exercise A (12).

For completion and possible applications (see, for instance, Sec. 6.6.2), we extend the argument principle (3.5.3.3) one step farther to the following

Generalized argument principle. Let $f(z)$ be meromorphic in a domain Ω with the zeros a_j of order n_j and the poles b_p of order m_p. Then

$$\frac{1}{2\pi i} \int_\gamma \frac{f'(z)}{f(z)} dz = \sum_j n_j n(\gamma; a_j) - \sum_p m_p n(\gamma; b_p)$$

for every cycle γ which is homologous to zero in Ω and does not pass through any of the zeros or poles.

(4.11.3.4)

In case γ is a piecewise smooth Jordan closed curve and γ *passes through* a *simple zero* a_j of $f(z)$, then (4.7.9) and Exercise A (12) below say that the corresponding coefficient $n(\gamma; a_j)$ of n_j, which is equal to 1 in this case, in (4.11.3.4) is obliged to be replaced by

$$\frac{\theta(a_j)}{2\pi}$$

(4.11.3.5)

where $\theta(a_j)$ is the interior angle of γ at a_j. Similar explanation goes to a *simple* pole lying on a γ, too.

The most direct and elementary application of the residue theorem is used to evaluate the complex line integrals, as we mentioned before. For more delicate usages, see Sec. 4.12, the argument principle in Sec. 3.5.3, and the Rouché's theorem in Sec. 3.5.4.

Example 1. Evaluate the following integrals.

(1) $\displaystyle\int_{|z|=3} \frac{e^{z^{-1}}}{z^2+4}\,dz.$

(2) $\displaystyle\int_{|z|=4} \frac{e^z}{\sinh z}\,dz.$

(3) $\displaystyle\int_{|z|=80} \frac{z^2}{(z^6+100)(z-1)}\,dz.$

(4) $\displaystyle\int_{|z|=2} \frac{z(z+3)}{z^n-1}\,dz \quad (n\geq 2).$

(5) $\displaystyle\int_{|z|=R} \frac{p(z)}{q(z)}\,dz$, where $p(z)$ and $q(z)$ are polynomials, relatively prime, of degrees n and m, respectively. It is required that all the zeros of $q(z)$ lie in the disk $|z|<R$.

(6) $\int_{|z|=R} \frac{p'(z)}{p(z)}\,dz$, where $p(z)$ is a polynomial of degree $n\geq 1$ and has all its zeros contained in $|z|<R$. Refer to the proof of Example 5(2) of Sec. 4.11.2.

Solution. In all these cases, use I to denote the integrals and $f(z)$ to denote the integrands.

(1) $f(z)$ has two simple poles at $z=\pm 2i$ and an essential singularity at $z=0$. Now

$$\operatorname{Re}s(f;2i)=\frac{e^{-\frac{i}{2}}}{4i}; \quad \operatorname{Re}s(f;-2i)=-\frac{e^{\frac{i}{2}}}{4i};$$

$$\operatorname{Re}s(f;0)=\text{the coefficient } a_{-1}\text{ of }\left(\sum_0^\infty \frac{(-1)^n z^{2n}}{2^{2n+2}}\right)\left(\sum_0^\infty \frac{1}{n!z^n}\right)$$

$$=\frac{1}{2}\sum_0^\infty \frac{(-1)^n\left(\frac{1}{2}\right)^{2n+1}}{(2n+1)!}=\frac{1}{2}\sin\frac{1}{2}$$

$$\Rightarrow I=2\pi i\{\operatorname{Re}s(f;2i)+\operatorname{Re}s(f;-2i)+\operatorname{Re}s(f;0)\}=0.$$

Alternatively, observing that $z=\infty$ is a zero of order 2, so $\operatorname{Re}s(f;\infty)=0$ (see the Note in (4.11.2.1)). Therefore, by (4.11.1.7), $I=-2\pi i\operatorname{Re}s(f;\infty)=0.$

(2) $I=2\pi i\{\operatorname{Re}s(f;0)+\operatorname{Re}s(f;-\pi i)+\operatorname{Re}s(f;\pi i)\}=2\pi i\{1+1+1\}=6\pi i.$

(3) Observe that f has seven simple poles lying in $|z| < 80$ and ∞ is a zero of order 5. By use of the Note in (4.11.2.1) and the identity in (4.11.1.7), $I = -2\pi i \operatorname{Re} s(f; \infty) = 0$.

(4) In case $n = 2$: Since $\operatorname{Re} s(f; 1) = 2$ and $\operatorname{Re} s(f; -1) = 1$, so $I = 2\pi i$ $(2 + 1) = 6\pi i$. In case $n = 3$: ∞ is a simple zero of f. The Laurent (or Taylor) series expansion of f at ∞ is

$$f(z) = \frac{z+3}{z^2} \cdot \frac{1}{1 - z^{-3}} = \left(\frac{3}{z^2} + \frac{1}{z}\right)\left(1 + \frac{1}{z^3} + \frac{1}{z^6} + \frac{1}{z^9} + \cdots\right),$$

$$1 < |z| \leq \infty$$

$$\Rightarrow \operatorname{Re} s(f; \infty) = -a_{-1} = -1.$$

Hence, $I = -2\pi i \operatorname{Re} s(f; \infty) = 2\pi i$. In case $n \geq 4$: ∞ is a zero of order $(n - 2)$ of f. By reasons as in (3), $I = -2\pi i \operatorname{Re} s(f; \infty) = 0$. Alternatively, choose $R > 2$, then

$$I = \int_{|z|=2} \frac{z(z+3)}{z^n - 1} dz = \int_{|z|=R} \frac{z(z+3)}{z^n - 1} dz$$

$$\Rightarrow |I| \leq \int_{|z|=R} \frac{R(R+3)}{R^n - 1}|dz| = \frac{2\pi R^2(R+3)}{R^n - 1} \to 0 \quad \text{as } R \to \infty.$$

It follows that $I = 0$.

(5) Let $p(z) = a_n z^n + \cdots + a_1 z + a_0$, $n \geq 1$ and $a_n \neq 0$, and $q(z) = b_m z^m + \cdots + b_1 z + b_0$, $m \geq 1$ and $b_m \neq 0$. If $m > n$, ∞ is a zero of $m - n$ of f; if $m \leq n$, let $r(z)$ be the remainder after dividing $p(z)$ by $q(z)$. Then

$$I = -2\pi i \operatorname{Re} s(f; \infty) = \begin{cases} \dfrac{2\pi i a_n}{b_m}, & \text{if } m = n + 1 \\ 0, & \text{if } m > n + 1 \\ \dfrac{2\pi i r^{(m-1)}(0)}{(m-1)! b_m}, & \text{if } m < n + 1 \end{cases}.$$

(6) By the argument principle, it is easily seen that $I = 2n\pi i$, even if $p(z)$ has multiple zeros in $|z| < R$. Alternatively, let $p(z) = a_n z^n + \cdots + a_1 z + a_0$, where $n \geq 1$ and $a_n \neq 0$. Then $p'(z) = n a_n z^{n-1} + \cdots + a_1$ and

$$f(z) = \frac{p'(z)}{p(z)} = \frac{n a_n z^{n-1} + \cdots + a_1}{a_n z^n + \cdots + a_1 z + a_0} = \frac{n}{z} + \cdots$$

$$\Rightarrow \operatorname{Re} s(f; \infty) = -n.$$

Therefore, $I = -2\pi i\,\mathrm{Re}\,s(f;\infty) = 2n\pi i$. Or, since $|z| = R$ is given by $z = Re^{i\theta},\ 0 \le \theta \le 2\pi$,

$$
I = \int_{|z|=R} f(z)dz = \int_0^{2\pi} \frac{na_n R^{n-1}e^{i(n-1)\theta} + \cdots + a_1}{a_n R^n e^{in\theta} + \cdots + a_1 Re^{i\theta} + a_0} Rie^{i\theta}d\theta
$$

$$
= i\int_0^{2\pi} \frac{na_n + (n-1)\frac{a_{n-1}}{R}e^{-i\theta} + \cdots + \frac{a_1}{R^n}e^{-in\theta}}{a_n + \frac{a_{n-1}}{R}e^{-i\theta} + \cdots + \frac{a_0}{R^n}e^{-in\theta}}d\theta
$$

$$
\to i\int_0^{2\pi} n\,d\theta = 2n\pi i
$$

as $R \to \infty$. Observe that the integrand converges to $\frac{na_n}{a_n} = n$ uniformly as $R \to \infty$.

Example 2. Evaluate the following integrals.

(1) $\displaystyle\int_{|z|=(n+\frac{1}{2})\pi} \cosh z \cot z\, dz, \quad n = 0, 1, 2, \ldots .$

(2) $\displaystyle\int_{|z|=r} \frac{z^2}{e^{2\pi i z^3} - 1} dz, \quad n < r^3 < n+1, \quad n = 1, 2, 3, \ldots .$

Solution.

(1) The integrand $f(z)$ has simple poles at $z_k = k\pi,\ k = 0, \pm 1, \pm 2, \ldots,$ and

$$
\mathrm{Re}\,s(f; z_k) = \frac{\cosh z_k \cos z_k}{\cos z_k} = \cosh z_k = \cosh k\pi.
$$

In case $|k| \le n$, then $|z_k| \le \left(n + \frac{1}{2}\right)\pi$ holds. Therefore, the integral

$$
I = 2\pi i \sum_{k=-n}^{n} \mathrm{Re}\,s(f; z_k) = 2\pi i \left(1 + 2\sum_{k=0}^{n} \cosh k\pi\right)
$$

$$
= 2\pi i \cdot \frac{\sinh\left(n + \frac{1}{2}\right)\pi}{\sinh \frac{1}{2}\pi}.
$$

(2) $e^{2\pi i z^3} - 1 = 0 \Leftrightarrow 2\pi i z^3 = 2k\pi i$ or $z^3 = k,\ k = 0, \pm 1, \pm 2, \ldots .$ Then f has a simple poles at $z = 0$ and at each of the points $\pm k^{\frac{1}{3}},\ \pm \omega k^{\frac{1}{3}},$ $\pm \omega^2 k^{\frac{1}{3}}$ (where $\omega = e^{\frac{2\pi i}{3}}$), $k = 1, 2, 3, \ldots .$ By computation,

$$
\mathrm{Re}\,s(f; 0) = \lim_{z\to 0} z \cdot \frac{z^2}{e^{2\pi i z^3} - 1} = \frac{1}{2\pi i};
$$

$$
\mathrm{Re}\,s(f; z_0) = \lim_{z\to z_0} \frac{(z - z_0)z^2}{e^{2\pi i z^3} - 1} = \frac{1}{6\pi i}, \quad \text{where } z_0 \ne 0 \text{ and } e^{2\pi i z_0^3} = 1.
$$

In case $n < r^3 < n+1$,

$$I = 2\pi i \left\{ \operatorname{Re} s(f;0) + \sum_{k=1}^{n} \left[\operatorname{Re} s(f; \pm k^{\frac{1}{3}}) + \operatorname{Re} s(f; \pm w k^{\frac{1}{3}}) \right. \right.$$

$$\left. \left. + \operatorname{Re} s(f; \pm w^2 k^{\frac{1}{3}}) \right] \right\}$$

$$= 2\pi i \left\{ \frac{1}{2\pi i} + 6n \cdot \frac{1}{6\pi i} \right\} = 2n + 1.$$

Example 3. Evaluate the following integrals.

(1) Let γ denote the vertical line $z = \alpha + iy$, $0 < \alpha < 1$, and $-\infty < y < \infty$, oriented in the direction of increasing y.

$$\frac{1}{2\pi i} \int_{\gamma} \frac{dz}{a^z \sin \pi z} \quad (a > 0 \text{ and } a \neq 1) = ?$$

(2) $\frac{1}{2\pi i} \int_{\gamma} \frac{e^z dz}{\cos z} = ?$ where γ is shown in Fig. 4.59.

Solution.

(1) Here, we consider the principal branch $e^{z \operatorname{Log} a}$, where $\operatorname{Log} a = \log a$ (real logarithm), of a^z. Let Γ be the boundary of the rectangle with vertices at $\alpha - i\beta$, $2 + \alpha - i\beta$, $2 + \alpha + i\beta$, and $\alpha + i\beta$, as shown in Fig. 4.60, where $\beta > 0$. The integrand $f(z)$ has simple poles at 1 and 2. Now

$$\operatorname{Re} s(f;1) = \lim_{z \to 1} (z-1)f(z) = \left. \frac{(a^z)^{-1}}{\pi \cos \pi z} \right|_{z=1} = \frac{-1}{a\pi} \quad \text{and}$$

$$\operatorname{Re} s(f;2) = \lim_{z \to 2} (z-2)f(z) = \frac{1}{a^2 \pi}.$$

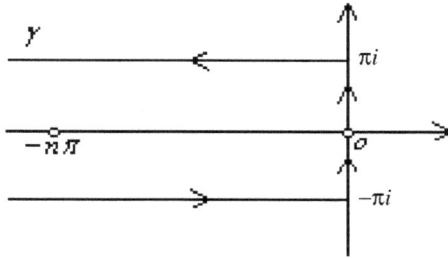

Fig. 4.59

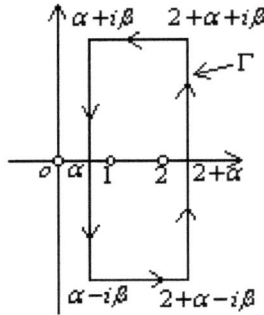

Fig. 4.60

$$\Rightarrow \frac{1}{2\pi i} \int_\Gamma f(z)dz = \frac{-1}{a\pi} + \frac{1}{a^2\pi} = \frac{1}{a\pi}\left(-1 + \frac{1}{a}\right). \qquad (*_1)$$

The integration along Γ can be divided in to four parts:

$$\int_{2+\alpha-i\beta}^{2+\alpha+i\beta} f(z)dz = \int_{\alpha-i\beta}^{\alpha+i\beta} \frac{dz}{a^{2+z}\sin\pi(2+z)} = \frac{1}{a^2}\int_{\alpha-i\beta}^{\alpha+i\beta} f(z)dz,$$

$$\int_{\alpha+i\beta}^{\alpha-i\beta} f(z)dz = -\int_{\alpha-i\beta}^{\alpha+i\beta} f(z)dz;$$

while, along $z = x-i\beta, \alpha \le x \le 2+\alpha, |a^z| = a^x \ge \min\{a^\alpha, a^{2+\alpha}\} = M,$ and $|\sin\pi z| \ge \sinh\pi\beta$ (see (3)4 in (2.6.2.6)), therefore

$$\left| \int_{\alpha-i\beta}^{2+\alpha+i\beta} f(z)dz \right| \le \frac{2}{M\sinh\pi\beta} \to 0 \quad \text{as } \beta \to \infty,$$

and, similarly,

$$\lim_{\beta\to\infty} \int_{2+\alpha-i\beta}^{\alpha-i\beta} f(z)dz = 0.$$

Substituting these results into $(*_1)$ and letting $\beta \to \infty$, we obtain

$$\left(\frac{1}{a^2} - 1\right)\frac{1}{2\pi i}\int_\gamma f(z) = \frac{1}{a\pi}\left(\frac{1}{a} - 1\right)$$

$$\Rightarrow \frac{1}{2\pi i}\int_\gamma f(z)dz = \frac{1}{\pi(1+a)}.$$

(2) Inside Fig. 4.59, let Γ be the boundary of the rectangle with vertices at $-n\pi - \pi i$, $-\pi i$, πi, and $-n\pi + \pi i$, $n = 1, 2, \ldots$. The integrand $f(z)$ has simple poles at $-\frac{\pi}{2}, -\frac{3\pi}{2}, \ldots, -n\pi + \frac{\pi}{2} = -\frac{(2n-1)}{2}\pi$. The corresponding residues are

$$\operatorname{Re} s\left(f; \frac{-(2k-1)\pi}{2}\right) = -\frac{e^{-\frac{(2k-1)\pi}{2}}}{\sin \frac{-(2k-1)\pi}{2}} = (-1)^{k-1}e^{-k\pi}e^{\frac{\pi}{2}}, \ 1 \le k \le n.$$

$$\Rightarrow \frac{1}{2\pi i}\int_\Gamma f(z)dz = \sum_{k=1}^n (-1)^{k-1}e^{-k\pi}e^{\frac{\pi}{2}} = e^{-\frac{\pi}{2}}\sum_{k=0}^{n-1}(-1)^k e^{-k\pi}. \quad (*_2)$$

Along the vertical segment $z = -n\pi + iy$, $-\pi \le y \le \pi$, $|e^z| = e^{-n\pi}$, and $|\cos z| = \cosh y \ge 1$, so

$$\left|\int_{-n\pi+i\pi}^{-n\pi-i\pi} f(z)dz\right| \le 2\pi e^{-n\pi} \to 0, \quad \text{as } n \to \infty.$$

Finally, letting $n \to \infty$ in $(*_2)$, its follows that

$$\frac{1}{2\pi i}\int_\gamma f(z)dz = e^{-\frac{\pi}{2}}\sum_{k=0}^\infty (-1)^k e^{-k\pi} = \frac{e^{-\frac{\pi}{2}}}{1+e^\pi}.$$

Example 4. Designate $\sqrt{(z-a)(z-b)} = z\sqrt{\left(1-\frac{a}{z}\right)\left(1-\frac{b}{z}\right)}$, $z \neq 0$, in which two branches of $\sqrt{\left(1-\frac{a}{z}\right)\left(1-\frac{b}{z}\right)}$ can be defined in $\mathbf{C}^* - \gamma$ where γ is a Jordan curve connecting a and b. Choose the branch determined by $\sqrt{1} = 1$ at $z = \infty$. Evaluate the following integrals: If R is larger enough,

(1) $\displaystyle\int_{|z|=R} \sqrt{(z-a)(z-b)}dz = -\frac{1}{4}\pi i(a-b)^2;$

(2) $\displaystyle\int_{|z|=R} \frac{z}{\sqrt{(z-a)(z-b)}}dz = \pi i(a+b);$ and

(3) $\displaystyle\int_{|z|=R} \frac{1}{\sqrt{(z-a)(z-b)}}dz = 2\pi i.$

Solution. Refer to Fig. 4.61. In $\sqrt{\left(1-\frac{a}{z}\right)\left(1-\frac{b}{z}\right)} = \left(1-\frac{a}{z}\right)^{\frac{1}{2}}\left(1-\frac{b}{z}\right)^{\frac{1}{2}}$, both $\left(1-\frac{a}{z}\right)^{\frac{1}{2}}$ and $\left(1-\frac{b}{z}\right)^{\frac{1}{2}}$ are chosen to be the branches determined by $\sqrt{1} = 1$ at $z = \infty$. Via the binomial series expansions (see Example 2 in

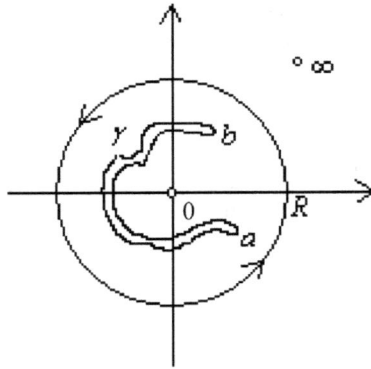

Fig. 4.61

Sec. 3.3.2) and the Cauchy product process, we have the Laurent series expansion

$$\sqrt{\left(1-\frac{a}{z}\right)\left(1-\frac{b}{z}\right)} = \left\{\sum_{n=0}^{\infty} C_n^{\frac{1}{2}}\left(-\frac{a}{z}\right)^n\right\}\left\{\sum_{n=0}^{\infty} C_n^{\frac{1}{2}}\left(-\frac{b}{z}\right)^n\right\},$$

$$R < |z| \leq \infty$$

$$= 1 - \frac{1}{2}(a+b)\frac{1}{z} + \sum_{n=1}^{\infty} a_{-n}\frac{1}{z^{n+1}}, \quad \text{where}$$

$$a_{-(n-1)} = (-1)^n \sum_{k=0}^{n} C_k^{\frac{1}{2}} C_{n-k}^{\frac{1}{2}} a^k b^{n-k}, \quad n \geq 2.$$

(1) The integrand $f(z)$ has a simple pole at ∞ and the corresponding residue is

$$\operatorname{Re} s(f;\infty) = -a_{-1} = \frac{1}{8}(a-b)^2.$$

$$\Rightarrow \int_{|z|=R} f(z)dz \quad \text{(in counterclockwise direction)}$$

$$= -\int_{|z|=R} f(z)dz \quad \text{(in clockwise direction)}$$

$$= -2\pi i \operatorname{Re} s(f;\infty) = -\frac{\pi i}{4}(a-b)^2.$$

(2) The integrand $f(z)$ is analytic at ∞ and the Taylor series expansion is

$$f(z) = \frac{z}{z\sqrt{\left(1 - \frac{a}{z}\right)\left(1 - \frac{b}{z}\right)}} = \frac{1}{1 - \frac{1}{2}(a+b)\frac{1}{z} + \sum_1^\infty a_{-n}\frac{1}{z^{n+1}}}$$

$$= 1 + \frac{1}{2}(a+b)\frac{1}{z} + \sum_1^\infty b_{-n}\frac{1}{z^n}, \quad R < |z| \le \infty,$$

$$\Rightarrow \operatorname{Re} s(f;\infty) = -\frac{1}{2}(a+b),$$

$$\Rightarrow \int_{|z|=R} f(z)dz = -2\pi i \cdot \operatorname{Re} s(f;\infty) = (a+b)\pi i.$$

(3) The integrand $f(z)$ is analytic at ∞ and $\operatorname{Re} s(f;\infty) = -1$. The result follows obviously.

It is well-known in Calculus that

$$\int_0^\infty x^{p-1}e^{-ax}dx = \frac{1}{a^p}\Gamma(p), \quad a > 0, \ p > 0$$

where $\Gamma(p)$ is the Gamma function of p (see Sec. 5.6). By invoking the *formal* change of variables $x = it$, where $t > 0$ is a real variable, we have

$$\int_0^\infty (it)^{p-1}e^{-ait}idt = \frac{1}{a^p}\Gamma(p). \tag{$*_3$}$$

Suppose $(it)^{p-1} = i^{p-1}t^{p-1}$ holds. And *choose* $i^{p-1} = e^{(p-1)\operatorname{Log} i} = e^{(p-1)\cdot\frac{\pi i}{2}} = -ie^{\frac{p\pi i}{2}}$. Substituting this into the above integrand, then

$$\int_0^\infty t^{p-1}e^{-ait}dt = \frac{\Gamma(p)}{a^p}e^{-\frac{p\pi}{2}i},$$

$$\Rightarrow \text{(by equating both sides their real and imaginary parts)}$$

$$\int_0^\infty t^{p-1}\cos at\,dt = \frac{\Gamma(p)}{a^p}\cos\frac{p\pi}{2},$$

$$\int_0^\infty t^{p-1}\sin at\,dt = \frac{\Gamma(p)}{a^p}\sin\frac{p\pi}{2}, \quad a > 0, \ p > 0. \tag{$*_3$}'$$

These two integrals are correct, even though the process leading to them is nonrigorous. Well, the residue method helps us justify their validities. This is the following example.

Example 5. (See Ref. [76], pp. 107–108). Let γ be composed of the following curves: the circle arc $C_r : z = re^{i\theta}$, $0 \le \theta \le \frac{\pi}{2}$; the segment $[r, R]$;

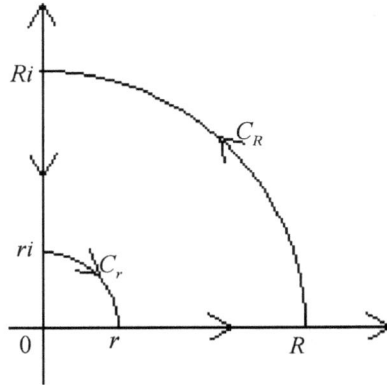

Fig. 4.62

the circle arc $C_R : z = \mathrm{Re}^{i\theta}$, $0 \le \theta \le \frac{\pi}{2}$, and the segment $[ir, iR]$, where $0 < r < R$, as shown in Fig. 4.62. Try to evaluate

$$\int_\gamma z^{p-1} e^{-az} dz, \quad a > 0, \ p > 0$$

and then, let $r \to 0$ and $R \to \infty$, to justify $(*_3)'$.

Solution. Let $f(z) = z^{p-1} e^{-az}$, where $z^{p-1} = e^{(p-1)\mathrm{Log}\,z}$, the principal branch, is analytic on $\overline{\mathrm{Int}\,\gamma}$. Hence,

$$\int_\gamma f(z)dz = \int_{C_r} f(z)dz + \int_r^R f(z)dz + \int_{C_R} f(z)dz + \int_{iR}^{ir} f(z)dz = 0. \quad (*_4)$$

Since $p > 0$, $\lim_{z \to 0} z f(z) = \lim_{z \to 0} z^p e^{-az} = 0$, and thus,

$$\left| \int_{C_r} f(z)dz \right| \le \int_0^{\frac{\pi}{2}} |f(re^{i\theta})| |rie^{i\theta}| d\theta$$

$$= \int_0^{\frac{\pi}{2}} r|f(re^{i\theta})| d\theta \to 0 \quad \text{as } r \to \infty$$

(see also Exercise A (28)(b), (c) of Sec. 2.9 or (4.12.3.3)). Since $a > 0$, $\lim_{z \to \infty} z f(z) = \lim_{z \to \infty} z^p e^{-az} = 0$, and thus

$$\left| \int_{C_R} f(z)dz \right| \le \int_0^{\frac{\pi}{2}} |f(Re^{i\theta})| |Rie^{i\theta}| d\theta$$

$$= \int_0^{\frac{\pi}{2}} R|f(Re^{i\theta})| d\theta \to 0 \quad \text{as } R \to \infty.$$

In $(*_4)$, letting $r \to 0$ and $R \to \infty$, then

$$\int_0^\infty f(x)dx + \int_\infty^0 f(iy)idy = 0$$

$$\Rightarrow \int_0^\infty f(ix)idx = \int_0^\infty (ix)^{p-1}e^{-aix}idx = \int_0^\infty x^{p-1}e^{-ax}dx = \frac{\Gamma(p)}{a^p}.$$

This justifies the validity of $(*_3)$ and the results follow as the argument before $(*_3)'$.

Example 6. Evaluate

$$P.V. \int_{|z|=1} \frac{\sin z}{z^2(z-1)(z+i)}dz.$$

Note that, the integrand has simple poles at 1 and $-i$ which lie on the path of integration, the unit circle $|z| = 1$, and hence, we can only evaluate the principal value of the integral (refer to Sec. 4.7 for details; in particular, Example 3 there). For general setting in this direction, see Exercise A (12).

Solution. Let $f(z)$ denote the integrand. Since $f(z)$ is analytic in $\mathbf{C} - \{0, 1, -i\}$, we can evaluate this integral in a couple of ways. Firstly, the corresponding residues are

$$\operatorname{Re} s(f;0) = i; \quad \operatorname{Re} s(f;1) = \frac{\sin 1}{1+i} = \frac{(1-i)\sin 1}{2};$$

$$\operatorname{Re} s(f;-i) = \frac{\sin(-i)}{1+i} = -\frac{i\sinh 1}{1+i} = -\frac{(1+i)\sinh 1}{2}.$$

Secondly, consider the integral

$$\int_\gamma f(z)dz,$$

where the path of integration γ is chosen to be the dented contour shown either in Fig. 4.63(a) or in (b), obtained by attaching outward or inward circular arcs of $|z-1| = r$ and $|z+i| = r$ to $|z| = 1$, respectively.

In Fig. 4.63(a): we have

$$\int_\gamma f(z)dz = 2\pi i\{\operatorname{Re} s(f;0) + \operatorname{Re} s(f;1) + \operatorname{Re} s(f;-i)\}. \tag{$*_5$}$$

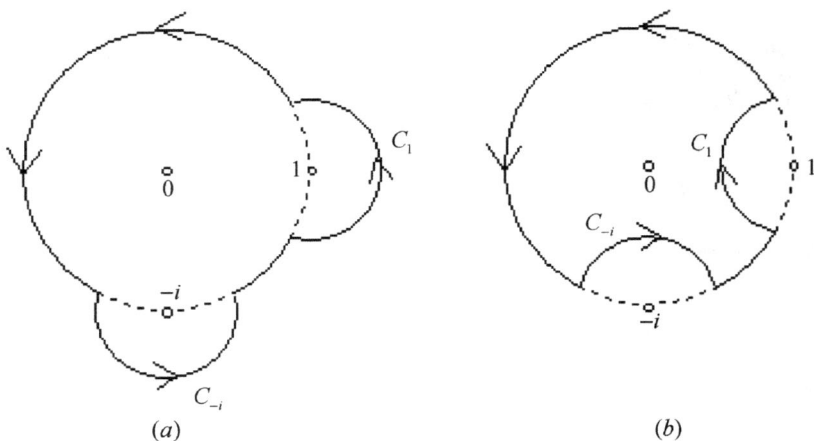

(a)   (b)

Fig. 4.63

Along the circular arc C_1, $z = 1 + re^{i\theta}$, $\alpha \le \theta \le \beta$: as the limit $r \to 0$, $\alpha \to -\frac{\pi}{2}$ and $\beta \to \frac{\pi}{2}$. Hence, as $r \to 0$

$$\int_{C_1} f(z)dz = \int_{\alpha}^{\beta} \frac{\sin(1 + re^{i\theta})}{(1 + re^{i\theta})^2 re^{i\theta}(1 + i + re^{i\theta})} rie^{i\theta} d\theta$$

$$= i \int_{\alpha}^{\beta} \frac{\sin(1 + re^{i\theta})}{(1 + re^{i\theta})^2(1 + i + re^{i\theta})} d\theta$$

$$\to i \int_{-\frac{\pi}{2}}^{\frac{\pi}{2}} \frac{\sin 1}{1 + i} d\theta = \pi i \cdot \operatorname{Re} s(f; 1)$$

(see also (4.7.9) or (4.12.3.3)).

While, along the circular arc C_{-i}, $z = -i + re^{i\theta}$, $\alpha \le \theta \le \beta$: as the limit $r \to 0$, $\alpha \to \pi$ and $\beta \to 2\pi$. Hence, as $r \to 0$,

$$\int_{C_{-i}} f(z)dz = \int_{\alpha}^{\beta} \frac{\sin(-i + re^{i\theta})}{(-i + re^{i\theta})^2(-1 - i + re^{i\theta})re^{i\theta}} rie^{i\theta} d\theta,$$

$$= i \int_{\alpha}^{\beta} \frac{\sin(-i + re^{i\theta})}{(-i + re^{i\theta})^2(-1 - i + re^{i\theta})} d\theta,$$

$$\to i \int_{\pi}^{2\pi} \frac{\sin(-i)}{(-i)^2(-1 - i)} d\theta = \pi i \cdot \operatorname{Re} s(f; -i)$$

(see also (4.7.9)).

Substituting these results in $(*_5)$ in the limit $r \to 0$, we obtain

$$\text{P.V.} \int_{|z|=1} f(z)dz + \pi i \left\{ \operatorname{Re} s(f;1) + \operatorname{Re} s(f;-i) \right\},$$

$$= 2\pi i \left\{ \operatorname{Re} s(f;0) + \operatorname{Re} s(f;1) + \operatorname{Re} s(f;-i) \right\},$$

$$\Rightarrow \text{P.V.} \int_{|z|=1} f(z)dz = 2\pi i \operatorname{Re} s(f;0) + \pi i \left\{ \operatorname{Re} s(f;1) + \operatorname{Re} s(f;-i) \right\},$$

$$= -2\pi + \frac{\pi}{2}[(1+i)\sin 1 + (1-i)\sinh 1]. \qquad (*_6)$$

In Fig. 4.63(b): We have, in this case,

$$\int_\gamma f(z)dz = 2\pi i \operatorname{Re} s(f;0). \qquad (*_7)$$

Along C_1: as the limit $r \to 0$, $\alpha \to -\frac{\pi}{2}$ and $\beta \to -\frac{3\pi}{2}$. Therefore, as $r \to \infty$,

$$\int_{C_1} f(z)dz \to i \int_{-\frac{\pi}{2}}^{-\frac{3\pi}{2}} \frac{\sin 1}{1+i} d\theta = -\pi i \operatorname{Re} s(f;1)$$

$$\text{(see also (4.7.9) or (4.12.3.3)).}$$

While, along C_{-i}: as the limit $r \to 0$, $\alpha \to -\pi$ and $\beta \to -2\pi$. Hence, as $r \to 0$,

$$\int_{C_{-i}} f(z)dz \to i \int_{-\pi}^{-2\pi} \frac{\sin(-i)}{(-i)^2(-1-i)} d\theta = -\pi i \operatorname{Re} s(f;-i)$$

$$\text{(see also (4.7.9)).}$$

Substituting these results in $(*_7)$, we obtain the same result as in $(*_6)$.

Exercises A

(1) Evaluate the integral

$$\int_\gamma \frac{1}{z^4+1} dz$$

according to the paths $\gamma : x^2 + y^2 = 2x$ and $\gamma : x^2 - xy + y^2 + x + y = 0$, respectively, where $z = x + iy$.

(2) Let $f(z) = (z-a)^{-n}(z-b)^{-m}$, where $a \neq b$, and m and n are positive integers. Show that $\operatorname{Re} s(f;a) = -\operatorname{Re} s(f;b) = (-1)^n C_{m-1}^{m+n-2}$ $(b-a)^{-m-n+1}$. Use this to deduce that

  (i) if $|a| < R < |b|$, $\frac{1}{2\pi i} \int_{|z|=R} f(z)dz = (-1)^n C_{m-1}^{m+n-2}(b-a)^{-m-n+1}$; and

  (ii) if $|a| < 1 < |b|$ and $m = n$, $\frac{1}{2\pi i} \int_{|z|=1} f(z)dz = \frac{(-1)^{n-1}(2(n-1))!}{((n-1)!)^2(a-b)^{2n-1}}$.

(3) Try to use (4.11.1.7) to evaluate the following integrals (refer to Example 1 (3)–(6)).

(a) $\displaystyle\int_{|z|=3} \frac{z^{21}}{(z^2-1)^4(z^4-2)^3}\,dz,$

(b) $\displaystyle\int_{|z|=R} \frac{z^{2n+3m-1}}{(z^2+a)^n(z^3+b)^m}\,dz,$ where $ab \neq 0$, m and n are positive integers, and $R > \max\{|a|^{\frac{1}{2}}, |b|^{\frac{1}{3}}\}$, and

(c) $\displaystyle\int_{|z|=2} \frac{1}{(z-3)(z^n-1)}\,dz, \quad n = 1, 2, 3, \ldots .$

(4) Use the residue theorem to evaluate:

(a) $\displaystyle\int_{|z|=1} \frac{\operatorname{Re} z}{z-\alpha}\,dz \quad (0 < |\alpha| < 1),$

(b) $\displaystyle\int_{|z|=1} \frac{|dz|}{|z-\alpha|^4} \quad (|\alpha| \neq 1),$ and

(c) $\displaystyle\int_{|z|=r} \frac{z^2 + z^{-2}}{(\bar{z}-a)(\bar{z}-b)}\,dz \quad (0 < |a| < r < |b|).$

(5) Use the residue theorem to evaluate the following integrals.

(a) $\displaystyle\int_{|z|=2} \frac{dz}{(z-3)(z^5-1)},$

(b) $\displaystyle\int_{|z|=1} \sin\frac{1}{z}\,dz,$

(c) $\displaystyle\int_{|z|=1} \sin^2\frac{1}{z}\,dz,$

(d) $\displaystyle\frac{1}{2\pi i}\int_{|z|=r} z^n e^{\frac{2}{z}}\,dz \quad (n \text{ is an integer}),$

(e) $\displaystyle\int_{|z|=R} \frac{p(z)}{z^{n+1}(z-a)}\,dz \quad (p(z) \text{ is a polynomial of degree at most equal to } n \text{ and } |a| < R),$

(f) $\displaystyle\int_{|z|=1} \frac{\sin^6 z}{(z-\frac{\pi}{6})^3}\,dz,$

(g) $\displaystyle\int_{|z|=n} \frac{z}{\cos \pi z}\,dz \quad (n \text{ is a positive integer}),$

(h) $\displaystyle\int_{|z|=2} \frac{dz}{\sin^2 z \cos z},$

(i) $\displaystyle\int_\gamma \frac{dz}{(z^2-1)^2(z-3)^2}$ (γ is the curve $x^{\frac{2}{3}}+y^{\frac{2}{3}}=2^{\frac{2}{3}}$, where $z = x+iy$),

(j) $\displaystyle\int_{|z|=2} \frac{z^3 e^{z^{-1}}}{1+z}dz,$

(k) $\displaystyle\int_{|z|=2} \frac{3z^7+4}{z^8-1}dz,$ and

(l) $\displaystyle\int_{|z|=2} \frac{\sin z}{z^3+z^2+z+1}dz.$

(6) Evaluate the following integrals (note that one should consider each branch of the integrands).

(a) $\displaystyle\int_{|z|=r} z^n \log\frac{z-a}{z-b}dz,\; r > \max\{|a|,|b|\},$ where n is a positive integer,

(b) $\displaystyle\int_{|z|=2} \sqrt{\frac{z}{z+1}}dz,$

(c) $\displaystyle\int_{|z|=1} \frac{dz}{\sqrt{4z^2+4z+3}},$ and

(d) $\displaystyle\int_{|z|=2} \frac{z^n}{\sqrt{1+z^2}}dz,$ where n is a positive integer.

(7) Let γ be the boundary of the rectangle with vertices at $\pi(\pm1\pm i)$. Show that

$$\operatorname{Log} w = \frac{1}{2\pi i}\int_\gamma \frac{ze^z}{e^z-w}dz,\; w\in \operatorname{Int}\gamma.$$

(8) Let γ be a rectifiable Jordan closed curve whose interior contains the origin 0. Suppose f and g are analytic in $\operatorname{Int}\gamma$ and continuous on $\overline{\operatorname{Int}\gamma}$. In case g has only simple zeros at $a_1,\dots,a_n$ in $\operatorname{Int}\gamma-\{0\}$, find

$$\frac{1}{2\pi i}\int_\gamma \frac{f(z)}{zg(z)}dz.$$

(9) Let Ω be a domain containing a neighborhood $R < |z| < \infty$ of ∞. Suppose f is analytic in Ω except finitely many isolated singularities and γ is a rectifiable Jordan closed curve in Ω, not passing any isolated singularities of f. In case the exterior $\operatorname{Ext}\gamma$ contains the isolated singularities $a_1,\dots,a_k$ $(a_k\neq\infty, 1\le k\le n)$ of f, then

$$\int_\gamma f(z)dz = 2\pi i\sum_{k=1}^n \operatorname{Re}s(f;a_k) + 2\pi i\operatorname{Re}s(f;\infty)\qquad (4.11.3.6)$$

where γ is endowed with the clockwise direction. Use this to evaluate

$$\int_{|z|=r} \frac{\sin\left(\frac{a}{z}\right)}{z^2 + b^2} dz, \quad a, b \neq 0 \text{ and } 0 < r \neq |b|$$

(in the counterclockwise direction).

(10) Let γ be a rectifiable Jordan closed curve. Suppose f is analytic in Int γ except the isolated singularities $a_1, \ldots, a_n$, and continuous on $\overline{\text{Int } \gamma} - \{a_1, \ldots, a_n\}$. Show that, if $z \in \text{Int } \gamma - \{a_1, \ldots, a_n\}$, then

$$\frac{1}{2\pi i} \int_\gamma \frac{f(\zeta)}{\zeta - z} d\zeta = f(z) - \sum_{k=1}^n \text{Re } s\left(\frac{f(\zeta)}{\zeta - z}; \zeta = a_k\right). \qquad (4.11.3.7)$$

In particular, if $a_1, \ldots, a_n$ are poles of f and $\varphi_k(z)$ is the principal part of f at a_k for $1 \leq k \leq n$, then

$$f(z) = \frac{1}{2\pi i} \int_\gamma \frac{f(\zeta)}{\zeta - z} d\zeta + \sum_{k=1}^n \varphi_k(z), \quad z \in \text{Int } \gamma - \{a_1, \ldots, a_n\}.$$
$$(4.11.3.8)$$

If Int γ is replaced by the exterior Ext γ, then (4.11.3.6) is replaced by

$$f(z) = -\sum_{k=1}^n \text{Re } s\left(\frac{f(\zeta)}{\zeta - z}; \zeta = a_k\right) - \text{Re } s\left(\frac{f(\zeta)}{\zeta - z}; \zeta = \infty\right),$$
$$z \in \text{Ext } \gamma - \{a_1, \ldots, a_n\}. \qquad (4.11.3.9)$$

(11) Let $p(z)$ be a polynomial of degree k. Let $r > 0$ but $r \neq 1, 2, \ldots, k$. Show that

$$\int_{|z|=r} p(z) \left(\sum_{j=0}^l \exp \frac{1}{z-j}\right) dz = 2\pi i \sum_{j=0}^l \sum_{m=0}^k \frac{p^{(m)}(j)}{m!(m+1)!},$$

where l is a nonnegative integer satisfying: if $r < k$, $l < r < l+1$; if $r > k$, $l = k$.

(12) Let γ: $z(t)$, $a \leq t \leq b$, be a rectifiable or piecewise smooth Jordan closed curve. Suppose f is analytic on $\overline{\text{Int } \gamma}$, except finitely many isolated singularities $a_1, \ldots, a_n$ in Int γ and finitely many simple poles $b_1, \ldots, b_m$, lying on γ at each of which γ is regular, namely, γ has a

nonzero tangent vector $z'(b_l) \neq 0$ for $1 \leq l \leq m$ (see Sec. 4.7). Then

$$P.V. \int_\gamma f(z)dz = 2\pi i \sum_{k=1}^n \mathrm{Re}\, s(f; a_k) + \pi i \sum_{l=1}^m \mathrm{Re}\, s(f; b_l).$$

$$(4.11.3.10)$$

Note that, this formula is still valid if the principal part of f at *each* b_l, $1 \leq l \leq m$, is of the form $p_l(z) = \sum_{p=0}^{q_l} \frac{a_{-(2p+1)}^{(l)}}{(z-z_l)^{2p+1}}$, namely, $p_l(z)$ contains only *odd powers* of $(z - z_l)^{-1}$. What happens if this constrained condition is discarded?

(13) Prove (4.11.3.4) and (4.11.3.5) in detail.

4.12.   The Applications of the Residue Theorem in Evaluating the Integrals

In calculus, to evaluate integrals $\int_a^b f(x)dx(-\infty \leq a < b \leq \infty)$ in which the primitives of the integrands $f(x)$ cannot be expressed as elementary functions, one usually adopts the method of integration with parameters. In general, this method is performed case-by-case and it is not easy to track down its original idea behind the whole process. While, the residue theorem provides a more unified and effective way of evaluating this kind of integrals, even though it might be not easier in the amount of computation.

Roughly speaking, the *residue method* (or *method of residues*) is as follows:

Step one. Extend the interval of integration $[a, b]$ or (a, b) as part of a rectifiable Jordan closed curve γ in the complex plane **C**.

Step two. Replace the integrand $f(x)$ by a suitably chosen analytic or meromorphic (auxiliary) function $F(z)$, so that $F(z)$ has finitely many isolated singularities in Int γ and is continuous on γ. In many cases, some auxiliary functions are indispensable.

Step three. Apply the residue theorem to $F(z)$ in $\overline{\mathrm{Int}\, \gamma}$.

The main difficulties in this method seem to come from how to choose suitable path of integration γ and the integrand $F(z)$ so that they will cooperate nicely in $\int_\gamma F(z)dz$. One needs to practice more before acquiring familiarity.

Here, we just illustrate some basic examples to see how powerful the residue method is. There are many known variants of this method and complicated examples, which are far from being exhausted at the present time. For more and better information concerned, refer to Refs. [7, 80], and [60] Vol. I, Vol. II, and the references and literatures mentioned therein.

4.12.1. $\int_0^{2\pi} f(\cos\theta, \sin\theta)d\theta$, where $f(x, y)$ is a function
in x and y, etc.

How to evaluate the integral

$$\int_0^{\frac{\pi}{2}} \frac{d\theta}{a + \sin^2\theta} = \frac{1}{4}\int_0^{2\pi} \frac{d\theta}{a + \sin^2\theta}(a > 0)?$$

Let $z = e^{i\theta}$, $0 \le \theta \le 2\pi$. Then

$$dz = ie^{i\theta}d\theta = izd\theta \Rightarrow d\theta = \frac{dz}{iz}, \quad \text{and}$$

$$\sin^2\theta = \left[\frac{1}{2i}(e^{i\theta} - e^{-i\theta})\right]^2 = \left[\frac{1}{2i}(z - z^{-1})\right]^2 = \frac{z^4 - 2z^2 + 1}{-4z^2},$$

$$\Rightarrow \frac{1}{4}\int_0^{2\pi} \frac{d\theta}{a + \sin^2\theta} = \int_{|z|=1} \frac{iz}{z^4 - (2 + 4a)z^2 + 1}dz.$$

The problem reduces to how to evaluate the above right integral. Now $z^4 - (2 + 4a)z^2 + 1 = 0$ has four zeros $\pm\sqrt{1 + 2a \pm 2\sqrt{a(1 + a)}}$. Among then, only $z_1 = \sqrt{1 + 2a - 2\sqrt{a(1 + a)}}$ and $z_2 = -z_1$ lie in the disk $|z| < 1$. The integrand $F(z)$ has simple poles at z_1 and $-z_1$, and the residues are

$$\operatorname{Re} s(F; z_1) = \frac{-i}{8\sqrt{a^2 + a}} = \operatorname{Re} s(F; -z_1)$$

$$\Rightarrow \int_0^{\frac{\pi}{2}} \frac{d\theta}{a + \sin^2\theta} = \int_{|z|=1} F(z)dz$$

$$= 2\pi i\{\operatorname{Re} s(F; z_1) + \operatorname{Re} s(F; -z_1)\} = \frac{\pi}{2\sqrt{a^2 + a}}.$$

The method introduced above can be summarized as

The evaluation of $\int_0^{2\pi} f(\cos\theta, \sin\theta)d\theta$. By invoking the change of variables $z = e^{i\theta}$, $0 \le \theta \le 2\pi$ (and then, $d\theta = \frac{dz}{iz}$), if

$$F(z) = f\left(\frac{z + z^{-1}}{2}, \frac{z - z^{-1}}{2i}\right)\frac{1}{iz}$$

is analytic on $|z| \le 1$ except isolated singularities $z_1, \ldots, z_n$ in the interior $|z| < 1$, then

$$\int_0^{2\pi} f(\cos\theta, \sin\theta)d\theta = \int_{|z|=1} F(z)dz = 2\pi i\sum_{k=1}^n \operatorname{Re} s(F; z_k). \qquad (4.12.1.1)$$

Usually $f(\cos\theta, \sin\theta)$ is a rational function in $\cos\theta$ and $\sin\theta$; in this case, $F(z)$ is a rational function of z.

Remark. When changing the real reariable θ, $0 \le \theta \le 2\pi$, to the complex variable z, $|z| = 1$ via $z = e^{i\theta}$, one needs to use the following relations:

$$\cos n\theta = \frac{1}{2}(z^n + z^{-n}) \quad \text{and} \quad \sin n\theta = \frac{1}{2i}(z^n - z^{-n}), \quad n = 1, 2, 3, \dots,$$

$$\text{and} \quad d\theta = \frac{dz}{iz}. \tag{4.12.1.2}$$

Since $f(\cos\theta, \sin\theta)$ is a function of period 2π, so

$$\int_a^{a+2\pi} f(\cos\theta, \sin\theta)d\theta = \int_0^{2\pi} f(\cos\theta, \sin\theta)d\theta \tag{4.12.1.3}$$

can be evaluated by the same method.

For integrals of the form

$$\int_0^{2\pi} f(\cos\theta, \sin\theta) \begin{Bmatrix} \cos n\theta \\ \sin n\theta \end{Bmatrix} d\theta, \quad \text{where } n \text{ is a positive integer,} \tag{4.12.1.4}$$

they can be combined into a single one integral as

$$\int_0^{2\pi} f(\cos\theta, \sin\theta)e^{in\theta} d\theta. \tag{4.12.1.5}$$

And then, use the same method as in (4.12.1.1) to evaluate it. It is worth to summarize as

The evaluation of $\int_0^{2\pi} f(\cos\theta, \sin\theta)g(e^{i\theta})d\theta$. Let $f(\cos\theta, \sin\theta)$ and $F(z)$ be as in (4.12.1.1). If $g(z)$ is analytic on $|z| \le 1$, then

$$\int_0^{2\pi} f(\cos\theta, \sin\theta)g(e^{i\theta})d\theta = \int_{|z|=1} F(z)g(z)dz = 2\pi i \sum_{k=1}^{n} \text{Re}\,s(F(z)g(z); z_k).$$

$$\tag{4.12.1.6}$$

$\square$

In what follows, we give some related examples.

Example 1. Evaluate the integral

$$\int_0^{2\pi} e^{-\cos\theta} \cos(n\theta + \sin\theta)d\theta, \quad n = 1, 2, 3, \dots.$$

Solution. Substitute $z = e^{i\theta}$ into

$$e^{-\cos\theta}\cos(n\theta + \sin\theta) - ie^{-\cos\theta}\sin(n\theta + \sin\theta) = e^{-e^{i\theta}}e^{-in\theta} = e^{-z}z^{-n},$$

$$\Rightarrow \int_0^{2\pi} e^{-e^{i\theta}}e^{-in\theta}d\theta = \int_{|z|=1} e^{-z}z^{-n}\frac{1}{iz}dz = -i\int_{|z|=1}\frac{e^{-z}}{z^{n+1}}dz$$

$$= -i \cdot 2\pi i \operatorname{Re} s\left(\frac{e^{-z}}{z^{n+1}};0\right) = 2\pi \cdot \frac{(-1)^n}{n!}$$

$$\Rightarrow \text{The original integral} = \frac{2(-1)^n\pi}{n!},$$

$$\int_0^{2\pi} e^{-\cos\theta}\sin(n\theta + \sin\theta)d\theta = 0, \quad n = 1,2,3,\ldots.$$

Observe that the integrand of the last integral is an odd function in $[-\pi,\pi]$.

Example 2. Evaluate the integral

$$\int_0^\pi \cot(x+a)dx, \quad \text{where } a \in \mathbf{C} \text{ is a constant with } \operatorname{Im} a \neq 0.$$

Solution. Since cot is a function with period π, then

$$\int_0^\pi \cot(x+a)dx = \frac{1}{2}\int_0^{2\pi}\cot(x+a)dx.$$

Let $a = \alpha + i\beta$, where $\beta \neq 0$. Then

$$\cot(x+a) = i\frac{e^{2i(x+\alpha)} + e^{2\beta}}{e^{2i(x+\alpha)} - e^{2\beta}} = i\frac{z^2 + e^{2\beta}}{z^2 - e^{2\beta}} \quad \text{if } e^{i(x+\alpha)} = z$$

$$\Rightarrow \int_0^{2\pi}\cot(x+a)dx = \int_\alpha^{2\pi+\alpha}\cot(x+a)dx$$

$$= i\int_{|z|=1}\frac{z^2 + e^{2\beta}}{z^2 - e^{2\beta}}\cdot\frac{1}{iz}dz = \int_{|z|=1}\frac{z^2 + e^{2\beta}}{z(z^2 - e^{2\beta})}dz.$$

The integrand $F(z)$ has simple poles at 0, e^β, and $-e^\beta$ with the corresponding residues

$$\operatorname{Re} s(F;0) = -1, \quad \operatorname{Re} s(F;e^\beta) = \operatorname{Re} s(F;-e^\beta) = 1,$$

$$\Rightarrow \int_0^{2\pi}\cot(x+a)dx = 2\pi i\begin{cases} -1, & \text{if } \beta > 0 \\ 1, & \text{if } \beta < 0 \end{cases} = -2\pi i \operatorname{sgn}\operatorname{Im} a.$$

$$\Rightarrow \int_0^\pi \cot(x+a)dx = -\pi i \operatorname{sgn}\operatorname{Im} a.$$

Example 3. (Adamović; 1959) In case $b > a > -1$, show that

$$\int_0^{\frac{\pi}{2}} \cos^a t \cos bt\, dt = \frac{\Gamma(a+1)\Gamma\left(\frac{b-a}{2}\right)}{2^{a+1}\Gamma\left(1+\frac{a+b}{2}\right)} \sin\frac{\pi(b-a)}{2},$$

where Γ is Euler's gamma function (see Sec. 5.6).

Solution. Set $z = e^{it}$, $0 \le t \le 2\pi$. Then

$$\cos^a t \cos bt + i\cos^a t \sin bt$$
$$= \cos^a t\, e^{ibt} = 2^{-a}(z+z^{-1})^a\, z^b = 2^{-a}(z^2+1)^a\, z^{b-a}.$$

$$\Rightarrow \int_0^{\frac{\pi}{2}} \cos^a t \cos bt\, dt = \frac{1}{2}\int_{-\frac{\pi}{2}}^{\frac{\pi}{2}} \cos^a t \cos bt\, dt = \frac{1}{2}\text{Re}\int_{-\frac{\pi}{2}}^{\frac{\pi}{2}} e^{ibt}\cos^a t\, dt$$

$$= \frac{1}{2}\text{Re}\frac{1}{2^a i}\int_{\substack{|z|=1 \\ \text{Re } z \ge 0}} (z^2+1)^a\, z^{b-a-1}dz.$$

The function $(z^2+1)^a\, z^{b-a-1}$ has branches in the domain $\mathbf{C}-(\infty,0]-[-i,i]$. We choose the branch $F(z)$, determined uniquely by $(1^2+1)^a\, 1^{b-a-1} = 2^a$ in this domain, as the integrand. In this case, $\text{Arg}(1+i) = \frac{\pi}{4}$, $\text{Arg}(1-i) = \frac{-\pi}{4}$, and $\text{Arg}\,1 = 0$; and, if z lies on the right side of the cut $[-i,i]$, $\text{Arg}(z^2+i) = \text{Arg}(z+i) + \text{Arg}(z-i) = \frac{\pi}{2} - \frac{\pi}{2} = 0$, and $\text{Arg}\,z = \frac{\pi}{2}$ in $\text{Im } z > 0$, $= -\frac{\pi}{2}$ in $\text{Im } z < 0$.

Consider the dented contour in Fig. 4.64, where $0 < r < \frac{1}{2}$. Let $\theta(r)$ denote the acute angle between OA and the positive imaginary axis. The

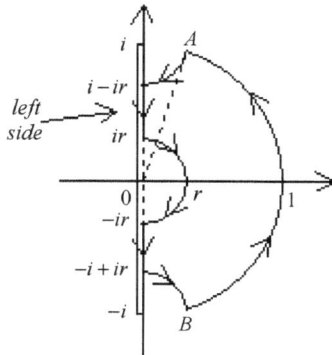

Fig. 4.64

path of integration γ is composed of the following curves: the segments $[i - ir, ir]$, $[-ir, -i + ir]$, and

$$\gamma_0 : z = re^{i\theta}, \quad -\frac{\pi}{2} \le \theta \le \frac{\pi}{2};$$

$$\gamma_1 : z = i + re^{i\theta}, \quad -\frac{\pi}{2} \le \theta \le \mathrm{Arg}(A - i);$$

$$\gamma_2 : z = -i + re^{i\theta}, \quad \mathrm{Arg}(B + i) \le \theta \le \frac{\pi}{2};$$

$$\gamma_3 : z = e^{i\theta}, \quad -\frac{\pi}{2} + \theta(r) \le \theta \le \frac{\pi}{2} - \theta(\gamma).$$

Cauchy's integral theorem says that

$$\int_\gamma F(z)\,dz = 0. \tag{$*_1$}$$

Wherein, as $r \to 0$,

$$\int_{\gamma_3} F(z)\,dz = \int_{-\frac{\pi}{2}+\theta(r)}^{\frac{\pi}{2}-\theta(r)} (e^{2it} + 1)^a e^{i(b-a-1)t} i e^{it}\,dt$$

$$= 2^a i \int_{-\frac{\pi}{2}+\theta(r)}^{\frac{\pi}{2}-\theta(r)} e^{ibt} \cos^a t\,dt$$

$$\to 2^a i \int_{-\frac{\pi}{2}}^{\frac{\pi}{2}} e^{ibt} \cos^a t\,dt;$$

$$\int_{[i(1-r),ir]} F(z)\,dz = \int_{1-r}^{r} (1 - y^2)^a (iy)^{b-a-1} i\,dy$$

$$= -i e^{\frac{\pi}{2}(b-a-1)i} \int_{r}^{1-r} (1 - y^2)^a y^{b-a-1}\,dy$$

$$\to -i e^{\frac{\pi}{2}(b-a-1)i} \int_{0}^{1} (1 - y^2)^a y^{b-a-1}\,dy$$

and

$$\int_{[-ir,-i(1-r)]} F(z)\,dz = \int_{r}^{1-r} (1 - y^2)^a (-iy)^{b-a-1}(-i\,dy),$$

$$\to -i e^{-\frac{\pi}{2}(b-a-1)i} \int_{0}^{1} (1 - y^2)^a y^{b-a-1}\,dy.$$

On the other hand, since $b > a > -1$, therefore, as $r \to 0$ (compare to Exercise A (28)(b) of Sec. 2.9 or (4.12.3.3)),

$$\left|\int_{\gamma_0} F(z)dz\right| \leq \left|\int_{\gamma_0} |F(z)|\,|dz|\right| \leq \int_{-\frac{\pi}{2}}^{\frac{\pi}{2}} (1+r^2)^a r^{b-a-1} \cdot r d\theta$$

$$= \int_{-\frac{\pi}{2}}^{\frac{\pi}{2}} (1+r^2)^a r^{b-a} d\theta \to 0;$$

$$\left|\int_{\gamma_1} F(z)dz\right| \leq \int_{-\frac{\pi}{2}}^{\text{Arg}(A-i)} r^a (1+r)^a (1+r)^{b-a-1} \cdot r d\theta$$

$$= \int_{-\frac{\pi}{2}}^{\text{Arg}(A-i)} r^{a+1}(1+r)^{b-1} dr \to 0,$$

and, similarly, $\left|\int_{\gamma_2} F(z)dz\right| \to 0$.

Letting $r \to 0$ in $(*_1)$ and substituting these results in it, then

$$\left[-e^{\frac{\pi}{2}(b-a)i} + e^{-\frac{\pi}{2}(b-a)i}\right] \int_0^1 (1-y^2)^a y^{b-a-1} dy + 2^a i \int_{-\frac{\pi}{2}}^{\frac{\pi}{2}} e^{ibt} \cos^a t\, dt$$

$$= 0 \Rightarrow \int_{-\frac{\pi}{2}}^{\frac{\pi}{2}} e^{ibt} \cos^a t\, dt$$

$$= \frac{1}{2^a i} \left[e^{\frac{\pi}{2}(b-a)i} - e^{-\frac{\pi}{2}(b-a)i}\right] \int_0^1 (1-y^2)^a y^{b-a-1} dy$$

$$= \frac{1}{2^{a-1}} \sin\frac{\pi}{2}(b-a) \cdot \frac{\Gamma(a+1)\Gamma\left(\frac{b-a}{2}\right)}{2\Gamma\left(a+1+\frac{b-a}{2}\right)}$$

and the result follows, where, in the last step, we have used the gamma and beta functions in

$$\int_0^1 (1-y^2)y^{b-a-1} dy$$

$$= \frac{1}{2} \int_0^1 (1-t)^a t^{\frac{b-a}{2}} dt$$

$$= \frac{1}{2} B\left(a+1, \frac{b-a}{2}\right) = \frac{\Gamma(a+1)\Gamma\left(\frac{b-a}{2}\right)}{2\Gamma\left(a+1+\frac{b-a}{2}\right)} \quad \text{(see Sec. 5.6)}.$$

Example 4. Show that

$$\int_0^\pi \log\sin\theta d\theta = -\pi\log 2.$$

Solution. Observe that $\sin\theta \geq 0$ on $0 \leq \theta \leq \pi$. Then

$$\log\sin\theta = \log|\sin\theta| = \log\left|\frac{e^{i\theta} - e^{-i\theta}}{2i}\right| = \log|e^{2i\theta} - 1| - \log 2$$

$$\Rightarrow \int_0^\pi \log\sin\theta d\theta = \int_0^\pi \log|e^{2i\theta} - 1|d\theta - \pi\log 2$$

$$= \int_0^{2\pi} \log|e^{i\theta} - 1|d\theta - \pi\log 2.$$

The problem reduces to show that $\int_0^{2\pi} \log|e^{i\theta} - 1|d\theta = 0$. To see this, we try to set $z = e^{i\theta}$, $0 \leq \theta \leq 2\pi$, and recall that $d\theta = \frac{dz}{iz}$. Consequently, consider the function $F(z) = \frac{\log(1-z)}{iz}$ which has branches in $\mathbf{C} - [1, \infty)$. We designate $F(z)$ the branch uniquely determined by $\log 1 = 0$. In this case, $F(z)$ has a removable singularity at $z = 0$ and

$$\mathrm{Re}\,s(F; 0) = \lim_{z\to 0} zF(z) = \lim_{z\to 0} \frac{\mathrm{Log}(1 - z)}{i} = 0$$

$$\Rightarrow \int_\gamma F(z)dz = 2\pi i \cdot 0 = 0 \qquad (*_2)$$

where $\gamma = \gamma_0 + \gamma_1$ is shown in Fig. 4.65 with $\gamma_0 : z = e^{i\theta}$, $\theta(r) \leq \theta \leq 2\pi - \theta(r)$ and $\gamma_1: z = 1 + re^{it}$, $-\frac{\pi}{2} - \varphi(r) \leq t \leq \frac{\pi}{2} + \varphi(r)$ in which both $\theta(r)$ and $\varphi(r)$ approach 0 as $r \to 0$. Along γ_1, we have, as $r \to 0$,

$$\int_{\gamma_1} F(z)dz = \int_{-\frac{\pi}{2}-\varphi(r)}^{\frac{\pi}{2}+\varphi(r)} \frac{\mathrm{Log}(-re^{it})}{i(1 + re^{it})} rie^{it}dt$$

$$= \int_{-\frac{\pi}{2}-\varphi(r)}^{\frac{\pi}{2}+\varphi(r)} \frac{r(\log r + i(\pi + t))}{1 + re^{it}} e^{it}dt \to 0.$$

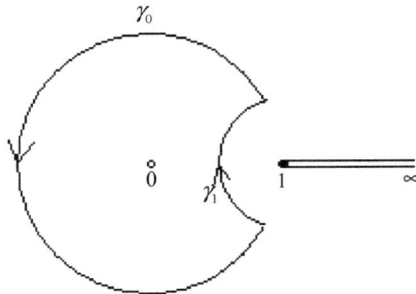

Fig. 4.65

Letting $r \to 0$ in $(*_2)$, we have then

$$\int_{|z|=1} \frac{\text{Log}(1-z)}{iz}dz = \int_0^{2\pi} \frac{\log|1-e^{i\theta}| + i\,\text{Arg}(1-e^{i\theta})}{ie^{i\theta}} ie^{i\theta}d\theta,$$

$$= \int_0^{2\pi} [\log|1-e^{i\theta}| + i\,\text{Arg}(1-e^{i\theta})]d\theta = 0,$$

$$\Rightarrow \int_0^{2\pi} \log|e^{i\theta}-1|d\theta = 0,$$

$$\int_0^{2\pi} \text{Arg}(1-e^{i\theta})d\theta = 0.$$

Note that this integral formula can also be justified easily as follows. Let $I = \int_0^\pi \log\sin\theta d\theta$. Then

$$I = 2\int_0^{\frac{\pi}{2}} \log\sin\theta d\theta$$

$$= 2\int_0^{\frac{\pi}{2}} \log\cos\theta d\theta = \int_0^\pi \log\left(2\sin\frac{\theta}{2}\cos\frac{\theta}{2}\right)d\theta = \pi\log 2 + 2I$$

and the result follows.

Exercises A

(1) Evaluate the following integrals.

(a) $\displaystyle\int_0^{2\pi} \frac{dx}{a\cos x + b\sin x + c}$ $\quad(a > 0,\, b > 0,\, c > 0 \text{ and } c^2 > a^2 + b^2)$

and $\displaystyle\int_0^{2\pi} \frac{1}{(a\cos x + b\sin x + c)^2}dx.$

(b) $\displaystyle\int_0^{\frac{\pi}{2}} \frac{dx}{a^2\sin^2 x + b^2\cos^2 x + c^2}$ $\quad(a > 0,\, b > 0,\, c > 0).$

(c) $\displaystyle\int_0^\pi \frac{\sin^2 x}{(1 - 2a\cos x + a^2)(1 - 2b\cos x + b^2)}dx$ $\quad(0 < a < b < 1).$

(d) $\displaystyle\int_0^{2\pi} \frac{d\theta}{a + b\sin\theta}$ $\quad(a > |b|)$ and $\displaystyle\int_0^{2\pi} \frac{d\theta}{(a + b\sin\theta)^2}.$

(e) $\displaystyle\int_0^{2\pi} \frac{(1 + 2\cos\theta)^n \cos n\theta}{5 - 4\cos\theta}d\theta$ $\quad(n = 0, 1, 2, \ldots).$

(f) $\displaystyle\int_0^\pi \frac{\cos\theta(1 + \cos\theta)}{(1 + \cos^2\theta)^2}d\theta.$

(g) $\displaystyle\int_{-\frac{\pi}{2}}^{\frac{\pi}{2}} \frac{d\theta}{1 - 2a\sin\theta + a^2}$, $0 < |a| < 1$.

(h) $\displaystyle\int_0^{2\pi} \frac{\sin 2\theta}{(a + \cos\theta)(a - \sin\theta)} d\theta$.

(2) Evaluate the following integrals.

(a) $\displaystyle\int_0^{2\pi} \frac{d\theta}{a + b\cos\theta}$ $(a > |b|)$ and $\displaystyle\int_0^{2\pi} \frac{d\theta}{(a + b\cos\theta)^2}$.

(b) $\displaystyle\int_0^{2\pi} \frac{\cos n\theta}{a + b\cos\theta} d\theta$ $(a, b \in \mathbf{R}, |a| > |b|$ and n is a positive integer).

(c) $\displaystyle\int_0^{2\pi} \frac{d\theta}{(a + b\cos^2\theta)^2}$ $(a > 0, b > 0)$.

(d) $\displaystyle\int_0^{2\pi} \frac{1 + \cos x}{(13 - 5\cos x)^2} dx$.

(e) $\displaystyle\int_0^{2\pi} \frac{1 + \cos x}{1 + \cos^2 x} dx$.

(f) $\displaystyle\int_0^{\frac{\pi}{2}} \frac{1 + 2\sin^2\theta}{1 + 2\cos^2\theta} d\theta$.

(g) $\displaystyle\int_0^{2\pi} \frac{\sin^2 x}{a - bx} dx = 2\int_{-1}^1 \frac{\sqrt{1 - x^2}}{a - bx} dx$ $(a > b > 0)$.

(3) Evaluate the following integrals, where $f(\theta) = 1 - 2a\cos\theta + a^2$.

(a) $\displaystyle\int_0^{2\pi} \frac{d\theta}{f(\theta)}$ $(|a| \neq 1)$.

(b) $\displaystyle\int_0^\pi \frac{\cos(n-1)\theta}{(f(\theta))^n} d\theta$ $(0 < a < 1$ and n is a positive integer).

(c) $\displaystyle\int_0^{2\pi} \frac{\cos^2\theta}{f(\theta - \varphi)} d\theta$ $(|a| \neq 1)$.

(d) $\displaystyle\int_0^{2\pi} \frac{\cos^3\theta}{f(\theta)} d\theta$ $(|a| < 1)$.

(e) $\displaystyle\int_0^{2\pi} \frac{\cos 2\theta}{f(\theta)} d\theta$ $(0 < |a| < 1)$.

(f) $\displaystyle\int_0^{2\pi} \frac{\cos^2 3\theta}{f(\theta)} d\theta$ $(a \neq \pm 1)$.

(g) $\displaystyle\int_0^{2\pi} \frac{\cos^2 3\theta}{1 + a^2 - 2a\cos 2\theta} d\theta$ $(0 < |a| < 1)$.

(h) $\displaystyle\int_0^{2\pi} \frac{\cos n\theta}{1 + a^2 + 2a\cos\theta} d\theta$ $(|a| < 1$ and $n = 0, 1, 2, \ldots)$.

(4) Evaluate the following integrals.

(a) $\int_0^{\pi} \tan(x+ia)dx$, where a is a real constant.

(b) $\int_0^{2\pi} \dfrac{dt}{a - ib\cos t}$, where $a, b \in \mathbf{R}$ and $a \neq 0$.

(5) Evaluate or verify the following integrals or formulae.

(a) $\int_0^{2\pi} e^{\cos\theta}\cos(\sin\theta - n\theta)d\theta = \dfrac{2\pi}{n!}$ and $\int_0^{2\pi} e^{\cos\theta}\sin(\sin\theta - n\theta)$
$d\theta = 0$ $n = 0, 1, 2, \ldots$.

(b) $\int_0^{2\pi} e^{2a\cos\theta}d\theta = 2\pi \sum\limits_{n=0}^{\infty} \left(\dfrac{a^n}{n!}\right)^2$, $a \in \mathbf{C}$.

(c) $\int_0^{2\pi} \cos^{2n}\theta d\theta = \int_0^{2\pi} \sin^{2n}\theta d\theta = \dfrac{2\pi(2n)!}{2^{2n}(n!)^2}$, $n = 0, 1, 2, \ldots$.

(d) $\int_0^{\frac{\pi}{2}} \cos^n x \cos nx dx$, $n = 0, 1, 2, \ldots$.

(e) $\int_0^{\pi} \cos^n x \cos mx dx$, $m, n = 0, 1, 2, \ldots$.

(f) $\int_0^{\pi} \sin^m x \cos(2n+1)x dx$, $m, n = 0, 1, 2, \ldots$.

(g) $\int_0^{\pi} (\cos nx)\log(1-2a\cos x + a^2)dx$ and $\int_0^{\pi} (\sin nx)\log(1-2a\cos x + a^2)dx$, $n = 0, 1, 2, \ldots$; $0 < a < 1$.

(6) Evaluate or verify the following integrals or formulae.

(a) $\int_0^{2\pi} (a\cos\theta + b\sin\theta)^{2n}d\theta = \dfrac{2\pi(2n)!\sqrt{a^2+b^2}}{2^{2n}(n!)^2}$, $n = 1, 2, \ldots; a, b \in \mathbf{R}$.

(b) $\int_0^{2\pi} \dfrac{(1+2\cos t)^n \cos nt}{1 - 2a\cos t - a}dt$, $n = 0, 1, 2, \ldots; -1 < a < \frac{1}{3}$.

(c) $\int_{-1}^{1} \dfrac{x^{2n}}{\sqrt{1-x^2}}dx = \dfrac{(2n-1)(2n-3)\cdots 1}{2n(2n-2)\cdots 2}\pi$, $n = 1, 2, 3, \ldots$.

(d) $\int_0^{\pi} \dfrac{\cos nx}{(\cos x + \cosh a)^2}dx = (-1)^n \dfrac{\pi e^{-\pi a}}{\sinh^3 a}(n\sinh a + \cosh a)$, $n = 1, 2, 3 \ldots; a \in \mathbf{R}$ and $a \neq 0$.

(e) $\int_0^{2\pi} \dfrac{(1 - 4\sin^2\theta)^n \cos 2\theta}{2 - \cos\theta}d\theta = \dfrac{2\pi(91 - 52\sqrt{3})}{\sqrt{3}}$, $n \geq 1$.

(f) $\int_0^{\frac{\pi}{2}} \log(a^2\sin^2 x + b^2\cos^2 x)dx$, $0 < a < b$.

(g) $\displaystyle\int_0^{2\pi} \log\left(\frac{5}{4} + \sin x\right) dx.$

(7) (Boas 1964) Suppose $P(z) = a_n z^n + \cdots + a_1 z + a_0 \neq 0$ for all $z \in \mathbf{C}$, where $n \geq 1$ and $a_n \neq 0$. Consider the integral (and verify it)

$$0 < \int_0^{2\pi} \frac{d\theta}{P(2\cos\theta)\overline{P(2\cos\theta)}} = \frac{1}{i}\int_{|z|=1} \frac{z^{2n-1}}{h(z)} dz = 0$$

where $h(z) = z^{2n} P\left(z + \frac{1}{z}\right) \overline{P}\left(z + \frac{1}{z}\right)$, $z \in \mathbf{C} - \{0\}$. This contradiction yields another proof of the fundamental theorem of algebra.

Exercises B

(1) (Mitrinović and Marsh, 1961, 1962) Suppose n and m are positive integers. If $|a| < 1$, show that

$$J(a, n, m) = \int_{-\pi}^{\pi} \frac{\cos mt}{(1 - 2a\cos t + a^2)^n} dt,$$

$$= \frac{2\pi|a|^m}{(1 - a^2)^n} \sum_{k=0}^{n-1} C_{n-1-k}^{n-1+m} C_k^{n-1+k} \frac{a^{2k}}{(1 - a^2)^k};$$

if $|a| > 1$, show that $J(a^{-1}, n, m) = a^{-2n} J(a, n, m)$. Observe that, in case $m \neq 0$, then $J(0, n, m) = 0$; while $J(0, n, 0) = 2\pi$.

(2) Show that

$$\int_{-\pi}^{\pi} \frac{x \sin x}{1 + a^2 - 2a\cos x} dx = \begin{cases} \dfrac{2\pi}{a} \log\dfrac{1+a}{a}, & a > 1 \\[2mm] 2\pi a \log(1 + a), & 0 < a < 1 \end{cases}$$

by integrating $f(z) = \frac{z}{a - e^{-iz}}$ $(a > 1)$ along γ shown in Fig. 4.66.

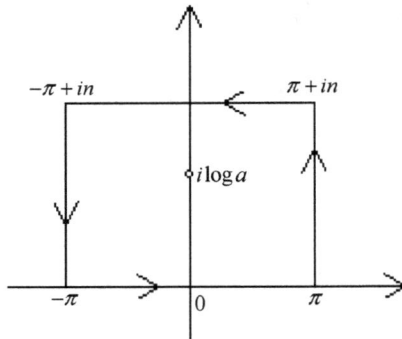

Fig. 4.66

(3) (Schmidt, 1903) Let n be a positive integer. Show that

$$\int_0^\pi \frac{dt}{(a + \sqrt{a^2 - 1}\cos t)^{n+1}} = \pm\frac{\pi}{2^n n!}\frac{d^n}{da^n}(a^2 - 1)^n,$$

where

$$P_n(x) = \frac{1}{2^n n!}\frac{d^n}{dx^n}(x^2 - 1)^n,$$

$$= \frac{(-1)^n}{2^n}\sum_{k=0}^n (-1)^k(C_k^n)^2(1 + x)^k(1 + x)^{n-k}$$

is the *Legendre polynomial of order* n, by integrating

$$F(z) = \frac{1}{z^{n+1}\sqrt{1 - 2az + z^2}}$$

along $\Gamma = \{|z| = R\} - \gamma_1 - (l_1 + l_2) - \gamma_2 - \gamma_0$ as shown in Fig. 4.67, in which $c_1 = a + \sqrt{a^2 - 1}$ and $c_2 = a - \sqrt{a^2 - 1}$.

(4) (Gilles and Casteren 1983) Show that

$$\int_0^{\frac{\pi}{2}} t^2\sqrt{\tan t}\sin 2t\, dt$$

$$= \frac{\sqrt{2}\pi}{32}\left[\frac{5\pi^2}{6} + 2\pi - 2\pi\log 2 - 4 + 4\log 2 - 2(\log 2)^2\right]$$

by integrating $F(z) = (\log z)^2\sqrt{\frac{1-z}{1+z}}$ along γ shown in Fig. 4.68 (compare to Example 3).

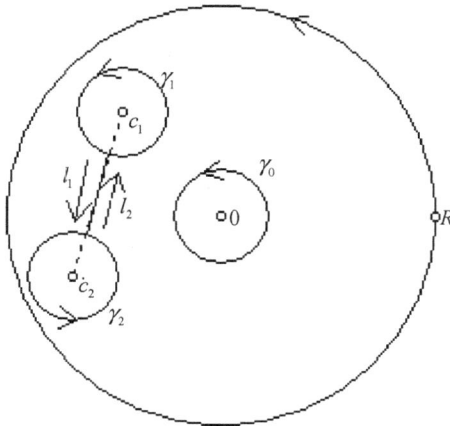

Fig. 4.67

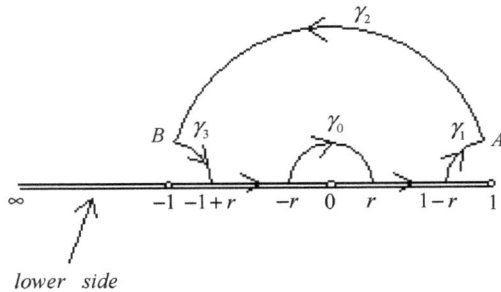

Fig. 4.68

4.12.2. $\int_0^1 x^{\alpha-1}(1-x)^{-\alpha}f(x)dx \ (0 < \alpha < 1)$, etc.

In this section, we try to evaluate improper integrals of the type $\int_a^b F(x)dx$, where $-\infty < a < b < \infty$. As a standard yet general setting in this direction, we have

The evaluation of $\int_0^1 x^{\alpha-1}(1-x)^{-\alpha}f(x)dx \ (0 < \alpha < 1)$. Suppose f is analytic in $\mathbf{C}^*$ except a finite number of isolated singularities $z_1, \ldots, z_n$ in $\mathbf{C}$ which do not lie in the real segment $[0,1]$ so that ∞ is a removable singularity of f. Then

$$\int_0^1 x^{\alpha-1}(1-x)^{-\alpha}f(x)dx = \frac{f(\infty)\pi}{\sin\alpha\pi} - \frac{\pi}{e^{\alpha\pi i}\sin\alpha\pi}\sum_{k=1}^{n}\operatorname{Re} s(F(z);z_k),$$

where $F(z) = z^{\alpha-1}(1-z)^{-\alpha}f(z)$. In particular, in case $f(z) = \frac{p(z)}{q(z)}$ is a rational function in which the degree of $p(z)$ is not larger than that of $q(z)$ and $q(x) \neq 0$ for all $x \in [0,1]$, then this formula is applicable.   (4.12.2.1)

Proof.   According to (3) in Example 5 of Sec. 2.7.3, $z^{\alpha-1}(1-z)^{-\alpha}$ has branches in the slit domain $\mathbf{C}^* - [0,1]$ with ∞ as a node point.

To choose a branch of $F(z)$ so that it is analytic in $\mathbf{C}^* - [0,1]$ except at $z_1, \ldots, z_n$ and is continuous to the upper side of the cut $(0,1)$ from the upper half-plane and continuous to the lower side of $(0,1)$ from the lower half-plane: Designate the value of $z^{\alpha-1}(1-z)^{-\alpha}$ to be positive *along the upper side* of $(0,1)$, namely,

$$\arg x = \arg(1-x) = 0,$$

and, *in this case,* $F(x) = x^{\alpha-1}(1-x)^{-\alpha}f(x)$ holds. If a moving point z starts from a point in the upper side and winds around the branch point 1 to a point in the lower side, both $\arg z$ and $\arg(1-z)$ keep changing continuously

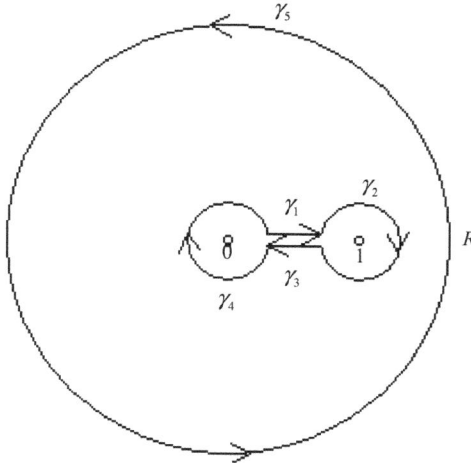

Fig. 4.69

and assume the final values

$$\arg x = 0 \quad \text{and} \quad \arg(1-x) = -2\pi,$$
$$\Rightarrow F(x) = e^{(\alpha-1)\log x}e^{-\alpha\log(1-x)}f(x)$$
$$= e^{(\alpha-1)[\log|x|+i\arg x]}e^{-\alpha[\log|1-x|+i\arg(1-x)]}f(x)$$
$$= e^{(\alpha-1)\log|x|}e^{-\alpha[\log|1-x|-2\pi i]}f(x) = x^{\alpha-1}(1-x)^{-\alpha}e^{2\alpha\pi i}f(x)$$

along the lower side of $(0,1)$.

$F(x)$ will attain the same value if z winds around the branch point 0.

Integrate $F(z)$ along the dented contour $\gamma = \gamma_5 + (\gamma_1+\gamma_2+\gamma_3+\gamma_4)$ shown in Fig. 4.69 where γ_1: $z = t, r \le t \le 1-r$ (the upper side), $0 < r < \frac{1}{2}$; γ_2: $z = 1+re^{it}, -\pi \le t \le \pi$; γ_3: $z = t, r \le t \le 1-r$ (the lower side); γ_4: $z = re^{it}$, $-2\pi \le t \le 0$, and γ_5: $z = Re^{it}, 0 \le t \le 2\pi, R > 1$. By the residue theorem,

$$\int_\gamma F(z)dz = \int_{\gamma_5} F(z)dz + \sum_{k=1}^4 \int_{\gamma_k} F(z)dz = 2\pi i\sum_{k=1}^n \mathrm{Re}\,s(F;z_k). \quad (*_1)$$

Inside $(*_1)$,

$$\int_{\gamma_1} F(z)dz = \int_r^{1-r} x^{\alpha-1}(1-x)^{-\alpha}f(x)dx;$$
$$\int_{\gamma_3} F(z)dz = \int_{1-r}^r x^{\alpha-1}(1-x)^{-\alpha}e^{2\alpha\pi i}f(x)dx$$
$$= -e^{2\alpha\pi i}\int_r^{1-r} x^{\alpha-1}(1-x)^{-\alpha}f(x)dx.$$

Along γ_2, $|(1-z)^{-\alpha}| = r^{-\alpha}$; both $z^{\alpha-1}$ and $f(z)$ are bounded, say, $|z^{\alpha-1}f(z)| \le M_1$. Then

$$\left| \int_{\gamma_2} F(z)dz \right| \le M_1 r^{1-\alpha} \to 0 \quad \text{as } r \to 0.$$

Along γ_4, $|z^{\alpha-1}| = r^{\alpha-1}$ and $|(1-z)^{-\alpha}f(z)| \le M_0$ (a constant). Then

$$\left| \int_{\gamma_4} F(z)dz \right| \le M_0 r^{\alpha} \to 0 \quad \text{as } r \to 0.$$

Along γ_5, $\lim_{z\to\infty} f(z) = f(\infty)$ exists since f is analytic at ∞. In what follows, we try to use result stated in Exercise A (28)(c) of Sec. 2.9 or (4.12.3.4): Observe that $z^{\alpha}(1-z)^{-\alpha}$ is analytic at ∞ and its modulus approaches 1 as $z \to \infty$. Fix a ray $\operatorname{Arg} z = \theta$. If z moves along $\operatorname{Arg} z = \theta$ to ∞, then

$$\lim_{z\to\infty} \operatorname{Arg} z = \theta \quad \text{and} \quad \lim_{z\to\infty} \operatorname{Arg}(1-z) = -\pi + \theta \quad \text{(see Fig. 4.70).}$$

$$\Rightarrow \lim_{z\to\infty} z^{\alpha}(1-z)^{-\alpha} = \lim_{z\to\infty} |z^{\alpha}(1-z)^{-\alpha}| e^{i[\alpha \operatorname{Arg} z + (-\alpha)\operatorname{Arg}(1-z)]}$$

$$= e^{i[\alpha\theta - \alpha(-\pi+\theta)]} = e^{\alpha\pi i},$$

$$\Rightarrow \lim_{z\to\infty} zF(z) = \lim_{z\to\infty} z^{\alpha}(1-z)^{-\alpha}f(z) = e^{\alpha\pi i}f(\infty),$$

$$\Rightarrow \text{(applying Exercise A (28)(c) of Sec. 2.9 to } \gamma_5) \lim_{R\to\infty} \int_{\gamma_5} F(z)dz$$

$$= e^{\alpha\pi i}f(\infty) \cdot 2\pi i.$$

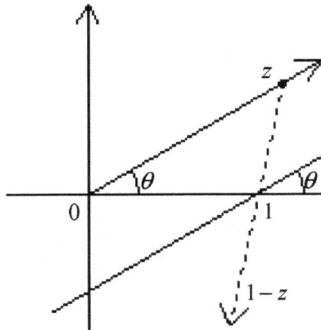

Fig. 4.70

Letting $r \to 0$ and $R \to \infty$ in $(*_1)$ and substituting the above results into it, then

$$(1 - e^{2\alpha\pi i})\int_0^1 F(x)dx + 2\pi i e^{\alpha\pi i}f(\infty) = 2\pi i \sum_{k=1}^n \operatorname{Re} s(F; z_k).$$

The required formula follows after simplification. $\qquad\square$

Example 1. Show that

$$\int_0^1 \frac{x^{1-p}(1-x)^p}{(1+x)^3}dx = \frac{2^{p-3}\pi p(1-p)}{\sin p\pi}, \quad -1 < p < 2.$$

Solution. One can verify this integral formula by using exactly the same process as in the proof of $(4.12.2.1)$. Or, we can use the established formula in $(4.12.2.1)$. Let $1 - p = \alpha - 1$, so $\alpha = 2 - p$ and $0 < \alpha < 1$ holds if $1 < p < 2$. In this case, choosing $f(x) = \frac{(1-x)^2}{(1+x)^3}$, then

$$\int_0^1 \frac{x^{1-p}(1-x)^p}{(1+x)^3}dx = \int_0^1 x^{\alpha-1}(1-x)^{-\alpha}f(x)dx$$

$$= -\frac{\pi}{e^{\alpha\pi i}\sin\alpha\pi}\operatorname{Re} s(F; -1)$$

$$= -\frac{\pi}{e^{(2-p)\pi i}\sin(2-p)\pi}e^{-p\pi i}2^{p-3}p(1-p)$$

$$= \frac{2^{p-3}\pi p(1-p)}{\sin p\pi}, \quad 1 < p < 2.$$

In case $0 < p < 1$, let $\alpha - 1 = -p$, so $0 < \alpha < 1$. Then choose $f(x) = \frac{(1-x)x}{(1+x)^3}$.
In case $-1 < p < 0$, let $\alpha = -p$, so $0 < \alpha < 1$. Then choose $f(x) = \frac{x^2}{(1+x)^3}$.
In case $p = 0$ or $p = 1$, use the limit $\frac{1}{8}$ or $\frac{1}{4}$ of $\frac{2^{p-3}\pi p(1-p)}{\sin p\pi}$ as $p \to 0$ or 1, respectively.

To practice more, we give another example.

Example 2. Show that

$$\int_0^1 \frac{1}{\sqrt[n]{1-x^n}}dx = \frac{\pi}{n\sin\frac{\pi}{n}}, \quad n = 2, 3, 4, \dots.$$

Solution. This is an easy consequence of $(4.12.2.1)$ by observing that, after setting $t = x^n$, then

$$\int_0^1 \frac{1}{\sqrt[n]{1-x^n}}dx = \int_0^1 (1-t)^{-\frac{1}{n}} \cdot \frac{1}{n}t^{\frac{1}{n}-1}dt = \frac{\pi}{n\sin\frac{\pi}{n}}.$$

On the other hand, this problem can also be solved as follows. The multiple-valued function

$$F(z) = \frac{1}{f(z)}, \quad \text{where } f(z) = \sqrt[n]{1 - z^n},$$

has n distinct branches in the simple connected domain $\mathbf{C}^* - \cup_{k=0}^{n-1} l_k$ in which $l_k = [0, e^{\frac{2k\pi i}{n}}] = \{z | z = te^{\frac{2k\pi i}{n}}, \, 0 \le t \le 1\}$ is a radius of the unit circle $|z| = 1$ for $0 \le k \le n - 1$. Note that the Riemann surface of $f(z)$ has an algebraic branch point of order $n - 1$ at each $e^{\frac{2k\pi i}{n}}$, $0 \le k \le n - 1$, and each l_k as a branch cut. *Note* that, after changing z to $\frac{1}{z}$, $f(z)$ still has n distinct branches in the simple connected domain $\mathbf{C}^* - \cup_{k=0}^{n-1} \widetilde{l_k}$, where

$$\widetilde{l_k} = [e^{\frac{2k\pi i}{n}}, \infty] = \{z \mid z = te^{\frac{2k\pi i}{n}}, 1 \le t \le \infty\} \quad \text{for } 0 \le k \le n - 1.$$

Along the interval $l_0 = [0, 1]$, designate

$$F(x) = \frac{1}{\sqrt[n]{1 - x^n}} > 0.$$

This uniquely defines a branch in $\mathbf{C}^* - \cup_{k=0}^{n-1} l_k$, still denoted as $F(z)$. What interests us is how this branch $F(z)$ takes values along l_k, $1 \le k \le n - 1$. For this purpose, let

$$z_k = e^{\frac{2k\pi i}{n}}, \quad 0 \le k \le n - 1$$

for simplicity, then $f(z) = \sqrt[n]{(z_0 - z)(z_1 - z) \cdots (z_{n-1} - z)}$.

If a moving point z starts from a point x $(0 < x < 1)$ on l_0 and moves along σ_1 to the point $xe^{\frac{2\pi i}{n}}$ on l_1 as shown in Fig. 4.71, the variations of

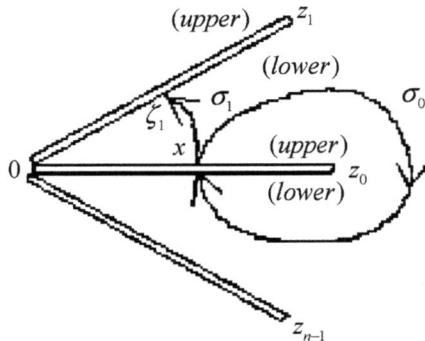

Fig. 4.71

$\mathrm{Arg}(z_k - z)$, $0 \le k \le n-1$ (see (2.7.1.4) and refer to Example 2, etc. in Sec. 2.7.2) are as follows:

(1) Along σ_0,

$$\Delta_{\sigma_0} \mathrm{Arg}(z_0 - z) = -2\pi - 0 = -2\pi;$$

$$\Delta_{\sigma_0} \mathrm{Arg}(z_k - z) = 0, \quad 1 \le k \le n-1;$$

(2) Along σ_1, letting $\zeta_1 = x e^{\frac{2\pi i}{n}}$, $0 < x < 1$, be the point where σ_1 reaches the lower side of l_1,

$$\Delta_{\sigma_1} \mathrm{Arg}(z_0 - z) = \mathrm{Arg}(z_0 - \zeta_1) - \mathrm{Arg}(z_0 - x)$$

$$= \mathrm{Arg}(z_0 - \zeta_1) = \mathrm{Arg}(z_0 - x e^{\frac{2\pi i}{n}});$$

$$\Delta_{\sigma_1} \mathrm{Arg}(z_k - z) = \mathrm{Arg}(z_k - \zeta_1) - \mathrm{Arg}(z_k - x),$$

$$2 \le k \le n-1.$$

Consequently,

(a) Along the lower side of l_0, $f(z)$ assumes the values

$$\sqrt[n]{(z_0 - x)e^{-2\pi i}(z_1 - x) \cdots (z_{n-1} - x)} = \sqrt[n]{1 - x^n} e^{\frac{-2\pi i}{n}}.$$

(b) Along the lower side of l_1, $f(z)$ assumes the values, in which $\zeta_1 = x e^{\frac{2\pi i}{n}}$, $0 < x < 1$,

$$f(x e^{\frac{2\pi i}{n}})$$

$$= \sqrt[n]{(z_0 - x e^{\frac{2\pi i}{n}})(z_1 - x e^{\frac{2\pi i}{n}})(z_2 - x e^{\frac{2\pi i}{n}}) \cdots (z_{n-1} - x e^{\frac{2\pi i}{n}})},$$

$$= e^{\frac{2\pi i}{n}} \sqrt[n]{(z_0 e^{-\frac{2\pi i}{n}} - x)(z_1 e^{-\frac{2\pi i}{n}} - x)(z_2 e^{-\frac{2\pi i}{n}} - x) \cdots (z_{n-1} e^{-\frac{2\pi i}{n}} - x)},$$

$$= e^{\frac{2\pi i}{n}} f(x).$$

Proceed in this manner in the counterclockwise direction, we obtain: for $0 \le k \le n-1$, $f(z)$ assumes $e^{\frac{2k\pi i}{n}} f(x)$ at $e^{\frac{2k\pi i}{n}} x$ ($0 < x < 1$) which lies on l_k.

Now consider $l_0 = [0, 1]$ as a branch cut with two sides. If a moving point z starts from a point x ($0 < x < 1$) on the upper side and moves along σ_0 to the same point on the lower side, then

$$f(x e^{\frac{2\pi i}{n}}) = e^{-\frac{2\pi i}{n}} f(x).$$

This process works for other cuts l_k, $1 \le k \le n-1$. Therefore, after circulating the point $e^{\frac{2k\pi i}{n}}$ ($0 \le k \le n-1$) in the clockwise direction, the values taken by $F(z)$ on the lower side of the cut are multiplied by $e^{-\frac{2\pi i}{n}}$.

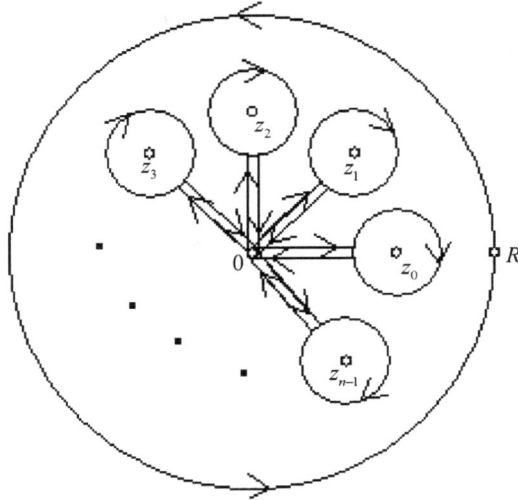

Fig. 4.72

Choose $R > 1$ and $r > 0$, sufficiently small. Construct the following curves: for $0 \le k \le n - 1$,

$$\gamma_k^- : z = te^{\frac{2k\pi i}{n}}, \quad 0 \le t \le 1 - r \ \text{(the lower side of } l_k\text{)};$$

$$\gamma_k^+ : z = te^{\frac{2k\pi i}{n}}, \quad 0 \le t \le 1 - r \ \text{(the upper side of } l_k\text{)};$$

$$\gamma_k : z = z_k + re^{it}, \quad 0 \le t \le 2\pi, \quad \text{and}$$

$$\gamma_n : z = Re^{it}, \quad 0 \le t \le 2\pi.$$

Integrate $F(z)$ along the curve $\gamma = \gamma_n + \sum_{k=0}^{n-1} (\gamma_k^+ + \gamma_k + \gamma_k^-)$ as shown in Fig. 4.72 and then, by the residue theorem,

$$\int_\gamma F(z)dz = \int_{\gamma_n} F(z)dz + \sum_{k=0}^{n-1} \left(\int_{\gamma_k^+} + \int_{\gamma_k} + \int_{\gamma_k^-} \right) F(z)dz = 0. \qquad (*2)$$

In $(*2)$, as $r \to 0$,

$$\int_{\gamma_k^+} F(z)dz = \int_0^{1-r} \frac{1}{\sqrt[n]{1 - t^n}} e^{\frac{-2k\pi i}{n}} \cdot e^{\frac{2k\pi i}{n}} dt \to \int_0^1 \frac{1}{\sqrt[n]{1 - x^n}} dx;$$

$$\int_{\gamma_k^-} F(z)dz = \int_{1-r}^0 \frac{1}{\sqrt[n]{1 - t^n}} e^{-2(k-1)\pi i/n} \cdot e^{\frac{2k\pi i}{n}} dt$$

$$\to -e^{\frac{2\pi i}{n}} \int_0^1 \frac{1}{\sqrt[n]{1 - x^n}} dx.$$

Along γ_k, $|1 - z^n| \geq 1 - |z|^n \geq 1 - (1-r)^n$, then, for $0 \leq k \leq n-1$,

$$\left| \int_{\gamma_k} F(z)dz \right| \leq \int_0^{2\pi} \frac{1}{\sqrt[n]{1-(1-r)^n}} |dz| = \frac{2\pi r}{\sqrt[n]{1-(1-r)^n}} \to 0, \quad \text{as } r \to 0.$$

Meanwhile,

$$\int_{\gamma_n} F(z)dz = -2\pi i \operatorname{Re} s(F; \infty).$$

To compute $\operatorname{Re} s(F; \infty)$, let $z = \frac{1}{w}$ and $G(w) = F\left(\frac{1}{w}\right) = \frac{w}{\sqrt[n]{w^n-1}} = \frac{e^{\frac{\pi i}{n}} w}{\sqrt[n]{1-w^n}}$.
Then

$$\operatorname{Re} s(F; \infty) = -\{\text{the coefficient of } w \text{ in the Laurent series expansion of}$$
$$G(w) \text{ at } w = 0\}$$
$$= -e^{\frac{\pi i}{n}},$$
$$\Rightarrow \int_{\gamma_n} F(z)dz = 2\pi i e^{\frac{\pi i}{n}}, \quad n = 2, 3, 4, \ldots.$$

Setting $r \to 0$ in $(*_2)$ and substituting above results into it, we obtain

$$2\pi i e^{\frac{\pi i}{n}} + n(1 - e^{\frac{2\pi i}{n}}) \int_0^1 \frac{1}{\sqrt[n]{1-x^n}} dx = 0,$$

$$\Rightarrow \int_0^1 \frac{dx}{\sqrt[n]{1-x^n}} = -\frac{2\pi i e^{\frac{\pi i}{n}}}{n(1 - e^{\frac{2\pi i}{n}})} = \frac{\pi}{n \sin \frac{\pi}{n}}, \quad n = 2, 3, 4, \ldots.$$

Exercises A

(1) Evaluate the following integrals, where (a) and (b) are chosen from Julia (1947), while (e)–(g) are chosen from Tisserand's book (1933).

(a) $\displaystyle\int_0^1 \frac{1}{\sqrt[3]{x^2(1-x)}} \cdot \frac{1}{1+x} dx.$

(b) $\displaystyle\int_0^1 \frac{\sqrt[3]{x^2(1-x)}}{(1+x)^3} dx.$

(c) $\displaystyle\int_0^1 \sqrt{\frac{(1-x)^3}{x}} \cdot \frac{1}{(1+x)^2} dx.$

(d) $\displaystyle\int_{-1}^1 \frac{\sqrt[4]{(1-x)^3(1+x)}}{1+x^2} dx.$

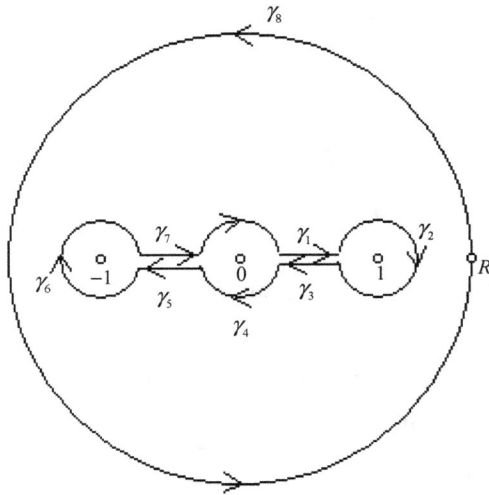

Fig. 4.73

(e) $\displaystyle\int_{-1}^{1}\frac{1}{\sqrt[3]{(1-x)(1+x)^2}}\,dx.$

(f) $\displaystyle\int_{0}^{1}\frac{x^{2n}}{\sqrt[3]{x(1-x^2)}}\,dx$ $(n=1,2,\ldots;$ see Fig. 4.73$)$.

(g) $\displaystyle\int_{0}^{1}\frac{1}{1+x^2}\log\left(x+\frac{1}{x}\right)\,dx.$

(2) Evaluate the following integrals, where $-1<p<2$.

(a) $\displaystyle\int_{0}^{1}\frac{x^{1-p}(1-x)^p}{1+x^2}\,dx.$

(b) $\displaystyle\int_{0}^{1}\frac{x^{1-p}(1-x)^p}{(1+x)^2}\,dx.$

(c) $\displaystyle\int_{-1}^{1}\frac{(1+x)^{1-p}(1-x)^p}{1+x^2}\,dx.$

(3) Designate $\sqrt{1-x^2}>0$ if $-1<x<1$. Let a be a complex number such that $a\neq\pm1$. Evaluate

$$\int_{-1}^{1}\frac{dx}{(x-a)\sqrt{1-x^2}}.$$

(4) Let b be a complex number such that $b\neq0,1$. Evaluate

$$\int_{0}^{1}\frac{x^{1-p}(1-x)^p}{b-x}\,dx,\qquad 0<p<1.$$

Exercises B

(1) (D. Kurepa and D. Mitrović) Let n be a positive integer. Show that

$$\int_0^1 \frac{x^{2n}}{\sqrt[7]{x^3(1-x^2)^2}}dx = \frac{\pi}{2\sin\frac{2\pi}{n}}\cdot\frac{2\cdot 9\cdots(7n-5)}{7\cdot 14\cdots(7n)}$$

by integrating $F(z) = \dfrac{z^{2n}}{\sqrt[7]{z^3(1-z^2)^2}}$ along the path shown in Fig. 4.73.

(2) Let n be a positive integer. Show that

$$\int_0^1 \frac{x^{2n}dx}{(1+x^2)\sqrt{1-x^2}} = \begin{cases} \dfrac{\pi}{2}\left(\dfrac{1}{\sqrt{2}}-S_n\right), & \text{if } n \text{ is even} \\[2ex] -\dfrac{\pi}{2}\left(\dfrac{1}{\sqrt{2}}-S_n\right), & \text{if } n \text{ is odd} \end{cases},$$

where

$$S_n = 1 + \sum_{k=1}^{n-1}(-1)^k\frac{(2k-1)!!}{(2k)!!},$$

$(2k)!! = 2k(2k-2)\cdots 2$ and $(2k-1)!! = (2k-1)(2k-3)\cdots 1$, by integrating $F(z) = \dfrac{z^{2n}}{(1+z^2)\sqrt{1-z^2}}$ along the path shown in Fig. 4.74.

H. G. Garnir et J. Gobert: *Fonctions d'une variable complexe. Louvain. Paris, 1965* extended (4.12.2.1) to

The evaluation of $\int_a^b(x-a)^p(x-b)^q f(x)dx$, where p, $q \in (-1,1)$ and $p+q = -1$, 0 or 1. Suppose $f(z)$ satisfies the following conditions:

(1) f is analytic in **C** except at finitely many isolated singularities $z_1,\ldots,z_n$ in **C**;

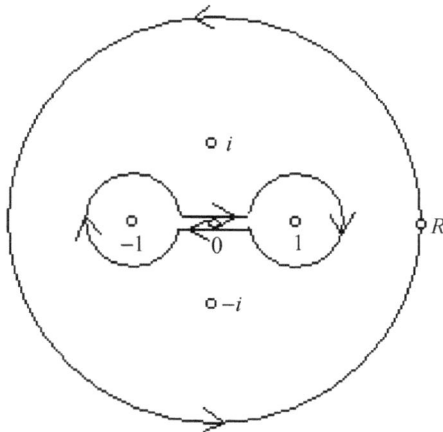

Fig. 4.74

(2) f is analytic on $[a, b]$; and

(3) $\lim_{z \to \infty} z^{p+q+1} f(z) = A \neq \infty$.

Then

$$\int_a^b (x-a)^p (b-x)^q f(x) dx$$

$$= \frac{\pi}{\sin q\pi} \left[\sum_{k=1}^n \operatorname{Re} s((z-a)^p (b-z)^q f(z); z_k) - A \right]. \qquad (4.12.2.2)$$

In case Condition 2 is changed to: f has simple poles $x_1, \dots, x_m$ in the open interval (a, b), then

$$\text{P.V.} \int_a^b (x-a)^p (b-x)^q f(x) dx$$

$$= \frac{\pi}{\sin q\pi} \left[\sum_{k=1}^n \operatorname{Re} s((z-a)^p (b-z)^q f(z); z_k) - A \right]$$

$$+ \pi \cot q\pi \sum_{l=1}^m \operatorname{Re} s((z-a)^p (b-z)^q f(z); x_l). \qquad (4.12.2.3)$$

If the Condition 3 is replaced by: $\lim_{z \to \infty} z^{p+q+1} f(z) = \infty$, then choose integer k large enough so that $\lim_{z \to \infty} \frac{z(z-a)^p (b-z)^q}{(1-\lambda z)^k} f(z)$ exists as a finite number. In this case,

$$\int_a^b (x-a)^p (b-x)^q f(x) dx$$

$$= \lim_{\lambda \to 0} \int_a^b \frac{(x-a)^p (b-x)^q}{(1-\lambda x)^k} f(x) dx. \qquad (4.12.2.4)$$

If either p or q is a positive integer, $f(z)$ may have a simple pole at a or b, respectively. Do the following problems.

(3) Note that, in (4.12.2.2), $(z-a)^p (b-z)^q = e^{p \log(z-a) + q(b-z)}$ has branches in $\mathbf{C} - [a, b]$. Try to integrate $F(z) = (z-a)^p (b-z)^q f(z)$ along a path similar to the one shown in Fig. 4.69 to prove (4.12.2.2)

(4) Prove (4.12.2.3) in detail.

(5) Evaluate the following integrals.

(a) $\displaystyle\int_a^b \sqrt{(x-a)(b-x)}\, dx.$

(b) $\displaystyle\int_a^b \frac{1}{\sqrt{(x-a)(b-x)}}\, dx.$

(c) $\displaystyle\int_{-a}^{b} \sqrt{\frac{b-x}{x+a}}\, dx.$

4.12.3.   *Improper integrals over $(-\infty, \infty)$*

To start with, we use concrete examples to give a general description of how to use the residue method to evaluate improper integrals over $(-\infty, \infty)$ or $(0, \infty)$. For beginners in complex analysis, this introductory content might provide some rough ideas in this direction so that they can handle similar yet elementary problems by themselves, without involving so deeply into the materials in subsections of Secs. 4.12.3 and 4.12.4.

Example 1. How to evaluate the improper integral

$$\int_{-\infty}^{\infty} \frac{1}{1+x^4}\, dx?$$

Solution. By calculus, just find a primitive of $f(x) = \frac{1}{1+x^4}$ as

$$\frac{1}{1+x^4} = \frac{-\frac{1}{2\sqrt{2}}x + \frac{1}{2}}{x^2 - \sqrt{2}x + 1} + \frac{\frac{1}{2\sqrt{2}}x + \frac{1}{2}}{x^2 + \sqrt{2}x + 1},$$

$$\Rightarrow \int \frac{1}{1+x^4}\, dx = \frac{1}{4\sqrt{2}} \log \frac{x^2 + \sqrt{2}x + 1}{x^2 - \sqrt{2}x + 1}$$

$$+ \frac{1}{2\sqrt{2}} [\tan^{-1}(\sqrt{2}x - 1) + \tan^{-1}(\sqrt{2}x + 1)].$$

Then, by the very definition of an improper integral,

$$\int_{-\infty}^{\infty} \frac{1}{1+x^4}\, dx = \lim_{R,R'\to\infty} \int_{-R'}^{R} \frac{dx}{1+x^4} = \frac{1}{2\sqrt{2}}(\pi + \pi) = \frac{\pi}{\sqrt{2}}.$$

Well, as we will see, the residue method provides a variety of ways to evaluate this integral in a completely different flavor.

The function $f(z) = \frac{1}{1+z^4}$ has simple poles at $z_k = e^{\frac{(2k+1)\pi i}{4}}$, $0 \le k \le 3$. The corresponding residues are

$$\mathrm{Re}\, s(f; z_0) = -\frac{1}{4}z_0, \quad \mathrm{Re}\, s(f; z_1) = \frac{1}{4}\overline{z_0},$$

$$\mathrm{Re}\, s(f; z_2) = -\frac{1}{4}\overline{z_0}, \quad \text{and} \quad \mathrm{Re}\, s(f; z_3) = \frac{1}{4}z_0.$$

If we choose $R > 1$, then by the residue theorem, $\int_{|z|=R} f(z)dz = -2\pi i\, \mathrm{Re}\, s(f; \infty) = 0$ which does not help in evaluating this integral since, eventually, $\int_{-\infty}^{\infty} \frac{1}{1+x^4} dx$ does not appear even letting $R \to \infty$.

Instead, we choose the path γ, composed of the segment $[-R, R]$, $R > 1$, and the upper half-circle $C_R : z = Re^{i\theta}$, $0 \le \theta \le \pi$, as shown in Fig. 4.75.

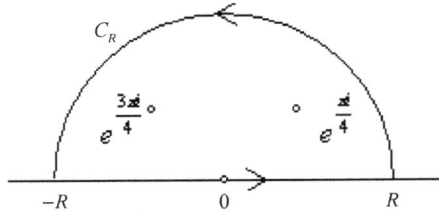

Fig. 4.75

In this case, only the poles z_0 and z_1 lie in Int γ. Then,

$$\int_\gamma f(z)dz = 2\pi i[\operatorname{Re} s(f;z_0) + \operatorname{Re} s(f;z_1)] = 2\pi i \cdot \frac{1}{4}\left(e^{-\frac{\pi}{4}i} - e^{\frac{\pi}{4}i}\right)$$

$$= \pi \sin\frac{\pi}{4} = \frac{\pi}{\sqrt{2}}. \tag{$*_1$}$$

Letting $R \to \infty$, then

$$\lim_{R\to\infty}\int_{-R}^R \frac{dx}{1+x^4} = \text{P.V.}\int_{-\infty}^\infty \frac{dx}{1+x^4} = \int_{-\infty}^\infty \frac{dx}{1+x^4}$$

appears. While, along C_R,

$$\left|\int_{C_R} f(z)dx\right| \le \int_0^\pi \frac{Rd\theta}{R^4-1} = \frac{\pi R}{R^4-1} \to 0 \quad \text{as } R\to\infty. \tag{$*_2$}$$

Finally, letting $R \to \infty$ in $(*_1)$ and using these two results, we obtain

$$\int_{-\infty}^\infty \frac{dx}{1+x^4} = \frac{\pi}{\sqrt{2}}.$$

Observe that $\frac{1}{1+x^4}$ is an even function in $(-\infty,\infty)$ and, along the positive imaginary axis $z = iy, 0 \le y \le \infty, 1+z^4 = 1+y^4 > 0$ holds. Also, $f(z)$ has a pole z_0 in the first quadrant. So we may choose the path of integration γ as a quarter circle plus two radii as shown in Fig. 4.76. Therefore,

$$\int_\gamma f(z)dz = \int_0^R \frac{dx}{1+x^4} + \int_0^{\frac{\pi}{2}} \frac{Rie^{i\theta}}{1+R^4e^{i4\theta}}d\theta + \int_R^0 \frac{idy}{1+y^4},$$

$$= 2\pi i\operatorname{Re} s(f;z_0) = -\frac{\pi i}{2}e^{\frac{\pi}{4}i}.$$

Letting $R \to \infty$, then it follows that

$$(1-i)\int_0^\infty \frac{dx}{1+x^4} = -\frac{\pi i}{2}\cdot\frac{1+i}{\sqrt{2}},$$

$$\Rightarrow \int_{-\infty}^\infty \frac{dx}{1+x^4} = 2\int_0^\infty \frac{dx}{1+x^4} = 2\cdot\frac{\pi}{2}\cdot\frac{1}{\sqrt{2}} = \frac{\pi}{\sqrt{2}}.$$

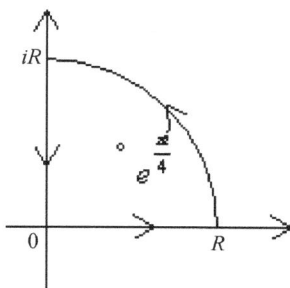

Fig. 4.76

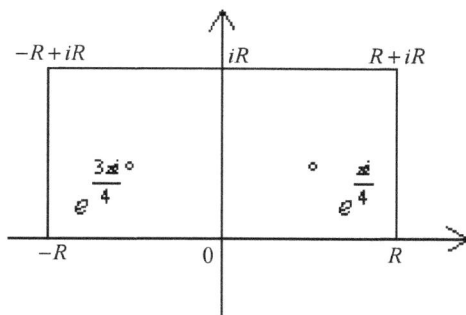

Fig. 4.77

Also, observe that, along the vertical segment $z = \pm R + iy$, $0 \leq y \leq R$, or the horizontal segment $z = x + iR$, $-R \leq x \leq R$, we always have $|1 + z^4| \geq |z|^4 - 1 \geq R^4 - 1$. Therefore, estimate similar to $(*_2)$ holds along these kinds of segments, namely, as $R \to \infty$,

$$\left| \int_0^R \frac{i\,dy}{1 + (\pm R + iy)^4} \right| \leq \frac{R}{R^4 - 1} \to 0 \quad \text{and}$$

$$\left| \int_{-R}^R \frac{dx}{1 + (x + iR)^4} \right| \leq \frac{2R}{R^4 - 1} \to 0.$$

Consequently, we can choose the path of integration γ the boundary of a rectangle with vertices at $\pm R$ and $\pm R + iR$ as shown in Fig. 4.77. Then

$$\int_\gamma f(z)dz = 2\pi i[\operatorname{Re} s(f; z_0) + \operatorname{Re} s(f; z_1)] = \frac{\pi}{\sqrt{2}}.$$

Letting $R \to \infty$, the result follows.   □

Since $\frac{x}{1+x^4}$ is odd in $(-\infty, \infty)$, then

$$\int_{-\infty}^{\infty} \frac{x}{1+x^4}\,dx = 0, \text{ but } \int_{0}^{\infty} \frac{x}{1+x^4}\,dx = \frac{\pi}{4}.$$

The second integral in the above follows either by a direct computation in calculus or by using contours shown in Figs. 4.75, 4.76, and 4.77. So is

$$\int_{-\infty}^{\infty} \frac{x^2}{1+x^4}\,dx = 2\int_{0}^{\infty} \frac{x^2}{1+x^4}\,dx = \frac{\pi}{\sqrt{2}}.$$

As an improper integral,

$$\int_{-\infty}^{\infty} \frac{x^3\,dx}{1+x^4} = \lim_{R,R'\to\infty} \int_{-R'}^{R} \frac{x^3\,dx}{1+x^4} = \lim_{R,R'\to\infty} \frac{1}{4}\log(1+x^4)|_{-R'}^{R}$$

does not converge to a definite finite number and hence, is divergent. If choosing $R' = R$, the *Cauchy principal value*

$$\text{P.V.} \int_{-\infty}^{\infty} \frac{x^3\,dx}{1+x^4} = \lim_{R\to\infty} \frac{1}{4}\log(1+x^4)|_{-R}^{R} = 0$$

does exist.

The appearance of $\sin x$ or $\cos x$ in an integrand has the effect of decreasing the amount of "largeness" if x goes beyond any finite bound. Hence, we pose the

Example 2. Is the improper integral

$$\int_{-\infty}^{\infty} \frac{x^3\sin x}{1+x^4}\,dx = 2\int_{0}^{\infty} \frac{x^3\sin x}{1+x^4}\,dx$$

convergent? If it is, how to evaluate it?

Solution. Since $\frac{x^3}{1+x^4}$ decreases to zero if $x > \sqrt[4]{3}$ and $\sin x$ is bounded in $(-\infty, \infty)$, by Dirichlet test in calculus, it does converge, conditionally but not absolutely.

If integrating $\frac{z^3\sin z}{1+z^4}$ along paths γ shown in Figs. 4.75–4.77, the main difficulty we might encounter lies on the fact that we cannot estimate its integration along the half-circle C_R because $\sin z$ is not bounded if $|z|$ is large (see (4) in (2.6.2.6)). Since $\operatorname{Im} e^{ix} = \sin x$ and $e^{iRe^{i\theta}} = e^{iR\cos\theta}e^{-R\sin\theta}$ is small enough in absolute value if R is large for $0 \le \theta \le \pi$, instead we consider the function

$$f(z) = \frac{z^3 e^{iz}}{1 + z^4}$$

and adopt the path γ in Fig. 4.75. Then, along $C_R : z = Re^{i\theta}$, $0 \le \theta \le \pi$,

$$\left| \int_{C_R} f(z)dz \right| = \left| \int_0^\pi \frac{R^3 e^{i3\theta} e^{iRe^{i\theta}}}{1 + R^4 e^{i4\theta}} Rie^{i\theta} d\theta \right|$$

$$\le \int_0^\pi \frac{R^4}{R^4 - 1} e^{-R\sin\theta} d\theta$$

$$= \frac{2R^4}{R^4 - 1} \int_0^{\frac{\pi}{2}} e^{-R\sin\theta} d\theta,$$

$$\le \frac{2R^4}{R^4 - 1} \int_0^{\frac{\pi}{2}} e^{-R \cdot \frac{2\theta}{\pi}} d\theta \quad \left(\text{using } \frac{2\theta}{\pi} \le \sin\theta \le \theta \text{ if } 0 \le \theta \le \frac{\pi}{2} \right),$$

$$= \frac{2R^4}{R^4 - 1} \cdot \frac{\pi}{2R} (1 - e^{-R}) \to 0 \quad \text{as } R \to \infty \qquad (*_3)$$

(referring to Example 9 in Sec. 2.9).

This is the estimate we need as in $(*_2)$. On the other hand,

$$\operatorname{Re} s(f; e^{\frac{\pi i}{4}}) = \frac{1}{4} e^{-\frac{1}{\sqrt{2}}} e^{\frac{i}{\sqrt{2}}} \quad \text{and} \quad \operatorname{Re} s(f; e^{\frac{3\pi i}{4}}) = \frac{1}{4} e^{-\frac{1}{\sqrt{2}}} e^{-\frac{i}{\sqrt{2}}}.$$

Finally, by the residue theorem,

$$\int_\gamma f(z)dz = \int_{-R}^R \frac{x^3 e^{ix}}{1 + x^4} dx + \int_{C_R} \frac{z^3 e^{iz}}{1 + z^4} dz,$$

$$= 2\pi i [\operatorname{Re} s(f; e^{\frac{\pi i}{4}}) + \operatorname{Re} s(f; e^{\frac{3\pi i}{4}})] = \pi i e^{-\frac{1}{\sqrt{2}}} \cos \frac{1}{\sqrt{2}}. \qquad (*_4)$$

Letting $R \to \infty$ in $(*_4)$, we obtain

$$\int_{-\infty}^\infty \frac{x^3 e^{ix}}{1 + x^4} dx = \pi i e^{-\frac{1}{\sqrt{2}}} \cos \frac{1}{\sqrt{2}}$$

$\Rightarrow$ (by equating the real and imaginary parts of both sides)

$$\int_{-\infty}^\infty \frac{x^3 \sin x}{1 + x^4} dx = \pi e^{-\frac{1}{\sqrt{2}}} \cos \frac{1}{\sqrt{2}} \quad \text{and} \quad \int_{-\infty}^\infty \frac{x^3 \cos x}{1 + x^4} dx = 0.$$

Readers are urged to adopt paths in Figs. 4.76 and 4.77 to justify the above integral formulas.

In case α is real, x^α is usually defined for $x > 0$ only. Then Example 2 can be extended one step farther as the Example 3.

Example 3. Is the improper integral

$$\int_0^\infty \frac{x^\alpha \sin x}{1+x^4}\,dx, \quad -1 < \alpha < 4,$$

convergent? If it is, how to evaluate it? In case α is an integer, the range of integration can be replaced by $(-\infty, \infty)$.

Solution. Dirichlet test says that the improper integral is convergent if $-1 < \alpha < 4$ holds. Our experience in handling Example 2 suggests that we are able to consider the function

$$f(z) = \frac{z^\alpha e^{iz}}{1+z^4}$$

as the integrand. Owing to $z^\alpha = e^{\alpha \log z}$, $f(z)$ is a multiple-valued function in $\mathbf{C} - \{0\}$. Yet z^α has branches in $\mathbf{C} - (\infty, 0]$. So we designate this $f(z)$ as the branch uniquely determined by $1^\alpha = 1$, namely, $f(z) = \frac{e^{\alpha \mathrm{Log}\, z} e^{iz}}{1+z^4}$, where $z \in \mathbf{C} - (\infty, 0]$. In this case, points in $(\infty, 0]$ are singular points of $f(z)$ (see (2.2.5)) but they are not isolated; in particular, this is the situation for the point 0.

How to choose a path of integration for this $f(z)$? Firstly, 0 should be out of the path because no matter which branch of z^α is chosen, it is not defined at 0. To do this, we construct a small circular arc to surround the point 0, say, the upper half-circle $|z| = \varepsilon$, $\mathrm{Im}\, z \geq 0$. Secondly, the branch $e^{\alpha \mathrm{Log}\, z}$ of z^α is discontinuous on the nonpositive real axis $(\infty, 0]$ but it is continuous downward to the upper side of the cut $(\infty, 0]$ from the upper half-plane. Hence, for $0 < \varepsilon < 1 < R$, we choose the path γ composed of the following curves: the segment $[\varepsilon, R]$, the upper half-circle $|z| = R$ with $\mathrm{Im}\, z \geq 0$, the upper side of the segment $[-R, -\varepsilon]$ and the upper half-circle $|z| = \varepsilon$ with $\mathrm{Im}\, z \geq 0$, as shown in Fig. 4.78. This γ is a variation of the path appeared in Fig. 4.75.

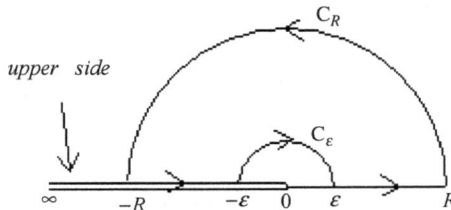

Fig. 4.78

Again, $f(z)$ has two simple poles $z_0 = e^{\frac{\pi i}{4}}$ and $z_1 = e^{\frac{3\pi i}{4}}$ inside this γ, and

$$\operatorname{Res}(f; z_0) = \lim_{z \to z_0} \frac{z^\alpha e^{iz}}{4z^3} = \frac{1}{4} e^{-\frac{1}{\sqrt{2}}} e^{\left[\frac{(\alpha-3)\pi}{4} + \frac{1}{\sqrt{2}}\right]i},$$

$$\operatorname{Res}(f; z_1) = \lim_{z \to z_1} \frac{z^\alpha e^{iz}}{4z^3} = \frac{1}{4} e^{-\frac{1}{\sqrt{2}}} e^{\left[\frac{(3\alpha-1)\pi}{4} - \frac{1}{\sqrt{2}}\right]i},$$

$$\Rightarrow \int_\gamma f(z)dz = 2\pi i[\operatorname{Res}(f; z_0) + \operatorname{Res}(f; z_1)],$$

$$= \left(\int_{[\varepsilon, R]} + \int_{C_R} + \int_{[-R, -\varepsilon]} + \int_{C_\varepsilon} \right) f(z)dz. \qquad (*_5)$$

Along $C_\varepsilon : z = \varepsilon e^{i\theta}$, $0 \leq \theta \leq \pi$,

$$\left| \int_{C_\varepsilon} f(z)dz \right| = \left| \int_\pi^0 \frac{\varepsilon^\alpha e^{i\alpha\theta} e^{i\varepsilon e^{i\theta}}}{1 + \varepsilon^4 e^{4i\theta}} i\varepsilon e^{i\theta} d\theta \right|$$

$$\leq \int_0^\pi \frac{\varepsilon^{\alpha+1} e^{-\varepsilon \sin \theta}}{1 - \varepsilon^4} d\theta \to 0 \quad \text{as } \varepsilon \to 0.$$

Along $C_R : z = Re^{i\theta}$, $0 \leq \theta \leq \pi$,

$$\left| \int_{C_R} f(z)dz \right| \leq \int_0^\pi \frac{R^{\alpha+1} e^{-R \sin \theta}}{R^4 - 1} d\theta \leq \frac{2R^{\alpha+1}}{R^4 - 1} \int_0^{\frac{\pi}{2}} e^{-R \cdot \frac{2\theta}{\pi}} d\theta \quad (\text{see } (*_3))$$

$$= \frac{2R^{\alpha+1}}{R^4 - 1} \cdot \frac{\pi}{2R}(1 - e^{-R}) = \frac{R^\alpha \pi}{R^4 - 1}(1 - e^{-R}) \to 0 \quad \text{as } R \to \infty.$$

On the other hand, as $\varepsilon \to 0$ and $R \to \infty$,

$$\int_{[\varepsilon, R]} f(z)dz = \int_\varepsilon^R \frac{x^\alpha e^{ix}}{1 + x^4} dx \to \int_0^\infty \frac{x^\alpha e^{ix}}{1 + x^4} dx,$$

$$\int_{[-R, -\varepsilon]} f(z)dz = \int_{-R}^{-\varepsilon} \frac{x^\alpha e^{ix}}{1 + x^4} dx = \int_\varepsilon^R \frac{(-x)^\alpha e^{-ix}}{1 + x^4} dx$$

$$= \int_\varepsilon^R \frac{e^{\alpha(\log|-x| + \pi i)} e^{-ix}}{1 + x^4} dx,$$

$$= e^{\alpha\pi i} \int_\varepsilon^R \frac{x^\alpha e^{-ix}}{1 + x^4} dx \to e^{\alpha\pi i} \int_0^\infty \frac{x^\alpha e^{-ix}}{1 + x^4} dx.$$

Letting $\varepsilon \to 0$ and $R \to \infty$ in $(*_5)$ and using the above results, we have

$$\int_0^\infty \frac{x^\alpha e^{ix}}{1+x^4}dx + e^{\alpha\pi i}\int_0^\infty \frac{x^\alpha e^{-ix}}{1+x^4}dx = 2\pi i[\mathrm{Re}\, s(f;z_0) + \mathrm{Re}\, s(f;z_1)].$$

By equating the real and imaginary parts of both sides,

$$(1+\cos\alpha\pi)\int_0^\infty \frac{x^\alpha \cos x}{1+x^4}dx + \sin\alpha\pi\int_0^\infty \frac{x^\alpha \sin x}{1+x^4}dx = 2\pi \,\mathrm{Re}\,\beta;$$

$$\sin\alpha\pi\int_0^\infty \frac{x^\alpha \cos x}{1+x^4}dx + (1-\cos\alpha\pi)\int_0^\infty \frac{x^\alpha \sin x}{1+x^4}dx = 2\pi \,\mathrm{Im}\,\beta,$$

where $\beta = i[\mathrm{Re}\, s(f;z_0) + \mathrm{Re}\, s(f;z_1)]$. Since the coefficient determinant of the left sides is equal to zero, we cannot solve out $\int_0^\infty \frac{x^\alpha \cos x}{1+x^4}dx$ and $\int_0^\infty \frac{x^\alpha \sin x}{1+x^4}dx$ simultaneously. This is the farthest point we can go in this direction. In case $\alpha = 3$, we do have the result as we obtained in Example 2 above. Probably we have to choose a somewhat different path of integration or different integrand (see Sec. 4.12.3.2).

Example 4. (See Ref. [1], p. 161). Evaluate

$$\int_0^\infty \frac{\log x}{1+x^2}dx.$$

Solution. Integrate $f(z) = \frac{\mathrm{Log}\, z}{1+z^2}$ over the path γ shown in Fig. 4.78, where $\mathrm{Log}\, z = \log|z| + i\,\mathrm{Arg}\, z$, $-\pi < \mathrm{Arg}\, z < \pi$. Then

$$\int_\gamma f(z)dz = 2\pi i\,\mathrm{Re}\, s(f;i) = 2\pi i \cdot \frac{\mathrm{Log}\, i}{2i} = \pi\left(\log|i| + \frac{\pi i}{2}\right) = \frac{\pi^2 i}{2}. \qquad (*_6)$$

Along C_ε and C_R, we have

$$\left|\int_{C_\varepsilon} f(z)dz\right| \leq \int_0^\pi \frac{\log\varepsilon + \pi}{1-\varepsilon^2}\cdot\varepsilon d\theta = \frac{\pi\varepsilon(\pi + \log\varepsilon)}{1-\varepsilon^2} \to 0 \quad \text{as } \varepsilon \to 0;$$

$$\left|\int_{C_R} f(z)dz\right| \leq \int_0^\pi \frac{\log R + \pi}{R^2-1}\cdot R d\theta = \frac{\pi R(\pi + \log R)}{R^2-1} \to 0 \quad \text{as } R \to \infty.$$

While, as $\varepsilon \to 0$ and $R \to \infty$,

$$\int_\varepsilon^R \frac{\log x}{1+x^2}dx \to \int_0^\infty \frac{\log x}{1+x^2}dx;$$

$$\int_{-R}^{-\varepsilon} \frac{\log x}{1+x^2}dx = \int_\varepsilon^R \frac{\log(-x)}{1+x^2}dx = \int_\varepsilon^R \frac{\log x + \pi i}{1+x^2}dx \to \int_0^\infty \frac{\log x + \pi i}{1+x^2}dx.$$

Letting $\varepsilon \to 0$ and $R \to \infty$ in $(*_6)$, we have then

$$\int_0^\infty \frac{\log x}{1+x^2}\,dx + \int_0^\infty \frac{\log x + \pi i}{1+x^2}\,dx = \frac{\pi^2 i}{2},$$

$$\Rightarrow \int_0^\infty \frac{\log x}{1+x^2}\,dx = 0, \quad \text{and} \quad \int_0^\infty \frac{1}{1+x^2}\,dx = \frac{\pi}{2}.$$

Note that

$$\int_0^\infty \frac{\log x}{1+x^2}\,dx = \int_0^1 \frac{\log x}{1+x^2}\,dx + \int_1^\infty \frac{\log x}{1+x^2}\,dx$$

$$= -\int_1^\infty \frac{\log x}{1+x^2}\,dx + \int_1^\infty \frac{\log x}{1+x^2}\,dx = 0,$$

while

$$\int_0^\infty \frac{1}{1+x^2}\,dx = \tan^{-1} x \big|_0^\infty = \frac{\pi}{2} - 0 = \frac{\pi}{2}.$$

If we define the principal branch of $\log z$ as $\operatorname{Log} z = \log |z| + i \operatorname{Arg} z$, $0 < \operatorname{Arg} z < 2\pi$, then the path of integration γ should be changed to one shown in Fig. 4.79. In this case,

$$\int_\gamma f(z)\,dz = 2\pi i [\operatorname{Re} s(f; i) + \operatorname{Re} s(f; -i)]$$

$$= 2\pi i \left(\frac{1}{2i} \cdot \frac{\pi i}{2} - \frac{1}{2i} \cdot \frac{3\pi i}{2} \right) = -\pi^2 i.$$

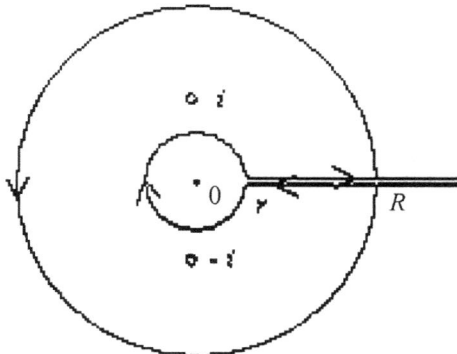

Fig. 4.79

Letting $\varepsilon \to 0$ and $R \to \infty$, we have

$$\int_0^\infty \frac{\log x}{1+x^2}\,dx - \int_0^\infty \frac{\log x + 2\pi i}{1+x^2}\,dx = -\pi^2 i,$$

$$\Rightarrow \int_0^\infty \frac{1}{1+x^2}\,dx = \frac{\pi}{2} \quad\text{and}\quad \int_0^\infty \frac{\log x}{1+x^2}\,dx \text{ disappears.}$$

Try to use $\frac{(\text{Log } z)^2}{1+z^2}$ to replace the original $\frac{\text{Log } z}{1+z^2}$ and you will find that it does work. Say precise reason why! Refer to (4.12.4.2.10).

Example 5. (See Ref. [1], p. 161) Evaluate

$$\int_0^\infty \frac{\log(1+x^2)}{x^{1+\alpha}}\,dx \quad (0 < \alpha < 2).$$

Solution. Via integration by parts,

$$\int_0^\infty \frac{\log(1+x^2)}{x^{1+\alpha}}\,dx = -\left.\frac{\log(1+x^2)}{\alpha x^\alpha}\right|_0^\infty - \int_0^\infty \frac{-1}{\alpha x^\alpha} \cdot \frac{2x}{1+x^2}\,dx$$

$$= \frac{2}{\alpha} \int_0^\infty \frac{x^{1-\alpha}}{1+x^2}\,dx.$$

The problem reduces to how to evaluate $\int_0^\infty \frac{x^{1-\alpha}}{1+x^2}\,dx$.

Integrate $f(z) = \frac{z^{1-\alpha}}{1+z^2}$ over the same path γ shown in Fig. 4.78, where $z^{1-\alpha} = e^{(1-\alpha)\text{Log } z}$ and $\text{Log } z = \log|z| + i \arg z$, $-\pi < \arg z < \pi$. Then

$$\int_\gamma f(z)dz = 2\pi i \operatorname{Re} s(f; i) = 2\pi i \cdot \frac{e^{(1-\alpha)\text{Log } i}}{2i} = \pi e^{(1-\alpha)\cdot\frac{\pi i}{2}}$$

$$= \pi i e^{-\frac{\alpha \pi i}{2}}. \tag{$*$7}$$

Along C_ε and C_R, we have

$$\left|\int_{C_\varepsilon} f(z)dz\right| = \left|\int_\pi^0 \frac{e^{(1-\alpha)(\log \varepsilon + i\theta)}}{1+\varepsilon^2 e^{i2\theta}} i\varepsilon e^{i\theta}\,d\theta\right|$$

$$\leq \int_0^\pi \frac{\varepsilon \cdot \varepsilon^{1-\alpha}}{1-\varepsilon^2}\,d\theta = \frac{\pi\varepsilon^{2-\alpha}}{1-\varepsilon^2} \to 0 \quad\text{as } \varepsilon \to 0 \quad\text{and}$$

$$\left|\int_{C_R} f(z)dz\right| = \left|\int_0^\pi \frac{e^{(1-\alpha)(\log R + i\theta)}}{1+R^2 e^{i2\theta}} i \operatorname{Re}^{i\theta}\,d\theta\right| \leq \frac{\pi R^{2-\alpha}}{R^2-1} \to 0 \quad\text{as } R \to \infty.$$

On the other hand, as $\varepsilon \to 0$ and $R \to \infty$,

$$\int_\varepsilon^R \frac{x^{1-\alpha}}{1+x^2}\,dx \to \int_0^\infty \frac{x^{1-\alpha}}{1+x^2}\,dx,$$

$$\int_{-R}^{-\varepsilon} \frac{x^{1-\alpha}}{1+x^2}\,dx = \int_\varepsilon^R \frac{(-x)^{1-\alpha}}{1+x^2}\,dx = \int_\varepsilon^R \frac{e^{(1-\alpha)(\log x + \pi i)}}{1+x^2}\,dx$$

$$= \int_\varepsilon^R \frac{x^{1-\alpha}e^{(1-\alpha)\pi i}}{1+x^2}\,dx, \quad \to -e^{-\alpha\pi i}\int_0^\infty \frac{x^{1-\alpha}}{1+x^2}\,dx.$$

In $(*_7)$, letting $\varepsilon \to 0$ and $R \to \infty$, hence we have

$$(1 - e^{-\alpha\pi i})\int_0^\infty \frac{x^{1-\alpha}}{1+x^2}\,dx = \pi i e^{-\frac{\alpha\pi i}{2}},$$

$$\Rightarrow \int_0^\infty \frac{x^{1-\alpha}}{1+x^2}\,dx = \pi i \frac{e^{-\frac{\alpha\pi i}{2}}}{1 - e^{-\alpha\pi i}} = \frac{\pi}{2\sin\frac{\alpha\pi}{2}} \quad (0 < \alpha < 2).$$

In particular, if $\alpha = \frac{2}{3}$ (see Ref. [1], p. 161),

$$\int_0^\infty \frac{x^{\frac{1}{3}}}{1+x^2}\,dx = \frac{\pi}{2\sin\frac{\pi}{3}} = \frac{\pi}{\sqrt{3}}.$$

Example 6. Evaluate the following integrals.

(1) $\displaystyle\int_0^\infty \frac{x}{e^x - 1}\,dx.$

(2) $\displaystyle\int_0^\infty \frac{x}{e^x + 1}\,dx.$

Solution.

(1) It seems naturally to let the integrand $f(z) = \frac{z}{e^z-1}$. Observe that $e^z - 1$ is a function with period $2\pi i$ and has simple poles at $z = 2n\pi i$ for $n = 0, \pm 1, \pm 2, \ldots$. And along a vertical line $z = R + iy$, it is easy to estimate $e^z - 1$ because $|e^z - 1| = |e^{R+iy} - 1| \ge e^R - 1$ if $R > 0$. Hence, a reasonable path of integration γ is as shown in Fig. 4.80, wherein dented quarter circles should be added to the path in order to avoid the poles 0 and $2\pi i$. When integrating along the horizontal sides $z = x$, $\varepsilon \le x \le R$ and $z = x + 2\pi i$, $\varepsilon \le x \le R$, we encounter the dilemma such as

$$\int_\varepsilon^R \frac{x}{e^x-1}\,dx + \int_R^\varepsilon \frac{x+2\pi i}{e^x-1}\,dx = \int_\varepsilon^R \frac{x}{e^x-1}\,dx - \int_\varepsilon^R \frac{x+2\pi i}{e^x-1}\,dx$$

$$= -\int_\varepsilon^R \frac{2\pi i}{e^x-1}\,dx$$

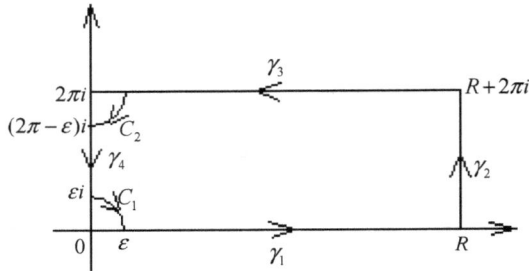

Fig. 4.80

in which the desired integral $\int_\varepsilon^R \frac{x}{e^x-1}dx$ disappears (compare to the latter part of Example 4). Consequently this integrand is not suitable to the chosen path γ. Fortunately, owing to the rapid increase to $+\infty$ of e^R as $R \to \infty$, we can replace $\frac{z}{e^z-1}$ by $\frac{z^2}{e^z-1}$. This will overcome the above claimed difficulty yet not harmless to the estimate of the integral along the vertical sides as we will see in what follows.

From now on, let $f(z) = \frac{z^2}{e^z-1}$ and γ be the path in Fig. 4.80. Then

$$\int_\gamma f(z)dz = \left(\int_{\gamma_1} + \int_{\gamma_2} + \int_{\gamma_3} + \int_{C_2} + \int_{\gamma_4} + \int_{C_1} \right) f(z)dz = 0. \qquad (*8)$$

Along $C_1 : z = \varepsilon e^{i\theta}$, $0 \le \theta \le \frac{\pi}{2}$, direct computation shows that

$$\int_{C_1} f(z)dz = \int_{\frac{\pi}{2}}^0 \frac{\varepsilon^2 e^{2i\theta}}{e^{\varepsilon e^{i\theta}}-1}\varepsilon i e^{i\theta}d\theta \to 0 \quad \text{as } \varepsilon \to 0.$$

Alternatively, we can apply Exercise A (28)(b) of Sec. 2.9 (see also (4.12.3.3)) to this case: Since $\lim_{z\to 0} zf(z) = 0$, $\lim_{\varepsilon\to 0}\int_{C_1} f(z)dz = 0(0 - \frac{\pi}{2})i = 0$. Similarly, either by direct computation or by applying the same fact, as $\varepsilon \to 0$,

$$\int_{C_2} f(z)dz \to i\int_0^{-\frac{\pi}{2}} (2\pi i)^2 d\theta$$

$$= 2\pi^3 i = \left(\lim_{z\to 2\pi i} (z - 2\pi i)f(z) \right)\left(-\frac{\pi}{2} - 0 \right)i,$$

$$= -\frac{1}{4} \cdot 2\pi i \operatorname{Re} s(f; 2\pi i).$$

Along the two horizontal sides γ_1 and γ_3:

$$\int_{\gamma_1} f(z)dz = \int_\varepsilon^R \frac{x^2}{e^x-1}dx,$$

$$\int_{\gamma_3} f(z)dz = \int_R^\varepsilon \frac{(x+2\pi i)^2}{e^x-1}dx = -\int_\varepsilon^R \frac{x^2 + 4\pi ix - 4\pi^2}{e^x-1}dx.$$

While, along the two vertical sides γ_2 and γ_4:

$$\left| \int_{\gamma_2} f(z) dz \right| \leq \int_0^{2\pi} \frac{|R + iy|^2}{|e^{R+iy} - 1|} |idy| \leq \int_0^{2\pi} \frac{2R^2}{e^R - 1} dy \quad (\text{if } R \geq 2\pi)$$

$$= \frac{4\pi R^2}{e^R - 1} \to 0 \quad \text{as } R \to \infty,$$

$$\int_{\gamma_4} f(z) dz = \int_{2\pi-\varepsilon}^{\varepsilon} \frac{(iy)^2}{e^{iy} - 1} idy = \int_{\varepsilon}^{2\pi-\varepsilon} \frac{iy^2}{e^{iy} - 1} dy.$$

Substituting these results into $(*_8)$, we have

$$\int_{\varepsilon}^{R} \frac{-4\pi ix + 4\pi^2}{e^x - 1} dx + 2\pi^3 i + \int_{\varepsilon}^{2\pi-\varepsilon} \frac{iy^2}{e^{iy} - 1} dy + o(1) = 0$$

$\Rightarrow$ (by equating the imaginary parts of both sides and noting that $\operatorname{Im} \frac{iy^2}{e^{iy}-1} = -\frac{1}{2} y^2$, and then letting $\varepsilon \to 0$ and $R \to \infty$)

$$-4\pi \int_0^{\infty} \frac{x}{e^x - 1} dx + 2\pi^3 - \frac{1}{2} \int_0^{2\pi} y^2 dy = 0,$$

$$\Rightarrow \int_0^{\infty} \frac{x}{e^x - 1} dx = \frac{\pi^2}{6}.$$

Upon changing $1 - e^x$ by x in the above integral, we have $\int_0^1 \frac{\log x}{x-1} dx = \frac{\pi^2}{6}$.

(2) By almost the same consideration as in (1), integrate now the function $f(z) = \frac{z^2}{e^z + 1}$ along the path γ shown in Fig. 4.81. In this case,

$$\int_{\gamma} f(z) dz = \int_0^{R} \frac{-4\pi ix + 4\pi^2}{e^x + 1} dx - \frac{1}{2} \cdot 2\pi i \cdot \operatorname{Re} s(f; \pi i)$$

$$+ \left(\int_0^{\pi-\varepsilon} + \int_{\pi+\varepsilon}^{2\pi} \right) \frac{iy^2}{e^{iy} + 1} dy + o(1) = 0.$$

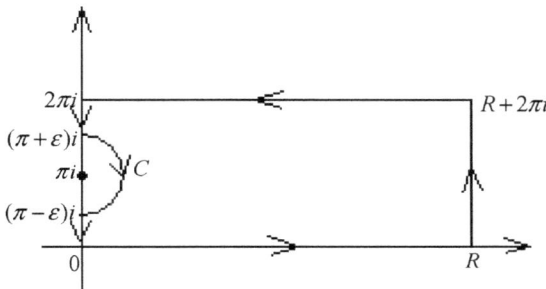

Fig. 4.81

Equating the imaginary parts of both sides and noting that $\operatorname{Re} s(f; \pi i) = \pi^2$ and $\operatorname{Im} \frac{iy^2}{e^{iy}+1} = \frac{y^2}{2}$, and then, letting $\varepsilon \to 0$ and $R \to \infty$, we have

$$-4\pi \int_0^\infty \frac{x}{e^x + 1} dx - \pi^3 + \frac{1}{2} \int_0^{2\pi} y^2 dy = 0$$

$$\Rightarrow \int_0^\infty \frac{x}{e^x + 1} dx = \frac{\pi^2}{12}.$$

Example 7. Evaluate the following integrals.

(1) P.V. $\displaystyle\int_{-\infty}^\infty \frac{dx}{x^3 - 1}$.

(2) P.V. $\displaystyle\int_0^\infty \frac{dx}{x^3 - 1}$.

Solution.

(1) $f(z) = \frac{1}{z^3 - 1}$ has simple poles at $z_k = e^{\frac{2k\pi i}{3}}$, $k = 0, 1, 2$. Let γ be the dented contour shown in Fig. 4.82. Then

$$\int_\gamma f(z)dz = 2\pi i \operatorname{Re} s(f; z_1) = 2\pi i \cdot \frac{1}{3} e^{\frac{2\pi i}{3}}. \qquad (*_9)$$

Since $\lim_{z \to z_0}(z - z_0)f(z) = \operatorname{Re} s(f; z_0) = \frac{1}{3}$ and $\lim_{z \to \infty} zf(z) = 0$, by using Exercises A (28)(b) and (c) of Sec. 2.9 (see also (4.12.3.4)) or just by direct computation, it follows that $\lim_{r \to 0} \int_{C_0} f(z)dz = \frac{i}{3}(0 - \pi) = -\frac{\pi i}{3}$ and $\lim_{R \to \infty} \int_{C_R} f(z)dz = 0 \cdot i(\pi - 0) = 0$. Therefore, letting $r \to 0$ and $R \to \infty$ in $(*_9)$, we have (see also (4.11.3.7))

$$\text{P.V.} \int_{-\infty}^\infty f(x)dx - \frac{\pi i}{3} = \frac{2\pi i}{3} e^{\frac{2\pi i}{3}},$$

$$\Rightarrow \text{P.V.} \int_{-\infty}^\infty \frac{dx}{x^3 - 1} = -\frac{\sqrt{3}}{3}\pi.$$

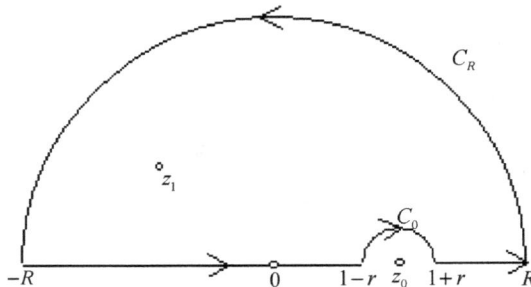

Fig. 4.82

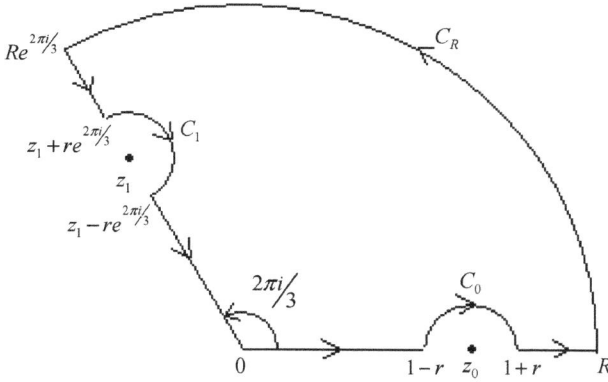

Fig. 4.83

(2) Integrate $f(z) = \frac{1}{z^3 - 1}$ along the path γ shown in Fig. 4.83, where $C_0 : z = 1 + re^{i\theta}$, $0 \le \theta \le \pi$; $C_R : z = Re^{i\theta}$, $0 \le \theta \le \frac{2\pi}{3}$ and $C_1 : z = z_1 + re^{i\theta}$, $-\frac{\pi}{3} \le \theta \le \frac{2\pi}{3}$. Then

$$\int_\gamma f(z)dz = 0$$

$$= \int_0^{1-r} \frac{dx}{x^3 - 1} + \int_{C_0} f(z)dz + \int_{1+r}^R \frac{dx}{x^3 - 1}$$

$$+ \int_{C_R} f(z)dz + \int_{Re^{\frac{2\pi i}{3}}}^{z_1 + re^{\frac{2\pi i}{3}}} f(z)dz$$

$$+ \int_{C_1} f(z)dz + \int_{z_1 - re^{\frac{2\pi i}{3}}}^0 f(z)dz. \qquad (*_{10})$$

Along C_0, C_1, and C_R, we have, by using Exercise A (28) of Sec. 2.9 (see also (4.12.2.3) and (4.12.3.4)) or by direct computation (see also (4.11.3.7)),

$$\lim_{r \to 0} \int_{C_0} f(z)dz = (0 - \pi)i \operatorname{Re} s(f; z_0) = -\frac{\pi i}{3},$$

$$\lim_{r \to 0} \int_{C_1} f(z)dz = \left(-\frac{\pi}{3} - \frac{2\pi}{3}\right) i \operatorname{Re} s(f; z_1) = -\frac{\pi i}{3} e^{\frac{2\pi i}{3}},$$

$$\lim_{R \to \infty} \int_{C_R} f(z)dz = \left(\frac{2\pi}{3} - 0\right) \cdot i \cdot 0 = 0;$$

on the other hand, as $r \to 0$ and $R \to \infty$,

$$\int_0^{1-r} \frac{dx}{x^3 - 1} + \int_{1+r}^R \frac{dx}{x^3 - 1} \to \text{P.V.} \int_0^\infty \frac{dx}{x^3 - 1},$$

$$\int_{\operatorname{Re}^{\frac{2\pi i}{3}}}^{z_1 + re^{\frac{2\pi i}{3}}} \frac{dz}{z^3 - 1} + \int_{z_1 - re^{\frac{2\pi i}{3}}}^0 \frac{dz}{z^3 - 1}$$

$$= \int_R^{1+r} \frac{e^{\frac{2\pi i}{3}} dz}{e^{2\pi i} z^3 - 1} + \int_{1-r}^0 \frac{e^{\frac{2\pi i}{3}} dz}{e^{2\pi i} z^3 - 1}$$

$$\to (-e^{\frac{2\pi i}{3}}) \text{P.V.} \int_0^\infty \frac{dx}{x^3 - 1}.$$

Letting $r \to 0$ and $R \to \infty$ in $(*10)$, we then have

$$(1 - e^{\frac{2\pi i}{3}}) \text{P.V.} \int_0^\infty \frac{dx}{x^3 - 1} = \frac{\pi i}{3}(1 + e^{\frac{2\pi i}{3}}),$$

$$\Rightarrow \text{P.V.} \int_0^\infty \frac{dx}{x^3 - 1} = -\frac{\pi}{3\sqrt{3}}.$$

Exercises A

(1) Show that

$$\int_{-\infty}^\infty \frac{dx}{1 + x^{2n}} = \frac{\pi}{n \sin(\frac{\pi}{2n})}, \quad n = 1, 2, 3, \ldots$$

by the following three methods: (a) Integrate $f(z) = \frac{1}{1+z^{2n}}$ along the path shown in Fig. 4.75; (b) along the rectangular path shown in Fig. 4.77; (c) along the curve shown in Fig. 4.84.

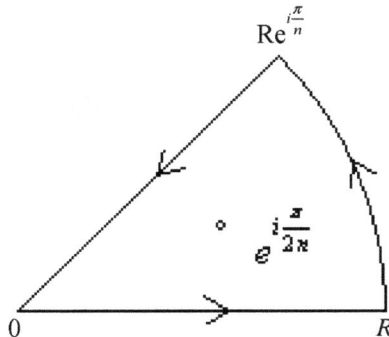

Fig. 4.84

(2) Use the three methods in Exercise (1) to show that, if m and n are integers,

$$\int_0^\infty \frac{x^{m-1}}{1+x^n}dx = \frac{\pi}{n \sin\left(\frac{m\pi}{n}\right)}, \qquad n > m > 0.$$

When adopting methods (a) and (b), it is understood that m is an odd integer.

(3) Integrate $f(z) = \frac{e^{az}}{1+e^z}$ along the path shown in Fig. 4.85 to show that

$$\int_{-\infty}^\infty \frac{e^{ax}}{1+e^x}dx = \frac{\pi}{\sin \pi a}, \qquad 0 < a < 1.$$

Note that $f(z + 2\pi i) = e^{2\pi ai}f(z)$. Set $e^x = t$ in the above integral. Then it turns out to be

$$\int_0^\infty \frac{t^{a-1}}{1+t}dt = \frac{\pi}{\sin \pi a}, \qquad 0 < a < 1.$$

Try to prove this directly by integrating $g(z) = \frac{z^{a-1}}{1+z} = \frac{e^{(a-1)\mathrm{Log}\, z}}{1+z}$ along the path shown in Fig. 4.79. Observe that $\lim_{z\to 0} |z|\,|g(z)| = \lim_{z\to\infty} |z|\,|g(z)| = 0$. Again, let $t = x^n$ and $a = \frac{m}{n}$ where m and n are positive integers and $n > m$, then the above integral becomes the one in Exercise (2).

(4) Integrate $f(z) = \frac{e^{iz}}{z}$ over a path similar to the one in Fig. 4.78 to show that

$$\int_0^\infty \frac{\sin x}{x}dx = \frac{\pi}{2}.$$

Instead, if we set $f(z) = \frac{\sin z}{z}$, does this work over the same path? Why?

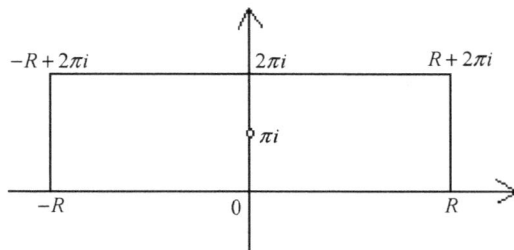

Fig. 4.85

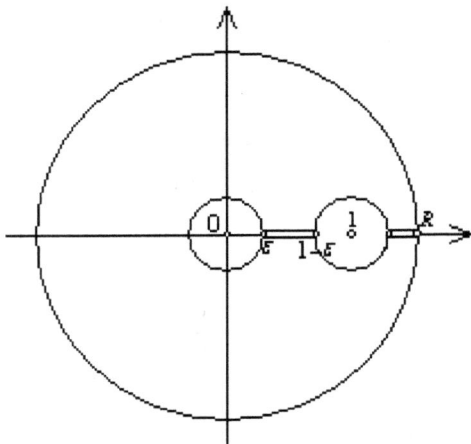

Fig. 4.86

(5) Integrate $f(z) = \frac{1-e^{2i\alpha z}}{2z^2}$ over a path as in Fig. 4.78 to show that

$$\int_0^\infty \frac{\sin^2 \alpha x}{x^2}\, dx = \frac{\alpha\pi}{2}, \quad \alpha > 0.$$

(6) Integrate $f(z) = \frac{z^{a-1}}{1-z}$ over the path γ shown in Fig. 4.86 to show that

$$\text{P.V.} \int_0^\infty \frac{x^{\alpha-1}}{1-x}\, dx = \pi \cot \alpha\pi, \quad 0 < \alpha < 1.$$

(7) Show that $\int_0^\infty \frac{\log x}{(1+x)^2}\, dx = 0$ by integrating $f(z) = \frac{(\log z)^2}{(1+z)^2}$ along the path in Fig. 4.79. Refer to Example 4. Also, note that $\int_0^\infty \cdots dx = \int_0^1 \cdots dx + \int_1^\infty \cdots dx = 0$.

(8) (See Ref. [1], p. 160) Note that $1 - e^{2iz} = -2ie^{iz}\sin z$ is real and negative only if $x = n\pi$, $y \leq 0$ where $z = x + iy$ and n is any integer. Integrate $f(z) = \log(1 - e^{2iz})$ along the contour γ shown in Fig. 4.87 to show that $\int_0^\pi \log(-2ie^{ix}\sin x)dx = 0$ by letting $r \to 0$ and $R \to \infty$. Choose $\log e^{ix} = ix$ and $\log(-i) = \frac{-\pi i}{2}$, show that

$$\int_0^\pi \log \sin x\, dx = -\pi \log 2.$$

Refer to Example 4 in Sec. 4.12.1.

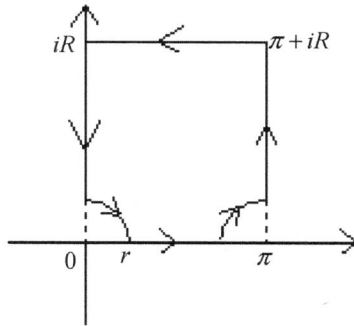

Fig. 4.87

(9) Show that

$$\int_0^\infty \frac{\sin mx}{x(x^2+b^2)^2}\,dx = \frac{\pi}{2b^4}\left[1 - \frac{e^{-mb}}{2}(mb+2)\right], \quad b > 0,\ m > 0,$$

by integrating $f(z) = \frac{e^{\mathrm{Im}\,z}}{z(z^2+b^2)}$ over the path shown in Fig. 4.78.

(10) Show that, via Fig. 4.79

$$\int_0^\infty \frac{x^\alpha}{(x^2+1)^2}\,dx = \frac{\pi(1-\alpha)}{4\cos\left(\frac{\alpha\pi}{2}\right)}, \quad -1 < \alpha < 3.$$

(11) Evaluate P.V. $\int_0^\infty \frac{dx}{1+x^3}$.

(12) By integrating $f(z) = \frac{e^{iaz}}{\sinh z}$ along the contour shown in Fig. 4.88, show that

$$\text{P.V.} \int_0^\infty \frac{\sin ax}{\sinh x}\,dx = \frac{\pi}{2}\tanh\frac{a\pi}{2}, \quad a > 0.$$

Exercises B

Gauss integral:

$$\int_0^\infty e^{-x^2}\,dx = \frac{\sqrt{\pi}}{2}.$$

This formula can be easily obtained by considering the double integral

$$\left(\int_0^\infty e^{-x^2}\,dx\right)^2 = \int_0^\infty e^{-x^2}\,dx \int_0^\infty e^{-y^2}\,dy = \int_0^\infty\int_0^\infty e^{-(x^2+y^2)}\,dx\,dy$$

$$= \int_0^{\frac{\pi}{2}}\int_0^\infty e^{-r^2}r\,dr\,d\theta = \frac{\pi}{4}.$$

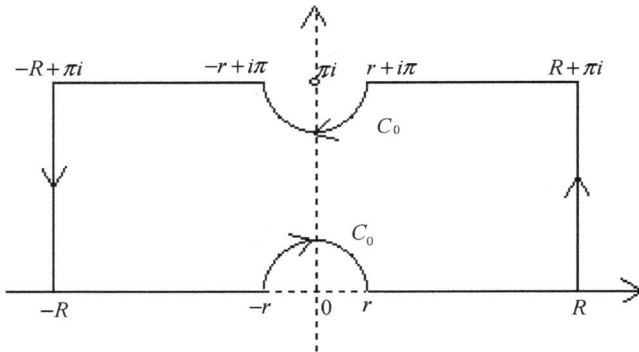

Fig. 4.88

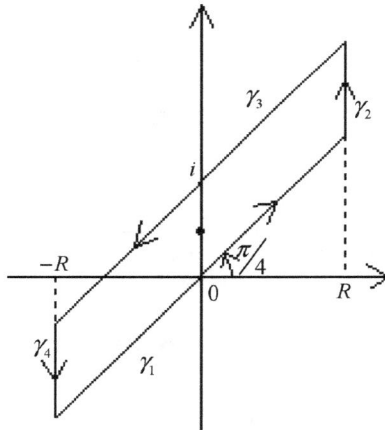

Fig. 4.89

Well, there are several ways to evaluate this integral by the method of residues. Some of them are given in the following Exercises (1)–(8). Among them, only the last one seems to be a direct proof.

(1) Integrate $f(z) = \dfrac{e^{\pi i z^2 + 2\pi z}}{e^{2\pi z}+1}$ over the path shown in Fig. 4.89. Note that

$$\operatorname{Re} s\left(f; \frac{i}{2}\right) = \frac{1}{2\pi}e^{-\frac{\pi i}{4}} \quad \text{and} \quad f(z) - f(z+i) = e^{\pi i z^2}.$$

This method will lead to $\int_{-\infty}^{\infty} e^{-\pi r^2}\, dr = ie^{-\frac{\pi i}{2}} = 1$. Then, set $x = \sqrt{\pi}\, r$.

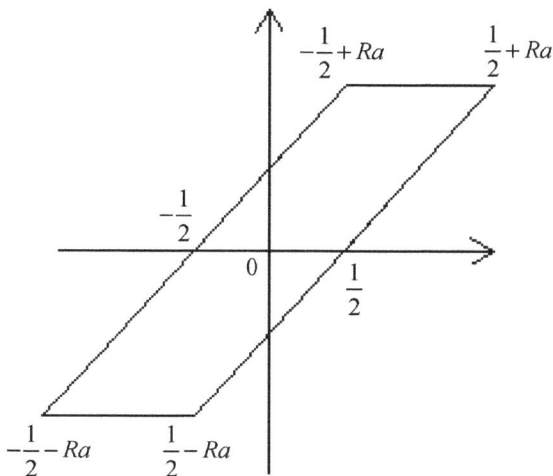

Fig. 4.90

(2) (Mirsky, 1947) Integrate $f(z) = \dfrac{e^{iaz^2}}{\sin \pi z}$, where $a = e^{\frac{\pi i}{4}}$, along the path shown in Fig. 4.90. Note that

$$f\left(ax + \frac{1}{2}\right) - f\left(ax - \frac{1}{2}\right) = 2e^{\frac{\pi i}{4}}e^{-\pi x^2}, \quad x \in \mathbf{R}, \quad \text{and}$$

$$\operatorname{Re} s(f; 0) = \frac{1}{\pi}.$$

Along the horizontal side,

$$|f(z)| \leq \frac{2e^{-R(R \pm \sqrt{2}t)}}{e^{\frac{\pi R}{\sqrt{2}}} - e^{\frac{-\pi R}{\sqrt{2}}}}, \quad -\frac{1}{2} \leq t \leq \frac{1}{2}.$$

(3) (Pólya, 1946). Integrate $f(z) = e^{\pi i z^2} \tan \pi z$ along the path γ in Fig. 4.91. Let $\mathbf{R} \to \infty$. Then

$$\int_{-(1+i)\infty}^{(1+i)\infty} [e^{\pi i (z+1)^2} - e^{\pi i z^2}] \tan \pi z \, dz = -2ie^{\frac{\pi i}{4}} = \frac{2}{i}\int_{-(1+i)\infty}^{(1+i)\infty} e^{\pi i z^2} \, dz$$

$$= \frac{2}{i}e^{\frac{\pi i}{4}}\int_{-\infty}^{\infty} e^{-\pi t^2} \, dt.$$

(4) (Cadwell, 1947). Integrate e^{-z^2} along the path in Fig. 4.92 to obtain

$$\int_0^\infty \cos x^2 \, dx = \int_0^\infty \sin x^2 \, dx = \frac{1}{\sqrt{2}}\int_0^\infty e^{-x^2} \, dx.$$

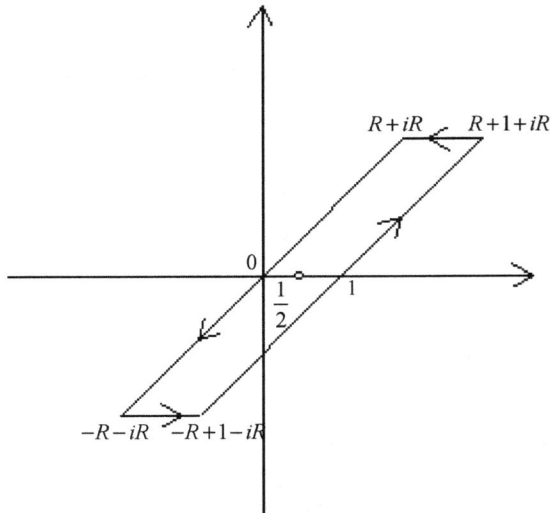

Fig. 4.91

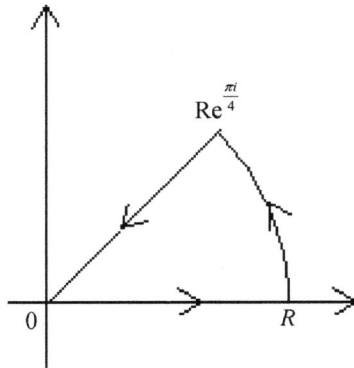

Fig. 4.92

Then, integrate $f(z) = \frac{e^{iz^2}}{\sin\sqrt{\pi z}}$ along the path γ in Fig. 4.93 to obtain $\int_0^\infty \cos x^2 dx = \int_0^\infty \sin x^2 dx = \frac{1}{2}\sqrt{\frac{\pi}{2}}$, the Fresnel integral.

(5) Integrate $f(z) = \frac{e^{-z^2}}{1+e^{-2az}}$, $a = (1+i)\sqrt{\frac{\pi}{2}}$, along the path γ shown in Fig. 4.94. Note that $\operatorname{Re} s(f; \frac{a}{2}) = -\frac{i}{2\sqrt{\pi}}$ and $f(z) - f(z+a) = e^{-z^2}$. Be aware that a is a period of e^{-2az}.

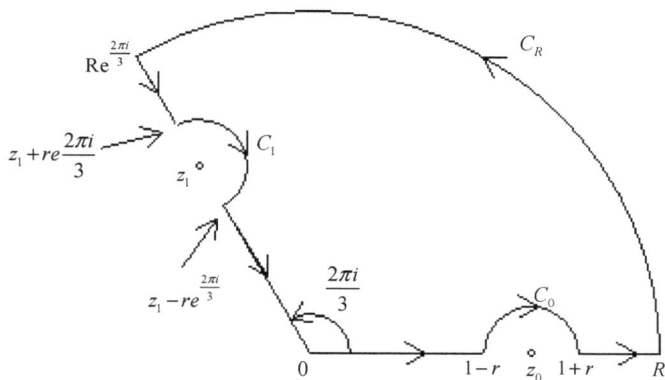

Fig. 4.93

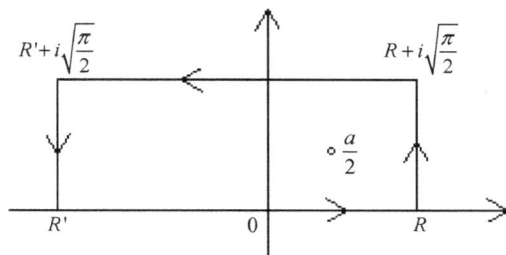

Fig. 4.94

(6) Integrate $f(z) = e^{-z^2}$ along the path γ in Fig. 4.95 and then, let $R \to \infty$ to obtain the *Poisson's integral*

$$\int_0^\infty e^{-x^2} \cos 2bx \, dx = \frac{\sqrt{\pi}}{2} e^{-b^2}, \quad b > 0.$$

Finally, let $b \to 0^+$.

(7) Integrate $f(z) = e^{-z^2}$ along the boundary of the right half-rectangle shown in Fig. 4.95, namely, the one with the vertices 0, R, $R + ib$, and ib. Show that

$$\int_0^\infty e^{-x^2} \cos 2bx \, dx = \frac{e^{-b^2} \sqrt{\pi}}{2} \quad \text{and}$$

$$\int_0^\infty e^{-x^2} \sin 2bx \, dx = e^{-b^2} \int_0^\infty e^{-x^2} dx.$$

And, then, let $b \to 0$.

(8) (Srinivasa Rao, 1974) Integrate $f(z) = \dfrac{e^{\frac{iz^2}{2}}}{e^{-\sqrt{\pi}z} - 1}$ along the path γ shown in Fig. 4.96, where $A = -R - iR$, $B = -r - ir$, $C = r + ir$,

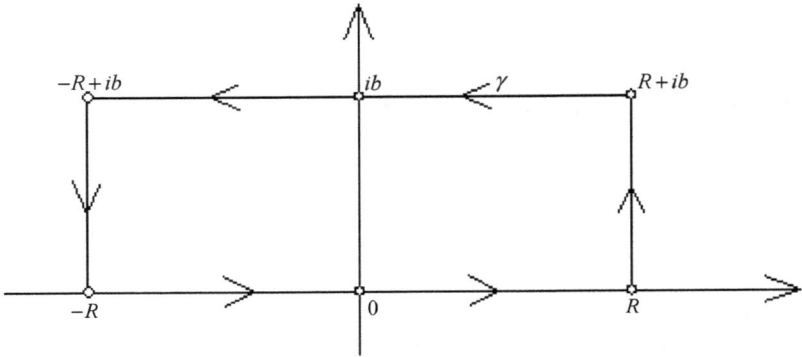

Fig. 4.95

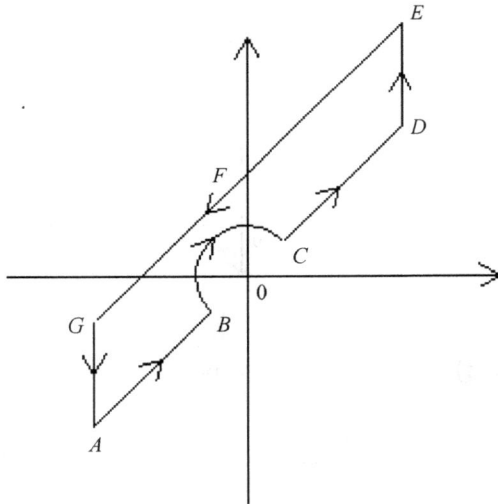

Fig. 4.96

$D = R + iR$, $E = R + i(R + \sqrt{\pi})$, $F = \sqrt{\pi}i$, and $G = -R - i$
$(R - \sqrt{\pi})$. This will lead *directly* to $-2i \int_0^\infty e^{-x^2} = -\sqrt{\pi}i$ if $r \to 0$
and $R \to \infty$.

In the sequel, we list some applications of the Gauss integral.

(9) Integrate e^{-z^2} along the boundary of the rectangle with vertices $-R$, R, $R + ia$, and $-R + ia$, where $R > 0$ and a is real, to show that

$$\int_{-\infty}^{\infty} e^{-(x+ia)^2} dx = \int_{-\infty+ia}^{\infty+ia} e^{-x^2} dx = \sqrt{\pi}, \quad a \in \mathbf{R}.$$

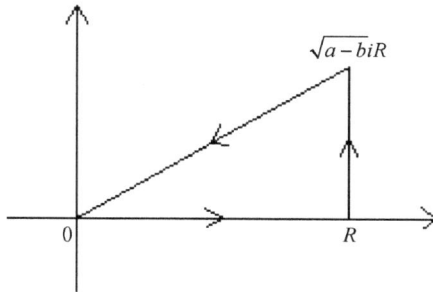

Fig. 4.97

(10) Let a and b be real numbers, where $a > 0$. Let $\sqrt{a - bi}$ denote the square root of $a - bi$ with positive real part (see Exercise B (1) of Sec. 1.3). Integrate e^{-z^2} along the triangular path in Fig. 4.97 to show that

$$\int_0^\infty e^{-(a-bi)x^2}\,dx = \frac{\sqrt{\pi}}{2\sqrt{a - bi}},$$

$$\Rightarrow \int_0^\infty e^{-ax^2}\cos bx^2\,dx = \frac{\sqrt{2\pi}}{4\rho}\sqrt{\rho + a}, \quad \text{and}$$

$$\int_0^\infty e^{-ax^2}\sin bx\,dx = -(\operatorname{sgn} b)\frac{\sqrt{2\pi}}{4\rho}\sqrt{\rho - a}.$$

where $\rho = \sqrt{a^2 + b^2}$.

(11) Integrate e^{-z^2} along the path γ shown in Fig. 4.84 with $\frac{\pi}{n}$ replaced by α, where $0 \le \alpha \le \frac{\pi}{4}$, to show that

$$\int_0^\infty e^{-x^2\cos 2\alpha}\cos(x^2\sin 2\alpha)\,dx = \frac{\sqrt{\pi}}{2}\cos\alpha,$$

$$\int_0^\infty e^{-x^2\cos 2\alpha}\sin(x^2\sin 2\alpha)\,dx = \frac{\sqrt{\pi}}{2}\sin\alpha.$$

In case $\alpha = \frac{\pi}{4}$, we have the *Fresnel integrals* in Exercise (4). Note that, in the estimation of the integral along the circular arc in Fig. 4.84, one might need the inequality $1 - \frac{2\theta}{\pi} \le \cos\theta$ if $0 \le \theta \le \frac{\pi}{2}$.

(12) Observe that $ax^2 + 2bxi = a\left(x + \frac{b}{a}i\right)^2 + \frac{b^2}{a}$. Integrate e^{-z^2} along the boundary of the rectangle with vertices $\pm R$, $\pm R + \frac{b}{a}i$

$\left(\text{or } \pm R, \pm R + \frac{b}{2\sqrt{a}}\right)$ to show that

$$\int_0^\infty e^{-ax^2} \cos 2bx\, dx = \frac{\sqrt{\pi}}{2\sqrt{a}} e^{-\frac{b^2}{a}}, \quad a > 0 \quad \text{and} \quad b \in \mathbf{R}.$$

This is *Poisson's integral* mentioned in Exercise (6).

As we have experienced so far, there are some *commonly used inequalities and integrals* in the process of evaluating or estimating line integrals along curves. They appeared in Example 9 of Sec. 2.9 and in Exercises A ranged from (24) to (28) there. For convenience to reach them, we collect them here for reference:

(I) *Jordan's inequalities.*

    (1) $\dfrac{2\theta}{\pi} \le \sin\theta \le \pi$,   if $0 \le \theta \le \dfrac{\pi}{2}$;

    (2) $1 - \dfrac{2\theta}{\pi} \le \cos\theta$,   if $0 \le \theta \le \dfrac{\pi}{2}$.   (4.12.3.1)

(II) *Jordan's lemma* (1913). If f is continuous in the sector domain $\bar{\Omega}$: $0 < R_0 \le |z| < \infty$, $0 \le \alpha \le \operatorname{Arg} z = \theta \le \beta \le \pi$ and $\lim_{z\in\bar{\Omega}\to\infty} f(z) = 0$, then, for any positive constant m,

$$\lim_{R\to\infty} \int_{C_R} e^{\operatorname{Im} z} f(z)\, dz = 0, \quad \text{where } C_R: z = Re^{i\theta}, \quad \alpha \le \theta \le \beta.$$

          (4.12.3.2)

(III) Suppose f is continuous in the sector domain $\bar{\Omega}$: $0 < |z - z_0| \le r_0$, $0 \le \alpha \le \operatorname{Arg}(z - z_0) \le \beta \le 2\pi$ and $\lim_{z\in\bar{\Omega}\to z_0}(z - z_0)f(z) = A \ne \infty$. Then

$$\lim_{r\to 0} \int_{C_r} f(z)\, dz = Ai(\beta - \alpha),$$

where

$$C_r : z = z_0 + re^{i\theta}, \quad 0 < r < r_0, \alpha \le \theta \le \beta.$$

In case z_0 is a simple pole of f, then

$$\lim_{r\to 0} \int_{C_r} f(z)\, dz = i(\beta - \alpha)\operatorname{Re} s(f; z_0). \quad\quad (4.12.3.3)$$

(IV) Suppose f is continuous in the sector domain $\bar{\Omega}$: $0 < R_0 \le |z| < \infty$, $0 \le \alpha \le \operatorname{Arg}(z - z_0) \le \beta \le 2\pi$ and $\lim_{z\in\bar{\Omega}\to\infty}(z - z_0)f(z) = A \ne \infty$. Then

$$\lim_{R\to\infty} \int_{C_R} f(z)\, dz = Ai(\beta - \alpha),$$

where

$$C_R : z = Re^{i\theta}, \quad R_0 < R, \alpha \le \theta \le \beta.$$

In case ∞ is a simple pole of f, then

$$\lim_{R \to \infty} \int_{C_R} f(z)dz = -i(\beta - \alpha)\operatorname{Re} s(f; \infty).$$

In particular, if f is analytic at ∞ and ∞ is a zero of order $m \ge 2$, then

$$\lim_{R \to \infty} \int_{C_R} f(z)dz = 0.$$

This is the case if $f(z) = \frac{p(z)}{q(z)}$, where $p(z)$ and $q(z)$ are relatively prime polynomials and $q(z)$ has degree at least two larger than that of $p(z)$.   (4.12.3.4)

(V) If f is analytic in a domain Ω which contains a real segment $[a, b]$ in which f has a simple pole at $x_0 \in (a, b)$, then the Cauchy principal value

$$\text{P.V.} \int_a^b f(x)dx = \lim_{\varepsilon \to 0} \left(\int_a^{x_0 - \varepsilon} f(x)dx + \int_{x_0 + \varepsilon}^b f(x)dx \right)$$

$$= A \log \frac{b - x_0}{x_0 - a} + \int_a^b \varphi(x)dx$$

exists, where $f(z) = \frac{A}{z - x_0} + \varphi(z)$, $0 < |z - x_0| < r$, and $\varphi(z)$ is analytic in $|z - x_0| < r$.   (4.12.3.5)

All proofs concerned are left to the readers.

Based on these previous contents — examples, exercises and (4.12.3.1)–(4.12.3.5), we will turn to more theoretical and general treatment on how to apply the method of residues in the evaluation of improper integrals over $(-\infty, \infty)$ and $(0, \infty)$ (see Sec. 4.12.4). As we proceed, we will sketch the main ideas and leave the details to readers.

Three basic types of improper integrals are discussed in the remaining of this section.

4.12.3.1. $\int_{-\infty}^{\infty} f(x)dx$

Here, $f(z)$ is always supposed to be analytic in the upper half-plane $\operatorname{Im} z > 0$ *except finitely many isolated singularities* $z_1, \ldots, z_n$ *(in* $\operatorname{Im} z > 0$*).*

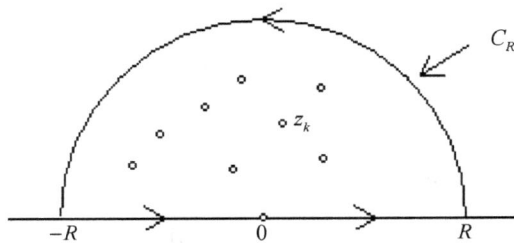

Fig. 4.98

We separate into two cases.

Case 1. Suppose $f(z)$ is continuous on $\{\operatorname{Im} z \geq 0\} - \{z_1, \ldots, z_n\}$ so that $f(z)$ does not have singularities along the real axis.

Choose $R > 0$ large enough so that $z_1, \ldots, z_n$ are contained in $|z| < R$. Integrate $f(z)$ over the path γ shown in Fig. 4.98. Then, by the residue theorem,

$$\int_\gamma f(z)dz = \int_{-R}^R f(x)dx + \int_{C_R} f(z)dz = 2\pi i \sum_{k=1}^n \operatorname{Re} s(f; z_k).$$

We want to have $\int_{-R}^R f(x)dx$ converge and $\int_{C_R} f(z)dz$ approach zero as $R \to \infty$. A sufficient condition is given by $\lim_{z \to \infty} zf(z) = 0$. This follows obviously by (4.12.3.4). In this case, $\int_{-\infty}^\infty f(x)dx$ exists as a principal value.

Another stronger condition is: There is a number $p > 1$ so that $f(z) = O\left(\frac{1}{|z|^p}\right)$ if $|z|$ is large enough. There are three aspects concerned. This condition guarantees that $f(z)$ can only have finitely many singularities in the upper half-plane. And, since $|f(x)| \leq \frac{M}{|x|^p}$ if $|x|$ is large, hence

$$\int_{-\infty}^\infty f(x)dx = \lim_{R,R' \to \infty} \int_{-R'}^R f(x)dx$$

converges absolutely; in particular, P.V. $\int_{-\infty}^\infty f(x)dx = \int_{-\infty}^\infty f(x)dx$ holds. Finally, of course, $\lim_{R \to \infty} \int_{C_R} f(z)dz = 0$ holds, too.

Summarize the above as

The evaluation of $\int_{-\infty}^\infty f(x)dx$ (no real poles). Suppose $f(z)$ is analytic in $\operatorname{Im} z > 0$ except $z_1, \ldots, z_n$ (in $\operatorname{Im} z > 0$) and continuous in $\operatorname{Im} z \geq 0$ except $z_1, \ldots, z_n$. If $\lim_{z \to \infty} zf(z) = 0$ holds, then

$$\text{P.V.} \int_{-\infty}^\infty f(x)dx = 2\pi i \sum_{k=1}^n \operatorname{Re} s(f; z_k).$$

If $f(z) = O\left(\frac{1}{|z|^p}\right)$ for some $p > 1$ as $z \to \infty$, then $\int_{-\infty}^{\infty} f(x)dx$ converges absolutely and is equal to P.V. $\int_{-\infty}^{\infty} f(x)dx$. In particular, if $f(z) = \frac{p(z)}{q(z)}$ is rational, where $p(z)$ and $q(z)$ are relatively prime polynomials and the degree of $q(z)$ is at least two larger than that of $p(z)$, then the result is applicable once $q(x) \neq 0$ on the real axis. $\hspace{2cm}$ (4.12.3.1.1)

Note that, if $f(x)$ is an even function of x, then $\int_0^{\infty} f(x)dx = \frac{1}{2}\int_{-\infty}^{\infty} f(x)dx$ can be evaluated by using (4.12.3.1.1) if applicable; if not, refer to Sec. 4.12.4 for $\int_0^{\infty} f(x)dx$.

For instance, see Example 1 in Sec. 4.12.2 and the following:

$$\int_{-\infty}^{\infty} \frac{1}{(1+x^2)^{n+1}}dx = \frac{(2n-1)!!}{(2n)!!}\pi,$$

where n is a positive integer and $(2n-1)!! = (2n-1)(2n-3)\cdots 3\cdot 1$, etc.;

$$\int_{-\infty}^{\infty} \frac{x^2 + 2x + 1}{x^4 + 8x + 16}dx = \frac{5\pi}{16};$$

$$\int_{-\infty}^{\infty} \frac{1}{(x^4+1)(x^2+a^2)^2}dx = \frac{5a^4 + 1 + a^3(a^4 - 2a^2 - 1)\sqrt{2}}{2a^3(1+a^4)^2}\pi,$$

where $a > 0$.

Case 2. Suppose $f(z)$ is analytic in $\{\operatorname{Im} z \geq 0\} - \{z_1, \ldots, z_n\}$ except finitely many *simple poles* $x_1, \ldots, x_m$ along the real axis.

In this case, we have to use the dented contour γ shown in Fig. 4.99 in order to avoid these poles, where the half-circles all have the same radius r. Then

$$\int_{\gamma} f(z)dz = 2\pi i \sum_{k=1}^{n} \operatorname{Re} s(f; z_k),$$

$$= \int_{\Gamma_R} + \sum_{l=1}^{m} \int_{\gamma_l} f(z)dz + \sum_{l=0}^{m} \int_{\tau_l} f(x)dx.$$

In case $\lim_{z \to \infty} zf(z) = A \neq \infty$, then, by (4.12.3.4), $\lim_{R \to \infty} \int_{\Gamma_R} f(z)dz = \pi i A$; while, by (4.12.3.3), $\lim_{r \to \infty} \int_{\gamma_l} f(z)dz = i(0 - \pi)\operatorname{Re} s(f; x_l) = -\pi i \operatorname{Re} s(f; x_l)$ for $1 \leq l \leq m$.

Summarize the above as

The evaluation of $\int_{-\infty}^{\infty} f(x)dx$ *(with real poles).* Suppose $f(z)$ satisfies the following conditions:

(1) $f(z)$ is analytic in $\operatorname{Im} z \geq 0$ except finitely many isolated singularities $z_1, \ldots, z_n$ in $\operatorname{Im} z > 0$ and finitely many *simple poles* $x_1, \ldots, x_m$ on the *real axis* and

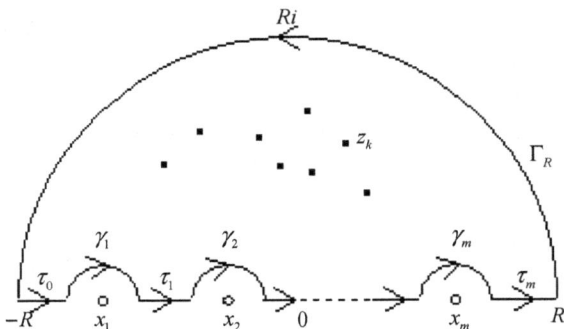

Fig. 4.99

(2) $\lim_{z \to \infty} z f(z) = A \neq \infty$ (a constant). For instance, $f(z) = \frac{A}{z} + o(\frac{1}{z})$ is the case.

Then,

$$\text{P.V.} \int_{-\infty}^{\infty} f(x)dx = -\pi i A + 2\pi i \sum_{k=1}^{n} \text{Re } s(f; z_k) + \pi i \sum_{l=1}^{m} \text{Re } s(f; x_l).$$

In particular, if $f(z) = \frac{p(z)}{q(z)}$ where $p(z)$ and $q(z)$ are relatively prime polynomials and the degree of $q(x)$ is at least one larger than that of $q(z)$, the result is applicable.   (4.12.3.1.2)

See Exercise A (12) of Sec. 4.11.3 for a generalization.
For instance, see Example 7 in Sec. 4.12.3 and the following:

$$\text{P.V.} \int_{0}^{\infty} \frac{x^4}{x^6 - 1}dx = \frac{1}{2}\text{P.V.} \int_{-\infty}^{\infty} \frac{x^4}{x^6 - 1}dx$$

$$= \frac{1}{2}\left\{ 2\pi i \cdot \frac{1}{6} \left(\frac{1 - \sqrt{3}i}{2} + \frac{-1 - \sqrt{3}i}{2} \right) \right.$$

$$\left. + \pi i \left(\frac{1}{6} - \frac{1}{6} \right) \right\} = \frac{\pi}{2\sqrt{3}};$$

$$\text{P.V.} \int_{-\infty}^{\infty} \frac{x^2 + 2}{x^6 + 4x^4 - x^2 - 4}dx = 2\pi i \left(\frac{i}{12} + \frac{i}{30} \right) + \pi i \left(\frac{3}{20} - \frac{3}{20} \right) = \frac{-7\pi}{30};$$

$$\text{P.V.} \int_{-\infty}^{\infty} \frac{\cos x}{a^2 - x^2}dx = \frac{\pi \sin a}{a} \quad (a > 0).$$

Exercises A

(1) Evaluate the following integrals.

(a) $\displaystyle\int_{-\infty}^{\infty} \frac{x^2 - x + 2}{x^4 + 10x^2 + 9}\,dx.$

(b) $\displaystyle\int_{0}^{\infty} \frac{x^2}{x^4 + 6x^2 + 13}\,dx.$

(c) $\displaystyle\int_{0}^{\infty} \frac{x^2}{(x^2 + a^2)^3}\,dx \quad (a > 0).$

(d) $\displaystyle\int_{0}^{\infty} \frac{1}{(x^2 + a^2)^2(x^2 + b^2)}\,dx \quad (a > 0, \, b > 0).$

(e) $\displaystyle\int_{0}^{\infty} \frac{dx}{(x^2 + a^2)^{n+1}} \quad (a > 0).$

(f) $\displaystyle\int_{0}^{\infty} \frac{x^2 + 1}{(x^2 - 2x \cos a + 1)(x^2 - 2x \cos b + 1)}\,dx \quad \left(0 < a < b < \frac{\pi}{2}\right).$

(g) $\displaystyle\int_{0}^{\infty} \frac{x^6}{x^8 + 1}\,dx.$

(h) $\displaystyle\int_{0}^{\infty} \frac{dx}{(x^2 + a^2)^2(x^2 + b^2)^2(x^2 + c^2)^2} \quad (a > b > c > 0).$

(i) $\displaystyle\int_{-\infty}^{\infty} \frac{dx}{(x^2 + a^2)^m(x^2 + b^2)^n} \quad (a > b > 0; \, m, \, n \text{ positive integers}).$

(j) $\displaystyle\int_{0}^{\infty} \frac{x^2}{(a^4 + x^4)^3}\,dx \quad (a > 0).$

(k) $\displaystyle\lim_{a \to \infty} \int_{0}^{\infty} \frac{1}{1 + \left(a - x + \frac{1}{2x}\right)^2}\,dx.$

(2) Evaluate the following integrals.

(a) P.V. $\displaystyle\int_{-\infty}^{\infty} \frac{dx}{(x - 1)(x^2 + 1)}.$

(b) P.V. $\displaystyle\int_{-\infty}^{\infty} \frac{x^2 + x + 2}{x^4 - 5x + 4}\,dx.$

(c) P.V. $\displaystyle\int_{-\infty}^{\infty} \frac{x^2 + 4}{x^6 + x^4 - x^2 - 1}\,dx.$

(d) P.V. $\displaystyle\int_{-\infty}^{\infty} \frac{dx}{x^3 + a^3} \quad (a > 0); \text{ P.V. } \int_{0}^{\infty} \frac{dx}{x^3 + a^3}.$

(e) P.V. $\displaystyle\int_{-\infty}^{\infty} \frac{x + 5}{x^4 - 10x^2 + 9}\,dx.$

(f) P.V. $\displaystyle\int_{-\infty}^{\infty} \frac{1-x}{1+x^3}\,dx.$

(3) Let $m = 2, 3, 4, \ldots$. Show that

$$\int_0^1 \frac{(1-x^2)^m}{1-(-x^2)^m}\,dx = \frac{2^m\pi}{8m}\sum_{|k|\le \frac{m}{2}}(-1)^k\left(\cos\frac{k\pi}{n}\right)^{m-1}.$$

4.12.3.2. $\int_{-\infty}^{\infty} f(x)e^{imx}\,dx$ $(m \in R)$

Based on the contour in Fig 4.98, we have an extension of (4.12.3.1.1).

The evaluation of $\int_{-\infty}^{\infty} f(x)e^{\mathrm{Im}\,x}\,dx$ $(m \in \mathbf{R}$, no real poles). Suppose f satisfies:

(1) f is analytic in $\mathrm{Im}\,z > 0$ except finitely many isolated singularities $z_1, \ldots, z_n$ in $\mathrm{Im}\,z > 0$;
(2) f is continuous in $\{\mathrm{Im}\,z \ge 0\} - \{\,z_1, \ldots, z_n\}$; and
(3) $f(z) = O\left(\frac{1}{|z|^p}\right)$, $p > 0$, as $z \to \infty$.

Then

$$\int_{-\infty}^{\infty} f(x)e^{\mathrm{Im}\,x}\,dx = 2\pi i\sum_{k=1}^{n}\mathrm{Re}\,s(f(z)e^{\mathrm{Im}\,z}; z_k).$$

In particular, this is the case if $f(z) = \frac{p(z)}{q(z)}$, where $p(z)$ and $q(z)$ are relatively prime polynomials, $q(x) \ne 0$ for real x, and the degree of $q(x)$ is at least one larger than that of $p(x)$.   (4.12.3.2.1)

Since $p > 0$, Dirichlet test says that the improper integral $\int_{-\infty}^{\infty} f(x)e^{\mathrm{Im}\,x}\,dx$ converges conditionally in general. The same proof of (4.12.3.1.1) is applicable here but, now, Jordan's lemma (4.12.3.2) is needed in the process of proving $\lim_{R\to\infty}\int_{C_R} f(z)e^{\mathrm{Im}\,z}\,dz = 0$. By equating the real and imaginary parts of both sides, (4.12.3.2.1) can be used to evaluate the improper integrals

$$\int_{-\infty}^{\infty} f(x)\cos mx\,dx \quad\text{and}\quad \int_{-\infty}^{\infty} f(x)\sin mx\,dx.$$

For another proof, see Exercise A (1).

For instance, see Example 2 in Sec. 4.12.3 and the following:

$$\int_{-\infty}^{\infty} \frac{\cos x}{(x^2 + 4)(x^2 + 9)}\,dx = \frac{\pi}{5}\left(\frac{e^{-2}}{2} - \frac{e^{-3}}{3}\right);$$

$$\int_{-\infty}^{\infty} \frac{\sin x}{(x^2 + 4)(x^2 + 9)}\,dx = 0;$$

$$\int_{0}^{\infty} \frac{\cos mx}{x^4 + x^2 + 1}\,dx = \frac{1}{2}\int_{-\infty}^{\infty} \frac{\cos mx}{x^4 + x^2 + 1}\,dx$$

$$= \frac{\pi}{\sqrt{3}}e^{-\frac{\sqrt{3}m}{2}}\sin\left(\frac{m}{2} + \frac{\pi}{6}\right), \quad m \geq 0;$$

$$\int_{0}^{\infty} \frac{\cos ax}{\cosh^2 x}\,dx = \frac{\pi a}{2\sinh(\frac{\pi a}{2})}, \quad a > 0.$$

The third integral above can be evaluated by integrating $\frac{e^{iaz}}{\cosh^2 z}$ along the boundary of the rectangle with vertices at $\pm R$ and $\pm R + i$.

While, corresponding to (4.12.3.1.2), comes

The evaluation of $\int_{-\infty}^{\infty} f(x)e^{\mathrm{Im}\,x}dx$ *($m \in \mathbf{R}$, with real poles).* Suppose f satisfies:

(1) Condition 1 in (4.12.3.1.2); and
(2) $\lim_{z \to \infty} f(z) = 0$.

Then, if $m > 0$,

$$\text{P.V.} \int_{-\infty}^{\infty} f(x)e^{\mathrm{Im}\,x}dx = 2\pi i \sum_{k=1}^{n} \mathrm{Re}\,s(f(z)e^{\mathrm{Im}\,z}; z_k)$$

$$+ \pi i \sum_{l=1}^{m} \mathrm{Re}\,s(f(z)e^{\mathrm{Im}\,z}; x_l). \qquad (4.12.3.2.2)$$

In case $m < 0$, we have

$$\text{P.V.} \int_{-\infty}^{\infty} f(x)e^{\mathrm{Im}\,x}dx = -2\pi i \sum_{k=1}^{P} \mathrm{Re}\,s(f(z)e^{\mathrm{Im}\,z}; \widetilde{z}_k)$$

$$+ \pi i \sum_{l=1}^{m} \mathrm{Re}\,s(f(z)e^{\mathrm{Im}\,z}; x_l), \qquad (4.12.3.2.3)$$

where $\widetilde{z}_1, \ldots, \widetilde{z}_p$ are isolated singularities of f in the *lower half-plane* $\mathrm{Im}\,z < 0$, if any. Equating the real and imaginary parts of both sides, (4.12.3.2.2) can be used to evaluate

$$\text{P.V.} \int_{-\infty}^{\infty} f(x) \cos mx\, dx \quad \text{and} \quad \text{P.V.} \int_{-\infty}^{\infty} f(x) \sin mx\, dx.$$

For another proof, see Exercise A (2).

For instance, see Example 7 in Sec. 4.12.3 and the following:

$$\int_{0}^{\infty} \frac{\sin x}{x} dx = \frac{1}{2} \text{Im} \int_{-\infty}^{\infty} \frac{e^{ix}}{x} dx = \frac{1}{2} \text{Im} \left\{ \pi i \cdot \text{Re}\, s \left(\frac{e^{iz}}{z}; 0 \right) \right\} = \frac{\pi}{2};$$

$$\int_{0}^{\infty} \frac{\sin^2 x}{x^2} dx = \frac{1}{2} \int_{-\infty}^{\infty} \frac{\sin^2 x}{x^2} dx = \frac{1}{4} \int_{-\infty}^{\infty} \frac{1 - \cos 2x}{x^2} dx$$

$$= \frac{1}{4} \pi i \cdot \text{Re}\, s \left(\frac{1 - e^{2iz}}{z}; 0 \right) = \frac{\pi}{2};$$

$$\int_{0}^{\infty} \frac{\sin^3 x}{x^3} dx = \frac{1}{8} \text{Im} \int_{-\infty}^{\infty} \frac{(1 - e^{3ix}) - 3(1 - e^{ix})}{x^3} dx$$

$$= \frac{1}{8} \text{Im} \left\{ \pi i \cdot \text{Re}\, s \left(\frac{(1 - e^{3iz}) - 3(1 - e^{iz})}{z^3}; 0 \right) \right\} = \frac{3\pi}{8};$$

$$\int_{0}^{\infty} \frac{\sin^4 x}{x^4} dx = \frac{\pi}{3}.$$

Note that the integrands $\frac{\sin^n z}{z^n}$, for $1 \le n \le 4$, all have a removable singularity at $z = 0$. So the symbol P.V. is not needed before each of the above integrals. Also, if $a > 0$ and $m \ge 0$,

$$\text{P.V.} \int_{0}^{\infty} \frac{\sin mx}{x(x^2 - a^2)} dx = \frac{1}{2} \text{Im} \left\{ \pi i \left(-\frac{1}{a^2} + \frac{1}{2a^2} e^{\text{Im}\, a} + \frac{1}{2a^2} e^{-\text{Im}\, a} \right) \right\}$$

$$= \frac{\pi}{2a^2} (\cos ma - 1);$$

$$\text{P.V.} \int_{0}^{\infty} \frac{\cos mx}{x^2 - a^2} dx = -\frac{\pi}{a} \sin ma.$$

Try to justify how to deduce the second integral from the first by differentiation process.

Exercises A

(1) Prove (4.12.3.2.1) by integrating $f(z)e^{\text{Im}\, z}$ along a rectangular path with vertices at $R_2, R_2 + iR_3, -R_1 + iR_3$ and $-R_1$, where $R_1 > 0, R_2 > 0$, and $R_3 > 0$.

(2) Prove (4.12.3.2.2) by integrating $f(z)e^{\text{Im}\, z}$ along the path shown in Fig. 4.100. In this case, condition 2 is replaced by the *assumption*

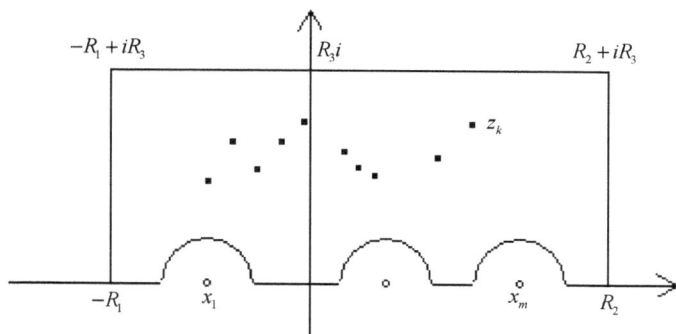

Fig. 4.100

that $f(z) = O\left(\frac{1}{|z|}\right)$ as $z \to \infty$. Also, show that the improper integral $\int_{-\infty}^{\infty} f(x)e^{\mathrm{Im}\, x}dx$ does converge if $f(z)$ does not have real simple poles.

(3) Evaluate the following integrals.

(a) $\displaystyle\int_{-\infty}^{\infty} \frac{x \cos x}{x^2 - 2x + 5}dx$ and $\displaystyle\int_{-\infty}^{\infty} \frac{x \sin x}{x^2 - 2x + 5}dx.$

(b) $\displaystyle\int_{-\infty}^{\infty} \frac{a \cos x + x \sin x}{x^2 + a^2}dx$ $(a > 0)$.

(c) $\displaystyle\int_{-\infty}^{\infty} \frac{x \sin mx}{x^4 + a^4}dx$ $(m > 0, a > 0)$ and $\displaystyle\int_{0}^{\infty} \frac{x^3 \sin mx}{x^4 + a^4}dx.$

(d) $\displaystyle\int_{-\infty}^{\infty} \frac{\cos mx}{(x^2 + a^2)(x^2 + b^2)}dx$ $(a > 0, b > 0, a \neq b$ and $m \geq 0)$ and
$\displaystyle\int_{0}^{\infty} \frac{\cos mx}{(x^2 + a^2)^2}dx.$

(e) $\displaystyle\int_{-\infty}^{\infty} \frac{\cos mx}{x^4 + a^4}dx$ $(m > 0, a > 0)$. Anything to do with (c) above?

(f) $\displaystyle\int_{0}^{\infty} \frac{\cos x}{(1 + x^2)^n}dx$ $(n = 1, 2, \ldots)$.

(g) $\displaystyle\int_{0}^{\infty} \frac{\cos ax}{1 + x^2 + x^4}dx$ $(a > 0)$.

(h) $\displaystyle\int_{0}^{\infty} \frac{\tanh\left(\frac{\pi x}{2}\right) \sin x}{1 + x^2}dx.$

(i) $\displaystyle\int_{0}^{\infty} \frac{\cos(ax^2) - \sin(ax^2)}{x^4 + 1}dx$ $(a > 0)$.

(j) $\displaystyle\int_{-\infty}^{\infty} \frac{e^{\frac{ax}{(x^2+1)}}}{x^2+1} \cos\frac{a}{x^2+1}\,dx$ and $\displaystyle\int_{0}^{\infty} \frac{e^{\frac{ax}{(x^2+1)}}}{x^2+1} \sin\frac{a}{x^2+1}\,dx$
$(a>0)$.

(4) Evaluate the following integrals.

(a) P.V. $\displaystyle\int_{-\infty}^{\infty} \frac{\sin\pi x}{x^5-x}\,dx$.

(b) $\displaystyle\int_{0}^{\infty} \frac{\sin^2 ax}{x^2(x^2-b^2)}\,dx$ $(a>0, b>0)$.

(c) $\displaystyle\int_{-\infty}^{\infty} \frac{\sin\pi x\cos\pi x}{x(2x-1)}\,dx$.

(d) $\displaystyle\int_{0}^{\infty} \frac{2\sin ax\cos\beta x}{x(x^2+a^2)}\,dx$ $(\alpha>0, \beta\geq 0, a>0)$.

(e) P.V. $\displaystyle\int_{-\infty}^{\infty} \frac{x\cos x}{x^2-5x+6}\,dx$.

(f) P.V. $\displaystyle\int_{-\infty}^{\infty} \frac{\cos ax}{1+x^3}\,dx$ $(a\geq 0)$.

(g) $\displaystyle\int_{0}^{\infty} \frac{x^2-b^2}{x^2+b^2}\frac{\sin ax}{x}\,dx$ $(a,b\in\mathbf{R})$

(h) $\displaystyle\int_{-\infty}^{\infty} \frac{\sin m(x-a)}{x-a}\frac{\sin n(x-b)}{x-b}\,dx$ $(m>n>0,$ integers; $a,b\in\mathbf{R}$
and $a\neq b$).

(i) $\displaystyle\int_{-\infty}^{\infty} \frac{x^{2m}-x^{2n}}{1-x^{2p}}\,dx$ $(m,n,p,$ positive integers and $m<n<p$).

4.12.3.3. $\int_{-\infty}^{\infty} f(x)dx$ (with periodic $f(x)$)

If the integrand contains a periodic function, say, of period $bi(b>0)$, then it is reasonable to choose a rectangular path of height b as the path of integration.

Firstly, we have

The evaluation of $\int_{-\infty}^{\infty} f(x)g(x)dx$ (with periodic $f(x)$ and no real simple poles). Suppose f and g satisfy:

(1) f and g are analytic in the parallel strip domain $0<\operatorname{Im}z<b$ except finitely many isolated singularities $z_1,\ldots,z_n$ there and are continuous on the closed strip $0\leq\operatorname{Im}z\leq b$, except at $z_1,\ldots,z_n$.
(2) $f(z+bi)=f(z)$, $0\leq\operatorname{Im}z\leq b$.
(3) $\lim_{z\to\infty} f(z)g(z)=0$ uniformly in $0\leq\operatorname{Im}z\leq b$.

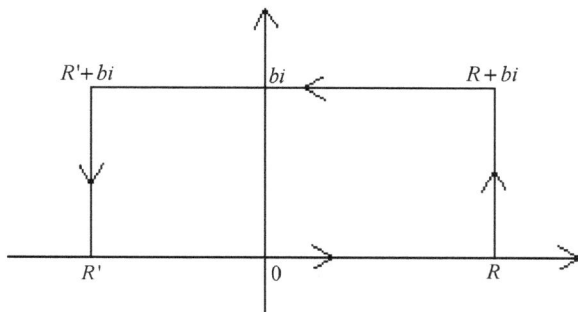

Fig. 4.101

Then

$$\int_{-\infty}^{\infty} f(x)[g(x) - g(x+bi)]dx = 2\pi i \sum_{k=1}^{n} \operatorname{Re} s(f(z)g(z); z_k). \qquad (4.12.3.3.1)$$

The path of integration is shown in Fig. 4.101. The proof is left to the readers. *In particular*, if f is analytic in $\mathbf{C}$ except at finitely many isolated singularities $z_1, \ldots, z_n$ so that $0 < \operatorname{Arg} z_k < 2\pi$, for $1 \le k \le n$, and g is analytic in the closed strip $0 \le \operatorname{Im} z \le 2\pi$, and if $\lim_{z \to \infty} f(e^z)g(z) = 0$ uniformly in $0 \le \operatorname{Im} z \le 2\pi$, then

$$\int_{-\infty}^{\infty} f(e^x)[g(x) - g(x+2\pi i)]dx = 2\pi i \sum_{k=1}^{n} \operatorname{Re} s(f(e^z)g(z); \operatorname{Log} z_k).$$
$$(4.12.3.3.2)$$

This is the case if $f(x) = \frac{p(x)}{q(x)}$ and $g(x) = e^{\lambda x}$ ($\lambda \in \mathbf{C}$, a constant) satisfy:

(1) $p(x)$ and $q(x)$ are relatively prime polynomials of degrees n and m, respectively;

(2) $q(x) \ne 0$ if $x > 0$ and $\frac{p(x)}{q(x)} = O(\frac{1}{x^l})$ as $x \to 0^+$; and

(3) $-l < \operatorname{Re} \lambda < m - n$,

then

$$\int_{-\infty}^{\infty} \frac{p(e^x)}{q(e^x)} e^{\lambda x} dx = \frac{2\pi i}{1 - e^{2\lambda \pi i}} \sum_{k} \operatorname{Re} s \left(\frac{p(e^z)}{q(e^z)} e^{\lambda z}; z_k \right), \qquad (4.12.3.3.3)$$

where z_k are singular points of $\frac{p(e^z)}{q(e^z)} e^{\lambda z}$ in the strip $0 < \operatorname{Im} z < 2\pi$. Note that the condition $-l < \operatorname{Re} \lambda < m - n$ will assure that the improper integral is absolutely convergent. This formula can also be obtained by integrating the integrand along the half circular path shown in Fig. 4.98.

For instance, see Exercise A (3) of Sec. 4.12.3 and the following integrals:

$$\int_{-\infty}^{\infty} \frac{e^{ax}}{\cosh bx} dx = \frac{\pi}{b} \cdot \frac{1}{\cos \frac{a\pi}{2b}}, \quad -b < a < b \left(f(z) = \frac{1}{\cosh bz}, g(z) = e^{az} \right);$$

$$\int_{-\infty}^{\infty} \frac{\cosh ax}{\cosh bx} dx = \frac{\pi}{2b} \cdot \frac{1}{\cos \frac{a\pi}{2b}} \text{ (using the above integral)};$$

$$\int_{0}^{\infty} \frac{x^2}{\cosh x} dx = \frac{\pi^3}{8} \left(f(z) = \frac{1}{\cosh z}, g(z) = z^2 \right). \text{ Does } g(z) = z^3 \text{work?}$$

Readers are urged to carry out the third integral above completely.

If real simple poles are permitted, then (4.12.3.3.1) can be extended to

The evaluation of $\int_{-\infty}^{\infty} f(x)dx$ (with real simple poles). Suppose f satisfies:

(1) $f(z)$ is analytic in the closed parallel strip $0 \leq \operatorname{Im} z \leq a \ (a > 0)$ except finitely many isolated singularities $z_1, \ldots, z_n$ in $0 < \operatorname{Im} z < a$ and simple poles $x_1, \ldots, x_m$ on the real axis $\operatorname{Im} z = 0$.

(2) $f(z + ai) = bf(z)$ for some constant $b \neq 1$. Hence, $x_1 + ai, \ldots, x_m + ai$ are simple poles of f on $\operatorname{Im} z = a$.

(3) $f(z) = c + o(1)$ as $z \to \infty$ on $0 \leq \operatorname{Im} z \leq a$.

Then,

$$\text{P.V.} \int_{-\infty}^{\infty} f(x)dx = \frac{2\pi i}{1-b} \sum_{k=1}^{n} \operatorname{Re} s(f; z_k) + \frac{(1+b)\pi i}{1-b} \sum_{l=1}^{m} \operatorname{Re} s(f; x_l).$$

$$(4.12.3.3.4)$$

In particular, if the assumptions about (4.12.3.3.3) are still valid for $f(x) = \frac{p(x)}{q(x)}$ except the condition "$q(x) \neq 0$ if $x > 0$" is replaced by the condition "$q(x)$ has *positive* simple zeros $x_1, \ldots, x_m$", then (4.12.3.3.3) turns out to be

$$\text{P.V.} \int_{-\infty}^{\infty} \frac{p(e^x)}{q(e^x)} e^{\lambda x} dx = \frac{2\pi i}{1 - e^{2\lambda \pi i}} \sum_{k=1}^{n} \operatorname{Re} s \left(\frac{p(e^z)}{q(e^z)} e^{\lambda z}; z_k \right)$$

$$+ \frac{(1 + e^{2\lambda \pi i})\pi i}{1 - e^{2\lambda \pi i}} \sum_{l=1}^{m} \operatorname{Re} s \left(\frac{p(e^z)}{q(e^z)} e^{\lambda z}; \log x_l \right)$$

$$(4.12.3.3.5)$$

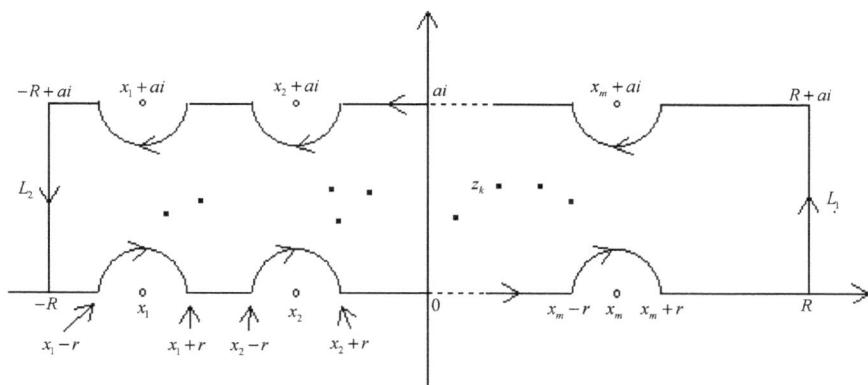

Fig. 4.102

To **prove** (4.12.3.3.4), choose $r > 0$ small enough and $R > 0$ large enough so that $z_1, \ldots, z_n$ all lie in the interior of a rectangle with vertices $\pm R$ and $\pm R + ai$; $|x_l| < R$ for $1 \le l \le m$ (say $x_1 < \cdots < x_m$); and $-R < x_1 - r < x_1 + r < x_2 - r < \cdots < x_m - r < x_m + r < R$. Let γ be the dented contour, shown in Fig 4.102, with half-circles $C_l : z = x_l + re^{i\theta}$, $0 \le \theta \le \pi$, and $\widetilde{C_l} : z = x_l + ai + re^{i\theta}$, $-\pi \le \theta \le 0$, $1 \le l \le m$. If $R > 0$ is large enough, then $|f(z) - c| < \varepsilon$ holds for $|z| \ge R$ and $0 \le \operatorname{Im} z \le a$, where $\varepsilon > 0$ is preassigned. Then along the two vertical sides L_1 and L_2,

$$\left| \int_{L_1} f(z)dz + \int_{L_2} f(z)dz \right|$$

$$= \left| \int_0^b [(f(R + iy) - c) - (f(-R + iy) - c)]i\,dy \right| \le 2\varepsilon b,$$

$$\Rightarrow \lim_{R \to \infty} \left(\int_{L_1} f(z)dz + \int_{L_2} f(z)dz \right) = 0.$$

Also, recall (see (4.12.3.4)) that, for $1 \le l \le m$,

$$\lim_{r \to \infty} \left(\int_{C_l} f(z)dz + \int_{\widetilde{C_l}} f(z)dz \right)$$

$$= (0 - \pi)i \operatorname{Re} s(f; x_l) + b(-\pi - 0)i \operatorname{Re} s(f; x_l)$$

$$= -(1 + b)\pi i \operatorname{Re} s(f; x_l).$$

The remaining details are left to the readers.

For instance, Exercise A (12) of Sec. 4.12.3 can be evaluated directly as

$$\text{P.V.} \int_0^\infty \frac{\sin ax}{\sinh x}dx = \frac{1}{2}\int_{-\infty}^\infty \frac{\sin ax}{\sinh x}dx = \frac{1}{2}\text{Im}\int_{-\infty}^\infty \frac{e^{iax}}{\sinh x}dx$$

$$= \frac{1}{2}\cdot\pi i\cdot\frac{1-e^{-a\pi}}{1+e^{-a\pi}}\text{Re}\,s\left(\frac{e^{iaz}}{\sinh z};0\right)$$

$$= \frac{\pi}{2}\tanh\frac{a\pi}{2}\left(f(z) = \frac{e^{iaz}}{\sinh z}, f(z+\pi i)\right)$$

$$= -e^{-a\pi}f(z)\text{ with } b = -e^{-a\pi}\bigg).$$

Also,

$$\text{P.V.}\int_{-\infty}^\infty \frac{e^{\lambda x}}{e^{2x}-1}dx = \frac{2\pi i}{1-e^{2\pi\lambda i}}\text{Re}\,s(f(z);\pi i)+\pi i\frac{1+e^{2\pi\lambda i}}{1-e^{2\pi\lambda i}}\text{Re}\,s(f(z);0)$$

$$= -\frac{\pi}{2}\cot\frac{\pi\lambda}{2}, 0<\text{Re}\,\lambda<2\left(f(z)=\frac{e^{\lambda z}}{e^{2z}-1},\right.$$

$$f(z+2\pi i) = e^{2\pi\lambda i}f(z)\text{ with } b = e^{2\pi\lambda i}\bigg);$$

$$\text{P.V.}\int_{-\infty}^\infty \frac{e^{\lambda x}}{1+e^x}dx = \frac{\pi}{\sin\lambda\pi},\quad 0<\text{Re}\,\lambda<1.$$

Exercises A

(1) Prove (4.12.3.3.1)–(4.12.3.3.3) in detail.
(2) Prove (4.12.3.3.4)–(4.12.3.3.5) in detail.
(3) Evaluate the following integrals.

(a) $\int_{-\infty}^\infty \frac{e^x}{e^{2x}+e^{2a}}\cdot\frac{1}{x^2+\pi^2}dx$ $(a\in\mathbf{R})$.

(b) $\int_{-\infty}^\infty \frac{x^2}{(x^2+\pi^2)\cosh x}dx.$

(c) $\int_0^\infty \frac{\cos ax}{\cosh^2 x}dx$ $(a>0)$.

(d) $\int_0^\infty \frac{\cosh ax}{\cosh x}dx$ $(-1<a<1)$.

(e) $\int_0^\infty \frac{\cos ax}{\cosh x}dx$ $(-1<a<1)$.

(4) Evaluate the following integrals by the following two methods.

   (i) Try to use (4.12.3.3.4)

   (ii) Try to imitate Exercise A (12) of Sec. 4.12.3 and Fig. 4.88.

   (a) $\int_{-\infty}^{\infty} \frac{\sin x}{x} dx$ (for method 2, consider a rectangular path with vertices $\pm R$, $\pm R + iR$, with a half circular detour at 0).

   (b) $\int_{-\infty}^{\infty} \frac{e^{tx}}{\cosh x} dx$ $(|\operatorname{Re} t| < 1)$.

   (c) $\int_{-\infty}^{\infty} \frac{\cosh tx}{\cosh x} dx$ $(|\operatorname{Re} t| < 1)$.

   (d) $\int_{0}^{\infty} \frac{x}{\sinh x} dx$.

   (e) $\int_{0}^{\infty} \frac{x \cos ax}{\sinh x} dx$ $(a \in \mathbf{R})$.

   (f) $\int_{0}^{\infty} \frac{\sin ax}{e^{2x} - 1} dx$ ($a > 0$; for method 2, consider a rectangular path with vertices $0, \pi, \pi + iR, iR$ ($R > 0$), with two quadrant circular detours at 0 and π).

(5) Integrate Log sin z along a rectangular path with vertices $0, \pi, \pi + iR, iR$ ($R > 0$), with two quadrant circular detours at 0 and π, to show that

$$\int_{0}^{\pi} \operatorname{Log} \sin x \, dx = -\pi \log 2.$$

Compare to Example 4 of Sec. 4.12.1 and Exercise A (8) of Sec. 4.12.3.

(6) Integrate $\frac{1}{(1+z^2)\cosh \frac{\pi z}{2}}$ along a rectangular path with vertices $\pm N$, $\pm N + 2iN$ (N is a positive integer) to show that

$$\int_{0}^{\infty} \frac{1}{(1 + x^2)\cosh\left(\frac{\pi x}{2}\right)} dx = \log 2.$$

(7) Use (4.12.3.3.3) to show that

$$\int_{-\infty}^{\infty} \frac{e^{ax}}{1 + e^{x} + e^{2x}} dx = \int_{0}^{\infty} \frac{t^{a-1}}{1 + t + t^2} dt$$

$$= \frac{2\pi}{\sqrt{3}} \frac{\sin \frac{1}{3}\pi(1 - a)}{\sin a\pi}, \quad 0 < a < 2.$$

Try to show this directly by integrating $e^{az}(1 + e^{z} + e^{2z})^{-1}$ along a rectangular path with vertices $\pm R$, $\pm R + 2\pi i$ ($R > 0$).

(8) Integrate $\dfrac{z}{a-e^{-iz}}$ $(a > 0)$ along a rectangular path with vertices $\pm\pi$, $\pm\pi + Ri$ $(R > 0)$ to show that

$$\int_0^\infty \frac{x\sin x}{1+a^2-2a\cos x}\,dx = \begin{cases} \dfrac{\pi}{a}\log(1+a), & 0 < a < 1 \\[2mm] \dfrac{\pi}{a}\log\dfrac{1+a}{a}, & a > 1 \end{cases}.$$

4.12.4.   *Improper integrals over* $(0,\infty)$

In this section, we mainly consider improper integrals:

$$\int_0^\infty f(x)dx, \quad \int_0^\infty f(x)\log x\,dx, \quad \text{and} \quad \int_0^\infty x^\alpha f(x)dx.$$

Once again, readers are suggested to read Ref. [7], especially Chap. 8 there, and Ref. [60] for much more complicated types of improper integrals which can be evaluated by the method of residues.

4.12.4.1. $\int_0^\infty f(x)dx$

Suppose f is analytic in $\mathbf{C}$ except finitely many isolated singularities $z_1, \ldots, z_n$ which *do not* lie on *the nonnegative real axis* $[0, \infty]$.

Designate $\operatorname{Log} z = \log|z| + i\operatorname{Arg} z$, $0 \le \operatorname{Arg} z < 2\pi$, as the principal branch of $\log z$.

Consider the *auxiliary function*

$$F(z) = f(z)\operatorname{Log} z.$$

$F(z)$ has the same isolated singularities $z_1, \ldots, z_n$ as $f(z)$ does. Meanwhile, points in the nonnegative real axis $[0, \infty]$ are singular points of $F(z)$; in particular, $z = 0$ is a nonisolated singularity of $\operatorname{Log} z$, as well as, of $F(z)$. Yet $F(z)$ is *continuous* downward to the upper side of *the cut* $(0, \infty)$ from the upper half-plane and is *continuous* upward to the lower side of the cut $(0, \infty)$ from the lower half-plane (refer to (2.2.5) and Example 1 of Sec. 2.2).

Choose $R > 0$ large enough and $r > 0$ smaller so that $r < z_k < R$ for $1 \le k \le r$. Consider the closed curve γ shown in Fig. 4.103, composed of the following subcurves:

$$\begin{aligned}
&C_r : z = re^{it}, && 0 \le t \le 2\pi; \\
&[r, R] : z = t, && r \le t \le R \quad \text{(upper side)}; \\
&C_R : z = Re^{it}, && 0 \le t \le 2\pi, \quad \text{and} \\
&[r, R]^- : z = t, && r \le t \le R \quad \text{(lower side)}.
\end{aligned}$$

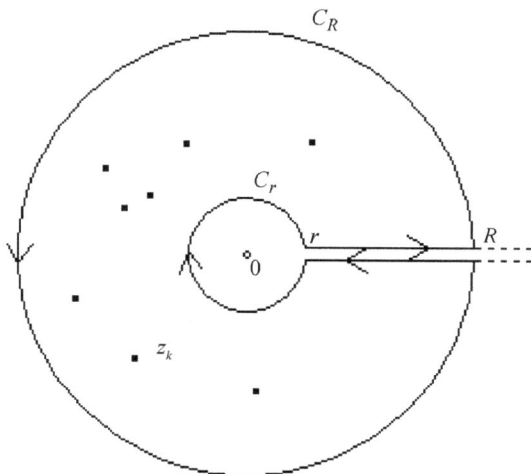

Fig. 4.103

Namely $\gamma = C_r + [r, R] + C_R + [r, R]^-$, oriented as indicated in Fig. 4.103.
By the residue theorem,

$$\int_\gamma F(z)dz = 2\pi i \sum_{k=1}^n \operatorname{Re} s(F(z); z_k). \qquad (*_1)$$

Along the upper side $[r, R]$ and the lower side $[r, R]^-$:
$\int_{[r,R]} F(z)dz = \int_r^R f(x)\operatorname{Log} x\,dx$ (by continuity of $\operatorname{Log} z$ from the upper
half-plane);

$$\int_{[r,R]^-} F(z)dz = \int_R^r f(x)[\operatorname{Log} x + 2\pi i]dx$$

(by continuity of $\operatorname{Log} z$ from the lower half-plane),

$$= -\int_r^R f(x)\operatorname{Log} x\,dx - 2\pi i\int_r^R f(x)dx,$$

$$\Rightarrow \int_{[r,R]} F(z)dz + \int_{[r,R]^-} F(z)dz = -2\pi i\int_r^R f(x)dx$$

$$\to -2\pi i\int_0^\infty f(x)dx \quad \text{if exists, as } r \to \infty \text{ and } R \to \infty.$$

Note that $\int_r^R f(x)\operatorname{Log} x\,dx$ cancels out in this sum process and this is the
main reason why we introduce $\operatorname{Log} z$ as part of the auxiliary function $F(z)$.

Along C_r and C_R: We do hope that $\lim_{r \to 0} \int_{C_r} F(z)dz = 0$ and $\lim_{R \to \infty} \int_{C_R} F(z)dz = 0$ hold, so that $\lim_{\substack{r \to 0 \\ R \to \infty}} \int_r^R f(x)dx = \int_0^\infty f(x)dx$ *really* exists as a finite number. To achieve this purpose, firstly observe that

$$\int_{C_r} F(z)dz = -\int_0^{2\pi} f(re^{i\theta})(\log r + i\theta)rie^{i\theta}\,d\theta,$$

$$\Rightarrow \left| \int_{C_r} F(z)dz \right| \le \int_0^{2\pi} |f(re^{i\theta})|[|\log r| + 2\pi]rd\theta \to 0 \quad \text{as } r \to 0.$$

This is because f is analytic at 0 and hence, is bounded in a neighborhood of 0, and that $\lim_{r \to 0} r|\log r| = 0$ holds. On the other hand, similar to above estimate, we have

$$\left| \int_{C_R} F(z)dz \right| \le \int_0^{2\pi} |f(Re^{i\theta})|[|\log R| + 2\pi]Rd\theta \quad (R > 1).$$

Suppose now that there is a constant $p > 1$ so that $\lim_{z \to \infty} z^p f(z) = 0$ holds true. Then

$$\left| \int_{C_R} F(z)dz \right| \le \frac{(\log R + 2\pi)}{R^{p-1}} \int_0^{2\pi} R^p |f(Re^{i\theta})|d\theta \to 0 \quad \text{as } R \to \infty.$$

Substituting these results in $(*_1)$ under the limit processes $r \to 0$ and $R \to \infty$, then

$$-2\pi i \int_0^\infty f(x)dx = 2\pi i \sum_{k=1}^n \operatorname{Re} s(f(z)\mathrm{Log}\, z; z_k).$$

After canceling $2\pi i$ out from both sides, the desired integral formula follows.

We summarize the above as

The evaluation of $\int_0^\infty f(x)dx$ (without positive simple poles). Suppose f satisfies:

(1) f is analytic in $\mathbf{C}$ except finitely many isolated singularities $z_1, \ldots, z_n \in \mathbf{C} - [0, \infty)$, where $[0, \infty)$ denotes the nonnegative real axis.

(2) $\lim_{z \to \infty} z^p f(z) = 0$ for some constant $p > 1$ or $\lim_{z \to \infty} zf(z)\mathrm{Log}\, z = 0$.

Then

$$\int_0^\infty f(x)dx = -\sum_{k=1}^n \operatorname{Re} s(f(z)\mathrm{Log}\, z; z_k)$$

where $\mathrm{Log}\, z = \log|z| + i\operatorname{Arg} z$, $0 \le \operatorname{Arg} z < 2\pi$, is the principal branch of $\log z$.   (4.12.4.1.1)

In particular, if $f(z) = \frac{p(z)}{q(z)}$, where $p(z)$ and $q(z)$ are relatively prime polynomials and the degree of $q(z)$ is at least two larger than that of $p(z)$, and

$q(x) \neq 0$ for $x \geq 0$, then

$$\int_0^\infty \frac{p(x)}{q(x)} dx = -\sum_{k=1}^n \operatorname{Re} s\left(\frac{p(z)}{q(z)} \operatorname{Log} z; z_k\right) \qquad (4.12.4.1.2)$$

where $z_1, \ldots, z_n$ are the *complex* or *negative* zeros of $q(z)$. For some theoretical extensions, see Exercise B.

For instance,

$$\int_0^\infty \frac{x}{1+x^6} dx = -\sum_{k=1}^5 \operatorname{Re} s\left(\frac{z \operatorname{Log} z}{1+z^6}; e^{\frac{(2k+1)\pi i}{6}}\right)$$

$$= -\sum_{k=0}^6 \frac{(2k+1)\pi i}{36} e^{-\frac{2(2k+1)\pi i}{3}} = \frac{\sqrt{3}\pi}{9};$$

$$\int_{-\infty}^0 \frac{x^3}{x^5-1} dx = \int_0^\infty \frac{x^3}{x^5+1} dx = -\sum_{k=0}^4 \operatorname{Re} s\left(\frac{z^3 \operatorname{Log} z}{z^5+1}; e^{\frac{(2k+1)\pi i}{5}}\right)$$

$$= \frac{\pi}{5}\left(2\sqrt{10-2\sqrt{5}} + \sqrt{10+2\sqrt{5}}\right).$$

Try to integrate $\frac{z}{1+z^6}$ along the boundary of the sector domain: $0 \leq |z| \leq R$ and $0 \leq \operatorname{Arg} \leq \frac{\pi}{3}$, to justify the first integral formula above.

In addition to the Condition 1 in (4.12.4.1.1), suppose $f(z)$ has a *simple* pole at $x_1 > 0$. Choose the closed curve γ shown in Fig. 4.104, composed of the following curves:

C_r and C_R are as in Fig. 4.103;
L_1 (upper side) and L_4 (lower side): $z = t, r \leq t \leq x_1 - r$;
L_2 (upper side) and L_3 (lower side): $z = t, r_1 + r \leq t \leq R$;
C_{1r} (upper half-circle): $z = x_1 + re^{it}, 0 \leq t \leq \pi$;
C_{1r}^* (lower half-circle): $z = x_1 e^{2\pi i} + re^{it}, \pi \leq t \leq 2\pi$
(*Note* that $x_1 e^{2\pi i}$ is considered as a point in the lower side of the cut $(0, \infty)$).

Then, as usual,

$$\int_\gamma f(z) \operatorname{Log} z \, dz = 2\pi i \sum_{k=1}^n \operatorname{Re} s(f(z)\operatorname{Log}; z_k). \qquad (*_2)$$

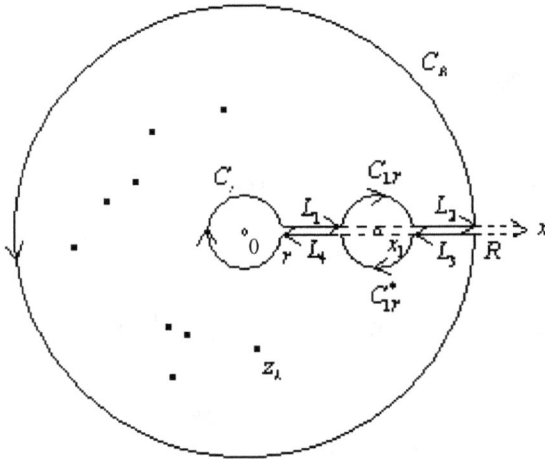

Fig. 4.104

As in the proof of (4.12.4.1.1), we still have

$$\lim_{r\to\infty}\int_{C_r} f(z)\mathrm{Log}\,z\,dz = 0 \quad\text{and}\quad \lim_{R\to\infty}\int_{C_R} f(z)\mathrm{Log}\,z\,dz = 0,$$

and

$$\int_{L_1} f(z)\mathrm{Log}\,z\,dz + \int_{L_4} f(z)\mathrm{Log}\,z\,dz = -2\pi i \int_r^{x_1-r} f(x)dx,$$

$$\int_{L_2} f(z)\mathrm{Log}\,z\,dz + \int_{L_3} f(z)\mathrm{Log}\,z\,dz = -2\pi i \int_{x_1+r}^{R} f(x)dx,$$

$\Rightarrow$ (adding together side by side)   as $r \to 0$ and $R \to \infty$

$$- 2\pi i \text{ P.V.} \int_0^\infty f(x)dx, \text{ if exists.}$$

Finally, along C_{1r} and C_{1r}^*: Along C_{1r}, by (4.12.3.3),

$$\lim_{r\to 0}\int_{C_{1r}} f(z)\mathrm{Log}\,z\,dz = (0 - \pi)i\,\mathrm{Re}\,s(f(z)\mathrm{Log}\,z; x_1)$$

$$= -\pi i\,\mathrm{Re}\,s(f(z)\mathrm{Log}\,z; x_1). \tag{$*_3$}$$

Be careful about the integration along C_{1r}^*. Because in the point of view of $\mathrm{Log}\,z$, the center of C_{1r}^* is a point on the lower side of the branch cut $(0, \infty)$, namely it is $x_1 e^{2\pi i}$ instead of x_1, yet this does not matter with respect to $f(z)$. Under this circumstance, by (4.12.3.3),

$$\lim_{r \to 0} \int_{C_{1r}^*} f(z) \text{Log}\, z \, dz = -(2\pi - \pi) i \, \text{Re}\, s(f(z) \text{Log}\, z; x_1 e^{2\pi i})$$

$$= -\pi i \, \text{Re}\, s(f(z) \text{Log}\, z e^{2\pi i}; x_1)$$

$$\text{(by the definition of a residue at a simple pole)}$$

$$= -\pi i [\text{Re}\, s(f(z) \text{Log}\, z; x_1) + 2\pi i \, \text{Re}\, s(f(z); x_1)]. \quad (*_4)$$

This can also be seen *by direct computation* of $\int_{C_{1r}^*} f(z) \text{Log}\, z \, dz$, where $C_{1r}^* : z = x_1 e^{2\pi i} + re^{it},\ \pi \le t \le 2\pi$: if $r > 0$ is small enough, then $f(z) = \frac{\varphi(z)}{z - x_1}$ where $\varphi(z)$ is analytic at $x_1 = x_1 e^{2\pi i}$ and $\varphi(x_1) \ne 0$. Since $\text{Re}\, s(f(z); x_1) = \lim_{z \to x_1} (z - x_1) f(z) = \varphi(x_1)$, then, as $r \to 0$,

$$\int_{C_{1r}^*} f(z) \text{Log}\, z \, dz = \int_{C_{1r}^*} \frac{\varphi(z)}{z - x_1} \text{Log}\, z \, dz$$

$$= \int_{2\pi}^{\pi} \frac{\varphi(x_1 + re^{it})}{re^{it}} \text{Log}(x_1 e^{2\pi i} + re^{it}) \cdot rie^{it} dt,$$

$$\to i \int_{2\pi}^{\pi} \varphi(x_1) \text{Log}\, x_1 e^{2\pi i} dt = -\pi i \varphi(x_1) \text{Log}\, x_1 e^{2\pi i},$$

$$= -\pi i [\varphi(x_1) \text{Log}\, x_1 + 2\pi i \varphi(x_1)]$$

and the claimed result follows. Combining $(*_3)$ and $(*_4)$ together, we have

$$\lim_{r \to \infty} \left(\int_{C_{1r}} f(z) \text{Log}\, z \, dz + \int_{C_{1r}^*} f(z) dz \right) = -2\pi i \, \text{Re}\, s(f(z) \text{Log}\, z; x_1)$$

$$-2\pi i \cdot \pi i \, \text{Re}\, s(f(z); x_1).$$

Putting these results in $(*_2)$ in the processes of $r \to 0$ and $R \to \infty$, then

$$-2\pi i \, \text{P.V.} \int_0^\infty f(x) dx - 2\pi i \, \text{Re}\, s(f(z) \text{Log}\, z; x_1) - 2\pi i \cdot \pi i \, \text{Re}\, s(f(z); x_1)$$

$$= 2\pi i \sum_{k=1}^n \text{Re}\, s(f(z) \text{Log}\, z; z_k)$$

$$\Rightarrow \text{P.V.} \int_0^\infty f(x) dx = -\sum_{k=1}^n \text{Re}\, s(f(z) \text{Log}\, z; z_k)$$

$$- \text{Re}\, s(f(z) \text{Log}\, z; x_1) - \pi i \, \text{Re}\, s(f(z); x_1).$$

We summarize the above as

The evaluation of $\int_0^\infty f(x)dx$ (with positive simple poles). Suppose f satisfies:

(1) f is analytic in $\mathbf{C}$ except finitely many isolated singularities $z_1, \ldots, z_n$ in $\mathbf{C} - [0, \infty)$ and finitely many simple poles $x_1, \ldots, x_m$ in $(0, \infty)$.
(2) $\lim_{z \to \infty} z f(z) \mathrm{Log}\, z = 0$.

Then

$$\text{P.V.} \int_0^\infty f(x)dx = -\sum_{k=1}^n \mathrm{Re}\, s(f(z)\mathrm{Log}\, z; z_k)$$

$$-\sum_{l=1}^m \mathrm{Re}\, s(f(z)\mathrm{Log}\, z; x_l) - \pi i \sum_{l=1}^m (f(z); x_l),$$

where $\mathrm{Log}\, z = \log|z| + i\,\mathrm{Arg}\, z$, $0 \le \mathrm{Arg}\, z < 2\pi$, is the principal branch of $\log z$.   (4.12.4.1.3)

And the *counterpart* of (4.12.4.1.2) is

$$\text{P.V.} \int_0^\infty \frac{p(x)}{q(x)}dx = -\sum_{k=1}^n \mathrm{Re}\, s\left(\frac{p(z)}{q(z)}\mathrm{Log}\, z; z_k\right) - \sum_{l=1}^m \mathrm{Re}\, s\left(\frac{p(z)}{q(z)}\mathrm{Log}\, z; x_l\right)$$

$$-\pi i \sum_{l=1}^m \left(\frac{p(z)}{q(z)}; x_l\right),$$ (4.12.4.1.4)

where $x_1, \ldots, x_m$ are the positive real simple zeros of $q(z)$; $z_1, \ldots, z_n$ are the complex or negative zeros of $q(z)$; and $p(z)$ and $q(z)$ have no common zeros so that the degree of $q(z)$ is at least two larger than that of $p(z)$.

For instance,

$$\text{P.V.} \int_0^\infty \frac{x^2}{x^4 - 1}dx$$

$$= -\left[\mathrm{Re}\, s\left(\frac{z^2 \mathrm{Log}\, z}{z^4 - 1}; i\right) + \mathrm{Re}\, s\left(\frac{z^2 \mathrm{Log}\, z}{z^4 - 1}; -i\right) + \mathrm{Re}\, s\left(\frac{z^2 \mathrm{Log}\, z}{z^4 - 1}; -1\right)\right]$$

$$- \mathrm{Re}\, s\left(\frac{z^2 \mathrm{Log}\, z}{z^4 - 1}; 1\right) - \pi i\, \mathrm{Re}\, s\left(\frac{z^2}{z^4 - 1}; 1\right)$$

$$= -\left[\frac{\pi}{8} - \frac{3\pi}{8} - \frac{\pi i}{4}\right] - 0 - \pi i \cdot \frac{1}{4} = \frac{\pi}{4};$$

$$\text{P.V.} \int_0^\infty \frac{1}{(x-2)(x^2 + 4)}dx = -\frac{\pi}{16}.$$

Exercises A

(1) Evaluate the following integrals.

(a) $\displaystyle\int_0^\infty \frac{x^n}{1+x^6}dx$ $(n = 1, 2, 3, 4)$.

(b) $\displaystyle\int_0^\infty \frac{x^3 - 2}{(x^4 + x + 1)(x^2 + 1)}dx$.

(c) $\displaystyle\int_0^\infty \frac{x^n}{(1 + x^2)^2}dx$ $(n = 0, 1, 2,)$.

(2) Evaluate the following integrals.

(a) P.V. $\displaystyle\int_0^\infty \frac{x^n}{x^4 - 1}dx$ $(n = 0, 1)$.

(b) P.V. $\displaystyle\int_0^\infty \frac{x}{x^3 - 6x^2 + 4x - 6}dx$.

(c) P.V. $\displaystyle\int_0^\infty \frac{x^2 + 1}{x^5 - 1}dx$.

(3) Integrate $F(z) = \frac{e^{iz}}{e^z - 1}$ along the rectangular path with vertices $\pm R$, $\pm R + 2\pi i$ with dented half-circles centered at 0 and $2\pi i$ to evaluate $\int_0^\infty \frac{\sin x}{e^x - 1}dx$.

(4) Integrate $F(z) = \frac{e^{iz} - e^{-z}}{z}$ along a path of a large quadrant circle, lying in the first quadrant and having a dented quadrant small circle centered at 0, to show that

$$\int_0^\infty \frac{\cos x - e^x}{x}dx = 0.$$

Exercises B

Extensions of (4.12.4.1.1) and (4.12.4.1.3).

Via linear fractional transformations, it is possible to map a line segment or a circular arc univalently onto the ray $[0, \infty]$ along the real axis. Owing to the conformal invariance of residues (see (4.11.1.4)), integrals of the form $\int_a^b f(x)dx$ over a finite interval $[a, b]$ can be transformed into the type $\int_0^\infty F(x)dx$ and be evaluated by the residues as indicated in (4.12.4.1.1) and (4.12.4.1.3).

P. R. Boas, Jr. and L. Schoenfeld: *Indefinite Integration by Residues*, *SIAM Review* **8** (1966), pp. 173–183 obtained the following results in Exercises (1) and (2).

(1) Suppose f satisfies:

(i) f is analytic in $\mathbf{C}^*$ except finitely many isolated singularities $z_1, \ldots, z_n$ which do *not* lie on the real segment (a, b).

(ii) f has simple poles $x_1, \ldots, x_m$ on the *open* interval (a, b).

(iii) f is analytic at both a and b.

Then ($\operatorname{Log} w = \log|w| + i\operatorname{Arg} w,\ 0 \le \operatorname{Arg} w < 2\pi$)

$$\int_a^b f(x)dx = -\left[\sum_{k=1}^n \operatorname{Re} s\left(f(z)\operatorname{Log}\frac{z-a}{b-z}; z_k\right)\right.$$
$$\left.+ \sum_{l=1}^m \operatorname{Re} s\left(f(z)\operatorname{Log}\frac{z-a}{b-z}; x_l\right)\right]$$
$$- \pi i \sum_{l=1}^m \operatorname{Re} s(f(z); x_l). \qquad (4.12.4.1.5)$$

To **prove** this, the change $t = \frac{x-a}{b-x}$ of variables transforms

$$\int_a^b f(x)dx = \int_0^\infty \frac{b-a}{(1+t)^2} f\left(\frac{a+bt}{1+t}\right) dt.$$

Consequently, apply (4.12.4.1.3) to $F(z) = \frac{b-a}{(1+z)^2} f\left(\frac{a+bz}{1+z}\right)$, $z \in \mathbf{C}$, and then, rewrite the result in terms of f. In this last step, observe that, if $w = \frac{a+bz}{1+z}$, then

$$F(z)\operatorname{Log} z = \frac{dw}{dz} f(w)\operatorname{Log}\frac{w-a}{b-w} \quad \text{or}$$
$$[F(z^{-1}(w))\operatorname{Log} z^{-1}(w)]\frac{dz}{dw} = f(w)\operatorname{Log}\frac{w-a}{b-w}$$

$$\Rightarrow (\text{by } (4.11.1.4))\operatorname{Re} s(F(z)\operatorname{Log} z; \widetilde{z_k})$$
$$= \operatorname{Re} s\left(f(w)\operatorname{Log}\frac{w-a}{b-w}; w_k\right), \quad w_k = w(\widetilde{z_k}).$$

When replacing w by z, w_k should be changed to z_k, etc. *Note*: In case f *does* not have simple poles on (a, b), then

$$\int_a^b f(x)dx = -\sum_{k=1}^n \operatorname{Re} s\left(f(z)\operatorname{Log}\frac{z-a}{b-z}; z_k\right)$$

Since $\operatorname{Log}\left(\frac{z-a}{b-z}\right) = (2p+1)\pi i + \log\frac{z-a}{z-b}$, where $\log\frac{z-a}{z-b}$ is *any chosen branch* in $\mathbf{C} - \widehat{ab}$ (see Example 4 in Sec. 2.7.3) and p is some integer,

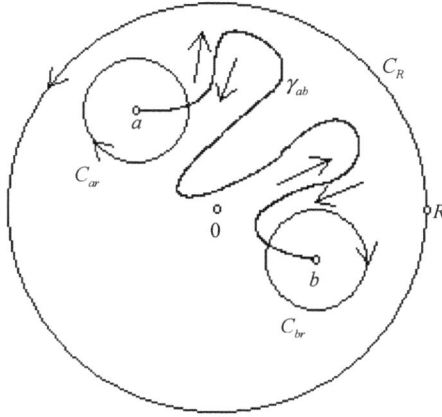

Fig. 4.105

and owing to $\sum_{k=1}^{n} \operatorname{Re} s(f; z_k) = 0$ (see (4.11.1.7)), it follows that

$$\int_a^b f(x)dx = -\sum_{k=1}^{n} \operatorname{Re} s \left(f(z) \log \frac{z-a}{z-b}; z_k \right).$$

Therefore, we try to *extend* (4.12.4.1.5) to the following: Let γ_{ab} be a rectifiable Jordan curve with end points a and b (see Fig. 4.105). Suppose f is analytic in $\mathbf{C}^* - \{a, b\}$ except finitely many isolated singularities $z_1, \ldots, z_n$ in $\mathbf{C}^* - \gamma$. Then

$$\int_{\gamma_{ab}} f(z)dz = -\sum_{k=1}^{n} \operatorname{Re} s \left(f(z) \log \frac{z-a}{z-b}; z_k \right), \qquad (4.12.4.1.6)$$

where $\log \frac{z-a}{z-b}$ is any fixed branch in $\mathbf{C}^* - \gamma$. Show that

$$\int_a^b \frac{dx}{1+x^4} = \frac{1}{4\sqrt{2}} \left(\log \frac{1 - \sqrt{2}a + a^2}{1 + \sqrt{2}a + a^2} + \log \frac{1 + \sqrt{2}b + b^2}{1 - \sqrt{2}b + b^2} \right)$$

$$+ \frac{1}{2\sqrt{2}} \left(\tan^{-1} \frac{b-a}{\sqrt{2} - a - b + \sqrt{2}ab} + \tan^{-1} \frac{b-a}{\sqrt{2} + a + b + \sqrt{2}ab} \right).$$

Also, show that, by real variable method and residues, respectively,

$$\int_1^2 e^{\frac{1}{x}} dx = \log 2 + 1 + \sum_{n=2}^{\infty} \frac{1}{n!} \cdot \frac{1 - 2^{1-n}}{n-1}.$$

(2) Let γ be the circular arc $|z| = 1$, $\alpha < \operatorname{Arg} z < \beta$, where $\beta - \alpha < 2\pi$. Suppose f satisfies:

(i) f is analytic in $\mathbf{C}^* - \{e^{i\alpha}, e^{i\beta}\}$ except finitely many isolated singularities $z_1, \ldots, z_n$ in $\mathbf{C}^* - \gamma$.

(ii) f has simple poles $a_1, \ldots, a_m$ in the open circular arc $\gamma - \{e^{i\alpha}, e^{i\beta}\}$.

Then $(\mathrm{Log}\, w = \log |w| + i \, \mathrm{Arg}\, w, \; 0 \le \mathrm{Arg}\, w < 2\pi$, and $\mu = \frac{1}{2}(\beta - \alpha))$

$$
\int_\alpha^\beta f(e^{ix})\,dx = i \left[\sum_{k=1}^n \mathrm{Re}\, s \left(\frac{f(z)}{z} \mathrm{Log}\left(-e^{i\mu}\frac{z - e^{i\alpha}}{z - e^{i\beta}}; z_k \right) \right) \right.
$$

$$
\left. + \sum_{l=1}^m \mathrm{Re}\, s \left(\frac{f(z)}{z} \mathrm{Log}\left(-e^{i\mu}\frac{z - e^{i\alpha}}{z - e^{i\beta}}; a_l \right) \right) \right] - \pi \sum_{l=1}^m \mathrm{Re}\, s \left(\frac{f(z)}{z}; a_l \right).
$$

$$(4.12.4.1.7)$$

To prove this, use $z = -e^{i\mu}\frac{w - e^{i\alpha}}{w - e^{i\beta}}$ to map $|z| = 1$ onto the real axis so that γ is mapped onto $[0, \infty]$. Observe that $\frac{dz}{dx} = \frac{\sin \mu}{2 \sin^2 \frac{\beta - x}{2}} > 0$ if $w = e^{ix}$, $\alpha < x < \beta$. Then

$$
\int_\alpha^\beta f(e^{ix})\,dx = \int_0^\infty F(x)\,dx, \quad \text{where}
$$

$$
F(z) = \frac{2 \sin \mu}{(z + e^{i\mu})(z + e^{-i\mu})} f\left(e^{i\beta}\frac{z + e^{-i\mu}}{z + e^{i\mu}} \right).
$$

In the last step, letting $w = g(z) = e^{i\beta}(z + e^{-i\mu})(z + e^{i\mu})^{-1}$, we have

$$
[F(g^{-1}(w))\mathrm{Log}\, g^{-1}(w)]\frac{dz}{dw} = -i\frac{f(w)}{w}\mathrm{Log}\left(-e^{i\mu}\frac{w - e^{i\alpha}}{w - e^{i\beta}} \right)
$$

and then, use (4.11.1.4). Try to show that

$$
\int_0^{\frac{\pi}{2}} e^{\cos x} \cos(\sin x)\,dx = \frac{\pi}{2} + \sum_{n=1}^\infty \frac{(-1)^{n-1}}{(2n-1)!(2n-1)}.
$$

(3) Suppose $f(z)$ satisfies:

(i) f is analytic in $\mathbf{C}$ except finitely many isolated singularities $z_1, \ldots, z_n$ which do not lie in the real interval $[a, \infty)$.

(ii) $\lim_{R\to\infty} \int_{|z|=R} f(z)\mathrm{Log}(a - z)\,dz = 0$, where $\mathrm{Log}(a - z)$ is the principal branch of $\log(a - z)\,dz$ in $\mathbf{C} - [a, \infty)$.

Then

$$
\int_a^\infty f(x)\,dx = -\sum_{k=1}^n \mathrm{Re}\, s(f(z)\mathrm{Log}(a - z); z_k), \quad a > 0. \quad (4.12.4.1.8)
$$

Since $\int_a^\infty f(x)\,dx = \int_0^\infty f(x+a)\,dx$, this follows easily from (4.12.4.1.1). Try to prove this integral directly without recourse to (4.12.4.1.1).

For instance,

$$\int_a^\infty \frac{1}{x^4+1}\,dx = \frac{1}{2\sqrt{2}}\operatorname{Arc\,tan}\frac{\sqrt{2}a}{a^2-1} - \operatorname{Arc\,tanh}\frac{\sqrt{2}a}{a^2+1}, \quad a > 0;$$

$$\int_a^\infty \frac{1}{x^2-2x\cos p+1}\,dx = \frac{1}{\sin p}\operatorname{Arc\,tan}\frac{\sin p}{a-\cos p}, \quad a > 0, 0 < p < \frac{\pi}{2}.$$

4.12.4.2. $\int_0^\infty f(x)(\log x)^p dx$ (*p*, positive integers)

The following integral is due to Copson (see Ref. [20], p. 154).

The evaluation of $\int_0^\infty f(x)\log x\,dx$, *etc.* Suppose f satisfies:

(1) $f(z)$ is analytic in $\mathbf{C}$ except finitely many isolated singularities $z_1, \ldots, z_n$ which all lie in the *open upper* half-plane $\operatorname{Im} z > 0$.
(2) $\lim_{z\to 0} z f(z)\log z = \lim_{z\to\infty} z f(z)\log z = 0$ (the former seems superfluous since f is analytic at 0).

Then

$$\int_0^\infty [f(x)+f(-x)]\log x\,dx + \pi i \int_0^\infty f(-x)\,dx = 2\pi i \sum_{k=1}^n \operatorname{Re} s(f(z)\operatorname{Log} z; z_k),$$

where $\operatorname{Log} z = \log|z| + i\operatorname{Arg} z$, $0 \le \operatorname{Arg} z < 2\pi$, is the chosen principal branch of $\log z$ (see the *Remark* 1 below).   (4.12.4.2.1)

Proof. Choose $r > 0$, $R > 0$ so that $r < |z_k| < R$ for $1 \le k \le n$. Choose the path γ shown in Fig. 4.106, where L_2 is the upper side of the cut $[r, R]$.

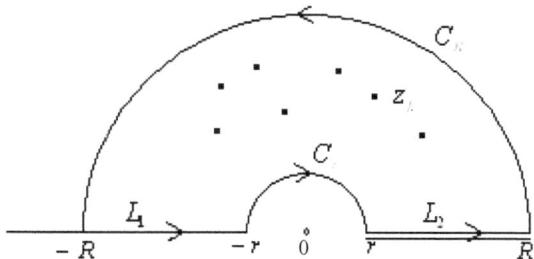

Fig. 4.106

Then

$$\int_{\gamma} f(z)\text{Log } z dz = 2\pi i \sum_{k=1}^{n} \text{Re } s(f(z)\text{Log } z; z_k). \qquad (*_1)$$

Along L_1 and L_2:

$$\int_{L_1} f(z)\text{Log } z dz = \int_{-R}^{-r} f(x)\text{Log } x dx = \int_{r}^{R} f(-x)\text{Log}(-x) dx$$

$$= \int_{r}^{R} f(-x)[\log x + \pi i] dx;$$

$$\int_{L_2} f(z)\text{Log } z dz = \int_{r}^{R} f(x)\text{Log } x dx.$$

Along C_r and C_R, by Condition 2 and (4.12.3.4),

$$\lim_{r \to 0} \int_{C_r} f(z)\text{Log } z dz = 0 \quad \text{and} \quad \lim_{R \to \infty} \int_{C_R} f(z)\text{Log } z dz = 0.$$

Putting these in $(*_1)$ if $r \to 0$ and $R \to 0$, the result follows. $\qquad \square$

Remark 1.
(a) In case f has additional *real simple poles* $x_1, \ldots, x_m$, all not equal to zero, then

$$\text{P.V.} \int_{0}^{\infty} [f(x) + f(-x)] \log x dx + \text{P.V.} \, \pi i \int_{0}^{\infty} f(-x) dx$$

$$= 2\pi i \sum_{k=1}^{n} \text{Re } s(f(z)\text{Log } z; z_k) + \pi i \sum_{l=1}^{m} \text{Re } s(f(z)\text{Log } z; x_l),$$

$$(4.12.4.2.2)$$

where $\text{Log } z = \log|z| + i \text{ Arg } z$, $0 \leq \text{Arg } z < 2\pi$.

If f is also an *odd* function, namely $f(-z) = -f(z)$, then (4.12.4.2.2) reduces to

$$\text{P.V.} \int_{0}^{\infty} f(x) dx = -2 \sum_{k=1}^{n} \text{Re } s(f(z)\text{Log } z; z_k) - \sum_{l=1}^{m} \text{Re } s(f(z)\text{Log } z; x_l).$$

$$(4.12.4.2.3)$$

Note that, in the case of odd functions, z_0 is an isolated singularity of f if and only if *so* is $-z_0$, and $\text{Re } s(f; z_0) = \text{Re } s(f; -z_0)$ holds (see Exercise A (4)(a) of Sec. 4.11.2). For a negative real simple pole x_0 of f, $\text{Re } s(f(z)\text{Log } z; x_0) = \text{Re } s(f(z)\text{Log } z; -x_0) + \pi i(f(z); -x_0)$. Under these circumstance, letting $x_1, \ldots, x_{m'}$, $m = 2m'$, be the *positive* simple poles of f, then (4.12.4.2.3) reduces to

$$\text{P.V.} \int_0^\infty f(x)dx = -2 \sum_{k=1}^n \text{Re } s(f(z)\text{Log } z; z_k)$$

$$- 2 \sum_{l=1}^{m'} \text{Re } s(f(z)\text{Log } z; x_l) - \pi i \sum_{l=1}^{m'} \text{Re } s(f(z); x_l).$$

This is coincident with (4.12.4.1.3).

Furthermore, suppose $f(z)$ assumes *real* value if z is real. Equating the real and imaginary parts of both sides in (4.12.4.2.1), we obtain

$$\int_0^\infty [f(x) + f(-x)] \log x\, dx = -2\pi \text{ Im} \left(\sum_{k=1}^n \text{Re } s(f(z)\text{Log } z; z_k) \right);$$

$$\int_0^\infty f(-x)dx = 2 \text{ Re} \left(\sum_{k=1}^n \text{Re } s(f(z)\text{Log } z; z_k) \right).$$

$$(4.12.4.2.4)$$

In case $f(z)$ is also an *even* function, then

$$\int_0^\infty f(x) \log x\, dx = -\pi \text{ Im} \left(\sum_{k=1}^n \text{Re } s(f(z)\text{Log } z; z_k) \right);$$

$$\int_0^\infty f(x)dx = \int_0^\infty f(-x)dx = 2 \text{ Re} \left(\sum_{k=1}^n \text{Re } s(f(z)\text{Log } z; z_k) \right).$$

$$(4.12.4.2.5)$$

Similar comments are also valid for (4.12.4.2.2) and (4.12.4.2.3).

(b) If Condition 2 in (4.12.4.2.1) is replaced by "$\lim_{z \to 0} z f(-z) \log z = \lim_{z \to \infty} z f(-z) \log z = \infty$", then integrate $f(-z)\text{Log } z$ around a similar contour as in Fig. 4.106 but now L_1 is the upper side of the cut $[-R, -r]$ and L_2 is the segment $[r, R]$. (4.12.4.2.1) is thus adjusted to

$$\int_0^\infty [f(x) + f(-x)] \log x\, dx + \pi i \int_0^\infty f(x)dx$$

$$= 2\pi i \sum_{k=1}^n \text{Re } s(f(-z)\text{Log } z; z_k) \qquad (4.12.4.2.6)$$

where $\text{Log } z = \log |z| + i \text{ Arg } z$, $-\pi < \text{Arg } z \leq \pi$, is now chosen as the principal branch of $\log z$.

Of course, (4.12.4.2.2)–(4.12.4.2.5) have corresponding counterparts in this case.

(c) If integrating $f(-z)(\mathrm{Log}\, z)^p$ along the same path as in (b), then

$$\int_0^\infty f(-x)(\log x)^p dx + \int_0^\infty f(x)[\log x + \pi i]^p dx$$

$$= 2\pi i \sum_{k=1}^n \mathrm{Re}\, s(f(-z)(\mathrm{Log}\, z)^p; z_k) \qquad (4.12.4.2.7)$$

and we have (due to Glasser (1964))

The evaluation of $\int_0^\infty f(x)(\log x)^p dx$ (p, positive integers). Suppose f satisfies:

(1) Condition 1 in (4.12.4.2.1);
(2) $\lim_{z \to 0} zf(-z)(\log z)^p = 0$.

Then ($\mathrm{Log}\, z = \log|z| + i\,\mathrm{Arg}\, z, \;\; -\pi < \mathrm{Arg}\, z \leq \pi$),

$$\int_0^\infty [f(x) + f(-x)](\log x)^p dx + \sum_{q=1}^p C_q^p (\pi i)^q \int_0^\infty f(x)(\log x)^{p-q} dx$$

$$= 2\pi i \sum_{k=1}^n \mathrm{Re}\, s(f(-z)(\mathrm{Log}\, z)^p; z_k).$$

This gives a recursive formula for computing $I_j = \int_0^\infty f(x)(\log x)^{j-1} dx$, $1 \leq j \leq p+1$:

(i) If $f(z)$ is an odd function,

$$I_p = \frac{1}{p}\left[2R_p - \sum_{q=2}^p C_q^p (\pi i)^{q-1} I_{p+1-q}\right];$$

(ii) if $f(z)$ is an even function,

$$I_{p+1} = \frac{\pi i}{2}\left[2R_p - \sum_{q=1}^p C_q^p (\pi i)^{q-1} I_{p+1-q}\right],$$

where

$$R_p = \sum_{k=1}^n \mathrm{Re}\, s(f(-z)(\mathrm{Log}\, z)^p; z_k). \qquad (4.12.4.2.8)$$

If, in addition, $f(z)$ has real simple poles $x_1, \ldots, x_m$, all not equal to zero, then (4.12.4.2.7) turns out to be

$$\text{P.V.} \int_0^\infty f(-x)(\log x)^p dx + \text{P.V.} \int_0^\infty f(x)[\log x + \pi i]^p dx$$

$$= 2\pi i \sum_{k=1}^n \text{Re } s(f(-z)(\text{Log } z)^p; z_k) + \pi i \sum_{l=1}^m \text{Re } s(f(-z)(\text{Log } z)^p; x_l).$$

$$(4.12.4.2.9)$$

This is a generalization of (4.12.4.2.2). □

For instance,

Example 1. Evaluate the integral

$$\int_0^\infty \frac{\log x}{(x^2 + a^2)^2 \sqrt{x}} dx, \quad a \neq 0.$$

Solution. $f(z) = \frac{\log z}{(z^2 + a^2)^2 \sqrt{z}}$ has branches in $\mathbf{C} - [0, \infty)$. We designate the branch determined uniquely by $\sqrt{1} = 1$, namely, $\sqrt{z} = e^{\frac{1}{2} \text{Log } z}$ with $\text{Log } z = \log |z| + i \text{ Arg } z$, $0 \leq \text{Arg } z < 2\pi$. Integrate this $f(z)$ along the path γ shown in Fig. 4.106. Owing to

$$\int_{L_1} f(z) dz = \int_{-R}^{-r} \frac{\log x}{(x^2 + a^2)^2 \sqrt{x}} dx = \int_r^R \frac{\log(-x)}{(x^2 + a^2)^2 \sqrt{-x}} dx$$

$$= \int_r^R \frac{\log x + \pi i}{(x^2 + a^2)^2 e^{\frac{1}{2}(\log x + \pi i)}} dx$$

$$= -i \int_r^R \frac{\log x + \pi i}{(x^2 + a^2)^2 \sqrt{x}} dx \to -i \int_0^\infty \frac{\log x + \pi i}{(x^2 + a^2)^2 \sqrt{x}} dx$$

$$\text{as } r \to 0 \quad \text{and} \quad R \to \infty.$$

$$\Rightarrow -i \int_0^\infty \frac{\log x + \pi i}{(x^2 + a^2)^2 \sqrt{x}} dx + \int_0^\infty \frac{\log x}{(x^2 + a^2)^2 \sqrt{x}} dx = 2\pi i \text{ Re } s(f(z); ai)$$

$$= 2\pi i \cdot \frac{-1}{8a^3 \sqrt{a}} \left(2 - 3 \log a - \frac{3\pi i}{2} \right) e^{-\frac{3\pi i}{4}}, \quad \text{if } a > 0.$$

$$\Rightarrow \int_0^\infty \frac{\log x}{(x^2 + a^2)^2 \sqrt{x}} dx = \frac{\pi}{2|a|^3 \sqrt{2|a|}} \left(\frac{3}{2} \log |a| - 1 - \frac{3\pi}{4} \right), \quad a \neq 0.$$

What is $\int_0^\infty \frac{1}{(x^2 + a^2) \sqrt{x}} dx$?

Example 2. Evaluate the integral

$$I_p = \int_0^\infty \frac{(\log x)^{p-1}}{x^2 + a^2} dx, \quad p = 1, 2, 3, \ldots, \quad \text{and} \quad a > 0.$$

Solution. Let $f(z) = \frac{1}{z^2 + a^2}$. Then

$$\operatorname{Re} s(f(-z)\operatorname{Log} z; ai) = \frac{\operatorname{Log} ai}{2ai} = -i\frac{\log a + \frac{\pi}{2}i}{2a}.$$

By (4.12.4.2.5),

$$I_1 = \int_0^\infty f(x)dx = 2\operatorname{Re}(\operatorname{Re} s(f(-z)\operatorname{Log} z; ai)) = 2\operatorname{Re}\frac{\frac{\pi}{2} - i\log a}{2a} = \frac{\pi}{2a};$$

$$I_2 = \int_0^\infty f(x)\log x dx = -\pi\operatorname{Im}(\operatorname{Re} s(f(-z)\operatorname{Log} z; ai)) = \frac{\pi}{2a}\log a.$$

According to (ii) in (4.12.4.2.8),

$$I_3 = \frac{\pi i}{2}[2R_2 - C_1^2 I_2 - C_2^2(\pi i)I_1].$$

Since

$$R_2 = \operatorname{Re} s(f(-z)(\operatorname{Log} z)^2; ai) = \frac{(\operatorname{Log} ai)^2}{2ai} = \frac{-i}{2a}\left[(\log a)^2 + \pi i\log a - \frac{\pi^2}{4}\right]$$

$$\Rightarrow I_3 = \frac{\pi i}{2}\left[-\frac{i}{a}(\log a)^2 + \frac{\pi^2}{4a}i + \frac{\pi}{a}\log a - \frac{\pi}{a}\log a - \frac{\pi}{2a}\cdot\pi i\right]$$

$$= \frac{\pi}{2}\left[\frac{1}{a}(\log a)^2 + \frac{\pi^2}{4a}\right].$$

The method introduced in Fig. 4.106 and the resulted formula (4.12.4.2.1) are especially beneficial to odd and even functions, but fail in the evaluation of integrals such as $\int_0^\infty \frac{\log x}{(x-1)(x+2)}dx$. An alternative method to evaluate integrals $\int_0^\infty f(x)(\log x)^m dx$ is the one we suggested near the end of Example 4 of Sec. 4.12.3, using the contour shown in Fig. 4.79, and Exercise A (7) there.

We formally state as

The evaluation of $\int_0^\infty f(x)\log x dx$ via Fig. 4.79 *(without positive simple poles).* Suppose f satisfies:

(1) $f(z)$ is analytic in $\mathbf{C}^*$ except finitely many isolated singularities $z_1, \ldots, z_n$ which do *not* lie in the nonnegative real axis $[0, \infty)$.

(2) $f(z)$ assumes *real* values on the *real* axis.

(3) $z = \infty$ is a zero of order at least 2 of $f(z)$, namely, $f(z) = O(\frac{1}{|z|^p})$, $p \geq 2$, as $z \to \infty$.

Then

$$\int_0^\infty f(x) \log x dx = -\frac{1}{2} \operatorname{Re} \left\{ \sum_{k=1}^n \operatorname{Re} s(f(z)(\operatorname{Log} z)^2; z_k) \right\},$$

$$\int_0^\infty f(x) dx = -\frac{1}{2\pi} \operatorname{Im} \left\{ \sum_{k=1}^n \operatorname{Re} s(f(z)(\operatorname{Log} z)^2; z_k) \right\},$$

where $\operatorname{Log} z = \log |z| + i \operatorname{Arg} z$, $0 \leq \operatorname{Arg} z < 2\pi$, is the chosen principal branch of $\log z$ (see *Remark 2* below).   (4.12.4.2.10)

Sketch of Proof

Along the upper side $[r, R]$ and the lower side $[r, R]^-$ of the cut in Fig. 4.103:

$$\int_r^R f(z)(\operatorname{Log} z)^2 dz = \int_r^R f(x)(\operatorname{Log} x)^2 dx,$$

$$\int_R^r f(z)(\operatorname{Log} z)^2 dz = -\int_r^R f(x)(\log x + 2\pi i)^2 dx.$$

Along $C_r : z = re^{i\theta}$, $0 \leq \theta \leq 2\pi$. Since f is analytic at 0, as $r \to 0$,

$$\left| \int_{C_r} f(z)(\operatorname{Log} z)^2 dz \right| = \left| \int_{2\pi}^0 f(re^{i\theta})(\log r + i\theta)^2 rie^{i\theta} d\theta \right|$$

$$\leq \int_0^{2\pi} |f(re^{i\theta})| r(\log r + \theta)^2 d\theta \to 0.$$

Along $C_R : z = Re^{i\theta}$, $0 \leq \theta \leq 2\pi$. Letting $|f(z)| \leq \frac{M}{|z|^p}$ ($M > 0$ is a constant), then as $R \to \infty$,

$$\left| \int_{C_R} f(z)(\operatorname{Log} z)^2 dz \right| = \left| \int_0^{2\pi} f(Re^{i\theta})(\log R + i\theta)^2 Rie^{i\theta} d\theta \right|$$

$$\leq \frac{M}{R^p} 4(\log R)^2 \cdot 2\pi R \to 0. \qquad \square$$

Remark 2.

(a) If f has a positive simple pole at x_1, then integrate $f(z) \operatorname{Log} z$ along the path γ shown in Fig. 4.104. Just like $(*_3)$ and $(*_4)$ in the proof of

(4.12.4.1.3), we have

$$\lim_{r \to 0} \int_{C_{1r}} f(z)(\text{Log } z)^2 dz = (0 - \pi)i \, \text{Re } s(f(z)(\text{Log } z)^2; x_1)$$

$$= -\pi i \, \text{Re } s(f(z)(\text{Log } z)^2; x_1);$$

$$\lim_{r \to 0} \int_{C_{1r}^*} f(z)(\text{Log } z)^2 dz = -(2\pi - \pi)i \, \text{Re } s(f(z)(\text{Log } z)^2; x_1 e^{2\pi i}),$$

$$= -\pi i \, \text{Re } s(f(z)(\text{Log } z + 2\pi i)^2; x_1).$$

Letting $r \to 0$ and $R \to \infty$, then

$$\text{P.V.} \int_0^\infty f(x)(\text{Log } x)^2 dx - \text{P.V.} \int_0^\infty f(x)(\text{Log } x + 2\pi i)^2 dx$$

$$-\pi i[\text{Res}(f(z)(\text{Log } z)^2; x_1] - \pi i[\text{Re } s(f(z)(\text{Log } z + 2\pi i)^2; x_1]$$

$$= 2\pi i \sum_{k=1}^n \text{Re } s(f(z)(\text{Log } z)^2; z_k)$$

$$\Rightarrow -4\pi i \, \text{P.V.} \int_0^\infty f(x) \text{Log } x dx + 4\pi^2 \text{P.V.} \int_0^\infty f(x) dx,$$

$$= 2\pi i \sum_{k=1}^n \text{Re } s(f(z)(\text{Log } z)^2; z_k) + \pi i \, \text{Re } s(f(z)[(\text{Log } z)^2$$

$$+ (\text{Log } z + 2\pi i)^2]; x_1).$$

Evaluating the real and imaginary parts of both sides, we get the desired formulas.

Summarize the above as

The evaluation of $\int_0^\infty f(x) \log x dx$ (with positive simple poles). In addition to Conditions 1–3 in (4.12.4.2.10), suppose $f(z)$ has positive simple poles $x_1, \ldots, x_m$. Then

$$2 \, \text{P.V.} \int_0^\infty f(x) \log x dx + 2\pi i \, \text{P.V.} \int_0^\infty f(x) dx,$$

$$= -\sum_{k=1}^n \text{Re } s(f(z)(\text{Log } z)^2; z_k)$$

$$-\frac{1}{2} \sum_{l=1}^m \text{Re } s(f(z)[(\text{Log } z)^2 + (\text{Log } z + 2\pi i)^2]; x_l),$$

where $\text{Log } z = \log |z| + i \text{Arg } z, 0 \le \text{Arg } z < 2\pi.$ (4.12.4.2.11)

(b) As a general setting, (4.12.4.2.10) can be extended as follows. Let

$$I_j = \int_0^\infty f(x)(\log x)^{j-1} dx, \quad j = 1, 2, 3, \ldots . \qquad (4.12.4.2.12)$$

Then, for $p = 2, 3, \ldots$,

$$\sum_{j=0}^{p-1} C_j^p (2\pi i)^{p-1-j} I_{j+1} = -\sum_{k=1}^n \operatorname{Re} s(f(z)(\operatorname{Log} z)^p; z_k). \qquad (4.12.4.2.13)$$

While corresponding to (4.12.4.2.11), $p = 2, 3, \ldots$,

$$\sum_{j=0}^{p-1} C_j^p (2\pi i)^{p-1-j} \, \text{P.V.} \, I_{j+1} = -\sum_{k=1}^n \operatorname{Re} s(f(z)(\operatorname{Log} z)^p; z_k)$$

$$-\frac{1}{2} \sum_{l=1}^m \operatorname{Re} s(f(z)[(\operatorname{Log} z)^p + (\operatorname{Log} z + 2\pi i)^p]; x_l). \qquad (4.12.4.2.14)$$

$\square$

Proofs are left to the readers.
    For instance,

Example 3. Evaluate the integrals:

(1) P.V. $\displaystyle\int_0^\infty \frac{\log x}{(x-1)(x+a)} dx \quad (a > 0)$;

(2) P.V. $\displaystyle\int_0^\infty \frac{\log x}{x^2 - 1} dx$;

(3) $\displaystyle\int_0^\infty \log\left(\frac{e^x + 1}{e^x - 1}\right) dx$.

Since $\lim_{x \to 1} \frac{\log x}{x-1} = 1$, the sign P.V. can be deleted from the above two integrals.

Solution.

(1) Let $f(z) = \frac{1}{(z-1)(z+a)}$. Then

$$\operatorname{Re} s(f(z)(\operatorname{Log} z)^2; -a) = \frac{(\operatorname{Log}(-a))^2}{-a-1} = -\frac{(\log a)^2 - \pi^2 + 2\pi i \log a}{1+a};$$

$$\operatorname{Res}(f(z)[(\log z)^2 + (\operatorname{Log} z + 2\pi i)^2]; 1) = \frac{(\operatorname{Log} 1)^2 + (\operatorname{Log} 1 + 2\pi i)^2}{1+a}$$

$$= -\frac{4\pi^2}{1+a}.$$

According to (4.12.4.2.11),

$$2\,\text{P.V.} \int_0^\infty f(x)\log x\,dx + 2\pi i\,\text{P.V.} \int_0^\infty f(x)\,dx$$

$$= \frac{(\log a)^2 - \pi^2 + 2\pi i\log a}{1+a} + \frac{1}{2}\cdot\frac{4\pi^2}{1+a}$$

$$\Rightarrow \text{P.V.} \int_0^\infty \frac{\log x}{(x-1)(x+a)}\,dx = \frac{(\log a)^2 + \pi^2}{2(1+a)};$$

$$\text{P.V.} \int_0^\infty \frac{1}{(x-1)(x+a)}\,dx = \frac{\log a}{1+a}.$$

(2) By direct computation suggested by (4.12.4.2.11) or just letting $a \to 1$ in (1), we have

$$\text{P.V.} \int_0^\infty \frac{\log x}{x^2-1}\,dx = \frac{\pi^2}{4}.$$

Another way to evaluate this integral (including the one in (1)) is to integrate $f(z) = \frac{\text{Log}\,z}{z-1}$ along the dented path shown in Fig. 4.107. Now

$$\int_{C_{0r}} f(z)\,dz + \int_r^{1-r} f(x)\,dx + \int_{C_{1r}} f(z)\,dz$$

$$+ \int_{1+r}^R f(x)\,dx + \int_{C_R} f(z)\,dz + \int_{Ri}^{ri} f(z)\,dz = 0. \qquad (*_2)$$

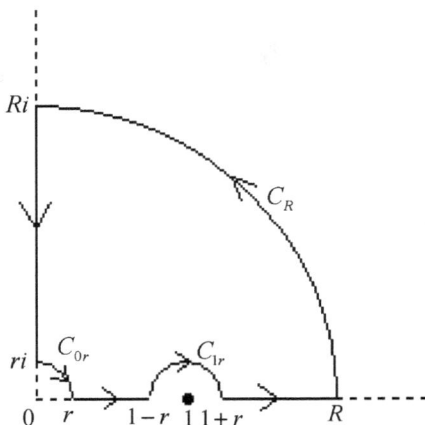

Fig. 4.107

Along the circular arcs: as $r \to 0$ and $R \to \infty$,

$$\int_{C_{or}} f(z)dz \to 0, \qquad \int_{C_R} f(z)dz \to 0, \quad \text{and}$$

$$\int_{C_{1r}} f(z)dz \to (0 - \pi)i \operatorname{Re} s(f(z); 1) = -\pi i \frac{\operatorname{Log} 1}{2} = 0.$$

While along the segment $[ir, iR]$: as $r \to 0$ and $R \to \infty$,

$$\int_{iR}^{ir} f(z)dz = \int_{R}^{r} \frac{\operatorname{Log}(iy)}{-y^2 - 1}(idy) = i \int_{r}^{R} \frac{\log y + \frac{\pi}{2}i}{y^2 + 1} dy$$

$$= i \int_{r}^{R} \frac{\log y}{y^2 + 1} dy - \frac{\pi}{2} \int_{r}^{R} \frac{1}{y^2 + 1} dy,$$

$$\to i \int_{0}^{\infty} \frac{\log y}{y^2 + 1} dy - \frac{\pi}{2} \int_{0}^{\infty} \frac{1}{y^2 + 1} dy$$

$$= i \cdot 0 - \frac{\pi}{2} \cdot \frac{\pi}{2} = -\frac{\pi^2}{4},$$

where we have used the fact that $\int_{0}^{\infty} \frac{\log y}{y^2+1} dy = 0$ (see Example 2 or just by direct computation via change of variables $y \to \frac{1}{y}$). Finally, putting these results into $(*_2)$ in the processes of letting $r \to 0$ and $R \to \infty$, the result follows.

(3) Perform the change of variables $t = \frac{e^x + 1}{e^x - 1}$. Note that $e^x = \frac{t+1}{t-1}$ and $dx = \frac{-2}{(t-1)(t+1)} dt$. Hence

$$\int_{0}^{\infty} \log\left(\frac{e^x + 1}{e^x - 1}\right) dx = 2 \int_{1}^{\infty} \frac{\log t}{(t-1)(t+1)} dt = \int_{0}^{\infty} \frac{\log t}{t^2 - 1} = \frac{\pi^2}{4}.$$

Exercises A

(1) Prove (4.12.4.2.8) and (4.12.4.2.9) in detail.

(2) Prove (4.12.4.2.13) and (4.12.4.2.14) in detail.

(3) By Example 2 and by direct computation, respectively, show that

$$\int_{0}^{\infty} \frac{(\log x)^{2k-1}}{x^2 + 1} dx = 0, \quad k = 1, 2, 3, \ldots.$$

(4) Evaluate the following integrals by using (4.12.4.2.8) and (4.12.4.2.13), respectively, if possible.

(a) $\displaystyle\int_{0}^{\infty} \frac{\log x}{x^2 + 2x + 2} dx.$

(b) $\displaystyle\int_{0}^{\infty} \frac{\sqrt{x} \log x}{x^2 + 1} dx.$

(c) $\displaystyle\int_0^\infty \frac{\log x}{(x^2+1)^2\sqrt{x}}\,dx.$

(d) $\displaystyle\int_0^\infty \frac{(1-x^2)\log x}{(1+x^2)^2}\,dx.$

(e) $\displaystyle\int_0^\infty \frac{\log x}{(x^2+a^2)(1+b^2x^2)}\,dx \quad (a>0, b>0).$

(f) $\displaystyle\int_0^\infty \frac{x^2\log x}{(a^2+b^2x^2)(1+x^2)}\,dx \quad (ab>0).$

(g) $\displaystyle\int_0^\infty \frac{(\log x)^p}{(x^2+a^2)^2}\,dx \quad (a>0;$ try to use Example 2, too), $p = 0, 1, 2, 3.$

(h) $\displaystyle\int_0^\infty \frac{(\log x)^p}{x^2+x+1}\,dx, \quad p = 0, 1, 2, 3.$

(i) $\displaystyle\int_0^\infty \frac{(1+x^2)(\log x)^p}{1+x^4}\,dx, \quad p = 0, 1, 2, 3.$

(j) $\displaystyle\int_0^\infty \frac{2\cos x\log x - \pi\sin x}{x^2+4}\,dx.$

(k) $\displaystyle\int_0^\infty \frac{x\log x}{(1+x^2)^3}\,dx.$

(l) $\displaystyle\int_0^\infty \frac{\log x}{(1+x)^n}\,dx, \quad n \geq 2.$

(m) $\displaystyle\int_0^\infty \frac{\sqrt{x}\log x}{(1+x)^2}\,dx.$

(5) Evaluate the following integrals by using (4.12.4.2.9) and (4.12.4.2.14), respectively, if possible.

(a) $\displaystyle\int_0^\infty \frac{(\log x)^p}{(x-1)(x+a)}\,dx \quad (a>0;$ what happens if $a<0$), $p = 2, 3.$

(b) $\displaystyle\int_0^\infty \frac{(x+1)(\log x)^p}{(x-2)(x^2+4)}\,dx, \quad p = 0, 1, 2, 3.$

(6) Integrate $(1+z^4)^{-1}\operatorname{Log} z$ along the path shown in Fig. 4.108 to show that

$$\int_0^\infty \frac{\log x}{1+x^4}\,dx = -\frac{\pi^2}{8\sqrt{2}}; \qquad \int_0^\infty \frac{x^2\log x}{1+x^4}\,dx = \frac{\pi^2}{8\sqrt{2}}.$$

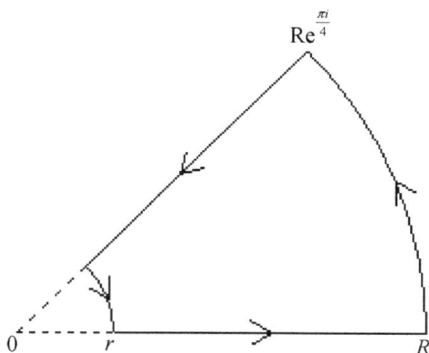

$\mathrm{Re}^{\frac{\pi i}{4}}$

Fig. 4.108

(7) Integrate $(1+z^2)^{-1} \operatorname{Log}(z+i)$ along the upper half of the circle $|z| = R$ (including the diameter) to show that

$$\int_0^\infty \frac{\log(1 + x^2)}{1 + x^2} dx = \pi \log 2.$$

Then, deduce that $\int_0^{\frac{\pi}{2}} \log \sin \theta d\theta = \int_0^\infty \log \cos \theta = -\frac{\pi}{2} \log 2$ and

$$\int_0^1 \log\left(x + \frac{1}{x}\right) \frac{dx}{1 + x^2} = \frac{\pi}{2} \log 2.$$

(8) Integrate $\left(\log \frac{1-z}{z}\right)^2 \frac{1}{1+z}$ along the path shown in Fig. 4.69 to evaluate

$$\int_0^1 \log \frac{1 - x}{x} \cdot \frac{dx}{1 + x}.$$

Exercises B

(1) By integrating $F(z) = \frac{1}{(z^2+a^2)(\operatorname{Log} z - \pi i)}$ along the path γ shown in Fig. 4.103, show that

$$\int_0^\infty \frac{1}{(x^2 + a^2)((\log x)^2 + \pi^2)} dx = \frac{\pi}{2a} \cdot \frac{1}{(\operatorname{Log} a)^2 + \frac{\pi^2}{4}} - \frac{1}{1 + a^2}, \quad a > 0.$$

Recall that $\operatorname{Log} z = \log |z| + i \operatorname{Arg} z$, $0 \le \operatorname{Arg} z < 2\pi$. Based on this experience, try to show that, for $a > 0$ and $n = 0, 1, 2, 3 \ldots$,

$$\int_0^\infty \frac{dx}{(x^2 + a^2)((\log x)^2 + (2n + 1)^2 \pi^2)}$$

$$= \frac{1}{2n + 1} \left\{ \frac{\pi}{2a} \sum_{k=0}^{2n} (-1)^k \frac{2k + 1}{(\log a)^2 + \left(k + \frac{1}{2}\right)^2 \pi^2} - \frac{1}{1 + a^2} \right\};$$

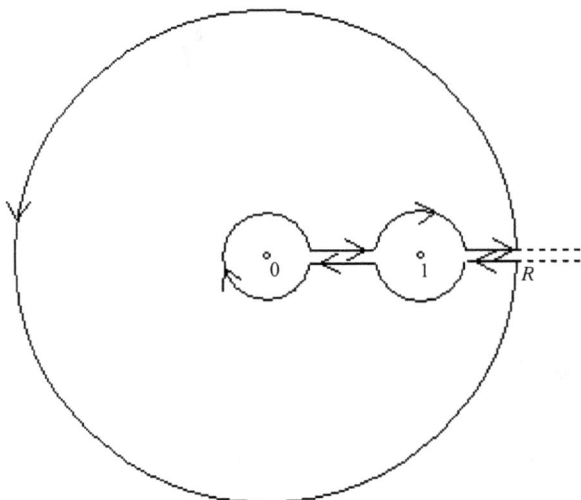

Fig. 4.109

$$\int_0^\infty \frac{dx}{(x^2 + a^2)((\log x)^2 + 4n^2\pi^2)}$$

$$= \frac{1}{2n} \left\{ \frac{\pi}{2a} \sum_{k=0}^{2n-1} (-1)^{k+1} \frac{2k+1}{(\log a)^2 + \left(k + \frac{1}{2}\right)^2 \pi^2} + \frac{1}{1 + a^2} \right\}.$$

When evaluating, adopt Fig. 4.103 for the first integral; while for the second integral, adopt the path γ shown in Fig. 4.109 but say the reason why.

As a general setting (due to Gheorghiu (1957)): Suppose f satisfies:

(i) f is analytic in **C** except finitely many isolated singularities which do not lie on the nonnegative real axis $[0, \infty)$.

(ii) $f(z) = O\left(\frac{1}{|z|}\right)$ as $z \to \infty$.

Then, for $n = 0, 1, 2, \ldots$,

$$\int_0^\infty \frac{f(x)}{(\log x + 2n\pi i)^2 + \pi^2} dx = \sum_{k=1}^n \operatorname{Re} s \left(\frac{f(z)}{\operatorname{Log} z + (2n-1)\pi i}; z_k \right),$$

where $\operatorname{Log} z = \log |z| + i \operatorname{Arg} z$, $0 \le \operatorname{Arg} z < 2\pi$, and $z_1, \ldots, z_n$ are the total of the isolated singularities of

$$\frac{f(z)}{\operatorname{Log} z + (2\pi - 1)\pi i}. \tag{4.12.4.2.15}$$

What happens if $\frac{f(z)}{\operatorname{Log} z + (2\pi - 1)\pi i}$ has simple poles along $[0, \infty)$?

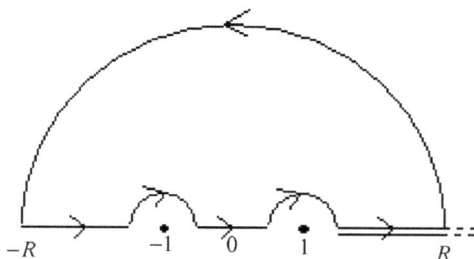

Fig. 4.110

(2) Integrate $F(z) = \frac{z \operatorname{Log}(z-1)}{(z+1)(z^2+1)}$ along the path γ shown in Fig. 4.110 to show that

$$\text{P.V.} \int_{-\infty}^{\infty} \frac{x \log|x-1|}{(x+1)(x^2+1)} dx + \pi i \,\text{P.V.} \int_{-\infty}^{1} \frac{x dx}{(x+1)(x^2+1)}$$

$$= 2\pi i \operatorname{Re} s(F; i) + \pi i \operatorname{Re} s(F; -1) = 2\pi i \cdot \frac{(1-i)\left(\frac{1}{2}\log 2 + \frac{3\pi}{4} i\right)}{4}$$

$$+ \pi i \cdot \frac{-\log 2 - \pi i}{2},$$

$$\Rightarrow \text{P.V.} \int_{-\infty}^{\infty} \frac{x \log|x-1|}{(x+1)(x^2+1)} dx = \frac{2\pi \log 2 + \pi^2}{8};$$

$$\text{P.V.} \int_{-\infty}^{1} \frac{x dx}{(x+1)(x^2+1)} = \frac{3\pi - 2\log 2}{8}.$$

In general, suppose f satisfies:

(i) Suppose f is analytic in the closed upper half-plane $\operatorname{Im} z \geq 0$ except finitely many isolated singularities $z_1, \ldots, z_n$ in the *open* half-plane $\operatorname{Im} z > 0$ and finitely many *real* simple poles $x_1, \ldots, x_m$, where $x_l \neq a$ (a *fixed* real number) for $1 \leq l \leq m$.

(ii) $\lim_{r \to \infty} \int_{C_r} f(z) \operatorname{Log}(z-a) dz = \lim_{R \to \infty} \int_{C_R} f(z) \operatorname{Log}(z-a) dz = 0$, where, $C_r : z = a + re^{i\theta}$, $0 \leq \theta \leq \pi$ and $C_R : z = Re^{i\theta}$, $0 \leq \theta \leq \pi$.

Then $(\operatorname{Log}(z-a) = \log|z-a| + i \operatorname{Arg}(z-a)$ with $0 \leq \operatorname{Arg}(z-a) < 2\pi)$

$$\text{P.V.} \int_{-\infty}^{\infty} f(x) \log|x-a| dx + \pi i \,\text{P.V.} \int_{-\infty}^{a} f(x) dx$$

$$= 2\pi i \sum_{k=1}^{n} \operatorname{Re} s(f(z) \operatorname{Log}(z-a); z_k)$$

$$+ \pi i \sum_{l=1}^{m} \operatorname{Re} s(f(z) \operatorname{Log}(z-a); x_l). \qquad (4.12.4.2.16)$$

This is the case if $f(z) = \frac{p(z)}{q(z)}$, where $p(z)$ and $q(z)$ are polynomials which do not have common zeros, and the degree of $q(z)$ is at least two larger than that of $p(z)$. Some remarks are as follows.

(a) Note that the symbol P.V. is not needed before the integrals if $f(z)$ does not have real simple poles. In this case, there is no second summation in the right.

(b) By equating the real and imaginary parts of both sides, one can evaluate

$$\text{P.V.} \int_{-\infty}^{\infty} f(x) \log |x - a| dx \quad \text{and} \quad \text{P.V.} \int_{-\infty}^{a} f(x) dx$$

if $f(x)$ assumes real values for real variable x.

(c) In case $f(z)$ is an even function and $f(z)$ is analytic at 0, then one can evaluate

$$\text{P.V.} \int_{-\infty}^{\infty} f(x) \log |x - a| dx = \text{P.V.} \int_{0}^{\infty} f(x) \log |x^2 - a^2| dx.$$

For instance,

$$\int_{-\infty}^{\infty} \frac{\log |x - 2|}{(x^2 + 4)(x^2 + 9)} dx = \int_{0}^{\infty} \frac{\log |x^2 - 4|}{(x^2 + 4)(x^2 + 9)} dx$$

$$= \frac{\pi}{20} \log 8 - \frac{\pi}{30} \log 13.$$

(d) Let $a > 0$. Then $\log(z^2 - a^2)$ has branches in $\mathbf{C} - [-a, \infty)$. Designate $\text{Log}(z^2 - a^2) = \log |z^2 - a^2| + i\text{Arg}(z^2 - a^2)$ where both $\text{Arg}(z+a)$ and $\text{Arg}(z-a)$ lie between 0 and 2π. Replace a by $-a$ in (4.12.4.2.16) and add the resulted formula to (4.12.4.2.16). Then, we obtain

$$\text{P.V.} \int_{-\infty}^{\infty} f(x) \log |x^2 - a^2| dx$$

$$+ \pi i \, \text{P.V.} \int_{-\infty}^{-a} f(x) dx + \pi i \, \text{P.V.} \int_{-\infty}^{a} f(x) dx$$

$$= 2\pi i \sum_{k=1}^{n} \text{Re } s(f(z)\text{Log}(z^2 - a^2); z_k)$$

$$+ \pi i \sum_{l=1}^{m} \text{Re } s(f(z)\text{Log}(z^2 - a^2); x_l), \qquad (4.12.4.2.17)$$

where $x_l \neq \pm a$ for $1 \leq l \leq m$. In case $f(x)$ is real if x is real, then P.V. $\int_{-\infty}^{\infty} f(x) \log |x^2 - a^2| dx$ = the real part of the right side of (4.12.4.2.17).   (4.12.4.2.18)

Of course, there are comments similar to (a)–(c) in this case.

(3) Integrate $F(z) = \frac{z}{z^4-1} \mathrm{Log}(i+z)$, where $f(z) = \frac{z}{z^4-1}$, along the path γ shown in Fig. 4.111 with branch cut $[-i\infty, -i]$ and use the identity $\mathrm{Arc\,tan}\, x = \frac{1}{2i}[\mathrm{Log}(i-x) - \mathrm{Log}(i+x)]$, $x \in \mathbf{R}$, to show that $\left(\mathrm{Log}(i+z) = \log|i+z| + i\,\mathrm{Arg}(i+z),\ -\frac{\pi}{2} \leq \mathrm{Arg}(i+z) < \frac{3\pi}{2}\right)$

$$\text{P.V.} \int_{-\infty}^{\infty} f(x)\mathrm{Arc\,tan}\, x dx = i\ \text{P.V.} \int_{-\infty}^{\infty} f(x)\mathrm{Log}(i+x)dx$$

$$= -2\pi \mathrm{Re}\, s(f(z)\mathrm{Log}(i+z); i) - \pi[\mathrm{Re}\, s(f(z)\mathrm{Log}(i+z); -1)$$

$$+ \mathrm{Re}\, s(f(z)\mathrm{Log}(i+z); 1)]$$

$$= -2\pi \cdot \frac{-\log 2 - \frac{\pi}{2} i}{4} - \pi \left(\frac{\frac{1}{2}\log 2 + \frac{\pi}{4} i}{4} + \frac{\frac{1}{2}\log 2 + \frac{3\pi}{4} i}{4} \right)$$

$$= \frac{\pi \log 2}{4},$$

$$\Rightarrow \text{P.V.} \int_0^{\infty} \frac{x}{x^4-1} \mathrm{Arc\,tan}\, x dx$$

$$= \frac{1}{2}\text{P.V.} \int_{-\infty}^{\infty} \frac{x}{x^4-1} \mathrm{Arc\,tan}\, x dx = \frac{\pi \log 2}{8}.$$

Alternatively, integrate $G(z) = \frac{z}{z^4-1} \mathrm{Log}(1-iz)$ along the same path γ as in Fig. 4.111 but now with branch cut $[-i-i\infty, -i]$, a ray parallel

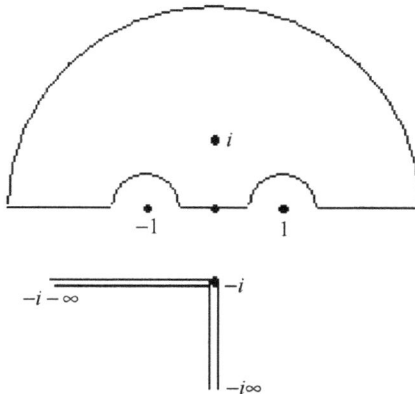

Fig. 4.111

to the nonpositive real axis. Observe that, in the case, $\text{Log}(1 - ix) = \log(1 + x^2)^{\frac{1}{2}} - i \operatorname{Arc}\tan x$ for $x \in \mathbf{R}$. Show that

$$\text{P.V.} \int_{-\infty}^{\infty} \frac{x}{x^4 - 1} \left[\log(1 + x)^{\frac{1}{2}} - i \operatorname{Arc}\tan x\right] dx,$$

$$= 2\pi i \operatorname{Re} s(f(z)\text{Log}(1 - iz); i) + \pi i [\operatorname{Re} s(f(z)\text{Log}(1 - iz); -1)$$

$$+ \operatorname{Re} s(f(z)\text{Log}(1 - iz); 1)]$$

$$= 2\pi i \cdot \frac{-\log 2}{4} + \pi i \left(\frac{\frac{1}{2}\log 2 + \frac{3}{4}\pi i}{4} + \frac{\frac{1}{2}\log 2 + \frac{1}{4}\pi i}{4}\right)$$

$$= -\frac{\pi^2}{4} - \frac{\pi \log 2}{4} i$$

$$\Rightarrow \text{P.V.} \int_{-\infty}^{\infty} \frac{x}{x^4 - 1} \log(x^2 + 1)^{\frac{1}{2}} dx = -\frac{\pi^2}{4},$$

$$\text{P.V.} \int_{-\infty}^{\infty} \frac{x}{x^4 - 1} \operatorname{Arc}\tan x\, dx = \frac{\pi \log 2}{4}.$$

In general, suppose f satisfies:

(1) As Condition 1 in (4.12.4.2.16), disregarding $x_l \neq a$ for $1 \leq l \leq m$.
(2) $f(z) = O(\frac{1}{z \log |z|})$ as $|z| \to \infty$.
(3) $f(x)$ is real if x is real.

Then $(\text{Log}(i + z) = \log|i + z| + i\text{Arg}(i + z)$ with $-\pi < \text{Arg}(1 - iz) \leq \pi)$,

$$\text{P.V.} \int_{-\infty}^{\infty} f(x)\operatorname{Arc}\tan x\, dx = -2\pi \sum_{k=1}^{n} \operatorname{Re} s(f(z)\text{Log}(i + z); z_k)$$

$$-\pi \sum_{l=1}^{m} \operatorname{Re} s(f(z)\text{Log}(i + z); x_l),$$

or $(\text{Log}(1 - iz) = \log|1 - iz| + i\text{Arg}(1 - iz)$ with $-\pi < \text{Arg}(1 - iz) \leq \pi)$,

$$\text{P.V.} \int_{0}^{\infty} f(x)\operatorname{Arc}\tan x\, dx = -2\pi \sum_{k=1}^{n} \operatorname{Res}(f(z)\text{Log}(1 - iz); z_k)$$

$$-\pi \sum_{l=1}^{m} \operatorname{Re} s(f(z)\text{Log}(1 - iz); x_l).$$

$$(4.12.4.2.19)$$

If, in addition, $f(x)$ is odd, namely $f(-x) = -f(x)$, then

$$\text{P.V.} \int_0^\infty f(x)\text{Arc}\tan x dx = \frac{1}{2}\text{P.V.} \int_{-\infty}^\infty f(x)\text{Arc}\tan x dx.$$

As usual, the symbol P.V. is not needed before the integral if f does not have real simple poles, and then, the second summation in the right of (4.12.4.2.19) should be deleted. Try to evaluate the following integrals:

(a) $\displaystyle\int_{-\infty}^\infty \frac{x}{x^4 + 1}\text{Arc}\tan x dx.$

(b) $\displaystyle\int_0^\infty \frac{x}{x^4 + x^2 + 1}\text{Arc}\tan x dx.$

(c) $\displaystyle\int_0^\infty \frac{x^n}{x^6 + 1}\text{Arc}\tan x dx \quad (n = 1, 3).$

(d) $\displaystyle\int_0^\infty \frac{1}{x(1 + x^2)}\text{Arc}\tan p x dx \quad (p > 0).$

(e) $\displaystyle\int_{-\infty}^\infty \frac{x}{(x^2 + 4x + 20)(x^2 + 1)}\text{Arc}\tan x dx.$

(4) Integrate $F(z) = f(z)\text{Log}(z + ai)$ along the upper half-circle (including the diameter) with the branch cut $[-\infty i, -ai]$, $a > 0$, to show the following *general* result. Suppose f satisfies:

(i) $f(z)$ is analytic in the closed upper half-plane $\text{Im } z \geq 0$ except finitely many isolated singularities $z_1, \ldots, z_n$ in the *open* upper half-plane $\text{Im } z > 0$.

(ii) $\lim_{R\to\infty} \int_{C_R} f(z)\text{Log}(z + ai)dz = 0$, where $C_R : z = \text{Re}^{i\theta}, 0 \leq \theta \leq \pi$.

Then, for $a > 0$,

$$\int_{-\infty}^\infty f(x)\log(x^2 + a^2)^{\frac{1}{2}} dx + i\int_{-\infty}^\infty f(x)\text{Arg}(x + ai)dx$$

$$= 2\pi i \sum_{k=1}^n \text{Re } s(f(z)\text{Log}(z + ai); z_k),$$

where (see Remark in Sec. 1.2)

$$\text{Arg}(x + ai) = \begin{cases} \text{Arc}\tan \frac{a}{x}, & x \geq 0 \\ \text{Arc}\tan \frac{a}{x} + \pi, & x < 0 \end{cases}.$$

In case $f(x)$ assumes real values if x is real, then

$$\int_{-\infty}^{\infty} f(x)\log(x^2 + a^2)dx = 4\pi \operatorname{Re} i \sum_{k=1}^{n} \operatorname{Re} s(f(z)\operatorname{Log}(z + ai); z_k);$$

$$\int_{-\infty}^{\infty} f(x)\operatorname{Arc tan}\frac{a}{x}dx$$

$$= 2\pi \operatorname{Im} i \left\{ \sum_{k=1}^{n} \operatorname{Re} s(f(z)\operatorname{Log}(z + ai); z_k) - \sum_{k=1}^{n} \operatorname{Re} s(f(z)\operatorname{Log} z; z_k) \right\},$$

$$(4.12.4.2.20)$$

where, in the second formula, $\int_{-\infty}^{0} f(x)dx = 2\operatorname{Im} i \sum_{k=1}^{n} \operatorname{Re} s(f(z); z_k)$ has been used (see (4.12.4.2.16)). What happens if $f(z)$ has finitely many real simple poles $x_1, \ldots, x_m$? Try to evaluate:

$$\int_{-\infty}^{\infty} \frac{x^2}{x^4 + 1} \log(x^2 + 2)dx \quad \text{and} \quad \int_{-\infty}^{\infty} \frac{x^2}{x^4 + 1} \operatorname{Arc tan}\frac{\sqrt{2}}{x}dx.$$

(5) Let a and b be real such that $a < b$. With $[a, \infty]$ as branch cut, designate the branch $\operatorname{Log}(z - a)\operatorname{Log}(z - b)$ of $\log(z - a)\log(z - b)$ as

$$\operatorname{Log}(x-a)\operatorname{Log}(x-b) = \begin{cases} (\log|x - a| + \pi i)(\log|x - b| + \pi i), & x < a \\ (\log|x - a|)(\log|x - b| + \pi i), & a < x < b \,. \\ (\log|x - a|)(\log|x - b|), & x > b \end{cases}$$

Integrate $F(z) = f(z)\operatorname{Log}(z - a)\operatorname{Log}(z - b)$ along the path γ shown in Fig. 4.112. Show the following *general* result. Suppose f satisfies:

(i) Condition 1 as in (4.12.4.2.20).
(ii) $\lim_{r \to 0} \int_{C_{ar}} f(z)\operatorname{Log}(z - a)\operatorname{Log}(z - b)dz = \lim_{r \to 0} \int_{C_{br}} f(z)\operatorname{Log}(z - a) (z - b)dz = 0$, and $\lim_{R \to \infty} \int_{C_R} f(z)\operatorname{Log}(z - a)\operatorname{Log}(z - b)dz = 0$.

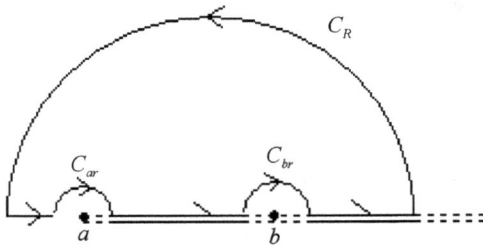

Fig. 4.112

Then,

$$\int_{-\infty}^{a} f(x)(\log|x-a| + \pi i)(\log|x-b| + \pi i)dx$$

$$+ \int_{a}^{b} f(x)(\log|x-a|)(\log|x-b| + \pi i)dx$$

$$+ \int_{b}^{\infty} f(x)(\log|x-a|)(\log|x-b|)dx$$

$$= 2\pi i \sum_{k=1}^{n} \operatorname{Re} s(f(z)\operatorname{Log}(z-a)\operatorname{Log}(z-b); z_k).$$

In case $f(x)$ assumes real value if x is real, then

$$\int_{-\infty}^{\infty} f(x) \log|x-a| \log|x-b| dx$$

$$= \operatorname{Re}\left\{2\pi i \sum_{k=1}^{n}\operatorname{Re} s(f(z)[\operatorname{Log}(z-a)\operatorname{Log}(z-b) - \pi i \operatorname{Log}(z-b)]; z_k)\right\}$$

$$(4.12.4.2.21)$$

Try to set up an integral formula in case f has finitely many real simple poles $x_1, \ldots, x_m$. Evaluate the integrals:

$$\int_{-\infty}^{\infty} \frac{x \log|x| \log|x-1|}{(x^2+2)(x^2+5)} dx; \quad \text{P.V.} \int_{-\infty}^{\infty} \frac{\log|x-2| \log|x-4|}{(x^2-1)(x^2+2x+5)} dx.$$

(6) By integrating $F(z) = \frac{z}{z^2+a^2}\operatorname{Log}(z+1)$, where $a > 0$ and $\operatorname{Log}(z+1)$ is the principal branch of $\log(z+1)$ in $\mathbf{C} - (\infty, -1]$, along a path γ similar to the one shown in Fig. 4.106, show that

$$\int_{0}^{\infty} \frac{x}{x^2+a^2} \log\left|\frac{x+1}{x-1}\right| dx = \pi \operatorname{Arc tan} \frac{1}{a}.$$

Find the limit if $a \to 0^+$. Also, show that

$$\int_{0}^{\infty} \frac{x(1-x^2)}{(x^2+a^2)^2} \log\left|\frac{x+1}{x-1}\right| dx = \pi \left(\frac{1}{2a} - \operatorname{Arc tan} \frac{1}{a}\right), \quad a > 0.$$

4.12.4.3. $\int_{0}^{\infty} x^{\alpha} f(x)dx$ $(\alpha \in \mathbf{C})$

Recall (see (2.7.3.4)) that $z^{\alpha} = e^{\alpha \log z}$, where α is a complex constant, has branches in $\mathbf{C} - [0, \infty)$ with 0 and ∞ as its logarithmic branch points. We designate its *principal branch* as

$$z^\alpha = e^{\alpha \operatorname{Log} z}, \quad \operatorname{Log} z = \log|z| + i \operatorname{Arg} z \quad \text{and} \quad 0 \le \operatorname{Arg} z < 2\pi$$

$$(4.12.4.3.1)$$

throughout the whole section unless otherwise stated.

Firstly, we have

The evaluation of $\int_0^\infty x^\alpha f(x)dx$ (without positive simple poles). Suppose f satisfies:

(1) f is analytic in $\mathbf{C}^*$ except finitely many isolated singularities $z_1, \ldots, z_n$ which do not lie on the nonnegative real axis $[0, \infty)$.
(2) There is a complex constant α with $|\operatorname{Re}\alpha| < 1$ or a nonintegral real number α such that

$$\lim_{z \to 0} z^{\alpha+1} f(z) = \lim_{z \to \infty} z^{\alpha+1} f(z) = 0.$$

Then

$$\int_0^\infty x^\alpha f(x)dx = \frac{2\pi i}{1 - e^{2\pi i \alpha}} \sum_{k=1}^n \operatorname{Re} s(z^\alpha f(z); z_k)$$

$$= -\frac{\pi e^{-\alpha \pi i}}{\sin \alpha \pi} \sum_{k=1}^n \operatorname{Re} s(z^\alpha f(z); z_k).$$

This is the case if $f(z) = \frac{p(z)}{q(z)}$, where the polynomials $p(z)$ and $q(z)$ do not have common zeros, $q(x) \ne 0$ if $x \ge 0$, $p(0) \ne 0$, and $-1 < \alpha < m - n - 1$ with n the degree of $p(z)$ and m the degree of $q(z)$. $(4.12.4.3.2)$

Note that Condition 2 guarantees that f can only have a finite number of isolated singularities in $\mathbf{C}^*$.

Sketch of Proof

Integrate $F(z) = z^\alpha f(z)$ along the path γ shown in Fig. 4.103. Then

$$\int_\gamma z^\alpha f(z)dz = 2\pi i \sum_{k=1}^n \operatorname{Re} s(z^\alpha f(z); z_k). \qquad (*_1)$$

Along the upper side $[r, R]$ and the lower side $[r, R]^-$:

$$\int_{[r,R]} z^\alpha f(z)dz = \int_r^R x^\alpha f(x)dx;$$

$$\int_{[r,R]^-} z^\alpha f(z)dz = \int_R^r e^{\alpha(\log x + 2\pi i)} f(x)dx = -e^{2\alpha \pi i} \int_r^R x^\alpha f(x)dx.$$

According to Condition 2,

$$\lim_{r \to \infty} \int_{C_r} z^{\alpha} f(z) dz = 0 \quad \text{and} \quad \lim_{R \to \infty} \int_{C_R} z^{\alpha} f(z) dz = 0.$$

Putting these results in $(*_1)$ in the process of letting $r \to 0$ and $R \to \infty$, the result follows.

Remark.

(a) Suppose $0 < \operatorname{Re} \alpha < \beta$ and $\lim_{x \to \pm\infty} x^z e^{\alpha x} f(e^{\beta x}) = 0$ holds. Via the change of variables $e^{\beta x} = t$, (4.12.4.3.2) can be used to evaluate integrals of the type

$$\int_{-\infty}^{\infty} e^{\alpha x} f(e^{\beta x}) dx = \frac{1}{\beta} \int_0^{\infty} t^{\frac{\alpha}{\beta}-1} f(t) dt. \tag{4.12.4.3.3}$$

For instance,

$$\int_{-\infty}^{\infty} \frac{e^{\alpha x}}{e^x + 1} dx = \int_0^{\infty} \frac{t^{\alpha-1}}{1+t} dt = \frac{2\pi i}{1 - e^{2\pi i (\alpha-1)}} \operatorname{Re} s \left(\frac{z^{\alpha-1}}{1+z}; -1 \right)$$

$$= \frac{-2\pi i e^{\alpha \pi i}}{1 - e^{2\alpha\pi i}} = \frac{\pi}{\sin \alpha \pi}, \quad 0 < \alpha < 1;$$

$$\int_{-\infty}^{\infty} \frac{e^{\alpha x}}{1 + e^x + e^{2x}} dx = \int_0^{\infty} \frac{t^{\alpha-1}}{1 + t + t^2} dt$$

$$= \frac{2\pi i}{1 - e^{2\pi i (\alpha-1)}} \left[\operatorname{Re} s \left(\frac{z^{\alpha-1}}{1 + z + z^2}; \frac{-1 + \sqrt{3}i}{2} \right) \right.$$

$$\left. + \operatorname{Re} s \left(\frac{z^{\alpha-1}}{1 + z + z^2}; \frac{-1 - \sqrt{3}i}{2} \right) \right]$$

$$= \frac{2\pi}{\sqrt{3} \sin \alpha \pi} \sin \frac{\pi(1 - \alpha)}{3}, \quad 0 < \operatorname{Re} \alpha < 2.$$

Suppose $0 < \operatorname{Re} \alpha < \beta$ and n is a positive integer. If $\lim_{x \to \pm\infty} f(x) = 0$, then from (4.12.4.3.3) we can evaluate

$$\int_{-\infty}^{\infty} x^n e^{\alpha x} f(e^{\beta x}) dx = \frac{\partial^n}{\partial \alpha^n} \int_{-\infty}^{\infty} e^{\alpha x} f(e^{\beta x}) dx. \tag{4.12.4.3.4}$$

(b) Let α be a real number so that $\lim_{x \to 0} x^{\alpha+1} f(x) = \lim_{x \to \infty} x^{\alpha+1} f(x) = 0$. If m is a positive integer, then via (4.12.4.3.2),

$$\int_0^{\infty} x^{\alpha} f(x) (\log x)^m dx = \frac{\partial^m}{\partial \alpha^m} \int_0^{\infty} x^{\alpha} f(x) dx. \tag{4.12.4.3.5}$$

$\square$

Example 1.

$$\int_0^\infty \frac{x^\alpha}{(x+a)^2}\,dx = \frac{2\pi i}{1-e^{2\alpha\pi i}}\lim_{z\to -a}\frac{d}{dz}(z+a)^2\frac{z^\alpha}{(z+a)^2}$$

$$= a^{\alpha-1}\frac{\pi\alpha}{\sin\pi\alpha},\quad a>0,\ -1<\alpha<1;$$

in particular,

$$\int_0^\infty \frac{dx}{(x+a)^2} = \frac{1}{a}\quad(\alpha=0);\quad \int_0^\infty \frac{\sqrt{x}}{(x+a)^2}\,dx = \frac{\pi}{2\sqrt{a}}\quad\left(\alpha=\frac{1}{2}\right);$$

$$\int_0^\infty \frac{dx}{\sqrt{x}(x+a)^2} = \frac{\pi}{2a^{\frac{3}{2}}}\quad\left(\alpha=-\frac{1}{2}\right);$$

$$\int_0^\infty \frac{x^\alpha\log x}{(x+a)^2}\,dx = \frac{\pi}{a}\frac{\partial}{\partial\alpha}\left\{\frac{a^\alpha\alpha}{\sin\pi\alpha}\right\} = \pi a^{\alpha-1}\left\{\frac{1+\alpha\log a}{\sin\pi\alpha} - \frac{\pi\alpha\cos\pi\alpha}{\sin^2\pi\alpha}\right\}.$$

Also,

$$\int_0^\infty \frac{x^\alpha}{(x^2+1)^2}\,dx = \frac{2\pi i}{1-e^{2\alpha\pi i}}[\mathrm{Re}\,s(z^\alpha f(z);i) + \mathrm{Re}\,s(z^\alpha f(z);-i)]$$

$$= \frac{\pi(1-\alpha)}{4\cos\frac{\alpha\pi}{2}},\quad -1<\alpha<3.$$

What is

$$\int_0^\infty \frac{x^\alpha\log x}{(x^2+1)^2}\,dx = ?$$

Secondly, we have

The evaluation of $\int_0^\infty x^\alpha f(x)dx$ (with positive simple poles). Suppose f satisfies:

(1) In addition to Condition 1 of (4.12.4.3.2), f has finitely many positive simple poles $x_1,\ldots,x_m$.
(2) Condition 2 of (4.12.4.3.2).

Then

$$\mathrm{P.V.}\int_0^\infty x^\alpha f(x)dx = -\frac{\pi e^{-\alpha\pi i}}{\sin\alpha\pi}\sum_{k=1}^n \mathrm{Re}\,s(z^\alpha f(z);z_k)$$

$$-\pi\cot\alpha\pi\sum_{l=1}^m \mathrm{Re}\,s(z^\alpha f(z);x_l).$$

This is the case if $f(z) = \frac{p(z)}{q(z)}$, where the polynomials $p(z)$ and $q(z)$ do not have common zeros, $z_1,\ldots,z_m$ are zeros of $q(z)$ in $\mathbf{C}-[0,\infty)$, $x_1,\ldots,x_m$

are simple zeros of $q(z)$ in $(0, \infty)$, $p(0) \neq 0$, and $-1 < \alpha < m - n - 1$ with m the degree of $q(z)$ and n the degree of $p(z)$.   (4.12.4.3.6)

Sketch of Proof

Integrate $F(z) = z^\alpha f(z)$ along the path γ shown in Fig. 4.104. Then

$$\int_\gamma z^\alpha f(z)dz = 2\pi i \sum_{k=1}^n \operatorname{Re} s(z^\alpha f(z); z_k). \qquad (*_2)$$

As before in (4.12.4.3.2),

$$\lim_{r \to 0} \int_{C_r} z^\alpha f(z)dz = \lim_{R \to \infty} \int_{C_R} z^\alpha f(z)dz = 0.$$

Along L_1, L_4, and L_2, L_3:

$$\int_{L_1} z^\alpha f(z)dz + \int_{L_4} z^\alpha f(z)dz = (1 - e^{2\alpha\pi i}) \int_r^{x_1 - r} x^\alpha f(x)dx,$$

$$\int_{L_2} z^\alpha f(z)dz + \int_{L_3} z^\alpha f(z)dz = (1 - e^{2\alpha\pi i}) \int_{x_1 + r}^R x^\alpha f(x)dx,$$

$$\Rightarrow \text{(adding both sides and letting } r \to 0, R \to \infty)$$

$$(1 - e^{2\alpha\pi i}) \text{P.V.} \int_0^\infty x^\alpha f(x)dx, \text{ if exists.}$$

Along C_{1r} and C_{1r}^*: By (4.12.3.4), and referring to $(*_3)$ and $(*_4)$ in the proof of (4.12.4.1.3),

$$\lim_{r \to 0} \int_{C_{1r}} z^\alpha f(z)dz = (0 - \pi)i \operatorname{Re} s(z^\alpha f(z); x_1) = -\pi i \operatorname{Re} s(z^\alpha f(z); x_1);$$

$$\lim_{r \to 0} \int_{C_{1r}^*} z^\alpha f(z)dz = (\pi - 2\pi)i \operatorname{Re} s(z^\alpha f(z); x_1 e^{2\pi i})$$

$$= -\pi i \operatorname{Re} s(e^{\alpha \operatorname{Log} ze^{2\pi i}} f(ze^{2\pi i}); x_1)$$

$$= -\pi i \operatorname{Re} s((z^\alpha e^{2\alpha\pi i})f(z); x_1)$$

$$= -\pi i e^{2\alpha\pi i} \operatorname{Re} s(z^\alpha f(z); x_1).$$

Substituting these into $(*_2)$ in the process of letting $r \to 0$ and $R \to \infty$, we obtain

$$(1 - e^{2\alpha\pi i}) \text{P.V.} \int_0^\infty x^\alpha f(x)dx - \pi i(1 + e^{2\alpha\pi i})\operatorname{Re} s(z^\alpha f(z); x_1)$$

$$= 2\pi i \sum_{k=1}^n \operatorname{Re} s(z^\alpha f(z); z_k),$$

$$\Rightarrow \text{P.V.} \int_0^\infty x^\alpha f(x)\,dx = \frac{2\pi i}{1 - e^{2\alpha\pi i}} \sum_{k=1}^n \text{Re}\, s(z^\alpha f(z); z_k)$$

$$+ \frac{\pi i(1 + e^{2\alpha\pi i})}{1 - e^{2\alpha\pi i}} \text{Re}\, s(z^\alpha f(z); x_1).$$

And the result follows.   □

Example 2.

$$\text{P.V.} \int_0^\infty \frac{x^{\alpha-1}}{x-a}\,dx = -\pi \cot(\alpha - 1)\pi \,\text{Re}\, s\left(\frac{z^{\alpha-1}}{z-a}; a\right)$$

$$= -\pi a^{\alpha-1} \cot \alpha\pi, \quad 0 < \alpha < 1, \ a > 0.$$

$$\text{P.V.} \int_0^\infty \frac{x^{\alpha-1}}{x^2-1}\,dx = -\frac{\pi e^{-(\alpha-1)\pi i}}{\sin(\alpha-1)\pi} \,\text{Re}\, s\left(\frac{z^{\alpha-1}}{z^2-1}; -1\right)$$

$$-\pi \cot(\alpha-1)\pi \text{Re}\, s\left(\frac{z^{\alpha-1}}{z^2-1}; 1\right),$$

$$= -\frac{\pi}{2}\left(\frac{1}{\sin \alpha\pi} + \cot \alpha\pi\right), \quad 0 < \alpha < 2.$$

$$\text{P.V.} \int_0^\infty \frac{\sin(\alpha \log x)}{x^2-1}\,dx = \text{Im}\left(\text{P.V.} \int_0^\infty \frac{e^{i\alpha \log x}}{x^2-1}\,dx\right)$$

$$= \text{Im}\left(\text{P.V.} \int_0^\infty \frac{x^{i\alpha}}{x^2-1}\,dx\right),$$

$$= \text{Im}\left[-\frac{\pi e^{-\alpha\pi i}}{\sin \alpha\pi} \,\text{Re}\, s\left(\frac{z^{i\alpha}}{z^2-1}; -1\right)\right.$$

$$\left. - \pi \cot \alpha\pi \text{Re}\, s\left(\frac{z^{i\alpha}}{z^2-1}; 1\right)\right],$$

$$= \frac{\pi}{2} \tanh \frac{\alpha\pi}{2}.$$

Exercises A

(1) Evaluate the following integrals.

(a) $\displaystyle\int_0^\infty \frac{x^{\alpha-1}}{x+e^{i\beta}}\,dx$ $(0 < \alpha < 1, \ -\pi < \beta < \pi)$.

(b) $\displaystyle\int_0^\infty \frac{x^\alpha\,dx}{(x+a)(x+b)}$ $(-1 < \alpha < 1, \ a > 0, \ b > 0 \text{ and } a \neq b)$. What happens if $b \to a$?

(c) $\int_0^\infty \dfrac{x^{\alpha-1}}{x^2+a^2}\,dx$ $(0 < \alpha < 2, a > 0)$ and $\int_0^\infty \dfrac{x^\alpha}{(x^2+a^2)^2}\,dx$ $(-1 < \alpha < 3,$ $a > 0)$.

(d) $\int_0^\infty \dfrac{x^\alpha}{x^4+1}\,dx$ $(-1 < \alpha < 3)$.

(e) $\int_0^\infty \dfrac{x^\alpha}{1+2x\cos\theta+x^2}\,dx$ $(-1 < \alpha < 1, \ -\pi < \theta < \pi)$.

(f) $\int_0^\infty \dfrac{x^{\alpha-1}}{x^2+x+1}\,dx$ $(0 < \alpha < 2)$.

(g) $\int_0^\infty \dfrac{x^\alpha}{x^3+a^3}\,dx$ $(-1 < \alpha < 2, \ a > 0)$.

(h) $\int_0^\infty \dfrac{x^{1/6}\log x}{(x^2+1)^2}\,dx$.

(i) $\int_0^\infty \dfrac{\sqrt{x}\log x}{(1+x)^2}\,dx$ and $\int_0^\infty \dfrac{\log x}{(1+x)^2\sqrt{x}}\,dx$.

(j) $\int_0^\infty \dfrac{\sqrt{x}}{x^5+1}\,dx$.

(k) $\int_0^\infty \dfrac{\log x}{(1+x^3)x^{1/4}}\,dx$.

(l) $\int_0^\infty \dfrac{\log(1+x)}{x^{\alpha+1}}\,dx$ $(0 < \alpha < 1)$.

(2) Show that $\displaystyle\int_0^{\frac{\pi}{2}} (\tan x)^\alpha dx = \dfrac{\pi}{2}\sec\dfrac{\alpha\pi}{2}$, $-1 < \alpha < 1$.

(3) Show that $\displaystyle\int_0^1 \dfrac{x^{1-\alpha}(1-x)^\alpha}{(1+x)^3}\,dx = \dfrac{\pi\alpha(1-\alpha)}{2^{3-\alpha}\sin\alpha\pi}$, $-1 < \alpha < 2$.

(4) Verify the following formulas, where $A = \pi\sin\dfrac{\alpha\pi}{2} + 2\log a \cdot \cos\dfrac{\alpha\pi}{2}$ in (b)–(d).

(a) $\displaystyle\int_0^\infty \dfrac{x^\alpha \log x}{x+a}\,dx = \dfrac{\pi a^\alpha}{\sin^2\alpha\pi}(\pi\cot\alpha\pi - \sin\alpha\pi \cdot \log a)$, $-1 < \alpha < 1, \ a > 0$.

(b) $\displaystyle\int_0^\infty \dfrac{x^\alpha \log x}{x^2+a^2}\,dx = \dfrac{\pi a^{\alpha-1}}{4\cos^2\left(\frac{\alpha\pi}{2}\right)}A$, $-1 < \alpha < 1, \ a > 0$.

(c) $\displaystyle\int_0^\infty \dfrac{x^\alpha \log x}{(x^2+a^2)^2}\,dx = \dfrac{\pi a^{\alpha-3}}{8\cos^2\left(\frac{\alpha\pi}{2}\right)}\left[(1-\alpha)A - 2\cos\dfrac{\alpha\pi}{2}\right]$, $-1 < \alpha < 3, \ b > 0$.

(d) $\displaystyle\int_0^\infty \frac{x^\alpha (\log x)^2}{(x^2+a^2)^2}dx = \frac{\pi a^{\alpha-3}}{16\cos^3(\alpha\pi/2)}\left[(1-\alpha)(\pi^2+A^2)-4A\cos\frac{\alpha\pi}{2}\right],$
$-1 < \alpha < 3,\ b > 0.$

(5) Evaluate the following integrals.

(a) P.V. $\displaystyle\int_0^\infty \frac{x^\alpha}{(x^2-a^2)(x^2-b^2)}dx,\quad -1 < \alpha < 3,\ a > 0,\ b > 0\quad$ and
$a \neq b.$

(b) P.V. $\displaystyle\int_0^\infty \frac{x^\alpha}{x^2-a^2}dx,\quad -1 < \alpha < 1,\ a > 0.$

(c) P.V. $\displaystyle\int_0^\infty \frac{x^\alpha}{a-x}dx,\quad -1 < \alpha < 0,\ a > 0.$

(d) P.V. $\displaystyle\int_{-\infty}^\infty \frac{e^{\alpha x}}{1-e^x}dx,\quad 0 < \alpha < 1.$

Exercises B

Frullani's integral

Suppose $f : (0,\infty) \to \mathbf{R}$ is continuous and satisfies that $\lim_{x\to 0+} f(x) = f(0)$ and $\lim_{x\to\infty} f(x) = f(\infty)$ both exist as finite numbers. In case $a > 0$ and $b > 0$, it can be shown in calculus that

$$\int_0^\infty \frac{f(ax)-f(bx)}{x}dx = (f(0)-f(\infty))\log\frac{a}{b}$$

which is called *Frullani's integral.* E. B. Elliot and G. H. Hardy extended this result, in 1900, to the following general case.

The evaluation of $\int_0^\infty \frac{f(ae^{i\alpha}x)-f(be^{i\beta}x)}{x}dx$ $(a > 0, b > 0, 0 \le \alpha - \beta < 2\pi)$.
Suppose f satisfies:

(1) f is analytic in the closed sector domain $\bar\Omega = \{z \mid \alpha \le \operatorname{Arg} z \le \beta\}$ (including 0) except finitely many poles $z_1, \ldots, z_n$ which do not lie on the rays $\operatorname{Arg} z = \alpha$ and $\operatorname{Arg} z = \beta$.
(2) In case $z \in \Omega = \{z \mid \alpha < \operatorname{Arg} z < \beta\}$, both $\lim_{z\to 0} f(z) = f(0)$ and $\lim_{z\to\infty} f(z) = f(\infty)$ exist as finite complex numbers.

Then

$$\int_0^\infty \frac{f(ae^{i\alpha}x)-f(be^{i\beta}x)}{x}dx = [f(\infty)-f(0)]\left[\log\frac{a}{b}+i(\alpha-\beta)\right]$$

$$+2\pi i\sum_{k=1}^n \operatorname{Re} s\left(\frac{f(z)}{z};z_k\right).\qquad (4.12.4.3.7)$$

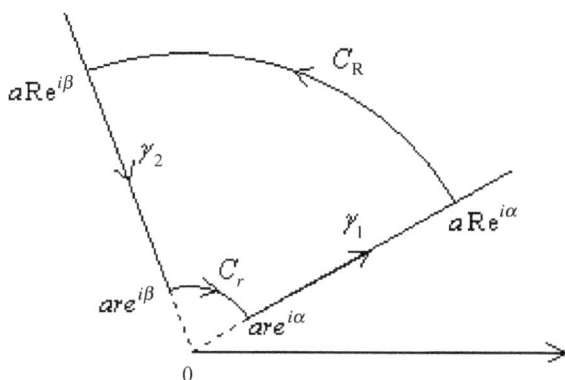

Fig. 4.113

To prove this, integrate $F(z) = \frac{f(z)}{z}$ along the path shown in Fig. 4.113.
Then

$$\int_{\gamma} \frac{f(z)}{z} dz = 2\pi i \sum_{k=1}^{n} \operatorname{Re} s\left(\frac{f(z)}{z}; z_k\right). \tag{*3}$$

Along γ_1 and γ_2:

$$\int_{\gamma_1} \frac{f(z)}{z} dz = \int_r^R \frac{f(ae^{i\alpha}t)}{t} dt \quad \text{and}$$

$$\int_{\gamma_2} \frac{f(z)}{z} dz = -\int_{are^{i\beta}}^{bre^{i\beta}} \frac{f(z)}{z} dz - \int_{bre^{i\beta}}^{bRe^{i\beta}} \frac{f(z)}{z} dz - \int_{bRe^{i\beta}}^{aRe^{i\beta}} \frac{f(z)}{z} dz$$

in which, as $r \to 0$ and $R \to \infty$,

$$\int_{are^{i\beta}}^{bre^{i\beta}} \frac{f(z)}{z} dz = \int_{are^{i\beta}}^{bre^{i\beta}} \frac{f(z) - f(0)}{z} dz + \int_{are^{i\beta}}^{bre^{i\beta}} \frac{f(0)}{z} dz$$

$$\to 0 + f(0) \log \frac{b}{a} = f(0) \log \frac{b}{a};$$

$$\int_{aRe^{i\beta}}^{bRe^{i\beta}} \frac{f(z)}{z} dz \to 0 + f(\infty) \log \frac{b}{a} = f(\infty) \log \frac{b}{a}.$$

Along $C_r : z = re^{i\theta}$, $\alpha \le \theta \le \beta$, and $C_R : z = Re^{i\theta}$, $\alpha \le \theta \le \beta$: as $r \to 0$ and $R \to \infty$,

$$\int_{C_r} \frac{f(z)}{z} dz = -i \int_\alpha^\beta f(re^{i\theta}) d\theta = -i \int_\alpha^\beta [f(re^{i\theta}) - f(0)] d\theta$$

$$-i f(0) \int_\alpha^\beta d\theta \to i f(0)(\alpha - \beta);$$

$$\int_{C_R} \frac{f(z)}{z} dz \to -i f(\infty)(\alpha - \beta).$$

Putting these results in ($*_3$) as $r \to 0$ and $R \to \infty$, the integral formula follows.

For instance, apply (4.12.4.3.7) to $f(z) = e^{-z}$ with $a > 0$, $b = 1$ and $\beta = 0 \le \text{Arg } z < \frac{\pi}{2} = \alpha$ and we have

$$\int_0^\infty \frac{e^{-aix} - e^{-x}}{x} dx = (0 - 1) \left[\log a + i \left(\frac{\pi}{2} - 0 \right) \right] = -\log a - \frac{\pi}{2} i,$$

$$\Rightarrow \int_0^\infty \frac{\sin ax}{x} dx = \frac{\pi}{2},$$

$$\int_0^\infty \frac{\cos ax - e^{-x}}{x} dx = -\log a, \quad a > 0.$$

If choosing $b = a$, $\alpha = \frac{\pi}{2}$ and $\beta = 0$, $f(z) = e^{-z}$, then

$$\int_0^\infty \frac{\cos bx - e^{-x}}{x} dx - \log b,$$

$$\Rightarrow \int_0^\infty \frac{\cos ax - \cos bx}{x} dx = -\log a - (-\log b) = \log \frac{b}{a}, \quad a > 0, \ b > 0.$$

If choosing $\alpha = \beta = 0$ and $f(z) = e^{-z}$, then (also, follows from the above two integrals)

$$\int_0^\infty \frac{e^{-ax} - e^{-bx}}{x} dx = (0 - 1) \left[\log \frac{a}{b} + i(0 - 0) \right] = \log \frac{b}{a}, \quad a > 0, \ b > 0$$

$$\Rightarrow \text{(by integration by parts)} \int_0^\infty \frac{e^{-ax} - e^{-bx} - (b - a)xe^{-bx}}{x^2} dx$$

$$= b - a - a \log \frac{b}{a}, \quad a > 0, \ b > 0.$$

Try to use Exercise B (4) of Sec. 1.5 and the above results to show that, for $n = 1, 2, 3, \ldots$,

$$\int_0^\infty \frac{\cos^{2n-1} ax - \cos^{2n-1} bx}{x} dx = \log \frac{b}{a}, \quad a > 0, b > 0, \quad \text{and}$$

$$\int_0^\infty \frac{\cos^{2n} ax - \cos^{2n} bx}{x} dx = \left[1 - \frac{(2n)!}{(n!)^2} \cdot \frac{1}{2^{2n}}\right] \log \frac{b}{a}, \quad a > 0, b > 0.$$

4.13. The Integral $\int_{x_0-i\infty}^{x_0+i\infty} f(z)dz$ along a Line $\operatorname{Re} z = x_0$

Given a fixed real number x_0. If a complex-valued function $f(z)$ is integrable (see (2.9.5)) on each finite segment of the line $\operatorname{Re} z = x_0$ and the limit

$$\lim_{R, R' \to \infty} \int_{x_0-iR'}^{x_0+iR} f(z)dz \tag{4.13.1}$$

exists as a finite complex number, denoted as $\int_{x_0-i\infty}^{x_0+i\infty} f(z)dz$, then $f(z)$ is said to be *integrable* along the line $\operatorname{Re} z = x_0$, oriented upward. And the *Cauchy principal integral value* of $f(z)$ along $\operatorname{Re} z = x_0$ is defined as

$$\text{P.V.} \int_{x_0-i\infty}^{x_0+i\infty} f(z)dz = \lim_{R \to \infty} \int_{x_0-iR}^{x_0+iR} f(z)dz \tag{4.13.2}$$

if the right limit exists as a finite number. Equation (4.13.1), of course, implies (4.13.2) but vice versa is not true. The one we introduced in (4.7.10) will be coincident with the above one if γ is the vertical line $\operatorname{Im} z = x_0$. Reader should review (4.7.11) and Example 4 in Sec. 4.7.

Remark 1. In case f is analytic in a closed vertical strip $\alpha_1 \leq \operatorname{Re} z \leq \alpha_2$ and $f(z) \to 0$ uniformly on the strip as $z \to \infty$, then, for each x, $\alpha_1 \leq x \leq \alpha_2$,

$$\int_{x-i\infty}^{x+i\infty} f(z)dz \tag{4.13.3}$$

exists as a constant, independent of the choice of such x. This follows easily by integrating $f(z)$ along a rectangular path with vertices $\alpha_1 - iR'$, $x - iR'$, $x + iR$, $\alpha_1 + iR$.

For our main purpose in what follows, we have

A fundamental integral formula. Suppose $F(z)$ satisfies:

(1) $F(z)$ is analytic in the closed half-plane $\operatorname{Re} z \leq x_0$ except finitely many isolated singularities $z_1, \ldots, z_n$ in the open half-plane $\operatorname{Re} z < x_0$ and finitely many simple poles $a_1, \ldots, a_m$ on the line $\operatorname{Re} z = x_0$.

(2) $\lim_{z \to \infty} F(z) = 0$ on $\operatorname{Re} z = x_0$.

Then, in case $a \geq 1$,

$$\text{P.V.} \int_{x_0-i\infty}^{x_0+i\infty} a^z F(z)dz = 2\pi i \sum_{k=1}^{n} \text{Re}\, s(a^z F(z); z_k) + \pi i \sum_{l=1}^{m} \text{Re}\, s(a^z F(z); a_l),$$

where $a^z = e^{z \log a}$, $\log a \geq 0$. $\hspace{4cm}$ (4.13.4)

According to (4.13.2),

$$\text{P.V.} \int_{x_0-i\infty}^{x_0+i\infty} a^z F(z)dz = \lim_{R\to\infty} \int_{x_0-iR}^{x_0+iR} e^{z \log a} F(z)dz$$

$$= \lim_{R\to\infty} \int_{-R}^{R} e^{(x_0+iy)\log a} F(x_0+iy)i\,dy,$$

$$= ia^{x_0} \lim_{R\to\infty} \int_{-R}^{R} e^{iy \log a} F(x_0+iy)dy,$$

$$= ia^{x_0} \text{P.V.} \int_{-\infty}^{\infty} e^{iy \log a} F(x_0+iy)dy. \hspace{1cm} (4.13.5)$$

Consequently all we need to do is to show that the right sides of (4.13.4) and (4.13.5) are coincident. For a direct yet another proof of (4.13.4), see Exercise A (1).

Sketch of Proof

Let $b_l = a_l - x_0$, $1 \leq l \leq m$, and $\widetilde{z_k} = z_k - x_0$, $1 \leq k \leq n$. As usual, integrate $a^{z+x_0} F(x_0+z) = a^{x_0} \cdot e^{z \log a} F(x_0+z)$ along the dented path shown in Fig. 4.114, where $|\widetilde{z_k}| < R$, $1 \leq k \leq n$, and $|b_l| < R$, $1 \leq l \leq m$. Then

$$\int_{\gamma} a^{z+x_0} F(x_0+z)dz = a^{x_0} \int_{\gamma} e^{z \log a} F(x_0+z)dz,$$

$$= a^{x_0} \cdot 2\pi i \sum_{k=1}^{n} \text{Re}\, s(e^{z \log a} F(x_0+z); \widetilde{z_k}). \hspace{1cm} (*)$$

Along $C_R : z = \text{Re}^{i\theta}$, $\frac{\pi}{2} \leq \theta \leq \frac{3\pi}{2}$:

$$\left| \int_{C_R} e^{z \log a} F(x_0+z)dz \right| \leq \int_{\frac{\pi}{2}}^{\frac{3\pi}{2}} e^{R \log a \cos\theta} |F(x_0+\text{Re}^{i\theta})| R\,d\theta,$$

$$\leq 2R \max_{|z|=R} |F(x_0+\text{Re}^{i\theta})|$$

$$\times \int_{0}^{\frac{\pi}{2}} e^{-R \log a \sin\theta} d\theta \to 0 \quad \text{as } R \to \infty.$$

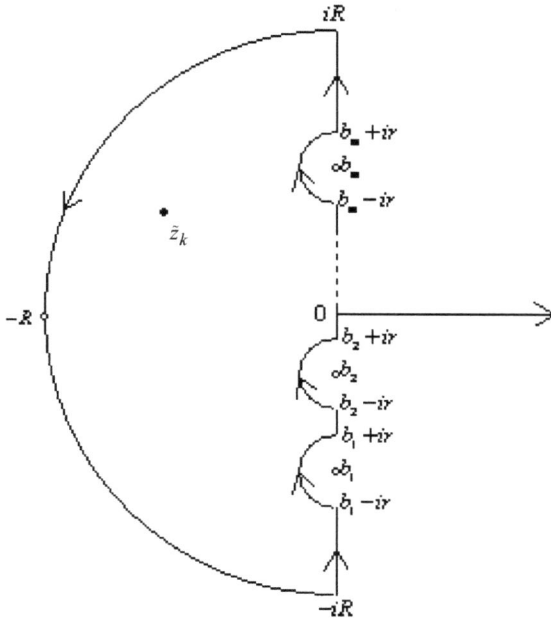

Fig. 4.114

Along $C_{lr} : z = b_l + re^{i\theta}$, $-\frac{3\pi}{2} \le \theta \le -\frac{\pi}{2}$ for $1 \le l \le m$:

$$\lim_{r \to 0} \int_{C_r} e^{z \log a} F(x_0 + z)dz = \left[-\frac{3\pi}{2} - \left(-\frac{\pi}{2}\right)\right] i \,\mathrm{Re}\, s(e^{z \log a} F(x_0 + z); b_l),$$

$$= -\pi i \,\mathrm{Re}\, s(e^{z \log a} F(x_0 + z); b_l).$$

Along the segments $L_0 = [-Ri, b_1 - ri]$, $L_l = [b_l + ir, b_{l+1} - ir]$ for $1 \le l \le m-1$ and $L_m = [b_m + ir, iR]$ (here, we suppose that $\mathrm{Im}\, b_1 < \cdots < \mathrm{Im}\, b_m$):

$$\lim_{\substack{r \to 0 \\ R \to \infty}} \sum_{l=0}^{m} \int_{L_l} e^{z \log a} F(x_0 + z)dz = \mathrm{P.V.} \int_{-\infty}^{\infty} e^{z \log a} F(x_0 + z)dz.$$

Letting $r \to 0$ and $R \to \infty$ in $(*)$ and using these results, we obtain

$$a^{x_0} \cdot \mathrm{P.V.} \int_{-\infty}^{\infty} e^{z \log a} F(x_0 + z)dz$$

$$= i a^{x_0} \cdot \mathrm{P.V.} \int_{-\infty}^{\infty} e^{iy \log a} F(x_0 + iy)dy$$

$$= a^{x_0} \cdot \left\{ 2\pi i \sum_{k=1}^{n} \operatorname{Re} s(e^{z \log a} F(x_0 + z); \widetilde{z_k}) \right.$$

$$\left. + \pi i \sum_{l=1}^{m} \operatorname{Re} s(e^{z \log a} F(x_0 + z); b_l) \right\}.$$

Observe that

$$a^{x_0} \operatorname{Re} s(e^{z \log a} F(x_0 + z); \widetilde{z_k}) = \operatorname{Re} s(e^{(x_0 + z) \log a} F(x_0 + z); \widetilde{z_k})$$

$$= \operatorname{Re} s(e^{z \log a} F(z); \widetilde{z_k} + x_0),$$

$$= \operatorname{Re} s(e^z F(z); z_k), \ 1 \le k \le n;$$

$$a^{x_0} \operatorname{Re} s(e^{z \log a} F(x_0 + z); b_l) = \operatorname{Re} s(a^z F(z); a_l), \quad 1 \le l \le m$$

and the final result follows.   □

Remark 2.

(a) If $\operatorname{Re} z \le x_0$ is changed to the closed right half-plane $\operatorname{Re} z \ge x_0$ (and Conditions 1 and 2 remain unchanged) in (4.13.4), then, in case $0 < a \le 1$,

$$\text{P.V.} \int_{x_0 - i\infty}^{x_0 + i\infty} a^z F(z) dz = - \left\{ 2\pi i \sum_{k=1}^{n} \operatorname{Re} s(a^z F(z); z_k) \right.$$

$$\left. + \pi i \sum_{l=1}^{m} \operatorname{Re} s(a^z F(z); a_l) \right\}. \qquad (4.13.6)$$

(b) If $F(z)$ is *analytic* along the line $\operatorname{Re} z = x_0$ and $a \ge 1$ is replaced by e^t, $t \ge 0$, then (4.13.4) turns out to be

$$\text{P.V.} \int_{x_0 - i\infty}^{x_0 + i\infty} e^{tz} F(z) dz = 2\pi i \sum_{k=1}^{n} \operatorname{Re} s(e^{tz} F(z); z_k), \quad t \ge 0. \qquad (4.13.7)$$

Section 4.13.2 will use this formula to compute the *inversion formula for Laplace transforms*. Similar comment holds for (4.13.6).

(c) Equation (4.13.7) can be further generalized as follows. Suppose $F(z)$ satisfies:

(1) $F(z)$ is analytic in $\operatorname{Im} z \le x_0$ except countably infinitely many isolated singularities $z_1, z_2, \ldots, z_n, \ldots$ in $\operatorname{Im} z < x_0$.

(2) There is a sequence R_n, $0 \le |x_0| < R_1 < R_2 < \cdots < R_n < \cdots$, satisfying $\lim_{n \to \infty} R_n = \infty$, so that the circles

$$C_{R_n} : z = R_n e^{i\theta}, \quad -\mathrm{Arg}\Big(x_0 - \sqrt{R_n^2 - x_0^2}\,i\Big)$$

$$\le \theta \le \mathrm{Arg}\Big(x_0 + \sqrt{R_n^2 - x_0^2}\,i\Big), \quad n \ge 1$$

do not pass through the isolated singularities of $F(z)$ and

$$\lim_{n \to \infty} \int_{C_{R_n}} e^{tz} F(z) dz = 0, \quad t > 0 \text{ being treated as a constant.}$$

(This is the case if $F(z) = O\big(\frac{1}{|z|^\alpha}\big)$, $\alpha > 0$, $z \to \infty$). See Fig. 4.115.

(3) As an infinite series, $\sum_{k=1}^{\infty} \mathrm{Re}\,s(e^{tz}F(z); z_k)$ converges. Then

$$\text{P.V.}\, \frac{1}{2\pi i} \int_{x_0-i\infty}^{x_0+i\infty} e^{tz} F(z) dz = \sum_{k=1}^{\infty} \mathrm{Re}\,s(e^{tz}F(z); z_k). \qquad (4.13.8)$$

For a concrete illustration, see Example 4(1) of Sec. 4.13.2. Similar generalization holds, too, for (4.13.6).   □

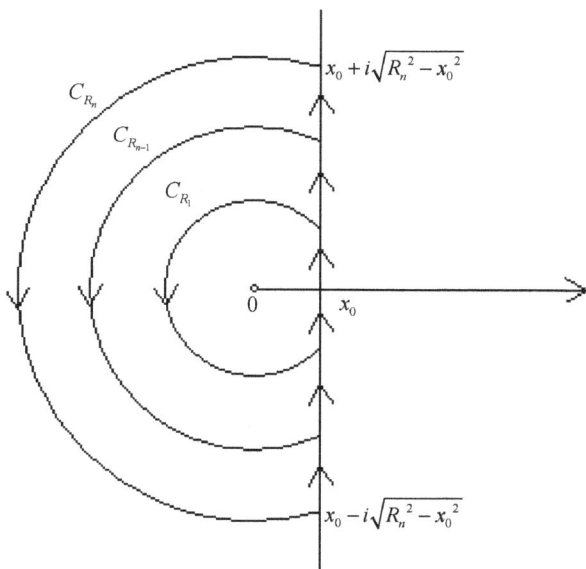

Fig. 4.115

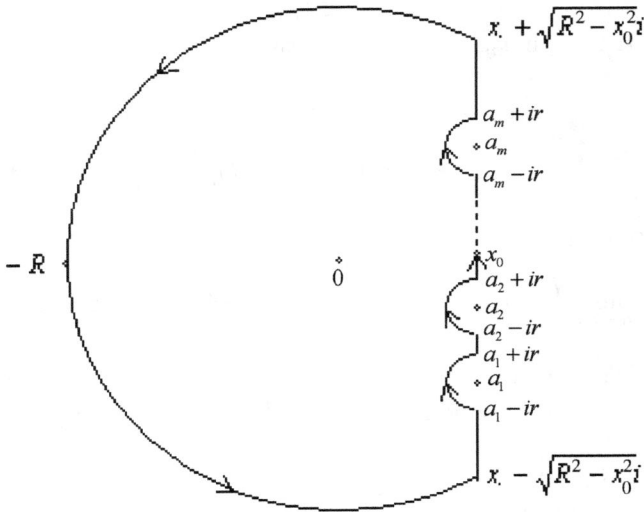

Fig. 4.116

Exrecises A

(1) Try to use path shown in Fig. 4.116 to reprove (4.13.4).

(2) Prove (4.13.6) in detail.

(3) Prove (4.13.8) in detail.

(4) Integrate $F(z) = \frac{e^{tz-x\sqrt{z}}}{z}$ ($t > 0$ and $x > 0$ are constants) along the path shown in Fig. 4.117 to show that

$$\frac{1}{2\pi i} \int_{x_0-i\infty}^{x_0+i\infty} \frac{e^{tz-x\sqrt{z}}}{z} dz = 1 - \frac{2}{\sqrt{\pi}} \int_0^{\frac{x}{2\sqrt{t}}} e^{-u^2} du.$$

(5) Let $x_0 > 0$ and $a > 0$. Show that, by changing the order of integration,

$$\int_0^\infty dt \int_{x_0-i\infty}^{x_0+i\infty} \frac{e^{\frac{z-at}{z}}}{z^2} dz = \frac{2\pi i}{a}.$$

(6) Let $x_0 > 0$ and $a > 0$, and b be a real number. Try to use Example 4 ($t < 0$) of Sec. 4.7 to show that

$$\frac{1}{2\pi i} \int_{x_0-i\infty}^{x_0+i\infty} \frac{e^{bz}}{z} dz \int_0^\infty e^{-az\cosh x} dx = \begin{cases} 0, & b < 0 \\ \cosh^{-1}\frac{b}{a}, & b > 0 \end{cases}.$$

Two subsections are divided.

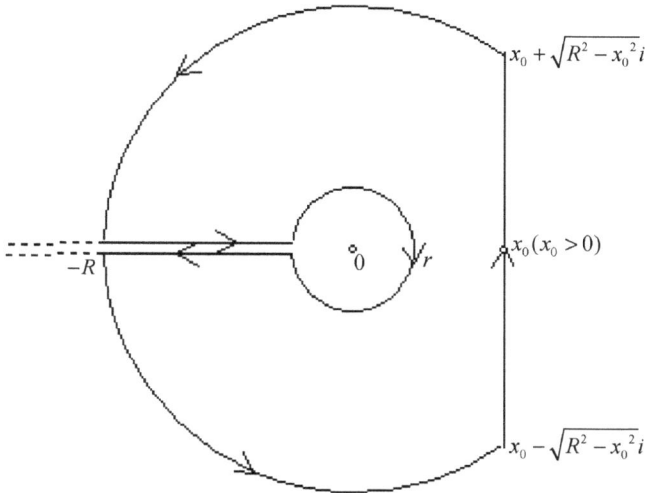

Fig. 4.117

Section 4.13.2 mainly concerns with the inversion formula for the Laplace transforms. As a preparatory work in this direction, Sec. 4.13.1 introduces the inversion formula for the Fourier transforms. We just sketch both transforms without going deeply into the theory. For details, refer to Ref. [43].

4.13.1. *Fourier transforms*

A real-valued function $\varphi(t)$, defined on $\mathbf{R} = (-\infty, \infty)$, is said to be piecewisely continuously differentiable on $\mathbf{R}$, or simply a *piecewise C^1 function* on $\mathbf{R}$, if there exist $t_1, \ldots, t_N$, $-\infty < t_1 < \cdots < t_N < \infty$, such that

(1) both $\varphi(t)$ and $\varphi'(t)$ are continuous on each of the subintervals $(-\infty, t_1), (t_1, t_2), \ldots, (t_{N-1}, t_N)$ and (t_N, ∞);
(2) $\varphi(t)$ and $\varphi'(t)$ have the left and right finite limits at each of the points $t_1, \ldots, t_N$. \hfill (4.13.1.1)

If, in addition, suppose that

$$\int_{-\infty}^{\infty} |\varphi(t)| dt < \infty,$$

then the *Fourier transform* of $\varphi(t)$ on the real line $(-\infty, \infty)$ is defined by

$$\widehat{\varphi}(x) = \int_{-\infty}^{\infty} \varphi(t) e^{-ixt} dt, \quad x \in (-\infty, \infty). \tag{4.13.1.2}$$

$\widehat{\varphi}(x)$ is well-defined in $(-\infty, \infty)$ because its integral representation converges absolutely (and uniformly) on $(-\infty, \infty)$.

Residues can be used to compute the Fourier transforms.

Example 1. For $x \in \mathbf{R}$,

(1) $\widehat{\varphi}(x) = \int_{-\infty}^{\infty} \varphi(t) e^{-ixt} dt = \dfrac{2 \sin \alpha x}{x}$,   where $\varphi(t) = \begin{cases} 1, & |t| \le \alpha \\ 0, & |t| > \alpha \end{cases}$,

$\alpha > 0$;

(2) $\widehat{\varphi}(x) = \int_{-\infty}^{\infty} e^{-t^2} e^{-ixt} dt = \sqrt{\pi} \exp\left(-\dfrac{x^2}{4} \right)$;

(3) $\widehat{\varphi}(x) = \int_{-\infty}^{\infty} \dfrac{1}{1+t^2} e^{-ixt} dt = \pi e^{-|x|}$.

Note that (1) follows by direct computation. For (2), observing that $e^{-t^2} e^{-ixt} = e^{-\left(\frac{t+ix}{2} \right)^2} e^{\frac{x^2}{4}}$, integrate $f(z) = e^{-z^2}$ along a rectangular path with vertices $\pm R$ and $\pm R + \frac{xi}{2}$ ($x > 0$ is fixed), apply the residue theorem and then let $R \to \infty$ (refer to Sec. 4.12.3). As for (3), integrate $f(z) = \frac{e^{-ixz}}{1+z^2}$ along the upper half-circle $|z| = R$, $\operatorname{Im} z > 0$, plus the diameter $[-R, R]$ or the lower half-circle according to $x > 0$ or $x < 0$, respectively.

Some basic properties are listed in Exercises A (2)–(6) for reference. What we really care is the following point of view, even though informal and nonrigorous. Represent $\varphi(t)$ sinusoidally by its Fourier series over an interval $\left[-\frac{l}{2}, \frac{l}{2} \right]$, $l > 0$, of length l as

$$\varphi(t) = \sum_{n=-\infty}^{\infty} C_n e^{\frac{i2n\pi t}{l}},   \text{where}$$

$$C_n = \frac{1}{l} \int_{-\frac{l}{2}}^{\frac{l}{2}} \varphi(t) e^{-\frac{i2n\pi t}{l}} dt,   n = 0, \pm 1, \pm 2 \ldots$$

$$= \sum_{n=-\infty}^{\infty} \widehat{\varphi}_n e^{\frac{i2n\pi t}{l}} \frac{[(n+1) - n]2\pi}{l},   \text{where}$$

$$\widehat{\varphi}_n = \frac{1}{2\pi} \int_{-\frac{l}{2}}^{\frac{l}{2}} \varphi(t) e^{-\frac{i2n\pi t}{l}} dt,   n = 0, \pm 1, \ldots$$

$$= \sum_{n=-\infty}^{\infty} \widehat{\varphi}_n(x_n) e^{ix_n t} (x_{n+1} - x_n),   \text{where}$$

$$x_n = \frac{2n\pi}{l}   \text{and}   \widehat{\varphi}_n(x) = \frac{1}{2\pi} \int_{-\frac{l}{2}}^{\frac{l}{2}} \varphi(t) e^{-ixt} dt.$$

As $l \to \infty$, it is reasonable to expect that the above relations will eventually lead to

$$\varphi(t) = \int_{-\infty}^{\infty} \widehat{\varphi}(x) e^{ixt} dx, \quad \text{where} \quad \widehat{\varphi}(x) = \frac{1}{2\pi} \int_{-\infty}^{\infty} \varphi(t) e^{-ixt} dt.$$

$$(4.13.1.3)$$

This integral might be considered as a generalized "sum" of sinusoids, summed over a continuum of frequencies x. We formally state as

The inversion formula for Fourier transform. The Fourier transform $\widehat{\varphi}(x)$ of $\varphi(t)$ satisfies the integral identity, called the *inversion formula*,

$$\varphi(t) = \frac{1}{2\pi} \int_{-\infty}^{\infty} \widehat{\varphi}(x) e^{ixt} dx$$

at point t where φ is continuous ($\varphi(t)$ should be replaced by $\frac{1}{2}[\varphi(t-) + \varphi(t+)]$ if φ is discontinuous at t). Moreover,

$$2\pi \int_{-\infty}^{\infty} |\varphi(t)|^2 dt = \int_{-\infty}^{\infty} |\widehat{\varphi}(x)|^2 dx,$$

which is called *Parseval's identity*; in general,

$$2\pi \int_{-\infty}^{\infty} \varphi(t) \overline{\psi(t)} dt = \int_{-\infty}^{\infty} \widehat{\varphi}(x) \overline{\widehat{\psi}(x)} dx. \qquad (4.13.1.4)$$

See Ref. [77] for a rigorous proof or see Exercises B (1) and (2).

Example 2. Show that

$$(1) \int_{-\infty}^{\infty} \frac{\sin \alpha x}{x} e^{ixt} dt = \begin{cases} \pi, & |t| \le \alpha \\ 0, & |t| > \alpha \end{cases}, \quad \text{where } \alpha > 0.$$

$$(2) \int_{-\infty}^{\infty} \frac{(1 - \cos x)^2}{x^4} dx = \frac{\pi}{3}.$$

Solution.

(1) According to (1) in Example 1, the Fourier transform of $\varphi(t)$ there is $\widehat{\varphi}(x) = \frac{2 \sin \alpha x}{x}$. While by the inversion formula (4.13.1.4),

$$\frac{1}{2\pi} \int_{-\infty}^{\infty} \frac{2 \sin \alpha x}{x} e^{ixt} dt = \varphi(t).$$

(2) The Fourier transform of $\varphi(t) = 1 + t$, $-1 \leq t \leq 0$; $1 - t$, $0 \leq t \leq 1$; 0, $|t| > 1$ is $\widehat{\varphi}(x) = \frac{2(1-\cos x)}{x^2}$. The result follows by using Parseval's identity

$$\int_{-\infty}^{\infty} \frac{4(1-\cos x)^2}{x^4} dx = 2\pi \int_{-\infty}^{\infty} |\varphi(t)|^2 dt = \frac{4\pi}{3}.$$

To extend the Fourier transform (4.13.1.2) to the complex variable z:

Still suppose $\varphi(t)$ is a piecewise C^1 function along the real axis. We want to pick up a horizontal domain $\Omega : y_1 < \text{Im } z < y_2$ on which the Fourier transform of $\varphi(t)$ is defined. Observe that

$$|\varphi(t)e^{-izt}| = |\varphi(t)|e^{(\text{Im } z)t}.$$

In case $-\infty < t \leq 0$: Suppose there are a constant $M_1 > 0$ and a real constant y_1 so that

$$|\varphi(t)| \leq M_1 e^{-y_1 t}, \quad t \leq 0,$$
$$\Rightarrow |\varphi(t)e^{-izt}| \leq M_1 e^{t(y-y_1)}, \quad t \leq 0 \quad \text{and} \quad y = \text{Im } z, \qquad (*_1)$$
$$\Rightarrow \int_{-\infty}^{0} \varphi(t)e^{-izt} dt \quad \text{converges absolutely if} \quad y = \text{Im } z > y_1.$$

In case $0 \leq t < \infty$: Suppose there are another constant $M_2 > 0$ and a real constant y_2 so that

$$|\varphi(t)| \leq M_2 e^{-y_2 t}, \quad t \geq 0,$$
$$\Rightarrow |\varphi(t)e^{-izt}| \leq M_2 e^{t(y-y_2)}, \quad t \geq 0 \quad \text{and} \quad y = \text{Im } z,$$
$$\Rightarrow \int_{0}^{\infty} \varphi(t)e^{-izt} dt \quad \text{converges absolutely if} \quad y = \text{Im } z < y_2. \quad (*_2)$$

In case $y_1 < y_2$, then the improper integral

$$\widehat{\varphi}(z) = \int_{-\infty}^{\infty} \varphi(t)e^{-izt} dt, \quad y_1 < \text{Im } z < y_2 \qquad (4.13.1.5)$$

converges absolutely and is called the *Fourier transform* of $\varphi(t)$ on $\Omega : y_1 < \text{Im } z < y_2$.

Moreover, $\widehat{\varphi}(z)$ is analytic in Ω. To see this, choose any rectangular path ∂R shown in Fig. 4.118. Then

$$\int_{\partial R} e^{-izt} dz = 0, \quad t \in \mathbf{R}$$

$$\Rightarrow \int_{\partial R} \widehat{\varphi}(z) dz = \int_{\partial R} \left\{ \int_{\infty}^{-\infty} \varphi(t) e^{-izt} dt \right\} dz,$$

$$= \int_{-\infty}^{\infty} \left\{ \int_{\partial R} e^{-izt} dz \right\} dt \quad \text{(by Fubini's theorem)} = 0.$$

Morera's theorem (see (3.4.2.13)) says that $\widehat{\varphi}(z)$ is thus analytic in Ω. For another proof, see Exercise A (1).

Fix y, $y_1 < y < y_2$. Observe that, now, both $(*_1)$ and $(*_2)$ hold simultaneously. Let $\psi(t) = \varphi(t) e^{ty}$ for the moment. Then

$$\begin{aligned}
\widehat{\psi}(x) &= \int_{-\infty}^{\infty} \psi(t) e^{-ixt} dt \\
&= \int_{-\infty}^{\infty} \varphi(t) e^{-i(x+iy)t} dt = \widehat{\varphi}(x + iy), \quad x \in \mathbf{R},
\end{aligned}$$

$$\begin{aligned}
\Rightarrow \varphi(t) &= \psi(t) e^{-ty} = e^{-ty} \cdot \frac{1}{2\pi} \int_{-\infty}^{\infty} \widehat{\psi}(x) e^{itx} dx \quad \text{(by (4.13.1.4))} \\
&= \frac{1}{2\pi} \int_{-\infty}^{\infty} \widehat{\varphi}(x + iy) e^{it(x+iy)} dx = \frac{1}{2\pi} \int_{-\infty+iy}^{\infty+iy} \widehat{\varphi}(z) e^{itz} dz \\
&\underset{\text{(def.)}}{=} \frac{1}{2\pi} \lim_{R \to \infty} \int_{-R+iy}^{R+iy} \widehat{\varphi}(z) e^{itz} dz. \quad\quad (4.13.1.6)
\end{aligned}$$

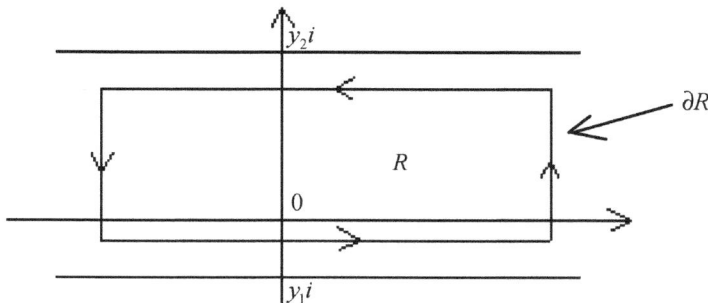

Fig. 4.118

We summarize the above as

The complex Fourier transform and its inversion formula. Suppose $\varphi(t)$: $\mathbf{R} \to \mathbf{R}$ (or $\mathbf{C}$) is a piecewise $\mathbf{C}^1$ function (see (4.13.1.1)) satisfying the condition that, there exist a constant $M > 0$ and two real constants y_1, y_2 with $y_1 < y_2$, such that

$$|\varphi(t)| \leq \begin{cases} Me^{-y_1 t}, & t \leq 0 \\ Me^{-y_2 t}, & t \geq 0 \end{cases}.$$

Then the *Fourier transform* of $\varphi(t)$ on the parallel strip domain $\Omega : y_1 < \text{Im } z < y_2$:

$$\widehat{\varphi}(z) = \int_{-\infty}^{\infty} \varphi(t) e^{-izt} dt$$

is analytic in Ω. For any fixed y, $y_1 < y < y_2$, the *inversion formula* of $\widehat{\varphi}(z)$ is

$$\varphi(t) = \frac{1}{2\pi} \int_{-\infty + iy}^{\infty + iy} \widehat{\varphi}(z) e^{itz} dz, \quad y_1 < \text{Im } z = y < y_2 \quad \text{and} \quad t \in \mathbf{R}.$$

Moreover,

$$\widehat{\varphi}^{(n)}(z) = (-i)^n \int_{-\infty}^{\infty} \varphi(t) t^n e^{-izt} dt, \quad n = 1, 2, 3, \ldots.$$

(see Exercise A (1)). $\hspace{6cm}$ (4.13.1.7)

Remark. There are variations in the definition of Fourier transforms. The factor $\frac{1}{\sqrt{2\pi}}$ may be added before the integral and, in the corresponding inversion formula, it is $\frac{1}{\sqrt{2\pi}}$, instead of $\frac{1}{2\pi}$, before the integral. In case $z = x$ is real and $\varphi(t)$ is real for real t, the real and imaginary parts form the *Fourier cosine* and *sine transform* of φ:

$$\int_{-\infty}^{\infty} \varphi(t) \cos xt \, dt = \text{Re } \widehat{\varphi}(x) \quad \text{and} \quad \int_{-\infty}^{\infty} \varphi(t) \sin xt \, dt = -\text{Im } \widehat{\varphi}(x).$$

These integrals can be evaluated by using methods mentioned in Sec. 4.12.3.2. $\hspace{6cm}$ □

Example 3. Show that

(1) $\widehat{\varphi}(z) = \displaystyle\int_{-\infty}^{\infty} e^{-|t|} e^{-izt} dt = \dfrac{2}{1 + z^2}$, $\quad -1 < \text{Im } z < 1$.

(2) If $\widehat{\varphi}(z) = \dfrac{1}{z^2}$, then $\varphi(t) = \begin{cases} 0, & t > 0 \\ t, & t \leq 0 \end{cases}$ or $\begin{cases} t, & t \geq 0 \\ 0, & t < 0 \end{cases}$.

Solution.

(1) Direct computation shows that, for $-1 < \operatorname{Im} z < 1$,

$$\lim_{R \to \infty} \int_{-R}^{R} e^{-|t|} e^{-izt} dt$$

$$= \lim_{R \to \infty} \left\{ \int_{-R}^{0} e^{t(1-iz)} dt + \int_{0}^{R} e^{t(-1-iz)} dt \right\}$$

$$= \lim_{R \to \infty} \left\{ \frac{1}{1-iz} - \frac{e^{-R(1-iz)}}{1-iz} + \frac{e^{-R(1+iz)}}{-1-iz} - \frac{1}{-1-iz} \right\}$$

$$= \frac{2}{1+z^2}.$$

Or, by (3) in Example 1 and the inversion formula in (4.13.1.4),

$$\frac{1}{2\pi} \int_{-\infty}^{\infty} \pi e^{-|x|} e^{ixt} dx = \frac{1}{1+t^2},$$

$$\Rightarrow \int_{-\infty}^{\infty} e^{-|x|} e^{ixt} dx = \frac{2}{1+t^2},$$

$$\Rightarrow \text{(interchanging } x \text{ and } t) \quad \int_{-\infty}^{\infty} e^{-|t|} e^{ixt} dt = \frac{2}{1+x^2}, \quad x \in \mathbf{R}.$$

Suppose we know beforehand that $\widehat{\varphi}(z)$ is analytic in $\Omega : |\operatorname{Im} z| < 1$, and $\widehat{\varphi}(x) = \frac{2}{1+x^2}$ holds identically on $\mathbf{R}$, then by the interior uniqueness theorem (see (4) in (3.4.2.9)), $\widehat{\varphi}(z) = \frac{2}{1+z^2}$ should hold throughout Ω. *Note:* $\psi(t) = 2(1+t^2)^{-1}$ is *not* of exponential order as required in the assumption of (4.13.1.7). The resulted Fourier transform $\widehat{\psi}(z) = \int_{-\infty}^{\infty} \psi(t) e^{-izt} dt = 2\pi e^{-|\operatorname{Re} z|}$, $z \in \mathbf{C}$, is *not* analytic in $\mathbf{C}$.

(2) By the inversion formula,

$$\varphi(t) = \frac{1}{2\pi} \int_{-\infty+iy}^{\infty+iy} \frac{e^{izt}}{z^2} dz.$$

Since $\widehat{\varphi}(z) = \frac{1}{z^2}$ is analytic in $\mathbf{C} - \{0\}$, so y in the above integral may be chosen to be positive.

In case $t \geq 0$, integrate $f(z) = \frac{e^{izt}}{z^2}$ along the path shown in Fig. 4.119(a); if $t < 0$, along the one shown in Fig. 4.119(b). By residue theorem, it is easy to show that

$$\varphi(t) = \begin{cases} 0, & t > 0 \\ t, & t \leq 0 \end{cases}.$$

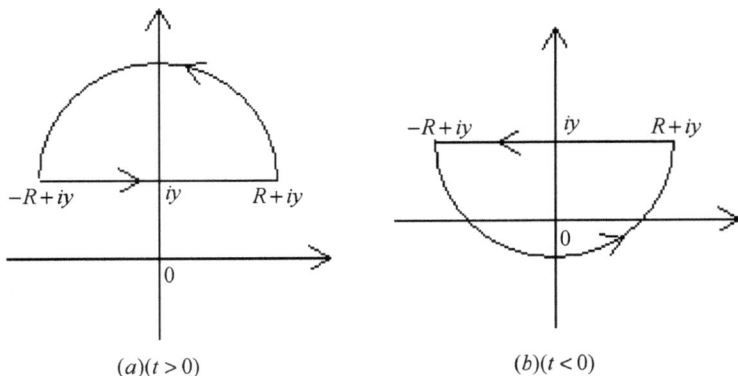

$(a)(t > 0)$ $(b)(t < 0)$

Fig. 4.119

In this case, choose $y_1 = 0$ and $y_2 = \infty$ in (4.13.1.7) and then $\widehat{\varphi}(z)$ is analytic in $0 < \operatorname{Im} z < \infty$. Instead, we can choose $y < 0$ to start with. Then $\widehat{\varphi}(z)$ is analytic in $-\infty < \operatorname{Im} z < 0$.

Exrecises A

(1) Let $\varphi(t)$ be as in (4.13.1.7). Define $f_n(z) = \int_{-n}^{n} \varphi(t)e^{-izt}dt$ for $n \geq 1$ and $\Omega : y_1 < \operatorname{Im} z < y_2$. Show that

    (a) $f_n(z)$ is analytic in Ω and converges uniformly on every compact subset of Ω to $\widehat{\varphi}(z)$. Give precise reasons why $\widehat{\varphi}(z)$ is analytic in Ω, too.

    (b) $f_n'(z) = -i \int_{-n}^{n} \varphi(t)te^{-izt}dt$ converges uniformly on every compact subset of Ω to $-i \int_{-\infty}^{\infty} \varphi(t)te^{-izt}dt$ which coincides with $\widehat{\varphi}'(z)$ on Ω. Refer to (4.7.4) and (5.3.1.1) if necessary.

(2) Basic properties for Fourier transform. Let $\varphi(t)$, $\varphi_1(t)$, and $\varphi_2(t)$ be as in (4.13.1.7).

    (a) *Linearity.* $(\lambda_1\varphi_1 + \lambda_2\varphi_2)^\wedge(z) = \lambda_1\widehat{\varphi_1}(z) + \lambda_2\widehat{\varphi_2}(z)$, $\lambda_1, \lambda_2 \in \mathbf{C}$.

    (b) *Shifting property.* Let $a \in \mathbf{R}$ and $\psi(t) = \varphi(a + t)$, $t \in \mathbf{R}$. Then $\widehat{\psi}(z) = e^{iaz}\widehat{\varphi}(z)$.

    (c) *Stretching property.* Let $a \neq 0$ be real and $\psi(t) = \varphi(at)$, $t \in \mathbf{R}$. Then $\widehat{\psi}(z) = \frac{1}{|a|}\widehat{\varphi}\left(\frac{z}{a}\right)$.

    (d) *The Fourier transform of the derivative.* Suppose $\varphi(t)$ is continuously differentiable in $\mathbf{R}$ such that $\varphi'(t)$ is also of exponential order as $\varphi(t)$ did. Then $\widehat{\varphi}'(z) = iz\widehat{\varphi}(z)$.

Note: It is sufficient to suppose that $\varphi(t)$ is continuously differentiable on $\mathbf{R}$ and both

$$\int_{-\infty}^{\infty} |\varphi(t)|dt < \infty \quad \text{and} \quad \int_{-\infty}^{\infty} |\varphi'(t)|dt < \infty$$

hold, then $\widehat{\varphi}'(x) = ix\widehat{\varphi}(x), \quad x \in \mathbf{R}.$

(e) *The derivative of a Fourier transform.*

$$\frac{d}{dz}\widehat{\varphi}(z) \underset{\text{(def.)}}{=} \widehat{\varphi}^{(1)}(z) = -i(t\varphi(t))^\wedge(z).$$

Note: If $\varphi(t)$ is piecewise C^1 so that $\int_{-\infty}^{\infty} |\varphi(t)|dt < \infty$ and $\int_{-\infty}^{\infty} |t\varphi(t)|dt < \infty$, then

$$\frac{d}{dx}\widehat{\varphi}(x) = -i(t\varphi(t))^\wedge(x), \quad x \in \mathbf{R}.$$

(3) In (4.13.1.7), suppose $\varphi(t) = 0$ if $|t| \geq t_0 > 0$. Show that

(i) $\widehat{\varphi}(z)$ is an entire function;

(ii) $\widehat{\varphi}(z)$ is bounded on the real axis; and

(iii) $|\widehat{\varphi}(z)| \leq 2t_0 \max_{|t|\leq t_0} |\varphi(t)| \cdot e^{t_0 \operatorname{Im} z}, z \in \mathbf{C}.$

(4) Let $\varphi(t)$ be continuous on $\mathbf{R}$ such that $\int_{-\infty}^{\infty} |\varphi(t)|dt < \infty$. Define $\widehat{\varphi}(x) = \int_{-\infty}^{\infty} \varphi(t)e^{-ixt}dt, x \in \mathbf{R}$. Show that $\widehat{\varphi}(x)$ is continuous on $\mathbf{R}$, and that $\widehat{\varphi}(x) \equiv 0$ on $\mathbf{R}$ if and only if $\varphi(t) \equiv 0$ on $\mathbf{R}$. Similar result is still valid for $\varphi(t)$ in (4.13.1.7).

(5) Let $\varphi(t)$ be piecewisely continuous on $\mathbf{R}$ such that $\int_{-\infty}^{\infty} |\varphi(t)|dt < \infty$. Prove the *Riemann–Lebesgue lemma*: Defining $\widehat{\varphi}(x) = \int_{-\infty}^{\infty} \varphi(t)e^{-ixt}dt$, then

$$\lim_{|x|\to\infty} \widehat{\varphi}(x) = 0.$$

Try the following steps:

(i) $\widehat{\varphi}(x) = -\int_{-\infty}^{\infty} \varphi\left(t + \frac{\pi}{x}\right)e^{-ixt}dt, x \neq 0.$

(ii) $\widehat{\varphi}(x) = \int_{-\infty}^{\infty} \frac{1}{2}\left[\varphi(t) - \varphi\left(t + \frac{\pi}{x}\right)\right]e^{-ixt}dt, x \neq 0.$

(iii) $|\widehat{\varphi}(x)| \leq \left\{\int_{-R}^{R} + \int_{-\infty}^{-R} + \int_{R}^{\infty}\right\}\psi(t)dt$, where $\psi(t) = \frac{1}{2}|\varphi(t) - \varphi(t + \frac{\pi}{x})|$. Try to show that $|\widehat{\varphi}(x)| < \varepsilon$ if $|x|$ is large enough for fixed $R > 0$.

Try to extend this result to $\widehat{\varphi}(z)$ in (4.13.1.7).

(6) Let $\varphi(t)$ and $\psi(t)$ be piecewisely continuous on $\mathbf{R}$ such that $\int_{-\infty}^{\infty} |\varphi(t)| dt < \infty$ and $\int_{-\infty}^{\infty} |\psi(t)| dt < \infty$. The *convolution* $\varphi * \psi$, of $\varphi(t)$ and $\psi(t)$, is defined by

$$(\varphi * \psi)(t) = \int_{-\infty}^{\infty} \varphi(s)\psi(t-s)ds, \ t \in \mathbf{R}.$$

   (a) Suppose $\varphi(t) = \psi(t) = 1$, if $|t| < \sigma$; $= 0$, if $|t| > \sigma$. Compute $\varphi * \psi$.

   *(b) Suppose $u(t)$ is a bounded continuous function on $\mathbf{R}$. Let $K_y(t) = \frac{y}{\pi(t^2+y^2)}$, $y > 0$. The *convolution* of $u(t)$ and $K_y(t)$ is still defined as

$$U(x,y) = (u * K_y)(x) = \frac{1}{\pi} \int_{-\infty}^{\infty} u(t) \frac{y}{(x-t)^2 + y^2} dt.$$

   Show that $U(x,y)$ is harmonic in Im $z = y > 0$ (see Sec. 3.4.3) and bounded there, and assumes the boundary values $u(t)$, namely,

$$\lim_{x+iy \to t} U(x,y) = u(t), \ t \in \mathbf{R}.$$

   This is *Schwarz's theorem* (see (6.3.2.8) in Chapter 6).

   (c) *Basic properties* of convolutions:

      (i) $\varphi * \psi = \psi * \varphi$.
      (ii) $\varphi * (\psi_1 + \psi_2) = \varphi * \psi_1 + \varphi * \psi_2$.
      (iii) $(\varphi * \psi)^\wedge(x) = \widehat{\varphi}(x)\widehat{\psi}(x)$, $x \in \mathbf{R}$.

   Try to extend these properties to $\widehat{\varphi}(z)$ and $\widehat{\psi}(z)$ defined in (4.13.1.7).

(7) Use Example 1(3) to evaluate the Fourier transforms $\widehat{\varphi}(x)$ of $\varphi(t)$.

   (a) $\varphi(t) = \dfrac{1}{t^2 + 4t + 5}$.

   (b) $\varphi(t) = \dfrac{1}{t^2 + 8t + 20}$.

   (c) $\varphi(t) = \dfrac{1}{a^2 + b^2 t^2}$ $(a > 0, b > 0)$.

(8) Use Example 1(2) to evaluate the Fourier transforms $\widehat{\varphi}(x)$ of $\varphi(t)$.

   (a) $\varphi(t) = 16e^{-4t^2}$.

   (b) $\varphi(t) = e^{-\frac{t^2}{2}}$.

   (c) $\varphi(t) = \begin{cases} e^{-t^2}, & |t| > t_0 \\ e^{-t^2} - e^{-t_0^2}, & |t| < t_0 \end{cases}$, where $t_0 > 0$ is a constant.

(9) (a) Evaluate the Fourier transforms $\widehat{\varphi}(x)$ of $\varphi(t)$:

$$\varphi(t) = \begin{cases} -1, & t_0 < t < 0 \\ 1, & 0 < t < t_0 \\ 0, & |t| > t_0 \end{cases}, \quad \text{where } t_0 > 0 \text{ is a constant,}$$

and then, use (4.13.1.4) to show that

$$\int_{-\infty}^{\infty} \frac{(\sin^2 ax)(\sin^2 bx)}{x^2} dx = \frac{\pi}{2} \min\{a, b\}, \text{ where } a > 0 \text{ and } b > 0.$$

What is $\int_{-\infty}^{\infty} (\frac{1-\cos ax}{x})^2 dx$?

(b) Use (a) to evaluate the Fourier transform of $\varphi(t) = 1 + t$, $-1 \le t \le 0$; $1 - t$, $0 \le t \le 1$; 0, $|t| \ge 1$.

(c) Use (b) to evaluate the Fourier transform of $\varphi(t) = (1 + t)^2$, $-1 \le t \le 0$; $(1 - t)^2$, $0 \le t \le 1$; 0, $|t| \ge 1$. And then, try to evaluate $\int_{-\infty}^{\infty} \frac{(x-\sin x)^2}{x^6} dx$.

(10) Let $[a, b]$ be an interval and $a < p < b$. The Dirac δ_p-function on $[a, b]$ is defined as

$$\delta_p(x) = \begin{cases} \infty, & x = p \\ 0, & x \ne p \text{ and } x \in [a, b] \end{cases}.$$

Show that, for any continuous function $f(x)$ on $[a, b]$,

$$\int_a^b f(x)\delta_p(x)dx = f(p).$$

(11) Define the *Heaviside function*

$$H(x) = \begin{cases} 0, & a \le x < p \\ 1, & p < x \le b \end{cases}.$$

Try to use integration by parts to show that $H'(x) = \delta_p(x)$.

Exercises B

(1) Prove the inversion formula in (4.13.1.4) by the following steps.

(a) Show that

$$\int_{-R}^{R} \widehat{\varphi}(s)e^{ixs} ds = 2\int_{-\infty}^{\infty} \varphi(t)\frac{\sin R(x-t)}{x-t} dt$$

$$= 2\int_{-\infty}^{\infty} \varphi(x+t)\frac{\sin Rt}{t} dt \quad \text{for } R > 0.$$

(b) Use the Riemann–Lebesgue lemma (see Exercise A (5)) to show that: for each $x \in \mathbf{R}$ and $\delta > 0$,

$$\lim_{R \to \infty} \int_\delta^\infty \varphi(x+t) \frac{\sin Rt}{t} dt = \lim_{R \to \infty} \int_{-\infty}^{-\delta} \varphi(x+t) \frac{\sin Rt}{t} dt = 0.$$

(c) Choose $\delta > 0$ small enough so that $\varphi(x+t) - \varphi(x+)$ is continuous on $[0, \delta]$ and $\varphi(x+t) - \varphi(x-)$ is continuous on $[-\delta, 0]$. Show that

$$\lim_{R \to \infty} \int_0^\delta [\varphi(x+t) - \varphi(x+)] \frac{\sin Rt}{t} dt$$

$$= \lim_{R \to \infty} \int_{-\delta}^0 [\varphi(x+t) - \varphi(x-)] \frac{\sin Rt}{t} dt = 0.$$

(d) By (b) and (c), show that

$$\lim_{R \to \infty} \int_{-R}^R \varphi(x+t) \frac{\sin Rt}{t} dt = [\varphi(x+) + \varphi(x-)] \int_0^\infty \frac{\sin x}{x} dx$$

$$= \frac{\pi}{2} [\varphi(x+) + \varphi(x-)].$$

Then, try to finish the proof.

(2) Try the following steps to prove the Parseval's identity in (4.13.1.4).

(a) For any fixed $\varepsilon > 0$, let $\psi_\varepsilon(t) = \overline{\widehat{\varphi}(t)} e^{-\varepsilon^2 t^2}$, $w_\varepsilon(t) = \frac{1}{2\sqrt{\pi}\varepsilon} e^{-\frac{t^2}{4\varepsilon^2}}$, and $p(t) = \overline{\varphi(-t)}$. Show that $\overline{\widehat{\varphi}(t)} = \widehat{p}(t)$ and (see Exercise A (6)(c)) $\psi_\varepsilon(t) = \widehat{p}(t)\widehat{w_\varepsilon}(t) = (p * w_\varepsilon)^\wedge(t)$.

(b) Use the inversion formula to show that

$$\frac{1}{2\pi} \widehat{\psi_\varepsilon}(x) = (p * w_\varepsilon)(-x) = \int_{-\infty}^\infty p(-x-s) w_\varepsilon(s) ds$$

$$= \frac{1}{\sqrt{\pi}} \int_{-\infty}^\infty p(-x - 2\varepsilon t) e^{-t^2} dt$$

$$= \frac{1}{\sqrt{\pi}} \int_{-\infty}^\infty \overline{\varphi(x + 2\varepsilon t)} e^{-t^2} dt.$$

(c) Use Fubini's theorem to show that, for general $\varphi(t)$ and $\psi(t)$ in (4.13.1.4),

$$\int_{-\infty}^\infty \widehat{\varphi}(x)\psi(x) e^{isx} dx = \int_{-\infty}^\infty \varphi(t+s)\widehat{\psi}(t) dt.$$

Then, letting $s = 0$ and $\psi = \psi_s$, show that

$$\int_{-\infty}^{\infty} |\widehat{\varphi}(x)|^2 e^{-\varepsilon^2 x^2}\,dx$$

$$= 2\sqrt{\pi}\int_{-\infty}^{\infty} e^{-t^2}\left\{\int_{-\infty}^{\infty} \varphi(x)\overline{\varphi(x+2\varepsilon t)}dx\right\}dt.$$

Finally, let $\varepsilon \to 0$ to finish the proof.

4.13.2.   Laplace transforms

Let $\varphi(t) : [0, \infty) \to \mathbf{R}$ or $\mathbf{C}$ be a function. The *Laplace transform* of φ

$$(L\varphi)(z) = \int_0^\infty \varphi(t)e^{-zt}\,dt \tag{4.13.2.1}$$

is defined for these $z \in \mathbf{C}$ for which the improper integral converges. For instance,

$$\int_0^\infty e^{-zt}\,dt = \frac{1}{z}, \quad \mathrm{Re}\,z > 0;$$

$$\int_0^\infty t^k e^{-zt}\,dt = \frac{k!}{z^{k+1}} \quad (k \text{ is a positive integer}), \quad \mathrm{Re}\,z > 0;$$

$$\int_0^\infty e^{at} e^{-zt}\,dt = \frac{1}{z-a} \quad (a \text{ is a complex constant}), \quad \mathrm{Re}\,z > \mathrm{Re}\,a;$$

$$\int_0^\infty \delta(t-t_0)e^{-zt}\,dt = e^{-zt_0}, \quad \mathrm{Re}\,z > 0, \quad \text{where } \delta(x) = \begin{cases} 0, & x \neq 0 \\ \infty, & x = 0 \end{cases} \text{ is}$$

the *Dirac δ-function* and $t_0 \in \mathbf{R}$;

$$\int_0^\infty H(t-t_0)e^{-zt}\,dt = \begin{cases} \dfrac{1}{z}(1-e^{-zt_0}), & t_0 > 0 \\ 0, & t_0 \le 0 \end{cases}, \quad \mathrm{Re}\,z > 0, \quad \text{where}$$

$$H(x) = \begin{cases} 0, & x \le 0 \\ 1, & x > 0 \end{cases} \text{ is the *Heaviside function*.}$$

Four subsections are divided in what follows.

Section (1) The existence (or convergence) and uniqueness theorems

For meaningful result, we impose the following conditions on $\varphi(t)$:

(1) $\varphi(t)$ is integrable on $[0, R]$ for any $R > 0$; i.e., φ is locally integrable on $[0, \infty)$.

(2) $\varphi(t)$ is of exponential order, i.e., there are constants $M > 0$, $\alpha \in \mathbf{R}$ and $t_0 \ge 0$ such that $|\varphi(t)| \le Me^{\alpha t}$, $t \ge t_0$. $\hfill (4.13.2.2)$

In case $\varphi(t)$ is piecewisely continuous, then Condition 1 is automatically valid, not to say a piecewise C^1 function on $[0, \infty)$ (see (4.13.1.1)). Also, any polynomial and the exponential functions $e^{at}(a \in \mathbf{C})$ satisfy the Condition 2.

Start from the Condition 2. For any fixed z with $\operatorname{Re} z > \alpha$, in case $t \geq t_0$,

$$|\varphi(t)e^{-zt}| \leq Me^{\alpha t}e^{-(\operatorname{Re} z)t} = Me^{-(\operatorname{Re} z - \alpha)t},$$

$$\Rightarrow \int_{t_0}^{\infty} |\varphi(t)e^{-zt}|dt \leq M \int_{t_0}^{\infty} e^{-(\operatorname{Re} z - \alpha)t}dt = M\frac{e^{-(\operatorname{Re} z - \alpha)t_0}}{\operatorname{Re} z - \alpha} < \infty. \qquad (*_1)$$

This means the improper integral $\int_0^{\infty} \varphi(t)e^{-zt}dt$ converges *absolutely* on $\operatorname{Re} z > \alpha$ and then, $(L\varphi)(z)$ is defined for $\operatorname{Re} z > \alpha$.

Secondly, in case $\int_0^{\infty} \varphi(t)e^{-zt}dt$ converges at z_0, then it will converge *uniformly* in the sector domain $S(z_0; \theta) = \{z \in \mathbf{C} | |\operatorname{Arg}(z - z_0)| \leq \theta\}$ for any θ, $0 \leq \theta < \frac{\pi}{2}$. This means that, for every $\varepsilon > 0$, there is a $R_0 = R_0(\varepsilon) > 0$ so that, for any $R_2 \geq R_1 \geq R_0$,

$$\left|\int_{R_1}^{R_2} \varphi(t)e^{-zt}dt\right| < \varepsilon \quad \text{for all } z \in S(z_0; \theta). \qquad (*_2)$$

To do this, rewrite

$$\int_{R_1}^{R_2} \varphi(t)e^{-zt}dt = \int_{R_1}^{R_2} [\varphi(t)e^{-z_0 t}]e^{-(z-z_0)t}dt$$

$$= \int_{R_1}^{R_2} e^{-(z-z_0)t}df(t), \quad \text{where } f(x)$$

$$= \int_0^x \varphi(t)e^{-z_0 t}dt - \int_0^{\infty} \varphi(t)e^{-z_0 t}dt$$

$$= e^{-(z-z_0)R_2}f(R_2) - e^{-(z-z_0)R_1}f(R_1)$$

$$+ (z - z_0)\int_{R_1}^{R_2} e^{-(z-z_0)t}f(t)dt$$

(integration by parts). $\qquad (*_3)$

To estimate this last expression, observing that $\lim_{x\to\infty} f(x) = 0$, hence there is a $R_0 > 0$ so that $x \geq R_0$ always implies that $|f(x)| < \varepsilon$. Now, if

$z \in S(z_0; \theta)$, then $\operatorname{Re} z_0 = x_0 < \operatorname{Re} z = x$, and

(1) $|e^{-(z-z_0)R_2} f(R_2)| \leq e^{-(x-x_0)R_2} |f(R_2)| \leq |f(R_2)| < \varepsilon$ if $R_2 > R_0$;

(2) $|e^{-(z-z_0)R_1} f(R_1)| < \varepsilon$ if $R_1 > R_0$, and then

(3)
$$\left| (z - z_0) \int_{R_1}^{R_2} e^{-(z-z_0)t} f(t) dt \right| \leq |z - z_0| \varepsilon \int_{R_1}^{R_2} e^{-(x-x_0)t} dt$$

$$= \varepsilon |z - z_0| \frac{e^{-(x-x_0)R_1} - e^{-(x-x_0)R_2}}{x - x_0} \leq 2\varepsilon \frac{|z - z_0|}{|x - x_0|} = \frac{2\varepsilon}{\cos \theta}.$$

Putting these estimates into $(*_3)$, we obtain that, if $R_2, R_1 \geq R_0$, then for all $z \in S(z_0; \theta)$,

$$\left| \int_{R_1}^{R_2} \varphi(t) e^{-zt} dt \right| < 2 \left(1 + \frac{1}{\cos \theta} \right) \varepsilon.$$

This is the same as $(*_2)$ after a suitable adjustment of ε there.

Consequently, we define

$$\sigma = \inf \left\{ \operatorname{Re} z \, \middle| \, \int_0^\infty \varphi(t) e^{-zt} dt \text{ converges} \right\} \qquad (4.13.2.3)$$

From what we obtained in the last paragraph, if $(L\varphi)(z_0)$ converges, then it follows that $(L\varphi)(z)$ converges *uniformly* on every compact subsets of $\operatorname{Re} z > \operatorname{Re} z_0$. On the other hand, in case $\operatorname{Re} z > \sigma$, then there is a z_0 with $\sigma < \operatorname{Re} z_0 < \operatorname{Re} z$ such that $(L\varphi)(z_0)$ converges and, thus, $(L\varphi)(z)$ converges on $\operatorname{Re} z \geq \operatorname{Re} z_0 > \sigma$; in particular, at this particularly chosen z. In case $\operatorname{Re} z < \sigma$, choose any z_0 so that $\operatorname{Re} z < \operatorname{Re} z_0 < \sigma$. If $(L\varphi)(z)$ converges, so does $(L\varphi)(z_0)$ and hence, $\sigma \leq \operatorname{Re} z_0$, a contradiction. Then, this means that $(L\varphi)(z)$ does not converge if $\operatorname{Re} z < \sigma$.

To prove the analyticity of $(L\varphi)(z)$ on $\operatorname{Re} z > \sigma$: Let $f_n(z) = \int_0^n \varphi(t) e^{-zt} dt$. By (4.7.4), $f_n(z)$ is analytic in $\operatorname{Re} z > \sigma$ for each n and $f'_n(z) = -\int_0^n \varphi(t) t e^{-zt} dt$, $n \geq 1$. Just like in the proof of $(*_2)$, $f_n(z)$ converges uniformly to $(L\varphi)(z)$ on compact subsets of $\operatorname{Re} z > \sigma$. So does $f'_n(z)$ to $-\int_0^\infty \varphi(t) t e^{-zt} dt$. By Weierstrass's theorem (see (5.3.1.1) in Chapter Five) or imitating Exercise A (1) of Sec. 4.13.1, it follows that $(L\varphi)(z)$ is analytic in $\operatorname{Re} z > \sigma$ and $(L\varphi)'(z) = -\int_0^\infty \varphi(t) t e^{-zt} dt$ holds on $\operatorname{Re} z > \sigma$.

We summarize the above as

The existence (or convergence) theorem for Laplace transforms. Suppose $\varphi(t)$ satisfies the two conditions stated in (4.13.2.2).

(1) There is a unique number σ (see (4.13.2.3)), $-\infty \leq \sigma < \infty$, such that the improper integral

$$(L\varphi)(z) = \int_0^\infty \varphi(t)e^{-zt}dt$$

converges absolutely and uniformly on every compact subset of the half-plane $\operatorname{Re} z > \sigma$ to the analytic function $(L\varphi)(z)$ and diverges on $\operatorname{Re} z < \sigma$ (if it exists).

(2) Moreover, for $\operatorname{Re} z > \sigma$,

$$(L\varphi)^{(n)}(z) = \int_0^\infty (-1)^n\varphi(t)t^n e^{-zt}dt, \quad n = 1,2,3,\dots.$$

This $\sigma = \sigma(\varphi)$ is called the *abscissa of convergence* of φ and

$$\sigma(\varphi) \leq \rho = \inf\{\alpha \mid \text{there are constants } M > 0 \text{ and } t_0 \geq 0 \text{ such that}$$

$$|\varphi(t)| \leq Me^{\alpha t} \text{ for } t \geq t_0\}.$$

Also, $\lim_{\operatorname{Re} z \to \infty}(L\varphi)(z) = 0$ as $(*_1)$ indicated.   (4.13.2.4)

No information can be obtained, in general, about the convergence of $(L\varphi)(z)$ along the vertical line $\operatorname{Re} z = \sigma$.

We pose a converse problem: if $(L\varphi)(z) = 0$ on $\operatorname{Re} z > \sigma$, does $\varphi(t) \equiv 0$ on $[0,\infty)$? The answer is in the affirmation in case $\varphi(t)$ is continuous on $[0,\infty)$. In other words, $(L\varphi)(z)$ is the Laplace transform of at most one function because $\varphi(t) \to (L\varphi)(z)$ is linear in the sense that $L(\lambda_1\varphi_1 + \lambda_2\varphi_2)(z) = \lambda_1(L\varphi_1)(z) + \lambda_2(L\varphi_2)(z)$, for $\lambda_1, \lambda_2 \in \mathbf{C}$.

To see this, fix $x_0 > \sigma$. Letting $z = x_0 + n$, $n = 0, 1, 2, \dots$,

$$(L\varphi)(x_0 + n) = 0 = \int_0^\infty \varphi(t)e^{-(x_0+n)t}dt = \int_0^\infty \varphi(t)e^{-x_0 t}e^{-nt}dt,$$

$$= \int_1^0 u^n u^{x_0}\varphi(-\log u)\left(-\frac{1}{u}\right)du \quad \text{(putting } u = e^{-t}),$$

$$= \int_0^1 u^n[u^{x_0-1}\varphi(-\log u)]du$$

$\Rightarrow$ (by Weierstrass approximation theorem)

$$\int_0^1 p(u)[u^{x_0-1}\varphi(-\log u)]du = 0 \quad \text{for any continuous function } p(u) \text{ on } [0,1].$$

$\Rightarrow$ (choosing $p(u) = \overline{u^{x_0-1}\varphi(-\log u)}$) $\displaystyle\int_0^1 |u^{x_0-1}\varphi(-\log u)|^2 du = 0,$

$\Rightarrow$ (by continuity of $\varphi(t)$) $u^{x_0-1}\varphi(-\log u) = e^{-t(x_0-1)}\varphi(t)$

$$= 0 \quad \text{on} \quad [0,\infty),$$

$\Rightarrow \varphi(t) = 0$ on $[0,\infty)$.

We summarize the above as

The uniqueness theorem for Laplace transforms. Let $\varphi(t)$ be continuous and satisfy (4.13.2.2). If $(L\varphi)(z) = 0$ for $\operatorname{Re} z > \sigma_0$ for some $\sigma_0 \geq \sigma(\varphi)$, then $\varphi(t) \equiv 0$ on $[0,\infty)$. Namely, for such $\varphi(t)$ and $\psi(t)$,

$$(L\varphi)(z) = (L\psi)(z) \quad \text{on} \quad \operatorname{Re} z > \max(\sigma(\varphi), \sigma(\psi))$$
$$\Leftrightarrow \varphi(t) = \psi(t) \quad \text{on} \quad [0,\infty). \tag{4.13.2.5}$$

Section (2) Basic properties

Let $\varphi(t) : [0,\infty) \to \mathbf{R}$ or $\mathbf{C}$ be as in (4.13.2.2). Redefine $\varphi(t)$ so that $\varphi(t) = 0$ for $t < 0$. Then the Fourier transform $\hat{\varphi}(\zeta)$ of $\varphi(t)$, as defined in (4.13.1.7), exists on $-\infty < \operatorname{Im}\zeta < -\sigma$ and

$$(L\varphi)(z) = \hat{\varphi}(-iz), \quad \operatorname{Re} z > \sigma.$$

See Fig. 4.120. Under this circumstance, basic properties listed in Exercises A (2)–(6) for the Fourier transforms can be easily restated for the Laplace transforms.

However, we list them in the following

Basic properties for Laplace transforms. Let $\varphi(t)$, $\psi(t)$, $\varphi_1(t)$, and $\varphi_2(t)$ be as in (4.13.2.2).

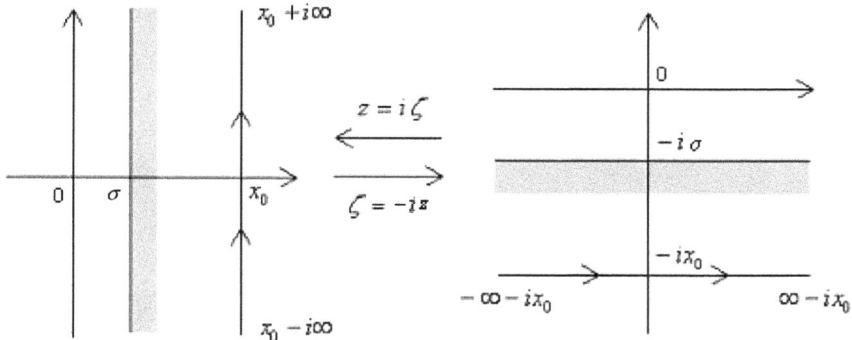

Fig. 4.120

(1) *Linearity.* $L(\lambda_1\varphi_1 + \lambda_2\varphi_2)(z) = \lambda_1(L\varphi_1)(z) + \lambda_2(L\varphi_2)(z)$,

$$\operatorname{Re} z > \max(\sigma(\varphi_1), \sigma(\varphi_2)), \quad \lambda_1, \lambda_2 \in \mathbf{C}.$$

(2) *Shifting property.* Let $a \in \mathbf{C}$ and $\psi(t) = e^{-at}\varphi(t)$. Then

$$(L\psi)(z) = (L\varphi)(z+a), \quad \operatorname{Re} z > \sigma(\varphi) - \operatorname{Re} a.$$

(3) *Shifting property.* Let $H(t) = \begin{cases} 0, \, t < 0 \\ 1, \, t \geq 0 \end{cases}$ be the Heaviside function. Fix $a \geq 0$ and let $\psi(t) = \varphi(t-a)H(t-a)$. Then

$$(L\psi)(z) = e^{-az}(L\varphi)(z), \quad \operatorname{Re} z > \sigma(\varphi).$$

Note: In other words, let $\psi(t) = \varphi(t)H(t-a)$ and $\tilde{\varphi}(t) = \varphi(t+a)$, then $(L\psi)(z) = e^{-az}(L\tilde{\varphi})(z)$.

(4) *The Laplace transform of the integral.* Let $\phi(x) = \int_0^x \varphi(t)dt$, $x \geq 0$. Then

$$(L\psi)(z) = \frac{1}{z}(L\varphi)(z), \quad \operatorname{Re} z > \max(0, \sigma(\varphi)).$$

(5) *The Laplace transform of the derivative.* Suppose $\varphi(t)$ is continuous and piecewise C^1 (see (4.13.1.1)) on $[0, \infty)$. Then,

$$(L\varphi')(z) = -\varphi(0) + z(L\varphi)(z), \quad \operatorname{Re} z > \rho \quad (\rho \text{ is in } (4.13.2.4)).$$

If φ is piecewise C^2, then

$$(L\varphi'')(z) = -\varphi'(0) - z\varphi(0) + z^2(L\varphi)(z), \quad \operatorname{Re} z > \rho.$$

Note: As a contrast, for $n = 1$ in (2) of (4.13.2.4),

$$(L\varphi)'(z) = (L(-t\varphi(t)))(z), \quad \operatorname{Re} z > \sigma(f).$$

(6) *The convolution.* The convolution of $\varphi(t)$ and $\psi(t)$ is defined by

$$(\varphi * \psi)(t) = \int_0^\infty \varphi(s)\psi(t-s)ds, \quad t > 0$$

where, under the condition that $\psi(x) = 0$ if $x < 0$. Then

   (i) $\varphi * \psi = \psi * \varphi$.
   (ii) $\varphi * (\psi_1 + \psi_2) = \varphi * \psi_1 + \varphi * \psi_2$.
   (iii) $L(\varphi * \psi)(z) = (L\varphi)(z) \cdot (L\psi)(z)$, $\operatorname{Re} z > \max(\sigma(\varphi), \sigma(\psi))$.

(7) $\lim_{z \to \infty}(L\varphi)(z) = 0$ on $\operatorname{Re} z > \sigma(\varphi)$.   (4.13.2.6)

The important inversion formula will be given in Section (3). The proofs, independent of the Fourier transforms, are left to the readers as Exercise A (1).

Section (3) The inversion formula

Suppose $\varphi(t)$ satisfies Conditions 1 and 2 in (4.13.2.2). Let the Laplace transform $(L\varphi)(z)$ be analytic in **C** except a finite number of isolated singularities $z_1, \ldots, z_n$ in $\operatorname{Re} z < \sigma$. In case $\lim_{z \to \infty}(L\varphi)(z) = 0$ on $\operatorname{Re} z \le x_0$, where $x_0 > \sigma$. Then, by (4.13.7),

$$\text{P.V.} \int_{x_0 - i\infty}^{x_0 + i\infty} (L\varphi)(z)e^{zt}dz = 2\pi i \sum_{k=1}^{n} \operatorname{Re}(e^{zt}(L\varphi)(z); z_k),$$

$$\underset{\text{(def.)}}{=} 2\pi i \psi(t), \quad t \ge 0. \tag{$*_4$}$$

Choose $R > 0$ large enough so that $|z_k| < R$, $1 \le k \le n$. Consider the circle $C_R : |z| = R$. Let the segment $l : z = x_0 + iy$, $-\sqrt{R^2 - x_0^2} \le y \le \sqrt{R^2 - x_0^2}$, divide C_R into the left part γ_1 and the right part γ_2 as shown in Fig. 4.121. As in the proof of (4.13.4), we have

$$2\pi i \psi(t) = \int_{\gamma_1 + l} (L\varphi)(z)e^{zt}dz = \int_{C_R} (L\varphi)(z)e^{zt}dz, \quad t \ge 0. \tag{$*_5$}$$

Two further considerations will be imposed on $(*_5)$.

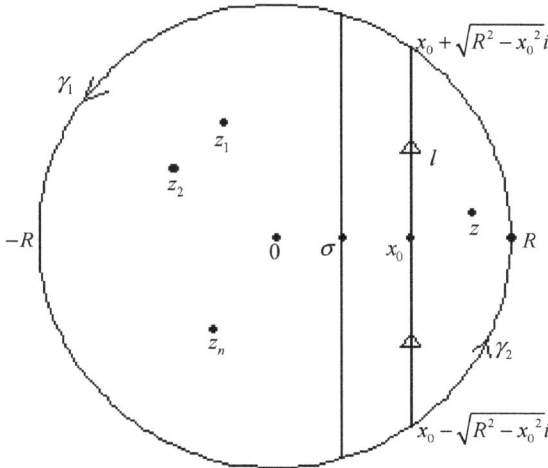

Fig. 4.121

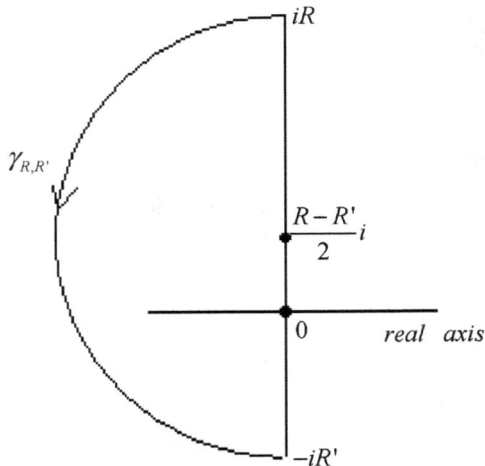

Fig. 4.122

One: In case $|(L\varphi)(z)| \leq \frac{M}{|z|^\alpha}$, *for some constants* $M > 0$ *and* $\alpha > 0$, *holds for all sufficiently large* $|z|$. For the sake of easier estimation, we deform the path γ_1 to the one $\gamma_{R,R'}$ shown in Fig. 4.122, where $R > 0$ and $R' > 0$ are independently large numbers. Then, if $t > 0$,

$$\left| \int_{\gamma_{R,R'}} e^{(x_0+z)t}(L\varphi)(x_0 + z)dz \right|$$

$$\leq \frac{R+R'}{2} e^{x_0 t} \frac{M}{(\min(R,R'))^\alpha} \int_{\frac{\pi}{2}}^{\frac{3\pi}{2}} e^{t(\frac{R+R'}{2})\cos\theta} d\theta,$$

$$= \frac{R+R'}{2} e^{x_0 t} \frac{M}{(\min(R,R'))^\alpha} \int_0^\pi e^{-t(\frac{R+R'}{2})\sin\theta} d\theta,$$

$$\leq \frac{e^{x_0 t}}{t} \frac{M\pi}{(\min(R,R'))^\alpha} [1 - e^{-(\frac{R+R'}{2})t}] \to 0,$$

as $R, R' \to \infty$ independently. This shows that, by the residue theorem,

$$\int_{x_0-i\infty}^{x_0+i\infty} (L\varphi)(z)e^{zt} dz$$

converges as an *improper Riemann integral* when the integral is taken along the vertical line $\operatorname{Re} z = x_0$. This is also true for $t = 0$. Consequently, $(*_5)$

and (*₄) can be rewritten as

$$\psi(t) = \frac{1}{2\pi i} \int_{x_0-i\infty}^{x_0+i\infty} (L\varphi)(z)e^{zt}dt, \quad t \geq 0$$

and the "P.V." sign is not needed now before the integral.

The other: We try to show that $(L\psi)(z) = (L\varphi)(z)$ on $\text{Re }z > \sigma$, and thus, by (4.13.2.5), $\psi(t) = \varphi(t)$ holds on $[0,\infty)$. To see this, by (*₅),

$$2\pi i(L\psi)(z) = \int_0^\infty e^{-zt}2\pi i\psi(t)dt,$$

$$= \lim_{a\to\infty} \int_0^a e^{-zt}\left[\int_{\gamma_1+l}(L\varphi)(\zeta)e^{\zeta t}d\zeta\right]dt,$$

$$= \lim_{a\to\infty} \int_{\gamma_1+l}\left[\int_0^a e^{(\zeta-z)t}dt\right](L\varphi)(\zeta)d\zeta,$$

$$= \lim_{a\to\infty} \int_{\gamma_1+l}[e^{(\zeta-z)a}-1]\frac{(L\varphi)(\zeta)}{\zeta-z}d\zeta.$$

Fix any point z with $\text{Re }z > x_0$, then $e^{(\zeta-z)a} \to 0$ as $a \to \infty$ if ζ lies on the closed path $\gamma_1 + l$ and then, the integrand converges uniformly to $-\frac{(L\varphi)(\zeta)}{\zeta-z}$ on $\gamma_1 + l$ as $a \to \infty$. Hence, by Fig. 4.121,

$$2\pi i(L\psi)(z) = -\int_{\gamma_1+l}\frac{(L\varphi)(\zeta)}{\zeta-z}d\zeta = \int_{\gamma_2-l}\frac{(L\varphi)(\zeta)}{\zeta-z}d\zeta - \int_{C_R}\frac{(L\varphi)(\zeta)}{\zeta-z}d\zeta$$

$$= 2\pi i(L\varphi)(z) - \int_{C_R}\frac{(L\varphi)(\zeta)}{\zeta-z}d\zeta$$

provided that $R > 0$ is large enough such that z lies in $\gamma_2 - l$. Suppose furthermore that $(L\varphi)(z) \to 0$ as $|z| \to \infty$. Then

$$\left|\int_{C_R}\frac{(L\varphi)(\zeta)}{\zeta-z}d\zeta\right| \leq \frac{2\pi R}{R-|z|}\max_{C_R}|(L\varphi)(\zeta)| \to 0 \quad \text{as } R \to \infty,$$

$$\Rightarrow 2\pi i(L\psi)(z) = 2\pi i(L\varphi)(z) \quad \text{or} \quad (L\psi)(z) = (L\varphi)(z), \quad \text{Re }z > x_0.$$

Since $x_0 > \sigma$ is arbitrary, it follows that $(L\psi)(z) = (L\varphi)(z)$, $\text{Re }z > \sigma$.

Finally, in case $(L\varphi)(z)$ has a singularity along the line $\text{Re }z = \sigma$, we will show that the abscissa of convergence of φ (see (4.13.2.4)) is σ, namely, $\sigma(\varphi) = \sigma$. By the last two paragraphs, $\sigma(\varphi) \leq \sigma$ holds since $(L\varphi)(z)$ converges on $\text{Re }z > \sigma$. If $\sigma(\varphi) < \sigma$ holds, by (4.13.2.4), $(L\varphi)(z)$ is analytic in $\text{Re }z > \sigma(\varphi)$; in particular, $(L\varphi)(z)$ is analytic along the line $\text{Re }z = \sigma$, contradicting to the assumption that $(L\varphi)(z)$ has a singularity there. Thus, $\sigma(\varphi) = \sigma$ should hold.

We summarize the above as

The inversion formula for Laplace transforms. Let $\varphi(t)$ be continuous and satisfy Conditions 1 and 2 in (4.13.2.2). Suppose its Laplace transform $(L\varphi)(z)$ is analytic in $\mathbf{C}$ except finitely many isolated singularities $z_1, \ldots, z_n$ so that $(L\varphi)(z)$ is analytic in $\operatorname{Re} z > \sigma$ for some real σ. Also, suppose that $\lim_{z \to \infty}(L\varphi)(z) = 0$ holds.

(1) Then (by (4.13.7)), for any fixed $x_0 > \sigma$,

$$\varphi(t) = \frac{1}{2\pi i} \, \text{P.V.} \int_{x_0 - i\infty}^{x_0 + i\infty} (L\varphi)(z) e^{zt} dz$$

$$= \sum_{k=1}^{n} \operatorname{Re} s(e^{zt}(L\varphi)(z); z_k), \quad t \geq 0,$$

$$\underset{\text{(def.)}}{=} (L^{-1}(L\varphi)(z))(t).$$

In particular, if there exist constants $M > 0$ and $\alpha > 0$ so that $|(L\varphi)(z)| \leq \frac{M}{|z|^\alpha}$ for all sufficiently large $|z|$, then the sign "P.V." can be dropped before the integral and the integral converges as an improper Riemann integral when the integration is taken along the vertical line $\operatorname{Re} z = x_0$.

(2) In case $(L\varphi)(z)$ has a singularity along the line $\operatorname{Re} z = \sigma$, then $\sigma = \sigma(\varphi)$, the abscissa of convergence of φ (see (4.13.2.4)).   (4.13.2.7)

As an easy application, if $(L\varphi)(z) = \frac{p(z)}{q(z)}$, where $p(z)$ and $q(z)$ are relatively prime polynomials and the degree of $q(z)$ is at least one larger than that of $p(z)$, and if the denominator $q(z)$ has only simple zeros, say at $z_1, \ldots, z_n$, then

$$\varphi(t) = \sum_{k=1}^{n} e^{z_k t} \frac{p(z_k)}{q'(z_k)}, \quad \sigma(\varphi) = \max_{1 \leq k \leq n} \operatorname{Re} z_k, \qquad (4.13.2.8)$$

which is called the *Heaviside expansion formula*. What happens to this formula if $q(z)$ has zeros of orders greater than one? Refer to Ref. [30], Chap. VII, for a concise introduction to the Laplace transform.

Section (4) Examples

Example 1. Use the following functions $\varphi(t)$, $t \geq 0$, to justify (4.13.2.4) and (4.13.2.7).

(1) $\varphi(t) = \cos at$ (a, a nonzero constant).

(2) $\varphi(t) = e^t \sin e^t$.

(3) $\varphi(t) = e^{-e^t}$.

Solution.

(1) By direct computation (recalling $\cos at = \frac{1}{2}(e^{iat} + e^{-iat})$) or integration by parts,

$$(L\varphi)(z) = \int_0^\infty e^{-zt} \cos at \ dt = \frac{z}{z^2 + a^2}, \quad \text{Re} \, z > |\text{Im} \, a|.$$

Since $\frac{z}{z^2+a^2}$ has poles at $\pm ai$, the abscissa of convergence of φ is $\sigma(\varphi) = |\text{Im} \, a|$. On the other hand, $|\cos at| \le e^{|\text{Im} \, a|t}$ holds for $t \ge 0$. Therefore ρ in (4.13.2.4) is not greater than $|\text{Im} \, a|$. It follows that, in this case, $\sigma(\varphi) = |\text{Im} \, a| = \rho$.

Conversely, by (4.13.2.8), for $x_0 > |\text{Im} \, a|$,

$$\frac{1}{2\pi i} \int_{x_0-i\infty}^{x_0+i\infty} \frac{ze^{zt}}{z^2 + a^2} dz = e^{ait} \frac{ai}{2ai} + e^{-ait} \frac{-ai}{-2ai}$$

$$= \frac{1}{2}(e^{ait} + e^{-ait}) = \cos at.$$

This is the inversion transform of $(L\varphi)(z) = \frac{z}{z^2+a^2}$, $\text{Re} \, z > \sigma(\varphi) = |\text{Im} \, a|$.

(2) Observe that $|\varphi(t)| \le e^t$ for all $t \ge 0$. Hence $\sigma(\varphi) \le \rho = 1$ in (4.13.2.4). In case $\text{Re} \, z > 0$, then by integration by parts,

$$\int_0^\infty e^{-zt} e^t \sin e^t \ dt = \int_0^\infty e^{-zt}(-d\cos e^t)$$

$$= -e^{-zt} \cos e^t \big|_0^\infty - z \int_0^\infty e^{-zt} \cos e^t dt$$

$$= \cos 1 - z \int_0^\infty e^{-zt} \cos e^t dt.$$

Since $|\int_0^\infty e^{-zt} \cos e^t dt| \le \int_0^\infty e^{-(\text{Re} \, z)t} dt < \infty$, so $(L\varphi)(z)$ is convergent for $\text{Re} \, z > 0$ and hence, is well-defined there. At $z = 0$, $\int_0^\infty e^t \sin e^t dt$ does not converge. Therefore $\sigma(\varphi) = 0$ holds.

This example shows that, in general, $\sigma(\varphi) < \rho$ in (4.13.2.4).

(3) For any $\alpha < 0$, we try to show that $|\varphi(t)| = e^{-e^t} \le Me^{\alpha t}$ for some constant $M > 0$ and for all $t \ge 0$. This shows that $\rho = -\infty$ in (4.13.2.4) and thus, $\sigma(\varphi) = -\infty$, in this case. Now,

$$e^{-e^t} \le Me^{\alpha t}, \quad t \ge t_0 \ge 0$$

$$\Leftrightarrow -e^t \le \log M + \alpha t, \quad t \ge t_0 \quad \text{if} \quad M > 0.$$

Consider $f(t) = e^t + \alpha t + \log M$. Then $f'(t) = e^t + \alpha \geq 0$ if $t \geq \log(-\alpha)$. We want to choose $M > 0$ so that, in case $t_0 = \log(-\alpha)$, $f(\log(-\alpha)) = -\alpha + \alpha \log(-\alpha) + \log M \geq 0$ holds, namely, let $M \geq e^{\alpha - \alpha \log(-\alpha)}$. Under this circumstance, $f(t) \geq f(t_0) \geq 0$, $t \geq t_0$ and the result follows.

Example 2.

(1) Compute the Laplace transform of $\varphi(t) = \sin at$, $a \neq 0$, a complex constant.

(2) Use (a) to find the inverse transform of $\frac{ae^{-z}}{z^2 + a^2}$.

(3) Use (a) and (b) to find the inverse transform of $\frac{(L\varphi)(z)}{z^2 + a^2}$ for a given $\varphi(t)$.

Solution.

(1) By direct computation or integration by parts,

$$\int_0^\infty e^{-zt} \sin at \, dt = \frac{a}{z^2 + a^2}$$

with $|\operatorname{Im} a|$ as its abscissa of convergence.

(2) Observe that, if $\operatorname{Re} z > |\operatorname{Im} a|$ and $x_0 > |\operatorname{Im} a|$ is any fixed number,

$$\frac{1}{2\pi i} \int_{x_0 - i\infty}^{x_0 + i\infty} \frac{a}{z^2 + a^2} e^{zt} dz = \sin at, \quad t \geq 0,$$

$$\Rightarrow \frac{1}{2\pi i} \int_{x_0 - i\infty}^{x_0 + i\infty} \frac{a}{z^2 + a^2} e^{z(t-1)} dz$$

$$= \frac{1}{2\pi i} \int_{x_0 - i\infty}^{x_0 + i\infty} e^{zt} \frac{ae^{-z}}{z^2 + a^2} dz = \sin a(t-1), \quad t \geq 1.$$

According to (3) in (4.13.2.6) with $a = 1$ there, the inverse transform is

$$\varphi(t) = \begin{cases} 0, & t < 1 \\ \sin a(t-1), & t \geq 1 \end{cases}$$

which is equal to $\sin a(t-1) \cdot H(t-1)$, where $H(x)$ is the Heaviside function.

(3) Observe that, if $\operatorname{Re} z > \max(\operatorname{Im} a, \sigma(\varphi))$ and x_0 is fixed,

$$\frac{1}{2\pi i} \int_{x_0 - i\infty}^{x_0 + i\infty} e^{zt} \frac{(L\varphi)(z)}{z^2 + a^2} dz = \frac{1}{2\pi i} \int_{x_0 - i\infty}^{x_0 + i\infty} \frac{e^{zt}}{z^2 + a^2} \left[\int_0^\infty e^{-zs} \varphi(s) ds \right] dz,$$

$$= \int_0^\infty \varphi(s) \left[\frac{1}{2\pi i} \int_{x_0 - i\infty}^{x_0 + i\infty} \frac{e^{z(t-s)}}{z^2 + a^2} dz \right] ds,$$

$$= \int_0^\infty \varphi(s) \sin a(t-s) \cdot H(t-s) ds,$$

$$= \int_0^t \varphi(s) \sin a(t-s) ds = (\varphi * \sin a(\cdot))(t),$$

where we have used (6) in (4.13.2.6).

Example 3.

(1) Compute the Laplace transform of $\frac{1}{\sqrt{t}}$, $t > 0$.
(2) Find the inverse transform of $\frac{1}{\sqrt{z}}$, $z \in \mathbf{C} - (\infty, 0]$.

Solution.

(1) Firstly, notice that

$$\int_0^\infty \frac{e^{-zt}}{\sqrt{t}} dt, \quad t > 0$$

does converge on $\operatorname{Re} z > 0$ and diverges at $z = 0$, so its abscissa of convergence is 0.

For any fixed z with $\operatorname{Re} z > 0$, invoke the successive changes of variables: $t \to \zeta = tz \to \eta = \sqrt{\zeta} = \sqrt{tz}$, $0 < t < \infty$. See Fig. 4.123. Then, for $\operatorname{Re} z > 0$,

$$\left(L\left(\frac{1}{\sqrt{t}}\right)\right)(z) = \int_0^\infty \frac{e^{-zt}}{\sqrt{t}} dt = \int_0^\infty e^{-\zeta} \frac{\sqrt{z}}{\sqrt{\zeta}} \frac{d\zeta}{z}$$

$$= \frac{1}{\sqrt{z}} \int_0^\infty e^{-\zeta} \zeta^{-\frac{1}{2}} d\zeta = \frac{2}{\sqrt{z}} \int_0^\infty e^{-\eta^2} d\eta$$

$$= \frac{2}{\sqrt{z}} \cdot \frac{\sqrt{\pi}}{2} = \sqrt{\frac{\pi}{z}}, \quad z \in \mathbf{C} - (\infty, 0]$$

where we have used $\int_0^\infty e^{-\eta^2} d\eta = \frac{\sqrt{\pi}}{2}$ (see Exercise B of Sec. 4.12.3).

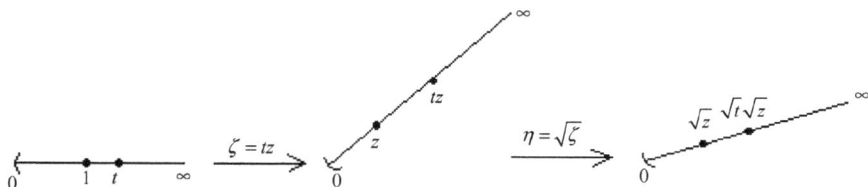

Fig. 4.123

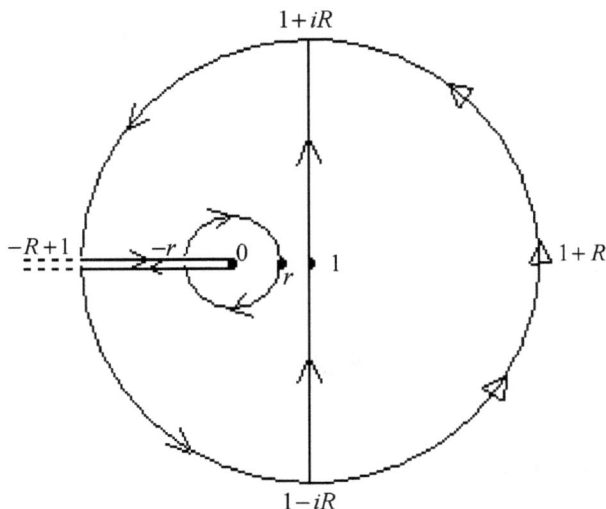

Fig. 4.124

(2) Notice that $\frac{1}{\sqrt{z}}$ is the single-valued branch determined by $\sqrt{1} = 1$ in $\mathbf{C} - (\infty, 0]$. For $t > 0$, integrate $\frac{e^{zt}}{\sqrt{z}}$ along the dented half-circle γ in the left part of Fig. 4.124. Then

$$\int_\gamma \frac{e^{zt}}{\sqrt{z}} dz = 0. \qquad (*6)$$

Along the circular arc $\gamma_1 : z = 1 + Re^{i\theta}$, $\frac{\pi}{2} \leq \theta \leq \pi$:

$$\left| \int_{\gamma_1} \frac{e^{tz}}{\sqrt{z}} dz \right| \leq \left| \int_{\pi/2}^{\pi} \frac{e^{t(1+Re^{i\theta})}}{\sqrt{1 + Re^{i\theta}}} \cdot Re i\theta d\theta \right| \leq \frac{Re^t}{\sqrt{R-1}} \int_0^{\frac{\pi}{2}} e^{-tR\sin\theta} d\theta$$

$$\leq \frac{\pi e^t}{2t\sqrt{R-1}}(1 - e^{-Rt}) \to 0 \quad \text{as } R \to \infty;$$

similarly, the integral along the circular arc $z = 1 + Re^{i\theta}$, $\pi \leq \theta \leq \frac{3\pi}{2}$, will approach zero, too, as $R \to \infty$. Along the small circle $C_r : z = re^{i\theta}$, $-\pi \leq \theta \leq \pi$:

$$\left| \int_{C_r} \frac{e^{tz}}{\sqrt{z}} dz \right| \leq \left| \int_{\pi}^{-\pi} \frac{e^{tre^{i\theta}}}{\sqrt{re^{i\theta}}} \cdot re^{i\theta} d\theta \right| \leq \sqrt{r} \cdot 2\pi \cdot e^{tr} \to 0 \quad \text{as } r \to 0.$$

While, along the upper side of the cut $[-R+1, -r]$ and the lower side:

$$\int_{-R+1}^{-r} \frac{e^{tx}}{\sqrt{x}} dx = \int_{-R+1}^{-r} \frac{e^{tx}}{\sqrt{-x}e^{\frac{\pi i}{2}}} dx = -i \int_{r}^{R-1} \frac{e^{-tx}}{\sqrt{x}} dx;$$

$$\int_{-r}^{-R+1} \frac{e^{tx}}{\sqrt{x}} dx = \int_{-r}^{-R+1} \frac{e^{tx}}{\sqrt{-x}e^{-\frac{\pi i}{2}}} dx = -i \int_{r}^{R-1} \frac{e^{-tx}}{\sqrt{x}} dx.$$

Now, in $(*_6)$, letting $r \to 0$ and $R \to \infty$, we have

$$\int_{1-i\infty}^{1+i\infty} \frac{e^{tz}}{\sqrt{z}} dz = 2i \int_{0}^{\infty} \frac{e^{-tx}}{\sqrt{x}} dx = 2i \cdot \int_{0}^{\infty} e^{-y^2} \frac{\sqrt{t}}{y} \cdot \frac{2y}{t} dy,$$

$$= \frac{4}{\sqrt{t}} i \int_{0}^{\infty} e^{-y^2} dy,$$

$$= \frac{4}{\sqrt{t}} i \cdot \frac{\sqrt{\pi}}{2} = 2\sqrt{\frac{\pi}{t}} i,$$

$$\Rightarrow \frac{1}{2\pi i} \int_{1-i\infty}^{1+i\infty} \frac{e^{tz}}{\sqrt{z}} dz = \frac{1}{2\pi i} \cdot 2\sqrt{\frac{\pi}{t}} i = \frac{1}{\sqrt{\pi t}}, \quad t > 0.$$

For $t < 0$, integrate $\frac{e^{zt}}{\sqrt{z}}$ along the right half-circle in Fig. 4.124. Since

$$\left| \int_{-\frac{\pi}{2}}^{\frac{\pi}{2}} \frac{e^{t(1+Re^{i\theta})}}{\sqrt{1+Re^{i\theta}}} \cdot Re^{i\theta} d\theta \right| \leq \frac{\pi R}{\sqrt{R-1}} e^{t(1+R)} \to 0 \quad \text{as } R \to \infty,$$

by Cauchy's integral theorem or residue theorem,

$$\frac{1}{2\pi i} \int_{1-i\infty}^{1+i\infty} \frac{e^{tx}}{\sqrt{z}} dz = 0, \quad t < 0.$$

Example 4. Evaluate the following integrals: $\alpha > 0$, $t > 0$.

(1) $\dfrac{1}{2\pi i} \displaystyle\int_{\alpha-i\infty}^{\alpha+i\infty} \dfrac{e^{tz} \sinh r\sqrt{z}}{rz \sinh a\sqrt{z}} dz \quad (a > r > 0)$, where $\sqrt{1} = 1$.

(2) $\dfrac{1}{2\pi i} \displaystyle\int_{\alpha-i\infty}^{\alpha+i\infty} \dfrac{e^{tz} \log(1+z)}{z} dz$, where $\log 1 = 0$.

(3) $\dfrac{1}{2\pi i} \displaystyle\int_{\alpha-i\infty}^{\alpha+i\infty} \dfrac{t^z}{z(z+1)\cdots(z+n)} dz \quad (n$, a positive integer$)$.

Solution.

(1) Let

$$F(z) = \frac{\sinh r\sqrt{z}}{rz \sinh a\sqrt{z}} = \frac{e^{r\sqrt{z}}(1 - e^{-2r\sqrt{z}})}{rze^{a\sqrt{z}}(1 - e^{-2a\sqrt{z}})},$$

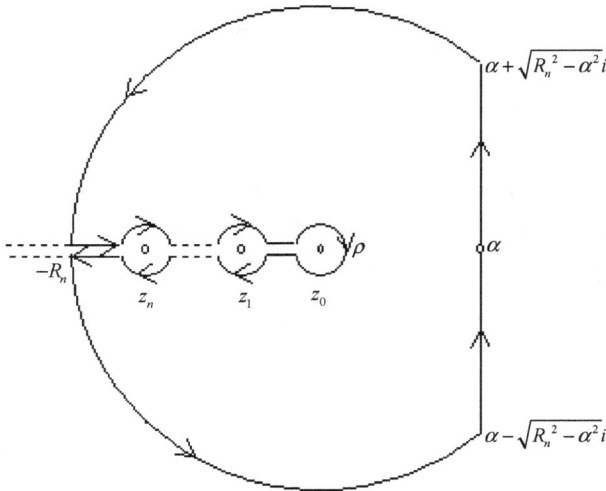

Fig. 4.125

where $\sqrt{1} = 1$. $F(z)$ is single-valued and analytic in $\mathbf{C} - (\infty, 0]$. $F(z)$ has a simple pole at $z_0 = 0$, and $\sinh a\sqrt{z} = 0$ produces other simple poles $z_n = -\frac{n^2\pi^2}{a^2}$, $n = 1, 2, \ldots$, for $F(z)$. Choose $R_n > 0$ satisfying $\frac{n\pi}{a} < \sqrt{R_n} < \frac{(n+1)\pi}{a}$, $n = 1, 2, \ldots$, and $\rho > 0$ small enough. Consider a dented contour γ as shown in Fig. 4.125, where the small circles are of the same radius ρ. By the residue theorem,

$$\int_\gamma F(z)e^{tz}\,dz = 0. \qquad (*_7)$$

Along C_{R_n} : $z = R_n e^{i\theta}$, $\theta_n \leq \theta \leq 2\pi - \theta_n$, where $\theta_n = \mathrm{Arg}(\alpha + i\sqrt{R_n^2 - \alpha^2})$:

$$|F(z)| \leq \frac{e^{(r-a)\sqrt{R_n}\cos\frac{\theta}{2}}}{rR_n} \cdot \frac{1 + e^{-2r\sqrt{R_n}\cos\frac{\theta}{2}}}{|1 - e^{-2r\sqrt{R_n}\cos\frac{\theta}{2}}|} \to 0 \quad \text{as } R_n \to \infty$$

$$\Rightarrow \lim_{n\to\infty} \int_{C_{R_n}} F(z)e^{tz}\,dz = 0 \quad \text{(refer to problem (2) if necessary).}$$

Along the upper side L_n and the lower side L_n^* of the cut $[-R_n, z_n - \rho]$:

$$\int_{L_n} F(z)e^{tz}\,dz + \int_{L_n*} F(z)e^{tz}\,dz$$

$$= -\int_{-z_n+\rho}^{R_n} \frac{e^{-tx} \cdot i \cdot \sin r\sqrt{x}}{rx \cdot i \cdot \sin a\sqrt{x}}\,dx + \int_{-z_n+\rho}^{R_n} \frac{e^{-tx}(-i)\sin r\sqrt{x}}{rx(-i)\sin a\sqrt{x}}\,dx$$

$$= 0.$$

Along the upper side L_0 and the lower side L_0^* of the cut $[z_1 + \rho, -\rho]$, the upper side L_l and the lower side L_l^* of the cut $[z_{l+1} + \rho, z_l - \rho]$, $1 \le l \le n - 1$:

$$\int_{L_l} F(z)e^{tz}dz + \int_{L_l*} F(z)e^{tz}dz = 0, \quad 0 \le l \le n - 1.$$

Along the small circle $C_{0\rho} : z = \rho e^{i\pi}$, $-\pi \le \theta \le \pi$. By (4.12.3.3),

$$\int_{C_{0\rho}} F(z)e^{tz}dz \rightarrow \lim_{z \to 0}(ze^{tz}F(z)) \cdot i(-\pi - \pi)$$

$$= \frac{1}{r} \cdot \frac{r}{a} \cdot (-2\pi i) = -\frac{2\pi i}{a} \quad \text{as } \rho \to 0.$$

Along $C_{l\rho} : z = z_l + \rho e^{i\theta}$, $0 \le \theta \le \pi$ and $C_{l\rho}* : z = z_l + \rho e^{i\pi}$, $-\pi \le \theta \le 0$. By (4.12.3.3),

$$\lim_{\rho \to 0}\left\{\int_{C_{l\rho}} F(z)e^{tz}dz + \int_{C_{l\rho}*} F(z)e^{tz}dz\right\}$$

$$= (-\pi - \pi)i \operatorname{Re} z(F(z)e^{tz}; z_l)$$

$$= -2\pi i \cdot \frac{2(-1)^l}{l\pi r}e^{-\frac{l^2\pi^2 t}{a^2}} \sin \frac{l\pi r}{a}, \quad 1 \le l \le n.$$

Substituting these results in $(*_7)$, we have

$$\frac{1}{2\pi i} \int_{\alpha-iR_n}^{\alpha+iR_n} F(z)e^{tz}dz = \frac{1}{a} + 2\sum_{l=1}^{n} \frac{(-1)^l}{l\pi r}e^{-\frac{l^2\pi^2 t}{a^2}} \sin \frac{l\pi r}{a},$$

$$n = 1, 2, \dots.$$

The finite sum in the right does converge if $n \to \infty$. Letting $n \to \infty$, then

$$\frac{1}{2\pi i} \int_{\alpha-i\infty}^{\alpha+i\infty} F(z)e^{tz}dz = \frac{1}{a} + 2\sum_{l=1}^{\infty} \frac{(-1)^l}{l\pi r}e^{-\frac{l^2\pi^2 t}{a^2}} \sin \frac{l\pi r}{a}.$$

Note: As a matter of fact,

$$F(z) = \frac{e^{tz}\left(\sum_{n=1}^{\infty} \frac{1}{(2n-1)!}r^n z^n\right)}{rz\left(\sum_{n=1}^{\infty} \frac{1}{(2n-1)!}a^n z^n\right)}, \quad 0 < |z| < \infty.$$

Hence $F(z)$ is a *single-valued* analytic function in $\mathbf{C}-\{0\}$, except having simple poles $z_n = -\frac{n^2\pi^2}{a^2}$, $n = 1, 2, \dots$. Hence we may apply (4.13.8)

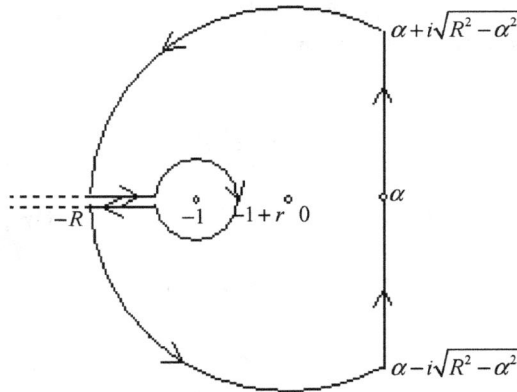

Fig. 4.126

directly to compute this integral without recourse to the dented contour shown in Fig. 4.125.

(2) Note that $z = 0$ is a removable singularity of $F(z) = \frac{\log(1+z)}{z}$. Adopt the dented contour γ shown in Fig. 4.126. Then

$$\int_{\gamma} F(z)e^{tz}dz = 0. \tag{$*_8$}$$

Along $C_R : z = Re^{i\theta_0}$, $\theta_0 \le \theta \le \pi$, where $\theta_0 = \operatorname{Arg}(\alpha + i\sqrt{R^2 - \alpha^2})$:

$$\int_{C_R} F(z)e^{tz}dz = \left\{ \int_{\theta_0}^{\frac{\pi}{2}} + \int_{\frac{\pi}{2}}^{\pi} \right\} F(z)e^{tz}dz,$$

wherein

$$\left| \int_{\theta_0}^{\frac{\pi}{2}} F(z)e^{tz}dz \right|$$

$$\le \int_{\theta_0}^{\frac{\pi}{2}} \frac{\left[(\log\sqrt{1 + R^2 + 2R\cos\theta})^2 + \left(\tan^{-1}\frac{R\sin\theta}{1+R\cos\theta}\right)^2 \right]^{\frac{1}{2}}}{R} e^{tR\cos\theta} R\,d\theta$$

$$\le 2\log R \cdot e^{t\alpha} \left(\frac{\pi}{2} - \tan^{-1}\sqrt{\frac{R^2 - \alpha^2}{\alpha}} \right)$$

$$= 2e^{t\alpha} \cdot \log R \cdot \tan^{-1}\sqrt{\frac{\alpha}{R^2 - \alpha^2}} \to 0 \quad \text{as } R \to \infty;$$

$$\left| \int_{\pi/2}^{\pi} F(z)e^{tz}dz \right| \to 0 \quad \text{as } R \to \infty \text{ (by applying (4.12.3.1))}.$$

$$\Rightarrow \lim_{R\to\infty} \int_{C_R} F(z)e^{tz}dz = 0; \quad \text{and, similarly,}$$

$$\lim_{R\to\infty} \int_{C_R^*} F(z)e^{tz}dz = 0, \quad \text{where } C_R^* : z = \text{Re}^{i\theta}, \quad \pi \le \theta \le 2\pi - \theta_0.$$

Along L and L^*, the upper and the lower side of the cut $[-R, -1-r]$:

$$\int_L F(z)e^{tz}dz + \int_{L^*} F(z)e^{tz}dz = \int_{-R}^{-1-r} \frac{e^{tx}[\log(-1-x) + \pi i]}{x}dx$$

$$+ \int_{-1-r}^{-R} \frac{e^{tx}[\log(-1-x) - \pi i]}{x}dx$$

$$= -2\pi i \int_{r+1}^{R} \frac{e^{-tx}}{x}dx.$$

Putting these results in $(*_8)$ when letting $r \to 0$ and $R \to \infty$, we obtain

$$\frac{1}{2\pi i} \int_{\alpha-i\infty}^{\alpha+i\infty} F(z)e^{tz}dz = \int_1^{\infty} \frac{e^{-tx}}{x}dx = -\int_{-\infty}^{-t} \frac{e^x}{x}dx, \quad t > 0.$$

(3) Let $F(z) = [z(z+1)\cdots(z+n)]^{-1}$. Recall that $t^z = e^{z\log t}$, where $t > 0$. In case $t \ge 1$: Integral $F(z)t^z$ along the path shown in Fig. 4.127 or apply (4.13.7) directly. Hence,

$$\frac{1}{2\pi i} \int_{\alpha-i\infty}^{\alpha+i\infty} F(z)t^z dz = \sum_{k=0}^{n} \text{Re } s(F(z)t^z; -k)$$

$$= \frac{1}{n!} \sum_{k=0}^{n} \frac{(-1)^k}{k!(n-k)!} \cdot \frac{n!}{t^k} = \frac{1}{n!}\left(1 - \frac{1}{t}\right)^n, \quad t \ge 1.$$

In case $0 < t \le 1$: Integral $F(z)t^z$ along the closed path shown in Fig. 4.128. Then,

$$\frac{1}{2\pi i} \int_{\gamma} F(z)t^z dz = 0. \tag{$*_9$}$$

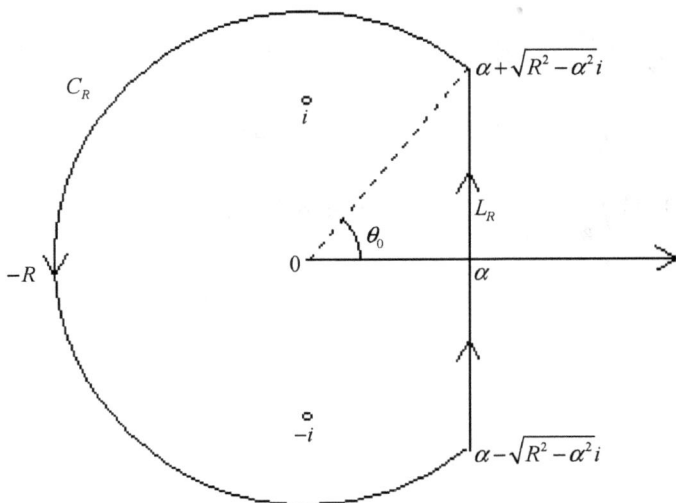

Fig. 4.127

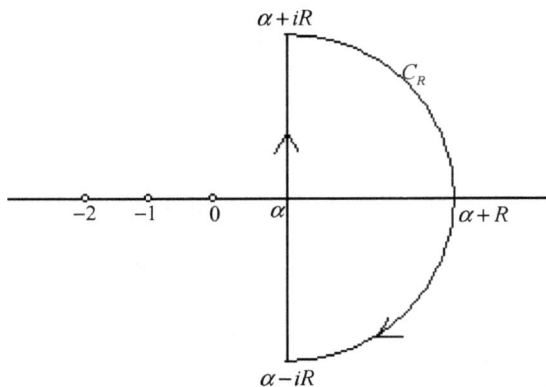

Fig. 4.128

Along the half-circle $C_R : z = \alpha + \mathrm{Re}^{i\theta}$, $-\frac{\pi}{2} \le \theta \le \frac{\pi}{2}$:

$$\left| \int_{C_R} F(z) t^z dz \right| \le \int_{-\frac{\pi}{2}}^{\frac{\pi}{2}} \frac{t^\alpha e^{R\cos\theta \log t}}{(R-\alpha)(R+1-\alpha)\cdots(R+n-\alpha)} R d\theta$$

$$\le \frac{Rt^\alpha}{(R-\alpha)(R+1-\alpha)\cdots(R+n-\alpha)} \to 0 \quad \text{as } R \to \infty.$$

Letting $R \to \infty$ in $(*_9)$, we have

$$\frac{1}{2\pi i} \int_{\alpha-i\infty}^{\alpha+i\infty} F(z)t^z dz = 0, \quad 0 < t \le 1.$$

Example 5. Use the Laplace transform to solve the following differential equations with initial values.

(1) $\dfrac{d^2 f}{dt^2} + 6\dfrac{df}{dt} - 7f = 0, \quad f(0) = 1 \quad \text{and} \quad f'(0) = 0.$

(2) $\dfrac{d^2 f}{dt^2} - \dfrac{df}{dt} - 2f = e^{-t} \sin 2t, \quad f(0) = 0 \quad \text{and} \quad f'(0) = 2.$

Solution.

(1) Taking the Laplace transform of the equation and using the linear property (1) in (4.13.2.6), we have

$$(Lf'' + 6Lf' - 7Lf)(z) = 0$$

$\Rightarrow$ (by using (5) in (4.13.2.6))

$$- f'(0) - zf(0) + z^2(Lf)(z) + 6[-f(0) + z(Lf)(z)]$$
$$- 7(Lf)(z) = 0,$$

$\Rightarrow$ (by using $f(0) = 1$ and $f'(0) = 0$) $\quad (z^2 + 6z - 7)(Lf)(z) = z + 6,$

$$\Rightarrow (Lf)(z) = \frac{z+6}{z^2 + 6z - 7} = \frac{7}{8} \cdot \frac{1}{z-1} + \frac{1}{8} \cdot \frac{1}{z+7},$$

$$\Rightarrow f(t) = \frac{7}{8}L^{-1}\left(\frac{1}{z-1}\right)(t) + \frac{1}{8}L^{-1}\left(\frac{1}{z+7}\right)(t) = \frac{7}{8}e^t + \frac{1}{8}e^{-7t}.$$

In the last step, try to use (4.13.2.7) or see the table in (4.13.2.11).

(2) Taking the Laplace transform of the equation, then

$$(Lf'' - Lf' - 2Lf)(z) = L(e^{-t}\sin 2t)(z)$$

$$= \frac{2}{(z+1)^2 + 2^2}(\operatorname{Re} z > -1)$$

$$\Rightarrow -f'(0) - zf(0) + z^2(Lf)(z) - [-f(0) + z(Lf)(z)] - 2(Lf)(z)$$

$$= \frac{2}{(z+1)^2 + 4}$$

$$\Rightarrow (z^2 - z - 2)(Lf)(z) - 1 = \frac{2}{(z+1)^2 + 4}, \quad \text{or}$$

$$(Lf)(z) = \frac{2}{z^2 - z - 2} + \frac{2}{(z^2 - z - 2)(z^2 + 2z + 5)},$$

$$\Rightarrow f(t) = L^{-1}\left(\frac{2}{z^2 - z - 2}\right)(t) + L^{-1}\left(\frac{2}{(z^2 - z - 2)(z^2 + 2z + 5)}\right)(t).$$

Wherein,

$$L^{-1}\left(\frac{2}{z^2 - z - 2}\right) = L^{-1}\left(-\frac{2}{3} \cdot \frac{1}{z+1} + \frac{2}{3} \cdot \frac{1}{z-2}\right) = -\frac{2}{3}e^{-t} + \frac{2}{3}e^{2t}.$$

On the other hand, by direct computation using (4.13.2.7) or by the table in (4.13.2.11),

$$\frac{2}{(z^2 - z - 2)(z^2 + 2z + 5)} = -\frac{1}{6} \cdot \frac{1}{z+1} + \frac{2}{39}\frac{1}{z-2} + \frac{\frac{3}{26}(z+1) - \frac{2}{13}}{(z+1)^2 + 2^2},$$

$$\Rightarrow L^{-1}\left(\frac{2}{(z^2 - z - 2)(z^2 + 2z + 5)}\right)(t)$$

$$= -\frac{1}{6}L^{-1}\left(\frac{1}{z+1}\right)(t) + \frac{2}{39}L^{-1}\left(\frac{1}{z-2}\right)(t)$$

$$+ \frac{3}{26}L^{-1}\left(\frac{z+1}{(z+1)^2 + 2^2}\right)(t) - \frac{1}{13}L^{-1}\left(\frac{2}{(z+1)^2 + 2^2}\right)(t)$$

$$= -\frac{1}{6}e^{-t} + \frac{2}{39}e^{2t} + \frac{3}{26}e^{-t}\cos 2t - \frac{1}{13}e^{-t}\sin 2t.$$

The solution is thus

$$f(t) = \frac{28}{29}e^{2t} - \frac{5}{6}e^{-t} + \frac{3}{26}e^{-t}\cos 2t - \frac{1}{13}e^{-t}\sin 2t.$$

Remark. How to use the Laplace transform to solve ordinary differential equations with constant coefficients?

Given the differential equation with the initial values:

$$\frac{d^n f}{dt^n} + a_{n-1}\frac{d^{n-1}f}{dt^{n-1}} + \cdots + a_1\frac{df}{dt} + a_0 f = \varphi(t),$$

$$f(0) = \alpha_0, f'(0) = \alpha_1, \ldots, f^{(n-1)}(0) = \alpha_{n-1}. \qquad (4.13.2.9)$$

Suppose f is C^{n-1} on $[0, \infty)$ on which $f^{(n)}$ exists and is piecewisely continuous (refer to (5) in (4.13.2.6)). Also suppose that there exist a positive constant M and a constant α so that $|f^{(k)}(t)| \le Me^{\alpha t}$, $t \ge 0$, for $0 \le k \le n$. Then the Laplace transforms $(Lf^{(k)})(z)$ exist for $\operatorname{Re} z > \alpha$, $0 \le k \le n$, and

$$(Lf^k)(z) = -f^{(k-1)}(0)$$
$$- zf^{(k-2)}(0) - \cdots - z^{k-1}f(0) + z^k(Lf)(z), \quad 1 \le k \le n.$$
$$(4.13.2.10)$$

To start with, take the Laplace transform of Eq. (4.13.2.9) to obtain

$$(Lf^{(n)})(z) + a_{n-1}(Lf^{(n-1)})(z) + \cdots + a_1(Lf')(z) + a_0(Lf)(z) = (L\varphi)(z)$$

and then, use (4.13.2.10) and the initial values to reduce it to

$$(Lf)(z) = \frac{(L\varphi)(z)}{z^n + a_{n-1}z^{n-1} + \cdots + a_1 z + a_0}$$

$$+ \frac{p(z)}{z^n + a_{n-1}z^{n-1} + \cdots + a_1 z + a_0}, \quad \text{where } p(z) \text{ is a polynomial.}$$

Finally, try to use the inversion formula (4.13.2.7) to solve $f(t)$. Occasionally, the following *table of Laplace transforms* is helpful in computation: $(L\varphi)(z)$ is written as $L(\varphi)$, for short.

$$L(e^{at}) = \frac{1}{z - a}, \quad \operatorname{Re} z > \operatorname{Re} a \text{ (a, a complex constant)}.$$

$$L(\cos at) = \operatorname{Re} L(e^{iat}) = \frac{z}{z^2 + a^2}, \quad \operatorname{Re} z > 0 \ (a \in \mathbf{R}, \text{a real constant}).$$

$$L(\sin at) = \operatorname{Im} L(e^{iat}) = \frac{a}{z^2 + a^2}, \quad \operatorname{Re} z > 0.$$

$$L(\cosh at) = L(\cos iat) = \frac{z}{z^2 - a^2}, \quad \operatorname{Re} z > |a| \ (a, \text{a real constant}).$$

$$L(\sinh at) = L(-i\sin iat) = \frac{a}{z^2 - a^2}, \quad \operatorname{Re} z > |a|.$$

$$L\left(e^{-bt}\left\{\begin{matrix} \cos at \\ \sin at \end{matrix}\right\}\right) = \left\{\begin{matrix} \operatorname{Re} \\ \operatorname{Im} \end{matrix} L(e^{(-b+ia)t})\right.$$

$$= \left\{\begin{matrix} \dfrac{z+b}{(z+b)^2 + a^2} \\[2mm] \dfrac{a}{(z+b)^2 + a^2} \end{matrix}\right., \quad \operatorname{Re} z > -b (a, b, \text{real}).$$

$$L\left(t\left\{\begin{array}{l}\cos at\\\sin at\end{array}\right\}\right)=\left\{\begin{array}{l}\mathrm{Re}\\\mathrm{Im}\end{array}\right.L(te^{iat})=\left\{\begin{array}{l}\dfrac{z^2-a^2}{(z^2+a^2)^2}\\[2mm]\dfrac{2az}{(z^2+a^2)^2}\end{array}\right.,\qquad \mathrm{Re}\,z>0\ (a,\ \mathrm{real}).$$

$$L(t^n e^{at})=\frac{n!}{(z-a)^{n+1}},\qquad \mathrm{Re}\,z>\mathrm{Re}\,a\ (a,\ \text{a complex constant}).$$

$$L(t^a)=\frac{\Gamma(a+1)}{z^{a+1}},\qquad \mathrm{Re}\,z>0\ (a>-1,\ \text{a constant};$$

$$\Gamma \text{ is the Gamma function (see Sec. 5.6)).}\qquad\qquad (4.13.2.11)$$

Exercises A

(1) Prove (4.13.2.6) in detail.

(2) Prove the table of Laplace transforms shown in (4.13.2.11) in detail.

(3) Find the Laplace transform $L\varphi$ for each of the following φ and determine its abscissa of convergence $\sigma(\varphi)$:

(a) $\varphi(t)=\begin{cases}1, & 0<t_1<t<t_2\\ 0, & \text{otherwise}\end{cases}$.

(b) $\varphi(t)=\begin{cases}t, & 0<t<t_0\\ 0, & t>t_0\end{cases}$.

(c) $\varphi(t)=\begin{cases}1, & 0<t<t_0\\ -1, & t_0<t<2t_0\\ 0, & t>2t_0\end{cases}$.

(d) $\varphi(t)=\displaystyle\int_0^t H(s)H(t-s)ds,$ where $H(s)$ is the Heaviside function.

(e) $\varphi(t)=\displaystyle\int_0^t \cos(t-s)\sin s\,ds.$

(f) $\varphi(t)=\begin{cases}0, & t<0\\ \dfrac{\sin t}{t}, & t>0\end{cases}$.

(g) $\varphi(t)=\begin{cases}0, & t\le 0\\ \dfrac{1}{\sqrt{t}}e^{-\frac{c}{t}}, & t>0\end{cases},$ where $c>0$.

(4) Let $t_0>0$. Suppose φ is piecewise C^1 on $(0,\infty)$ and $\varphi(t)=0$ if $t\ge t_0$. Show that $(L\varphi)(z)$ is entire and has its Taylor series expansion

at $z = 0$ as

$$(L\varphi)(z) = \sum_{n=0}^{\infty} \left\{ \frac{(-1)^n}{n!} \int_0^{t_0} \varphi(t)t^n \, dt \right\} z^n.$$

(5) Let φ be piecewise C^1 on $(0, \infty)$ and be period with period $\sigma \neq 0$, namely, $\varphi(t) = \varphi(t + \sigma)$ for all $t \geq 0$. Show that

$$(L\varphi)(z) = \frac{1}{1 - e^{-\sigma z}} \int_0^{\sigma} \varphi(t)e^{-zt} \, dt, \quad \mathrm{Re}\, z > 0.$$

In case

$$\varphi(t) = \begin{cases} 1, & 0 < t < 1 \\ 0, & 1 < t < 2 \end{cases} \quad \text{and} \quad \varphi(t) = \varphi(t + 2) \quad \text{for } t \geq 0,$$

show that

$$(L\varphi)(z) = \frac{1}{z} \cdot \frac{1 - e^{-z}}{1 - e^{-2z}}.$$

(6) Evaluate the following integrals, where $x_0 > 0$ and n is a positive integer. Try to use the basic properties in (4.13.2.6), if possible.

(a) $\dfrac{1}{2\pi i} \displaystyle\int_{x_0-i\infty}^{x_0+i\infty} \dfrac{e^{zt}(1 - z^{-az})^2}{z} \, dz.$

(b) $\dfrac{1}{2\pi i} \displaystyle\int_{x_0-i\infty}^{x_0+i\infty} \dfrac{e^{zt}}{z\sqrt{1+z}} \, dz \quad (\sqrt{1} = 1).$

(c) $\dfrac{1}{2\pi i} \displaystyle\int_{x_0-i\infty}^{x_0+i\infty} \dfrac{t^z}{z^{n+1}} \, dz.$

(d) $\dfrac{1}{2\pi i} \displaystyle\int_{x_0-i\infty}^{x_0+i\infty} \dfrac{e^{zt}}{z^2(z^2+1)} \, dz.$

(e) $\dfrac{1}{2\pi i} \displaystyle\int_{x_0-i\infty}^{x_0+i\infty} \dfrac{e^{zt}}{\sqrt{1+z}} \, dz.$

(f) $\dfrac{1}{2\pi i} \displaystyle\int_{x_0-i\infty}^{x_0+i\infty} \dfrac{e^{zt}}{\sqrt{i+z}} \, dz.$

(g) $\dfrac{1}{2\pi i} \displaystyle\int_{x_0-i\infty}^{x_0+i\infty} \dfrac{e^{zt}}{1-iz} \, dz.$

(h) $\dfrac{1}{2\pi i} \displaystyle\int_{x_0-i\infty}^{x_0+i\infty} \dfrac{e^{zt}}{(z+1)\sqrt{z+2}} \, dz.$

(i) $\dfrac{1}{2\pi i}\displaystyle\int_{x_0-i\infty}^{x_0+i\infty}\dfrac{e^{zt}}{z(1-e^{-az})}dz$ $(a>0)$.

(j) $\dfrac{1}{2\pi i}\displaystyle\int_{x_0-i\infty}^{x_0+i\infty}e^{zt}\log\left(\dfrac{z+1}{z-1}\right)dz,\ \log(-1)=\dfrac{\pi}{2}i$.

(k) $\dfrac{1}{2\pi i}\displaystyle\int_{x_0-i\infty}^{x_0+i\infty}\dfrac{z^2e^{zt}}{(z^2+4)^2}dz$.

(l) $\dfrac{1}{2\pi i}\displaystyle\int_{x_0-i\infty}^{x_0+i\infty}\dfrac{e^{zt}}{(z-1)^4}dz$.

(7) Imitate Example 4(1) to prove the following: If $F(z)=(L\varphi)(z)$, then denote

$$\varphi(t)=L^{-1}(F(z))=(L^{-1}F)(z)=\dfrac{1}{2\pi i}\int_{x_0-i\infty}^{x_0+i\infty}F(z)e^{zt}dz.$$

(a) $L^{-1}\left(\dfrac{\sinh az}{z^2\cosh az}\right)=a+\dfrac{8}{\pi^2}\sum_{n=1}^{\infty}\dfrac{(-1)^n}{(2n-1)^2}\cos\frac{1}{2}(2n-1)\pi t\sin\frac{1}{2}(2n-1)a\pi$, $0<a<1$.

(b) $L^{-1}\left(\dfrac{1}{z\cosh z}\right)=1-\dfrac{4}{\pi}\sum_{n=1}^{\infty}\dfrac{(-1)^{n-1}}{2n-1}\cos\frac{1}{2}(2n-1)\pi t$.

(c) $L^{-1}\left(\dfrac{1}{z^2\sinh z}\right)=\dfrac{t^2}{2}+\dfrac{2}{\pi^2}\sum_{n=1}^{\infty}\dfrac{(-1)^n}{n^2}(1-\cos n\pi t)$.

(d) $L^{-1}\left(\dfrac{1}{z^2\cosh z}\right)=t+\dfrac{8}{\pi^2}\sum_{n=1}^{\infty}\dfrac{(-1)^n}{(2n-1)^2}\sin\frac{1}{2}(2n-1)\pi t$.

(e) $L^{-1}\left(\dfrac{\cosh az}{z^3\cosh bz}\right)=\frac{1}{2}(t^2+a^2-b^2)-\dfrac{16b^2}{\pi^3}\sum_{n=1}^{\infty}\dfrac{(-1)^n}{(2n-1)^3}\cos\dfrac{(2n-1)\pi a}{2b}\cos\dfrac{(2n-1)\pi t}{2b}$, $0<a<b$.

(8) Solve the following differential equations via the Laplace transforms.

(a) $\dfrac{d^2f}{dt^2}-2\dfrac{df}{dt}+3f=\varphi(t)$, where $\varphi(t)$ is the one in Exercise (3)(c) with $t_0=1$, $f(0)=0$, $f'(0)=1$.

(b) $\dfrac{d^2f}{dt^2}+f=t\sin t$, $f(0)=1$, $f'(0)=0$.

4.14. Asymptotic Function and Expansion of Functions Defined by Integrals with a Parameter

Suppose $f(z)$ is defined in a neighborhood of ∞. Let $0\le\alpha\le\beta\le2\pi$. If there is another function $\varphi(z)$ such that, for $\alpha\le\operatorname{Arg}z\le\beta$,

$$\lim_{z\to\infty}\dfrac{f(z)}{\varphi(z)}=1,\quad\text{denoted by }f(z)\sim\varphi(z)\quad\text{as }z\to\infty\text{ in }\alpha\le\operatorname{Arg}z\le\beta,$$

$$(4.14.1)$$

then $\varphi(z)$ is called an *asymptotic function* of $f(z)$ as $z \to \infty$ in $\alpha \le$ Arg $z \le \beta$. In most cases, if $|z|$ becomes larger and larger, then $f(z)$ turns to be more complicated while $\varphi(z)$ turns out to be more simpler.

In case there is a series $\sum_{n=0}^{\infty} \frac{a_n}{z^n}$ (not necessarily convergent) such that, for $n = 0, 1, 2, \ldots$, and, for $\alpha \le$ Arg $z \le \beta$,

$$\lim_{n\to\infty} z^n \left[f(z) - \sum_{k=0}^{n} \frac{a_k}{z^k} \right] = 0,$$

$$\text{denoted as } f(z) \sim \sum_{n=0}^{\infty} \frac{a_n}{z^n} \quad \text{as } z \to \infty \quad \text{in } \alpha \le \text{Arg } z \le \beta, \quad (4.14.2)$$

then the series $\sum_{n=0}^{\infty} \frac{a_n}{z^n}$ is called an *asymptotic expansion* of $f(z)$ as $z \to \infty$ in $\alpha \le$ Arg $z \le \beta$.

Here, we just illustrate some examples and then state three general main theoretical results (4.14.3), (4.14.4), and (4.14.7). For details, refer to Refs. [22 and 29].

Section (1) $f(x) = \frac{1}{2\pi i} \int_{x_0-i\infty}^{x_0+i\infty} F(z) e^{zt} dz, \ t > 0$

Suppose z_0 is a pole of order k of F. Let $F(z) = \frac{\varphi(z)}{(z-z_0)^k}$, where φ is analytic at z_0 and $\varphi(z_0) \ne 0$. Then

$$\text{Re } s(e^{zt}F(z); z_0) = \frac{1}{(k-1)!} \lim_{z\to z_0} \frac{d^{k-1}}{dz^{k-1}} \{(z-z_0)^k e^{zt} F(z)\}$$

$$= \frac{1}{(k-1)!} \lim_{z\to z_0} \frac{d^{k-1}}{dz^{k-1}} \{e^{zt}\varphi(z)\}$$

$$= \frac{1}{(k-1)!} e^{tz_0} \{\varphi(z_0)t^{k-1} + \cdots + \varphi^{(k-1)}(z_0)\}$$

$$= \frac{1}{(k-1)!} e^{tz_0} p(t), \quad (*_1)$$

where $p(t) = \varphi(z_0)t^{k-1} + \cdots + \varphi^{(k-1)}(z_0)$ is a polynomial of order $k-1$.

In case Re $z_0 > 0$: Then

$$|\text{Re } s(e^{zt}F(z); z_0)| = \frac{1}{(k-1)!} e^{t\text{Re } z_0} |p(t)| \to \infty \quad \text{as } t \to \infty.$$

In case Re $z_0 < 0$: Then

$$|\text{Re } s(e^{zt}F(z); z_0)| \to 0 \quad \text{as } t \to \infty.$$

In case $z_0 = iy_0 \neq 0$: Then

$\operatorname{Re} s(e^{zt} F(z); iy_0)$

$\quad = \dfrac{1}{(k-1)!} e^{ity_0} p(t)$

$\quad = \begin{cases} e^{ity_0} \cdot \varphi(z_0), & \text{which is bounded and oscillating as } t \to \infty, \\ & \text{if } k = 1 \\[2ex] \dfrac{1}{(k-1)!} e^{ity_0} p(t), & \text{which is unbounded and oscillating as } t \to \infty, \\ & \text{if } k \geq 2 \end{cases}$

In case $z_0 = 0$: Then

$\quad \operatorname{Re} s(e^{zt} F(z); z_0)$

$\quad = \dfrac{1}{(k-1)!} p(t) = \begin{cases} \varphi(z_0), & \text{if } k = 1 \\[2ex] \dfrac{1}{(k-1)!} p(t) \to \infty & \text{as } t \to \infty, \quad \text{if } k \geq 2 \end{cases}$

These results describe how the term $\operatorname{Re} s(e^{zt} F(z); z_0)$ behaves as $t \to \infty$.

Based on this argument, it is easy to obtain

The asymptotic function of $f(t) = \frac{1}{2\pi i} \int_{x_0 - i\infty}^{x_0 + i\infty} e^{zt} F(z) dz$ as $t \to \infty$. Suppose $F(z)$ is analytic in $\operatorname{Re} z \leq x_0$ ($x_0 > 0$) except finitely many poles in $\operatorname{Re} z < x_0$. Also suppose that $F(z) \to 0$ as $z \to \infty$ in $\operatorname{Re} z \leq x_0$. Then

$$f(t) = \text{P.V.} \frac{1}{2\pi i} \int_{x_0 - i\infty}^{x_0 + i\infty} e^{zt} F(z) dz \sim \sum_{\operatorname{Re} z_k \geq 0} \operatorname{Re} s(e^{zt} F(z); z_k), \quad \text{as } t \to \infty$$

where the sum is taken over these poles z_k with $\operatorname{Re} z_k \geq 0$. (4.14.3)

For instance,

$$\frac{1}{2\pi i} \int_{x_0 - i\infty}^{x_0 + i\infty} \frac{z e^{zt - \sqrt{z^2 + 2az}}}{(z - bi)\sqrt{z^2 + 2az}} dz \sim \operatorname{Re} s(e^{zt} F(z); bi)$$

$$= \frac{bi}{\sqrt{2abi - b^2}} e^{ibt - \sqrt{2abi - b^2}} \quad \text{as } t \to \infty,$$

where $a > 0$, $b > 0$, $x_0 > 0$ and $\sqrt{z^2 + 2az} > 0$ if $z > 0$;

$$\frac{1}{2\pi i} \int_{x_0 - i\infty}^{x_0 + i\infty} \frac{e^{zt}}{z^2(z + a)^3} dz \sim \operatorname{Re} s(e^{zt} F(z); 0)$$

$$= \frac{at - 3}{a^4} \quad \text{as } t \to \infty,$$

where $\operatorname{Re} a > 0$ and $x_0 > 0$.

Section (2) Asymptotic series expansions

We illustrate two examples.

Example 1. In case $\operatorname{Re} z \neq 0$, show that, as $z \to \infty$,

$$\int_0^z e^{z^2 - t^2}\, dt \sim \pm \frac{\sqrt{\pi}}{2} e^{z^2} - \left[\frac{1}{2z} + \sum_{n=2}^{\infty}(-1)^{n-1}\frac{1 \cdot 3 \cdot 5 \cdots (2n-3)}{2^n} \cdot \frac{1}{z^{2n-1}}\right],$$

where the plus or minus sign depends on $\operatorname{Re} z > 0$ or $\operatorname{Re} z < 0$, respectively. In case $\operatorname{Re} z = 0$, the term $\pm\frac{\sqrt{\pi}}{2} e^{z^2}$ should be omitted.

Solution. In case $\operatorname{Re} z > 0$: Integrate $e^{z^2 - t^2}$ (z fixed while t varies) along the path γ shown in Fig. 4.129. By Cauchy's integral theorem, we have

$$\int_\gamma e^{z^2 - t^2}\, dt = 0 = -\int_0^{t_0} e^{z^2 - t^2}\, dt + \int_0^z e^{z^2 - t^2}\, dt$$

$$+ \int_z^{t_0 + i\operatorname{Im} z} e^{z^2 - t^2}\, dt + \int_{t_0 + i\operatorname{Im} z}^{t_0} e^{z^2 - t_0^2}\, dt.$$

Since $e^{z^2 - t^2} \to 0$ uniformly for fixed z on $\operatorname{Re} z > 0$ as $t \to \infty$, let $t_0 \to \infty$ in the above relation, and we obtain

$$\int_0^z e^{z^2 - t^2}\, dt = \int_0^\infty e^{z^2 - t^2}\, dt - \int_x^\infty e^{z^2 - (t+iy)^2}\, dt,$$

$$z = x + iy \quad \text{with } \operatorname{Re} z = x > 0. \tag{$*_1$}$$

To estimate the right side of $(*_1)$: Notice that

$$\int_0^\infty e^{z^2 - t^2}\, dt = e^{z^2}\int_0^\infty e^{-t^2}\, dt = \frac{\sqrt{\pi}}{2} e^{z^2}.$$

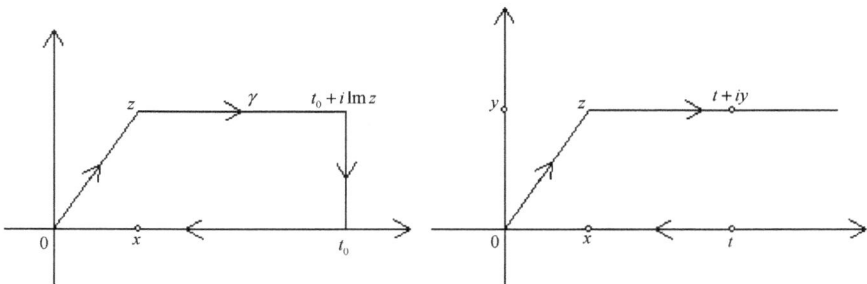

Fig. 4.129

Remind again that $z = x + iy$ is a fixed constant in the sequel. Then

$$\int_x^\infty e^{z^2 - (t+iy)^2} \, dt$$

$$= -\frac{1}{2} \int_x^\infty \frac{1}{t+iy} d\{e^{z^2 - (t+iy)^2}\}$$

$$= \frac{1}{2z} - \frac{1}{2} \int_x^\infty \frac{1}{(t+iy)^2} e^{z^2 - (t+iy)^2} \, dt \quad \text{(by integration by parts)}$$

$$= \frac{1}{2z} - \frac{1}{2^2 z^3} + \frac{3}{2} \int_x^\infty \frac{1}{(t+iy)^4} e^{z^2 - (t+iy)^2} \, dt$$

$$\text{(by integration by parts)}$$

$$= \cdots$$

$$= \frac{1}{2z} + \sum_{k=2}^n (-1)^{k-1} \frac{1 \cdot 3 \cdot 5 \cdots (2k-3)}{2^k z^{2k-1}}$$

$$+ (-1)^n \frac{1 \cdot 3 \cdot 5 \cdots (2n-1)}{2^n} \int_x^\infty \frac{e^{z^2 - (t+iy)^2}}{(t+iy)^{2n}} \, dt. \tag{$*_2$}$$

Wherein the remainder can be estimated as follows:

$$\left| \frac{(-1)^n \cdot 1 \cdot 3 \cdot 5 \cdots (2n-1)}{2^n} \int_x^\infty \frac{e^{z^2 - (t+iy)^2}}{(t+iy)^{2n}} \, dt \right|$$

$$\leq \frac{1 \cdot 3 \cdot 5 \cdots (2n-1)}{2^n} \int_x^\infty \frac{e^{x^2 - t^2}}{|t+iy|^{2n}} \, dt;$$

$$\int_x^\infty \frac{e^{x^2 - t^2}}{|t+iy|^{2n}} \, dt = -\frac{1}{2} \int_x^\infty \frac{1}{t(t^2 + y^2)^n} d\{e^{x^2 - t^2}\}$$

$$= \frac{1}{2x|z|^{2n}} - \frac{1}{2} \int_x^\infty \frac{(2n+1)t^2 + y^2}{t^2(t^2+y^2)^{n+1}} e^{x^2 - t^2} \, dt < \frac{1}{2x|z|^{2n}}.$$

Substituting this estimate into $(*_2)$ and then, into $(*_1)$, we obtain

$$|z^{2n-1}| \left| \int_0^z e^{z^2 - t^2} \, dt - \frac{\sqrt{\pi}}{2} e^{z^2} + \left[\frac{1}{2z} + \sum_{k=2}^n (-1)^{k-1} \frac{1 \cdot 3 \cdot 5 \cdots (2k-3)}{2^k z^{2k-1}} \right] \right|$$

$$\leq |z|^{2n-1} \frac{1 \cdot 3 \cdot 5 \cdots (2n-1)}{2^n} \cdot \frac{1}{2x|z|^{2n}}$$

$$= \frac{1 \cdot 3 \cdot 5 \cdots (2n-1)}{2^{n+1} x |z|} \to 0 \quad \text{as } z \to \infty.$$

This proves the case that $\operatorname{Re} z > 0$.

In case Re $z < 0$: Integrate $e^{z^2-t^2}$ along the path shown in Fig. 4.130. Then

$$\int_0^z e^{z^2-t^2}\,dt = -\int_0^\infty e^{z^2-t^2}\,dt - \int_x^\infty e^{z^2-(t+iy)^2}\,dt.$$

What remains is the same as in the previous case.

In case Re $z = 0$: Integrate $e^{z^2-t^2}$ along the path shown in Fig. 4.131. We have, in this case,

$$\int_0^z e^{z^2-t^2}\,dt = \int_0^\infty e^{z^2-t^2}\,dt - \int_0^\infty e^{z^2-(t+iy)^2}\,dt, \quad \text{where } z = iy$$

$$= \frac{\sqrt{\pi}}{2}e^{z^2} - \left[\frac{1}{2z} + \sum_{k=2}^n (-1)^{k-1}\frac{1\cdot 3\cdot 5\cdots(2k-3)}{2^k z^{2k-1}} \right.$$

$$\left. +(-1)^n\frac{1\cdot 3\cdot 5\cdots(2n-1)}{2^n}\int_0^\infty \frac{e^{z^2-(t+iy)^2}}{(t+iy)^{2n}}\,dt \right]$$

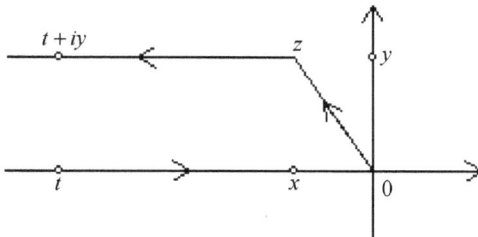

Fig. 4.130

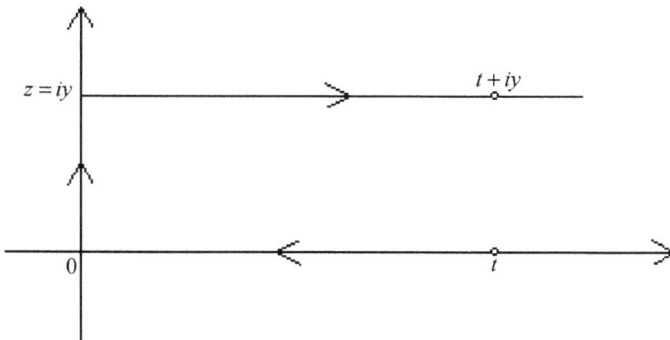

Fig. 4.131

wherein

$$\left| \int_0^\infty \frac{e^{z^2-(t+iy)^2}}{(t+iy)^{2n}} dt \right| \le \int_0^\infty \frac{e^{-t^2}}{y^{2n}} dt = \frac{\sqrt{\pi}}{2y^{2n}} = \frac{\sqrt{\pi}}{2|z|^{2n}}.$$

Since $\lim_{z=iy\to\infty} z^{2n-1}e^{z^2} = 0$, therefore

$$\left| z^{2n-1} \left[\int_0^z e^{z^2-t^2} dt + \left\{ \frac{1}{2z} + \sum_{k=2}^n (-1)^{k-1} \frac{1\cdot 3\cdot 5 \cdots (2k-3)}{2^k z^{2k-1}} \right\} \right] \right|$$

$$\le |z|^{2n-1} \cdot \frac{\sqrt{\pi}}{2} |e^{z^2}| + |z|^{2n-1} \frac{1\cdot 3\cdot 5 \cdots (2n-1)}{2^n} \cdot \frac{\sqrt{\pi}}{2|z|^{2n}}$$

$$\to 0 \quad \text{as } z = iy \to \infty.$$

And the required result follows.

Example 2. Let $\alpha > 0$. Designate $z^{\frac{3}{2}} > 0$ if $z = x > 0$. Show that, as $t \to \infty$,

$$f(t) = \frac{1}{2\pi i} \int_{\alpha-i\infty}^{\alpha+i\infty} \frac{e^{zt}}{z(1+z^{\frac{3}{2}})} dz \sim 1 + \frac{1}{\pi} \sum_{n=0}^\infty (-1)^n \frac{\Gamma\left(3n+\frac{3}{2}\right)}{t^{3n+\frac{3}{2}}},$$

where $\Gamma(x) = \int_0^\infty e^{-t} t^{x-1} dt$, $x > 0$. In case $\alpha > 1$ and $t > 0$ is small enough,

$$f(t) = \frac{t^{\frac{3}{2}}}{\Gamma\left(\frac{5}{2}\right)} - \frac{t^3}{\Gamma(4)} + \frac{t^{\frac{9}{2}}}{\Gamma\left(\frac{11}{2}\right)} - \cdots \simeq \frac{t^{\frac{3}{2}}}{\Gamma\left(\frac{5}{2}\right)}$$

$$= \frac{4t^{\frac{3}{2}}}{3\sqrt{\pi}} \quad \text{(approximate value).}$$

Solution. $z^{\frac{3}{2}}$ denotes the branch on $\mathbf{C} - (\infty, 0]$ determined uniquely by $z^{\frac{3}{2}} > 0$ in case $z > 0$ is real. Observe that $1 + z^{\frac{3}{2}} = 0$ has solutions $z = e^{\frac{2(\pi i + 2k\pi i)}{3}}$, where $k = 0, 1, 2$ (see Example 1 in Sec. 2.7.2), in which the one $e^{2\pi i} = 1$ corresponding to $k = 1$ should be omitted. As a consequence, $F(z) = \frac{1}{z(1+z^{\frac{3}{2}})}$ has simple poles at $z_1 = e^{\frac{2\pi i}{3}}$ and $z_2 = e^{\frac{4\pi i}{3}}$. Note that $z = 0$ is *not* a pole but is a nonisolated singular point of $F(z)$ (see (2.2.5)). Direct computation shows that

$$\text{Re } z(e^{zt}F(z); z_1) = \frac{e^{tz_1}}{z_1 \cdot \frac{3}{2} \cdot z_1^{\frac{1}{2}}} = -\frac{2}{3} e^{tz_1};$$

$$\text{Re } z(e^{zt}F(z); z_2) = \frac{e^{tz_2}}{z_2 \cdot \frac{3}{2} \cdot z_2^{\frac{1}{2}}} = -\frac{2}{3} e^{tz_2}.$$

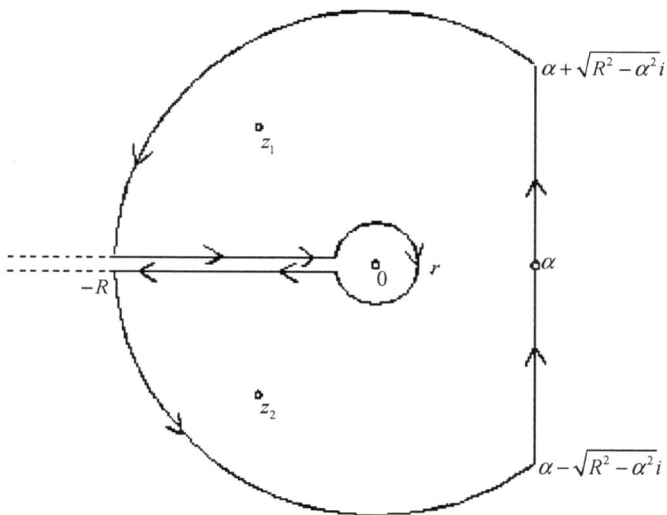

Fig. 4.132

By integrating $e^{zt}F(z)$ along the dented path γ shown in Fig. 4.132, we have

$$\frac{1}{2\pi i}\int_{\gamma} e^{zt}F(z)dz = -\frac{2}{3}(e^{tz_1}+e^{tz_2}) = -\frac{4}{3}e^{-\frac{t}{2}}\cos\frac{\sqrt{3}}{2}t. \qquad (*_3)$$

Along C_R : $z = Re^{i\theta}$, $\theta_0 \le \theta \le 2\pi - \theta_0$, where $\theta_0 = \mathrm{Arg}(\alpha + \sqrt{R^2 - \alpha^2}i)$: As usual,

$$\lim_{R\to\infty}\int_{C_R} e^{zt}F(z)dz = 0.$$

Along C_{0r} : $z = re^{i\theta}$, $-\pi \le \theta \le \pi$: As $r \to 0$,

$$\int_{C_{0r}} e^{zt}F(z)dz = \int_{\pi}^{-\pi}\frac{e^{tre^{i\theta}}}{re^{i\theta}(1 + r^{\frac{3}{2}}e^{\frac{3i\theta}{2}})}re^{i\theta}d\theta \to -2\pi i.$$

Along the upper and lower sides of the cut $[-R, 0]$: As $R \to \infty$,

$$\int_{\text{upper side}} + \int_{\text{lower side}}$$

$$= \int_{-R}^{0}\frac{e^{tx}}{x[1 + |x|^{\frac{3}{2}}e^{\frac{3\pi i}{2}}]}dx + \int_{0}^{-R}\frac{e^{tx}}{x[1 + |x|^{\frac{3}{2}}e^{-\frac{3\pi i}{2}}]}dx$$

$$= -2i\int_{0}^{R}\frac{x^{1/2}e^{-tx}}{1 + x^3}dx \to -2i\int_{0}^{\infty}\frac{x^{1/2}e^{-tx}}{1 + x^3}dx.$$

Substituting these results into $(*_3)$ if letting $r \to 0$ and $R \to \infty$, then

$$\frac{1}{2\pi i}\int_{\alpha-i\infty}^{\alpha+i\infty} e^{zt}F(z)dz$$

$$= -\frac{4}{3}e^{-\frac{t}{2}}\cos\frac{\sqrt{3}}{2}t + 1 + \frac{1}{\pi}\int_0^\infty \frac{x^{\frac{1}{2}}e^{-tx}}{1+x^3}dx, \quad t > 0. \qquad (*_4)$$

To estimate the behavior of the integral in the right of $(*_4)$ as $t \to \infty$:
Since

$$\frac{1}{1+x^3} = \sum_{k=0}^{n}(-1)^k x^{3k} + (-1)^{n+1}\frac{x^{3n+3}}{1+x^3}$$

$$\Rightarrow \int_0^\infty \frac{x^{\frac{1}{2}}e^{-tx}}{1+x^3}dx = \sum_{k=0}^{n}(-1)^k \int_0^\infty x^{3k+\frac{1}{2}}e^{-tx}dx$$

$$+ (-1)^{n+1}\int_0^\infty \frac{x^{3n+\frac{7}{2}}e^{-tx}}{1+x^3}dx$$

$$= \sum_{k=0}^{n}(-1)^k \frac{\Gamma\left(3k+\frac{3}{2}\right)}{t^{3k+\frac{3}{2}}} + R_n(t),$$

where $R_n(t) = (-1)^{n+1}\int_0^\infty \frac{x^{3n+\frac{7}{2}}e^{-tx}}{1+x^3}dx$. Now,

$$|R_n(t)| \le \int_0^\infty x^{3n+\frac{7}{2}}e^{-tx}dx$$

$$= \frac{1}{t^{3n+\frac{9}{2}}}\int_0^\infty x^{3n+\frac{7}{2}}e^{-x}dx = \frac{\Gamma\left(3n+\frac{9}{2}\right)}{t^{3n+\frac{9}{2}}}.$$

$$\Rightarrow \left|t^{3n+\frac{3}{2}}\left[-\frac{4}{3}e^{-\frac{t}{2}}\cos\frac{\sqrt{3}}{2}t + \frac{1}{\pi}R_n(t)\right]\right|$$

$$\le \frac{4}{3}t^{3n+\frac{3}{2}}e^{-\frac{t}{2}} + \frac{1}{\pi}t^{3n+\frac{3}{2}}\cdot\frac{\Gamma\left(3n+\frac{9}{2}\right)}{t^{3n+\frac{9}{2}}} \to 0 \quad \text{as } t \to \infty.$$

Putting this result into $(*_4)$ and by definition (4.14.2), we obtain the asymptotic series expansion as $t \to \infty$.

As for $t > 0$ is sufficiently small: Choose $\alpha > 1$. Then $1 + z^{\frac{3}{2}}$ is analytic in $\operatorname{Re} z > \alpha$ and is not equal to zero there. Then

$$\frac{1}{1+z^{\frac{3}{2}}} = z^{-\frac{3}{2}}(1+z^{-\frac{3}{2}})^{-1} = z^{-\frac{3}{2}}\sum_{n=0}^{\infty}(-1)^n z^{-\frac{3n}{2}}$$

$$= \sum_0^{\infty}(-1)^n z^{-\frac{(3n+3)}{2}}, \quad |z| > 1$$

$$\Rightarrow f(t) = \sum_{n=0}^{\infty} (-1)^n \frac{1}{2\pi i} \int_{a-i\infty}^{a+i\infty} \frac{e^{tz}}{z^{(3n+3)/2+1}} dz$$

$$= \sum_{n=0}^{\infty} (-1)^n \frac{t^{(3n+3)/2}}{\Gamma\left(\frac{3n+5}{2}\right)} \quad \text{(see (5.6.1.8))} \quad \simeq \frac{t^{\frac{3}{2}}}{\Gamma\left(\frac{5}{2}\right)} = \frac{4t^{\frac{3}{2}}}{3\sqrt{\pi}}.$$

The approximate value means that $f(t) - \frac{t^{\frac{3}{2}}}{\Gamma\left(\frac{5}{2}\right)} = o(t^3)$ as $t \to \infty$.

However, there is a general formula for

The asymptotic series expansion of the Laplace transform. Suppose $\varphi(z)$ is analytic in a domain containing the positive real axis $[0, \infty)$. Also suppose that either $\varphi(z)$ is bounded on the positive real axis or φ and all of its derivatives are of exponential order, namely, there are constants $M_n > 0$ and α_n such that $|\varphi^{(n)}(z)| \le M_n e^{\alpha_n t}$ for $t \ge 0$ and $n \ge 0$. Then

$$(L\varphi)(x) = \int_0^{\infty} e^{-xt} \varphi(t) dt \sim \sum_{n=0}^{\infty} \frac{\varphi^{(n)}(0)}{x^{n+1}} \quad \text{as } x \to \infty. \tag{4.14.4}$$

Note that this result is still valid for C^{∞} function φ on $[0, \infty)$ such that each $\varphi^{(n)}(t)$ is of exponential order. This formula can be proved either by the Taylor series expansion $\sum_0^{\infty} a_n z^n$ of $\varphi(z)$ at 0 or by repeated usage of integration by parts. The readers are invited to present a detailed proof in Exercise A (2).

Section (3) Approximate function of $\int_a^{\infty} \varphi(t) e^{-zt} dt$ as Re $z \to \infty$

The function $\Gamma(z+1)$ defined through the integral with a real variable t

$$\Gamma(z+1) = \int_0^{\infty} t^z e^{-t} dt \tag{4.14.5}$$

is called the *gamma function* of the complex variable z. It is analytic in Re $z > -1$. We take it for granted here. For formal and detailed account about the gamma function, see Sec. 5.6.

Example 3 (Stirling's formula).

$$\Gamma(z+1) = \sqrt{2\pi} z^{z+\frac{1}{2}} e^{-z} \left(1 + O\left(\frac{1}{z}\right)\right), \quad \text{or}$$

$$\Gamma(z+1) \sim \sqrt{2\pi} z^{z+\frac{1}{2}} e^{-z}$$

as $z \to \infty$ in Re $z > -1$. $\tag{4.14.6}$

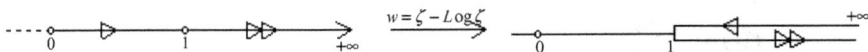

Fig. 4.133

Proof.   Fix a point z in $\operatorname{Re} z > -1$. Applying the change of variables $t = z\zeta$ where ζ denotes temporarily a real variable, (4.14.5) turns out to be

$$\Gamma(z+1) = z^{z+1} \int_0^\infty e^{-z(\zeta - \log \zeta)} d\zeta. \tag{$*_5$}$$

From now on, treat ζ as a complex variable and consider the transform $w = \zeta - \log \zeta$. Choose the principal branch $\operatorname{Log} \zeta$ in $\mathbf{C} - (\infty, 0]$. Then $w = \zeta - \operatorname{Log} \zeta$ is single-valued and analytic in $\mathbf{C} - (\infty, 0]$. Note that $\frac{dw}{d\zeta} = 1 - \frac{1}{\zeta} \neq 0$ if $\zeta \neq 1$. Therefore $w = \zeta - \operatorname{Log} \zeta$ is conformal everywhere in $\mathbf{C} - (\infty, 0]$ except at $\zeta = 1$. It maps the interval $(0,1)$ onto the interval $(1, \infty)$ strictly decreasing, and maps $(1, \infty)$ onto $(1, \infty)$ strictly increasing. See Fig. 4.133. As a whole, $w = \zeta - \operatorname{Log} \zeta$ maps $(0, \infty)$ onto $[1, \infty)$ in 2-to-1 manner except at $\zeta = 1$, where w assumes $w = 1$ twice because $w(1) = 1, w'(1) = 0$, and $w''(1) \neq 0$.

Let $\zeta = \zeta_1(w)$ be the unique inverse function of $w = w(\zeta) : (0,1) \to (1, \infty)$, and $\zeta = \zeta_2(w) : (1, \infty) \to (1, \infty)$. Observe that $0 < \zeta_1(w) < 1$ and $\lim_{w \to \infty} \zeta_1(w) = 0$, while $1 < \zeta_2(w) < \infty$ and $\lim_{w \to \infty} \zeta_2(w) = \infty$. Under these circumstance, $(*_5)$ is subject to the following change

$$z^{-z-1}\Gamma(z+1) = \int_0^1 e^{-z(\zeta - \operatorname{Log} \zeta)} d\zeta + \int_1^\infty e^{-z(\zeta - \operatorname{Log} \zeta)} d\zeta$$

$$= \int_\infty^1 e^{-zw} \zeta_1'(w) dw + \int_1^\infty e^{-zw} \zeta_2'(w) dw$$

$$= \int_1^\infty e^{-zw} [\zeta_2'(w) - \zeta_1'(w)] dw. \tag{$*_6$}$$

What remains is to consider the behavior of $\zeta_1'(w)$ and $\zeta_2'(w)$ on $(1, \infty)$, and as $w \to 1$ and ∞, respectively.

Since $\zeta_k(w) - \operatorname{Log} \zeta_k(w) = w$ on $(1, \infty)$, for $k = 1, 2$,

$$\zeta_k'(w) - \frac{\zeta_k'(w)}{\zeta_k(w)} = 1$$

$$\Rightarrow \zeta_k'(w) = \frac{\zeta_k(w)}{\zeta_k(w) - 1}, \quad k = 1, 2.$$

It follows obviously that $\lim_{w \to 1} \zeta_k'(w) = \infty$ for $k = 1, 2$, and $\lim_{w \to \infty} \zeta_1'(w) = 0$ and $\lim_{w \to \infty} \zeta_2'(w) = 1$. Moreover, $w = \zeta - \operatorname{Log} \zeta = 1 + \eta(\zeta)^2$, where $\eta(\zeta) = (\zeta - 1)\{\sum_{n=2}^\infty \frac{(-1)^n}{n}(\zeta - 1)^{n-2}\}^{\frac{1}{2}}$ with $\{\ \}^{\frac{1}{2}}$ assuming

$\frac{1}{\sqrt{2}}$ or $\frac{-1}{\sqrt{2}}$ at $\zeta = 1$. According to (3) in (3.5.1.8), there is a univalent function $\zeta = g(\eta)$ in an open neighborhood of $\eta = 0$ such that $g(0) = 1$, $g'(0) = -\sqrt{2}$ or $\sqrt{2}$ (subject to $\{\ \}^{\frac{1}{2}}$ assuming $\frac{-1}{\sqrt{2}}$ or $\frac{1}{\sqrt{2}}$ at $\zeta = 1$), and the double valued inverse function $\zeta = \zeta(w)$ of $w = \zeta - \operatorname{Log}\zeta$ near $\zeta = 1$ can be expressed as

$$\zeta = \zeta(w) = g(\sqrt{w-1}),$$
$$\Rightarrow \zeta_1(w) = 1 - \sqrt{2}\sqrt{w-1} + \cdots, \quad w > 1, \ \text{if } g'(0) = -\sqrt{2},$$
$$\zeta_2(w) = 1 + \sqrt{2}\sqrt{w-1} + \cdots, \quad w > 1, \ \text{if } g'(0) = \sqrt{2},$$
$$\Rightarrow \zeta_2'(w) - \zeta_1'(w) = \frac{\sqrt{2}}{\sqrt{w-1}} + \varphi(w), \quad w > 1,$$

where $|\varphi(w)| \le M\sqrt{|w-1|}$, $w > 1$, for some constant $M > 0$. Returning to $(*_6)$,

$$z^{-z-1}\Gamma(z+1) = \sqrt{2}\int_1^\infty (w-1)^{-\frac{1}{2}}e^{-zw}dw + \int_1^\infty \varphi(w)e^{-zw}dw,$$

where

$$\int_1^\infty (w-1)^{-\frac{1}{2}}e^{-zw}dw = \frac{e^{-z}}{\sqrt{z}}\int_0^\infty t^{-\frac{1}{2}}e^{-t}dt$$
$$= \frac{2e^{-z}}{\sqrt{z}}\int_0^\infty e^{-t^2}dt = \sqrt{\frac{\pi}{z}}e^{-z};$$

$$\left|\int_1^\infty \varphi(w)e^{-zw}dw\right| \le M\int_1^\infty |w-1|^{\frac{1}{2}}e^{-(\operatorname{Re}z)w}dw$$
$$\le \frac{Me^{-\operatorname{Re}z}}{(\operatorname{Re}z)^{\frac{3}{2}}}\int_0^\infty t^{\frac{1}{2}}e^{-t}dt$$
$$\le M_1 \cdot \frac{e^{-\operatorname{Re}z}}{(\operatorname{Re}z)^{\frac{3}{2}}},$$

where $M_1 = M\int_1^\infty t^{\frac{1}{2}}e^{-t}dt$ is a constant.

Note that, in case $z > -1$ is real, one can use z to replace $\operatorname{Re}z$ in the above expression. Consequently,

$$\Gamma(z+1) = \sqrt{2\pi}z^{z+\frac{1}{2}}e^{-z}\left(1 + O\left(\frac{1}{\operatorname{Re}z}\right)\right) \quad \text{as } \operatorname{Re}z > -1 \quad \text{and} \quad z \to \infty.$$

$\square$

The key points in the process of proving Stirling's formula in Example 3 are twofold: Transform $\int_0^\infty t^z e^{-t}dt$ into $\int_1^\infty \left\{\frac{c}{\sqrt{w-1}} + \varphi(w)\right\}e^{-zw}dw$ via the change of variables, where $|\varphi(w)| \le M\sqrt{|w-1|}$, and then try to estimate the approximate value of the latter integral. This method is of general interest and can be used to prove the following

approximate function of $\int_0^\infty \varphi(t)e^{-zt}dt$ *as* Re $z \to \infty$. Suppose $\varphi(t)$ satisfies:

(1) $|\varphi(t)| \leq t$,   if $t - a \geq \rho > 0$; and
(2) $\varphi(t) = \sum_{k=0}^\infty c_k(t-a)^{\alpha+\beta k}$, $0 < t - a < \rho$.

Then, as Re $z \to \infty$,

$$\int_a^\infty \varphi(t)e^{-zt}dt = \sum_{k=0}^{n-1} c_k\Gamma(\alpha+\beta k)z^{-\alpha-\beta k-1} + O\left(\frac{1}{(\text{Re }z)^{\alpha+\beta n+1}}\right)$$

where $n \geq 1$ is any fixed integer.   (4.14.7)

   The proof is left as Exercise A (3). The readers are invited to visit Ref. [29], pp. 159–165 for details.

Exercises A

(1) Prove (4.14.3) in detail.
(2) Prove (4.14.4) in detail.
(3) Prove (4.14.7) in detail.
(4) Suppose $\varphi(z)$ is analytic in a domain containing the real axis. Show that

$$\int_{-\infty}^\infty e^{-\frac{xt^2}{2}}\varphi(t)dt \sim \frac{\sqrt{2\pi}}{\sqrt{x}}$$

$$\times \left(a_0 + \frac{a_2}{x} + \cdots + \frac{a_{2n}\cdot 1\cdot 3\cdots(2n-1)}{x^n} + \cdots\right)$$

as x (real) $\to \infty$, where $\varphi(z) = \sum_{n=0}^\infty a_n z^n$ near 0. Justify that

$$\int_{-\infty}^\infty e^{-\frac{xt^2}{2}}\cos t\, dt \sim \frac{\sqrt{2\pi}}{\sqrt{x}}$$

$$\times \left(1 - \frac{1}{2x} + \frac{1}{2^2\cdot 2!\cdot x^2} - \cdots\right) \quad \text{as } x \text{ (real)} \to \infty, \quad \text{and}$$

$$\int_{-\infty}^\infty e^{-\frac{xt^2}{2}}\sin t\, dt \sim \frac{\sqrt{2\pi}}{\sqrt{x}}$$

$$\times \left(\frac{1}{x} - \frac{1\cdot 3\cdot 5}{3!}\frac{1}{x^3} + \frac{1\cdot 3\cdot 5\cdot 7\cdot 9}{5!}\frac{1}{x^5} - \cdots\right) \quad \text{as } x \text{ (real)} \to \infty.$$

Try to write out the general terms of both series on the right.

(5) Show the following asymptotic expansions.

(a) $\displaystyle \int_x^\infty \frac{e^{-t}}{t}\,dt \sim e^{-x}\sum_{n=0}^\infty (-1)^n \frac{n!}{x^{n+1}}$, $x > 0$. What is $\int_x^\infty \frac{e^{x-t}}{t}\,dt \sim$?

(b) $\displaystyle \int_0^\infty \frac{e^{-xt}}{1+t^2}\,dt \sim \frac{1}{x} + \sum_{n=1}^\infty (-1)^n \frac{(2n)!}{x^{2n+1}}$, $x > 0$.

(c) P.V. $\displaystyle \int_{-x}^\infty \frac{e^{-x-t}}{t}\,dt \sim -\sum_{n=1}^\infty \frac{(n-1)!}{x^n}$, $x > 0$.

(d) P.V. $\displaystyle \int_{-\infty}^\infty \frac{e^{-t^2}}{t-x}\,dx \sim \sqrt{\pi}\sum_{n=0}^\infty \frac{(2n)!}{2^{2n}\cdot n!}\frac{1}{x^{2n+1}}$, $x \in \mathbf{R}$.

(e) $\displaystyle \int_0^\infty \frac{t^\alpha e^{-t\beta}}{t-x}\,dx \sim \frac{1}{\beta}\sum_{n=1}^\infty \Gamma(\frac{\alpha+n}{\beta})\frac{1}{x^n}$, $\alpha > -1$, $\beta > 0$, $x \in \mathbf{R}$.

Note that, in case $x > 0$, the integral is taken as its principal value.

(6) Imitate Example 2 to show that, as $t \to \infty$,

$$ f(t) = \frac{1}{2\pi i}\int_{\alpha-i\infty}^{\alpha+i\infty} \frac{e^{zt}\sqrt{z}}{z^2+a^2}\,dz \sim \frac{1}{a}\sin\left(at+\frac{\pi}{4}\right) $$

$$ -\frac{1}{a^2\pi}\sum_{n=0}^\infty \frac{\Gamma\left(2n+\frac{3}{2}\right)}{a^{2n}t^{2n+\frac{3}{2}}}, \qquad \alpha > 0,\ a > 0. $$

For $t > 0$ small, show that

$$ f(t) = 2\sqrt{\frac{t}{\pi}}\left[1 - \frac{(2at)^2}{1\cdot 3\cdot 5} + \frac{(2at)^4}{1\cdot 3\cdot 5\cdot 7\cdot 9} - \cdots\right] \sim 2\sqrt{\frac{t}{\pi}}. $$

(7) Let $\alpha > 0$. Show that, as $t \to \infty$,

$$ \frac{1}{2\pi i}\int_{\alpha-i\infty}^{\alpha+i\infty} \frac{e^{zt}}{\sqrt{z}(z^2+1)}\,dz - \sin\left(t-\frac{\pi}{4}\right) \sim -\frac{2}{\sqrt{\pi}} $$

$$ \times \sum_{n=0}^\infty (-1)^n \frac{(4n)!}{(2n)!}\left(\frac{1}{2\sqrt{t}}\right)^{4n+1}. $$

4.15.   The Summation of Series by Residues

The method of residues can also be used to evaluate the sum of series such as

$$ \sum_{-\infty}^\infty a_n; \qquad \sum_{-\infty}^\infty (-1)^n a_n. $$

One needs only to choose an analytic function $f(z)$ on $\mathbf{C}$, except *countably many* isolated singularities z_k in $\mathbf{C}$, such that

$$f(n) = a_n, \quad n = 0, \pm 1, \pm 2, \ldots;$$

and another analytic function $g(z)$ on $\mathbf{C}$, except *simple poles* at each of $n = 0, \pm 1, \pm 2, \ldots$, such that

$$\mathrm{Re}\,s(g(z); n) = 1 \text{ or } (-1)^n, \ n = 0, \pm 1, \pm 2, \ldots .$$

Then choose a rectifiable Jordan closed curve γ, not passing the integers and the singularities z_k of f. By residue theorem,

$$\int_\gamma f(z)g(z)dz = 2\pi i \left\{ \sum_{n\in\text{Int }\gamma} f(n) + \sum_{z_k\in\text{Int }\gamma} \mathrm{Re}\,s(f(z)g(z); z_k) \right\} \quad \text{or}$$

$$2\pi i \left\{ \sum_{n\in\text{Int }\gamma} (-1)^n f(z) + \sum_{z_k\in\text{Int }\gamma} \mathrm{Re}\,s(f(z)g(z); z_k) \right\}.$$

$$(4.15.1)$$

Try to construct a sequence of such curves γ_m, expanding outwardly and tending to ∞ as $m \to \infty$, so that $\lim_{m\to\infty} \int_{\gamma_m} f(z)g(z) = 0$. If this is the case, we obtain

$$\sum_n f(n) \text{ or } \sum_n (-1)^n f(n) = -\sum_{z_k} \mathrm{Re}\,s(f(z)g(z); z_k). \qquad (4.15.2)$$

The applicability and availability of this process depends mainly on how easy it is to compute the residues and to show that $\int_\gamma f(z)g(z)dz \to 0$ as $\gamma \to \infty$ outwardly.

As for $g(z)$, usually we choose

$$g(z) = \pi \cot \pi z \quad \text{or} \quad \frac{2\pi i}{e^{2\pi i z} - 1} \quad \text{or} \quad \frac{-2\pi i}{e^{-2\pi i z} - 1}, \qquad (4.15.3)$$

which has simple poles at each integer n with $\mathrm{Re}\,s(g(z); n) = 1$ for $n = 0, \pm 1, \pm 2, \ldots$; and

$$g(z) = \frac{\pi}{\sin \pi z}, \qquad (4.15.4)$$

which has simple poles at each integer n with $\mathrm{Re}\,s(g(z); n) = (-1)^n$, $n = 0, \pm 1, \pm 2, \ldots .$

While the choice of the path γ of integration depends actually on what the functions $f(z)$ and $g(z)$ are. Take, for instance, the sequence path γ_n with vertices $(n + \frac{1}{2})(\pm 1 \pm i)$, for $n = 1, 2, \ldots$ (see Fig. 4.134). In this case, along the lower and upper horizontal sides $z = x \pm (n + \frac{1}{2})i$, $|x| \le (n + \frac{1}{2})$,

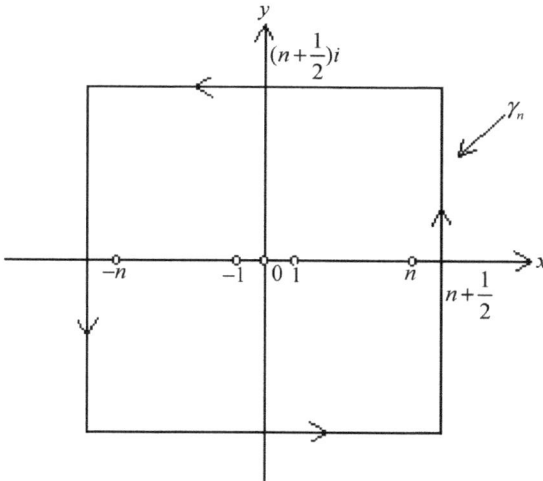

Fig. 4.134

we have

$$\left|\pi \cot \pi z\right| \leq \pi \sqrt{\frac{\sinh^2 \pi y + 1}{\sinh^2 \pi y}} < 2\pi, y = n + \frac{1}{2}, \quad \text{and}$$

$$\left|\frac{\pi}{\sin \pi z}\right| \leq \frac{2\pi}{\left|e^{-\pi y} - e^{\pi y}\right|} \leq 2\pi e^{-n\pi}, y = n + \frac{1}{2}; \qquad (4.15.5)$$

along the left and right vertical sides of $z = \pm(n + \frac{1}{2}) + iy$, $|y| \leq n + \frac{1}{2}$,

$$\left|\pi \cot \pi z\right| = \pi \sqrt{\frac{\sinh^2 \pi y}{1 + \sinh^2 \pi y}} < \pi;$$

$$\left|\frac{\pi}{\sin \pi z}\right| \leq \frac{\pi}{\cosh \pi y}. \qquad (4.15.6)$$

These estimates might be helpful in general. Remember that the absolute values of both $\pi \cot \pi z$ and $\frac{\pi}{\sin \pi z}$ along γ_n are less than 2π, a constant independent of n.

We summarize this process as a universal method in

The series summation. Suppose $g(z)$ is analytic in $\mathbf{C}$ except isolated singularities α_k so that $\lim_{k \to \infty} \alpha_k = \infty$ and $f(z)$ is analytic in $\mathbf{C}$ except isolated singularities z_l. Suppose there is a sequence γ_n of rectifiable Jordan closed curves satisfying:

(1) $0 \in \text{Int } \gamma_1$ and $\gamma_n \subseteq \text{Int } \gamma_{n+1}$ for $n \geq 0$, and each γ_n does not pass all such α_k and z_l.

(2) The distance $d_n = \text{dist}(0, \gamma_n) \to \infty$ as $n \to \infty$.

(3) $\dfrac{\text{the length } l_n \text{ of } \gamma_n}{d_n} \leq M$, a positive constant, for $n \geq 0$.

(4) $\lim_{n \to \infty} \int_{\gamma_n} f(z)g(z)dz = 0$.

Then

$$\sum_k \text{Re}\, s(f(z)g(z); \alpha_k) = -\sum_l \text{Re}\, s(f(z)g(z); z_l),$$

where the summation $\sum_l$ in the right is taken over by these $z_l \neq \alpha_k$ for all k.

$$(4.15.7)$$

As an easy consequence of the residue theorem, the details are left to the readers.

Two sections are divided in the sequel.

Section 4.15.1 establishes formulas mainly for $\sum_{-\infty}^{\infty} f(n)$, while Sec. 4.5.2 for $\sum_{-\infty}^{\infty} (-1)^n f(n)$. Readers are advised to refer to Ref. [7], Chap. 10, and the references therein for much more information concerned with the series summation.

4.15.1. $\sum_{-\infty}^{\infty} f(n)$

To start with

$\sum_{-\infty}^{\infty} f(n)$ (Lindelöf, 1905, 1947). Suppose $f(z)$ is analytic in $\mathbf{C}$ except finitely many isolated singularities $z_1, \ldots, z_m$, all of them different from the integers $0, \pm 1, \pm 2 \ldots$. In case $\lim_{z \to \infty} zf(z) = 0$, then $\lim_{n \to \infty} \sum_{k=-n}^{n} f(k)$ exists and

$$\sum_{-\infty}^{\infty} f(n) = -\sum_{k=1}^{m} \text{Re}\, s(f(z) \cdot \pi \cot \pi z; z_k).$$

This is the case if

(1) either $f(z) = O(\frac{1}{z^2})$ as $z \to \infty$;

(2) or $f(z) = \frac{p(z)}{q(z)}$, where the polynomial $q(z)$ has degree at least 2 larger than that of $p(z)$. $\qquad (4.15.1.1)$

Sketch of Proof

Adopt Fig. 4.134. Choose n large enough so that $z_1, \ldots, z_m$ all lie in Int γ_n. Then, letting $g(k) = \pi \cot \pi z$ and by (4.15.1),

$$\frac{1}{2\pi i} \int_{\gamma_n} f(z) \cdot \pi \cot \pi z\, dz = \sum_{k=-n}^{n} f(k) + \sum_{k=1}^{m} \text{Re}\, s(f(z) \cdot \pi \cot \pi z; z_k).$$

For any $\varepsilon > 0$, there is an $R > 0$ so that $|zf(z)| < \varepsilon$ if $|z| \geq R$. Choose n large enough so that $n \geq R$. By using (4.15.5) and (4.15.6),

$$\left| \int_{\gamma_n} f(z) \cdot \pi \cot \pi z dz \right| \leq 2\pi\varepsilon \int_{\gamma_n} \frac{|dz|}{|z|} \leq 2\pi\varepsilon \cdot \frac{8(n + \frac{1}{2})}{n + \frac{1}{2}} = 16\pi\varepsilon,$$

$$\Rightarrow \lim_{n \to \infty} \int_{\gamma_n} f(z) \cdot \pi \cot \pi z dz = 0$$

and the result follows. □

Remark 1. The assumption about $f(z)$ can be relaxed a little bit to grant that some of its singular points happen to be integers. Say, $z_1 = l$, an integer. Then

$$\sum_{\substack{n=-\infty \\ n \neq l}}^{\infty} f(n) = - \sum_{k=1}^{m} \operatorname{Re} s(f(z) \cdot \pi \cot \pi z; z_k), \tag{4.15.1.2}$$

where the residue at $z_1 = l$ is still included in the right-handed sum.

In case $f(z)$ is an even function, then the formula turns out to be the form

$$\frac{1}{2}f(0) + \sum_{n=1}^{\infty} f(n) = -\frac{1}{2} \sum_{k=1}^{m} \operatorname{Re} s(f(z) \cdot \pi \cot \pi z; z_k). \tag{4.15.1.3}$$

On the other hand, it is possible that $f(z)$ has countably infinitely many isolated singularities $z_1, \ldots, z_m \ldots$. In this case,

$$\sum_{n=-\infty}^{\infty} f(n) = - \sum_{k=1}^{\infty} \operatorname{Re} s(f(z) \cdot \pi \cot \pi z; z_k). \tag{4.15.1.4}$$

Series on both sides might be divergent; if they are convergent, then the series in the left becomes the series in the right, which might converge faster. For instance, $f(z) = \frac{z}{e^{\pi a z} - e^{-\pi a z}}$ ($a \neq 0$ is real) has simple poles at $z = \frac{ki}{a}$, $k = \pm 1, \pm 2, \ldots$ ($z = 0$ is a removable singularity), and then,

$$\sum_{n=1}^{\infty} (-1)^n \frac{n}{e^{n\pi a} - e^{-n\pi a}} = -\frac{1}{4\pi a} - \frac{1}{a^2} \sum_{k=1}^{\infty} (-1)^k \frac{k}{e^{k\pi/a} - e^{-k\pi/a}}.$$

Note that the series in the right does converge faster if $|a| < 1$.

Example 1. Show that

$$\sum_{n=1}^{\infty} \frac{1}{n^2 + a^2} = \frac{1}{2a} \left(\pi \coth \pi a - \frac{1}{a} \right), \quad a > 0.$$

Deduce that (or prove independently) $\sum_{n=1}^{\infty} \frac{1}{n^2} = \frac{\pi}{6}$. What is $\sum_{n=1}^{\infty} \frac{1}{(n^2+a^2)^2} = ?$

Solution. $f(z) = \frac{1}{z^2+a^2}$ has simple poles at $\pm ai$. According to (4.15.1.3),

$$\sum_{n=1}^{\infty} \frac{1}{n^2 + a^2} = -\frac{1}{2a^2} - \frac{1}{2}\left[\operatorname{Res}\left(\frac{\pi \cot \pi z}{z^2 + a^2}; ai\right) + \operatorname{Res}\left(\frac{\pi \cot \pi z}{z^2 + a^2}; -ai\right)\right]$$

$$= -\frac{1}{2a^2} - \frac{\pi}{2}\left[\frac{\cot \pi ai}{2ai} + \frac{\cot \pi ai}{2ai}\right]$$

$$= \frac{1}{2a}\left(\pi \coth \pi a - \frac{1}{a}\right), \quad \text{by using } \cot \pi ai = -i \coth \pi a. \quad (*_1)$$

Note that the series $\sum_{n=1}^{\infty} \frac{1}{n^2+a^2}$ converge uniformly on $0 \le a < \infty$ and each term $\frac{1}{n^2+a^2}$ is a continuous function in a. Consequently its sum is a continuous function of a and

$$\sum_{n=1}^{\infty} \frac{1}{n^2} = \sum_{n=1}^{\infty} \left(\lim_{a \to 0} \frac{1}{n^2 + a^2}\right) = \lim_{a \to 0} \sum_{n=1}^{\infty} \frac{1}{n^2 + a^2} = \lim_{a \to 0} \frac{\pi a \coth \pi a - 1}{2a^2}$$

$$= \frac{1}{2} \lim_{a \to 0} \frac{\pi a \cosh \pi a - \sinh \pi a}{a^2 \sinh \pi a}$$

$$= \frac{1}{2} \lim_{a \to 0} \frac{\pi \cosh \pi a + \pi^2 a \sinh \pi a - \pi \cosh \pi a}{2a \sinh \pi a + \pi a^2 \cosh \pi a} \quad \text{(by L'Hospital's rule)}$$

$$= \frac{1}{2} \lim_{a \to 0} \frac{\pi^3 \cosh \pi a}{3\pi \cosh \pi a + \pi a^2 \sinh \pi a} = \frac{\pi^2}{6}.$$

Observe that $\sum_{n=1}^{\infty} \frac{1}{(n^2+a^2)^2}$ converges uniformly on $0 \le a < \infty$. Hence, it is permissible to differentiate $(*_1)$ termwise with respect to a and we have

$$\sum_{n=1}^{\infty} \frac{-2a}{(n^2 + a^2)^2} = \frac{d}{da} \frac{\pi a \coth \pi a - 1}{2a^2} = \frac{2 - \pi a \coth \pi a - \pi^2 a^2 \operatorname{csch}^2 \pi a}{2a^3}$$

$$\Rightarrow \sum_{n=1}^{\infty} \frac{1}{(n^2 + a^2)^2} = \frac{\pi a \coth \pi a + \pi^2 a^2 \operatorname{csch}^2 \pi a - 2}{4a^4}, \quad a > 0.$$

Try to compute $\sum_{n=1}^{\infty} \frac{1}{n^4} = ?$

Example 2. Try to sum the following series:

(1) $\sum_{n=-\infty}^{\infty} \frac{1}{n^3+a^3}$, where $a > 0$ and $a \neq 1, 2, 3, \ldots$.

(2) $\sum_{n=1}^{\infty} \frac{1}{n^4+a^4}$, where a is real.

(3) $\sum_{n=1}^{\infty} \frac{1}{n^4-a^4}$, where a is real and $a \neq \pm n, \pm ni$ for $n = 0, 1, 2, \ldots$.

Solution.

(1) Choose $f(z) = \frac{1}{z^3+a^3}$. $f(z)$ has simple poles at $z_k = ae^{\frac{(2k+1)\pi i}{3}}$, $k = 0, 1, 2$. Then

$$\sum_{n=-\infty}^{\infty} f(n) = -\pi \sum_{k=0}^{3} \operatorname{Re} s(f(z) \cot \pi z; z_k)$$

$$= -\frac{\pi}{3a^2}\left[-\cot \pi a + \frac{\cot \pi a e^{\frac{\pi i}{3}}}{e^{2\pi i/3}} + \frac{\cot \pi a e^{\frac{5\pi i}{3}}}{e^{10\pi i/3}} \right]$$

$$= \frac{\pi}{3a^2}\left[\cot \pi a - 2\operatorname{Re}\frac{\cot \pi a e^{\frac{\pi i}{3}}}{e^{\frac{2\pi i}{3}}} \right].$$

(2) $f(z) = \frac{1}{z^4+a^4}$ has simple poles $z_k = \frac{1}{\sqrt{2}a}(\pm 1 \pm i)$, $1 \leq k \leq 4$. Hence

$$\operatorname{Re} s(f(z); z_k) = -\frac{1}{4}a^{-4}z_k, \quad k = 1, 2, 3, 4,$$

$$\Rightarrow \sum_{n=-\infty}^{\infty} f(n) = -\sum_{k=1}^{4} \operatorname{Re} s(f(z) \cdot \pi \cot \pi z; z_k)$$

$$= \frac{\pi}{4}a^{-4}\sum_{k=1}^{4} z_k \cot \pi z_k$$

$$= \frac{\pi}{2^{\frac{3}{2}}}a^{-3}\left[(1+i)\cot \pi a \cdot \frac{1+i}{\sqrt{2}} + (1-i)\cot \pi a \cdot \frac{1-i}{\sqrt{2}} \right]$$

$$= \frac{\pi}{a^3\sqrt{2}} \cdot \frac{\sinh \pi a\sqrt{2} + \sin \pi a\sqrt{2}}{\cosh \pi a\sqrt{2} - \cos \pi a\sqrt{2}}.$$

By observing $\sum_{-\infty}^{\infty} = 2\sum_{n=1}^{\infty} + \frac{1}{a^4}$ (see (4.15.1.3)), the result follows.

(3) The function $f(z) = \frac{1}{z^4-a^4}$ has simple poles at $z_k = \pm a, \pm ai$, $1 \leq k \leq 4$. Then

$$\operatorname{Re} s(f(z); z_k) = \frac{1}{4a^4}z_k, \quad k = 1, 2, 3, 4,$$

$$\Rightarrow \sum_{n=-\infty}^{\infty} f(n) = -\sum_{k=1}^{4} \operatorname{Re} s(f(z) \cdot \pi \cot \pi z; z_k)$$

$$= -\frac{\pi}{4a^4}\sum_{k=1}^{4} z_k \cot \pi z_k$$

$$= -\frac{\pi}{2a^4}[ai \cot \pi ai - a \cot \pi(-a)]$$

$$= -\frac{\pi}{2a^3}(\cot \pi a + \coth \pi a)$$

$$\Rightarrow \sum_{n=1}^{\infty} \frac{1}{n^4 - a^4} = \frac{1}{2a^4} - \frac{\pi}{4a^3}(\cot \pi a + \coth \pi a).$$

Example 3. Let z be a complex number which is not equal to any integer. Then

$$\sum_{n=-\infty}^{\infty} \frac{1}{(z-n)^2} = \frac{\pi^2}{\sin^2 \pi z}.$$

Solution. For fixed $z \neq 0, \pm 1, \pm 2, \ldots$, $f(w) = \frac{1}{(z-w)^2}$ has a pole of order 2 at $w = z$. Hence

$$\mathrm{Re}\, s(f(w) \cdot \pi \cot \pi w; z) = \lim_{w \to z} \frac{d}{dw}\left\{(w-z)^2 \cdot \frac{1}{(z-w)^2} \cdot \pi \cot \pi w\right\}$$

$$= -\frac{\pi^2}{\sin^2 \pi z}$$

$$\Rightarrow \sum_{k=-\infty}^{\infty} f(n) = -\mathrm{Re}\, s(f(w) \cdot \pi \cot \pi w; z) = \frac{\pi^2}{\sin^2 \pi z}.$$

Example 4. Show that

$$\sum_{n=1}^{\infty} \frac{1}{n^{2k}} = (-1)^{k-1} 2^{2k-1} \frac{B_{2k}}{(2k)!} \pi^{2k}, \quad k = 1, 2, 3, \ldots$$

where B_{2k} are the Bernoulli's number defined by $\frac{z}{e^z - 1} = 1 - \frac{z}{2} + \sum_{n=1}^{\infty} \frac{B_{2n}}{(2n)!} z^{2n}$, $|z| < 2\pi$ (see Example 5 in Sec. 4.8).

Solution. The function $f(z) = \frac{1}{z^{2k}}$ has a pole of order $2k$ at $z = 0$. According to Exercise A (5)(a) of Sec. 4.8,

$$\pi z \cot \pi z = \sum_{k=0}^{\infty} (-1)^k \frac{B_{2k}}{(2k)!}(2\pi z)^{2k}, \quad |z| < 1,$$

$$\Rightarrow \operatorname{Re} s(f(z) \cdot \pi \cot \pi z; 0)$$

$$= \text{The coefficient of } z^{2k-1} \text{ in the}$$

$$\text{Laurent series expansion of } \pi \cot \pi z \text{ at } z = 0$$

$$= (-1)^k \cdot \frac{B_{2k}}{(2k)!} (2\pi)^{2k},$$

$$\Rightarrow \text{(by (4.15.1.2))} \quad \sum_{k=1}^{\infty} \frac{1}{n^{2k}} = \frac{1}{2} \sum_{\substack{n=-\infty \\ n \neq 0}}^{\infty} f(n)$$

$$= -\frac{1}{2} \operatorname{Re} s(f(z) \cdot \pi \cot \pi z; 0)$$

$$= (-1)^{k-1} 2^{2k-1} \frac{B_{2k}}{(2k)!} \pi^{2k}.$$

Next comes a useful formula about Fourier series.

The sum of the Fourier series $\sum_{n=-\infty}^{\infty} \frac{p(n)}{q(n)} e^{in\theta}$. *Let* $f(z) = \frac{p(z)}{q(z)}$ *be a rational function such that* $z_1, \ldots, z_m$ *are all distinct zeros of* $q(z)$ *and no* z_k $(1 \leq k \leq m)$ *is equal to an integer. Then*

$$\sum_{n=-\infty}^{\infty} \frac{p(n)}{q(n)} e^{in\theta} = -\sum_{k=1}^{m} \operatorname{Re} s\left(\frac{p(z)}{q(z)} e^{i(\theta - \pi)z} \frac{\pi}{\sin \pi z}; z_k\right)$$

uniformly on $0 \leq \theta \leq 2\pi$ if $\frac{p(n)}{q(n)} = O(\frac{1}{z^2})$ as $z \to \infty$; uniformly on compact subsets of $0 < \theta < 2\pi$ if $\frac{p(n)}{q(n)} = O(\frac{1}{z})$ as $z \to \infty$. (4.15.1.5)

Note that $\frac{p(n)}{q(n)} = O(\frac{1}{z^2})$, $z \to \infty$, says that the degree, say m, of $q(z)$ is at least two larger than that, say l, of $p(z)$, namely, $m \geq l + 2$. In this case, Weierstrass test (see (2.3.5) of Sec. 2.3) guarantees that the series $\sum_{n=-\infty}^{\infty} \frac{p(n)}{q(n)} e^{in\theta}$ will converge uniformly on $[0, 2\pi]$. While, $\frac{p(n)}{q(n)} = O(\frac{1}{z})$, $z \to \infty$, says that $m \geq l + 1$, and Abel or Dirichlet test (see (2.3.5) of Sec. 2.3) will justify that the series converges locally uniformly in the *open* interval $(0, 2\pi)$.

Proof. Adopt Fig. 4.134. The key step is to show if

$$\lim_{n \to \infty} \int_{\gamma_n} f(z) e^{i\theta z} \cdot \pi \cot \pi z \, dz = 0 \tag{$*_2$}$$

holds. To see this, we present the following preliminary estimation: Along the four sides of γ_n, as for $e^{i\theta z}$:

$$\left| e^{i\theta[x+i(n+\frac{1}{2})]} \right| = e^{-\theta(n+\frac{1}{2})} \quad \text{(the upper side)};$$

$$\left| e^{i\theta[x-i(n+\frac{1}{2})]} \right| = e^{\theta(n+\frac{1}{2})} \quad \text{(the lower side)};$$

$$\left| e^{i\theta[-(n+\frac{1}{2})+iy]} \right| = e^{-\theta y}, \, -\left(n+\frac{1}{2} \right) \leq y \leq n+\frac{1}{2} \quad \text{(the left side)};$$

$$\left| e^{i\theta[(n+\frac{1}{2})+iy]} \right| = e^{-\theta y}, \, -\left(n+\frac{1}{2} \right) \leq y \leq n+\frac{1}{2} \quad \text{(the right side)}; \quad (*_3)$$

as for $\pi \cot \pi z$: letting $z = x + iy$,

$$\lim_{y \to -\infty} \pi \cot \pi z = \lim_{y \to -\infty} \pi i \cdot \frac{1 + e^{-2\pi i x} e^{2\pi y}}{1 - e^{-2\pi i x} e^{2\pi y}} = \pi i \quad \text{(uniformly in } x\text{)}. \quad (*_4)$$

Even though $f(z) = O(\frac{1}{z^2})$ as $z \to \infty$, we still cannot prove the validity of $(*_2)$ via the above estimates.

Owing to $(*_4)$ we consider, instead of the original $\pi \cot \pi z$, the function

$$t(z) = \pi \cot \pi z - \pi i = \frac{2\pi i}{e^{2\pi i z} - 1} = \frac{\pi e^{-i\pi z}}{\sin \pi z}.$$

This $t(z)$ still has simple poles at the integers $n = 0, \pm 1, \pm 2, \ldots$ with residue 1 at each pole. The main advantage of $t(z)$ over $\pi \cot \pi z$ lies on the fact that $|t(z)|$ will turn out to be small if $|e^{2\pi i z}|$ becomes large enough. As a matter of fact, along the sides of γ_n,

$$|t(z)| \leq \begin{cases} \dfrac{2\pi}{|1 - e^{-2\pi y}|} \leq \dfrac{2\pi}{1 - e^{-\pi(2n+1)}} & \text{(the upper side)} \\[3mm] \dfrac{2\pi}{|e^{-2\pi y} - 1|} \leq \dfrac{2\pi}{e^{\pi(2n+1)} - 1} & \text{(the lower side)} \end{cases} ;$$

$$|t(z)| \leq \left| \frac{2\pi}{1 - e^{-\pi y}} \right|, \, -\left(n+\frac{1}{2} \right) \leq y \leq n+\frac{1}{2} \text{ (the left and right sides)}.$$

$$(*_5)$$

Combing $(*_3)$ and $(*_5)$, there is a constant $M_1 > 0$ so that, for all sufficiently large n,

$$\sup_{\gamma_n} |t(z) e^{i\theta z}| \leq 2\pi M_1, \ 0 \leq \theta \leq 2\pi.$$

In case $f(z) = O(\frac{1}{z^2})$ as $z \to \infty$: There is a constant $M_2 > 0$ so that, for all large $|z|$, $|f(z)| \le M_2 \frac{1}{|z|^2}$.

Remember now that $\pi \cot \pi z$ in $(*_2)$ shall be replaced by $t(z)$. Then

$$\left| \int_{\gamma_n} f(z) e^{i\theta z} t(z) dz \right| \le 2\pi M_1 M_2 \int_{\gamma_n} \frac{|dz|}{|z|^2}$$

$$\le 2\pi M_1 M_2 \frac{8(n+\frac{1}{2})}{(n+\frac{1}{2})^2} \to 0 \quad \text{as } n \to \infty \qquad (*_6)$$

$$\Rightarrow \lim_{n \to \infty} \sum_{k=-n}^{n} f(k) e^{ik\theta} + \sum_{k=1}^{m} \text{Re}\, s\left(f(z) e^{i\theta z} \cdot \frac{2\pi i}{e^{2\pi i z} - 1}; z_k \right) = 0. \qquad (*_7)$$

And the result follows.

In case $f(z) = O(\frac{1}{z})$ as $z \to \infty$: In this case, $(*_6)$ fails to be true and more subtle estimates for $t(z)$ than those shown in $(*_5)$ are needed. We proceed as follows. If $|z|$ is large enough, the integral $f(z) e^{i\theta z} t(z)$ can be estimated as, letting $z = x + iy$ and for some constant $M > 0$,

$$|f(z) e^{i\theta z} t(z)| = \left| f(z) e^{i(\theta - \pi) z} \frac{\pi}{\sin \pi z} \right| \le \frac{e^{(\pi - \theta) y}}{\sqrt{x^2 + y^2}} \cdot \frac{M}{\sqrt{\sinh^2 \pi y + \sin^2 \pi x}}.$$

Consequently, along the upper side $L_n^+ : z = x + il_n$, $l_n = n + \frac{1}{2}$, $-l_n \le x \le l_n$:

$$\left| \int_{L_n^+} f(z) e^{i\theta z} t(z) dz \right| \le \int_{-l_n}^{l_n} \frac{e^{(\pi - \theta) l_n}}{\sqrt{x^2 + l_n^2}} \frac{M}{\sqrt{\sinh^2 \pi l_n + \sin^2 \pi x}} dx$$

$$= \int_{-1}^{1} \frac{e^{(\pi - \theta) l_n}}{\sqrt{x^2 + 1}} \frac{M dx}{\sqrt{\sinh^2 \pi l_n + \sin^2 \pi l_n x}}$$

$$\text{(using } l_n x \text{ to replace } x\text{)}$$

$$\le \int_{-1}^{1} \frac{e^{(\pi - \theta) l_n}}{\sqrt{x^2 + 1}} \frac{M dx}{\sinh \pi l_n} = \frac{2M e^{-\theta l_n}}{1 - e^{-2\pi l_n}} \int_{-1}^{1} \frac{dx}{\sqrt{x^2 + 1}}$$

$$\to 0 \quad \text{as } n \to \infty, \quad \text{if } \theta > 0.$$

Along the lower side $L_n^- : z = x - il_n$, $-l_n \le x \le l_n x$:

$$\left| \int_{L_n^-} f(z) e^{i\theta z} t(z) dz \right| \le \frac{2M e^{\theta l_n}}{e^{2\pi l_n} - 1} \int_{-1}^{1} \frac{dx}{\sqrt{x^2 + 1}} \to 0 \quad \text{as } n \to \infty, \quad \text{if } \theta < 2\pi.$$

Along the right side $U_n^+ : z = l_n + iy,\ -l_n \le y \le l_n$:

$$\left| \int_{U_n^+} f(z)e^{i\theta z}t(z)dz \right|$$

$$\le \int_{-l_n}^{l_n} \frac{Me^{(\pi-\theta)y}}{\sqrt{l_n^2+y^2}\sqrt{\sinh^2\pi y + \sin^2\pi l_n}}dy$$

$$= \int_{-l_n}^{l_n} \frac{Me^{(\pi-\theta)y}}{\sqrt{l_n^2+y^2}\cosh\pi y}dy = \int_{-1}^{1} \frac{Me^{(\pi-\theta)l_n y}}{\sqrt{y^2+1}\cosh\pi l_n y}dy$$

$$= \int_{-1}^{1} \frac{2Me^{-\theta l_n y}}{\sqrt{y^2+1}(1+e^{-2\pi l_n y})}dy \to 0 \quad \text{as } n \to \infty, \quad \text{if } \theta > 0.$$

Similarly, along the left side $U_n^- : z = -l_n + iy,\ -l_n \le y \le l_n$:

$$\left| \int_{U_n^-} f(z)e^{i\theta z}t(z)dz \right| \le \int_{-1}^{1} \frac{2Me^{\theta l_n y}}{\sqrt{y^2+1}(e^{2\pi l_n y}+1)}dy \to 0$$

$$\text{as } n \to \infty, \quad \text{if } \theta < 2\pi.$$

As a whole, if $0 < \theta < 2\pi$, then

$$\lim_{n\to\infty} \int_{\gamma_n} f(z)e^{i\theta z}t(z)dz = 0$$

and thus, $(*_7)$ holds.   □

Remark 2. Firstly, (4.15.1.2) is still valid in the case (4.15.1.5). See the note at the end of Example 5.

Equating the real and imaginary parts of (4.15.1.5), we obtain

$$\sum_{n=-\infty}^{\infty} \frac{p(n)}{q(n)}\cos\pi\theta = -\operatorname{Re}\sum_{k=1}^{m}\operatorname{Res}\left(\frac{p(z)}{q(z)}e^{i(\theta-\pi)z}\frac{\pi}{\sin\pi z}; z_k\right); \quad (4.15.1.6)$$

$$\sum_{n=-\infty}^{\infty} \frac{p(n)}{q(n)}\sin\pi\theta = -\operatorname{Im}\sum_{k=1}^{m}\operatorname{Res}\left(\frac{p(z)}{q(z)}e^{i(\theta-\pi)z}\frac{\pi}{\sin\pi z}; z_k\right). \quad (4.15.1.7)$$

In case $\frac{p(n)}{q(n)}$ is even in (4.15.1.6), then

$$\sum_{n=1}^{\infty} \frac{p(n)}{q(n)}\cos n\theta = -\frac{p(0)}{2q(0)} - \frac{\pi}{2}\operatorname{Re}\sum_{k=1}^{m}\operatorname{Res}\left(\frac{p(z)}{q(z)}\frac{e^{i(\theta-\pi)z}}{\sin\pi z}; z_k\right). \quad (4.15.1.8)$$

In case $\frac{p(n)}{q(n)}$ is odd in (4.15.1.7), then

$$\sum_{n=1}^{\infty} \frac{p(n)}{q(n)}\sin n\theta = -\frac{\pi}{2}\operatorname{Im}\sum_{k=1}^{m}\operatorname{Res}\left(\frac{p(z)}{q(z)}\frac{e^{i(\theta-\pi)z}}{\sin\pi z}; z_k\right). \quad (4.15.1.9)$$

Remark 3. Similarly, by considering the integral

$$\int_{\gamma_n} \frac{p(z)}{q(z)} e^{i\theta z} \cdot \frac{e^{-\frac{\pi}{2}z}}{\cos \frac{\pi z}{2}} dz \quad \text{with } \gamma_n \text{ the square with vertices at}$$

$$\pm n\pi(1+i), n \geq 1,$$

where $0 \leq \theta \leq \pi$ if $\frac{p(n)}{q(n)} = O(\frac{1}{z^2})$ as $z \to \infty$, and $0 < \theta < \pi$ if $\frac{p(n)}{q(n)} = O(\frac{1}{z})$ as $z \to \infty$, we have

$$\sum_{n=-\infty}^{\infty} \frac{p(2n+1)}{q(2n+1)} e^{i(2n+1)\theta} = \frac{\pi i}{2} \sum_{k=1}^{m} \operatorname{Re} s \left(\frac{p(z)}{q(z)} \frac{e^{i(\theta-\frac{\pi}{2})z}}{\cos \frac{\pi z}{2}}; z_k \right). \quad (4.15.1.10)$$

where $z_1, \ldots, z_m$ are all distinct zeros of the denominator $q(z)$. The readers are invited to prove (4.15.1.10) in detail in Exercise A (1). Of course, there are formulas, in this case, corresponding to (4.15.1.6)–(4.15.1.9).

Example 5. Show that

$$\sum_{n=1}^{\infty} \frac{\cos n\theta}{n^2 + a^2} = \frac{\pi}{2a} \cdot \frac{\cosh a(\pi - \theta)}{\sinh a\pi} - \frac{1}{2a^2}, \quad a > 0, \ 0 \leq \theta \leq 2\pi.$$

Then, deduce that

$$\sum_{n=1}^{\infty} \frac{\cos n\theta}{n^2} = \frac{1}{12}(2\pi^2 - 6\pi\theta + 3\theta^2), \quad 0 \leq \theta \leq 2\pi;$$

$$\sum_{n=1}^{\infty} \frac{n \sin n\theta}{n^2 + a^2} = \frac{\pi}{2} \cdot \frac{\sinh a(\pi - \theta)}{\sinh a\pi}, \quad a > 0, \ 0 < \theta < 2\pi.$$

Also, sum the series:

$$\sum_{n=1}^{\infty} \frac{\sin n\theta}{n}, \quad 0 < \theta < 2\pi; \quad \sum_{n=1}^{\infty} \frac{\sin n\theta}{n^3}, \quad 0 \leq \theta \leq 2\pi, \quad \text{and}$$

$$\sum_{n=1}^{\infty} \frac{\cos n\theta}{n^4}, \quad 0 \leq \theta \leq 2\pi.$$

Solution. The function $f(z) = \frac{1}{z^2 + a^2}$ has simple poles at $\pm ai$ and

$$\operatorname{Re} s\left(f(z)\frac{\pi e^{i(\theta - \pi)z}}{\sin \pi z}; ai\right) = -\frac{\pi}{a} \cdot \frac{e^{-a\theta}}{1 - e^{-2\pi a}};$$

$$\operatorname{Re} s\left(f(z)\frac{\pi e^{i(\theta - \pi)z}}{\sin \pi z}; -ai\right) = -\frac{\pi}{a} \cdot \frac{e^{a\theta}}{e^{2\pi a} - 1}.$$

$$\Rightarrow \sum_{n=-\infty}^{\infty} \frac{e^{in\theta}}{z^2 + a^2} = -\left\{-\frac{\pi}{a} \cdot \frac{e^{-a\theta}}{1 - e^{-2\pi a}} - \frac{\pi}{a} \cdot \frac{e^{a\theta}}{e^{2\pi a} - 1}\right\}$$

$$= \frac{\pi}{a}\frac{\cosh a(\pi - \theta)}{\sinh a\pi}.$$

By equaling the real parts of both sides, we obtain

$$\sum_{n=1}^{\infty} \frac{\cos n\theta}{n^2 + a^2} = \frac{\pi}{2a} \cdot \frac{\cosh a(\pi - \theta)}{\sinh a\pi} - \frac{1}{2a^2}, \quad a > 0, \ 0 \le \theta \le 2\pi.$$

This series converges uniformly either in $0 \le \theta \le 2\pi$ or in $0 < a < \infty$. By repeated use of L'Hospital's rule three times,

$$\sum_{n=1}^{\infty} \frac{\cos n\theta}{n^2}$$

$$= \lim_{a \to 0} \sum_{n=1}^{\infty} \frac{\cos n\theta}{n^2 + a^2} = \lim_{a \to 0} \frac{\pi a \cosh a(\pi - \theta) - \sinh a\pi}{2a^2 \sinh a\pi}$$

$$= \lim_{a \to 0} \frac{3\pi(\pi - \theta)^2 \cosh a(\pi - \theta) + a\pi(\pi - \theta)^3 \sinh a(\pi - \theta) - \pi^3 \cosh a\pi}{12\pi \cosh a\pi + 12a\pi^2 \sinh a\pi + 2a^2\pi^3 \cosh a\pi}$$

$$= \frac{3\pi(\pi - \theta)^2 - \pi^3}{12\pi} = \frac{1}{12}(2\pi^2 - 6\theta\pi + 3\theta^2), \quad 0 \le \theta \le 2\pi.$$

A direct derivation of this sum is given in the *Note* at the end of this Example. On the other hand, by Abel or Dirichlet test (see (2.3.5) of Sec. 2.3), the series $\sum_{n=1}^{\infty} \frac{n \sin n\theta}{n^2 + a^2}$ converges locally uniformly in $0 < \theta < 2\pi$. Hence, by termwise differentiation in θ of $\sum_{n=1}^{\infty} \frac{\cos n\theta}{n^2 + a^2}$, we have

$$\sum_{n=1}^{\infty} \frac{n \sin n\theta}{n^2 + a^2} = \frac{d}{d\theta}\left\{\frac{\pi}{2a} \cdot \frac{\cosh a(\pi - \theta)}{\sinh a\pi} - \frac{1}{2a^2}\right\} = \frac{\pi}{2} \cdot \frac{\sinh a(\pi - \theta)}{\sinh a\pi},$$

$$a > 0, 0 < \theta < 2\pi.$$

Alternatively, by (4.15.1.9)

$$\sum_{n=1}^{\infty} \frac{n \sin n\theta}{n^2 + a^2} = -\frac{\pi}{2} \operatorname{Im} \left[\operatorname{Re} s \left(\frac{z}{z^2 + a^2} \cdot \frac{e^{i(\theta - \pi)z}}{\sin \pi z}; ai \right) \right.$$

$$\left. + \operatorname{Re} s \left(\frac{z}{z^2 + a^2} \cdot \frac{e^{i(\theta - \pi)z}}{\sin \pi z}; -ai \right) \right]$$

$$= -\frac{\pi}{4} \frac{1}{\sinh a\pi} \operatorname{Im} \left[\frac{1}{i} (e^{(\pi - \theta)a} - e^{-(\pi - \theta)a}) \right]$$

$$= \frac{\pi}{2} \frac{\sinh a(\pi - \theta)}{\sinh a\pi}.$$

Similarly, $\sum_{n=1}^{\infty} \frac{\sin n\theta}{n}$ converges locally uniformly in $0 < \theta < 2\pi$, and then, termwise differentiation of $\sum_{n=1}^{\infty} \frac{\cos n\theta}{n^2}$ or the limit of $\sum_{n=1}^{\infty} \frac{n \sin n\theta}{n^2 + a^2}$ as $a \to 0$ will lead to

$$\sum_{n=1}^{\infty} \frac{\sin n\theta}{n} = \frac{1}{2}(\theta - \pi), \quad 0 < \theta < 2\pi.$$

According to (4.15.1.5) and (4.15.1.2), this can also be evaluated by

$$\sum_{\substack{n=-\infty \\ n \neq 0}}^{\infty} \frac{e^{in\theta}}{n} = -\operatorname{Re} s \left(\frac{1}{z} e^{i(\theta - \pi)z} \frac{\pi}{\sin \pi z}; 0 \right) = -\lim_{z \to 0} \frac{d}{dz} \frac{e^{i(\theta - z)\pi} \cdot \pi z}{\sin \pi z}$$

$$= -[-i(\theta - \pi)] = i(\theta - \pi),$$

$$\Rightarrow \sum_{n=1}^{\infty} \frac{\sin n\theta}{n} = \frac{1}{2}(\theta - \pi), \quad 0 < \theta < 2\pi.$$

By integrating $\sum_{n=1}^{\infty} \frac{\cos n\theta}{n^2}$ termwise from $\theta = 0$ to θ, where $0 \leq \theta \leq 2\pi$, we obtain

$$\sum_{n=1}^{\infty} \frac{\sin n\theta}{n^3} = \int_0^\theta \frac{1}{12} (2\pi^2 - 6\theta\pi + 3\theta^2) d\theta$$

$$= \frac{1}{12} (2\pi^2 \theta - 3\pi\theta^2 + \theta^3), \quad 0 \leq \theta \leq 2\pi.$$

Integrating termwise again, we obtain

$$-\sum_{n=1}^{\infty} \frac{\cos n\theta}{n^4} + \sum_{n=1}^{\infty} \frac{1}{n^4} = \frac{1}{12} \left(\pi^2 \theta^2 - \pi\theta^3 + \frac{1}{4}\theta^4 \right),$$

$$\Rightarrow \text{(by Example 4)} \quad \sum_{n=1}^{\infty} \frac{\cos n\theta}{n^4} = \frac{\pi^4}{90} - \frac{1}{12} \left(\pi^2 \theta^2 - \pi\theta^3 + \frac{1}{4}\theta^4 \right),$$

$$0 \leq \theta \leq 2\pi.$$

The readers are asked to derive directly, via (4.15.1.8) and (4.15.1.9), these last two formulas.

Note: An illustration of (4.15.1.2) for (4.15.1.5).
   The function

$$F(z) = \frac{e^{i\theta z}}{z^2} \frac{2\pi i}{e^{2\pi i z} - 1} = \frac{\pi}{z^2} \cdot \frac{e^{i(\theta - \pi)z}}{\sin \pi z}$$

has a simple pole at each of $\pm 1, \pm 2, \ldots$ and a pole of order 3 at $z = 0$. Now,

$\operatorname{Re} s(F(z); 0) = $ the coefficient of z^2 in the Taylor series expansion of

$$\frac{2\pi i e^{i\theta z}}{z^{-1}(e^{2\pi i z} - 1)} \quad \text{(by long division)}$$

$$= -\frac{1}{2}\theta^2 + \pi\theta - \frac{1}{6}\pi^2;$$

$$\operatorname{Re} s(F(z); k) = \frac{i e^{i\theta k}}{k^2}, \quad k = \pm 1, \pm 2, \ldots,$$

$$\Rightarrow \text{(adopting Fig. 4.134)} \quad \sum_{\substack{k=-n \\ k\neq 0}}^{n} \frac{e^{i\theta k}}{k^2} - \left(\frac{1}{2}\theta^2 - \pi\theta + \frac{1}{6}\pi^2\right)$$

$$= \int_{\gamma_n} F(z)dz \to 0 \quad \text{as } n \to \infty,$$

$$\Rightarrow \sum_{n=1}^{\infty} \frac{\cos n\theta}{n^2} = \frac{1}{2}\lim_{n\to\infty} \sum_{\substack{k=-n \\ k\neq 0}}^{n} \frac{e^{i\theta k}}{k^2} = \frac{1}{2}\left(\frac{1}{2}\theta^2 - \pi\theta + \frac{1}{6}\pi^2\right)$$

$$= \frac{1}{12}(3\theta^2 - 6\pi\theta + 2\pi^2).$$

Example 6. Show that

$$\sum_{n=0}^{\infty} \frac{\sin(2n+1)\theta}{2n+1} = \begin{cases} \frac{\pi}{4}, & 0 < \theta < \pi \\ 0, & \theta = 0, \pi \end{cases}.$$

Then, try to sum the following series:

$$\sum_{n=1}^{\infty} \frac{\cos(2n+1)\theta}{(2n+1)^2}, \quad 0 \le \theta \le \pi, \quad \text{and} \quad \sum_{n=1}^{\infty} \frac{\sin(2n+1)\theta}{(2n+1)^3}, \quad 0 \le \theta \le \pi.$$

Solution. According to (4.15.1.10), for $0 < \theta < \pi$,

$$\sum_{n=-\infty}^{\infty} \frac{e^{i(2n+1)\theta}}{2n+1} = \frac{\pi i}{2} \operatorname{Re} s \left(\frac{1}{z} \cdot \frac{e^{i(\theta - \frac{\pi}{2})z}}{\cos \frac{\pi z}{2}} ; 0 \right) = \frac{\pi i}{2}$$

$\Rightarrow$ (by equating the imaginary parts) $\quad \displaystyle\sum_{n=-\infty}^{\infty} \frac{\sin(2n+1)\theta}{2n+1} = \frac{\pi}{2}$

and the result follows by observing $\sum_{-\infty}^{-1} = \sum_0^{\infty}$. This series converges locally uniformly in $0 < \theta < \pi$. By integrating termwise from 0 to θ, where $0 < \theta < \pi$, of the previous series, one has

$$\sum_{n=0}^{\infty} \frac{1}{(2n+1)^2} [1 - \cos(2n+1)\theta] = \frac{\pi}{4}\theta.$$

Setting $\theta = \frac{\pi}{2}$, we obtain

$$\sum_{n=0}^{\infty} \frac{1}{(2n+1)^2} = \frac{\pi^2}{8}$$

$$\Rightarrow \sum_{n=0}^{\infty} \frac{\cos(2n+1)\theta}{(2n+1)^2} = \frac{\pi^2}{8} - \frac{\pi}{4}\theta, \quad 0 \le \theta \le \pi.$$

Similar argument shows that

$$\sum_{n=0}^{\infty} \frac{\sin(2n+1)\theta}{(2n+1)^3} = \frac{\pi^2}{8}\theta - \frac{\pi}{8}\theta^2, \quad 0 \le \theta \le \pi.$$

Exercises A

(1) Prove (4.15.1.10) in detail. Also, derive the following formulas corresponding to (4.15.1.6)–(4.15.1.9), in case $\frac{p(x)}{q(x)}$ is real if x is real:

(a) By separating the real and imaginary parts of (4.15.1.10), then

$$\sum_{n=-\infty}^{\infty} \frac{p(2n+1)}{q(2n+1)} \cos(2n+1)\theta$$

$$= \frac{\pi}{2} \operatorname{Re} \left\{ i \sum_{k=1}^{m} \operatorname{Re} s \left(\frac{p(z)}{q(z)} \cdot \frac{e^{i(\theta - \frac{\pi}{2})z}}{\cos \frac{\pi z}{2}} \right) ; z_k \right\};$$

$$\sum_{n=-\infty}^{\infty} \frac{p(2n+1)}{q(2n+1)} \sin(2n+1)\theta$$

$$= \frac{\pi}{2} \operatorname{Im} \left\{ i \sum_{k=1}^{m} \operatorname{Re} s \left(\frac{p(z)}{q(z)} \cdot \frac{e^{i(\theta - \frac{\pi}{2})z}}{\cos \frac{\pi z}{2}} \right) ; z_k \right\}.$$

where, as before, $0 \le \theta \le \pi$ if $\frac{p(z)}{q(z)} = O(\frac{1}{z^2})$ as $z \to \infty$; $0 < \theta < \pi$ if $\frac{p(z)}{q(z)} = O(\frac{1}{z})$ as $z \to \infty$.

(b) If $\frac{p(z)}{q(z)}$ is even in z, then

$$\sum_{n=0}^{\infty} \frac{p(2n+1)}{q(2n+1)} \cos(2n+1)\theta$$

$$= \frac{\pi}{4} \mathrm{Re} \left\{ i \sum_{k=1}^{m} \mathrm{Re}\, s \left(\frac{p(z)}{q(z)} \cdot \frac{e^{i\left(\theta - \frac{\pi}{2}\right)z}}{\cos \frac{\pi z}{2}} \right) ; z_k \right\};$$

if $\frac{p(z)}{q(z)}$ is odd in z, then

$$\sum_{n=0}^{\infty} \frac{p(2n+1)}{q(2n+1)} \sin(2n+1)\theta$$

$$= \frac{\pi}{4} \mathrm{Im} \left\{ i \sum_{k=1}^{m} \mathrm{Re}\, s \left(\frac{p(z)}{q(z)} \cdot \frac{e^{i\left(\theta - \frac{\pi}{2}\right)z}}{\cos \frac{\pi z}{2}} \right) ; z_k \right\}.$$

(2) Sum the following series.

(a) $\displaystyle \sum_{n=1}^{\infty} \frac{n^2}{n^4 + a^4}$ $(a > 0)$.

(b) $\displaystyle \sum_{n=1}^{\infty} \frac{n^2}{n^4 - a^4}$ $(a \ne \pm n, \pm ni, \text{ for } n = \pm 1, \pm 2, \ldots)$.

(c) $\displaystyle \sum_{n=1}^{\infty} \frac{1}{(n-a)(n-b)}$ (both a and b are not integers).

(d) $\displaystyle \sum_{n=-\infty}^{\infty} \frac{1}{(a+n)^4}$ (a is real but is not an integer), and try to show
that $\sum_{n=1}^{\infty} \frac{1}{n^4} = \frac{\pi^4}{90}$.

(e) $\displaystyle \sum_{n=1}^{\infty} \frac{1}{n^{2k} + a^{2k}}$ ($a > 0$ and k is a positive integer).

(f) $\lim_{n \to \infty} \sum_{k=-n}^{n} \frac{1}{k-a}$ (a is not an integer), and deduce that

$$\sum_{n=-\infty}^{\infty} \frac{1}{(n-a)^2} = \frac{\pi^2}{\sin^2 a\pi}; \quad \sum_{n=-\infty}^{\infty} \frac{1}{(n-a)^3} = -\frac{\pi^3 \cos a\pi}{\sin^3 a\pi}.$$

(3) Suppose a is not an integer. Show that

$$\sum_{n=-\infty}^{\infty} \frac{1}{n^2 - a^2} = -\frac{\pi \cot a\pi}{a}; \quad \sum_{n=1}^{\infty} \frac{2a}{a^2 - n^2} = \pi \cot a\pi - \frac{1}{a},$$

either directly or by using Example 1. Then, deduce that

$$\sum_{n=1}^{\infty} \frac{1}{n^2} = \frac{\pi^2}{6}, \quad \text{and} \quad \sum_{n=1}^{\infty} \frac{1}{(2n+1)^2} = \frac{\pi^2}{8},$$

and (refer to Example 1 in Sec. 5.5.2)

$$\sin \pi z = \pi z \prod_{n=1}^{\infty} \left(1 - \frac{z^2}{n^2}\right), \quad n = \pm 1, \pm 2, \ldots.$$

(4) Try to sum $\sum_{n=-\infty}^{\infty} \frac{1}{(n-a)^3}$ without recourse to Exercise (2)(f), where a is a complex number but is not an integer.

(5) Sum the series $\sum_{n=-\infty}^{\infty} \frac{e^{in a\theta}}{(n-a)^2}$, $0 \le \theta \le 2\pi$, where $a \ne 0, \pm 1, \pm 2, \ldots$.

(6) Sum the series $\sum_{n=-\infty}^{\infty} \frac{e^{in a\theta}}{1+n+n^2+\cdots n^{2k}}$, $0 \le \theta \le 2\pi$, for $k = 1, 2, 3, \ldots$.

(7) Integrate $f(z) = \pi z^{-7} \cot \pi z \coth \pi z$ along the path γ_n shown in Fig. 4.134 to show that

$$\sum_{n=1}^{\infty} \frac{\coth n\pi}{n^7} = \frac{19\pi^7}{56{,}700}.$$

(8) The *Bernoulli polynomial* $B_k(z)$ of order k, for $k \ge 0$, is defined by

$$\frac{\zeta e^{z\zeta}}{e^\zeta - 1} = \sum_{k=0}^{\infty} \frac{B_k(z)}{k!} \zeta^k, \quad |\zeta| < 2\pi.$$

(see the *Note* below). Integrate $f_k(z) = \frac{1}{\zeta^{k+1}} \cdot \frac{\zeta e^{z\zeta}}{e^\zeta - 1}$, $k \ge 1$, along $\gamma_n : |\zeta| = (2n+1)\pi$ for $n \ge 1$, to show that, for $0 \le \theta \le 1$ and $k = 1, 2, 3, \ldots$,

$$\sum_{n=1}^{\infty} \frac{\cos 2n\pi\theta}{n^{2k}} = (-1)^{k+1} \frac{(2\pi)^{2k}}{2(2k)!} B_{2k}(\theta);$$

$$\sum_{n=1}^{\infty} \frac{\sin 2n\pi\theta}{n^{2k}} = (-1)^{k+1} \frac{(2\pi)^{2k}}{2(2k+1)!} B_{2k+1}(\theta).$$

Note: As a matter of fact,

$$B_k(z) = \sum_{n=0}^{k} C_n^k B_n z^{n-k} = z^k - \frac{k}{2} z^{n-1} + \cdots + B_k, \quad k > 1$$

where $B_k = B_k(0)$, $0 \le k \le n$, are the Bernoulli's number (defined by $\frac{\zeta}{e^\zeta - 1} = \sum_{n=0}^{\infty} \frac{B_n z^n}{n!}$, see Example 5 in Sec. 4.8). For instance,

$$B_0(z) = 1, \quad B_1(z) = z - \frac{1}{2}, \quad B_2(z) = z^2 - z + \frac{1}{6},$$

$$B_3(z) = z^3 - \frac{3}{2}z^2 + \frac{1}{2}z, \quad B_4(z) = z^4 - 2z^3 + z^2 - \frac{1}{30}, \cdots.$$

They satisfy the following properties: for $k \ge 1$,

(1) $B_k'(z) = kB_{k-1}(z)$;
(2) (difference equation) $B_k(z+1) - B_k(z) = kz^{k-1}$;
(3) $B_k(1-z) = (-1)^k B_k(z)$.

Exercises B

Suppose $f(u, v)$ is a rational function of u and v, and a is not an integer. By integrating $f(z, \sqrt{z^2 + a^2})\pi \cot \pi z$ along the path γ_n shown in Fig. 4.135, one can evaluate the sum of the series $\sum_{n=-\infty}^{\infty} f(n, \sqrt{n^2 + a^2})$. Refer to Example 5 in Sec. 4.15.2.

(1) Sum the series $\sum_{n=-\infty}^{\infty} \frac{1}{\sqrt{n^4 + a^4}}$.

(2) Sum the series $\sum_{n=-\infty}^{\infty} \frac{1}{\sqrt{n^4 - a^4}}$.

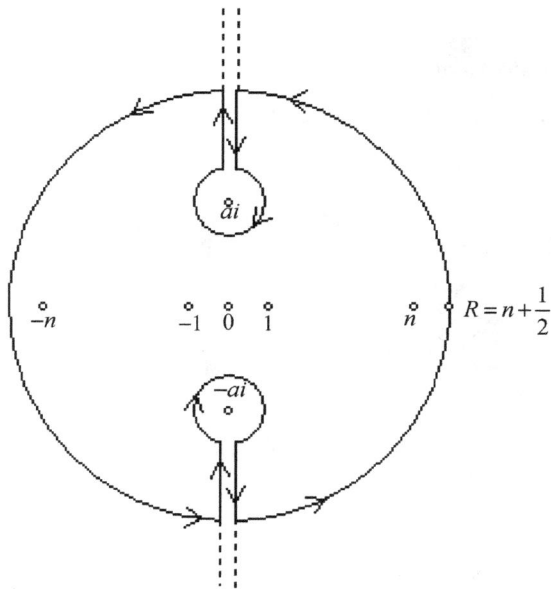

Fig. 4.135

4.15.2. $\sum_{-\infty}^{\infty} (-1)^n f(n)$

The theory concerned is parallel to that of Sec. 4.15.1. We just summarize the main results and invite the readers to give detailed proofs for each of them.

Corresponding to (4.15.1.1),

$\sum_{-\infty}^{\infty} (-1)^n f(n)$. Let $f(z)$ be as in (4.15.1.1). Then

$$\lim_{n \to \infty} \sum_{k=-n}^{n} (-1)^n f(n) \text{ exists and}$$

$$\sum_{n=-\infty}^{\infty} (-1)^n f(n) = -\sum_{k=1}^{n} \operatorname{Re} s \left(f(z) \cdot \frac{\pi}{\sin \pi z}; z_k \right). \quad (4.15.2.1)$$

Also, (4.15.1.2) and (4.15.1.3) in Remark 1 of Sec. 4.15.1 both have counterparts in this case, too.

Corresponding to (4.15.1.5),

The sum of the Fourier series $\sum_{n=-\infty}^{\infty} (-1)^n \frac{p(n)}{q(n)} e^{in\theta}$. *Let* $f(z) = \frac{p(z)}{q(z)}$ *be as in* (4.15.1.5). *Then*

$$\sum_{n=-\infty}^{\infty} (-1)^n \frac{p(n)}{q(n)} e^{in\theta} = -\sum_{k=1}^{m} \operatorname{Re} s \left(\frac{p(z)}{q(z)} e^{in\theta} \frac{\pi}{\sin \pi z}; z_k \right)$$

uniformly on $-\pi \le \theta \le \pi$ if $\frac{p(z)}{q(z)} = O(\frac{1}{z^2})$ as $z \to \infty$; uniformly on compact subsets of $-\pi < \theta < \pi$ if $\frac{p(z)}{q(z)} = O(\frac{1}{z})$ as $z \to \infty$. $\quad (4.15.2.2)$

(4.15.1.6)–(4.15.1.9) in Remark 2 of Sec. 4.15.1 all have corresponding statements in this case. *Note that, as in* (4.15.1.5), *as long as* $(f(z)$, *in general; instead of* $\frac{p(z)}{q(z)}$, *in particular*)

$$\int_{\gamma_n} f(z) \csc \pi z \, dz \to 0 \quad \text{as } n \to \infty$$

holds, where γ_n is as in Fig. 4.134, then (4.15.2.2) is still true.

Corresponding to (4.15.1.10),

The sum of the series $\sum_{n=-\infty}^{\infty} (-1)^n \frac{p(2n+1)}{q(2n+1)} e^{i(2n+1)\theta}$. *Let* $f(z) = \frac{p(z)}{q(z)}$ *be as in* (4.15.1.5). *Then*

$$\sum_{n=-\infty}^{\infty} (-1)^n \frac{p(2n+1)}{q(2n+1)} e^{i(2n+1)\theta} = \frac{\pi}{2} \sum_{k=1}^{m} \operatorname{Re} s \left(\frac{p(z)}{q(z)} e^{i\theta z} \frac{1}{\cos \frac{\pi z}{2}}; z_k \right)$$

uniformly on $-\frac{\pi}{2} \le \theta \le \frac{\pi}{2}$ if $\frac{p(z)}{q(z)} = O(\frac{1}{z^2})$ as $z \to \infty$; uniformly on compact subsets of $-\frac{\pi}{2} < \theta < \frac{\pi}{2}$ if $\frac{p(z)}{q(z)} = O(\frac{1}{z})$ as $z \to \infty$. $\quad (4.15.2.3)$

There are counterparts, in this case, as in Exercise A (1) of Sec. 4.15.1.

In what follows, we present a series of examples.

Example 1. Show that

(1) $\displaystyle\sum_{n=1}^{\infty} \frac{(-1)^n}{n^4 - a^4} = \frac{1}{2a^4} - \frac{\pi}{4a^3}\left(\frac{1}{\sin \pi a} + \frac{1}{\sinh \pi a}\right)$, $a \neq \pm n, \pm ni$ for $n =$
$0, 1, 2, \ldots$.

(2) $\displaystyle\sum_{n=-\infty}^{\infty} \frac{(-1)^n}{n^4 + a^4} = \frac{\pi}{a^3\sqrt{2}} \cdot \frac{\sin\frac{\pi a}{\sqrt{2}}\cosh\frac{\pi a}{\sqrt{2}} + \cos\frac{\pi a}{\sqrt{2}}\sinh\frac{\pi a}{\sqrt{2}}}{\sin^2\frac{\pi a}{\sqrt{2}} + \sinh^2\frac{\pi a}{\sqrt{2}}}$, $a \neq \frac{n}{\sqrt{2}}$
$(\pm 1 \pm i)$ for $n = 0, 1, 2, \ldots$.

Solution.

(1) $f(z) = \frac{1}{z^4 - a^4}$ has simple poles at $z_k = \pm a, \pm ai$, $1 \le k \le 4$, and

$$\mathrm{Re}\,s\left(f(z)\frac{\pi}{\sin \pi z}; z_k\right) = \frac{z_k}{4a^4} \cdot \frac{\pi}{\sin \pi z_k}, \quad k = 1, 2, 3, 4.$$

$$\Rightarrow \sum_{n=-\infty}^{\infty} (-1)^n f(n)$$

$$= -\frac{\pi}{4a^4}\left\{\frac{a}{\sin \pi a} - \frac{a}{\sin \pi(-a)} + \frac{ai}{\sin \pi ai} - \frac{ai}{\sin \pi(-ai)}\right\}$$

$$= -\frac{\pi}{4a^4}\left\{\frac{2}{\sin \pi a} + \frac{2}{\sinh \pi a}\right\}.$$

And the result follows.

(2) $f(z) = \frac{1}{n^4 + a^4}$ has simple poles at $z_k = ae^{\frac{(2k+1)\pi i}{4}}$, $k = 0, 1, 2, 3$, and

$$\mathrm{Re}\,s\left(f(z)\frac{\pi}{\sin \pi z}; z_k\right) = -\frac{z_k}{4a^4} \cdot \frac{\pi}{\sin \pi z_k}, \quad k = 0, 1, 2, 3.$$

What remains is left to the readers.

Example 2. Show that, for $a \neq 0, \pm 1, \pm 2, \ldots$,

$$\sum_{n=1}^{\infty} \frac{(-1)^n \cos n\theta}{a^2 - n^2} = -\frac{1}{2a^2} + \frac{\pi}{2a}\frac{\cos a\theta}{\sin a\pi}, \quad -\pi \le \theta \le \pi;$$

$$\sum_{n=1}^{\infty} \frac{(-1)^n n \sin n\theta}{a^2 - n^2} = \frac{\pi}{2} \cdot \frac{\sin a\theta}{\sin a\pi}, \quad -\pi < \theta < \pi.$$

Then, deduce that

$$\sum_{n=1}^{\infty} \frac{\sin n\theta}{n} = \frac{\pi - \theta}{2}, \quad 0 < \theta < 2\pi; \qquad \sum_{n=1}^{\infty} \frac{(-1)^{n-1}}{2n - 1} = \frac{\pi}{4}.$$

Also, try to sum the following series: for $a \neq 0, \pm i, \pm 2i, \dots$,

$$\sum_{n=1}^{\infty} \frac{(-1)^n \cos n\theta}{a^2 + n^2}; \quad \sum_{n=1}^{\infty} \frac{(-1)^n n \sin n\theta}{a^2 + n^2}.$$

Solution. The function $f(z) = \frac{1}{a^2 - z^2}$ has simple poles at $z = \pm a$. Then

$$\operatorname{Re} s\left(\frac{e^{i\theta z}}{a^2 - z^2} \cdot \frac{\pi}{\sin \pi z}; a\right) = -\frac{\pi e^{ia\theta}}{2a \sin a\pi};$$

$$\operatorname{Re} s\left(\frac{e^{i\theta z}}{a^2 - z^2} \cdot \frac{\pi}{\sin \pi z}; -a\right) = -\frac{\pi e^{-ia\theta}}{2a \sin a\pi},$$

$$\Rightarrow \sum_{n=-\infty}^{\infty} (-1)^n f(n) e^{in\theta} = \frac{\pi}{2a \sin a\pi} (e^{ia\theta} + e^{-ia\theta})$$

$$= \frac{\pi \cos a\theta}{a \sin a\pi} \quad \text{(by (4.15.2.2))}$$

$$\Rightarrow \sum_{n=1}^{\infty} \frac{(-1)^n \cos n\theta}{a^2 - n^2} = -\frac{1}{2a^2} + \frac{\pi}{2a} \frac{\cos a\theta}{\sin a\pi},$$

$$-\pi \leq \theta \leq \pi. \tag{$*_1$}$$

Similarly, if $f(z) = \frac{z}{a^2 - z^2}$, then

$$\sum_{n=-\infty}^{\infty} (-1)^n f(n) e^{in\theta}$$

$$= -\left\{\operatorname{Re} s\left(f(z)\frac{\pi e^{i\theta z}}{\sin \pi z}; a\right) + \operatorname{Re} s\left(f(z)\frac{\pi e^{i\theta z}}{\sin \pi z}; -a\right)\right\}$$

$$= -\frac{\pi}{2 \sin a\pi}(-e^{ia\theta} + e^{-ia\theta}) = \frac{\pi \sin a\theta}{\sin a\pi} i$$

$$\Rightarrow \sum_{n=1}^{\infty} \frac{(-1)^n n \sin n\theta}{a^2 - n^2} = \frac{\pi}{2} \cdot \frac{\sin a\theta}{\sin a\pi}, \quad -\pi < \theta < \pi. \tag{$*_2$}$$

This series converges locally uniformly in $-\pi < \theta < \pi$ and, thus, can be obtained by differentiating $(*_1)$ termwise.

Let $a \to 0$ in $(*_2)$. We obtain

$$-\sum_{n=1}^{\infty} (-1)^n \frac{\sin n\theta}{n} = \frac{\theta}{2}, \quad -\pi < \theta < \pi$$

$$\Rightarrow \text{(replacing θ by $\pi - \theta$)} \sum_{n=1}^{\infty} \frac{\sin n\theta}{n} = \frac{\pi - \theta}{2}, \quad 0 < \theta < 2\pi.$$

Choosing $\theta = \frac{\pi}{2}$, then it follows that $\sum_{n=1}^{\infty} \frac{(-1)^{n-1}}{2n-1} = \frac{\pi}{4}$.

Replace a by ai in $(*_1)$ and $(*_2)$, we obtain

$$\sum_{n=1}^{\infty} \frac{(-1)^n \cos nx}{n^2 + a^2} = -\frac{1}{2a^2} + \frac{\pi}{2a} \frac{\cosh a\theta}{\sinh a\pi}, \quad -\pi \le \theta \le \pi;$$

$$\sum_{n=1}^{\infty} \frac{(-1)^n n \sin nx}{n^2 + a^2} = -\frac{\pi}{2} \cdot \frac{\sinh a\theta}{\sinh a\pi}, \quad -\pi < \theta < \pi.$$

Example 3. Suppose a, b, and t are real numbers, and $|a| < |b|$. Show that

$$\sum_{n=-\infty}^{\infty} (-1)^n \frac{bt}{n^2\pi^2 + b^2t^2} \cos \frac{n\pi a}{b} = \frac{\cosh at}{\sinh bt};$$

$$\sum_{n=-\infty}^{\infty} (-1)^n \frac{bt}{n^2\pi^2 + b^2t^2} \sin \frac{n\pi a}{b} = 0.$$

Solution. In (4.15.2.2), set $p(z) = bt$, $q(z) = \pi^2 z^2 + b^2t^2$, and $\theta = \frac{\pi a}{b}$. Consider

$$F(z) = \frac{bt}{\pi^2 z^2 + b^2t^2} e^{i\frac{\pi a}{b} z} \cdot \frac{\pi}{\sin \pi z},$$

$$\Rightarrow \operatorname{Re} s\left(F(z); \frac{bt}{\pi} i\right) = \frac{bt}{2\pi^2 \cdot \frac{bt}{\pi} i} e^{i\frac{\pi a}{b} \frac{bt}{\pi} i} \frac{\pi}{\sin \pi \frac{bt}{\pi} i} = -\frac{1}{2\sinh bt} e^{-at};$$

$$\operatorname{Re} s\left(F(z); -\frac{bt}{\pi} i\right) = -\frac{1}{2\sinh bt} e^{at}$$

$$\Rightarrow \sum_{n=-\infty}^{\infty} (-1)^n F(n) = \frac{1}{\sinh bt} \cdot \frac{1}{2}(e^{at} + e^{-at}) = \frac{\cosh at}{\sinh bt}.$$

The results follow by equating the real and imaginary parts.

Example 4. Show that

$$\sum_{n=1}^{\infty} (-1)^n \frac{\cos(2n+1)\theta}{2n+1} = \frac{\pi}{4}, \quad -\frac{\pi}{2} < \theta < \frac{\pi}{2} = 0, \quad \text{if } \theta = \pm\frac{\pi}{2},$$

either directly or by using Example 6 in Sec. 4.15.1.

Solution. According to (4.15.2.3),

$$\sum_{n=-\infty}^{\infty} (-1)^n \frac{\cos(2n+1)\theta}{2n+1} = \frac{\pi}{2} \operatorname{Re} s\left(\frac{1}{z} \cdot \frac{e^{i\theta z}}{\cos \frac{\pi z}{2}}; 0\right) = \frac{\pi}{2},$$

$$\Rightarrow \sum_{n=1}^{\infty} (-1)^n \frac{\cos(2n+1)\theta}{2n+1} = \frac{\pi}{4}, \quad -\frac{\pi}{2} < \theta < \frac{\pi}{2}.$$

Or, in Example 6 of Sec. 4.15.1, replace θ by $\theta + \frac{\pi}{2}$ and the result follows immediately.

By the way, integrate this Fourier series termwise and we have

$$\sum_{n=-\infty}^{\infty} (-1)^n \frac{\sin(2n+1)\theta}{(2n+1)^2} = \frac{\pi}{4}\theta, \quad -\frac{\pi}{2} \le \theta \le \frac{\pi}{2}.$$

Similar process leads to

$$\sum_{n=-\infty}^{\infty} (-1)^n \frac{1 - \cos(2n+1)\theta}{(2n+1)^3} = \frac{\pi}{8}\theta^2, \quad -\frac{\pi}{2} \le \theta \le \frac{\pi}{2},$$

$$\Rightarrow \left(\text{setting } \theta = \frac{\pi}{2}\right) \quad \sum_{n=-\infty}^{\infty} (-1)^n \frac{1}{(2n+1)^3} = \frac{\pi^3}{32},$$

$$\Rightarrow \sum_{n=-\infty}^{\infty} (-1)^n \frac{\cos(2n+1)\theta}{(2n+1)^3} = \frac{\pi^3}{32} - \frac{\pi}{8}\theta^2, \quad -\frac{\pi}{2} \le \theta \le \frac{\pi}{2}.$$

Try to derive these formulas from Example 6 in Sec. 4.15.1.

Example 5. Show that

$$\sum_{n=-\infty}^{\infty} \frac{(-1)^n}{\sqrt{n^2 + a^2}} = 2 \int_a^\infty \frac{dx}{\sinh \pi x \cdot \sqrt{x^2 - a^2}}, \quad a > 0.$$

Solution. Adopt the path γ_n shown in Fig. 4.135, where $R = n + \frac{1}{2}$ and $\varepsilon > 0$ is small enough:

$C_R : z = Re^{i\theta}, \quad 0 \le \theta \le 2\pi;$

$C_\varepsilon : z = ai + \varepsilon e^{i\theta}, \quad -\frac{3\pi}{2} \le \theta \le \frac{\pi}{2}; \quad C_\varepsilon^* : z = -ai + \varepsilon e^{i\theta}, \quad -\frac{\pi}{2} \le \theta \le \frac{3\pi}{2};$

$L_{\varepsilon,R} :$ the right and left sides of the segment $[ai + \varepsilon i, iR]$.

$L_{\varepsilon,R}^* :$ the left and right sides of the segment $[-ai - \varepsilon i, -iR]$.

Setting $f(z) = \frac{\pi \csc \pi z}{\sqrt{z^2 + a^2}}$, by the residue theorem,

$$\frac{1}{2\pi i} \int_{\gamma_n} f(z)dz = \frac{1}{2\pi i} \left\{ \int_{C_R} + \int_{L_{\varepsilon,R}} + \int_{C_\varepsilon} + \int_{L_{\varepsilon,R}^*} + \int_{C_\varepsilon^*} \right\}$$

$$= \sum_{k=-n}^{n} \text{Re}\, s(f(z); k) = \sum_{k=-n}^{n} \frac{(-1)^k}{\sqrt{k^2 + a^2}}. \tag{$*3$}$$

Along C_R: Observe that $\sin \pi z \neq 0$ on C_R. Then

$$\left| \int_{C_R} f(z) dz \right| \le \frac{2\pi R}{\sqrt{R^2 - a^2}} \int_0^\pi \frac{d\theta}{\sqrt{\sinh^2(R\cos\theta) + \sin^2(R\sin\theta)}} \to 0$$

$$\text{as } R \to \infty.$$

Along C_ε and C_ε^*:

$$\int_{C_\varepsilon} f(z) dz = \int_{\frac{\pi}{2}}^{-\frac{3\pi}{2}} \frac{\pi \csc(ai + \varepsilon e^{i\theta})}{\sqrt{2ai + \varepsilon e^{i\theta}} \cdot \varepsilon^{\frac{1}{2}} e^{\frac{i\theta}{2}}} \cdot \varepsilon i e^{i\theta} d\theta \to 0 \quad \text{as } \varepsilon \to 0;$$

$$\int_{C_\varepsilon^*} f(z) dz \to 0 \quad \text{as } \varepsilon \to 0.$$

Along $L_{\varepsilon,R}$: On the right side, $\sqrt{z^2 + a^2} = |yi - ai|^{\frac{1}{2}} e^{\frac{\pi i}{4}} \cdot |yi + ai| e^{\frac{\pi i}{4}} = i\sqrt{y^2 - a^2}$, $a + \varepsilon \le y \le R$; on the left side, $\sqrt{z^2 + a^2} = |yi - ai|^{\frac{1}{2}} e^{-\frac{3\pi i}{4}} \cdot |yi + ai| e^{\frac{\pi i}{4}} = -i\sqrt{y^2 - a^2}$, $a + \varepsilon \le y \le R$. Hence,

$$\int_{L_{\varepsilon,R}} f(z) dz = \int_R^{a+\varepsilon} \frac{\pi \csc(iy)}{i\sqrt{y^2 - a^2}} i \, dy + \int_{a+\varepsilon}^R \frac{\pi \csc(iy)}{-i\sqrt{y^2 - a^2}} i \, dy$$

$$= 2\pi i \int_{a+\varepsilon}^R \frac{1}{\sinh \pi y \sqrt{y^2 - a^2}} dy$$

$$\to 2\pi i \int_a^\infty \frac{1}{\sinh \pi y \sqrt{y^2 - a^2}} dy \quad \text{as } \varepsilon \to 0 \text{ and } R \to \infty.$$

Similarly,

$$\int_{L_{\varepsilon,R}^*} f(z) dz \to 2\pi i \int_a^\infty \frac{1}{\sinh \pi y \sqrt{y^2 - a^2}} dy \quad \text{as } \varepsilon \to 0 \text{ and } R \to \infty.$$

Substituting these results in $(*_3)$ as $\varepsilon \to 0$ and $R \to \infty$, the result follows.

Exercises A

(1) Prove (4.15.2.1) and its counterparts of (4.15.1.2)–(4.15.1.3) in detail.

(2) Prove (4.15.2.2) and its counterparts of (4.15.1.6)–(4.15.1.9) in detail. Also, try to derive it directly from (4.15.1.5) by noting that $(-1)^n = e^{in\pi}$ and replacing θ by $\theta + \pi$.

(3) Prove (4.15.2.3) and its counterparts of Exercise A (1) in Sec. 4.15.1 in detail. Also, try to derive it directly from (4.5.1.10) by observing that $(-1)^n = \frac{1}{i} e^{\frac{i(2n+1)\pi}{2}}$ and replacing θ by $\theta + \frac{\pi}{2}$.

(4) Compare to Example 1 in Sec. 4.15.1. Show that, if $a \neq \pm ni$ for $n = 1, 2, 3, \ldots$,

$$\sum_{n=1}^{\infty} \frac{(-1)^n}{n^2 + a^2} = -\frac{1}{2a^2} + \frac{\pi}{2a \sinh \pi a};$$

$$\sum_{n=1}^{\infty} \frac{(-1)^n}{(n^2 + a^2)^2} = \frac{1}{2a} \left[-\frac{1}{a^3} + \frac{\pi}{2a^2 \sinh \pi a} + \frac{\pi^2 \cosh \pi a}{2a \sinh^2 \pi a} \right].$$

Then, try to deduce that

$$\sum_{n=1}^{\infty} \frac{(-1)^n}{n^2} = -\frac{\pi^2}{12}; \quad \sum_{n=1}^{\infty} \frac{(-1)^n}{n^4} = -\frac{7\pi^4}{720}.$$

Also, prove these last two identities directly by considering $-\mathrm{Re}\, s \left(\frac{\pi}{z^2 \sin \pi z}; 0 \right)$ and $-\mathrm{Re}\, s \left(\frac{\pi}{z^4 \sin \pi z}; 0 \right)$, respectively. Similarly,

$$\sum_{n=1}^{\infty} \frac{(-1)^n}{n^6} = -\frac{1}{2} \mathrm{Re}\, s \left(\frac{\pi}{z^6 \sin \pi z}; 0 \right) = -\frac{31\pi^6}{30,240}.$$

(5) Sum the following series.

(a) $\displaystyle\sum_{n=-\infty}^{\infty} \frac{(-1)^n}{(n + a)^2}$ $(a \neq 0, \pm 1, \pm 2, \ldots)$. What happens if $a = 0$?

(b) $\displaystyle\sum_{n=-\infty}^{\infty} \frac{(-1)^n n^2}{n^4 + a^4}$ $\left(a \neq \frac{n}{\sqrt{2}} (\pm 1 \pm i) \text{ for } n = 0, 1, 2, \ldots \right)$.

(c) $\displaystyle\sum_{\substack{n=-\infty \\ n \neq 0}}^{\infty} (-1)^n \left(\frac{1}{z - n} + \frac{1}{n} \right)$ $(z \neq \pm 1, \pm 2, \ldots)$.

(6) In (4.15.2.2), let $p(z) = 1$, $q(z) = z^{2k}$, and $\theta = 0$. Show that, for $k = 1, 2, \ldots$,

$$\sum_{n=1}^{\infty} (-1)^n \frac{1}{n^{2k}} = -\frac{1}{2} \cdot \frac{1}{(2k)!} \lim_{z \to 0} \frac{d^{2k}}{dz^{2k}} \left(\frac{\pi}{\sin \pi z} \right).$$

Recall that $\frac{z}{\sin z} = \sum_{n=0}^{\infty} (-1)^{n-1} \frac{(4^n - 2) B_{2n}}{(2n)!} z^{2n}$, $|z| < \pi$ (see Exercise A (5)(f) of Sec. 4.8).

(7) Integrate $f(z) = \frac{\pi \sin \theta z}{z^3 \sin \pi z}$, $-\pi < \theta < \pi$, along the path γ_n shown in Fig. 4.134 and show that

(a) $\displaystyle\sum_{n=1}^{\infty} (-1)^{n+1} \frac{\sin n\theta}{n^3} = \frac{\theta}{12} (\pi^2 - \theta^2)$, $\quad -\pi < \theta < \pi$.

(b) $\displaystyle\sum_{n=0}^{\infty} \frac{(-1)^n}{(2n + 1)^3} = \frac{\pi^3}{32}$.

(8) Let z be a complex number which is not an integer and $-\pi < \theta < \pi$. Show that

$$\lim_{n\to\infty} \int_{\gamma_n} \frac{\cos\theta\zeta}{(z^2 - \zeta^2)\sin\pi\zeta}\,d\zeta = 0,$$

where γ_n is as in Fig. 4.134, and hence,

$$\frac{\pi\cos\theta z}{\sin\pi z} = \frac{1}{z} + 2z\sum_{n=1}^{\infty}(-1)^n\frac{\cos n\theta}{z^2 - n^2}, \qquad -\pi < \theta < \pi.$$

(9) Let z and θ be as in Exercise (8). Show that, in case γ_n is as in Fig. 4.134,

$$\lim_{n\to\infty} \int_{\gamma_n} \frac{\zeta\sin\theta\zeta}{(z^2 - \zeta^2)\sin\pi\zeta}\,d\zeta = 0.$$

Then, deduce that:

(a) $\dfrac{\pi}{2}\cdot\dfrac{\sin\theta z}{\sin\pi z} = \displaystyle\sum_{n=1}^{\infty}(-1)^n\frac{n\sin n\theta}{z^2 - n^2}$, $\quad -\pi < \theta < \pi$, $\quad z$ is not an integer.

(b) $\dfrac{\pi}{\cos\pi z} = \displaystyle\sum_{n=0}^{\infty}(-1)^n\frac{2n+1}{(n+\frac{1}{2})^2 - z^2}$, $\quad z \neq \pm\dfrac{1}{2},\pm\dfrac{3}{2},\pm\dfrac{5}{2},\ldots$.

(c) $\dfrac{\pi}{\cosh\pi z} = \displaystyle\sum_{n=0}^{\infty}(-1)^n\frac{2n+1}{z^2 + (n+\frac{1}{2})^2}$, $\quad z \neq \pm\dfrac{1}{2}i,\pm\dfrac{3}{2}i,\pm\dfrac{5}{2}i,\ldots$.

(10) Sum the following series:

$$\sum_{n=2}^{\infty}(-1)^n\frac{\cos n\theta}{n(n-1)},\ 0\le\theta\le 2\pi;\qquad \sum_{n=2}^{\infty}(-1)^n\frac{\sin n\theta}{n(n-1)},\ 0\le\theta\le 2\pi.$$

(11) Sum the following series:

$$\sum_{n=-\infty}^{\infty}\frac{(-1)^n}{\sqrt{n^4 + a^4}}\,(a>0);\qquad \sum_{n=-\infty}^{\infty}\frac{(-1)^n}{\sqrt{n^4 - a^4}}\quad (a\text{ is not an integer}).$$

(12) Let $f(z)$ be an entire function such that $|f(z)| \le Me^{|\operatorname{Im} z|}$, $z \in \mathbf{C}$, for some constant $M > 0$. Suppose $F(z) = \frac{f(z)}{z^2\cos z}$, $z \neq 0$, $(n+\frac{1}{2})\pi$ for $n = 0,\pm1,\pm2,\ldots$. Let σ_n be the square path with vertices at $n\pi(\pm1\pm i)$, $n = 1, 2, 3, \ldots$.

(a) Show that $\frac{1}{2\pi i}\int_{\sigma_n}F(z)dz = f'(0) - \sum_{k=-n}^{n-1}(-1)^k\frac{f((k+\frac{1}{2})n)}{(k+\frac{1}{2})^2\pi^2}$.

(b) Show that $|\cos z| \geq \frac{1}{4} e^{|\operatorname{Im} z|}$ and $|F(z)| \leq \frac{4M}{|z|^2}$ along σ_n.

(c) Use (a) and (b) to sow that $f'(0) = \sum_{n=-\infty}^{\infty} (-1)^n \frac{f((n+\frac{1}{2})n)}{(n+\frac{1}{2})^2 \pi^2}$.

(d) Apply (c) to $f(z) = \sin z$ and show that $\sum_{n=-\infty}^{\infty} \frac{1}{(2n+1)^2} = \frac{\pi^2}{4}$.

(e) Use (c) and (d) to show that $|f'(0)| \leq \sup_{x \in \mathbf{R}} |f(x)|$.

(f) Bernstein. Let $g(z)$ be an entire function such that $|g(z)| \leq M e^{c|\operatorname{Im} z|}$, $z \in \mathbf{C}$, $M > 0$, $c > 0$. Show that

$$|g'(x)| \leq c \sup_{x \in \mathbf{R}} |g(x)|, \quad x \in \mathbf{R}.$$

Appendix

Appendix A. The Real Number System R

Readers of this book are supposed to be familiar with basic properties of the real number system **R**. For the sake of reference and comparison, we list some of them barely needed in the study of the complex analysis.

(I) **R** *is a field.*

For each pair of elements $a, b \in \mathbf{R}$, there are unique elements $a + b$ (called the *sum* of a and b) and ab (called the *product* of a and b) in **R** for which the following conditions hold for all $a, b, c \in \mathbf{R}$:

(1) Addition

    (a) $a + b = b + a$.
    (b) $(a + b) + c = a + (b + c)$.
    (c) (zero element 0) $\quad a + 0 = a$.
    (d) (inverse element $-a$) $\quad a + (-a) = a - a = 0$.

(2) Multiplication

    (a) $ab = ba$.
    (b) $(ab)c = a(bc)$.
    (c) (identity element 1) $\quad 1a = a$.
    (d) (inverse element a^{-1} if $a \neq 0$) $\quad aa^{-1} = 1$.

(3) Addition and Multiplication: $a(b + c) = ab + ac$.

0 and 1 are unique. We call **R** a *number system* just because it owns the properties (1)(a), (b); (2)(a), (b), and (3). **R**, in turn, contains the *rational number system* **Q**, the *integral number system* **Z**, and the *natural number system* **N**. **R** can be interpreted geometrically as a *real line* on which a real number is then called a *real point*, and conversely.

(II) R *is an ordered field.*

For each pair of elements $a, b \in \mathbf{R}$, designate $a > b$ if $a - b > 0$; $a < b$ if $a - b < 0$, and $a = b$ if $a - b = 0$. Then the relation "$<$" *orders* **R** in the following sense: For any two elements $a, b \in \mathbf{R}$, one and only one of the statements $a < b$, $a = b$, and $a > b$ can hold; and satisfies the following three properties: for any $a, b, c \in \mathbf{R}$,

(1) $a < b$ and $b < c \Rightarrow a < c$.
(2) $a < b \Rightarrow a + c < b + c$.
(3) $a < b$ and $c > 0 \Rightarrow ac < bc$.

Note that $a \le b$ means $a < b$ or $a = b$. And $a^2 \ge 0$ always holds for any $a \in \mathbf{R}$.

(III) *Basic point sets.*

For $a, b \in \mathbf{R}$ with $a < b$. Designate the

    open interval $(a, b) = \{x \in \mathbf{R} \mid a < x < b\}$ and the

    closed interval $[a, b] = \{x \in \mathbf{R} \mid a \le x \le b\}$, where $a = b$ is permitted.

The *extended real number system* $\mathbf{R}^*$ is the one obtained from **R** by adjoining to it the plus infinite $+\infty$ and the minus infinite $-\infty$, namely, $\mathbf{R}^* = [-\infty, +\infty] = \mathbf{R} \cup \{-\infty, +\infty\}$ with $-\infty < x < +\infty$ for each $x \in \mathbf{R}$. Of course, $\mathbf{R}^*$ can also be realized as *the extended real line*.

    Let S be a nonempty subset of **R**. S is said to be *bounded above* if there is a number x such that $a \le x$ for each $a \in S$, and then x is called an *upper bound* of S. Suppose x is an upper bound of S, and for each $\varepsilon > 0$, there is an element $a \in S$ satisfying $x - \varepsilon < a \le x$, then such an x is called the *least upper bound* of S and is simply denoted as sup S. Designate sup $S = +\infty$ if S is not bounded above.

    Correspondingly, define inf $S = -\sup(-S)$, where $-S = \{-a \mid a \in S\}$.

    A point $a \in \mathbf{R}^*$ is called a *limit point* of a sequence $a_n \in \mathbf{R}$, $n \ge 1$, if this sequence has a subsequence a_{n_k} converging to a, if $a \in \mathbf{R}$; or diverging to a, if $a = +\infty$ or $-\infty$. Let S be the set of all limit points of the sequence $a_n, n \ge 1$. The *upper limit* is defined and denoted by

$$\varlimsup_{n \to \infty} a_n = \sup S$$

$$= \inf_{n \ge 1} \sup_{k \ge n} a_k,$$

and can be characterized as: setting $L = \varlimsup_{n \to \infty} a_n$.

(1) If $L \in \mathbf{R}$: then for each $\varepsilon > 0$, there is an integer N such that $a_n < L+\varepsilon$ for each $n \geq N$; and, for each integer $k \geq 1$, there is an integer $n_k \geq k$ such that $L - \varepsilon < a_{n_k}$.
(2) If $L = +\infty$: then there is a subsequence $a_{n_k} \to L = +\infty$ if $k \to \infty$.
(3) If $L = -\infty$: then the sequence a_n itself diverges to $-\infty$.

Correspondingly, the *lower limit* is

$$\lim_{n \to \infty} a_n = - \overline{\lim_{n \to \infty}} (-a_n).$$

(IV) *Completeness of* $\mathbf{R}$.

The following statements are true and equivalent:

(1) Every nonempty subset of $\mathbf{R}$, bounded above, has a least upper bound.
(2) Every Cauchy sequence in $\mathbf{R}$ converges. *Note*: A sequence $x_n, n \geq 1$, is *Cauchy* if for each $\varepsilon > 0$, there is an integer $N \geq 1$ so that $|x_n - x_m| < \varepsilon$ if any $m, n \geq N$.
(3) (*Nested intervals theorem*) Let $[a_n, b_n], n \geq 1$, be a sequence of intervals in $\mathbf{R}$ satisfying
  (i) $[a_n, b_n] \supseteq [a_{n+1}, b_{n+1}]$ for $n \geq 1$; and
  (ii) $b_n - a_n \to 0$ as $n \to \infty$,
  then there is a unique point $x_0 \in \mathbf{R}$ such that $\bigcap_{n=1}^{\infty}[a_n, b_n] = \{x_0\}$.
(4) (*Bolzano–Weierstrass*) Any bounded infinite set in $\mathbf{R}$ has at least one limit point. *Note*: A set S is *bounded* if $|a| \leq M$ for each $a \in A$ and for some constant $M > 0$. x_0 is called a *limit point* of an infinite set S if S contains a sequence consisting of distinct terms and converging to x_0.
(5) (*Bolzano–Weierstrass*) Any bounded sequence in $\mathbf{R}$ has a convergent subsequence.
(6) (*Heine–Borel*) Let $[a, b]$ be a closed interval in $\mathbf{R}$. Then, for any open covering $\{I_\lambda\}$ of $[a, b]$ (where each I_λ is an open interval or open set), there is a *finite* open subcovering, namely, there are finitely many open sets $I_{\lambda_1}, \ldots, I_{\lambda_n}$ from $\{I_\lambda\}$ such that

$$[a, b] \subseteq \bigcup_{k=1}^{n} I_{\lambda_k}.$$

Prove these.

References

1. Ahlfors LV (1979). *Complex Analysis*, 3rd Ed. McGraw-Hill.
2. Ahlfors LV (1973). *Conformal Invariants, Topics in Geometric Function Theory.* McGraw-Hill.
3. Ahlfors LV (1966). *Lectures on Quasiconformal Mappings.* Van Nostrand.
4. Ahlfors LV (1981). *Möbius Transformations in Several Dimensions.* School of Mathematics, University of Minnesota.
5. Ahlfors LV (1938). An extension of Schwarz's Lemma. *Transaction of the American Mathematical Society*, **43**, 359–364.
6. Ahlfors LV and Sario L (1960). *Riemann Surfaces.* Princeton University Press.
7. Antimirov MY, Kolyshkin AA and Vaillancourt R (1998). *Complex Variables.* Academic Press.
8. Beardon AF (1989). *Complex Analysis, The Angument Principle in Analysis and Topology.* John-Wiley & Sons.
9. Berenstein CA and Gay R (1991). *Complex Variables, An Introduction.* Springer-Verlag.
10. Bers L (1957–1958). *Riemann Surfaces.* New York University.
11. Bochner S and Martin WT (1948). *Several Complex Variables.* Princeton University Press.
12. Burckel RB (1979). *An Introduction to Classical Complex Analysis.* Birkhäuser Verlag.
13. Campbell R (1966). *Les Intégrales Eulériennes et Leurs Applications.* Dunod.
14. Carathéodory C (1954). *Theory of Functions of a Complex Variable*, Vols. I, II. Chelsea Pub. Co.
15. Carleson L and Gamelin TW (1993). *Complex Dynamics.* Springer-Verlag.
16. Cartan H (1963). *Elementary Theory of Analytic Functions of One or Several Complex Variables.* Adiwes International Series.
17. Chen X, Lu P and Tian G (2005). *A Note on Uniformization of Riemann Surfaces by Ricci Flow*, arXiv: Math.DG/0505163 V2 27 May 2005.
18. Chern SS. Collected Papers.
19. Collingwood EF and Lohwater AJ (1966). *The Theory of Cluster Sets.* Cambridge University Press.
20. Copson ET (1955). *An Introduction to the Theory of Functions of a Complex Variable.* Oxford University Press.

21. de Branges L (1985). A proof of the bieberbach conjecture, *Acta Mathematica*, **154**, 137–152.
22. de Bruijn NG (1958). *Asymptotic Methods in Analysis*. Interscience, Wiley.
23. Dieudonné J (1960). *Foundations of Modern Analysis*. Academic Press.
24. Dineen S (1989). *The Schwarz Lemma*. Oxford: Clarendon Press.
25. Dinghas A (1961). *Vorlesungen über Funktionentheorie*. Springer-Verlag.
26. Dixion JD (1971). A brief proof of Cauchy's integral theorem, *Proceedings of the American Mathematical Society*, **29**, 625–626.
27. Dugundji J (1966). *Topology*. Allyn and Bacon.
28. Duren PL (1983). *Univalent Functions*. Springer-Verlag.
29. Evgrafov MA (1962). *Asymptotic Estimates and Entire Functions*. Gordon and Breach.
30. Evgrafov MA (1966). *Analytic Functions*. Sanders Math. Books.
31. Evgrafov MA, Sidorov IV, Fedoryuk MV, Shabunin MI and Bezanov KA (1969). *Collection of Problems on the Theory of Analytic Functions*. Moscow: Nauka (Russian).
32. Farkas HM and Kra I (1980). *Riemann Surfaces*. Springer-Verlag.
33. Forster O (1981). *Lectures on Reimann Surfaces*. Springer-Verlag.
34. Fuchs BA and Shabat BV (1964). *Functions of a Complex Variables and Some of their Application*, Vol. II. Addision-Wesley.
35. Goluzin GM (1969). *Geometric Theory of a Complex Variable*. AMS.
36. Gonzáles MO (1992). *Classical Complex Analysis*, Vols. 1, 2. Marcel Dekker.
37. Grauert H and Reckziegel H (1956). Hermiteschen metriken and normale familien holomorpher abbildungen, *Math. Zeltsch*, **89**, 108–125.
38. Gunning RC (1966). *Lectures on Riemann Surfaces*. Princeton University Press.
39. Guning RC and Rossi H (1965). *Analytic Functions of Several Complex Variables*. Prentice-Hall, Inc.
40. Hayman WK (1964). *Meromorphic Functions*. Oxford University Press.
41. Hayman WK and Kennedy PB (1976). *Subharmonic Functions*. Academic Press.
42. Heins M (1949). The conformal mapping of simply connected Riemann surfaces, *Annals of Mathematics*, **50**, 686–690.
43. Henrici P (1974). *Applied and Computational Complex Analysis*, Vol. I. John Wiley & Sons; Vol. II (1977), Vol. III (1987).
44. Hille E (1962). *Analytic Function Theory*, Vol. I. Ginn and Co.; Vol. II (1962).
45. Hörmander L (1973). *An Introduction to Complex Analysis in Several Variables*. North-Holland.
46. Hua LK (1985).
47. Kline M (1972). *Mathematical Thought: From Ancient to Modern Times*. Oxford University Press.
48. Knopp K (1947). *Theory of Functions*, Vol. II. Dover.
49. Kobayashi Z (1970). *Hyperbolic Manifolds and Holomorphic Mappings*. Marcel Dekkar.
50. Kober H (1957). *Dictionary of Conformal Representations*. Dover Pub. Inc.
51. Krantz SG (1990). *Complex Analysis: The Geometric Viewpoint*. MAA.

52. Krzyż JG (1971). *Problems in Complex Variable Theory*. Elsevier Pub. Inc.
53. Lehto O (1987). *Univalent Functions and Teichmüller Spaces*. Springer-Verlag.
54. Lehto O and Virtanen KI (1970). *Quasiconformal Mappings in the Plane*. Springer-Verlag.
55. Li Z (1988).
56. Lin IH (2005). *Geometric Linear Algebra*, Vol. 1. World Scientific; Vol. 2 (2008).
57. Massey WS (1991). *A First Course in Algebraic Topology*. Springer-Verlag.
58. Markushevich AI (1969). *Theory of Functions of a Complex Variable*, Vol. I. Prentice-Hall; Vol. II (1965), Vol. III (1968, 1970); Chelsea, 1977.
59. Minda D and Schober G (1983). Another elementary approach to the theorems of Landau, Montel, Picard and Schottky, *Complex Variables*, **2**, 157–164.
60. Mitrinovič DS and Kečkić JD (1984). *The Cauchy Method of Residues, Theory and Applications*, Vol. I. D. Reidel Publ. Co.; Vol. II (1993) Kluwer Academic Press.
61. Narasimhan R (1985). *Complex Analysis in One Variable*. Birkhäuser.
62. Needham T (1997). *Visual Complex Analysis*. Clarendon Press.
63. Nehari Z (1952). *Conformal Mapping*. McGraw-Hill Book Co.
64. Nevanlinna R (1970). *Analytic Functions*. Springer-Verlag.
65. Newman MHA (1951). *Elements of the Topology of Plane Sets of Points*. Cambridge University Press.
66. Palka BP (1991). *An Introduction to Complex Function Theory*. Springer-Verlag.
67. Pommerenke C (1992). *Boundary Behaviour of Conformal Maps*. Springer-Verlag.
68. Ratcliffe IG (1994). *Foundations of Hyperbolic Manifolds*. Springer-Verlag.
69. Robinson RM (1939). A generalization of Picard's and related theorems, *Duke Mathematical Journal*, **5**, 118–132.
70. Rudin W (1966). *Real and Complex Analysis*. McGraw-Hill.
71. Sansone G and Gerretsen J (1960). *Lectures on the Theory of Functions of a Complex Variable*. Noordhoff.
72. Sario L and Nakai M (1970). *Classification Theory of Riemann Surfaces*. Springer-Verlag.
73. Sario L and Noshiro K (1966). *Value Distribution Theory*. D. Van Nostramd Co. Inc.
74. Spanier EH (1966). *Algebraic Topology*. McGraw-Hill.
75. Springer G (1957). *Introduction to Riemann Surfaces*. Addison-Wesley.
76. Titchmarsh EC (1939). *The Theory of Functions*, 2nd Ed. Oxford University Press.
77. Titchmarsh EC (1937). *The Fourier Transform*. Oxford University Press.
78. Tsuji M (1959). *Potential Theory in Modern Function Theory*. Tokyo: Maruzen Co. LTD.
79. Väisälä J (1971). *Lectures on n-Dimensional Quasiconformal mappings*, Lectures Notes in Mathematics 229. Springer-Verlag.

80. Volkovyskii LV, Lunts GL and Aramanovich IG (1965). *A Collection of Problems on Complex Analysis*. Pergamon Press.
81. Weyl H (1923). *Die Idee der Riemannschen Flächen*, 2nd Ed. Leipzig: Teubner; New York: Chelsea, 1951.
82. Yang L (1982).
83. Zalcman L (1975). Heuristic principle in complex function theory, *The American Mathematical Monthly*, **82**, 813–817.
84. Zhuang QT (1982).

Index of Notations

z^*	the symmetric (or reflection) point of z w.r.t. a circle $\|z - z_0\| = r : z_0 + \frac{r^2}{\bar{z} - \bar{z_0}}$; or, w.r.t. a line $z = a + bt : a + b^2(\bar{z} - \bar{a})$ 53
$\sqrt[n]{z}$ or $z^{\frac{1}{n}}$ $(n \geq 2)$	nth root of z; nth root function of z 57, 214
∞	the point at infinity or the infinite point 67
$\mathbf{C}^*$	the extended complex plane: $\mathbf{C} \cup \{\infty\}$ 67
$\{z_n\}_{n=1}^{\infty}$ or z_n, $n \geq 1$ or z_n	a sequence with general term z_n 76
$\lim_{n \to \infty} z_n = z_0$ or $\lim z_n = z_0$ or $z_n \to z_0$ (as $n \to \infty$)	the sequence z_n converging to the limit z_0 as $n \to \infty$ 76, 77
$\overline{\lim}_{n \to \infty} a_n$ or $\overline{\lim} a_n$	the upper limit of the *real* sequence $a_n : \inf_{n \geq 1} \sup_{k \geq n} a_k$ 1037
$\underline{\lim}_{n \to \infty} a_n$ or $\underline{\lim} a_n$	the lower limit of the *real* sequence $a_n : \sup_{n \geq 1} \inf_{k \geq n} a_k$ 1038
$\sum_{n=1}^{\infty} z_n$ or $\sum z_n$	the series with general term z_n 83
$D_{\varepsilon}(a)$	the open disk with center a and radius $\varepsilon : \|z - a\| < \varepsilon$; an open ε-neighborhood of a 86
A'	derived set of a set A: the set of all the limit points of A 87
$\bar{A}$	closure of $A : A \cup A'$ 87
$A^{\sim}$	complement of A (in a larger set): $\mathbf{C} - A$ 88
Int A or A°	the interior of a set A 89
Bdry A or ∂A	the boundary of A 89
Ext A or $\bar{A}^{\sim}$	the exterior of A 89
$f : A \to \mathbf{C}$ or $w = f(z) : A \to \mathbf{C}$ or $f(z)$ or f	complex-valued function of a complex variable $z \in A \subseteq \mathbf{C}$ 104
Re $f(z)$ or $u(z)$	real part of $f(z) = u(z) + iv(z)$ 105
Im $f(z)$ or $v(z)$	imaginary part of $f(z) = u(z) + iv(z)$ 105
$\lim_{z \in A \to z_0} f(z) = w_0$ or $\lim_{z \to z_0} f(z) = w_0$ or $f(z) \to w_0$ as $z \to z_0$	the limit w_0 of $f(z)$ as z approaches z_0, where z_0 is usually a limit point of the set A 107

$\frac{\partial f}{\partial x}$ or f_x partial derivative of $f = u + iv$ w.r.t.
$x : u_x + iv_x$ 273

$\frac{\partial f}{\partial y}$ or f_y $u_y + iv_y$ 273

$\frac{\partial}{\partial z}$ or ∂_z or ∂ complex differential operator w.r.t.
$z : \frac{1}{2}(\frac{\partial}{\partial x} - i\frac{\partial}{\partial y})$;
$\frac{\partial f}{\partial z} = \partial_z f = \partial f = f_z = \frac{1}{2}(f_x - if_y)$ 273

$\frac{\partial}{\partial \bar{z}}$ or $\partial_{\bar{z}}$ or $\bar{\partial}$ complex differential operator w.r.t.
$\bar{z} : \frac{1}{2}(\frac{\partial}{\partial x} + i\frac{\partial}{\partial y})$;
$\frac{\partial f}{\partial \bar{z}} = \partial_{\bar{z}} f = \bar{\partial} f = f_{\bar{z}} = \frac{1}{2}(f_x + if_y)$ 273, 275

$D_\theta f(z)$ the directional derivative of f at z in the direction
$e^{i\theta} : e^{-i\theta} df_z(e^{i\theta}) = f_z(z) + f_{\bar{z}}(z)e^{-2i\theta}$ 273, 275

$C^k(O)$ the vector space of all k-times continuously differentiable functions (in the real sense) on an open set O 275

$dxdy$ absolute area element 276

$dx \wedge dy$ oriented area element 276

$dz \wedge d\bar{z}$ fundamental complex differential form of order 2: $-2idx \wedge dy = -d\bar{z} \wedge dz$ 277

df a complex differential form of order 1:
$f_x dx + f_y dy = f_z dz + f_{\bar{z}} d\bar{z} = D_\theta f(z)dz$ 277

$d(fdz + gd\bar{z})$ $= (\frac{\partial g}{\partial z} - \frac{\partial f}{\partial \bar{z}})dz \wedge d\bar{z}$ 277

Δ the Laplacian operator:
$\frac{\partial^2}{\partial x^2} + \frac{\partial^2}{\partial y^2} = r\frac{\partial}{\partial r}(r\frac{\partial}{\partial r}) + \frac{\partial^2}{\partial \theta^2}$;
$\Delta u = \frac{\partial^2 u}{\partial x^2} + \frac{\partial^2 u}{\partial y^2} = u_{xx} + u_{yy}$, etc. 280

$\int_\gamma f(z)dz$ complex integral of f along a curve γ w.r.t. z 281

$\int_\gamma f(z)d\bar{z}$ complex integral of f along a curve γ w.r.t. $\bar{z}$ 285

$\int_\gamma f(z)|dz|$ complex integral of f along a curve γ w.r.t. arc length $|dz|$ 286

C_n^λ generalized binomial coefficient: 1, if $n = 0$;
$\frac{\lambda(\lambda-1)\cdots(\lambda-n+1)}{n!}$, if $n \geq 1$ 368

$\mathcal{S}$ the family of the univalent analytic functions $f(z)$ in $|z| < 1$ with the property that $f(0) = 0$ and $f'(0) = 1$ 611

$\gamma \sim 0 (\bmod \Omega)$ a cycle γ homologous to zero in a domain Ω 664

Index

W
Weierstrass's factorization theorem 594
winding number 208, 211, 514
   integral representation 525
   geometric interpretation 208, 211, 525
   basic properties 214, 528
   homotopic invariance 214, 528

Y
Young inequality 34